Lecture Notes in Computer Science   3496

Commenced Publication in 1973
Founding and Former Series Editors:
Gerhard Goos, Juris Hartmanis, and Jan van Leeuwen

D0529020

Jun Wang   Xiaofeng Liao   Zhang Yi (Eds.)

Advances in Neural Networks – ISNN 2005

Second International Symposium on Neural Networks
Chongqing, China, May 30 - June 1, 2005
Proceedings, Part I

 Springer

Volume Editors

Jun Wang
The Chinese University of Hong Kong
Department of Automation and Computer-Aided Engineering
Shatin, New Territories, Hong Kong
E-mail: jwang@acae.cuhk.edu.hk

Xiaofeng Liao
Chongqing University, School of Computer Science and Engineering
Chongqing, 400044, China
E-mail: xfliao@cqu.edu.cn

Zhang Yi
University of Electronic Science and Technology of China
School of Computer Science and Engineering
Chengdu, Sichuan, China
E-mail: zhangyi@uestc.edu.cn

Library of Congress Control Number: 2005926239

CR Subject Classification (1998): F.1, F.2, D.1, G.2, I.2, C.2, I.4-5, J.1-4

ISSN   0302-9743
ISBN-10   3-540-25912-0 Springer Berlin Heidelberg New York
ISBN-13   978-3-540-25912-1 Springer Berlin Heidelberg New York

Springer is a part of Springer Science+Business Media

springeronline.com

© Springer-Verlag Berlin Heidelberg 2005
Printed in Germany

Typesetting: Camera-ready by author, data conversion by Olgun Computergrafik
Printed on acid-free paper   SPIN: 11427391   06/3142   5 4 3 2 1 0

Preface

This book and its sister volumes constitute the proceedings of the 2nd International Symposium on Neural Networks (ISNN 2005). ISNN 2005 was held in the beautiful mountain city Chongqing by the upper Yangtze River in southwestern China during May 30–June 1, 2005, as a sequel of ISNN 2004 successfully held in Dalian, China. ISNN emerged as a leading conference on neural computation in the region with increasing global recognition and impact. ISNN 2005 received 1425 submissions from authors on five continents (Asia, Europe, North America, South America, and Oceania), 33 countries and regions (Mainland China, Hong Kong, Macao, Taiwan, South Korea, Japan, Singapore, Thailand, India, Nepal, Iran, Qatar, United Arab Emirates, Turkey, Lithuania, Hungary, Poland, Austria, Switzerland, Germany, France, Sweden, Norway, Spain, Portugal, UK, USA, Canada, Venezuela, Brazil, Chile, Australia, and New Zealand). Based on rigorous reviews, 483 high-quality papers were selected by the Program Committee for presentation at ISNN 2005 and publication in the proceedings, with an acceptance rate of less than 34%. In addition to the numerous contributed papers, 10 distinguished scholars were invited to give plenary speeches and tutorials at ISNN 2005.

The papers are organized into many topical sections under 20 coherent categories (theoretical analysis, model design, learning methods, optimization methods, kernel methods, component analysis, pattern analysis, signal processing, image processing, financial analysis, system modeling, control systems, robotic systems, telecommunication networks, incidence detection, fault diagnosis, power systems, biomedical applications, and industrial applications, and other applications) spanning all major facets of neural network research and applications. ISNN 2005 provided an international forum for the participants to disseminate new research findings and discuss the state of the art. It also created a pleasant opportunity for the participants to interact and exchange information on emerging areas and future challenges of neural network research.

Many people made significant efforts to ensure the success of this event. The ISNN 2005 organizers are grateful to Chongqing University, Southwest Normal University, Chongqing University of Posts and Telecommunications, Southwest Agricultural University, and Chongqing Education College for their sponsorship; grateful to the National Natural Science Foundation of China for the financial support; and to the Asia Pacific Neural Network Assembly, the European Neural Network Society, the IEEE Computational Intelligence Society, and the IEEE Circuits and Systems Society for their technical co-sponsorship. The organizers would like to thank the members of the Advisory Committee for their spiritual support, the members of the Program Committee for reviewing the papers, and the members of the Publication Committee for checking the papers. The organizers would particularly like to thank the publisher, Springer, for their cooperation in publishing the proceedings as three volumes of the Lecture Notes in Computer Science series. Last but not least, the organizers would like to thank all the authors for contributing their papers to ISNN 2005. Their enthusiastic contributions and participation were essential parts of the symposium with which the organizers were proud to be involved.

May 2005

Jun Wang
Xiaofeng Liao
Zhang Yi

ISNN 2005 Organization

ISNN 2005 was organized and sponsored by Chongqing University, Southwest Normal University, Chongqing University of Posts and Telecommunications, Southwest Agricultural University, and Chongqing Education College in cooperation with the Chinese University of Hong Kong. It was technically cosponsored by the Asia Pacific Neural Network Assembly, the European Neural Network Society, the IEEE Circuits and Systems Society, and the IEEE Computational Intelligence Society. It was financially supported by the National Natural Science Foundation of China and K.C. Wong Education Foundation of Hong Kong.

General Chair

Jun Wang, Hong Kong, China

Advisory Committee Co-chairs

Shun-ichi Amari, Tokyo, Japan

Jacek M. Zurada, Louisville, USA

Advisory Committee Members

Zheng Bao, X'ian, China

Ruwei Dai, Beijing, China

Walter J. Freeman, Berkeley, USA

Kunihiko Fukushima, Tokyo, Japan

Zhenya He, Nanjing, China

Frank L. Lewis, Fort Worth, USA

Erkki Oja, Helsinki, Finland

Shoujue Wang, Beijing, China

Bo Zhang, Beijing, China

Guoliang Chen, Hefei, China

Chunbo Feng, Nanjing, China

Toshio Fukuda, Nagoya, Japan

Aike Guo, Shanghai, China

Okyay Kaynak, Istanbul, Turkey

Yanda Li, Beijing, China

Tzyh-Jong Tarn, St. Louis, USA

Youshou Wu, Beijing, China

Nanning Zheng, Xi'an, China

Steering Committee Chairs

Xiaohong Li, Chongqing, China

Yixin Zhong, Beijing, China

Steering Committee Members

Wlodzislaw Duch, Torun, Poland

Max Q.H. Meng, Hong Kong, China

Yuhui Qiu, Chongqing, China

DeLiang Wang, Columbus, USA

Zongben Xu, Xi'an, China

Fuliang Yin, Dalian, China

Yinguo Li, Chonqing, China

Marios M. Polycarpou, Cincinnati, USA

Zhengqi Sun, Beijing, China

Zhongfu Wu, Chongqing, China

Gary G. Yen, Stillwater, USA

Juebang Yu, Chengdu, China

Program Committee Co-chairs

Xiaofeng Liao, Chongqing, China   *Zhang Yi*, Chengdu, China

Program Committee Members

Shigeo Abe, Kobe, Japan
Amit Bhaya, Rio de Janeiro, Brazil

Jinde Cao, Nanjing, China
Ke Chen, Manchester, UK
Tianping Chen, Shanghai, China
Yiu Ming Cheung, Hong Kong, China
Hyungsuk Cho, Dae Jeon, Korea
Shuang Cong, Hefei, China
Meng Joo Er, Singapore
Jun Gao, Hefei, China
Ping Guo, Beijing, China
Baogang Hu, Beijing, China
Jinglu Hu, Fukuoka, Japan
Licheng Jiao, Xi'an, China
Hon Keung Kwan, Windsor, Canada
Cees van Leeuwen, Tokyo, Japan
Yangmin Li, Macau, China
Yanchun Liang, Changchun, China
Chin-Teng Lin, Hsingchu, Taiwan
Qing Liu, Wuhan, China
Hongtao Lu, Shanghai, China
Zhiwei Luo, Nagoya, Japan
Satoshi Matsuda, Narashino, Japan
Stanislaw Osowski, Warsaw, Poland
Rudy Setiono, Singapore
Daming Shi, Singapore
Jianbo Su, Shanghai, China
Fuchun Sun, Beijing, China
Johan Suykens, Leuven, Belgium
Ying Tan, Hefei, China
Lipo Wang, Singapore
Wei Wu, Dalian, China
Hong Yan, Hong Kong, China
Wen Yu, Mexico City, Mexico
Huaguang Zhang, Shenyang, China
Liqing Zhang, Shanghai, China

Sabri Arik, Istanbul, Turkey
Abdesselam Bouzerdoum, Wollongong, Australia
Laiwan Chan, Hong Kong, China
Luonan Chen, Osaka, Japan
Yen-Wei Chen, Kyoto, Japan
Zheru Chi, Hong Kong, China
Andrzej Cichocki, Tokyo, Japan
Chuanyin Dang, Hong Kong, China
Mauro Forti, Siena, Italy
Chengan Guo, Dalian, China
Zengguang Hou, Beijing, China
Dewen Hu, Changsha, China
Danchi Jiang, Hobart, Australia
Nikola Kasabov, Auckland, New Zealand
Irwin King, Hong Kong, China
Xiaoli Li, Birmingham, UK
Yuanqing Li, Singapore
Lizhi Liao, Hong Kong, China
Ju Liu, Jinan, China
Baoliang Lu, Shanghai, China
Fa-Long Luo, San Jose, USA
Qing Ma, Kyoto, Japan
Tetsuo Nishi, Fukuoka, Japan
Paul S. Pang, Auckland, New Zealand
Yi Shen, Wuhan, China
Peter Sincak, Kosice, Slovakia
Changyin Sun, Nanjing, China
Ron Sun, Troy, USA
Ah Hwee Tan, Singapore
Dan Wang, Singapore
Wanliang Wang, Hangzhou, China
Michel Verleysen, Louvain, Belgium
Mao Ye, Chengdu, China
Zhigang Zeng, Hefei, China
Liming Zhang, Shanghai, China
Chunguang Zhou, Changchun, China

Special Sessions Chair

Derong Liu, Chicago, USA

Organizing Chairs

Guoyin Wang, Chongqing, China   *Simon X. Yang*, Guelph, Canada

Finance Chairs

Guangyuan Liu, Chongqing, China   *Qingyu Xiong*, Chongqing, China
Yu Wu, Chongqing, China

Publication Co-chairs

Yi Chai, Chongqing, China   *Hujun Yin*, Manchester, UK
Jianwei Zhang, Hamburg, Germany

Publicity Co-chairs

Min Han, Dalian, China   *Fengchun Tian*, Chongqing, China

Registration Chairs

Yi Chai, Chongqing, China   *Shaojiang Deng*, Chongqing, China

Local Arrangements Chairs

Wei Zhang, Chongqing, China   *Jianqiao Yu*, Chongqing, China

Secretariat and Webmaster

Tao Xiang, Chongqing, China

Table of Contents, Part I

1 Theoretical Analysis

2   Model Design

3   Learning Methods

4   Optimization Methods

5   Kernel Methods

Table of Contents, Part II

7 Pattern Analysis

8   System Modeling

9   Signal Processing

10   Image Processing

11   Financial Analysis

11 Financial Analysis

Table of Contents, Part III

12 Control Systems

15   Incidence Detection

16 Fault Diagnosis

17 Power Systems

18   Biomedical Applications

Information-geometric concepts play a fundamental role in this decomposition. We finally show that higher-order correlations are necessary for neural synchronization.

3   Bayesian Posterior and Predictive Distributions

Let us assume the ordinary statistical framework of parametric family $p(x|\theta)$ of probability distributions. Let the prior distribution be given by $\pi(\theta)$. Given independent data $D = \{x_1, \cdots, x_n\}$, its distribution is written as $p(D|\theta) = \prod p(x_i|\theta)$.

Then, the Bayes posterior distribution of θ given D is written as

$$p(\theta|D) = \frac{\pi(\theta)p(D|\theta)}{p(D)},\tag{2}$$

where

$$p(D) = \int \pi(\theta)p(D|\theta)d\theta.\tag{3}$$

In the case of population coding, θ represents a stimulus s and D is the neural activity r. It is believed that information of a stimulus s is kept in the form of Bayes posterior distribution $p(s|r)$ in the brain.

The Bayes predictive distribution given data D is written as

$$p(x|D) = \int p(x|\theta)p(\theta|D)d\theta.\tag{4}$$

This is also a useful concept. This is the mixture of probability distributions $p(x|\theta)$ whose weights are given by the Bayes posterior distribution $p(\theta|D)$.

4   Integration of Two Probability Distributions

Let us consider that there are two probability distributions $p_1(\theta)$ and $p_2(\theta)$, concerning the stimulus θ, which are given, for example, from sensors of different modalities. How to combine them? A simple idea is the arithmetic mean (or weighted arithmetic mean),

$$p(\theta) = \frac{1}{2}\{p_1(\theta) + p_2(\theta)\}, \quad p(\theta) = cp_1(\theta) + (1-c)p_2(\theta).\tag{5}$$

Another idea is the geometric mean (weighted geometric mean),

$$p(\theta) = c\sqrt{p_1(\theta)p_2(\theta)}, \quad p(\theta) = c\sqrt{\{p_1(\theta)\}^a \{p_2(\theta)\}^{1-a}}\tag{6}$$

We can generalize this in the following way. Let us introduce the α-function

$$f_\alpha(u) = \frac{2}{1-\alpha}u^{\frac{1-\alpha}{2}}\tag{7}$$

and denote the α-representation of probability density $p(\theta)$,

$$l_\alpha(\theta) = f_\alpha\{p(\theta)\}. \tag{8}$$

For $\alpha = -1$, we have

$$l_{-1}(\theta) = p(\theta) \tag{9}$$

When $\alpha = 1$, by taking the limit, we have

$$l_1(\theta) = \lim_{\alpha \to 1} f_\alpha\{p(\theta)\} = \log p(\theta) \tag{10}$$

The α-mean of the probability distributions is given by

$$\tilde{p}_\alpha(\theta) = f_\alpha^{-1}\left[\frac{1}{2}\{f_\alpha(p_1(\theta)) + f_\alpha(p_2(\theta))\}\right]. \tag{11}$$

When $\alpha = 1$, this gives the geometric mean. When $\alpha = -1$, it is the arithmetic mean. When $\alpha = 0$, it is the mean of the arithmetic and geometric means, and is called the Hellinger mean. When $\alpha = -\infty$, we have

$$\tilde{p}_\infty(\theta) = \max\{p_1(\theta), p_2(\theta)\} \tag{12}$$

while for $\alpha = \infty$,

$$\tilde{p}_{-\infty}(\theta) = \min\{p_1(\theta), p_2(\theta)\} \tag{13}$$

These two show the optimistic mean and pessimistic mean, respectively.

Each mean has its own nice property. We show some optimality results in the next.

5 α-Divergence of Probability Distributions

The KL divergence of two probability distributions $p(\theta)$ and $q(\theta)$ is written as

$$KL[p : q] = \int p(\theta) \log \frac{p(\theta)}{q(\theta)} d\theta \tag{14}$$

This is not symmetric with respect to p and q, so that we may consider the reversed KL divergence $KL[q : p]$. Another well known divergence is the Hellinger distance,

$$H[p : q] = 2\int \left(\sqrt{p(\theta)} - \sqrt{q(\theta)}\right)^2 d\theta. \tag{15}$$

This is the square of the true distance satisfying the axiom of distance.

Information geometry gives the α-divergence defined by

$$D_\alpha[p : q] = \frac{4}{1-\alpha^2}\left[1 - \int p(\theta)^{\frac{1-\alpha}{2}} q(\theta)^{\frac{1+\alpha}{2}} d\theta\right] \tag{16}$$

most one of them is gaussian. The problem is further formalized by Comon [7] under the name ICA. Although ICA has been studied from different perspectives, such as the minimum mutual information (MMI) [1, 4] and maximum likelihood (ML) [5], in the case that W is invertible, all such approaches are equivalent to minimizing the following cost function

$$D(W) = \int p(y; W) \ln \frac{p(y, W)}{\prod_{i=1}^{n} q(y_i)} dy, \tag{1}$$

where $q(y_i)$ is the pre-determined model probability density function (pdf), and $p(y, W)$ is the distribution on $y = Wx$. With each model pdf $q(y_i)$ prefixed, however, this approach works only for the cases that the components of y are either all sub-Gaussians [1] or all super-Gaussians [4].

To solve this problem, it is suggested that each model pdf $q(y_i)$ is a flexibly adjustable density that is learned together with W, with the help of either a mixture of sigmoid functions that learns the cumulative distribution function (cdf) of each source [24, 26] or a mixture of parametric pdfs [23, 25], and a so-called learned parametric mixture based ICA (LPMICA) algorithm is derived, with successful results on sources that can be either sub-Gaussian or super-Gaussian, as well as any combination of both types. The mixture model was also adopted in a so called context-sensitive ICA algorithm [17], although it did not explicitly target at separating the mixed sub- and super-Gaussian sources.

On the other hand, it has also been found that a rough estimate of each source pdf or cdf may be enough for source separation. For instance, a simple sigmoid function such as $tanh(x)$ seems to work well on the super-Gaussian sources [4], and a mixture of only two or three Gaussians may be enough already [23] for the mixed sub- and super-Gaussian sources. This leads to the so-called one-bit-matching conjecture [22], which states that "all the sources can be separated as long as there is an one-to-one same sign- correspondence between the kurtosis signs of all source pdf's and the kurtosis signs of all model pdf's." In past years, this conjecture has also been implicitly supported by several other ICA studies [10, 11, 14, 19]. In [6], a mathematical analysis was given for the case involving only two sub-Gaussian sources. In [2], stability of an ICA algorithm at the correct separation points was also studied via its relation to the nonlinearity $\phi(y_i) = d \ln q_i(y_i)/dy_i$, but without touching the circumstance under which the sources can be separated.

Recently, the conjecture on multiple sources has been proved mathematically in a weak sense [15]. When only sources' skewness and kurtosis are considered with $Es = 0$ and $Ess^T = I$, and the model pdf's skewness is designed as zero, the problem $\min_W D(W)$ by eq.(1) is simplified via pre-whitening into the following problem

$$\max_{RR^T=I} J(R), \ J(R) = \sum_{i=1}^{n} \sum_{j=1}^{n} r_{ij}^4 \nu_j^s k_i^m, \ n \geq 2, \tag{2}$$

where $R = (r_{ij})_{n \times n} = WA$ is an orthonormal matrix, and ν_j^s is the kurtosis of the source s_j, and k_i^m is a constant with the same sign as the kurtosis ν_i^m of the model $q(y_i)$. Then, it is further proved that the global maximization of eq. (2) can only be reachable by setting R a permutation matrix up to certain sign indeterminacy. That is, the one-bit-matching conjecture is true when the global minimum of $D(W)$ in eq.(1) with respect to W is reached. However, this proof still can not support the successes of many existing iterative ICA algorithms that typically implement gradient based local search and thus usually converge to one of local optimal solutions.

In the next section of this paper, all the local maxima of eq.(2) are investigated via a special convex-concave programming on a polyhedral set, from which we prove the one-bit-matching conjecture in a strong sense that it is true when anyone of local maxima by eq.(2) or equivalently local minima by eq.(1) is reached in help of investigating convex-concave programming on on a polyhedral-set. Theorems have also been provided on separation of a part of sources when there is a partial matching between the kurtosis signs, and on an interesting duality of maximization and minimization. Moreover, corollaries are obtained from theorems to state that the duality makes it possible to get super-gaussian sources via maximization and sub-gaussian sources via minimization. Another corollary is also to confirm the symmetric orthogonalization implementation of the kurtosis extreme approach for separating multiple sources in parallel, which works empirically but in a lack of mathematical proof [13].

In section 3, we further discuss that eq. (2) with R being a permutation matrix up to certain sign indeterminacy becomes equivalent to a special example of the following combinatorial optimization:

$$\min_{V} E_o(V), \; V = \{v_{ij}, i = 1, \cdots, N, j = 1, \cdots, M\}, \quad subject \; to$$

$$C^c : \; \sum_{i=1}^{N} v_{ij} = 1, j = 1, \cdots, M, \quad C^r : \; \sum_{j=1}^{M} v_{ij} = 1, i = 1, \cdots, N;$$

$$C^b : \; v_{ij} \; takes \; either \; 0 \; or \; 1. \tag{3}$$

This connection suggests to investigate combinatorial optimization from a perspective of gradient flow searching within the Stiefel manifold , with algorithms that guarantee convergence and constraint satisfaction.

2   One-Bit-Matching Theorem and Extension

2.1   An Introduction on Convex Programming

To facilitate mathematical analysis, we briefly introduce some knowledge about convex programming. A set in R^n is said to be *convex*, if $x_1 \in S, x_2 \in S$, we have $\lambda x_1 + (1 - \lambda)x_2 \in S$ for any $0 \leq \lambda \leq 1$. Shown in Fig.1 are examples of convex sets. As an important special case of convex sets, a set in R^n is called a *polyhedral* set if it is the intersection of a finite number of closed half-spaces, that is, $S = \{x : a_i^t x \leq \alpha_i, \; for \; i = 1, \cdots, m\}$, where a_i is a nonzero vector and α_i is a scalar for $i =, \cdots, m$. The second and third ones in Fig.1 are two examples. Let S be a nonempty convex set, a vector $x \in S$ is called an extreme

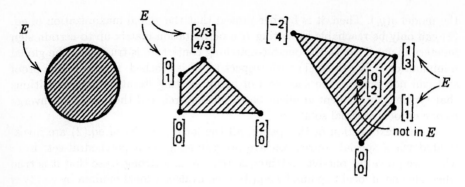

Fig. 1. Convex set and polyhedral set.

point of S if $x = \lambda x_1 + (1 - \lambda)x_2$ with $x_1 \in S, x_2 \in S$, and $0 < \lambda < 1$ implies that $x = x_1 = x_2$. We denote the set of extreme point by E and illustrate them in Fig.1 by dark points or dark lines as indicated.

Let $f : S \to R$, where S is a nonempty convex set in R^n. As shown in Fig.1, the function f is said to be convex on S if

$$f(\lambda x_1 + (1 - \lambda)x_2) \leq \lambda f(x_1) + (1 - \lambda)f(x_2) \tag{4}$$

for $x_1 \in S, x_2 \in S$ and for $0 < \lambda < 1$. The function f is called *strictly convex* on S if the above inequality is true as a strict inequality for each distinct $x_1 \in S, x_2 \in S$ and for $0 < \lambda < 1$. The function f is called concave (strictly concave) on S if $-f$ is convex (strict convex) on S.

Considering an optimization problem $\min_{x \in S} f(x)$, if $\bar{x} \in S$ and $f(x) \geq f(\bar{x})$ for each $x \in S$, then $\bar{x}$ is called a global optimal solution. If $\bar{x} \in S$ and if there exists an ε-neighborhood $N_\varepsilon(\bar{x})$ around $\bar{x}$ such that $f(x) \geq f(\bar{x})$ for each $x \in S \cap N_\varepsilon(\bar{x})$, then $\bar{x}$ is called a local optimal solution. Similarly, if $\bar{x} \in S$ and if $f(x) > f(\bar{x})$ for all $x \in S \cap N_\varepsilon(\bar{x}), x \neq \bar{x}$, for some ε, then $\bar{x}$ is called a strict local optimal solution. Particularly, an optimization problem $\min_{x \in S} f(x)$ is called a *convex programming problem* if f is a convex function and S is a convex set.

Lemma 1

(a) Let S be a nonempty open convex set in R^n, and let $f : S \to R$ be twice differentiable on S. If its Hessian matrix is positive definite at each point in S, the f is strictly convex.

(b) Let S be a nonempty convex set in R^n, and let $f : S \to R$ be convex on S. Consider the problem of $\min_{x \in S} f(x)$. Suppose that $\bar{x}$ is a local optimal solution to the problem. Then (i) $\bar{x}$ is a global optimal solution. (ii) If either $\bar{x}$ is a strict local minimum or if f is strictly convex, then $\bar{x}$ is the unique global optimal solution.

(c) Let S be a nonempty compact polyhedral set in R^n, and let $f : S \to R$ be a strict convex function on S. Consider the problem of $\max_{x \in S} f(x)$. All the local maxima are reached at extreme points of S.

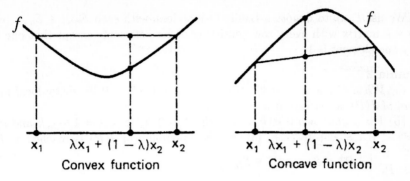

Fig. 2. Convex and concave function.

The above (a)(b) are basically known from a foundation course on mathematics during a undergraduate study. Though the statement (c) may not be included, it is not difficult to understand. Assume $\bar{x}$ is a local maximum but not an extreme point, we may find $x_1 \in N_\varepsilon(\bar{x}), x_2 \in N_\varepsilon(\bar{x})$ such that $\bar{x} = \lambda x_1 + (1 - \lambda)x_2$ for $0 < \lambda < 1$. It follows from eq.(4) that $f(\bar{x}) < \lambda f(x_1) + (1 - \lambda)f(x_2) \leq \max[f(x_1), f(x_2)]$, which contradicts to that $\bar{x}$ is a local maximum, while at an extreme point x of S, $x = \lambda x_1 + (1 - \lambda)x_2$ with $x_1 \in S, x_2 \in S$ and $0 < \lambda < 1$ implies that $x = x_1 = x_2$, which does not contradict the definition of a strict convex function made after eq.(4). That is, a local maximum can only be reached at one of the extreme points of S.

Details of the above knowledge about convex programming are referred to one of textbooks on nonlinear programming, e.g., [3].

2.2 One-Bit-Matching Theorem

For the problem by eq. (2), neither the set $RR^T = I$ is convex nor $J(R)$ is always convex. To use the knowledge given in the previous section as a tool, we let $p_{ij} = r_{ij}^2$ and considering $RR^T = I$ via keeping the part of normalization conditions but ignoring the part of orthogonal conditions, then we can relax the problem by eq. (2) as follows:

$$\max_{P \in S} J(P), \ J(P) = \sum_{i=1}^{n}\sum_{j=1}^{n} p_{ij}^2 \nu_j^s k_i^m, \quad P = (p_{ij})_{n \times n}, \ n \geq 2$$

$$S = \{p_{ij}, i, j = 1, \cdots, n : \sum_{j=1}^{n} p_{ij} = 1, \ for \ i = 1, \cdots, n, \ and \ p_{ij} \geq 0\}, \quad (5)$$

where ν_j^s and k_i^m are same as in eq. (2), and S become a convex set or precisely a polyhedral set. Moreover, we stack P into a vector $vec[P]$ of n^2 elements and compute the Hessian H_P with respect to $vec[P]$, resulting in that

$$H_P \text{ is a } n^2 \times n^2 \text{ diagonal matrix with each diagonal element being } \nu_j^s k_i^m. \ (6)$$

Thus, whether $J(P)$ is convex can be checked simply via all the signs of $\nu_j^s k_i^m$.

We use $\mathcal{E}_{n \times k}$ to denote a family of matrices, with each $E_{n \times k} \in \mathcal{E}_{n \times k}$ being a $n \times k$ matrix with every row consisting of zero elements except that one and only one element is 1.

Lemma 2

(a) When either $\nu_i^s > 0, k_i^m > 0, \forall i$ or $\nu_i^s < 0, k_i^m < 0, \forall i$, every local maximum of $J(P)$ is reached at a $P \in \mathcal{E}_{n \times n}$.

(b) For a unknown $0 < k < n$ with $\nu_i^s > 0, k_i^m > 0, i = 1, \cdots, k$ and $\nu_i^s < 0, k_i^m < 0, i = k+1, \cdots, n$, every local maximum of $J(P)$ is reached at $P = \begin{bmatrix} P_1^+ & 0 \\ 0 & P_2^- \end{bmatrix}$, $P_1^+ \in \mathcal{E}_{k \times k}$, $P_2^- \in \mathcal{E}_{(n-k) \times (n-k)}$.

Proof. (a) In this case, we have every $\nu_j^s k_i^m > 0$ and thus it follows from eq. (6) and Lemma 1(a) that $J(P)$ is strictly convex on the polyhedral set S. It further follows from Lemma 1 (c) that all the local maxima of $J(P)$ are reached at the polyhedral set's extreme points that satisfy $\sum_{j=1}^n p_{ij} = 1$, $for\ i = 1, \cdots, n$, i.e., each local maximum $P \in \mathcal{E}_{n \times n}$.

(b) Notice that the constraint $\sum_{j=1}^n p_{ij} = 1$ effects only on the i-th row, and $J(P)$ is additive, we see that the task by eq. (5) is solved by separately considering the following two tasks:

$$T_1 : \max_{P_1} J(P_1), \ J(P_1) = \sum_{i=1}^k \sum_{j=1}^n p_{ij}^2 \nu_j^s k_i^m,$$

$$P_1 = (p_{ij})_{i=1,\cdots,k,j=1,\cdots,n} \ with\ every\ p_{ij} \geq 0,$$

$$Subject\ to\ \sum_{j=1}^n p_{ij} = 1, \ for\ i = 1, \cdots, k. \tag{7}$$

$$T_2 : \max_{P_2} J(P_2), \ J(P_2) = \sum_{i=k+1}^N \sum_{j=1}^n p_{ij}^2 \nu_j^s k_i^m,$$

$$P_2 = (p_{ij})_{i=k+1,\cdots,n,j=1,\cdots,n} \ with\ every\ p_{ij} \geq 0,$$

$$Subject\ to\ \sum_{j=1}^n p_{ij} = 1, \ for\ i = k+1, \cdots, N. \tag{8}$$

First, we consider T_1. Further let $J(P_1) = J_+^+(P_1^+) + J_+^-(P_1^-)$ with

$$J_+^+(P_1^+) = \sum_{i=1}^k \sum_{j=1}^k p_{ij}^2 \nu_j^s k_i^m, \quad P_1^+ = (p_{ij})_{i=1,\cdots,k,j=1,\cdots,k},$$

$$J_+^-(P_1^-) = \sum_{i=1}^k \sum_{j=k+1}^n p_{ij}^2 \nu_j^s k_i^m, \quad P_1^- = (p_{ij})_{i=1,\cdots,k,j=k+1,\cdots,n}, \tag{9}$$

we see that J_+^+ and J_+^- are decoupled if ignoring the constraints $\sum_{j=1}^n p_{ij} = 1$, $for\ i = 1, \cdots, k$. So, the key point is considering the roles of the constraints.

Without the constraints $\sum_{j=1}^{n} p_{ij} = 1$, for $i = 1, \cdots, k$, $J_+^-(P_1^-) \leq 0$ is strictly concave from Lemma 1(a) by observing $\nu_j^s k_i^m < 0$ for every term, and thus has only one maximum at $P_1^- = \mathbf{0}$. Then, the constraints can be re-taken in consideration via written as $\sum_{j=k+1}^{n} p_{ij} = c_i$, for $i = 1, \cdots, k$ with a unknown $c_i = 1 - \sum_{j=1}^{k} p_{ij}$. For $c_i > 0$, the boundary $\sum_{j=k+1}^{n} p_{ij} = c_i$ is inactive and will not affect that $J_+^- \leq 0$ reaches its only maximum at $P_1^- = \mathbf{0}$. For $c_i = 0$, $J_+^- \leq 0$ reaches its only maximum at the boundary $\sum_{j=k+1}^{n} p_{ij} = 0$ which is still $P_1^- = \mathbf{0}$. Thus, all the local maxima of $J(P_1)$ are reached at $P_1 = [P_1^+, P_1^-] = [P_1^+, \mathbf{0}]$ and thus determined by all the local maxima of $J_+^+(P_1^+)$ on the polyhedral set of $p_{ij} \geq 0, i = 1, \cdots, k, j = 1, \cdots, k$ and $\sum_{j=1}^{k} p_{ij} = 1$, for $i = 1, \cdots, k$ (because $p_{ij} = 0$, for $i = 1, \cdots, k, j = k+1, \cdots, n$). It follows from Lemma 1(b) that $J_+^+(P_1^+)$ is strictly convex on this polyhedral set since $\nu_j^s k_i^m > 0$ for every term. Similar to the above (a), each of the local maxima of $J_+^+(P_1^+)$ is $P_1^+ \in \mathcal{E}_{k \times k}$.

Second, we can consider T_2 in a same way and have $J(P_2) = J_-^+(P_2^+) + J_-^-(P_2^-)$ with

$$J_-^+(P_2^+) = \sum_{i=k+1}^{n} \sum_{j=1}^{k} p_{ij}^2 \nu_j^s k_i^m, \quad P_2^+ = (p_{ij})_{i=k+1,\cdots,n,j=1,\cdots,k},$$

$$J_-^-(P_2^-) = \sum_{i=k+1}^{n} \sum_{j=k+1}^{n} p_{ij}^2 \nu_j^s k_i^m, \quad P_2^- = (p_{ij})_{i=k+1,\cdots,n,j=k+1,\cdots,n}. \quad (10)$$

Now, J_-^+ is strictly concave and J_-^- is strictly convex. As a result, all the local maxima of $J(P_2)$ are reached at $P_2 = [P_2^+, P_2^-] = [\mathbf{0}, P_2^-]$ with $P_2^- \in \mathcal{E}_{(n-k) \times (n-k)}$.   **Q.E.D.**

Further considering $p_{ij} = r_{ij}^2$ and the part of orthogonal conditions in $RR^T = I$, we get

Theorem 1. *Every local maximum of $J(R)$ on $RR^T = I$ by eq. (2) is reached at R that is an permutation matrix up to sign indeterminacy at its nonzero elements, as long as there is a one-to-one same-sign-correspondence between the kurtosis of all source pdf's and the kurtosis of all model pdf's.*

Proof. From $p_{ij} = r_{ij}^2$ and Lemma 2, we have $r_{ij} = 0$ for $p_{ij} = 0$ and either $r_{ij} = 1$ or $r_{ij} = -1$ for $p_{ij} = 1$. All the other choices of P in Lemma 2(a) or of P_1^+ and P_2^- in Lemma 2(b) can not satisfy the part of orthogonal conditions in $RR^T = I$ and thus should be discarded, except that P is a $n \times n$ permutation matrix for Lemma 2(a) or P_1^+ is a $k \times k$ permutation matrix and P_2^- is a $(n-k) \times (n-k)$ permutation matrix for Lemma 2(b). That is, R should be an permutation matrix up to sign indeterminacy at its nonzero elements. On the other hand, any other R on $RR^T = I$ with the corresponding P being not a local maximum of $J(P)$ is also not a local maxima of $J(R)$ on $RR^T = I$. Thus, we get the theorem proved by noticing that k_i^m has the same sign as the kurtosis ν_i^m of the model density $q_i(y_i)$.   **Q.E.D.**

The above theorem is obtained from eq. (2) that is obtained from eq. (1) by approximately only considering the skewness and kurtosis and with the model pdfs without skewness. Thus, in such an approximative sense, all the sources can also be separated by a local searching ICA algorithm (e.g., a gradient based algorithm) obtained from eq.(1) as long as there is a one-to-one same-sign-correspondence between the kurtosis of all source pdf's and the kurtosis of all model pdf's.

Though how seriously such an approximation will affect the separation performance by an ICA algorithm obtained from eq.(1) is unclear yet, this approximation can be removed by an ICA algorithm obtained directly from eq. (2). Under the one-to-one kurtosis sign matching assumption, we can derive a local search algorithm that is equivalent to maximize the problem by eq.(2) directly. A pre-whitening is made on observed samples such that we can consider the samples of x with $Ex = 0, Exx^T = I$. As a results, it follows from $I = Exx^T = AEss^T A^T$ and $Ess^T = I$ that $AA^T = I$, i.e., A is orthonormal. Thus, an orthonormal W is considered to let $y = Wx$ become independent among its components via

$$\max_{WW^T=I} J(W), \quad J(W) = \sum_{i=1}^{n} k_i^m \nu_i^y, \tag{11}$$

where $\nu_i^y = Ey_i^4 - 3, i = 1, \cdots, n$, $\nu_j^x = Ex_j^4 - 3, j = 1, \cdots, n$, and $k_i^m, i = 1, \cdots, n$ are pre-specified constants with the same signs as the kurtosis ν_i^m. We can derive its gradient $\nabla_W J(W)$ and then project it onto $WW^T = I$, which results in an iterative updating algorithm for updating W in a way similar to eq.(19) and eq.(20) at the end of the next section. Such an ICA algorithm actually maximizes the problem by eq.(2) directly by considering $y = Wx = WAs = Rs, R = WA, RR^T = I$, and thus

$$\nu_i^y = \sum_{j=1}^{n} r_{ij}^4 \nu_j^s, i = 1, \cdots, n. \tag{12}$$

That is, the problem by eq.(11) is equivalent to the problem by eq.(2). In other words, under the one-to-one kurtosis sign matching assumption, it follows from Theorem 1 that all the sources can be separated by an ICA algorithm not in an approximate sense, as long as eq.(12) holds.

However, Theorem 1 does not tell us how such a kurtosis sign matching is built, which is attempted via eq.(1) through learning each model pdf $q_i(y_i)$ together with learning W [23, 24, 26] as well as further advances either given in [14, 19] or given by eqn. (103) in [20]. Still, it remains an open problem whether these efforts or the possibility of developing other new techniques can guarantee such an one-to-one kurtosis sign matching surely or in certain probabilistic sense, which deserves future investigations.

2.3   Cases of No Matching and Partial Matching

Next, we consider what happens when one-to-one kurtosis-sign-correspondence does not hold. We start at the extreme situation via the following Lemma.

Lemma 3. (no matching case)

When either $\nu_i^s > 0, k_i^m < 0, \forall i$ or $\nu_i^s < 0, k_i^m > 0, \forall i$, $J(P)$ has only one maximum that is reached usually not in $\mathcal{E}_{n \times n}$.

Proof. From eq.(6) and Lemma 1(a) that $J(P)$ is strictly concave since $\nu_j^s k_i^m < 0$ for every term. Thus, it follows from Lemma 1(b) that it has only one maximum usually at an interior point in S (thus not in $\mathcal{E}_{n \times n}$) instead of at the extreme points of S.   **Q.E.D.**

Lemma 4. (partial matching case)

Given two unknown integers k, m with $0 < k < m < n$, and provided that $\nu_i^s > 0, k_i^m > 0, i = 1, \cdots, k$, $\nu_i^s k_i^m < 0, i = k+1, \cdots, m$, and $\nu_i^s < 0, k_i^m < 0, i = m+1, \cdots, n$, every local maximum of $J(P)$ is reached either at $P = \begin{bmatrix} P_1^+ & 0 \\ 0 & P_2^- \end{bmatrix}$, where either $P_1^+ \in \mathcal{E}_{k \times m}, P_2^- \in \mathcal{E}_{(n-k) \times (n-m)}$ when $\nu_i^s > 0, k_i^m < 0, i = k+1, \cdots, m$ or $P_1^+ \in \mathcal{E}_{m \times k}, P_2^- \in \mathcal{E}_{(n-m) \times (n-k)}$ when $\nu_i^s < 0, k_i^m > 0, i = k+1, \cdots, m$.

Proof. The proof is made similar to proving Lemma 2. The difference is that both P_1^+ and P_2^- are not square matrices.   **Q.E.D.**

Theorem 2. Given two unknown integers k, m with $0 < k < m < n$, and provided that $\nu_i^s > 0, k_i^m > 0, i = 1, \cdots, k$, $\nu_i^s k_i^m < 0, i = k+1, \cdots, m$, and $\nu_i^s < 0, k_i^m < 0, i = m+1, \cdots, n$, every local maximum of $J(R)$ on $RR^T = I$ by eq.(2) is reached at $R = \begin{bmatrix} \Pi & 0 \\ 0 & \bar{R} \end{bmatrix}$ subject to a 2×2 permutation, where Π is a $(k+n-m) \times (k+n-m)$ permutation matrix up to sign indeterminacy at its nonzero elements; while $\bar{R}$ is a $(m-k) \times (m-k)$ orthonormal matrix with $\bar{R}\bar{R}^T = I$, but usually not a permutation matrix up to sign indeterminacy.

Proof. By Lemma 2, putting $p_{ij} = r_{ij}^2$ in P we can directly select a $(k+n-m) \times (k+n-m)$ sub-matrix Π that is of full rank in both row and column, also automatically with $\Pi \Pi^T = I$ satisfied. The remaining part in P must be linear dependent of Π with $RR^T = I$ still satisfied. Thus, the entire R should be the above form with $\bar{R}\bar{R}^T = I$. As a result, $\max_{RR^T=I} J(R)$ in eq. (2) is decoupled with $\bar{R}$ maximized via $\max_{\bar{R}\bar{R}^T=I} J(\bar{R})$, $J(\bar{R}) = \sum_{i=k+n-m+1}^{n} \sum_{j=k+n-m+1}^{n} \bar{r}_{ij}^4 \nu_j^s k_i^m$ with every $\nu_j^s k_i^m < 0$, which is a situation similar to Lemma 3. That is, $\bar{R}$ is usually not a permutation matrix up to sign indeterminacy. On the other hand, if the second row of R is not $[\mathbf{0}, \bar{R}]$ but in a form $[A, B]$ with both A, B being nonzero and $[A, B][A, B]^T = I$, the first row of R will non longer be $[\Pi, \mathbf{0}]$ and the resulting P deviates from a local maximum of $J(P)$. Thus, the corresponding R is not a local maxima of $J(R)$ on $RR^T = I$.   **Q.E.D.**

In other words, there will be $k + n - m$ sources that can be successfully separated in help of a local searching ICA algorithm when there are $k + n - m$ pairs of matching between the kurtosis signs of source pdf's and of model pdf's. However, the remaining $m - k$ sources are not separable. Suppose that the kurtosis sign of each model is described by a binary random variable ξ_i with 1 for + and 0 for −, i.e., $p(\xi_i) = 0.5^{\xi_i} 0.5^{1-\xi_i}$. When there are k sources with their kurtosis signs in positive, there is still a probability $p(\sum_{i=1}^{n} \xi_i = k)$ to have an one-to-one kurtosis-sign-correspondence even when model pdf's are prefixed without knowing the kurtosis signs of sources. Moreover, even when an one-to-one kurtosis-sign-correspondence does not hold for all the sources, there will still be $n - |\ell - k|$ sources recoverable with a probability $p(\sum_{i=1}^{n} \xi_i = \ell)$. This explains not only why those early ICA studies [1, 4], work in some case while fail in other cases due to the pre-determined model pdf's, but also why some existing heuristic ICA algorithms can work in this or that way.

2.4 Maximum Kurtosis vs Minimum Kurtosis

Interestingly, it can be observed that changing the maximization in eq. (2), eq. (5) and eq. (11) into the minimization will lead to similar results, which are summarized into the following Lemma 5 and Theorem 3.

Lemma 5

(a) When either $\nu_i^s > 0, k_i^m > 0, \forall i$ or $\nu_i^s < 0, k_i^m < 0, \forall i$, $J(P)$ has only one minimum that is reached usually not in $\mathcal{E}_{n \times n}$.

(b) When either $\nu_i^s > 0, k_i^m < 0, \forall i$ or $\nu_i^s > 0, k_i^m < 0, \forall i$, every local minimum of $J(P)$ is reached at a $P \in \mathcal{E}_{n \times n}$.

(c) For a unknown $0 < k < n$ with $\nu_i^s > 0, k_i^m > 0, i = 1, \cdots, k$ and $\nu_i^s < 0, k_i^m < 0, i = k+1, \cdots, n$, every local minimum of $J(P)$ is reached at $P = \begin{bmatrix} 0 & P_1^- \\ P_2^+ & 0 \end{bmatrix}$, $P_1^- \in \mathcal{E}_{k \times (n-k)}$, $P_2^+ \in \mathcal{E}_{(n-k) \times k}$.

(d) For two unknown integers k, m with $0 < k < m < n$ with $\nu_i^s > 0, k_i^m > 0, i = 1, \cdots, k$, $\nu_i^s k_i^m < 0, i = k+1, \cdots, m$, and $\nu_i^s < 0, k_i^m < 0, i = m+1, \cdots, n$, every local minimum of $J(P)$ is reached either at $P = \begin{bmatrix} 0 & P_1^- \\ P_2^+ & 0 \end{bmatrix}$, where either $P_1^- \in \mathcal{E}_{k \times (n-m)}, P_2^+ \in \mathcal{E}_{(n-k) \times m}$ when $\nu_i^s > 0, k_i^m < 0, i = k+1, \cdots, m$ or $P_1^+ \in \mathcal{E}_{m \times (n-k)}, P_2^- \in \mathcal{E}_{(n-m) \times k}$ when $\nu_i^s < 0, k_i^m > 0, i = k+1, \cdots, m$.

Proof. The proof can be made similar to those in proving Lemma 2, Lemma 3, and Lemma 4. The key difference is shifting our focus from the maximization of a convex function on a polyhedral set to the minimization of a concave function on a polyhedral set, with switches between 'minimum' and 'maximum', 'maxima' and 'minima', 'convex' and 'concave', and 'positive' and 'negative', respectively. The key point is that Lemma 1 still remains to be true after these switches. **Q.E.D.**

Similar to Theorem 2, from the above lemma we can get

Theorem 3

(a) When either $\nu_i^s k_i^m < 0, i = 1, \cdots, n$ or $\nu_i^s > 0, k_i^m > 0, i = 1, \cdots, k$ and $\nu_i^s < 0, k_i^m < 0, i = k+1, \cdots, n$ for a unknown $0 < k < n$, every local minimum of $J(R)$ on $RR^T = I$ by eq. (2) is reached at R that is an permutation matrix up to sign indeterminacy at its nonzero elements.

(b) For two unknown integers k, m with $0 < k < m < n$ with $\nu_i^s > 0, k_i^m > 0, i = 1, \cdots, k$, $\nu_i^s k_i^m < 0, i = k+1, \cdots, m$, and $\nu_i^s < 0, k_i^m < 0, i = m+1, \cdots, n$, every local minimum of $J(R)$ on $RR^T = I$ by eq. (2) is reached at $R = \begin{bmatrix} \Pi & 0 \\ 0 & \bar{R} \end{bmatrix}$ subject to a 2×2 permutation. When $m + k \geq n$, Π is a $(n - m + n - k) \times (n - m + n - k)$ permutation matrix up to sign indeterminacy at its nonzero elements, while $\bar{R}$ is a $(m + n - k) \times (m + n - k)$ orthonormal matrix with $\bar{R}\bar{R}^T = I$, but usually not a permutation matrix up to sign indeterminacy. When $m + k < n$, Π is a $(k + m) \times (k + m)$ permutation matrix up to sign indeterminacy at its nonzero elements, while $\bar{R}$ is a $(n - k - m) \times (n - k - m)$ orthonormal matrix with $\bar{R}\bar{R}^T = I$, but usually not a permutation matrix up to sign indeterminacy.

In a comparison of Theorem 2 and Theorem 3, when $m + k \geq n$, comparing $n - m + n - k$ with $k + n - m$, we see that more source can be separated by minimization than maximization if $k < 0.5n$ while maximization is better than minimization if $k > 0.5n$. When $m + k < n$, comparing $k + m$ with $k + n - m$, we see that more source can be separated by minimization than maximization if $m > 0.5n$ while maximization is better than minimization if $m < 0.5n$.

We further consider a special case that $k_i^m = 1, \forall i$. In this case, eq. (2) is simplified into

$$J(R) = \sum_{i=1}^{n} \sum_{j=1}^{n} r_{ij}^4 \nu_j^s, \ n \geq 2, \tag{13}$$

From Theorem 2 at $n = m$, we can easily obtain

Corollary 1. For a unknown integer $0 < k < n$ with $\nu_i^s > 0, i = 1, \cdots, k$ and $\nu_i^s < 0, i = k + 1, \cdots, n$, every local maximum of $J(R)$ on $RR^T = I$ by eq. (13) is reached at $R = \begin{bmatrix} \Pi & 0 \\ 0 & \bar{R} \end{bmatrix}$ subject to a 2×2 permutation, where Π is a $k \times k$ permutation matrix up to sign indeterminacy at its nonzero elements, while $\bar{R}$ is a $(n - k) \times (n - k)$ orthonormal matrix with $\bar{R}\bar{R}^T = I$, but usually not a permutation matrix up to sign indeterminacy.

Similarly, from Theorem 3 we also get

Corollary 2. For a unknown integer k with $0 < k < n$ with $\nu_i^s > 0, i = 1, \cdots, k$ and $\nu_i^s < 0, i = k + 1, \cdots, n$, every local minimum of $J(R)$ on $RR^T = I$ by eq.(2) is reached at $R = \begin{bmatrix} \bar{R} & 0 \\ 0 & \Pi \end{bmatrix}$ subject to a 2×2 permutation, where Π is a $(n - k) \times (n - k)$ permutation matrix up to sign indeterminacy at its nonzero elements, while $\bar{R}$ is a $k \times k$ orthonormal matrix with $\bar{R}\bar{R}^T = I$, but usually not a permutation matrix up to sign indeterminacy.

It follows from Corollary 1 that k super-gaussian sources can be separated by $\max_{RR^T=I} J(R)$, while it follows from Corollary 2 that $n - k$ sub-gaussian sources can be separated by $\min_{RR^T=I} J(R)$. In implementation, from eq. (11) we get

$$J(W) = \sum_{i=1}^{n} \nu_i^y, \tag{14}$$

and then make $\max_{WW^T=I} J(W)$ to get k super-gaussian source and make $\min_{WW^T=I} J(W)$ to get $n-k$ sub-gaussian source. Thus, instead of learning an one-to-one kurtosis sign matching, the problem can also be equivalently turned into a problem of selecting super-gaussian components from $y = Wx$ with W obtained via $\max_{WW^T=I} J(W)$ and of selecting sub-gaussian components from $y = Wx$ with W obtained via $\min_{WW^T=I} J(W)$. Though we know neither k nor which of components of y should be selected, we can pick those with positive signs as super-gaussian ones after $\max_{WW^T=I} J(W)$ and pick those with negative signs as sub-gaussian ones after $\min_{WW^T=I} J(W)$. The reason comes from $\nu_i^y = \sum_{j=1}^{n} r_{ij}^4 \nu_j^s$ and the above corollaries. By Corollary 1, the kurtosis of each super-gaussian component of y is simply one of $\nu_j^s > 0, j = 1, \cdots, k$. Though the kurtosis of each of the rest components in y is a weighted combination of $\nu_j^s < 0, j = k+1, \cdots, n$, the kurtosis signs of these rest components will all remain negative. Similarly, we can find out those sub-gaussian components according to Corollary 2.

Anther corollary can be obtained from eq.(11) by considering a special case that $k_i^m = sign[\nu_i^y], \forall i$. That is, eq.(11) becomes

$$\max_{WW^T=I} J(W), \ J(W) = \sum_{i=1}^{n} |\nu_i^y|. \tag{15}$$

Actually, this leads to what is called kurtosis extreme approach and extensions [8, 13, 16], where studies were started at extracting one source by a vector w and then extended to extracting multiple sources by either sequentially implementing the one vector algorithm such that the newly extracted vector is orthogonal to previous ones or in parallel implementing the one vector algorithm on all the vectors of W separately together with a symmetric orthogonalization made at each iterative step. In the literature, the success of using one vector vector w to extract one source has been proved mathematically and the proof can be carried easily to sequentially extracting a new source with its corresponding vector w being orthogonal to the subspace spanned by previous. However, this mathematical proof is not applicable to implementing the one vector algorithm in parallel on all the vectors of W separately together with a symmetric orthogonalization, as suggested in Sec.8.4.2 of [13] but with no proof. Actually, what was suggested there can only ensure a convergence of such a symmetric orthogonalization based algorithm but is not able to guarantee that this local searching featured iterative algorithm will surely converge to a solution that can separate all the sources, though experiments usually turned out with successes.

When $\nu_i^y = \sum_{j=1}^{n} r_{ij}^4 \nu_j^s$ holds, from eq.(15) we have $\min_{RR^T=I} J(R), \ J(R) = \sum_{i=1}^{n} \sum_{j=1}^{n} r_{ij}^4 |\nu_j^s|$, which is covered by Lemma 2(a) and Theorem 2. Thus, we can directly prove the following corollary:

Corollary 3. *As long as $\nu_i^y = \sum_{j=1}^{n} r_{ij}^4 \nu_j^s$ holds, every local minimum of the above $J(R)$ on $RR^T = I$ is reached at a permutation matrix up to sign indeterminacy.*

Actually, it provides a mathematical proof on the success of the above symmetric orthogonalization based algorithm on separating all the sources.

The last but not least, it should be noticed that the above corollaries are true only when the relation $\nu_i^y = \sum_{j=1}^{n} r_{ij}^4 \nu_j^s, i = 1, \cdots, n$ holds, which is true only when there is a large size of samples such that the pre-whitening can be made perfectly.

3   Combinatorial Optimization in Stiefel Manifold

The combinatorial optimization problem by eq.(3) has been encountered in various real applications and still remains a hard task to solve. Many efforts have also been made on in the literature of neural networks since Hopfield and Tank [12]. As summarized in [21], these efforts can be roughly classified according to the features on dealing with C_e^{col}, C_e^{row} and C_b. Though having a favorable feature of being parallel implementable, almost all the neural network motivated approaches share one unfavorable feature that these intuitive approaches have no theoretical guarantees on convergence to even a feasible solution. Being different from several existing algorithms in the literature, a general LAGRANGE-enforcing iterative procedure is proposed firstly in [27] and further developed in the past decade, and its convergence to even a feasible solution is guaranteed. Details are referred to [21].

Interestingly, focusing at local maxima only, both eq.(2) and eq.(5) can be regarded as special examples of the combinatorial optimization problem by eq.(3) simply via regarding p_{ij} or r_{ij} as v_{ij}. Though such a linkage is not useful for ICA since we need not to seek a global optimization for making ICA, linking from eq.(3) reversely to eq.(2) and even eq.(1) leads to one motivation. That is, simply let $v_{ij} = r_{ij}^2$ and then use $RR^T = I$ to guarantee the constraints C_e^{col}, C_e^{row} as well as a relaxed version of C_b (i.e., $0 \le v_{ij} \le 1$). That is, the problem eq.(3) is relaxed into

$$\min_{RR^T = I \ for \ N \le M} E_o(\{r_{ij}^2\}_{i=1,j=1}^{i=N,j=M}), \ R = \{r_{ij}\}_{i=1,j=1}^{i=N,j=M}. \qquad (16)$$

We consider the problems with

$$\frac{\partial^2 E_o(V)}{\partial vec[V]\partial vec[V]^T} = \mathbf{0}, \quad \frac{\partial^2 E_o(V)}{\partial vec[V]\partial vec[V]^T} \text{ is negative definite,} \qquad (17)$$

or $E_o(V)$ in a form similar to $J(P)$ in eq.(5), i.e.,

$$E_o(V) = -\sum_{i=1}^{n}\sum_{j=1}^{n} v_{ij}^2 a_j b_i, \qquad (18)$$

with $a_i > 0, b_i > 0, i = 1, \cdots, k$ and $a_i < 0, b_i < 0, i = 1, \cdots, k$ after an appropriate permutation on $[a_1, \cdots, a_n]$ and on $[b_1, \cdots, b_n]$. Similar to the study of eq.(5), maximizing $E_o(V)$ under the constraints C_e^{col}, C_e^{row} and $v_{ij} \ge$ will imply the satisfaction of C_b. In other words, the solutions of eq.(16) and of eq.(3)

are same. Thus, we can solve the hard problem of combinatorial optimization by eq.(3) via a gradient flow on the Stiefel manifold $RR^T = I$ to maximize the problem by eq.(16). At least a local optimal solution of eq.(3) can be reached, with all the constraints C_e^{col}, C_e^{row}, and C_b guaranteed automatically.

To get an appropriate updating flow on the Stiefel manifold $RR^T = I$, we first compute the gradient $\nabla_V E_o(V)$ and then get $G_R = \nabla_V E_o(V) \circ R$, where the notation $\circ$ means that

$$\begin{bmatrix} a_{11} & a_{12} \\ a_{21} & a_{22} \end{bmatrix} \circ \begin{bmatrix} b_{11} & b_{12} \\ b_{21} & b_{22} \end{bmatrix} = \begin{bmatrix} a_{11}b_{11} & a_{12}b_{12} \\ a_{21}b_{21} & a_{22}b_{22} \end{bmatrix}.$$

Given a small disturbance δ on $RR^T = I$, it follows from $RR^T = I$ that the solution of $\delta RR^T + R\delta R^T = 0$ must satisfy

$$\delta R = ZR + U(I - R^T R), \tag{19}$$

where U is any $m \times d$ matrix and $Z = -Z$ is an asymmetric matrix.

From $Tr[G_R^T \delta R] = Tr[G_R^T(ZR+UI-R^T R)] = Tr[((G_R R^T)^T Z] + Tr[(G_R(I-R^T R))^T U]$, we get

$$Z = G_R R^T - R G_R^T, \quad U = G_R(I - R^T R), \quad \delta R = \begin{cases} U(I - R^T R) = U, & \text{(a)}, \\ ZR, & \text{(b)}, \\ ZR + U, & \text{(c)}. \end{cases}$$

$$R^{new} = R^{old} + \gamma_t \delta R. \tag{20}$$

That is, we can use anyone of the above three choices of δR as the updating direction of R. A general technique for optimization on the Stiefel manifold was elaborately discussed in [9], which can also be adopted for implementing our problem by eq.(16).

3.1   Concluding Remarks

The one-to-one kurtosis sign matching conjecture has been proved in a strong sense that every local maximum of $\max_{RR^T=I} J(R)$ by eq.(2) is reached at a permutation matrix up to certain sign indeterminacy if there is an one-to-one same-sign-correspondence between the kurtosis signs of all source pdf's and the kurtosis signs of all model pdf's. That is, all the sources can be separated by a local search ICA algorithm. Theorems have also been proved not only on partial separation of sources when there is a partial matching between the kurtosis signs, but also on an interesting duality of maximization and minimization on source separation. Moreover, corollaries are obtained from the theorems to state that seeking a one-to-one same-sign-correspondence can be replaced by a use of the duality, i.e., super-gaussian sources can be separated via maximization and sub-gaussian sources can be separated via minimization. Furthermore, a corollary is also obtained to provide a mathematical proof on the success of symmetric orthogonalization implementation of the kurtosis extreme approach.

Due to the results, the open problem of the one-to-one kurtosis sign matching conjecture [22] can be regarded as closed. However, there still remain problems to be further studied. First, the success of those eq.(1) based efforts along this direction [14, 19, 20, 23, 24, 26] can be explained as their ability of building up an one-to-one kurtosis sign matching. However, we still need a mathematical analysis to prove that such a matching can be achieved surely or in certain probabilistic sense by these approaches. Second, as mentioned at the end of Sec. 2.4, a theoretical guarantee on either the kurtosis extreme approach or the approach of extracting super-gaussian sources via maximization and sub-gaussian sources via minimization is true only when there is a large size of samples such that the pre-whitening can be made perfectly. In practice, usually with only a finite size of samples, it remains to be further studied on comparison of the two approaches as well as of those eq.(1) based approaches. Also, comparison may deserve to made on convergence rates of different ICA algorithms.

The last but not the least, the linkage of the problem by eq. (3) to eq.(2) and eq.(5) leads us to a Stiefel manifold perspective of combinatorial optimization with algorithms that guarantee convergence and satisfaction of constraints, which also deserve further investigations.

References

1. Amari, S. I., Cichocki, A., Yang, H.: A New Learning Algorithm for Blind Separation of Sources. Advances in Neural Information Processing. Vol. 8. MIT Press, Cambridge, MA (1996) 757-763
2. Amari, S., Chen, T.-P., & Cichocki, A.: Stability analysis of adaptive blind source separation, Neural Networks, **10** (1997) 1345-1351
3. Bazaraa, M.S., Sherall, H.D., Shetty, C.M.: Nonlinear Programming: Theory and Algorithms, John Wileys & Sons, Inc., New York (1993)
4. Bell, A., Sejnowski, T.: An Information-maximization Approach to Blind Separation and Blind Deconvolution. Neural Computation, **7** (1995) 1129-1159
5. Cardoso, J.-F. Blind signal separation: Statistical Principles, Proc. of IEEE, **86** (1998) 2009-2025
6. Cheung, C. C., Xu, L.: Some Global and Local Convergence Analysis on the Information-theoretic Independent Component Analysis Approach. Neurocomputing, **30** (2000) 79-102
7. Comon, P.: Independent component analysis - a new concept ? Signal Processing **36** (1994) 287-314
8. Delfosse, N., Loubation, P.: Adaptive Blind Separation of Independent Sources: A Deflation Approach. Signal Processing, **45** (1995) 59-83
9. Edelman, A., Arias, T.A., Smith, S.T.: The Geometry of Algorithms with Orthogonality Constraints, SIAM J. Matrix Anal. APPL., **20** (1998) 303-353
10. Everson, R., Roberts, S.: Independent Component Analysis: A Flexible Nonlinearity and Decorrelating Manifold Approach. Neural Computation, **11** (1999) 1957-1983
11. Girolami, M.: An Alternative Perspective on Adaptive Independent Component Analysis Algorithms. Neural Computation, **10** (1998) 2103-2114
12. Hopfield, J. J. & Tank, D. W.: Neural computation of decisions in optimization problems, Biological Cybernetics 52, 141-152 (1985).
13. Hyvarinen, A., Karhunen, J., Oja, A.: Independent Component Analysis, John Wileys, Sons, Inc., New York (2001)

14. Lee, T. W., Girolami, M., Sejnowski, T. J.: Independent Component Analysis Using an Extended Infomax Algorithm for Mixed Subgaussian and Supergaussian Sources. Neural Computation, **11** (1999) 417-441
15. Liu, Z.Y., Chiu, K.C., Xu, L.: One-Bit-Matching Conjecture for Independent Component Analysis, Neural Computation, **16** (2004) 383-399
16. Moreau, E., Macchi, O.: High Order Constrasts for Self-adaptive Source Separation. International Journal of Adaptive Control and Signal Processing, **10** (1996) 19?6
17. Pearlmutter, B. A., Parra, L. C.: A Context-sensitive Genaralization of ICA. In Proc. of Int. Conf. on Neural Information Processing. Springer-Verlag, Hong Kong (1996)
18. Tong, L., Inouye, Y., Liu, R.: Waveform-preserving Blind Estimation of Multiple Independent Sources. Signal Processing, **41** (1993) 2461-2470
19. Welling, M., Weber, M.: A Constrained EM Algorithm for Independent Component Analysis. Neural Computation, **13** (2001) 677-689
20. Xu, L.: Independent Component Analysis and Extensions with Noise and Time: A Bayesian Ying-Yang Learning Perspective, Neural Information Processing Letters and Reviews, **1** (2003) 1-52
21. Xu, L.: Distribution Approximation, Combinatorial Optimization, and Lagrange-Barrier, Proc. of International Joint Conference on Neural Networks 2003 (IJCNN '03), July 20-24, Jantzen Beach, Portland, Oregon, (2003) 2354-2359
22. Xu, L., Cheung, C. C., Amari, S. I.: Further Results on Nonlinearity and Separtion Capability of a Liner Mixture ICA Method and Learned LPM. In C. Fyfe (Ed.), Proceedings of the I&ANN?8 (pp39-45) (1998a)
23. Xu, L., Cheung, C. C., & Amari, S. I.: Learned Parametric Mixture Based ICA Algorithm. Neurocomputing, **22** 69-80 (1998b)
24. Xu, L., Cheung, C. C., Yang, H. H., Amari, S. I.: Independent component analysis by the information-theoretic approach with mixture of density. Proc. of 1997 IEEE Intl. Conf on Neural Networks, Houston, TX. **3** 1821-1826 (1997)
25. Xu, L. Bayesian Ying-Yang Learning Based ICA Models, Proc. 1997 IEEE Signal Processing Society Workshop on Neural Networks for Signal Processing **VI** , Florida, 476-485 (1997).
26. Xu, L., Yang, H. H., Amari, S. I.: Signal Source Separation by Mixtures: Accumulative Distribution Functions or Mixture of Bell-shape Density Distribution Functions. Rresentation at FRONTIER FORUM. Japan: Institute of Physical and Chemical Research, April (1996)
27. Xu, L.: Combinatorial optimization neural nets based on a hybrid of Lagrange and transformation approaches, Proc. of World Congress on Neural Networks, San Diego, 399-404 (1994).

Dynamic Models for Intention (Goal-Directedness) Are Required by Truly Intelligent Robots

Walter J. Freeman

Department of Molecular and Cell Biology University of California
Berkeley CA 94720-3206, USA
wfreeman@sulcus.berkeley.edu
http://sulcus.berkeley.edu

Abstract. Intelligent behavior is characterized by flexible and creative pursuit of endogenously defined goals. Intentionality is a key concept by which to link brain dynamics to goal-directed behavior, and to design mechanisms for intentional adaptations by machines. Evidence from vertebrate brain evolution and clinical neurology points to the limbic system as the key forebrain structure that creates the neural activity which formulate goals as images of desired future states. The behavior patterns created by the mesoscopic dynamics of the forebrain take the form of hypothesis testing. Predicted information is sought by use of sense organs. Synaptic connectivity of the brain changes by learning from the consequences of actions taken. Software and hardware systems using coupled nonlinear differential equations with chaotic attractor landscapes simulate these functions in free-roving machines learning to operate in unstructured environments.

1 Introduction

1.1 Neurodynamics of Intentionality in the Process of Observation

The first step in pursuit of an understanding of intentionality is to ask, what happens in brains during an act of observation? This is not a passive receipt of information from the world. It is an intentional action by which an observer directs the sense organs toward a selected aspect of the world and interprets the resulting barrage of sensory stimuli. The concept of intentionality has been used to describe this process in different contexts, since its first proper use by Aquinas 700 years ago. The three salient characteristics of intentionality as it is treated here are (i) intent, (ii) unity, and (iii) wholeness [1]. These three aspects correspond to use of the term in psychology with the meaning of purpose, in medicine with the meaning of mode of healing and integration of the body, and in analytic philosophy with the meaning of the way in which beliefs and thoughts are connected with ("about") objects and events in the world.

(i) Intent is directedness of behavior toward some future state or goal; it comprises the endogenous initiation, construction, and direction of actions into the world. It emerges from brains by what is known as the "action-perception-assimilation cycle" of Maurice Merleau-Ponty and Jean Piaget. Humans and animals select their own goals, plan their own tactics, and choose when to begin, modify, and stop sequences of action. Intent is the creative process by which images of future states are constructed, in accordance with which the actions of search, observation, and hypothesis testing are directed. (ii) Unity appears in the combining of input from all sensory modalities into *Gestalts*, the coordination of all parts of the body, both musculoskeletal and autonomic, into adaptive, flexible, and effective movements, and the focused preparation for action that is experienced in consciousness. (iii) Wholeness is the construction of a life history, by which all of experience is sifted, catalogued, and organized in support of future contingencies in the ongoing adaptation to the environment. The aim of this report is to describe the dynamics by which animals prefigure their own goal-directed actions and predict the sensory information they need to perform their actions and achieve their goals, and to sketch briefly the simulation of these processes in simple robots.

1.2   The Limbic System Is the Organ of Intentional Behavior

Brain scientists have known for over a century that the necessary and sufficient part of the vertebrate brain to sustain minimal intentional behavior is the ventral forebrain, including those components that comprise the external shell of the phylogenetically oldest part of the forebrain, the paleocortex, and the deeper lying nuclei with which the cortex is connected. Intentional behavior is severely altered or absent after major damage to the basal forebrain, as manifested most clearly in Alzheimer's disease.

Phylogenetic evidence comes from observing intentional behavior in simpler animals, particularly (for vertebrates} that of salamanders, which have the simplest of the existing vertebrate forebrains [2]. The three parts are sensory (which, also in small mammals, is predominantly olfactory), motor, and associational. The associational part contains the primordial hippocampus with its septal, amygdaloid and striatal nuclei. It is identified in higher vertebrates as the locus of the functions of spatial orientation (the "cognitive map") and of temporal integration in learning (organization of long term and short term memories). These processes are essential, because intentional action takes place into the world, and even the simplest action, such as searching for food or evading predators, requires an animal to know where it is with respect to its world, where its prey or refuge is, and what its spatial and temporal progress is during sequences of attack and escape.

The three parts are schematized in Fig.1. All sensory systems in mammals send their output patterns to the entorhinal cortex, which is the gateway to the hippocampus and also a main target of hippocampal output. Multisensory signals are integrated in transit through the hippocampus and then transmitted back to the entorhinal cortex (the "time loop") and then to all the sensory cortices (the "reafference loop"), so that within a third of a second of receipt of new information in any sensory area, all sensory areas are apprised and up-dated. The hippocampal and entorhinal outputs also go to the musculoskeletal, autonomic and neuroendocrine motor systems with feedback inside the brain ("control loop") and through the body (the "proprioceptive loop"). Feedback through the environment to the sensory receptors (the "motor loop") constitutes the action-perception cycle, and the learning that follows completes the process of assimilation.

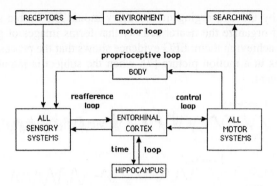

Fig. 1. Schematic flow diagram of mechanism of intentional (goal-directed) action. The limbic system is schematized as the interaction between entorhinal cortex where sensory inputs are combined and the hippocampus where time-space integration takes place.

2 Neurodynamics of Intentionality

The question for neuroscientists and engineers alike is, how do brains create the patterns of neural activity that sustain intentional behavior? Answers are provided by studies of the patterns of mesoscopic electrical activity of the primary sensory cortices of animals that have been trained to discriminate conditioned stimuli [3]. While cortical neurons are selectively excited by sensory receptor input and generate microscopic action potential [4], the relevance to intentionality appears only in the mass activity formed by interactions among the cortical neurons populations that "enslave" their action potentials into mesoscopic patterns as "order parameters" [5].

The brain activity patterns that are seen in electroencephalograms (EEGs) reveal the mesoscopic brain states that are induced, not evoked, by the arrival of stimuli. These brain states are not representations of stimuli, nor are they the simple effects caused by stimuli. Each learned stimulus serves to elicit the construction of a pattern that is shaped by the synaptic modifications among cortical neurons from prior learning, and by neurohormones of the brain stem nuclei. It is a dynamic action pattern that creates and enacts the meanings of stimuli for the subject [6]. It reflects the individual history, present context, and imaging of future possible states, corresponding to the wholeness of the intentionality. The patterns created in each cortex are unique to each subject. All sensory cortices transmit their patterns in the same basic form into the limbic system, where they are integrated with each other and organized in temporal sequences corresponding to the unity of intention. The resultant integrated pattern is transmitted back to the cortices by reafference, which consists in the processes of selective attending, expectancy, and the prediction of future sensory inputs. Animals know what they are looking for before they find it.

The same kinds of patterns of EEG activity (Fig. 2) as those found in the sensory and motor cortices are found in various parts of the limbic system: amplitude modulation of chaotic carrier waves. This discovery indicates that the limbic system employs the same creative dynamics as the sensory cortices to create its own spatiotemporal patterns of neural activity. The patterns are created from memories of past experience that are shaped by the converging multisensory input. The limbic system serves to generate and self-organize the neural activity that forms images of goals and directs behavior toward achieving them. EEG evidence shows that the process occurs in rapid steps, like frames in a motion picture [7], when the subject is intentionally engaged with its environment.

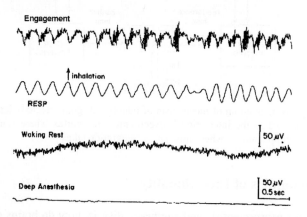

Fig. 2. EEGs in deep anesthesia (open loop, bottom, 4th trace), resting (3rd trace), and aroused (top). Bursts of oscillation result from destabilization of the system with each inhalation (upward in 2nd trace).

The top trace in Fig. 2 shows with each inhalation a brief oscillation that results from destabilization of the bulb by action potentials from sensory receptors in the nose. Each oscillation carries an amplitude pattern that is formed through a dynamic state transition, in which the bulbar assembly of neuron populations jumps suddenly from one spatiotemporal pattern to the next, as the behavior evolves. Being intrinsically unstable, both the bulb and the limbic system continually transit across states that emerge, spread into other parts of the brain, and then dissolve to give rise to new ones. Limbic output controls the brain stem neurohumoral nuclei that serve to regulate its excitability levels, implying that it regulates its own neurohumoral context, enabling it to respond with equal facility to changes that call for arousal and adaptation or rest and recreation, both in the body and the environment. The neurodynamics of the limbic system, assisted by other parts of the forebrain such as the frontal lobes, initiates the creative actions seen in search by trial and error.

3 Hierarchy of Neural Populations Generating Intentional Behavior

The proposed neural basis for a theory of intentionality is a hierarchy of nonlinear ordinary differential equations, K-sets, having noninteractive neural populations (KO) near equilibrium at its base [8]. By synaptic transmission KO sets create interactive populations of excitatory (KIe) or inhibitory (KIi) neurons with nonequilibrium point attractors. KI populations of olfactory excitatory or inhibitory neurons comprise the KII set having also a limit cycle attractor used to model the olfactory bulb, olfactory nucleus, or the olfactory cortex. The KIII set incorporates 3 KII sets to model the EEG with chaotic attractors that govern the olfactory EEG [9] (see Fig. 6), and to simulate the olfactory capacity for classification of incomplete patterns embedded in complex backgrounds with facilitation by noise [10]. The KIV model of the primordial vertebrate forebrain comprising its limbic system is currently under development as an interactive network of 3 KIII sets used by autonomous vehicles for navigation [11].

The basis for the K-set hierarchy, the KO set, represents a noninteractive collection of neurons with a linear operator, a 2nd order ordinary differential equation in dendritic potential, v, that represents dendritic integration; a static nonlinear operator, G(p), that represents the operation that transforms dendritic current density to pulse density output; and a synaptic weight, k_{ij}, that represents the transformation of pulse density to wave density at synapses where learning takes place. The linear dendritic response to impulse input, $\delta(t)$, gives rates of rise, a, and decay, b, from fitting the sum of two exponential terms that are the solution of the 2nd order ODE. The coupling coefficient, k_{ij}, representing the conversion at synapses of axonal input to dendritic synaptic current, v, is measured from the ratio of output amplitude to input amplitude.

$$d^2v/dt^2 + (a+b)dv/dt + ab\, v = k_{ij}G(p) \qquad (1)$$

The static nonlinearity in the system, G(p), is located at the trigger zones of the neurons in the populations, where dendritic currents serve to modulate the firing rates of the neurons. The collective property gives the operator the form of an asymmetric sigmoid curve, such that input not only increases output but also the gain of the interacting neurons (Fig. 3, equation 2). When a threshold is reached for a regenerative gain, the neurons are no longer driven preferentially by the sensory input but instead by each other. The gain, k_{ij}, of synapses by which they interact are modified by learning, so that the formation of cortical output patterns draws on past experience to retrieve memories. The static nonlinearity is derived from the Hodgkin-Huxley equations, giving an asymmetric function for wave-to-pulse conversion:

$$G(p)=p_0\,(1+\{1-\exp[-(e^v-1)/Q_m]\}), \quad v > -u_0 \qquad (2)$$

$$dp/dv = u_0 \exp[v-(e^v-1)/Q_m] \qquad (3)$$

The KI_e set of excitatory neurons is the source of background" spontaneous" activity in the brain [7] by mutual excitation, in which the connectivity exceeds unity, in the sense that each neuron maintains enough synapses with other neurons that it receives back more impulses than it transmits. However, it cannot exceed a steady state because of the refractory periods and thresholds, which provide for homeostatic stability in all parts of the brain. This property is reflected in the fact that the distribution amplitudes in histograms (Fig. 3, lower left frame) is on the descending part of the nonlinear gain curve. In contrast, the amplitude histograms of EEGs from KII sets are located on the rising part, meaning that input to the population not only increases its output; it also increases its internal gain. This is that basis for the input-dependent instability that is shown by the bursts of oscillation with inhalation in the top trace of Fig. 2, whereby the cortex (including the bulb) can jump from one state to another in a state transition.

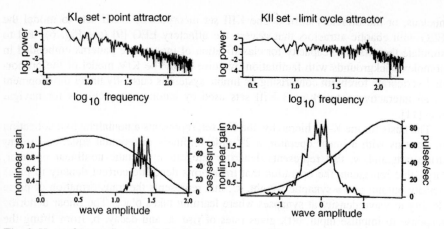

Fig. 3. Upper frames: Spectra of simulated activity of a KI_e set of mutually excitatory neurons (left) and of a KII set of interactive excitatory and inhibitory neurons (right) in the olfactory bulb. Lower frames: amplitude histograms of the same signals as shown in Fig. 2. The gain curves are from equation (3), the derivative of the sigmoid function in equation (2). From [9].

4 Characteristics of Brain States

The "state" of the brain is a description of what it is doing in some specified time period. A state transition occurs when the brain changes and does something else. For example, locomotion is a state, within which walking involves large parts of brain, spinal cord, muscles and bones. The entire neuromuscular system changes almost instantly with the transition to a pattern of jogging or running. Similarly, a sleeping state can be divided into a sequence of slow wave and REM stages. Transit to a waking state can occur in a fraction of a second, whereby the entire brain and body change in concert. The state of a neuron can be described as active and firing or as silent, with sudden changes in the firing manifesting state transitions. Populations of neurons also have a range of states, such as slow wave, fast activity, seizure, or silence. Neurodynamics is used to study these states and their transitions by measuring the spatial patterns of the EEG recorded from high-density electrode arrays.

The most critical question to ask about a state is its degree of stability or resistance to change. Evaluation is by perturbing an object or a system [8],[12]. A person standing on a moving bus and holding on to a railing is stable, but someone walking is not. If a person regains posture after each perturbation, no matter in which direction the displacement occurred, that state is regarded as stable, meaning that an attractor governs it. This is a metaphor to say that the system goes ("is attracted") to the state through an interim state of transience. The range of displacement from which recovery can occur defines the basin of attraction, in analogy to a ball rolling to the bottom of a bowl. If the perturbation is so strong that it causes concussion or a broken leg, and the person cannot stand up again, then the system has been placed outside the basin of attraction, and it goes to a new state with its own attractor and basin.

Stability is always relative to the time duration of observation and the criteria for what variables are chosen for observation. In the perspective of a lifetime, brains appear to be highly stable, in their numbers of neurons, their architectures and major patterns of connection, and in the patterns of behavior they produce, including the character and identity of the individual that can be recognized and followed for many years. Brains undergo repeated transitions from waking to sleeping and back again, coming up refreshed with a good night or irritable with insomnia, but still, giving the same persons as the night before. Personal identity is usually quite stable. But in the perspective of the short term, brains are highly unstable. Measurements of their internal states of neural activity reveal patterns that are more like hurricanes than the orderly march of symbols in a computer. Brain states and the states of populations of neurons that interact to give brain function, are highly irregular in spatial form and time course. They emerge, persist for about a tenth of a second (see Fig. 2, top trace), then disappear and are replaced by other states.

The neurodynamic approach to the problem used here is to define three kinds of stable state, each with its type of attractor [12]. The point attractor: the neurons are in a steady state unless perturbed, and they return to that steady state when allowed to do so. The steady state may be at zero firing rate under anesthesia, but in the normal state the neurons are firing randomly. The reason for randomness is that each neuron interacts with thousands of others through one, two, many serial synapses, so that the return path can be modeled with a 1-dimensional diffusion equation [8]. Examples of point attractors are silent neurons or populations that have been isolated from the brain or depressed by a strong anesthetic, to the point where the EEG has flattened (Fig. 2, lowest trace; Figure 3, left frames). This is the "open loop" state in which the time and space constants of populations are measured.

A special case is a point attractor perturbed by steady-state noise. This state is observed in populations of neurons in the brain of a subject at rest, with no evidence of overt behavior (Fig. 2, 3rd trace from top). The neurons fire continually but not in concert with each other. Their pulses occur in long trains at irregular times. Knowledge about the prior pulse trains from each neuron and those of its neighbors up to the present fails to support the prediction of when the next pulse will occur. The state of noise has continual activity with no history of how it started, and it gives only the expectation that its amplitude and other statistical properties will persist unchanged.

A system that gives periodic behavior is said to have a limit cycle attractor. The classic example is the clock. When it is viewed in terms of its ceaseless motion, it is regarded as unstable until it winds down, runs out of power, and goes to a point attractor. If it resumes its regular beat after it is re-set or otherwise perturbed, it is stable as long as its power lasts. Its history is limited to one cycle, after which there is no retention of its transient approach in its basin to its attractor. Neurons and populations rarely fire periodically, and when they appear to do so, close inspection shows that the activities are in fact irregular and unpredictable in detail, and when periodic activity does occur, it is either intentional, as in rhythmic drumming, or pathological, as in nystagmus and Parkinsonian tremor.

The third type of attractor gives aperiodic oscillations of the kind that are observed in recordings of EEGs. The power spectral densities (PSDs) are $1/f$ as in the upper frames of Fig. 3. There is no one or small number of frequencies at which the system oscillates. The system behavior is therefore unpredictable, because performance can only be projected far into the future for periodic behavior. This type was first called "strange"; it is now widely known as "chaotic". The existence of this type of oscillation was known to Poincaré a century ago, but systematic study was possible only recently after the full development of digital computers. The best-known systems with deterministic chaotic attractors have few degrees of freedom, as for example, the double-hinged pendulum, and the dripping faucet. They are autonomous and noise-free. Large, noisy, open systems such as neurons and neural populations are capable of high-dimensional, nondeterministic "stochastic" chaos, but proof is not yet possible at the present level of developments in mathematics.

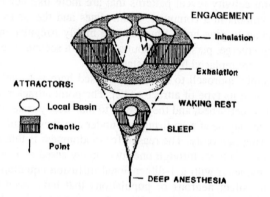

Fig. 4. A portrait is shown of the state space of the olfactory system. In a motivated state the system oscillates between chaotic and limit cycle attractors recreating and dissolving the landscape rapidly enough to allow repeated sampling of odors (see Fig. 5 for a different view of the attractor landscape). Adapted from [13].

The discovery of chaos has profound implications for the study of brain function [13]. A chaotic system has the capacity to create novel and unexpected patterns of activity. It can jump instantly from one mode of behavior to another, which manifests the facts that it has a landscape of attractors, each with its basin, and that it can move from one basin to another in an itinerant trajectory [14]. It retains in its pathway across its basins its history, which fades into its past, just as its predictability into its future decreases with distance. Transitions between chaotic states constitute the dynamics we need to understand how brains perform intentional behavior.

5 The Cortical Phase Transition Is the Basic Step of Intentionality

Systems such as neurons and brains that have multiple chaotic attractors also have point and limit attractors. A system that is in the basin of one of its chaotic attractors is legendary for the sensitivity to what are called the "initial conditions". This refers to the way in which a simple system is placed into the basin of one of its attractors. If the basin is that of a point or a limit cycle attractor, the system proceeds predictably to the same end state. If the basin leads to a chaotic attractor, the system goes into ceaseless fluctuation, as long as its energy lasts. If the starting point is identical on repeated trials, which can only be assured by simulation of the dynamics on a digital computer, the same aperiodic behavior appears. This is why chaos is sometimes called "deterministic". If the starting point is changed by an arbitrarily small amount, although the system is still in the same basin, the trajectory is not identical. If the difference in starting conditions is too small to be originally detected, it can be inferred from the unfolding behavior of the system, as the difference in trajectories becomes apparent.

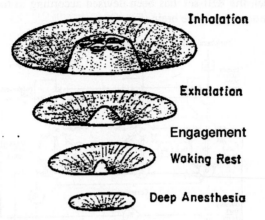

Inhalation

Exhalation

Engagement

Waking Rest

Deep Anesthesia

Fig. 5. A perspective projection of the state space from Fig. 4 shows the learned attractor landscape that forms with each inhalation, giving the basis for rapid selection by a stimulus, and that then dissolves, clearing the entire landscape to receive and classify the next sample when the subject is engaged with the environment. Adapted from [13].

This observation shows that a chaotic system has the capacity to create information in the course of continually constructing its own trajectory into the future. In each sensory cortex there are multiple basins corresponding to previously learned classes of stimuli, and also the unstimulated state. Together they form an attractor landscape that repeatedly forms and dissolves at each new stimulus frame. The chaotic prestimulus state in "exhalation" (Fig. 5) establishes the sensitivity of the cortex, so that the very small number of sensory action potentials evoked by an expected stimulus can carry the cortical trajectory into the basin of an appropriate attractor. The stimulus is selected by the limbic brain through control of the sensory receptors by sniffing, looking, and listening. The basins of attraction are shaped by reafferent limbic action that sensitizes reception to desired classes of stimuli. The web of synaptic connections modified by learning forms basins and attractors that are selected by the input.

Access to attractor basins is by destabilization of the cortical neural populations, manifested in a 1st order state transition. The attractor by which the system is then governed determines the spatial pattern of the oscillatory output by imposing amplitude modulation on the shared wave form, which is transmitted by axons in parallel using pulse frequency modulation and time multiplexing to an array of targets, where the signal is extracted after smoothing by spatial integration imposed by a divergent-convergent axon projection of cortical output [8].

The selection of the basin of attraction is done by the stimulus given to the receptors feeding the primary receiving cortex. The basins and their attractors are shaped by Hebbian and non-Hebbian learning, as well as by reafferent barrages that bias and shape the attractor landscape. The trajectory manifested in the EEG is higher in dimension than the landscape, so that every basin is immediately accessible from every point of the trajectory with minimal need for search. Capture by one of the attractors reduces the dimension of activity below that of the landscape.

6 Noise-Stabilized Chaotic Dynamics of KIII Neural Populations

A dynamic model, the KIII set, has been devised according to these principles [10] and the olfactory anatomy in the brain (Fig. 6):

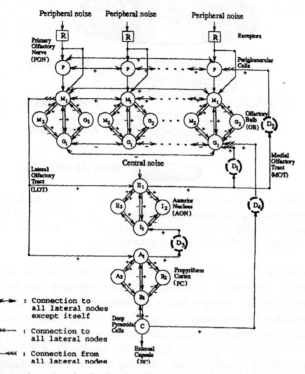

Fig. 6. Chaotic dynamics arises by feedback with delays among 3 coupled KII oscillators having incommensurate characteristic frequencies. The landscape is stabilized by additive noise. Each node is governed by equations (1) and (2).

The model embodies use of dynamic attractor landscapes for memory storage, with recall by input that acts without need for tree search or gradient descent, and that can learn new patterns in very few trials without significant loss of past acquisitions. The device has proven to provide a versatile and flexible interface between the infinite complexity of real world environments and finite state devices such as digital computers, which can operate effectively to determine the hypotheses to be tested and interpret the results. An important aspect to be discussed is the use of additive noise for the stabilization of chaotic attractors in the digital platforms that have been tested thus far [9], [10]. The use of rational numbers to implement the solutions of difference equations imposes coarse graining by truncation into the attractor landscapes of digital embodiments, so that the landscape resembles the fur of mammals, and the system is unstable due to the small size of the basins from attractor crowding. Basin diameters approach the digitizing step size. Low-level additive noise modeled on biological properties of neural action potentials empirically solves this aspect of the problems of attractor crowding while improving the classification efficacy of the KIII model, to be discussed, somewhat in the manner of stochastic resonance. Additive noise also enables precise simulation of the temporal (Fig. 7) and spectral properties (Fig. 3) of brain activity.

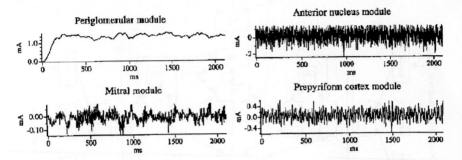

Fig. 7. Simulations by the KIII model are shown of activity patterns observed in 4 parts of the olfactory system in the rest state (Fig. 2, 3rd trace).

7   Construction of a KIV Set to Provide for Intentional Navigation

The unity of intentionality requires that all messages from all sensory modalities be converged into a central location and there be integrated. In simpler vertebrates this is done by direct connections among all KIII sets, including those for exteroception (the state of the outer world), interoception (the inner state, such as fuel resources, needs for rest and repair), and proprioception (the state of the motor system, including whether the actions that have been commanded have been taken, and how well the intended actions conform to the expected actions. The requirement for associational structures (including the components of the hippocampus) is imposed by the necessity for the subject (or robot) to place each action and resulting sensation in a space-time context. In a search for food or fuel, for example, the scent of prey or of gasoline has no value to a search for its location or source unless the observer can retain in short term memory a record of its strength and the time and place at which the observation was made. Then the observer must move in some direction, take another sample, and compare it with the first sample. Depending on whether the second sample is stronger or weaker, the observer must continue forward or reverse course, while at the same time building a central conception of the trajectory of search. This is the wholeness of intentionality, and it explains the role of the hippocampal formation in maintaining what is called the 'cognitive map' as well as the central mechanisms for short term memory. All sensory systems require this collection of mechanisms, which explains why they operate only after convergence and integration of the several sensory streams. Thus any robot that is expected to engage in autonomous, self-directed exploration of its environment must be equipped with some form of associational modules.

Acknowledgments
This work was funded in part by research grants from NIMH (MH06686), ONR (N63373 N00014-93-1-0938), NASA (NCC 2-1244), and NSF (EIA-0130352).

32   Walter J. Freeman

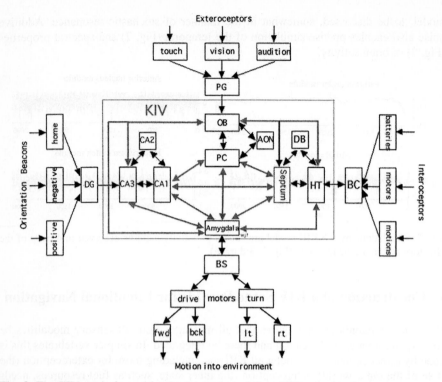

Fig. 8. The KIV set is formed by the interaction of three KIII sets plus the motor apparatus required for locomotion and execution of intentional commands. From Kozma, Freeman and Erdí [11].

References

1. Freeman, W.J.: Societies of Brains. A Study in the Neuroscience of Love and Hate. Lawrence Erlbaum Associates, Hillsdale NJ (1995)
2. Herrick, C. J.: The Brain of the Tiger Salamander. Univ. Chicago Press, IL (1948)
3. Barrie, J.M., Freeman, W.J. and Lenhart, M.D.: Spatiotemporal Analysis of Prepyriform, Visual, Auditory and Somesthetic Surface EEG in Trained Rabbits. J. Neurophysiol,**76** (1996) 520-539
4. Singer, W., Gray, C.M.: Visual Feature Integration and the Temporal Correlation Hypothesis. Annual Review of Neuroscience, **18** (1995) 555-586
5. Haken, H.: Synergetics: An Introduction. Springer-Verlag, Berlin (1983).
6. Freeman, W.J.: A Neurobiological Theory of Meaning in Perception. Part 1. Information and Meaning in Nonconvergent and Nonlocal Brain Dynamics. Int. J. Bif. Chaos, **13** (2003) 2493-2511
7. Freeman, W.J.: (2005) Origin, Structure and Role Of Background EEG activity. Part 3. Neural frame classification. Clin. Neurophysiol (2005)
8. Freeman, W.J.: Mass Action in the Nervous System. Academic Press, New York (2005) Available in electronic form on http://sulcus.berkeley.edu
9. Freeman, W.J., Chang, H.-J., Burke, B.C., Rose, P.A., Badler, J.: Taming Chaos: Stabilization of Aperiodic Attractors by Noise. IEEE Trans. Circuits Systems, **44** (1997) 989-996

10. Kozma, R., Freeman, W.J.: Chaotic Resonance: Methods and Applications For Robust Classification Of Noisy and Variable Patterns. Int. J. Bif. Chaos, **10** (2001) 2307-2322
11. Kozma, R., Freeman, W.J., Erdí, P.: The KIV Model - Nonlinear Spatiotemporal Dynamics of the Primordial Vertebrate Forebrain. Neurocomputing, **52** (2003) 819-826
12. Freeman, W.J.: Neurodynamics. An Exploration of Mesoscopic Brain Dynamics. London UK: Springer-Verlag (2000) Translated into Chinese by Prof. Gu Fanji, Springer-Verlag (2004)
13. Skarda, C. A., Freeman, W. J.: How Brains Make Chaos in Order to Make Sense of the World. Behavioral and Brain Sciences, **10** (1987) 161-195
14. Tsuda, I.: Toward An Interpretation of Dynamical Neural Activity in Terms of Chaotic Dynamical Systems. Behav. Brain Sci, **24** (2001) 793-847

Differences and Commonalities
Between Connectionism and Symbolicism

Shoujue Wang and Yangyang Liu

Laboratory of Artificial Neural Networks, Institute of Semiconductors,
Chinese academy of Sciences 100083, Beijing, China
{wsjue,liuyangyang}@red.semi.ac.cn

Abstract. The differences between connectionism and symbolicism in artificial intelligence (AI) are illustrated on several aspects in details firstly; then after conceptually decision factors of connectionism are proposed, the commonalities between connectionism and symbolicism are tested to make sure, by some quite typical logic mathematics operation examples such as "parity"; At last, neuron structures are expanded by modifying neuron weights and thresholds in artificial neural networks through adopting high dimensional space geometry cognition, which give more overall development space, and embodied further both commonalities.

1 Introduction

After Turing machines theory, Artificial Intelligence (AI) researches spread out to two branches of connectionism and symbolicism.

Symbolicism, developed enlightened algorithm – experts system – knowledge engineering theories and techniques, and had flourished in 80's. Symbolicism is still the mainstream of AI by now.

From the first artificial neuron model built, until Rumelhart proposed many layer back propagation (BP) algorithm. Connectionism, from theories analyzing to the engineering carrying out, lay the foundation for NN computer to head on the market. In AI research symbolicism and connectionism swiftly and violently develop at long intervals, but their divergences more and more clear. For expanding novel NN analytical angle, this paper will set out from illustration on differences and commonalities between symbolicism and connectionism, and validate connectionism to realize logic functions, adopt high dimensional space reasoning geometry cognition, and its analysis means to expand NN structure.

2 Thought Rifts of Connectionism and Symbolicism

There is some thought divergences between symbolic AI and connection AI, which incarnate theories, calculating process etc. a few aspects.

From theories aspect, symbolicism's principles lay on mainly physical symbolic system assumption and limited rationality theory. Connectionism, its principle is neural network (NN),even the learning algorithms and connection mechanism in it.

J. Wang, X. Liao, and Z. Yi (Eds.): ISNN 2005, LNCS 3496, pp. 34–38, 2005.

Symbolicism carries on researches from the intelligence phenomenons represented by outside of mankind's brain. But the characteristics of connection is to imitate the mankind's physiological neural network structure.

In the core problem aimed to AI calculation process, the symbolic AI calculation particularly emphasizes on problem solutions and inductive methods. But connection AI calculation emphasizes the training process.

Symbol system could realize characteristics of serial, linearity, accurateness, concision, and easy expression. In contrast to symbolicism, connectionism flourish on Large-scale parallel processing and distributing information storages[1], good at both self-adaptivity, self-organization,and so on.

So, whether connectionism exists commonalities together with symbolicism or not? whether the connection function could expand symbolicism realm or not? Holding upper problems, some further discussion differences and commonalities between connectionism and symbolicism are here.

3 Commonalities Between Connectionism and Symbolicism

Speaking essentially, connectionism calculation is to connect a great deal of same operation units mutually to form a kind of paralleled computing method without procedures.

3.1 Decision Factors of Connectionism

So artificial NN calculation process decided by three aspects as follows:

(1) connection between neurons;
(2) values of weights of neuron;
(3) the structure (mathematical model) of neurons.

And NN development, whatever to approach iteratively the expectation result from modifying weights, or still to reckon the neurons connection and weights from the expectation result, traditional connectionism adopted methods had all neglected seriously the importance of neuron mathematical model in NN development trend.

3.2 Connectionism Functions to Expand

After analyzed "parity" operation process and neurons structure, one NN structure designed such as Fig.1show, the single-weighted and single-thresholded hard limiter NN of Fig.1 needs only 15 neurons, the numbers in circles are thresholds of neurons. then could carry out respectively logic operation function of 8 binary input "parity"[2]. but the same "parity" operation have to be done with 35 "NAND Gate" at least.

Another typical logic example which usually is used to compare NN models and learning algorithms is Bi-level logic operation. Fig.2 adopted single-weighted and single-thresholded hard limiter NN to carry out the function of distinguishing 7's whole multiple from 6 binary digits.

Supposed NN doesn't limit at single thresholded value, change monotonic activation function to non-monotonic function, for example, to carry out respectively six,

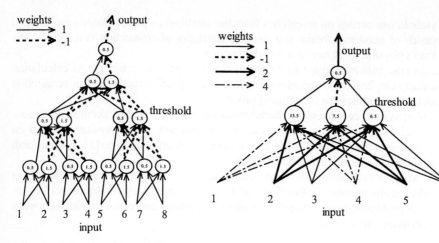

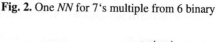

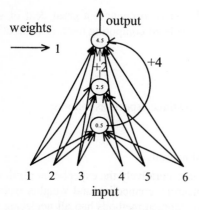

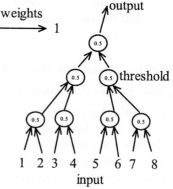

Fig. 1. One *NN* for 8 binary input "parity"   **Fig. 2.** One *NN* for 7's multiple from 6 binary

Fig. 3. One *NN* for *six* input of "parity"   **Fig. 4.** One *NN* for *eight* input of "parity"

eight input of "parity" by taking multi-thresholded hard limiter NN, such as Fig.3 and 4 showed[2], the traditional NN structure is changed. All thresholded value partition within figures is $\theta_1 = 1$, numbers showed in circles is values for θ_0.

But by the analysis and verification above, multi-thresholded NN of Fig. 3 avoid to increase algorithms complexity, and is obviously more simple when compared with Fig.1 single-thresholded NN, and must be easier than the 35 "NAND gate" structure.

After breaking tradational limitness in Radius Basis Function structure NN (RBF), multi-thresholded RBF neurons could also carry out eight numbers of "parity" such as Fig.5.

It could be seen, while performing the same logic operation function, NN structure is easier than symbolism, especially multi-thresholded NN.Then further probing into the neuron structure, if the structure is multi-weighted and multi-thresholded, whether NN function would have to expand or not? Does connectionism would unify more with the symbolism in a specific way?

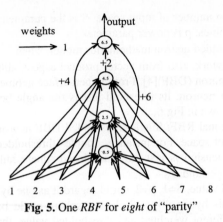

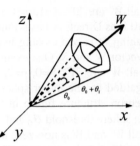

Fig. 5. One *RBF* for *eight* of "parity"

Fig. 6. 3D input space of *double-weighted DBF*

4 Discusses on Neuron Structure

From broadly sense, one neuron could be treated as an basic operation rule as follows:

$$Y = F[\Phi(W,X) - \theta] \quad . \tag{1}$$

and then traditional BP network of mathematics model is:

$$Y = f(\sum_{i=0}^{n} w_i x_i - \theta) . \tag{2}$$

$$\Phi(W,X) = \sum_{i=0}^{n} w_i x_i \quad . \tag{3}$$

so in traditional RBF network structure

$$Y = f(\sum_{i=0}^{n} (w_i - x_i)^2 - \theta^2) \quad . \tag{4}$$

$$\Phi(W,X) = \sum_{i=0}^{n} (w_i - x_i)^2 \quad . \tag{5}$$

if building one novel double-weighted neuron, $\Phi(W,W',...,X)$, to make its ability both have BP and RBF neuron structure, and realize multi-weighted and multi-thresholded neuron structure, therefore the novel basic calculation model should be [3]:

$$Y = f[\sum_{j=1}^{n} (\frac{W_{ji}}{|W_{ji}|})^s |W_{ji}(X_j - W'_{ji})|^P - \theta] \quad . \tag{6}$$

$$\Phi(W,W',X) = \sum_{j=1}^{n} (\frac{W_{ji}}{|W_{ji}|})^s |W_{ji}(X_j - W'_{ji})|^P \quad . \tag{7}$$

by Eq.(6,7):where Y is neuron output; f is inactivation function; θ is neuron threshold; Wji and W'ji is the two weights from no. j input connection to no. i neuron; Xj is

No. j input (positive value); n is dimension number of input space; S is the parameter to decide monomial sign of positive and minus; p is power parameter.

The novel multi-weighted multi- thresholded neuron mathematics model, when

(1) all W' are "0", S=1, p=1, is BP network, also from vector product aspect, able to regard it as Direction-basis-Function Neuron (DBF)[4]. From hypersurface geometry meaning, it is one closing hyper-taper neuron; its threshold just is the angle between vector W and X, One example is shown in Fig.6;

(2) all W' are "1", S=0, p=2, is traditional RBF. Single-thresholded DBF neuron was regarded as one hyper-sphere in input space. For instance, double- thresholded DBF neuron structure figure in three dimension is just like one closed sphere hull between sphere threshold θ_0 and sphere threshold $\theta_0 + \theta_1$;

(3) all W' and W is non-zero numerical value, S=1, p=2, could regard it as one hyper-ellipsoid neuron structure. Further this hyper-ellipsoid closing sphere could be changed by the numerical value of p; different weighted and thresholded value, the mulit-dimension space shape what it materialized certainly correspond to change;

(4) hyper-sausage neuron could be seen as double or multi-thresholded neuron structure from the hyper-sphere geometry theories[5],of course, its mathematical model also is establish as the basis of Eq. (6).

Obviously hyperplane or hypersurface geometry concept for helping understanding to NN behavior and analysis is very valid. The analytical theories using multi-dimension space helps the NN development, certainly will consumedly boost the applied space of NN, make NN applied accuracy raised greatly.

5 Conclusion

It could be seen: NN usually could carry out the logic operation function of symbolicism more simply; and because the agility and diversity in connectionism structure, the different structure, different weighted and thresholded NN certainly will hold vaster and more overall development space.

In short, symbolicism and connectionism both have the special features each other, one ideal intelligence system and its expressed intelligence behavior should be the result what both is organic to combine mutually.

References

1. Lau, C. (ed.): Neural Networks: Theoretical Foundations and Analysis. IEEE PRESS, A Selected Reprints. Neural Networks Council, Sponsor (1992)
2. Wang, S.: Multi-valued Neuron (MVN) and Multi-Thresholded Neuron (MTV), Their Combination and Applications. Chinese Journal of Electronics, **24** (1996) 1-6
3. Wang, S., Li, Z. Chen, X., Wang, B.: Discussion on the Basic Mathematical Models of Neurons in General Purpose Neurocomputer. Chinese Journal of Electronics, **29** (2001) 577-580
4. Wang, S., Shi, J., Chen, C., Li, Y.: Direction-basis-Function Neural Network. # 58 Session: 4.2, IJCNN'99 (1999)
5. Wang, S., Zhao, X.: Biomimetic Pattern Recognition Theory and Its Applications. Chinese Journal of Electronics, **13** (2004) 372-377

Pointwise Approximation for Neural Networks*

Feilong Cao[1,2], Zongben Xu[2], and Youmei Li[3]

[1] Department of Information and Mathematics Sciences, College of Science,
China Jiliang University, Hangzhou, Zhejiang 310018, China
flcao@263.net
[2] Institute for Information and System Sciences, Faculty of Science,
Xi'an Jiaotong University, Xi'an, Shanxi 710049, China
zbxu@mail.xjtu.edu.cn
[3] Department of Computer Science, Shaoxing College of Arts and Sciences,
Shaoxing, Zhejiang 312000, China
liyoumei@sohu.com

Abstract. It is shown in this paper by a constructive method that for any $f \in C^{(m)}[a, b]$, the function and its m order derivatives can be simultaneously approximated by a neural network with one hidden layer in the pointwise sense. This approach naturally yields the design of the hidden layer and the estimate of rate of convergence. The obtained results describe the relationship among the approximation degree of networks, the number of neurons in the hidden layer and the input sample, and reveal that the approximation speed of the constructed networks depends on the smoothness of approximated function.

1 Introduction

Artificial forward neural networks are nonlinear parametric expressions representing numerical functions. In connection with such paradigms there arise mainly three problems: a density problem, a complexity problem and an algorithmic problem. The density problem was satisfactorily solved in the late 1980's and in the early 1990's.

More recently, the complexity problem which deals with the relationship between the size of an expression (i.e., the number of neurons) and its approximation capacity was discussed in many literatures, such as [6], [7], [1], [8] and [9]. The conclusions in these papers provide answers of Jackson's type to the question: how closely can a continuous function be uniformly approximated by a network with one hidden layer of a given size? or, in other words, how many hidden units are sufficient to approximate such a function uniformly with an error not exceeding ϵ? On the other hand, the simultaneous approximation of a function and its derivatives was also studied by some scholars in recent years (see [4], [5], etc.). In particular, Li [5] proved that any function and its derivatives can be simultaneously and uniformly approximated by a network. The method in [5],

* Supported by the Nature Science Foundation of China (60473034) and the Science Foundation of Past Doctor of China (20040350225).

J. Wang, X. Liao, and Z. Yi (Eds.): ISNN 2005, LNCS 3496, pp. 39–44, 2005.

different from previous ones, is based on the Bernstein operators and therefore, is constructive in a certain sense. The obtained results, however, are still only concerned with the density of simultaneous approximation. Such type of results are important and useful in theory, but nothing on the topology specification, i.e., providing no solution on how many hidden units are needed to perform the expected simultaneous approximation.

In this paper, we try to introduce some approximation methods of linear operators into the investigation of simultaneous approximation for neural networks in the pointwise sense. A pointwise estimate of approximation rate for feedforward neural networks with one input and one hidden layer simultaneously approximating the continuous functions is established by using a constructive method. Meanwhile, the relationship among the sample point, the approximation error, and the number of neurons in the hidden layer is described, and the approximation behavior for the network is characterized in term of the smoothness of approximated functions.

2　Main Result

We use the following notations. The symbols N, N_0 and R stand for the sets of positive negative integers, nonnegative integers, and real numbers, respectively. For $m \in N_0$, $C^{(m)}[a, b]$ denotes the space of m order continuous differentiable real-valued functions defined on $[a, b]$, it is equaled by the supremum norm $\| \cdot \|$.

Let f be a real-valued function defined on $[a, b]$. We will use the modulus of continuity of f, $\omega(f, \delta)$, defined by $\omega(f, \delta) := \sup_{0 < h \leq \delta} \| f(\cdot + h) - f(\cdot) \|$. We have $\lim_{\delta \to 0} \omega(f, \delta) = 0$, and for any real number $\lambda \geq 0$,

$$\omega(f, \lambda \delta) \leq (\lambda + 1)\omega(f, \delta).$$

The modulus of continuity is usually considered as the measure of the smoothness of function and the approximation degree in approximation theory. The function f is called Lipschitz $\alpha, 0 < \alpha \leq 1$ continuous and is written as $f \in \text{Lip } \alpha$, if there exists a constant $C(f)$ such that $\omega(f, \delta) \leq C(f)\delta^\alpha$, here and in the following $C(a)$ denotes the positive constant dependents only on a, its value may be different at different occurrence.

Our main result is as follows.

Theorem 1. *Let φ be any continuous and bounded function defined on R, which has up to $r+1$ order continuous derivatives $\varphi^{(k)}, k = 1, 2, \ldots, r+1$, the derivative functions $\varphi^{(k)}, k = 1, 2, \ldots, r+1$ all have bounded range, and there is a common point, say θ in R, such that $\varphi^{(k)}(\theta) \neq 0$ for all $1 \leq k \leq r+1$ (such functions are clearly abundant, as substantiated by $\sigma(x) = (1 + e^{-\alpha x})^{-1}$ with any $\alpha > 0$). Then for any $f \in C^{(m)}[a, b], 0 \leq m \leq r$, there is a feedforword network with one input, $r - m + 1$ neurons in the hidden layer and the activation function φ:*

$$N_n(x) = \sum_{j=1}^{n} c_j \varphi(w_j x + \theta), \quad n = r - m + 1, x \in [a, b], c_j, w_j \in R,$$

such that

$$\left| N_n^{(m)}(x) - f^{(m)}(x) \right|$$

$$\leq C(m) \left(\frac{1}{n} \left| f^{(m)}(x) \right| + \omega \left(f^{(m)}, \left(\frac{(x-a)(b-x)}{n} + \frac{1}{n^2} \right)^{1/2} \right) \right), \qquad (2.1)$$

and for $m < \alpha < m + 1$

$$\left| N_n^{(m)}(x) - f^{(m)}(x) \right| \leq C(m) \left(\frac{(x-a)(b-x)}{n} + \frac{1}{n^2} \right)^{(\alpha-m)/2}, \qquad (2.2)$$

if and only if $f^{(m)} \in Lip\,(\alpha - m)$.

3   Proof of Main Result

To facilitate the following discussion, we assume that r is a fixed integer, and let

$$H_r(x) = a_0 + a_1 x + \cdots + a_r x^r \qquad (3.1)$$

denote the univariate polynomial of degree r defined on $[a, b]$, where $a_i \in R$. To prove Theorem 1.1, we first use the following result of simultaneous approximation of H_r and its derivatives by networks as an intermediate step, its proof is similar to Theorem 2 in [9] where the result of special case $m = 0$ and its proof are given. We here omit the details of proof.

Theorem 2. *Suppose that* φ *satisfies the conditions of Theorem 1,* $H_r(x)$ *is a univariate polynomial given in (3.1), and* m *is a integer and* $0 \leq m \leq r$. *Then for any* $\epsilon > 0$, *we can construct a neural network with one hidden layer, one input and* $(r - m + 1)$ *units in the hidden layer:*

$$N_n(x) = \sum_{i=1}^{n} c_i \varphi(w_i x + \theta), \quad c_i, w_i \in R, \quad x \in [0,1], \quad n = r - m + 1,$$

such that

$$\left| N_n^{(m)}(x) - H_r^{(m)}(x) \right| < \epsilon.$$

Secondly, we use simultaneous approximation results of the Bernstein operators. Since the interval $[a, b]$ can be translated to the unit interval $[0, 1]$ by a linear transformation, it is sufficient to prove the theorem 1 on $[0, 1]$.

Let $f \in C[0, 1]$, the Bernstein operators associated with an integer $n \in N$ are defined by $B_n(f, x) := \sum_{k=0}^{n} P_{n,k}(x) f\left(\frac{k}{n}\right)$, $x \in [0, 1]$, where $P_{n,k}(x) := \binom{n}{k} x^k (1-x)^{n-k}$. We have for $m < n$

$$B_n^{(m)}(f, x) = \frac{n!}{(n-m)!} \sum_{k=0}^{n-m} P_{n-m,k}(x)$$

$$\times \int_0^{1/n} \cdots \int_0^{1/n} f^{(m)} \left(\frac{k}{n} + \sum_{i=1}^{m} u_i \right) du_1 \cdots du_m.$$

For description, we introduce the linear operators

$$B_{n,m}(g,x) := \frac{n!}{(n-m)!} \sum_{k=0}^{n-m} P_{n-m,k}(x) \int_0^{1/n} \cdots \int_0^{1/n} g\left(\frac{k}{n} + \sum_{i=1}^{m} u_i\right) du_1 \cdots du_m.$$

Then

$$B_n^{(m)}(f,x) = B_{n,m}\left(f^{(m)},x\right). \tag{3.2}$$

Now, by using the methods of [2] and [3], we can get the following lemma for the operators $B_{n,m}$.

Lemma 1. *Let* $m \in N_0$, $0 \le m \le n$ *and* $x \in [0,1]$, *we have*

$$B_{n,m}\left((t-x)^2, x\right) \le C(m)\left(\frac{x(1-x)}{n} + \frac{1}{n^2}\right); \tag{3.3}$$

$$\left|B'_{n,m}(f,x)\right| \le 2n\|f\|, \quad f \in C[0,1]; \tag{3.4}$$

$$\left|B'_{n,m}(f,x)\right| \le \|f'\|, \quad f \in C^{(1)}[0,1]; \tag{3.5}$$

$$\left|\sqrt{x(1-x)}B'_{n,m}(f,x)\right| \le 4n^{1/2}\|f\|, \quad f \in C^{(1)}[0,1]. \tag{3.6}$$

We now begin with the proof of Theorem 1. We define, for any $f \in C[0,1]$, a K-functional as $K(f,t) := \inf_{g \in C^{(1)}[0,1]} \{\|f-g\| + t\|g'\|\}$. It was shown in [3] that the K-functional is equivalent to the modulus of continuity $\omega(f,t)$, i.e., there exist the constants C_1 and C_2, such that

$$C_1\omega(f,t) \le K(f,t) \le C_2\omega(f,t). \tag{3.7}$$

Using Schwarz's inequality, (3.3) and the fact

$$|B_{n,m}(g,x) - g(x)| \le |g(x)(B_{n,m}(1,x)-1)| + \left|B_{n,m}\left(\int_x^t g'(u)du, x\right)\right|$$

$$\le C(m)\frac{1}{n}|g(x)| + \|g'\|\, B_{n,m}(|t-x|,x).$$

we obtain

$$|B_{n,m}(g,x) - g(x)| \le C(m)\frac{1}{n}|g(x)| + \|g'\|\left(B_{n,m}(|t-x|^2,x)\right)^{1/2}$$

$$\le C(m)\left(\frac{1}{n}\left|f^{(m)}(x)\right| + \left\|f^{(m)} - g\right\| + \|g'\|\left(\frac{x(1-x)}{n} + \frac{1}{n^2}\right)\right),$$

which implies from Theorem 2 and (3.2) that

$$\left|N_n^{(m)}(x) - f^{(m)}(x)\right|$$

$$\le \left|N_n^{(m)}(x) - B_n^{(m)}(f,x)\right| + \left|B_n^{(m)}(f,x) - f^{(m)}(x)\right|$$

$$\le \epsilon + B_{n,m}\left(|f^{(m)} - g|, x\right) + \left\|f^{(m)} - g\right\| + |B_{n,m}(g,x) - g(x)|$$

$$\le \epsilon + C(m)\left(\frac{1}{n}\left|f^{(m)}(x)\right| + \left\|f^{(m)} - g\right\| + \|g'\|\left(\frac{x(1-x)}{n} + \frac{1}{n^2}\right)\right).$$

Thus, from the definition of K-functional and (3.7) it follows that

$$\left| N_n^{(m)}(x) - f^{(m)}(x) \right|$$

$$\leq \epsilon + C(m) \left(\frac{1}{n} \left| f^{(m)}(x) \right| + \omega \left(f^{(m)}, \left(\frac{x(1-x)}{n} + \frac{1}{n^2} \right)^{1/2} \right) \right).$$

Let $\epsilon \to 0$, then (2.1) holds.

If $f^{(m)} \in \text{Lip}\,(\alpha - m)$, then from (2.1) it follows that

$$\left| N_n^{(m)}(x) - f^{(m)}(x) \right| \leq C(m) \left(\frac{x(1-x)}{n} + \frac{1}{n^2} \right)^{(\alpha-m)/2},$$

which shows that (2.2) holds. Inversely, if (2.2) holds, then

$$\left| f^{(m)}(x+t) - f^{(m)}(x) \right| \leq \left| B_n^{(m)}(f, x+t) - N_n^{(m)}(x+t) \right|$$

$$+ \left| N_n^{(m)}(f, x+t) - f^{(m)}(x+t) \right| + \left| B_n^{(m)}(f, x) - N_n^{(m)}(x) \right|$$

$$+ \left| N_n^{(m)}(f, x) - f^{(m)}(x) \right| + \left| \int_x^{x+t} B_n^{(m+1)}(f, u) du \right|$$

$$\leq 2\epsilon + \left| \int_x^{x+t} B_{n,m}'(f^{(m)}, u) du \right|.$$

Using the methods of [10], we can imply that $f^{(m)} \in \text{Lip}\,(\alpha - m)$. We omit the details. This completes the proof of Theorem 1.

4   Conclusion

We have constructed a class of neural networks with one hidden layer to realize simultaneous and pointwise approximation for any smooth function and its existing derivatives. By making use of the Bernstein operators as a tool, an upper bound estimation on simultaneous approximation accuracy and a characterization of approximation order in the pointwise sense are established in term of the modulus of smoothness of approximated function.

From the upper bound estimation we imply the density or feasibility of simultaneously approximating any smooth function and its existing derivatives in the pointwise sense. Actually, by the monotonically decreasing property of modulus of smoothness, we get from (2.1)

$$\left| N_n^{(m)}(x) - f^{(m)}(x) \right| \to 0 \quad \text{as } n \to \infty$$

which holds for any $0 \leq m \leq r$. This shows that for any $f \in C^{(m)}[a, b]$ the constructed networks, at every sample point $x \in [a, b]$, can simultaneously approximate $f \in C^{(m)}[a, b]$ and its derivatives $f^{(m)}(x)$ arbitrarily well as long as the number of hidden units is sufficiently large.

The obtained results clarify the relationship among the approximation speed of the constructed networks, the number of hidden units and smoothness of the approximated functions. From (2.1), we can conclude in particular that the approximation speed of the constructed networks not only depend on the number of the hidden units used, and also depend on the input sample x. In general, the nearer the distance of the sample point x to the ends a or b is, the faster the approximation speed is. As shown in (2.2) for the Lipschitz function class $\text{Lip}(\alpha - m)$ the approximation speed of networks is positively proportional both to the number of hidden units and to the smoothness of the approximated function. For the given number of hidden units n and the sample x, the better construction properties of approximated function has, the faster approximation speed of the constructed networks has, and inversely, it is also true.

References

1. Chen, X.H., White, H.: Improve Rates and Asymptotic Normality Normality for Nonparametric Neural Network Estimators. IEEE Trans. Information Theory, **49** (1999) 682-691
2. Ditzian, Z.: A Global Inverse Theorem for Combinations of Bernstein Polynomials. J. Approx. Theory, **26** (1979) 277-292
3. Ditzian, Z., Totik, V. (ed): Moduli of Smoothness. Springer-Verlag, Berlin Heidelberg New York (1987)
4. Gallant, A.R., White, H.: On Learning the Derivatives of an Unknown Mapping with Multilayer Feedforward Networks. Neural Networks, **5** (1992) 129-138
5. Li, X.: Simultaneous Approximations of Multivariate Functions and their Derivatives by Neural Networks with one Hidden Layer. Neuocomputing, **12** (1996) 327-343
6. Mhaskar, H.N., Micchelli, C.A.: Degree of Approximation by Neural Networks with a Single Hidden Layer. Adv. Applied Math. **16** (1995) 151-183
7. Maiorov, V., Meir, R.S.: Approximation Bounds for Smooth Functions in R^d by Neural and Mixture Networks. IEEE Trans. Neural Networks, **9** (1998) 969-978
8. Suzuki, Shin.: Constructive Function Approximation by Three-layer Artificial Neural Networks. Neural Networks, **11** (1998) 1049-1058
9. Xu, Z.B., Cao, F.L.: The Essential Order of Approximation for Neural Networks. Science in China (Ser. F), **47** (2004) 97-112
10. Zhou, D.X.: On Smoothness Characterized by Bernstein Type Operators. J. Approx. Theory, **81** (1995) 303-315

On the Universal Approximation Theorem
of Fuzzy Neural Networks
with Random Membership Function Parameters

Lipo Wang[1,2], Bing Liu[1], and Chunru Wan[1]

[1] School of Electrical and Electronic Engineering, Nanyang Technology University,
Block S1, 50 Nanyang Avenue 639798, Singapore
[2] College of Information Engineering, Xiangtan University,
Xiangtan, Hunan, China
{elpwang,liub0002,ecrwan}@ntu.edu.sg

Abstract. Lowe [1] proposed that the kernel parameters of a radial basis function (RBF) neural network may first be fixed and the weights of the output layer can then be determined by pseudo-inverse. Jang, Sun, and Mizutani (p.342 [2]) pointed out that this type of two-step training methods can also be used in fuzzy neural networks (FNNs). By extensive computer simulations, we [3] demonstrated that an FNN with randomly fixed membership function parameters (FNN-RM) has faster training and better generalization in comparison to the classical FNN. To provide a theoretical basis for the FNN-RM, we present an intuitive proof of the universal approximation ability of the FNN-RM in this paper, based on the orthogonal set theory proposed by Kaminski and Strumillo for RBF neural networks [4].

1   Introduction

Due to its ability to approximate nonlinear functions, the FNN has attracted extensive research interests in the area of function approximation and pattern classification [2, 5–11]. Traditionally, the FNN is trained by adjusting all system parameters with various optimization methods [12, 13]. However the gradient methods are usually slow, which forms a bottle-neck in many applications. To overcome such problems, an approach is to fix the membership functions first and only adjust the consequent part by pseudo-inverse method [2], which was first proposed by Lowe [1] for training the RBF network [15–18]. By extensive computer simulations, we [3] demonstrated that such an FNN with randomly fixed membership function parameters (FNN-RM) has faster training and better generalization in comparison to the classical FNN. To provide a theoretical basis for the FNN-RM, here we present an intuitive proof for the universal approximation ability of the FNN-RM, based on the orthogonal set theory proposed by Kaminski and Strumillo for RBF neural networks [4]. The key idea in [4] is to transform the RBF kernels into an orthonormal set of functions by using the Gram-Schmidt orthogonalization.

J. Wang, X. Liao, and Z. Yi (Eds.): ISNN 2005, LNCS 3496, pp. 45–50, 2005.

Our paper is organized as follows. In Section 2, we prove the universal approximation capability of the FNN-RM. Finally, discussions and conclusions are given in Section 3.

2   An Intuitive Proof for the Universal Approximation Theorem for the FNN-RM

Let $L^2(X)$ be a space of function f on X such that $|f|^2$ are integrable. Similar to Kwok and Yeung [14], for $u, v \in L^2(X)$, the inner product $< u, v >$ is defined by

$$< u, v > = \int_X u(\boldsymbol{x})v(\boldsymbol{x})d\boldsymbol{x} \qquad (1)$$

The norm in $L^2(X)$ space will be denoted as $\| \cdot \|$, and the closeness between network function f_n and the target function f is measured by the L^2 distance:

$$\|f_n - f\| = \left[\int_X |f_n(\boldsymbol{x}) - f(\boldsymbol{x})|^2 d\boldsymbol{x} \right]^{1/2} \qquad (2)$$

Let $\{g_1, g_2, ..., g_n\}$ be a function sequence and g_i is continuous at a compact subset $X \subset R^d$.

Definition 1.1 [14]: *A function sequence $\{g_1, g_2, ..., g_n\}$ in $L^2(X)$ is linearly independent if $\sum_{i=1}^n \alpha_i g_i = 0$ implies $\alpha_i = 0$ for all $i = 1, 2, ..., n$. Otherwise, a nontrivial combination of the g_i is zero and $g_1, g_2, ..., g_n$ is said to be linearly dependent.*

Definition 1.2 [14]: *A function sequence $\{u_1, u_2, ..., u_n\}$ in $L^2(X)$ is orthonormal if $< u_i, u_j >= 0$ and $< u_i, u_i >= 1$ whenever $i \neq j$.*

We prove the universal approximation ability of the FNN-RM as follows. Firstly we use the principle proposed in [4] to prove the membership functions can be transformed into an orthonormal set of functions as long as they are linearly independent. Then we prove that the membership functions are linearly independent if the locations of these functions are distinct. Based on the two lemmas stated above, we prove the final theorem using the generalized form of Parseval's theorem [14]. The generalized form of Parseval's theorem is an important property of the generalized Fourier transform, which states that the power computed in either frequency domain equals to the power in the other domain.

Lemma 1. *Given a vector $\boldsymbol{\gamma} = [\gamma_1, \gamma_2, ...\gamma_n]^T$ and activations $g_1, g_2, ..., g_n$, ($g_i \in L^2(X), g_i \neq 0, i = 1, 2, ..., n$), $g_1, g_2, ..., g_n$ are linearly independent, there exists an orthonormal function sequence $\{h_1, h_2, ..., h_n\}$ ($h_i \in L^2(X), h_i \neq 0, i = 1, 2, ..., n$), and a vector $\boldsymbol{\beta} = [\beta_1, \beta_2, ..., \beta_n]$, such that*

$$\sum_{i=1}^n \beta_i g_i = \sum_{i=1}^n \gamma_i h_i. \qquad (3)$$

Proof. Following the method in [4], we apply the Gram-Schmidt orthonormalization algorithm to activation functions $g_i(x)$ to obtain an orthonormal set of basis functions $h_i(x)$ as follows:

$$h_1 = \frac{g_1}{\|g_1\|}$$
$$h_2 = \frac{g_2 - <g_2,h_1>h_1}{\|g_2 - <g_2,h_1>h_1\|}$$
$$\cdots$$
$$h_i = \frac{g_i - \sum_{j=1}^{i-1}<g_i,h_j>h_j}{\|g_i - \sum_{j=1}^{i-1}<g_i,h_j>h_j\|}$$
$$(4)$$

where $i = 1, 2, ..., n$. It is noted that functions $h_i(x)$ can be expressed as linear combinations of $g_j(x)$ $(j = 1, 2, ...k)$. Hence we have

$$h_i(x) = \sum_{j=1}^{i} c_{ij}g_j(x), \; i = 1, 2, ..., n \qquad (5)$$

where c_{ik} are calculated from equation set (4). Then we have

$$\sum_{i=1}^{n} \gamma_i h_i = \sum_{i=1}^{n} \gamma_i \sum_{j=1}^{i} c_{ij}g_j = \sum_{i=1}^{n} \sum_{j=1}^{i} \gamma_i c_{ij}g_j = \sum_{i=1}^{n} \beta_i g_i. \qquad (6)$$

where

$$\beta_i = \sum_{k=i}^{n} \gamma_j c_{jk}. \qquad (7)$$

Hence the proof is completed. $\square$

Lemma 1 shows that a network with rule layer output $g_1, g_2, ..., g_n$, which are linearly independent and continuous on a compact subspace X, can be equivalent to another network with orthonormal continuous rule layer output $h_1, h_2, ..., h_n$.

According to [14], a network sequences $f_n = \sum_{i=1}^{n} \beta_i u_i$ can converge to any continuous function, where $u_1, u_2, ..., u_n$ are orthonormal continuous activation functions. Therefore, we have the following theorem.

Theorem 1. *Given continuous activation functions $g_i(x, \vartheta_i) \in L^2(X)$, ϑ_i is the parameter vector for g_i, $i = 1, 2, ..., n$, and $g_1, g_2, ..., g_n$ are linearly independent, for any continuous function f, there exists a vector $\beta = [\beta_1, \beta_2, ..., \beta_n]^T$ and a network sequence $f_n = \sum_{i=1}^{n} \beta_i g_i$, such that $\lim_{n\to\infty} \|f - f_n\| = 0$.*

Proof. Assume $\{h_1, h_2, ..., h_n\}$ is an orthonormal function sequence obtained by (5), according to Lemma 1, there exists a vector $\beta = [\beta_1, \beta_2, ..., \beta_n]^T$ such that

$$f_n = \sum_{i=1}^{n} \beta_i g_i = \sum_{i=1}^{n} \gamma_i h_i \qquad (8)$$

When $\gamma_i = <f, h_i>$, $(i = 1, 2, ..., n)$, we have

$$\lim_{n\to\infty} \|f - \sum_{i=1}^{n} \gamma_i h_i\| = 0. \qquad (9)$$

Hence

$$\lim_{n \to \infty} \left\| f - \sum_{i=1}^{n} \beta_i g_i \right\| = 0. \tag{10}$$

The Proof is completed. $\square$

Theorem 1 shows that the incremental network f_n can converge to any continuous target function f when rule layer output functions are linearly independent. Next we will show when fuzzy neural networks choose Gaussian function as membership function, the function sequence $g_1, g_2, ..., g_n$ are linearly independent for any $\boldsymbol{a}_k \neq \boldsymbol{a}_j$ ($k \neq j$), where $\boldsymbol{a}_k = [a_{1k}, a_{2k}, ..., a_{dk}]^T$.

Lemma 2. *Given membership functions* $p_{ik}(x_i, a_{ik}, \sigma_{ik}) = \exp\left(-\frac{(x_i - a_{ik})^2}{2\sigma_{ik}^2}\right)$, $i = 1, 2, ..., d$, $k = 1, 2, ..., n$, $n \geq 2$, $g_k(x_1, x_2, ..., x_d) = \prod_{i=1}^{d} p_{ik}$. *Assume* $\boldsymbol{a}_k = [a_{1k}, ..., a_{dk}]^T$, *if* $\boldsymbol{a}_j \neq \boldsymbol{a}_k$, *for any* $j \neq k$, *then* $g_1, g_2, ..., g_n$ *are linearly independent.*

Proof. When $n = 2$, suppose g_1, g_2 are linearly dependent, according to Definition 1.1, there exists a constant vector $\boldsymbol{\phi} = [\phi_1, \phi_2]^T \neq \boldsymbol{0}$, such that

$$\phi_1 g_1 + \phi_2 g_2 = 0. \tag{11}$$

Hence

$$\phi_1 \frac{\partial g_1}{\partial x_i} + \phi_2 \frac{\partial g_2}{\partial x_i} = 0 \tag{12}$$

where $\frac{\partial g_k}{\partial x_i}$ is partial differential form. Since

$$\frac{\partial g_k}{\partial x_i} = -\frac{(x_i - a_{ik})}{\sigma_{ik}^2} g_k, \tag{13}$$

we obtain

$$\phi_1 \frac{(x_i - a_{i1})}{\sigma_{i1}^2} g_1 + \phi_2 \frac{(x_i - a_{i2})}{\sigma_{i2}^2} g_2 = 0 \tag{14}$$

Assume $\phi_1 \neq 0$, when $x_i = a_{i2}$, we have

$$\phi_1 \frac{(a_{i2} - a_{i1})}{\sigma_{i1}^2} g_1 = 0, \tag{15}$$

Since $g_1 = \prod_i^d p_{i1} \neq 0$ and $\phi_1 \neq 0$, we obtain $a_{i2} = a_{i1}$, $i = 1, 2, ..., d$. Therefore $\boldsymbol{a}_1 = \boldsymbol{a}_2$. However, all $\boldsymbol{a}_k$ are different. This presents a contradiction, and hence g_1, g_2 are linearly independent.

Assume $g_1, g_2, \ldots, g_{n-1}$ are linearly independent, we will show that $g_1, g_2, \ldots,$ g_n are also linearly independent. Suppose $g_1, g_2, ..., g_n$ are linearly dependent, according to Definition 1.1, there exists a constant vector $\boldsymbol{\phi} = [\phi_1, \phi_2, ..., \phi_n]^T \neq \boldsymbol{0}$, such that

$$\phi_1 g_1 + \phi_2 g_2 + ... + \phi_n g_n = 0. \tag{16}$$

Hence, we have

$$\phi_1 \frac{\partial g_1}{\partial x_i} + \phi_2 \frac{\partial g_2}{\partial x_i} + \ldots + \phi_{n-1} \frac{\partial g_{n-1}}{\partial x_i} + \phi_n \frac{\partial g_n}{\partial x_i} = 0 \qquad (17)$$

Furthermore,

$$\phi_1 \frac{(x_i - a_{i1})}{\sigma_{i1}^2} g_1 + \phi_2 \frac{(x_i - a_{i2})}{\sigma_{i2}^2} g_2 + \ldots + \phi_{n-1} \frac{(x_i - a_{i,n-1})}{\sigma_{i,n-1}^2} g_{n-1} + \phi_n \frac{(x_i - a_{in})}{\sigma_{i2}^2} g_n = 0 \qquad (18)$$

Assume $\phi_1 \neq 0$, when $x_i = a_{in}$, we obtain

$$\phi_1 \frac{(a_{in} - a_{i1})}{\sigma_{i1}^2} g_1 + \phi_2 \frac{(a_{in} - a_{i2})}{\sigma_{i2}^2} g_2 + \ldots + \phi_{n-1} \frac{(a_{in} - a_{i,n-1})}{\sigma_{i,n-1}^2} g_{n-1} = 0 \qquad (19)$$

Since $g_1, g_2, \ldots, g_{n-1}$ are linearly independent, and $\phi_1 \neq 0$, we have $a_{i1} = a_{in}$ for $i = 1, 2 \ldots, d$. Therefore $\boldsymbol{a}_1 = \boldsymbol{a}_n$. However, all $\boldsymbol{a}_k$ are different. This presents a contradiction, and hence $g_1, g_2, \ldots, g_n$ are linearly independent, which completes the proof. □

Theorem 2. *Given membership functions* $p_{ik}(x_i, a_{ik}, \sigma_{ik}) = \exp\left(-\frac{(x_i - a_{ik})^2}{2\sigma_{ik}^2}\right)$, $i = 1, 2, \ldots, d$, $k = 1, 2, \ldots, n$, $n \geq 2$, $g_k(x_1, x_2, \ldots, x_d) = \prod_{i=1}^{d} p_{ik}$. *Assume* $\boldsymbol{a}_k = [a_{1k}, \ldots, a_{dk}]^T$, *and when* $k \neq j$, $\boldsymbol{a}_k \neq \boldsymbol{a}_j$, *for any continuous function* f, *there exists a vector* $\boldsymbol{\beta} = [\beta_1, \beta_2, \ldots, \beta_n]^T$ *and a network sequence* $f_n = \sum_{k=1}^{n} \beta_k g_k$, *such that* $\lim_{n \to \infty} \|f - f_n\| = 0$.

Proof. According to Lemma 3, when $\boldsymbol{a}_j \neq \boldsymbol{a}_k$ for any $j \neq k$, $g_1, g_2, \ldots, g_n$ are linearly independent. Since $g_k(x_1, x_2, \ldots, x_d)$ is a continuous function and $g_k \in L^2(X)$, then based on Theorem 1, there exists a vector $\boldsymbol{\beta} = [\beta_1, \beta_2, \ldots, \beta_n]^T$ and a network sequence $f_n = \sum_{k=1}^{n} \beta_k g_k$, such that $\lim_{n \to \infty} \|f - f_n\| = 0$. Hence, we completes the proof. □

  If centers are randomly chosen in an FNN, then with probability 1, all the centers $\boldsymbol{a}_i$ are different. Hence, $g_1, g_2, \ldots, g_n$ will be linearly independent with probability 1. According to Theorem 2, this shows that the network will be a universal function approximator with probability 1. When the FNN chooses other nonlinear continuous functions, such as the generalized bell membership function, the sigmoidal membership function [2], sine and cosine, we can similarly prove that the incremental network f_n with different parameters, i.e. arbitrary parameters, converges to any continuous target function f.

3   Discussion and Conclusion

In this paper, we presented an intuitive proof for the universal approximation ability of an FNN with random membership function parameters. This theorem provided a theoretical basis for the FNN-RM algorithm [1–3]. According to this theorem, as long as the centers of the membership functions are different from each other and the number of rules is sufficiently large, the FNN is able to approximate any continuous function.

References

1. Lowe, D.: Adaptive Radial Basis Function Nonlinearities, and the Problem of Generalisation. Proc. First IEE International Conference on Artificial Neural Networks, (1989) 29–33
2. Jang, J.S.R., Sun, C.T., Mizutani, E.: Neuro-fuzzy and Soft Computing. Prentice Hall International Inc (1997)
3. Wang, L., Liu, B., Wan, C.R.: A Novel Fuzzy Neural Network with Fast Training and Accurate Generalization. International Symposium Neural Networks, (2004) 270–275
4. Kaminski, W., Strumillo, P.: Kernel Orthogonalization in Radial Basis Function Neural Networks. IEEE Trans. Neural Networks, 8 (1997) 1177–1183
5. Frayman, Y., Wang, L.: Torsional Vibration Control of Tandem Cold Rolling Mill Spindles: a Fuzzy Neural Approach. Proc. the Australia-Pacific Forum on Intelligent Processing and Manufacturing of Materials, 1 (1997) 89–94
6. Frayman, Y., Wang, L.: A Fuzzy Neural Approach to Speed Control of an Elastic Two-mass System. Proc. 1997 International Conference on Computational Intelligence and Multimedia Applications, (1997) 341–345
7. Frayman, Y., Wang, L.: Data Mining Using Dynamically Constructed Recurrent Fuzzy Neural Networks. Proc. the Second Pacific-Asia Conference on Knowledge Discovery and Data Mining, 1394 (1998) 122–131
8. Frayman, Y., Ting, K. M., Wang, L.: A Fuzzy Neural Network for Data Mining: Dealing with the Problem of Small Disjuncts. Proc. 1999 International Joint Conference on Neural Networks, 4 (1999) 2490–2493
9. Wang, L., Frayman, Y.: A Dynamically-generated Fuzzy Neural Network and Its Application to Torsional Vibration Control of Tandem Cold Rolling Mill Spindles. Engineering Applications of Artificial Intelligence, 15 (2003) 541–550
10. Frayman, Y., Wang, L., Wan, C.R.: Cold Rolling Mill Thickness Control Using the Cascade-correlation Neural Network. Control and Cybernetics, 31 (2002) 327–342
11. Frayman Y., Wang, L.: A Dynamically-constructed Fuzzy Neural Controller for Direct Model Reference Adaptive Control of Multi-input-multi-output Nonlinear Processes. Soft Computing, 6 (2002) 244-253
12. Jang, J.S.R.: Anfis: Adaptive-network-based Fuzzy Inference Systems. IEEE Trans. Syst., Man, Cybern. B, 23 (1993) 665–685
13. Shar, S., Palmieri, F., Datum, M.: Optimal Filtering Algorithms for Fast Learning in Feedforward Neural Networks. Neural Networks, 5 (1992) 779–787
14. Korner, T.W.: Fourier Analysis. Cambridge University Press (1988)
15. Wang, L., Fu, X.J.: Data Mining with Computational Intelligence. Springer. Berlin (2005)
16. Fu, X.J., Wang, L.: Data Dimensionality Reduction with Application to Simplifying RBF Network Structure and Improving Classification Performance. IEEE Trans. Syst., Man, Cybern. B, 33 (2003) 399-409
17. Fu, X.J., Wang, L.: Linguistic Rule Extraction from a Simplified RBF Neural Network. Computational Statistics, 16 (2001) 361-372
18. Fu, X.J., Wang, L., Chua, K.S., Chu, F.: Training RBF Neural Networks on Unbalanced Data. In: Wang, L. et al (eds.): Proc. 9th International Conference on Neural Information Processing (ICONIP 2002), 2 (2002) 1016-1020
19. Kwok, T.Y., Yeung, D.Y.: Objective Functions for Training New Hidden Units in Constructive Neural Networks. IEEE Trans. Neural Networks, 18 (1997) 1131–1148

A Review: Relationship Between Response Properties of Visual Neurons and Advances in Nonlinear Approximation Theory

Shan Tan, Xiuli Ma, Xiangrong Zhang, and Licheng Jiao

National Key Lab for Radar Signal Processing and Institute of Intelligent
Information Processing, Xidian University, Xi'an 710071, China
tanshan5989@yahoo.com.cn

Abstract. In this review, we briefly introduce the 'sparse coding' strategy employed in the sensory information processing system of mammals, and reveal the relationship between the strategy and some new advances in nonlinear approximation theory.

1 Introduction

Some researches indicated that there exists a ubiquitous strategy employed in the sensory information processing system of mammals. This strategy, referred to as 'sparse coding', represents information only using a relatively small number of simultaneously active neurons out of a large population [0].

Neuroscience is usually studied together with mathematics. For example, some works focused on the relations between the 'sparse coding' strategy and the theory of coding [2], or the static of natural scene [3]. In this review, we will reveal the similarity between the 'sparse coding' strategy and nonlinear approximation theory.

The fundamental problem of approximation theory is to resolve a possibly complicated function, called the target function, by simpler, easier to compute functions called the approximants. Nonlinear approximation means that the approximants do not come from linear spaces but rather from nonlinear manifolds. The central question of nonlinear approximation theory is to study the rate of approximation which is the decrease in error versus the number of parameters in the approximant [4]. Essentially, 'sparse coding' strategy is to employ relatively small number simultaneously active neurons to represents information. Such, there should exists intrinsic relationship between the 'sparse coding' strategy and the nonlinear approximation theory. We believe that the advances in the two fields, namely, the neuroscience and approximation theory should shed light on each other and better understanding can be got when we consider the two fields together.

This review is organized as follows. In Section 2, 'sparse coding' strategy in sensory information processing system is briefly reviewed. And some advance in nonlinear approximation is presented in Section 3, where we also analysis the relationship between the 'sparse coding' strategy and these advances. Finally, concluding remarks are given in Section 4.

J. Wang, X. Liao, and Z. Yi (Eds.): ISNN 2005, LNCS 3496, pp. 51–56, 2005.

2 'Sparse Coding' Strategy
in Sensory Information Processing System

In the visual system of mammals, the images that fall upon the retina when viewing the natural world have a relatively regular statistical structure, which arises from the contiguous structure of objects and surfaces in the environment. The sensory system has long been assumed that neurons are adapted, at evolutionary, developmental, and behavioral timescales, to the signals to which they are exposed. The spatial receptive fields of simple cells in mammalian striate cortex have been reasonably well described physiologically and can be characterized as being localized, oriented, and band-pass [3], namely, the simple cells in the primary visual cortex process incoming visual information with receptive fields localized in space and time, band-pass in spatial and temporal frequency, tuned in orientation, and commonly selective for the direction of movement. It is the special spatial structure of the receptive fields of simple cells that produce sparse representation of information.

Some work have examined the 'sparse coding' strategy employed in the V1 [5–8], among which, maybe the most well known one is Olshausen and Field's experiment in [5]. In [5], the relationship between simple-cell receptive fields and sparse coding was reexamined. Olshausen and Field created a model of images based on a linear superposition of basis functions and adapted these functions so as to maximize the sparsity of the representation while preserving information in the images. The set of functions that emerges after training on hundreds of thousands of image patches randomly extracted from natural scenes, starting from completely random initial conditions, strongly resemble the spatial receptive field properties of simple cells, i.e. they are spatially localized, oriented, and band-pass in different spatial frequency bands. Example basis functions derived using sparseness criterion in [5] are shown in Fig. 1.

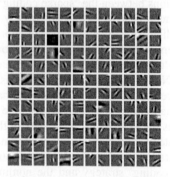

Fig. 1. Example basis functions derived using sparseness criterion in [5]

In [6], Hateren and Schaaf compared properties of the receptive fields of simple cells in macaque cortex with the properties of independent component filters generated by independent component analysis (ICA) on a large set of natural images. Their results showed that the two kinds of receptive field properties match well, according to the histograms of spatial frequency bandwidth, orientation tuning bandwidth, aspect ratio and length of the receptive fields. This indicates that simple cells are well tuned to the expected statistics of natural stimuli.

In [7], Bell and Sejnowski showed that an unsupervised learning algorithm based on information maximization, a nonlinear infomax network, when applied to an ensemble of natural scenes produces sets of visual filters that are localized and oriented. Some of these filters are Gabor-like and resemble those produced by the sparseness-maximization network. In addition, the outputs of these filters are as independent as possible, since this infomax network performs ICA, for sparse component distributions. They also resembled the receptive fields of simple cells in visual cortex, which suggests that these neurons form a natural, information-theoretic coordinate system for natural images.

In [8], Hyvarinen and Hoyer demonstrated that the principle of independence maximization, which is similar with that used in Olshausen and Field's experiment in [5], could explain the emergence of phase- and shift-invariant features, similar to those found in complex cells.

To summarize, the work mentioned above, with the purpose to find the sparse component or sparse representation of natural images derived using optimization algorithm, all showed that the resulting basic functions or the response of the corresponding filter having the visually same receptive field as those of primary cortex neurons. These works, in fact, as well known to researcher in the neuroscience, can be viewed as proofs of the 'sparse coding' strategy employed in the sensory system in mammals. Below, we shall show a new kind of proofs that are from another field, the nonlinear approximation theory used in computationally harmonic analysis or image processing. Different from the works above, these new proofs are along a contrary line, namely, they first intimidate the receptive field of neurons in V1 rather than to obtain sparse representation using some optimization algorithm.

3 New Advances in Nonlinear Approximation Theory and Their Relation with 'Sparse Coding' Strategy

To some degree, the sparse representation of a transform to signals or functions, roughly, is the same as the 'sparse coding' strategy employed in V1 when one consider that V1 functions as if it takes a transform that maps the sensory input into the combination of states of neurons: active or not. Hence, more or less, the advances in the theory of nonlinear approximation should shed light on the understanding the mechanism of sensory information processing system.

Wavelet analysis has achieved tremendous success in many fields of contemporary science and technology, especially in signal and image processing application. The success of wavelet mainly arises from its ability to provide optimal sparse representation for broad function classes, e.g., those smooth away from point singularities.

However, it is well known that the commonly used 2-D separable wavelet system fails at efficiently representing the bivariate functions with straight and curvilinear singularities. The separable wavelet system is the tensor products of 1-D wavelet. It is this special kind of structure that makes the 2-D separable wavelet has only few orientations, i.e. horizontal, vertical and diagonal. And as a result, the approximation accuracy of functions using separable wavelet in 2-D is contaminated by straight and curvilinear singularities.

Whether a system can efficiently represent the underlying functions or not, is always crucial in signal and image processing application such as denoising and compression. Motivated by the failure of wavelet at efficiently representing functions with straight and curvilinear singularities, much work has been done to look for new systems superior to separable wavelet system.

Recently, several new systems for function analysis and image processing have been proposed, which can provide more orientation selectiveness than separable wavelet system hence can effectively deal with straight and curved edges in images.

In [9], Candes developed a new system, ridgelet analysis, and showed how they can be applied to solve important problems such as constructing neural networks, approximating and estimating multivariate functions by linear combinations of ridge functions. In a following paper [10], Donoho constructed orthonormal ridgelet, which provides an orthonormal basis for $L^2(R^2)$. It was shown that both ridgelet analysis and orthonormal ridgelet is optimal to represent functions that are smooth away from straight singularity. In paper [11], it was shown that both ridgelet analysis and orthonormal ridgelet are optimal to represent functions that are smooth away from straight singularity. For example, for the function mentioned above, g, the number of orthonormal ridgelet coefficients with amplitude exceeding $1/M$ grows with M more slowly than any fractional power of M, namely, $\#\{\alpha_\lambda^R \| \alpha_\lambda^R |> 1/M\} \sim O(M^{-s})$ for $\forall s \in Z^+$, where α_λ^R is the orthonormal ridgelet expansion coefficient of g with index λ. By comparison, when decomposing function g into wavelets series, we only have $\#\{\alpha_\lambda^W \| \alpha_\lambda^W |> 1/M\} \sim O(M^{-1})$. It is exciting that the higher approximation rate of g is achieved by orthonormal ridgelet than that by wavelet system. In fact, orthonormal ridgelet can optimally represent functions smooth away from straight singularity in the sense that nor orthonormal system achieves higher approximation rate than orthonormal ridgelet do. The key idea of orthonormal ridgelet is that it first transforms the straight singularity in spatial space into point singularity in Radon domain, then deal with the resulting point singularity using wavelet system. As a result, in effect, it has 'anisotropic' basis functions as shown in Fig. 2. As an extension version of orthonormal ridgelet, ridgelet frame and dual ridgelet frame was proposed in paper [12], [13], both of which also can effectively deal with straight edges in images. Though the multi-resolution and localization property, rigorously, are not introduced into these system, they can provide the effectively sparse representation for images with straight edges. We suggest that the multi-orientation property, maybe, plays a more important role than others, i.e. band-pass and localization property, in the 'sparse coding' strategy in V1, considering that edges are dominating features in natural images.

Based on a localization principle and subband decomposition, monoscale ridgelet and Curvelet were proposed [14], both of which are derived from Candes' ridgelet analysis and can efficiently deal with smooth images with smooth edges (including both straight and curved ones). Besides band-pass and localized, the basis functions of Curvelet exhibit very high direction sensitivity and are highly anisotropic, as shown in Fig. 3.

Fig. 2. From left to right: example basis function of ridgelet analysis, orthonormal ridgelet and ridgelet frame

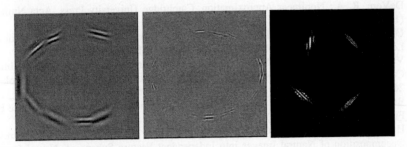

Fig. 3. Example basis functions of Curvelet (the first two) and Countourlet (the last one)

Suppose we have an object supported in $[0,1]^2$ which has a discontinuity across a nice curve Γ, and which is otherwise smooth. The error of m-term nonlinear approximation using Curvelet can achieve the optimal approximation rate: $\left\| f - f_M^C \right\|^2 \sim O(M^{-2}(\log M)), M \to \infty$. Whereas using a standard wavelet representation, the error of n-term nonlinear approximation only satisfies $\left\| f - f_M^W \right\|^2 \sim O(M^{-1}), M \to \infty$.

Minh N. Do and Vetterli developed a new 'true' two-dimensional representation for images that can capture the intrinsic geometrical structure of pictorial information [15]. The new system, called countourlet transform, provides a flexible multi-resolution, local and directional expansion for images. Several basis functions of countourlet transform are shown in Fig. 3. The countourlet transform also provides a sparse representation for two-dimension piecewise smooth signals, namely, in such case, the error of m-term nonlinear approximation using countourlet can achieve the optimal approximation rate: $\left\| f - f_M^C \right\|^2 \sim O(M^{-2}(\log M^3)), M \to \infty$.

4 Discussion

We have reveal the relationship between 'sparse coding' strategy employed in the sensory information processing system of mammals and some new advances in nonlinear approximation theory. Several newly-proposed function analysis or image representation systems, whose basis functions imitated the response properties of neurons in V1, namely, band-pass (multi-resolution), localized and multi-orientation, interestingly, can provide sparse representation for some special classes of images dominated by edges. The relationship allows us to understand both fields better.

References

1. Olshausen, B. A., Field, D. J.: Sparse Coding of Sensory Inputs. Current Opinion in Neurobiology, **14** (2004) 481-487
2. Olshausen, B. A., Field, D. J.: Sparse Coding With An Overcomplete Basis Set: a Strategy Employed By V1?. Vision Res, **37** (1997) 3311-3325
3. Simoncelli, E. P., Olshausen, B. A.: Natural Image Statistics and Neural Representation. Annu. Rev. Neurosci, **24** (2001) 1193-1216
4. DeVore, R. A.: Nonlinear Approximation. Cambridge University Press. Acta Numerica (1998)
5. Olshausen, B. A., Field, D.J.: Emergence of Simple-cell Receptive Field Properties By Learning A Sparse Code for Natural Images. Nature, **381** (1996) 607–609
6. Van Hateren, J. H, Van Der Schaaf A.: Independent Component Filters of Natural Images Compared with Simple Cells in Primary Visual Cortex. Proc R. Soc. Lond B Biol. Sci, **265** (1998) 359-366
7. Bell, A. J., Sejnowski, T. J.: The 'Independent Components' of Natural Scenes Are Edge Filters. Vision Res, **37** (1997) 3327-3338
8. Hyvarinen, A., Hoyer, P. O.: Emergence of Phase and Shift Invariant Features by Decomposition of Natural Images into Independent Feature Subspaces. Neural Comput, **12** (2000) 1705-1720
9. Candès, E.J.: Harmonic Analysis of Neural Networks. Appl. Comput. Harmon. Anal, **6** (1999) 197–218
10. Donoho, D.L.: Orthonormal Ridgelet and Straight Singularities. SIAM J. Math Anal, **31** (2000) 1062–1099
11. Candès, E. J.: On the Representation of Mutilated Sobolev Functions. SIAM J. Math. Anal., **1** (1999) 2495-2509
12. Tan, Jiao, L. C., Feng, X. C.: Ridgelet Frame. In: Campilho, Aurélio, Kamel, Mohamed (eds.): Proc. Int. Conf. Image Analysis and Recognition, Springer Lecture Notes in Computer Science (LNCS) series, (2004) 479-486
13. Tan, S., Zhang, X. R, Jiao, L. C.: Dual Ridgelet Frame Constructed Using Biorthonormal Wavelet Basis. In: IEEE International Conference on Acoustics, Speech, and Signal Processing, (2005)
14. Candès, E. J., Donoho, D. L.: Curvelet-A Surprisingly Effective Nonadaptive Representation for Objects With Edges. In: Cohen, A., Rabut, C., Schumaker, L. L. (eds.): Curve and Surface Fitting: Saint-Malo. Nashville, Van-derbilt Univ. Press, TN (1999)
15. Do, M. N., Vetterli, M.: Contourlets. In: Stoeckler, J. Welland, G.V. (eds.): Beyond Wavelet, Academic Press (2002)

Image Representation in Visual Cortex and High Nonlinear Approximation

Shan Tan, Xiangrong Zhang, Shuang Wang, and Licheng Jiao

National Key Lab for Radar Signal Processing and Institute of Intelligent
Information Processing, Xidian University, Xi'an 710071, China
tanshan5989@yahoo.com.cn

Abstract. We briefly review the "sparse coding" principle employed in the sensory information processing system of mammals and focus on the phenomenon that such principle is realized through over-complete representation strategy in primary sensory cortical areas (V1). Considering the lack of quantitative analysis of how many gains in sparsenality the over-complete representation strategy brings in neuroscience, in this paper, we give a quantitative analysis from the viewpoint of nonlinear approximation. The result shows that the over-complete strategy can provide sparser representation than the complete strategy.

1 Introduction

At any given moment, our senses are receiving vast amount of information about the environment in the form of light intensities, changes in sound pressure, deformations of skin, stimulation of taste and olfactory receptors and more. How the brain makes sense of this flood of time-varying information and forms useful internal representations for mediating behavior remains one of the outstanding mysteries in neuroscience.

Much work has shown that "sparse coding" is a common underlying principle involved in the sensory information processing system of mammals, namely that information is represented by a relatively small number of simultaneously active neurons out of a large population [1]. The "sparse coding" principle could possibly confer several advantages. Early work on associative memory models showed that sparse representations are most effective for storing patterns as they maximize memory capacity because of the fact there are fewer collisions between patterns [2]. Barlow's work indicated that in many sensory nervous systems, neurons at later stages of processing are generally less active than those at earlier stages, which suggested that nervous system was attempting to form neural representations with higher degrees of specificity and thus make it possible for higher areas to learn structure in the data [3]. Other work showed that, given the actual energy constraints of the mammalian cortex, sparse coding would seem to be a necessity [4] [5].

In this paper, we focus on how the mammals realize the "sparse coding" principle in V1 in section 2, and in section 3 we give closely correlated facts emerged in the field of nonlinear approximation and commonly referred as "high nonlinear approximation", which provide quantitative analysis of the gains in sparsenality brought by over-complete representation. Finally, concluding marks are given in section 4.

J. Wang, X. Liao, and Z. Yi (Eds.): ISNN 2005, LNCS 3496, pp. 57–62, 2005.

2 Realization of "Sparse Coding" Principle Through Over-Complete Representation in V1

For convenience below, we first give a simplified model according to the standard notions of linear coding. A typical form of image-coding principle is to apply a linear transform to the image by taking the inner-product of the image, $I(\vec{x})$, with a set of spatial weighting functions, ψ:

$$b_i = \sum_{\vec{x}_j} \psi_i(\vec{x}_j) I(\vec{x}_j) .$$
(1)

Alternatively, we may write this operation in vector notation as

$$b = WI ,$$
(2)

where I is the vector with components $I_i = I(\vec{x}_j)$ and W is the matrix with components $W_{i,j} = \psi_i(\vec{x}_j)$.

Generally, the goal in such a standard coding principle is to find an invertible weight matrix W that transforms the input so that some criterion of optimality on the output activities is met (e.g., de-correlation, sparseness, etc.). This is the basis of coding strategies such as Discrete Cosine Transform (used in JPEG image compression), or orthonormal wavelet transform.

It is known that, different from the standard image coding, the coding principle employed in V1 need not be invertible exactly and usually it cannot be modeled actually using linear models such as (1). However, how the V1 realize the "sparse coding" does remain one of the most challenging questions in neuroscience. Many researchers have focused on such a challenging question [1] [6] [7] [8].

To realize the "sparse coding" principle, a ubiquitous strategy employed in V1 is that they over-complete represent the sensory inputs many times over [6]. For example, in cat V1, there exists 25:1 expansion ration in terms of the number of axons projecting from layers 2/3 on to higher areas relative to the number of inputs from the Lateral Geniculate Nucleus (LGN).

Specialists in neuroscience suggested and several evidences have indicated that the over-complete representation is utilized to produce an even higher degree of sparsity among neurons by making them more selective to specific patterns of input. Some work showed that natural images lie along a continuous curved surface or "manifold" embedded in the high-dimensional state space of images [7] [8] [9]. The surface represents the smooth changes that follow from the transformations that are likely to occur in natural scenes (e.g., translation, scale, rotation, etc.). For example, an object that moves across the pixel array of an image gives rise to a series of different spatial patterns. Each of these spatial patterns corresponds to a point in the state space, and the set of all points resulting from this motion would form a smooth, curved trajectory in the state space, and the set of all points resulting from this motion would form a smooth, curved trajectory in the state space. So how can the visual system represent this curved surface of probable images? One possibility is that this is achieved through the over-complete representation of the state space, in which the number of neurons used to represent the image is potentially much greater than the dimensionality of the input (number of pixels) [10] [11]. In this coding scheme, each neuron would have a preferred pattern in the input space and a neuron would become active only when the input pattern is sufficiently close to its preferred pattern. It was suggested that the advantage of such a coding scheme is that the manifold then becomes flattened out (less curvature) in the higher-dimensional space defined by the neural response, thus making it easier for higher areas to learn structure in the data.

In the other side, namely, from the mathematic and physical viewpoint, one obvious reason for V1 using the over-complete coding strategy is that it possesses greater robustness in the face of noise and other forms of degradation. And a more pertinent reason is that it will allow greater flexibility in matching the generative model to the input structure. In addition, over-complete codes allow for small translation or scaling of local image features to result in a smooth and graceful change in the distribution of activity among coefficients. By contrast, in a critically sampled code, local image changes typically result in fairly global and drastic undulations among coefficients values. Such instabilities would be undesirable for doing pattern analysis, and also in terms of maintaining a sparse image representation.

3   Quantitative Analysis of Gains Brought by Over-Complete Representation Strategy in V1 Using Nonlinear Approximation Theory

In this section, we shall quantitatively analyze the gain, if any, in terms of the "sparsity" brought by the over-complete representation strategy in V1 through technology in another field – nonlinear approximation.

The fundamental problem of approximation is to resolve a possibly complicated function, called the target function, by simpler, easier to compute function called approximants. Nonlinear approximation means that the approximants do not come from linear spaces but rather from nonlinear manifolds. Recently, a new setting, namely, the so-called "high nonlinear approximation", has received more and more attention in the field of data analysis and nonlinear approximation [12] [13] [14] [15]. The high nonlinear approximation is essentially a kind of over-complete representation for function classes of some kind. Reminiscing the strategy of representation employed in the visual sensing system of mammals, we believe that the high nonlinear approximation in fact makes it available to understanding the essence of the over-complete representation in V1. In other words, we can understand mathematically what and how much gains the strategy in V1 provides in terms of the theory of nonlinear approximation.

In this paper, we compare the *n-term* nonlinear approximation in two cases, namely, the case of orthonormal basis representation and the case of over-complete representation.

Consider Hilbert space H and let $\Sigma_n := span\ \{\eta_k : 1 \le k \le n\}$ with constraint $\#\Sigma_n \le n$, here $n \le \dim(H)$ and $\eta_k\ (k \le \dim(H))$ is an orthonormal system for H. Then, for all $f \in H$ the error of *n-term* nonlinear approximation by $\eta_k\ (k \le \dim(H))$ can be express as

$$\sigma_n(f)_H := \inf_{S \in \Sigma_n} \|f - S\|_H .$$ (3)

The essential question of nonlinear approximation is that given a real number $\alpha > 0$, for which elements $f \in H$ do we have, $\sigma_n(f)_H \leq Mn^{-\alpha}$, $n=1, 2, \ldots$ for some constant $M > 0$? In other words, which functions are approximated at a given rate like $O(n^{-\alpha})$? Usually, we use approximation space to characterizes the approximation efficiency of a given η_k ($k \leq \dim(H)$). And the approximation space is defined as

$$\mathbf{A}_q^\alpha := \{ f \in H \mid (\sum_{n=1}^{\infty} [n^\alpha \sigma_n(f)]^q \frac{1}{n})^{\frac{1}{q}} < \infty \}, \tag{4}$$

here, the approximation space $\mathbf{A}_q^\alpha$ gathers functions together under one roof, which have a common approximation order.

Commonly, in the theory of nonlinear approximation, the broader the approximation space is, we say that the higher nonlinear approximation efficiency is obtained by the associating basis, hence the more sparsely such basis represents a given functions. In the standard notions of linear coding (1), for example, the orthonormal wavelet system is more preferred than the Discrete Cosine Transform for most of still images in that the former has much broader approximation space than the latter.

Now, consider the case of high nonlinear approximation. Call an arbitrary subset of H, D, dictionary. For example, D can be the collection of different orthonormal bases whose element has different waveform. Usually, D is over-complete for H, namely, the family of functions used in the approximation process is highly redundant and there are many more functions in the dictionary than needed to approximation any target function f. The hope is that the redundancy will increase the efficiency of approximation. As in the case of visual research in neuroscience, results on highly nonlinear approximation are quite fragmentary and a cohered theory still needs to be developed. However, some original work on the highly nonlinear approximation has emerged recently. These primary results, we believe, shed light, more or less, on the understanding the over-complete strategy used in V1.

Let $\Sigma_n := \Sigma_n(D)$ denote the collection of all functions in H which can be expressed as a linear combination of at most n elements of D, namely, $\#\Sigma_n \leq n$. Then, define

$$S = \sum_{g_k \in \Lambda} \beta_k g_k, \tag{5}$$

here $\Lambda \subset D$, $\#\Lambda \leq n$ and $\beta_k \in R$. Then, for $\forall f \in H$, the error of *n-term* nonlinear approximation by D can be written as

$$\sigma_n(f) := \sigma_n(f, D)_H := \inf_{S \in \Sigma_n(D)} \| f - S \|_H. \tag{6}$$

As mentioned above, $B := \{ \eta_k, k \leq \dim(H) \}$ is an orthonormal basis for H and the approximation error is defined in (3). It follows that $f \in A_\infty^\alpha(H)$ if and only if the expansion coefficients series $f_k := < f, \eta_k >$ is in weak $l_{\tau(\alpha),\infty}$, here $\tau(\alpha) := (\alpha + \frac{1}{2})^{-1}$. Note that $f \in A_\infty^\alpha(\mathcal{H})$ is equivalent to $E_n(f)_X = O(n^{-\alpha})$. In other words, let $\gamma_n(f, B)$ be the nth largest of the absolute values of the expansion coefficients series, we have $\| (\gamma_n(f, B)) \|_{l_{\tau,\infty}} \sim |f|_{A_\infty^\alpha}$, here the semi-norm $|f|_{A_\infty^\alpha}$ is defined by (4). Now, for the purpose to compare the approximation efficiency of over-complete dictionary $\mathbb{D}$ with that of orthonormal basis, consider a simply case in which the dictionary $\mathbb{D}$ consists of a family of orthonormal system $\mathcal{L} = \{\mathbf{B}\}$, here $\mathcal{L}$ is the collection of several orthonormal system such as B. Then, it is obvious that the error of *n-term* nonlinear approximation satisfy

$$\sigma_n^{\mathbb{D}}(f)_H := \inf_{B \in \mathcal{L}} \sigma_n(f, B)_H . \tag{7}$$

Using the relation $\left\| (\gamma_n(f,B)) \right\|_{l_{\tau,\infty}} \sim |f|_{\mathcal{A}^{\alpha}}$, it is not difficult to obtain

$$\sigma_n^{\mathbb{D}}(f)_H \leq Cn^{-\alpha} \inf_B \left\| (\gamma_n(f,B)) \right\|_{l_{\tau,\infty}} , \tag{8}$$

with C an absolute constant. Moreover, for any $\alpha > 0$, we have

$$\cap_B \mathcal{A}_\infty^\alpha(H, \ B) \subset \mathcal{A}_\infty^\alpha(H, \ \mathbb{D}) . \tag{9}$$

The relation (9) can be essentially interpreted that higher approximation efficiency is obtained from dictionary $\mathbb{D}$, compared with that from individual orthonormal basis such as B. Note the gain is resulted from the introduction of over-completeness. And in the nonlinear approximation frame the over-complete representation is superior to orthonormal one in terms of the *n*-term nonlinear approximation, which also can be interpreted roughly as sparseness. It has long been assumed that neurons in the sensing processing system of mammals are adapted, at evolutionary, developmental, and behavioral timescales, to the signals to which they are exposed. Interestingly, to obtain the sparse representation of input information, these neurons actually develop an over-complete representation strategy, which we can prove that it is preferred to sparser representation of input information using the theory of nonlinear approximation.

4 Conclusion

In this paper we have focused on the over-complete representation strategy in V1 of mammals and analyzed the advantages of this strategy from the viewpoint of nonlinear approximation theory. The relationship has been illuminated, which showed quantitatively that the over-complete strategy is more preferable to "sparse coding" employed in V1 compared with the complete representation such as orthonormal basis.

References

1. Olshausen, B.A., Field, D.J.: Sparse Coding of Sensory Inputs. Current Opinion in Neurobiology, **14** (2004) 481–487
2. Willshaw, D.J., Buneman, O.P., Longuet-Higgins, H. C.: Nonholographic Associative Memory. Nature, **222** (1969) 960–962
3. Barlow, H.B.: Possible Principles Underlying the Transformation of Sensory Messages. In: Rosenblith, W.A. (eds.), Sensory Communication, Cambridge, MA: MIT Press, (1961) 217–234
4. Attwell, D., Laughlin, S.B.: An Energy Budget for Signaling in The Grey Matter of The Brain. J Cereb Blood Flow Metab, **21** (2001) 1133–1145

5. Lennie, P.: The Cost of Cortical Computation. Curr. Biol., **13** (2003) 493–497
6. Olshausen, B.A.: Principles of Image Representation in Visual Cortex. In: Chalupa, L.M., Werner, J.S., Boston, M.A. (eds.), The Visual Neurosciences. MIT Press, (2003) 1603–1615
7. Lee, K.S., Pedersen, D., Mumford: The Nonlinear Statistics of High-contrast Patches in Natural Images. Int J. Comput Vis. **54**(2003) 83–103
8. Roweis, S.T., Saul, L.L.: Nonlinear Dimensionality Reduction By Locally Linear Embedding. Science, **290** (2000) 2323–2326
9. Wiskott, L., Sejnowski, L.: Slow Feature Analysis: Unsupervised Learning of Invariances. Neural Comput, **14** (2002) 715–770
10. Olshausen, B.A. Field, D.J.: Sparse Coding With An Overcomplete Basis Set: A Strategy Employed By V1? Vision Res, **37** (1997) 3311–3325
11. Lewicki, M.S., Sejnowski, T.J.: Learning Overcomplete Representations. Neural Comput, **12** (2000) 337–365
12. DeVore, R.A.: Nonlinear Approximation. Acta Numerica, Cambridge University Press, (1998)
13. D. L. Donoho and M. Elad, Optimally Sparse Representation in General (Non-orthogonal) Dictionaries Via Minimization, Proc. Nat. Aca. Sci. **100** (2003) 2197–2202
14. Feuer, Nemirovsky, A.: On Sparse Representations in Pairs of Bases. IEEE Trans. Inform. Theory, **49** (2003) 1579-1581
15. Starck, J.L., Elad, M., Donoho, D. L.: Image Decomposition Via the Compination of Sparse Representations And A Variational Approach, Submitted to IEEE. IP., (2004)

Generalization and Property Analysis of GENET*

Youmei Li[1], Zongben Xu[2], and Feilong Cao[3]

[1] Department of Computer Science, Shaoxing College of Arts and Sciences,
Shaoxing 312000, China
li_youmei@zscas.edu.cn
[2] Institute for Information and System Sciences, Faculty of Science,
Xi'an Jiaotong University, Xi'an, Shaan'xi 710049, China
[3] Department of Information and Mathematics Sciences, College of Science,
China Jiliang University, Hangzhou 310018, China

Abstract. GENET model has attracted much attention for its special feature in solving constraint satisfaction problems. However, the parameter setting problems have not been discussed in detail. In this paper, the convergent behavior of GENET is thoroughly analyzed and its learning strategy is generalized. The obtained results can shed light on choosing parameter values and exploiting problem specific information.

1 Introduction

Consider a binary constraint satisfaction problem (CSP): (Z, D, C) with

- $Z = \{\xi_1, \xi_2, ..\xi_n)$ being a finite set of variables;
- $D = \cup D_j$ and $D_j = (\eta_{1j}, \eta_{2j}, ..., \eta_{mj})$ being the domain of variable ξ_j (i.e., each ξ_j takes any value in D_j, $j = 1, 2, ..., n$). Without loss of generality, we assume that the scale of all D_j is the same.
- C being a set of constraints of any form, that restricts the values each pair of variables may take simultaneously.

The task is then to assign one value η_{ij} $(i = 1, 2, ..., m)$ to each variable ξ_j $(j = 1, 2, ..., n)$ satisfying all the constraints C.

CSPs are well-known to be NP-hard in general. In many situations, stochastic search methods are needed (see [1],[2],[3]). GENET, proposed by Davenport A. et al. , belongs to this kind of methods. GENET consists of three components: a network architecture, local search and a reinforcement learning strategy.

Network Architecture: A GENET network $\mathcal{N}$ representing CSP: (Z, D, C) consists of a set of label nodes and $|C|$ connections. A label node is a tuple (ξ_i, η_{xi}) where ξ_i is a variable and η_{xi} is a value in the domain of ξ_i. Each label node (ξ_i, η_{xi}) (in short,(x, i)) is associated with an output v_{xi}, which is 1 (active) if η_{xi} is assigned to variable ξ_i, and 0 (inactive) otherwise. The label nodes for the same variable are grouped into a cluster. In each cluster, only one label node is active.

* Supported by the Nature Science Foundation of China under Grant 60473034 and the Youth Science Foundation of Shanxi Province under Grant 20031028.

J. Wang, X. Liao, and Z. Yi (Eds.): ISNN 2005, LNCS 3496, pp. 63–68, 2005.

A connection is made between each pair of label nodes $(x, i), (y, j), i \neq j$ that represent two incompatible labels between two different variables. Associated with each connection is a weight $w_{xi,yj}$, which is a negative value indicating the strength of an inhibitory relationship between the two connected nodes. All weights are initialized to -1 at start.

Local Search: By representing CSP as above, each label node (x, i) receives an input u_{xi}

$$u_{xi} = \sum_{(y,j) \in A(\mathcal{N}, (x,i))} w_{xi,yj} v_{yj} \tag{1}$$

where $A(\mathcal{N}, (x, i))$ is the set of all label nodes connected to (x, i). A solution state of network $\mathcal{N}$ has the inputs of all active label nodes being zero. The initial state of network $\mathcal{N}$ can be set by randomly selecting a label node in each cluster to be active and other label nodes to be inactive. During local search process, the label nodes for each cluster repeatedly competes for activation, and the node that receives the maximum input will become active. GENET iteratively update the network state cluster by cluster asynchronously till the network settles in a stable state. Since all weights are negative, this state update rule means minimizing the number of weighted constraint violations $- \sum_{((x,i)(y,j)) \in C} w_{xi,yj} v_{xi} v_{yj} = - \sum_{(x,i)} v_{xi} u_{xi}$.

Reinforcement Learning: When the network has converged to a local minimum, the weight associated with the connection between active nodes are reduced by 1, that is

$$w_{xi,yj} = w_{xi,yj} - v_{xi} v_{yj}. \tag{2}$$

In this way, the local minimum is removed by decreasing the weights of violated connections, and reducing the possibility of any local minimum appearing again.

However, there is two reasons that greatly influence the efficiency of GENET. First, since the initial weights were set to -1, the problem specific information does not be well exploited. Second, in learning process, for a violated connection, its weight is punished by decreasing 1 unit, thus in order to satisfy this constraint, the punishment process will be implemented several times. So we hope to determine a suitable punishment unit to improve the speed of GENET. In the following section, we will separately discuss this two problems.

2   The Generalization of Connection Weights

In this section, we assume that between any two nodes there is a connection and associated connection weight, and we will give a convergence theorem in order to guide the determination of connection weights of GENET.

A CSP can be represented as the following combinatorial optimization problem:

$$\min E(V) = -\frac{1}{2} \sum_{x=1}^{m} \sum_{y=1}^{m} \sum_{i=1}^{n} \sum_{j=1}^{n} w_{xi,yj} v_{xi} v_{yj} \tag{3}$$

$$subject \ to \ V = (v_{xi})_{m \times n} \in \Omega_c \tag{4}$$

where Ω_c is defined as follows:

$$\Omega_c = \left\{ V = (v_{x,i})_{m \times n} : v_{x,i} \in \{0,1\} \ and \ \sum_{x=1}^{m} v_{x,i} = 1, \ for \ i = 1,2,...,n \right\} \quad (5)$$

The set Ω_c could be referred to as the feasible set associated with CSP. We need introducing several useful notion and notations.

Definition 1. (WTA Operator). (i) We call a mapping $WTA:\mathbf{R}^m \to \{0,1\}^m$ a WTA (Winner Takes All) operator if for any $x = (x_1, x_2, ..., x_m)^T \in R^m$,

$$WTA(x) = ([WTA(x)]_1, [WTA(x)]_2, ..., [WTA(x)]_m)^T \quad (6)$$

with

$$[WTA(x)]_i = \begin{cases} 1, \ if \ x_i = \max_{j=1,2,...,m}\{x_j\} \\ 0, \ otherwise \end{cases} i = 1,2,...,m \quad (7)$$

where it is regulated that only one component of $WTA(x)$ takes 1 even when more than one component in vector x takes the maximum value.

(ii) A mapping $WTA_c:\mathbf{R}^{mn} \to \{0,1\}^m$ is called a column-WTA operator if for any $U = (u_{xi}) \in \mathbf{R}^{mn}$, write $U = (U^{(\cdot,1)}, U^{(\cdot,2)}, ...U^{(\cdot,n)})$, where $U^{(\cdot,i)}$ denote the column of the matrix U, and we have

$$WTA_c(U) = (WTA(U^{(\cdot,1)}), WTA(U^{(\cdot,2)}), ...WTA(U^{(\cdot,n)})). \quad (8)$$

Let U denote the input matrix of all label node (x,i), and $U^{(\cdot,i)} = (u_{1i}, u_{2i}, ..., u_{mi})^T$ be the ith column. By Definition 1, the output of node (x,i) is computed by

$$v_{xi} = [WTA(U^{(\cdot,i)})]_x \quad (9)$$

The state of a CSP network then can be represented by $V = WTA_c(U)$ accordingly, where $U = (U^{(\cdot,1)}, U^{(\cdot,2)}, ...U^{(\cdot,n)})$ are whole inputs of the network.

With the above notion and notations, the GENET local search can then be formulated as

$$V(t+1) = WTA_c(U(t)) = WTA_c(WV(t)), t = 1,2,..., \quad (10)$$

where $WV(t) = ([WV(t)]_{xi})_{m \times n} = (u_{xi})_{m \times n}, u_{xi} = \sum_{y=1}^{m} \sum_{j=1}^{n} w_{xi,yj} v_{yj}$.

In GENET, the states of the network are updated column by column in a predefined or random order, and if the ith column of the network is chosen to be updated at time t, the state of the network at time $t+1$ is determined by

$$[V(t+1)]^{(\cdot,k)} = \begin{cases} WTA(U(t)^{(\cdot,i)}), \ if \ k = i \\ V(t)^{(\cdot,k)}, \quad\quad\quad if \ k \neq i \end{cases} \quad (11)$$

Remark 1. GENET local search can be regulated *in* the following way: whenever there are more than one nodes in the ith column which receive the maximal input, we regulate that a node (x,i) wins the competition if either the node *wins* the competition at the previous time instant (i.e., $v_{x,i}(t) = 1$) or x is the minimal index such that $u_{x,i}(t) \geq u_{yi}(t)$ for any $y \in \{1, 2, ..., m\}$. This regulation could be called the Old-Winner-Least-Index ($OWLI$) priority regulation.

Let the set of all local minimizers of $E(V)$ be denoted by $\Omega(E)$, and the set of all stable states of system (10) be denoted by $\Omega(N)$. The convergence of the GENET local search (11) is given in the following theorem.

Theorem 1. *Assume* $W = (w_{xi,yj})_{nm \times nm}$ *is symmetric, i.e.,* $w_{xi,yj} = w_{yj,xi}$, *the OWLI priority regulation is made, and* $\{V(t)\}$ *is the mutual different sequence defined by the GENET local search (10) with the sequential operation mode. For any* $x, y \in \{1, 2, ..., m\}$, $k \in \{1, 2, ..., n\}$, *let*

$$C_{x,y}^{(k)} = w_{xk,xk} + w_{yk,yk} - 2w_{xk,yk} \tag{12}$$

Then

(i) $V(t)$ *converges to a stable state of (11) within a finite number of steps, from any initial value* $V(0)$ *in* Ω_c, *if* $C_{x,y}^{(k)} \geq 0$, *for any* $x, y \in \{1, 2, ..., m\}$, $k \in \{1, 2, ..., n\}$ *and* $x \neq y$;

(ii) Any stable state of (11) is a local minimizer of (3) if $C_{x,y}^{(k)} \leq 0$, $\forall\ x, y \in \{1, 2, ..., m\}$, $k \in \{1, 2, ..., n\}$;

(iii) $V(t)$ *converges to a local minimizer of (11) within a finite number of steps if* $C_{x,y}^{(k)} = 0$, *for any* $x, y \in \{1, 2, ..., m\}$, $k \in \{1, 2, ..., n\}$ *and* $x \neq y$.

The proof is omitted.

Remark 2. (i) When applied to CSP, one normally sets $w_{xk,xk} = 0$ and $w_{xk,yk} = w_{yk,xk} = 0$, which meets the condition $C_{xy}^{(k)} = 0$ naturally. So, from Theorem 1, the convergence of GENET local search to a local minimum solution of CSP can always be guaranteed. (The same is true if we set $w_{xk,xk} + w_{yk,yk} = -2w_{xk,yk}$ for any $x, y \in \{1, 2, ..., m\}$ and $k \in \{1, 2, ..., n\}$).

(ii) If allowing the convergent limit of $\{V(t)\}$ not necessarily is a local minimizer of E (that is maybe needed in the learning process in order to escape from local minimizer), Theorem 1 (i) says that $C_{xy}^{(k)} \geq 0$ is sufficient for convergence of the GENET local search (11).

(iii) The problem information can be classified into two classes. One class demands that some constraints are hard and others are soft. As for this kind of information, we can combine it into the connection weights, that is, the hard constraints should correspond to absolutely large weight. The other demands that some nodes may be more desired to be active. We can also exploit this information through assigning a positive value to self-feedback connection $w_{xk,xk}$.

3   The Generalization of Learning Strategy

We illustrate in this section that how those suitable learning factors are chosen. For clarity, we treat the state variable $V = (v_{xi})_{m \times n}$ as a long $m \times n$ vector aligned as

$$V = (v_{11}, ..., v_{1n} \vdots v_{21}, ... v_{2n} \vdots ...... \vdots v_{m1}, ..., v_{mn})^T \tag{13}$$

and, the tensor $W = (w_{xi,yj})_{mn \times mn}$ should be understood as a $mn \times mn$ matrix whose element in $xi - th$ row and $yj - th$ column is given by $w_{xi,yj}$.

Let us suppose that $V^*(k)$ is yielded through minimization of the kth step objective function

$$E^{(k)}(V) = -\frac{1}{2} \sum_{x=1}^{m} \sum_{y=1}^{m} \sum_{i=1}^{n} \sum_{j=1}^{n} w_{xi,yj}^{(k)} v_{xi} v_{yj} \tag{14}$$

under the constraint $V \in \Omega_c$, where $W^{(k)} = (w_{xi,yj}^{(k)})_{nm \times nm}$ is the strength tensor associated with the CSP network. Then $V^*(k)$ may be or may not be a global minimum solution, depending on if $E(V^*(k)) = 0$. If $V^*(k)$ is not a global minimum solution of the CSP, GENET suggests finding a further local minimum $V^*(k+1)$, starting from the kth local minimum $V^*(k)$ and through applying reinforcement learning strategy and implementing the GENET local search for the $(k+1)$-step objective function.

The GENET reinforcement learning rule (2) can be embedded into a more general form as follows

$$W^{(k+1)} = W^{(k)} - \alpha_k[(V^*(k))(V^*(k))^T - \beta_k I] \tag{15}$$

where I is the identity tensor that satisfies $i_{xi,xi} = 1$ and $i_{xi,yj} = 0$ whenever $(x,i) \neq (y,j)$, $\alpha_k > 0$, $\beta_k \in (0,1)$. It can be seen from (15) that, in principle, except for penalizing the two violated nodes (x,i) and (y,j), a certain amount of penalization is also imposed directly on node (x,i) itself in the new rule (15). With the update rule (15), The following facts can be observed:

(1) The modified $(k+1)th$ step objective function can be defined by

$$E^{(k+1)}(V) = -\frac{1}{2} \sum_{x=1}^{m} \sum_{y=1}^{m} \sum_{i=1}^{n} \sum_{j=1}^{n} w_{xi,yj}^{(k+1)} v_{xi} v_{yj}$$

$$= -\frac{1}{2} V^T [W^{(k)} - \alpha_k[(V^*(k))(V^*(k))^T - \beta_k I] V$$

$$= E^{(k)}(V) + \frac{\alpha_k}{2} \{[V^*(k)^T V]^2 - \beta_k[V^T V]\}$$

When restricted to feasible set Ω_c, one has $V^T V = n$. So we have

$$E^{(k+1)}(V) = E^{(k)}(V) + \frac{\alpha_k}{2} \{[V^*(k)^T V]^2 - n\beta_k\}, \forall V \in \Omega_c \tag{16}$$

$$\geq E^{(k)}(V), \quad whenever \frac{1}{2} d_H(V, V^*(k)) \leq n - \sqrt{n\beta_k}. \tag{17}$$

where $d_H(V, V^*(k))$ is the Hamming distance between V and $V^*(k)$. The inequality (17) shows that

$$E^{(k+1)}(V^*(k)) > E^{(k)}(V^*(k)) \tag{18}$$

This shows that GENET escape from $V^*(k)$ by increasing the value of $E^{(k)}(V)$ whenever V is restricted to the neighborhood of $V^*(k)$.

(2) With the update rule (15), for any $a \neq b$, it can be observed that

$$C_{ab}^{(I)}(k+1) = w_{aI,aI}^{(k+1)} + w_{bI,bI}^{(k+1)} - 2w_{aI,bI}^{(k+1)}$$

$$= C_{ab}^{(I)}(k) + \begin{cases} \alpha_k(2\beta_k - 1), & if \ v_{aI}^*(k) \ or \ v_{bI}^*(k) = 1 \\ 2\alpha_k\beta_k, & if \ v_{aI}^*(k) \ and \ v_{bI}^*(k) = 0 \end{cases} \quad (19)$$

which shows that $C_{ab}^{(I)}(k+1) \geq 0$, if $\beta_k \geq 1/2, \alpha_k \geq 0$, and $C_{ab}^{(I)}(k) \geq 0$.

Remark 3. (i) It is first observed from (16)(17) that β_k can be used to control the scope in Ω_c of points, at which the objective value will be increased. In original GENET model, $\beta_k = 0$ means the objective value of every points in feasible domain will be increased. When $\beta_k \neq 0$, for those points far from $V^*(k)$, their objective values will be decreased, thus they has more possibility to be a stable state of $(k+1)th$ step local search. The larger β_k is, the nearer $V^*(k+1)$ is from $V^*(k)$. Parameter α_k can be used to adjust the increment amount of each point. Since we do not want $E^{(k+1)}(V)$ far from original objective function, the size of α_k must not be too large.

(ii) It can also be observed that, whenever $\beta_k \geq 1/2, \alpha_k \geq 0$, Theorem 1 and (19) implies that the GENET local search (11) can be employed to yield a stable state or a local minimum of objective function $E^{(k+1)}(V)$. This conclusion guarantees the applicability of learning strategy (15).

4   Conclusion

How to escape from local minimizer and find a better solution always is a main issue in optimization. In this paper, we theoretically demonstrate the convergent behavior of GENET model, generalize the learning strategy and analyze the function of each learning factor. The obtained results shed light on the setting of initial connection weights and learning factors. In future lots of simulation experiments will be realized in order to provide more accurate and detailed results.

References

1. Milis, P., .Tsang, E.P.K.: Guided Local Search for Solving SAT and Weighted MAX-SAT Problems. Journal of Automated Reasoning, **24** (2000) 205–223
2. Tsang, E.P.K., Wang, C.-J., Davenport, A., Voudouris, C., Lau, T.-L.: A Family of Stochastic Methods for Constraint Satisfaction and Optimization, The First International Conference on The Practical Application of Constraint Technologies and Logic Programming, London, April (1999) 359–383
3. Voudouris, C., Tsang, E.P.K.: Guided Local Search and Its Application to the Traveling Salesman Problem. European Journal of Operational Research, **113** (1999) 469–499

On Stochastic Neutral Neural Networks

Yumin Zhang, Lei Guo, Lingyao Wu, and Chunbo Feng

Research Institute of Automation, Southeast University,
Nanjing, Jiangshu 210096, China
zhyminus@sohu.com, l.guo@seu.edu.cn

Abstract. A new type of neutral neural network (NNN) model is established and the corresponding stability analysis is studied in this paper. By introducing the neutral term into the classical neural network model, the inspiration and associate memory phenomenon can be well described and explained. The stochastic Hopfield NNN model (HNNN) is investigated, respectively. Some criteria for mean square exponential stability and asymptotic stable are provided.

1 Introduction

Artificial neural networks theory has been received considerable attention in the past two decades, since the Hopfield neural network (NN) was introduced in [1] (see [2], [3] for surveys). Various approaches have been provided to study the dynamics of different Hopfield NN models, where time-delay has also been widely considered (see [3,4,8-10,18-19]). On the other hand, along with the development of the neutron sciences, some new type neural networks models have been established base on different behaviour of the neural cells such as the Bolzmann machines model bases on simulated annealing (SA) algorithm [5], the adaptive resonance theory (ART) model bases on adaptive resonance theory [6], and the cellular neural network in case that all of the neural cells are not connected entirely [7], [17], *et al.*

However, due to the complicated dynamic properties of the neural cells, in many cases the existing NN models cannot characterize the properties of a neural reaction process precisely. A natural and important problem is how to further describe and model the dynamic behavior for such complex neural reactions. This leads to the motivation of this paper, where a new type of NN models named neutral neural network (NNN) is established and analyzed.

There are three reasons to introduce the neutral-type operation into the conventional NN models. Firstly, based on biochemistry experiments, neural information may transfer across chemical reactivity, which results in a neutral-type process [11]. Secondly, in view of electronics, it has been shown that neutral phenomena exist in large-scale integrated (LSI) circuits [12], [13], [14], [16]. And the key point is that cerebra can be considered as a super LSI circuit with chemical reactivity, which reasonably implies that the neutral dynamic behaviors should be included in neural dynamic systems. However, to our best knowledge, there is no sufficient research up to date concentrating on neural network models using neutral dynamics. It is an important character of this paper.

J. Wang, X. Liao, and Z. Yi (Eds.): ISNN 2005, LNCS 3496, pp. 69–74, 2005.

2 Main Results

2.1 Motivation of the NNN Model

It is known that homo-information is distributed in lots of different neurons. The memory function of information can be denoted by the synapse connection strength among different neurons with one another. When the synapse connection strength changes, the information memory will be modified, transferred and processed, while in the meantime, a new information memory mode is completed. Based on the above idea and the influence of time delays for the information propagation, the neutral characteristic of neural networks can be investigated as follows.

Consider a neural network $\mathcal{A}$, which includes a main network $N^{a1} = \left(N_1^{a1}, N_2^{a1}, \ldots, N_{N^a}^{a1} \right)$ and a backup network $N^{a2} = \left(N_1^{a2}, N_2^{a2}, \ldots, N_{N^a}^{a2} \right)$. Both of the networks have N^a neurons. Suppose that N_k^{a1} and N_k^{a2} memorize the same information that generated from the similar stimulation. On the other hand, it is noted that the cerebrum can still remember or recover the existing information even if some correlative neurons are hurt or changed. Thus, in network $\mathcal{A}$, N^{a1} and N^{a2} can be supposed to have a (weak) coupling relationship. In this case, the dynamics of N^{a2} need not be changed immediately when N^{a1} is modified, which is the virtue of the so-called distribution or backup of the memory. Based on such observations, the model modification of network $\mathcal{A}$ is required to be further investigated when its partial information in local network changes. In the following, the modified dynamics of $\mathcal{A}$ will be concerned along with the changes of either N^{a2} or N^{a1}, where two kinds of modifications are considered.

The first case is resulted from the change of neurons in N^{a1} or N^{a2}, which includes the signal attenuation or amplification due to the pathological changes of neurons. We assume that the modified information with respect to N^{a1} is modeled by $y = g_1(x)$, as a function of the original information x. Thus, although both N^{a1} and N^{a2} can memorize the information generated from the same x, the reaction information of N^{a1} is different from that of N^{a2} because of the neuron changes in N^{a1}. Hence, there are two types of information carrying on the network $\mathcal{A}$: the original information and the information of N^{a2} (N^{a1}) attenuated or amplified by N^{a1} (N^{a2}). The procedures can be considered as a reconstruction (or simply, an addition) for the dynamics of N^{a1} and N^{a2}. Considering the delay of signal transmitting, the modified information of N^{a2} can be characterized as $z = x + g_2(y_t)$, where y_t is denoted as the delayed information of y resulting from N^{a1}, and $g_2(y_t)$ represents the degree of received information from N^{a1}. In general, differential dynamic equations can be used to describe the dynamics of N^{a1}, which leads to a neutral differential form as $\dfrac{d}{dt}z = \dfrac{d}{dt}\left(x + g_2\left(y_t\right)\right)$. On the other hand, similarly N^{a1} is also influenced by N^{a2}, and the modified dynamics of N^{a1} can be defined by $w = y + g_3\left(x_t\right)$, where x_t is denoted as the delayed dynamics of x from N^{a2}.

Hence, the differential dynamic equation is related to $\dfrac{d}{dt}w = \dfrac{d}{dt}\left(y + g_3\left(x_t\right)\right)$, where the neutral form also appears. Denoting that $\overline{x} = \begin{bmatrix} x & y \end{bmatrix}^T$, $\begin{bmatrix} g_2(y_t) & g_3(x_t) \end{bmatrix}^T = -G\left(\overline{x}_t\right)$, it can be verified that

$$\frac{d}{dt}\overline{x} = \frac{d}{dt}\left(\overline{x} - G\left(\overline{x}_t\right)\right) \tag{1}$$

where it is shown the neutral term exists in the dynamical equation of the entire network with connections and local modifications.

The second case originated from the so-called associate memory phenomena. Consider two neural network $\mathcal{A}$ (including N^{a1} and N^{a2}) and $\mathcal{B}$ (including N^{b1} and N^{b2}) with a (weak) connection between them. Constructing $\mathcal{A}$ and $\mathcal{B}$ to a new NN denoted as $\mathcal{C}$. By using $\mathcal{C}$, the associate memory phenomena can be described in a reasonable way. Similarly to the above analysis on N^{a1} and N^{a2}, the combined dynamics of $\mathcal{C}$ can also be formulated in terms of the neutral differential equations (see equation (1)).

Furthermore, by using of the conventional formulations studied in the classical stochastic NN model, the following NNN model can be introduced

$$d\left(x(t) - G\left(x_t\right)\right) = \rho\left(t, x, x_t\right)dt + \sigma\left(t, x, x_t\right)dw(t), \quad t \geq 0 \tag{2}$$

with $x_0 = \xi\left(-\tau \leq t \leq 0\right)$. Where $G: C\left(\left[-\tau, 0\right); R^n\right) \to R^n$, $\rho: R^n \times C\left(\left[-\tau, 0\right); R^n\right) \to R^n$, and $\sigma: R^n \times C\left(\left[-\tau, 0\right); R^n\right) \to R^n$ are continuous functionals. $w(t)$ represents the stochastic property in associate process. $x_t = \left\{x(t+\theta): -\tau \leq \theta \leq 0\right\}$ denotes the $C\left(\left[-\tau, 0\right]; R^n\right)$ -valued stochastic process, and $\xi = \left\{\xi(\theta): -\tau \leq \theta \leq 0\right\} \in C^b_{F_0}\left(\left[-\tau, 0\right]; R^n\right)$.

2.2 Stability Analysis for Stochastic HNNN

Consider a special case of system (2) as follows,

$$d\left(x - G(x_t)\right) = \left[-Ax + Wf(x) + W_\tau f(x_t)\right]dt + \sigma(t, x, x_t)\,dw(t) \quad (3)$$

with initial value $x(s) = \xi(s)$ $(-\tau \le s \le 0)$. Where $A = \mathrm{diag}(a_i)_{n\times n} \in R^{n\times n}$, $W = (w_{ij})_{n\times n} \in R^{n\times n}$, $W_\tau = (w_{ij}^\tau)_{n\times n} \in R^{n\times n}$ are weighted matrices. The other signs are same with ones in references [4] and [13]. In Hopfield networks, the nonlinear activation function is usually chosen as differentiable sigmoid function. Throughout this paper, we assume that $\sigma(t,\cdot,\cdot)$ and $f(\cdot)$ are Lipschitz continuous and satisfy the nonlinear incremental conditions:

Assumption 1: There exist constants L_j , $0 \le L_j < \infty$, $j = 1, 2, \cdots, n$, such that

$$f_j : R \to R \text{ satisfies } 0 \le \left|f_j(x_j) - f_j(y_j)\right| \le L_j \left|x_j - y_j\right|, \forall y_j \ne x_j \in R.$$

Assumption 2: There exist constant b_{ik} and c_{ik} , $i = 1, 2, \cdots, n$, $k = 1, 2, \cdots, m$ such that $\sigma(t,\cdot,\cdot)$ satisfies $\sigma_{ik}^2\left(t, x_i(t), x_i(t-\tau_i)\right) \le b_{ik} x_i^2(t) + c_{ik} x_i^2(t-\tau_i)$.

With Assumption 1-2, it is known (see conference [15]) that system (4) has a unique global solution on $t \ge 0$, denoted by $x(t;\xi)$. Moreover, it is assumed that $\sigma(t,0,0) \equiv 0$ so that system (4) admits an equation solution $x(t;0) \equiv 0$. For convenience, let $L = \mathrm{diag}(L_i)_{n\times n}$ and $b = \max\left\{\sum_{k=1}^{m} b_{ik} : i = 1, 2, \cdots, n\right\}$,

$c = \max\left\{\sum_{k=1}^{m} c_{ik} : i = 1, 2, \cdots, n\right\}$. $G(\cdot)$ satisfies the following assumption.

Assumption 3: There is $k \in (0,1)$ such that

$$E\left|G(\varphi)\right|^2 \le k \sup_{-\tau \le \theta \le 0} E\left|\varphi(\theta)\right|^2, \quad \varphi \in L_{F_\infty}^2\left([-\tau, 0]; R^n\right) \quad (4)$$

Denote $C^{1,2}\left([-\tau, \infty) \times R^n, R_+\right)$ as the family of all functions $V(t,x)$ $:[-\tau, \infty) \times R^n \to R_+$, which are continuously twice differentiable in x and once in t .

In this subsection, it is supposed that Assumptions 1-3 hold.

Theorem 1: For a definite positive matrix Q and a Lyapunov candidate $V(t, \varphi) = \varphi^T Q \varphi$. Suppose that for all $t \ge 0$ and $\varphi \in L_{F_t}^2\left([-\tau, 0]; R^n\right)$ satisfying $\left|\varphi(0)\right| \le \left|\varphi(\theta)\right|$, and the following inequality

$$E\left|\varphi(\theta)\right|^2 \leq EV\left(t,\varphi(0)-G(\varphi)\right)q/\lambda_{\min}(Q), \ -\tau \leq \theta \leq 0 \qquad (5)$$

holds, then for all $\xi \in C_{F_0}^b\left([-\tau,0];R^n\right)$, we have

$$\limsup_{t\to\infty}\frac{1}{t}\log\left(E\left|x(t,\xi)\right|^2\right) \leq -\frac{1}{\tau}\log\frac{q}{\left[1+(kq)^{1/2}\right]^2} := -\gamma \qquad (6)$$

That is, the zero-solution of HNNN system (4) is mean square exponentially stable. In (6) and (7), q is the unique positive root of the following algebraic equation

$$-\lambda_{\min}\left(Q^{\frac{1}{2}}AQ^{-\frac{1}{2}}+Q^{-\frac{1}{2}}A^TQ^{\frac{1}{2}}\right)+2\sqrt{\frac{kq\|Q\|}{\lambda_{\min}(Q)}}\|A\|+2\sqrt{\frac{q\|Q\|}{\lambda_{\min}(Q)}}\|W\|\|L\|$$

$$+2\sqrt{\frac{q\|Q\|}{\lambda_{\min}(Q)}}\|W_\tau\|\|L\|+\frac{q\|Q\|(b+c)}{\lambda_{\min}(Q)}=-\gamma \qquad (7)$$

and satisfies $q > \left(1-k^{1/2}\right)^{-2}$.

Corollary 1: If rewrite '$=-\gamma$' as '<0' in (7), system (3) is asymptotic stable.

Remark 1: If cite $Q = I$ in Theorem 1, Theorem 1 will be simplified.

Remark 2: If $\sigma(\cdot,\cdot,\cdot) \equiv 0$, system (3) turns to be a deterministic model of HNNN and corresponding results to Theorem 1 and Corollary 1are obtained.

Remark 3: If $G(\cdot) \equiv 0$, the neutral term is eliminated and model (3) is reduced to the conventional Hopfield NN model (see 3,4, 8-10), then some existed results can be generalized by the approach in Theorem 1and Corollary 1.

3 Conclusions

A novel type of stochastic NNN model is studied in this paper. Following the recent development on the conventional HNN systems with time delays, some new criteria on HNNN are provided for the mean square exponential stability as well as the asymptotical stability. Further researches include more elegant stability analysis approaches with less conservativeness and some control strategies for the neutral differential equations.

Acknowledgement
This work is supported partially by the Leverhulm Trust, UK, the NSF of China, the China Post-Doctoral Foundation and the Jiangsu Post-Doctoral Foundation of China.

References

1. Hopfield, J. J.: Neural Networks and Physical Systems with Emergent Collect Computational Abilities. Proc. Natl. Acad. Sci., USA 79 (1982) 2554-2558
2. De Wilde, P.: Neural Networks Models, An Analysis. Springer-Verlag (1996)
3. Liao, X.: Theory and Applications of Stability for Dynamical Systems. National Defensive Industrial Press, Beijing (2000)
4. Zhang, Y., Shen, Y., Liao, X., et al.: Novel Criteria Stability for Stochastic Interval Delayed Hopfield Neural Networks. Advances in Syst. Sci. and Appl., 2 (2002) 37-41
5. Ackley, D.H., Hinton, G.E., Sejinowski. T.J.: A Lining Algorithm for Bolzmann Machines. Cognitive Science, 9 (1985) 147-169
6. Grossberg, S.: Compectitive Learning: From Interactive Activation To Adaptive Resonance. Conitive Science, 11 (1987) 23-63
7. Chua, L.O., Yang, L.: Cellular Neural Network: Theory. IEEE Trans. Circuits Syst., 35 (1988) 1257-1272
8. Cao, J.: Global Stability Analysis in Delayed Cellular Neural Networks. Phys. Rev., 59 (1999) 5940-5944
9. Zeng, Z., Wang, J., Liao, X.: Stability Analysis of Delayed Cellular Neural Networks Described Using Cloning Templates. IEEE Transactions on Circuits and Systems I: Fundamental Theory and Applications, 51 (2004) 2313-2324
10. Liao, X., Yu, J., et al.: Robust Stability Criteria for Interval Hopfield Neural Networks with Time Delay. IEEE Trans. Neural Networks, 9 (1998) 1042-1046
11. Curt, W.: Reactive Molecules: the Neutral Reactive Intermediates in Organic Chemistry. Wiley Press, New York (1984)
12. Salamon, D.: Control and Observation of Neutral Systems. Pitman Advanced Pub. Program, Boston (1984)
13. Zhang, Y., Shen, Y., Liao, X.: Robust Stability of Neutral Stochastic Interval Systems. Journal of Control Theory and Applications, 2 (2004) 82-84
14. Wang, Z., Lam, J., Burnham, K.J.: Stability Analysis and Observer Design for Delay System. IEEE Trans. on Auto. Contr., 47 (2002) 478-483
15. Mao, X.: Exponential Stability of Stochastic Differential Equations. Marcel Dekker, New York (1994)
16. Shen, Y., Liao, X.: Razumikhin-type Theorems on Exponential Stability of Neutral Stochastic Functional Differential Equations. Chinese Science Bulletin, 44 (1999) 2225-2228
17. Shen, Y., Liao, X.: Dynamic Analysis for Generalized Cellular Networks with Delay. ACTA Electronica Sinica., 27 (1999) 62-644
18. Zhang, Y., Shen, Y., Liao, X.: Exponential Stability for Stochastic Interval Delayed Hopfieild Neural Networks. Control Theory & Applications, 20 (2003) 746-748
19. Sun, C.: A Comment on 'Global Stability Analysis in Delayed Hopfield Neural Network Models. Neural Networks, 15 (2002) 1299-1300

Eigenanalysis of CMAC Neural Network

Chunshu Zhang

Electrical and Computer Engineering Department, University of New Hampshire
03824 Durham, NH, USA
czhang@cisunix.unh.edu

Abstract. The CMAC neural network is by itself an adaptive processor. This paper studies the CMAC algorithm from the point of view of adaptive filter theory. Correspondingly, the correlation matrix is defined and the Wiener-Hopf equation is obtained for the CMAC neural network. It is revealed that the trace (i.e., sum of eigenvalues) of the correlation matrix is equal to the generalization parameter of the CMAC neural network. Using the tool of eigenanalysis, analytical bounds of the learning rate of CMAC neural network are derived which guarantee convergence of the weight vector in the mean. Moreover, a simple formula of estimating the misadjustment due to the gradient noise is derived.

1 Introduction

The Cerebellar Model Arithmetic Computer (CMAC), an associative memory neural network in that each input maps to a subset of weights whose values are summed to produce outputs, was first introduced by James Albus [1] in early 1970's to approximate the information processing characteristics of the human cerebellum. A practical implementation of the CMAC neural network that could be used in the real-time control applications was developed by Miller [2]. The built-in algorithm of CAMC results in such advantages [3] as: a) fast learning property, b) rapid generalization capability, c) no local-minima problem, and, d) modeling or learning abilities for nonlinear plants as well as linear plants. Evidently since mid-1980's, study on CMAC has made significant progresses [4, 5] and applications have been found in fields such as real-time robotics [6], vibration control [7, 8], pattern recognition [9], and signal processing [10].

Impressive advantages as it has, however, CMAC has been conspicuously absent from many major literatures in related fields. Perhaps part of the reason is due to the lack of systematic analytical study on CMAC, which is resulted from the complexity of organization of its receptive fields. Or, simply not enough efforts have been made to bridge the CMAC study to other fields concerning adaptive process.

Look beyond the details of CMAC algorithms, CMAC neural network is by itself an adaptive processor. This suggests that it is possible to study CMAC within a general adaptation context which has been studied by such disciplines as adaptive signal processing and adaptive control. In this paper, a basic tool of analysis in the study of digital signal processing – eigenanalysis [11] is used to study the analytical properties of the CMAC neural network. As will be seen, the eigenanalysis involves a useful decomposition of the correlation matrix in terms of its eigenvalues and associated eigenvectors.

J. Wang, X. Liao, and Z. Yi (Eds.): ISNN 2005, LNCS 3496, pp. 75–80, 2005.

2 Performance Function and Correlation Matrix

Assume at time step k, a pair of data (x_k, d_k) is presented, in which x_k is the input and d_k is the desired output (target). The output of CMAC corresponding to x_k is:

$$y_k = \mathbf{s}_k^T \bullet \mathbf{w} = \mathbf{w}^T \bullet \mathbf{s}_k . \tag{1}$$

where $\mathbf{w}$ is the weight vector of size M (memory size) and $\mathbf{s}_k$ is the excitation (selection) vector determined by x_k. For a conventional CMAC neural network whose generalization parameter is set to be ρ, $\mathbf{s}_k$ is a vector with ρ elements of one and M-ρ elements of zero.

The error between the desired output d_k and the estimated output y_k is:

$$e_k = d_k - y_k = d_k - \mathbf{s}_k^T \bullet \mathbf{w} . \tag{2}$$

The goal of adaptation is to minimize the following performance function (MSE):

$$J(\mathbf{w}) = E[e_k^2] = E[(d_k - y_k)^2] = E[(d_k - \mathbf{s}_k^T \bullet \mathbf{w})^2] . \tag{3}$$

Take the derivative of $J(\mathbf{w})$,

$$\frac{\partial}{\partial \mathbf{w}} J(\mathbf{w}) = (-2) \times \{E[\mathbf{s}_k d_k] - E[\mathbf{s}_k \mathbf{s}_k^T]\mathbf{w}\} . \tag{4}$$

Set $\dfrac{\partial}{\partial \mathbf{w}} J(\mathbf{w})|_{\mathbf{w}=\mathbf{w}^*} = 0$,

$$E[\mathbf{s}_k \mathbf{s}_k^T]\mathbf{w} = E[\mathbf{s}_k d_k] . \tag{5}$$

Let $\mathbf{R}$ denote the M-by-M correlation matrix of the excitation vector $\mathbf{s}_k$ of the CMAC neural network:

$$R \equiv E[\mathbf{s}_k \mathbf{s}_k^T] = \begin{bmatrix} E[\mathbf{s}_k(1)\mathbf{s}_k(1)] & \cdots & E[\mathbf{s}_k(1)\mathbf{s}_k M)] \\ \vdots & \ddots & \vdots \\ E[\mathbf{s}_k(M)\mathbf{s}_k(1)] & \cdots & E[\mathbf{s}_k(M)\mathbf{s}_k(M)] \end{bmatrix} . \tag{6}$$

Let $\mathbf{p}$ denote the M-by-1 cross-correlation vector between the excitation vector and the desired response d_k:

$$\mathbf{p} \equiv E[\mathbf{s}_k d_k] = \begin{bmatrix} E[\mathbf{s}_k(1)d_k] \\ \vdots \\ E[\mathbf{s}_k(M)d_k] \end{bmatrix} . \tag{7}$$

Then equation (5) becomes:

$$\mathbf{R}\mathbf{w}^* = \mathbf{p} . \tag{8}$$

Equation (8) is the Wiener-Hopf equation in matrix form, which gives the optimal weight vector:

$$\mathbf{w}^* = \mathbf{R}^{-1}\mathbf{p} . \tag{9}$$

under the assumption that $\mathbf{R}^{-1}$ exists.

The properties of the correlation matrix **R** are very important in the study of adaptive filter theory, which will be explored in the next section.

3 Properties of Correlation Matrix

In this section, several useful properties of the correlation matrix are discussed. The first two properties apply to general correlation matrix [11]. Properties 3 and 4 apply to the CMAC neural network only.

Property 1. The correlation matrix **R** is always nonnegative definite (or positive semidefinite).

Property 2. Let $\lambda_1, \lambda_2, ..., \lambda_M$ be the eigenvalues of the correlation matrix **R**. Then all these eignevalues are real and nonnegative.

Property 3. The trace of correlation matrix **R** of the CMAC neural network is equal to the generalization parameter (ρ) of the CMAC neural network. That is

$$trace(\mathbf{R}) = \rho . \tag{10}$$

Proof: Let r_{ij} denote the product of the ith element and j^{th} element of the excitation vector s_k, i.e.,

$$r_{ij} = \mathbf{s}_k(i) \bullet \mathbf{s}_k(j) .$$

The value of r_{ii} may be determined by the following equation:

$$r_{ii} = [\mathbf{s}_k(i)]^2 = \begin{cases} 1 & if \ i^{th} \ element \ of \ \mathbf{s}_k \ is \ 1 \\ 0 & f \ i^{th} \ element \ of \ \mathbf{s}_k \ is \ 0 \end{cases} .$$

Since s_k is the excitation vector that has ρ elements of one and M-ρ elements of zero,

$$\sum_{i=1}^{M} r_{ii} = \rho .$$

Hence,

$$trace(\mathbf{R}) = \sum_{i=1}^{M} E[r_{ii}] = E[\sum_{i=1}^{M} r_{ii}] = E[\rho] = \rho .$$

Property 4. Let $\lambda_1, \lambda_2, ..., \lambda_M$ be the eigenvalues of the correlation matrix **R**. Then,

$$\sum_{i=1}^{M} \lambda_i = \rho . \tag{11}$$

Proof: This follows directly from Property 3.

4   Convergence and Misadjustment of CMAC Algorithm

The LMS algorithm uses the instantaneous gradients at each step, $\hat{\nabla}_k = \dfrac{\partial e_k^2}{\partial w} = 2e_k \dfrac{\partial e_k}{\partial w} = -2e_k s_k$, as the guide to adjust the weight vector:

$$w_{k+1} = w_k + \mu(-\hat{\nabla}_k) = w_k + 2\mu \cdot e_k s_k .\tag{12}$$

The conventional weight-updating algorithm of CMAC is obtained by substituting $\mu = \alpha/(2\rho)$,

$$w_{k+1} = w_k + \frac{\alpha}{\rho}(d_k - y_k)s_k .\tag{13}$$

Taking the expected value, we get

$$E[w_{k+1}] = E[w_k] + \frac{\alpha}{\rho}\{E[d_k s_k] - E[s_k s_k^T w_k]\} .\tag{14}$$

To continue on, we need to make an assumption that the excitation vector s_k is independent of the weight vector w_k. The independence can be interpreted as the result of slow adaptation. Assume that the learning rate μ is small (or the generalization parameter ρ is big), the adaptive weight vector depends on many past input samples, and its dependence on the present excitation vector is negligible. Furthermore, when the training process is stabilized, the weight vector will remain unchanged while the excitation vector will still respond to the input at every tick of time. A similar assumption was first made by Widrow in 1970 and then again in 1996 for the study of convergence of adaptive filters [12]. For CMAC neural networks with hashing, another layer of independence is added. It follows that

$$E[w_{k+1}] = E[w_k] + \frac{\alpha}{\rho}\{E[d_k s_k] - E[s_k s_k^T]E[w_k]\}$$

$$= E[w_k] + \frac{\alpha}{\rho}\{p - RE[w_k]\}\tag{15}$$

$$= (I - \frac{\alpha}{\rho}R)E[w_k] + \frac{\alpha}{\rho}p .$$

Substituting $w = v + w^*$, we get

$$E[v_{k+1}] = (I - \frac{\alpha}{\rho}R)E[v_k] .\tag{16}$$

By defining a new vector $v_k' = Q^{-1}(w_k - w^*)$, in which Q is such a matrix that $Q^{-1}RQ = \Lambda$ and $\Lambda = diag(\lambda_1, \lambda_2, \cdots, \lambda_M)$, the following equation is derived:

$$E[\mathbf{v}_{k+1}'] = (\mathbf{I} - \frac{\alpha}{\rho}\mathbf{\Lambda})^{k+1} E[\mathbf{v}_0'] .\tag{17}$$

For the stability of Eq. (17), it is necessary that

$$\left| 1 - \frac{\alpha}{\rho} \lambda_i \right| < 1 \ , \qquad i = 1, 2, \ldots, M. \tag{18}$$

This leads directly to the following theorem:

Theorem 1. For a CMAC neural network trained by Eq. (13), a necessary and sufficient condition for convergence of the weight vector in the mean is

$$\frac{2\rho}{\lambda_{max}} > \alpha > 0 \ . \tag{19}$$

Corollary 1. For a CMAC neural network trained by Eq. (13), a sufficient condition for convergence of the weight vector in the mean is

$$2 > \alpha > 0 \ . \tag{20}$$

The proof of Corollary 1 can be done by using Property 2 ~ 4 in Theorem 1.

Eq. (19) and Eq. (20) give two bounds of the learning rate of the CMAC neural network that guarantee convergence of the weight vector in the mean. Theorem 1 is a new conclusion about the convergence of CMAC neural networks. Other authors [5] [8] presented conditions of convergence similar to Corollary 1, with different approach. While it is difficult to calculate the bound given by Eq. (19), it points out the theoretical bound is bigger than two. For example, if the maximum eigenvalue of the correlation matrix R is half the sum of all eigenvalues, the maximum bound of the learning rate will be four.

Applying Property 2, 3, and 4 to the concept of misadjustment due to gradient noise presented by Widrow [12] leads to another useful conclusion.

Theorem 2. For a CMAC neural network trained with Eq. (13), the misadjustment due to gradient noise after adaptive transients die out may be estimated by:

$$Misadjustment = \frac{\alpha}{2} \ . \tag{21}$$

Theorem 2 gives us a quick way to select the parameter of CMAC neural network to meet certain design specification. For example, typically an experienced designer would expect no more than 10% misadjustment, so one can select a learning rate (α) less than 0.2. The tradeoff is that the adaptation time will be longer when α decreases.

5 Conclusion

An analytical research on the properties of the CMAC neural network was conducted. The theoretical results contribute to the development of the CMAC neural network and help improve the general understanding of the CMAC neural network. In this paper, the research was conducted from the viewpoint of adaptive filter theory, which not only revealed some conclusions regarding the convergence and performance of the CMAC neural network but also provides a new perspective of studying the CMAC neural network.

References

1. Albus, J.S.: Theoretical and Experimental Aspects of a Cerebellar Model. Ph.D Dissertation. University of Maryland (1972)
2. Miller, W. T, Sutton, R. S, Werbos, P. J.: Neural Networks for Control. MIT Press, Cambridge, MA (1990)
3. Miller, W. T., Glanz, F. H., Kraft, L. G.: CMAC: An Associative Neural Network Alternative to Backpropagation. Proceedings of the IEEE, Special Issue on Neural Networks, **II** (1990) 1561-1567
4. Lee, H.-M., Chen, V.-M., Lu, Y.-F.: A Self-organizing HCMAC Neural-Network Classifier. IEEE Trans. Neural Networks, (2003) 15-26
5. Lin. C.-S., Chiang, C.-T.: Learning Convergence of CMAC Technique. IEEE Trans. Neural Networks, (1997) 1281-1292
6. Miller, W.T.: Real Time Application of Neural Networks for Sensor-Based Control of Robots with Vision. IEEE SMC, (1989) 825-831
7. Zhang, C., Canfield, J., Kraft, L.G., Kun, A.: A New Active Vibration Control Architecture Using CMAC Neural Networks. Proc. of 2003 IEEE ISIC, (2003) 533-536
8. Canfield, J., Kraft, L. G., Latham, P., Kun, A.: Filtered-X CMAC: An Efficient Algorithm for Active Disturbance Cancellation in Nonlinear Dynamical Systems. Proc. of 2003 IEEE ISIC, (2003) 340-345
9. Herold, D., Miller, W.T., Glanz, F.H., Kraft, L.G.: Pattern Recognition Using a CMAC Based Learning System. Proc. SPIE, Automated Inspection and High Speed Vision Architectures **II**, (1989) 84-90
10. Glanz, F.H., Miller, W.T.: Deconvolution and Nonlinear Inverse Filtering Using a CMAC Neural Network. Intl. Conf. on Acoustics and Signal Processing (1989) 2349-2352
11. Haykin, S.: Adaptive Filter Theory, 3rd ed. Prentice Hall, NJ (1996)
12. Widrow, B., Walach, E.: Adaptive Inverse Control, Englewood Cliffs, Prentice-Hall, NJ (1996)

A New Definition of Sensitivity for RBFNN and Its Applications to Feature Reduction*

Xizhao Wang and Chunguo Li

Faculty of Mathematics and Computer Science, Hebei University, Baoding, Hebei, China
{wangxz,licg}@mail.hbu.edu.cn

Abstract. Due to the existence of redundant features, the Radial-Basis Function Neural Network (RBFNN) which is trained from a dataset is likely to be huge. Sensitivity analysis technique usually could help to reduce the features by deleting insensitive features. Considering the perturbation of network output as a random variable, this paper defines a new sensitivity formula which is the limit of variance of output perturbation with respect to the input perturbation going to zero. To simplify the sensitivity expression and computation, we prove that the exchange between limit and variance is valid. A formula for computing the new sensitivity of individual features is derived. Numerical simulations show that the new sensitivity definition can be used to remove irrelevant features effectively.

1 Introduction

As a sort of neural networks, the Radial-Basis Function Neural Network (RBFNN) is usually used to approximating a very complex and smooth function. In its basic form, the structure of RBFNN involves three layers, i.e., the input layer, the hidden layer and the output layer, with entirely different roles. The input layer accepts the information from the environment; the second layer (the only hidden layer) applies a nonlinear transformation to the accepted information; and the output layer supplies the response of the network. The input layer is made up of sensory units, the hidden layer nonlinear neurons, and the output layer pure linear neurons [1].

RBFNNs are able to approximate any smooth function within the required accuracy. According to RBFNN's training procedure, the hidden layer keeps adding neurons one by one until the required accuracy is reached. The network obtained is likely to be very huge. It is due to the high dimension of the hidden space when the training data have redundant information. If the redundant information could be deleted before training, the network size and performance would be improved. Sensitivity analysis of the network consequently arises and becomes one of the most important means for redundant information removal of neural networks [2–5].

The sensitivity analysis of neural networks has been investigated for over 30 years. During this period, a number of useful methodologies were put forward to investigate the sensitivity of Multilayer Perceptrons (MLP) [3–6]. One of the most popular techniques is to delete redundant inputs of MLPs using partial derivative [3, 4]. Another

* This research work is supported by NSFC (60473045) and Natural Science Foundation of Hebei Province (603137).

J. Wang, X. Liao, and Z. Yi (Eds.): ISNN 2005, LNCS 3496, pp. 81–86, 2005.

popular technique is to consider the weight perturbation [5, 6]. For example, Piché [6] investigated the effects of weight errors upon a statistical model of an ensemble of Madalines instead of the effects upon specific known networks.

Recently, the investigation of sensitivity analysis for RBFNNs has been found [7]. Sensitivity analysis plays an extremely important role in removing irrelevant features and improving the performance of RBFNNs. Due to the simplicity and the good approximation property of RBFNNs, the study on RBFNN's sensitivity has attracted more and more research fellows [8].

This paper aims at giving a new definition of sensitivity of RBFNNs. Considering the perturbation of network output as a random variable, this paper defines a new sensitivity formula which is the limit of variance of output perturbation with respect to the input perturbation going to zero. The exchange of limit and variance is proved here to be valid. A formula for computing the new sensitivity of individual features is derived. Numerical simulations show that the new sensitivity definition can be used to remove irrelevant features effectively.

2   A New Sensitivity Definition

Here we focus on a new definition of sensitivity: sensitivity based on variance of output perturbation. Statistically, the variance of one random variable measures the deviation from its center.

A trained RBFNN can be expressed as follows:

$$f(x_1, x_2, \cdots, x_n) = \sum_{j=1}^{m} w_j \left(\exp \left(\frac{\sum_{i=1}^{n} (x_i - u_{ij})^2}{-2v_j^2} \right) \right) \tag{1}$$

where n is the number of features, m the number of centers, $(u_{1j}, u_{2j}, \cdots, u_{nj})$ the j-th center, v_j the spread of the j-th center, and w_j the weight of the output layer, $j = 1, 2, \cdots, m$.

Definition. The magnitude of sensitivity of the first feature is defined

$$S(x_1) = \lim_{\Delta x \to 0} \left(Var \left(\frac{f(x_1 + \Delta x, x_2, \cdots, x_n) - f(x_1, x_2, \cdots, x_n)}{\Delta x} \right) \right) \tag{2}$$

where $x_1, x_2, \cdots, x_n$ are n independent random variables with a joint distribution $\Phi(x_1, x_2, \cdots, x_n)$, $Var(y)$ means the variance of random variable y, and the input perturbation Δx is a real variable (not a random variable).

Similarly, the sensitivity definition for other features can be given.

If the limit works before the integral (i.e., Var) in Eq. (2), the definition can be simplified by using the partial derivative expression:

$$\frac{\partial f}{\partial x_1} = \lim_{\Delta x \to 0} \frac{f(x_1 + \Delta x, x_2, \cdots, x_n) - f(x_1, x_2, \cdots, x_n)}{\Delta x} \tag{3}$$

Mathematically, it is a very important Limit-Integral exchange problem.

Theorem. The limit operation and the variance operation in Eq.(2) can be exchanged, that is,

$$\underset{\Delta x \to 0}{Lim}\left(Var\left(\frac{f(x_1 + \Delta x, x_2, \cdots, x_n) - f(x_1, x_2, \cdots, x_n)}{\Delta x} \right) \right) = Var\left(\frac{\partial f}{\partial x_1} \right) \qquad (4)$$

where the function f is defined as Eq.(1) and $x = (x_1, x_2, \cdots, x_n)$ is a random vector with the density function $\varphi(x)$. The first and second order moments of the random vector x are supposed to be finite.

Proof. Consider a series of numbers $\Delta_k \to 0 (k \to \infty)$ to replace the continuous $\Delta x \to 0$. Let

$$g_k = \frac{f(x_1 + \Delta_k, x_2, \cdots, x_n) - f(x_1, x_2, \cdots, x_n)}{\Delta_k} \qquad (5)$$

According to Eq.(1), the function g_k can be expressed as

$$g_k = \sum_{j=1}^{m} w_j \left(\exp\left(\frac{\sum_{i=1}^{n}(x_i - u_{ij})^2}{-2v_j^2} \right) \right)\left(\frac{1}{\Delta_k}\left(\exp\left(\frac{2\Delta_k(x_1 - u_{1j}) + \Delta_k^2}{-2v_j^2} \right) - 1 \right) \right) \qquad (6)$$

We now first prove

$$\underset{k \to \infty}{Lim} \int g_k \varphi(x)\, dx = \int \left(\underset{k \to \infty}{Lim}\, g_k \right)\varphi(x)\, dx \qquad (7)$$

Consider the last term in Eq.(6) and let

$$h_k = \left(\frac{1}{\Delta_k}\left(\exp\left(\frac{2\Delta_k(x_1 - u_{1j}) + \Delta_k^2}{-2v_j^2} \right) - 1 \right) \right), \qquad (8)$$

We have

$$\underset{k \to \infty}{\lim}\, h_k = \underset{\Delta \to 0}{\lim}\left(\frac{\dfrac{d}{d\Delta}\exp\left(\dfrac{2\Delta(x_1 - u_{1j}) + \Delta^2}{-2v_j^2} \right)}{\dfrac{d}{d\Delta}(\Delta)} \right)$$

$$= \underset{\Delta \to 0}{\lim}\left(\exp\left(\frac{2\Delta(x_1 - u_{1j}) + \Delta^2}{-2v_j^2} \right) \cdot \frac{(x_1 - u_{1j}) + \Delta}{-v_j^2} \right) = \frac{(x_1 - u_{1j})}{-v_j^2} \qquad (9)$$

It implies that $|h_k| \le \left| \dfrac{(x_1 - u_{1j})}{-v_j^2} \right| + C$ holds well for all k where C is a constant. There-fore, from Eq.(6), we can obtain that

$$\left|g_k\varphi(x)\right|=\left|\sum_{j=1}^{m}w_j\left(\exp\left(\frac{\sum_{i=1}^{n}(x_i-u_{ij})^2}{-2v_j^2}\right)\right)h_k\varphi(x)\right|$$

$$\leq\sum_{j=1}^{m}\left|w_j\right|\left(\left|\frac{(x_1-u_{1j})}{-v_j^2}\right|+C\right)\varphi(x) \tag{10}$$

Noting that the right of Eq.(10) is independent of k and the first order moments of x are supposed to exist, we have

$$\int\left(\sum_{j=1}^{m}\left|w_j\right|\left(\left|\frac{(x_1-u_{1j})}{-v_j^2}\right|+C\right)\varphi(x)\right)dx<\infty \tag{11}$$

which results in the correctness of Eq. (7), according to the integral convergence theorem.

Similar to the proof of Eq. (7), we can prove that

$$\underset{k\to\infty}{Lim}\int g_k^2\varphi(x)\,dx=\int\left(\underset{k\to\infty}{Lim}\,g_k^2\right)\varphi(x)\,dx \tag{12}$$

Further, paying attention to the following fundamental equalities on random variables,

$$Var(g_k)=E\left(g_k^2\right)-\left(E(g_k)\right)^2=\int g_k^2\varphi(x)\,dx-\left(\int g_k\varphi(x)\,dx\right)^2 \tag{13}$$

we obtain from Equations (7), (12) and (13) that

$$\begin{aligned}
\underset{k\to\infty}{Lim}\,Var(g_k)&=\underset{k\to\infty}{Lim}\left(E\left(g_k^2\right)-\left(E(g_k)\right)^2\right)\\
&=\underset{k\to\infty}{Lim}\int g_k^2\varphi(x)\,dx-\left(\underset{k\to\infty}{Lim}\int g_k\varphi(x)\,dx\right)^2\\
&=\int\left(\underset{k\to\infty}{Lim}\,g_k\right)^2\varphi(x)\,dx-\left(\int\left(\underset{k\to\infty}{Lim}\,g_k\right)\varphi(x)\,dx\right)^2\\
&=E\left(\underset{k\to\infty}{Lim}\,g_k\right)^2-\left(E\left(\underset{k\to\infty}{Lim}\,g_k\right)\right)^2\\
&=Var(\underset{k\to\infty}{Lim}\,g_k)
\end{aligned} \tag{14}$$

Noting that

$$\underset{k\to\infty}{Lim}\,g_k=\frac{\partial f}{\partial x_1} \tag{15}$$

and $\{\Delta_k\}$ is an arbitrary sequence going to zero, we complete the proof of the theorem.

3 The Computational Formula of Sensitivity for RBFNNs

To numerically calculate the new sensitivity for each feature, the following equations are followed. Noting that

$$\frac{\partial f}{\partial x_1} = \sum_{j=1}^{m} w_j \left(\frac{(x_1 - u_{1j})}{-v_j} \exp\left(\frac{\sum_{i=1}^{n}(x_i - u_{ij})^2}{-2v_j} \right) \right) \tag{16}$$

and   Eq.(4), we have

$$Var\left(\frac{\partial f}{\partial x_1} \right) = Var\left(\sum_{j=1}^{m} w_j \left(\frac{(x_1 - u_{1j})}{-v_j^2} \exp\left(\frac{\sum_{i=1}^{n}(x_i - u_{ij})^2}{-2v_j^2} \right) \right) \right) \tag{17}$$

Eq.(17) can also be expressed as

$$Var\left(\frac{\partial f}{\partial x_1} \right) = \sum_{j=1}^{m} \frac{w_j^2}{v_j^4} \left\{ \left(\frac{\sigma_1^2 v_j^3 (2\sigma_1^2 + v_j^2) + v_j^5 (\mu_1 - u_{1j})^2}{(2\sigma_1^2 + v_j^2)^{5/2}} \right) \exp\left(\frac{\sum_{i=2}^{n} \left(\sigma_i^2 (\sigma_i^2 - v_j^2) + (\mu_i - u_{ij})^2 (2\sigma_i^2 - v_j^2) \right)}{v_j^4} - \frac{2(\mu_1 - u_{1j})^2}{(2\sigma_1^2 + v_j^2)^2} \right) \right. $$
$$\left. - \left(\frac{v_j^6 (\mu_1 - u_{1j})^4}{(\sigma_1^2 + v_j^2)^3} \right) \exp\left(\frac{\sum_{i=2}^{n} \left(\sigma_i^2 (\sigma_i^2 - 2v_j^2) + 2(\mu_i - u_{ij})^2 (\sigma_i^2 - v_j^2) \right)}{2v_j^4} - \frac{(\mu_1 - u_{1j})^2}{(\sigma_1^2 + v_j^2)^2} \right) \right\} \tag{18}$$

where $x_1, x_2, \cdots, x_n$ are supposed to be independent normal distributions with means $\mu_1, \mu_2, \cdots, \mu_n$ and variances $\sigma_1, \sigma_2, \cdots, \sigma_n$ respectively. Similarly, the computational formula for computing other features sensitivity can be derived.

For irrelevant feature deletion, we need the following formula

$$E\left(\frac{\partial f}{\partial x_1} \right) = \sum_{j=1}^{m} \frac{w_j}{-v_j^2} \left\{ \left(\frac{v_j^3 (\mu_1 - u_{1j})^2}{(\sigma_1^2 + v_j^2)^{3/2}} \right) \exp\left(\frac{\sum_{i=2}^{n} \left(\sigma_i^2 (\sigma_i^2 - 2v_j^2) + 2(\mu_i - u_{ij})^2 (\sigma_i^2 - v_j^2) \right)}{4v_j^4} - \frac{(\mu_1 - u_{1j})^2}{2(\sigma_1^2 + v_j^2)^2} \right) \right\} \tag{19}$$

It is the expectation of output's partial derivative with respect to individual features. The feature deletion policy takes the following steps. First, sort the features by sensitivity in descending order. Second, choose the features with the smallest sensitivity constituting a set A. Third, select the features with the smallest expectation from A to eliminate.

A number of simulations are conducted on some UCI databases. The simulation result shows that the training and testing accuracy can be kept if features with small sensitivity are deleted. Without losing any accuracy, Table 1 shows the results of deleting features according to our feature deletion policy. It is worth noting that, for the irrelevant feature deletion, we use both the variance (i.e., the sensitivity) and the expectation of the partial derivative. One question is whether or not the expectation is important for the irrelevant feature deletion? It remains to be studied further.

Table 1. The sensitivity simulations using UCI databases for input feature selection.

Database	Total Feature Number	Eliminated Feature Number
Sonar	60	12
Wine	13	3
Zoo	13	4
Ionosphere	34	5

4   Conclusions

A new sensitivity definition for RBFNNs is given and a formula for computing the new sensitivity is derived in this paper. The simulation result shows that the training and testing accuracy can be kept if features with stable and small expectation are deleted for most selected datasets.

References

1. Haykin, S.: Neural Networks: A Comprehensive Foundation (Second Edition). Prentice-Hall, Incorporation 2001
2. Winter, R.: Madaline Rule II: A New Method for Training Networks of Adalines. Ph. Thesis, Stanford University, Stanford, CA 1989
3. Zurada, J.M.: Perturbation Method for Deleting Redundant Inputs of Perceptron Networks. Neurocomputingc **14** (1997) 177-193
4. Choi, J.Y., Choi, C.: Sensitivity Analysis of Multilayer Perceptron with Differentiable Activation Functions. IEEE Transactions on Neural Networks, 3 (1992) 101-107
5. Zeng, X., Yeung, D.S.: Sensitivity Analysis of Multilayer Perceptron to Input and Weight Perturbations. IEEE Transactions on Neural Networks, **12** (2001) 1358-1366
6. Stephen, W., Piché.: The Selection of Weight Accuracies for Madalines. IEEE Transactions on Neural Networks, **6** (1995) 432-445
7. Karayiannis, N.B.: Reformulated Radial Basis Neural Networks Trained by Gradient Decent. IEEE Transactions on Neural Networks, **10(3)** (1999) 657-671
8. Karayiannis, N.B.: New Development in the Theory and Training of Reformulated Radial Basis Neural Networks. IEEE International Joint Conference on Neural Network, (2000) 614-619

Complexity of Error Hypersurfaces
in Multilayer Perceptrons
with General Multi-input and Multi-output Architecture

Xun Liang[1,2]

[1] Institute of Computer Science and Technology, Peking University, Beijing 100817, China
liangxun@icst.pku.edu.cn
[2] Department of Management Science, Stanford University, Palo Alto, CA 94305, USA

Abstract. For the general multi-input and multi-output architecture of multi-layer perceptrons, the issue of classes of congruent error hypersurfaces is converted into the issue of classes of congruent pattern sets. By finding the latter number which is much smaller than the total number of error hypersurfaces, the complexity of error hypersurfaces is reduced. This paper accomplishes the remaining work left by [4] which only addresses multi-input and single-output architecture. It shows that from the input side, group $G(N)$ includes all the possible orthogonal operations which make the error hypersurfaces congruent. In addition, it extends the results from the case of single output to the case of multiple outputs by finding the group $S(M)$ of orthogonal operations. Also, the paper shows that from the output side, group $S(M)$ includes all the possible orthogonal operations which make the error hypersurfaces congruent. The results in this paper simplify the complexity of error hypersurfaces in multilayer perceptrons.

1 Introduction

The shape of error hypersurface in multilayer perceptrons has been studied for many years [1][4][7]. However, most of papers assume a fixed training pattern set, namely, a fixed χ mapping from N-dimensional space to M-dimensional space [1][7]. This paper treats all the possible pattern sets as a whole, and studies whether different mappings χ can result in similar error hypersurfaces. The topic in this paper is therefore wider.

Consider the binary pattern pairs. Clearly, for the N-dimensional input, there are totally 2^N input patterns. For an M-dimensional output, the output pattern for each output m ($m=1,\dots,M$) could be -1 or 1. Hence there are totally $\left(2^M\right)^{2^N} = 2^{M2^N}$ possible pattern sets. Because different pattern sets lead to different error hypersurfaces, intuitively there are 2^{M2^N} error hypersurfaces. However, noticing that some pattern sets coincide after operations of rotations, may we find their corresponding error hypersurfaces similar? If yes, after grouping similar error hypersurfaces into the same class (see Figs. 1 and 2), might we obtain a smaller number of classes of similar error hypersurfaces? The results could simplify the understanding on classes of error hypersurfaces that actually exist with difference. This is the motivation of the paper.

J. Wang, X. Liao, and Z. Yi (Eds.): ISNN 2005, LNCS 3496, pp. 87–94, 2005.
© Springer-Verlag Berlin Heidelberg 2005

A binary input-output pattern pair can be written as $(X_p, Y_p) \in \{-1,+1\}^N$ $\times \{-1,+1\}^M$, $p = 1, \cdots, 2^N$, where $X_p = (x_{1p}, \ldots, x_{Np})^{\tau}$ and $Y_p = (y_{1p}, \ldots, y_{Mp})^{\tau}$, $(\bullet)^{\tau}$ denotes the transpose of $(\bullet)$. The input pattern set is $X = \{X_p | p = 1, \ldots, 2^N\}$ and the output pattern set $Y = \{Y_p | p = 1, \ldots, 2^N\}$. The N-dimensional cube $\{-1,+1\}^N$ with 2^N vertices forms the input pattern space. On the 2^N vertices, there are patterns (X_p, Y_p), $p = 1, \cdots, 2^N$ with the input pattern $X_p \in \{-1,+1\}^N$ as the coordinates of the vertex and output pattern $Y_p \in \{-1,+1\}^M$ as the "vector-value" of the vertex. We use

$$e(w) = \sum_{p=1}^{2^N} \|\xi_w(X_p) - Y_p\|^2 \tag{1}$$

as the error function where $\|\cdot\|$ is the Euclidean norm, $\xi_w(X_p)$ is a mapping from N-dimension space to M-dimension space performed by the multilayer perceptron, and w is the set composed of all weights in the multilayer perceptron. There exist other types of error functions that can be used, although this paper will not consider them. If $M=1$, the "vector-value" becomes a scalar quantity, and takes two possible values, -1 and +1.

The activation function ϕ is assumed to be a generic odd function,

$$\phi(-x) = -\phi(x) \tag{2}$$

The remaining of this paper is organized as follows. Section 2 gives the examples of congruent pattern sets for $N=1$, 2 and $M=1$. Section 3 shows that $G(N)$ includes all possible orthogonal operations. Section 4 extends the case of $M=1$ to the general case and forms orthogonal operation group $S(M)$. Section 5 calculates the numbers of congruent error hypersurfaces. Section 6 concludes the paper.

2 Examples of Congruent Pattern Sets for $N = 1$ and 2, and $M=1$

For $N=1$ and $M=1$ (see Fig.1), we have totally the total number $2^{2^1} = 4$ pattern sets. By examining the input pattern features based on reflections and rotations, we obtain 3 different groups (see Fig.2).

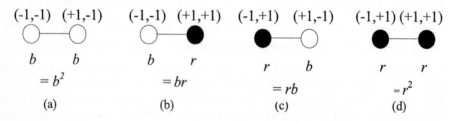

Fig. 1. All pattern sets for $N=1$ and $M=1$. $X_p=-1$ is for the left node, and $X_p=+1$ is for the right node. $y_{mp}=-1$ is denoted by O or b (blue). $y_{mp}=+1$ is denoted by ● or r (red)

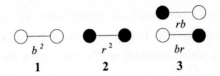

Fig. 2. All coloring schemes for $N=1$, $M=1$. The members of each group are still pairwise congruent

Definition 1. For a single output pattern y_{mp}, $m=1,\cdots,M$, $p=1,\cdots,P$, the operation $y_{mp}'=-y_{mp}$ is called the operation of *output pattern inversion*.

If we furthermore apply the output pattern inversion, groups 1 and 2 coincide each other. Denote by $Q(N,M)$ the number of congruent error hypersurfaces for the multilayer perceptron with N inputs and M outputs. We know that the number of congruent pattern sets is always equal to the number of congruent error hypersurfaces [4]. As a result, 4 pattern sets can be grouped into $Q(1,1)=2$ congruent pattern sets (see Fig.3).

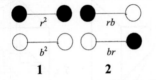

Fig. 3. Classes of congruent pattern sets for $N=1$ and $M=1$

Definition 2. In the 2^{M2^N} pattern sets, two pattern sets are congruent if they coincide after applying any of the operations of rotations, reflections, and output pattern inversions, or a combination of them.

For $N=2$ and $M=1$, there are $2^{2^2} = 16$ pattern sets, we obtain $Q(2,1)=4$ classes of congruent pattern sets (see Fig.4).

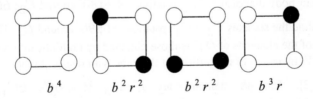

Fig. 4. Representatives for 4 classes of congruent pattern sets for $N=2$ and $M=1$

3   Operation Group on Inputs, $G(N)$

[4] proved that congruent pattern sets lead to congruent error hypersurfaces. Hence, in the following, we only need to discuss the congruent pattern sets.

For the operations on input patterns, firstly [4] has recursively constructed the set $G(N)$ and then proves it is a group. In this section, we repeat the approach briefly first, and then show that $G(N)$ includes all of the possible N-dimensional orthogonal matrices we are interested.

For an N-dimensional cube, with $2N$ hyperplanes $X_i = \{(x_1,\cdots,x_N)\mid x_k = d\}$, $k = 1,\cdots,N$, $d=\pm1$, each made by fixing the value on one coordinate, providing that we already have $G(N)$ which performs on the vertices of the N-dimensional cube, then for any operation $f\in G(N)$,

$$f(x_1,\cdots,x_N) = (y_1,\cdots,y_{k-1},y_{k+1},\cdots,y_{N+1}) \tag{3}$$

where $x_1,\ldots,x_N$ and $y_1,\ldots,y_{k-1},y_{k+1},\ldots,y_{N+1}$ are the coordinates of vertices on the N-dimensional cube, and let

$$g:\quad (\ 1,x_1,\cdots,x_N)\mid \to (y_1,\cdots,y_{k-1},\ d,y_{k+1},\cdots y_{N+1}), \tag{3a}$$

$$(-1,x_1,\cdots,x_N)\mid \to (y_1,\cdots,y_{k-1},-d,y_{k+1},\cdots,y_{N+1}), \tag{3b}$$

then the operations g form $G(N+1)$.

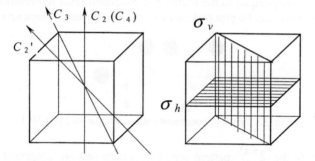

Fig. 5. The rotations for $N=3$ **Fig. 6.** The reflections for $N=3$

Particularly, $G(3)$ is the O group in the symmetry group theory [2][3] with 24 elements in the O group (see Figs. 9 and 10), I : identity; $8C_3$: rotations of $\pm120°$ around the four body diagonals; $3C_2$: rotations of $+180°$ around C_2; $6C_2'$: rotations of $+180°$ around the six axes C_2'; $6C_4$: rotations of $\pm90°$ around C_4. The group O_h is composed of the elements of O plus those obtained by multiplication with inversion in the origin V, inversion of all vertices of the cube with respect to the origin, i.e., O_h $=O\times\{e,V\}$ [2]. The 48 elements are I, $8C_3$, $3C_2$, $6C_2'$, $6C_4$, V, $8S_6$, $3\sigma_h$, $6\sigma_v$, $6S_4$ where σ_h is the reflection in a horizontal plane which runs through the middle points of the opposite four edges of the cube and perpendicular to C_4 (see Figs. 5 and 6); σ_v is the reflection in a vertical plane containing the opposite edges of the cube (see Figs. 9 and 10); S_n is the rotary reflection (adjunction of reflection [2], a combination of a rotation by an angle of $\dfrac{2\pi}{n}$ together with a reflection in a plane perpendicular to the axis), where $VC_4 = S_4{}^3$, $VC_3 = S_6{}^5$.

Definition 3. If all the elements in an orthogonal matrix are -1, or 0, or 1, it is called the *simple orthogonal matrix*.

Theorem 1. The group $G(N)$ includes all of the N-dimensional simple orthogonal matrices.

Proof. We proceed with induction. First, $G(1) = \{$ [-1], [+1] $\}$. Clearly, [-1] and [+1] are all the 1-dimensional simple orthogonal matrices. Assume that $G(N)$ consists of all N-dimensional simple orthogonal matrices. For an $(N+1)$-dimensional simple orthogonal matrix, every row has only one -1 or 1 and N 0's, and every column has only one -1 or 1 and N 0's. As a result, (3a) and (3b) formulate all possible $(N+1)$-dimensional simple orthogonal matrix in $G(N+1)$.

Clearly, from the input side, only the operations in group $G(N)$ can make the error hypersurfaces congruent. This paper therefore accomplishes the study of congruent error hypersurfaces from the input side of multilayer perceptrons.

4 Operation Group on Outputs, $S(M)$

Definition 4. For the M-dimensional output pattern vector Y_p, $p = 1, \cdots, P$, the operation $S(M)$ is called the *output pattern inversion matrix*, if after applying it, for at least one m, $m = 1, \cdots, M$, an output pattern inversion for y_{mp} is performed.

For the M-dimensional outputs, the output pattern inversion matrix $S(M)$ is operated on the M-dimensional space. If $M=1$, $S(M) = [-1]$ or $[+1]$. If $M>1$,

$$
\begin{bmatrix} \vdots \\ y_{mp}{}' \\ \vdots \end{bmatrix} = \begin{bmatrix} S(m-1) & & 0 \\ & -1 & \\ 0 & & S(M-m) \end{bmatrix} \begin{bmatrix} \vdots \\ y_{mp} \\ \vdots \end{bmatrix} \tag{4}
$$

where $S(m-1)$ and $S(M-m)$ are $(m-1)$-dimensional and $(M-m)$-dimensional *output pattern inversion* matrices, respectively.

The combinations of -1 and 1 in $S(M)$ result in $\binom{0}{M} + \binom{1}{M} + \binom{2}{M} + \cdots \binom{M}{M} = 2^M$ different $S(M)$. Clearly, all $S(M)$ are orthogonal matrices. Let

$$
S(M) = \{ \text{ all output pattern inversion matrices, } S(M) \} \tag{5}
$$

Then we have the number of elements in $S(M)$, $|S(M)|=2^M$.

Theorem 2. The operations $S_i(M)$ ($i=1, \ldots, |S(M)|$) compose a group $S(M)$.

Proof. Unit element, associate law: obvious. Inverse element: If $S_j(M) \in S(M)$, $S_j^{-1}(M) = S_j(M) \in S(M)$. Closure: $S_j(M) \in S(M)$, $S_k(M) \in S(M)$, $S_j(M)S_k(M) = S_i(M) \in S(M)$.

Clearly, $S(M)$ is orthogonal and $S(M)$ is a special orthogonal matrix group.

[4] proved that any operation in $G(N)$ and $S(1)$ can result in the congruent error hypersurfaces. Next we show it is also true for the general case of $S(M)$ with $M \geq 1$.

Theorem 3. Any operation in $S(M)$ can result in the congruent error hypersurfaces.

Theorem 4. The set $S(M)$ includes all of the M-dimensional inversion operations.

Proof. By Definition 4 and Theorem 3, this is straightforward.

Clearly, from the output side, only the operations in group $S(M)$ can make the error hypersurfaces congruent. This paper therefore accomplishes the study of congruent error hypersurfaces from the output side of multilayer perceptrons.

5 Number of Congruent Error Hypersurfaces, $Q(N,M)$

[4] gave $Q(N,M)$ for $N \geq 1$, $M=1$. Next we calculate $Q(N,M)$ for $N \geq 1$, $M \geq 1$.

[4] derived the polynomials based on Polya Theorem [5]. We use the same polynomials and obtain the coefficients we need in this paper. Since we are concerning the case that the patterns abound in all the input space in this paper as opposed to any subset of the whole pattern set as in [4], we are only interested in the terms without y.

We directly list the polynomial results. The entire derivation can be found in [4].

For $N=1$, we have

$$\left(b^2 + y^2\right) + \left(br + by\right) \tag{7}$$

Summarizing the coefficients of the terms without y leading to $Q(1,1)=2$, which is consistent with the result given by the diagram-based method in Fig.3.

For $N=2$, we have

$$\left(b^4 + y^4\right) + 2\left(b^2 r^2 + b^2 y^2\right) + \left(b^3 r + b^3 y + by^3\right) + 2\left(b^2 ry + bry^2\right) \tag{8}$$

Summarizing the coefficients of the terms without y leading to $Q(2,1)=4$, which is also consistent with the result given by the diagram-based method in Fig.4.

For $N=3$, we have

$$\left(b^8 + y^8\right) + \left(b^7 r + b^7 y + by^7\right) + 3\left(b^6 r^2 + b^6 y^2 + b^2 y^6\right) + 3\left(b^5 r^3 + b^5 r^3 + b^3 r^5\right)$$
$$+ 6\left(b^4 r^4 + b^4 y^4\right) + 3\left(b^6 ry + bry^6\right) + 6\left(b^5 r^5 y + b^5 ry^2 + b^2 ry^5\right)$$
$$+ 10\left(b^4 r^3 y + b^4 ry^3 + b^3 ry^4\right) + 16\left(b^4 r^2 y^2 + b^2 r^2 y^4\right) + 17\left(b^3 r^3 y^2 + b^3 r^2 y^3\right) \tag{9}$$

Summarizing the coefficients of the terms without y leading to $Q(3,1)=20$.

For $N=4$, we have

$$\frac{1}{384}[51(b^2 + r^2 + y^2)^8 + 96(b^2 + r^2 + y^2)^2(b^6 + r^6 + y^6)^2 + 48(b^8 + r^8 + y^8)^2$$
$$+ 84(b^4 + r^4 + y^4)^4 + 32(b+r+y)^4(b^3 + r^3 + y^3)^4$$
$$+ 48(b+r+y)^2(b^2 + r^2 + y^2)(b^4 + r^4 + y^4)^3 + 12(b+r+y)^8(b^2 + r^2 + y^2)^4$$
$$+ 12(b+r+y)^4(b^2 + r^2 + y^2)^6 + (b+r+y)^{16}]$$
$$= b^{16} + b^{15}r + 4b^{14}r^2 + 6b^{13}r^3 + 19b^{12}r^4 + 27b^{11}r^5 + 50b^{10}r^6 + 56b^9 r^7 + 74b^8 r^8$$
$$+ 56b^7 r^9 + 50b^6 r^{10} + 27b^5 r^{11} + 19b^4 r^{12} + 6b^3 r^{13} + 4b^2 r^{14} + br^{15} + r^{16} + (...)y^k \tag{10}$$

where $k \geq 1$ and $(...)$ is a polynomial. The formula is derived by the software tool MATHEMATICA. Summarizing the coefficients of the first 17 terms leading to $Q(4,1)=402$.

For $N=5$, we have

$$\frac{1}{3840}[231(b^2 + r^2 + y^2)^{16} + 720(b^2 + r^2 + y^2)^4(b^6 + r^6 + y^6)^4$$
$$+ 240(b+r+y)^4(b^2 + r^2 + y^2)^2(b^4 + r^4 + y^4)^6 + 384(b+r+y)^2(b^5 + r^5 + y^5)^6$$
$$+ 520(b^4 + r^4 + y^4)^8 + 480(b^8 + r^8 + y^8)^4 + 384(b^2 + r^2 + y^2)(b^{10} + r^{10} + y^{10})^3$$

$$+ 240(b^2 + r^2 + y^2)^4(b^4 + r^4 + y^4)^6 + 60(b + r + y)^8(b^2 + r^2 + y^2)^{12}$$
$$+ 160(b + r + y)^4(b^2 + r^2 + y^2)^2(b^3 + r^3 + y^3)^4(b^6 + r^6 + y^6)^2$$
$$+ 320(b^4 + r^4 + y^4)^2(b^{12} + r^{12} + y^{12})^2 + 80(b + r + y)^8(b^3 + r^3 + y^3)^8$$
$$+ 20(b + r + y)^{16}(b^2 + r^2 + y^2)^8 + (b + r + y)^{32}] = \sum_{i=0}^{32} \beta_i b^{32-i} r^i + (\ldots) y^k \qquad (11)$$

where $k \geq 1$, $(\ldots)$ is a polynomial, and β_i, $i=1,\ldots,32$, are 1, 1, 5, 10, 47, 131, 472, 1326, 3779, 9013, 19963, 38073, 65664, 98804, 133576, 158658, 169112, 158658, 133576, 98804, 65664, 38073, 19963, 9013, 3779, 1326, 472, 131, 47, 10, 5, 1, 1. The formula is also derived by the software tool MATHEMATICA. Summarizing the coefficients of the first 33 terms leading to $Q(5,1)=1228113$.

For $M>1$, we can think that we have $Q(N,1)$ different colors, $c_1, c_2, \ldots, c_{L(N,1)}$, and we want to pick out one and assign it to each of the M outputs. Any color c_i $(i=1,\ldots,Q(N,1))$ can be picked out repeatedly. The above process is equivalent to computing the following generating function for the term c^M

$$\left(c_1^0 + c_1^1 + c_1^2 + \cdots + c_1^M\right)\left(c_2^0 + c_2^1 + c_2^2 + \cdots + c_2^M\right)\cdots\left(c_{Q(N,1)}^0 + c_{Q(N,1)}^1 + c_{Q(N,1)}^2 + \cdots + c_{Q(N,1)}^M\right). \qquad (12)$$

The summation of the coefficients of term $\sum_{i+j+\cdots+k=M} c_1^i c_2^j \cdots c_{Q(N,1)}^k$ is $Q(N,M)$. Alternatively, by combinatorics, it is easy to see that the above number is C_{M+Q-1}^M, given in Table 1 for different N and M. Noticing that $Q(N,M)$ is polynomial by Polya Theorem [5], we know that $Q(N,M)/2^{M2^N} \to 0$ as $N,M \to \infty$ at an exponential rate, given in Table 2 for different N and M.

Table 1. $Q(N,M)$, number of congruent error hypersurfaces

M \ N	1	2	3	4	5	...
1	2	4	20	402	1228113	...
2	3	10	210	81003	754131384441	...
3	4	20	1540	10908404	308720021734252000	...
4	5	35	8855	1104475905	9478599955304590000000	...
5	6	56	42504	89683443486	232816594826176000000000000	...
...	...	...	...	...	...	...

Table 2. $Q(N,M)/2^{M2^N}$ decreases fast as N and M increase

M \ N	1	2	3	4	5	...
1	0.5	0.25	0.078	0.0061	0.00029	...
2	0.19	0.039	0.0032	0.000019	0.000000041	...
3	0.06	0.0049	0.00009	0.00000004	0.000000000004	...
4	0.02	0.0005	0.000002	0.00000000006	0.0000000000000003	...
5	0.01	0.0001	0.00000004	0.0000000000001	0.00000000000000000002	...
...	...	...	...	...	...	...

6 Conclusions

This paper improves the analysis in [4] by proving that group $G(N)$ includes all the possible orthogonal operations that make the error hypersurfaces congruent, extending the results from $M=1$ to any possible integer of M, and forming $S(M)$ with a proof that $S(M)$ includes all possible orthogonal operations.

The result shows that $Q(N,M)$ is much smaller than the total number of error hypersurfaces, and with N and M growing, $Q(N,M)$ increases at a much slower rate than the total number. The result simplifies the understanding on classes of error hypersurfaces which actually exist with difference.

References

1. Chen, A.M., Lu, H., Hecht-Nielsen, R.: On the Geometry of Feedforward Neural Network Error Surfaces. Neural Computation **5** (1993) 910-927
2. Hamermesh, M.: Group Theory and Its Application to Physical Problems. Addison-Wesley, Berkeley (1964) 59-60
3. Sternberg, S.: Group Theory and Physics Cambridge University Press, New York (1994)
4. Liang, X.: Complexity of Error Hypersurfaces in Multilayer Perceptrons with Binary Pattern Sets. Int. J. of Neural Systems **14** (2004) 189-200
5. Polya, G., Read, R.C.: Combinatorial Enumeration of Groups, Graphs, and Chemical Compounds Springer-Verlag, Berlin Heidelberg New York (1987)
6. Spence, L.E., Friedberg, S., Insel, A.J., Friedberg, S.H.: Elementary Linear Algebra Prentice Hall, New Jersey (1999)
7. Sussmann, H.J.: Uniqueness of the Weights for Minimal Feedforward Nets with a Given Input-output Map. Neural Networks **5** (1992) 589-593

Nonlinear Dynamical Analysis on Coupled Modified Fitzhugh-Nagumo Neuron Model

Deepak Mishra, Abhishek Yadav, Sudipta Ray, and Prem K. Kalra

Department of Electrical Engineering, IIT Kanpur, India-208 016
{dkmishra,ayadav,sray,kalra}@iitk.ac.in

Abstract. In this work, we studied the dynamics of modified FitzHugh-Nagumo (MFHN) neuron model. This model shows how the potential difference between spine head and its surrounding medium vacillates between a relatively constant period called the silent phase and large scale oscillation reffered to as the active phase or bursting. We investigated bifurcation in the dynamics of two MFHN neurons coupled to each other through an electrical coupling. It is found that the variation in coupling strength between the neurons leads to different types of bifurcations and the system exhibits the existence of fixed point, periodic and chaotic attractor.

1 Introduction

Determining the dynamical behavior of an ensemble of coupled neurons is an important problem in computational neuroscience. The primary step for understanding this complex problem is to understand the dynamical behavior of individual neurons. Commonly used models for the study of individual neurons which display spiking/bursting behavior include (a) Integrate-and-fire models and their variants [1, 2] (b) FitzHugh-Nagumo model [3], (c) Hindmarsh-Rose model [13], (d) Hodgkin-Huxley model [4, 7] and (e) Morris-Lecar model [5]. A short review of models in neurobiology is provided by Rinzel in [6, 9, 12]. The study of Type I neuronal models is more important as pyramidal cells in the brain [11] exhibits this type of behavior. Biophysical models such as the Hodgkin-Huxley(HH) model and Morris-Lecar(ML) model have been observed to display Type I neural excitability [2]. Mathematical techniques to study Type I neurons were developed by Ermentrout and Kopell [1] and the individual behavior of Type I neurons was fairly well understood. Bifurcation phenomena in individual neuron models including the Hodgkin-Huxley, Morris-Lecar and FitzHugh-Nagumo have been investigated in the literature [9–11, 13, 14]. Rinzel and Ermentrout [9] studied bifurcations in the Morris-Lecar (ML) model by treating the externally applied direct current as a bifurcation parameter. It is also important to note that the choice of system parameters in these neuron models can influence the type of excitability [9]. Rinzel, proposed a neuron model which produces a Type III burst is studied here [12]. The study of coupled neuron models is one of the fundamental problem in computational neuroscience that

J. Wang, X. Liao, and Z. Yi (Eds.): ISNN 2005, LNCS 3496, pp. 95–101, 2005.

helps in understanding the behavior of neurons in a network. From dynamical point of view it is of crucial importance to investigate the effect of variation in coupling strength to its dynamical behavior. In this work, we presented general conditions under which a system of two coupled neurons shows different types of dynamical behavior like converging, oscillatory and chaotic. For our analysis we considered coupling strength as a parameter. Our studies are based on modified FitzHugh-Nagumo system [12] which is an extension of FitzHugh-Nagumo system.

In section 2, we discuss the three dimensional mathematical model of modified FitzHugh-Nagumo type neurons. Nonlinear dynamical analysis of this model is presented in section 2. In section 3, we study a mathematical model of coupled neural oscillators and its bifurcation diagram. Finaly, in section 4, we concluded our work.

2   The Modified FitzHugh-Nagumo Neuron Model

The modified FitzHugh-Nagumo equations are a set of three simple ordinary differential equations which exhibit the qualitative behavior observed in neurons, viz quiescence, excitability and periodic behavior [12]. The system can be represented as

$$\dot{v} = -v - \frac{v^3}{3} - w + y + F(t) \tag{1}$$

$$\dot{w} = \phi(v + a - bw) \tag{2}$$

$$\dot{y} = \epsilon(-v + c - dy) \tag{3}$$

The function $F(t)$ represents the external stimulus. From biological point of view, variable v represents the potential difference between the dendritic spine head and its surrounding medium, w is recovery variable and y represents the slowly moving current in the dendrite. In this model, v and w together make up a fast subsystem relative to y. The equilibrium point (v^*, w^*, y^*) is calculated by substituting $\dot{v} = \dot{w} = \dot{y} = 0$ in equations (1), (2) & (3). The jacobian matrix at this point is found to be

$$J = \begin{bmatrix} -1 - v^{*2} & -1 & 1 \\ \phi & -b\phi & 0 \\ -\epsilon & 0 & -\epsilon d \end{bmatrix} \tag{4}$$

The three eigenvalues λ_1, λ_2 and λ_3 are the roots of equation $det(J - \lambda I) = 0$. If at a neighborhood of a particular value μ_0 of the parameter μ, there exists a pair of eigenvalues of $J(\mu)$ of the form $\alpha(\mu) \pm i\beta(\mu)$ such that $\alpha(\mu) = 0$, $\beta(\mu) \neq 0$, then no other eigenvalue of $A(\mu_0)$ will be an integral multiple of $i\beta(\mu_0)$. Thus $A(\mu_0)$ has a pair of pure imaginary eigenvalues. This provides the information about the bifurcation in the system.

2.1 Nonlinear Dynamical Analysis
   of Modified FitzHugh-Nagumo Model

The numerical simulation for this model shows a sudden disappearance of two (stable and unstable) periodic orbits, via different bifurcations associated with the homoclinic orbits of the saddle.

Phase Plane Analysis of the Model by Considering ϕ as Bifurcation Parameter. The time responses and phase portraits for the model are drawn in (Fig. 1). In (Fig. 1(a)) and (Fig. 1(c)) the model shows bursting behavior. In (Fig. 1(b)) and (Fig. 1(d)) the system moves to oscillatory state and there apears a periodic orbit (limit cycle). We also found that if $\phi > 0.16$ the system shows converging response. It is observed here that the model shows change in dynamical behavior at some parametric values.

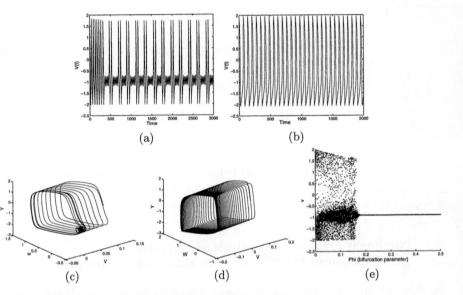

(a)   (b)

(c)   (d)   (e)

Fig. 1. Time responses, three dimensional phase portraits and bifurcation diagram for modified FHN model with following parameters $a = 0.7, b = 0.8, c = -0.775, d = 1, \epsilon = 0.001$ & $F(t) = 0.3125$. (a) Time response at $\phi = 0.08$ (b) Time response at $\phi = 0.9$ (c) Three dimensional phase portrait at $\phi = 0.08$ (d) Three dimensional phase portrait at $\phi = 0.9$ (e) Bifurcation diagram with ϕ as bifurcation parameter.

Bifurcation Diagram of the Model by Considering ϕ as Bifurcation Parameter. As the parameter is varied, the behavior of the system changes qualitatively from a stable fixed point to a limit cycle. At some point, the fixed point looses its stability, implying the real part of at least one of the eigenvalues changes from negative to positive. In other words, the real part passes through zero. The bifurcation diagram for the model is shown in (Fig. 1(e)). Model shows converging (stable) dynamics for $\phi > 0.15$ and multiple limit cycles attractor for

$\phi < 0.15$. It can be observed from this figure that the model exhibits bifurcation as the parameter ϕ is changed.

3   Coupled Modified FitzHugh-Nagumo Neuron Model

This study is designed to describe the behavior of two neurons linked with electrical coupling (coupling via the flow of ions through the gap junctions between neurons). We choose this coupling to be a constant times the difference in the voltages of two cells. This leads to a set of six coupled nonlinear ordinary differential equations:

$$\dot{v}_1 = -v_1 - \frac{v_1^3}{3} - w_1 + y_1 + \gamma_{12}(v_1 - v_2) + F_1(t) \tag{5}$$

$$\dot{w}_1 = \phi_1(v_1 + a_1 - b_1 w_1) \tag{6}$$

$$\dot{y}_1 = \epsilon_1(-v_1 + c_1 - d_1 y_1) \tag{7}$$

$$\dot{v}_2 = -v_2 - \frac{v_2^3}{3} - w_2 + y_2 + \gamma_{21}(v_2 - v_1) + F_2(t) \tag{8}$$

$$\dot{w}_2 = \phi_2(v_2 + a_2 - b_2 w_2) \tag{9}$$

$$\dot{y}_2 = \epsilon_2(-v_2 + c_2 - d_2 y_2) \tag{10}$$

For our analysis, we consider the behavior when two coupled neurons posses identical values of parameters *i.e.* $a = a_1 = a_2$, $b = b_1 = b_2$, $c = c_1 = c_2$, $d = d_1 = d_2$, $\gamma = \gamma_{12} = \gamma_{21}$, $\phi = \phi_1 = \phi_2$ and $\epsilon = \epsilon_1 = \epsilon_2$.

3.1   Nonlinear Dynamical Analysis of Coupled MFHN Neuron Model with γ as Bifurcation Parameter

In this section, we investigate nonlinear dynamics of coupled MFHN neuron model. We consider coupling variable γ as the bifurcation parameter for analysis. We have shown time response for both neurons at different values of bifurcation parameter. We calculated the eigenvalues for the model at different values of γ. The bifurcation diagrams are plotted for the coupled model. The leading Lyapunov exponent plot is also drawn in order to verify the presence of chaos.

Eigenvalue Analysis for Coupled MFHN Neuron Model. Eigenvalues for the linearised model at different values of bifurcation parameter are shown in (Table 1). It is observed that for $\gamma = 3.0$, real parts of all eigenvalues of the linearized model are negative and therefore the system response is stable. Values of other parameters used for our analysis are $a = 0.7$, $b = 0.8$, $c = -0.775$, $d = 1.0$, $\phi = 0.08$ and $\epsilon = 0.1$. Eigenvalues at $\gamma = 4.4$ and $\gamma = 4.95$ are complex. We found that the system shows hopf bifurcation at $\gamma = 3.64$.

Phase Plane Analysis for Coupled MFHN Model with γ as Bifurcation Parameter. In this section, time responses for the model are drawn for different values of γ for neuron I and neuron II (Fig. 2). This system shows converging behavior at $\gamma = 3.95$ as depicted in (Fig. 2(a,d)). We found oscillatory response shown in (Fig. 2(b,e)) at $\gamma = 4.4$. It shows chaotic behavior at $\gamma = 4.95$, shown in (Fig. 2(c,f)). Three regimes of dynamics i.e. stable, periodic and chaotic are observed.

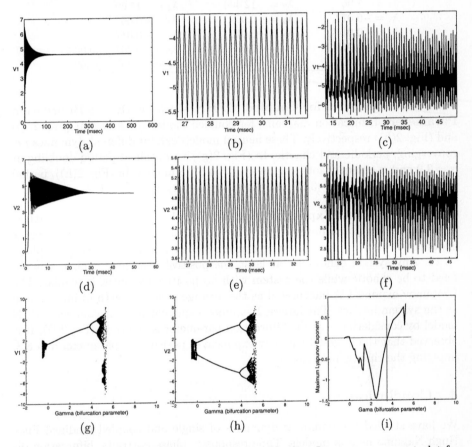

Fig. 2. Time responses, bifurcation diagrams and leading Lyapunov exponent plot for coupled modified FHN model. (a) Time response for coupled neuron I at $\gamma = 3.95$ (b) Time response for neuron I at $\gamma = 4.4$ (c) Time response for neuron I at $\gamma = 4.95$ (d) Time response for coupled neuron II at $\gamma = 3.95$ (e) Time response for coupled neuron II at $\gamma = 4.4$ (f) Time response for coupled neuron II at $\gamma = 4.95$ (g) Bifurcation diagram for neuron I, the variable $v_1(t)$ is on the vertical axis, and the bifurcation parameter γ is on the horizontal axis (h) Bifurcation diagram for neuron II, the variable $v_2(t)$ is on the vertical axis, and the bifurcation parameter γ is on the horizontal axis (i) Plot for leading Lyapunov exponent.

Table 1. Eigenvalues of Linearized Model at Different Values of Parameter, γ.

$At(\gamma = 3.0)$	$At(\gamma = 4.4)$	$At(\gamma = 4.95)$
Equilibrium Points		
$v_1^* = 3.66$	$v_1^* = 4.44$	$v_1^* = 5.692$
$v_2^* = -3.87$	$v_2^* = -4.77$	$v_2^* = -5.035$
Eigenvalues		
$\lambda_1 = -13.97$	$\lambda_1 = -21.27$	$\lambda_1 = -20.25$
$\lambda_2 = -7.96$	$\lambda_2 = -12.406$	$\lambda_2 = -17.60$
$\lambda_3 = -0.072$	$\lambda_3 = -0.069$	$\lambda_3 = -0.66$
$\lambda_4 = -0.014$	$\lambda_4 = -0.015$	$\lambda_4 = -0.017$
$\lambda_5 = -0.013$	$\lambda_5 = -0.008 - j0.002$	$\lambda_5 = -0.007 - j0.003$
$\lambda_6 = -0.008$	$\lambda_6 = -0.008 + j0.002$	$\lambda_6 = -0.007 - j0.003$

Bifurcation Diagram for Coupled MFHN Model with γ as Bifurcation Parameter. Bifurcation diagrams for coupled model are shown in (Fig. 2(g)) and (Fig. 2(h)) respectively. These neuron models exhibit different dynamics *i.e.* converging, periodic and chaotic. In (Fig. 2(g)) we obtained period doubling at $\gamma = 3.9$ and it enters chaotic region at $\gamma = 4.67$. Similarly, in (Fig. 2(h)), period doubling occured at $\gamma = 3.87$ and chaotic region starts at $\gamma = 4.51$.

Leading Lyapunov Exponent Plot for Coupled MFHN Model with γ as Bifurcation Parameter. Lyapunov exponents are related to the average exponential rates of divergence or convergence of neighboring orbits in phase space. Any system containing at least one positive Lyapunov exponent is defined to be chaotic while the system with no positive exponent is regular. The Lyapunov exponent is interpreted as the average rate of growth of information as the system evolves. The largest Lyapunov exponent plot for coupled MFHN model by considering γ as the bifurcation parameter is shown in (Fig 2(i)). It is observed that the largest Lyapunov exponent is positive for parameter $\gamma > 3.8$, implying the chaotic behavior.

4   Conclusion

We have studied the nonlinear dynamics of single and coupled modified Fitz-Hugh-Nagumo neuron models. Time responses, phase portraits, bifurcation diagrams and leading Lyapunov exponent plots are drawn in order to analyze the behavior at different parametric values. We found that in a single MFHN neuron there exist point attractors as well as mutiple limit cycles at different values of parameters. We also found that in coupled MFHN neuron model, the transition from stable to chaotic regime occurs with the variation of coupling strength.

References

1. Ermentrout, G.B., Kopell, N.: Parabolic Bursting in an Excitable System Coupled with a Slow Oscillation. SIAM Journal on Applied Mathematics, **46** (1986) 233–253

2. Ermentrout, G.B.: Type I Membranes, Phase Resetting Curves and Synchrony. Neural Computing, **8** (1996) 979–1001
3. Fitzhugh, R.: Impulses and Physiological States in Models of Nerve Membrane. Biophysical Journal, **1** (1961) 445–466
4. Hodgkin, A.L., Huxley, A.F.: A Quantitative Description of Membrane Current and Application to Conduction and Excitation in Nerve. Journal of Physiology, **117** (1954) 500–544
5. Morris, C., Lecar, H.: Voltage Oscillations in the Barnacle Giant Muscle Fiber. Journal of Biophysics, **35** (1981) 193–213
6. Rinzel, J.: Models in neurobiology. Nonlinear Phenomena in Physics and Biology, Plenum Press, New York (1981) 345–367
7. Hodgkin, A.L.: The Local Changes Associated with Repetitive Action in a Non-Modulated Axon. Journal of Physiology, **107** (1948) 165–181
8. Izhikevich, E.M.: Class 1 Neural Excitability, Conventional Synapses, Weakly Connected Networks and Mathematical Foundations of Pulse Coupled Models. IEEE Transactions on Neural Networks, **10** (1999) 499–507
9. Rinzel, J., Ermentrout, G.B.: Analysis of Neural Excitability and Oscillations. Methods in Neuronal Modeling, MIT press, Cambridge MA (1989)
10. Ehibilik, A.I., Borisyuk, R.M., Roose, D.: Numerical Bifurcation Analysis of a Model of Coupled Neural Oscillators. International Series of Numerical Mathematics, **104** (1992) 215–228
11. Hoppensteadt, F.C., Izhikevich, E.M.: Weakly Connected Neural Networks. Springer-Verlag (1997)
12. Rinzel, J.: A Formal Classification of Bursting Mechanisms in Excitable Systems, in Mathematical Topics in Population Biology, Morphogenesis and Neurosciences. Lecture Notes in Biomathematics. Springer-Verlag, New York, **71** (1987) 267–281
13. Izhikevich, E.M.: Neural Excitability, spiking and bursting. International Journal of Bifurcation and Chaos, **10** (2000) 1171–1266.
14. Mishra, D., Yadav, A., Kalra, P.K.: Chaotic Behavior in Neural Network and FitzHugh-Nagumo Neuronal Model. LNCS 3316 ICONIP-2004, (2004) 868–873

Stability of Nonautonomous Recurrent Neural Networks with Time-Varying Delays*

Haijun Jiang[1,2], Jinde Cao[1], and Zhidong Teng[2]

[1] Department of Mathematics, Southeast University, Nanjing 210096, China
[2] College of Mathematics and System Sciences, Xinjiang University,
Urumqi 830046, China
jianghai@xju.edu.cn, jdcao@seu.edu.cn

Abstract. The paper studies the nonautonomous delayed recurrent neural networks. By applying Lyapunov functional method and utilizing the technique of inequality analysis, we obtain the sufficient condition to ensure the globally asymptotic stability and globally exponential stability. The results given in this paper are new and useful.

1 Introduction

Recently, dynamical neural networks have attracted increasing interest in both theoretical studies and engineering applications. Many important results on the existence and uniqueness of equilibrium point, global asymptotic stability and global exponential stability have been established (see [1-9] and the references cited therein) and successfully applied to signal processing system, especially in static image treatment, and to solve nonlinear algebraic equations, such application rely on the qualitative properties of stability.

In this paper, we will consider the following recurrent neural networks with time-varying delays.

$$\frac{dx_i(t)}{dt} = -c_i(t)x_i(t) + \sum_{j=1}^n a_{ij}(t)f_j(x_j(t)) + \sum_{j=1}^n b_{ij}(t)f_j(x_j(t-\tau_j(t))) + u_i(t), \quad (1)$$

or rewritten as

$$\dot{x}(t) = -C(t)x(t) + A(t)f(x(t)) + B(t)f(x(t-\tau(t))) + u(t). \quad (2)$$

Where $x(t) = (x_1(t), \cdots, x_n(t))^T$ is the state vector the neural networks. $C(t) = \text{diag}(c_1(t), \cdots, c_n(t))$, $A(t) = (a_{ij}(t))_{n \times n}$ is weight matrix. $B(t) = (b_{ij}(t))_{n \times n}$ is the delayed weight matrix. $u(t) = (u_1(t), \cdots, u_n(t))^T$ is the input vector.

* This work was supported by the 973 Program of China under Grant 2003CB316904, the National Natural Science Foundation of China under Grants 60373067 and 10361004, the Natural Science Foundation of Jiangsu Province, China under Grants BK2003053 and BK2003001, and The Natural Science Foundation of Xinjiang University.

J. Wang, X. Liao, and Z. Yi (Eds.): ISNN 2005, LNCS 3496, pp. 102–107, 2005.
© Springer-Verlag Berlin Heidelberg 2005

$f(x(t)) = (f_1(x_1(t)), \cdots, f_n(x_n(t)))^T$ and $f(x(t - \tau(t))) = (f_1(x_1(t - \tau_1(t))), \cdots,$
$f_n(x_n(t - \tau_n(t))))^T$.

In this paper, we introduce the assumptions as follows.

(H_1) $c_i(t)$, $a_{ij}(t)$, $b_{ij}(t)$ and $u_i(t)$ $(i, j = 1, 2, \cdots, n)$ are bounded continuous functions defined on R_+.

(H_2) $f_i(u)$ is bounded and monotonic nondecreasing on $R = (-\infty, \infty)$ and satisfies global Lipschitz condition, that is, there exist real number $k_j > 0$ such that $|f_j(u_1) - f_j(u_2)| \le k_j|u_1 - u_2|$ for all $u_1, u_2 \in R$.

(H_3) $\tau_i(t)(i = 1, 2, \cdots, n)$ is nonnegative, bounded and continuous differentiable defined on $R_+ = [0, \infty)$, and derivative $\dot{\tau}_i(t)$ is also uniformly continuous on R_+ and $\inf_{t \in R_+}\{1 - \dot{\tau}_i(t)\} > 0$.

Let $\tau = \sup\{\tau_i(t) : t \in [0, +\infty), i = 1, 2, ..., n\}$. We denoted by $C[-\tau, 0]$ the Banach space of continuous functions $x(\cdot) = (x_1(\cdot), \cdots, x_n(\cdot))^T : [-\tau, 0] \to R^n$ with the following norm $\|\phi\| = \sup_{-\tau \le \theta \le 0}|\phi(\theta)|$ with $|\phi(\theta)| = [\sum_{i=1}^{n} \phi_i^2(\theta)]^{\frac{1}{2}}$. In this paper we always assume that all solutions of system (1) satisfy the following initial conditions

$$x_i(\theta) = \phi_i(\theta), \quad \theta \in [-\tau, 0], \ i = 1, 2, ..., n, \tag{3}$$

where $\phi = (\phi_1, \phi_2, \cdots, \phi_n) \in C[-\tau, 0]$. It is well known that by the fundamental theory of functional differential equations, system (1) has a unique solution $x(t) = (x_1(t), x_2(t), \cdots, x_n(t))$ satisfying the initial condition (3).

Definition 1. System (1) is said to be globally asymptotically stable, if for any two solutions $x(t) = (x_1(t), \cdots, x_n(t))$ and $y(t) = (y_1(t), \cdots, y_n(t))$ of system (1), one has

$$\lim_{t \to \infty} |x_i(t) - y_i(t)| = 0.$$

Definition 2. System (1) is said to be globally exponentially stable, if there are constants $\epsilon > 0$ and $M \ge 1$ such that for any two solutions $x(t) = (x_1(t), \cdots, x_n(t))$ and $y(t) = (y_1(t), \cdots, y_n(t))$ of system (1) with the initial conditions $\phi \in C[-\tau, 0]$ and $\psi \in C[-\tau, 0]$, respectively, one has

$$\|x(t) - y(t)\| \le M\|\phi - \psi\| \exp(-\epsilon t).$$

2 Global Asymptotic Stability

Theorem 1. Suppose that (H_1)-(H_3) hold and if there exist a diagonal matrix $P = \text{diag}(p_1, p_2, \cdots, p_n)$ with positive numbers p_i $(i = 1, 2, \cdots, n)$, such that

$$-C(t)PK^{-1} + \frac{PA(t) + A^T(t)P}{2} + \frac{\|PB(t)\|}{2}I + \frac{PB(\psi^{-1}(t))}{2}Q(t)$$

is negative definite for all $t \ge 0$, where $K = \text{diag}(k_1^{-1}, \cdots, k_n^{-1})$, $Q(t) = \text{diag}(\frac{1}{1 - \dot{\tau}_1(\psi_1^{-1}(t))}, \cdots, \frac{1}{1 - \dot{\tau}_n(\psi_n^{-1}(t))})$ and $\psi^{-1}(t) = (\psi_1^{-1}(t), \cdots, \psi_n^{-1}(t))$, $\psi_i^{-1}(t)$ is the inverse function of $\psi_i(t) = t - \tau_i(t)$, $B(\psi^{-1}(t)) = (b_{ij}(\psi_i^{-1}(t)))_{n \times n}$. Then system (1) is globally asymptotically stable.

Proof. Let $x(t)$ and $x^*(t)$ are two solutions of system (2). By applying the transformation $y(t) = x(t) - x^*(t)$, system (2) can be rewritten as

$$y(t) = -C(t)y(t) + A(t)g(y(t) + B(t)g(y(t - \tau(t))), \tag{4}$$

where $y(t) = (y_1(t), \cdots, y_n(t))^T$, $g(y(t)) = (g_1(y_1(t), \cdots, g_n(y_n(t))^T$, $g(y(t - \tau(t))) = (g_1(y_1(t - \tau_1(t)), \cdots, g_n(y_n(t - \tau_n(t)))^T$ and $g_j(y_j(t)) = f_j(x_j(t)) - f_j(x_j^*(t))$ $(j = 1, 2, \cdots, n)$. According to the properties of f_j, g_j possess the following properties:

(A_1) $|g_j(y_j(t))| \leq k_j|y_j(t)|$,

(A_2) g_j is bounded and monotonic nondecreasing and $g_j(0) = 0$,

(A_3) $y_jg(y_j) \geq 0$, $g_j^2(y_j) \leq k_jy_jg_j(y_j)$ and $y_jg_j(y_j) \leq k_jy_j^2$ for any y_j.

We choose the Lyapunov function as follows:

$$V(t) = \sum_{i=1}^{n} p_i \int_0^{y_i} g_i(s)ds + \sum_{i=1}^{n} \int_{t-\tau_i(t)}^{t} \frac{\|PB(\psi^{-1}(s))\|}{2(1 - \dot{\tau}_i(\psi_i^{-1}(s))} g_i^2(y_i(s))ds.$$

Differentiating $V(t)$ with respect to time t along the solution of system (4), we obtain

$$\begin{aligned}
\frac{dV(t)}{dt} &= \sum_{i=1}^{n} p_i g_i(y_i(t))[-c_i(t)y_i(t) + \sum_{i=1}^{n} a_{ij}(t)g_i(y_i(t)) \\
&\quad + \sum_{j=1}^{n} b_{ij}(t)g_j(y_j(t - \tau_j(t)))] \\
&\quad + \sum_{i=1}^{n} \frac{\|PB(\psi^{-1}(s))\|}{2(1 - \dot{\tau}_i(\psi_i^{-1}(s))} g_i^2(y_i(t)) \\
&\quad - \frac{1}{2} \sum_{i=1}^{n} \|PB(t)\|g_i^2(y_i(t - \tau_i(t))) \\
&\leq - \sum_{i=1}^{n} p_i g_i(y_i(t))c_i(t)y_i(t) \\
&\quad + \sum_{i=1}^{n}\sum_{j=1}^{n} p_i g_i(y_i(t)a_{ij}(t)g_j(y_j(t)) \\
&\quad + |g^T(y_i(t))|\|PB(t)\||g(y(t - \tau(t)))| \\
&\quad + \sum_{i=1}^{n} \frac{\|PB(\psi^{-1}(t))\|}{2(1 - \dot{\tau}_i(\psi_i^{-1}(t)))} g_i^2(y_i(t)) \\
&\quad - \frac{1}{2} \sum_{i=1}^{n} \|PB(t)\|g_i^2(y_i(t - \tau_i(t))) \\
&\leq - \sum_{i=1}^{n} p_i g_i(y_i(t))c_i(t)y_i(t) \\
&\quad + \sum_{i=1}^{n}\sum_{j=1}^{n} p_i g_i(y_i(t)a_{ij}(t)g_j(y_j(t)) \\
&\quad + \frac{\|PB(t)\|}{2}[|g^T(y_i(t))|^2 + |g(y(t - \tau(t)))|^2]
\end{aligned}$$

$$+ \sum_{i=1}^{n} \frac{\|PB(\psi^{-1}(t))\|}{2(1 - \dot{\tau}_i(\psi_i^{-1}(t)))} g_i^2(y_i(t))$$

$$- \frac{1}{2} \sum_{i=1}^{n} \|PB(t)\| g_i^2(y_i(t - \tau_i(t)))$$

$$= - \sum_{i=1}^{n} \frac{c_i(t)}{k_i} p_i g_i^2(y_i(t)) + g(y(t)) PA(t) g(y(t)) \tag{5}$$

$$+ g^T(y(t)) \frac{\|PB(t)\|}{2} g(y(t)) + g^T(y(t))$$

$$\times \frac{\|PB(\psi^{-1}(t))\|}{2} Q(t) g(y(t))$$

$$= g(y(t)[-C(t)PK^{-1} + \frac{PA(t) + A^T(t)P}{2} + \frac{\|PB(t)\|}{2} I$$

$$+ \frac{1}{2} \|PB(\psi^{-1}(t))\| Q(t)] g(y(t))$$

$$< 0.$$

This implies that system (1) is globally asymptotically stable. This complete the proof.

3   Global Exponential Stability

Theorem 2. Suppose that (H_1)-(H_3) hold and if there exist n positive numbers $p_1, p_2, \cdots, p_n$ and $\sigma > 0$ such that

$$\frac{c_i(t)}{k_i} - \sum_{j=1}^{n} p_j |a_{ji}(t)| - \sum_{j=1}^{n} p_j |b_{ji}(\psi_i^{-1}(t))| \frac{1}{1 - \dot{\tau}_i(\psi_i^{-1}(t))} > \sigma \tag{6}$$

for all $t \geq 0$. Then the system (1) is globally exponentially stable.

Proof. If the condition (6) holds, then there exists a positive number $\epsilon > 0$ with $\epsilon < c_i(t)$ such that for all $t \in R_+ = [0, \infty)$ we have

$$\frac{-\epsilon + c_i(t)}{k_i} - \sum_{j=1}^{n} p_j |a_{ji}(t)| - \sum_{j=1}^{n} p_j |b_{ji}(\psi_i^{-1}(t))| \frac{e^{\epsilon \tau_i(\psi_i^{-1}(t))}}{1 - \dot{\tau}_i(\psi_i^{-1}(t))} > \sigma \tag{7}$$

Let $y(t) = (y_1(t), \cdots, y_n(t))^T$ is solution of system (4) through initial condition $y_i(\theta) = \phi_i(\theta)$, we define Lyapunov function as follows:

$$V(t) = \sum_{i=1}^{n} p_i \{ |y_i(t)| e^{\epsilon t}$$

$$+ \sum_{j=1}^{n} \int_{t-\tau_j(t)}^{t} \frac{|b_{ij}(\psi_j^{-1}(s))|}{1 - \dot{\tau}_j(\psi_j^{-1}(s))} |g_j(y_j(s))| e^{\epsilon(s + \tau_j(\psi_j^{-1}(s)))} ds \}. \tag{8}$$

Calculating the upper right Dini derivative of $V(t)$ along the solution of system (4), we obtain

$$
\begin{aligned}
D^+V(t) &\leq \sum_{i=1}^{n} p_i \Big\{ \frac{\epsilon - c_i(t)}{k_i} |y_i(t)| e^{\epsilon t} + \sum_{j=1}^{n} |a_{ij}(t)| |g_j(y_j(t))| e^{\epsilon t} \\
&\quad + \sum_{j=1}^{n} |b_{ij}(t)| |g_j(y_j(t - \tau_j(t)))| e^{\epsilon t} \\
&\quad + \sum_{j=1}^{n} \frac{|b_{ij}(\psi_j^{-1}(t))|}{1 - \dot{\tau}_j(\psi_j^{-1}(t))} |g_j(y_j(t))| e^{\epsilon(t + \tau_j(\psi_j^{-1}(t)))} \\
&\quad - \sum_{j=1}^{n} |b_{ij}(t)| |g_j(y_j(t - \tau_j(t)))| e^{\epsilon t} \Big\} \\
&\leq \sum_{i=1}^{n} \Big\{ p_i \frac{\epsilon - c_i(t)}{k_i} + \sum_{j=1}^{n} p_j |a_{ji}(t)| \\
&\quad + \sum_{j=1}^{n} p_j \frac{|b_{ji}(\psi_i^{-1}(t))|}{1 - \dot{\tau}_i(\psi_i^{-1}(t))} e^{\epsilon(\tau_i(\psi_i^{-1}(t)))} \Big\} |g_i(y_i(t))| e^{\epsilon t} \\
&\leq -\sigma \sum_{i=1}^{n} |g_i(y_i(t))| e^{\epsilon t} < 0.
\end{aligned}
\tag{9}
$$

From (9) we have

$$
V(t) \leq V(0) \quad \text{for all} \quad t \geq 0.
\tag{10}
$$

From (8), we obtain

$$
\min_{1 \leq i \leq n} \{p_i\} e^{\epsilon t} \sum_{i=1}^{n} |y_i(t)| \leq \sum_{i=1}^{n} p_i |y_i(t)| e^{\epsilon t} \leq V(t).
\tag{11}
$$

According to (9), we obtain that

$$
\begin{aligned}
V(0) &= \sum_{i=1}^{n} p_i \Big\{ |\phi_i(0)| \\
&\quad + \sum_{j=1}^{n} \int_{-\tau_j(t)}^{0} \frac{b_{ij}(\psi_j^{-1}(s))}{1 - \dot{\tau}_j(\psi_j^{-1}(s))} |g_j(\phi_j(s))| e^{\epsilon(s + \tau_j(\psi_j^{-1}(s)))} ds \Big\} \\
&\leq \max_{1 \leq i \leq n} \{p_i\} \sum_{i=1}^{n} |\phi_i(0)| \\
&\quad + \sum_{i=1}^{n} \sum_{j=1}^{n} \sup_{s \in [-\tau, 0]} \Big\{ p_i k_j \frac{b_{ij}(\psi_j^{-1}(s))}{1 - \dot{\tau}_j(\psi_j^{-1}(s))} e^{\epsilon \tau_j(\psi_j^{-1}(s))} \Big\} \\
&\quad \times \int_{-\tau_j(t)}^{0} |\phi_j(s)| ds \\
&\leq \max_{1 \leq i \leq n} \{p_i\} \|\phi\| + \max_{1 \leq i \leq n} \Big\{ \sum_{i=1}^{n} L_{ij} \Big\} \int_{-\tau}^{0} |\phi_j(s)| ds \\
&\leq \Big\{ \max_{1 \leq i \leq n} \{p_i\} + \max_{1 \leq i \leq n} \Big\{ \sum_{i=1}^{n} L_{ij} \tau \Big\} \Big\} \|\phi\|,
\end{aligned}
\tag{12}
$$

where $L_{ij} = \sup_{s \in [-\tau, 0]} \{p_i k_j \frac{b_{ij}(\psi_j^{-1}(s))}{1 - \dot{\tau}_j(\psi_j^{-1}(s))} e^{\epsilon \tau_j(\psi_j^{-1}(s))}\}$. Combining (10), (11) and (12), we finally obtain

$$\sum_{i=1}^{n} |y_i(t)| \leq M \|\phi\| e^{-\epsilon t},$$

where

$$M = \max\{(\min_{1 \leq i \leq n} \{p_i\})^{-1}(\max_{1 \leq i \leq n} \{p_i\} + \max_{1 \leq j \leq n} \{\sum_{i=1}^{n} L_{ij}\} \tau)\} \geq 1.$$

This completes the proof of Theorem 2.

4   Conclusions

In this paper, we considered the nonautonomous neural networks with time-varying delays. By constructing suitable Lyapunov function and applying some inequality analysis technique, we have obtained new criteria for checking the global asymptotical stability and global exponential stability. The method given in this paper may extend to discuss more complicated system such as Hopfield neural networks.

References

1. Arik, S.: Global Asymptotic Stability of a Larger Class of Neural Networks with Constant Time Delay. Phys. Lett. A, **311** (2003) 504-511
2. Gopalsamy, K., He, X.: Stability in Asymmetric Hopfield Nets with Transmission Delays. Phys. D, **76** (1994) 344-358
3. Lu, H.: Stability Criteria for Delay Neural Networks. Phys. Rev. E, **64** (2001) 1-13
4. Cao, J.: New Results Concerning Exponential Stability and Periodic Solutions of Delayed Cellular Neural Networks. Phys. Lett. A, **307** (2003) 136-147
5. Cao, J., Wang, J.: Global Asymptotic Stability of a Delays. IEEE Trans. Circuits and Systems-I, **50** (2003) 34-44
6. Huang, H., Cao, J.: On Global Asymptotic Stability of Recurrent Neural Networks with Time-varying Delays. Appl. Math. Comput. **142** (2003) 143-154
7. Peng, J., Qiao, H., Xu, Z.: A New Approach to Stability of Neural Networks with Time-varying Delays. Neural Networks, **15** (2002) 95-103
8. Jiang, H., Teng, Z.: Global Exponential Stability of Cellular Neural Networks with Time-varying Coefficients and Delays. Neural Networks, **17** (2004) 1415-1435
9. Zhou, J., Liu, Z., Chen, G.: Dynamics of Periodic Delayed Neural Networks. Neural Networks, **17** (2004) 87-101

Global Exponential Stability of Non-autonomous Neural Networks with Variable Delay

Minghui Jiang[1], Yi Shen[1], and Meiqin Liu[2]

[1] Department of Control Science and Engineering, Huazhong University of Science and Technology, Wuhan, Hubei 430074, China
[2] College of Electrical Engineering, Zhejiang University, Hangzhou 310027, China

Abstract. The aim of this work is to discuss the exponential stability of the non-autonomous neural networks with delay. By applying the generalized Halanay inequality and Lyapunov second method, several sufficient conditions are obtained ensuring the global exponential stability of the non-autonomous neural networks. Some previous results are improved and extended in this paper. In the end, a example is also given for illustration.

1 Introduction

Recently, the dynamics of autonomous neural networks have attracted increasing interest in both theoretical studies and engineering applications, and many important results on the global exponential stability of equilibrium have been obtained ([1]-[7],[9]-[11]). However, from the view point of reality ,it should also be taken into account evolutionary processes of some practical systems as well as disturbances of external influence such as varying environment of biological systems and so on. Therefore, the study of the non-autonomous neural networks with variable delay have more important significance than the autonomous one. Up to now, to the best of our knowledge, less work has addressed for the non-autonomous neural networks with variable delay. In this paper, we will consider the following generalized neural networks with variable delays.

$$\frac{\mathrm{d}x_i(t)}{\mathrm{d}t} = -c_i(t)x_i(t) + \sum_{j=1}^{n} a_{ij}(t)f_j(x_j(t))$$

$$+ \sum_{j=1}^{n} b_{ij}(t)f_j(x_j(t - \tau_j(t))) + \mu_i(t)), i = 1, 2, \cdots, n. \qquad (1)$$

where $n \geq 2$ is the number of neurons in the networks, x_i denotes the state variable associated with the ith neuron, and $c_i(\cdot)$ is an appropriately behaved function. The connection matrix $A(t) = (a_{ij}(t))_{n \times n}$ tells how the neurons are connected in the network, and the activation function $f_j(\cdot)$ shows how the jth neuron reacts to the input. matrix $B(t) = (b_{ij}(t))_{n \times n}$ represents the interconnections with delay and $\mu_i(t), i = 1, 2, \cdots, n$ denote the inputs at times t from outside of the network(1).

J. Wang, X. Liao, and Z. Yi (Eds.): ISNN 2005, LNCS 3496, pp. 108–113, 2005.

Throughout this paper, we assume that

(H1) Each function $c_i(t), a_{ij}(t), b_{ij}(t), \tau_j(t), i, j = 1, 2, \cdots, n$, is continuous and bounded, i.e. there exist positive numbers $\bar{c}, \bar{a}, \bar{b}, \tau$ such that $0 < c_i(t) \leq \bar{c}, |a_{ij}(t)| \leq \bar{a}, |b_{ij}(t)| \leq \bar{b}, 0 < \tau_j(t) \leq \tau$ for all $t \in [0, \infty)$.

(H2) Each function $f_i(x), i = 1, 2, \cdots, n$ is bounded and Lipschitz continuous with Lipschitz constants $\alpha_i > 0, i = 1, 2, \cdots, n$.

The initial conditions associated with (1) is assumed to be the form

$$x_i(s) = \phi_i(s), s \in [-\tau, 0], i = 1, 2, \cdots, n.$$

where $\phi_i(s) : [-\tau, 0] \to R$ are assumed to be continuous functions.

Obviously, under the hypotheses (H1)-(H2), the right-hand side of the equation (1) satisfies the local Lipschitz condition. We know (see [8]) that the network (2) with the initial condition has a unique continuous solution on $t \geq -\tau$. Our purpose of this paper is to investigate the global exponential stability of the network (1) by using the generalized Halanay inequality and constructing suitable Lyapunov function. In this paper we will establish a new criteria for the global exponential stability .

2   Preliminaries

In this section, we shall give the following definition and Lemma1.

Definition 1. The network(1) is said to be globally exponentially stable, if there are constants $\varepsilon > 0$ and $M \geq 1$ such that for any two solution $x(t)$ and $y(t)$ with the initial functions $\phi \in C^1[-\tau, 0]$ and $\varphi \in C^1[-\tau, 0]$, respectively, for all $t \geq t_0$ one has

$$|x(t) - y(t)| \leq M\|\phi - \varphi\|e^{-\varepsilon(t-t_0)},$$

here $\|\phi - \varphi\| = \sup_{t_0-\tau \leq s \leq t_0} \sum_{j=1}^{n} |\phi_j(s) - \varphi_j(s)|.$

The following Lemma 1 generalizes the famous Halanay inequality.

Lemma 1. [12] (Generalized Halanay Inequality) Assume $p(t)$ and $q(t)$ be continuous with $p(t) > q(t) \geq 0$ and $\inf_{t \geq t_0} \frac{p(t)-q(t)}{1+1.5\tau q(t)} \geq \eta > 0$ for all $t \geq 0$, and $y(t)$ is a continuous function on $t \geq t_0$ satisfying the following inequality for all $t \geq t_0$

$$D^+y(t) \leq -p(t)y(t) + q(t)\overline{y}(t), \tag{2}$$

where $\overline{y}(t) = \sup_{t-\tau \leq s \leq t}\{|y(s)|\}$, then, for $t \geq t_0$, we have

$$y(t) \leq \overline{y}(t_0)e^{-\lambda^*(t-t_0)}, \tag{3}$$

in which λ^* is defined as

$$\lambda^* = \inf_{t \geq t_0}\{\lambda(t) : \lambda(t) - p(t) + q(t)e^{\lambda(t)\tau} = 0\}. \tag{4}$$

3   Exponential Stability

By using the above generalized Halanay inequality and the Lyapunov method, we give two algebraic criterion on the exponential stability of the network(1).

Theorem 1. If (H1)-(H2)hold, and moreover, there exist positive numbers $\eta, w_i, i = 1, 2, \cdots, n$ such that the network (1) satisfies the following condition

(1) $p(t) = \min\limits_{1 \le i \le n} \{w_i c_i(t) - \alpha_i \sum\limits_{j=1}^{n} w_j |a_{ji}(t)|\} > q(t) = \max\limits_{1 \le i \le n} \{\alpha_i \sum\limits_{j=1}^{n} w_j |b_{ji}(t)|\}$

$> 0,$

(2) $\inf\limits_{t \ge t_0} \{\dfrac{p(t) - q(t)}{1 + 1.5\tau q(t)}\} \ge \eta > 0, \eta := \text{const.},$

then the network (1) is globally exponentially stable .

Proof. Let $x(t)$ is any solution of the network (1) with any initial function $\phi(s), s \in [-\tau, 0]$, and $x^*(t)$ denotes an solution of the network (1) with the initial function $\varphi(s), s \in [-\tau, 0]$.

Set $y_i(t) = x_i(t) - x_i^*(t), i = 1, 2, \cdots, n$. Substituting $y_i(t) = x_i(t) - x_i^*(t), i = 1, 2, \cdots, n$. into the network(1) leads to

$$\frac{dy_i(t)}{dt} = - c_i(t)y_i(t) + \sum_{j=1}^{n} a_{ij}(t)(f_j(y_j(t) + x_j^*(t)) - f_j(x_j^*(t))) +$$

$$\sum_{j=1}^{n} b_{ij}(t)(f_j(y_j(t - \tau_j(t)) + x_j^*(t - \tau_j(t))) - f_j(x_j^*(t - \tau_j(t)))),$$

$$i = 1, 2, \cdots, n. \tag{5}$$

Consider the following Lyapunov function $V(y) = \sum_{i=1}^{n} w_i |y_i(t)|$, where $w_i, i = 1, 2, \cdots, n$.

Calculating the Dini derivative D^+V of $V(y)$ along the solution of the equation(5), we get

$$D^+V = \sum_{i=1}^{n} w_i \text{sgn}(y_i(t))\dot{y}_i(t)$$

$$\le \sum_{i=1}^{n} \{w_i[-c_i(t)|y_i(t)| + \sum_{j=1}^{n} \alpha_j |a_{ij}(t)||y_j(t)|$$

$$+ \sum_{j=1}^{n} \alpha_j |b_{ij}(t)||y_j(t - \tau_j(t))|]\}$$

$$\le - \min\limits_{1 \le i \le n} \{w_i c_i(t) - \alpha_i \sum\limits_{j=1}^{n} w_j |a_{ji}(t)|\} \sum\limits_{i=1}^{n} |y_i(t)|$$

$$+ \max\limits_{1 \le i \le n} \{\alpha_i \sum\limits_{j=1}^{n} w_j |b_{ji}(t)|\} \sum\limits_{j=1}^{n} |\overline{y}_j(t)|$$

$$\le -p(t)V + q(t)\overline{V}. \tag{6}$$

According to Lemma1, we have

$$\sum_{i=1}^{n} |y_i(t, t_0, y_0)| = V(t) \leq \overline{V}(t_0)e^{-\lambda^*(t-t_0)} \tag{7}$$

in which λ^* is defined as (4). It follows from (7) that the network (1) is globally exponentially stable.This completes the proof.

Remark 1. While $c_i \equiv c, a_{ij} \equiv a, b_{ij} \equiv b$, the network (1) has a unique equilibrium point. then the theorem 1 reduces to theorem 1 in [2]. Particularly, when the network parameters and input stimuli are varied periodically in time, the periodical solution is exponentially stable in [4].

Theorem 2. If assumption (H1)-(H2)hold, and there exist a positive definite matrix Q, two functions $p(t) > q(t) > 0$ such that

(1) $S^2 - q(t)Q < 0$,

(2) $-(QC(t) + C(t)Q) + QA(t)A^T(t)Q + S^2 + QB(t)B^T(t)Q + p(t)Q < 0$,

(3) $\inf_{t \geq t_0} \{ \dfrac{p(t) - q(t)}{1 + 1.5\tau q(t)} \} \geq \eta > 0, \eta := \text{const.}$,

in which $S = \text{diag}(\alpha_1, \alpha_2, \cdots, \alpha_n), C(t) = \text{diag}(c_1(t), c_2(t), \cdots, c_n(t)), A(t) = (a_{ij}(t)), B(t) = (b_{ij}(t))$, then the network (1) is globally exponentially stable .

Proof. The Eq.(5) can be rewritten as the following form.

$$\dot{y}(t) = -C(t)y(t) + A(t)(f_j(y_j(t) + x_j^*(t)) - f_j(x_j^*(t)))$$
$$+ B(t)(f_j(y_j(t - \tau_j(t)) + x_j^*(t - \tau_j(t)))) - f_j(x_j^*(t - \tau_j(t)))). \tag{8}$$

Consider the Lyapunov function $V(t) = y^T(t)Qy(t)$. Calculating the derivative $V(t)$ of along the solution of equation (8), we get

$$\dot{V}(t) = y^T(-QC(t) - C(t)Q)y(t) + 2y^T QA(t)[f_j(y_j(t) + x_j^*(t)) - f_j(x_j^*(t))]$$
$$+ 2y^T QB(t)[f_j(y_j(t - \tau_j(t)) + x_j^*(t - \tau_j(t)))) - f_j(x_j^*(t - \tau_j(t)))]. \tag{9}$$

By the inequality $2a^T b \leq a^T a + b^T b$ for any $a, b \in R^n$, we have

$$2y^T(t)QA(t)[f_j(y_j(t) + x_j^*(t)) - f_j(x_j^*(t))] \leq y^T(t)QA(t)A^T(t)Qy(t)$$
$$+ [f_j(y_j(t) + x_j^*(t)) - f_j(x_j^*(t))]^T[f_j(y_j(t) + x_j^*(t)) - f_j(x_j^*(t))]$$
$$\leq y^T(t)(QA(t)A^T(t)Q + S^2)y(t), \tag{10}$$

and

$$2y^T QB(t)[f_j(y_j(t - \tau_j(t)) + x_j^*(t - \tau_j(t)))) - f_j(x_j^*(t - \tau_j(t)))]$$
$$\leq y^T(t)QB(t)B^T(t)Qy(t) + [f_j(y_j(t - \tau_j(t)) + x_j^*(t - \tau_j(t))) -$$
$$f_j(x_j^*(t - \tau_j(t)))]^T \times [f_j(y_j(t - \tau_j(t)) + x_j^*(t - \tau_j(t))) - f_j(x_j^*(t - \tau_j(t)))]$$
$$\leq y^T(t)QB(t)B^T(t)Qy(t) + \overline{y}^T(t)S^2\overline{y}(t). \tag{11}$$

By substituting (10) and (11) into (9), and using the condition (1) and (2) of the Theorem 2, we have

$$
\begin{aligned}
\dot{V}(t) &\leq y^T(t)[-(QC(t)+C(t)Q)+QA(t)A^T(t)Q+S^2+QB(t)B^T(t)Q]y(t) \\
&\quad +\overline{y}^T S^2 \overline{y}(t) \\
&= y^T(t)[-(QC(t)+C(t)Q)+QA(t)A^T(t)Q+S^2+QB(t)B^T(t)Q \\
&\quad +p(t)Q]y(t)+\overline{y}^T(S^2-q(t)Q)\overline{y}(t)-y^T(t)p(t)Qy(t)+q(t)\overline{y}^T Q\overline{y}(t) \\
&\leq -p(t)V(t)+q(t)\overline{V}(t).
\end{aligned}
\tag{12}
$$

By Lemma 1, we obtain

$$
\lambda_{\min}(Q)\|y(t)\|^2 \leq V(t) \leq \overline{V}(t_0)e^{-\lambda^*(t-t_0)} \leq \lambda_{\max}(Q)\|\overline{y}(t_0)\|^2 e^{-\lambda^*(t-t_0)}.
$$

Therefore,

$$
\|y(t)\| = \|x(t)-x^*(t)\| \leq \frac{\lambda_{\max}(Q)}{\lambda_{\min}(Q)}\|\phi-\varphi\|e^{-\frac{\lambda^*(t-t_0)}{2}}.
$$

The proof is completed.

Remark 2. If $C(t) \equiv C, A(t) \equiv A, B(t) \equiv B$ (A, B, C are constant matrix), the system (1) has a unique equilibrium point, then the result reduces to corollary 1 in [1].

4   Example

In this section, a example will be given to show the validity of our results.

Example. Consider a two dimensional non-autonomous neural network with delay.

$$
\dot{x}(t) = -C(t)x(t)+A(t)(f_j(x_j(t)))+B(t)(f_j(x_j(t-\tau_j(t))))+\mu(t), \tag{13}
$$

where $C(t) = \begin{pmatrix} 2+0.25\sin(t) & 0 \\ 0 & 3+0.5\cos(t) \end{pmatrix}, \mu(t) = (e^{-t},e^{-t})^T, \tau_i(t)=1, A(t)$
$$
= \begin{pmatrix} 0.5-0.5\cos(t) & -0.5+0.2\sin(t) \\ -0.4+0.6\cos(t) & -0.7+0.3\sin(t) \end{pmatrix}, f_i(x)=\tfrac{1}{2}(|x+1|-|x-1|), i=1,2, B(t)
$$
$$
= \begin{pmatrix} -0.05+0.02\sin(t) & -0.2+0.5\cos(t) \\ 0.1+0.8\cos(t) & -0.06+0.04\cos(t) \end{pmatrix}.
$$

It is easy to verify that the network(13) satisfies the assumption (H1)-(H2) with $\alpha_1 = \alpha_2 = 1, S = I$, Let $p(t) = 1.1+\sin^2(t), q(t) = 1+\sin^2(t), Q = 2I$, we get

(1) $S^2 - q(t)Q = \begin{pmatrix} -1-\sin^2(t) & 0 \\ 0 & -1-\sin^2(t) \end{pmatrix} < 0,$

(2) $-(QC(t)+C(t)Q)+QA(t)A^T(t)Q+S^2+QB(t)B^T(t)Q+p(t)Q$
$$
= \begin{pmatrix} h_{11} & h_{12} \\ h_{21} & h_{22} \end{pmatrix} < 0, \text{ where } h_{11} = -6.63974-0.8080\sin(t)-2.0800\cos(t)+
$$

$2.8384 \cos^2(t), h_{12} = h_{21} = 0.8680 - 1.1520 \sin(t) + 1.6880 \cos(t) + 0.0320 \sin(2t) - 1.3600 \cos^2(t), h_{22} = -4.1852 - 1.2992 \cos(t) - 1.6800 \sin(t) + 0.3564 \sin^2(t),$

(3) $\inf_{t \geq t_0} \{ \frac{p(t) - q(t)}{1 + 1.5 \tau q(t)} \} \geq 0.03000 > 0.$

Therefore, all the conditions of the theorem 2 are satisfied. According to the theorem 2, the network (13) is globally exponentially stable.

Acknowledgments

The work was supported by Natural Science Foundation of Hubei (2004ABA055) and National Natural Science Foundation of China (60074008).

References

1. Zhang, Q., Wei, X.P., Xu, J.: Globally Exponential Convergence Analysis of Delayed Neural Networks with Time-varying Delays. Physics Letters A, **318** (2003) 537-544
2. Zhou, D, Cao, J.: Globally Exponential Stability Conditions for Cellular Neural Networks with Time-varying Delays. Applied Mathematics and Computation, **131**(2002) 487-496
3. Liao, X.X. : Stability of the Hopfield Neural Networks. Sci. China, **23**(1993) 523-532
4. Cao, J.: On Exponential Stability and Periodic Solution of CNN's with Delay. Physics Letters A, **267** (2000) 312-318
5. Cao, J., Liang, J.: Boundedness and Stability for Cohen-Grossberg Neural Networks with Time-varying Delays. J. Math. Anal. Appl., **296** (2004) 665-685
6. Liao, X.F.: Novel Robust Stability Criteria for Interval-delayed Hopfield Neural Networks. IEEE Trans. Circuits and Systems, **48** (2001) 1355-1359
7. Wang, J., Gan, Q., Wei, Y.: Stability of CNN with Opposite-sign Templates and Nonunity Gain Output Functions. IEEE Trans Circuits Syst I, **42** (1995) 404-408
8. Kolmanovskii, V., Myshkis, A.: Introducion to the Theory and Applications of Functional Differential Equations. Kluwer Academic Publishers, London (1999)
9. Tian, H.: The Exponential Asymptotic Stability of Singularly Perturbed Delay Differential Equations with A Bounded Lag. J. Math. Anal. Appl., **270** (2002) 143-149
10. Sun, C., Zhang, K., Fei, S., Feng, C.B.: On Exponential Stability of Delayed Neural Networks with a General Class of Activation Functions. Physics Letters A, **298** (2002) 122-132
11. Yucel, E., Arik, S.: New Exponential Stability Results for Delayed Neural Networks with Time Varying Delays. Physica D, **191** (2004) 314-322
12. Jiang, M., Shen, Y., Liao, X.: On the Global Exponential Stability for Functional Differential Equations. Communications in Nonlinear Science and Numerical Simulation, **10** (2005) 705-713

A Generalized LMI-Based Approach to the Global Exponential Stability of Recurrent Neural Networks with Delay

Yi Shen, Minghui Jiang, and Xiaoxin Liao

Department of Control Science and Engineering,
Huazhong University of Science and Technology, Wuhan, Hubei 430074, China

Abstract. A new theoretical result on the global exponential stability of recurrent neural networks with delay is presented. It should be noted that the activation functions of recurrent neural network do not require to be bounded. The presented criterion, which has the attractive feature of possessing the structure of linear matrix inequality (LMI), is a generalization and improvement over some previous criteria. A example is given to illustrate our results.

1 Introduction

In recent years, recurrent neural networks are widely studied in [1-2], because of their immense potentials of application perspective. The Hopfield neural networks and cellular neural networks are typical representative recurrent neural networks among others, and have been successfully applied to signal processing, especially in image processing, and to solving nonlinear algebraic and transcendental equations. Such applications rely heavily on the stability properties of equilibria. Therefore, the stability analysis of recurrent neural networks is important from both theoretical and applied points of view[1-5]. In this paper, a new theoretical result on the global exponential stability of recurrent neural networks with delay is presented. It should be noted that the activation functions of recurrent neural network do not require to be bounded. The presented criterion, which has the attractive feature of possessing the structure of linear matrix inequality (LMI), is a generalization and improvement over some previous criteria in Refs.[3-5]. A example is given to illustrate our results.

The model of recurrent neural network to be considered herein is described by the state equation

$$\dot{x}(t) = -Dx(t) + Af(x(t)) + Bf(x(t-\tau)) + u, \tag{1}$$

where $x(t) = [x_1(t), \cdots, x_n(t)]^T$ denotes the state vector, $f(x(t)) = [f_1(x_1(t)), \cdots, f_n(x_n(t))]^T$ is the output vector, $D = \text{diag}(d_1, \cdots, d_n)$ is the self feedback matrix, $A = (a_{ij})_{n \times n}$ is the feedback matrix, $B = (b_{ij})_{n \times n}$ is the delayed feedback matrix, $u = [u_1, \cdots, u_n]^T$ is the constant vector, and τ is the delay. The activation function $f_i(\cdot)$ satisfies the following condition

J. Wang, X. Liao, and Z. Yi (Eds.): ISNN 2005, LNCS 3496, pp. 114–119, 2005.
© Springer-Verlag Berlin Heidelberg 2005

(**H**) There exist $k_i > 0$ such that for any $\theta, \rho \in R$

$$0 \le f_i(\theta) - f_i(\rho)(\theta - \rho)^{-1} \le k_i. \tag{2}$$

It is easy to see that these activation functions satisfied (**H**) are not necessarily bounded. However, some usual sigmoidal functions and piecewise linear functions which are employed in [3-5] are bounded on R.

Throughout this paper, unless otherwise specified, we let $A \ge 0$ denote non-negative definite matrix, $A > 0$ denote positive definite symmetric matrix and etc. $A^T, A^{-1}, \lambda_{min}(A), \lambda_{max}(A)$, denotes, respectively, the transpose of, the inverse of, the minimum eigenvalue of, the maximum eigenvalue of a square matrix A. Let $|\cdot|$ and $\|A\|$ denote the Euclidean norm in R^n and the 2-norm in $R^{n\times n}$, respectively.

2   Main Results

Theorem 2.1. If there exist positive definite symmetric matrices $P = [p_{ij}] \in R^{n\times n}$ and $G = [g_{ij}] \in R^{n\times n}$ and a positive definite diagonal matrix $Q = \mathrm{diag}(q_1, \cdots, q_2)$ such that

$$M = \begin{pmatrix} PD + D^T P & -PA & -PB \\ -A^T P & 2QDK^{-1} - QA - A^T Q - G & -QB \\ -B^T P & -B^T Q & G \end{pmatrix} > 0, \tag{3}$$

where $K = \mathrm{diag}(k_1, \ldots, k_n)$ (see (2)), then the network (1) has the unique equilibrium point x^* and it is globally exponentially stable.

Proof. By (3), it follows

$$\begin{pmatrix} 2QDK^{-1} - QA - A^T Q - G & -QB \\ -B^T Q & G \end{pmatrix} > 0. \tag{4}$$

From the Schur Complement and (4), we obtain

$$2QDK^{-1} - QA - A^T Q - G - QBG^{-1}B^T Q > 0. \tag{5}$$

By (5) and $G + QBG^{-1}B^T Q - B^T Q - QB \ge 0$, we get

$$2QDK^{-1} - Q(A + B) - (A + B)^T Q > 0. \tag{6}$$

Set $J(x) = -Dx + (A+B)f(x) + u$, then J is injective on R^n. Otherwise, there is $x \ne y, x, y \in R^n$ such that $J(x) = J(y)$, that is to say

$$D(x - y) = (A + B)(f(x) - f(y)). \tag{7}$$

It is obvious that $f(x) \ne f(y)$, otherwise, $x = y$ by (7). Multiplying the both sides of (7) by $2(f(x) - f(y))^T Q$, we obtain

$$2(f(x) - f(y))^T QD(x - y) = 2(f(x) - f(y))^T Q(A + B)(f(x) - f(y)). \tag{8}$$

By (2), we have

$$2(f(x) - f(y))^T QD(x - y) \geq 2(f(x) - f(y))^T QDK^{-1}(f(x) - f(y)). \quad (9)$$

Therefore, by (8) and (9), we get

$$(f(x) - f(y))^T (2QDK^{-1} - Q(A + B) - (A + B)^T Q)(f(x) - f(y)) \leq 0. \quad (10)$$

Since $f(x) - f(y) \neq 0$, by (6), we have

$$(f(x) - f(y))^T (2QDK^{-1} - Q(A + B) - (A + B)^T Q)(f(x) - f(y)) > 0. \quad (11)$$

It is obvious that (10) contradicts (11). Therefore, J is injective.

Now one can prove $\lim_{|x| \to \infty} |J(x)| = \infty$. Because u and $f(0)$ are constants, it only needs to prove $\lim_{|x| \to \infty} |\overline{J}(x)| = \infty$, where $\overline{J}(x) = -Dx + (A + B)(f(x) - f(0))$. If $|f(x)|$ is bounded, then $\lim_{|x| \to \infty} |\overline{J}(x)| = \infty$. Assume that $|f(x)|$ is unbounded, by (2), $f_i(x)$ is monotonic and non-decreasing, then $\lim_{|x| \to \infty} |f(x) - f(0)| = \infty$. And by (2), we have

$$\begin{aligned}
2(f(x) - f(0))^T Q\overline{J}(x) \\
\leq -(f(x) - f(0))^T (2QDK^{-1} - Q(A + B) - (A + B)^T Q)(f(x) - f(0)) \\
\leq -\mu|f(x) - f(0)|^2, \quad (12)
\end{aligned}$$

where $\mu = \lambda_{min}(2QDK^{-1} - Q(A + B) - (A + B)^T Q)$, it follows from (6) that $\mu > 0$. So, by (12), we get

$$\mu|f(x) - f(0)|^2 \leq 2\|Q\||f(x) - f(0)||\overline{J}(x)|. \quad (13)$$

$\lim_{|x| \to \infty} |f(x) - f(0)| = \infty$ together with (13) implies $\lim_{|x| \to \infty} |\overline{J}(x)| = \infty$. By Ref.[2], it follows that $J(x)$ is a homeomorphism from R^n to itself. Therefore, the equation $J(x) = 0$ has a unique solution; i.e., the system (1) has a unique equilibrium x^*.

One will prove that the unique equilibrium $x^* = (x_1^*, \cdots, x_n^*)^T$ of the system (1) is globally exponentially stable. Since x^* is the equilibrium of the system (1), $-Dx^* + (A + B)f(x^*) + u = 0$, by (1), we have

$$\dot{y}(t) = -Dy(t) + Ag(y(t)) + Bg(y(t - \tau)), \quad (14)$$

where $y(\cdot) = [y_1(\cdot), \cdots, y_n(\cdot)]^T = x(\cdot) - x^*$ is the new state vector, $g(y(\cdot)) = [g_1(y_1(\cdot)), \cdots, g_n(y_n(\cdot))]^T$ represents the output vector of the transformed system, and $g_i(y_i(\cdot)) = f_i(y_i(\cdot) + x_i^*) - f_i(x_i^*), i = 1, 2, \cdots, n$. Obviously, $g(0) = 0, y^* = 0$ is a unique equilibrium of the system (14). Therefore the stability of the equilibrium x^* of the system (1) is equivalent to the one of the trivial solution of the system (14).

Choose the following Lyapunov functional

$$V(y(t)) = y^T(t)Py(t) + 2\sum_{i=1}^n q_i \int_0^{y_i(t)} g_i(s)\mathrm{d}s + \int_{t-\tau}^t g(y(s))^T Gg(y(s))\mathrm{d}s,$$

The time derivative of $V(y(t))$ along the trajectories of (14) turns out to be

$$\dot{V}(y(t)) = -(y^T(t), g^T(y(t)), g^T(y(t-\tau)))M \begin{pmatrix} y(t) \\ g(y(t)) \\ g(y(t-\tau)) \end{pmatrix}$$

$$-2g^T(y(t))QDy(t) + 2g^T(y(t))QDK^{-1}g(y(t))$$

$$\leq -\lambda|y(t)|^2 - \lambda|g(y(t))|^2 - \lambda|g(y(t-\tau))|^2$$

$$\leq -\lambda|y(t)|^2, \tag{15}$$

where $\lambda = \lambda_{min}(M) > 0$, (15) has been obtained by (2). Set $h(\varepsilon) = -\lambda + \lambda_{max}(P)\varepsilon + \|Q\|\|K\|\varepsilon + \|G\|\|K\|^2 e^{\varepsilon\tau}\tau\varepsilon$. Owing to $h'(\varepsilon) > 0, h(0) = -\lambda < 0, h(+\infty) = +\infty$, there exists a unique ε such that $h(\varepsilon) = 0$; i.e., there is $\varepsilon > 0$ satisfies

$$-\lambda + \lambda_{max}(P)\varepsilon + \|Q\|\|K\|\varepsilon + \|G\|\|K\|^2 e^{\varepsilon\tau}\tau\varepsilon = 0. \tag{16}$$

From the definition of V and (2), it follows

$$|V(y(t))| \leq (\lambda_{max}(P) + \|Q\|\|K\|)|y(t)|^2 + \|G\|\|K\|^2 \int_{t-\tau}^{t} |y(s)|^2 ds. \tag{17}$$

For $\varepsilon > 0$ satisfying (16), by (15) and (17), we get

$$(e^{\varepsilon t}V(y(t)))' = \varepsilon e^{\varepsilon t}V(y(t)) + e^{\varepsilon t}\dot{V}(y(t))$$

$$\leq e^{\varepsilon t}(-\lambda + \lambda_{max}(P)\varepsilon + \|Q\|\|K\|\varepsilon)|y(t)|^2$$

$$+\varepsilon e^{\varepsilon t}\|G\|\|K\|^2 \int_{t-\tau}^{t} |y(s)|^2 ds. \tag{18}$$

Integrating the both sides of (18) from 0 to an arbitrary $t \geq 0$, we can obtain

$$e^{\varepsilon t}V(y(t)) - V(y(0))$$

$$\leq \int_{0}^{t} e^{\varepsilon s}(-\lambda + \lambda_{max}(P)\varepsilon + \|Q\|\|K\|\varepsilon)|y(s)|^2 ds$$

$$+ \int_{0}^{t} \varepsilon e^{\varepsilon s}\|G\|\|K\|^2 ds \int_{s-\tau}^{s} |y(r)|^2 dr. \tag{19}$$

And

$$\int_{0}^{t} \varepsilon e^{\varepsilon s}\|G\|\|K\|^2 ds \int_{s-\tau}^{s} |y(r)|^2 dr$$

$$\leq \int_{-\tau}^{t} \varepsilon\|G\|\|K\|^2 \tau e^{\varepsilon(r+\tau)}|y(r)|^2 dr. \tag{20}$$

Substituting (20) into (19) and applying (16), we have

$$e^{\varepsilon t}V(y(t)) \leq C, t \geq 0. \tag{21}$$

where $C = V(y(0)) + \int_{-\tau}^{0} \varepsilon \|G\| \|K\|^2 \tau e^{\varepsilon \tau} |y(r)|^2 dr$. By the definition of V and (21), we get

$$|y(t)| \leq \frac{C}{\lambda_{min}(P)} e^{-\varepsilon t}, t \geq 0. \tag{22}$$

This implies the trivial solution of (14) is globally exponentially stable and its exponential convergence rate ε is decided by (16).

Remark 1. In Theorem 2.1, we do not suppose that the activation function is bounded . However, in [3], several asymptotic stability results are derived by the bounded piecewise linear function while our result ensures the globally exponential stability of the system (1) in similar conditions. In addition, the network (1) is even more general system than the one in [3].

Corollary2.2. Assume that $2DK^{-1} - I > 0$, then, neural network (1) has the unique equilibrium point and it is globally exponentially stable. If there exists a positive constant β such that

$$A + A^T < 0, \quad \|B\| \leq \sqrt{\lambda_{min}(2DK^{-1} - I)}$$

or

$$\begin{cases} A + A^T + \beta I < 0, \|B\| \leq \sqrt{2\beta \lambda_{min}(DK^{-1})} & \text{if } \beta > 1; \\ A + A^T + \beta I < 0, \|B\| \leq \sqrt{(1+\beta)\lambda_{min}(2DK^{-1} - I)} & \text{if } 0 < \beta \leq 1. \end{cases}$$

where I denotes the identity matrix.

The proof is omitted.

Remark 2. Assume that $D = I, f_i(x_i) = \frac{1}{2}(|x_i + 1| - |x_i - 1|)$ in (1), then Corollary 2.2 reduces to the main results in [3-5]. However, Refs.[3-5] only prove the asymptotic stability of the network (1).

3 Illustrative Example

Example 3.1. Assume that the network parameters of system (1) are chosen as:
$\tau = 2, u_1 = 1, u_2 = -1, D = \begin{pmatrix} 1 & 0 \\ 0 & 1 \end{pmatrix}, A = \begin{pmatrix} 0 & 1 \\ -1 & -1 \end{pmatrix}, B = \begin{pmatrix} 0.5 & 0.5 \\ 1 & 0 \end{pmatrix},$
and $f_i(x_i) = \frac{4}{3}x_i, i = 1, 2$ with $K = \text{diag}(\frac{4}{3}, \frac{4}{3})$. In this case, all of the criteria in [3-5] cannot be applied. On the other hand, there exists at least one feasible solution (computation by LMI control toolbox), i.e., $P = \begin{pmatrix} 178.3947 & 34.3916 \\ 34.3916 & 326.4909 \end{pmatrix},$
$G = \begin{pmatrix} 872.0375 & 115.3044 \\ 115.3044 & 1386.9807 \end{pmatrix}, Q = \begin{pmatrix} 887.0318 & 0 \\ 0 & 1557.0953 \end{pmatrix}$, of the LMI (3) in Theorem 2.1. Therefore, the unique equilibrium point $x^* = [0.5, -0.5]$ of the above system is globally exponentially stable, and its behavior is shown in Fig.1.

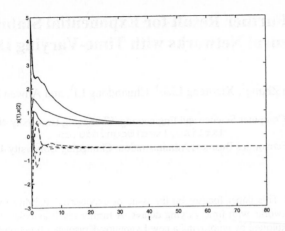

Fig. 1. The convergence behavior of the network in Example 3.1. The states x, with initial value $x_0 = (-1,3); (1,1); (5,-3)$, respectively, ultimately converge to the equilibrium $x^* = [0.5, -0.5]$. x(1)-solid line, x(2)-dashed line.

4   Conclusion

In this paper, we present theoretical results on the global exponential stability of a general class of recurrent neural networks with delay. These conditions obtained here are easy to be checked in practice, and are of prime importance and great interest in the design and the applications of network. The criterions are more general than the respective criteria reported in [3-5]. Finally, an illustrative example is also given to compare the new results with existing ones.

Acknowledgments

The work was supported by Natural Science Foundation of Hubei (2004ABA055) and National Natural Science Foundation of China (60274007, 60074008).

References

1. Cao, J., Wang, J.: Global Asymptotic Stability of Recurrent Neural Networks with Lipschitz-continuous Activation Functions and Time-Varying Delays. IEEE Trans. Circuits Syst. I, **50** (2003) 34-44
2. Zeng, Z., Wang, J., Liao, X.: Global Exponential Stability of a General Class of Recurrent Neural Networks with Time-Varying Delays. IEEE Trans. Circuits Syst., **50** (2003) 1353-1359
3. Singh, V.: A Generalized LMI-Based Approach to the Global Asymptotic Stability of Delayed Cellular Neural Networks. IEEE Trans. Neural Networks, **15** (2004) 223-225
4. Arik, S.: An Analysis of Global Asymptotic Stability of Delayed Cellular Neural Networks. IEEE Trans. Neural Networks, **13** (2002) 1239-1242
5. Liao, T., Wang, F.: Global Stability for Cellular Neural Networks with Time Delay. IEEE Trans. Neural Networks , **11** (2000) 1481-1484

A Further Result for Exponential Stability
of Neural Networks with Time-Varying Delays

Jun Zhang[1], Xiaofeng Liao[1], Chuandong Li[1], and Anwen Lu[2]

[1] College of Computer Science and Engineering, Chongqing University 400030, China
{xfliao,licd}@cqu.edu.cn
[2] College of Economics Business Administration, Chongqing University 400030, China

Abstract. This paper focuses on the issue of exponential stability for delayed neural networks with time-varying delays. A further result associated with this issue is exploited by employing a new Lyapunov-Krasovskii functional together with linear matrix inequality technique. Moreover, an approach to estimate the degree of exponential convergence is formulated.

1 Introduction

During past decade, the problem of exponential stability of DNN has recently been extensively studied and many exponential stability results have also been proposed [1–4, 6–11]. In [1], for example, two sufficient conditions for the exponential stability of delayed neural networks with time-varying delays have been derived, which were shown to be less restrictive than those reported recently. In this paper, we exploit an improved exponential stability criterion for DNN with time-varying delays with less restriction and less conservatism.

The delayed neural network model considered in this paper is defined by the following delayed differential equation:

$$\frac{du(t)}{dt} = -Au(t) + W_0 g(u(t)) + W_1 g(u(t-\tau(t))) + I, \tag{1}$$

where $u = [u_1, u_2, \cdots, u_n]^T$ is the state vector of system (1), $A = \text{diag}(a_i)$ is a positive diagonal matrix, $\tau(t)$ the transmission delay, $W_0 = \left(w_{ij}^0\right)_{n \times n}$ and $W_1 = \left(w_{ij}^1\right)_{n \times n}$ are the interconnection matrices representing the weight coefficients of the neurons, $I = [I_1, I_2, \cdots, I_n]^T$ is the constant external input vector, and the $g(u) = [g_1(u_1), g_2(u_2), \cdots, g_n(u_n)]^T$ denotes the neuron activation functions with the following assumption:

(H) For each $\xi_1, \xi_2 \in R$, $\xi_1 \neq \xi_2$, there exists $\sigma_j > 0$ such that

$$0 \leq \frac{g_j(\xi_1) - g_j(\xi_2)}{\xi_1 - \xi_2} \leq \sigma_j, \quad j = 1, 2, \cdots, n.$$

J. Wang, X. Liao, and Z. Yi (Eds.): ISNN 2005, LNCS 3496, pp. 120–125, 2005.

Let $u^* = [u_1^*, u_2^*, \cdots, u_n^*]^T$ be the equilibrium point of system (1). As usual, we shift the equilibrium point u^* into the origin by the transformation $x(t) = u(t) - u^*$, which puts system (1) into the following form:

$$\frac{dx(t)}{dt} = -ax(t) + W_0 f(x(t)) + W_1 f(x(t - \tau(t))),$$

$$x(s) = \phi(s), \quad s \in [-\tau(t), 0)$$

(2)

where $f_j(x_j) = g_j(x_j + u_j^*) - g_j(u_j^*)$ with $f_j(0) = 0$ and $x = [x_1, x_2, \cdots, x_n]^T$ is the state vector of the transformed system.

2   Improved Exponential Stability Result

The main result in this paper is as follows:

Theorem 1. Suppose $\dot{\tau}(t) \leq \eta < 1$ and $\Sigma = diag(\sigma_i)$. If the condition (H) is satisfied and there exist positive definite matrix P, positive diagonal matrix $D = diag(d_i > 0)$, $Q = diag(q_i > 0)$, and positive constant k such that the following LMI holds:

$$M \equiv \begin{bmatrix} \Delta_1 & -PW_0 - Q & -PW_1 \\ -W_0^T P - Q & \Delta_2 & -DW_1 \\ -W_1^T P & -W_1^T D & 2(1 - \dot{\tau}(t))e^{-2k\tau(t)}Q\Sigma^{-1} \end{bmatrix} > 0$$

(3)

where $\Delta_1 = PA + AP - 2kP - 4k\Sigma D$, $\Delta_2 = 2DA\Sigma^{-1} - (DW_0 + W_0^T D)$. Then, the origin of system (2) is exponentially stable.

Proof.   Consider the following Lyapunov-Krasovskii functional:

$$V(x(t)) = e^{2kt} x^T(t) Px(t) + 2e^{2kt} \sum_{i=1}^{n} d_i \int_0^{x_i(t)} f_i(s) ds$$

$$+ 2 \int_{t - \tau(t)}^{t} e^{2k\xi} x^T(\xi) Qf(x(\xi)) d\xi$$

(4)

where $P = P^T > 0$, $D = diag(d_i > 0)$, $Q = diag(q_i > 0)$ are the solutions of LMI (3). The time derivative of the functional along the trajectories of system (2) is calculated and estimated as follows:

$$\dot{V}(x(t)) = e^{2kt} \Big\{ x^T(t)[2kP - (PA + Ap)]x(t) + 2x^T(t)(PW_0 + Q)f(x(t))$$

$$+ 2x^T(t)PW_1 f(x(t - \tau(t))) + 4k\sum_{i=1}^{n} d_i \int_0^{x_i(t)} f_i(s) ds - 2f^T(x(t))DAx(t)$$

$$+ 2f^T(x(t))DW_0 f(x(t)) + 2f^T(x(t))PW_1 f(x(t-\tau(t)))$$

$$- 2(1-\dot{\tau}(t))e^{-2k\tau(t)} f^T(x(t-\tau(t)))Qx(t-\tau(t))\ \Big\}$$

$$\le e^{2kt}\Big\{\ x^T(t)[2kP - (PA + Ap)]x(t) + 2x^T(t)(PW_0 + Q)f(x(t))$$

$$+ 2x^T(t)PW_1 f(x(t-\tau(t))) + 4kx^T(t)\Sigma Dx(t) - 2f^T(x(t))DA\Sigma^{-1}f(x(t))$$

$$+ 2f^T(x(t))DW_0 f(x(t)) + 2f^T(x(t))PW_1 f(x(t-\tau(t)))$$

$$\dot{-}\, 2(1-\dot{\tau}(t))e^{-2k\tau(t)} f^T(x(t-\tau(t)))Q\Sigma^{-1}f(x(t-\tau(t)))\ \Big\}$$

$$= -e^{2kt}\Big[x^T(t)\quad f^T(x(t))\quad f^T(x(t-\tau(t)))\Big]M\begin{bmatrix} x(t) \\ f(x(t)) \\ f(x(t-\tau(t))) \end{bmatrix}.$$

Since $M > 0$, $\dot{V}(x(t)) \le 0$. Therefore,

$$e^{2kt}\lambda_m(P)\|x(t)\|^2 \le V(x(t)) \le V(x(0)).$$

Note that

$$V(x(0)) = x^T(0)Px(0) + 2\sum_{i=1}^{n} d_i \int_0^{x_i(0)} f_i(s)\,ds$$

$$+ 2\int_{-\tau(0)}^{0} e^{2k\xi}x^T(\xi)Qf(x(\xi))\,d\xi$$

$$\le \lambda_M(P)\|\phi\|^2 + 2\max_{1\le i\le n}\{d_i\sigma_i\}\|\phi\|^2 + 2\max_{1\le i\le n}\{q_i\sigma_i\}\|\phi\|^2 \int_{-\tau(t)}^{0} e^{2k\xi}\,d\xi$$

$$\le \Big[\lambda_M(P) + 2\max_{1\le i\le n}\{d_i\sigma_i\} + \frac{1}{k}\max_{1\le i\le n}\{q_i\sigma_i\}\Big]\|\phi\|^2,$$

where $\|\phi\| = \sup_{-\tau(t)\le\theta\le0}\|x(\theta)\|$. Hence,

$$e^{2kt}\lambda_m(P)\|x(t)\|^2 \le \Big[\lambda_M(P) + 2\max_{1\le i\le n}\{d_i\sigma_i\} + \frac{1}{k}\max_{1\le i\le n}\{q_i\sigma_i\}\Big]\|\phi\|^2.$$

This follows

$$\|x(t)\| \le \left[\frac{\lambda_M(P) + 2\max\limits_{1\le i\le n}\{d_i\sigma_i\} + \max\limits_{1\le i\le n}\{q_i\sigma_i\}/k}{\lambda_m(P)}\right]^{\frac{1}{2}}\|\phi\|\,e^{-kt}.$$

The proof is thus completed.

Remark 1. When system (2) is exponentially stable, we can also estimate the corresponding maximum value of degree of exponential convergence by solving the following optimization problem:

$$\begin{cases} \max\{k\} \\ \\ s.t. \quad \text{The LMI (3) is satisfied.} \end{cases} \quad (5)$$

3   A Numerical Example

To demonstrate the advantage of our result, a numerical example is presented in this section. Without loss of generality, we assume that the charging matrix A in system (2) be an identity matrix. The activation functions are chosen as $f_i(\alpha) = 0.5\left[\,|\alpha+1|-|\alpha-1|\,\right]$, which implies $\Sigma = E$ (identity matrix). We select respectively the initial values and the delay function as $x(\theta) = [0.2, -0.5]^T$, $\theta \in [-\overline{\tau}, 0]$ and $\tau(t) = \left(1 - e^{-t}\right)\big/\left(1 + e^{-t}\right)$. Noting that $0 \le \tau(t) < 1$ and $0 < \dot{\tau}(t) \le 0.5$.

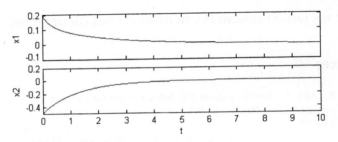

Fig. 1. The time response curves of states in Example 1.

Example 1.   Consider system (1) with

$$W_0 = \begin{bmatrix} -4 & 0.3 \\ - & \\ 0.2 & 0.13 \end{bmatrix}, \quad W_1 = \begin{bmatrix} 2.6 & -0.001 \\ -0.001 & 0.1 \end{bmatrix}.$$

In this case, the matrix inequalities of Theorems 1-2 in Ref. [1] have no feasible solutions for any $k>0$. However, the LMI (3) proposed above has at least one feasible solution:

$$P = \begin{bmatrix} 0.29165 & 0.03763 \\ 0.03763 & 0.79038 \end{bmatrix}, \quad Q = \begin{bmatrix} 1.13616 & 0 \\ 0 & 0.7 \end{bmatrix},$$

$$D = \begin{bmatrix} 0.12452 & 0 \\ 0 & 0.88804 \end{bmatrix}.$$

It follows from Theorem 1 that the considered system in this example is exponentially stable, as shown in Fig.1. Furthermore, by solving optimization problem (5), we estimate the degree of exponential convergence, i.e., $\max\{k\} = 0.0534$.

4   Conclusions

In this paper, we have presented a novel result for the exponential stability of neural networks with time-varying delays, which is derived by employing a new Lyapunov-Krasovskii functional. From the theoretical analysis and numerical illustration, we find that the proposed result possesses the following advantages in comparison with those reported in [1]: (i) Theorem 1 imposes less restriction on network parameters than Theorems 1-2 in Ref. [1]; (ii) The exponential stability condition is represented in the form of linear matrix inequality (LMI), which allows us to verify these conditions fast and easily by interior-point algorithms; (iii) Based on the proposed exponential stability conditions, we also estimate the degree of exponential convergence via solving an optimization problem.

Acknowledgements

This work was partially supported by the NNSF of China (Grant No. 60271019).

References

1. Yuce, E., Arik, S.: New Exponential Stability Results for Delayed Neural Networks with Time Varying Delays. Physica D, **191** (2004) 314-322
2. Liao, X.F., Chen, G., Sanchez, E. N.: Delay-dependent Exponential Stability Analysis of Delayed Neural Networks: an LMI Approach. Neural Networks, **15** (2002) 855-866
3. Zeng, Z., Wang, J, Liao X.: Global Exponential Stability of a General Class of Recurrent Neural Networks with Time-varying Delays. IEEE Trans. Circuits and Systems-I 50, (2003) 1353 - 1358
4. Liao, X. X., Wang, J.: Algebraic Criteria for Global Exponential Stability of Cellular Neural Networks With Multiple Time Delays. IEEE Trans. Circuits and Systems -I 50, (2003) 268-274.
5. Boyd, S. Ghaoui, L. Feron, E. and Balakrishnan, V.: Linear Matrix Inequality in Systems and Control Theory. SIAM: Philadelphia PA (1994)
6. Liao, X. F., Wong, K.-W., Wu, Z.: Asymptotic Stability Criteria for a Two-neuron Network with Different Time Delays. IEEE Trans. Neural Networks, **14** (2003) 222-227
7. Liao, X.F., Chen, G., Sanchez, E. N.: LMI-based Approach for Asymptotically Stability Analysis of Delayed Neural Networks. Neural Networks, **49** (2001) 1033-1039
8. Liao, X.F., Wong, K.-W., W,u Z., Chen, G.: Novel Robust Stability Criteria for Interval-delayed Hopfield Neural Networks. IEEE Trans. Circuits and Systems –I, **48** (2001) 1355-1359
9. Liao, X.F., Wong, K.-W., Yu, J.: Stability Switches and Bifurcation Analysis of a Neural Network with Continuously Delay. IEEE TSMC-A, **29** (1999) 692-696
10. Liao X. F., Yu J.: Robust stability for Interval Hopfield Neural Networks with Time Delay. IEEE Trans. Neural Networks, **9** (1998)1042-1045
11. Cao J., Wang L.: Exponential Sstability and Periodic Oscillatory Solution in Bam Networks with Delays. IEEE Trans. Neural Networks, **13** (2002) 457-463
12. Chuandong Li, Xiaofeng Liao, Rong Zhang.: Global Robust Asymptotical Stability of Multi-delayed Interval Neural Networks: an LMI Approach. Physics Letters A, **328** (2004) 452-462

13. Chuandong Li, Xiaofeng Liao.: New Algebraic Conditions for Global Exponential Stability of Delayed Recurrent Neural Networks. Neurocomputing, **64C** (2005)319-333
14. Chuandong Li, Xiaofeng Liao.: Delay-dependent Exponential Stability Analysis of Bidirectional Associative Memory Neural Networks: an LMI Approach, Chaos, Solitons & Fractals, **24** (2005) 1119-1134
15. Xiaofeng Liao, Chuandong Li.: An LMI Approach to Asymptotical Stability of Multidelayed Neural Networks. Physica D, **200** (2005) 139-155
16. Changyin Sun, Chun-Bo Feng.: Exponential Periodicity and Stability of Delayed Neural Networks. Mathematics and Computers in Simulation, **66** (2004) 469-478
17. Changyin Sun, Chun-Bo Feng.: Discrete-time Analogues of Integrodifferential Equations Modeling Neural Networks. Physics Letters A, **334** (2005) 180-191
18. Changyin Sun, Kanjian Zhang, Shumin Fei, Chun-Bo Feng.: On Exponential Stability of Delayed Neural Networks with A General Class of Activation Functions. Physics Letters A, **298** (2002) 122-132

Improved Results for Exponential Stability of Neural Networks with Time-Varying Delays

Deyin Wu[1], Qingyu Xiong[2], Chuandong Li[1], Zhong Zhang[1], and Haoyang Tang[3]

[1] College of Mathematics and Physics, Chongqing University 400030, China
cd_licqu@163.com
[2] College of Automation, Chongqing University 400030, China
[3] College of Economics Business Administration, Chongqing University 400030, China

Abstract. This paper presents several exponential stability criteria for delayed neural networks with time-varying delays and a general class of activation functions, which are derived by employing Lyapyunov-Krasovskii functional approach and linear matrix inequality technique. The proposed results are shown theoretically and numerically to be less restrictive and more easily verified than those reported recently in the open literature. In addition, an approach to estimate the degree of exponential convergence is also formulated.

1 Introduction

During past decades, the stability issue on the delayed neural networks (DNN) has been gained increasing interest for the potential applications in the signal processing, image processing, optimal programming and so on, and many global and/or local stability criteria for the equilibrium point of DNN have been reported in the literature [1-17]. It is worth noting that it is important that the designed neural networks be globally exponentially stable for some intended applications of DNN such as real-time computation, optimization problems and pattern formations. As we know, the problem of exponential stability of DNN has recently been extensively studied and many exponential stability results have also been proposed [1-4]. In [1], several sufficient conditions for the exponential stability of delayed neural networks with time-varying delays have been derived, which were shown to be less restrictive than those reported recently. In this paper, we present several improved exponential stability criteria for DNN with time-varying delays in comparison with [1].

The delayed neural network model considered in this paper is defined by the following delayed differential equation:

$$\frac{du(t)}{dt} = -Au(t) + W_0 g(u(t)) + W_1 g(u(t - \tau(t))) + I, \tag{1}$$

where $u = [u_1, u_2, \cdots, u_n]^T$ is the neuron state vector, $A = \text{diag}\,(a_i)$ is a positive diagonal matrix, $\tau(t)$ the transmission delay, $W_0 = \left(W_{ij}^0\right)_{n \times n}$ and $W_1 = \left(W_{ij}^1\right)_{n \times n}$ are the interconnection matrices representing the weight coefficients of the neurons, $I = [I_1, I_2, \cdots, I_n]^T$ is the constant external input vector, and the $g(u) = [g(u_1), g(u_2), \cdots, g(u_n)]^T$ denotes the neuron activation functions with the following assumption:

J. Wang, X. Liao, and Z. Yi (Eds.): ISNN 2005, LNCS 3496, pp. 126–131, 2005.
© Springer-Verlag Berlin Heidelberg 2005

(H) For each $\xi_1, \xi_2 \in R$, $\xi_1 \neq \xi_2$, there exists $\sigma_j > 0$ such that

$$0 \le \frac{g_j(\xi_1) - g_j(\xi_2)}{\xi_1 - \xi_2} \le \sigma_j, \quad j = 1, 2, \cdots, n.$$

Let $u^* = \begin{bmatrix} u_1^*, & u_2^*, & \cdots, & u_n^* \end{bmatrix}^T$ be the equilibrium point of system (1). We always shift the equilibrium point u^* into the origin of system (1), which puts system (1) into the following form:

$$\frac{dx(t)}{dt} = -ax(t) + W_0 f(x(t)) + W_1 f(x(t - \tau(t))),$$

$$x(s) = \phi(s), \quad s \in [-\tau(t), 0)$$

(2)

where $x = [x_1, \quad x_2, \quad \cdots, \quad x_n]^T$ is the state vector of the transformed system, and $f_j(x_j) = g_j(x_j + u_j^*) - g_j(u_j^*)$ with $f_j(0) = 0$.

2   Improved Exponential Stability Results

Corresponding to Theorem 1 and 2 in [1], the improved results are as follows.

Theorem 1. Suppose $\dot{\tau}(t) \le \eta < 1$, $\Sigma = diag(\sigma_i)$. If the condition (H) is satisfied and there exist positive definite matrices P and Q, the diagonal matrices $D = diag(d_i > 0)$ and $R = diag(r_i \ge 0)$, and positive constants β and k such that

$$\Omega_1^* = PA + AP - 2kP - P - 4k\beta\Sigma D$$

$$- (1 - \dot{\tau}(t))^{-1} e^{2k\tau(t)} PW_1 Q^{-1} W_1^T P - \Sigma R\Sigma > 0,$$

(3)

$$\Omega_2^* = 2\beta DA\Sigma^{-1} + R - \beta(DW_0 + W_0^T D) - W_0^T PW_0$$

$$- 2Q - \beta^2 (1 - \dot{\tau}(t))^{-1} e^{2k\tau(t)} DW_1 Q^{-1} W_1^T D \ge 0.$$

(4)

Then, the origin of system (2) is exponentially stable.

Proof. From Eq. (9) in Ref. [1], we have

$$\dot{V}(x(t)) \le -e^{2kt} \left[x^T (t)\Omega_1 x(t) + f^T (x(t))\Omega_2 f(x(t)) \right]$$

$$- e^{2kt} \left[-f^T (x(t))Rf(x(t)) + f^T (x(t))Rf(x(t)) \right]$$

$$= -e^{2kt} \left[x^T (t)[\Omega_1 - \Sigma R\Sigma]x(t) + f^T (x(t))[\Omega_2 + R]f(x(t)) \right]$$

$$= -e^{2kt} \left[x^T (t)\Omega_1^* x(t) + f^T (x(t))\Omega_2^* f(x(t)) \right]$$

(5)

Example. Consider system (2) with

$$W_0 = \begin{bmatrix} 0.8 & -0.05 \\ 1 & 0.6 \end{bmatrix}, \; W_1 = \begin{bmatrix} 0.1 & -0.02 \\ -0.5 & 0.1 \end{bmatrix}.$$

By calculating the LMIs in Theorems 1-2, we find that, for any $k \geq 0$, both have no feasible solution. However, the conditions of Corollaries 1-2 are satisfied, as shown in Table 1. The time response curve of the states in this example is shown in Fig. 1.

Table 1. Feasible solutions and estimates of degree of exponential convergence.

	Feasible solutions (k=0.001)	Max$\{k\}$
Corollary 1	P = [1065.52144, -64.43705; -64.43705 118.09761], Q = [296.57229 -37.22869; -37.22869 43.67397], D = [1422.99724 0; 0 280.01617], R = [892.50023 0; 0 68.70048].	0.01999
Corollary 2	P = [562.12244 -44.3431; -44.3431689 55.17457], Q = [151.959 0; 0 17.79], D = [756.81 0; 0 136.16], R = [144.12843 0; 0 0.41260].	0.0017

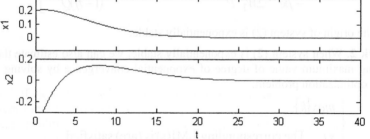

Fig. 1. The time response curves of states in Example 1.

4   Conclusions

In this paper, we have presented two new results for the exponential stability of neural networks with time-varying delays, which are derived by rearranging the analysis process given in [1]. Furthermore, we have also proposed an approach to estimate the maximum degree of exponential convergence.

Acknowledgements

This work was partially supported by the National Natural Science Foundation of China (Grant No.60271019, 60375024).

References

1. Yuce, E., Arik, S.: New Exponential Stability Results for Delayed Neural Networks with Time Varying Delays, Physica D., **191** (2004) 314-322
2. Liao, X. F., Chen, G.R., Sanchez E. N.: Delay-Dependent Exponential Stability Analysis of Delayed Neural Networks: an LMI Approach. Neural Networks, **15** (2002) 855-866
3. Zeng, Z., Wang, J. Liao, X.X.: Global Exponential Stability of a General Class of Recurrent Neural Networks with Time-Varying Delays. IEEE TCAS-I, **50** (2003) 1353-1358
4. Liao, X.F., Wang J.: Algebraic Criteria for Global Exponential Stability of Cellular Neural Networks with Multiple Time Delays. IEEE TCAS-I, **50** (2003) 268-274
5. Boyd, S., Ghaoui, L., Feron E., Balakrishnan, V.: Linear Matrix Inequality in Systems and Control Theory. SIAM: Philadelphia PA (1994)
6. Liao, X.F., Wong, K.W., Wu, Z.: Asymptotic Stability Criteria for a Two-Neuron Network with Different Time Delays. IEEE TNN, **14** (2003) 222-227
7. Liao, X.F., Chen, G.R., Sanchez, E.N.: LMI-based Approach for Asymptotically Stability Analysis of Delayed Neural Networks. **49** (2001) 1033-1039
8. Liao, X.F., Wong, K.W., Wu, Z., Chen, G.R.: Novel Robust Stability Criteria for Interval-Delayed Hopfield Neural Networks. IEEE TCAS-I, **48** (2001) 1355-1359
9. Liao, X.F., Wong, K.W., Yu, J.: Stability Switches and Bifurcation Analysis of a Neural Network with Continuously Delay. IEEE TSMC-A, **29** (1999) 692-696
10. Liao, X.F., Yu, J.: Robust Stability for Interval Hopfield Neural Networks with Time Delay. IEEE TNN, **9** (1998) 1042-1045
11. Cao, J., Wang, L.: Exponential Stability and Periodic Oscillatory Solution in Bam Networks with Delays. IEEE TNN, **13** (2002) 457-463
12. Sun, C., Feng, C,.: Exponential Periodicity of Continuous-time and Discrete-Time Neural Networks with Delays. Neural Processing Letters, **19** (2004) 131-146
13. Sun, C., Feng, C.: On Robust Exponential Periodicity of Interval Neural Networks with Delays. Neural Processing Letters, **20** (2004) 53-61
14. Li, C., Liao, X., Zhang, R.: Global Robust Asymptotical Stability of Multi-delayed Interval Neural Networks: an LMI approach. Physics Letters A, **328** (2004) 452-462
15. Li, C., Liao, X.F.: New Algebraic Conditions for Global Exponential Stability of Delayed Recurrent Neural Networks. Neurocomputing, **64C** (2005) 319-333.
16. Li, C., Liao, X.F.: Delay-dependent Exponential Stability Analysis of Bi-directional Associative Memory Neural Networks: an LMI Approach. Chaos, Solitons & Fractals, **24** (2005) 1119-1134
17. Liao, X.F., Li, C..: An LMI Approach to Asymptotical Stability of Multi-Delayed Neural Networks. Physica D, **200** (2005) 139-155

Global Exponential Stability of Recurrent Neural Networks with Infinite Time-Varying Delays and Reaction-Diffusion Terms

Qiankun Song[1,2], Zhenjiang Zhao[1], and Xuedong Chen[1]

[1] Department of Mathematics, Huzhou University, Huzhou, Zhejiang 313000, China
sqk@hutc.zj.cn
[2] Department of Computation Science, Chengdu University of Information Technology, Chengdu 610041, China
qiankunsong@163.com

Abstract. The global exponential stability is discussed for a general class of recurrent neural networks with infinite time-varying delays and reaction-diffusion terms. Several new sufficient conditions are obtained to ensure global exponential stability of the equilibrium point of recurrent neural networks with infinite time-varying delays and reaction-diffusion terms. The results extend the earlier publications. In addition, an example is given to show the effectiveness of the obtained results.

1 Introduction

The global stability of recurrent neural networks is a basic requirement in some practical applications of neural networks, such as in solving optimization problems [1]. In hardware implementation, time delays inevitably occur due to the finite switching speed of the amplifiers and communication time. What is more, to process moving images, one must introduce time delays in the signals transmitted among the cell[2]. Time delays may lead to oscillation, divergence, or instability which may be harmful to a system[3]. Therefore, study of neural dynamics with consideration of the delayed problem becomes extremely important to manufacture high quality neural networks. The stability of recurrent neural networks with delays has been studied, e.g. Refs.[1]-[23]. But the discussions of the neural networks with infinite time-varying delay are very few[24]. However, strictly speaking, diffusion effects cannot be avoided in the neural networks model when electrons are moving in asymmetric electromagnetic fields, so we must consider that the activations vary in space as well as in time. Refs.[25]-[30] have considered the stability of neural networks with diffusion terms, which are expressed by partial differential equations. It is also common to consider the neural networks with delays and reaction-diffusion terms. In this paper, we will derive a set of new sufficient conditions of the global exponential stability by using some analytic techniques. The work will have significance impact on the design and applications of globally exponentially stable recurrent neural networks with infinite time-varying delays and diffusion terms, and are of great interest in many applications.

J. Wang, X. Liao, and Z. Yi (Eds.): ISNN 2005, LNCS 3496, pp. 132–141, 2005.

In this paper, we consider the following model:

$$
\begin{cases}
\frac{\partial u_i(t,x)}{\partial t} = \sum_{k=1}^{m} \frac{\partial}{\partial x_k}\left(a_{ik}\frac{\partial u_i(t,x)}{\partial x_k}\right) - b_i u_i(t,x) \\
\qquad\qquad + \sum_{j=1}^{n} c_{ij} f_j(u_j(t,x)) + \sum_{j=1}^{n} d_{ij} g_j(u_j(t-\tau(t),x)) + I_i, \; x \in \Omega, \quad (1)\\
\frac{\partial u_i(t,x)}{\partial \bar{n}} = 0, \; t \geq t_0 \geq 0, x \in \partial\Omega, \\
u_i(t_0 + s, x) = \phi_i(s,x), -\infty < s \leq 0,
\end{cases}
$$

where $\frac{\partial u_i}{\partial \bar{n}} =: \left(\frac{\partial u_i}{\partial x_1}, \frac{\partial u_i}{\partial x_2}, \cdots, \frac{\partial u_i}{\partial x_m}\right)^T$, $i = 1, 2, \ldots, n$, and n corresponds to the number of units in a neural network; $u_i(t,x)$ corresponds to the state of the ith unit at time t and in space x; $f_j(u_j(t,x))$, $g_j(u_j(t,x))$ denotes the activation functions of the jth unit at time t and in space x; I_i denotes the external bias on the ith unit; $b_i > 0, c_{ij}, d_{ij}$ are constants, b_i represents the rate with which the ith unit will reset its potential to the resting state in isolation when disconnected from the networks and external inputs, c_{ij} denotes the strength of the jth unit on the ith unit at time t and in space x, d_{ij} denotes the strength of the jth unit on the ith unit at time $t - \tau(t)$ and in space x; time-varying delay $\tau(t)$ is continuous and $\tau(t) \geq 0$. Smooth function $a_{ik} = a_{ik}(t,x) \geq 0$ corresponds to the transmission diffusion operator along the ith unit; $x_i(i = 1, 2, \ldots, m)$ corresponds to the ith coordinate in the space; Ω is a compact set with smooth boundary and mes$\Omega > 0$ in space R^m. The initial condition is given by

$$
\Phi(s,x) = (\phi_1(s,x), \phi_2(s,x), \cdots, \phi_n(s,x))^T,
$$

where $\phi_i \in C((-\infty, 0] \times R^m, R)$, $i = 1, 2, \ldots, n$.

We assume that the activation functions $f_i, g_i (i = 1, 2, \ldots, n)$ satisfy the following properties:

(H) There exist positive constants k_i, l_i such that

$$
0 \leq (v_1 - v_2)(f_i(v_1) - f_i(v_2)) \leq k_i(v_1 - v_2)^2,
$$

$$
0 \leq (v_1 - v_2)(g_i(v_1) - g_i(v_2)) \leq l_i(v_1 - v_2)^2,
$$

for any $v_1, v_2 \in R$, and $i = 1, 2, .., n$.

For convenience, we introduce several notations. For an $n \times n$ matrix $A = (a_{ij})_{n\times n}$, $|A|$ denotes the absolute-value matrix given by $|A| = (|a_{ij}|)_{n\times n}$, $A^+ = (a_{ij}^+)_{n\times n}$, where $a_{ii}^+ = max(0, a_{ii}), a_{ij}^+ = |a_{ij}|(i \neq j)$.

In model (1), let $B = diag(b_1, b_2, \cdots, b_n)$, $C = (c_{ij})_{n\times n}$, $D = (d_{ij})_{n\times n}$, $K = diag(k_1, k_2, \cdots, k_n)$, $L = diag(l_1, l_2, \cdots, l_n)$, $C_K = (c_{ij}k_j)_{n\times n}$, $D_L = (d_{ij}l_j)_{n\times n}$. $L^2(\Omega)$ denotes the real Lebesgue measurable function space. For any $u = (u_1, u_2, ..., u_n)^T \in R^n$, define

$$
\|u_i\|_2 = [\int_\Omega |u_i(x)|^2 dx]^{\frac{1}{2}}, \; i = 1, 2, \ldots n.
$$

Definition 1 *The equilibrium point $u^* = (u_1^*, u_2^*, \cdots, u_n^*)^T$ of model (1) is said to be globally exponentially stable, if there exist constants $\delta > 0$ and $M \geq 1$ such* *that*

$$\sum_{i=1}^{n} ||u_i(t,x) - u_i^*||_2 \leq M \sup_{-\infty < s \leq 0} \sum_{i=1}^{n} ||\phi_i(s,x) - u_i^*||_2 e^{-\delta(t-t_0)}$$

for all $t \geq t_0$, where $u(t,x) = (u_1(t,x), u_2(t,x), ..., u_n(t,x))^T$ is a solution of model (1) with initial values $u_i(t_0+s,x) = \phi_i(s,x)$, $s \in (-\infty, 0]$, $i = 1, 2, ..., n$.

Lemma 1 *Let A be an $n \times n$ real matrix, if λ is an eigenvalue of matrix $\begin{pmatrix} 0 & A \\ A^T & 0 \end{pmatrix}$, then $-\lambda$ is also an eigenvalue of the matrix $\begin{pmatrix} 0 & A \\ A^T & 0 \end{pmatrix}$.*

Proof. Because $\lambda E(-A^T) = (-A^T)\lambda E$, then

$$det \begin{pmatrix} \lambda E & -A \\ -A^T & \lambda E \end{pmatrix} = det(\lambda^2 E - A^T A).$$

Since λ is an eigenvalue of matrix $\begin{pmatrix} 0 & A \\ A^T & 0 \end{pmatrix}$, we have $det \begin{pmatrix} \lambda E & -A \\ -A^T & \lambda E \end{pmatrix} = 0$, so $det(\lambda^2 E - A^T A) = 0$, which means $-\lambda$ is also an eigenvalue of matrix $\begin{pmatrix} 0 & A \\ A^T & 0 \end{pmatrix}$. The proof is completed.

2   Main Results

Theorem 1 *Assume that the assumption* (**H**) *is satisfied, $W = B - C^+ K - |D|L$ is an M-matrix, and for $\omega_i > 0$, $i = 1, 2, ..., n$, if there exist $M \geq 1$, $\lambda > 0$ such that*

$$\alpha + \beta e^{-\alpha \tau(t)} \leq h(t), \qquad (2)$$

where $\alpha = \max_{1 \leq i \leq n} \{-b_i + \frac{k_i}{\omega_i} \sum_{j=1}^{n} \omega_j c_{ij}^+\} \leq 0, \beta = \max_{1 \leq i \leq n} \{\frac{l_i}{\omega_i} \sum_{j=1}^{n} \omega_j |d_{ij}|\}$, the continuous function $h(t)$ satisfies

$$e^{\int_{t_0}^{t} h(s)ds} \leq M e^{-\lambda(t-t_0)}, \qquad (3)$$

then the unique equilibrium point of model (1) is globally exponentially stable.

Proof. For model (1), since the assumption (**H**) is satisfied, and $W = B - C^+ K - |D|L$ is an M-matrix, according to the results in [26], we know that model (1) has one unique equilibrium point $u^* = (u_1^*, u_2^*, \cdots, u_n^*)^T$.

Suppose $u(t,x) = (u_1(t,x), u_2(t,x), \cdots, u_n(t,x))^T$ is a solution of model (1). For convenience, in the following sections, we note $u_i = u_i(t,x), i = 1, 2, \cdots, n$, and rewrite model (1) as

$$\frac{\partial(u_i - u_i^*)}{\partial t} = \sum_{k=1}^{m} \frac{\partial}{\partial x_k}(a_{ik} \frac{\partial(u_i - u_i^*)}{\partial x_k}) - b_i(u_i - u_i^*) + \sum_{j=1}^{n} c_{ij}(f_j(u_j) - f_j(u_j^*))$$

$$+ \sum_{j=1}^{n} d_{ij}(g_j(u_j(t - \tau(t), x)) - g_j(u_j^*)),$$

multiply both sides of the equation above with $u_i - u_i^*$, and integrate

$$\frac{1}{2}\frac{d}{dt}\int_\Omega (u_i - u_i^*)^2 dx = \sum_{k=1}^{m}\int_\Omega (u_i - u_i^*)\frac{\partial}{\partial x_k}(a_{ik}\frac{\partial(u_i - u_i^*)}{\partial x_k})dx$$

$$-b_i\int_\Omega (u_i - u_i^*)^2 dx$$

$$+\sum_{j=1}^{n}\int_\Omega (u_i - u_i^*)[c_{ij}(f_j(u_j) - f_j(u_j^*))]dx$$

$$+\sum_{j=1}^{n}\int_\Omega (u_i - u_i^*)[d_{ij}(g_j(u_j(t - \tau(t), x)) - g_j(u_j^*))]dx.$$

From the boundary condition of model (1) and the proof of theorem in [28], we get

$$\sum_{k=1}^{m}\int_\Omega (u_i - u_i^*)\frac{\partial}{\partial x_k}(a_{ik}\frac{\partial(u_i - u_i^*)}{\partial x_k})dx = -\sum_{k=1}^{m}\int_\Omega a_{ik}(\frac{\partial(u_i - u_i^*)}{\partial x_k})^2 dx.$$

From the two equations above, assumption (**H**) and Cauchy inequality, we have

$$\frac{d}{dt}\|u_i - u_i^*\|_2^2 \le -2b_i\|u_i - u_i^*\|_2^2 + 2\sum_{j=1}^{n}c_{ij}^+ k_j\|u_i - u_i^*\|_2\|u_j - u_j^*\|_2$$

$$+2\sum_{j=1}^{n}|d_{ij}|l_j\|u_i - u_i^*\|_2\|u_j(t - \tau(t), x) - u_j^*\|_2,$$

i.e.

$$\frac{d\|u_i - u_i^*\|_2}{dt} \le -b_i\|u_i - u_i^*\|_2 + \sum_{j=1}^{n}c_{ij}^+ k_j\|u_j - u_j^*\|_2$$

$$+\sum_{j=1}^{n}|d_{ij}|l_j\|u_j(t - \tau(t), x) - u_j^*\|_2. \qquad (4)$$

For $\omega_i > 0, i = 1, 2, \ldots, n$, let $V(t) = \sum_{i=1}^{n}\omega_i\|u_i - u_i^*\|_2$, then

$$D^+ V(t) \le \alpha V(t) + \beta V(t - \tau(t)).$$

Let $y(t) = V(t)e^{-\alpha(t-t_0)}$, $\overline{y}(t_0) = \sup_{-\infty < s \le t_0} V(s)$, then

$$D^+ y(t) \le \beta y(t - \tau(t))e^{-\alpha\tau(t)},$$

and for any $t \ge t_0$,

$$y(t) \le y(t_0) + \int_{t_0}^{t}\beta y(s - \tau(s))e^{-\alpha\tau(s)}ds. \qquad (5)$$

Theorem 4 *Assume that the assumption* (**H**) *is satisfied,* $W = B - C^+ K - |D|L$ *is an M-matrix, if there exist* $M \geq 1$, $\lambda > 0$, *such that*

$$\alpha + \beta e^{-\alpha \tau(t)} \leq h(t),$$

where $\alpha = 2 \max\limits_{1 \leq i \leq n} \{-b_i\} + \lambda_3 + \lambda_2 \leq 0$, $\beta = \lambda_2$, *and the continuous function* $h(t)$ *satisfies*

$$e^{\int_{t_0}^t h(s)ds} \leq M e^{-\lambda(t-t_0)},$$

then the unique equilibrium point of model (1) is globally exponentially stable.

Proof. Let $V(t) = \sum\limits_{i=1}^n \|u_i - u_i^*\|_2^2$, then

$$D^+ V(t) \leq (2 \max\limits_{1 \leq i \leq n} \{-b_i\} + \lambda_3 + \lambda_2) V(t) + \lambda_2 V(t - \tau(t)).$$

The rest part of the proof is similar to the proof of Theorem 1. The proof is completed.

Theorem 5 *Assume that the assumption* (**H**) *is satisfied,* $W = B - C^+ K - |D|L$ *is an M-matrix, and for any* $\omega_i > 0$, $q_{ij}, r_{ij}, q_{ij}^*, r_{ij}^* \in R$, $i,j = 1,2,\cdots,n$, *if there exist* $M > 1$, $\lambda > 0$, *such that*

$$\alpha + \beta e^{-\alpha \tau(t)} \leq h(t),$$

where

$$\alpha = \max\limits_{1 \leq i \leq n} \left\{ -2b_i + \sum_{j=1}^n \left((c_{ij}^+)^{2-q_{ij}} k_j^{2-r_{ij}} + |d_{ij}|^{2-q_{ij}^*} l_j^{2-r_{ij}^*} + \frac{\omega_j}{\omega_i} k_i^{r_{ji}} (c_{ji}^+)^{q_{ji}} \right) \right\} \leq 0,$$

$$\beta = \max\limits_{1 \leq i \leq n} \{ \frac{1}{\omega_i} \sum_{j=1}^n \omega_j l_i^{r_{ji}^*} |d_{ji}|^{q_{ji}^*} \}, \text{ and the continuous function } h(t) \text{ satisfies}$$

$$e^{\int_{t_0}^t h(s)ds} \leq M e^{-\lambda(t-t_0)},$$

then the unique equilibrium point of model (1) is globally exponentially stable.

Proof. By the process of proof of Theorem 1, we know

$$\frac{d\|u_i - u_i^*\|_2}{dt} \leq -b_i \|u_i - u_i^*\|_2 + \sum_{j=1}^n c_{ij}^+ k_j \|u_j - u_j^*\|_2$$

$$+ \sum_{j=1}^n |d_{ij}| l_j \|u_j(t - \tau(t), x) - u_j^*\|_2.$$

Let $V(t) = \sum\limits_{i=1}^n \omega_i \|u_i(t,x) - u_i^*\|_2^2$, then

$$D^+ V(t) \leq \alpha V(t) + \beta V(t - \tau(t)).$$

The rest part of the proof is similar to the proof of Theorem 1. The proof is completed.

3 Example

Consider the neural network with infinite time-varying delays and diffusion terms:

$$
\begin{cases}
\frac{\partial u_1(t,x)}{\partial x} = \frac{\partial}{\partial x}\left(\sin^2(tx)\frac{\partial u_1(t,x)}{\partial x}\right) - 3u_1(t,x) + \frac{1}{2}\sin(u_1(t,x)) + \frac{1}{3}\cos(u_2(t,x)) \\
\qquad\qquad + \frac{1}{2}\sin(u_1(t-\tau(t),x)) + \frac{1}{3}\cos(u_2(t-\tau(t),x)) + 1, \\[2mm]
\frac{\partial u_2(t,x)}{\partial x} = \frac{\partial}{\partial x}\left(\cos^4(tx)\frac{\partial u_2(t,x)}{\partial x}\right) - 3u_2(t,x) + \frac{1}{2}\sin(u_1(t,x)) + \frac{2}{3}\cos(u_2(t,x)) \\
\qquad\qquad + \frac{1}{2}\sin(u_1(t-\tau(t),x)) + \frac{2}{3}\cos(u_2(t-\tau(t),x)) + 2,
\end{cases}
$$

where $\tau(t) = \frac{1}{2}\ln(1 + \triangle(t))$, for positive integral m,

$$
\triangle(t) = \begin{cases}
t, & t = m, \\[1mm]
\pi(t), \; t \in (m - \frac{1}{m}\frac{1}{2^m}, m) \cup (m, m + \frac{1}{m}\frac{1}{2^m}), \\[1mm]
e^{-t}, \; t \in (m + \frac{1}{m}\frac{1}{2^m}, m+1 - \frac{1}{m+1}\frac{1}{2^{m+1}}), \\[1mm]
e^{-\frac{3}{2}}, \; t \in (0, \frac{1}{2}],
\end{cases}
$$

$0 \le \pi(t) \le t$, such that $\triangle(t)$ is continuous. Obviously $B = \begin{pmatrix} 3 & 0 \\ 0 & 3 \end{pmatrix}$, $C = D = \begin{pmatrix} 1/2 & 1/3 \\ 1/2 & 2/3 \end{pmatrix}$, $K = L = \begin{pmatrix} 1 & 0 \\ 0 & 1 \end{pmatrix}$. $B - C^+K - |D|L = \begin{pmatrix} 2 & -2/3 \\ -1 & 5/3 \end{pmatrix}$ is an M-matrix. Take $\omega_1 = \omega_2 = 1$, since $\alpha = \max\limits_{1 \le i \le 2}\{-b_i + \frac{k_i}{\omega_i}\sum\limits_{j=1}^{2}\omega_j c_{ji}^+\} = -2, \beta = \max\limits_{1 \le i \le 2}\{\frac{l_i}{\omega_i}\sum\limits_{j=1}^{2}\omega_j |d_{ji}|\} = 1$, $\alpha + \beta e^{-\alpha\tau(t)} = -1 + \triangle(t)$. take $h(t) = -1 + \triangle(t)$, then for any $t \ge t_0$, $e^{\int_{t_0}^{t} h(s)ds} \le e^3 e^{-(t-t_0)}$. Form Theorem 1 we know the equilibrium point of this networks is globally exponentially stable.

It is worth noting that all of the references are not applicable to ascertain the stability of such a neural network because the time-varying delay is infinite.

4 Conclusion

The recurrent neural networks model with infinite time-varying delays and reaction-diffusion terms was studied, Several new sufficient conditions are obtained to ensure global exponential stability of the equilibrium point of reaction-diffusion recurrent neural networks with infinite time-varying delays. The results extend and improve the earlier publications. In addition, an example is given to show the effectiveness of the obtained result. The work will have significance impact on the design and applications of globally exponentially stable recurrent neural networks with infinite time-varying delays and diffusion terms, and are of great interest in many applications.

Acknowledgements

The work was supported by the National Natural Science Foundation of China under Grant 10272091, the key project, Ministry of Education of China under Grant 03051, the Natural Science Foundation of Zhejiang Province under Grant M103087, the Natural Science Foundation of Huzhou City, Zhejiang Province, China under Grant 2004SZX0703 and the Scientific and Technological Project of Huzhou City, Zhejiang Province, China under Grant 2004KYZ1026 and 2004KYZ1019.

References

1. Lu, H.T., Chung, F.L., He, Z.Y.: Some Sufficient Conditions for Global Exponential Stability of Hopfield Neural Networks. Neural Networks, **17** (2004) 537–544
2. Civalleri, P.P., Gilli, L.M., Pabdolfi, L.: On Stability of Cellular Neural Networks with Delay. IEEE Transactions on Circuits and Systems, **40** (1993) 157–164
3. Baldi, P., Atiga, A.F., How Delays Affect Neural Dynamics and Learning. IEEE Transactions on Neural Networks, **5** (1994) 612–621
4. Chen, T.P.: Global Exponential Stability of Delayed Hopfield Neural Networks. Neural Networks, **14** (2001) 977–980
5. Cao, J.D., Wang, J.: Global Asymptotic Stability of a General Class of Recurrent Neural Networks with Time-varying Delays. IEEE Transactions on Circuits and Systems I, **50** (2003) 34–44
6. Cao, J.D., Wang, L.: Exponential Stability and Periodic Oscillatory Solution in BAM Networks with Delays. IEEE Transactions on Neural Networks, **13** (2002) 457–463
7. Cao, J.D.: Global Stability Conditions for Delayed CNNs. IEEE Transactions on Circuits and Systems-I, **48** (2001) 1330–1333
8. Ca,o J.D.: A Set of Stability Criteria for Delayed Cellular Neural Networks. IEEE Transactions on Circuits and Systems-I, **48** (2001) 494–498
9. Chen, T.P., Rong, L.B.: Delay-independent Stability Analysis of Cohen-Grossberg Neural Networks. Physics Letters A, **317** (2003) 436–449
10. Cao, J.D.: An Estimation of the Domain of Attraction and Convergence Rate for Hopfield Continuous Feedback Neural Networks. Physics Letters A, **325** (2004) 370–374
11. Zhang, J.Y.: Global Exponential Satbility of Neural Networks with Varible Delays. IEEE Transactions on Circuits and Systems I, **50** (2003) 288–291
12. Liao, X., Chen, G., Sanches, E.: Delay-dependent Exponential Stability Analysis of Delayed Neural Networks: an LMI approach. Neural Networks, **15** (2002) 855–866
13. Arik, S.: An Analysis of Global Asymptotic Stability of Delayed Cellular Networks. IEEE Transactions on Neural Networks, **13** (2002) 1239–1242

14. Cao, J.D., Ho, D.W.C.: A General Framework for Global Asymptotic Stability Analysis of Delayed Neural Networks Based on LMI approach. Chaos, Solitons & Fractals, **24** (2005) 1317–1329
15. Zhao, H.Y.: Global Asymptotic Stability of Hopfield Neural Network Involving Distributed Delays. Neural Networks, **17** (2004) 47–53
16. Arik, S.: Global Asymptotic Stability of a Large Class of Neural Networks with Constant Time Delay. Physics Letters A, **311** (2003) 504–511
17. Sun, C.Y., Feng, C.B.: Global Robust Exponential Stability of Interval Neural Networks with Delays. Neural Process Letters, **17** (2003) 107–115
18. Sun, C.Y., Feng, C.B.: Exponential Periodicity and Stability of Delayed Neural Networks. Mathematics and Computers in Simulation, **66** (2004) 469–478
19. Cao, J.D., Huang, D.S., Qu, Y.Z.: Global Robust Stability of Delayed Recurrent Neural Networks. Chaos, Solitons & Fractals, **23** (2005) 221–229
20. Cao, J.D., Liang, J.L., Lam, J.: Exponential Stability of High-order Bidirectional Associative Memory Neural Networks with Time Delays. Physica D: Nonlinear Phenomena, **199** (2004) 425–436
21. Cao, J.D., Liang, J.L.: Boundedness and Stability for Cohen-Grossberg Neural Networks with Time-varying Delays. Journal of Mathematical Analysis and Applications, **296** (2004) 665–685
22. Cao, J.D., Wang, J.: Absolute Exponential Stability of Recurrent Neural Networks with Time Delays and Lipschitz-continuous Activation Functions. Neural Networks, **17** (2004) 379–390
23. Cao, J.D.: New Results Concerning Exponential Stability and Periodic Solutions of Delayed Cellular Neural Networks. Physics Letters A, **307** (2003) 136–147
24. Zeng, Z.G., Fu, C.J., Liao, X.X.: Stability Analysis of Neural Networks with Infinite Time-varying Delay. Journal of Mathematics, **22** (2002) 391–396
25. Liao, X.X., Fu, Y.L., Gao, J., Zhao, X.Q.: Stability of Hopfield Neural Networks with Reaction-diffusion Terms. Acta Electronica Sinica, **28** (2000) 78–80(in chinese)
26. Wang, L.S., Xu, D.Y.: Global Exponential Stability of Hopfield Reaction-diffusion Neural Networks with Time-varying Delays. Sci. China Ser. F, **46** (2003) 466–474
27. Liao, X.X., Li, J.: Stability in Gilpin-Ayala Competition Models with Diffusion. Nonliear Analysis, **28** (1997) 1751–1758
28. Liang, J.L., Cao, J.D.: Global Exponential Stability of Reaction-diffusion Recurrent Neural Networks with Time-varying Delays, Physics Letters A, **314** (2003) 434–442
29. Song, Q.K., Zhao, Z.J., Li, Y.M.: Global Exponential Stability of BAM Neural Networks with Distributed Delay and Reaction-diffusion Terms. Physics Letters A, **335** (2005) 213–225
30. Song, Q.K., Cao, J.D.: Global Exponential Stability and Existence of Periodic Solutions in BAM Networks with Delays and Reaction-diffusion Terms. Chaos, Solitons & Fractals, **23** (2005) 421–430

Exponential Stability Analysis of Neural Networks with Multiple Time Delays

Huaguang Zhang[1], Zhanshan Wang[1,2], and Derong Liu[3]

[1] Northeastern University, Shenyang, Liaoning 110004, China
hg_zhang@21cn.com
[2] Shenyang Ligong University, Shenyang, Liaoning 110045, P.R. China
[3] University of Illinois, Chicago, IL 60607-7053, USA
dliu@ece.uic.edu

Abstract. Without assuming the boundedness, strict monotonicity and differentiability of the activation function, a result is established for the global exponential stability of a class of neural networks with multiple time delays. A new sufficient condition guaranteeing the uniqueness and global exponential stability of the equilibrium point is established. The new stability criterion imposes constraints, expressed by a linear matrix inequality, on the self-feedback connection matrix and interconnection matrices independent of the time delays. The stability criterion is compared with some existing results, and it is found to be less conservative than existing ones.

1   Introduction

It is well known that stability and convergence properties are important in the design and application of recurrent neural networks. Basically, the stability properties of neural networks depend on the intended applications. For instance, in solving optimization problems, neural networks must be designed to have only one equilibrium point and this equilibrium point must be globally asymptotically stable, so as to avoid the risk of having spurious equilibriums and being trapped in local minima [1–7]. However, in some applications of neural networks, such as associative memories, the existence of multiple equilibriums is a necessary feature to be utilized and the exponential stability of the global pattern formation is important [1, 4, 7–11]. Moreover, the exponential stability property is particularly important when the exponential convergence rate is used to determine the speed of neural computation and the convergence to the memory points in associative memories. Thus, it is important to determine the exponential stability of dynamical neural networks.

In hardware implementation, time delays inevitably occur due to finite switching speed of amplifiers and communication time. Neural networks with time delays have much more complicated dynamics due to the incorporation of delays. A single time delay is first introduced into recurrent neural networks in [12]. Multiple time delays in different communication channels are considered in [13] for neural networks. For the delay model studied in [13], its modified model has been

J. Wang, X. Liao, and Z. Yi (Eds.): ISNN 2005, LNCS 3496, pp. 142–148, 2005.
© Springer-Verlag Berlin Heidelberg 2005

extensively studied, e.g., in [2, 3, 5, 8, 14–18], and some stability criteria related to the global asymptotic stability and global exponential stability independent of the delays have been obtained.

Usually, a single fixed time delay in the model of delayed neural networks can serve as good approximation in simple circuits having a small number of cells. However, a neural network usually has a spatial nature due to the presence of parallel pathways and layers, which produces the time nature of neural networks. It is desirable to model them by introducing multiple time delays. In this paper, we establish a sufficient condition for the uniqueness and global exponential stability of the equilibrium point for a class of neural networks with multiple time delays. The condition obtained can be expressed by a linear matrix inequality (LMI). The main advantages of using the LMI include: (i) the stability condition contains all information about the neural network, and eliminate the differences between excitatory and inhibitory effects of the neural networks; and (ii) the stability condition can efficiently be verified via the LMI Toolbox in Matlab. In many existing results [6, 17–20], the exponential convergence conditions consist of a set of algebraic inequalities or equalities including some scalar parameters. Although the suitability of the criteria is improved due to these parameters, it is not easy to verify the inequalities or to make the equalities hold by efficiently choosing these parameters because we have no systematic approach to tune these parameters in practice. Moreover, the condition obtained here is less restrictive, and is a generalization of some previous works.

The present paper is organized as follows. In Section 2, we give the model description and establish a lemma to show the existence and uniqueness of the model's equilibrium. In Section 3, we conduct a global exponential stability analysis of the delayed neural networks and provide some remarks, which are used to compare our result with existing ones. In Section 4, we conclude the paper.

2   Problem Description and Uniqueness of the Equilibrium Point

Consider the following neural networks with multiple time delays

$$\frac{du(t)}{dt} = -Au(t) + W_0 g(u(t)) + \sum_{i=1}^{N} W_i g(u(t - \tau_i)) + U \tag{1}$$

where $u(t) = [u_1(t), u_2(t), \ldots, u_n(t)]^T$ is the neuron state vector; n is the number of neurons; $A = \mathrm{diag}(a_1, a_2, \ldots, a_n)$ is a diagonal matrix with positive diagonal entries; $W_0 \in \Re^{n \times n}$ and $W_i \in \Re^{n \times n}$ $(i = 1, 2, \ldots, N)$ are the connection weight matrix and delayed connection weight matrices, respectively; N denotes the number of time delayed vectors; $\tau_i = [\tau_{i1}, \tau_{i2}, \ldots, \tau_{in}]$ with $\tau_{ij} > 0$ denotes the transmission delay of the jth state vector in the ith delayed connection matrix $(i = 1, 2, \ldots, N, \ j = 1, 2, \ldots, n)$; $U = [U_1, U_2, \ldots, U_n]^T$ denotes the external constant input vector; and $g(u(t)) = [g_1(u_1(t), g_2(u_2(t)), \ldots, g_n(u_n(t))]^T$ denotes the neuron activations function.

Remark 1. (i) When $W_0 = 0$, $N = 1$ and $g_i(u_i)$ is the sigmoidal function, system (1) becomes a Hopfield neural network with transmission delay [2, 12, 16].
(ii) When $W_0 = 0$, $N = 1$, and $W_1 = \begin{bmatrix} 0 & W_{12} \\ W_{21} & 0 \end{bmatrix}$, where W_{12} and W_{21} are
$(n/2) \times (n/2)$ matrices, respectively, system (1) is reduced to a BAM (bidirectional associative memory) network with delay studied in [13]. (iii) When $N = 1$ and $g_i(u_i)$ is bounded, system (1) changes into delayed cellular neural networks studied in [15, 21]. (iv) When $N = 1$, system (1) becomes the delayed neural networks investigated in [1, 8].

Throughout the paper, we need the following notation and assumptions.

Let B^T, B^{-1}, $\lambda_M(B)$, $\lambda_m(B)$ and $\|B\| = \sqrt{\lambda_M(B^T B)}$ denote the transpose, the inverse, the largest eigenvalue, the smallest eigenvalue, and the Euclidean norm of a square matrix B, respectively. $B > 0$ ($B < 0$) denotes a positive (negative) definite symmetric matrix. Let $\rho = \max\{\rho_i\}$, where $\rho_i = \max_{1 \le j \le n}\{\tau_{ij}\}$. We use a continuous function $\phi_{\tau i}$ to represent the segment of $u(\theta)$ on $[t - \rho_i, \ t]$, and $\|\phi_\tau\| = \sup_{t-\rho \le \theta \le t} \|u(\theta)\|$.

Assumption A. The activation function $g_j(u_j)$, $j = 1, 2, \ldots, n$, satisfies the following condition

$$0 \le \frac{g_j(\xi) - g_j(\zeta)}{\xi - \zeta} \le \sigma_j \qquad (2)$$

for arbitrary $\xi, \zeta \in \Re$ and $\xi \ne \zeta$, and for some positive constant $\sigma_j > 0$.

Let $\Delta = \mathrm{diag}(\sigma_1, \sigma_2, \ldots, \sigma_n)$ denote a positive diagonal matrix.

Remark 2. As pointed out in [19], the capacity of an associative memory model can be remarkably improved by replacing the usual sigmoid activation functions with non-monotonic activation functions. Hence, it seems that in some cases, non-monotonic functions might be better candidates for neuron activation function in designing and implementing neural networks. Many popular activation functions satisfy the Assumption A, for example, $\frac{e^u - e^{-u}}{e^u + e^{-u}}$, $\frac{1 - e^{-u}}{1 + e^{-u}}$, $\frac{1}{1 + e^{-u}}$ and the piecewise linear saturation function $\frac{|u+1| - |u-1|}{2}$, the linear function $g(u) = u$, etc. As see from these functions, the functions under the Assumption A may be bounded or unbounded. At the same time, we can see that Δ is a positive definite diagonal matrix.

Assumption B. The equilibrium point set of system (1) is a non-empty set when $\tau_i = 0$, $i = 1, \ldots, N$.

Remark 3. As pointed out in [4], for neural networks with bounded activation functions, there exists at least one equilibrium point. But for the case of unbounded activation functions, there is no such a conclusion. To make the present investigation meaningful, Assumption B is necessary.

The following lemma and definition are needed.

Lemma 1 ((see [22])). For any given symmetric matrix $Q = \begin{bmatrix} Q_{11} & Q_{12} \\ Q_{12}^T & Q_{22} \end{bmatrix} < 0$, where Q_{ii} is an $r_i \times r_i$ nonsingular matrix, $i = 1, 2$, the following conditions are equivalent:

(1) $Q < 0$;

(2) $Q_{11} < 0$ and $Q_{22} - Q_{12}^T Q_{11}^{-1} Q_{12} < 0$;

(3) $Q_{22} < 0$ and $Q_{11} - Q_{12} Q_{22}^{-1} Q_{12}^T < 0$.

Definition. Consider the system defined by (1), if there exist positive constants $k > 0$ and $\gamma > 0$ such that $\|u(t)\| \le \gamma e^{-kt} \sup_{-\rho \le \theta \le 0} \|u(\theta)\|$, $\forall t > 0$, then the equilibrium point of system (1) is exponentially stable, where k is called as the exponential convergence rate.

We first provide the following lemma without proof to establish the existence and uniqueness of the equilibrium for system (1).

Lemma 2. Suppose that $g(u)$ satisfies Assumption A, and there exist positive diagonal matrix P and positive diagonal matrices Q_i $(i = 0, 1, \ldots, N)$ such that

$$- PA\Delta^{-1} - \Delta^{-1}AP + \sum_{i=0}^{N} PW_i Q_i^{-1} W_i^T P + \sum_{i=0}^{N} Q_i < 0, \tag{3}$$

then for each $U \in \Re^n$, system (1) has a unique equilibrium point.

3   Global Exponential Stability Result

By Lemma 2, system (1) has a unique equilibrium point, namely, $u^* = [u_1^*, u_2^*, \ldots, u_n^*]^T$. In the following, we will shift the equilibrium point of system (1) to the origin. The transformation $x(\cdot) = u(\cdot) - u^*$ converts system (1) into the following form

$$\frac{dx(t)}{dt} = -Ax(t) + W_0 f(x(t)) + \sum_{i=1}^{N} W_i f(x(t - \tau_i)) \tag{4}$$

$$x(t) = \phi(t), \ t \in [-\rho, \ 0)$$

where $x(t) = [x_1(t), x_2(t), \ldots, x_n(t)]^T$ is the state vector of the transformed system, $f_j(x_j(t)) = g_j(x_j(t) + u_j^*) - g_j(u_j^*)$ with $f_j(0) = 0$, $j = 1, 2, \ldots, n$. By Assumption A, we can see that $0 \le \frac{f_j(x_j(t))}{x_j(t)} \le \sigma_j$, $f_j(x_j(t))x_j(t) \ge 0$ and $|f_j(x_j(t))| \le \sigma_j |x_j(t)|$. Clearly, the equilibrium point u^* of system (1) is globally exponentially stable if and only if the zero solution of system (4) is globally exponentially stable.

We are now in a position to provide our main theorem. Its proof is omitted due to space limitations.

Theorem 1. If $g(u)$ satisfies Assumption A and equation (3) holds, then system (1) is globally exponentially stable, independent of the time delays. Moreover,

$$\|u(t) - u^*\| \le \sqrt{\frac{\Pi}{N+1}} \|\phi\| e^{\frac{-k}{2}t} \tag{5}$$

where $\Pi = (N+1) + \alpha \lambda_M(P)\lambda_M(\Delta) + \sum_{i=1}^{N} \lambda_M(Q_i)(\alpha + \beta_i)\frac{e^{k\rho_i} - 1}{k}\lambda_M^2(\Delta)$, $\alpha > 0, \beta_i > 0$.

15. Lu, H.: On Stability of Nonlinear Continuous-Time Neural Networks with Delays. Neural Networks, **13** (2000) 1135–1143
16. Zhang, J., Jin, X.: Global Stability Analysis in Delayed Hopfield Neural Network Models. Neural Networks, **13** (2000) 745–753
17. Liao, X., Yu, J.: Robust Stability for Interval Hopfield Neural Networks with Time Delay. IEEE Trans. Neural Networks, **9** (1998) 1042–1046
18. Cao, J., Wang, J.: Global Asymptotic Stability of a General Class of Recurrent Neural Networks with Time-Varying Delays. IEEE Trans. Circuits and Systems-I **50** (2003) 34–44
19. Morita, M.: Associative Memory with Nonmonotone Dynamics. Neural Networks, **6** (1993) 115–126
20. Zhang, Y., Heng, P. A., Fu, A.W.C.: Estimate of Exponential Convergence Rate and Exponential Stability for Neural Networks. IEEE Trans. Neural Networks, **10** (1999) 1487–1493
21. Arik, S.: An Improved Global Stability Result for Delayed Cellular Neural Networks. IEEE Trans. Circuits and Systems-I, **49** (2002) 1211–1214
22. Boyd, S., El Ghaoui, L., Feron, E., Balakrishnan, V.: Linear Matrix Inequalities in System and Control Theory, SIAM Press, Philadelphia, PA (1994)

Exponential Stability
of Cohen-Grossberg Neural Networks with Delays

Wei Zhang[1,3] and Jianqiao Yu[2]

[1] Department of Computer Science and Engineering, Chongqing University,
Chongqing 400044, China
[2] College of Information, Southwest Agricultural University 400716,
Chongqing, China
[3] Department of Computer and Modern Education Technology,
Chongqing Education College, Chongqing 400067, China

Abstract. The exponential stability characteristics of the Cohen-Grossberg neural networks with discrete delays are studied in this paper, without assuming the symmetry of connection matrix as well as the monotonicity and differentiability of the activation functions and the self-signal functions. By constructing suitable Lyapunov functionals, the delay-independent sufficient conditions for the networks converge exponentially toward the equilibrium associated with the constant input are obtained. It does not doubt that our results are significant and useful for the design and applications of the Cohen-Grossberg neural networks.

1 Introduction

Cohen-Grossberg had proposed a large class of neural networks, which can function as stable content addressable memories or CAMs [1],[2]. These Cohen-Grossberg networks were designed to include additive neural networks, later studied by Hopfield [3],[4], and shunting neural networks. In original analysis, Cohen and Grossberg assumed that the weight matrix was symmetric. Meanwhile, the activation functions are assumed to be continuous, differentiable, monotonically increasing and bounded, such as the sigmoid-type function. However, for some application purpose of networks, non-monotonic and not necessarily smooth functions might be better candidates for neuron activation functions in designing and implementing an artificial neural network. We also note that in many electronic circuits, amplifiers that have neither monotonically increasing nor continuously differentiable, input-output functions are frequently adopted (i.e., piecewise linear functions) [17],[18]. Ye et al.[10] introduced discrete delays into the Cohen-Grossberg model. Furthermore, their global stability needs to satisfy the requirements that the connection should possess certain amount of symmetry and the discrete delays are sufficiently small. Hence, the work of Ye et al. [10] could not tell what would happen when the delays are increased. Moreover, the large delay could destroy the stability of an equilibrium in a network.

For the delayed Hopfield networks [5],[6],[7],[10],[11],[13],[14],[19],[22],[23], [24],cellular neural networks [17],[18] as well as BAM networks [8],[12],[15],[16], some delay-independent criteria for the global asymptotic stability are established

J. Wang, X. Liao, and Z. Yi (Eds.): ISNN 2005, LNCS 3496, pp. 149–155, 2005.
© Springer-Verlag Berlin Heidelberg 2005

without assuming the monotonicity and the differentiability of the activation functions and also the symmetry of the connection. Wang and Zou[9] also studied the Cohen-Grossberg model with time delays. The global stability criteria of the Cohen-Grossberg neural network with delays were also obtained by constructing appropriate Lyapunov functionals, or Lyapunov functions combined with the Rezumikhin technique. All of these criteria are independent of the magnitudes of the delays, and therefore, delays are harmless in a network with the structure satisfying one of the criteria. However, they did not involve in the convergence or exponential stability of Cohen-Grossberg neural networks with time delays. Actually, the global exponential stability implies global asymptotic stability, results leading to global exponential stability can provide relevant estimates on how fast such networks perform during real time computations.

In this paper, the amplification functions are required to be continuous, positive and bounded, the self-signal functions are not assumed to be differentiable, by contraries, only require to satisfy condition (H₂). At the same time, we do not confine ourselves to the symmetric connections and not assume monotonicity and differentiability for the activation functions. The rest of this paper is organized as follows. In Section II, the Cohen-Grossberg neural network with time delays and some preliminary analyses are given. By constructing Lyapunov functionals, some global exponential stability criteria for the network are presented in Section III. Finally, conclusions are drawn in Section IV.

2   The Cohen-Grossberg Model and Some Preliminaries

In this paper, we consider the following Cohen-Grossberg neural networks with multiple delays:

$$\dot{x}_i(t) = -a_i(x_i)\left[b_i(x_i) - \sum_{k=0}^{K}\sum_{j=1}^{n} t_{ij}^{(k)} s_j(x_j(t-\tau_k)) + J_i \right], \quad i=1,2,\cdots,n \ . \tag{1}$$

where $J_i, i=1,2,\cdots,n$, denote the constant inputs from outside of the system and the matrix $T_k = \left(t_{ij}^{(k)} \right)_{n\times n}$ represents the interconnections associated with delay τ_k of which $\tau_k, k=0,1,2,\cdots,K$, are arranged such that $0=\tau_0 < \tau_1 < \cdots < \tau_K$. $a_i(x_i)$ and $b_i(x_i), i=1,2,\cdots,n$, are the amplification functions and the self-signal functions, respectively, while $s_j(x_j), j=1,2,\cdots,n$, are the activation functions.

The initial conditions associated with (1) are given in the form:

$$x_i(s) = \phi_i(s) \in C[(-\tau,0],R), i=1,2,\cdots,n, \ . \tag{2}$$

where $\tau = \max_{0\le k\le K}\{\tau_k\} = \tau_K$.

Let R denote the set of real numbers and if $x \in R^n$, then $x^T = (x_1,x_2,\cdots,x_n)$ denotes the transpose of x. Let $R^{n\times n}$ denote the set of $n\times n$ real matrices. For functions $a_i(x_i)$ and $b_i(x_i), i=1,2,\cdots,n$, we state the assumptions used in this paper.

(H$_1$) $a_i(.), i = 1,2,\cdots,n$, are continuous and there exist positive constants $\underline{\alpha}_i$ and $\overline{\alpha}_i$ such that

$$\underline{\alpha}_i \leq a_i(u_i) \leq \overline{\alpha}_i, \quad \text{for all } u \in R^n, i = 1,2,\cdots,n.$$

(H$_2$) b_i and $b_i^{-1}, i = 1,2,\cdots,n$, are locally Lipschiz continuous and there exist $\gamma_i > 0, i = 1,2,\cdots,n$, such that $u[b_i(u+x) - b_i(x)] \geq \gamma_i u^2, i = 1,2,\cdots,n.$

Remark 1. In [9], the authors need $b_i'(x) > 0$. However, our condition (H$_2$) does not need that $b_i(x), i = 1,2,\cdots,n$, are continuous and differentiable. Moreover, our results are less conservative and restrictive than that given in [9].

In this paper, we assume that the activation functions $s_i, i = 1,2,\cdots,n$, satisfy either (S$_1$) or (S$_2$):

(S$_1$) (i) $s_i, i = 1,2,\cdots,n$, are bounded in R, i.e.,

$$|s_i(x)| \leq M_i, x \in R \text{ for some constannt } M_i > 0$$

(ii) $0 < \dfrac{s_i(x+u) - s_i(x)}{u} \leq L_i, i = 1,2,\cdots,n.$

(S$_2$) (i) $s_i, i = 1,2,\cdots,n$, are bounded in R, i.e.,

$$|s_i(x)| \leq M_i, x \in R \text{ for some constannt } M_i > 0$$

(ii) $|s_i(x+u) - s_i(x)| \leq L_i|u|, i = 1,2,\cdots,n.$

Remark 2. Note that unlike in [10], $T = \sum\limits_{k=0}^{K} T_k$ is not required to be symmetric in this paper, which means that our results will be applicable to networks with much boarder connection structures.

3 Main Results

Note that x* is an equilibrium of system (1) if and only if $x^* = (x_1^*, x_2^*, \cdots, x_n^*)^T$ is a solution of the following equations

$$b_i(x_i^*) - \sum_{k=0}^{K} \sum_{j=1}^{n} t_{ij}^{(k)} s_j(x_j^*) + J_i = 0, \quad i = 1,2,\cdots,n \ . \tag{3}$$

Let x^* be an equilibrium of system (1) and $u(t) = x(t) - x^*$. Substituting $x(t) = u(t) + x^*$ into system (1) leads to

$$\dot{u}_i(t) = -a_i(u_i(t) + x_i^*) \left[b_i(u_i(t) + x_i^*) - \sum_{k=0}^{K} \sum_{j=1}^{n} t_{ij}^{(k)} s_j(u_j(t-\tau_k) + x_j^*) + J_i \right]. \tag{4}$$

for $i = 1,2,\cdots,n$. By (3), system (4) can be rewritten as

$$\dot{u}_i(t) = -\alpha_i(u_i(t)) \left[\beta_i(u_i(t)) - \sum_{k=0}^{K} \sum_{j=1}^{n} t_{ij}^{(k)} g_j(u_j(t-\tau_k)) \right], \quad i = 1,2,\cdots,n, \ . \tag{5}$$

where

$$
\begin{cases}
\alpha_i(u_i(t)) = a_i(u_i(t) + x_i^*) \\
\beta_i(u_i(t)) = b_i(u_i(t) + x_i^*) - b_i(x_i^*) \\
g_j(u_j(t)) = s_j(u_j(t) + x_j^*) - s_j(x_j^*)
\end{cases} \quad . \tag{6}
$$

It is obvious that x^* is globally exponentially stable for system (1) if and only if the trivial solution $u=0$ of system (5) is globally exponentially stable.

Theorem 1. Consider the delayed system (1) and assume that conditions (H_1)-(H_2) and (S_1) or (S_2) are satisfied. Suppose the following conditions hold

$$
\frac{\underline{\alpha}_i}{\bar{\alpha}_i} \gamma_i > L_i \sum_{k=0}^{K} \sum_{j=1}^{n} \left| t_{ji}^{(k)} \right|, \quad i = 1, 2, \cdots, n, \; . \tag{7}
$$

then the equilibrium point x^* for system (1) is globally exponentially stable. This implies that there exist constants $C_1 \geq 1$ and $\sigma_1 > 0$ such that

$$
\sum_{i=1}^{n} \left| x_i(t) - x_i^* \right| \leq C_1 e^{-\sigma_1 t} \sum_{i=1}^{n} \left(\sup_{s \in [-\tau, 0]} \left| x_i(s) - x_i^* \right| \right), \quad \text{for } t > 0 \; . \tag{8}
$$

Proof. The proof of the existence and uniqueness of the equilibrium x^* of system (1) is similar to that of [9]. In the following, we establish the global exponential stability of the equilibrium x^* of system (1), i.e., the global exponential stability of the equilibrium solution of system (5). Let $u(t)$ denote an arbitrary solution of (5). From (5) and by using (S_1) or (S_2) we obtain the following inequalities

$$
\frac{d^+ |u_i(t)|}{dt} \leq -\gamma_i \underline{\alpha}_i |u_i(t)| + \bar{\alpha}_i \sum_{k=0}^{K} \sum_{j=1}^{n} \left| t_{ij}^{(k)} \right| L_j |u_j(t - \tau_k)|, \quad i = 1, 2, \cdots, n, t > 0 \; . \tag{9}
$$

We note from the condition (7) that

$$
\underline{\alpha}_i \gamma_i - \bar{\alpha}_i L_i \sum_{k=0}^{K} \sum_{j=1}^{n} \left| t_{ji}^{(k)} \right| \geq \eta, i = 1, 2, \cdots, n, \; . \tag{10}
$$

where

$$
\eta = \min_{1 \leq i \leq n} \left\{ \underline{\alpha}_i \gamma_i - \bar{\alpha}_i L_i \sum_{k=0}^{K} \sum_{j=1}^{n} \left| t_{ji}^{(k)} \right| \right\} > 0 \; . \tag{11}
$$

We consider functions $F_i(.)$ defined by

$$
F_i(\xi_i) = \underline{\alpha}_i \gamma_i - \xi_i - \bar{\alpha}_i L_i \sum_{k=0}^{K} \sum_{j=1}^{n} \left| t_{ji}^{(k)} \right| e^{\xi_i \tau_k}, \quad \text{for } \xi_i \in [0, \infty), i = 1, 2, \cdots, n \; . \tag{12}
$$

We have from (11) and (12) that $F_i(0) \geq \eta > 0, i = 1, 2, \cdots, n$, and hence by the continuity of $F_i(.)$ on $[0, \infty)$, there exists a constant $\sigma_1 > 0$ such that

$$
F_i(\sigma_1) = \underline{\alpha}_i \gamma_i - \sigma_1 - \bar{\alpha}_i L_i \sum_{k=0}^{K} \sum_{j=1}^{n} \left| t_{ji}^{(k)} \right| e^{\sigma_1 \tau_k}, \quad i = 1, 2, \cdots, n \; . \tag{13}
$$

Next we consider functions

$$z_i(t) = e^{\sigma_1 t}|u_i(t)|, \, i = 1,2,\cdots,n, \, t \in [-\tau,\infty) \ . \tag{14}$$

By using (9) and (14) we obtain

$$\frac{d^+ z_i(t)}{dt} \le -(\underline{\alpha}_i \gamma_i - \sigma_1) z_i(t) + \overline{\alpha}_i \sum_{k=0}^{K}\sum_{j=1}^{n}|t_{ij}^{(k)}|L_j e^{\sigma_1 \tau_k} z_j(t-\tau_k), \, t > 0, \, i = 1,2,\cdots,n \ \cdot \tag{15}$$

We define a Lyapunov functional $V(t)$ as follows

$$V(t) = \sum_{i=1}^{n}\left(z_i(t) + \sum_{k=0}^{K}\sum_{j=1}^{n}|t_{ij}^{(k)}|L_j e^{\sigma_1 \tau_k}\int_{-\tau_k} z_j(s)ds\right), \, t > 0 \ . \tag{16}$$

Calculating the rate of change of $V(t)$ along (15) we have

$$\frac{d^+ V(t)}{dt} \le -\sum_{i=1}^{n}\left(\underline{\alpha}_i \gamma_i - \sigma_1 - \overline{\alpha}_i L_i \sum_{k=0}^{K}\sum_{j=1}^{n}|t_{ji}^{(k)}|e^{\sigma_1 \tau_k}\right)z_i(t), \qquad t > 0. \tag{17}$$

By using (13) in (17) we have $d^+/dt \le 0$, for $t > 0$ implying that $V(t) \le V(0)$ for $t > 0$. We obtain from (16) that

$$\sum_{i=1}^{n}z_i(t) \le \sum_{i=1}^{n}\left(z_i(0) + L_i\sum_{k=0}^{K}\sum_{j=1}^{n}|t_{ji}^{(k)}|e^{\sigma_1 \tau_k}\int_{-\tau_k}^{0}z_j(s)ds\right), \quad t > 0$$

and by using (16) in the above inequalities

$$\sum_{i=1}^{n}|x_i(t) - x_i^*| \le e^{-\sigma_1 t}\sum_{i=1}^{n}\left(1 + L_i\sum_{k=0}^{K}\sum_{j=1}^{n}|t_{ji}^{(k)}|e^{\sigma_1 \tau_k}\tau_k\right)\left(\sup_{s\in[-\tau,0]}|x_i(s) - x_i^*|\right) \le C_1 e^{-\sigma_1 t}\left(\sup_{s\in[-\tau,0]}|x_i(s) - x_i^*|\right)$$

where

$$C_1 = \max_{1 \le i \le n}\left\{1 + L_i\sum_{k=0}^{K}\sum_{j=1}^{n}|t_{ji}^{(k)}|e^{\sigma_1 \tau_k}\tau_k\right\} \ge 1$$

Since $\sigma_1 > 0$ and $x(t)$ denotes an arbitrary solution of the network (1), we conclude that the equilibrium x^* of system (1) is globally exponentially stable and hence the proof is complete.

Remark 3. We note that from the estimate in (8) that the constant $C_1 \ge 1$ can play a role in approximating the convergence time of the network (2). We have from (8) that the constant C_1 is highly dependent that the estimated time required for the network to converge towards the equilibrium for the certain accuracy can become longer.

4   Conclusions

In this paper, the criteria for the global exponential stability of Cohen-Grossberg neural networks with discrete delays have been derived. Analyses have also shown that the neuronal input-output activation function and the self-signal function only need to satisfy, respectively, conditions (S1), (S2) and (H2) given in this paper, but do not need to be continuous, differentiable, monotonically increasing and bounded, as usually required by other analyzing methods. Novel stability conditions are stated in simple algebraic forms so that their verification and applications are straightforward and convenient. The criteria obtained in this paper are independent of the magnitudes of the delays, and delays are harmless in a network with the structure satisfying the criteria.

Acknowledgements

The work described in this paper was supported by a grant from the National Natural Science Foundation of China (No.60271019), the Doctorate Foundation of the Ministry of Education of China (No.20020611007), Chongqing Education Committee (No.011401), Chongqing Science and Technology Committee (No.2001-6878).

References

1. Cohen, M.A., Grossberg S.: Absolute Stability and Global Pattern Formation and Parallel Memory Storage by Competitive Neural Networks. IEEE Transactions on Systems, Man and Cybernetics, SMC-13, (1981) 815-821
2. Grossberg, S.: Nonlinear Neural Networks: Principles, Mechanisms, and Architectures. Neural Networks, 1 (1988) 17-61
3. Hopfield, J.J.: Meuronal Networks and Physical Systems with Emergent Collective Computational Abilities. Proceedings of the National Academy of Sciences, 79 (1982) 2554-2558
4. Hopofield, J.J.: Neurons with Graded Response Have Collective Computational Properties Like Those of Two-state Neurons. Proceeding of the National Academy of Sciences, .81 (1984) 3058-3092
5. Vander, Driessche P., Zou X.: Global Attractivity in Delayed Hopfield Neural Network Models. SIAM Journal of on Applied Mathematics, 58 (1998) 1878-1890
6. Marcus, C.M., Westervelt, R.M.: Stability of Analog Neural Networks with Delay. Physical Review A, 39 (1989) 347-359
7. Gopalsamy, K., He, X.: Stability in Asymmetric Hopfield Nets with Transmission Delays. Physica D, 76 (1994) 344-358
8. Gopalsamy, K., He, X.: Delay-independent Stability in Bi-directional Associative Memory Networks. IEEE Transactions on Neural Networks, 5 (1994) 998-1002
9. Wang, L., Zou, X.: Harmless Delays in Cohen-Grossberg Neural Networks. Physical D, 170 (2002) 162-173
10. Ye, H., Michel, A.N., Wang, K.: Qualitative Analysis of Cohen-Grossberg Neural Networks with Multiple Delays. Physical Review E, 51 (1995) 2611-2618
11. Joy, M.: Results Concerning the Absolute Stability of Delayed Neural Networks. Neural Networks, 13 (2000) 613-616
12. V. Sree Hari Rao, Bh. R. M. Phaneendra: Global Dynamics of Bi-directional Associative Memory Neural Networks Involving Transmission Delays and Dead Zones. Neural Networks, 12 (1999) 455-465
13. 13 Liao, X.F., Yu, J.B.: Robust Stability for Interval Hopfield Neural Networks with Time Delays. IEEE Transactions on Neural Networks, 9 (1998) 1042-1046
14. Liao, X.F., Wong, K.W., Wu, Z.F., Chen, G.: Novel Robust Stability Criteria for Interval Delayed Hopfield Neural Networks. IEEE Transactions on CAS-I, 48 (2001) 1355-1359

15. Liao, X.F., Yu, J.B.: Qualitative Analysis of Bi-directional Associative Memory Networks with Time Delays. Int. J. Circuit Theory and Applicat., **26** (1998) 219-229
16. Liao, X.F., Yu, J.B., Chen, G.: Novel Stability Criteria for Bi-directional Associative Memory Neural Networks with Time Delays. Int. J. Circuit Theory and Application,**30** (2002) 519-546
17. Liao, X.F., Wong, K.W., Yu, J.B.: Novel Stability Conditions for Cellular Neural Networks with Time Delay. Int. J. Bifur. And Chaos, **11** (2002) 1853-1864
18. Liao, X.F., Wu, Z.F., Yu, J.B.: Stability Analyses for Cellular Neural Networks with Continuous Delay. Journal of Computational and Applied Math, **143** (2002) 29-47
19. Liao, X.F., Chen, G., Sanchez, E.N.: LMI-based Approach for Asymptotically Stability Analysis of Delayed Neural Networks. IEEE Transactions on CAS-I, **49** (2002) 1033-1039
20. Liao, X.F., Chen, G., Sanchez, E.N.: Delay-dependent Exponential Stability Analysis of Delayed Neural Networks: an LMI Npproach. Neural Networks, **15** (2002) 855-866
21. Gopalsamy, K.: Stability and Oscillations in Delays Differential Equations of Population Dynamics. Kluwer, Dordrecht (1992)
22. Cao, J., Liang, J.: Boundness and Stability for Cohen-Grossberg Neural Networks with Time-varying Delays. Journal of Mathematical Analysis and Applications, **296** (2004) 665-685
23. Yuan, K., Cao, J.: Global Exponential Stability of Cohen-Grossberg Neural Networks with Multiple Time-varying Delays. Lecture Notes in Computer Science, ISNN 2004, **3173** (2004) 78-83
24. Cao, J., Wang, J.: Global Asymptotic Stability of a General Class of Recurrent Neural Networks with Time-varying Delays. IEEE Trans. Circuits Syst.-I, **50** (2003) 34-44

Global Exponential Stability of Cohen-Grossberg Neural Networks with Time-Varying Delays and Continuously Distributed Delays

Yi Shen, Minghui Jiang, and Xiaoxin Liao

Department of Control Science and Engineering, Huazhong University
of Science and Technology, Wuhan, Hubei 430074, China

Abstract. In this paper, we investigate the global exponential stability for a class of Cohen-Grossberg neural networks (CGNN) with time-varying delays and continuously distributed delays. CGNN herein is a general neural networks model which includes some well-known neural networks as its special cases. Firstly, applying the homeomorphism theory, we establish the new sufficient condition of existence and uniqueness of the equilibrium point to CGNN. Then, the sufficient criteria of global exponential stability of CGNN, which are easy to verify, are given by M-matrix. Our results imply and generalize some existed ones in previous literature. Furthermore, it is convenient to estimate the exponential convergence rates of the neural networks by using the criteria. Compared with the previous methods, our method does not resort to any Lyapunov functions or functionals. Finally, a example is given to illustrate the effective of our results.

1 Introduction

Since Cohen and Grossberg proposed a class of neural networks in 1983 [1], this model has received increasing interest due to its promising potential for applications in classification, parallel computing,etc. Such applications rely on the existence of equilibrium points or of a unique equilibrium point, and the qualitative properties of stability. Thus, the qualitative analysis of these dynamic behaviors is a prerequisite step for the practical design and application of neural networks. In neural processing and signal transmission, significant time delays may occur. This needs introducing delays into communication channels which leads to delayed CGNN model. To date, most research on delayed CGNN has been restricted to simple cases of constant fixed delays or time-varying delay [1-3]. Since a neural network usually has a spatial nature due to the presence of an amount of parallel pathways of a variety of axon sizes and lengths, it is desired to model them by introducing continuously distributed delays over a certain duration of time such that the distant past has less influence compared to the recent behavior of the state [4]. But discussion about CGNN with the continuously distributed delays are few [3,4]. In this paper, we will consider a class of CGNN with time-varying delays and continuously distributed delays.

J. Wang, X. Liao, and Z. Yi (Eds.): ISNN 2005, LNCS 3496, pp. 156–161, 2005.
© Springer-Verlag Berlin Heidelberg 2005

To our knowledge, this is the first paper for studying CGNN with time-varying delays and distributed delays, which is described by the following equations

$$\dot{u}_i(t) = -\bar{a}_i(u_i(t))(\bar{b}_i(u_i(t)) - \sum_{j=1}^{n} c_{ij}\bar{f}_j(u_j(t)) - \sum_{j=1}^{n} d_{ij}\bar{g}_j(u_j(t - \tau_{ij}(t)))$$

$$-\sum_{j=1}^{n} e_{ij}\int_{-\infty}^{t} K_{ij}(t-s)\bar{h}(u_j(s))\mathrm{d}s + I_i), \quad i = 1, 2, \cdots, n, \qquad (1)$$

where $u(t) = [u_1(t), \cdots, u_n(t)]^T$ denotes the state vector, matrix $C = (c_{ij})_{n\times n}$, $D = (d_{ij})_{n\times n}, E = (e_{ij})_{n\times n}$ are the connection weight matrix, time-varying delay connection weight matrix and continuously distributed delay connection weight matrix, respectively. $\bar{a}_i(u_i(t))(i = 1, \cdots, n)$ represent an amplification functions, $\bar{b}_i(u_i(t))(i = 1, \cdots, n)$ are an appropriately behaved function. $\bar{f}_i, \bar{g}_i, \bar{h}_i$ $(i = 1, \cdots, n)$ are activation functions. $0 \leq \tau_{ij} \leq \tau(i, j = 1, \cdots, n)$ are the time delay, $K_{ij} : R_+ \to R_+, i, j = 1, \cdots, n$, the kernels of distributed delays, are continuous on R_+, $I_i(i = 1, \cdots, n)$ are the bias.

It can be easily seen that CGNN (1) is a general neural networks model which includes some well-known neural networks as its special cases [1-6].

To establish the main results, some of the following assumptions will apply:

($\mathbf{H}_1$) $\int_0^\infty K_{ij}(s)\mathrm{d}s = 1, \int_0^\infty e^{\rho s}K_{ij}(s)\mathrm{d}s = \kappa_{ij} < \infty, i, j = 1, \cdots, n$,
in which ρ is a small positive constant.

($\mathbf{H}_2$) Each function $\bar{a}_i(u)$ is positive, continuous, and $\bar{a}_i(u) \geq \alpha_i > 0$.

($\mathbf{H}_3$) Each function $\bar{b}_i(u) \in C^1(R, R)$ and $\bar{b}'_i(u) \geq \beta_i > 0$.

($\mathbf{H}_4$) Each function $\bar{f}_i, \bar{g}_i, \bar{h}_i : R \to R$ satisfied the Lipschitz condition with Lipschitz constants $k_i, l_i, \eta_i > 0$, respectively, i.e.,

$|\bar{f}_i(\gamma) - \bar{f}_i(\delta)| \leq k_i|\gamma - \delta|, \quad |\bar{g}_i(\gamma) - \bar{g}_i(\delta)| \leq l_i|\gamma - \delta|, \quad |\bar{h}_i(\gamma) - \bar{h}_i(\delta)| \leq$
$\eta_i|\gamma - \delta|$, for all $\gamma, \delta \in R, i = 1, \cdots, n$.

Assume that the network (1) is supplemented with initial conditions of the form $u_i(s) = \phi_i(s), \quad i = 1, \cdots, n, \quad s \leq 0$, in which $\phi_i(s)$ is bounded, and continuous on $(-\infty, 0]$.

Throughout this paper, unless otherwise specified, we denote $|A|$ and $|x|$ as the absolute-value in $A \in R^{n\times n}$ and in $x \in R^n$, respectively.

2   Existence and Uniqueness of the Equilibrium Point

Theorem 2.1. Assume that $A \overset{\triangle}{=} B - |C|K - |D|L - |E|N$ is a nonsingular M-matrix, then (1) has a unique equilibrium point u^*, where $B = \mathrm{diag}(\beta_1, \cdots, \beta_n)$ is decided by ($\mathbf{H}_3$), $K = \mathrm{diag}(k_1, \cdots, k_n), L = \mathrm{diag}(l_1, \cdots, l_n), N = \mathrm{diag}(\eta_1, \cdots, \eta_n)$ is defined by ($\mathbf{H}_4$).

Proof. By assumptions ($\mathbf{H}_1$) and ($\mathbf{H}_2$), in order to prove existence and uniqueness of equilibrium point of the networks (1), we only need to prove that the following equation (2) has a unique solution.

$$\bar{b}_i(u_i) - \sum_{j=1}^{n} c_{ij}\bar{f}_j(u_j) - \sum_{j=1}^{n} d_{ij}\bar{g}_j(u_j) - \sum_{j=1}^{n} e_{ij}\bar{h}_j(u_j)$$
$$+I_i = 0, \quad i = 1, 2, \cdots, n \qquad (2)$$

Set $\bar{B}(u) = (\bar{b}_1(u_1), \cdots, \bar{b}_n(u_n))^T, \bar{f}(u) = (\bar{f}_1(u_1), \cdots, \bar{f}_n(u_n))^T, \bar{g}(u) = (\bar{g}_1(u_1), \cdots, \bar{g}_n(u_n))^T, \bar{h}(u) = (\bar{h}_1(u_1), \cdots, \bar{h}_n(u_n))^T, I = (I_1, \cdots, I_n)^T$, then Eq.(2) can be rewritten

$$\bar{B}(u) - C\bar{f}(u) - D\bar{g}(u) - E\bar{h}(u) + I = 0. \qquad (3)$$

Let $J(u) = \bar{B}(u) - C\bar{f}(u) - D\bar{g}(u) - E\bar{h}(u) + I$, then J is injective on R^n. Otherwise, on the one hand, there is $u \neq v, u, v \in R^n$ such that $J(u) = J(v)$; i.e., $\bar{B}(u) - \bar{B}(v) = C(\bar{f}(u) - \bar{f}(v)) + D(\bar{g}(u) - \bar{g}(v)) + E(\bar{h}(u) - \bar{h}(v))$. By $(\mathbf{H}_3)$ and $(\mathbf{H}_4)$, we have $B|u - v| \leq |C|K|u - v| + |D|L|u - v| + |E|N|u - v|$. Hence

$$A|u - v| \leq 0. \qquad (4)$$

On the other hand, since A is a nonsingular M-matrix, there is an n-dimensional vector $\xi > 0$ such that $\xi^T A > 0$, then $\xi^T A|u-v| > 0$. This contradicts (4), thus, J is injective on R^n. In addition, since A is a nonsingular M-matrix, for sufficiently small $\varepsilon > 0$, there exists a positive definite diagonal matrix $Q = \text{diag}(q_1, \cdots, q_n)$ such that $-QA - A^T Q \leq -2\varepsilon I_{n \times n}$. Thus, by $(\mathbf{H}_3)$ and $(\mathbf{H}_4)$, we obtain

$$2(Qu)^T(\bar{B}(0) - \bar{B}(u) + C(\bar{f}(u) - \bar{f}(0)) + D(\bar{g}(u) - \bar{g}(0)) + E(\bar{h}(u) - \bar{h}(0)))$$
$$\leq -|u|^T(QA + A^T Q)|u| \leq -2\varepsilon\|u\|^2$$

By the Schwartz inequality, we have

$$\varepsilon\|u\|^2 \leq \|Q\|\|u\|\|\bar{B}(0) - \bar{B}(u) + C(\bar{f}(u) - \bar{f}(0)) + D(\bar{g}(u) - \bar{g}(0)) + E(\bar{h}(u) - \bar{h}(0))\|.$$

Then, if $\|u\| \to \infty$, then $\|\bar{B}(0) - \bar{B}(u) + C(\bar{f}(u) - \bar{f}(0)) + D(\bar{g}(u) - \bar{g}(0)) + E(\bar{h}(u) - \bar{h}(0))\| \to \infty$. Thus, $\|J\| \to +\infty$. Therefore, J is a homeomorphism of R^n onto itself, so there exists a unique $u^* \in R^*$ such that $J(u^*) = 0$, i.e., (1) has a unique equilibrium point.

3   Global Exponential Stability

Theorem 3.1. Assume that $\bar{A} \triangleq B - |C|K - |D|L - |\bar{E}|N$, where $\bar{E} = (e_{ij}\kappa_{ij})_{n \times n}$, ($\kappa_{ij}$ see $(\mathbf{H}_1)$), is a nonsingular M-matrix, then (1) has a unique equilibrium point u^* and it is globally exponentially stable.

Proof. Because $\bar{A}$ is a nonsingular M-matrix, there exists a positive definite diagonal matrix $P = \text{diag}(p_1, \ldots, p_n)$, such that

$$p_i\beta_i - \sum_{j=1}^{n} |c_{ij}|k_j p_j - \sum_{j=1}^{n} |d_{ij}|l_j p_j - \sum_{j=1}^{n} |e_{ij}|\kappa_{ij}\eta_j p_j > 0, i = 1, \ldots, n. \qquad (5)$$

By $(\mathbf{H_1})$, it is obvious that $\kappa_{ij} \geq 1$, therefore (5) reduces to

$$p_i\beta_i - \sum_{j=1}^{n}|c_{ij}|k_jp_j - \sum_{j=1}^{n}|d_{ij}|l_jp_j - \sum_{j=1}^{n}|e_{ij}|\eta_jp_j > 0. \tag{6}$$

By (6), the definition of M-matrix, and Theorem 2.1, (1) has a unique equilibrium $u^* = (u_1^*, \ldots, u_n^*)^T$. Let $x(t) = (x_1(t), \ldots, x_n(t))^T = (u_1(t)-u_1^*, \ldots, u_n(t)-u_n^*)^T$, $f_i(x_i(t)) = \bar{f}_i(x_i(t)+u_i^*)-\bar{f}_i(u_i^*)$, $g_i(x_i(t)) = \bar{g}_i(x_i(t)+u_i^*)-\bar{g}_i(u_i^*)$, $h_i(x_i(t)) = \bar{h}_i(x_i(t) + u_i^*) - \bar{h}_i(u_i^*)$, $a_i(x_i(t)) = \bar{a}_i(x_i(t) + u_i^*)$, $b_i(x_i(t)) = \bar{b}_i(x_i(t) + u_i^*) - \bar{b}_i(u_i^*)$. Then the neural networks (1) can be rewritten as

$$\dot{x}_i(t) = -a_i(x_i(t))[b_i(x_i(t)) - \sum_{j=1}^{n}c_{ij}f_j(x_j(t)) - \sum_{j=1}^{n}d_{ij}g_j(x_j(t - \tau_{ij}(t)))$$

$$- \sum_{j=1}^{n}e_{ij}\int_{-\infty}^{t}K_{ij}(t-s)g_j(x_j(s))ds], \quad i = 1, \cdots, n. \tag{7}$$

Let $U_i(\lambda) = \alpha_i^{-1}p_i\lambda - \beta_ip_i + \sum_{j=1}^{n}|c_{ij}|k_jp_j + \sum_{j=1}^{n}|d_{ij}|l_jp_j\exp\{\lambda\tau_{ij}(t)\} + \sum_{j=1}^{n}|e_{ij}|\kappa_{ij}\eta_jp_j$, where α_i is determined by $(\mathbf{H_2})$. Then, (5) implies $U_i(0) < 0$, while $U_i(\beta_i\alpha_i) \geq 0, dU_i(\lambda)/d\lambda > 0$, thus there exists $\lambda_i \in (0, \beta_i\alpha_i]$ such that $U_i(\lambda_i) = 0$, and when $\lambda \in (0, \lambda_i), U_i(\lambda) < 0$. Taking $\lambda_{\min} = \min_i \lambda_i, U_i(\lambda_{\min}) \leq 0$ for all $i \in \{1, 2, \ldots, n\}$. Setting $0 < \theta < \min\{\lambda_{\min}, \rho\}$, where ρ see $(\mathbf{H_1})$. Thus

$$U_i(\theta) = \alpha_i^{-1}p_i\theta - \beta_ip_i + \sum_{j=1}^{n}|c_{ij}|k_jp_j + \sum_{j=1}^{n}|d_{ij}|l_jp_j\exp\{\theta\tau_{ij}(t)\}$$

$$+ \sum_{j=1}^{n}|e_{ij}|\kappa_{ij}\eta_jp_j < 0, \quad i = 1, \cdots, n. \tag{8}$$

Let

$$V_i(t) = p_i^{-1}|x_i(t)| - \bar{x}(0)\exp(-\theta t), \tag{9}$$

where $\bar{x}(0) = \sum_{i=1}^{n}\sup_{-\infty<s\leq0}|p_i^{-1}x_i(s)|$. Without loss of generality, we suppose that $\bar{x}(0) > 0$, then for all $i \in \{1, 2, \ldots, n\}, t \geq 0, V_i(t) \leq 0$. Otherwise, on the one hand, for all $i \in \{1, 2, \ldots, n\}, \forall \ s \in (-\infty, 0], V_i(s) \leq 0$, then there exists $i \in \{1, 2, \ldots, n\}, t_1 > 0$, such that

$$V_i(t_1) = 0, \tag{10}$$

$$D^+V_i(t_1) \geq 0, \tag{11}$$

and for $j = 1, 2, \ldots, n, t \in (-\infty, t_1]$

$$V_j(t) \leq 0 \tag{12}$$

On the other hand, by(7),(9),(10), (12), and (8), one have

$$D^+V_i(t_1) \le p_i^{-1}a_i(x_i(t_1))[\sum_{j=1}^{n}|c_{ij}|k_j|x_j(t_1)| + \sum_{j=1}^{n}|d_{ij}|l_j|x_j(t_1 - \tau_{ij}(t_1))|$$

$$+ \sum_{j=1}^{n}|e_{ij}|\eta_j \int_{-\infty}^{t_1} K_{ij}(t_1 - s)|x_j(s)|ds - \beta_i|x_i(t_1)|] + \bar{x}(0)\theta\exp(-\theta t_1)$$

$$\le \{p_i^{-1}a_i(x_i(t_1))[\sum_{j=1}^{n}|c_{ij}|k_j p_j + \sum_{j=1}^{n}|d_{ij}|l_j p_j \exp(\theta\tau_{ij}(t_1))$$

$$+ \sum_{j=1}^{n}|e_{ij}|\kappa_{ij}\eta_j p_j - \beta_i p_i] + \theta\}\bar{x}(0)\exp(-\theta t_1)$$

$$\le p_i^{-1}a_i(x_i(t_1))U_i(\theta)\bar{x}(0)\exp(-\theta t_1) < 0.$$

This contradicts (11), thus $|x_i(t)| \le p_i\bar{x}(0)\exp(-\theta t)$, for all $i \in \{1, 2, \ldots, n\}, t \ge -\infty$, i.e., (1) is globally exponentially stable.

Remark 1. Theorem 3.1 generalizes and improves the corresponding result in [5,6]. Denote $c_{ii}^+ = \max\{0, c_{ii}\}$. When $\bar{f}_i$ is a monotone increasing function, the results in Theorem 2.1 and Theorem 3.1 still hold if c_{ii} is substituted by c_{ii}^+. In addition, Refs.[2,3] deduced stability results on networks (1)without distributed delay under the condition that the activation functions are bounded, or delays $\tau_{ij}'(t) \le 1$. But Theorem 3.1 cancels these restrictions.

4   Simulation Result

Example 4.1. Consider a two dimensional Cohen-Grossberg neural networks (1) whose parameters are chosen as $\tau = 5+0.1\sin(t), \rho = 0.5, \bar{a}_1(u_1(t)) = 1+\frac{1}{1+u_1^2(t)}$, $\bar{a}_2(u_2(t)) = 2 + \frac{1}{1+u_2^2(t)}, \bar{b}_1(u_1(t)) = 2u_1(t),\ \bar{b}_2(u_2(t)) = 1.5u_2(t), \bar{f}_i(u_i(t)) = \bar{g}_i(u_i(t)) = \sin(u_i(t)), \bar{h}_i(u_i(t)) = \tanh(u_i(t)), i = 1, \cdots, n,\ \rho = \frac{1}{2}, K_{11}(t) = K_{12}(t) = K_{22}(t) = \exp(-t), K_{21}(t) = 2\exp(-2t),\ C = \begin{pmatrix} 0 & -0.5 \\ 0.2 & 0 \end{pmatrix}, D = \begin{pmatrix} -0.4 & 0 \\ 0 & -0.3 \end{pmatrix}, E = \begin{pmatrix} -0.2 & 0.3 \\ 0.2 & 0.3 \end{pmatrix}$. Obviously, $\bar{A} = \begin{pmatrix} 1.2 & -0.9 \\ -0.6 & 0.6 \end{pmatrix}$. It is easy to verify$\bar{A}$ is a non-singular M-matrix. Therefore, by Theorem 3.1, the above CGNN exponentially converge to the unique equilibrium point $x^* = (-2.9045, 1.0967)$. Fig.1 demonstrates the networks ultimately exponentially converge to the unique equilibrium point with different initial values.

5   Conclusion

In this paper, we present theoretical results on the global exponential stability of a class of Cohen-Grossberg neural networks with time-varying delays and continuously distributed delays. These stability conditions are mild and some of they are easy to verify by using the connection weights of the neural networks.

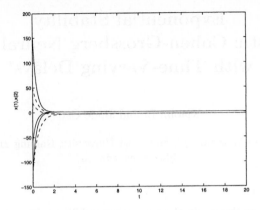

Fig. 1. The states x, with initial value $x_0 = (-120, 160); (2., -60); (50, -120)$, respectively, ultimately converge to the equilibrium $x^* = (-2.9045, 1.0967)$. x(1)-dashed line, x(2)-solid line.

In additions, the rate of exponential convergence can be estimated by means of simple computation based on the stability results herein. Our results imply and generalize some existed ones in previous literature. Furthermore, it is convenient to estimate the exponential convergence rates of the neural networks by using the criteria. Compared with the previous methods, our method does not resort to any Lyapunov functions or functionals.

Acknowledgments

The work was supported by Natural Science Foundation of Hubei (2004ABA055) and National Natural Science Foundation of China (60274007, 60074008).

References

1. Cohen, M., Grossberg, S.: Absolute Stability and Global Pattern Formation and Parallel Memory Storage by Competitive Neural Networks. IEEE Trans Syst Man Cybernet, **13** (1983) 815-821
2. Chen, T., Rong, L.: Robust Global Exponential Stability of Cohen-Grossberg Neural Networks. IEEE Trans on Neural Networks, **15** (2004) 203-205.
3. Cao, J., Liang, J.: Boundedness and Stability for Cohen-Grossberg Neural Networks with Time-varying Delays. J. Math. Anal. Appl. **296** (2004) 665-685
4. Gopalsamy, K.: Stability in Asymmetric Hopfield Nets with Transmission Delays. Physica D, **76** (1994) 344-358
5. Liao, X., Wang, J.: Algebraic Criteria for Global Exponential Stability of Cellular Neural Networks with Multiple Time Delays. IEEE Trans. Circuits Syst., **50** (2003) 268-274
6. Zhang, Y., Pheng, A.H., and Kwong, S.L.: Convergence Analysis of Cellular Neural Networks with Unbounded Delay. IEEE Trans. Circuits Syst., **48** (2001) 680-687

Exponential Stability
of Stochastic Cohen-Grossberg Neural Networks
with Time-Varying Delays*

Xiaolin Li and Jinde Cao

Department of Mathematics, Southeast University, Nanjing 210096, China
jdcao@seu.edu.cn

Abstract. In this paper, the exponential stability is discussed for a class of stochastic Cohen-Grossberg neural networks with time-varying delays. A set of novel sufficient conditions on exponential stability are given for the considered system by using the well-known Razumikhin-type theorem. A numerical example is also given to illustrate the effectiveness of our results.

1 Introduction

The Cohen-Grossberg neural network is introduced by Cohen and Grossberg in 1983 [1]. This model has attracted increasing interest due to its potential applications in classification, parallel computation, associative memory, especially in solving some optimization problems. Some results on Cohen-Grossberg neural networks have been obtained(see [2-5]).

However, a real system is usually affected by external perturbations which in many cases are of great uncertainty and hence may be treated as random, as pointed out by Haykin [10] that in real nervous systems, synaptic transmission is a noisy process brought on by random fluctuations form the release of neurotransmitters and other probabilistic causes. Therefore, it is of significant importance to consider stochastic effects to the stability of neural networks. On the other hand, time delays inevitably exist in biological and artificial neural networks due to the finite switching speed of neurons and amplifiers. They are often the source of oscillation and instability in neural networks. For this reason, it is important to study the stability of the stochastic neural networks with delays. Although various stability of neural networks has been extensively investigated by many authors in the past two decades, the problem of stochastic effects on the stability has been not studied until recent years(see [6-9],[13-14]). In [6,7], the authors studied the mean square exponential stability and instability of cellular neural networks. In [8], the almost sure exponential stability for

---

* This work was jointly supported by the National Natural Science Foundation of China under Grant 60373067, the Natural Science Foundation of Jiangsu Province, China under Grants BK2003053, Qing-Lan Engineering Project of Jiangsu Province, China.

J. Wang, X. Liao, and Z. Yi (Eds.): ISNN 2005, LNCS 3496, pp. 162–167, 2005.

a class of stochastic cellular neural networks with discrete delays was discussed. In [14], stochastic Cohen-Grossberg neural networks without time delays was investigated. However, to the best of our knowledge, few authors have considered stochastic Cohen-Grossberg neural networks with time-varying delays. In this paper, we shall study the exponential stability of stochastic Cohen-Grossberg neural network with time-varying delays

$$dx(t) = -\alpha(x(t))[\beta(x(t)) - Ag(x(t)) - Bg(x(t - \tau(t)))]dt$$
$$+\sigma(t, x(t), x(t - \tau(t)))d\omega(t) \qquad (1)$$

where $x(t) = (x_1(t), \cdots, x_n(t))^T$ is the neuron states vector, $\alpha(x(t)) = diag(a_1(x_1(t)), \cdots, a_n(x_n(t)))$, $\beta(x(t)) = (\beta_1(x_1(t)), \cdots, \beta_n(x_n(t)))^T$, $A = (a_{ij})_{n \times n}$ and $B = (b_{ij})_{n \times n}$ are the connection weight matrix and delayed connection weight matrix, respectively, $g(x(t)) = (g_1(x_1(t)), \cdots, g_n(x(t)))^T$ is the activation functions vector, $g(x(t - \tau(t))) = (g_1(x_1(t - \tau(t))), \cdots, g_n(x_n(t - \tau(t)))^T$, $0 < \tau(t) \le \tau$, $\omega(t) = (\omega_1(t), \cdots, \omega_n(t))^T$ is an n-dimensional Brown motion defined on a complete probability space $(\Omega, \mathcal{F}, P)$ with a natural filtration $\{\mathcal{F}\}_{t \ge 0}$ (i.e. $\mathcal{F}_t = \sigma\{\omega(s) : 0 \le s \le t\}$), and $\sigma : R_+ \times R^n \times R^n \to R^{n \times n}$, i.e. $\sigma = (\sigma_{ij})_{n \times n}$ is the diffusion coefficient matrix.

The initial conditions for system (1) are $x(s) = \xi(s)$, $-\tau \le s \le 0$, $\xi \in L^2_{\mathcal{F}_0}([-\tau, 0], R^n)$, here $L^2_{\mathcal{F}_0}([-\tau, 0], R^n)$ is regarded as a R^n-valued stochastic process $\xi(s)$, $-\tau \le s \le 0$, moreover, $\xi(s)$ is $\mathcal{F}_0$ measurable, $\int_{-\tau}^0 E|\xi(s)|^2 ds < \infty$.

Throughout this paper we always assume that $\alpha(x(t))$, $g(x(t))$ and $\sigma(t, x(t), x(t - \tau(t)))$ satisfy the local Lipschitz condition and the linear growth condition. It is known ([10] or [11]) that given any initial value $\xi \in R^n$, system (1) has a unique global solution on $t \ge 0$ and we denote the solution by $x(t; \xi)$. We will also assume that $\beta(0) = 0, g(0) = 0, \sigma(t, 0, 0) = 0$ for the stability of this paper. So system (1) admits a zero solution or trivial solution $x(t, 0) = 0$.

Let $C^{2,1}([-\tau, \infty) \times R_n; R_+)$ denote the family of all nonnegative functions $V(t, x)$ on $[-\tau, \infty) \times R_n$ which are continuous twice differentiable in x and once differentiable in t. If $V \in C^{2,1}([-\tau, \infty) \times R_n; R_+)$, define an operator $\mathcal{L}V$ associated with (1) as

$$\mathcal{L}V(t, x) = V_t(t, x) + V_x(t, x)(-\alpha(x(t))[\beta(x(t)) - Ag(x(t)) - Bg(x(t - \tau(t)))])$$
$$+ \tfrac{1}{2}trace[\sigma^T V_{xx}(t, x)\sigma],$$

where

$$V_t(t, x) = \frac{\partial V(t,x)}{\partial t}, V_x(t, x) = \left(\frac{\partial V(t,x)}{\partial x_1}, \cdots, \frac{\partial V(t,x)}{\partial x_n}\right), V_{xx}(t, x) = \left(\frac{\partial^2 V(t,x)}{\partial x_i x_j}\right)_{n \times n}.$$

We assume the following conditions are satisfied:

(H_1) For each $i \in \{1, 2, \cdots, n\}$, $a_i(x)$ is bounded, positive and locally Lipschitz continuous, furthermore $0 < \underline{\alpha}_i \le a_i(x) \le \bar{\alpha}_i$.

(H_2) For each $i \in \{1, 2, \cdots, n\}$, $x_i\beta_i(x_i(t)) \ge \gamma_i x_i^2(t)$.

(H_3) For each $i \in \{1, 2, \cdots, n\}$, there exist constant $G_i > 0$ such that
$$|g_i(x) - g_i(y)| \le G_i|x - y|, \quad \forall x, y \in R.$$

(H_4) There are nonnegative constants v_i, μ_i such that
$$trace[\sigma^T(t, x, y)\sigma(t, x, y)] \le \sum_{i=1}^n (v_i x_i^2 + \mu_i y_i^2), \forall (t, x, y) \in R_+ \times R^n \times R^n.$$

2 Main Results

In this section, we will apply the Razumikhin-type theorem [9] to deal with the exponential stability of stochastic Cohen-Grossberg neural networks with time-varying delays.

Theorem 1. *Under assumptions* $(H_1) - (H_4)$, *if there exist constant* $r \in [0,1]$ *and positive diagonal matrix* $Q = diag(q_1, \cdots, q_n)$ *such that*

$$\lambda_1 > \lambda_2, \tag{2}$$

where

$$\lambda_1 = \min_{1 \le i \le n} \left[2\underline{\alpha}_i \gamma_i - \sum_{j=1}^n \bar{\alpha}_i |a_{ij}| G_j^{2r} - \frac{1}{q_i} \sum_{j=1}^n \bar{\alpha}_j q_j |a_{ji}| G_i^{2(1-r)} - \sum_{j=1}^n \bar{\alpha}_i |b_{ij}| G_j^{2r} - \frac{v_i}{q_i} \max_{1 \le i \le n} q_i \right],$$

$$\lambda_2 = \max_{1 \le i \le n} \left[\frac{\mu_i}{q_i} \max_{1 \le i \le n} q_i + \frac{1}{q_i} \sum_{j=1}^n |b_{ji}| G_i^{2(1-r)} \bar{\alpha}_j q_j \right],$$

then for all $\xi \in L_{\mathcal{F}_0}^2([-\tau, 0]; R^n)$, *the trivial solution of system (1) is exponentially stable in mean square and also almost surely exponentially stable.*

Proof: Consider the following Lyapunov function

$$V(t, x) = x^T(t) Q x(t).$$

Then the operator $\mathcal{L}V$ associated with system (1) has the form

$$\mathcal{L}V(t, x) = 2x^T Q \left\{ -\alpha(x) \Big[\beta(x) - Ag(x) - Bg(x(t - \tau(t))) \Big] \right\} + trace(\sigma^T Q \sigma)$$

$$= -2 \sum_{i=1}^n \alpha_i(x_i(t)) q_i x_i(t) \beta_i(x_i(t))$$

$$+ 2 \sum_{i=1}^n \alpha_i(x_i(t)) q_i x_i(t) \sum_{j=1}^n a_{ij} g_j(x_j(t))$$

$$+ 2 \sum_{i=1}^n \alpha_i(x_i(t)) q_i x_i(t) \sum_{j=1}^n b_{ij} g_j(x_j(t - \tau(t))) + trace(\sigma^T Q \sigma)$$

$$\le -2 \sum_{i=1}^n \underline{\alpha}_i q_i \gamma_i x_i^2(t) + 2 \sum_{i=1}^n \bar{\alpha}_i q_i |x_i(t)| \sum_{j=1}^n |a_{ij}| G_j |x_j(t)|$$

$$+ 2 \sum_{i=1}^n \bar{\alpha}_i q_i |x_i(t)| \sum_{j=1}^n |b_{ij}| G_j |x_j(t - \tau(t))|$$

$$+ \max_{1 \le i \le n} q_i \sum_{i=1}^n \left[v_i x_i^2(t) + \mu_i x_i^2(t - \tau(t)) \right]$$

$$\le -2 \sum_{i=1}^n \underline{\alpha}_i q_i \gamma_i x_i^2(t) + \sum_{i=1}^n \bar{\alpha}_i q_i \sum_{j=1}^n |a_{ij}| \left[G_j^{2r} x_i^2(t) + G_j^{2(1-r)} x_j^2(t) \right]$$

$$+ \sum_{i=1}^{n} \bar{\alpha}_i q_i \sum_{j=1}^{n} |b_{ij}| \left[G_j^{2r} x_i^2(t) + G_j^{2(1-r)} x_j^2(t - \tau(t)) \right]$$

$$+ \max_{1 \leq i \leq n} q_i \sum_{i=1}^{n} \left[v_i x_i^2(t) + \mu_i x_i^2(t - \tau(t)) \right]$$

$$\leq - \sum_{i=1}^{n} q_i \left[2\underline{\alpha}_i \gamma_i - \sum_{j=1}^{n} \bar{\alpha}_i |a_{ij}| G_j^{2r} - \frac{1}{q_i} \sum_{j=1}^{n} \bar{\alpha}_j q_j |a_{ji}| G_i^{2(1-r)} \right.$$

$$\left. - \sum_{j=1}^{n} \bar{\alpha}_i |b_{ij}| G_j^{2r} - \frac{v_i}{q_i} \max_{1 \leq i \leq n} q_i \right] x_i^2(t) + \sum_{i=1}^{n} q_i \left[\frac{\mu_i}{q_i} \max_{1 \leq i \leq n} q_i \right.$$

$$\left. + \frac{1}{q_i} \sum_{j=1}^{n} |b_{ji}| G_i^{2(1-r)} \bar{\alpha}_j q_j \right] x_i^2(t - \tau(t))$$

$$\leq -\lambda_1 \sum_{i=1}^{n} q_i x_i^2(t) + \lambda_2 \sum_{i=1}^{n} q_i x_i^2(t - \tau(t)),$$

by (2), there exists $q > 1$ such that

$$-\lambda^* = -\lambda_1 + q\lambda_2 < 0,$$

therefore, for any $t \geq 0$ and $x_t = \{x(t + \theta) : -\tau \leq \theta \leq 0\} \in L^2_{\mathcal{F}_t}([-\tau, 0]; R^n)$ satisfying

$$EV(t + \theta, x(t + \theta)) < qEV(t, x(t)), \quad -\tau \leq \theta \leq 0,$$

we have

$$E\mathcal{L}V(t, x) \leq (-\lambda_1 + q\lambda_2)EV(t, x) = -\lambda^* EV(t, x),$$

by the Razuminkhin-type theorem in [9], for all $\xi \in L^2_{\mathcal{F}_0}([-\tau, 0]; R^n)$, the trivial solution of system (1) is exponentially stable in mean square and also almost surely exponentially stable.

In Theorem 1, if we take $r = \frac{1}{2}$, $Q = E$(identity matrix), we can easily obtain the following corollary.

Corollary 1. *Under assumptions* $(H_1) - (H_4)$, *if*

$$\lambda_1 > \lambda_2, \tag{3}$$

where
$$\lambda_1 = \min_{1 \leq i \leq n} \left[2\underline{\alpha}_i \gamma_i - \sum_{j=1}^{n} \bar{\alpha}_i |a_{ij}| G_j - \sum_{j=1}^{n} \bar{\alpha}_j |a_{ji}| G_i - \sum_{j=1}^{n} \bar{\alpha}_i |b_{ij}| G_j - v_i \right],$$

$$\lambda_2 = \max_{1 \leq i \leq n} \left[\mu_i + \sum_{j=1}^{n} |b_{ji}| G_i \bar{\alpha}_j \right],$$

then for all $\xi \in L^2_{\mathcal{F}_0}([-\tau, 0]; R^n)$, *the trivial solution of system (1) is exponentially stable in mean square and also almost surely exponentially stable.*

Remark 1. For system (1), when $a_i(x_i(t)) \equiv 1$, $\beta_i(x_i(t)) = c_i x_i(t)$, $a_{ij} = 0$, then it turns out to be following stochastic delayed Hopfield neural networks

$$dx(t) = [-Cx(t) + Bg(x(t - \tau(t)))]dt + \sigma(t, x(t), x(t - \tau(t)))d\omega(t) \qquad (4)$$

For system (4), by Theorem 1, we have the following results.

Corollary 2. *Under assumptions (H_3) and (H_4), if there exist constant $r \in [0, 1]$ and positive diagonal matrix $Q = diag(q_1, \cdots, q_n)$ such that*

$$\min_{1 \le i \le n} \left[2c_i - \sum_{j=1}^{n} |b_{ij}| G_j^{2r} - \frac{v_i}{q_i} \max_{1 \le i \le n} q_i \right] \qquad (5)$$

$$> \max_{1 \le i \le n} \left[\frac{\mu_i}{q_i} \max_{1 \le i \le n} q_i + \frac{1}{q_i} \sum_{j=1}^{n} |b_{ji}| G_i^{2(1-r)} q_j \right],$$

then for all $\xi \in L^2_{\mathcal{F}_0}([-\tau, 0]; R^n)$, the trivial solution of system (4) is exponentially stable in mean square and also almost surely exponentially stable.

Remark 2. For system (1), when $\sigma(t, x(t), x(t - \tau(t))) = 0$, then it turns to be the following Cohen-Grossberg neural networks with time-varying delays

$$\frac{dx(t)}{dt} = -\alpha(x(t))[\beta(x(t)) - Ag(x(t)) - Bg(x(t - \tau(t)))] \qquad (6)$$

For system (6), by Theorem 1, we have the following results.

Corollary 3. *Under assumptions $(H_1) - (H_3)$, if there exist constant $r \in [0, 1]$ and positive diagonal matrix $Q = diag(q_1, \cdots, q_n)$ such that*

$$\lambda_1 > \lambda_2, \qquad (7)$$

where

$$\lambda_1 = \min_{1 \le i \le n} \left[2\underline{\alpha}_i \gamma_i - \sum_{j=1}^{n} \bar{\alpha}_i |a_{ij}| G_j^{2r} - \frac{1}{q_i} \sum_{j=1}^{n} \bar{\alpha}_j q_j |a_{ji}| G_i^{2(1-r)} - \sum_{j=1}^{n} \bar{\alpha}_i |b_{ij}| G_j^{2r} \right],$$

$$\lambda_2 = \max_{1 \le i \le n} \left[\frac{1}{q_i} \sum_{j=1}^{n} |b_{ji}| G_i^{2(1-r)} \bar{\alpha}_j q_j \right],$$

then the trivial solution of system (6) is globally exponentially stable.

3   An Illustrative Example

Consider the following stochastic Cohen-Grossberg neural networks with time-varying delays

$$dx(t) = - \begin{pmatrix} 2 + \sin(x_1(t)) & 0 \\ 0 & 2 + \cos(x_2(t)) \end{pmatrix} \left[\begin{pmatrix} 10 & 0 \\ 0 & 10 \end{pmatrix} \begin{pmatrix} x_1(t) \\ x_2(t) \end{pmatrix} \right.$$

$$\left. - \begin{pmatrix} 1 & -0.2 \\ 1.5 & 1 \end{pmatrix} \begin{pmatrix} \tanh(x_1(t)) \\ \tanh(x_2(t)) \end{pmatrix} - \begin{pmatrix} 0.5 & 0 \\ -0.5 & 1 \end{pmatrix} \begin{pmatrix} \tanh(x_1(t - \tau(t))) \\ \tanh(x_2(t - \tau(t))) \end{pmatrix} \right]$$

$$+ \sigma(t, x(t), x(t - \tau(t)))d\omega(t).$$

Moreover, $\sigma : R_+ \times R^2 \times R^2 \to R^{2 \times 2}$ satisfies

$$trace[\sigma^T(t, x, y)\sigma(t, x, y)] \leq 2x_1^2 + 0.4x_2^2 + 0.5y_1^2 + 0.8y_2^2.$$

Obviously, we have
$$\underline{\alpha_i} = 1, \quad \bar{\alpha}_i = 3, \quad \gamma_i = 10, \quad G_i = 1, \quad i = 1, 2.$$
By simple computation, we can easily get that

$$\min_{1 \leq i \leq 2}[2\underline{\alpha_i}\gamma_i - \sum_{j=1}^{2} \bar{\alpha}_i|a_{ij}|G_j - \sum_{j=1}^{2} \bar{\alpha}_j|a_{ji}|G_i - \sum_{j=1}^{2} \bar{\alpha}_i|b_{ij}|G_j - v_i] = 4$$

$$\max_{1 \leq i \leq 2}[\mu_i + \sum_{j=1}^{2} |b_{ji}|G_i\bar{\alpha}_j] = 3.8$$

Thus, it follows Corollary 1 that system (1) is exponentially stable in mean square and also almost surely exponentially stable.

References

1. Cohen, M.A., Grossberg, S.: Absolute Stability and Global Pattern Formation and Parallel Memory Storage by Competitive Neural Networks. IEEE Trans. Systems, Man and Cybernetics, **13** (1983) 815-821
2. Cao, J., Liang, J.: Boundedness and Stability for Cohen-Grossberg Neural Network with Time-varying Delays. J. Math. Anal. Appl., **296** (2004) 665-685
3. Wang, L., Zou, X.: Harmless Delays in Cohen-Grossberg Neural Networks. Physica D, **170** (2002) 162-173
4. Wang, L., Zou, X.: Exponential Stability of Cohen-Grossberg Neural Networks. Neural Networks, **15** (2002) 415-422
5. Chen, T., Rong, L.: Delay-independent Stability Analysis of Cohen-Grossberg Neural Networks. Phys. Lett. A, **317** (2003) 436-449
6. Liao, X., Mao, X.: Exponential Stability and Instability of Stochastic Neural Networks. Stochast. Anal. Appl, **14** (1996a) 165-185.
7. Liao, X., Mao, X.: Stability of Stochastic Neural Networks. Neural. Parallel Sci. Comput, **14** (1996b) 205-224
8. Blythe, S., Mao, X., Liao, X.: Stability of Stochastic Delay Neural Networks. Journal of the Franklin Institute, **338** (2001) 481-495
9. Mao, X.: Razumikhin-type Theorems on Exponential Stability of Stochastic Functional Differential Equations. Stochastic Process. Appl, **65** (1996) 233-250
10. Haykin, S.: Neural Networks, Prentice-Hall, NJ (1994)
11. Mao, X.: Exponential Stability of Stochastic Differential Equations. New York: Marcel Dekker (1994)
12. Mohammed, S-E.A.: Stochastic Functional Differential Equations. Longman Scientific and Technical (1986)
13. Mao, X., Shah, A.: Exponential Stability of Stochastic Differential Delay Equations. Stochastics and Stochastics Reports, **60** (1997) 135-153
14. Wang, L.: Stability of Stochastic Cohen-Grossberg Neural Networks. ISNN 2004, LNCS 3173, (2004) 84-89

Exponential Stability of Fuzzy Cellular Neural Networks with Unbounded Delay

Tingwen Huang[1] and Linhua Zhang[2]

[1] Texas A&M University at Qatar, Doha, P. O. Box 5825, Qatar
tingwen.huang@qatar.tamu.edu
[2] Department of Computer and Engineering, Chongqing University
Chongqing 400044, China

Abstract. In this paper, we investigate the exponential stability of fuzzy cellular networks with unbounded delay. An easily verified sufficient condition is obtained. Moreover, we get the exponential convergent rate for a broad class of the unbounded delayed fuzzy cellular networks.

1 Introduction

There are lots of results on the analysis of the stability of cellular neural networks since the stability and convergence are prerequisites for the designing neural networks. At the same time, there is another type of fundamental neural networks, fuzzy cellular neural networks (FCNN), introduced by T. Yang and L.B. Yang [11], combining the fuzzy logic with the traditional CNN. FCNN can be applied to the image processing and pattern recognition. Like the traditional CNN, the stability is very important for the design of neural networks. T. Yang et al. in [11], [12],[13] have studied the existence and uniqueness of equilibrium point and the exponential stability of equilibrium point of FCNN without delay. Y. Liu et al. in [7] have obtained the exponential stability of FCNN with constant delay and time varying delay. In this paper, we would like to study FCNN with distributed delay.

$$\frac{dx_i}{dt} = -d_i x_i(t) + \sum_{j=1}^{n} b_{ij}\mu_j + I_i$$

$$+ \bigwedge_{j=1}^{n} \alpha_{ij} \int_{-\infty}^{t} k_{ij}(t-s)f_j(x_j(s))ds + \bigwedge_{j=1}^{n} T_{ij}\mu_j$$

$$+ \bigvee_{j=1}^{n} \beta_{ij} \int_{-\infty}^{t} k_{ij}(t-s)f_j(x_j(s))ds + \bigvee_{j=1}^{n} H_{ij}\mu_j \qquad (1)$$

where $\alpha_{ij}, \beta_{ij}, T_{ij}$ and H_{ij} are elements of fuzzy feedback MIN template, fuzzy feedback MAX template, fuzzy feed forward MIN template and fuzzy feed forward MAX template respectively; b_{ij} are elements of feed forward template; $\bigwedge$ and $\bigvee$ denote the fuzzy AND and fuzzy OR operation respectively; x_i, μ_i and I_i

J. Wang, X. Liao, and Z. Yi (Eds.): ISNN 2005, LNCS 3496, pp. 168–173, 2005.

denote state, input and bias of the ith neurons respectively; f_i is the activation function; $k_{ij}(s) \geq 0$ is the feedback kernel, defined on the interval $[0, \infty)$, Kernels satisfy

$$\int_0^\infty k_{ij}(s)ds = 1, \qquad \int_0^\infty e^{\tau s} k_{ij}(s)ds = k_{ij}, \qquad i = 1, \cdots, n. \qquad (2)$$

where τ is a positive constant. Without loss of generality, we assume that $\tau \leq \min_{1 \leq i \leq n} d_i$, where $d_i, i = 1, \cdots, n$, are the constants in system (1); it is obvious that if $\int_0^\infty e^{\tau s} k_{ij}(s)ds$ is finite for $\tau \geq \min_{1 \leq i \leq n} d_i$, $\int_0^\infty e^{\tau s} k_{ij}(s)ds$ is finite when we limit $\tau < \min_{1 \leq i \leq n} d_i$.

The initial conditions of (1) are of the form $x_i(t) = \varphi_i(t), -\infty \leq t \leq 0, i = 1, \cdots, n$, where φ_i bounded and continuous on $[-\infty, 0]$.

In this paper, we assume that

H: f_i is a bounded function defined on R and satisfies

$$|f_i(x) - f_i(y)| \leq l_i |x - y|, \qquad i = 1, \cdots, n. \qquad (3)$$

for any $x, y \in R$.

Definition 1. *The equilibrium point x^* of (1) is said to be globally exponential stable if there exist constants $\lambda > 0$ and $M > 0$ such that*

$$|u_i(t) - x_i^*| \leq M \max_{1 \leq i \leq n} \| \varphi_i - x_i^* \| e^{-\lambda t} \qquad (4)$$

for all $t \geq 0$, where $\| \varphi_i - x_i^* \| = \sup_{s \in (-\infty, 0]} |\varphi_i(s) - x_i^*|, i = 1, \cdots, n$.

For the convenience, we give the matrix notations here. For $A, B \in R^{n \times n}$, $A \leq B(A > B)$ means that each pair of the corresponding elements of A and B satisfies the inequality $' \leq' ('>')$.

2   Main Results

In order to get the main result regarding the exponential stability of FCNN with distributed delay, we would like to cite two lemmas first.

Lemma 1. *([6]). If $M \geq 0$ and $\rho(M) < 1$, then $(I - M)^{-1} \geq 0$, where I denotes the identity matrix and $\rho(M)$ denotes the spectral radius of a square matrix M.*

Lemma 2. *([13]). For any $a_{ij} \in R$, $x_j, y_j \in R$, $i, j = 1, \cdots, n$, we have the following estimations,*

$$\left| \bigwedge_{j=1}^n a_{ij} x_j - \bigwedge_{j=1}^n a_{ij} y_j \right| \leq \sum_{1 \leq j \leq n} (|a_{ij}| \cdot |x_j - y_j|) \qquad (5)$$

and

$$\left| \bigvee_{j=1}^n a_{ij} x_j - \bigvee_{j=1}^n a_{ij} y_j \right| \leq \sum_{1 \leq j \leq n} (|a_{ij}| \cdot |x_j - y_j|) \qquad (6)$$

Now, we are ready to state and prove the main result regarding to the exponential stability of system (1).

Theorem 1. *If the spectral radius of the matrix $D^{-1}AL$ is less than 1, i.e., $\rho(D^{-1}AL) < 1$, where $D = diag(d_1, \cdots, d_n)$, $A = (a_{ij})$ is an $n \times n$ matrix with $a_{ij} = k_{ij}(|\alpha_{ij}| + |\beta_{ij}|)$, $L = diag(l_1, \cdots, l_n)$, then there is a unique equilibrium point of system (1), and the equilibrium point of the system is globally exponential stable. Moreover, the convergent rate of exponential stability of system (1) is $\lambda - \varepsilon$, where $\lambda = \min\{\eta|\tau \geq \eta > 0, \rho((D - \eta I)^{-1}AL) = 1\}$, ε is a positive number which is smaller than λ.*

Proof. The proof of the existence of the equilibrium point of system (1) can be done similarly as in [12] by using Brown fix point theorem, so it is omitted here. The uniqueness of the equilibrium point follows from the globally exponential stability of the equilibrium point.

In the following, we derive the exponential stability of the equilibrium point of system (1).

Let $x^* = (x_1^*, \cdots, x_n^*)^T$ be the equilibrium point. $y(t) = x(t) - x^* = (x_1(t) - x_1^*, \cdots, x_n(t) - x_n^*)^T$. Thus we have

$$
\begin{aligned}
\frac{dy_i(t)}{dt} = &-d_i y_i(t) + \bigwedge_{j=1}^{n} \alpha_{ij} \int_{-\infty}^{t} k_{ij}(t-s)f_j(y_j(s)+x_j^*)ds \\
&- \bigwedge_{j=1}^{n} \alpha_{ij} \int_{-\infty}^{t} k_{ij}(t-s)f_j(x_j^*)ds \\
&+ \bigvee_{j=1}^{n} \beta_{ij} \int_{-\infty}^{t} k_{ij}(t-s)f_j(y_j(s)+x_j^*)ds \\
&- \bigvee_{j=1}^{n} \beta_{ij} \int_{-\infty}^{t} k_{ij}(t-s)f_j(x_j^*)ds
\end{aligned}
\tag{7}
$$

By the results of *Lemma 2* and the assumption H, we have the following estimations

$$
\begin{aligned}
&\left| \bigwedge_{j=1}^{n} \alpha_{ij} \int_{-\infty}^{t} k_{ij}(t-s)f_j(y_j(s)+x_j^*)ds - \bigwedge_{j=1}^{n} \alpha_{ij} \int_{-\infty}^{t} k_{ij}(t-s)f_j(x_j^*)ds \right| \\
&\leq \sum_{j=1}^{n} |\alpha_{ij}| \cdot \left| \int_{-\infty}^{t} k_{ij}(t-s)f_j(y_j(s)+x_j^*)ds - \int_{-\infty}^{t} k_{ij}(t-s)f_j(x_j^*)ds \right| \\
&\leq \sum_{j=1}^{n} |\alpha_{ij}| l_j \int_{-\infty}^{t} k_{ij}(t-s)|y_j(s)|ds
\end{aligned}
\tag{8}
$$

and

$$
\left| \bigvee_{j=1}^{n} \beta_{ij} \int_{-\infty}^{t} k_{ij}(t-s)f_j(y_j(s)+x_j^*)ds - \bigvee_{j=1}^{n} \beta_{ij} \int_{-\infty}^{t} k_{ij}(t-s)f_j(x_j^*)ds \right|
$$

$$\leq \sum_{j=1}^{n} |\beta_{ij}| \cdot |\int_{-\infty}^{t} k_{ij}(t-s)f_j(y_j(s)+x_j^*)ds - \int_{-\infty}^{t} k_{ij}(t-s)f_j(x_j^*)ds|$$

$$\leq \sum_{j=1}^{n} |\beta_{ij}|l_j \int_{-\infty}^{t} k_{ij}(t-s)|y_j(s)|ds \qquad (9)$$

From (7)-(9), we get

$$\frac{d|y_i|}{dt} \leq -d_i|y_i(t)| + \sum_{j=1}^{n}(|\alpha_{ij}|l_j + |\beta_{ij}|l_j)\int_{-\infty}^{t} k_{ij}(t-s)|y_j(s)|ds \qquad (10)$$

Let $z(t) = (z_1(t), \cdots, z_n(t))^T$ be the solution of the following differential equation: For $i = 1, \cdots, n$,

$$\frac{dz_i}{dt} = -d_i z_i(t) + \sum_{j=1}^{n}(|\alpha_{ij}|l_j + |\beta_{ij}|l_j)\int_{-\infty}^{t} k_{ij}(t-s)z_j(s)ds, \quad t \geq 0.$$

$$z_i(t) = \|\varphi_i\|, \quad t \leq 0. \qquad (11)$$

By virtue of comparison principle of [10], $|y_i(t)| \leq z_i(t)$ for all $t \in R$, and $i = 1, \cdots, n$.

Now we define a function $f(\eta) = \rho((D - \eta I)^{-1}AL)$ on $[0, \min_{1 \leq i \leq n}(d_i))$. It is a continuous function. By the condition, we have $f(0) < 1$, and $f(\eta) \to \infty$ when η approaches to $\min_{1 \leq i \leq n}(d_i)$, so there exists an η_0 such that $f(\eta_0) = 1$.

Let $\lambda = \min\{\eta | \tau \geq \eta \geq 0, f(\eta) = 1\}$, ε is a positive number which is smaller than λ, $\lambda_0 = \lambda - \varepsilon$.

Let $\gamma(t) = (\gamma_1(t), \cdots, \gamma_n(t))^T$ be defined as $\gamma_i(t) = z_i(t)e^{\lambda_0 t}$, as $t \geq 0$, and $\gamma_i(t) = \|\varphi_i\|$, as $t \leq 0$.

From the definition of $\gamma(t)$, when $t \geq 0$, we have

$$\frac{d\gamma_i}{dt} = -(d_i - \lambda_0)\gamma_i(t) + e^{\lambda_0 t}\sum_{j=1}^{n}(|\alpha_{ij}|l_j + |\beta_{ij}|l_j)\int_{-\infty}^{t} k_{ij}(t-s)z_j(s)ds$$

$$\leq -(d_i - \lambda_0)\gamma_i(t) + \sum_{j=1}^{n}(|\alpha_{ij}|l_j + |\beta_{ij}|l_j)\int_{-\infty}^{t} e^{\lambda_0(t-s)}k_{ij}(t-s)\gamma_j(s)ds$$

$$\leq -(d_i - \lambda_0)\gamma_i(t) + \sum_{j=1}^{n}(|\alpha_{ij}|l_j + |\beta_{ij}|l_j)\int_{-\infty}^{t} e^{\lambda_0(t-s)}k_{ij}(t-s)ds \cdot \bar{\gamma}_j(t)$$

$$\leq -(d_i - \lambda_0)\gamma_i(t) + \sum_{j=1}^{n}(|\alpha_{ij}|l_j + |\beta_{ij}|l_j)\int_{-\infty}^{t} e^{\tau(t-s)}k_{ij}(t-s)ds \cdot \bar{\gamma}_j(t)$$

$$= -(d_i - \lambda_0)\gamma_i(t) + \sum_{j=1}^{n}(|\alpha_{ij}|l_j + |\beta_{ij}|l_j)k_{ij}\bar{\gamma}_j(t) \qquad (12)$$

where $\bar{\gamma}_j(t) = \sup_{-\infty < s \le t} \gamma_j(s)$. Integrating both sides of the above inequality from 0 to t ($t > 0$), we obtain

$$\gamma_i(t) \le \|\varphi_i\| e^{-(d_i - \lambda_0)t} + \sum_{j=1}^{n} (|\alpha_{ij}|l_j + |\beta_{ij}|l_j) k_{ij} \int_{-\infty}^{t} e^{-(d_i - \lambda_0)(t-s)} \bar{\gamma}_j(s) ds$$

$$\le \|\varphi_i\| e^{-(d_i - \lambda_0)t} + \sum_{j=1}^{n} (|\alpha_{ij}|l_j + |\beta_{ij}|l_j) k_{ij} \int_{-\infty}^{t} e^{-(d_i - \lambda_0)(t-s)} ds \cdot \bar{\gamma}_j(t)$$

$$\le \|\varphi_i\| + \sum_{j=1}^{n} \frac{(|\alpha_{ij}|l_j + |\beta_{ij}|l_j)}{d_i - \lambda_0} \bar{\gamma}_j(t) \tag{13}$$

Let $\bar{\gamma}(t) = (\bar{\gamma}_1(t), \cdots, \bar{\gamma}_n(t))^T, \|\varphi\| = (\|\varphi_1\|, \cdots, \|\varphi_n\|)^T$, we have the following matrix form inequality.

$\gamma(t) \le \|\varphi\| + (D - \lambda_0 I)^{-1} AL \bar{\gamma}(t)$, when $t \ge 0$.

It is clear that $\bar{\gamma}_i(t)$ is an increasing function on $[0, +\infty]$ by the definition. Thus, for $s \in [0, t]$,

$\gamma(s) \le \|\varphi\| + (D - \lambda_0 I)^{-1} AL \bar{\gamma}(t)$.

So, we have

$\bar{\gamma}(t) \le \|\varphi\| + (D - \lambda_0 I)^{-1} AL \bar{\gamma}(t)$, when $t \ge 0$.

From the above inequality, we get $(I - (D - \lambda_0 I)^{-1} AL) \bar{\gamma}(t) \le \|\varphi\|$.

It is clear that $(D - \lambda_0 I)^{-1} AL \ge 0$, and we have $\rho((D - \lambda_0 I)^{-1} AL) < 1$ by the definition of λ_0. So by *Lemma 1*, we have $(I - (D - \lambda_0 I)^{-1} AL)^{-1} \ge 0$. Thus, there is an $M > 0$, such that $\gamma_i(t) \le M \max_{1 \le i \le n} \|\varphi_i\|, i = 1, \cdots, n$. From the definition of $\gamma_i(t)$, we have

$z_i(t) \le M \max_{1 \le i \le n} \|\varphi_i\| e^{-\lambda_0 t}, \quad i = 1, \cdots, n$.

Since $|y_i(t)| \le z_i(t), t \ge 0$, we get

$|y_i(t)| \le M \max_{1 \le i \le n} \|\varphi_i\| e^{-\lambda_0 t}, \quad i = 1, \cdots, n$.

So far, we have completed the proof of the theorem.

3   Conclusion

In last section, we have obtained the stability of the equilibrium point of system (1) under the condition of $\rho(D^{-1}AL) < 1$ $(A \ge 0)$. This condition basically is equivalent to that $DL^{-1} - A$ is an M-Matrix by the result in [6]. It includes a broad class of neural networks. The condition is easy to be verified, and this is very important to the designing neural networks and applications of the neural networks.

Acknowledgements

The first author is grateful for the support of Texas A&M University at Qatar.

References

1. Arik, S.: Global Robust Stability of Delayed Neural Networks. IEEE Trans. Circ. Syst. I, **50** (2003) 156-160
2. Cao, J.: Global Stability Analysis in Delayed Cellular Neural Networks. Phys. Rev. E, **59** (1999)5940-5944
3. Cao, J., Wang, J.: Absolute Exponential Stability of Recurrent Neural Networks with Time Delays and Lipschitz-continuous Activation Functions. Neural Networks, **17** (2004) 379-390
4. Cao, J., Wang, J., Liao, X.: Novel Stability Criteria of Delayed Cellular Neural Networks. International Journal of Neural Systems, **13** (2003) 365-375
5. Forti, M., Manetti, S., Marini, M.: Necessary and Sufficient Condition for Absolute Stability of Neural Networks. IEEE Trans. Circuits Syst. **41** (1994) 491-494
6. Horn, R.A., Johnson, C.R.: Topics in Matrix Analysis, Cambridge University Press, Cambridge (1999)
7. Liu, Y., Tang, W.: Exponential Stability of Fuzzy Cellular Neural Networks with Constant and Time-varying Delays. Physics Letters A, **323** (2004) 224-233
8. Liao, X.F., Wu, Z.F., Yu, J.B.: Stability Analyses for Cellular Neural Networks with Continuous Delay, Journal of Computational and Applied Mathematics, **143** (2002) 29-47
9. Liao, X.F., Wong, K.W., Li, C.: Global Exponential Stability for a Class of Generalized Neural Networks with Distributed Delays. Nonlinear Analysis: Real World Applications, **5** (2004) 527-547
10. Michel, A.N., Miller, R.K.: Qualitative Analysis of Large-scale Dynamical Systems. Academic Press, New York (1977)
11. Yang, T., Yang, L.B., Wu, C.W., Chua, L.O.: Fuzzy Cellular Neural Networks: Theory. In Proc. of IEEE International Workshop on Cellular Neural networks and Applications, (1996) 181-186
12. Yang, T., Yang, L.B., Wu, C.W., Chua, L.O.: Fuzzy Cellular Neural Networks: Applications. In Proc. of IEEE International Workshop on Cellular Neural Networks and Applications, (1996) 225-230
13. Yang, T., Yang, L.B.: The Global Stability of Fuzzy Cellular Neural Network, Circuits and Systems I: Fundamental Theory and Applications, **43** (1996) 880-883
14. Zhang, Q., Wei, X., Xu, J.: Global Exponential Stability of Hopfield Neural Networks with Continuously Distributed Delays, Physics Letters A, **315** (2003) 431-436

Global Exponential Stability of Reaction-Diffusion Hopfield Neural Networks with Distributed Delays

Zhihong Tang[1], Yiping Luo[1,2], and Feiqi Deng[1]

[1] College of Automation Science and Engineering, South China University of Technology, Guangdong, Guangzhou 510640, China
[2] Hunan Institute of Engineering, Xiangtan, Hunan 411101, China
lyp8688@sohu.com

Abstract. The global exponential stability of reaction-diffusion Hopfield neural networks with distributed delays is studied. Without assuming the boundedness, monotonicity and differentiability of the activation functions, the sufficient conditions were obtained by utilizing Dini's derivative, F-function and extended Hanaly's inequality. These conditions are easy to check and apply in practice and can be regarded as an extension of existing results.

1 Introduction

There has recently been increasing interest in the potential applications of the dynamics of artificial neural networks in signal and image processing. In hardware implementation, time delays are inevitably present due to the finite switching speed of amplifiers, so many researchers investigated the dynamics neural networks with delay[1]-[18]. Fewer researchers study reaction-diffusion neural network with delay. However, diffusion effect cannot be avoided in the neural networks model when electrons are moving in asymmetric electromagnetic field, so we must consider the space is varying with the time. Refs.[15],[16],[17] have considered the stability of neural networks with diffusion terms, which are expressed by partial differential equations. But the models[15],[16] were considered the discrete time-delay. To the best of our knowledge, few authors have considered the global exponential stability (GES) for the reaction-diffusion recurrent neural network with distributed delay. In this paper, we consider neural networks with distributed delays and diffusion terms described by the following system of integral-differential equations

$$\frac{\partial u_i}{\partial t} = \sum_{k=1}^{m} \frac{\partial}{\partial x_k}(D_{ik}\frac{\partial u_i}{\partial x_k}) - a_i u_i + \sum_{j=1}^{n} b_{ij}(\int_0^T k_{ij}(s)f_j(u_j((t-s),x))ds + I_i \tag{1}$$

$$\frac{\partial u_i}{\partial n} = Col(\frac{\partial u_i}{\partial x_1},\cdots\frac{\partial u_i}{\partial x_n}) = 0, t \geq t_0 > 0, x \in \partial\Omega \tag{2}$$

$$u_i(t_0 + s, x) = \phi_i(s,x), x \in \partial\Omega, i = 1,\cdots n \tag{3}$$

where $D_{ik} = D_{ik}(t,x,u) \geq 0$ denotes diffusion operator. s is time delay, $a_i > 0$, b_{ij} is the weight coefficient between neurons i and j, u_i, x_i denote the state variable and spatial

J. Wang, X. Liao, and Z. Yi (Eds.): ISNN 2005, LNCS 3496, pp. 174–180, 2005.
© Springer-Verlag Berlin Heidelberg 2005

variable respectively, The kernel functions $k_{ij}(s): [0,+\infty) \rightarrow [0,+\infty)(i,j=1,\cdots,n)$ are continuous on $[0,+\infty)$ with $\int_0^{+\infty} k_{ij}(s)ds = k_{ij}$. I_i, $f_j(\cdot)$ are external input function and activation function respectively. $\frac{\partial u_i}{\partial n} = 0$, $t \geq t_0$, $x \in \partial\Omega$, $\phi_i(s,x)$ is the initial function. Ω is a compact set with smooth boundary. It is pointed out that the dynamics of the neural networks can be remarkably changed by incorporating nonlinear non-monotone activation functions [19], So we investigate the GES of the model (1-3) without assuming the boundedness, monotonicity and differentiability of the activation function. In order to convenience, we introduce the following conditions:

(H1) $\left| f_j(u_1) - f_j(u_2) \right| \leq L_j \left| u_1 - u_2 \right|$, $u_1, u_2 \in R$, $j=1,2,\ldots,n$

(H2) $W = A - LP^+$ is an M-matrix, where $A = diag(a_1, \cdots a_n)$, $a_i > 0$,

$L = diag(L_1, \cdots L_n)$, $B = \left(b_{ij} \right)_{n \times n}$, $P^+ = \left(\left| p_{ij} \right| \right)_{n \times n}$. $P = \left(p_{ij} \right)_{n \times n}$, $p_{ij} = b_{ij}k_{ij}$,

(H3) $D^+\left[u_i(t) - u_i^* \right]^L \leq -a_i\left[u_i(t) - u_i^* \right]^L + \sum_{j=1}^{n} L_j \cdot \left| p_{ij} \right| \left(\left[u_j(t) - u_j^* \right]^L \right)^s$ \hfill (4)

where $u^* = (u_1^*, \cdots u_n^*)$ is equilibrium of system (1),

$\left(\left[u(t) - u^* \right]^\dagger \right)^s = Col\left(\left\| u_1(t) - u_1^* \right\|_2^s, \cdots, \left\| u_n(t) - u_n^* \right\|_2^s \right).$

$\left\| u_j(t) - u_j^* \right\|_2^s = \underset{-T \leq s \leq 0}{Sup} \left\| u_j(t+s) - u_j^* \right\|_2 ,$

$[u(t) - u^*]^+ = Col\left(\left\| u_1(t) - u_1^* \right\|_2, \cdots, \left\| u_n(t) - u_n^* \right\|_2 \right).$ $i=1,2,\ldots,n.$

Using the theory of Dini's derivative, F-function and extended Hanalay's inequality, we can obtain sufficient conditions for globally exponential stability of systems (1-3)

In addition, we denote $L^2(\Omega)$ real Lebesgue measurable function space on Ω and its L_2—norm by

$$\left\| u \right\|_2 = \left[\int_\Omega \left| u(x) \right|^2 dx \right]^{1/2}.$$ \hfill (5)

which constructs a Banach space, where $|u|$ denotes Euclid-norm of vector $u \in R^n$ $u = Col(u_1, \cdots u_n)$.

2 Definitions and Lemmas

Definition 1. The equilibrium of system (1-3) is global exponential stable with respect to $\left\| \bullet \right\|_G$, if for arbitrary solution of system (1-3), $u(t, x)$, there exist positive constants $M > 0$ and $\delta > 0$ such as

$$\left\| u(t) - u^* \right\|_G \leq Me^{-\delta(t-t_0)}$$ \hfill (6)

Definition 2 [8]. Let $C = C([t-T,t], R^n), T \geq 0$, $F(t,x,y) \in C(R_+ \times R^n \times C, R^n)$,

$F(t,x,y) = Col(f_1(t,x,y), \cdots f_n(t,x,y))$ is called F-function, if

(C1)   For $\forall t \in R_+$, $x \in R^n, y^{(1)}, y^{(2)} \in C$, if $y^{(1)} \leq y^{(2)}$, then we have

$F(t,x,y^{(1)}) \leq F(t,x,y^{(2)})$, where $y^{(1)} = Col(y_1^{(1)}, \cdots y_n^{(1)})$, $y^{(2)} = Col(y_1^{(2)}, \cdots y_n^{(2)})$,

(C2)   For each component of function F, and $y \in C$, $t \geq t_0$ we have

$f_i(t, x^{(1)}, y) \leq f_i(t, x^{(2)}, y)$, if $x^{(1)} \leq x^{(2)}$ and for some $i, x_i^{(1)} = x_i^{(2)}$.

Lemma 1 [8] Assume vector functions

$x(t) = Col(x_1(t), \cdots x_n(t))$, $y(t) = Col(y_1(t), \cdots y_n(t))$,

$x^s := \underset{-T \leq s \leq 0}{Sup} \, x(t+s) = Col(x_1^s(t), \cdots x_n^s(t))$, $x_i^s = \underset{-T \leq s \leq 0}{Sup} \, x_i(t+s), i = 1, 2, \cdots n$

$y^s := \underset{-T \leq s \leq 0}{Sup} \, y(t+s) = Col(y_1^s(t), \cdots y_n^s(t))$, $y_i^s = \underset{-T \leq s \leq 0}{Sup} \, y_i(t+s), i = 1, 2, \cdots n$

satisfy the following conditions

(L1) $x(t) < y(t)$, $t \in [t_0 - T, \ t_0]$

(L2) if $D^+y(t) > F(t, y(t), y^s(t)), t \geq t_0 \geq 0$, $D^+x(t) \leq F(t, x(t), x^s(t)), t \geq t_0 \geq 0$,

where $F(t,x,y) = Col(f_1(t,x,y), \cdots f_n(t,x,y))$ is F-function, then we have

$$x(t) < y(t), \ t \geq t_0. \tag{7}$$

Lemma 1 is called extended Halany inequality.

Lemma 2. If conditions (H1-H3) satisfy, then there exist $r_j > 0, \alpha > 0$, such that for the solution of (H3), we have the following estimate

$$\left\| u_j(t) - u_j^* \right\|_2 \leq r_j \left\| u(t_0) - u^* \right\|_2^s e^{-\alpha(t-t_0)}, \ t \geq t_0. \tag{8}$$

where

$$\left\| u_j(t_0) - u_j^* \right\|_2^s = \underset{-T \leq s \leq 0}{Sup} \left\| u_j(t_0+s) - u_j^* \right\|_2, \left\| u(t_0) - u^* \right\|_2^s = \left(\sum_{j=1}^n \left(\left\| u_j(t_0) - u_j^* \right\|_2^s \right)^2 \right)^{\frac{1}{2}}.$$

Proof: From condition (H2), there exists a vector $d = Col(d_1, \cdots, d_n) > 0$ such that [20]

$$-a_i d_i + \sum_{j=1}^n |P_{ij}| L_j d_j < 0 \ i = 1, \cdots, n \tag{9}$$

Let $\delta \ll 1$, such that

$$\delta d_i - a_i d_i + \sum_{j=1}^n |P_{ij}| L_j d_j e^{\delta \tau} < 0 \tag{10}$$

Choose $R \gg 1$, such that

$$Rde^{-\delta \tau} > \overline{I}, \ t \in [t_0 - T, t_0]. \tag{11}$$

where $\overline{I} = Col(1, \cdots 1)$.

$\forall \varepsilon > 0$, Let

$$q_i(t) = Rd_i e^{-\delta(t-t_0)} \left(\left\| \phi(t_0) - u^* \right\|_2^s + \varepsilon \right), t \geq t_0.$$ (12)

By (10), (11) and (12), we can obtain

$$D^+ q_i(t) = -\delta d_i R e^{-\delta(t-t_0)} \left(\left\| \phi(t_0) - u^* \right\|_2^s + \varepsilon \right)$$

$$> [(-a_i d_i + \sum_{j=1}^n |p_{ij}| L_j d_j e^{\alpha T}) \cdot R \cdot \left(\left\| \phi(t_0) - u^* \right\|_2^s + \varepsilon \right) e^{-\alpha(t-T-t_0)}$$

$$= -a_i q_i(t) + \sum_{j=1}^n |p_{ij}| L_j d_j (\left\| u(t_0) \right\|_2^s + \varepsilon) R e^{-\delta(t-t_0-T)}$$

$$\geq -a_i q_i(t) + \sum_{j=1}^n |p_{ij}| L_j q_j^s(t) =: F_i(t, q(t), q^s(t)), i = 1, \cdots, n.$$ (13)

Where

$$q_i^s(t) = \underset{-T \leq s \leq 0}{Sup} Rd_i (\left\| \phi(t_0) - u^* \right\|_2^s + \varepsilon) e^{-\delta(t+s-t_0)}, \left\| \phi(t_0) - u^* \right\|^s = \left(\sum_{j=1}^n (|\phi_j(t_0) - u_j^*|^s)^2 \right)^{1/2}.$$

$$|\phi_j(t_0) - u_j^*|^s = \underset{-T \leq s \leq 0}{\sup} |\phi_j(s) - u_j^*|$$

We define the right of equality (13) as $F_i(t, q_i(t), q_i^s(t))$, $t \geq t_0$, namely,

$$F_i(t, q_i(t), q_i^s(t)) = -a_i q_i(t) + \sum_{j=1}^n |p_{ij}| L_j q_j^s(t).$$ (14)

It is easy to be verified $F(t, x, y) = (F_1(t, x, y), \cdots, F_n(t, x, y))^T$ is an F-function. In addition, from (11) we have

$$\left\| u_i(t) - u_i^* \right\| \leq \left\| u_i(t) - u_i^* \right\|^s < \left\| \phi(t_0) - u^* \right\|_2^s < Rd_i e^{-\delta(t-t_0)} \left\| \phi(t_0) - u^* \right\|_2^s$$

$$< Rd_i e^{-\delta(t-t_0)} \left(\left\| \phi(t_0) - u^* \right\|_2^s + \varepsilon \right) = q_i(t), \; t \in [t_0 - T, t_0].$$

Namely

$$[u - u^*]^+ < q(t), t \in [t_0 - T, t_0].$$ (15)

From (H3), Eqs.(15) and Lemma 1, we get

$$\left\| u_i(t) - u_i^* \right\|_2 < r_i e^{-\delta(t-t_0)} \left\| \phi(t_0) - u^* \right\|_2^s, \; t \geq t_0.$$ (16)

Where $r_i = Rd_i$. The proof is complete.

3 Main Results

Theorem 1. If conditions (H1), (H2) satisfy, then system (1-3) has a unique equilibrium, and it is global exponential stablility.

Proof. The uniqueness of equilibrium can be concluded from the global exponential stability of the system (1-3). So in the following we only prove the global exponential stability of the system (1-3). Let u^* is a equilibrium of systems (1), $u(t,x) = Col(u_1(t,x),\cdots,u_n(t,x))$ is the arbitrary solution of system (1), then systems (1) can be rewritten as

$$\frac{\partial(u_i - u_i^*)}{\partial t} = \sum_{k=1}^{m} \frac{\partial}{\partial x_k}\left(D_{ik}\frac{\partial(u_i - u_i^*)}{\partial x_k}\right) - a_i(u_i - u_i^*)$$

$$+\sum_{j=1}^{n} b_{ij}[\int_0^T k_{ij}(s)f(u_j(t-s,x))ds - \int_0^T k_{ij}(s)f(u_j^*)ds] \qquad (17)$$

Multiply $u_i - u_i^* = u_i(t,x) - u_i^*$ with equation (17) and integrate it, we have

$$\frac{1}{2}\frac{d}{dt}\int_\Omega(u_i - u_i^*)^2\, dx = \sum_{k=1}^{m}\int_\Omega(u_i - u_i^*)\frac{\partial}{\partial x_k}\left(D_{ik}\frac{\partial(u_i - u_i^*)}{\partial x_k}\right)dx \qquad (18)$$

$$-a_i\int_\Omega(u_i - u_i^*)^2\, dx + \sum_{j=1}^{n}\int_\Omega\left\{(u_i - u_i^*)b_{ij}\int_0^T k_{ij}(s)\left[f_j(u_j(t-s,x)) - f_j(u_j^*)\right]ds\right\}dx$$

Based on the Green formula and its initial value condition, we have

$$\sum_{k=1}^{m}\int_\Omega(u_i - u_i^*)\frac{\partial}{\partial x_k}\left(D_{ik}\frac{\partial(u_i - u_i^*)}{\partial x_k}\right)dx = -\sum_{k=1}^{m}\int_\Omega D_{ik}\left(\frac{\partial(u_i - u_i^*)}{\partial x_k}\right)^2 dx \qquad (19)$$

Using (H1), we have the following

$$\sum_{j=1}^{n}\int_\Omega\left\{(u_i - u_i^*)b_{ij}\int_0^T k_{ij}(s)\left[f_j(u_j(t-s)) - f_j(u_j^*)\right]ds\right\}dx$$

$$\le \sum_{j=1}^{n}\int_\Omega\left\{|u_i - u_i^*|\|b_{ij}\|\left|\int_0^T k_{ij}(s)\left[f_j(u_j(t-s)) - f_j(u_j^*)\right]ds\right|\right\}dx$$

$$\le \sum_{j=1}^{n}|b_{ij}|\cdot L_j\int_\Omega|u_i - u_i^*|\cdot\int_0^T k_{ij}(s)|u_j(t-s) - u_j^*|ds$$

$$= \sum_{j=1}^{n}|b_{ij}|\cdot L_j\int_0^T k_{ij}(s)ds\cdot\int_\Omega|u_i - u_i^*|\cdot|u_j(t-s) - u_j^*|dx$$

$$\le \sum_{j=1}^{n}|b_{ij}|\cdot L_j\left(\int_0^T k_{ij}(s)ds\cdot\int_\Omega|u_i - u_i^*|^2\, dx\cdot\int_0^T k_{ij}(s)ds\int_\Omega|u_j(t-s) - u_j^*|^2\, dx\right)^{\frac{1}{2}}$$

$$\le \sum_{j=1}^{n}|b_{ij}|L_j k_{ij}\|u_i - u_i^*\|_2\cdot\|u_j(t) - u_j^*\|_2^s \qquad (20)$$

where $k_{ij} = \int_0^T k_{ij}(s)ds$.

Substitute equations (20) and (19) into equation (18) and simplify it:

$$\frac{d\left\|u_i - u_i^*\right\|_2}{dt} \leq -a_i \left\|u_i - u_i^*\right\|_2 + \sum_{j=1}^{n} \left|p_{ij}\right| \cdot L_j \cdot \left\|u_j(t) - u_j^*\right\|_2^s$$

(21)

i.e: $D^+ \left\|u_i - u_i^*\right\|_2 \leq -a_i \left\|u_i - u_i^*\right\|_2 + \sum_{j=1}^{n} \left|p_{ij}\right| L_j \left\|u_j(t) - u_j^*\right\|_2^s$

So condition (H3) satisfies. From Lemma 2, we have:

$$\left\|u_i(t) - u_i^*\right\|_2 \leq r_j R \left\|u(t_0) - u^*\right\|_2^s e^{-\alpha(t-t_0)}$$

and

$$\left\|u(t) - u^*\right\|_2 = \left(\sum_{i=1}^{n} \left\|u_i(t) - u_i^*\right\|_2^2\right)^{\frac{1}{2}} \leq R\left|r\right| \left\|u(t_0) - u^*\right\|_2^s e^{-\alpha(t-t_0)} = Me^{-\alpha(t-t_0)} \to 0 .$$

as $t \to \infty$.

where $\left|r\right| = \max\{r_1, r_2, \cdots r_n\}$ or $\left|r\right| = (\sum_{i=1}^{n} r_i^2)^{\frac{1}{2}}$

So all solutions of the system (1-3) tend to u^*, as $t \to \infty$. System (1-3) has an unique and global exponential stable equilibrium with respect to L_2-norm.

Corollary 1. If conditions (H1), (H2) satisfy, then the following system

$$\frac{du_i}{dt} = -a_i u_i + \sum_{j=1}^{n} b_{ij} \int_0^T k_{ij}(s) f(u_j(t-s,x)) ds + I_i. \qquad i=1,2,\ldots, n.$$

(22)

is global exponential stable with respect to E-norm.

4 Conclusions

Some sufficient criteria, which are independent of the delay parameter, have been given ensuring the global exponential stability of distributed delayed Hopfield neural networks with reaction-diffusion by using an approach based on Dini's derivative and the extended Hanalay inequality. It has been shown that the results we have obtained are applicable to a larger class of activation functions than the class of the functions considered in previous works

Acknowledgments

The paper is supported by National Science Foundation of China under Grant 60374023 and Key Science Foundation of Educational Department of Hunan Province Grant 04A012.

References

1. Chua, L.O., Yang, L.: Cellular Neural Network: Theory. IEEE Trans CAS-I, **35** (1988) 1257-1272
2. Cao, J.D., Zhou, D.: Stability Analysis of Delayed Cellular Neural Networks. Neural Networks. **11** (1998) 1601-1605

2   Preliminaries

We consider the delayed impulsive Hopfield type neural networks described by

$$
\begin{cases}
C_i \dot{u}_i(t) = -u_i(t)/R_i + \sum_{j=1}^{n} T_{ij} g_j(u_j(t - \tau_j)) + I_i, t \neq t_k \\
\Delta u_i(t) = d_i u_i(t^-) + \sum_{j=1}^{n} W_{ij} h_j(u_j(t^- - \tau_j)), t = t_k
\end{cases}
, i = 1, 2, \cdots, n,
$$

$$(1)$$

where $\Delta u_i(t_k) = u_i(t_k) - u_i(t_k^-), u_i(t_k^-) = \lim\limits_{t \to t_k^-} u_i(t), k \in \mathcal{Z} = \{1, 2, \cdots\}$; the
time sequence $\{t_k\}$ satisfies $0 < t_0 < t_1 < t_2 < \cdots < t_k < t_{k+1} < \cdots$, and
$\lim\limits_{k \to \infty} t_k = \infty; C_i > 0, R_i > 0$, and I_i are, respectively, the capacitance, resistance,
and external input of the ith neuron; T_{ij} and W_{ij} are the synaptic weights of
the neural networks; and $\tau_i \geq 0$ is the transmission delay of the ith neuron.

The initial condition for system (1) is given by $u_i(s) = \psi_i(s)$, $s \in [t_0 - \tau, t_0]$,
$i = 1, 2, \cdots, n$ where $\psi_i : [t_0 - \tau, t_0] \to \Re$, $(i = 1, 2, \cdots, n)$, is a continuous
function, and $\tau = \max\limits_{1 \leq i \leq n} \{\tau_i\}$.

Throughout this paper, we assume that the neuron activation functions $g_i(u)$,
$h_i(u)$, $i = 1, 2, \cdots, n$, are continuous and satisfy the following conditions:

$$
|g_i(u_i)| \leq M_i, 0 \leq \frac{g_i(u_i) - g_i(v_i)}{u_i - v_i} \leq K_i, \forall \, u_i \neq v_i, u_i, v_i \in R, i = 1, 2, \cdots, n,
$$

$$(2)$$

$$
|h_i(u_i)| \leq N_i, 0 \leq \frac{h_i(u_i) - h_i(v_i)}{u_i - v_i} \leq L_i, \forall \, u_i \neq v_i, u_i, v_i \in R, i = 1, 2, \cdots, n. \quad (3)
$$

Let $u^* = (u_1^*, u_2^*, \cdots, u_n^*)^T$ be an equilibrium point of system (1), and set
$x_i(t) = u_i(t) - u_i^*, d_i u_i^* + \sum_{j=1}^{n} W_{ij} h_j(u_j^*) = 0, f_i(x_i(t - \tau_i)) = g_i(u_i(t - \tau_i)) -$
$g_i(u_i^*), \varphi_i(x_i(t - \tau_i)) = h_i(u_i(t - \tau_i)) - h_i(u_i^*), i = 1, 2, \cdots, n$. Then, for each
$i = 1, 2, \cdots, n$,

$$
|f_i(z)| \leq K_i|z|, \, z f_i(z) \geq 0, \, |\varphi_i(z)| \leq L_i|z|, \, z\varphi_i(z) \geq 0, \, \forall z \in \Re. \quad (4)
$$

System (1) may be rewritten as follows.

$$
\begin{cases}
C_i \dot{x}_i(t) = -x_i(t)/R_i + \sum_{j=1}^{n} T_{ij} f_j(x_j(t - \tau_j)), t \neq t_k \\
\Delta x_i(t) = d_i x_i(t^-) + \sum_{j=1}^{n} W_{ij} \varphi_j(x_j(t^- - \tau_j)), t = t_k
\end{cases}
, \, i = 1, 2, \cdots, n, \quad (5)
$$

Define $C = diag(C_1, C_2, \cdots, C_n)$, $D = diag(d_1, d_2, \cdots, d_n)$, $T = (T_{ij})_{n \times n}$,
$R = diag(R_1, R_2, \cdots, R_n)$, $K = diag(K_1, K_2, \cdots, K_n)$, $W = (W_{ij})_{n \times n}$,
$\varphi(x(t^- - \tau)) = \left(\varphi_1(x_1(t^- - \tau_1)), \varphi_2(x_2(t^- - \tau_2)), \cdots, \varphi_n(x_n(t^- - \tau_n)) \right)^T$,
$f(x(t - \tau)) = \left(f_1(x_1(t - \tau_1)), f_2(x_2(t - \tau_2)), \cdots, f_n(x_n(t - \tau_n)) \right)^T$,
$\Delta x = (\Delta x_1, \Delta x_2, \cdots, \Delta x_n)^T$, $x(t - \tau) = \left(x_1(t - \tau_1), x_2(t - \tau_2), \cdots, x_n(t - \tau_n) \right)^T$

System (5) is reduces to

$$\begin{cases} C\dot{x}(t) = -R^{-1}x(t) + Tf(x(t-\tau)), t \neq t_k \\ \Delta x(t) = Dx(t^-) + W\varphi(x(t^- - \tau)), t = t_k \end{cases} \tag{6}$$

The initial condition for system (6) is given by $x(t) = \phi(t)$, $t \in [t_0 - \tau, t_0]$, where $\phi(t) = (\phi_1(t), \phi_2(t), \cdots, \phi_n(t))^T$, and $\phi_i(t) = \psi_i(t) - u_i^*$, $t \in [t_0 - \tau, t_0]$, $i = 1, 2, \cdots, n$.

The following notations will be used throughout the paper: The notation $P > 0$, (respectively, $P < 0$) means that P is symmetric and positive definite (respectively, negative definite) matrix. We use P^T, P^{-1}, $\lambda_{\min}(P)$, and $\lambda_{\max}(P)$, to denote, respectively, the transpose of, the inverse of, the smallest and the largest eigenvalues of a square matrix P. The norm $\| \cdot \|$ is either the Euclidean vector norm or the induced matrix norm. For a function $U(t), t \in [t_0 - \tau, \infty)$, $\overline{U}(t)$ and $\overline{U}(t^-)$ are defined by $\overline{U}(t) = \sup\limits_{t-\tau \leq s \leq t} U(s)$ and $\overline{U}(t^-) = \sup\limits_{t-\tau \leq s < t} U(s)$.

Lemma 2.1. System (1) admits at least one equilibrium point.

The proof of Lemma 2.1 is similar to that given in [3, Theorem 1]. An additional difference is the consideration of the impulse effect.

3   Exponential Stability

In this section, we shall obtain some sufficient conditions for global exponential stability of delayed impulsive Hopfield type neural networks. If $u^* = (u_1^*, u_2^*, \cdots, u_n^*)^T$ is an equilibrium point of system (1), then $x = (0, 0, \cdots, 0)^T$ is an equilibrium point of system (5) and system (6).

Theorem 3.1. Assume that the following conditions are satisfied.

(i) There exists a $n \times n$ matrix $P > 0$ and constant $\varepsilon > 0$ such that

$$\Omega = \begin{bmatrix} -PC^{-1}R^{-1} - R^{-1}C^{-1}P & PC^{-1}T \\ T^T C^{-1}P & -\varepsilon I \end{bmatrix} < 0;$$

(ii) $a = \lambda_{\min}(\Psi)/\lambda_{\max}(P) > \varepsilon \max\limits_{1 \leq i \leq n} \{K_i^2\}/\lambda_{\min}(P) = b \geq 0$,

where $\Psi = PC^{-1}R^{-1} + R^{-1}C^{-1}P - \frac{1}{\varepsilon}PC^{-1}TT^T C^{-1}P$;

(iii) There exists a constant δ satisfying $\delta > \frac{\ln(\rho e^{\lambda\tau})}{\lambda\tau}$ such that $\inf\limits_{k \in Z}(t_k - t_{k-1}) > \tau\delta$, where $\lambda > 0$ is the unique solution of the equation

$$\lambda = a - be^{\lambda\tau}, \text{ and } \rho = \max\left\{1, 2\frac{\lambda_{\max}(P)}{\lambda_{\min}(P)}\left(\|I + D\|^2 + \max\limits_{1 \leq i \leq n}\{L_i^2\}\|W\|^2 e^{\lambda\tau}\right)\right\}.$$

Then, the equilibrium point u^* of system (1) is globally exponentially stable with convergence rate $\frac{1}{2}(\lambda - \frac{\ln(\rho e^{\lambda\tau})}{\delta\tau})$.

Proof. Let $V(t) = x^T(t)Px(t)$. Then, $V(t)$ is radially unbounded in $x(t)$ and $\lambda_{\min}(P)\|x(t)\|^2 \leq V(t) \leq \lambda_{\max}(P)\|x(t)\|^2$.

For $t \neq t_k$, compute the derivative of $V(t)$ along the trajectories of system (6), we have

$$\dot{V}(t)\big|_{(6)} = -x^T(t)(R^{-1}C^{-1}P + PC^{-1}R^{-1})x(t) + 2x^T(t)PC^{-1}Tf(x(t-\tau)).(7)$$

By $2u^Tv \le \frac{1}{\varepsilon}u^Tu + \varepsilon v^Tv, \forall\, u,v \in \Re^n$, $\varepsilon > 0$, (4) and (7), we obtain

$$\dot{V}(t)\big|_{(6)} \le -x^T(t)\Psi x(t) + \varepsilon f^T(x(t-\tau))f(x(t-\tau))$$

$$\le -\lambda_{\min}(\Psi)\|x(t)\|^2 + \varepsilon \max_{1\le i\le n}\{K_i^2\}\|x(t-\tau)\|^2 \le -aV(t) + b\overline{V}(t).(8)$$

Since $\Omega < 0$, it is clear from Schur complement that $\Psi > 0$. Thus $a > 0$. In view of system (6), it follows from (4) that

$$V(t_k) \le \lambda_{\max}(P)\big[\|(I+D)\|\|x(t_k^-)\| + \|W\|\|\varphi(x(t_k^- - \tau))\|\big]^2$$

$$\le 2\lambda_{\max}(P)\|(I+D)\|^2\|x(t_k^-)\|^2 + 2\lambda_{\max}(P)\max_{1\le i\le n}\{L_i^2\}\|W\|^2\|x(t_k^- - \tau)\|^2$$

$$\le 2\frac{\lambda_{\max}(P)}{\lambda_{\min}(P)}\|(I+D)\|^2 V(t_k^-) + 2\frac{\lambda_{\max}(P)}{\lambda_{\min}(P)}\|W\|^2 \max_{1\le i\le n}\{L_i^2\}\overline{V}(t_k^-). \qquad (9)$$

By assumptions (ii) and (iii), and Lemma 1 in [6], we obtain

$$V(t) \le \rho\overline{V}(t_0)\exp\left\{-\left(\lambda - \frac{\ln(\rho e^{\lambda\tau})}{\delta\tau}\right)(t-t_0)\right\}, \; \forall\, t \ge t_0.$$

It follows that

$$\|x(t)\| \le \sqrt{\frac{\rho\lambda_{\max}(P)}{\lambda_{\min}(P)}}\|\overline{x}(t_0)\|\exp\left\{-\frac{1}{2}\left(\lambda - \frac{\ln(\rho e^{\lambda\tau})}{\delta\tau}\right)(t-t_0)\right\}, \; \forall\, t \ge t_0.$$

This completes the proof.

Theorem 3.2. Assume that the following conditions are satisfied.

(i) $a = \min\limits_{1\le i\le n}\left\{\frac{1}{R_iC_i}\right\} > \max\limits_{1\le j\le n}\left\{\sum\limits_{i=1}^n \frac{K_j}{C_i}|T_{ij}|\right\} = b \ge 0$;

(ii) There exists a constant δ satisfying $\delta > \frac{\ln(\rho e^{\lambda\tau})}{\lambda\tau}$ such that $\inf\limits_{k\in Z}(t_k - t_{k-1}) > \tau\delta$, where $\lambda > 0$ is the unique solution of the equation $\lambda = a - be^{\lambda\tau}$, and $\rho = \max\left\{1, \max\limits_{1\le i\le n}\{|1+d_i|\} + \max\limits_{1\le j\le n}\left\{\sum\limits_{i=1}^n |W_{ij}|L_j\right\}e^{\lambda\tau}\right\}$.

Then, the equilibrium point u^* of system (1) is globally exponentially stable with convergence rate $\lambda - \frac{\ln(\rho e^{\lambda\tau})}{\delta\tau}$.

Proof. Construct a radially unbounded Lyapunov function $V(t)$ by $V(t) = \sum\limits_{i=1}^n |x_i(t)|$. Then, for $t \ne t_k$, we compute the Dini derivative of $V(t)$ along the trajectories of system (5), and by (4), giving

$$D^+V(t)\big|_{(5)} \le -\min_{1\le i\le n}\left\{\frac{1}{R_iC_i}\right\}\sum_{i=1}^n |x_i(t)| + \sum_{i=1}^n\sum_{j=1}^n \frac{K_j}{C_i}|T_{ij}||x_j(t-\tau_j)|$$

$$\le -a\sum_{i=1}^n |x_i(t)| + b\sum_{j=1}^n |x_j(t-\tau_j)| \le -aV(t) + b\overline{V}(t). \qquad (10)$$

In view of system (5), it follows from (4) that

$$V(t_k) \leq \sum_{i=1}^{n} |1 + d_i||x_i(t_k^-)| + \sum_{i=1}^{n} \sum_{j=1}^{n} |W_{ij}|L_j|x_j(t_k^- - \tau_j)|$$

$$\leq \max_{1 \leq i \leq n} \{|1 + d_i|\} V(t_k^-) + \max_{1 \leq j \leq n} \left\{ \sum_{i=1}^{n} |W_{ij}|L_j \right\} \overline{V}(t_k^-). \tag{11}$$

The remaining part of the proof of this theorem is similar to that of the proof of Theorem 1. The proof is complete.

Theorem 3.3. Assume that the following conditions are satisfied.

(i) $a = \lambda_{\min}(R^{-1}C^{-1}) > \mu = \sqrt{\lambda_{\max}(KA^TC^{-2}AK)}$, where $A = (|T_{ij}|)_{n \times n}$;

(ii) There exists a constant δ satisfying $\delta > \frac{\ln(\rho e^{2\lambda \tau})}{2\lambda \tau}$ such that $\inf_{k \in Z}(t_k - t_{k-1}) > \tau\delta$, where $\lambda > 0$ is the unique solution of the equation

$$\lambda = a - \frac{\mu}{2} - \frac{\mu}{2}e^{2\lambda \tau}, \text{ and } \rho = \max \left\{ 1, 2\|I + D\|^2 + 2 \max_{1 \leq i \leq n} \{L_i^2\}\|W\|^2 e^{\lambda \tau} \right\}.$$

Then, the equilibrium point u^* of system (1) is globally exponentially stable with convergence rate $\lambda - \frac{\ln(\rho e^{2\lambda \tau})}{2\delta \tau}$.

Proof. Construct a radially unbounded Lyapunov function $V(t)$ by $V(t) = \frac{1}{2} \sum_{i=1}^{n} x_i^2(t)$. Then, for $t \neq t_k$, computing the derivative of $V(t)$ along the trajectories of system (5), and by (4), we obtain

$$\dot{V}(t)\big|_{(5)} \leq -a \sum_{i=1}^{n} x_i^2(t) + \frac{1}{2} \left[\begin{matrix} |x(t)| \\ |x(t-\tau)| \end{matrix} \right]^T \left[\begin{matrix} 0 & C^{-1}AK \\ KA^TC^{-1} & 0 \end{matrix} \right] \left[\begin{matrix} |x(t)| \\ |x(t-\tau)| \end{matrix} \right]$$

$$\leq -(2a - \mu)V(t) + \mu\overline{V}(t), \tag{12}$$

where $|x(t)| = \left[|x_1(t)|, |x_2(t)|, \cdots, |x_n(t)| \right]^T$, and $|x(t-\tau)| = \left[|x_1(t-\tau_1)|, |x_2(t-\tau_2)|, \cdots, |x_n(t-\tau_n)| \right]^T$.

It follows from (9) that, $V(t_k) \leq 2\|I + D\|^2 V(t_k^-) + 2\|W\|^2 \max_{1 \leq i \leq n} \{L_i^2\} \overline{V}(t_k^-)$.

The remaining part of the proof follows from a argument similar to that used in the proof of Theorem 1. The proof is complete.

4   Numerical Example

In this section, we give an example to illustrate our results.

Example 1. Consider the following delayed impulsive Hopfield type neural networks

$$\begin{cases} C_i \dot{u}_i(t) = -u_i(t)/R_i + \sum_{j=1}^{3} T_{ij}g_j(u_j(t-\tau_j)), \ t \neq t_k \\ \Delta u_i(t) = d_i u_i(t^-) + \sum_{j=1}^{3} W_{ij}h_j(u_j(t^- - \tau_j)), \ t = t_k \end{cases} \quad i = 1, 2, 3, \tag{13}$$

where $g_1(u_1) = \tanh(0.63u_1), g_2(u_2) = \tanh(0.78u_2), g_3(u_3) = \tanh(0.46u_3)$, $h_1(u_1) = \tanh(0.09u_1), h_2(u_2) = \tanh(0.02u_2), h_3(u_3) = \tanh(0.17u_3), C = diag(0.89, 0.88, 0.53), R = diag(0.16, 0.12, 0.03), D = -diag(0.95, 0.84, 0.99),$

$$T = \begin{bmatrix} 0.19 & 0.35 & 1.29 \\ 0.31 & 0.61 & -0.25 \\ 0.07 & -0.37 & 0.44 \end{bmatrix}, W = \begin{bmatrix} -0.04 & -0.05 & 0.16 \\ 0.19 & -0.17 & -0.02 \\ 0.03 & 0.13 & 0.04 \end{bmatrix}, 0 \leq \tau_i \leq 0.5, i = 1, 2, 3.$$

In this case, $K = diag(0.63, 0.78, 0.46), L_1 = 0.09, L_2 = 0.02, L_3 = 0.17, \tau = 0.5$, and $u^* = (0, 0, 0)^T$ is an equilibrium point of neural network (13).

By direct computation, it follows that the matrix $P = diag(0.9, 0.7, 0.8)$ and constant $\varepsilon = 1$ such that $\Omega < 0$ in Theorem 1, and that $a = 11.9671 > 0.8691 = b$, and $\lambda = 4.3432$ is the unique solution of the equation $\lambda = a - be^{\lambda\tau}$, and $\delta = 1.01 > \frac{\ln(\rho e^{\lambda\tau})}{\lambda\tau} = 1$.

To use Theorem 2, we note that $a = 7.0225 > 1.392 = b$, and $\lambda = 2.4003$ is the unique solution of the equation $\lambda = a - be^{\lambda\tau}$, and $\delta = 1.01 > \frac{\ln(\rho e^{\lambda\tau})}{\lambda\tau} = 1$.

To use Theorem 3, we note that $a = 7.0225 > 1.0898 = \mu$, and that $\lambda = 2.0867$ is the unique solution of the equation $\lambda = a - \frac{\mu}{2} - \frac{\mu}{2}e^{2\lambda\tau}$, and $\delta = 1.01 > \frac{\ln(\rho e^{2\lambda\tau})}{2\lambda\tau} = 1$.

Hence, by Theorem 1-3, we see that the equilibrium point u^* of system (13) is globally exponentially stable for $\inf_{k \in \mathbb{Z}}\{t_k - t_{k-1}\} > 0.505$, and the convergence rate computed by Theorem 1-3 are, respectively, 0.0215, 0.0238 and 0.0207.

5   Conclusions

The problem of global exponential stability analysis for delayed impulsive Hopfield type neural networks was discussed in this paper. By means of Lyapunov functions, some global exponential stability criteria have been derived and the exponential convergence rate is also estimated. These criteria are easy to verify.

References

1. Liao, X. X., Liao, Y.: Stability of Hopfield-type Neural Networks (II). Science in China, (Series A), **40** (1997) 813-816
2. Sun, C. Y., Zhang, K. J., Fei, S. M., Feng, C. B.: On Exponential Stability of Delayed Neural Networks with a General Class of Activation Functions. Physics Letters A, **298** (2002) 122-132
3. Xu, B. J., Liu, X. Z., Liao, X. X.: Global Asymptotic Stability of High-Order Hopfield Type Neural Networks with Time Delays. Computers and Mathematics with Applications, **45** (2003) 1729-1737
4. Gopalsamy, K.: Stability of Atificial Neural Networks with Impulses. Applied Mathematics and Computation, **154** (2004) 783-813
5. Acka, H., Alassar, R., Covachev, V., et al.: Continuous-time Additive Hopfield-type Neural Networks with Impulses. Journal of Mathematical Analysis and Applications, **290** (2004) 436-451
6. Yue, D., Xu, S. F., Liu, Y. Q.: Differential Inequality with Delay and Impulse and Its Applications to Design Robust Control. Control Theory and Applications, **16** (1999) 519-524 (in Chinese)

Global Exponential Stability of Hopfield Neural Networks with Impulsive Effects

Zhichun Yang[1,2], Jinan Pei[3], Daoyi Xu[1], Yumei Huang[1], and Li Xiang[1]

[1] College of Mathematics, Sichuan University, Chengdu 610064, China
zhichy@yahoo.com.cn
[2] Basic Department, Chengdu Textile Institute, Chengdu 610023, China
[3] Department of Math., Chongqing Education College, Chongqing 400067, China

Abstract. A class of Hopfield neural network model involving variable delays and impulsive effects is considered. By applying idea of piecewise continuous vector Lyapunov function, the sufficient conditions ensuring the global exponential stability of impulsive delay neural networks are obtained. The results extend and improve some recent work.

1 Introduction

Stability of Hopfield neural networks is a prerequisite in the design and applications of the networks and have attracted considerable attention. In implementation of neural networks, time delays are unavoidably encountered because of the finite switching speed of amplifiers [1].On the other hand, artificial electronic networks are subject to instantaneous perturbations and experience abrupt change of the state, that is, do exhibit impulsive effects (see, [2],[3],[4],[5]). Furthermore, both delays and impulses can affect the dynamical behaviors of the system creating oscillatory and unstable characteristics. Therefore, it is necessary to investigate impulse and delay effects on the stability of Hopfield neural networks.

In this paper, we consider the stability of a class of Hopfield neural networks with variable delays and impulsive effects described by the following impulsive delay differential equations

$$
\begin{cases}
x_i'(t) = -a_i x_i(t) + \sum_{j=1}^{n} b_{ij} g_j(x_j(t - \tau_{ij}(t))) + J_i, \quad t \neq t_k, \ t \geq 0, \\
\triangle x_i(t_k) = x_i(t_k^+) - x_i(t_k^-), \quad i = 1, 2, \ldots, n, \ k \in N \overset{\triangle}{=} \{1, 2, \ldots\},
\end{cases}
\tag{1}
$$

where $a_i > 0$, x_i is the state of the neurons, g_i represents the activation function, b_{ij} denotes the weight coefficients, J_i is the constant input, transmission delay $\tau_{ij}(t)$ satisfies $0 \leq \tau_{ij}(t) \leq \tau$, impulsive moments $\{t_k, k \in N\}$ satisfy $t_0 = 0 < t_1 < t_2 < \ldots$, $\lim_{k \to \infty} t_k = \infty$, $\triangle x_i(t_k)$ corresponds to the abrupt change of the state at fixed impulsive moment t_k. If $\triangle x_i \equiv 0$, then the model (1) becomes continuous Hopfield neural networks (see, [6],[7], [8],[9],[10],[11],[12],[13],[14])

$$
x_i'(t) = -a_i x_i(t) + \sum_{j=1}^{n} b_{ij} g_j(x_j(t - \tau_{ij}(t))) + J_i, i = 1, \ldots, n, t \geq 0.
\tag{2}
$$

J. Wang, X. Liao, and Z. Yi (Eds.): ISNN 2005, LNCS 3496, pp. 187–192, 2005.

A piecewise continuous function $x(t) = (x_1(t), \ldots, x_n(t))^T : [-\tau, +\infty) \to R^n$ is called a solution of Eq.(1) with the initial condition

$$x(s) = \phi(s), \quad \phi \in C([-\tau, 0], R^n),$$

if $x(t)$ is continuous at $t \neq t_k$, $k \in N$, $x(t_k) = x(t_k^+)$ and $x(t_k^-)$ exists, $x(t)$ satisfies Eq.(1) for $t \geq 0$ under the initial condition. Especially, a point $x^* \in R^n$ is called an equilibrium point of Eq. (1), if $x(t) = x^*$ is a solution of (1).

Assume that g_i satisfies global Lipschitz condition and impulsive operator is viewed as perturbation of the equilibrium point x^* of continuous system (2), i.e.,

(A_1) $|g_i(s_1) - g_i(s_2)| \leq L_i|s_1 - s_2|, \forall s_1, s_2 \in R, i = 1, 2, \ldots, n.$
(A_2) $\triangle x_i(t_k) = I_{ik}(x_i(t_k^-) - x_i^*), I_{ik}(0) = 0, |s + I_{ik}(s)| \leq \beta_{ik}|s|, \forall s \in R, k \in N.$

For any initial function ϕ, (A_1) and (A_2) guarantee the existence and uniqueness of the solution of Eq. (1) [5]. If the continuous system (2) has exactly one equilibrium point x^*, then x^* is also an equilibrium point of (1) by (A_2).

For convenience, we introduce the following denotation.

For $x = (x_1, x_2, \ldots, x_n)^T \in R^n$, $|x| = (|x_1|, |x_2|, \ldots, |x_n|)^T$, $\|x\|$ denotes any norm in R^n. For $X, Y \in R^{m \times n}$ or $X, Y \in R^n$, $X \geq Y(X > Y)$ means that each pair of corresponding elements of X and Y satisfies the inequality "$\geq($ $>)$". $D = (d_{ij})_{n \times n} \in \mathcal{M}$ denotes the matrix D belongs to M-matrix class, i.e., $d_{ii} > 0$, $d_{ij} \leq 0$ for $i \neq j$, $i, j = 1, 2, \ldots, n$, and all the leading principle minors of D are positive. $g(x) = (g_1(x_1), g_2(x_2), \ldots, g_n(x_n))^T$, $J = (J_1, J_2, \ldots, J_n)^T$, $A = \text{diag}\{a_1, a_2, \ldots, a_n\}$, $L = \text{diag}\{L_1, L_2, \ldots, L_n\}$, $B = (b_{ij})$ and $|B| = (|b_{ij}|)$, E denotes an unit matrix.

2   Global Exponential Stability

In this section, we shall study the global exponential stability of the equilibrium point of impulsive delay system (1).

Theorem 1. Assume that, in addition to (A_1) and (A_2),

(A_3) there exists a number $\lambda > 0$ and a vector $z = (z_1, z_2, \ldots, z_n)^T > 0$ such that $(A - \lambda E - |B|Le^{\lambda \tau})z > 0$;
(A_4) let $\eta_k = \max\{1, \beta_{1k}, \beta_{2k}, \ldots, \beta_{nk}\}$ and $\eta \overset{\triangle}{=} \sup\limits_{k \in N}\{\frac{\ln \eta_k}{t_k - t_{k-1}}\} < \lambda.$

Then the equilibrium point x^* of (1) is unique and globally exponentially stable in the following sense:

$$|x_i(t) - x_i^*| \leq d_i e^{-(\lambda - \eta)t}\|\phi\|, t \geq 0,$$

where $d_i = \frac{z_i}{\min\limits_{1 \leq j \leq n}\{z_j\}}$ and $\|\phi\| = \sup\limits_{s \in [-\tau, 0]}\|x(s) - x^*\|, i = 1, 2, \ldots, n.$

Proof. From (A_2), the existence of the equilibrium point of (2) implies one of (1). The proof of the existence of the equilibrium point x^* of (2) can be found in [10] and we omit it here. Next, We shall show the stability of the equilibrium point x^*. Let $x(t)$ be any solution of Eq.(1) with the initial function ϕ. Calculating the upper right derivative along the solutions of Eq.(1), from Condition (A_1), we can get for $t \neq t_k, k \in N$

$$D^+|x_i(t) - x_i^*| \leq -a_i|x_i(t) - x_i^*| + \sum_{j=1}^{n} b_{ij}|g_j(x_j(t - \tau_{ij}(t))) - g_j(x_j^*)|$$

$$\leq -a_i|x_i(t) - x_i^*| + \sum_{j=1}^{n} b_{ij}L_j|x_j(t - \tau_{ij}(t)) - x_j^*|. \qquad (3)$$

In the following, we shall prove $(\eta_0 = 1)$

$$|x_i(t) - x_i^*| \leq \eta_0\eta_1 \ldots \eta_{k-1}d_ie^{-\lambda t}\|\phi\|, \ t_{k-1} \leq t < t_k, \ k \in N. \qquad (4)$$

Since $d_i \geq 1$ and $\lambda > 0$,

$$|x_i(t) - x_i^*| \leq d_ie^{-\lambda t}\|\phi\|, \quad -\tau \leq t \leq 0. \qquad (5)$$

Now, we claim that for any $\varrho > \|\phi\| \geq 0$

$$|x_i(t) - x_i^*| \leq \varrho d_ie^{-\lambda t} \overset{\triangle}{=} y_i(t), \ \ 0 \leq t < t_1, i = 1, 2, \ldots, n. \qquad (6)$$

If this is not true, from the continuity of $x_i(t), y_i(t)$ as $t \in [0, t_1)$, then there must be a $t^* \in (0, t_1)$ and some integer m such that

$$|x_i(t) - x_i^*| \leq y_i(t), \ t \leq t^*, \ \ i = 1, \ldots, n, \qquad (7)$$

$$|x_m(t^*) - x_m^*| = y_m(t^*), \ \ D^+|x_m(t^*) - x_m^*| \geq y_m'(t^*). \qquad (8)$$

By using (3), (7) and (8),

$$D^+|x_m(t^*) - x_m^*| \leq -a_m|x_m(t^*) - x_i^*| + \sum_{j=1}^{n} |b_{mj}|L_j|x_j(t^* - \tau_{mj}(t^*)) - x_j^*|$$

$$\leq -a_my_m(t^*) + \sum_{j=1}^{n} |b_{mj}|L_jy_j(t^* - \tau_{mj}(t^*))$$

$$\leq [-a_md_m + \sum_{j=1}^{n} |b_{mj}|L_je^{\lambda\tau}d_j]\varrho e^{-\lambda t^*}. \qquad (9)$$

From (A_3), $-a_md_m + \sum_{j=1}^{n} |b_{mj}|L_je^{\lambda\tau}d_j < -\lambda d_m$, and so

$$D^+|x_m(t^*) - x_m^*| < -\lambda\varrho d_me^{-\lambda t^*} = y_m'(t^*),$$

which contradicts the inequality in (8). That is, (6) holds for any $\varrho > \|\phi\| \geq 0$. Letting $\varrho \to \|\phi\|$, then the inequalities (4) hold for $t \in [t_0, t_1)$. Suppose that for all $l = 1, \ldots, k$ the following inequalities hold

$$|x_i(t) - x_i^*| \leq \eta_0 \ldots \eta_{l-1}d_ie^{-\lambda t}\|\phi\|, t_{l-1} \leq t < t_l. \qquad (10)$$

By Eq.(1) and Condition (A_2), we have

$$|x_i(t_k^+) - x_i^*| = |I_{ik}(x_i(t_k^-) - x_i^*) + x_i(t_k^-) - x_i^*|$$
$$\leq \beta_{ik} |x_i(t_k^-) - x_i^*|$$
$$\leq \eta_k |x_i(t_k^-) - x_i^*|, \ k \in N. \tag{11}$$

From (10), (11) and $\eta_k \geq 1$, we get

$$|x_i(t) - x_i^*| \leq \eta_0 \ldots \eta_{k-1} \eta_k d_i e^{-\lambda t} \|\phi\|, \ t_k - \tau \leq t \leq t_k. \tag{12}$$

In a similar way as the proof of (4) with $k = 1$, we can prove that (12) implies

$$|x_i(t) - x_i^*| \leq \eta_0 \ldots \eta_{k-1} \eta_k d_i e^{-\lambda t} \|\phi\|, \ t_k \leq t < t_{k+1}.$$

By a simple induction, the inequalities (4) hold for any $k \in N$. Since $\eta_k \leq e^{\eta(t_k - t_{k-1})}$,

$$|x_i(t) - x_i^*| \leq e^{\eta t_1} \ldots e^{\eta(t_{k-1} - t_{k-2})} d_i e^{-\lambda t} \|\phi\|$$
$$\leq d_i e^{-(\lambda - \eta)t} \|\phi\|, \ t_{k-1} \leq t < t_k, \ k \in N.$$

Accordingly, we obtain global exponential stability and the uniqueness of the equilibrium point x^*. The proof is complete.

Remark 1. Assumption (A_1) is equivalent to $A - |B|L \in \mathcal{M}$. In fact, if $A - |B|L \in \mathcal{M}$, there must be a vector $z > 0$ satisfies $(A - |B|L)z > 0$. By the continuity, there exists a $\lambda > 0$ such that (A_3) holds. The reverse is easily derived from the definition of M-Matrix.

Using Theorem 1, we easily obtain the following sufficient conditions independent of delays for global exponential stability of impulsive neural networks.

Corollary 1. In addition to (A_1), assume that

i) $A - |B|L \in \mathcal{M}$;
ii) $\triangle x_i(t_k) = I_{ik}(x_i(t_k^-) - x_i^*), |s + I_{ik}(s)| \leq s, \forall s \in R, i = 1, 2, \ldots, n, k \in N$.

Then the equilibrium point of impulsive delay neural networks (1) is unique and globally exponentially stable.

Remark 2. Akca et al. [3] have ever proved that the equilibrium point of impulsive delay neural network model (1) with $\tau_{ij}(t) \equiv \tau_{ij}$ is unique and globally exponentially stable under the following assumptions

i) g_i is bounded and (A_1) holds, $i = 1, 2, \ldots, n$;
ii) $A - |B|L$ is column strictly dominant diagonal;
iii) $\triangle x_i(t_k) = -\gamma_{ik}[x_i(t_k^-) - x_i^*]$ and $0 < \gamma_{ik} < 2$.

It is easily seen that these conditions are restrictive than ones given in Corollary 1. So, their result is a special case of Corollary 1(or Theorem 1).

Corollary 2. If (A_1) holds and $A - |B|L \in \mathcal{M}$, then the equilibrium point of continuous delay neural networks (2) (i.e., $I_{ik}(s) \equiv 0$) is unique and globally exponentially stable.

Remark 3. Corollary 2 extends or improves the corresponding results in [6],[7], [8],[9],[10],[11],[12],etc..

3 Example

Example 1. Consider delay Hopfield neural networks with impulsive effects

$$\begin{cases} x_1'(t) = -3x_1(t) + |x_1(t - \tau_{11}(t))| + 0.5|x_2(t - \tau_{12}(t))|, \ t \neq t_k, \\ x_2'(t) = -3x_2(t) - 2|x_1(t - \tau_{21}(t))| + |x_2(t - \tau_{22}(t))|, \ t \geq 0, \\ \triangle x_1(t_k) = x_1(t_k^+) - x_1(t_k^-) = I_{1k}(x_1(t_k^-)), \ t_k = k, \\ \triangle x_2(t_k) = x_2(t_k^+) - x_2(t_k^-) = I_{2k}(x_2(t_k^-)), \ k \in N, \end{cases} \tag{13}$$

where $\tau_{ij}(t) = |\cos(i+j)t|, \ i,j = 1,2$.

i) If $I_{1k}(x_1) = I_{2k}(x_2) = 0, k \in N$, then system (13) becomes continuous Hopfield delay neural networks. We easily observe that $\tau = 1, L = E$ and $A - |B|L \in \mathcal{M}$. By Corollary 2 the system (13) has exactly one globally exponentially stable equilibrium point x^*, which is actually the point $(0,0)^T$.

ii) If $I_{1k}(x_1) = 0.3x_1, I_{2k}(x_2) = 0.3x_2, k \in N$, then system (13) is delay neural networks with impulses. Taking $z = (1,2)^T$, $\lambda = 0.3$, $\beta_{1k} = \beta_{2k} = 1.3$, $\eta_k = 1.3$ and $\eta = \ln(1.3)$, we can verify that all the conditions in Theorem 1 are satisfied and so the equilibrium point is globally exponentially stable.

Taking initial values: $x_1(s) = \cos(s), x_2(s) = \sin(s), s \in [-1,0]$, Fig. 1 depicts the time responses of state variables $(x_1(t), x_2(t))^T$ in the above two case.

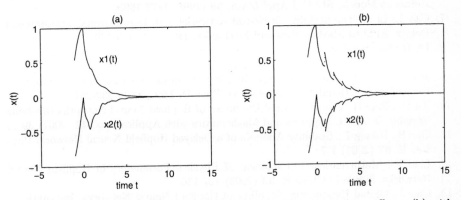

Fig. 1. Stability of delay neural networks (13): (a) without impulsive effects; (b) with impulsive effects.

4 Conclusion

In this paper, we have formulated a model of Hopfield neural network system with time-varying delays and impulsive effects. Sufficient conditions for global exponential stability of the hybrid system have been presented. The conditions are dependent on impulses and delays and so our results shows delay and impulsive effects on the stability of Hopfield neural networks. The criteria should have an important significance in the design of neural networks with global stability which can counteract the perturbations of both variable delays and impulsive effects.

Acknowledgments

The work is supported by the National Natural Science Foundation of China under Grant 10371083 and the Foundation of Technology Project of Chongqing Education Commission under Grant 041503.

References

1. Baldi P., Atiya A. F.: How Delays Affect Neural Dynamics and Learning. IEEE Trans. Neural Networks, **5** (1994) 612-621
2. Panas A. I., Yang T., Chua L. O.: Experimental Results of Impulsive Synchronization Between Two Chua's Circuits Int. J Bifurcation Chaos Appl Sci Eng., **8** (1998) 639-644.
3. Akca H., Alassar R., Covachev V., Covacheva Z., Al-Zahrani E.: Continuous-time Additive Hopfield-type Neural Networks with Impulses. J Math Anal Appl, **290** (2004) 436-451
4. Guan Z.-H., Chen G.: On Delayed Impulsive Hopfield Neural Networks. Neural Networks, **12** (1999) 273-280
5. Lakshmikantham V., Bainov D. D., Simeonov P. S.: Theory of Impulsive Differential Equations. World Scientific, Singapore (1989)
6. Driessche P. V. D., Zou X. F.: Global Attractivity in Delayed Hopfield Neural Networks Models. SIAM J Appl Math, **58** (1998) 1878-1890
7. Cao J., Li J.: The Stability in Neural Networks with Interneuronal Transmission Delays. Applied Mathematics and Mechanics, **19** (1998) 425-430
8. Lu H.: On stability of Nonlinear Continuous-time Neural Networks with Delays. Neural Networks, **13** (2000) 1135-1143
9. Mohamad S.: Global Exponential Stability of Continuous-time and Discrete-time Delayed Bidirectional Neural Networks. Phys D, **159** (2001) 233-251
10. Xu D., Zhao H., Zhu H.: Global Dynamics of Hopfield Neural Networks Involving Variable Delays. Computers and Mathematics with Applications, **42** (2001) 39-45
11. Guo S., Huang L.: Stability Analysis of a Delayed Hopfield Neural Network. Phys Rev E, **67** (2003) 1-7
12. Liao X.: Mathematical Connotation of Physical Parameters in Hopfield Neural Networks. Sience in China E, **33** (2003) 127-136
13. Cao J.: Global Exponential Stability of Hopfield Neural Networks, International Journal of Systems Science, **2** (2001) 233-236
14. Cao J., Wang J.: Absolute Exponential Stability of Recurrent Neural Networks with Time Delays and Lipschitz-continuous Activation Functions. Neural Networks, **3** (2004) 379-390

Global Exponential Stability of Discrete Time Hopfield Neural Networks with Delays*

Qiang Zhang[1], Wenbing Liu[2], and Xiaopeng Wei[1]

[1] University Key Lab of Information Science & Engineering, Dalian University,
Dalian 116622, China
zhangq26@hotmail.com
[2] School of Computer Science and Engineering, Wenzhou Normal College,
Wenzhou 325027, China

Abstract. By utilizing the Lyapunov function method to analyze stability of discrete time Hopfield neural networks with delays and obtain some new sufficient conditions for the global exponential stability of the equilibrium point for such networks. It is shown that the proposed conditions rely on the connection matrices and network parameters. The presented conditions are testable and less conservative than some given in the earlier references.

1 Introduction

Since Hopfield neural network were introduced in 1984, it has been successfully applied to many fields, such as signal processing, image processing, pattern recognition. In those applications, the stability plays an important role. In electronic circuits implementation, time delays are inevitable due to the finite switching speed of amplifiers and the inherent communication time of neurons, and this may result in an oscillation phenomenon or system instability. Therefore, it is significant to investigate stability conditions of Hopfield neural networks with delays. So far, many sufficient conditions have been presented ensuring global asymptotic stability and global exponential stability of the equilibrium point of continuous Hopfield neural network with delays, see, for example, [1]-[8] and references cited therein. On the other hand, a few results are given for global exponential stability of the equilibrium point for discrete time delayed Hopfield neural networks [9]-[10]. In this paper, we present a new sufficient condition for global exponential stability of the equilibrium point for discrete time delayed Hopfield neural networks.

2 Preliminaries

The dynamic behavior of a discrete time delayed Hopfield neural networks can be described by a system of difference equations with variable delays [9]

* The project supported by the National Natural Science Foundation of China (Grant Nos:60403001,60403002) and China Postdoctoral Science Foundation

J. Wang, X. Liao, and Z. Yi (Eds.): ISNN 2005, LNCS 3496, pp. 193–198, 2005.

$$u_i(k+1) = c_i u_i(k) + \sum_{j=1}^{n} a_{ij} f_j(u_j(k)) + \sum_{j=1}^{n} b_{ij} f_j(u_j(k-\tau_{ij})) + J_i \qquad (1)$$

where n corresponds to the number of units in a neural network; $u_i(k)$ is the activation of the ith neurons, $f(u(k)) = (f_1(u_1(k)), \cdots, f_n(u_n(k)))^T \in R^n$ denotes the activation function of the neurons; $A = (a_{ij})_{n \times n}$ is referred to as the feedback matrix, $B = (b_{ij})_{n \times n}$ represents the delayed feedback matrix, while J_i is an external bias vector, τ_{ij} is the transmission delay along the axon of the jth unit, c_i denotes the rate with which the cell i resets its potential to the resting state when isolated from other cells and inputs.

System (1) can be regarded as a discrete time analogue of the continuous time delayed Hopfield neural networks. Throughout this paper, we will use the following notations. For matrix $A = (a_{ij})_{n \times n}$, $|A|$ denotes absolute-value matrix given by $|A| = (|a_{ij}|)_{n \times n}$, $\rho(A)$ denotes its spectral radius, $A \geq 0$ means $a_{ij} \geq 0$ for all i,j and $||A||$ denotes its norm. I denotes the identity matrix with appropriate dimension.

In studying the stability of (1), we assume that $c_i > 0$, τ_{ij} are nonnegative integers with $\tau = \max_{1 \leq i,j \leq n} \tau_{ij}$ and each $f_i(\cdot)$ satisfies

$$|f_i(\xi_1) - f_i(\xi_2)| \leq L_i |\xi_1 - \xi_2|, \forall \xi_1, \xi_2. \qquad (2)$$

The system (1) is supplemented with an initial condition $x_i(l) = \varphi_i(l)$ for l is a nonnegative integer and $l \in [-\tau, 0]$.

We say that $u^* = (u_1^*, u_2^* \cdots, u_n^*)^T$ is an equilibrium point of (1) if

$$u_i^* = c_i u_i^* + \sum_{j=1}^{n} a_{ij} f_j(u_j^*) + \sum_{j=1}^{n} b_{ij} f_j(u_j^*) + J_i \qquad (3)$$

Definition 1. *Let u^* be an equilibrium point of system (1). If there exist real constants $\lambda > 1$ and $\nu \geq 1$ such that the solution $u(k) = (u_1(k), \cdots, u_n(k))^T$ of (1) satisfies*

$$\sum_{i=1}^{n} |u_i(k)| \leq \nu \sum_{i=1}^{n} \sup_{l \in [-\tau, 0]} |u_i(l)| \lambda^{-k} \qquad (4)$$

where l is a nonnegative integer. Then, the equilibrium point u^* is said to be globally exponentially stable and λ is the convergence rate.

Definition 2. *A real matrix $A = (a_{ij})_{n \times n}$ is said to be an M-matrix if $a_{ij} \leq 0, i,j = 1,2,\cdots,n, i \neq j$, and all successive principal minors of A are positive.*

In order to obtain the exponential stability conditions, we need the following lemma.

Lemma 1. *[11] Let $A = (a_{ij})_{n \times n}$ has nonpositive off-diagonal elements, then A is a nonsingular M matrix if one of the following conditions holds:*

1) A has all positive diagonal elements and there exist positive constants $d_j > 0$ such that

$$\sum_{j=1}^{n} d_j a_{ij} > 0 \text{ or } \sum_{j=1}^{n} d_j a_{ji} > 0, \ i = 1, 2, \cdots, n \qquad (5)$$

2) If A has the form $A = \alpha I - Q$, where $Q \geq 0$, then A is an M matrix if and only if $\alpha > \rho(Q)$.

3   Global Exponential Stability Analysis

In this section, we will use the above Lemma to establish the global exponential stability of system (1). Assume that $u^* = (u_1^*, u_2^* \cdots, u_n^*)^T$ is an equilibrium point of Eq. (1), one can derive from (1) that the transformation $x_i(k) = u_i(k) - u_i^*$ transforms system (1) into the following system

$$x_i(k+1) = c_i x_i(k) + \sum_{j=1}^{n} a_{ij} g_j(x_j(k)) + \sum_{j=1}^{n} b_{ij} g_j(x_j(k - \tau_{ij})) \qquad (6)$$

where $g_j(x_j(k)) = f_j(x_j(k) + u_j^*) - f_j(u_j^*)$. Obviously, to prove the stability of the equilibrium point of system (1), it is sufficient to prove the stability of the trivial solution of (6).

From (6) we can obtain

$$|x_i(k+1)| \leq c_i |x_i(k)| + \sum_{j=1}^{n} L_j |a_{ij}||x_j(k)| + \sum_{j=1}^{n} L_j |b_{ij}||x_j(k - \tau_{ij})| \qquad (7)$$

Theorem 1. *If the $\rho(\alpha) < 1$ where $\alpha = C + (|A| + |B|)L$, then the equilibrium point of system (1) is globally exponentially stable.*

Proof. Since $\rho(\alpha) < 1$ and $\alpha \geq 0$, by Lemma 1 we know that $I - \alpha = I - C - (|A| + |B|)L$ is an M matrix, and so there must exist constants $d_i > 0$ $(i = 1, 2, \cdots, n)$ satisfying

$$c_i + \sum_{j=1}^{n} \frac{d_j}{d_i} L_i |a_{ji}| + \sum_{j=1}^{n} \frac{d_j}{d_i} L_i |b_{ji}| < 1 \qquad (8)$$

Defining functions

$$F_i(\mu) = \mu c_i - 1 + \mu \sum_{j=1}^{n} \frac{d_j}{d_i} L_i |a_{ji}| + \sum_{j=1}^{n} \frac{d_j}{d_i} L_i |b_{ji}| \mu^{\tau_{ji}+1} \qquad (9)$$

for $\lambda \in [1, \infty)$.

We know that

$$F_i(1) = c_i - 1 + \sum_{j=1}^{n} \frac{d_j}{d_i} L_i |a_{ji}| + \sum_{j=1}^{n} \frac{d_j}{d_i} L_i |b_{ji}| < 0 \tag{10}$$

Due to the functions $F_i(\mu)$ being continuous on $[1, \infty)$, there exists a constant $\lambda > 1$ such that

$$\lambda c_i - 1 + \lambda \sum_{j=1}^{n} \frac{d_j}{d_i} L_i |a_{ji}| + \sum_{j=1}^{n} \frac{d_j}{d_i} L_i |b_{ji}| \lambda^{\tau_{ji}+1} < 0 \tag{11}$$

Let $y_i(k) = \lambda^k |x_i(k)|$. We will now show that the conditions given in Theorem 1 ensure the global exponential stability of the origin of (6). To this end, the following Lyapunov functional is used.

$$V(k) = \sum_{i=1}^{n} d_i \left\{ y_i(k) + \sum_{j=1}^{n} L_j |b_{ij}| \lambda^{\tau_{ij}+1} \sum_{l=k-\tau_{ij}}^{k-1} y_j(l) \right\} \tag{12}$$

Calculating the difference $\Delta V(k) = V(k+1) - V(k)$ along the solutions of (6), we get

$$\Delta V(k) = \sum_{i=1}^{n} d_i \left\{ y_i(k+1) + \sum_{j=1}^{n} L_j |b_{ij}| \lambda^{\tau_{ij}+1} \sum_{l=k-\tau_{ij}+1}^{k} y_j(l) \right\}$$

$$- \sum_{i=1}^{n} d_i \left\{ y_i(k) + \sum_{j=1}^{n} L_j |b_{ij}| \lambda^{\tau_{ij}+1} \sum_{l=k-\tau_{ij}}^{k-1} y_j(l) \right\}$$

$$= \sum_{i=1}^{n} d_i \left\{ \lambda^{k+1} |x_i(k+1)| + \sum_{j=1}^{n} L_j |b_{ij}| \lambda^{\tau_{ij}+1} \sum_{l=k-\tau_{ij}+1}^{k} y_j(l) \right\}$$

$$- \sum_{i=1}^{n} d_i \left\{ y_i(k) + \sum_{j=1}^{n} L_j |b_{ij}| \lambda^{\tau_{ij}+1} \sum_{l=k-\tau_{ij}}^{k-1} y_j(l) \right\} \tag{13}$$

By substituting (7) in the above, we can obtain

$$\Delta V(k) \leq \sum_{i=1}^{n} d_i \left\{ \lambda^{k+1} c_i |x_i(k)| + \lambda^{k+1} \sum_{j=1}^{n} L_j |a_{ij}| |x_j(k)| \right.$$

$$+ \lambda^{k+1} \sum_{j=1}^{n} L_j |b_{ij}| |x_j(k - \tau_{ij})| + \sum_{j=1}^{n} L_j |b_{ij}| \lambda^{\tau_{ij}+1} \sum_{l=k-\tau_{ij}+1}^{k} y_j(l) \right\}$$

$$- \sum_{i=1}^{n} d_i \left\{ y_i(k) + \sum_{j=1}^{n} L_j |b_{ij}| \lambda^{\tau_{ij}+1} \sum_{l=k-\tau_{ij}}^{k-1} y_j(l) \right\}$$

$$= \sum_{i=1}^{n} d_i \left\{ \lambda c_i y_i(k) + \lambda \sum_{j=1}^{n} L_j |a_{ij}| y_j(k) + \sum_{j=1}^{n} L_j |b_{ij}| \lambda^{\tau_{ij}+1} y_j(k - \tau_{ij}) \right.$$

$$\left. + \sum_{j=1}^{n} L_j |b_{ij}| \lambda^{\tau_{ij}+1} \sum_{l=k-\tau_{ij}+1}^{k} y_j(l) \right\}$$

$$- \sum_{i=1}^{n} d_i \left\{ y_i(k) + \sum_{j=1}^{n} L_j |b_{ij}| \lambda^{\tau_{ij}+1} \sum_{l=k-\tau_{ij}}^{k-1} y_j(l) \right\}$$

$$\leq \sum_{i=1}^{n} d_i \left\{ \lambda c_i - 1 + \lambda \sum_{j=1}^{n} \frac{d_j}{d_i} L_i |a_{ji}| + \sum_{j=1}^{n} \frac{d_j}{d_i} L_i |b_{ji}| \lambda^{\tau_{ji}+1} \right\} y_i(k)$$

$$(14)$$

By using (11) in the above inequality we have $\Delta V(k) \leq 0$ and so $V(k) \leq V(0)$. Since

$$\min_{1 \leq i \leq n} \{d_i\} \sum_{i=1}^{n} y_i(k) \leq \sum_{i=1}^{n} d_i y_i(k) \leq V(k) \qquad (15)$$

$$V(0) = \sum_{i=1}^{n} d_i \left\{ y_i(0) + \sum_{j=1}^{n} L_j |b_{ij}| \lambda^{\tau_{ij}+1} \sum_{l=-\tau_{ij}}^{-1} y_j(l) \right\}$$

$$\leq \max_{1 \leq i \leq n} \{d_i\} \sum_{i=1}^{n} \left\{ 1 + L_i \sum_{j=1}^{n} |b_{ji}| \lambda^{\tau_{ji}+1} \tau_{ji} \right\} \sup_{l \in [-\tau, 0]} |x_i(l)| \qquad (16)$$

Then we easily follow from the above

$$\sum_{i=1}^{n} |x_i(k)| \leq \frac{\max_{1 \leq i \leq n} \{d_i\}}{\min_{1 \leq i \leq n} \{d_i\}} \lambda^{-k} \sum_{i=1}^{n} \left\{ 1 + L_i \sum_{j=1}^{n} |b_{ji}| \lambda^{\tau_{ji}+1} \tau_{ji} \right\} \sup_{l \in [-\tau, 0]} |x_i(l)|$$

$$= \nu \sum_{i=1}^{n} \sup_{l \in [-\tau, 0]} |x_i(l)| \lambda^{-k} \qquad (17)$$

where

$$\nu = \frac{\max_{1 \leq i \leq n} \{d_i\}}{\min_{1 \leq i \leq n} \{d_i\}} \max_{1 \leq i \leq n} \left\{ 1 + L_i \sum_{j=1}^{n} |b_{ji}| \lambda^{\tau_{ji}+1} \tau_{ji} \right\} \geq 1$$

This completes the proof.

Since for any square matrix A one have $\rho(A) \leq ||A||$, it directly follows that the following corollary holds.

Corollary 1. *For any matrix norm $|| \cdot ||$, if $||C + (|A| + |B|)L|| < 1$, then the equilibrium point of system (1) is globally exponentially stable. In particular, any one of the following conditions ensures global exponential stability of (1).*

$$(1) \quad c_j + \sum_{i=1}^{n} L_j \left(|a_{ij}| + |b_{ij}| \right) < 1 \tag{18}$$

$$(2) \quad c_i + \sum_{j=1}^{n} L_j \left(|a_{ij}| + |b_{ij}| \right) < 1 \tag{19}$$

$$(3) \quad \sum_{i=1}^{n} \sum_{j=1}^{n} \left[c_i \delta_{ij} + L_j \left(|a_{ij}| + |b_{ij}| \right) \right]^2 < 1 \tag{20}$$

Proof. By applying the column norm, row norm and Frobenius norm of the matrix $C + (|A| + |B|)L$ respectively, combining with the inequality $|x + y| \leq |x| + |y|$ where x, y are any real constants, we can easily obtain the above corollary.

Remark 1. The condition in Corollary (2) corresponds to that of Theorem 3.1 in [9]. Therefore, condition of Theorem 3.1 in [9] can be included as a special case of our Theorem.

References

1. Arik, S.: An Improved Global Stability Result for Delayed Cellular Neural Networks. IEEE Trans.Circuits SystI, **49** (2002) 1211–1214
2. Cao, J., Wang, J.: Global Asymptotic Stability of a General Class of Recurrent Neural Networks with Time-Varying Delays. IEEE Trans.Circuits Syst.I, **50** (2003) 34–44
3. Liao, X., Chen, G., Sanchez, E.N.: LMI-Based Approach for Asymptotically Stability Analysis of Delayed Neural Networks. IEEE Trans.Circuits Syst.I, **49** (2002) 1033–1039
4. Zeng, Z., Wang, J., Liao, X.: Global Exponential Stability of a General Class of Recurrent Neural Networks with Time-Varying Delays. IEEE Trans.Circuits Syst.I, **50** (2003) 1353–1358
5. Zhang, J.: Globally Exponential Stability of Neural Networks with Variable Delays. IEEE Trans.Circuits Syst.I, **50** (2003) 288–290
6. Qiang, Z., Run, M., Chao, W., Jin, X.: On the Global Stability of Delayed Neural Networks. IEEE Trans.Automatic Control, **48** (2003) 794–797
7. Chen, T.: Global Exponential Stability of Delayed Hopfield Neural Networks. Neural Networks, **14** (2001) 977–980
8. Yi, Z., Heng, P.A., Leung, K.S.: Convergence Analysis of Cellular Neural Networks with Unbounded Delay. IEEE Trans. Circuits Syst.I, **48** (2001) 680–687
9. Mohamad, S., Gopalsamy, K.: Exponential Stability of Continuous-Time and Discrete-Time Cellular Neural Networks with Delays. Appl.Math.Comput. **135** (2003) 17–38
10. Guo, S., Huang, L., Wang, l.: Exponential Stability of Discrete-Time Hopfield Neural Networks. Computers and Mathematics with Applications, **47** (2004) 1249–1256
11. Berman, A., Plemmons, R.J.: Nonnegative Matrices in the Mathematical Science. Academic Press (1979)

Stability Analysis of Uncertain Neural Networks with Linear and Nonlinear Time Delays

Hanlin He[1], Zhongsheng Wang[2], and Xiaoxin Liao[3]

[1] College of Sciences, Naval University of Engineering,
Wuhan, Hubei 430033, China
hanlinhe62@yahoo.com.cn
[2] Department of Electrical Engineering, ZhongYuan Institute of Technology,
Zhengzhou, Henan 450007, China
[3] Department of Control Science and Engineering, Huazhong University of Science
and Technology,
Wuhan, Hubei 430074, China

Abstract. A method is proposed for stability analysis of neural networks with linear and nonlinear time delays. Given a neural network and the corresponding generalized algebraic Riccati equation with two unknown positive matrices, using the Razumikhin-type theory, the problem of insuring the globally asymptotic stability of the neural networks with linear and nonlinear time delays is obtained.

1 Introduction

Neural networks have attracted the attention of the scientists, due to their promising potential for the tasks of classification, associate memory and parallel computation, etc., and various results were reported [1-10]. However, in hardware implementation, uncertainty and time delays occur due to disturbance between the electric components and finite switching speeds of the amplifiers, can affect the stability of a network by creating oscillatory and unstable characteristics. It is important to investigate the dynamics of uncertain neural networks with time delays. Most papers discussed the stability of the neural networks with time delays and obtained some sufficient conditions for globally stability exponential stability, but there are few results of stability for the uncertain neural networks [10].

This paper continues the research line in [10] and extends the corresponding result to the case when the uncertain neural networks contain linear and nonlinear delays. By solving a generalized algebraic Riccati equation with two unknown positive matrices and by the using of Razumikhin-type theory, this paper achieves a sufficient criterion for globally asymptotic stability of the neural networks with linear and nonlinear time delays.

2 Main Result

Throughout this paper, the following notations are used. $\lambda_{\min}(P), \lambda_{\max}(P)$ denote the minimal and maximal eigenvalues of matrix P. $C_\tau = C([-\tau, 0], R^n)$

J. Wang, X. Liao, and Z. Yi (Eds.): ISNN 2005, LNCS 3496, pp. 199–202, 2005.

denotes the Banach space of continuous vector functions mapping the interval $[-\tau, 0]$ into R^n with the topology of uniform convergence. $\|.\|$ refers to either the Euclidean vector norm or the induced matrix 2-norm.

We consider the uncertain neural networks with linear and nonlinear time delays described by the differential-difference equation of the form

$$\begin{cases} \dot{x}(t) = [A + \Delta A(t)]x(t) + A_d x(t - \tau) + Bf(x(t - h(t))) \\ x(t) = \phi(t), t \in [-\max\{\tau, h\}, 0] \end{cases} \quad (1)$$

Where $x(t) \in R^n$ is the state of the neuron; A, B is real constant matrices with appropriate dimension; τ is constant time-delay; $h(t)$ is time-varying delay, satisfying $0 \le h(t) \le h$; $\phi(t)$ is a continuous vector-valued initial function defined on $[-\max\{\tau, h\}, 0]$; the norm of $\Delta A(t)$ is bounded by Δ; $f \in C(R^n, R^n)$ is bounded function, the boundedness is l, that is $\|f(x(t - h(t)))\| \le l \|x(t - h(t))\|$ and $f(0) = 0$;

Given symmetric and positive matrix Q, if there exist two symmetric and positive definite matrices P and S that are solutions of the following generalized algebraic Riccati equation

$$A^T P + PA + PA_d S^{-1} A_d^T P + S + Q = 0 \quad (2)$$

Then, we have the following theorem.

Theorem 1. Suppose P and S are the symmetric and positive definite matrix solutions of the Riccati equation (2). Then the zero solution of system (1) is globally uniformly asymptotically stable if

$$\Delta\|P\| + l\|PB\| < \frac{\lambda_{\min}(Q)}{2} \quad (3)$$

Proof. Take the following Lyapunov functional candidate:

$$V(x(t)) = x^T(t)Px(t) + \int_{t-\tau}^{t} x(\theta)^T Sx(\theta)d\theta \quad (4)$$

Clearly $V(t)$ satisfies the following inequalities

$$\lambda_{\min}(P)\|x(t)\|^2 \le V(x(t)) \le \{\lambda_{\max}(P) + \tau\lambda_{\max}(S)\}\|x_t\|_C^2$$

where $\|x_t\|_C^2 = \max_{-\tau \le \theta \le 0} \|x_t(\theta)\|$. The time derivative of $V(x(t))$ along the system (1) is given by:

$$\dot{V}(x(t)) = \dot{x}(t)^T Px(t) + x(t)^T P\dot{x}(t) + x(t)^T Sx(t) - x(t - \tau)^T Sx(t - \tau)$$
$$= x^T(t)[PA + A^T P]x(t) + 2x^T(t)P\Delta A(t)x(t)$$
$$+ x^T(t - \tau)A_d^T Px(t) + x^T(t)PA_d x(t - \tau)$$
$$+ x(t)^T Sx(t) - x(t - \tau)^T Sx(t - \tau) + 2x^T PBf(x(t - h(t)))$$

Then from (2), it follows

$$\dot{V}(x(t)) = -x(t)^T Qx(t) + 2x^T(t)P\Delta A(t)x(t) + 2x(t)^T PBf(x(t - h(t)))$$
$$- [x(t - \tau) - S^{-1}A_d^T Px(t)]^T S[x(t - \tau) - S^{-1}A_d^T Px(t)]$$

hence

$$\dot{V}(x(t)) \leq -\lambda_{\min}(Q)\|x(t)\|^2 + 2\|P\Delta A(t)\|\|x(t)\|^2$$
$$+2\|x(t)\|\|PBf(x(t-h(t))\|$$
$$\leq -(\lambda_{\min}(Q) - 2\Delta\|P\|)\|x(t)\|^2$$
$$+2l\|PB\|\|x(t)\|\|x(t-h(t))\|$$

From the use of the Razumikhin's Theorem [10-11], it is assumed that there exists a positive $q > 1$ such that

$$\|x(t-h(t))\| < q\|x(t)\|$$

Hence

$$\dot{V}(x(t)) \leq -(\lambda_{\min}(Q) - 2\Delta\|P\| - 2lq\|PB\|)\|x(t)\|^2$$
$$= -\alpha\|x(t)\|^2$$

where

$$\alpha = (\lambda_{\min}(Q) - 2\Delta\|P\| - 2lq\|PB\|)$$

When

$$\Delta\|P\| + l\|PB\| < \frac{\lambda_{\min}(Q)}{2}$$

there exists $q > 1$ small enough such that $\alpha > 0$. Applying Razumikhin Type Theorem, the zero solution of the system (1) is globally uniformly asymptotically stable.

Remark 1. Suppose $PA_dS^{-1}A_d^TP + S + Q > 0$, then from standard results on Lyapunov equations, A is stable iff $P \geq 0$. Hence, the necessary condition for the equation (2) has solutions is that A is stable.

Remark 2. There is a problem to be further studied: the solutions of equation (2) is not unique in general, for the neural networks (1), how to chose S and P to make the term $[x(t-\tau) - S^{-1}A_d^TPx(t)]^T S[x(t-\tau) - S^{-1}A_d^TPx(t)]$ small, then the inequality (3) will be satisfied for larger Δ and l.

3   Illustrative Example

Example. Suppose the system (1) with the following data

$$A = \begin{bmatrix} -22.333 & -6.232 \\ -2.0333 & -2.7732 \end{bmatrix}; A_d = \begin{bmatrix} 0 & -1 \\ 0 & 0 \end{bmatrix}; B = I$$

Then by choosing $Q = \begin{bmatrix} 7.4443 & 2.8041 \\ 2.8041 & 3.5978 \end{bmatrix}$ one gets

$$P = \begin{bmatrix} 0.2387 & -0.0537 \\ -0.0537 & 1.3102 \end{bmatrix}; S = \begin{bmatrix} 2.9810 & 0.0043 \\ 0.0043 & 2.9990 \end{bmatrix}$$

$\|P\| = \lambda_{\max}(P) = 1.3128, \lambda_{\min}(Q) = 2.1208$. Then, from Theorem 1 when

$$1.3128\Delta + 1.3128l < 2.1208 \tag{5}$$

the zero solution of neural networks (1) is globally uniformly asymptotically stable. For example, let

$$\Delta A(t) = \begin{bmatrix} 0 & r \\ 0 & 0 \end{bmatrix}; f(x(t - h(t))) = \begin{bmatrix} \sin x_2(t - h(t)) & 0 \\ 0 & \sqrt{1 - \cos x_1(x(t - h(t)))} \end{bmatrix}$$

where $|r| < 0.3$, then $\Delta = 0.3$, since $l = 1$, the inequality (5) is satisfied, the zero solution of neural networks (1) is globally uniformly asymptotically stable for the given data.

Acknowledgments

This work is supported by the Natural Science Foundation of China (60274007, 60474011), and Academic Foundation of Naval University of Engineering, China.

References

1. Liang, X. B., Wu, L. D.: Globally Exponential Stability of a Class of Neural Circuits. IEEE Trans. Circuits Syst. I, **46** (1999) 748-751
8. Liao, X. X., Xiao, D. M.: Globally Exponential Stability of Hopfield Neural Networks with Time-varying Delays. ACTA Electronica Sinica, **28** (2000) 1-4
3. Cao, J. D., Li, Q.: On the Exponential Stability and Periodic Solution of Delayed Cellular Neural Networks. J. Math. Anal. and Appl., **252** (2000) 50-64
4. Zhang, Y., Heng, P. A., Leung, K. S.: Convergence Analysis of Cellular Neural Networks with Unbounded Delay. IEEE Trans. Circuits Syst. I, **48** (2001) 680
5. Li, S. Y., Xu, D. Y.: Exponential Attractions Domain and Exponential Convergent Rate of Associative Memory Neural Networks with Delays. Control Theory and Applications, **19** (2002) 442-444
8. Liao, X. X., Wang, J.: Algebraic Criteria for Global Exponential Stability of Cellular Neural Networks with Multiple Time Delays. IEEE Trans. Circuits and Systems I., **50** (2003) 268-275
7. Zeng, Z. G., Wang, J., Liao, X. X.: Global Exponential Stability of a General Class of Recurrent Neural Networks with Time-varying Delays. IEEE Trans. Circuits and Syst. I, **50** (2003) 1353-1358
8. Sun, C., Feng, C.: Exponential Periodicity of Continuous-time and Discrete-time Neural Networks with Delays. Neural Processing Letters, **19** (2004) 131-146
9. Zeng, Z. G., Wang, J., Liao, X. X.: Stability Analysis of Delayed Cellular Neural Networks Described Using Cloning Templates. IEEE Trans. Circuits and Syst. I, **51** (2004) 2313-2324
10. Wang, Z. S., He, H. L., Liao, X. X.: Stability Analysis of Uncertain Neural Networks with Delay. Lecture Notes in Computer Science, **3173** (2004) 44-48
11. Xu, B. G., Liu, Y. Q.: An Improved Razumikhin Type Theorem and Its Applications. IEEE Transaction on Automatic Control, **39** (1994) 839-841

Robust Stability for Delayed Neural Networks with Nonlinear Perturbation*

Li Xie[1], Tianming Liu[2], Jilin Liu[1], Weikang Gu[1], and Stephen Wong[2]

[1] Department of Information and Electronic Engineering, Zhejiang University
310027 Hangzhou, P.R. China
Xiehan@Zju.edu.cn
[2] Center for Bioinformatics, HCNR, Harvard Medical School
02115 Boston, MA, USA
Tliu@Bwh.harvard.edu

Abstract. The problem of analysis of robust stability for time-delayed neural networks with nonlinear perturbation has been investigated via Lyapunov stability theory. The sufficient conditions for robust stability of neural networks with time delays have been developed. The exponential stability criterion for neural networks is also derived. The result includes the information on the state convergence degree of the neural networks. The robust stable criterion in this paper is presented in terms of linear matrix inequalities.

1 Introduction

The stability analysis for neural networks has received considerable attentions in recent years. It is well known that the stability of neural network is the prerequisite for its applications in either pattern recognition or optimization solving. There have been extensive research results presented about the stability analysis of neural network and its applications. In ref [1], matrix measure is used for the stability analysis of dynamical neural network. In ref [2], a new concept called nonlinear measure is introduced to quantify stability of nonlinear systems in the way similar to the matrix measure. In ref [3], absolute exponential stability of a general class of neural networks is analyzed based on Lyapunov stability theory. In general, time delays are natural components in various dynamic systems. There usually exist transmission delays among the neurons. It is recognized that time delay is frequently a major source of instability and poor performance in neural networks. The global exponential stability is analyzed for neural networks with time delays in ref [4]. In ref [5, 6], the authors introduced linear matrix inequality (LMI) theory for the stability analysis of neural networks.

In this paper, we will investigate the problem of analysis of robust stability for delayed neural networks with nonlinear perturbation. Some sufficient conditions for the delay independent robust stability of the neural networks will be developed. All of the results are presented in terms of linear matrix inequalities (LMIs). The remainder of

---

* This work was supported by Chinese National Science Foundation (60473129).

J. Wang, X. Liao, and Z. Yi (Eds.): ISNN 2005, LNCS 3496, pp. 203–208, 2005.

this paper is organized as follows: In section 2, the system model is described. Some necessary assumptions are given. In section 3, the robust stable criteria are developed. Two cases of time delay, i.e. single delay case and multiple delays case are discussed. Then, an exponential stability criterion for the considered neural networks is provided. Finally, conclusions are provided in section 4.

2 Systems Descriptions

Consider a delayed neural network with nonlinear perturbation, which is described by a set of functional differential equations as follows:

$$\dot{x}_i(t) = -a_i x_i(t) - a_{di} x_i(t-\tau) + \sum_{j=1}^{n} b_i \sigma_j[x_j(t)] + \sum_{j=1}^{n} b_{dij} \sigma_j[x_j(t-\tau)]$$
$$+ c_i f_i(x_i(t), x_i(t-\tau)). \tag{1}$$

or, the considered neural networks can be represented in the form of vector state space as follows

$$\dot{x}(t) = -Ax(t) - A_1 x(t-\tau) + B\sigma[x(t)] + B_1 \sigma[x(t-\tau)]$$
$$+ Cf(x(t), x(t-\tau)). \tag{2}$$

where $x(t) = (x_1(t), \ldots, x_n(t))^T$ is the state vector of the neural network, $x(t-\tau) = (x_1(t-\tau), \ldots, x_n(t-\tau))^T$ is the delayed state vector of the neural networks, and the delay $\tau \geq 0$. $f(x(t), x(t-\tau))$ is the exogenous perturbation with the form of $f(x(t), x(t-\tau)) = [f_1(x_1(t), x_1(t-\tau)), \ldots, f_n(x_n(t), x_n(t-\tau))]^T$, the activation function is $\sigma[x(t)] = \{\sigma_1[x_1(t)], \ldots, \sigma_n[x_n(t)]\}^T$.

Though out this paper, we assume that the activation function $\sigma_i[x_i(t)]$ and the perturbation function $f(x(t), x(t-\tau))$ satisfy the following conditions:

(A.1) If there exist positive constants k_i, $i = 1, \ldots, n$, such that

$$0 < \frac{\sigma_i(x_1) - \sigma_i(x_2)}{x_1 - x_2} \leq k_i, \ \forall x_1, x_2 \in R, \ x_1 \neq x_2, \ i = 1, \ldots, n. \tag{3}$$

(A.2) There exist positive constant matrices M and M_1, such that

$$\|f(x(t), x(t-\tau))\| \leq M\|x(t)\| + M_1\|x(t-\tau)\|. \tag{4}$$

3 Main Results

In this section, stability criteria for delayed neural networks with nonlinear perturbation are given.

Theorem 1. Consider the delayed neural networks with nonlinear perturbation (1), if there exist matrices $X > 0$, $W > 0$, $S > 0$, $S_1 > 0$ and constants $\xi_j > 0$ ($j = 1, 2$), satisfying the LMIs

(a). $\Omega = \begin{bmatrix} \Omega_{11} & XK^T & XM^T \\ KX & -S & 0 \\ MX & 0 & -\xi_1 I \end{bmatrix} < 0.$ $\qquad\qquad (5)$

where

$\qquad \Omega_{11} = -XA^T - AX + W + BSB^T + B_1 S_1 B_1^T + C(\xi_1 + \xi_2)C^T + 3X .$

(b). $\begin{bmatrix} X & XA_1^T \\ XA_1 & W \end{bmatrix} \geq 0 ; \begin{bmatrix} X & XK^T \\ XK & S_1 \end{bmatrix} \geq 0 ; \begin{bmatrix} X & XM_1^T \\ XM_1 & \xi_2 I \end{bmatrix} \geq 0 .$ $\qquad (6)$

then the system (1) is globally asymptotically stable. Here, $K = diag\{k_1, \ldots, k_n\}$.

Proof. Given $P > 0$, introduce a quadratic Lyapunov function as

$$V(x(t), t) = x^T(t) Px(t) . \qquad\qquad (7)$$

It is easy to obtain $0 \leq \varepsilon_1 \|x(t)\|^2 \leq V(x(t), t) \leq \varepsilon_2 \|x(t)\|^2$, where $\varepsilon_1 = \lambda_{\min}(P)$, $\varepsilon_2 = \lambda_{\max}(P)$.

The derivative of the Lyapunov function along the solution of system (1) is

$\dot{V}(x(t), t) = \dot{x}^T(t) Px(t) + x^T(t) P\dot{x}(t)$

$= \{-x^T(t) A^T - x^T(t-\tau) A_1^T + \sigma^T(x(t)) B^T + \sigma^T[x(t-\tau)] B_1^T \} Px(t)$

$\quad + x^T(t) P\{-Ax(t) - A_1 x(t-\tau) + B\sigma[x(t)] + B_1 \sigma[x(t-\tau)]\}$

$\quad + f^T(x(t), x(t-\tau)) C^T Px(t) + x(t)^T PCf(x(t), x(t-\tau))$

$\leq \{-x^T(t) A^T Px(t) - x^T(t) PAx(t)\} + x^T(t-\tau) A_1^T W^{-1} A_1 x(t-\tau)$

$\quad + x^T(t) PWPx(t) + \sigma^T[x(t)] S^{-1} \sigma[x(t)] + x^T(t) PBSB^T Px(t)$

$\quad + \sigma^T[x(t-\tau)] S_1^{-1} \sigma[x(t-\tau)] + x^T(t) PB_1 S_1 B_1^T Px(t)$

$\quad + 2\|x(t)^T PC\| \|f(x(t), x(t-\tau))\|$

$\leq \{-x^T(t) A^T Px(t) - x^T(t) PAx(t)\} + x^T(t-\tau) A_1^T W^{-1} A_1 x(t-\tau)$

$\quad + x^T(t) PWPx(t) + x^T(t) K^T S^{-1} Kx(t) + x^T(t) PBSB^T Px(t)$

$\quad + x^T(t-\tau) K^T S_1^{-1} Kx(t-\tau) + x^T(t) PB_1 S_1 B_1^T Px(t)$

$\quad + x^T(t) PC(\xi_1 + \xi_2) C^T Px(t) + \xi_1^{-1} x^T(t) M^T Mx(t)$

$\quad + \xi_2^{-1} x^T(t-\tau) M_1^T M_1 x(t-\tau) .$ $\qquad\qquad (8)$

To apply Razumikhin-type theorem, we assume that there exist a constant $\varepsilon > 1$ and a constant $\delta > 0$ such that

$$V(x(t+\theta),t+\theta) \le \varepsilon V(x(t),t), \ \forall \theta \in [-\tau,0].$$ (9)

let $\varepsilon = 1 + \delta$ and

$$\Xi_1 = -A^T P - PA + PWP + K^T S^{-1} K + PBSB^T P + PB_1 S_1 B_1^T P$$
$$+ PC(\xi_1 + \xi_2)C^T P + \xi_1^{-1} M^T M + (3 + 4\delta)P.$$ (10)

If $\Xi_1 < 0$, $P = X^{-1}$, from (6) we have

$$\dot{V}(x(t),t) \le x^T(t)\{-A^T P - PA + K^T S^{-1} K + PB_1 S_1 B_1^T P + PBSB^T P$$
$$+ PC(\xi_1 + \xi_2)C^T P + PWP + 3\varepsilon P + \xi_1^{-1} M^T M\}x(t)$$ (11)
$$\le -\delta V(x(t),t).$$

According to Razumikhin-type theorem, the considered neural network is globally asymptotically stable. By continuity,

$$\Xi < 0 \Leftrightarrow \Xi_1 < 0, \ \delta \to 0^+.$$ (12)

where

$$\Xi = -A^T P - PA + PWP + K^T S^{-1} K + PBSB^T P + PB_1 S_1 B_1^T P$$
$$+ PC(\xi_1 + \xi_2)C^T P + \xi_1^{-1} M^T M + 3P.$$ (13)

Pre- and post-multiply (13) with $X = P^{-1}$. By the Schur complement, $\Xi < 0$ if and only if inequality (5) holds.

This completes the proof.

Remark 1. Noting that the conditions of Theorem 1 do not include any information of the delay, that is, the theorem provides a delay-independent robust stability criterion for time-delayed neural networks with nonlinear perturbations in terms of LMIs. The above results can be extended to multi-delay case.

Consider the multi-delayed neural networks as follows

$$\dot{x}(t) = -Ax(t) - \sum_{i=1}^{l} A_i x(t-\tau_i) + B\sigma[x(t)] + \sum_{i=1}^{l} B_i \sigma[x(t-\tau_i)]$$
$$+ Cf(x(t), x(t-\tau_1), \ldots, x(t-\tau_l)).$$ (14)

where $x(t) = (x_1(t), \ldots, x_n(t))^T$ is the state vector of the neural network, $x(t-\tau_i) = (x_1(t-\tau_i), \ldots, x_n(t-\tau_i))^T$ is the delayed state vector of the neural network, $f(x(t), x(t-\tau_1), \ldots, x(t-\tau_i))$ is the exogenous perturbation, we assume that there exist positive constant matrices M, M_1, ..., M_l, satisfying the following condition:

$$\|f(x(t), x(t-\tau_1), \ldots, x(t-\tau_i))\| \le M\|x(t)\| + M_1\|x(t-\tau_1)\| + \ldots\ldots$$
$$+ M_l\|x(t-\tau_l)\|.$$ (15)

$\sigma[x(t)] = \{\sigma_1[x_1(t)], \ldots, \sigma_n[x_n(t)]\}^T$ denotes the neuron activation function, $\tau_i \geq 0$ are the delays, $i = 1, \ldots, I$.

Similarly, according to the proof of Theorem 1, we have the following result.

Theorem 2. Consider the multi-delayed neural networks with nonlinear perturbation (14), if there exist matrices $X > 0$, $S_0 > 0$, $W_i > 0$, $S_i > 0$ and constants $\xi_0 > 0$, $\xi_i > 0 (i = 1, \ldots, I)$, satisfying the LMIs

(a). $\Omega = \begin{bmatrix} \Omega_{11} & XK^T & \Omega_{13} \\ KX & -S_0 & 0 \\ \Omega_{31} & 0 & \Omega_{33} \end{bmatrix} < 0.$ 　　　　　　　　(16)

where

$$\Omega_{11} = -XA^T - AX + \sum_{i=1}^{I} W_I + \sum_{i=1}^{I} \xi_i CC^T + \sum_{i=1}^{I} B_I S_I B_I^T$$
$$+ BS_0 B^T + \xi_0 CC^T + 3X$$
$$\Omega_{13} = \Omega_{31}^T = [XM^T \quad XM_1^T \quad \cdots \quad XM_I^T]$$
$$\Omega_{33} = diag\{-\xi_0 I, -\xi_1 I, \ldots, -\xi_I I\}.$$ 　　　　　　　　(17)

(b). $\begin{bmatrix} X & XA_i^T \\ XA_i & W_i \end{bmatrix} \geq 0 ; \quad \begin{bmatrix} X & XK^T \\ XK & S_i \end{bmatrix} \geq 0 ; \quad \begin{bmatrix} X & XM_i^T \\ XM_i & \xi_i I \end{bmatrix} \geq 0$

$i = 1, \ldots, I$. 　　　　　　　　(18)

then the system (14) is globally asymptotically stable.

Remark 2. In Theorem 1 and Theorem 2, the criteria of delay-independent stability are developed. However, no information about the state convergence degree of the neural networks is given. Here, we investigate the problem of exponential stability analysis for delayed neural networks with nonlinear perturbation.

Theorem 3. Consider the time delayed neural networks with nonlinear perturbation (1), if there exist matrices $X > 0$, $W > 0$, $S > 0$, $S_1 > 0$ and constants $\xi_j > 0 (j = 1, 2)$, satisfying the LMIs

(a). $\Omega = \begin{bmatrix} \Omega_{11} & XK^T & XM^T \\ KX & -S & 0 \\ MX & 0 & -\xi_1 I \end{bmatrix} < 0.$ 　　　　　　　　(19)

Where

$$\Omega_{11} = -XA^T - AX + W + BSB^T + B_1 S_1 B_1^T + C(\xi_1 + \xi_2)C^T$$
$$+ (3 + 2\alpha)X$$

the global robust stability in a manner that they specify the size of perturbation Hopfield neural networks can endure when the structure of the network is given. On the other hand, from the viewpoint of system synthesis, our results can answer how to choose the parameters of neural networks to endure a given perturbation. To the best of our knowledge, there does not seem to exist robust stability criteria involving the use of linear matrix inequalities in the literature.

The present paper is organized as follows. In Section 2, the neural network model will be described. In Section 3, by constructing a suitable Lyapunov-Krasovskii functional and using the sector conditions, sufficient conditions for the robust stability of the present model will be established. Finally, Section 4 will conclude the present paper.

2   The Neural Network Model

We consider the following Hopfield neural network model with multiple delays described by equations of the form

$$\dot{x}(t) = -Cx(t) + T_0 S(x(t)) + \sum_{k=1}^{K} T_k S(x(t - \tau_k)) + I \tag{1}$$

where $x = (x_1, \cdots, x_n)^T \in \Re^n$ denotes the state variables, $C = \text{diag}[c_1, \cdots, c_n]$ with $c_i > 0$, $i = 1, \cdots, n$, denotes self-feedback connections of the neurons, $T_0 \in \Re^{n \times n}$ denotes the part of interconnecting structure with no delays, $T_k \in \Re^{n \times n}$ denotes the part of interconnecting structure that is associated with delay τ_k, $k = 1, \cdots, K$, $I = (I_1, \cdots, I_n)^T \in \Re^n$ is a constant vector representing bias terms, and $S(x) = (s_1(x_1), \cdots, s_n(x_n))^T$ denotes the activation functions. The initial condition is $x(s) = \varphi(s)$ for $s \in [-\tau_K, 0]$, where $\varphi \in C([-\tau_K, 0], \Re^n)$. Here, $C([-\tau_K, 0], \Re^n)$ denotes the Banach space of continuous vector-valued functions mapping the interval $[-\tau_K, 0]$ into $\Re^n$ with a topology of uniform convergence. We assume that $0 < \tau_1 < \cdots < \tau_K < +\infty$ and for $s_i(x_i)$, $i = 1, \cdots, n$, we assume that $s_i(0) = 0$, and $s_i(x_i)$ is bounded and satisfies the following sector condition

$$0 \le \frac{s_i(x_i)}{x_i} \le \sigma_i^M. \tag{2}$$

Considering the influence of uncertainties, (1) can be described as

$$\dot{x}(t) = -(C + \Delta C)x(t) + (T_0 + \Delta T_0)S(x(t)) + \sum_{k=1}^{K}(T_k + \Delta T_k)S(x(t - \tau_k)) + I \tag{3}$$

where $\Delta C = \text{diag}[\Delta c_1, \cdots, \Delta c_n] \in \Re^{n \times n}$ and $\Delta T_k \in \Re^{n \times n}$, $k = 0, \cdots, K$, are time-invariant matrices representing the norm-bounded uncertainties. The uncertainties ΔC and ΔT_k, $k = 0, \cdots, K$, are assumed to satisfy the following assumption.

Assumption. We assume that

$$[\Delta C \ \ \Delta T_0 \ \ \cdots \ \ \Delta T_K] = HF[A \ \ B_0 \ \ \cdots \ \ B_K] \tag{4}$$

where F is an unknown matrix representing parametric uncertainty which satisfies

$$F^T F \leq I \tag{5}$$

and $H, A, B_0, \cdots, B_K$ can be regarded as the known structural matrices of uncertainty with appropriate dimensions.

The uncertainty model of (4) and (5) has been widely adopted in robust control and filtering for uncertain systems [6–8].

Definition. The equilibrium point of system (1) is said to be globally robustly stable with respect to the uncertainties ΔC and ΔT_k, $k = 0, \cdots, K$, if the equilibrium point of system (3) is globally asymptotically stable.

Letting $\tilde{x} = x - x_e$, where x_e is an equilibrium point of network (3), and defining

$$\tilde{S}(\tilde{x}) = S(\tilde{x} + x_e) - S(x_e) \tag{6}$$

then the new description of the neural network (3) can be obtained as

$$\dot{\tilde{x}}(t) = -(C + \Delta C)\tilde{x}(t) + (T_0 + \Delta T_0)\tilde{S}(\tilde{x}(t)) + \sum_{k=1}^{K} (T_k + \Delta T_k)\tilde{S}(\tilde{x}(t - \tau_k)). \tag{7}$$

From equation (6), it follows that $\tilde{S}(0) = 0$ and the terms of $\tilde{S}(\tilde{x})$ satisfy the assumption (2) as well. As a result, the origin is an equilibrium point of (7). Thus, in order to study the global robust stability of the equilibrium point of system (1) with respect to parametric uncertainties and $\Delta C, \Delta T_k, k = 0, \cdots, K$, it suffices to investigate the globally asymptotic stability of the zero solution of system (7). Now, the interconnecting matrix $T = T_0 + \Delta T_0 + \sum_{k=1}^{K} (T_k + \Delta T_k)$ is nonsymmetric due to the influence of uncertainties ΔT_k, $k = 0, \cdots, K$.

3 Robust Stability Analysis Results

We are now in a position to establish the following robust stability analysis results. The proof of the theorem is sketched and the proofs of the corollaries are omitted due to space limitations.

Theorem. The equilibrium point $\tilde{x} = 0$ of system (7) is globally asymptotically stable for arbitrarily bounded delays τ_k if there exists a positive constant ε, a positive definite matrix P and positive diagonal matrices $\Lambda_k = \mathrm{diag}[\lambda_{k1}, \cdots, \lambda_{kn}]$, where $\lambda_{ki} > 0$, $i = 1, \cdots, n$, $k = 0, \cdots, K$, such that the following linear matrix inequality (LMI) holds

$$\begin{bmatrix} -C^T P - PC + \sum_{k=0}^{K} \Lambda_k + \varepsilon A^T A & PT_K E^M - \varepsilon A^T B_K E^M & \cdots \\ E^M T_K^T P - \varepsilon E^M B_K^T A & -\Lambda_K + \varepsilon E^M B_K^T B_K E^M & \cdots \\ \vdots & \vdots & \ddots \\ E^M T_0^T P - \varepsilon E^M B_0^T A & \varepsilon E^M B_0^T B_K E^M & \cdots \\ H^T P & 0 & \cdots \end{bmatrix}$$

$$\left.\begin{array}{cc} PT_0E^M - \varepsilon A^T B_0 E^M & PH \\ \varepsilon E^M B_K^T B_0 E^M & 0 \\ \vdots & \vdots \\ -\Lambda_0 + \varepsilon E^M B_0^T B_0 E^M & 0 \\ 0 & -\varepsilon I \end{array}\right] < 0 \tag{8}$$

where $E^M = \mathrm{diag}[\sigma_1^M, \cdots, \sigma_n^M]$.

Proof. By (2), we can rewrite (7) as

$$\dot{\tilde{x}}(t) = -(C + \Delta C)\tilde{x}(t) + (T_0 + \Delta T_0)E(\tilde{x}(t))\tilde{x}(t)$$

$$+ \sum_{k=1}^{K}(T_k + \Delta T_k)E(\tilde{x}(t - \tau_k))(\tilde{x}(t - \tau_k)) \tag{9}$$

where $E(\tilde{x}) = \mathrm{diag}[\sigma_1(\tilde{x}_1), \cdots, \sigma_n(\tilde{x}_n)]$, $\sigma_i(\tilde{x}_i) = s_i(\tilde{x}_i)/\tilde{x}_i$, $i = 1, \cdots, n$. We choose the following Lyapunov-Krasovskii functional

$$V(\tilde{x}_t) = \tilde{x}^T(t)P\tilde{x}(t) + \sum_{k=0}^{K}\int_{-\tau_k}^{0}\tilde{x}_t^T(\theta)\Lambda_k\tilde{x}_t(\theta)d\theta. \tag{10}$$

The derivative of $V(\tilde{x}_t)$ with respect to t along the trajectories of system (9) is given by

$$\dot{V}(\tilde{x}_t) = \dot{\tilde{x}}^T(t)P\tilde{x}(t) + \tilde{x}^T(t)P\dot{\tilde{x}}(t) + \sum_{k=1}^{K}\tilde{x}^T(t)\Lambda_k\tilde{x}(t) - \sum_{k=1}^{K}\tilde{x}^T(t-\tau_k)\Lambda_k\tilde{x}(t-\tau_k)$$

$$\leq -\tilde{x}^T(t)[(C + \Delta C)^T P + P(C + \Delta C)]\tilde{x}(t) + \sum_{k=0}^{K}\tilde{x}^T(t)\Lambda_k\tilde{x}(t)$$

$$+ \tilde{x}^T(t)P(T_0 + \Delta T_0)E(\tilde{x}(t))\Lambda_0^{-1}E^T(\tilde{x}(t))(T_0 + \Delta T_0)^T P\tilde{x}(t)$$

$$+ \sum_{k=1}^{K}\tilde{x}^T(t)P(T_k + \Delta T_k)E(\tilde{x}(t-\tau_k))\Lambda_k^{-1}E^T(\tilde{x}(t-\tau_k))(T_k + \Delta T_k)^T P\tilde{x}(t)$$

$$\leq \tilde{x}^T(t)\Big\{[(C + \Delta C)^T P + P(C + \Delta C)] + \sum_{k=0}^{K}\Lambda_k$$

$$+ \sum_{k=1}^{K}P(T_k + \Delta T_k)E^M\Lambda_k^{-1}E^M(T_k + \Delta T_k)^T P\Big\}\tilde{x}(t)$$

$$= \tilde{x}^T(t)S^M\tilde{x}(t). \tag{11}$$

From the Lyapunov stability theory of functional differential equations, we know that the equilibrium point $\tilde{x} = 0$ of system (7) is globally asymptotically stable when $S^M < 0$. Then, according to Schur Complement [9], $S^M < 0$ can be expressed by the following linear matrix inequality (LMI)

$$\begin{bmatrix} \alpha_{11} & P(T_K + \Delta T_K)E^M & \cdots & \cdots & P(T_0 + \Delta T_0)E^M \\ E^M(T_K + \Delta T_K)^T P & -\Lambda_K & 0 & \cdots & 0 \\ \vdots & 0 & \ddots & \ddots & \vdots \\ \vdots & \vdots & \ddots & \ddots & 0 \\ E^M(T_0 + \Delta T_0)^T P & 0 & \cdots & 0 & -\Lambda_0 \end{bmatrix} < 0$$

(12)

where $\alpha_{11} = -[(C + \Delta C)^T P + P(C + \Delta C)] + \sum_{k=0}^{K} \Lambda_k$.

By the assumption and the relevant matrix theory, (12) is equivalent to condition (8).

Corollary 1. When $\Delta C = \Delta T_k = 0$, $k = 0, \cdots, K$, the equilibrium point $\tilde{x} = 0$ of system (7) is globally asymptotically stable for arbitrarily bounded delays τ_k if there exists a positive definite matrix P, and positive definite diagonal matrices $\Lambda_k = \text{diag}[\lambda_{k1}, \cdots, \lambda_{kn}]$, $k = 0, \cdots, K$, such that the following LMI holds

$$\begin{bmatrix} -(C^T P + PC) + \sum_{k=0}^{K} \Lambda_k & PT_K E^M & \cdots & \cdots & PT_0 E^M \\ E^M T_K^T P & -\Lambda_K & 0 & \cdots & 0 \\ \vdots & 0 & \ddots & \ddots & \vdots \\ \vdots & \vdots & \ddots & \ddots & 0 \\ E^M T_0^T P & 0 & \cdots & 0 & -\Lambda_0 \end{bmatrix} < 0 \qquad (13)$$

where $E^M = \text{diag}[\sigma_1^M, \cdots, \sigma_n^M]$.

Corollary 2. The equilibrium point $\tilde{x} = 0$ of system (7) is globally asymptotically stable for arbitrarily bounded delays τ_k if there exists a positive definite matrix P, positive constants ε_k, $k = 0, 1, \cdots, K+1$, and positive definite diagonal matrices $\Lambda_k = \text{diag}[\lambda_{k1}, \cdots, \lambda_{kn}]$, such that the following LMI holds

$$\begin{bmatrix} \beta_{11} & PT_K E^M & \cdots & \cdots & PT_0 E^M & PH & \cdots & \cdots & PH \\ E^M T_K^T P & \beta_{22} & 0 & \cdots & 0 & 0 & \cdots & \cdots & 0 \\ \vdots & 0 & \ddots & \ddots & \vdots & \vdots & \ddots & \ddots & \vdots \\ \vdots & \vdots & \ddots & \ddots & \vdots & \vdots & \ddots & \ddots & \vdots \\ E^M T_0^T P & 0 & \cdots & 0 & -\Lambda_0 + \varepsilon_0 E^M B_0^T B_0 E^M & 0 & \cdots & \cdots & 0 \\ H^T P & 0 & \cdots & \cdots & 0 & -\varepsilon_{K+1}I & 0 & \cdots & 0 \\ \vdots & \vdots & \ddots & \ddots & \vdots & \vdots & \ddots & \ddots & \vdots \\ \vdots & \vdots & \ddots & \ddots & \vdots & \vdots & \ddots & \ddots & 0 \\ H^T P & 0 & \cdots & \cdots & 0 & 0 & \cdots & 0 & -\varepsilon_0 I \end{bmatrix} < 0$$

(14)

where $\lambda_{ki} > 0$, $i = 1, \cdots, n$, $k = 0, \cdots, K$, $\beta_{11} = -C^T P - PC + \sum_{k=0}^{K} \Lambda_k + \varepsilon_{K+1} A^T A$, and $\beta_{22} = -\Lambda_K + \varepsilon_K E^M B_K^T B_K E^M$.

Remark. The present criteria are expressed in the form of LMIs, and thus they can be verified conveniently by solving the linear matrix inequalities, for example, using the interior-point method [9].

4   Conclusions

During the implementation process of Hopfield neural networks by electronic circuits, time delays and perturbations are inevitable. The present paper analyzes the robust stability of a class of Hopfield neural network model with multiple time delays and uncertainties, and establish sufficient conditions for the global robust stability of equilibrium point under arbitrarily bounded delays τ_k, $k = 1, \cdots, K$. The results of this paper take the form of linear matrix inequalities and are very practical to the analysis and design of Hopfield neural networks with delays.

Acknowledgements

This work was supported in part by the National Natural Science Foundation of China under Grants 60274017 and 60325311 to H. Zhang.

References

1. Hopfield, J.J.: Neural Networks and Physical Systems with Emergent Collective Computational Abilities. Proceedings of the National Academy of Sciences USA, **79** (1982) 2554–2558
2. Huang, H., Cao, J.: On Global Asymptotic Stability of Recurrent Neural Networks with Time-Varying Delays. Applied Mathematics and Computation, **142** (2003) 143–154
3. Wang, R.L., Tang, Z., Cao, Q.P.: A Learning Method in Hopfield Neural Network for Combinatorial Optimization Problem. Neurocomputing, **48** (2002) 1021–1024
4. Liao, X., Wang, J., Cao, J.: Global and Robust Stability of Interval Hopfield Neural Networks with Time-Varying Delays. International Journal of Neural Systems, **13** (2003) 171–182
5. Cao, J., Huang, D., Qu, Y.: Global Robust Stability of Delayed Recurrent Neural Networks. Chaos, Solitons and Fractals, **23** (2005) 221–229
6. Du, C., Xie, L.: Stability Analysis and Stabilization of Uncertain Two-Dimensional Discrete Systems: An LMI Approach. IEEE Trans. on Circuits and Systems-I, **46** (1999) 1371–1374
7. Xu, S., Lam, J., Lin, Z., Galkowski, K.: Positive Real Control for Uncertain Two-Dimensional Systems. IEEE Trans. on Circuits and Systems-I, **49** (2002) 1659–1666
8. Yang, F., Hung, Y.S.: Robust Mixed H_2/H_∞ Filtering with Regional Pole Assignment for Uncertain Discrete-Time Systems. IEEE Trans. on Circuits and Systems-I, **49** (2002) 1236–1241
9. Boyd, S., El Ghaoui, L., Feron, E., Balakrishnan, V.: Linear Matrix Inequalities in System and Control Theory. Studies in Applied Mathematics, Vol. 15. SIAM, Philadelphia, PA (1994)

Robust Stability
of Interval Delayed Neural Networks

Wenlian Lu and Tianping Chen*

Laboratory of Nonlinear Science, Institute of Mathematics,
Fudan University, Shanghai 200433, China

Abstract. Recently, there are several papers discussing global robust stability of the equilibrium point for the interval delayed neural networks. However, we find these criteria are not accurate. In this paper, based on Linear Matrix Inequality (LMI) technique, we propose an algorithm to determine in which region the interval delayed system is globally robust stabile. This approach is much more powerful than the criteria given in previous papers. We also give a numerical example to illustrate the viability of our algorithm.

1 Introduction

Delayed recurrently connected neural networks can be modelled by the following delayed dynamical system:

$$\frac{du_i(t)}{dt} = -d_i u_i(t) + \sum_{j=1}^{n} a_{ij} g_j(u_j(t)) + \sum_{j=1}^{n} b_{ij} g_j(u_j(t-\tau)) + I_i, \quad i=1,\cdots,n$$

and activation functions $g_j(\cdot)$ satisfy

$$H: \qquad 0 \le \frac{g_j(\eta) - g_j(\zeta)}{\eta - \zeta} \le G_j \ for \ any \ \eta \ne \zeta.$$

where G_j are positive constants, $j = 1, \cdots, n$. We denote $diag\{G_1, \cdots, G_n\}$ by G. The initial conditions are $u_i(s) = \phi_i(s)$, $s \in [-\tau, 0]$, where $\phi_i(s) \in C([-\tau, 0])$, $i = 1, \cdots, n$. we can rewrite the system by the following matrix form:

$$\frac{du(t)}{dt} = -Du(t) + Ag(u(t)) + Bg(u(t-\tau)) + I, \tag{1}$$

where $u(t) = (u_1(t), u_2(t), \cdots, u_n(t))^{\top}$, $D = diag\{d_1, d_2, \cdots, d_n\}$, $A = (a_{ij})_{n \times n}$, $B = (b_{ij})_{n \times n}$, $g(u(t)) = (g_1(u_1(t)), g_n(u_n(t)), \cdots, g_n(u_n(t)))^{\top}$, and $I = (I_1, I_2, \cdots, I_n)^{\top}$.

Referring to [1, 2], we have the following theorem:

Theorem 1.   *If $g_j(\cdot)$ satisfy assumption H, $j = 1, \cdots, n$, and there exist a positive definite diagonal matrix P and a symmetric positive definite matrix Q such that*

---

* Corresponding author.

J. Wang, X. Liao, and Z. Yi (Eds.): ISNN 2005, LNCS 3496, pp. 215–221, 2005.

$$\begin{bmatrix} 2PDG^{-1} - PA - A^{\mathsf{T}}P - Q & -PB \\ -B^{\mathsf{T}}P & Q \end{bmatrix} \text{ is positive deifinite.} \qquad (2)$$

Then system (1) is global stable.

However, in practical implementation, constants d_i and connection coefficients a_{ij} and b_{ij} all depend on some which cause d_i, a_{ij} and b_{ij} may vary in some intervals because electricity quantities are all subject to uncertainty. Here, we assume the coefficients of system (1) satisfy:

$$[\underline{D}, \overline{D}] = \{D;\ \underline{D} \leq D \leq \overline{D}\}, \quad [\underline{A}, \overline{A}] = \{A;\ \underline{A} \leq A \leq \overline{A}\},$$
$$[\underline{B}, \overline{B}] = \{B;\ \underline{B} \leq B \leq \overline{B}\}, \qquad (3)$$

where $\underline{D} = diag\{\underline{d}_1, \cdots, \underline{d}_n\}$ $\overline{D} = diag\{\overline{d}_1, \cdots, \overline{d}_n\}$, $\underline{A} = (\underline{a}_{ij})$, $\overline{A} = (\overline{a}_{ij})$, $\underline{B} = (\underline{b}_{ij})$, $\overline{B} = (\overline{b}_{ij})$, and the symbol $\leq$ denotes the comparison of each components between matrices with the same size, i.e., $\underline{A} \leq A$ means $\underline{a}_{ij} \leq a_{ij}$, for all $i, j = 1, \cdots, n$. So it is with symbol $\geq, <, <$. Then we give a definition for robust stability of interval neural networks.

Definition 1. *Interval neural networks (1) are said to be globally robustly stable if for each coefficient matrices D, A, B satisfying interval condition(3), the system is globally stable.*

Recently, there are several papers investigating robust stability of interval delayed systems. In [3–6], authors presented criteria based on M matrix to guarantee robust stability of interval neural networks. In [7, 8], authors used a quadric Lyapunov functional to obtain robust stability criteria based on matrix inequalities.

It is clear that following theorem and corollary are direct consequences of Theorem 1.

Theorem 2. *Suppose $g_j(\cdot)$ satisfy assumption H, $j = 1, \cdots, n$. If for each D, A, B satisfying interval condition(3), there exist a positive definite diagonal matrix P and a symmetric positive definite matrix Q such that condition (2) is satisfied, then system (1) under interval condition (3) is globally robustly stable.*

Corollary 1. *Suppose $g_j(\cdot)$ satisfy assumption H, $j = 1, \cdots, n$. If there exist a positive definite diagonal matrix P and a symmetric positive definite matrix Q such that condition 2 holds for all D, A, B, then system (1) under interval condition (3) is globally robustly stable.*

It is clear that results in [7, 8] are some special cases of Corollary 1. Namely, they all provide criteria that there exist P and Q such that condition (2) holds for all interval matrices. However, there do exist cases that conditions of Theorem 2 can be satisfied despite that those of Corollary 1 can not . We will provide such counterexample in Section 3. In this paper, we give criteria based on Linear Matrix Inequality (LMI) technique for global robust stability of interval delayed neural networks (1).

Before the main results, we present some denotations for simplicity. Firstly, noticing the matrix form in condition (2), we can conclude that if with fixing $D = \underline{D}$ and $a_{ii} = \underline{a}_{ii}$, condition (2) can hold, then they will also hold for the whole interval condition (3). Hence, we can fix $D = \underline{D}$ and $a_{ii} = \underline{a}_{ii}$ in the following discussion. Secondly, we denote

$$S = \{(A, B); A \in [\underline{A}, \overline{A}], \ B \in [\underline{B}, \overline{B}\}, (of \ course \ we \ assume \ a_{ii} = \underline{a}_{ii} \ is \ fixed)$$

$$Z(A, B, P, Q) = \begin{bmatrix} 2PDG^{-1} - PA - A^\top P - Q & -PB \\ -B^\top P & Q \end{bmatrix},$$

$$\gamma(A, B, P, Q) = \min \lambda(Z(A, B, P, Q)),$$

$$L = \{(A, B); there \ exists \ P \ and \ Q \ such \ that \ condition \ (2) \ holds\},$$

$$N(P, Q) = \{(P, Q), P \ is \ a \ diagonal \ positive \ definite \ matrix \ with \ \|P\|_2 < 1$$

$$and \ Q \ is \ a \ symmetric \ positive \ definite \ matrix \ such \ that \ condition \ (2) \ holds\}.$$

2 Main Results

Theorem 2 is rewritten as the following proposition.

Proposition 1.   If $S \subset L$, then system (1) under interval condition (3) is globally robustly stable.

Let $A^c = (\underline{A} + \overline{A})/2$, $\Delta A = A^c - \underline{A} = (\Delta a_{ij})$, $B^c = (\underline{B} + \overline{B})/2$, and $\Delta B = B^c - \underline{B} = (\Delta b_{ij})$. We have

Theorem 3.   Suppose $g_j(\cdot)$ satisfy assumption H, $j = 1, \cdots, n$ and $(A^c, B^c) \in L$, i.e., there exist a positive definite diagonal matrix P and a symmetric positive definite matrix Q such that $Z(A^c, B^c, P, Q) - \gamma I_n$ is positive. The 2-norm of matrix is denoted by $\| \cdot \|_2$ and

$$\max_{i=1,\cdots,n} \left\{ \sum_{j=1, j \neq i}^{n} [\Delta a_{ij} + \Delta a_{ji}] + \sum_{j=1}^{n} \Delta b_{ij}, \sum_{j=1}^{n} \Delta b_{ji} \right\} < \frac{\gamma}{\|P\|_2}. \qquad (4)$$

Then system (1) under interval condition (3) is globally robustly stable.

Proof. Let $P = diag\{p_1, p_2, \cdots, p_n\}$, $\Delta A = (\Delta a_{ij})$, and $\Delta B = (\Delta b_{ij})$. For each $x, y \in R^n$, $\delta A = (\delta a_{ij}) \in [-\Delta A, \Delta A]$ and $\delta B = (\delta b_{ij}) \in [-\Delta B, \Delta B]$, we have

$$2x^\top P\delta Ax + 2x^\top P\delta By = \sum_{i \neq j} x_i[p_i \delta a_{ij} + p_j \delta a_{ji}]x_j + \sum_{i,j=1}^{n} x_i[p_i \delta b_{ij} + p_j \delta b_{ji}]y_j$$

$$\leq (\max_i p_i) \sum_{i=1}^{n} x_i^2 \sum_{j \neq i} (\Delta a_{ij} + \Delta a_{ji}) + \sum_{i=1}^{n} x_i^2 \sum_{j=1}^{n} \Delta b_{ij} + \sum_{i=1}^{n} y_i^2 \sum_{j=1}^{n} \Delta b_{ji}$$

$$\leq \|P\|_2 \max_{i=1,\cdots,n} \left\{ \sum_{j=1, j \neq i}^{n} [\Delta a_{ij} + \Delta a_{ji}] + \sum_{j=1}^{n} \Delta b_{ij}, \sum_{j=1}^{n} \Delta b_{ji} \right\} [x^\top x + y^\top y]$$

$$\leq \gamma[x^\top x + y^\top y],$$

which implies that for any $\delta A = (\delta a_{ij}) \in [-\Delta A, \Delta A]$ and $\delta B = (\delta b_{ij}) \in [-\Delta B, \Delta B]$, $Z(A^c, B^c, P, Q) - \begin{bmatrix} P\delta A + \delta A^\top P & P\delta B \\ \delta B^\top P & 0 \end{bmatrix}$ is positive definite. Hence, $(A + \delta A, B + \delta B) \in L$. The theorem is proven.

Without loss of generality, we can assume $\|P\|_2 < 1$, i.e., there exist some diagonal elements of P equal 1 and others are all less than 1.

Defining $v(A, B) = \sup_{(P,Q) \in N(A,B)} \{\gamma, Z(A, B, P, Q) - \gamma I_n \text{ is positive definite}\}$,

we have

Proposition 2. *1. If $v(A, B) < +\infty$, for all $(A, B) \in S$, then $v(A, B)$ is continuous on S;*

2. If $S \subset L$, then $\min_{(A,B) \in S} v(A, B) > 0$.

Proof.   Because item 2 is a direct corollary of item 1, we only need to prove item 1. For any $\epsilon > 0$, there exists $(P, Q) \in N(A, B)$ such that $\gamma(A, B, P, Q) \geq v(A, B) - \epsilon$. Therefore, according to the definitions of $v(\cdot, \cdot)$ and $\gamma(\cdot, \cdot, \cdot, \cdot)$, there must exist $\delta > 0$ such that for any $\|A_1 - A\|_2 < \delta$ and $\|B_1 - B\|_2 < \delta$, $v(A_1, B_1) \geq \gamma(A_1, B_1, P, Q) \geq v(A, B) - \frac{\epsilon}{2}$ holds, which implies $\lim\inf_{(A_1,B_1) \to (A,B)} v(A_1, B_1) \geq v(A, B)$. On the other hand, we will prove $\lim\sup_{(A_1,B_1) \to (A,B)} v(A_1, B_1) \leq v(A, B)$ by reduction to absurdity. Suppose $\lim\sup_{(A_1,B_1) \to (A,B)} v(A_1, B_1) > v(A, B)$. Then there exists $\epsilon_0 > 0$ such that for any $n \in N$, there exists a sequence (A_n, B_n) satisfying $\|A_n - A\|_2 < \frac{1}{n}$ and $\|B_n - B\|_2 < \frac{1}{n}$ such that $v(A_n, B_n) > v(A, B) + \epsilon_0$. Thus, there exists a sequence (P_n, Q_n) such that $\gamma(A_n, B_n, P_n, Q_n) > v(A, B) + \epsilon_0$. Then there must exist subsequences, still denoted by (A_n, B_n) and (P_n, Q_n), which are all convergent. Let $\lim_{n \to \infty}(P_n, Q_n) = (P_1, Q_1)$, where P_1 must be a nonnegative definite diagonal matrix and Q_1 must be a nonnegative definite symmetric matrix such that $\gamma(A, B, P_1, Q_1) \geq v(A, B) + \epsilon_0$. It can be seen that there exist a positive definite diagonal matrix P_2 and a positive definite symmetric matrix Q_2 such that $\gamma(A, B, P_2, Q_2) \geq v(A, B) + \frac{\epsilon_0}{2}$, which is contradict to $v(A, B) \geq \gamma(A, B, P_2, Q_2)$. Therefore, the proposition is proven.

Based on results of Theorem 3 and Proposition 2, we present the following algorithm to judge the global robust stability of interval delayed neural networks.

Algorithm

Step 1   *Let $A^c = (\underline{A} + \overline{A})/2$, $\Delta A = A^c - \underline{A} = (\Delta a_{ij})$, $B^c = (\underline{B} + \overline{B})/2$, and $\Delta B = B^c - \underline{B} = (\Delta b_{ij})$. We solve the following LMI Optimization problem:*

$$maximize \; \gamma$$

$$subject \; to \begin{cases} Z(A^c, B^c, P, Q) - \gamma I_n \text{ is positive definite,} \\ P \text{ is a positive definite diagonal matrix,} \\ I - P \text{ is a positive definite diagonal matrix,} \\ Q \text{ is a positive definite symmetric matrix;} \end{cases}$$

Step 2   *If $\gamma < 0$, then stop and say $S \subset L$ is not right; if $\gamma > 0$, then goto step 3;*

Step 3   *If condition (4) is satisfied, then stop and say $S \subset L$ is right; else goto step 4;*

Step 4   *Split set S into two subset S_1 and S_2. If for some i_0, j_0 such that $\Delta a_{i_0,j_0} = \max_{i,j} \max\{\Delta a_{ij}, \Delta b_{ij}\}$, then we construct S_1 and S_2 as follows:*

$$S_1 = \{(A,B), A \in [\underline{A}, \overline{A}^c], B \in [\underline{B}, \overline{B}],$$
$$S_2 = \{(A,B), A \in [\underline{A}^c, \overline{A}], B \in [\underline{B}, \overline{B}].$$

where

$$\overline{A}^c = \begin{cases} \overline{a}_{ij} & (i,j) \neq (i_0,j_0), \\ a^c_{i_0,j_0} & (i,j) = (i_0,j_0) \end{cases} \qquad \underline{A}^c = \begin{cases} \underline{a}_{ij} & (i,j) \neq (i_0,j_0) \\ a^c_{i_0,j_0} & (i,j) = (i_0,j_0) \end{cases}.$$

Similarly, if for some i_0, j_0 such that $\Delta b_{i_0,j_0} = \max_{i,j} \max\{\Delta a_{ij}, \Delta b_{ij}\}$, the construction of S_1 and S_2 is similar to above

$$S_1 = \{(A,B), A \in [\underline{A}, \overline{A}], B \in [\underline{B}, \overline{B}^c],$$
$$S_2 = \{(A,B), A \in [\underline{A}, \overline{A}], B \in [\underline{B}^c, \overline{B}],$$

where

$$\overline{B}^c = \begin{cases} \overline{b}_{ij} & (i,j) \neq (i_0,j_0) \\ b^c_{i_0,j_0} & (i,j) = (i_0,j_0) \end{cases}, \qquad \underline{B}^c = \begin{cases} \underline{b}_{ij} & (i,j) \neq (i_0,j_0) \\ b^c_{i_0,j_0} & (i,j) = (i_0,j_0) \end{cases}.$$

Then replace S by S_1 and S_2 respectively and goto step 1.

Proposition 3.   *If $S \subset L$, then algorithm above will succeed in verification in finite steps. In addition , if there exists an open subset of S in which condition of Theorem 2 can not hold, then the algorithm will succeed in verification in finite time.*

Proof.   If $S \subset L$, then $v_0 = \min_{(A,B) \in S} v(A,B) > 0$. If splitting of S leads to

$$\max_{i=1,\cdots,n} \left\{ \sum_{j=1,j\neq i}^{n} [\Delta a_{ij} + \Delta a_{ji}] + \sum_{j=1}^{n} \Delta b_{ij}, \sum_{j=1}^{n} \Delta b_{ji} \right\} < v_0.$$

then the algorithm succeeds in verifying $S \subset L$. This procession will last for only finite steps.

If there exists an open subset of S in which conditions of Theorem 2 do not hold, finite splitting will cause a subset S_1 will be contained in this open subset. Then by step 2, it will end with verifying $S \overline{\subset} L$.

3   Numerical Illustration

Here, we consider a two-dimensional interval delayed neural network system (1) with interval coefficient matrices :

$$D = G = \begin{bmatrix} 1 & 0 \\ 0 & 1 \end{bmatrix}, \quad A = \begin{bmatrix} -0.1425 & [0.4558, 3.1515] \\ -4.8381 & -3.5657 \end{bmatrix}, \quad B = \begin{bmatrix} 1 & 0 \\ 1 & 1 \end{bmatrix}. \quad (5)$$

If there exist P and Q such that $Z(A, B, P, Q) > 0$ under the interval condition(5), then it can be concluded that $P(DG^{-1} - A - B)$ must be positive definite under the interval condition(5). Without loss of generality, we assume P with the form $\begin{bmatrix} 1 & 0 \\ 0 & p \end{bmatrix}$. Taking the two vertexes, where $-DG^{-1} + A + B$ are respectively,

$$K_1 = \begin{bmatrix} -0.1425 & 0.4558 \\ -3.8381 & -3.5657 \end{bmatrix}, \quad K_2 = \begin{bmatrix} -0.1425 & 3.1515 \\ -3.8381 & -3.5657 \end{bmatrix}.$$

After some calculations, we can obtain that when $p \in (3.0015, 23.6320)$, $PK_1 + K_1^\top P$ is negative definite and when $p \in (0.8106, 1.8298)]$, $PK_2 + K_2^\top P$ is negative definite. It can be seen that there is no pubic range for p such that $PK_{1,2} + K_{1,2}^\top P$ are both negative definite. Therefore, there does not exist uniform P and Q such that $Z(A, B, P, Q)$ is positive definite for all A, B satisfying the interval condition (5). In other words, results in [7, 8] all fail to verify the robust stability.

Instead, we can use the algorithm presented in this paper to judge the robust stability. After 7 cycles are done, robust stability can be assured.

4   Conclusions

In this work, we propose a novel algorithm to determine whether the delayed system is robust stable.

Acknowledgements

This work is supported by National Science Foundation of China 69982003 and 60074005, and also supported by Graduate Student Innovation Foundation of Fudan University.

References

1. Lu, W., Rong, L., Chen, T.: Global Convergence of Delayed Neural Network Systems. Inter. J. Neural Syst., **13** (2003) 193-204
2. Arik, S.: Global Asymptotical Stability of a Large Class of Neural Networks with Constant time Delay. Phys. Lett. A, **311** (2003) 504-511
3. Chen, T., Rong, L.: Robust Global Exponential Stability of Cohen-Grossberg Neural Networks With Time Delays. IEEE Trans. Neural Networks, **15** (2004) 203-206

4. Liao, X., K. Wong, K.: Robust Stability of Interval Bidirectional Associative Memory Neural Network With Time Delays. IEEE Trans. Syst. Man Cybern. B, **34** (2004) 1142-1154
5. Huang, H., Cao, J., Qu, Y.: Global Robust Stability of Delayed Neural Networks with a Class of General Activation Functions. J. Compt. Syst. Scien., **69** (2004) 688-700
6. Li, X., Cao, J.: Global Exponential Robust Stability of Delayed Neural Networks. Inter. J. Bifur. Chaos, **14** (2004) 2925-2931
7. Cao, J., Huang, D.-S., Qu, Y.: Global Robust Stability of Delayed Recurrent Neural Networks. Chaos, Solitons and Fractals, **23** (2005) 221-229
8. Arik, S.: Global Robust Stability of Delayed Neural Networks. IEEE Trans. CAS-1, **50** (2003) 156-160

Impulsive Robust Control
of Interval Hopfield Neural Networks[*]

Yinping Zhang and Jitao Sun

Dept. of Applied Mathematics, Tongji University, Shanghai 200092, China
sunjt@sh163.net

Abstract. This paper discusses impulsive control and synchronization of interval Hopfield neural networks (HNN for short). Based on the matrix measure and new comparison theorem, this paper presents an impulsive robust control scheme of the interval HNN. We derive some sufficient conditions for the stabilization and synchronization of interval Hopfield neural networks via impulsive control with varying impulsive intervals. Moreover, the large upper bound of impulsive intervals for the stabilization and synchronization of interval HNN can be obtained.

1 Introduction

In the last three decades, neural network architectures have been extensively studied and developed [1]-[10]. HNN has been applied to associative memory, model identification, optimization problem etc. The most widely used neural networks today are classified into two groups: continuous and discrete networks. However, there are also many real world systems and natural processes which display some kind of dynamic behavior in a style of both continuous and discrete characteristics. These include for example many evolutionary processes, particularly some biological systems such as biological neural networks and bursting rhythm models in pathology. Other examples include optimal control models in economics, frequency-modulated signal processing systems, and flying object motions. All these systems are characterized by abrupt change of state at certain instants [11]-[14]. Moreover, examples of impulsive phenomena can also be found in other fields such as information science, electronics, automatic control systems, computer networks, artificial intelligence, robotics and telecommunication etc.[13]-[20].Many sudden and sharp changes occur instantaneously, in the form of impulses, which cannot be well described by using pure continuous or pure discrete models. Therefore, it is important and, in effect, necessary to introduce a new type of neural networks. Literatures [4],[5] have discussed the measure-type impulsive neural networks, Liu etc. in [6] derived some sufficient conditions for robust stability and robust asymptotic stability in the sense of HNN. The matrix measure is an important tool and calculation of matrix measure is convenient, in paper [22], the authors obtain some simple global synchronization criterions for some chaotic systems by using matrix measure.

---

[*] This work is supported by the NNSF(60474008) and the NSF of Shanghai City (03ZR14095), China

J. Wang, X. Liao, and Z. Yi (Eds.): ISNN 2005, LNCS 3496, pp. 222–228, 2005.
© Springer-Verlag Berlin Heidelberg 2005

In this paper, we consider the stabilization and synchronization of interval HNN via impulsive control with varying impulsive intervals by using new comparison theorem and matrix measure. Some sufficient conditions are derived under which the impulsively controlled interval HNN is asymptotically stable and two interval HNN is synchronize, these conditions for the stabilization and synchronization of interval HNN are less conservative. Moreover, the greater upper bound of impulsive intervals can be obtained.

This paper is organized as follows. In section 2, we design impulsive robust control law for interval HNN, and discuss the synchronization of interval HNN. Finally, concluding remarks are given in section 3.

2 Impulsive Stabilization and Synchronization of Interval HNN Systems

The continuous-time Hopfield neural network (HNN) is described by the following differential equation

$$
\begin{cases}
c_i \dfrac{du_i}{dt} = \displaystyle\sum_{j=1}^{n} T_{ij} v_j - \dfrac{u_i}{R_i} + I_i, \\
v_i = g_i(u_i), \quad i = 1, 2, \cdots n.
\end{cases}
\tag{1}
$$

where $c_i > 0$, $R_i > 0$ and I_i are capacity, resistance, and bias, respectively; u_i and v_i are in the input and output of the ith neuron, respectively; all the functions $\{g_i(\cdot)\}$ are nonlinear continuously differential functions, $g_i'(\cdot)$ is invertible [denoted $u_i = g_i^{-1}(v_i)$ $\triangleq G_i(v_i)$ below] and satisfies $0 < m_i \le g_i' \le M_i < \infty$ uniformly over the domain of g_i, $i = 1, 2, \cdots, n$. First, it follows from the local inversion theorem [21] that

$$
M_i^{-1} \le G_i' = (g_i^{-1})' = (g_i')^{-1} \le m_i^{-1}.
\tag{2}
$$

It is clear that system (1) is equivalent to

$$
c_i G_i'(v_i) \dfrac{dv_i}{dt} = \sum_{j=1}^{n} T_{ij} v_j - \dfrac{G_i(v_i)}{R_i} + I_i, \quad i = 1, 2, \cdots, n
\tag{3}
$$

if v^* is the equilibrium state of system (1) then it satisfies (3). Rewrite (3) as

$$
c_i G_i'(v_i) \dfrac{d(v_i - v_i^*)}{dt} = \sum_{j=1}^{n} T_{ij}(v_j - v_j^*) - \dfrac{G_i(v_i) - G_i(v_i^*)}{R_i}, \quad i = 1, 2, \cdots, n.
\tag{4}
$$

Letting $X^T = (v_1 - v_1^*, v_2 - v_2^*, \cdots, v_n - v_2^*)$, denote $p_{ij} = \begin{cases} m_i \dfrac{T_{ij}}{c_i}, & if \ T_{ij} \ge 0 \\[2mm] M_i \dfrac{T_{ij}}{c_i}, & if \ T_{ij} < 0 \end{cases}$,

Remark 3 Condition (10) implies $V(t, X)$ is only required to be non-increasing along an odd subsequence of switchings when $K(\tau_{2i+2}^+) = K(\tau_{2i+2})$.

Now, we study the impulsive synchronization of interval HNN systems. In an impulsive synchronization configuration, the driving system is given by

$$c_i \frac{dx_i}{dt} = \sum_{j=1}^{n} T_{ij} g(x_j) - \frac{x_i}{R_i} + I_i, i = 1, 2, \cdots, n. \tag{14}$$

while the driven system is given

$$c_i \frac{d\tilde{x}_i}{dt} = \sum_{j=1}^{n} T_{ij} g(\tilde{x}_j) - \frac{\tilde{x}_i}{R_i} + I_i, i = 1, 2, \cdots, n. \tag{15}$$

where $X = (x_1, x_2, \cdots x_n)^T$ is the state variable of the driving system, $\tilde{X} = (\tilde{x}_1, \tilde{x}_2, \cdots \tilde{x}_n)^T$ is the state variable of the driven system.

At discrete instants, $\tau_l, l = 1, 2, \cdots$, the state variable of the driving system are transmitted to the driven system are subjected to jumps at these instants. In this sense, the driven system is modeled by the following impulsive equations

$$\begin{cases} c_i \dfrac{d\tilde{x}_i}{dt} = \sum_{j=1}^{n} T_{ij} g(\tilde{x}_j) - \dfrac{\tilde{x}_i}{R_i} + I_i, \ i = 1, 2, \cdots n, \ t \neq \tau_l, \\ \Delta \tilde{X} \big|_{t=\tau_l} = -B_l e, \ l = 1, 2, \cdots n. \end{cases} \tag{16}$$

where $\{\tau_l : l = 1, 2, \cdots\}$ satisfy (7) and (8).

Let $e^T = (e_1, e_2, \cdots, e_n) = (x_1 - \tilde{x}_1, x_2 - \tilde{x}_2, \cdots, x_n - \tilde{x}_n)$ is the synchronization error, then the error system of the impulsive synchronization is given by

$$\begin{cases} \dot{e}(t) = \{N[P,Q] - \mathrm{diag}(\dfrac{1}{c_1 R_1}, \dfrac{1}{c_2 R_2}, \cdots, \dfrac{1}{c_n R_n})\}e, t \neq \tau_l, \\ \Delta e \big|_{t=\tau_l} = B_l e, \quad i = 1, 2, \cdots, n. \end{cases} \tag{17}$$

where $N[P, Q]$ is defined in (5), then, similar to the stabilization of interval HNN systems, we have the following result.

Theorem 2 Let $d(i) = \lambda_{\max}(I + B_i)^T (I + B_i)$, where $\lambda_{\max}(A)$ is the largest eigenvalue of matrix A and I is the identify matrix. Then the impulsive robust synchronization of two interval HNN systems, given in (17), is asymptotically stable, if for any $A \in N[P,Q] - \mathrm{diag}(\dfrac{1}{c_1 R_1}, \dfrac{1}{c_2 R_2}, \cdots, \dfrac{1}{c_n R_n})$, there exists an $r > 1$, differentiable at $t \neq \tau_i$ and non-increasing function $K(t) \geq m > 0$ which satisfies $-\dfrac{K'(t)}{K(t)} \leq \mu(A + A^T)$, and

$$[\mu(A+A^T)](\tau_{2i+3}-\tau_{2i+1}) \le \ln\frac{K(\tau_{2i+2}^+)K(\tau_{2i+1}^+)}{K(\tau_{2i+3})K(\tau_{2i+2})rd(2i+2)d(2i+1)}$$

or $$[\mu(A+A^T)](\tau_{i+1}-\tau_i) \le \ln\frac{K(\tau_i^+)}{K(\tau_{i+1})rd(i)}$$ hold.

Remark 4 From matrix measure, we can obtain some new comparison system and corollary of results in this paper. Moreover, the large upper bound of impulsive intervals for the stabilization and synchronization of interval HNN can be obtained.

3 Conclusion

In this paper, we have obtained an impulsive control scheme of the interval Hopfield neural network. First we use the theory of impulsive differential equation to find conditions under which impulsively robustly controlled interval HNN systems are asymptotically stable and synchronize of two interval HNN systems. Then we give the estimate of the upper bound of impulse interval for asymptotically robustly stable.

References

1. Hopfield, J.J.: Neural Networks and Physical Systems with Emergent Collective Computational Abilities. Proc. Nate. Acad. 9 (1982) 2554-2558
2. Hopfield, J.J.: Neurons with Graded Response Have Collective Computational Properties Like Those of Two-State Neurons. Proc. Nate. Acad. 81(1984) 3088-3092
3. Hopfield, J.J,Tank, DW.: Neural Computation of Decisions Optimization Problems. Biolcyhem. 52 (1985) 141-152
4. Guan, Z.H., Lam, J., Chen, G.R.: On Impulsive Autoassociative Neural Networks. Networks 13 (2000) 63-69
5. Guan, Z.H., Chen, GR., Qin, Y.: On Equilibria, Stability, and Instability of Hopfield Neural Networks. IEEE Trans. on Neural Networks 11 (2000) 534-540
6. Liu, B., Liu, X.Z., Liao, X.X.: Robust H-stability for Hopfield Neural Networks with Impulsive Effects. Proceeding of Dynamics of Continuous, Discrete and Impulsive Systems, 1 (2003)
7. Liao, X.X.: Stability of Hopfield-type Neural Networks (1). Science in China (A), 38 (1995) 407-418
8. Liao, X.X, Liao, Y.: Stability of Hopfield-type Neural Networks (2). Science in China (A), 40 (1997) 813-816
9. Fang, Y, Kincaid, T.G.: Stability Analysis of Dynamical Neural Networks. IEEE Trans. Neural Networks .7 (1996) 996-1006
10. Liang, X.B., Wu, L.D.: New Sufficient Conditions for Absolute Stability of Neural Networks. IEEE Trans. Circuits Syst. 45 (1998) 584-586
11. Bainov, D, Simeonov, P.S.: Systems With Impulse Effect: Stability, Theory and Applications. Halsted Press, New York (1989)
12. Lakshmikantham, V., Bainov, D.,Simeonov, P.S.: Theory of Impulsive Differential Equations. World Scientific, Singapore (1989)
13. Yang, T.: Impulsive Control. IEEE Trans. Autom. Control .44 (1999) 1081-1083

14. Yang, T.: Impulsive Systems and Control: Theory and Applications. Huntington, NY: Nova Science Publishers, Inc.,Sept (2001)
15. Li, Z.G., Wen, C.Y., Soh, Y.C.: Analysis and Design of Impulsive Control Systems. IEEE Trans. Autom. Control .46 (2001) 894-897
16. Sun, J.T., Zhang, Y.P.: Impulsive Control of a Nuclear Spin Generator. J. of Computational and Applied Mathematics.157 (2003) 235-242
17. Xie, W., Wen, C.Y., Li, Z.G.: Impulsive Control for the Stabilization and Synchronization of Lorenz Systems. Phys. Lett. A.275 (2000) 67-72
18. Yang, T.: Impulsive Control Theory, Spinger-Verlag, Berlin (2001)
19. Sun, J.T. , Zhang, Y.P., Wu, Q.D.: Less Conservative Conditions for Asymptotic Stability of Impulsive Control Systems. IEEE Trans. Autom. Control . 48 (2003) 829-831
20. Sun, J.T., Zhang, Y.P., Wu, Q.D.: Impulsive Control for the Stabilization and Synchronization of Lorenz Systems. Physics Letter A. 298 (2002) 153-160
21. Ambrosetti, A., Prodi, G.: A primer of Nonlinear Analysis. Cambridge University , New York (1993)
22. Sun, J.T., Zhang, Y.P.: Some Simple Global Synchronization Criterions for Coupled Time-varying Chaotic Systems. Chaos, Solitons & Fractals.19 (2003) 93-98

Global Attractivity of Cohen-Grossberg Model with Delays

Tao Xiang[1], Xiaofeng Liao[1], and Jian Huang[2]

[1] Department of Computer Science and Engineering
Chongqing University, Chongqing 400044, China
xfliao@cqu.edu.cn
[2] Department of Automation, Chongqing University

Abstract. In this paper, we have studied the global attractivity of the equilibrium of Cohen-Grossgerg model with both finite and infinite delays. Criteria for global attractivity are also derived by means of Lyapunov functionals. As a corollary, we show that if the delayed system is dissipative and the coefficient matrix is VL-stable, then the global attractivity of the unique equilibrium is maintained provided the delays are small. Estimates on the allowable sizes of delays are also given. Applications to the Hopfield neural networks with discrete delays are included.

1 Introduction

The purpose of this paper is to study the effect of delays on the global attractivity of the equilibrium of Cohen-Grossberg models[1-6] with finite and infinite delays. An effective technique to construct Lyapunov functionals for such models with both finite and infinite delays is presented. By means of Lyapunov functionals, we are able to establish the global attractivity of the equilibrium of such Cohen-Grossberg systems. As a corollary, we show that if the delayed system is dissipative and the coefficient matrix is VL-stable, then the global attractivity of the unique equilibrium is maintained provided the delays are small. Consequently, when the coefficient matrix is VL-stable, we confirm the common believe that small delays are harmless for the global attractivity of delayed Cohen-Grossberg systems provided the system has an equilibrium.

The organization of this paper is as follows. In the next section, we first describe our model in detail and give the main result by constructing suitable Lyapunov functionals. Finally, some conclusions are drawn in Section III.

2 Main Results

In this paper, we consider the following general autonomous Cohen-Grossberg system of pure-delay type:

$$
\dot{x}_i(t) = a_i(x_i(t))[b_i(x_i(t)) - \sum_{j=1}^{n} w_{ij} \int_{-\tau_{ij}}^{0} f_j(x_j(t+s))d\mu_{ij}(s)
$$
$$
- \sum_{j=1}^{n} v_{ij} \int_{-\infty}^{0} g_j(x_j(t+s))d\eta_{ij}(s) + J_i], \qquad i = 1, 2, \cdots, n
\tag{1}
$$

J. Wang, X. Liao, and Z. Yi (Eds.): ISNN 2005, LNCS 3496, pp. 229–234, 2005.
© Springer-Verlag Berlin Heidelberg 2005

$$V_{i3}(y_t) = \frac{1}{2}\sum_{j,k=1}^{n}[\bar{a}_iF_j\left|w_{ij}\right|\{\bar{a}_j^{-2}\alpha_{ij}F_k\left|w_{jk}\right|\int_{-\tau_{jk}}^{0}\int_{t+s}^{t}y_k^2(p)dpd\mu_{jk}(s)$$

$$+\bar{a}_j^{-2}\alpha_{ij}G_k\left|v_{jk}\right|\int_{-\infty}^{0}\int_{t+s}^{t}y_k^2(p)dpd\eta_{jk}(s)\}$$

$$+\bar{a}_iG_j\left|v_{ij}\right|\{\bar{a}_j^{-2}\beta_{ij}F_k\left|w_{jk}\right|\int_{-\tau_{jk}}^{0}\int_{t+s}^{t}y_k^2(p)dpd\mu_{jk}(s)$$

$$+\bar{a}_j^{-2}\beta_{ij}G_k\left|v_{jk}\right|\int_{-\infty}^{0}\int_{t+s}^{t}y_k^2(p)dpd\eta_{jk}(s)\}]$$

(8)

Then it turns out from (5)-(8) and (H_7) that for all large t,

$$\dot{V}_i(y_t) \le \sum_{j=1}^{n}m_{ij}y_i(t)y_j(t)+\frac{1}{2}\sum_{j=1}^{n}e_{ij}y_j^2(t)+\left|W_i(y_t)\right|$$

(9)

where

$$W_i(y_t) = -y_i(t)\sum_{j=1}^{n}\bar{a}_iG_j\left|v_{ij}\right|\int_{-\infty}^{t}\left[y_j(t)-y_j(t+s)\right]d\eta_{ij}(s), \quad i=1,2,\cdots,n.$$

From (H_6), we know that there exist $D = diag(d_1,\cdots,d_n)$ with $d_i > 0$ $\gamma_i > 0 (i\in I)$ such that

$$\frac{1}{2}x^T(DM+M^TD)x = \sum_{i,j=1}^{n}d_im_{ij}x_ix_j \le -\sum_{i=1}^{n}\gamma_ix_i^2 \qquad \text{for } x\in R^n$$

(10)

Let

$$V(y_t) = \sum_{i=1}^{n}d_iV_i(y_t), \quad \text{and} \quad W(y_t) = \sum_{i=1}^{n}d_i\left|W_i(y_t)\right|$$

(11)

with $V_i(y_t)$ defined by (5), Then, from (9)-(11),

$$\dot{V}(y_t) \le -\gamma_0\sum_{i=1}^{n}y_i^2(t)+W(y_t)$$

(12)

where $\gamma_0 = \min_{i\in I}\left\{\frac{1}{2}\left[\gamma_i-\sum_{j=1}^{n}d_je_{ij}\right]\right\}$. From ($H_9$), $\gamma_0 > 0$. Noting that

$$W(y_t) \le \sum_{i=1}^{n}d_i\sum_{j=1}^{n}\bar{a}_iG_j\left|v_{ij}\right|[\left|y_i(t)\right|\left|y_j(t)\right|+\frac{1}{2}(y_i^2(t)+\sup_{-\infty<s\le0}y_j^2(s))]\int_{-\infty}^{t}d\eta_{ij}(s)$$

$$\le \sum_{i,j=1}^{n}\frac{1}{2}d_i\bar{a}_iG_j\left|v_{ij}\right|[2y_i^2(t)+y_j^2(t)+\sup_{-\infty<s\le0}y_j^2(s))]\int_{-\infty}^{t}d\eta_{ij}(s)$$

(13)

By assumption (H_7), $\int_{-\infty}^{t}d\eta_{ij}(s) \to 0$ as $t\to\infty$. It then follows from (13) that there exist $0 < \delta < \gamma_0$ and a large $T>0$ such that $W(y_t) \le \delta\sum_{i=1}^{n}y_i^2(t)+Z(t)$ *for* $t\ge T$, where

$$Z(t) \equiv \sum_{i,j=1}^{n}\frac{1}{2}d_i\bar{a}_iG_j\left|v_{ij}\right|[\sup_{-\infty<s\le0}y_j^2(s))]\int_{-\infty}^{t}d\eta_{ij}(s)$$

From (H$_7$) and the boundedness of the initial functions, one can see that $Z(t) \in L_1[T, \infty]$. Therefore, we have from (12) and (13) that

$$\dot{V}(y_t) \le -(\gamma_0 - \delta) \sum_{i=1}^{n} y_i^2(t) + Z(t), \quad \text{for } t \ge T \tag{14}$$

Then one can show from (14) (see Kuang [16]) that $\lim_{t \to \infty} y(t) = 0$. This completes the proof of Theorem.

In our assumption (H$_9$), α_{ij} and β_{ij} $(i, j \in I)$ measure the size of the finite and infinite delays in (1), respectively. One can see that (H$_9$) is always satisfied provided the sizes of the delays are small. The estimates on the allowable sizes of delays can be obtained from (H$_9$). Therefore, from Theorem 1 we have the following corollary

Corollary 1. Assume the delay system (1) is dissipative and the coefficient matrix M is VL-stable. Then the global attractivity of the unique equilibrium is maintained provided the delays are (sufficiently) small.

In other words, when the coefficient matrix is VL-stable, small delays do not matter for the global attractivity of the unique equilibrium (if any) of Cohen-Grossberg systems. In practice, estimates on the allowable sizes of delays may be need.

3 Conclusions

Usually, there are general results for stability independent of delays, one may expect sharper, delay-dependent stability conditions. This is because the robustness of independent of delay properties is of course counterbalanced by very conservative conditions. In engineering practice, information on the delay range are generally available and delay-dependent stability criteria are likely to give better performances.

In this paper, we have shown that time delays are not negligible for the global attractivity of the delayed Cohen-Grossberg systems provided the coefficient matrix is VL-stable. Some criteria are also derived by means of Lyapunov functionals.

Acknowledgments

The work described in this paper was supported by a grant from the National Natural Science Foundation of China (No. 60271019) and the Doctorate Foundation of the Ministry of Education of China (No. 20020611007) and the Natural Science Foundation of Chongqing (No. 8509).

References

1. Cohen, M. A., & Grossberg, S.: Absolute Stability and Global Pattern Formation and Parallel Memory Storage by Competitive Neural Networks. IEEE Transactions on Systems, Man and Cybernetics, SMC-13, (1981) 815-821
2. Grossberg, S.: Nonlinear Neural Networks: Principles, Mechanisms, and Architectures. Neural Networks, 1 (1988) 17-61

3. Hopfield, J. J.: Neuronal Networks and Physical Systems with Emergent Collective Computational Abilities. Proceedings of the National Academy of Sciences, **79** (1982) 2554-2558

4. Hopfield, J. J.: Neurons with Graded Response Have Collective Computational Properties Like Those of Two-State Neurons. Proceeding of the National Academy of Sciences, **81** (1984) 3058-3092

5. Wang, L., & Zou, X.: Exponential Stability of Cohen-Grossberg Neural Network. Neural Networks, **15** (2002) 415-422

6. Ye, H., Michel, A. N., & Wang, K.: Qualitative Analysis of Cohen-Grossberg Neural Networks with Multiple Delays. Physical Review E, **51** (1995) 2611-2618

7. Liao, X. F., Yu, J. B.: Robust Stability for Interval Hopfield Neural Networks with Time Delays", IEEE Transactions on Neural Networks, **9** (1998) 1042-1046

8. Liao, X. F., Wong, K. W., Wu, Z. F., Chen, G.: Novel Robust Stability Criteria for Interval Delayed Hopfield Neural Networks. IEEE Transactions on CAS-I, **48** (2001) 1355-1359

9. Liao, X. F., Yu, J. B.: Qualitative Analysis of Bi-directional Associative Memory Networks with Time Delays. Int. J. Circuit Theory and Applicat., **26** (1998) 219-229

10. Liao, X. F., Yu, J. B., Chen, G.: Novel Stability Criteria for Bi-directional Associative Memory Neural Networks with Time Delays. Int. J. Circuit Theory and Applicat., **30** (2002) 519-546

11. Liao, X. F., Wong, K. W., Yu, J. B.: Novel Stability Conditions for Cellular Neural Networks with Time Delay. Int. J. Bifur. And Chaos, **11** (2001) 1853-1864

12. Liao, X. F., Wu, Z. F., Yu, J. B.: Stability Analyses for Cellular Neural Networks with Continuous Delay. Journal of Computational and Applied Math., **143** (2002) 29-47

13. Liao, X. F., Chen, G. Sanchez, E. N.: LMI-based Approach for Asymptotically Stability Analysis of Delayed Neural Networks. IEEE Transactions on CAS-I, **49** (2002a) 1033-1039

14. Liao, X. F., Chen, G. Sanchez, E. N.: Delay-dependent Exponential Stability Analysis of Delayed Neural Networks: an LMI Approach. Neural Networks, **15** (2002b) 855-866

15. Chen A., Cao J., and Huang L.: An Estimation of Upperbound of Delays for Global Asymptotic Stability of Delayed Hopfield Neural Networks, IEEE Transactions on CAS-I, **49** (2002) 1028-1032

16. Kuang Y., Delay Differential Equations with Applications in Population Dynamics, Academic Press, New York (1993)

High-Order Hopfield Neural Networks

Yi Shen, Xiaojun Zong, and Minghui Jiang

Department of Control Science and Engineering,
Huazhong University of Science and Technology, Wuhan, Hubei 430074, China

Abstract. In 1984 Hopfield showed that the time evolution of a symmetric Hopfield neural networks are a motion in state space that seeks out minima in the energy function (i.e., equilibrium point set of Hopfield neural networks). Because high-order Hopfield neural networks have more extensive applications than Hopfield neural networks, the paper will discuss the convergence of high-order Hopfield neural networks. The obtained results ensure that high-order Hopfield neural networks ultimately converge to the equilibrium point set. Our result cancels the requirement of symmetry of the connection weight matrix and includes the classic result on Hopfield neural networks, which is a special case of high-order Hopfield neural networks. In the end, A example is given to verify the effective of our results.

1 Introduction

Much of the current interest in artificial networks stems not only from their richness as a theoretical model of collective dynamics but also from the promise they have shown as a practical tool for performing parallel computation [1]. Theoretical understanding of neural networks dynamics has advanced greatly in the past fifteen years [2-13]. The neural networks proposed by Hopfield can be described by an ordinary differential equation of the form

$$C_i \dot{x}_i(t) = -\frac{1}{R_i} x_i(t) + \sum_{j=1}^{n} T_{ij} g_j(x_j(t)) + I_i, i = 1, 2, \cdots, n. \tag{1}$$

on $t \geq 0$. The variable $x_i(t)$ represents the voltage on the input of the ith neuron, and I_i is the external input current to the ith neuron. Each neuron is characterized by an input capacitance C_i and a transfer function g_j. The connection matrix element T_{ij} has a value $+1/R_{ij}$ when the non-inverting output of the jth neuron is connected to the input of the ith neuron through a resistance R_{ij}, and a value $-1/R_{ij}$ when the inverting output of the jth neuron is connected to the input of the ith neuron through a resistance R_{ij}. The parallel resistance at the input of each neuron is defined by $R_i = (\sum_{j=1}^{n} |T_{ij}|)^{-1}$. The nonlinear transfer function g_i is sigmoidal function. By defining

$$b_i = \frac{1}{C_i R_i}, \quad a_{ij} = \frac{T_{ij}}{C_i}, \quad c_i = \frac{I_i}{C_i},$$

J. Wang, X. Liao, and Z. Yi (Eds.): ISNN 2005, LNCS 3496, pp. 235–240, 2005.

Corollary 2.2. If the high-order Hopfield neural networks (6) satisfies

$$a_{ij} = a_{ji}, d_{ijk} = d_{jik} = d_{kji}, \quad i, j, k = 1, 2, \cdots, n, \tag{12}$$

then the high-order Hopfield neural networks (6) converge to M.

Proof. Set $P = I$ (I is an identity matrix), then Theorem 2.1 reduces to Corollary 2.2.

Remark 2. When $d_{ijk} = 0$, $i, j, k = 1, 2, \cdots, n$, the high-order Hopfield neural networks (6) becomes the Hopfield neural networks in general sense. Corollary 2.2 reduces to the classic result on the Hopfield neural networks in [1].

3   Simulation Results

Example 3.1. Consider a two dimensional high-order Hopfield neural networks

$$\begin{cases} \dot{x}_1(t) = -b_1 x_1(t) + \sum_{j=1}^{2} a_{1j} g_j(x_j(t)) + \sum_{j=1}^{2} \sum_{k=1}^{2} d_{1jk} g_j(x_j(t)) g_k(x_k(t)) + c_1 \text{ ;} \\ \dot{x}_2(t) = -b_2 x_2(t) + \sum_{j=1}^{2} a_{2j} g_j(x_j(t)) + \sum_{j=1}^{2} \sum_{k=1}^{2} d_{2jk} g_j(x_j(t)) g_k(x_k(t)) + c_2 \text{ .} \end{cases}$$

The network parameters are chosen as $b_1 = 0.1, b_2 = 0.2, a_{11} = 2, a_{22} = -1, a_{12} = 2, a_{21} = 1, d_{111} = d_{122} = 1, d_{112} = d_{121} = -1, d_{211} = -\frac{1}{2}, d_{222} = 1, d_{212} = d_{221} = \frac{1}{2}, c_1 = 1, c_2 = -1, g_i(x_i) = \frac{\exp(x_i) - \exp(-x_i)}{\exp(x_i) + \exp(-x_i)}, i = 1, 2$. When $p_1 = 1, p_2 = 2$, it is easy to verify the all conditions of Theorem 2.1 hold. Therefore, the above high-order Hopfield neural networks converge to M, which is defined by (7). Simulation results are depicted in Fig.1 and Fig.2. Since the equilibrium point set (7) has multi-points, Fig.1 and Fig.2 demonstrate the same network may converge to different equilibrium point with different initial values.

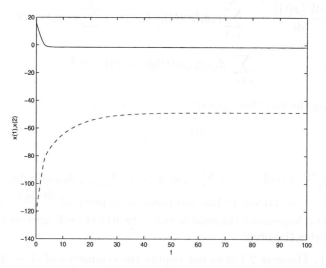

Fig. 1. The states x with initial value $x = (-122, 16)$ ultimately converge to the equilibrium $x^* = (-48.7679, -1.6245)$. x(1)-solid line, x(2)-dashed line.

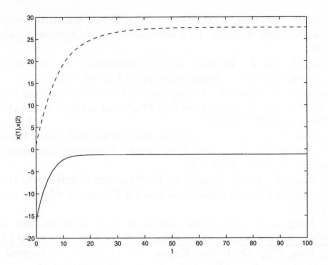

Fig. 2. The states x with initial value $x = (1, -16)$ ultimately converge to the equilibrium $x^* = (27.6633, -1.1678)$. x(1)-dashed line, x(2)-solid line.

4   Conclusion

In this paper, general high-order Hopfield neural network (6) are studied. The Theorem 2.1 and Corollary 2.2 ensure that the neural network (6) ultimately converge to the equilibrium point set M, i.e., the minimal point set of the energy function V. These results can be directly derived from the parameters of high-order Hopfield neural network (6), are very easy to verified. Hence, it is very convenience in application. Meanwhile, the results of this paper is a generalization and improvement over some previous criteria in [1]. Finally, simulations demonstrate the validity and feasibility of our proposed approach.

Acknowledgments

The work was supported by Natural Science Foundation of Hubei (2004ABA055) and National Natural Science Foundation of China (60274007, 60074008).

References

1. Hopfield, J J.: Neurons with Graded Response Have Collective Computational Properties Like those of Two-state Neurons. Porc.Natl Acad.Sci., **81** (1984) 3088-3092
2. Dembo, A., Farotimi, O., Kailath, T.: High-Order Absolutely Stable Neural Networks. IEEE Trans. Circ. Syst. II, **38** (1991) 57-65
3. Kosmatopoulos, E.B., Polycarpou, M.M., Christodoulou, M.A., et al.: High-Order Neural Networks Structures for Identification of Dynamical Systems. IEEE Trans. Neural Networks, **6** (1995) 422-431

4. Zhang, T., Ge, S.S., Hang, C.C.: Neural-Based Direct Adaptive Control for a Class of General Nonlinear Systems. International Journal of Systems Science, **28** (1997) 1011-1020

5. Su, J., Hu, A., He, Z.: Solving a Kind of Nonlinear Programming Problems via Analog Neural Networks. Neurocomputing, **18** (1998) 1-9

6. Stringera, S.M., Rollsa, E.T., Trappenbergb, T.P.: Self-organising Continuous Attractor Networks with Multiple Activity Packets, and the Representation of Space. Neural Networks, **17** (2004) 5-27

7. Xu, B.J., Liu, X.Z., Liao, X.X.: Global Asymptotic Stability of High-Order Hopfield Type Neural Networks with Time Delays. Computers and Mathematics with Applications, **45** (2003) 1729-1737

8. Sun, C., Zhang, K., Fei, S., Feng, C.B.: On Exponential Stability of Delayed Neural Networks with a General Class of Activation Functions. Physics Letters A, **298** (2002) 122-132

9. Sun, C., Feng, C.B.: Exponential Periodicity of Continuous-Time and Discrete-Time Neural Networks with Delays. Neural Processing Letters, **19** (2004)131-146

10. Sun, C., Feng, C.B.: On Robust Exponential Periodicity of Interval Neural Networks with Delays. Neural Processing Letters, **20** (2004) 53-61

11. Cao, J.: On Exponential Stability and Periodic Solution of CNN's with Delay. Physics Letters A, **267** (2000) 312-318

12. Liao, X., Chen, G., and Sanchez, E.: Delay-dependent Exponential Stability Analysis of Delayed Neural Networks: an LMI approach. Neural Networks, **15** (2002) 855-866

13. Cao, J., Wang, J.: Global Asymptotic Stability of Recurrent Neural Networks with Lipschitz-continuous Activation Functions and Time-Varying Delays. IEEE Trans. Circuits Syst. I, **50** (2003) 34-44

Stability Analysis of Second Order Hopfield Neural Networks with Time Delays

Jinan Pei[1,2], Daoyi Xu[2], Zhichun Yang[2,3], and Wei Zhu[2,4]

[1] Department of Math., Chongqing Education College, Chongqing 400067, China
peijn@sohu.com
[2] Institute of Mathematics, Sichuan University, Chengdu 610064, China
daoyixucn@yahoo.com
[3] Basic Department, Chengdu Textile Institute, Chengdu 610023, China
zhichy@yahoo.com.cn
[4] Institute for Nonlinear Systems,
Chongqing University of Posts and Telecommunications, 400065, Chongqing, China

Abstract. In this paper, some sufficient conditions of uniform stability and asymptotic stability about second order neural networks with time delays were obtained by the properties of nonnegative matrices and techniques of differential inequalities. In the results, it was assumed that the activation functions satisfy the Lipschitz condition and it was not required that the activation functions were bounded, differentiable and strictly increasing. Moreover, the symmetry of the connection matrix was not also necessary. Thus, we have improved some previous work of other researchers.

1 Introduction

As is well known,the delays are ubiquitous in biological and artificial neural networks,which not only control the dynamical action of networks,but also affect the stability of hardware neural networks.Thus it is very important to investigate the dynamics of delay neural networks.Hopfield neural networks with time delays(DHNN)have been extensively investigated over the years,and various sufficient conditions and applicable criterion for the stability of equilibrium point of this class of neural networks have been established.Second order neural networks have attracted considerable attention in recent years(see,e.g.,[1],[2],[3],[4],[5],[6]). This is due to the fact that second order neural networks have stronger approximation property,faster convergence rate,greater storage capacity,and higher fault tolerance than first order neural networks.However,there are very few results on the stability of the equilibrium point for second order Hopfield neural networks with time delays(SDHNN).Recently,some sufficient conditions for the global asymptotic stability of the equilibrium point of SDHNN system are obtained by utilizing Lyapunov functional or Razumikhin theorem in ([4],[5],[6]). These results are corresponded on boundedness of activation functions.The aims of this paper are to investigate stability of equilibrium point of SDHNN system by the properties of nonnegative matrices and techniques of inequalities,where the boundedness of activation functions is not required.

J. Wang, X. Liao, and Z. Yi (Eds.): ISNN 2005, LNCS 3496, pp. 241–246, 2005.
© Springer-Verlag Berlin Heidelberg 2005

2   Notations and Terminologies

Let $R^+ = [0, +\infty)$ and $C[X, Y]$ is the class of continuous mapping from the topological space X to the topological space Y. Especially, $C \triangleq C[[-\tau], R^n]$.

For $A, B \in R^n$, (or $A, B \in R^{n \times n}$), $A \leq B(A < B)$ means that each pair of corresponding elements of A and B satisfies this inequality " $\leq$ "(" $<$ "). Especially, A is called a nonnegative matrix if $A \geq 0$. The function matrix $A(x) \in C[\Omega, R^{n \times n}]$ is said to be monotonically non-decreasing in Ω, if for any $x', x'' \in \Omega \subset R^n$, $x' \leq x''$ implies $A(x') \leq A(x'')$.

For $u \in R^n$, we define $[u]^+ = (|u_1|, |u_2|, \cdots, |u_n|)^T$; for $\varphi \in C$, define $[\varphi]_\tau^+ = (\||\varphi_1\|_\tau, \|\varphi_2\|_\tau, \cdots, \|\varphi_n\|_\tau)^T$, where $\|\varphi_i\|_\tau = \sup_{-\tau \leq s \leq 0} |\varphi_i(t+s)|, i = 1, 2, \cdots, n$.

Consider second order Hopfield neural networks with time delays (SDHNN):

$$\frac{du_i(t)}{dt} = -b_i u_i(t) + \sum_{j=1}^{n} a_{ij} g_j(u_j(t - \tau_j))$$
$$+ \sum_{j=1}^{n} \sum_{k=1}^{n} c_{ijk} g_j(u_j(t - \tau_j)) g_k(u_k(t - \tau_k)) + J_i, i = 1, 2, \cdots, n, \tag{1}$$

where $t \geq 0$, $b_i > 0$ and c_{ijk} represent the first- and second-order synoptic weights of the neural networks; $u_i(t)$ and $0 \leq \tau_i \leq \tau, (\tau > 0)$ are, respectively, the output and time delay of the ith neuron; g_i the neuron input-output activation function, g_i are continuous function and $g_i(0) = 0, i = 1, 2, \cdots, n$.

Let $u_i(s) = \varphi_i(s), s \in [-\tau, 0]$ be the initial condition, for any $\varphi_i : [-\tau, 0] \to R$, a solution of (1) is a function $u_i : R^+ \to R$ satisfying for $t \geq 0$. We always assume that system (1) has a continuous solution denoted by $u_i(t, 0, \varphi)$ or simply $u_i(t)$ if no confusion should occur.

For the convenience, denote $B = \text{diag}\{b_i\}$, $A = (|a_{ij}|)_{n \times n}$, $L = \text{diag}\{l_i\}$ with $l_i > 0, J = (|J_1|, |J_2|, \cdots, |J_n|)^T, P = B^{-1}J, g(u) = (g_1(u_1), g_2(u_2), \cdots, g_n(u_n))^T$, $C(u) = (\sum_{k=1}^{n} |c_{ijk} + c_{ikj}|l_k u_k)_{n \times n}$ $(u \in R^n)$, $H(\cdot) = B^{-1}(A + C(\cdot))L$.

Clearly, $C(\cdot)$ is a monotonically non-decreasing matrix.

Throughout this paper, we always assume that

(H_1) $[g(u) - g(v)]^+ \leq L[u - v]^+$, for all $u, v \in R^n$;

(H_2) There exists a constant vector $K = (K_1, K_2, \cdots, K_n)^T > 0$, such that

$$\rho(H(K)) < 1 \quad \text{and} \quad K - H(K)K > P,$$

where the symbol $\rho(H(K))$ denotes the spectral radius of the matrix $H(K)$.

3   Main Results

We can write (1) as

$$\frac{du_i(t)}{dt} = -b_i u_i(t) + \sum_{j=1}^{n} a_{ij} g_j(u_j(t - \tau_j))$$
$$+ \frac{1}{2} \sum_{j=1}^{n} \sum_{k=1}^{n} (c_{ijk} + c_{ikj}) g_j(u_j(t - \tau_j)) g_k(u_k(t - \tau_k)) + J_i, t \geq 0. \tag{2}$$

Lemma 1. If (H_1) and (H_2) hold, then the set $\Omega = \{\varphi \in C | [\varphi]_\tau^+ \leq (I - H(K))^{-1}P < K\}$ is a invariable set of system (1).

Proof. Without loss of generality, we assume that $P > 0$. By Theorem 9.16 in [10] and (H_2), then $0 < N \overset{\Delta}{=} (I - H(K))^{-1}P < K$.

We first prove that for any $\varphi \in \Omega$ and for any given $\alpha \in (1, \beta)$,

$$[u(t)]^+ < \alpha N, \text{ for } t \geq 0 \tag{3}$$

where $\beta = \min_{1 \leq i \leq n} \{\frac{K_i}{N_i}\}$, K_i and N_i denote i^{th} component of vector K and N, respectively.

If (4) is not true, then there is a $l(1 \leq l \leq n)$ and $t_1 > 0$, such that

$$|u_l(t_1)| = \alpha N_l; \ |u_l(t)| \leq \alpha N_l, \text{ for } t < t_1; \tag{4}$$

and

$$[u(t)]_\tau^+ \leq \alpha N, \text{ for } 0 \leq t < t_1. \tag{5}$$

Let $E_l = (0, \cdots, 0, \underbrace{1}_{l}, 0, \cdots, 0)^T$. From (2) and (5), we have

$$
\begin{aligned}
|u_l(t_1)| &\leq e^{-b_l t_1} \|\phi_l\|_\tau + \int_0^{t_1} e^{-b_l(t_1-s)} \Big(\sum_{j=1}^n (|a_{lj}| \\
&\quad + \frac{1}{2} \sum_{k=1}^n l_k |c_{ljk} + c_{lkj}| |u_k(s)|) l_j |u_j(s)| + |J_l|) ds \\
&\leq E_l \{ e^{-Bt_1} [\varphi]_\tau^+ + \int_0^{t_1} e^{-B(t_1-s)} ((A + \frac{1}{2}C([u(s)]^+)) L[u(s)]^+ + J) ds \} \\
&\leq E_l \{ e^{-Bt_1} \alpha N + \int_0^{t_1} e^{-B(t_1-s)} ((A + C(K)) L\alpha N + J) ds \} \\
&= E_l \{ e^{-Bt_1} \alpha N + (I - e^{-Bt_1})(H(K)\alpha N + P) \} \\
&= E_l \{ e^{-Bt_1} (\alpha N - H(K)\alpha N - P) + (H(K)\alpha N + P) \} \\
&< E_l \{ \alpha N \} = \alpha N_l.
\end{aligned}
$$

This contradicts the first equality of (4), and so (3) holds. Letting $\alpha \to 1$, then $[u(t)]^+ \leq N$, for $t \geq 0$, and we complete the proof.

Let $u^* = (u_1^*, u_2^*, \cdots, u_n^*)^T \in \Omega$ be an equilibrium point of system (1), and $x = u - u^* = (x_1, x_2, \cdots, x_n)^T$, $f_i(x_i) = g_i(x_i + u_i^*) - g_i(u_i^*)$, then $|f_i(z)| \leq l_i|z|$, for all $z \in R$.

Using Taylor's theorem (see [4],[5],[6]), we can write (1) as

$$\frac{dx_i(t)}{dt} = -b_i x_i(t) + \sum_{j=1}^n (a_{ij} + \sum_{k=1}^n ((c_{ijk} + c_{ikj})\varsigma_k) f_j(x_j(t - \tau_j)), \text{ for } t \geq 0 \tag{6}$$

$$x_i(t) = \phi_i(t), \text{ for } 0 \leq t \leq -\tau, \tag{7}$$

where ς_k lies between $g_k(u_k(t - \tau_k))$ and $g_k(u_k^*)$; $\phi_i = \varphi_i - u_i^*$.

Clearly, the equilibrium point u^* of (1) is stable if and only if the equilibrium point 0 of (6) is stable.

Theorem 1. If (H_1) and (H_2) hold, then equilibrium point 0 of (6) is uniformly stable.

Proof. Let $E = (1, 1, \cdots, 1)^T$. We can find arbitrarily small constant $\varepsilon > 0$ and $\eta > 1$ such that

$$(I - H(K))^{-1}(P + E\eta\varepsilon) < K.$$

When $[\phi]_\tau^+ < (I - H(K))^{-1}E\eta\varepsilon = \eta V = \eta(V_1, V_2, \cdots, V_n)^T$, we have

$$[x(t)]^+ < \eta V, \text{ for } t \geq 0. \tag{8}$$

If (8) is not true, then there is a l $(1 \leq l \leq n)$ and $t_2 > 0$, such that

$$|x_l(t_2)| = \eta V_l; \ |x_l(t)| \leq \eta V_l, \text{ for } t < t_2; \tag{9}$$

and

$$[x(t)]_\tau^+ \leq \eta V, \text{ for } 0 \leq t \leq t_2. \tag{10}$$

Since $u^* \in \Omega$ and $[\phi]_\tau^+ < \eta V$,

$$[\varphi]_\tau^+ \leq [u^*]^+ + [\phi]_\tau^+ < (I - H(K))^{-1}(P + E\eta\varepsilon) < K.$$

By Lemma 1, then $[u(t)]_\tau^+ < K$, thus for $k = 1, 2, \cdots, n$

$$|\varsigma_k| \leq \max\{|g_k(u_k(t - \tau_k))|, |g_k(u_k^*)|\} \\ \leq l_k \max\{|u_k(t - \tau_k)|, |u_k^*|\} < l_k K_k. \tag{11}$$

From (6), (10) and (11), we have

$$|x_l(t_2)| \leq e^{-b_1 t_2}\|\phi_l\|_\tau + \int_0^{t_2} e^{-b_1(t_2-s)}(\sum_{j=1}^n (|a_{lj}|$$
$$+ \sum_{k=1}^n l_k|c_{ljk} + c_{lkj}|K_k)l_j|x_j(s)|)ds$$
$$\leq E_l\{e^{-Bt_2}[\phi]_\tau^+ + \int_0^{t_2} e^{-B(t_2-s)}((A + C(K))L[x(s)]^+)ds\}$$
$$\leq E_l\{e^{-Bt_2}\eta V + (I - e^{-Bt_2})H(K)\eta V\}$$
$$= E_l\{e^{-Bt_2}(\eta V - H(K)\eta V) + H(K)\eta V\}$$
$$< E_l\{\eta V\} = \eta V_l.$$

This contradicts the first equality of (9), and so (8) holds. So the equilibrium point 0 of (6) is uniformly stable, and the proof is completed.

Theorem 2. If (H_1) and (H_2) hold, then the equilibrium point 0 of (6) is asymptotically stable.

Proof. For any $\varphi \in \Omega$, $x(t)$ denote the solution of system (6), we claim that

$$\lim_{t \to +\infty} \sup[x(t)]^+ = 0. \tag{12}$$

If (12) is not true, from (8), there is a nonnegative constant vector $\sigma(\neq 0)$, such that

$$\lim_{t\to+\infty} \sup[x(t)]^+ = \sigma < \eta V. \tag{13}$$

Then for a sufficiently small constant $\delta > 0$ there is $t_3 > 0$, for any $t \geq t_3$,

$$[x(t)]^+ \leq \sigma + \delta E. \tag{14}$$

For the above $\delta > 0$, we can find a $T > 0$, satisfies

$$\int_T^{+\infty} e^{-Bs} BH(K)\eta V ds \leq \delta E, \tag{15}$$

According to the proof of theorem 1, by (14) and (15), we can get for any $t \geq \max\{t_3, T\}$,

$$
\begin{aligned}
[x(t)]^+ &\leq e^{-Bt}[\phi]_\tau^+ + \int_0^t e^{-B(t-s)} BH(K)[x(t)]_\tau^+ ds \\
&= e^{-Bt}[\phi]_\tau^+ + \{\int_0^{t-T} + \int_{t-T}^t\} e^{-B(t-s)} BH(K)[x(t)]^+ ds \\
&\leq e^{-Bt}[\phi]_\tau^+ + \int_T^{+\infty} e^{-Bs} BH(K)\eta V ds + \int_{t-T}^t e^{-B(t-s)} BH(K)(\sigma + \delta E) ds \\
&\leq e^{-Bt}\eta V + \delta E + (I - e^{-BT})H(K)(\sigma + \delta E) \\
&< e^{-Bt}\eta V + \delta E + H(K)(\sigma + \delta E).
\end{aligned}
$$

From (13), there exists monotone non-decreasing sequence of numbers $\{t_l\}$: $t_l \geq max\{t_3, T\}$, such that $lim_{t_l \to +\infty}[x(t_l)]^+ = \sigma$. Letting $t_l \to +\infty$, $\delta \to 0^+$, then $\sigma \leq H(K)\sigma$. Using Theorem 8.3.2 of [11], we have $\rho(H(K)) \geq 1$. This contradicts (H_2), and so (12) holds. We derive from the equilibrium point 0 of (6) is asymptotically stable.

4   Conclusions

In the present paper, the second order Hopfield neural networks with time delays (SDHNN) model are investigated. For the model, we obtain some sufficient conditions independent of delays that ensure the equilibrium point is asymptotically stable. Especially, we drop the condition that activation function $g_j(\cdot)$ are bounded, and present one such method which can be applied to neural networks with time delay system. The method,based on the properties of nonnegative matrices and inequality, yields weaker sufficient criteria on asymptotically stable being independent of delays.

Acknowledgments

The work is supported by the Foundation of Science and Technology project of Chongqing Education Commission 041503 and National Natural Science Foundation of China under Grant 10371083.

References

1. Kosmatopouloos, E. B., Polycarpou, M. M., Christodoulou, M . A., Ioannou, P. A.: High-order Neural Network Structures for Identification of Dynamical Systems. IEEE Trans. Neural Network, 6 (1995) 422-431
2. Zhang, T. Ge, S. S., Hang, C. C.: Neueal-based Direct Adaptive Control for a Class of General Nonlinear Systems. International Journal of Systems Science, 28 (1997) 1011-1020
3. Su, J.Z.: Solving a Kind of Nonlinear Programming Problems via Analog Neural Networks. International Journal of Systems Science, 18 (1998) 1-9
4. Xu, B., Liao, Xinzhi, Liao, X.: Global Asymptotic Stability of High-order Hopfield Type Neural Networks with Time Delay. Computers Mathematics Applic, 45 (2003) 1729-1737
5. Xu, B., Liao, X.: Stability of Higher-oeder Hopfield type Neural Networks with Time Delays. Journal of Circuits and Systems, 7 (2002) 9-12
6. Xu, B, Shen, Y., Liao, X., L. Xinzhi: Stability Anlysis of Second Order Hopfiel Neural with Time Delays. Systems Engineering and Electronics, 24 (2002) 77-81
7. Xu, D., Z. Hongyong, Z. Hong: Global Dynamics of Hopfield Neural Networks Invoing Variable Delays. Computers Mathematics Applic, 42 (2001) 39-45
8. Xu, D., L. Shuyong, P.Zhiling, G.Qingyi: Domain of Attraction of Nonlinear Discrete Systems with Delays. Computers Mathematics Applic, 38 (1999) 155-162
9. Xu, D.: Asymptptic Behavior of Hopfield Neural Networks with Delays. Differential Equations and Dynamical Systems, 9 (2001) 353-364
10. Lasalle, J.P.: The Stability of Dynamical System. SIAM Philadelphia (1976)
11. Horm, R.A., Johnson, C.R.: Matrix Analysis. Cambridge University Press (1985)
12. L. Shuyong, Xu, D., Z.Hongyong: Stability Region of Nonlinear Integrodifferential Equation. Applied Mathematics Letters, 13 (2000) 77-82
13. Yi, Z.: Robust Stabilization of Bilinear Uncertain Delay Systems. Journal of UEST of China, 22 (1993)
14. Kolmanovskii, V., Myshkis, A.: Introduction to the Theory and Applications of Functional Differential Equations. Kluwer Academic Pulishers (1999)

Convergence Analysis of Genetic Regulatory Networks Based on Nonlinear Measures

Hongtao Lu[1], Zhizhou Zhang[2], and Lin He[2]

[1] Department of Computer Science and Engineering,
Shanghai Jiao Tong University,
Shanghai 200030, China
lu-ht@cs.sjtu.edu.cn
[2] Bio-X Life Science Research Center,
Shanghai Jiao Tong University
Shanghai 200030, China

Abstract. In this paper, we propose a nonlinear 2-measure concept, and using it, together with the nonlinear 1-measure proposed earlier by other researchers, to analyze the global convergence of genetic regulatory networks. Two sufficient conditions are derived to ensure globally exponentially convergence of solutions of the networks. The derived conditions provide insight into the dynamical behavior of gene networks.

1   Introduction

In a biological organism genes and their products interact in a complicated way, this interaction behavior can be described by genetic regulatory network models, which usually take two forms: the Boolean model where the activity of each gene has only two states [1],[2] or the differential equations where continuous variables describe the concentrations of gene products such as mRNAs (message RNAs) or proteins [2],[3],[4],[5]. In this paper, we focus on the genetic regulatory network model described by the differential equations. To understand the various biological functions of a living organism, it is important to understand the complicated interaction between genes and proteins. In [2], Chen and Aihara studied the local stability and bifurcations of genetic regulatory networks with time delay by means of linearization approach. However, the global dynamics of gene networks is more important for understanding biological functions of living organisms. In this paper, with the help of concept of the nonlinear measures, we derive some conditions for global convergence of genetic regulatory networks.

The concept of nonlinear measure was initially proposed for stability analysis of neural networks by Qiao *et.al.* [6], where the proposed nonlinear measure is an extension of the matrix column measure, i.e. the $\mu_1(\cdot)$, which we refer to nonlinear 1-measure. In this paper, we first introduce a new nonlinear measure called nonlinear 2-measure for nonlinear mapping, which is an extension of the matrix μ_2-measure concept from linear mappings to nonlinear ones.

J. Wang, X. Liao, and Z. Yi (Eds.): ISNN 2005, LNCS 3496, pp. 247–252, 2005.

2   Nonlinear Measures

Matrix measures play an important role in the stability analysis of nonlinear systems. For neural networks many stability criteria have been derived, which are expressed in terms of the matrix measures or matrix norms of the connection weight matrix. The most widely used matrix norms and matrix measures are those induced by the vector l_p-norms with $p = 1, 2, \infty$, where for a vector $x = (x_1, \ldots, x_n)^T \in R^n$, the three vector norms are defined by $\|x\|_1 = \sum_i |x_i|$, $\|x\|_\infty = \max_i |x_i|$ and $\|x\|_2 = (\sum_i |x_i|^2)^{1/2}$, respectively. And for a $n \times n$ matrix $W = (w_{ij})$, the induced matrix measures are defined as:

$$\mu_1(W) \stackrel{\triangle}{=} \max_i (w_{ii} + \sum_{j \neq i} |w_{ji}|)$$

$$\mu_2(W) \stackrel{\triangle}{=} \lambda_{\max}\{(W + W^T)/2\}$$

$$\mu_\infty(W) \stackrel{\triangle}{=} \max_i (w_{ii} + \sum_{j \neq i} |w_{ij}|)$$

where $\lambda_{\max}(\cdot)$ represents the largest eigenvalue of a symmetrical matrix.

For a linear system described by a differential equation $\dot{x}(t) = Ax(t)$, where A is a $n \times n$ matrix, a well-known result is that if there is a matrix measure $\mu(\cdot)$ such that $\mu(A) < 0$, then the system is globally exponentially stable.

Due to the significance of the matrix measures in stability analysis, a natural question is: can we generalize the concept of matrix measures for linear systems to some quantities for nonlinear mappings that have similar properties as matrix measures in characterizing the stability of a nonlinear system. This question is first proposed and addressed by Qiao $et.al.$ [6], where the concept of the nonlinear 1-measure was present, which is only an extension of the matrix μ_1-measure. In this paper, we will define another nonlinear measure named nonlinear 2-measure which is an extension to the matrix μ_2-measure. It turns out that the nonlinear 2-measure established here is different from and independent of the nonlinear 1-measure in [6].

According to matrix theory, for any given vector norm denoted by $\| \cdot \|$, the induced matrix measure for any $n \times n$ matrix A is defined respectively by

$$\mu(A) \stackrel{\triangle}{=} \lim_{s \to 0+} \frac{\|I + sA\| - 1}{s} \tag{1}$$

For general vector norm, there is usually no explicit expression formula for the induced matrix measure, but for l_1, l_2, l_∞ vector norms, the induced matrix measures have simple expressions already given above.

In [6], by expressing the matrix μ_1-measure in a different yet equivalent way

$$\mu_1(A) = \sup_{x \in R^n, x \neq 0} \frac{< Ax, \text{sign}(x) >}{\|x\|_1} \tag{2}$$

where $< u, v >$ denotes the inner product of two vectors u and v, and $\text{sign}(\cdot)$ is the signum function, the authors introduced a nonlinear measure for the nonlinear mapping $F : \Omega \to R^n$ defined on an open set Ω in R^n as

$$m_1^\Omega(F) = \sup_{x,y \in \Omega, x \neq y} \frac{< F(x) - F(y), \text{sign}(x - y) >}{\|x - y\|_1} \tag{3}$$

The nonlinear measure so defined is obviously a generalization of the matrix μ_1-measure, so we call it the nonlinear 1-measure. For any nonlinear mapping F, it is obvious that

$$< F(x) - F(y), \text{sign}(x - y) > \leq m_1^\Omega(F) \cdot \|x - y\|_1, \forall x, y \in \Omega \tag{4}$$

For definition of new nonlinear measure, we observe that the matrix μ_2-measure has an alternative expression as

$$\mu_2(A) = \sup_{x \in R^n, x \neq 0} \frac{< Ax, x >}{\|x\|_2^2} = \sup_{x \in R^n, x \neq 0} \frac{< Ax, x >}{< x, x >} \tag{5}$$

This can be proved in the following way. On one hand

$$\mu_2(A) = \lim_{s \to 0^+} \frac{\|I + sA\|_2 - 1}{s}$$

$$= \lim_{s \to 0^+} \sup_{x \in R^n, x \neq 0} \frac{\|(I + sA)x\|_2 - \|x\|_2}{s\|x\|_2}$$

$$= \lim_{s \to 0^+} \sup_{x \in R^n, x \neq 0} \frac{\|(I + sA)x\|_2 \cdot \|x\|_2 - \|x\|_2^2}{s\|x\|_2^2}$$

$$\geq \lim_{s \to 0^+} \sup_{x \in R^n, x \neq 0} \frac{< (I + sA)x, x > - < x, x >}{s\|x\|_2^2}$$

$$= \sup_{x \in R^n, x \neq 0} \frac{< Ax, x >}{\|x\|_2^2} \tag{6}$$

On the other hand, if we let the eigenvector of $(A + A^T)/2$ associated to the largest eigenvalue $\lambda_{\max}\{(A+A^T)/2\}$ be x_0, i.e., $\frac{A+A^T}{2} x_0 = \lambda_{\max}\{(A+A^T)/2\}x_0$, then we have

$$\sup_{x \in R^n, x \neq 0} \frac{< Ax, x >}{\|x\|_2^2} \geq \frac{< Ax_0, x_0 >}{\|x_0\|_2^2} = \frac{< \frac{A+A^T}{2} x_0, x_0 >}{\|x_0\|_2^2}$$

$$= \frac{< \lambda_{\max}\{(A + A^T)/2\}x_0, x_0 >}{\|x_0\|_2^2}$$

$$= \frac{\lambda_{\max}\{(A + A^T)/2\}\|x_0\|_2^2}{\|x_0\|_2^2} = \lambda_{\max}\{(A + A^T)/2\} = \mu_2(A)$$

Now, based on (5) we define the nonlinear 2-measure as follows.

Definition 1. Suppose Ω is an open set of R^n, and $F : \Omega \to R^n$ is a nonlinear operator defined on Ω. The constant

$$m_2^\Omega(F) \triangleq \sup_{x,y\in\Omega, x\neq y} \frac{< F(x) - F(y), x - y >}{< x - y, x - y >}$$

$$= \sup_{x,y\in\Omega, x\neq y} \frac{< F(x) - F(y), x - y >}{\|x - y\|_2^2} \qquad (7)$$

is called the nonlinear 2-measure of F on Ω.

According to the definition of the nonlinear 2-measure, for any $x, y \in \Omega$, we have,

$$< F(x) - F(y), x - y > \le m_2^\Omega(F) \cdot \|x - y\|_2^2, \qquad \forall x, y \in \Omega \qquad (8)$$

Similar to the matrix μ_2 measure, it is easy to prove the following properties of the nonlinear 2-measure.

1. For any two operators F and G defined on Ω, $m_2^\Omega(F+G) \le m_2^\Omega(F) + m_2^\Omega(G)$;
2. For any nonnegative constant $\alpha \ge 0$, $m_2^\Omega(\alpha F) = \alpha m_2^\Omega(F)$.

Next, we prove an important role of the nonlinear 2-measure in characterizing the one-to-one property of a nonlinear operator.

Lemma 1. Suppose Ω is an open set in R^n, $F : \Omega \to R^n$ is an operator mapping Ω into R^n. If $m_2^\Omega(F) < 0$, the F is an injective (one-to-one mapping) from Ω to R^n. In addition, if $\Omega = R^n$, then F is a homeomorphism on R^n.

Proof. From the definition of the nonlinear 2-measure, for any $x, y \in R^n$,

$$< F(x) - F(y), x - y > \le m_2^\Omega(F)\|x - y\|_2^2 \qquad (9)$$

On the other hand, we have

$$| < F(x) - F(y), x - y > | \le \|F(x) - F(y)\|_2 \cdot \|x - y\|_2 \qquad (10)$$

So combining the two inequalities, we get

$$\|F(x) - F(y)\|_2 \cdot \|x - y\|_2 \ge - < F(x) - F(y), x - y > \ge -m_2^\Omega(F)\|x - y\|_2 \qquad (11)$$

When $m_2^\Omega(F) < 0$, from (11), if $F(x) = F(y)$, then there must have $\|x - y\|_2 = 0$, so $x = y$. This means that F is an injective. Next, if we fix y in (11), we then have $\|F(x)\|_2 \to +\infty$ whenever $\|x\|_2 \to +\infty$, thus F is a homeomorphism on R^n.

3   Genetic Regulatory Networks and Their Global Convergence

We consider a genetic regulatory network model proposed in [2], where the networks have time delay. Here, we just involve the model without time delay. The network can be described by a set of differential equations.

$$\dot{m}(t) = -K_m m(t) + c(p(t)) \qquad (12)$$

$$\dot{p}(t) = -K_p p(t) + d(m(t)) \qquad (13)$$

where $m(t) = (m_1(t), \ldots, m_n(t))^T \in R^n$ and $p(t) = (p_1(t), \ldots, p_n(t))^T \in R^n$ represent the concentrations of mRNAs and proteins, respectively. The positive diagonal matrices $K_m = \mathrm{diag}(k_{m1}, \ldots, k_{mn})$ and $K_p = \mathrm{diag}(k_{p1}, \ldots, k_{pn})$ represent the degradation rates of mRNAs and proteins, respectively. $c(p) = (c_1(p), \ldots, c_n(p))^T \in R^n$ and $d(m) = (d_1(m), \ldots, d_n(m))^T \in R^n$ are two nonlinear vector-valued functions describing the nonlinear effects of the biochemical reactions of transcription, splicing and translation processes.

If we let $x(t) = [(m(t))^T, (p(t))^T]^T \in R^{2n}$, and define a function f from R^{2n} to R^{2n} as $f(x(t)) = [(c(p(t)))^T, (d(m(t)))^T]^T$, then the network (12)-(13) can be rewritten as

$$\dot{x}(t) = -\begin{pmatrix} K_m & \\ & K_p \end{pmatrix} \cdot x(t) + f(x(t)) \qquad (14)$$

Theorem 1. If the nonlinear 2-measure of f satisfies $m_2^{R^{2n}}(f) < \min_{1 \leq i \leq n}(k_{mi}, k_{pi})$, then there exists unique equilibrium x^* for the genetic regulatory network (14) and any solution $x(t)$ of (14) with initial condition $x(t_0) = x^0 \in R^{2n}$ exponentially converges to x^* satisfying

$$\|x(t) - x^*\|_2 \leq e^{[2(m_2^{R^{2n}}(f) - \min_{1 \leq i \leq n}(k_{mi}, k_{pi})) \cdot (t - t_0)]} \cdot \|x^0 - x^*\|_2, \forall t \geq t_0 \quad (15)$$

Proof. We first prove the estimation (15) under the condition $m_2^{R^{2n}}(f) < \min_{1 \leq i \leq n}(k_{mi}, k_{pi})$. The existence of equilibrium is left for proof later. For this, we prove that any two solutions $x(t)$ and $y(t)$ with initial condition $x(t_0) = x^0$ and $y(t_0) = y^0$ satisfy

$$\|x(t) - y(t)\|_2 \leq e^{[2(m_2^{R^{2n}}(f) - \min_{1 \leq i \leq n}(k_{mi}, k_{pi})) \cdot (t - t_0)]} \cdot \|x^0 - y^0\|_2, \forall t \geq t_0 \quad (16)$$

Since $\|x(t) - y(t)\|_2^2 = \sum_{i=1}^n (x_i(t) - y_i(t))^2$, we have

$$\frac{1}{2} \frac{d\|x(t) - y(t)\|_2^2}{dt} = <\dot{x}(t) - \dot{y}(t), x(t) - y(t)>$$

$$= <-\begin{pmatrix} K_m & \\ & K_p \end{pmatrix}(x(t) - y(t)) + f(x(t)) - f(y(t)), x(t) - y(t)>$$

$$\leq -\min_{1 \leq i \leq n}(k_{mi}, k_{pi}) \cdot \|x(t) - y(t)\|_2^2 + <f(x(t)) - f(y(t)), x(t) - y(t)>$$

$$\leq \left\{ -\min_{1 \leq i \leq n}(k_{mi}, k_{pi}) + m_2^{R^{2n}}(f) \right\} \cdot \|x(t) - y(t)\|_2^2 \qquad (17)$$

Integrating both sides of the above differential inequality from t_0 to t, we obtain (16). (16) implies that the difference of any two solutions converges to 0 exponentially fast. If we let $y^0 = x(s)$ for some $s > t_0$, (16) implies that every solution $x(t)$ is a Cauchy sequence. Then based on the Cauchy convergence principle, $x(t)$ converges to an equilibrium x^*. This proves the existence of equilibrium, and the uniqueness is guaranteed by Lemma 1. The proof of Theorem 1 is completed.

The results in Theorem 1 are in terms of nonlinear 2-measure, the nonlinear 1-measure in [6] can also be used to characterize global convergence of the genetic regulatory networks.

Theorem 2. If the nonlinear 1-measure satisfies $m_1^{R^{2n}}(f) < \min_{1 \leq i \leq n}(k_{mi}, k_{pi})$, then there exists unique equilibrium x^* for the genetic regulatory network (14) and any solution $x(t)$ of (14) with initial condition $x(t_0) = x^0 \in R^{2n}$ exponentially converges to x^* satisfying

$$\|x(t) - x^*\|_1 \leq e^{[(m_1^{R^{2n}}(f) - \min_{1 \leq i \leq n}(k_{mi}, k_{pi})) \cdot (t-t_0)]} \cdot \|x^0 - x^*\|_1, \forall t \geq t_0 \quad (18)$$

Proof. The proof of this theorem is similar to that of Theorem 1. We first prove that any two solutions $x(t)$ and $y(t)$ of (14) with initial condition $x(t_0) = x^0$ and $y(t_0) = y^0$ satisfy the inequality

$$\|x(t) - y(t)\|_1 \leq e^{[(m_1^{R^{2n}}(f) - \min_{1 \leq i \leq n}(k_{mi}, k_{pi})) \cdot (t-t_0)]} \cdot \|x^0 - y^0\|_1, \forall t \geq t_0 \quad (19)$$

Since $\|x(t) - y(t)\|_1 = < x(t) - y(t), \text{sign}(x(t) - y(t)) >$, we have

$$\frac{d\|x(t) - y(t)\|_1}{dt} = < \dot{x}(t) - \dot{y}(t), \text{sign}(x(t) - y(t)) >$$

$$= < -\begin{pmatrix} K_m & \\ & K_p \end{pmatrix}(x(t) - y(t)) + f(x(t)) - f(y(t)), \text{sign}(x(t) - y(t)) >$$

$$\leq -\min_{1 \leq i \leq n}(k_{mi}, k_{pi}) \cdot \|x(t) - y(t)\|_1 + < f(x(t)) - f(y(t)), \text{sign}(x(t) - y(t)) >$$

$$\leq \left\{ -\min_{1 \leq i \leq n}(k_{mi}, k_{pi}) + m_2^{R^{2n}}(f) \right\} \cdot \|x(t) - y(t)\|_1 \quad (20)$$

Integrating both sides of the above differential inequality from t_0 to t, we obtain (19), and the remaining proof is same to Theorem 1.

Acknowledgments

This work is supported by Shuguang Program of Shanghai Education Development Foundation under grant 03SG11, and also supported by National Department of Education project (No. 03066) and BIO-X BDCC Fund (03DZ14025).

References

1. P. Smolen, D.A.B., Byrne, J.H.: Mathematical Modelling of Gene Networks Review. Neuron, **26** (2000) 567–580
2. Chen, L., Aihara, K.: Stability of Genetic Regulatory Networks with Time Delay. IEEE Trans. Circuits Syst. I, **49** (2002) 602–608
3. Elowitz, M.B., Leibler, S.: A Synthetic Oscillatory Network of Transcriptional Regulators. Nature, **403** (2000) 335–338
4. T. Chen, H.L.H., Church, G.M.: Modelling Gene Expression with Differential Equations. In: Proc. Pacific Symp. Biocomputing, (1999) 29–40
5. Sakamoto, E., Iba, H.: Inferring a System of Differential Equations for a Gene Regulatory Network by Using Genetic Programming. In Proc. Congress on Evolutionary Computation, (2001) 720–726
6. H. Qiao, J.P., Xu, Z.: Nonlinear measure: A New Approach to Exponential Stability Analyssi for Hopfield-type Neural Networks. IEEE Trans. Neural Networks, **12** (2001) 360–370

Stability Conditions for Discrete Neural Networks in Partial Simultaneous Updating Mode

Runnian Ma[1,2], Shengrui Zhang[3], and Sheping Lei[4]

[1] Telecommunication Engineering Institute, Air Force Engineering University,
Xi'an 710071, China
[2] University Key Lab of Information Sciences and Engineering,
Dalian University 111662, China
m314@163.com
[3] School of Highway, Chang'an University, Xi'an 710064, China
[4] School of Humanity Law and Economics , Northwestern Polytechnical University,
Xi'an 710072, China

Abstract. The stability analysis of discrete Hopfield neural networks not only has an important theoretical significance, but also can be widely used in the associative memory, combinatorial optimization, etc. The dynamic behavior of asymmetric discrete Hopfield neural network is mainly studied in partial simultaneous updating mode, and some new simple stability conditions of the networks are presented by using the Lyapunov method and some analysis techniques. Several new sufficient conditions for the networks in partial simultaneous updating mode converging towards a stable state are obtained. The results established here improve and extend the corresponding results given in the earlier references. Furthermore, we provide one method to analyze and design the stable discrete Hopfield neural networks.

1 Introduction

Discrete Hopfield neural network (DHNN) was proposed mainly as an associative memory model [1]. As it is known, the stability of DHNN is not only known to be one of the mostly basic problems, but also known to be bases of the network's various applications. Because of its importance, the stability analysis of the DHNN has attracted considerable interest. Many researchers have focused on the following three distinct updating modes: 1) serial mode, in which a neuron is chosen at random and then its value is updated, see, for example, [1-3,8] and references cited therein; 2) parallel (or fully parallel) mode, where all of the neurons are simultaneously updated, see, for example, [1-3,7] and references cited therein; 3) partial simultaneous updating (PSU) mode, in which a group of a fixed number of neurons are updated. Each of these groups is referred to as a "macroneuron", see, for example, [4-6].

This paper is organized as follows. Section two introduces some notations and definitions used in the paper. Section three investigates the stability of DHNN in PSU mode and gives some new conditions for DHNN converging to a stable state.

2 Basic Definitions

Discrete Hopfield neural networks consist of a large number of interconnected "artificial neurons". Mathematically, DHNN with n neurons can uniquely be defined by

J. Wang, X. Liao, and Z. Yi (Eds.): ISNN 2005, LNCS 3496, pp. 253–258, 2005.

$N=(W, \theta)$, where $W=(w_{ij})_{n \times n}$ is an $n \times n$ real matrix with w_{ij} denoting the connection strength between neuron i and neuron j, and w_{ii} denoting the strength of the self-feedback connection of neuron i; $\theta=(\theta_1, \theta_2, \cdots, \theta_n)^T$ is an n-dimensional column with θ_i representing the threshold of neuron i. There are two possible values for the state of each neuron i: 1 or -1. Denote the state of neuron i at time $t \in \{0,1,2,\cdots\}$ as $x_i(t)$, the vector $X(t)=(x_1(t), x_2(t), \cdots, x_n(t))^T$ is the state of the whole neurons at time t.

The updating mode of DHNN is determined by the following equations

$$x_i(t+1) = \operatorname{sgn}(\sum_{j=1}^{n} w_{ij} x_j(t) + \theta_i), \ i \in \{1,2,\cdots,n\}, \tag{1}$$

where $t \in \{0,1,2,\cdots\}$, and the sign function is defined as follows

$$\operatorname{sgn}(u) = \begin{cases} 1, & \text{if } u \geq 0 \\ -1, & \text{if } u < 0 \end{cases}. \tag{2}$$

Denote $I=\{1,2,\cdots,n\}$. For each state $X=(x_1,x_2,\cdots,x_n)^T$, denote

$$H_i(X) = \sum_{j \in I} w_{ij} x_j + \theta_i, \ i \in I. \tag{3}$$

If $H_i(X) \neq 0$ for each state X and each $i \in I$, then we say that DHNN (1) is strict. We rewrite equation (1) in the compact form

$$X(t+1)=\operatorname{sgn}(WX(t)+\theta). \tag{4}$$

If the state X^* satisfies the following condition

$$X^*=\operatorname{sgn}(WX^*+\theta), \tag{5}$$

then the state X^* is called a stable state (or an equilibrium point).

Let $N=(W,\theta)$ starting from any initial state $X(0)$. At any time $t \geq 1$, if one of the "macroneurons" $I_1, I_2, \cdots, I_m$ is chosen at random to update according to (1), where $I_1, I_2, \cdots, I_m$ satisfy the following conditions

$$\phi \neq I_i \subseteq I=\{1,2,\cdots,n\}, \ i=1,2,\cdots,m,$$

then the network is said to be operated in the partial simultaneous updating (PSU) mode. If there exists time $t_1 \in \{0,1,2,\cdots\}$ such that the updating sequence $X(0),X(1)$, $X(2),X(3),\cdots$ satisfies that $X(t)=X(t_1)$ for all $t \geq t_1$, then we call that the initial state $X(0)$ converges towards a stable state in the PSU mode. If the network always converges towards a stable state for each initial state $X(0)$, then we say the network being convergent or stable in the PSU mode.

If $Z^T A Z \geq 0$ for each $Z \in \{-2,0,2\}^n$, then matrix $A=(a_{ij})_{i,j \in I}$ is called to be nonnegative definite on the set $\{-2,0,2\}$, where matrix A is not necessarily symmetric.

For an $n \times n$ matrix $A=(a_{ij})_{i,j \in I}$, we define the corresponding matrix $A^*=(a_{ij}^*)_{i,j \in I}$ with

$$a_{ij}^* = \begin{cases} a_{ii} - \dfrac{1}{2}\sum_{k \in I} |a_{ki} - a_{ik}|, & i = j \\ a_{ij}, & i \neq j \end{cases}. \tag{6}$$

A matrix $A=(a_{ij})_{i,j\in I}$ is called to be row-diagonally dominant or column-diagonally dominant, if the matrix A respectively satisfies the following conditions

$$a_{ii} \ge \sum_{j\in I(j\ne i)}\left|a_{ij}\right|, \text{ or } a_{ii} \ge \sum_{j\in I(j\ne i)}\left|a_{ji}\right|, \; i\in I. \qquad (7)$$

The stability of DHNN in PSU mode is mainly investigated, and some results on the stability are given. The main contribution of this paper is expressed by the following results

3 The Stability in PSU Mode

Theorem 1. If matrix $A_{ss}^{*}=(a_{hk}^{*})_{h,k\in I_s}$ is nonnegative definite on the set $\{-2,0,2\}$ but not necessarily symmetric $s=1,2,\cdots,m$, matrix $B=(b_{ij})_{i,j\in I}$ is column-diagonally dominant, and matrix $C=(c_{ij})_{i,j\in I}$ is row-diagonally dominant, then $N=(A+B+C,\theta)$ is convergent in PSU mode, where the expression of a_{hk}^{*} as (6).

Proof. Without loss generality, we assume that $N=(W,\theta)$ is strict, where matrix $W=A+B+C$. Then we can easily prove that, whether $x_i(t+1)=1$ or -1, we have

$$x_i(t+1)H_i(X(t)) > 0, \; i\in I. \qquad (8)$$

where the expression of $H_i(X(t))$ as (3).

We define energy function (Lyapunov function) of DHNN (1) as follows.

$$E(X(t)) = -X^{T}(t)AX(t) - 2X^{T}(t)BX(t) - 2X^{T}(t)\theta. \qquad (9)$$

Then

$$\Delta E(X(t)) = E(X(t+1)) - E(X(t))$$

$$= \Delta X^{T}(t)AX(t) - X^{T}(t)A\Delta X(t) - \Delta X^{T}(t)A\Delta X(t) \qquad (10)$$

$$- 2\Delta X^{T}(t)[(A+B+C)X(t)+\theta]$$

$$- 2X^{T}(t+1)B\Delta X(t) + 2\Delta X^{T}CX(t) = -p(t) - q(t) - r(t).$$

Where

$$\Delta X(t) = X(t+1) - X(t),$$

$$p(t) = -\Delta X^{T}(t)AX(t) + X^{T}(t)A\Delta X(t) + \Delta X^{T}(t)A\Delta X(t),$$

$$q(t) = 2\Delta X^{T}(t)[(A+B+C)X(t)+\theta],$$

$$r(t) = 2X^{T}(t+1)B\Delta X(t) - 2\Delta X^{T}CX(t).$$

Now without loss generality, we assume that the "macroneuron" I_s is chosen to update at time t according to (1). So, if $i\notin I_s$, then $\Delta x_i(t)=0$. When $\Delta x_i(t)\ne 0$, then $\Delta x_i(t)=2x_i(t+1)=-2x_i(t)$.

In order to prove our results, let $I_{2s}(t)=\{i\in I_s|\Delta x_i(t)\neq 0\}$, $I_{1s}(t)=I\backslash I_{2s}(t)$. We consider two cases as follows.

Case 1 If $\Delta X(t)=0$, then $p(t)=q(t)=0$.

Case 2 If $\Delta X(t)\neq 0$, then there exists at least one neuron $i\in I_{2s}(t)$, such that $\Delta x_i(t)= 2x_i(t+1)=-2x_i(t)$. Based on (8), if $\Delta x_i(t)\neq 0$, then $q(t)>0$. We analyze $p(t)$ as follows.

Since matrix $B=(b_{ij})_{i,j\in I}$ is column-diagonally dominant and matrix $C=(c_{ij})_{i,j\in I}$ is row-diagonally dominant, we have

$$r(t) = 2X^T(t+1)B[X(t+1)-X(t)]-[X^T(t+1)-X^T(t)]CX(t)$$

$$= 2[X^T(t+1)-X^T(t)]B^T X(t+1) - 2[X^T(t+1)-X^T(t)]CX(t)$$

$$= 2\sum_{i\in I}[x_i(t+1)-x_i(t)]\sum_{j\in I}b_{ji}x_j(t+1) - 2\sum_{i\in I}[x_i(t+1)-x_i(t)]\sum_{j\in I}c_{ij}x_j(t)$$

$$= 4\sum_{i\in I_{2s}(t)}x_i(t+1)\sum_{j\in I}b_{ji}x_j(t+1) + 4\sum_{i\in I_{2s}(t)}x_i(t)\sum_{j\in I}c_{ij}x_j(t)$$

$$\geq 4\sum_{i\in I_{2s}(t)}(b_{ii}-\sum_{j\neq i}|b_{ji}|) + 4\sum_{i\in I_{2s}(t)}(c_{ii}-\sum_{j\neq i}|c_{ij}|) \geq 0. \tag{11}$$

Please refer to [5, theorem 2] for a detailed description, we have

$$p(t) = \sum_{i\in I_{2s}(t)}\Delta x_i(t)[\sum_{j\in I_{2s}(t)}a_{ij}\Delta x_j(t)-\sum_{j\in I}(a_{ij}-a_{ji})x_j(t)]$$

$$= \sum_{i\in I_{2s}(t)}\sum_{j\in I_{2s}(t)}\Delta x_i(t)a_{ij}^*\Delta x_j(t). \tag{12}$$

By assumption, matrix $A_{ss}^*=(a_{hk}^*)_{h,k\in I_s}$ is nonnegative definite on the set $\{-2,0,2\}$, so is any one of the principal submatrices of A_{ss}^*. This then implies $p(t)\geq 0$ in (12).

As proved above, note that (10), (11) and (12) implies $\Delta E(X(t))\leq 0$. Furthermore, if $\Delta X(t)=0$, then $\Delta E(X(t))=0$, if $\Delta X(t)\neq 0$, then $\Delta E(X(t))<0$. Consequently, the energy function is strict, and then theorem 1 holds. The proof is completed.

Corollary 1. Let matrix A is symmetric. If the principal submatrices $A_{ss}=(a_{hk})_{h,k\in I_s}$ are nonnegative definite on the set $\{-2,0,2\}$, $s=1,2,\ldots,m$, matrix $B=(b_{ij})_{i,j\in I}$ is column-diagonally dominant, and matrix $C=(c_{ij})_{i,j\in I}$ is row-diagonally dominant, then the network $N=(A+B+C,\theta)$ converges towards a stable state for each initial state $X(0)$ in the PSU mode.

Proof. By assumption, matrix A is symmetric, and the principal submatrices $A_{ss}=(a_{hk})_{h,k\in I_s}$ are nonnegative definite on the set $\{-2,0,2\}$, $s=1,2,\ldots,m$, we can easily prove that matrices $A_{ss}^*=(a_{hk}^*)_{h,k\in I_s}$ are nonnegative definite on the set $\{-2,0,2\}$. In fact, matrix $A_{ss}^*=A_{ss}$ because of matrix A being symmetric. So the conditions of theorem 1 are satisfied. By theorem 1, we know that corollary is true. The proof is completed.

Corollary 2. Let matrix $B=(b_{ij})_{i,j\in I}$ be column-diagonally dominant, and matrix $C=(c_{ij})_{i,j\in I}$ be row-diagonally dominant. If matrix $A=(a_{ij})_{i,j\in I}$ is satisfies the following conditions

$$a_{ss} \geq \frac{1}{2}\left(\sum_{j\in I}\left|a_{sj}-a_{js}\right| + \sum_{j\in I_s,(j\neq s)}\left|a_{sj}+a_{js}\right|\right), \; s\in I_s, s=1,2,\cdots,m. \tag{13}$$

then the network $N=(A+B+C,\theta)$ converges towards a stable state for each initial state $X(0)$ in the PSU mode.

Proof. By the expression (6) of $a_{ij}{}^*$, we rewrite (13) as

$$a_{ss}{}^* + a_{ss}{}^* \geq \sum_{j\in I_s,(j\neq s)}\left|a_{sj}{}^* + a_{js}{}^*\right|, \; s\in I_s, s=1,2,\cdots,m. \tag{14}$$

Based on (14), we know that matrix $(A_{ss}{}^* + (A_{ss}{}^*)^T) = (a_{hk}{}^* + a_{kh}{}^*)_{h,k\in Is}$ is both column-diagonally dominant and row-diagonally dominant. So, matrix $(A_{ss}{}^* + (A_{ss}{}^*)^T)$ is non-negative definite on the real number set R, of course, nonnegative definite on the set $\{-2,0,2\}$. This then implies that matrix $A_{ss}{}^*$ is nonnegative definite on the set $\{-2,0,2\}$. By assumption, matrix $B=(b_{ij})_{i,j\in I}$ being column-diagonally dominant, and matrix $C=(c_{ij})_{i,j\in I}$ being row-diagonally dominant, we know that conditions of theorem 1 are satisfied. This then implies that corollary 2 is true.

Remark. In theorem 1, corollary 1, and corollary 2, if there exists a positive diagonal matrix $D=\text{diag}(d_1,d_2,\dots,d_n)$ $(d_i>0, i=1,2,\dots,n)$ such that matrix DA, DB and DC respectively satisfy corresponding conditions of matrix A, B and C, then the network $N=(A+B+C,\theta)$ converges towards a stable state for each initial state $X(0)$ in PSU mode.

Example. Consider the stability of $N=(W,0)$ in the PSU mode($I_1=\{1,2\}$, $I_2=\{2,3\}$), where the expressions of matrices W and W^* are respectively in the following (we calculate the matrix W^* according to the definition of (6))

$$W = \begin{pmatrix} 4 & -1 & -3 \\ -3 & 3 & 1 \\ -6 & 1 & 1 \end{pmatrix}, \; W^* = \begin{pmatrix} 1.5 & -1 & -3 \\ -3 & 2 & 1 \\ -6 & 1 & -0.5 \end{pmatrix}. \tag{15}$$

Obviously, matrix W is not symmetric, and matrix W^* is not nonnegative definite. The stability of the network in PSU mode can not guaranteed by the previous results in references. However, matrix W can be decomposed into $W=A+B+C$, where

$$A = \begin{pmatrix} 1 & -1 & -3 \\ -1 & 1 & 1 \\ -3 & 1 & 1 \end{pmatrix}, \; B = \begin{pmatrix} 3 & 0 & 0 \\ 0 & 0 & 0 \\ -3 & 0 & 0 \end{pmatrix}, \; C = \begin{pmatrix} 0 & 0 & 0 \\ -2 & 2 & 0 \\ 0 & 0 & 0 \end{pmatrix}. \tag{16}$$

From (16), we know that matrix A is symmetric, matrix B is column-diagonally dominant and matrix C is row-diagonally dominant. The expressions of principal submatrices A_{11} and A_{22} of matrix A are respectively in the following

$$A_{11} = \begin{pmatrix} 1 & -1 \\ -1 & 1 \end{pmatrix}, \; A_{22} = \begin{pmatrix} 1 & 1 \\ 1 & 1 \end{pmatrix}. \tag{17}$$

Obviously, matrix A_{11} and A_{22} are nonnegative definite on the set $\{-2,0,2\}$. From all above, the conditions of corollary 1 are satisfied, then $N=(A+B+C,\theta)$ converges towards a stable state for each initial state $X(0)$ in the PSU mode. In fact, we can test that the network has two stable states $(-1,1,1)^T$ and $(1,-1,-1)^T$. The states $(-1,-1,1)^T$ and $(-1,1,-1)^T$ converge towards the stable state $(-1,1,1)^T$. The states $(1,1,1)^T$, $(1,1,-1)^T$ and $(1,-1,1)^T$ converge towards the stable state $(1,-1,-1)^T$. The state $(-1,-1,-1)^T$ converge towards either the stable state $(-1,1,1)^T$ or $(1,-1,-1)^T$.

Acknowledgments

The work described in this paper was supported by China Postdoctoral Science Foundation (No.2003033516), and partly supported by the Open Foundation of University Key Lab of Information Sciences and Engineering, Dalian University.

References

1. Hopfield, J.J.: NeuralNetworks, Physical Systems Emergent Collective Computational Abilities. Proc.Nat.Acad.Sci.USA, **79** (1982) 2554-2558
2. Xu, Z., Kwong, C.P.: Global Convergence and Asymptotic Stability of Asymmetrical Hopfield Neural Networks. J. Mathematical Analysis and Applications, **191** (1995) 405-426
3. Liao, X., Chang, L., Shen, Y.: Study on Stability of Discrete-time Hopfield Neural Networks. Acta Automatica Sinica, **25** (1999) 721-727
4. Cernuschi Frias B: Partial Simultaneous Updating in Hopfield Memories. IEEE Trans. Syst., Man, Cybern., **19** (1989) 887-888
5. Lee, D.: New Stability Conditions for Hopfield Neural Networks in Partial Simultaneous Update Mode. IEEE, Trans. Neural Networks, **10** (1999) 975-978
6. Ma, R., Zhang, Q., and Xu, J.: Convergence of Discrete-time Cellular Neural Networks. Chinese J. Electronics, **11** (2002) 352-356
7. Ma, R., Xi, Y., and Guo, J.: Dynamic Behavior of Discrete Hopfield Neural Networks. Asian J. Information Technology, **3** (2005) 9-15
8. Ma, R., Xi, Y., and Gao, H.: Stability of Discrete Hopfield Neural Networks with Delay in Serial Mode. Lecture Notes in Computer Science, **3173** (2004) 126-131

Dynamic Behavior Analysis of Discrete Neural Networks with Delay

Runnian Ma[1,2], Sheping Lei[3], and Shengrui Zhang[4]

[1] Telecommunication Engineering Institute, Air Force Engineering University,
Xi'an 710071, China
[2] University Key Lab of Information Sciences and Engineering,
Dalian University 111662, China
m314@163.com
[3] School of Humanity Law and Economics, Northwestern Polytechnic University,
Xi'an 710072, China
[4] School of Highway, Chang'an University, Xi'an 710064, China

Abstract. The stability of recurrent neural networks is known to be bases of successful applications of the networks. Discrete Hopfield neural networks with delay are extension of discrete Hopfield neural networks without delay. In this paper, the stability of discrete Hopfield neural networks with delay is mainly investigated. The method, which does not make use of energy function, is simple and valid for the dynamic behavior analysis of the neural networks with delay. Several new sufficient conditions for the networks with delay converging towards a limit cycle with length 2 are obtained. All results established here generalize the existing results on the stability of both discrete Hopfield neural networks without delay and with delay in parallel updating mode.

1 Introduction

Discrete Hopfield neural networks (DHNN) are one of the famous neural networks with a wide range of applications, such as content addressable memory, pattern recognition, and combinatorial optimization. Such applications heavily depend on the dynamic behavior of the networks. Therefore, the researches on the dynamic behavior are a necessary step for the design of the networks. Because the stability of DHNN is not only the foundation of the network's applications, but also the most basic and important problem, the researches on stability of DHNN have attracted considerable interest in Refs.[1-7]. Recently, discrete Hopfield neural network with delay (DHNND) is presented. We know that DHNND is an extension of DHNN. Of course, the stability of DHNND is an important problem. The stability of DHNND is investigated in serial updating mode and some results on stability of DHNND are given in Refs.[8-10]. Also, the stability of DHNND is mainly studied in parallel updating mode and some results on stability of DHNND are obtained in Refs.[10-13]. However, all previous researches on the network's stability assume that interconnection matrix W^0 is symmetric, quasi-symmetric or antisymmetric and interconnection matrix W^1 is row-diagonally dominant. In this paper, we improve the previous stability conditions of DHNND and obtain some new stability conditions for DHNND converging towards a limit cycle with length 2.

J. Wang, X. Liao, and Z. Yi (Eds.): ISNN 2005, LNCS 3496, pp. 259–264, 2005.

This paper is organized as follows. Section two introduces some notations and definitions used in the paper. Section three investigates the stability of DHNND and gives some new conditions for DHNND converging to a limit cycle with length 2. Section four concludes the paper.

2 Notations and Definitions

Discrete Hopfield neural networks with delay having n neurons can be determined by two $n \times n$ real matrices $W^0 = (w_{ij}^0)_{n \times n}$, $W^1 = (w_{ij}^1)_{n \times n}$, and an n-dimensional column vector $\theta = (\theta_1, \cdots, \theta_n)^T$, denoted by N=($W^0 \oplus W^1, \theta$). There are two possible values for the state of each neuron i: 1 or -1. Denote the state of neuron i at time $t \in \{0,1,2,\cdots\}$ as $x_i(t)$, the vector $X(t)=(x_1(t),\cdots,x_n(t))^T$ is the state of the whole neurons at time t.

The updating mode of DHNND is determined by the following equations

$$x_i(t+1) = \text{sgn}(\sum_{j=1}^n w_{ij}^0 x_j(t) + \sum_{j=1}^n w_{ij}^1 x_j(t-1) + \theta_i), \ i \in I = \{1,2,\cdots,n\}, \tag{1}$$

where $t \in \{0,1,2,\cdots\}$, and the sign function is defined as follows

$$\text{sgn}(u) = \begin{cases} 1, & \text{if } u \geq 0 \\ -1, & \text{if } u < 0 \end{cases}. \tag{2}$$

We rewrite equation (1) in the compact form

$$X(t+1)=\text{sgn}(W^0 X(t)+W^1 X(t-1)+\theta). \tag{3}$$

If the state X^* satisfies the following condition

$$X^*=\text{sgn}(W^0 X^*+W^1 X^*+\theta), \tag{4}$$

the state X^* is called a stable state (or an equilibrium point).

Let N=($W^0 \oplus W^1, \theta$) start from any initial states $X(0), X(1)$. For $t \geq 2$, if there exists time $t_1 \in \{0,1,2,\cdots\}$ such that the updating sequence $X(0), X(1)$, $X(2), X(3), \cdots$ satisfies that $X(t+T)=X(t)$ for all $t \geq t_1$, where T is the minimum value which satisfies the above condition, then we call that the initial states $X(0), X(1)$ converges towards a limit cycle with length T.

A matrix $A=(a_{ij})_{i,j \in I}$ is called to be row-diagonally dominant or column-diagonally dominant, if the matrix A respectively satisfies the following conditions

$$a_{ii} \geq \sum_{j \in I(j \neq i)} |a_{ij}|, \text{ or } a_{ii} \geq \sum_{j \in I(j \neq i)} |a_{ji}|, \ i \in I = \{1,2,\cdots,n\}. \tag{5}$$

If $Z^T A Z \geq 0$ for each $Z \in \{-2,0,2\}^n$, then matrix $A=(a_{ij})_{i,j \in I}$ is called to be nonnegative definite on the set $\{-2,0,2\}$, where matrix A is not necessarily symmetric.

A matrix A is called quasi-symmetric if there exists a positive diagonal matrix $D=\text{diag}(d_1, d_2, \ldots, d_n)$ $(d_i>0, i=1,2,\ldots,n)$ such that matrix DA is symmetric.

The main contribution of this paper is expressed by the following results.

3 Main Results

Theorem 1[12]. If matrix W^0 is symmetric and matrix W^1 is row-diagonally dominant, then the network N=$(W^0 \oplus W^1, \theta)$ converges towards a limit cycle with length at most 2 for all initial states $X(0), X(1)$ (theorem 3 in Ref.[12]).

Remark 1. This result, corresponding to the result of DHNN, is an extension of the result of DHNN.

Remark 2. If matrix W^0 is quasi-symmetric and matrix W^1 is row-diagonally dominant, then network N=$(W^0 \oplus W^1, \theta)$ converges towards a limit cycle with length at most 2 for the initial states $X(0), X(1)$.

Example 1. Consider N=$(W^0 \oplus W^1, \theta)$, the expressions of matrices W^0, W^1 and θ are respectively in the following

$$W^0 = \begin{pmatrix} 1 & 0 \\ 0 & 2 \end{pmatrix}, \ W^1 = \begin{pmatrix} 2 & 0 \\ 0 & 3 \end{pmatrix}, \ \theta = \begin{pmatrix} 0 \\ 0 \end{pmatrix}. \tag{6}$$

Obviously, the conditions of theorem 1 are satisfied. So, the network N=$(W^0 \oplus W^1, 0)$ converges towards a limit cycle with length at most 2. In fact, we can test that DHNND converges towards a stable state $X(0)$ for all initial states $X(0)=X(1)$, and DHNND converges towards limit cycle $(X(0), X(1))$ with length 2 for all initial states $X(0) \neq X(1)$.

Although matrix W^0 and W^1 are all positive diagonal matrix, i.e. matrix W^0 and W^1 satisfy very strong conditions, it is not guaranteed that network N=$(W^0 \oplus W^1, 0)$ convergences towards a stable state for the initial states $X(0)$, $X(1)$.

Generally, it is difficult for us to give one result on DHNND converging towards a stable state for the initial states $X(0), X(1)$. This point of DHNND does not correspond to that of DHNN.

Well then, what conditions can guarantee that DHNND converges towards limit cycle with length 2 for all initial states $X(0), X(1)$? In the following, we will give this question an answer. Some conditions for DHNND converging towards limit cycle with length 2 for all initial states $X(0), X(1)$ are obtained.

Theorem 2. Let matrix W^0 be symmetric, matrix W^1 be row-diagonally dominant. If there exists a neuron $i \in I$ such that the following condition is satisfied

$$w_{ii}^0 + w_{ii}^1 < - \sum_{j \in I (j \neq i)} |w_{ij}^0 + w_{ij}^1| - |\theta_i|, \tag{7}$$

then DHNND (1) converges towards a limit cycle with length 2 for any initial states $X(0), X(1)$ in parallel mode.

Proof. Based on theorem 1, we know that N=$(W^0 \oplus W^1, \theta)$ converges towards a limit cycle of length at most 2 for the initial states $X(0), X(1)$. In the following, we will prove that DHNND has no stable state. Reduction to absurdity, if there exists a stable state $X^* = (x_1^*, \cdots, x_n^*)^T$, then, by the equation (4), we have

$$x_i^* = \operatorname{sgn}(\sum_{j \in I} (w_{ij}^0 + w_{ij}^1) x_j^* + \theta_i) , \ i \in I. \tag{8}$$

By the equation (8), we can easily prove that

$$x_i^* (\sum_{j \in I} (w_{ij}^0 + w_{ij}^1) x_j^* + \theta_i) \geq 0 , \quad i \in I.$$

(9)

This then implies

$$w_{ii}^0 + w_{ii}^1 + \sum_{j \in I (j \neq i)} |w_{ij}^0 + w_{ij}^1| + |\theta_i| \geq w_{ii}^0 + w_{ii}^1 + \sum_{j \in I (j \neq i)} (w_{ij}^0 + w_{ij}^1) x_i^* x_j^* + \theta_i x_i^*) =$$

$$x_i^* (\sum_{j \in I} (w_{ij}^0 + w_{ij}^1) x_j^* + \theta_i) \geq 0$$

(10)

This conflicts (7), and means that N=($W^0 \oplus W^1, \theta$) has no stable state. By theorem 1, DHNND (1) converges towards a limit cycle with length 2 for any initial states $X(0), X(1)$ in parallel mode. The proof is completed.

Theorem 3. Let matrix W^0 be symmetric, matrix W^1 be row-diagonally dominant. If for any state X, the following conditions are satisfied

$$X^T (W^0 + W^1) X < -\sum_{i \in I} |\theta_i|,$$

(11)

then N=($W^0 \oplus W^1, \theta$) converges towards a limit cycle with length 2 for any initial states $X(0), X(1)$.

Proof. By condition (11), for any state X, we have

$$0 > X^T (W^0 + W^1) X + \sum_{i \in I} |\theta_i| \geq X^T ((W^0 + W^1) X + \theta)$$

(12)

If there exists a stable state $X^* = (x_1^*, \cdots, x_n^*)^T$, then condition (9) is satisfied. Condition (9) can be rewritten as

$$X^{*T} ((W^0 + W^1) X^* + \theta) \geq 0.$$

(13)

This conflicts (12), and means that N=($W^0 \oplus W^1, \theta$) has no stable state. By theorem 1, DHNND (1) converges towards a limit cycle with length 2 for any initial states $X(0), X(1)$ in parallel mode. The proof is completed.

Corollary. Let matrix W^0 be symmetric, matrix W^1 be row-diagonally dominant. If matrix $W^0 + W^1$ is negative definite on the set $\{-1, 1\}$, and $\theta = 0$, then N=($W^0 \oplus W^1, 0$) converges towards a limit cycle with length 2 for any initial states $X(0), X(1)$.

Proof. If matrix $W^0 + W^1$ is negative definite on the set $\{-1, 1\}$, and $\theta = 0$, then, for any X, we have

$$X^T (W^0 + W^1) X < 0.$$

(14)

This implies condition (11) is satisfied. So, corollary is true.

Remark 3. In theorem 2, theorem 3 and corollary, if the matrix W^0 is quasi-symmetric, and the rest conditions do not change, then the corresponding results holds either. This is because the theorem 1 holds when the matrix W^0 is quasi-symmetric.

Example 2. Consider $N=(W^0 \oplus W^1, \theta)$, where

$$W^0 = \begin{pmatrix} -4 & 1 \\ 2 & 2 \end{pmatrix}, \quad W^1 = \begin{pmatrix} 1 & 1 \\ -2 & 2 \end{pmatrix}, \quad \theta = \begin{pmatrix} 0 \\ 0 \end{pmatrix}. \tag{15}$$

Obviously, matrix W^0 is quasi-symmetric, matrix W^1 is row-diagonally dominant. Also, there exists a neuron $1 \in I$ such that the condition (7) is satisfied. By remark 3, $N=(W^0 \oplus W^1, \theta)$ converges towards a limit cycle with length 2. In fact, we can test that DHNND converges towards a limit cycle with length 2.

Remark 4. In corollary, if we replace matrix $W^0 + W^1$ being negative definite on the set $\{-1,1\}$ by matrix $W^0 + W^1$ being non-positive definite on the set $\{-1,1\}$, and the rest conditions are satisfied, then the corresponding results does not hold.

Example 3. Consider $N=(W^0 \oplus W^1, \theta)$, where

$$W^0 = \begin{pmatrix} -2 & 1 \\ 1 & -2 \end{pmatrix}, \quad W^1 = \begin{pmatrix} 1 & 0 \\ 0 & 1 \end{pmatrix}, \quad \theta = \begin{pmatrix} 0 \\ 0 \end{pmatrix}. \tag{16}$$

Obviously, matrix W^0 is symmetric, matrix W^1 is row-diagonally dominant. Furthermore, matrix $W^0 + W^1$ is non-positive definite on the set $\{-1,1\}$. However, we can test that DHNND has a stable state $X=(1,1)^T$. This then implies that remark 4 is true.

If matrix W^0 is asymmetric, we have the result as follows.

Theorem 4. If the following conditions are satisfied for each $i \in I$

$$\max\{w_{ii}^0 + w_{ii}^1, w_{ii}^0 - w_{ii}^1\} < -\sum_{j \neq i}\left(\left|w_{ij}^0\right| + \left|w_{ij}^1\right|\right) - \left|\theta_i\right|, \tag{17}$$

then $N=(W^0 \oplus W^1, \theta)$ converges towards a limit cycle with length 2 for any initial states $X(0), X(1)$.

Proof. For any initial states $X(0), X(1)$, if $x_i(0)=x_i(1)$, then, based on (1) and (17), we can calculate the state updating sequence of neuron i updating process, which can be interpreted as follows

$$x_i(0), \ x_i(0), \ -x_i(0), \ x_i(0), \ -x_i(0), \ x_i(0), \ -x_i(0), \ \ldots.$$

If $x_i(0) \neq x_i(1)$, then, based on (1) and (17) too, we can calculate the state updating sequence of neuron i updating process, which can be interpreted as follows

$$x_i(0), \ -x_i(0), \ x_i(0), \ -x_i(0), \ x_i(0), \ -x_i(0), \ x_i(0), \ \ldots.$$

As proved above, we know that $N=(W^0 \oplus W^1, \theta)$ converges towards a limit cycle with length 2 for any initial states $X(0), X(1)$.

4 Conclusion

This paper mainly studies the stability of DHNND and obtains some results on that of DHNND. The conditions for DHNND converging towards a stable state in serial mode are given. These results generalize the existing results. If matrix $W^1=0$, then DHNND is the same as DHNN. So, the results also generalize some results on stability of DHNN.

Acknowledgments

The work described in this paper was supported by China Postdoctoral Science Foundation (No.2003033516), and partly supported by the Open Foundation of University Key Lab of Information Sciences and Engineering, Dalian University.

References

1. Hopfield, J.J.: Neural Networks and Physical Systems Emergent Collective Computational Abilities. Proc.Nat.Acad.Sci.USA, **79** (1982) 2554-2558
2. Bruck, J.: On the Convergence Properties of the Hopfield Model. Proceedings IEEE, **78** (1990) 1579-1585
3. Xu, Z., Kwong, C.P.: Global Convergence and Asymptotic Stability of Asymmetrical Hopfield Neural Networks. J. Mathematical Analysis and Applications, **191** (1995) 405-426
4. Lee, D.: New Stability Conditions for Hopfield Neural Networks in Partial Simultaneous Update Mode. IEEE,Trans.Neural Networks, **10** (1999) 975-978
5. Xu, J., Bao, Z.: Stability Theory of Antisymmetric Discrete Hopfield Networks. Acta Electronica Sinica, **27** (1999) 103-107
6. Goles, E.: Antisymmetrical Neural Networks. Discrete Applied Mathematics, **13** (1986) 97-100.
7. Ma, R., Zhang, Q., Xu, J.: Convergence of Discrete-time Cellular Neural Networks. Chinese J. Electronics, **11** (2002) 352-356
8. Qiu, S., Xu, X., Liu, M., et al.: Convergence of Discrete Hopfield-type neural Network with Delay in Serial Mode. J. Computer Research and Development, **36** (1999) 546-552
9. Ma, R., Xi, Y., Gao, H.: Stability of Discrete Hopfield Neural Networks with Delay in Serial Mode. Lecture Notes in Computer Science, **3173** (2004) 126-131
10. Qiu, S., Tang, E.C.C., Yeung, D.S.: Stability of Discrete Hopfield Neural Networks with Time-delay. Proceedings of IEEE International Conference on system, Man Cybernetics, (2000) 2546-2552
11. Qiu, S., Xu, X., Li, C., et al: Matrix Criterion for Dynamic Analysis in Discrete Neural Network with Delay. J. Software, **10** (1999) 1108-1113
12. Ma, R., Zhang, Q., Xu, J.: Dynamic Analysis of Discrete Hopfield Neural Networks with Delay. J. Computer Research and Development, **40** (2003) 550-555
13. Ma, R., Liu, N., and Xu, J.: Stability Conditions for Discrete Hopfield Neural Networks with Delay. Acta Electronica Sinica, **32** (2004) 1674-1677

Existence and Stability of Periodic Solution in a Class of Impulsive Neural Networks

Xiaofan Yang[1], David J. Evans[2], and Yuanyan Tang[1,3]

[1] Department of Computer Science and Engineering, Chongqing University,
Chongqing 400044, China
xf_yang1964@yahoo.com
[2] Parallelism, Algorithms and Architectures Research Centre,
Department of Computer Science, Loughborough University,
Loughborough, Leicestershire, LE11 3TU, UK
[3] Department of Computer Science, Hong Kong Baptist University,
Kowloon Tong, Hong Kong

Abstract. In this paper, we initiate the study of a class of neural networks with impulses. A sufficient condition for the existence and global exponential stability of a unique periodic solution of the networks is established. Our condition does not assume the differentiability or monotonicity of the activation functions.

1 Introduction

Most neural networks can be classified as either continuous or discrete. However, there are many real-world systems and natural processes that behave in a piecewise continuous style interlaced with instantaneous and abrupt changes (impulses) [2, 10]. Motivated by this fact, several types of neural networks with impulses have recently been proposed and studied [1, 4, 6-7, 11-12]. In particular, Li and Hu [12] gave a criterion for the existence and global exponential stability of a periodic solution in a class of neural networks. In this paper, we initiate the study of periodic dynamics of a class of impulsive neural networks of the form

$$
\begin{cases}
\dfrac{du_i(t)}{dt} = -a_i u_i(t) + g_i\left(\sum_{j=1}^{n} b_{ij} u_j(t) + c_i(t) \right), i = 1, \cdots, n \\
\qquad \text{if } 0 \le t \ne t_k + r\omega \text{ for all } k \in \{1, \cdots, p\} \text{ and } r \in \{0,1,\cdots\}, \\
u_i(t+0) = \beta_{ik} u_i(t), i = 1, \cdots, n \\
\qquad \text{if } t = t_k + r\omega \text{ for some } k \in \{1, \cdots, p\} \text{ and } r \in \{0,1,\cdots\},
\end{cases}
\tag{1}
$$

where n is the number of neurons in the network, $u_i(t)$ is the state of the ith neuron at time t, $g_i(\cdot)$ is the activation function of the ith neuron, $a_i > 0$ is the rate at which the ith neuron reset the state when it is isolated from the system, b_{ij} is the connection strength from the jth neuron to the ith neuron, $c_i(\cdot)$ is an ω-periodic and Lipschitz continuous function that represents external input to the ith neuron, $0 = t_1 < t_2 < \ldots < t_n < \omega$, β_{ik} is a constant.

J. Wang, X. Liao, and Z. Yi (Eds.): ISNN 2005, LNCS 3496, pp. 265–270, 2005.

Letting $\beta_{ik} = 1$ for all $i \in \{1, \dots, n\}$ and $k \in \{1, \dots, p\}$, system (1) will reduce to the continuous neural networks

$$\frac{du_i(t)}{dt} = -a_i u_i(t) + g_i\left(\sum_{j=1}^{n} b_{ij} u_j(t) + c_i(t)\right), t \geq 0, i = 1, \dots, n. \tag{2}$$

Recently, Chen and Wang [3] investigated the existence and stability of a unique periodic solution of system (2). If the function $c_i(\cdot)$ is a constant c_i, then system (2) reduces to

$$\frac{du_i(t)}{dt} = -a_i u_i(t) + g_i\left(\sum_{j=1}^{n} b_{ij} u_j(t) + c_i\right), t \geq 0, i = 1, \dots, n. \tag{3}$$

The dynamics of system (3) has been studied extensively [5, 8-9, 13].

In this paper, we present a sufficient condition for the existence and global exponential stability of a periodic solution in system (1). The materials are organized as follows: In Section 2, the necessary knowledge is provided. The main result is established in Section 3, while an illustrative example is given in Section 4 to show the validity of our criterion. Some conclusions are drawn in Section 5.

2 Preliminaries

In this paper, we assume that for $i = 1, \dots, n$, $g_i(\cdot)$ is globally Lipschitz continuous with the Lipschitz constant L_i. That is, $|g_i(x) - g_i(y)| \leq L_i|x - y|$ for all $x, y \in R$.

Definition 1. A function $u : [0, \infty) \to R^n$ is said to be a solution of system (1) if the following two conditions are satisfied:

(a) u is piecewise continuous with first kind discontinuity at the points $t_k + r\Pi$, $k \in \{1, \dots, p\}$, $r \in \{0, 1, \dots\}$. Moreover, u is left continuous at each discontinuity point.

(b) u satisfies Eq. (1)

Definition 2. System (1) is said to be *globally exponentially periodic* if (a) it possesses a periodic solution w, and (b) w is globally exponentially stable. That is, there exist positive constants α and β such that every solution u of (1) satisfies

$$\|u(t) - w(t)\|_\infty \leq \alpha \|u(0) - w(0)\|_\infty e^{-\beta t} \text{ for all } t \geq 0.$$

In the next section, we will use the upper right Dini derivative of a continuous function $f \colon R \to R$. The following lemma follows directly from the definition of upper right Dini derivative.

Lemma 1. Let $f(\cdot)$ be a continuous function on R that is differentiable at t. Then

$$D^+|f(t)| = \begin{cases} \dot{f}(t) & \text{when } f(t) > 0, \\ -\dot{f}(t) & \text{when } f(t) < 0, \\ |\dot{f}(t)| & \text{when } f(t) = 0. \end{cases}$$

3 Main Results

The main result in this paper is stated below.

Theorem 2. System (1) is globally exponentially periodic if the following two conditions are satisfied:

(a) There exist positive numbers $\lambda_1, \ldots , \lambda_n$ such that $\lambda_i a_i > \sum_{j=1}^{n} \lambda_j L_j \left| b_{ji} \right|$, $i = 1$, 2, ..., n.

(b) $\left| \beta_{ik} \right| \leq 1$ for all $i \in \{1, \ldots , n\}$ and $k \in \{1, \ldots , p\}$.

As a direct result of Theorem 2, we have

Corollary 3. System (1) is globally exponentially periodic if the following two conditions are satisfied:

(a) $a_i > \sum_{j=1}^{n} L_j \left| b_{ji} \right|$, $i = 1, \ldots , n$.

(b) $\left| \beta_{ik} \right| \leq 1$ for all $i \in \{1, \ldots , n\}$ and $k \in \{1, \ldots , p\}$.

In the case $\beta_{ik} = 1$ for all $i \in \{1, \ldots , n\}$ and $k \in \{1, \ldots , p\}$, Theorem 2 will reduce to

Corollary 4. System (2) is globally exponentially periodic if there exist positive numbers $\lambda_1, \ldots , \lambda_n$ such that $\lambda_i a_i > \sum_{j=1}^{n} \lambda_j L_j \left| b_{ji} \right|$, $i = 1, \ldots , n$.

In order to prove Theorem 2, we need the following lemma.

Lemma 5. Let u and v be a pair of solutions of system (1). If the two conditions in Theorem 3.1 are satisfied, then there is a positive number ε such that

$$\left\| u(t) - v(t) \right\|_{\infty} \leq M \left\| u(0) - v(0) \right\|_{\infty} e^{-\varepsilon t} \text{ for all } t \geq 0 ,$$

where $M = \left(\sum_{i=1}^{n} \lambda_i \right) / \min_{1 \leq i \leq n} \{\lambda_i\}$.

Proof. From the first condition in Theorem 2, there is a small positive number ε such that

$$\lambda_i (\varepsilon - a_i) + \sum_{j=1}^{n} \lambda_j L_j \left| b_{ji} \right| < 0 , i = 1, \ldots , n. \tag{4}$$

Let $V(t) = e^{\varepsilon t} \sum_{i=1}^{n} \lambda_i \left| u_i(t) - v_i(t) \right|$. We now prove $V(t)$ is monotonically increasing by distinguishing two possibilities.

Case 1. $t \neq t_k + r\omega$ for all $k \in \{1, \ldots , p\}$ and $r \in \{0, 1, \ldots\}$. Then

$$D^+ V(t) = e^{\varepsilon t} \sum_{i=1}^{n} \lambda_i \left\{ \varepsilon \left| u_i(t) - v_i(t) \right| + D^+ \left| u_i(t) - v_i(t) \right| \right\} .$$

Observe that for $i = 1, \ldots , n$,

$$\dot{u}_i(t) - \dot{v}_i(t) = -a_i \left(u_i(t) - v_i(t) \right) + g_i \left(\sum_{j=1}^{n} b_{ij} u_j(t) + c_i(t) \right) - g_i \left(\sum_{j=1}^{n} b_{ij} v_j(t) + c_i(t) \right)$$

which plus Lemma 1 yields

$$D^+ \left| u_i(t) - v_i(t) \right| \le -a_i \left| u_i(t) - v_i(t) \right| + \left| g_i \left(\sum_{j=1}^{n} b_{ij} u_j(t) + c_i(t) \right) - g_i \left(\sum_{j=1}^{n} b_{ij} v_j(t) + c_i(t) \right) \right|$$

$$\le -a_i \left| u_i(t) - v_i(t) \right| + L_i \sum_{j=1}^{n} \left| b_{ij} \right| \left| u_j(t) - v_j(t) \right|$$

So
$$D^+ V(t) = e^{\varepsilon t} \sum_{i=1}^{n} \lambda_i (\varepsilon - a_i) \left| u_i(t) - v_i(t) \right| + e^{\varepsilon t} \sum_{i=1}^{n} \sum_{j=1}^{n} \lambda_i L_i \left| b_{ij} \right| \left| u_j(t) - v_j(t) \right|$$

$$= e^{\varepsilon t} \sum_{i=1}^{n} \lambda_i (\varepsilon - a_i) \left| u_i(t) - v_i(t) \right| + e^{\varepsilon t} \sum_{i=1}^{n} \sum_{j=1}^{n} \lambda_j L_j \left| b_{ji} \right| \left| u_i(t) - v_i(t) \right|$$

$$= e^{\varepsilon t} \sum_{i=1}^{n} \left\{ \lambda_i (\varepsilon - a_i) + \sum_{j=1}^{n} \lambda_j L_j \left| b_{ji} \right| \right\} \left| u_i(t) - v_i(t) \right| \le 0 .$$

Case 2. $t = t_k + r\omega$ for some $k \in \{1, \ldots, p\}$ and $r \in \{0, 1, \ldots\}$. Then

$$V(t+0) - V(t) = e^{\varepsilon t} \sum_{i=1}^{n} \lambda_i \left(\left| u_i(t+0) - v_i(t+0) \right| - \left| u_i(t) - v_i(t) \right| \right)$$

$$= e^{\varepsilon t} \sum_{i=1}^{n} \lambda_i \left(\left| \beta_{ik} \right| - 1 \right) \left| u_i(t) - v_i(t) \right| \le 0 .$$

Namely, $V(t + 0) \le V(t)$. Combining the above discussions, we obtain $V(t) \le V(0)$ for all $t \ge 0$. This plus the inspections that

$$V(t) \ge e^{\varepsilon t} \min_{1 \le i \le n} \{\lambda_i\} \sum_{i=1}^{n} \left| u_i(t) - v_i(t) \right| \ge e^{\varepsilon t} \min_{1 \le i \le n} \{\lambda_i\} \left\| u(t) - v(t) \right\|_{\infty}$$

and $V(0) = \sum_{i=1}^{n} \lambda_i \left| u_i(0) - v_i(0) \right| \le \sum_{i=1}^{n} \lambda_i \left\| u(0) - v(0) \right\|_{\infty}$ induces the desired inequality. □

Proof of Theorem 2. First, we prove that system (1) possesses a ω-periodic solution. For each solution u of (1) and each $t \ge 0$, we can define a point $u^t \in R^n$ in this way: $u_t = u(t)$. On this basis, we can define a mapping $P : R^n \to R^n$ by $Pu(0) = u(\omega) = u_\omega$.

Let u and v be an arbitrary pair of solutions of (1). Let ε be a positive number satisfying (4). Let $m \ge \ln(2M)/(\varepsilon\omega)$ be a positive integer. It follows from Lemma 5 that

$$\left\| P^m u(0) - P^m v(0) \right\|_{\infty} = \left\| u(m\omega) - u(m\omega) \right\|_{\infty} \le M \left\| u(0) - v(0) \right\|_{\infty} e^{-\varepsilon m\omega} \le (1/2) \left\| u(0) - v(0) \right\|_{\infty}$$

which shows that P^m is a contraction mapping on the Banach space R^n with the ∞-norm. According to the contraction mapping principle, P^m possesses a unique fixed point $w^* \in R^n$. Note that $P^m(Pw^*) = P(P^m w^*) = Pw^*$, which indicates that Pw^* is also a fixed point of P^m. It follows from the uniqueness of fixed point of P^m that $Pw^* = w^*$. Let w denote the solution of (1.1) with the initial condition w^*. Then $w(0) = w(\omega) = w^*$. So $w(t + \omega) = w(t)$ for all $t \ge 0$. Thus, w is ω-periodic. On the other hand, it follows from Lemma 5 that every solution u of (1) satisfies

$$\left\| u(t) - w(t) \right\|_{\infty} \le M \left\| u(0) - w(0) \right\|_{\infty} e^{-\varepsilon t} .$$

This shows that u exponentially tends to w. □

4 An Illustrative Example

Consider the impulsive neural network consisting of two neurons

$$\dot{u}_1(t) = -u_1(t) + \tanh\left(0.5u_1(t) + 0.2u_2(t) + \sin(t)\right),$$

$$\dot{u}_2(t) = -u_2(t) + \tanh\left(0.3u_1(t) + 0.4u_2(t) + \cos(t)\right),$$

where $t \geq 0$, $t \notin \{0.3 + 2\pi r, 0.6 + 2\pi r : r = 0, 1, \cdots\}$;

$u_1(0.3 + 2\pi r + 0) = -0.1u_1(0.3 + 2\pi r)$, $u_2(0.3 + 2\pi r + 0) = 0.7u_2(0.3 + 2\pi r)$,

$u_1(0.6 + 2\pi r + 0) = 0.4u_1(0.6 + 2\pi r)$, $u_2(0.6 + 2\pi r + 0) = -0.5u_2(0.6 + 2\pi r)$,

where $r = 0, 1, \ldots$.

Clearly, all the conditions in Corollary 3 are satisfied. Hence, this system has a unique 2π-periodic solution, which is globally exponentially stable.

5 Conclusions

We have introduced a class of neural networks with periodic impulses, and established a sufficient condition for the existence and global exponential stability of a periodic solution in the proposed neural networks.

In recent years, numerous results have been reported on the stability of discrete as well as continuous neural networks. It is worthwhile to introduce various impulsive neural networks and establish the corresponding stability results that include some known results for pure discrete/continuous neural networks as special cases.

Acknowledgements

This work is supported jointly by Chinese National Natural Science Funds (60271019) and Chongqing's Application-Oriented Fundamentals Research Funds (8028).

References

1. Akça, H., Alassar, R., Covachev, V., Covacheva, Z., Al-Zahrani, E.: Continuous-time Additive Hopfield-type Neural Networks with Impulses. Journal of Mathematical Analysis and Applications, **290** (2004) 436-451
2. Bainov, D.D., Simeonov, P.S.: Stability Theory of Differential Equations with Impulse Effects: Theory and Applications. Chichester: Ellis Horwood, (1989)
3. Chen, B.S., Wang, J.: Global Exponential Periodicity and Global Exponential Stability of a Class of Recurrent Neural Networks. Physics Letters A, **329** (2004) 36-48
4. Gopalsamy, K.: Stability of Artificial Neural Networks with Impulses. Applied Mathematics and Computation, **154** (2004) 783-813
5. Grossberg, S.: Nonlinear Neural Networks: Principles, Mechanisms and Architectures. Neural Networks, **1** (1988) 17-61

6. Guan, Z.H., Chen, G.R.: On Delayed Impulsive Hopfield Neural Networks. Neural Networks, **12** (1999) 273-280
7. Gua,n Z.H., Lam, J., Chen, G.R.: On Impulsive Autoassociative Neural Networks. Neural Networks, **13** (2000) 63-69
8. Hirsch, M.W.: Convergent Activation Dynamics in Continuous Time Networks. Neural Networks, **2** (1989) 331-349
9. Hu, S.Q., Wang, J.: Global Stability of a Class of Continuous-Time Recurrent Neural Networks. IEEE Transactions on Circuits and Systems-I, **49** (2002) 1334-1347
10. Lakshmikantham, V., Bainov D.D., Simeonov, P.S.: Theory of Impulsive Differential Equations. Singapore, World Scientific, (1989)
11. Li, YK: Global Exponential Stability of BAM Neural Networks with Delays and Impulses, Chaos, Solitons and Fractals, **24** (2005) 279-285
12. Li, YK, Hu, LH: Global Exponential Stability and Existence of Periodic Solution of Hopfield-type Neural Networks with Impulses. Physics Letters A, **333** (2004) 62-71
13. Liang, XB, Wang, J: A Recurrent Neural Network for Nonlinear Optimization with a Continuously Differentiable Objective Function and Bound Constraints. IEEE Transactions on Neural Networks, **11** (2000) 1251-1262

Globally Attractive Periodic Solutions of Continuous-Time Neural Networks and Their Discrete-Time Counterparts

Changyin Sun[1,2], Liangzhen Xia[2], and Chunbo Feng[2]

[1]College of Electrical Engineering, Hohai University, Nanjing 210098, China
[2]Research Institute of Automation, Southeast University, Nanjing 210096, China
cysun@seu.edu.cn

Abstract. In this paper, discrete-time analogues of continuous-time neural networks with continuously distributed delays and periodic inputs are investigated without assuming Lipschitz conditions on the activation functions. The discrete-time analogues are considered to be numerical discretizations of the continuous-time networks and we study their dynamical characteristics. By employing Halanay-type inequality, we obtain easily verifiable sufficient conditions ensuring that every solutions of the discrete-time analogue converge exponentially to the unique periodic solutions. It is shown that the discrete-time analogues inherit the periodicity of the continuous-time networks. The results obtained can be regarded as a generalization to the discontinuous case of previous results established for delayed neural networks possessing smooth neuron activation.

1 Introduction

There have been active investigations recently into the dynamics and applications of neural networks with delays because the dynamical properties of equilibrium points of neural systems play an important role in some practical problems [1-11], such as optimization solvers, associative memories, image compression, speed detection of moving objects, processing of moving images, and pattern classification. It is well known that an equilibrium point can be viewed as a special periodic solution of continuous-time neural networks with arbitrary period [6-8]. In this sense the analysis of periodic solutions of neural systems may be considered to be more general sense than that of equilibrium points. Among the most previous results of the dynamical analysis of continuous-time neural systems, a frequent assumption is that the activation functions are Lipschitz continuous or differentiable and bounded, such as the usual sigmoid-type functions in conventional neural networks [2]. But, in some evolutionary processes as well as optimal control models and flying object motions, there are many bounded monotone-nondecreasing signal functions which do not satisfy the Lipschitz condition. For instance, in the simplest case of the pulse-coded signal function which has received much attention in many fields of applied sciences and engineering, an exponentially weighted time average of sampled pulse is often used which does not satisfy the Lipschitz condition [11]. Since the Lipschitz continuous assumption is not always practical, it is necessary and important to investigate the dynamical properties

J. Wang, X. Liao, and Z. Yi (Eds.): ISNN 2005, LNCS 3496, pp. 271–275, 2005.

of the continuous-time neural systems in both theory and applications without assuming Lipschitz continuous of the activation functions.

On the other hand, in implementing the continuous-time neural system for simulation or computational purposes, it is essential to formulate a discrete-time system which is an analogue of the continuous-time system. A method which is commonly used in the formulation of a discrete-time analogue is by discretizing the continuous-time system. Certainly, the discrete-time analogue when derived as a numerical approximation of continuous-time system is desired to preserve the dynamical characteristics of the continuous-time system. Once this is established, the discrete-time analogue can be used without loss of functional similarity to the continuous-time system and preserving any physical or biological reality that the continuous-time system has. Though there exist a lot of numerical schemes (such as Euler scheme, Runge-Kutta scheme) that can be used to obtain discrete-time analogues of continuous-time neural networks. However, it is generally known [6,8,10] that these numerical schemes can exhibit spurious equilibrium and spurious asymptotic behavior, which are not present in the continuous-time counterparts. The existence of spurious equilibrium and asymptotic behavior in a discrete-time analogue can happen if one holds the positive discretization step size fixed and let the original parameters (i.e. parameters of the continuous-time systems) vary within the asymptotic parameter space of the continuous-time systems. As a consequence, one has to impose limitations either on the size of h or on the original parameters so as to avoid the existence of spurious equilibrium and spurious asymptotic behavior in the discrete-time analogues. With such limitations, the computational capability of the continuous-time network will not be achieved fully by the discrete-time analogues. In order to overcome this difficulty, Mohamad and Gopalsamy [10] have formulated the discrete-time analogues of continuous-time neural systems and addressed their stability properties. Sun and Feng [6-8] has extended the discrete-time analogues with constant inputs to the discrete-time analogues with periodic inputs, and further investigated their periodicity properties.

The results on dynamical analysis of continuous-time neural systems are mostly assumed that the time delays are discrete. However, neural networks usually have spatial extent due to the presence of a multitude of parallel pathways with a variety of axon sizes and lengths. Thus, there will be a distribution of propagation delays. This means that the signal propagation is no longer instantaneous and is better represented by a model with continuously distributed delays. In this paper, our focus is on periodicity of continuous-time neural networks with continuously distributed delays. By using the Halanay-type inequality, we will derive some easily checkable conditions to guarantee exponential periodicity of continuous-time neural networks with continuously distributed delays without assuming Lipschitz conditions on the activation functions. Discrete-time analogues of integrodifferential equations modeling neural networks with periodic inputs are introduced, and we will also study the exponential periodicity of the discrete-time analogues.

2   Continuous-Time Neural Network Model and Preliminaries

Consider the model of continuous-time neural networks described by the following integrodifferential equations of the form

$$\dot{x}_i(t)=-d_ix_i(t)+\sum_{j=1}^{n}a_{ij}g_j(x_j(t))+\sum_{j=1}^{n}b_{ij}g_j\left(\int_{0}^{\infty}K_{ij}(s)x_j(t-s)ds\right)+I_i(t),\ t\geq 0 \qquad (1)$$

for $i\in\{1,2,\cdots,n\}$, $x_i(t)$ is the state vector of the i th unit at time t, a_{ij}, b_{ij} are constants. For simplicity, let D be an $n\times n$ constant diagonal matrix with diagonal elements $d_i>0$, $i=1,2,...,n$, $A=(a_{ij})$ and $B=(b_{ij})$ are $n\times n$ constant interconnection matrices, $g_j(x_j(t))$ denotes the nonlinear activation function of the j th unit at time t, $I(t)=(I_1(t),I_2(t),..,I_n(t))^T\in R^n$ is an input periodic vector function with period ω, i.e., there exists a constant $\omega>0$ such that $I_i(t+\omega)=I_i(t)$ $(i=1,2,\cdots,n)$ for all $t\geq 0$, and $x=(x_1,x_2,...,x_n)^T\in R^n$. In this paper, we suppose that (H) for each $j\in\{1,2,\cdots,n\}$, g_j satisfy $(u-v)(g_j(u)-g_j(v))>0$ $(u-v\neq 0)$ and there exist $L_j\geq 0$ such that

$$\sup_{u-v\neq 0}\frac{g_j(u)-g_j(v)}{u-v}=L_j\ \text{for all}\ u,v\in R.$$

Note that the assumption (H) is weaker than the Lipschitz condition which is mostly used in previous literature. Some researchers have studied the pure-delay model (with $a_{ij}=0$). However, neural network model with periodic inputs in which instantaneous signaling as well as delayed signaling both occur (with $a_{ij}\neq 0$, $b_{ij}\neq 0$) has not been investigated yet. $K_{ij}(\cdot)$, $i,j=1,2,\cdots,n$ are the delay kernels which are assumed to satisfy the following conditions (i-iv) simultaneously:

(i) $K_{ij}:[0,\infty)\rightarrow[0,\infty)$;

(ii) K_{ij} are bounded and continuous on $[0,\infty]$;

(iii) $\int_{0}^{\infty}K_{ij}(s)ds=1$;

(iv) there exists a positive number μ such that $\int_{0}^{\infty}K_{ij}(s)e^{\mu s}ds<\infty$.

The literature [10] has given some examples to meet the above conditions. The initial conditions associated with the system (1) are given by

$$x_i(s)=\gamma_i(s),\ s\leq 0,i=1,2,\cdots,n,$$

where $\gamma_i(\cdot)$ is bounded and continuous on $(-\infty,0]$.

As a special case of neural system (1), the delayed neural networks with constant input vector $I=(I_1,I_2,...,I_n)^T\in R$ are described by the following functional differential equations

$$\dot{x}_i(t)=-d_ix_i(t)+\sum_{j=1}^{n}a_{ij}g_j(x_j(t))+\sum_{j=1}^{n}b_{ij}g_j\left(\int_{0}^{\infty}K_{ij}(s)x_j(t-s)ds\right)+I_i,\ t\geq 0 \qquad (2)$$

Let $\nu(t_0) = \max_{1 \le i \le n}\{\sup_{-\infty \le s \le t_0}\{y_i(s)\}\}$. Then for $\forall t \ge t_0$,

$$||y(t)|| \le \nu(t_0)\exp\{-\theta(t - t_0)\}. \tag{5}$$

Otherwise, there exist $t_2 > t_1 > t_0$, $q \in \{1, 2, \cdots, n\}$ and sufficiently small $\varepsilon > 0$ such that for $\forall s \in [t_0 - \tau, t_1]$, (5) holds, and

$$|y_i(s)| \le \nu(t_0)\exp\{-\theta(s - t_0)\} + \varepsilon, \ s \in (t_1, t_2], i \in \{1, 2, \cdots, n\}, \tag{6}$$

$$D^+|y_q(t_2)| + \theta\nu(t_0)\exp\{-\theta(t_2 - t_0)\} > 0. \tag{7}$$

But from (3), (4) and (6),

$$D^+|y_q(t_2)| + \theta\nu(t_0)\exp\{-\theta(t_2 - t_0)\} \le 0. \tag{8}$$

Hence, from this conclusion of absurdity, it shows that (5) holds.

Define $x_\phi^{(t)}(\theta) = x(t+\theta; t_0, \phi), \theta \in (-\infty, t_0]$. Define a mapping $H : C((-\infty, t_0], \Re^n) \to C((-\infty, t_0], \Re^n)$ by $H(\phi) = x_\phi^{(\omega)}$, then $H(C((-\infty, t_0], \Re^n)) \subset C((-\infty, t_0], \Re^n)$, and $H^m(\phi) = x_\phi^{(m\omega)}$.

From (5), $||x(t; t_0, \phi) - x(t; t_0, \varphi)||_{t_0} \le \frac{\max_{1 \le i \le n}\{\gamma_i\}}{\min_{1 \le i \le n}\{\gamma_i\}}||\phi, \varphi||_{t_0}\exp\{-\theta(t - t_0)\}$. Choose a positive integer m such that $\frac{\max_{1 \le i \le n}\{\gamma_i\}}{\min_{1 \le i \le n}\{\gamma_i\}}\exp\{-(m\omega - t_0)\} \le \alpha < 1$. Hence,

$$||H^m(\phi), H^m(\varphi)||_{t_0} = ||x(m\omega + \theta; t_0, \phi) - x(m\omega + \theta; t_0, \varphi)||_{t_0}$$
$$\le ||\phi, \varphi||_{t_0}\exp\{-(m\omega - t_0)\} \le \alpha||\phi, \varphi||_{t_0}.$$

Based on Lemma 1, there exists a unique fixed point $\phi^* \in C((-\infty, t_0], \Re^n)$ such that $H^m(\phi^*) = \phi^*$. In addition, for any integer $r \ge 1$, $H^m(H^r(\phi^*)) = H^r(H^m(\phi^*)) = H^r(\phi^*)$. This shows that $H^r(\phi^*)$ is also a fixed point of H^m, hence, by the uniqueness of the fixed point of the mapping H^m, $H^r(\phi^*) = \phi^*$, that is, $x_{\phi^*}^{(r\omega)} = \phi^*$. Let $x(t; t_0, \phi^*)$ be a state of the neural network (1) with initial condition ϕ^*. Then from (1), $\forall i = 1, 2, \cdots, n, \forall t \ge t_0$,

$$\frac{dx_i(t; t_0, \phi^*)}{dt} = -b_i(x_i(t; t_0, \phi^*)) -$$
$$\sum_{k=1}^{p}\sum_{j=1}^{n} c_{ij}^{(k)} f_j\left(\int_{-\infty}^{t} \Gamma_{kj}(t - s)x_j(s; t_0, \phi^*)ds\right) + u_i(t).$$

Hence, $\forall i = 1, 2, \cdots, n, \forall t + \omega \ge t_0$,

$$\frac{dx_i(t + \omega; t_0, \phi^*)}{dt} = -b_i(x_i(t + \omega; t_0, \phi^*)) -$$
$$\sum_{k=1}^{p}\sum_{j=1}^{n} c_{ij}^{(k)} f_j\left(\int_{-\infty}^{t+\omega} \Gamma_{kj}(t + \omega - s)x_j(s; t_0, \phi^*)ds\right) + u_i(t + \omega)$$
$$= -b_i(x_i(t + \omega; t_0, \phi^*)) -$$
$$\sum_{k=1}^{p}\sum_{j=1}^{n} c_{ij}^{(k)} f_j\left(\int_{-\infty}^{t} \Gamma_{kj}(t - s)x_j(s + \omega; t_0, \phi^*)ds\right) + u_i(t),$$

this implies $x(t + w; t_0, \phi^*)$ is also a state of the neural network (1) with initial condition (t_0, ϕ^*). $x_{\phi^*}^{(rw)} = \phi^*$ implies that $x(rw+\theta; t_0, \phi^*) = x((r-1)w+\theta; t_0, \phi^*)$. $\forall t \geq t_0$, there exist $\bar{r}$ and $\bar{\theta}$ such that $t = (\bar{r} - 1)w + \bar{\theta}$; i.e., $x(t + w; t_0, \phi^*) = x(t; t_0, \phi^*)$. Hence, $x(t; t_0, \phi^*)$ is a periodic orbit of the neural network (1) with period w.

From (5), it is easy to see that all other states of the neural network (1) converge to this periodic orbit as $t \to +\infty$. Hence, the periodic orbit $x(t; t_0, \phi^*)$ is globally attractive.

Consider the artificial neural network model

$$\frac{dx_i(t)}{dt} = -b_i(x_i(t)) - \sum_{k=1}^{p} \sum_{j=1}^{n} c_{ij}^{(k)} f_j(x_j(t - \tau_{kj}(t))) + u_i(t), \qquad (9)$$

where $i = 1, \cdots, n$, $x = (x_1, \cdots, x_n)^T \in \Re^n$ is the state vector, delay $\tau_{kj}(t) \leq \tau$ (constant), $c_{ij}^{(k)}$ is the interconnection associated with delay $\tau_{kj}(t)$.

Let $C([t_0 - \tau, t_0], \Re^n)$ be the space of continuous functions mapping $[t_0 - \tau, t_0]$ into $\Re^n$ with norm defined by $\|\phi\|_{t_0} = \max_{1 \leq i \leq n} \{\sup_{u \in (-\infty, t_0]} |\phi_i(u)|\}$, where $\phi(s) = (\phi_1(s), \phi_2(s), \cdots, \phi_n(s))^T$. The initial condition of neural network (9) is assumed to be $\phi(\vartheta) = (\phi_1(\vartheta), \phi_2(\vartheta), \cdots, \phi_n(\vartheta))^T$, where $\phi(\vartheta) \in C((-\infty, t_0], \Re^n)$.

Similar to the proof of Theorem 1, the following result can be obtained.

Theorem 2. If $\bar{B} - \Theta\mu$ is a nonsingular M matrix, then the neural network (9) has a periodic state which is globally attractive.

Consider delayed neural networks

$$\frac{dx_i(t)}{dt} = -x_i(t) + \sum_{j=1}^{n} c_{ij}^{(1)} f_j(x_j(t)) + \sum_{j=1}^{n} c_{ij}^{(2)} f_j(x_j(t - \tau_{2j}(t))) + u_i(t), \quad (10)$$

where $i = 1, \cdots, n$.

Corollary 1. If $E - (\sum_{k=1}^{2} |c_{ij}^{(k)}|)_{n \times n} \mu$ is a nonsingular M matrix, then the neural network (10) has a periodic state which is globally attractive.

Proof. Choose $a_i(x_i(t)) \equiv 1$, $b_i(x_i(t)) \equiv x_i(t)$, $p = 2$, $\tau_{1j}(t) \equiv 0, \tau_{2j}(t) \leq \tau$ (constant), according to Theorem 2, Corollary 1 holds.

3   Illustrative Examples

In this section, we give two examples to illustrate the new results.

Example 1. Consider a Cohen-Grossberg neural network with sigmoid activation function,

$$\begin{cases} \dot{x}_1(t) = -2(1 + \sin^2 x_1(t))x_1(t) + 10h(x_2(t)) - 2\sin t - 5; \\ \dot{x}_2(t) = -2(1 + \cos^2 x_2(t))x_2(t) - \frac{3}{10}h(x_1(t)) + \frac{1}{5}h(x_2(t)) - 2\cos t - 5, \end{cases} \qquad (11)$$

where $h(r) = (-\exp\{-r\} + \exp\{r\})/(\exp\{-r\} + \exp\{r\})$. Obviously,

$$\bar{B} = \begin{pmatrix} 2 & 0 \\ 0 & 2 \end{pmatrix}, \quad \left(c_{ij}^{(1)}\right)_{2 \times 2} = \begin{pmatrix} 0 & -10 \\ 3/10 & -1/5 \end{pmatrix}.$$

Globally Attractive Periodic State
of Discrete-Time Cellular Neural Networks
with Time-Varying Delays

Zhigang Zeng[1,2], Boshan Chen[3], and Zengfu Wang[2]

[1] School of Automation, Wuhan University of Technology,
Wuhan, Hubei 430070, China
zhigangzeng@163.com
[2] Department of Automation, University of Science and Technology of China
Hefei, Anhui 230026, China
[3] Department of Mathematics, Hubei Normal University,
Huangshi, Hubei 435002, China

Abstract. For the convenience of computer simulation, the discrete-time systems in practice are often considered. In this paper, Discrete-time cellular neural networks (DTCNNs) are formulated and studied in a regime where they act as a switchboard for oscillating inputs. Several sufficient conditions are obtained to ensure DTCNNs with delays have a periodic orbit and this periodic orbit is globally attractive using a method based on the inequality method and the contraction mapping principle. Finally, simulations results are also discussed via one illustrative example.

1   Introduction

Cellular neural networks have been found useful in areas of signal processing, image processing, associative memories, pattern classification (see for instance [1]). The existence of periodic orbits of CNNs and DCNNs is an interesting dynamic behavior. It is expected that it can be applied to association memory by storing targets in periodic orbits [2], [3]. In addition, an equilibrium point can be viewed as a special periodic orbit of neural networks with arbitrary period. In this sense the analysis of periodic orbits of neural networks could be more general than that of equilibrium points [4], [5], [6], [7], [8]. Recently, stability analysis and existence of periodic states have been widely researched for continuous-time cellular neural networks with and without delays in [9], [10]. However, we usually need to consider the discrete-time systems in practice such as computer simulation, etc [11], [12]. Motivated by the above discussions, our aim in this paper is to consider the globally attractive periodic state of DTCNNs.

This paper consists of the following sections. Section 2 describes some preliminaries. The main results are stated in Sections 3. Simulation results of one illustrative example are given in Section 4. Finally, concluding remarks are made in Section 5.

J. Wang, X. Liao, and Z. Yi (Eds.): ISNN 2005, LNCS 3496, pp. 282–287, 2005.

2   Preliminaries

In this paper, we always assume that $t \in \mathcal{N}$, where $\mathcal{N}$ is the set of all natural number.

Consider a class of DTCNNs described by the following difference equation: for $i = 1, 2, \cdots, n$,

$$\Delta x_i(t) = -c_i x_i(t) + \sum_{j=1}^{n} a_{ij} f(x_j(t)) + \sum_{j=1}^{n} b_{ij} f(x_j(t - \tau_{ij}(t))) + u_i(t) , \quad (1)$$

where $x = (x_1, \cdots, x_n)^T \in \Re^n$ is the state vector, $\Delta x_i(t) = x_i(t+1) - x_i(t)$, c_i is a positive constant that satisfies $c_i \in (0, 2)$, $A = (a_{ij})$ and $B = (b_{ij})$ are connection weight matrices that are not assumed to be symmetric, $u(t) = (u_1(t), u_2(t), \cdots, u_n(t))^T \in \Re^n$ is an input periodic vector function with period ω; i.e., there exits a constant $\omega > 0$ such that $u_i(t + \omega) = u_i(t)$, $\forall t \geq t_0, \forall i \in \{1, 2, \cdots, n\}$, $\tau_{ij}(t)$ is the time-varying delay that satisfies $0 \leq \tau_{ij}(t) \leq \tau = \max_{1 \leq i,j \leq n}\{\sup\{\tau_{ij}(t), t \in \mathcal{N}\}\}$, $\tau_{ij}(t)$ and τ are nonnegative integers, $f(\cdot)$ is the activation function defined by

$$f(r) = (|r+1| - |r-1|)/2 . \quad (2)$$

The initial value problem for DTCNNs (1) requires the knowledge of initial data $\{\phi(t_0 - \tau), \cdots, \phi(t_0)\}$. This vector is called initial string ϕ. For every initial string ϕ, there exists a unique state $x(t; t_0, \phi)$ of DTCNNs (1) that can be calculated by the explicit recurrence formula

$$x_i(t+1) = (1 - c_i)x_i(t) + \sum_{j=1}^{n} a_{ij} f(x_j(t)) + \sum_{j=1}^{n} b_{ij} f(x_j(t - \tau_{ij}(t))) + u_i(t) . (3)$$

Denote $\mathbf{z}$ as the set of all integers, $[a, b]_{\mathbf{z}} = \{a, a+1, \cdots, b-1, b\}$, where $a, b \in z, a \leq b$. Let $C([t_0 - \tau, t_0]_{\mathbf{z}}, \Re^n)$ be the Banach space of functions mapping $[t_0 - \tau, t_0]_{\mathbf{z}}$ into $\Re^n$ with norm defined by $\|\phi\|_{t_0} = \max_{1 \leq i \leq n}\{\sup_{r \in [t_0 - \tau, t_0]_{\mathbf{z}}} |\phi_i(r)|\}$, where $\phi(s) = (\phi_1(s), \phi_2(s), \cdots, \phi_n(s))^T$. Denote $\|x\| = \max_{1 \leq i \leq n}\{|x_i|\}$ as the vector norm of the vector $x = (x_1, \cdots, x_n)^T$.

Lemma 1 [13]. Let H be a mapping on complete metric space $(C([t_0 - \tau, t_0]_{\mathbf{z}}, \Re^n), \|\cdot\|_{t_0})$. If $H(C([t_0 - \tau, t_0]_{\mathbf{z}}, \Re^n)) \subset C([t_0 - \tau, t_0]_{\mathbf{z}}, \Re^n)$ and there exists a constant $\alpha < 1$ such that $\forall \phi, \varphi \in C([t_0 - \tau, t_0]_{\mathbf{z}}, \Re^n)$, $\|H(\phi), H(\varphi)\|_{t_0} \leq \alpha \|\phi, \varphi\|_{t_0}$, then there exists a unique fixed point $\phi^* \in C([t_0 - \tau, t_0]_{\mathbf{z}}, \Re^n)$ such that $H(\phi^*) = \phi^*$.

3   Globally Attractive Periodic State

For $i, j \in \{1, 2, \cdots, n\}$, let

$$T_{ij} = \begin{cases} c_i - |a_{ii}| - |b_{ii}|, & i = j, \\ -|a_{ij}| - |b_{ij}|, & i \neq j, \end{cases} \quad \tilde{T}_{ij} = \begin{cases} 2 - c_i - |a_{ii}| - |b_{ii}|, & i = j, \\ -|a_{ij}| - |b_{ij}|, & i \neq j . \end{cases}$$

Denote matrices $T_1 = (T_{ij})_{n \times n}, T_2 = (\tilde{T}_{ij})_{n \times n}$.

Theorem 3.1. If for $\forall i \in \{1, 2, \cdots, n\}$, $c_i \in (0, 1)$, T_1 is a nonsingular M-matrix, then DTCNNs (1) have a periodic state which is globally attractive.

Proof. Since T_1 is a nonsingular M-matrix, there exist positive constants $\gamma_1, \gamma_2, \cdots, \gamma_n$ such that for $\forall i \in \{1, 2, \cdots, n\}$,

$$\gamma_i c_i - \sum_{j=1}^{n} \gamma_j |a_{ij}| - \sum_{j=1}^{n} \gamma_j |b_{ij}| > 0 . \tag{4}$$

Let $\eta_i(\lambda) = \gamma_i \lambda^{\tau+1} - (\gamma_i(1 - c_i) + \sum_{j=1}^{n} \gamma_j |a_{ij}|)\lambda^{\tau} - \sum_{j=1}^{n} \gamma_j |b_{ij}|$, then $\eta_i(0) = -\sum_{j=1}^{n} \gamma_j |b_{ij}| \leq 0$, $\eta_i(1) > 0$. Hence, there exists $\lambda_{0i} \in (0, 1)$ such that $\eta_i(\lambda_{0i}) = 0$, and $\eta_i(\lambda) \geq 0$, $\lambda \in [\lambda_{0i}, 1)$.

In fact, we can choose the largest value of $\lambda \in (0, 1)$ satisfying $\eta_i(\lambda_{0i}) = 0$, since $\eta_i(\lambda)$ is a polynomial and it has at most $\tau + 1$ real roots. Choose $\lambda_0 = \max_{1 \leq i \leq n}\{\lambda_{0i}\}$, then for $\forall j \in \{1, 2, \cdots, n\}$,

$$\eta_j(\lambda_0) \geq 0 . \tag{5}$$

Denote $x(t; t_0, \phi)$ and $x(t; t_0, \varphi)$ as two states of DTCNN (1) with initial strings $\{\phi(t_0 - \tau), \cdots, \phi(t_0)\}$ and $\{\varphi(t_0 - \tau), \cdots, \varphi(t_0)\}$, where $\phi, \varphi \in C([t_0 - \tau, t_0]_{\mathbf{z}}, \Re^n)$. Let $z_i(t) = (x_i(t; t_0, \phi) - x_i(t; t_0, \varphi))/\gamma_i$, from (2) and (3), $\forall i = 1, 2, \cdots, n, \forall t \geq t_0$,

$$|z_i(t+1)| \leq (1 - c_i)|z_i(t)| + \sum_{j=1}^{n}(|a_{ij}|\gamma_j|z_j(t)| + |b_{ij}|\gamma_j|z_j(t - \tau_{ij}(t))|)/\gamma_i . \tag{6}$$

Let $\Upsilon = \max_{1 \leq i \leq n}\{\max\{|z_i(t_0)|, |z_i(t_0 - 1)|, \cdots, |z_i(t_0 - \tau)|\}\}$, then for natural number $t \geq t_0$, $|z_i(t)| \leq \Upsilon \lambda_0^{t-t_0}$. Otherwise, there exist $p \in \{1, 2, \cdots, n\}$ and natural number $q \geq t_0$ such that $|z_p(q)| > \Upsilon \lambda_0^{q-t_0}$, and for all $j \neq p, j \in \{1, 2, \cdots, n\}$, $|z_j(s)| \leq \Upsilon \lambda_0^{s-t_0}$, $-\tau \leq s \leq q$; $|z_p(s)| \leq \Upsilon \lambda_0^{s-t_0}$, $-\tau \leq s < q$.

If $|z_p(q)| > \Upsilon \lambda_0^{q-t_0}$, since $1 - c_p \geq 0$, from (6),

$$\Upsilon \lambda_0^{q-t_0} < |z_p(q)| \leq \Upsilon \lambda_0^{q-1-t_0}\{(1 - c_p) + [\sum_{j=1}^{n} \gamma_j |a_{pj}| + \sum_{j=1}^{n} \gamma_j |b_{pj}|\lambda_0^{-\tau}]/\gamma_p\} ;$$

i.e., $\gamma_p \lambda_0^{\tau+1} < [\gamma_p(1 - c_p) + \sum_{j=1}^{n} \gamma_j |a_{pj}|]\lambda_0^{\tau} + \sum_{j=1}^{n} \gamma_j |b_{pj}|$, this contradicts (5). Hence for natural number $t \geq t_0$,

$$|z_i(t)| \leq \Upsilon \lambda_0^{t-t_0} . \tag{7}$$

Define $x_\phi^{(t)}(\theta) = x(t + \theta; t_0, \phi)$, $\theta \in [t_0 - \tau, t_0]_{\mathbf{z}}$. Define a mapping $H : C([t_0 - \tau, t_0]_{\mathbf{z}}, \Re^n) \to C([t_0 - \tau, t_0]_{\mathbf{z}}, \Re^n)$ by $H(\phi) = x_\phi^{(\tau)}$, then

$$H(C([t_0 - \tau, t_0]_{\mathbf{z}}, \Re^n)) \subset C([t_0 - \tau, t_0]_{\mathbf{z}}, \Re^n),$$

and $H^m(\phi) = x_\phi^{(m\tau)}$. We can choose a positive integer m such that

$$\frac{\max_{1 \leq i \leq n}\{\gamma_i\}}{\min_{1 \leq i \leq n}\{\gamma_i\}}\lambda_0^{m\tau-\tau} \leq \alpha < 1.$$

Hence, from (7),

$$\|H^m(\phi), H^m(\varphi)\|_{t_0} \leq \max_{1 \leq i \leq n} \{ \sup_{\theta \in [t_0-\tau, t_0]_z} |x_i(m\tau + \theta; t_0, \phi) - x_i(m\tau + \theta; t_0, \varphi)| \}$$

$$\leq \|\phi, \varphi\|_{t_0} \frac{\max_{1 \leq i \leq n}\{\gamma_i\}}{\min_{1 \leq i \leq n}\{\gamma_i\}} \lambda_0^{m\tau + t_0 - \tau - t_0} \leq \alpha \|\phi, \varphi\|_{t_0} .$$

Based on Lemma 1, there exists a unique fixed point $\phi^* \in C([t_0 - \tau, t_0]_z, \Re^n)$ such that $H^m(\phi^*) = \phi^*$. In addition, for any integer $r \geq 1$, $H^m(H^r(\phi^*)) = H^r(H^m(\phi^*)) = H^r(\phi^*)$. This shows that $H^r(\phi^*)$ is also a fixed point of H^m. Hence, by the uniqueness of the fixed point of the mapping H^m, $H^r(\phi^*) = \phi^*$, that is, $x_{\phi^*}^{(r\tau)} = \phi^*$. Let $x(t; t_0, \phi^*)$ be a state of DTCNN (1) with initial initial strings $\{\phi^*(t_0 - \tau), \cdots, \phi^*(t_0)\}$. $x_{\phi^*}^{(r\tau)} = \phi^*$ implies that $x(r\tau + \theta; t_0, \phi^*) = x((r - 1)\tau + \theta; t_0, \phi^*) = \phi^*(\theta)$, $\forall \theta \in [t_0 - \tau, t_0]$. Hence, $x(t; t_0, \phi^*)$ is a periodic orbit of DTCNN (1).

From (7), it is easy to see that all other states of DTCNN (1) with initial string $\{\phi(t_0 - \tau), \cdots, \phi(t_0)\}$ ($\phi \in C([t_0 - \tau, t_0]_z, \Re^n)$) converge to this periodic orbit as $t \to +\infty$. Hence, the periodic orbit $x(t; t_0, \phi^*)$ is globally attractive.

Theorem 3.2. If for $\forall i \in \{1, 2, \cdots, n\}$, $c_i \in [1, 2)$, T_2 is a nonsingular M-matrix, then DTCNNs (1) have a periodic state which is globally attractive.

Proof. Since T_2 is a nonsingular M-matrix, there exist positive constants $\gamma_1, \gamma_2, \cdots, \gamma_n$ and $\lambda_0 \in (0, 1)$ such that for $\forall i \in \{1, 2, \cdots, n\}$,

$$\gamma_i \lambda_0^{\tau+1} - \gamma_i(c_i - 1)\lambda_0^\tau - \sum_{j=1}^n \gamma_j |a_{ij}| \lambda_0^\tau - \sum_{j=1}^n \gamma_j |b_{ij}| \geq 0 . \tag{8}$$

Denote $x(t; t_0, \phi)$ and $x(t; t_0, \varphi)$ as two states of DTCNN (1) with initial strings ϕ and φ, where $\phi, \varphi \in C([t_0 - \tau, t_0]_z, \Re^n)$. Let $z_i(t) = (x_i(t; t_0, \phi) - x_i(t; t_0, \varphi))/\gamma_i$, from (2) and (3), $\forall i = 1, 2, \cdots, n, \forall t \geq t_0$,

$$|z_i(t+1)| \leq (c_i - 1)|z_i(t)| + \sum_{j=1}^n (|a_{ij}|\gamma_j|z_j(t)| + |b_{ij}|\gamma_j|z_j(t - \tau_{ij}(t))|)/\gamma_i .$$

Similar to the proof of Theorem 3.1, based on Lemma 1, DTCNNs (1) have a periodic orbit with period ω, which is globally attractive.

Let $N_1 \bigcup N_2 = \{1, 2, \cdots, n\}$, $N_1 \bigcap N_2$ is empty. Similar to the proof of Theorem 3.1, we have the following result.

Theorem 3.3. If for $\forall i \in N_1$, $c_i \in (0, 1)$, $c_i - \sum_{p=1}^n |a_{ip}| - \sum_{p=1}^n |b_{ip}| > 0$; for $\forall j \in N_2$, $c_j \in [1, 2)$, $2 - c_j - \sum_{p=1}^n |a_{jp}| - \sum_{p=1}^n |b_{jp}| > 0$, then DTCNNs (1) have a periodic state which is globally attractive.

Proof. Choose $\gamma_1 = \gamma_2 = \cdots = \gamma_n = 1$. When $i \in N_1$, (4) holds; when $i \in N_2$, (8) holds. Similar to the proof of Theorem 3.1 and Theorem 3.2, Theorem 3.2 holds.

4   Simulation Result

In this section, we give one example to illustrate the new results.

Example 1. Consider a DTCNN:

$$\begin{cases} \Delta x_1(t) = -0.9x_1(t) + 0.4f(x_1(t-1)) + 0.5f(x_2(t-2)) + \sin(t); \\ \Delta x_2(t) = -0.9x_2(t) + 0.2f(x_1(t-1)) + 0.4f(x_2(t-2)) + \cos(t). \end{cases} \quad (9)$$

According to Theorem 3.1, (9) has a globally attractive periodic state. Simulation results with 18 random initial strings are depicted in Figures 1 and 2.

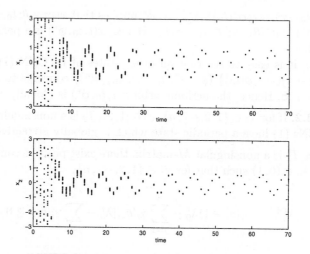

Fig. 1. The relation between t and x_i

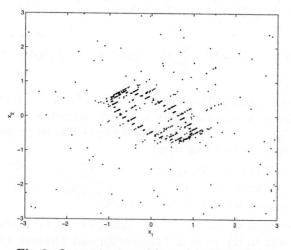

Fig. 2. Isometric view of (x_1, x_2) in Example 1

5   Concluding Remarks

In this paper, the obtained results showed that all trajectories of DTCNNs (1) converge to a periodic state when some sufficient conditions on weight matrices are satisfied. Conditions of these results can be directly derived from the parameters of DTCNNs, are very easy to verified. Hence, it is very convenience in application.

Acknowledgement

This work was supported by the Natural Science Foundation of China under Grant 60405002, the China Postdoctoral Science Foundation under Grant 2004035579 and the Young Foundation of Hubei Provincial Education Department of China under Grant 2003B001.

References

1. Grassi, G.: On Discrete-time Cellular Neural Networks for Associative Memories. IEEE Trans. Circuits Syst. I., **48** (2001) 107-111
2. Liu, D., Michel, A. N.: Sparsely Interconnected Neural Networks for Associative Memories with Applications to Cellular Neural Networks. IEEE Trans. Circ. Syst. II, **41** (1994) 295-307
3. Michel, A. N., Wang, K., Liu, D., Ye, H.: Qualitative Limitations Incurred in Implementations of Recurrent Neural Networks. IEEE Trans. Contr. Syst. Technol., **15** (1995) 52-65
4. Chua, L. O., Yang, L.: Cellular Neural Networks: Theory. IEEE Trans. Circuits Syst., **35** (1988) 1257-1272
5. Liao, X. X., Wang, J.: Algebraic Criteria for Global Exponential Stability of Cellular Neural Networks with Multiple Time Delays. IEEE Trans. Circuits and Systems I., **50** (2003) 268-275
6. Huang, D. S.: A Constructive Approach for Finding Arbitrary Roots of Polynomials by Neural Networks. IEEE Trans. Neural Networks, **15** (2004) 477-491
7. Zeng, Z. G., Wang, J., Liao, X. X.: Global Exponential Stability of A General Class of Recurrent Neural Networks with Time-varying Delays. IEEE Trans. Circuits and Syst. I, **50** (2003) 1353-1358
8. Zeng, Z. G., Wang, J., Liao, X. X.: Stability Analysis of Delayed Cellular Neural Networks Described Using Cloning Templates. IEEE Trans. Circuits and Syst. I, **51** (2004) 2313-2324
9. Sun, C., Feng, C.: Exponential Periodicity of Continuous-time and Discrete-time Neural Networks with Delays. Neural Processing Letters, **19** (2004) 131-146
10. Sun, C., Feng, C.: Exponential Periodicity and Stability of Delayed Neural Networks. Mathematics and Computers in Simulation, **66** (2004) 469-478
11. Huang, D. S.: The Local Minima Free Condition of Feedforward Neural Networks for Outer-supervised Learning. IEEE Trans. Systems, Man and Cybernetics, **28B** (1998) 477-480
12. Zeng, Z. G., Huang, D. S., Wang, Z. F.: Stability Analysis of Discrete-time Cellular Neural Networks. Lecture Notes in Computer Science, **3173** (2004) 114-119
13. Yosida, K.: Functional Analysis. Springer-Verlag, Berlin (1978)

An Analysis for Periodic Solutions of High-Order BAM Neural Networks with Delays[*]

Jianlong Qiu[1,2] and Jinde Cao[1]

[1] Department of Mathematics, Southeast University, Nanjing 210096, China
jdcao@seu.edu.cn
[2] Department of Mathematics, Linyi Normal University, Linyi 276005, China

Abstract. In this paper, by employing Lyapunov functional and LMI technique, a sufficient condition is derived for checking the existence and exponential stability of the periodic solution for high-order BAM networks. This criterion has important significance in the design and applications of periodic neural circuits for the high-order BAM networks.

1 Introduction

It is well known that neural networks play a great important role in various areas such as pattern recognition, associate memory, and combinatorial optimization. The dynamical characteristic of neural networks have been extensively studied in recent years, see [1-5]. The high-order bi-directional associative memory(BAM) networks is a special class of recurrent neural networks which can store and bipolar vector pairs. Recently, in Refs.[2,4], the authors discussed the problem of stability for high-order BAM networks. In fact the dynamical characteristic behavior of networks not only include stability but also periodic oscillatory, bifurcation, and chaos. Especially, the periodic oscillatory solutions is of great interest. To the best of our knowledge, few authors have considered the periodic oscillatory solutions of the high-order BAM networks. Motivated by the above discussion, the purpose of this paper is to investigate the existence and uniqueness of periodic oscillatory solution of the high-order BAM networks.

In this paper we consider the following second-order BAM neural networks with time-delays described by delayed differential equations:

$$x_i'(t) = -a_i x_i(t) + \sum_{j=1}^{n} b_{ij} \widetilde{g}_j(y_j(t-\tau))$$

$$+ \sum_{j=1}^{n} \sum_{l=1}^{n} e_{ijl} \widetilde{g}_j(y_j(t-\tau)) \widetilde{g}_l(y_l(t-\tau)) + I_i(t), \qquad (1)$$

[*] This work was jointly supported by the National Natural Science Foundation of China under Grant 60373067, the 973 Program of China under Grant 2003CB316904, the Natural Science Foundation of Jiangsu Province, China under Grants BK2003053 and BK2003001.

J. Wang, X. Liao, and Z. Yi (Eds.): ISNN 2005, LNCS 3496, pp. 288–293, 2005.

$$y_j'(t) = -d_j y_j(t) + \sum_{i=1}^{n} c_{ji} \widetilde{f}_i(x_i(t-\sigma))$$

$$+ \sum_{i=1}^{n} \sum_{l=1}^{n} s_{jil} \widetilde{f}_i(x_i(t-\sigma)) \widetilde{f}_l(x_l(t-\sigma)) + J_j(t), \qquad (2)$$

where $t \geq 0$; $X(t) = (x_1(t), x_2(t), \cdots, x_n(t))^T, Y(t) = (y_1(t), y_2(t), \cdots, y_n(t))^T$; a_i, d_j are positive constants, time delays τ, σ are non-negative constants, b_{ij}, c_{ji}, e_{ijl}, s_{jil} are the first- and second-order connection weights of the neural network, respectively; external inputs $I_i : R^+ \to R$ and $J_j : R^+ \to R$, are continuously periodic functions with period ω, i.e., $I_i(t+\omega) = I_i(t)$, $J_j(t+\omega) = J_j(t)$. The activation functions $\widetilde{f}_i(\cdot)$, $\widetilde{g}_j(\cdot)$ possess some of the following properties:

(H_1) There exist numbers $N_i > 0, M_j > 0$ such that $|\widetilde{f}_i(x)| \leq N_i$, $|\widetilde{g}_j(x)| \leq M_j$ for all $x \in R$ $(i = 1, 2, \ldots, n; j = 1, 2, \ldots, n)$.

(H_2) There exist numbers $L_i > 0, K_j > 0$ such that $|\widetilde{f}_i(x) - \widetilde{f}_i(y)| \leq L_i|x - y|$, $|\widetilde{g}_j(x) - \widetilde{g}_j(y)| \leq K_j|x - y|$ for all $x, y \in R$ $(i = 1, 2, \ldots, n; j = 1, 2, \ldots, n)$.

In the following of the paper, $X \in R^n$, its norm is defined as $\|X\| = \sqrt{X^T X}$. $A \geq 0$ means that matrix A is real symmetric and nonnegative definite. $\lambda_{\max}(A)$ and $\lambda_{\min}(A)$ represent the maximum and minimum eigenvalue of matrix A respectively. To obtain our main results, we need the following elementary lemmas:

Lemma 1. $i)^{[5]}$ *Suppose* W, U *are any matrices,* ϵ *is a positive number and matrix* $D > 0$, *then* $W^T U + U^T W \leq \epsilon W^T D W + \epsilon^{-1} U^T D^{-1} U$.
$ii)^{[6]}$ *(Schur complement) The following LMI*

$$\begin{bmatrix} Q(x) & S(x) \\ S^T(x) & R(x) \end{bmatrix} > 0,$$

where $Q(x) = Q^T(x)$, $R(x) = R^T(x)$, *and* $S(x)$ *depend affinely on* x, *is equivalent to* $R(x) > 0$, $Q(x) - S(x)R^{-1}(x)S^T(x) > 0$.

2   Periodic Oscillatory Solutions

Theorem 1. *Under the assumptions* (H_1)–(H_2), *if there exist positive definite matrices* P, Q, W, *and* T *such that*

$$\begin{bmatrix} AP + PA - LWL & P & PB \\ P & \frac{1}{M^*}I_{n \times n} & 0 \\ B^T P & 0 & I_{n \times n} \end{bmatrix} > 0, \qquad (3)$$

$$\begin{bmatrix} QD + DQ - KTK & Q & QC \\ Q & \frac{1}{N^*}I_{n \times n} & 0 \\ C^T Q & 0 & I_{n \times n} \end{bmatrix} > 0 \qquad (4)$$

$$W - I_{n \times n} - S^T S \geq 0, \quad T - I_{n \times n} - E^T E \geq 0. \tag{5}$$

then there exist exactly one ω-periodic solution of system (1)-(2) and all other solutions of system (1)-(2) converge exponentially to it as $t \to +\infty$.

Proof. Let $C = C\left(\left(\begin{matrix} [-\sigma, 0] \\ [-\tau, 0] \end{matrix}\right), R^{2n}\right)$ be the Banach space of continuous functions which map $\left(\begin{matrix} [-\sigma, 0] \\ [-\tau, 0] \end{matrix}\right)$ into R^{2n} with the topology uniform convergence. For any $\left(\begin{matrix} \Phi_X \\ \Phi_Y \end{matrix}\right) \in C$, we define $\left\| \left(\begin{matrix} \Phi_X \\ \Phi_Y \end{matrix}\right) \right\| = \sup_{s \in [-\sigma, \, 0]} |\Phi_X(s)| + \sup_{s \in [-\tau, \, 0]} |\Phi_Y(s)|$ where $|\Phi_X(s)| = \sum\limits_{i=1}^{n} (\phi_{xi}(s))^2, |\Phi_Y(s)| = \sum\limits_{j=1}^{n} (\phi_{yj}(s))^2$. For $\left(\begin{matrix} \Phi_X \\ \Phi_Y \end{matrix}\right), \left(\begin{matrix} \Psi_X \\ \Psi_Y \end{matrix}\right) \in C$, denote the solutions through $\left(\left(\begin{matrix} 0 \\ 0 \end{matrix}\right), \left(\begin{matrix} \Phi_X \\ \Phi_Y \end{matrix}\right)\right), \left(\left(\begin{matrix} 0 \\ 0 \end{matrix}\right), \left(\begin{matrix} \Psi_X \\ \Psi_Y \end{matrix}\right)\right)$ as

$$X(t, \Phi_X) = (x_1(t, \phi_{x1}), \cdots, x_n(t, \phi_{xn}))^T, Y(t, \Phi_Y) = (y_1(t, \phi_{y1}), \cdots, y_n(t, \phi_{yn}))^T,$$
$$X(t, \Psi_X) = (x_1(t, \psi_{x1}), \cdots, x_n(t, \psi_{xn}))^T, Y(t, \Psi_Y) = (y_1(t, \psi_{y1}), \cdots, y_n(t, \psi_{yn}))^T.$$

Define $X_t(\Phi_X) = X(t+s, \Phi_X), s \in [-\sigma, 0]; Y_t(\Phi_Y) = Y(t+s, \Phi_Y), s \in [-\tau, 0];$ then $\left(\begin{matrix} X_t(\Phi_X) \\ Y_t(\Phi_Y) \end{matrix}\right) \in C$; and $u_i(t, \phi_i, \psi_i) = x_i(t, \phi_{xi}) - x_i(t, \psi_{xi}), v_j(t, \phi_j, \psi_j) = y_j(t, \phi_{yj}) - y_j(t, \psi_{yj}), f_i(u_i(t - \sigma, \phi_i, \psi_i)) = \tilde{f}_i(x_i(t - \sigma, \phi_{xi})) - \tilde{f}_i(x_i(t - \sigma, \psi_{xi})), g_j(v_j(t - \tau, \phi_j, \psi_j)) = \tilde{g}_j(y_j(t - \tau, \phi_{yj})) - \tilde{g}_j(y_j(t - \tau, \psi_{yj})),$ for $\forall t \geq 0, i, j = 1, 2, \cdots, n$. Thus system (1)-(2) is transformed into

$$u_i'(t, \phi_i, \psi_i) = -a_i u_i(t, \phi_i, \psi_i) + \sum_{j=1}^{n} b_{ij} g_j(v_j(t, \phi_j, \psi_j))$$

$$+ \sum_{j=1}^{n} \sum_{l=1}^{n} (e_{ijl} + e_{ilj}) \xi_l g_j(v_j(t - \tau, \phi_j, \psi_j)), \tag{6}$$

$$v_j'(t, \phi_j, \psi_j) = -d_j v_j(t, \phi_j, \psi_j) + \sum_{i=1}^{n} c_{ji} f_i(u_i(t, \phi_i, \psi_i))$$

$$+ \sum_{i=1}^{n} \sum_{l=1}^{n} (s_{jil} + s_{jli}) \eta_l f_i(u_i(t - \sigma, \phi_i, \psi_i)), \tag{7}$$

where $\xi_l = e_{ijl}/(e_{ijl}+e_{ilj})\tilde{g}_l(y_l(t-\tau, \phi_{yl}))+e_{ilj}/(e_{ijl}+e_{ilj})\tilde{g}_l(y_l(t-\tau, \psi_{yl}))$ when $e_{ijl}+e_{ilj} \neq 0$, it lies between $\tilde{g}_l(y_l(t-\tau, \phi_{yl}))$ and $\tilde{g}_l(y_l(t-\tau, \psi_{yl}))$; otherwise $\xi_l = 0$. Similarly, $\eta_l = s_{jil}/(s_{jil}+s_{jli})\tilde{f}_l(x_l(t-\sigma, \phi_{xl}))+s_{jli}/(s_{jil}+s_{jli})\tilde{f}_l(x_l(t-\sigma, \psi_{xl}))$ when $s_{jil} + s_{jli} \neq 0$, it lies between $\tilde{f}_l(x_l(t - \sigma, \phi_{xl}))$ and $\tilde{f}_l(x_l(t - \sigma, \psi_{xl}))$; otherwise $\eta_l = 0$.

If we denote

$$U(t, \Phi, \Psi) = (u_1(t, \phi_1, \psi_1), u_2(t, \phi_2, \psi_2), \cdots, u_n(t, \phi_n, \psi_n))^T,$$

$$V(t, \Phi, \Psi) = (v_1(t, \phi_1, \psi_1), v_2(t, \phi_2, \psi_2), \cdots, v_n(t, \phi_n, \psi_n))^T,$$
$$F(U(t, \Phi, \Psi)) = (f_1(u_1(t, \phi_1, \psi_1)), f_2(u_2(t, \phi_2, \psi_2)), \cdots, f_n(u_n(t, \phi_n, \psi_n)))^T,$$
$$G(V(t, \Phi, \Psi)) = (g_1(v_1(t, \phi_1, \psi_1)), g_2(v_2(t, \phi_2, \psi_2)), \cdots, g_n(v_n(t, \phi_n, \psi_n)))^T,$$

and $A = \operatorname{diag}(a_1, a_2, \cdots, a_n)$, $D = \operatorname{diag}(d_1, d_2, \cdots, d_n)$; $B = (b_{ij})_{n \times n}$, $C = (c_{ji})_{n \times n}$; $E = (E_1 + E_1^T, E_2 + E_2^T, \cdots, E_n + E_n^T)^T$, where $E_i = (e_{ijl})_{n \times n}$, $S = (S_1 + S_1^T, S_2 + S_2^T, \ldots, S_n + S_n^T)^T$, where $S_j = (s_{jil})_{n \times n}$; $\Gamma = \operatorname{diag}(\xi, \xi, \cdots, \xi)_{n \times n}$, where $\xi = [\xi_1, \xi_2, \ldots, \xi_n]^T$, $\Theta = \operatorname{diag}(\eta, \eta, \ldots, \eta)_{n \times n}$, where $\eta = [\eta_1, \eta_2, \ldots, \eta_n]^T$; system (6)-(7) can be rewritten in the following vector-matrix form

$$\frac{dU(t, \Phi, \Psi)}{dt} = -AU(t, \Phi, \Psi) + BG(V(t - \tau, \Phi, \Psi)) + \Gamma^T EG(V(t - \tau, \Phi, \Psi)), \quad (8)$$

$$\frac{dV(t, \Phi, \Psi)}{dt} = -DV(t, \Phi, \Psi) + CF(U(t - \sigma, \Phi, \Psi)) + \Theta^T SF(U(t - \sigma, \Phi, \Psi)). \quad (9)$$

We consider the following Lyapunov functional

$$V(t) = e^{2kt} U^T(t, \Phi, \Psi) PU(t, \Phi, \Psi) + e^{2kt} V^T(t, \Phi, \Psi) QV(t, \Phi, \Psi)$$

$$+ \int_{t-\sigma}^{t} e^{2k(s+\sigma)} F^T(U(s, \Phi, \Psi)) WF(U(s, \Phi, \Psi)) ds$$

$$+ \int_{t-\tau}^{t} e^{2k(s+\tau)} G^T(V(s, \Phi, \Psi)) TG(V(s, \Phi, \Psi)) ds.$$

Calculate the derivative of $V(t)$ along the solutions of (8)-(9) and we obtain

$$\dot{V}(t)|_{(8)-(9)}$$
$$= e^{2kt} \{ U^T(t, \Phi, \Psi)[2kP - AP - PA] U(t, \Phi, \Psi)$$
$$+ V^T(t, \Phi, \Psi)[2kQ - DQ - QD] V(t, \Phi, \Psi)$$
$$+ G^T(V(t - \tau, \Phi, \Psi)) B^T PU(t, \Phi, \Psi) + U^T(t, \Phi, \Psi) PBG(V(t - \tau, \Phi, \Psi))$$
$$+ G^T(V(t - \tau, \Phi, \Psi)) E^T \Gamma PU(t, \Phi, \Psi) + U^T(t, \Phi, \Psi) P\Gamma^T EG(V(t - \tau, \Phi, \Psi))$$
$$+ F^T(U(t - \sigma, \Phi, \Psi)) C^T QV(t, \Phi, \Psi) + V^T(t, \Phi, \Psi) QCF(U(t - \sigma, \Phi, \Psi))$$
$$+ F^T(U(t - \sigma, \Phi, \Psi)) S^T \Theta QV(t, \Phi, \Psi) + V^T(t, \Phi, \Psi) Q\Theta^T SF(U(t - \sigma, \Phi, \Psi))$$
$$+ e^{2k\sigma} F^T(U(t, \Phi, \Psi)) WF(U(t, \Phi, \Psi)) - F^T(U(t - \sigma, \Phi, \Psi)) WF(U(t - \sigma, \Phi, \Psi))$$
$$+ e^{2k\tau} G^T(V(t, \Phi, \Psi)) TG(V(t, \Phi, \Psi)) - G^T(V(t - \tau, \Phi, \Psi)) TG(V(t - \tau, \Phi, \Psi)) \}.$$

By Lemma 1 and the assumptions, we have

$$G^T(V(t - \tau, \Phi, \Psi)) B^T PU(t, \Phi, \Psi) + U^T(t, \Phi, \Psi) PBG(V(t - \tau, \Phi, \Psi))$$
$$\leq U^T(t, \Phi, \Psi) PBB^T PU(t, \Phi, \Psi) + G^T(V(t - \tau, \Phi, \Psi)) G(V(t - \tau, \Phi, \Psi)), \quad (10)$$
$$G^T(V(t - \tau, \Phi, \Psi)) E^T \Gamma PU(t, \Phi, \Psi) + U^T(t, \Phi, \Psi) P\Gamma^T EG(V(t - \tau, \Phi, \Psi))$$
$$\leq U^T(t, \Phi, \Psi) P\Theta^T \Theta PU(t, \Phi, \Psi) + G^T(V(t - \tau, \Phi, \Psi)) E^T EG(V(t - \tau, \Phi, \Psi)),$$

$$(11)$$

$$F^T(U(t-\sigma,\Phi,\Psi))C^T QV(t,\Phi,\Psi) + V^T(t,\Phi,\Psi)QCF(U(t-\sigma,\Phi,\Psi))$$
$$\leq V^T(t,\Phi,\Psi)QCC^T QV(t,\Phi,\Psi) + F^T(U(t-\sigma,\Phi,\Psi))F(U(t-\sigma,\Phi,\Psi)), \quad (12)$$
$$F^T(U(t-\sigma,\Phi,\Psi))S^T \Theta QV(t,\Phi,\Psi) + V^T(t,\Phi,\Psi)Q\Theta^T SF(U(t-\sigma,\Phi,\Psi))$$
$$\leq V^T(t,\Phi,\Psi)Q\Theta^T \Theta QV(t,\Phi,\Psi) + F^T(U(t-\sigma,\Phi,\Psi))S^T SF(U(t-\sigma,\Phi,\Psi)),$$

$$\tag{13}$$

$$F^T(U(t,\Phi,\Psi))WF(U(t,\Phi,\Psi)) \leq U^T(t,\Phi,\Psi)LWLU(t,\Phi,\Psi), \tag{14}$$
$$G^T(V(t,\Phi,\Psi))TG(V(t,\Phi,\Psi)) \leq V^T(t,\Phi,\Psi)KTKV(t,\Phi,\Psi), \tag{15}$$
$$\Gamma^T \Gamma \leq M^*, \Theta^T \Theta \leq N^*, \tag{16}$$

where $M^* = \sum\limits_{j=1}^{n} M_j^2$, $N^* = \sum\limits_{i=1}^{n} N_i^2$. $K = \mathrm{diag}(K_1, K_2, \ldots, K_n)$, $L = \mathrm{diag}(L_1, L_2, \ldots, L_n)$.

From Lemma 1 and condition (3)-(4), there exists a scalar $k > 0$ such that

$$AP + PA - 2kP - e^{2k\sigma}LWL - M^*P^2 - PBB^T P \geq 0,$$
$$QD + DQ - 2kQ - e^{2k\tau}KTK - N^*Q^2 - QCC^T Q \geq 0,$$

And considering (10)-(16), we have $\dot{V}(t)|_{(8)-(9)} \leq 0$, $t \geq 0$, which means $V(t) \leq V(0)$, $t \geq 0$.

Since

$$V(t) \geq e^{2kt}[\lambda_{\min}(P)\|U(t,\Phi,\Psi)\|^2 + \lambda_{\min}(Q)\|V(t,\Phi,\Psi)\|^2],$$
$$V(0) \leq \lambda_{\max}(P)\|\Phi_X - \Psi_X\|^2 + \lambda_{\max}(Q)\|\Phi_Y - \Psi_Y\|^2$$
$$+ \frac{1}{2k}(e^{2k\sigma} - 1)\|L\|^2 \|W\| \sup_{s\in[-\sigma,0]} \|\Phi_X - \Psi_X\|^2$$
$$+ \frac{1}{2k}(e^{2k\tau} - 1)\|K\|^2 \|T\| \sup_{s\in[-\tau,0]} \|\Phi_Y - \Psi_Y\|^2.$$

Then we easily follow that

$$\|X(t,\Phi_X) - X(t,\Psi_X)\| \leq \gamma e^{-k(t-\sigma)}(\|\Phi_X - \Psi_X\| + \Phi_Y - \Psi_Y),$$
$$\|Y(t,\Phi_Y) - Y(t,\Psi_Y)\| \leq \gamma e^{-k(t-\tau)}(\|\Phi_X - \Psi_X\| + \Phi_Y - \Psi_Y),$$

for $\forall t \geq 0$, where $\gamma \geq 1$ is a constant. We can choose a positive integer m such that

$$\gamma e^{-k(m\omega-\sigma)} \leq \frac{1}{8}, \qquad \gamma e^{-k(m\omega-\tau)} \leq \frac{1}{8}.$$

Define a Poincare $P : C \to C$ by $P\begin{pmatrix} \Phi_X \\ \Phi_Y \end{pmatrix} = \begin{pmatrix} X_\omega(\Phi_X) \\ Y_\omega(\Phi_Y) \end{pmatrix}$. Then we can derive from BAM (1)-(2) that

$$\left\| P^m\begin{pmatrix} \Phi_X \\ \Phi_Y \end{pmatrix} - P^m\begin{pmatrix} \Psi_X \\ \Psi_Y \end{pmatrix} \right\| \leq \frac{1}{4} \left\| \begin{pmatrix} \Phi_X \\ \Phi_Y \end{pmatrix} - \begin{pmatrix} \Psi_X \\ \Psi_Y \end{pmatrix} \right\|.$$

This implies that P^m is a contraction mapping, hence there exist a unique fixed point $\begin{pmatrix} \Phi_X^* \\ \Phi_Y^* \end{pmatrix} \in C$ such that $P^m \begin{pmatrix} \Phi_X^* \\ \Phi_Y^* \end{pmatrix} = \begin{pmatrix} \Phi_X^* \\ \Phi_Y^* \end{pmatrix}$. Note that $P^m \left(P \begin{pmatrix} \Phi_X^* \\ \Phi_Y^* \end{pmatrix} \right) =$

$P \left(P^m \begin{pmatrix} \Phi_X^* \\ \Phi_Y^* \end{pmatrix} \right) = P \begin{pmatrix} \Phi_X^* \\ \Phi_Y^* \end{pmatrix}$. This shows that $P \begin{pmatrix} \Phi_X^* \\ \Phi_Y^* \end{pmatrix} \in C$ is also a fixed

point of P^m, and so $P \begin{pmatrix} \Phi_X^* \\ \Phi_Y^* \end{pmatrix} = \begin{pmatrix} \Phi_X^* \\ \Phi_Y^* \end{pmatrix}$, i.e., $\begin{pmatrix} X_\omega(\Phi_X^*) \\ Y_\omega(\Phi_Y^*) \end{pmatrix} = \begin{pmatrix} \Phi_X^* \\ \Phi_Y^* \end{pmatrix}$.

Let $\begin{pmatrix} X(t, \Phi_X^*) \\ Y(t, \Phi_Y^*) \end{pmatrix}$ be the solution of system (1)-(2) through $\left(\begin{pmatrix} 0 \\ 0 \end{pmatrix}, \begin{pmatrix} \Phi_X^* \\ \Phi_Y^* \end{pmatrix} \right)$,

then $\begin{pmatrix} X(t+\omega, \Phi_X^*) \\ Y(t+\omega, \Phi_Y^*) \end{pmatrix}$ is also a solution of BAM (1)-(2). Obviously,

$$\begin{pmatrix} X(t+\omega, \Phi_X^*) \\ Y(t+\omega, \Phi_Y^*) \end{pmatrix} = \begin{pmatrix} X_{t+\omega}(\Phi_X^*) \\ Y_{t+\omega}(\Phi_Y^*) \end{pmatrix} = \begin{pmatrix} X_t(X_\omega(\Phi_X^*)) \\ Y_t(Y_\omega(\Phi_Y^*)) \end{pmatrix} = \begin{pmatrix} X_t(\Phi_X^*) \\ Y_t(\Phi_Y^*) \end{pmatrix}.$$

This shows that $\begin{pmatrix} X(t, \Phi_X^*) \\ Y(t, \Phi_Y^*) \end{pmatrix}$ is exactly one ω-periodic solution of BAM (1)-(2), and it easy to see that all other solutions of BAM (1)-(2) converge exponentially to it as $t \to +\infty$. This complete the proof.

3   Conclusion

In this paper, we studied the high-order BAM nerual networks with delays. A sufficient condition is given to ensure the existence and uniqueness of periodic oscillatory solution by constructing Lyapunov functional and LMI technique. And we also prove all other solutions of the high-order BAM networks converge exponentially to the unique periodic solution as $t \to +\infty$.

References

1. Kosmatopoulos, E.B., Christodoulou, M.A.: Structural Properties of Gradient Recurrent High-order Neural Networks. IEEE Trans. Circuits Syst. II, **42** (1995) 592-603
2. Cao, J., Liang, J., Lam, J.: Exponential Stability of High-order Bidirectional Associative Memory Neural Networks with Time Delays. Physica D, **199** (2004) 425-436
3. Cao, J., Jiang, Q.: An Analysis of Periodic Solutions of Bi-directional Associative Memory Networks with Time-varying Delays. Phys. Lett. A, **330** (2004) 203-213
4. Xu, B., Liu, X., Liao, X.: Global Asymptotic Stability of High-order Hopfield Type Neural Networks with Time Delays. Comput. Math. Appl., **45** (2003) 1729-1737
5. Cao, J. Ho, D. W. C.: A General Framework for Global Asymptotic Stability Analysis of Delayed Neural Networks Based on LMI Approach. Chaos, Solitons and Fractals, **24** (2005) 1317-1329
6. Boyd, S., Ghaoui, L.E.I., Feron, E., Balakrishnan, V.: Linear Matrix Inequalities in System and Control Theory. SIAM, Philadelphia (1994)

Periodic Oscillation and Exponential Stability of a Class of Competitive Neural Networks

Boshan Chen

Dept. of Mathematics, Hubei Normal University, Huangshi, Hubei 435002, China
chenbs@hbnu.edu.cn

Abstract. In this paper, the periodic oscillation and the global exponential stability of a class of competitive neural networks are analyzed. The competitive neural network considered includes the Hopfield networks, Cohen-Grossberg networks as its special cases. Several sufficient conditions are derived for ascertaining the existence, uniqueness and global exponential stability of the periodic oscillatory state of the competitive neural networks with periodic oscillatory input by using the comparison principle and the theory of mixed monotone operator and mixed monotone flow. As corollary of results on the global exponential stability of periodic oscillation state, we give some results on the global exponential stability of the network modal with constant input, which extend some existing results. In addition, we provide a new and efficacious method for the qualitative analysis of various neural networks.

1 Introduction

There are many results on the dynamic behavior analysis of the neural networks with and without delays in the past decade. For example, results on the existence, uniqueness and global stability of the equilibria of the neural networks with and without time delay are discussed in [1],[2]. On the other hands, periodic oscillation in the neural networks is an interesting dynamic behavior. It has been found applications in associative memories [3], pattern recognition [4],[5], learning theory [6],[7], and robot motion control [8]. An equilibrium point can viewed as a special case of periodic state with an arbitrary period or zero amplitude. In this sense, the analysis of periodic oscillation of neural networks is more general than the stability analysis of equilibrium points. In recent years, studies of the periodic oscillation of various neural networks such as the Hopfield network, cellular neural networks, and bidirectional associative memories are reported in [9],[10]. In particular, the periodicity and stability of a general class of recurrent neural networks without time delays are analyzed in [10]. Several algebraic criteria are derived to ascertain the global exponential periodicity and stability of the recurrent neural networks model by using the comparison principle and the method of mixed monotone operator. In 1983, Cohen and Grossberg [11] first propose and study a class of competitive neural networks, which includes a number of modal from neurobiology, population biology, evolutionary theory. The competitive neural network also includes the former Hopfield neural

J. Wang, X. Liao, and Z. Yi (Eds.): ISNN 2005, LNCS 3496, pp. 294–301, 2005.
© Springer-Verlag Berlin Heidelberg 2005

networks as special case. Recently, exponential stability of the Cohen-Grossberg neural networks with and without delay are analyzed in [12]. However, to the best of our knowledge, a few result has been reported on the periodic oscillation of the competitive neural network.

In this paper, we analyze the periodic oscillation an global exponential stability of a class of competitive neural networks with periodic input. As corollary of results on the global exponential stability of periodic oscillation state , we give some results on the global exponential stability of the modal with constant input, which extend some results in [12].

The reminder of the paper is organized in four sections. Section 2 provides preliminary information. Section 3 presents the criteria for global exponential periodicity of the competitive neural networks with periodic oscillation input and for global exponential stability of the competitive neural networks with constant input. In Section 4, a numerical example is given to illustrate the results. Finally, Section 5 concludes the paper.

2   Preliminaries

Consider a class of competitive neural network (NN) described by the following the differential equations:

$$\dot{x}_i(t) = -a_i(x_i(t)) \left[b_i(x_i) - \sum_{j=1}^{n} c_{ij} d_j(x_j(t)) - I_i(t) \right], \ i = 1, \cdots, n \qquad (1)$$

where x_i denotes the state variable associated with the ith neuron, $\dot{x}_i$ denotes the derivative of x_i with respect to time t; $a_i(\cdot)$ represents an amplification function, and $b_i(\cdot)$ is an appropriately behaved function. $(c_{ij}) \in R^{n \times n}$ are connection weight matrix , $d_i(\cdot)$ are activation function and $I(t) = (I_1(t), \cdots, I_n(t))^T$ is an input periodic vector function with period ω; i.e., there exits a constant $\omega > 0$ such that $I_i(t+\omega) = I_i(t), \forall t \geq t_0, \forall i \in \{1, 2, \cdots, n\}$. NN (1) represents a general class of recurrent neural networks, which including the Hopfield networks, Cohen-Grossberg neural networks as special cases.

Throughout of the paper, we assume that the function $b_i(\cdot)$, the activation functions $d_i(\cdot)$ and amplification function $a_i(\cdot)$ possess the following properties:
Assumption A: There exist $k_i > 0$ and $l_i > 0$ such that

$$k_i(\theta - \rho)^2 \leq (\theta - \rho)(b_i(\theta) - b_i(\rho)) \text{ for } \forall \theta, \rho \in R, i = 1, \cdots, n \qquad (2)$$

and

$$0 < (\theta - \rho)(d_i(\theta) - d_i(\rho)) \leq l_i(\theta - \rho)^2 \text{ for } \forall \theta, \rho \in R, i = 1, \cdots, n \qquad (3)$$

and there exist two constant $\underline{a} \leq \bar{a}$ such that

$$0 < \underline{a} \leq a_i(\theta) \leq \bar{a} \qquad (4)$$

$v_{0i} = \min\{x_i(0), x_i^*(0)\}$. Clearly, $v_0 \leq x(0)$, $x^*(0) \leq u_0$ and $||x(0) - x^*(0)|| = ||u_0 - v_0||$. From Lemmas 1 and Lemma 3, it follows that

$$||x(t) - x^*(t)|| \leq ||u(t, u_0, v_0) - v(t, u_0, v_0)||$$
$$\leq M||u_0 - v_0||e^{-\mu t} = M||x(0) - x^*(0)||e^{-\mu t}, \; \forall t \geq 0. \qquad (12)$$

The proof is complete.

Now, we will give some simple and practical algebraic criteria for ascertaining the global exponential convergence of the periodic oscillatory state of NN (1) by giving the existence conditions for the equilibrium of the system (7).

Theorem 2. Suppose that the assumption A holds. If for each $i = 1, \cdots, n$, there are constants $M_i > 0$ such that

$$|d_i(\theta)| \leq M_i \text{ for } \forall \theta \in R, \qquad (13)$$

and the condition (9) holds, then NN (1) has a unique ω-periodic oscillatory state $x^*(t)$, which is globally exponentially stable.

We do not require the boundedness of activation functions in the following theorems.

Theorem 3. Suppose that the assumption A holds. If for each $i = 1, \cdots, n$,

$$d_i(-\theta) = -d_i(\theta) \text{ for } \forall \theta \in R, \; b_i(0) = 0 \qquad (14)$$

and the condition (9) holds, then NN (1) has a unique ω-periodic oscillatory state $x^*(t)$, which is globally exponentially stable.

Theorem 4. Suppose that the assumption A holds. If there exist constants $\alpha_i > 0$, $i = 1, \cdots, n$ such that

$$- k_i \alpha_i + l_i c_{ii}^+ \alpha_i + \sum_{j=1, j \neq i}^{n} \alpha_j l_j |c_{ij}| < 0 \text{ for } i = 1, \cdots, n, \qquad (15)$$

and

$$b_i(0) = 0, d_i(0) = 0 \text{ for } i = 1, \cdots, n, \qquad (16)$$

then NN (1) has a unique ω-periodic oscillatory state $x^*(t)$, which is globally exponentially stable.

NN (1) with constant input can be rewrite as following form:

$$\dot{x}_i(t) = -a_i(x_i(t)) \left[b_i(x_i) - \sum_{j=1}^{n} c_{ij} d_j(x_j(t)) - I_i \right], \; i = 1, \cdots, n. \qquad (17)$$

When $I(t) \equiv I$ constant (i.e., periodic ω is any real number), $x^*(t) \equiv x^*$. According to Theorems 1, 2, 3 and 4, we give the following results of the global exponential stability of the competitive neural network with constant input.

Corollary 1. Suppose that the assumption A and the condition (9) hold. For any constant input $I = (I_1, \cdots, I_n)^T$, NN (17) is globally exponentially stable if it has an equilibrium x^*.

Corollary 2. If the assumption A and one of the following conditions hold:
(i) the conditions (9) and (13) hold;
(ii) the conditions (9) and (14) hold;
(iii) the conditions (15) and (16) hold,
then for any constant input $I = (I_1, \cdots, I_n)^T$, NN (17) has a globally exponentially stable equilibrium.

Remark 1. It is worth pointing out that the conditions (3.10) and (3.11) in [12] are equivalent to the conditions (3) and (9) above, respectively. But, we do not require the boundedness of activation functions in Corollary 2.

4   Illustrative Example

In this section, we give a numerical example to illustrate the validity of the results.

Example. Consider the following competitive neural network:

$$\dot{x}_1(t) = -(2 + \sin x_1(t))[x_1(t) + \tfrac{1}{2}d_1(x_1(t)) - d_2(x_2(t)) + I_1(t)]$$
$$\dot{x}_2(t) = -(2 + \cos x_2(t))[x_2(t) + \tfrac{3}{4}d_1(x_1(t)) - \tfrac{1}{4}d_2(x_2(t)) + I_2(t)] \tag{18}$$

where the activation functions $d_i(\cdot)(i = 1, 2)$ satisfy the assumption A with $l_1 = l_2 = 1$, which include sigmoidal functions $\arctan \sigma$, $\frac{\exp(\sigma)-\exp(-\sigma)}{\exp(\sigma)+\exp(-\sigma)}$, the linear saturation function $\frac{(|\sigma+1|-|\sigma-1|)}{2}$, and unbounded odd function $\frac{p}{q}\sigma^{\frac{q}{p}}$, both p and q odd and $0 < q \le p$. In this example, since

$$-k_1 + l_1c_{11} + l_1|c_{12}| = -\tfrac{1}{2}, -k_2 + l_2c_{22} + l_2|c_{21}| = -\tfrac{1}{2};$$

and

$$-k_1 + l_1c_{11} + l_1|c_{12}| = -\tfrac{1}{2}, -k_2 + l_2c_{22} + l_2|c_{21}| = -\tfrac{1}{2}.$$

According to Theorem 3, for any 2π-periodic input vector $I(t) = (I_1(t), I_2(t))^T$, the neural network (18) is globally exponentially convergent to a unique 2π-periodic state $x^*(t)$. It can be seen from Fig. 1 that the periodic oscillation of the neural network.

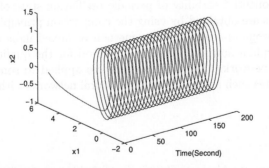

Fig. 1. Periodic oscillation of NN with $d_1(\sigma) = d_2(\sigma) = \tfrac{1}{3}\sigma^3$, and $I_1(t) = \sin t, I_2(t) = \cos t$ in the Example

In addition, according to Corollary 2, for any constant vector input; i.e, $I(t) \equiv (I_1, \ I_2)^T$, the neural network (18) globally exponentially convergent to a unique equilibrium x^*. It can be seen from Fig. 2.

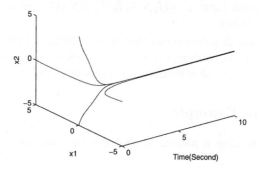

Fig. 2. Global exponential stability of NN with $d_1(\sigma) = d_2(\sigma) = \frac{1}{3}\sigma^3$, and $I_1(t) = I_2(t) \equiv 1$ in the Example

Remark 2. It is worth pointing out that the results above can not be obtained by using the theorems in [12], since the activation functions $d_i(\cdot)(i = 1, 2)$ can be unbounded.

5   Conclusion

This paper presents theoretical results on periodic oscillation of a class of competitive neural networks with periodic oscillating input. The neural network considered herein includes a number of modal from neurobiology, population biology, evolutionary theory. And the competitive neural network includes the former Hopfield neural networks as special case. In addition, the property of global exponential periodicity includes that of global exponential stability as a special case. Several algebraic criteria for ascertaining the existence, uniqueness, and global exponential stability of periodic oscillation state of the competitive neural networks are obtained by using the comparison principle and the theory of mixed monotone operator and the method of mixed monotone flow. In addition, we provide a new and efficacious method for the qualitative analysis of various neural networks. The methods may be applied to some more complex nonlinear systems such as the competitive neural networks with time delays and so on.

References

1. Liao, X., Wang, J.: Algebraic Criteria for Global Exponential Stability of Cellular Neural Networks with Multiple Time Delays. IEEE Trans. Circuits and Systems I., **50** (2003) 268-275

2. Zeng, Z., Wang, J., Liao, X.: Global Exponential Stability of A General Class of Recurrent Neural Networks with Time-varying Delays. IEEE Trans. Circuits and Syst. I, **50** (2003) 1353-1358

3. Nishikawa, T., Lai, Y. C., Hoppensteadt, F. C.: Capacity of Oscillatory Associative-memory Networks with Error-free Retrieval. Physical Review Letters, **92** (2004) 108101

4. Wang, D. L.: Emergent Synchrony in Locally Coupled Neural Oscillators. IEEE Trans. Neural Networks, **6** (1995) 941-948

5. Chen, K., Wang, D. L., Liu, X.: Weight Adaptation and Oscillatory Correlation for Image Segmentation. IEEE Trans. Neural Networks, **11** (2000) 1106-1123

6. Ruiz, A., Owens, D. H., Townley, S.: Existence, Learning, and Replication of Periodic Motion in Recurrent Neural Networks. IEEE Trans. Neural Networks, **9** (1998) 651-661

7. Townley, S., Ilchmann, A., Weiss, M. G., Mcclements, W., Ruiz, A. C., Owens, D. H., Pratzel-Wolters, D: Existence and Learning of Oscillations in Recurrent Neural Networks. IEEE Trans. Neural Networks, **11** (2000) 205-214

8. Jin, H., Zacksenhouse, M.: Oscillatory Neural Networks for Robotic Yo-yo Control. IEEE Trans. Neural Networks, **14** (2003) 317-325

9. Yang, H., Dillon, T. S.: Exponential Stability and Oscillation of Hopfield Graded Response Neural Network. IEEE Trans. Neural Networks, **5** (1994) 719-729

10. Chen, B., Wang, J.: Global Exponential Periodicity and Global Exponential Stability of A Class of Recurrent Neural Networks. Physics Letters A, **329** (2004) 36-48

11. Cohen, M. A., Grossberg, S.: Absolute Stability and Global Pattern Formation and Parallel Memory Storage by Competitive Neural Networks. IEEE Trans. on Systems, Man and Cybernetics, **13** (1983) 815-821

12. Wang, L., Zou, X.: Exponential Stability of Cohen-Grossberg Neural Networks. Neural Networks, **15** (2002) 415-422

13. Chen, B.: Existence and Attraction of the Bounded State for the Non-autonomous n-competing Species Systems. J. Sys. Math Scis(in Chinese), **16** (1996) 113-118

Synchronous Behaviors of Two Coupled Neurons*

Ying Wu[1], Jianxue Xu[1], and Wuyin Jin[1, 2]

[1] School of Architectural Engineering & Mechanics,
Xi'an Jiaotong University, Xi'an 710049, China
Wying36@163.com
[2] College of Mechano-Electronic Engineering,
Lanzhou University of Technology,
Lanzhou 730050, China

Abstract. We study the synchronization phenomena in a pair of Hindmarsh-Rose(HR) neurons that connected by electrical coupling and chemical coupling and combinations of electrical and chemical coupling. We find that excitatory synapses can antisynchronize two neurons and enough strong inhibition can foster phase synchrony. Investigating the affection of combination of chemical and electrical coupling on network of two HR neurons shows that combining chemical coupling and positive electrical coupling can promotes phase synchrony, and conversely, combining chemical coupling and negative electrical coupling can promotes antisynchrony.

1 Introduction

Synchronous oscillatory activity has been suggested that his activity may be important for cognition and sensory information processing. Synchronization of nonlinear oscillators has been widely study recently [1]-[5]. Especially, the affection of electrical and chemical coupling on synchrony of coupling neurons has attracted lots of attention.

In Ref. [6], the experimental studies of synchronization phenomena in a pair of biological neurons interacted through electrical coupling were reported. In Ref. [7], the synchronization phenomena in a pair of analog electronic neurons with both direct electrical connections and excitatory and inhibitory chemical connections was studied. Traditionally, it has been assumed that inhibitory synaptic coupling pushes neurons towards antisynchrony. In fact, this is the case for sufficiently rapid synaptic dynamics. If the time scale of the synapses is sufficiently slow with respect to the intrinsic oscillation period of the individual cells, inhibition can act to synchronize oscillatory activity [8]. Electrical coupling is usually thought to synchronize oscillation activity, however it has been shown that electrical coupling can induce stable antisynchronous activity in some cases [8].

In this paper, first, we investigate dynamics of network of two HR neurons with chemical synapses or electrical coupling alone, used models were given in Ref. [9]. The results show that excitatory synapses can antisynchronize two neurons and

---

*· Project supported by the National Natural Science Foundation of China (Grant Nos. 10172067 and 10432010).

J. Wang, X. Liao, and Z. Yi (Eds.): ISNN 2005, LNCS 3496, pp. 302–307, 2005.

enough strong inhibition can foster phase synchronization, which are different from the tradition views. Secondly, we investigate the affection of combined chemical and electrical coupling on network of two HR neurons. Results show that electrical coupling is predominant and chemical synapses can boost synchronous activity up in combined coupling, namely means that combining chemical synapses can promote two neurons with positive electrical coupling to be phase synchronization, and can promote two neurons with negative electrical coupling to be antiphase synchronization.

2 Hindmarsh-Rose Models with Electrical and Chemical Synaptic Connections

Consider two identical HR models with both electrical and reciprocal synaptic connections. The differential equations of the coupled systems are given as [9]:

$$\dot{x}_i = y_i + bx_i^2 - ax_i^3 - z_i + I_{dc} + e_e(x_j - x_i) + e_s(\frac{x_i + V_c}{1 + \exp\frac{x_j - X_0}{Y_0}})$$

$$\dot{y}_i = c - dx_i^2 - y_i \qquad\qquad (1)$$

$$\dot{z}_i = r[S(x_i - \chi) - z_i]$$

Where $i = 1, 2$, $j = 1, 2$, $i \neq j$

The HR neuron is characterized by three time-dependent variables: the membrane potential x, the recovery variable y, and a slow adaptation current z, in the simulation,   let $a = 1.0, b = 3.0, c = 1.0, d = 5.0, s = 4.0, r = 0.006, \chi = -1.56$, $I_{dc} = 3.0$, I_{dc} denotes the input constant current, e_e is the strength of the electrical coupling, e_s is the strength of the synaptic coupling, and $V_c = 1.4$ is synaptic reverse potential which is selected so that the current injected into the postsynaptic neuron are always negative for inhibitory synapses and positive for excitatory synapses. Since each neuron must receive an input every time the other neuron produces a spike, we set $Y_0 = 0.01$ and $X_0 = 0.85$ [9]. In numerical simulation, the double precision fourth-order Runge-kutta method with integration time step 0.01 was used, the initial condition is (0.1,1.0,0.2,0.1,0.2,0.3).

3 Synchronizing Two HR Neurons with Electrical Coupling Alone

To compare with the results of the Ref. [6],[7] the effects of positive and negative electrical coupling on systems are investigated all. The synchronized courses of two models are presented in Fig.1. When $e_e > 0$, two models will be realized from partial phase synchronization to full synchronization with the coupling strength increasing gradually (Fig.1 (a,b,c)). When $e_e < 0$, the oscillation modes of two systems will vary from partial antiphase to full antiphase (Fig.1 (d,e,f)). The results are in agreement with Ref. [6],[7]. But In Ref. [7], the average burst length decrease as the negative coupling becoming stronger, which isn't seen in our results.

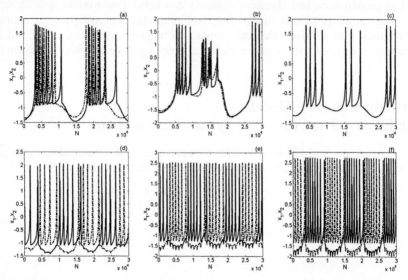

Fig. 1. Time courses of membrane potentials of two neurons with electrical coupling. (a,b,c) two neurons are partial phase synchrony for e_e=0.05, and phase synchrony for e_e=0.4 , and full synchrony for e_e=0.5, respectively; (d,e,f) two neurons are partial antiphase synchrony for e_e=-0.2 and e_e=-0.5, and full antiphase synchrony for e_e=-1.0

4 Synchronizing Two HR Neurons with Synaptic Connection Alone

4.1 Excitatory Synapse

The chemical synapse is excitatory for $e_s > 0$ and is inhibition for $e_s < 0$. The results show that two neurons will be irregular oscillation with small excitatory coupling strength, and will be in full antiphase for enough coupling strength, such as Fig.2. It's interesting that these results aren't agreement with which of Ref. [7], and are contrary to traditional view.

4.2 Inhibitory Synapse

When two systems are interacted by inhibitory synapses alone, The results are given in Fig.3. Two neurons oscillation are irregular for little coupling strength, and the phase difference of two neurons will increase gradually with coupling strength increasing, till $e_s = -0.45$ the phase difference of two neurons are biggest, which means that two neurons are full antiphase synchrony. Continuing to increase inhibition-coupling strength, the phase difference of two neurons will decrease, till $e_s = -0.9$ two neurons are full synchronous periodic oscillation without spikes.

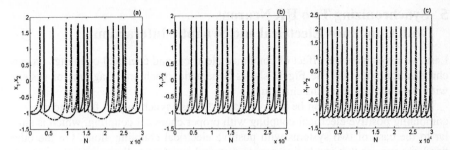

Fig. 2. Time courses of membrane potential of two neurons for excitatory synapse. (a,b,c) two neurons are irregular activity for $e_s = 0.03$, and period 2 antiphase synchrony for $e_s = 0.1$, and period 1 antiphase synchrony for $e_s = 0.3$, respectively

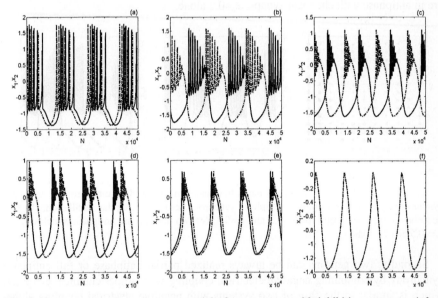

Fig. 3. Time courses of membrane potential of two neurons with inhibition synapse. (a,b,c) two neurons are irregular activity for $e_s = -0.05$, and partial antiphase synchrony for $e_s = -0.2$, and full antiphase synchrony for $e_s = -0.45$, respectively; (d,e,f) two neurons are partial phase synchrony for $e_s = -0.5$, and phase synchrony for $e_s = -0.6$, and full synchrony periodic oscillation for $e_s = -0.9$

Traditionally, it has been thought that inhibitory synaptic coupling pushes neurons towards antisynchrony [7]. In Ref. [8], intrinsic oscillation frequency of the individual cell was increased by increasing export stimulating current, and the systems with inhibition were evolved to synchronous state. In our paper, it has been seen that enough strong inhibition can synchronize two neurons to periodic oscillations without spikes.

5 Synchronizing Two HR Neurons with Combined Electrical and Synaptic Interaction

Last, we investigate the effect of combined electrical and chemical couplings on synchronization patterns. The Fig.4a shows Schematic synchrony diagrams of systems with two coupling strengths changing.

Firstly, two neurons can be always realized phase synchrony or full synchrony by combining positive electrical coupling with proper chemical coupling, which is similar to the case of two systems with positive electrical coupling alone. On the other hand, we can see that combining proper chemical coupling will promote neurons with positive electrical coupling to be phase synchronized. For example, two neurons with electrical coupling e_e=0.3, can be only in partial phase synchrony, but which can be full synchrony by combining chemical synapse e_s=0.3 (Fig.4b), although two neurons are in antiphase with chemical synapse e_s=0.3 alone.

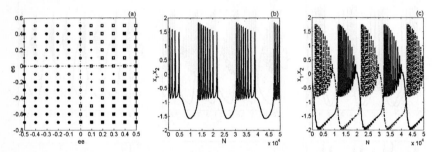

Fig. 4. (a) Schematic synchrony pattern diagram for two neurons with e_e and e_s, dark squares correspond to full synchrony, open squares to phase synchrony, dark circle to full antiphase synchrony, open circle to partial antiphase synchrony, cross to irregular oscillation; (b) two neurons are full synchrony for e_e=0.3 and e_s=0.3; (c) two neurons are antiphase synchrony for e_e=-0.5 and e_s=-0.8

Secondly, two neurons can be always realized partial antiphase or full antiphase synchrony by combining negative electrical coupling with proper chemical coupling, which is similar to the case of two systems with negative electrical coupling alone. And combining proper chemical synapse can promote antiphase activities of two systems with negative electrical coupling. Even we can see the interesting phenomenon that two neurons with synapses coupling e_s=-0.8 alone are phase synchrony, but after combining negative electrical coupling two neurons are full antiphase synchrony. (Fig.4c)

6 Conclusion

In this paper, we investigate two HR models that interact by electrical coupling and chemical coupling alone and combinations of electrical and chemical coupling. The results show that excitatory synapses can antisynchronize two neurons and enough strong inhibition can foster phase synchrony. Investigating the affection of combined chemical and electrical coupling on two HR neurons shows that two neurons with positive electrical coupling can be always realized phase synchronization or full synchronization by combining proper chemical coupling, and two neurons with negative electrical coupling can be always realized partial antiphase or full antiphase activities by combining proper chemical synapse. The electrical coupling is dominant in combined coupling, but chemical synapse can boost synchronous activity up by coupling with electrical coupling.

References

1. He, D.H., Shi, P.L., Stone, L.: Noise-induced Synchronization in Realistic Models. Phys. Rev. E, **67** (2003) 027201
2. Wu, Y., Xu, J.X., He, D.H., et al: Synchronization in Two Uncoupled Chaotic Neurons. LNCS, **3173** (2004) 183-189
3. Wu, Y., Xu, J.X., He, D.H., et al: Generalized Synchronization Induced by Noise and Parameter Mismatching in Hindmarsh-Rose neurons. Chaos Solitons & Fractals, **23** (2005) 1605-1611
4. Jin, W.Y., Xu, J.X., Wu, Y., et al: Rate of Afferent Stimulus Dependent Synchronization and Coding in Coupled Neurons System. Chaos Solitons & Fractals, **21** (2004) 1221-1229
5. Tamas, G., Eberhard H.B., Lörinz, A., et al: Proximally Targeted GABAergic Synapses and Gap Junctions Synchronize Cortical Interneurous. Nat. Neurosci, **3** (2000) 366-371
6. Elson, R.C., Selverston, A.I., Romon, H., et al: Synchronous Behavior of Two Coupled Biological Neurons. Phys. Rev. lett. **81** (1998) 5692-5695
7. Pinto, R.D., Varona, P., Volkovskii, A.R., et al: Synchronous Behavior of Two Coupled Electronic Neurons. Phys. Rev. E, **62** (2000) 2644-2656
8. Timothy, J.L., John, R.: Dynamics of Spiking Neurons Connected by Both Inhibitory and Electrical Coupling. Journal of computation neuroscience, **14** (2003) 283-309
9. Romon, H., Mikhail, I.R.: Spike-train Bifurcation in Two Coupled Chaotic Neurons. Phys. Rev. E, **55** (1997) R2108

Adaptive Synchronization of Delayed Neural Networks Based on Parameters Identification

Jin Zhou[1,2], Tianping Chen[1], and Lan Xiang[2]

[1] Laboratory of Nonlinear Science, Institute of Mathematics,
Fudan University, Shanghai 200433, China
{Jinzhou,Tchen}@fudan.edu.cn
[2] Department of Applied Mathematics and Physics,
Hebei University of Technology, Tianjin 300130, China
Xianglan.htu@eyou.com.cn

Abstract. By combining the adaptive control and linear feedback with the updated laws, an approach of adaptive synchronization and parameters identification of recurrently delayed neural networks with all the parameters unknown is proposed based on the invariance principle of functional differential equations. This approach supplies a systematic and analytical procedure for adaptive synchronization and parameters identification of such uncertain networks, and it is also simple to implement in practice. Theoretical proof and numerical simulation demonstrate the effectiveness and feasibility of the proposed technique.

1   Introduction

Synchronization of coupled chaotic systems and its potential application in engineering are currently a field of great interest (see [1–5] and references cited therein). A wide variety of approaches have been proposed for the synchronization of chaotic systems which include linear and nonlinear feedback control, time-delay feedback control, adaptive design control, impulsive control method, and so on [3, 4]. However, most of the developed methods are valid only for the chaotic systems whose parameters are precisely known. But in practical situation, the parameters of some systems cannot be exactly known in priori, the effect of these uncertainties will destroy the synchronization and even break it [3–5]. Therefore, synchronization of chaotic systems in the presence of unknown parameters is essential.

Recently, there has been increasing interest in applications of the dynamical properties of recurrently delayed neural networks such as delayed Hopfield neural networks and delayed cellular neural networks (CNN). Most of previous studies are predominantly concentrated on the stability analysis and periodic oscillations of such kind of networks [6, 7]. However, it has been shown that recurrently delayed neural networks not only can exhibit some complicated dynamics and even chaotic behaviors, but also synchronization of such coupled networks have potential applications in many fields including secure communication, parallel image processing, biological systems, information science, etc

J. Wang, X. Liao, and Z. Yi (Eds.): ISNN 2005, LNCS 3496, pp. 308–313, 2005.
© Springer-Verlag Berlin Heidelberg 2005

[6–9]. Therefore, the investigation of synchronization dynamics of recurrently delayed neural networks is an important step for practical design and applications of neural networks.

This paper proposes a novel method of adaptive synchronization and parameters identification for recurrently delayed neural networks with all the parameters unknown based on the invariance principle of functional differential equations. By this method, one can not only achieve global synchronization of such networks, but also identify all the unknown parameters dynamically. To this end, the theoretical results will be illustrated by numerical simulations on a typical chaotic delayed Hopfield neural networks.

2   Problem Formulations

First, we consider a class of recurrently delayed neural networks, which is described by the following set of differential equations with delays [6–9]:

$$\dot{x}_i(t) = -c_i h_i(x_i(t)) + \sum_{j=1}^{n} a_{ij} f_j(x_j(t)) + \sum_{j=1}^{n} b_{ij} g_j(x_j(t - \tau_{ij})) + u_i, i = 1, 2, \cdots, n.$$
(1)

or, in a compact form,

$$\dot{x}(t) = -Ch(x(t)) + Af(x(t)) + Bg(x(t - \tau)) + u,$$
(1)′

where $x(t) = (x_1(t), \cdots, x_n(t))^\top \in R^n$ is the state vector of the neural network, $C = \text{diag}(c_1, \ldots, c_n)$ is a diagonal matrix with $c_i > 0$, $i = 1, 2, \cdots, n$, $A = (a_{ij})_{n \times n}$ is a weight matrix, $B = (b_{ij})_{n \times n}$ is the delayed weight matrix, $u = (u_1, \cdots, u_n)^\top \in R^n$ is the input vector, $\tau(r) = (\tau_{rs})$ with the delays $\tau_{ij} \geq 0$, $i, j = 1, 2, \cdots, n$, $h(x(t)) = [h_1(x_1(t)), \cdots, h_n(x_n(t))]^\top$, $f(x(t)) = [f_1(x_1(t)), \cdots, f_n(x_n(t))]^\top$ and $g(x(t)) = [g_1(x_1(t)), \cdots, g_n(x_n(t))]^\top$. The initial conditions of (1) or (1)′ are given by $x_i(t) = \phi_i(t) \in C([-\tau, 0], R)$ with $\tau = \max_{1 \leq i,j \leq n} \{\tau_{ij}\}$, where $C([-\tau, 0], R)$ denotes the set of all continuous functions from $[-\tau, 0]$ to R.

Next, we list some assumptions which will be used in the main results of this paper [6–9]:

   (A_0) $h_i : R \to R$ is differentiable and $\eta_i = \inf_{x \in R} h_i'(x) > 0, h_i(0) = 0$, where $h_i'(x)$ represents the derivative of $h_i(x)$, $i = 1, 2, \cdots, n$.

Each of the activation functions in both $f_i(x)$ and $g_i(x)$ is globally Lipschitz continuous, i.e., either (A_1) or (A_2) is satisfied:

   (A_1) There exist constants $k_i > 0, l_i > 0$, $i = 1, 2, \cdots, n$, for any two different $x_1, x_2 \in R$, such that $0 \leq \frac{f_i(x_1) - f_i(x_2)}{x_1 - x_2} \leq k_i$, $|g_i(x_1) - g_i(x_2)| \leq l_i|x_1 - x_2|$, $i = 1, 2, \cdots, n$.

   (A_2) There exist constants $k_i > 0, l_i > 0$, $i = 1, 2, \cdots, n$, for any two different $x_1, x_2 \in R$, such that $|f_i(x_1) - f_i(x_2)| \leq k_i|x_1 - x_2|$, $|g_i(x_1) - g_i(x_2)| \leq l_i|x_1 - x_2|$, $i = 1, 2, \cdots, n$.

Now we consider the master (or drive) system in the form of the recurrently delayed neural networks (1) or (1)′, which may be a chaotic system. We also introduce an auxiliary variables $y(t) = (y_1(t), \cdots, y_n(t))^{\top} \in R^n$, the slave (or response) system is given by the following equation

$$\dot{y}(t) = -\bar{C}h(y(t)) + \bar{A}f(y(t)) + \bar{B}g(y(t-\tau)) + u, \qquad (2)$$

which has the same structure as the master system but all the parameters $\bar{C} = \text{diag}(\bar{c}_1, \ldots, \bar{c}_n), \bar{A} = (\bar{a}_{ij})_{n \times n}$ and $\bar{B} = (\bar{b}_{ij})_{n \times n}$ are completely unknown, or uncertain. In practical situation, the output signals of the master system (1) can be received by the slave system (2), but the parameter vector of the master system (1) may not be known a priori, even waits for identifying. To estimate all unknown parameters, by adding the controller U to the slave system (2), we have the following controlled slave system

$$\dot{y}(t) = -\bar{C}h(y(t)) + \bar{A}f(y(t)) + \bar{B}g(y(t-\tau)) + u + U(t : x(t), y(t)). \qquad (3)$$

Therefore, the goal of control is to design and implement an appropriate controller U for the slave system and an adaptive law of the parameters $\bar{C}, \bar{A}$ and $\bar{B}$, such that the controlled slave system (3) could be synchronous with the master system (1), and all the parameters $\bar{C} \to C, \bar{A} \to A$ and $\bar{B} \to B$ as $t \to +\infty$.

3   Main Results

Theorem 1   *Let the controller $U(t : x(t), y(t)) = \varepsilon(y(t) - x(t)) = \varepsilon e(t)$, where the feedback strength $\varepsilon = \text{diag}(\varepsilon_1, \ldots, \varepsilon_n)$ with the following update law*

$$\dot{\varepsilon}_i = -\delta_i e_i^2(t) \exp(\mu t), \quad i = 1, 2, \cdots, n, \qquad (4)$$

and the synchronization error $e(t) = (e_1(t), \cdots, e_n(t))^{\top}$, and an adaptive law of the parameters $\bar{C} = \text{diag}(\bar{c}_1, \ldots, \bar{c}_n), \bar{A} = (\bar{a}_{ij})_{n \times n}$ and $\bar{B} = (\bar{b}_{ij})_{n \times n}$ can be chosen as below:

$$\begin{cases} \dot{\bar{c}}_i = \gamma_i \, e_i(t) h_i(y_i(t)) \exp(\mu t), \quad i = 1, 2, \cdots, n. \\ \dot{\bar{a}}_{ij} = -\alpha_{ij} \, e_i(t) f_j(y_j(t)) \exp(\mu t), \quad i, j = 1, 2, \cdots, n. \\ \dot{\bar{b}}_{ij} = -\beta_{ij} \, e_i(t) g_j(y_j(t-\tau_{ij})) \exp(\mu t), \quad i, j = 1, 2, \cdots, n. \end{cases} \qquad (5)$$

in which $\mu \geq 0$ is a enough small number properly selected, $\delta_i > 0, \gamma_i > 0 \, (i = 1, 2, \cdots, n)$ and $\alpha_{ij} > 0, \beta_{ij} > 0 \, (i, j = 1, 2, \cdots, n)$ are arbitrary constants, respectively. If one of the following conditions is satisfied, then the controlled slave system (3) is globally synchronous with the master system (1) and satisfies

$$|e(t)|_2 = \left[\sum_{i=1}^{n} (y_i(t) - x_i(t))^2 \right]^{\frac{1}{2}} = O(\exp(-\frac{\mu}{2}t)). \qquad (6)$$

Moreover, $\lim_{t \to +\infty} (\bar{c}_i - c_i) = \lim_{t \to +\infty} (\bar{a}_{ij} - a_{ij}) = \lim_{t \to +\infty} (\bar{b}_{ij} - a_{ij}) = 0$ for all $i, j = 1, 2, \cdots, n$.

1) Assume that (A_0) and (A_1) hold, and that there exist n positive numbers $p_1, \cdots, p_n$ and two positive numbers $r_1 \in [0, 1], r_2 \in [0, 1]$. Let $\alpha_i \overset{\text{def}}{=} -c_i \eta_i p_i +$

$$\frac{1}{2}\sum_{\substack{j=1\\j\neq i}}^{n}\left(p_i|a_{ij}|k_j^{2r_1}+p_j|a_{ji}|k_i^{2(1-r_1)}\right)\;+\;\frac{1}{2}\sum_{j=1}^{n}\left(p_i|b_{ij}|l_j^{2r_2}+p_j|b_{ji}|l_i^{2(1-r_2)}\right),$$

$i=1,2,\cdots,n.$ *such that*

$$((a_{ii})^{+}k_i-\rho_i)p_i+\alpha_i<0,\quad i=1,2,\cdots,n. \tag{7}$$

where $(a_{ii})^{+}=\max\{a_{ii},0\}$ *and* $\rho_i\geq-\lim_{t\to+\infty}\varepsilon_i(t)=-\varepsilon_{i0}>0$, $i=1,2,\cdots,$
$n,$ *are constants properly selected.*

    2) Assume that (A_0) *and* (A_2) *hold, and*

$$(|a_{ii}|k_i-\rho_i)p_i+\alpha_i<0,\quad i=1,2,\cdots,n. \tag{8}$$

where $\rho_i\geq-\varepsilon_{i0}>0$, $i=1,2,\cdots,n,$ *are constants properly selected.*

Brief Proof. 1)   Let $e(t)=y(t)-x(t)$ be the synchronization error between the controlled slave system (3) and the master system (1), one can get the error dynamical system as follows:

$$\dot{e}(t)=-C\bar{h}(e(t))+A\bar{f}(e(t))+B\bar{g}(e(t-\tau))-(\bar{C}-C)h(y(t))$$
$$+(\bar{A}-A)f(y(t))+(\bar{B}-B)g(y(t-\tau))+\varepsilon e(t). \tag{9}$$

where $\bar{h}(e(t))=h(x(t)+e(t))-h(x(t)),\bar{f}(e(t))=f(x(t)+e(t))-f(x(t))$ and $\bar{g}(e(t))=g(x(t)+e(t))-g(x(t)).$

    Now construct a Lyapunov functional of the following form

$$V(t)=\frac{1}{2}\sum_{i=1}^{n}p_i\left\{e_i^2(t)\exp(\mu t)+2\sum_{j=1}^{n}|b_{ij}|l_j^{2(1-r_2)}\int_{t-\tau_{ij}}^{t}e_j^2(s)\exp(\mu(s+\tau_{ij}))\,ds\right.$$

$$\left.+\frac{1}{\gamma_i}(\bar{c}_i-c_i)^2+\sum_{j=1}^{n}\frac{1}{\alpha_{ij}}(\bar{a}_{ij}-a_{ij})^2+\sum_{j=1}^{n}\frac{1}{\beta_{ij}}(\bar{b}_{ij}-b_{ij})^2+\frac{1}{\delta_i}(\varepsilon_i+l_i)^2\right\}.$$

Differentiating V with respect to time along the solution of (9), from (A_0), (A_1) and (7), then by some elementary but tedious computations, one has $\dot{V}(t)\leq\sum_{i=1}^{n}\left(((a_{ii}^0)^{+}k_i-\rho_i)p_i+\alpha_i'\right)e_i^2(t)\exp(\mu t)\leq 0$, in which $\alpha_i'\overset{\text{def}}{=}\left(-c_i\eta_i+\frac{\mu}{2}\right)p_i+\frac{1}{2}\sum_{\substack{j=1\\j\neq i}}^{n}\left(p_i|a_{ij}|k_j^{2r_1}+p_j|a_{ji}|k_i^{2(1-r_1)}\right)+\frac{1}{2}\sum_{j=1}^{n}\left(p_i|b_{ij}|l_j^{2r_2}+p_j|b_{ji}|l_i^{2(1-r_2)}\right.$
$\left.\exp(\mu\tau_{ji})\right),i=1,2,\cdots,n.$ It is obvious that $\dot{V}=0$ if and only if $e_i=0,i=1,2,\cdots,n.$ Note that the construction of the Lyapunov functional implies the boundedness of all the $e_i(t)$. According to the well-known Liapunov-LaSall type theorem for functional differential equations [10], the trajectories of the error dynamical system (9), starting with arbitrary initial value, converges asymptotically to the largest invariant set E contained in $\dot{V}=0$ as $t\to\infty$, where the set $E=\{e=0\,|\,\bar{C}=C,\bar{A}=A,\bar{B}=B,\varepsilon=\varepsilon_0\in R^n\}$, the feedback coupling strength $\varepsilon_0=\text{diag}(\varepsilon_{10},\ldots,\varepsilon_{n0})$ depends on the initial value of (9), and the unknown parameters $\bar{C},\bar{B}$ and $\bar{A}$ with arbitrary initial values will approximate asymptotically the parameters identification values A,B and C of the master system, respectively. By the same arguments as in the proof of Theorem

1 in [7], one can also derive the inequality $|e(t)|_2 = \left[\sum_{i=1}^{n}(y_i(t) - x_i(t))^2\right]^{\frac{1}{2}} \le$ $N\|\phi\|_2 \exp(-\frac{\mu}{2}t))$, $N > 0$, which implies clearly (6). 2) The proof of 2) is precisely the same as that for 1) by employing the same Lyapunov functional as above from (A_0) and (A_2). In summary, this completes the proof of Theorem 1.

It can be seen that Theorems 1 supplies a systematic and analytical procedure for determining adaptive synchronization based on parameters identification of some well-known recurrently delayed neural networks such as delayed Hopfield neural networks and delayed cellular neural networks (CNNs). In particularly, the Lyapunov functional technique employed here can guarantee the global exponential stability of the synchronization error system, and also gives the corresponding exponential convergence rate. So just as stated in [5] that this estimation approach is not only robust against the effect of noise, but also able to response dynamically to changes in identifying parameters of the master system. It is useful to point out that our method can be applied to almost all recurrently delayed neural networks with the uniform Lipschitz activation functions. Therefore, the approach developed here is very convenient to implement in practice.

Example 1. In order to verify the effectiveness of the proposed method, let the master output signals are from the delayed neural networks (1), i.e., $\dot{x}(t) = -Ch(x(t)) + Af(x(t)) + Bg(x(t-\tau))$, in which $x(t) = (x_1(t), x_2(t))^\top, h(x(t)) = x(t)$, $f(x(t)) = g(x(t)) = (\tanh(x_1(t)), \tanh(x_2(t)))^\top, \tau = (1), C = \begin{bmatrix} 1 & 0 \\ 0 & 1 \end{bmatrix}, A = \begin{bmatrix} 2.0 & -0.1 \\ -5.0 & 3.0 \end{bmatrix}$ and $B = \begin{bmatrix} -1.5 & -0.1 \\ -0.2 & -2.5 \end{bmatrix}$. It should be noted that the networks is actually a typical chaotic delayed Hopfield neural networks [9]. For simplicity, we assume that only the four parameters a_{11}, a_{22}, b_{11} and b_{22} will be identified. By taking $\eta_i = k_i = l_i = p_i = 1, r_i = \dfrac{1}{2}, i = 1, 2$, it is easy to verify that if $\rho_i > 10$, $i = 1, 2$, then the conditions of Theorem 1 are satisfied. According to Theorem 1, one can easily construct the controlled slave system (3) with the feedback strength update law (4) and the parameters adaptive law (5), in which $\mu = 0.0035, \delta_i = \alpha_{ii} = \beta_{ii} = 8, i = 1, 2$, respectively. Let the initial condition of the feedback strength and the unknown parameters of the controlled slave system as follows: $(\varepsilon_1(0), \varepsilon_2(0))^\top = (-6.0, -6.0)^\top$, $(\bar{a}_{11}(0), \bar{a}_{22}(0), \bar{b}_{11}(0), \bar{b}_{22}(0))^\top = (-2.0, 5.0, -1.0, 0.8)^\top$, respectively. Numerical simulation shows that the parameters identification and adaptive synchronization are achieved successfully (see *Fig. 1*).

4   Conclusions

In this paper, we introduce an adaptive synchronization and parameters identification method for uncertain delayed neural networks. It is shown that the approach developed here further extends the ideas and techniques presented in the recent literature, and it is also simple to implement in practice. Numerical experiment shows the effectiveness of the proposed method.

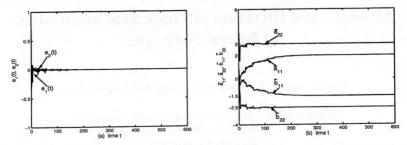

Fig. 1. Adaptive synchronization and parameters identification of the coupled delayed Hopfield neural networks (3) and (1) based on the feedback strength update law (4) and the parameters adaptive law (5) in time interval $[0, 600]$. (a) Graphs of synchronization errors varying with time, where $e_i(t) = y_i(t) - x_i(t)$, $i = 1, 2$. (b) Graphs of parameters identification results: $\bar{a}_{11}, \bar{a}_{22}, \bar{b}_{11}, \bar{b}_{22}$.

Acknowledgements

This work was supported by the National Science Foundation of China (Grant no. 60474071), the China Postdoctoral Science Foundation (Grant no. 20040350121) and the Science Foundation of Education Commission of Hebei Province (Grant no. 2003013).

References

1. Pecora, L. M., Carrol, T. L.: Synchronization in Chaotic Systems. Phys. Rev. Lett. **64** (1990) 821-823
2. Pecora, L. M., Carrol, T. L., Johnson, G. A.: Fundamentals of Synchronization in Chaotic Systems, Concepts, and Applications. Chaos, **7** (1998) 520-543
3. Chen, G., Dong, X.: From Chaos to Order: Methodologies, Perspectives, and Applications. World Scientific Pub. Co, Singapore (1998)
4. Chen, S., Hu, J., Wang, C., Lü, J.: Adaptive Synchronization of Uuncertain Rössler Hyperchaotic System Based on Parameter Identification. Phys. Lett. A. **321** (2004) 50-55
5. Huang, D.: Synchronization Based Estimation of All Parameters of Chaotic Systems From Time Series, Phys. Rev. E. **69** (2004) 067201
6. Cao, J., Wang, J.: Global Asymptotic Stability of A General Class of Recurrent Neural Networks with Time-varying Delays. IEEE Trans. CAS-I. **50** (2003) 34-44
7. Zhou, J., Liu, Z., Chen, G.: Dynamics of Periodic Delayed Neural Networks. Neural Networks, **17** (2004) 87-101
8. Zhou, J., Chen, T., Xiang, L.: Robust Synchronization of Coupled Delayed Recurrent Neural Networks. In: Yin, F., Wang, J., Guo, C. (eds.): Advances in Neural Networks - ISNN 2004. Lecture Notes in Computer Science, Vol. 3173. Springer-Verlag, Berlin Heidelberg New York (2004) 144-149
9. Chen, G., Zhou, J., Liu, Z.: Global Synchronization of Coupled Delayed Neural Networks and Applications to Chaotic CNN Model. Int. J. Bifur. Chaos, **14** (2004) 2229-2240
10. Kuang, Y.: Delay Differential Equations with Application in Population Dynamics, Academic Press, INC. New York (1993)

Strength and Direction of Phase Synchronization of Neural Networks

Yan Li, Xiaoli Li, Gaoxiang Ouyang, and Xinping Guan

Institute of Electrical Engineering, Yanshan University,
Qinhuangdao, 066004, China
xlli@ysu.edu.cn

Abstract. This paper studies the strength and direction of phase synchronization among Neural Networks (NNs). First, a nonlinear lumped-parameter cerebral cortex model is addressed and used to generate epileptic surrogate EEG signals. Second, a method that can be used to calculate the strength and direction of phase synchronization among NNs is described including the phase estimation, synchronization index and phase coupling direction. Finally, simulation results show the method addressed in this paper can be used to estimate the phase coupling direction among NNs.

1 Introduction

Phase synchronization is a universal concept of nonlinear sciences [1]. Phase synchronization phenomena are abundant in science, nature, engineering, and social life. In particular, phase synchronization is becoming a very useful analysis tool in life science. Many methods have been proposed to detect the phase coupling strength direction between two time series without a prior knowledge of the system [2],[3],[4]. For instance, the interaction of cardiorespiratory system is studied by above methods [5],[6]. [7]-[10] used synchronization definition to predict epileptic seizures. In neuroscience, phase synchronization method often only describe the phase coupling strength between two EEGs, but do not pay attention to the direction of phase coupling of both EEGs. To find out the localization of epileptic seizures, we hope to gain a new method to identify the direction of phase synchronization among NNs.

In this paper, we study the phase coupling direction and strength of two surrogate EEG signals, which are generated from a nonlinear lumped-parameter cerebral cortex model proposed by Lopes Da Silva et al. [11]. Simulation result shows the evolution map approach (EMA) proposed by [2],[5] can be applied to detect the phase coupling direction.

2 Phase Synchronization Strength and Direction

Prior to the analysis of phase relations, we have to estimate phases from a time series $s(t)$. Often, continuous wavelet transform and Hilbert transform (HT) can be used to estimate the phase of a time series [12]. In this paper, HT is applied to estimate the

J. Wang, X. Liao, and Z. Yi (Eds.): ISNN 2005, LNCS 3496, pp. 314–319, 2005.

phase of EEG data. The complex analytic signal $s_H(t)$ of the time series $s(t)$ and the estimated phase are below:

$$s_H(t) = \pi^{-1} P.V. \int_{-\infty}^{\infty} \frac{s(\tau)}{t-\tau}, \quad \phi(t) = \arctan \frac{s_H(t)}{s(t)}. \tag{1}$$

where $s_n(t)$ is HT of $s(t)$; $P.V.$ denotes the Cauchy principal value.

Often, mean phase couple method is applied to calculate the synchronization index for estimating the strength of phase coupling. The advantage of this measure is the absence of parameters [7], namely we only use the phases of two time series to estimate the phase synchronization strength and direction without using other parameters. The equation is below:

$$\rho(1,2) = \sqrt{\langle \cos(\phi_1 - \phi_2) \rangle^2 + \langle \sin(\phi_1 - \phi_2) \rangle^2}. \tag{2}$$

Where ϕ_1, ϕ_2 is the phase of time series 1 and 2, respectively. $\langle \bullet \rangle$ denote average over time. The value of $\rho(1,2)$ is between [0,1], if the value is zero, it means no locking. If the value of $\rho(1,2)$ is 1, it means there is a perfect phase synchronization between time series.

There exist some methods to calculate phase coupling direction [2],[3],[4],[7]. Such as, Palus [7] used an information-theoretic approach to detect the direction of phase coupling. In this paper, EMA proposed by Rosenblum and Pikovsky [2],[5] is applied to calculate the phase coupling direction of two EEGs. This method is based on empirical construction of model maps to describe phase dynamics of the two subsystems. Two original time series $\{x_{1,2}(t_i)\}_{i=1}^{N_x}$, $t_i = i\Delta t, i = 1,...,N_x$; Δt is a sampling interval. Their phase can be estimated by HT; $\{\phi_{1,2}(t_i)\}_{i=1}^{N_x}$. A global model map, which characterizes the dependence of phase increments (over a time interval $\tau\Delta t$) on the phases of subsystems' oscillations, is constructed as :

$$\Delta_{1,2} = \phi_{1,2}(t + \tau\Delta t) - \phi_{1,2}(t) = F_{1,2}(\phi_{2,1}(t), \phi_{1,2}(t), \mathbf{a}_{1,2}). \tag{3}$$

where t is a positive integer, $F_{1,2}$ are trigonometric polynomials, and $\mathbf{a}_{1,2}$ are vectors of their coefficients. The coefficients $\mathbf{a}_{1,2}$ can be obtained from by the least-squares routine (LSR). The intensity of influence of the second NN on the first one, c_1, is determined by the steepness of the dependence of F_1 on the phase of the second NN. Same principle, the intensity of influence of the first NN on the second one can be defined as c_2. The estimator of c_1 and c_2 are given as follows:

$$c_{1,2}^2 = \iint_0^{2\pi} (\frac{\partial F_{1,2}}{\partial \phi_{2,1}})^2 d\phi_1 d\phi_2 = \sum_{i=1}^{L_1} n_i^2 (a_{1i,2i})^2. \tag{4}$$

one computes intensities of influence of the second subsystem on the first one $(2 \to 1) c_1$ and of the first subsystem on the second one $(1 \to 2) c_2$ and directionality index $d = (c_2 - c_1)/(c_2 + c_1)$. Since $c_{1,2} > 0$, d takes the values within the interval $[-1,1]$ only: d=1 or d=-1 corresponds to unidirectional coupling $1 \to 2$ or $2 \to 1$, respectively, and $d = 0$ for ideally symmetric coupling.

3 Nonlinear Lumped-Parameter NN Models

The nonlinear lumped-parameter NN model is used in this paper to generate epileptic surrogate EEG signals. The model was first proposed by Lopes Da Silva et al. [11] and has been improved and extended later by Jansen et al. [13],[14] and by Wendling et al. [15] to generate EEG signals. The lumped-parameters of excitatory and inhibitory neuron ratio A/B and the connective strength are used to modulate the balance of excitation and inhibition, the degree and direction of couplings between different NNs. In each NN model, two dynamic linear transfer functions $h_e(t)$ and $h_i(t)$ are used to simulate excitatory and neurons, respectively, which transfers the pre-synaptic action potentials to the postsynaptic membrane potentials [15],[16]. When the neuron has been fired up, a static nonlinear sigmoid function [15] used to transfer the average membrane potential into the average pulse density of potentials, which will be conduct to other neurons along the axon of the firing neuron.

In this paper, three coupled neural networks are shown in Fig. 1. The combination of three NNs may simulate the propagation of electrical signals among neuron networks located at different areas of the brain.

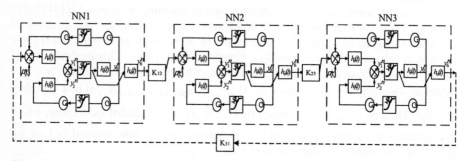

Fig. 1. The multiple coupled NNs model. K_{12} and K_{23} are the connectivity between NN1 and NN2, and NN2 and NN3 respectively. One-directional situation without K_{31}, circle situation if join K_{31}. K_{ij}: Connectivity constant associated with the connection between NNi and NNj. Each of NN is discribed by excitation function $h_e(t) = u(t)Aate^{-at}$. and restrain function $h_i(t) = u(t)Bbte^{-bt}$.

4 Results

In this paper, the surrogate EEG data are generated from three NN lumped-parameter models that are connected each other. We mainly study two different types EEG signals, one is from a unidirectional mode (NN1→NN2→NN3), another is from a loop mode NN1→NN2→NN3→NN1. The main aim of this paper is to investigate if the method proposed can estimate the phase coupling direction.

The ratio of A/B control the degree of excitability of neurons. The exciting level grows proportionally with the increasing of the ratio. In this paper, we only change the parameter A of NN 1, it grows from 3.25 (0-150s) up to 3.585. The parameters B=22 mv, K_{12}=200, K_{23} is a normal Gaussian distributed random coefficient (mean=130,variance=50), the other parameters of the model are adapted from [13,

15]. Fig. 2 (c) shows three EEG data without K_{13}. Seeing Fig 2.(a), NN1 generate spike first, then the NN 1 lead to the NN 2, the NN2 lead to the NN3, NN3 occur a seizure at last, there is a time delay among the NN1, NN2, NN3. The simulated data is analyzed using a moving-window of the length of 20 s with an overlap of 50%. When the value of A is 3.25 at the duration from 1s to 150s, all value d_{12} (d_{23}, d_{13}) fluctuate at the around of 0. That's because all parameters in the NNs are the same. When the value of A in the NN 1 grows up to 3.585 from 151s to the end, the value d_{12} (d_{23}, d_{13}) decrease down at the around of -1, that's mean the NN1 lead to NN2 and NN3, NN2 lead to NN3. Fig. 2 (e) shows the synchronization index among NNs. It is found that the value of ρ_{12} (ρ_{23}, ρ_{13}) is equal to 1 from 1s to 150s, because all of the NNs generate same EEG signals. From 151s to 400s, all the value of ρ_{12} (ρ_{23}, ρ_{13}) decreases to 0.8. This is due to the fact that the parameter A of NN 1graw up to 3.585, the synchronization among NNs decreases.

Do not change all of parameters, we add a reciprocal couple from NN 3 to NN 1 since 151s by using $K_{31}=200$. It causes the model to produce a different activity, as shown in Fig. 2 (d). Seeing Fig.2 (b), it is found that the model begins to generate sustained discharges of spikes, unlike sporadic spike in Fig. 2 (a). As can seen from Fig.2. (f), the value of synchronization index increased from 0.45 to 0.8. The value of directional index d_{12}(d_{23}, d_{13}) is at the around of zero and changes randomly before 150 s. It increases down -1 from 151s to 400s.

To further study the phase coupling direction among NNS. A value of B in NN1 often is 22mv, we try to decrease the value of B to 17.5mv from 150s. The NN model begins to generate sustained discharges of spikes. Three phase coupling direction indexes are at the round 1 when the parameter B decreases. That's mean the NN1 lead to NN2 and NN3. The final test is to change the connective strength K_{23} to be a normal Gaussian distributed random coefficient (mean=130,variance=50). It is found that the critical value of connect strength is 130, when K_{23} is less than 130, the model didn't generate spikes in the next NN. When K_{23} is greater than 200, the excitation activity will lead to the next NNs. This experiment result shows when we choose Gaussian function instead of a fix number, the EMA method can also work well also, the phase coupling direction among NNs can be identified. the better. The reason of this phenomenon is when we use Gaussian function instead of a fix number. It decreased the synchronization degree among the NNs.

5   Conclusions

In this paper, we study the direction and strength of phase synchronization among NNs. Three nonlinear lumped-parameter cerebral cortex models are used to simulate epileptic EEG signals with changing some important parameters A, B and K of the model. It is found that these epileptic surrogate signals have different strength and direction of phase synchronization by using EMA method. The limitation of the EMA method is that it can not accurately identify the phase coupling direction when the synchronization index is rather high (greater than 0.6). Thus, this method still needs to be improved or combine with other algorithm that performs better in the high coupling situation. It especially needs to be improved for short and noisy time series, such as actual human EEG signals.

318   Yan Li et al.

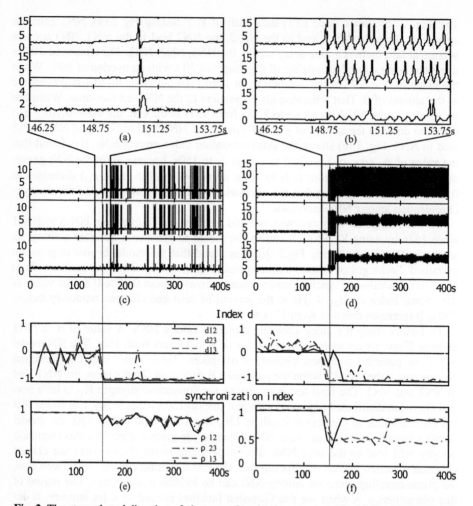

Fig. 2. The strength and direction of phase synchronization among NNs. (c) The EEG of unidirectional model(NN1→NN2→NN3). (top) EEG data of NN1; (middle) EEG data of NN2; (down) EEG data of NN3. (d) The EEG of back-loop model (NN1→NN2→NN3→NN1). (top) EEG data of NN1; (middle) EEG data of NN2; (down) EEG data of NN3. (e) the direction and strength of phase synchronization of unidirectional model. (top) The direction of phase synchronization; (down) the strength of phase synchronization. (f) the direction and strength of phase synchronization of back-loop model. (top) The direction of phase synchronization; (down) the strength of phase synchronization.

Acknowledgments

The support of National Science Foundation of China (60274023) and Important Subject Foundation of Hebei Province is gratefully acknowledged.

References

1. Pikovsky, A., Rosenblum, G.M., Kurths, J.: Synchronization, A Universal Concept in Nonlinear Sciences. Cambridge University Press. Cambridge (2001)
2. Rosenblum, G.M., Pikovsky, A.: Detecting Direction of Coupling in Interacting Oscillators Phys. Rev. E, **64** (2001)045202
3. Palus, M.: Detecting Phase Synchronization in Noisy Systems. Phys.Lett. A, **235** (1997) 341-351
4. Smirnov, D.A., Bezruchko, B.P.: Estimation of Interaction Strength and Direction from Short and Noisy Time Series. Phys. Rev. E, **68** (2003)
5. Rosenblum, G.M.: Identification of Coupling Direction: Application to Cardiorespiratory Interaction. Phys. Rev. E, **65** (2002)
6. Schafer, C., Rosenblum, G.M., Abel,H.H., Kurths, J.: Synchronization in the Human Cardiorespiratory System, Phys. Rev. E, **60** (1999) 857-870
7. Mormann, F., Andrzejak, R.G., Kreuz, T., Rieke, Ch., David, P., Elger, Ch. E., Lehnertz, K.: Automated Detection of a Preseizure State Based on a Decrease in Synchronization in Intracranial Electroencephalogram Recordings from Epilepsy Patients, Phys. Rev. E, **67** (2003)
8. Li, X., Li, J., Yao, X.: Complexity and Synchronization of EEG during Epileptic Seizures. IEEE Workshop on Life Science Data Mining (ICDM 2004 UK) (2004)
9. Li, X., Ouyang, G., Yao, X., Guan, X.: Dynamical Characteristics of Pre-epileptic Seizures in Rats with Recurrence Quantification Analysis. Phys. Lett. A, **333** (2004) 164-171
10. Li, X., Guan, X., Ru, D.: Using Damping Time for Epileptic Seizures Detection in EEG. Modelling and Control in Biomedical Systems. Elsevier Ltd (2003) 255-258
11. Lopes da Silva, F.H., Hoek, A., Smith, H., Zetterberg, L. H.: Model of Brain Rythmic Activity. Kybernetic (1974) 15-27
12. Rosenblum, G.M., Pikovsky, A., Kurths, J., Tass, P.A.: Phase Synchronization: From Theory to Data Analysis, in Neuro-informatics and Neural Modeling, volume 4 of Handbook of Biological Physics. Elsevier. Amsterdam (2001) 279-321
13. Jansen, B., Rit, V.G.: Electroencephalogram and Visual Evoked Potential Generation in a Mathermatical Model of Coupled Cortical Columns. Biol. Cybern (1995) 73-357
14. Jansen, B.H., Zouridakis, G., Brandt, M.E.: A Neuro-physiologically-based Mathematical Model of Flash Visual Evoked Potentials. Biol. Cybern (1993) 68-275
15. Wendling, F., Bellanger, J.J., Bartolomei, F., Chauvel, P.: Relevance of Nonlinear lumped-parameter Models in the Analysis of Depth-eeg Epileptic Signals. Biol. Cybern (2000) 83-367
16. Nicholls, J.G., Martin, A.R., Wllace, B.G., Fuchs, P.A.: From Neuron to Brain (2001)

Hopf Bifurcation in a Single Inertial Neuron Model: A Frequency Domain Approach

Shaorong Li[1], Shaowen Li[2,3], Xipeng Sun[2], and Jie Li[2]

[1] School of Opto-Electronic Information,
University of Electronic Science and Technology of China, Chengdu, Sichuan 610054, China
lsrxt@uestc.edu.cn
[2] Department of Mathematics, Southwestern University of Finance and Economics,
Chengdu, Sichuan 610074, China
lisw@swufe.edu.cn
[3] College of Electronic Engineering, University of Electronic Science and Technology of China,
Chengdu, Sichuan 610054, China

Abstract. In this paper, a single inertial neuron system with distributed delays for the weak kernel is investigated. By applying the frequency domain approach, the existence of bifurcation parameter point is determined. The direction and stability of the bifurcating periodic solutions are determined by the Nyquist criterion and the graphical Hopf bifurcation theorem. Some numerical simulations for justifying the theoretical analysis are also given.

1 Introduction

Neural networks [1] are complex and large-scale nonlinear dynamical systems. For simplicity, Schieve et al. [2] developed the notion of effective neurons. So, many researchers have directed their attention to the simple models only involving the effective neurons [3-7]. Wheeler and Schieve [3] considered an inertial neuron model:

$$M_1 \ddot{U}_1 = -\eta_1 \dot{U}_1 - K_1 U_1 + J_{11} \tanh(U_1) \tag{1}$$

where U_1 denotes the mean soma potential of the effective neuron, M_1, η_1, K_1 and J_{11} are non-zero real numbers. In this paper, an inertial neuron system with distributed delays and a weak kernel will be discussed:

$$\ddot{U}_1 = -\eta \dot{U}_1 - K U_1 + a \tanh[U_1 - b \int_0^{+\infty} F(\tau;\mu)U_1(t-\tau)d\tau] \tag{2}$$

where η, K, a and b are positive real numbers. Moreover, the memory function $F(\tau;\mu) = \mu e^{-\mu\tau}$ $(\mu > 0)$ is a weak kernel.

The organization of this paper is as follows. In Section 2, the existence of Hopf bifurcation of the system (2) with the weak kernel is determined and Hopf bifurcation occurs when the bifurcation parameter exceeds a critical value. In Section 3, by means of the frequency domain approach proposed in [8],[9],10], the direction of Hopf bifurcation and the stability of the bifurcating periodic solutions are analyzed. Some numerical simulation results and the frequency graph are presented in Section 4. Finally, some conclusions are made in Section 5.

J. Wang, X. Liao, and Z. Yi (Eds.): ISNN 2005, LNCS 3496, pp. 320–326, 2005.

2 Existence of Hopf Bifurcation

In this paper, system (2) with the weak kernel is considered. Define

$$x_1 = U_1 - b \int_0^{+\infty} F(\tau;\mu)U_1(t-\tau)d\tau ,$$

and let $x_2 = \dot{x}_1$, (2) become as:

$$\begin{cases} \dot{x}_1 = x_2 \\ \dot{x}_2 = -\eta x_2 - K x_1 + a\tanh(x_1) - ab \int_0^{+\infty} F(\tau;\mu)\tanh[x_1(t-\tau)]d\tau \end{cases} \tag{3}$$

The origin $(0,0)$ is the unique fixed points if and only if $K \geq a(1-b)$. Then, one discusses Hopf bifurcation at the origin $(0,0)$ when $K > a(1-b)$. Let

$$x_3 = \int_0^{+\infty} F(\tau;\mu)\tanh[x_1(t-\tau)]d\tau = \int_0^{+\infty} \mu e^{-\mu\tau}\tanh[x_1(t-\tau)]d\tau . \tag{4}$$

The non-linear system (3) can be changed to an ODE system:

$$\begin{cases} \dot{x}_1 = x_2 \\ \dot{x}_2 = -\eta x_2 - K x_1 + a\tanh(x_1) - abx_3 \\ \dot{x}_3 = -\mu x_3 + \mu\tanh(x_1) \end{cases} \tag{5}$$

To rewrite it in a matrix form as:

$$\begin{cases} \dot{x} = A(\mu)x + B(\mu)g(y;\mu) \\ y = -C(\mu)x \end{cases} , \tag{6}$$

where

$$x = \begin{pmatrix} x_1 \\ x_2 \\ x_3 \end{pmatrix}, A(\mu) = \begin{pmatrix} 0 & 1 & 0 \\ -K & -\eta & -ab \\ 0 & 0 & -\mu \end{pmatrix}, B(\mu) = \begin{pmatrix} 0 \\ a \\ \mu \end{pmatrix}, C(\mu) = (-1 \quad 0 \quad 0),$$

$$g(y;\mu) = \tanh(y) .$$

Next, taking a Laplace transform on (6) yields a standard transfer matrix of the linear part of the system

$$G(s;\mu) = C(\mu)[sI - A(\mu)]^{-1}B(\mu) = \frac{-a[s+(1-b)\mu]}{(s+\mu)(s^2+\eta s+K)} , \tag{7}$$

This feedback system is linearized about the equilibrium $y = 0$, then the Jacobian is given by $J(\mu) = \partial g / \partial y|_{y=0} = 1$. Then, by applying the generalized Nyquist stability criterion, with $s = i\omega$, the following results can be established:

Lemma 1. *[10] If an eigenvalue of the corresponding Jacobian of the nonlinear system, in the time domain, assumes a purely imaginary value $i\omega_0$ at a particular $\mu = \mu_0$, then the corresponding eigenvalue of the matrix $[G(i\omega_0;\mu_0)J(\mu_0)]$ in the frequency domain must assume the value $-1+i0$ at $\mu = \mu_0$.*

Let $\hat{\lambda}(i\omega;\mu)$ be the eigenvalue of $[G(i\omega;\mu)J(\mu)]$ that satisfies $\hat{\lambda}(i\omega_0;\mu_0) = -1+i0$. Then

$$\det[(-1)I - G(i\omega_0;\mu_0)J(\mu_0)] = -1 + \frac{a[i\omega_0 + (1-b)\mu_0]}{(i\omega_0 + \mu_0)(-\omega_0^2 + i\eta\omega_0 + K)} = 0. \tag{8}$$

One has

$$\mu_0^2 + (\eta - ab/\eta)\mu_0 + (K - a) = 0, \text{ and } \omega_0^2 = \eta\mu_0 + K - a. \tag{9}$$

Hence, the two roots of (9a) are

$$\mu_\pm = 1/2[-(\eta - ab/\eta) \pm \sqrt{(\eta - ab/\eta)^2 - 4(K - a)}]. \tag{10}$$

Considering that μ is the positive real number, we discuss four cases as follows:

Theorem 1. (*Existence of Hopf Bifurcation*)
   (i) If $-ab < K - a < 0$, then $\mu_+ > 0$ and $\mu_- < 0$. So, $\mu_0 = \mu_+$ is the unique Hopf bifurcation of system (3).
   (ii) If $K - a = 0$, then $\mu_+ = 0$ or $\mu_- = 0$. So, $\mu_0 = \mu_+$ is the unique Hopf bifurcation of system (3) when $\eta < \sqrt{ab}$, and the Hopf bifurcation of system (3) doesn't exist when $\eta \geq \sqrt{ab}$.
   (iii) If $0 < K - a < 1/4 \cdot (\eta - ab/\eta)^2$, then $\mu_+ > 0$, $\mu_- > 0$, or $\mu_+ < 0$, $\mu_- < 0$. So, then $\mu_0 = \mu_+$ and $\mu_0 = \mu_-$ are the Hopf bifurcations of system (3) when $\eta < \sqrt{ab}$, and the Hopf bifurcation of system (3) doesn't exist when $\eta \geq \sqrt{ab}$.
   (iv) If $K - a \geq 1/4 \cdot (\eta - ab/\eta)^2$, then μ_+ and μ_- are complex conjugates or $\mu_+ = \mu_-$. So, the Hopf bifurcation of system (3) doesn't exist.

3   Stability of Bifurcating Periodic Solutions and the Direction of Bifurcation

In order to study stability of bifurcating periodic solutions, the frequency-domain formulation of Moiola and Chen [10] is applied in this section. We first define an auxiliary vector

$$\xi_1(\tilde{\omega}) = -w^T[G(i\tilde{\omega};\tilde{\mu})]p_1/(w^T v), \tag{11}$$

where $\tilde{\mu}$ is the fixed value of the parameter μ, $\tilde{\omega}$ is the frequency of the intersection between the characteristic locus $\hat{\lambda}(i\omega; \tilde{\mu})$ and the negative real axis closet to the point $(-1 + i0)$, w^T and v are the left and right eigenvectors of $[G(i\tilde{\omega}; \tilde{\mu})J(\tilde{\mu})]$, respectively, associated with the value $\hat{\lambda}(i\tilde{\omega}; \tilde{\mu})$, and

$$p_1 = D_2 (V_{02} \otimes v + 1/2 \cdot \overline{v} \otimes V_{22}) + 1/8 \cdot D_3 v \otimes v \otimes \overline{v}, \tag{12}$$

where $\otimes$ is the tensor product operator, and

$$D_2 = \partial^2 g(y; \tilde{\mu}) / \partial y^2 \big|_{y=0}, \quad D_3 = \partial^3 g(y; \tilde{\mu}) / \partial y^3 \big|_{y=0},$$
$$V_{02} = -1/4 \cdot [I + G(0; \tilde{\mu}) J(\tilde{\mu})]^{-1} G(0; \tilde{\mu}) D_2 v \otimes \overline{v}, \tag{13}$$
$$V_{22} = -1/4 \cdot [I + G(2i\tilde{\omega}; \tilde{\mu}) J(\tilde{\mu})]^{-1} G(2i\tilde{\omega}; \tilde{\mu}) D_2 v \otimes v.$$

Now, the following Hopf Bifurcation Theorem formulated in the frequency domain can be established:

Lemma 2. *[10] Suppose that the locus of the characteristic function $\hat{\lambda}(s)$ intersects the negative real axis at $\hat{\lambda}(i\tilde{\omega})$ that is closest to the point $(-1 + i0)$ when the variable s sweeps on the classical Nyquist contour. Moreover, suppose that $\xi_1(\tilde{\omega})$ is nonzero and the half-line L_1 starting from $(-1 + i0)$ in the direction defined by $\xi_1(\tilde{\omega})$ first intersects the locus of $\hat{\lambda}(i\omega)$ at $\hat{\lambda}(i\hat{\omega}) = \hat{P} = -1 + \xi_1(\tilde{\omega})\theta^2$, where $\theta = O(|\mu - \mu_0|^{1/2})$. Finally, suppose that the following conditions are satisfied that the eigenlocus $\hat{\lambda}$ has nonzero rate of change with respect to its parameterization at the criticality (ω_0, μ_0), and the intersection is transversal. Then, the system (3) has a periodic solution $y(t)$ of frequency $\omega = \hat{\omega} + O(\hat{\theta}^4)$. Moreover, by applying a small perturbation around the intersection $\hat{P}$ and using the generalized Nyquist stability criterion, the stability of the periodic solution $y(t)$ can be determined.*

According to Lemma 2, ones can determine the direction of Hopf bifurcation and the stability of the bifurcating periodic solution by drawing the figure of the half-line L_1 and the locus $\hat{\lambda}(i\omega)$. As $g(y; \mu) = \tanh(y)$ and $J(\mu) = 1$, one has

$$G(i\tilde{\omega}; \tilde{\mu}) = \hat{\lambda}(i\tilde{\omega}; \tilde{\mu}), \text{ and } D_2 = 0, \; D_3 = -2. \tag{14}$$

Furthermore, the left and right eigenvectors for the Jacobian $[G(i\tilde{\omega}; \tilde{\mu})J(\tilde{\mu})]$ associated with the eigenvalue $\hat{\lambda}(i\tilde{\omega}; \tilde{\mu})$ for any fixed $\tilde{\mu}$ is $w = v = 1$ in this scalar case. So, one obtains:

$$\xi_1(\tilde{\omega}) = -w^T [G(i\tilde{\omega}; \tilde{\mu})] p_1 / w^T v = 1/4 \cdot \hat{\lambda}(i\tilde{\omega}; \tilde{\mu}). \tag{15}$$

If the bifurcation parameter μ be perturbed slightly from μ_0 to $\tilde{\mu}$, then $\tilde{\lambda} = \lambda(i\tilde{\omega}; \tilde{\mu})$ is a real number near -1. So, the half-line L_1 is in the direction of the negative real axis. As $\bar{\omega}$ is the frequency of the intersection between the characteristic locus $\hat{\lambda}(i\omega\,;\bar{\mu})$ and the negative real axis closet to the point $(-1+i0)$, then the bifurcating periodic solution exists if $\tilde{\lambda}<-1$, and the direction of the Hopf bifurcation is +1 (resp. −1) if $\tilde{\lambda}<-1$ when $\bar{\mu}>\mu_0$ (resp. $\bar{\mu}<\mu_0$). Noticing $\lambda(i\omega_0;\mu_0)=-1$, the following results can be obtained:

Theorem 2. If $\left.d\tilde{\lambda}/d\mu\right|_{\mu=\mu_0}<0$ (resp. $\left.d\tilde{\lambda}/d\mu\right|_{\mu=\mu_0}>0$), the direction of the Hopf bifurcation of the system (2) is +1 (resp. −1), i.e., $\mathrm{sgn}\left(-d\tilde{\lambda}/d\mu\big|_{\mu=\mu_0}\right)$.

By means of (8), one has

$$\lambda(s) = -\frac{a[s+(1-b)\mu]}{(s+\mu)(s^2+\eta s+K)}.\qquad(16)$$

So, the number of the poles of $\lambda(s)$ that have positive real parts is zero.

Theorem 3. Let k be the total number of anticlockwise encirclements of $\hat{\lambda}(i\omega;\bar{\mu})$ to the point $P_1=\hat{P}+\varepsilon\xi_1(\bar{\omega})$ for a sufficiently small $\varepsilon>0$. the bifurcating periodic solutions of system (3) is stable if $k=0$, and unstable if $k\neq0$.

4   Numerical Examples and the Frequency Graph

In this section, some numerical simulations for justifying the theoretical analysis are also given. The direction of the Hopf bifurcation and the stability of the bifurcating periodic solutions is determined by the graphical Hopf bifurcation theorem.

If $a=4$, $b=0.75$, $K=4.5$, $\eta=1$, one can easily calculate $\mu_-=0.2929$, $\mu_+=1.7071$. Set $\mu_0=0.2929$, then the bifurcation of the Hopf bifurcation is +1. The cases of $\mu=0.2$ and $\mu=0.4$ are simulated as Fig. 1 and 2.

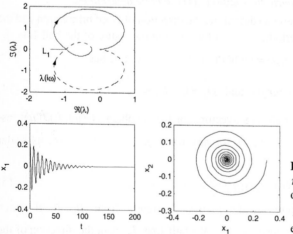

Fig. 1. $a=4$, $b=0.75$, $K=4.5$, $\eta=1$, $\mu=0.2$. The half-line L_1 does not intersect the locus $\hat{\lambda}(i\omega)$, so no periodic solution exists.

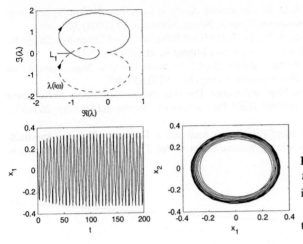

Fig. 2. $a = 4$, $b = 0.75$, $K = 4.5$, $\eta = 1$, $\mu = 0.4$. The half-line L_1 intersects the locus $\hat{\lambda}(i\omega)$, and $k = 0$, so a stable periodic solution exists.

5 Conclusions

An inertial neuron system with distributed delays usually provides very rich dynamical behaviors. For this kind of system, by using the average time delay as a bifurcation parameter, we have proved that a Hopf bifurcation occurs when this parameter passes through a critical value, showing that a family of periodic orbits bifurcates from the origin. The stability and direction of the bifurcating periodic orbits have also been analyzed by the frequency-domain methodology. It is deemed that the frequency-domain approach has great potential in bifurcation analysis of various delayed nonlinear systems.

Acknowledgements

The work described in this paper was supported by the Scientific Research Fund of Southwestern University of Finance and Economics (No. 04X16).

References

1. Hopfield J. J.: Neurons with Graded Response Have Collective Computational Properties Like Those of Two-State Neurons. Proceedings of the National Academy of Sciences of the USA, **81** (1984) 3088–3092
2. Schieve, W. C., Bulsara, A. R., Davis, G. M.: Single Effective Neurons. Physical Review A, **43** (1991) 2613-2623
3. Liao, X. F., Wong, K. W., Wu, Z. F.: Bifurcation Analysis on A Two-neuron System with Distributed Delays. Physica D, **149** (2001) 123-141
4. Diek, W. Wheeler, Schieve, W. C.: Stability and Chaos in An Inertial Ttwo Neuron System. Physica D, **105** (1997) 267-284
5. Liao, X. F., Li, S. W., Wong, K. W.: Hopf Bifurcation on A Two-Neuron System with Distributed Delays: A Frequency Domain Approach. Nonlinear Dynamics, **31** (2003) 299-326

6. Liao, X. F., Li, S. W., Chen, G. R.: Bifurcation Analysis on A Two-Neuron System with Distributed Delays in the Frequency Domain. Neural Networks, **17** (2004) 545–561
7. Li, C. G.., Chen G.. R., Liao, X. F., Yu, J. B.: Hopf Bifurcation and Chaos in a Single Inertial Neuron Model with Time Delay. Physics Journal B, **41** (2004) 337-343
8. Allwright, D. J.: Harmonic Balance and the Hopf Bifurcation Theorem. Mathematical Proceedings of the Cambridge Philosophical Society, **82** (1977) 453-467
9. Mees, A. I., Chua, L. O.: The Hopf Bifurcation Theorem and Its Applications to Nonlinear Oscillations in Circuits and Systems. IEEE Transactions on Circuits and Systems, **26** (1979) 235-254
10. Moiola, J. L., Chen, G. R.: Hopf Bifurcation Analysis: A Frequency Domain Approach. World Scientific Series on Nonlinear Science Series A, Vol. 21, World Scientific, Singapore (1996)

Hopf Bifurcation in a Single Inertial Neuron Model with a Discrete Delay

Shaowen Li[1,2] and Shaorong Li[3]

[1] Department of Mathematics, Southwestern University of Finance and Economics,
Chengdu, Sichuan 610074, China
lisw@swufe.edu.cn
[2] College of Electronic Engineering, University of Electronic Science and Technology of China,
Chengdu, Sichuan 610054, China
[3] School of Opto-Electronic Information,
University of Electronic Science and Technology of China,
Chengdu, Sichuan 610054, China
lsrxt@uestc.edu.cn

Abstract. In this paper, a single inertial neuron system with distributed delays for the weak kernel is investigated. By applying the frequency domain approach, the existence of bifurcation parameter point is determined. The direction and stability of the bifurcating periodic solutions are determined by the Nyquist criterion and the graphical Hopf bifurcation theorem. Some numerical simulations for justifying the theoretical analysis are also given.

1 Introduction

Neural networks [1] are complex and large-scale nonlinear dynamical systems. To simplify the analysis and computation, many researchers have directed their attention to the study of simple systems [2-7]. Li et al. [2] considered a delayed differential equation modelling a single neuron with inertial term subject to time delay:

$$\ddot{x} = -a\dot{x} - bx + cf(x - hx(t - \tau)). \tag{1}$$

where constants $a, b, c > 0$, $h \geq 0$, and $\tau > 0$ is the time delay. Hopf bifurcation is studied by using the normal form theory of retarded functional differential equations. When adopting a nonmonotonic activation function, chaotic behavior is observed.

In this paper, an inertial neuron system with distributed delays will be discussed:

$$\ddot{x}_1 = -\eta\dot{x}_1 - Kx_1(t - \mu) + L\tanh[x_1(t - \mu)]. \tag{2}$$

where $\eta, K, L > 0$, and $\mu > 0$ is the time delay. Hopf bifurcation of the system (2) is analyzed by using the frequency domain approach.

The organization of this paper is as follows. In Section 2, the existence of Hopf bifurcation of the system (2) with the weak and strong kernel is determined and Hopf bifurcation occurs when the bifurcation parameter exceeds a critical value. In Section 3, by means of the frequency domain approach proposed in [8], [9], [10], the direction of Hopf bifurcation and the stability of the bifurcating periodic solutions are analyzed. Some numerical simulation results and the frequency graph are presented in Section 4. Finally, some conclusions are made in Section 5.

J. Wang, X. Liao, and Z. Yi (Eds.): ISNN 2005, LNCS 3496, pp. 327–333, 2005.
© Springer-Verlag Berlin Heidelberg 2005

2   Existence of Hopf Bifurcation

In this paper, system (2) is considered. Let $x_2 = \dot{x}_1$, (2) becomes as:

$$\begin{cases} \dot{x}_1 = x_2 \\ \dot{x}_2 = -\eta x_2 - Kx_1(t-\mu) + L\tanh[x_1(t-\mu)] \end{cases}.$$ (3)

The origin $(0, 0)$ is the unique fixed points if and only if $K \geq L$. Then, one discusses Hopf bifurcation at the origin $(0, 0)$ when $K > L$. First, rewrite (3) in a matrix form as:

$$\begin{cases} \dot{x} = A_0 x + A_1 x(t-\mu) + Bg[y(t-\mu)] \\ y = -Cx \end{cases},$$ (4)

Where

$$x = \begin{pmatrix} x_1 \\ x_2 \end{pmatrix}, A_0 = \begin{pmatrix} 0 & 1 \\ 0 & -\eta \end{pmatrix}, A_1 = \begin{pmatrix} 0 & 0 \\ -K & 0 \end{pmatrix}, B = \begin{pmatrix} 0 \\ L \end{pmatrix}, C = (-1 \ \ 0), g(y) = \tanh(y).$$

Next, taking a Laplace transform on (4) yields a standard transfer matrix of the linear part of the system

$$G(s; \mu) = C[sI - A_0 - A_1 e^{-s\mu}]^{-1} B = -L/(s^2 + \eta s + Ke^{-s\mu}).$$ (5)

This feedback system is linearized about the equilibrium $y = 0$, then the Jacobian is given by $J(\mu) = \partial g / \partial y|_{y=0} = 1$. Then, by applying the generalized Nyquist stability criterion, with $s = i\omega$, the following results can be established:

Lemma 1. [10] *If an eigenvalue of the corresponding Jacobian of the nonlinear system, in the time domain, assumes a purely imaginary value $i\omega_0$ at a particular $\mu = \mu_0$, then the corresponding eigenvalue of the matrix $[G(i\omega_0; \mu_0)J(\mu_0)e^{-i\omega_0\mu_0}]$ in the frequency domain must assume the value $-1 + i0$ at $\mu = \mu_0$.*

Let $\hat{\lambda}(i\omega; \mu)$ be the eigenvalue of $[G(i\omega; \mu)J(\mu)e^{-i\omega\mu}]$ that satisfies $\hat{\lambda}(i\omega_0; \mu_0) = -1 + i0$. Then

$$\det[(-1)I - G(i\omega_0; \mu_0)J(\mu_0)e^{-i\omega_0\mu_0}] = -1 + \frac{Le^{-i\omega_0\mu_0}}{-\omega_0^2 + i\eta\omega_0 + Ke^{-i\omega_0\mu_0}} = 0.$$ (6)

One has

$$\omega_0^2 = (K-L)\cos(\omega_0\mu_0) \text{ and } \eta\omega_0 = (K-L)\sin(\omega_0\mu_0).$$ (7)

Since $\omega_0 > 0$, $\eta > 0$ and $K - L > 0$, one obtains

$$\begin{cases} \omega_0 = \sqrt{[\sqrt{\eta^4 + 4(K-L)^2} - \eta^2]/2}, \\ \mu_n = 1/\omega_0 \cdot \{2n\pi + \arccos[\omega_0^2/(K-L)]\}, \quad (n = 0,1,2,\cdots). \end{cases}$$ (8)

The number of the roots of $s^2 + \eta s + Ke^{-s\mu} = 0$ that has positive real parts is zero if $0 < \omega_K \mu \le \theta$, two if $\theta < \omega_K \mu \le 2\pi + \theta$, four if $2\pi + \theta < \omega_K \mu \le 4\pi + \theta$, and so on.

Theorem 3. *Let* k *be the total number of anticlockwise encirclements of* $\hat{\lambda}(i\omega, \tilde{\mu})$ *to the point* $P_1 = \hat{P} + \varepsilon \xi_{1d}(\tilde{\omega})$ *for a sufficiently small* $\varepsilon > 0$. *Note:*

$$c = \lceil (\omega_K \mu - \theta)/(2\pi) \rceil. \tag{17}$$

where $\lceil a \rceil$ *denotes the minimal integer which is not less than the number* a.

(i) If $k = 2c$, *the bifurcating periodic solutions of system (4) is stable;*
(ii) If $k \ne 2c$, *the bifurcating periodic solutions of system (4) is unstable.*

4 Numerical Examples and the Frequency Graph

In this section, some numerical simulations for justifying the theoretical analysis are also given. The Hopf bifurcation μ_0 of system (4) can be obtained if they exist. The direction of the Hopf bifurcation and the stability of the bifurcating periodic solutions is determined by the graphical Hopf bifurcation theorem.

If $\eta = 1, K = 2, L = 1$, one can easily calculate $\mu_0 = 1.1506, \mu_1 = 9.1429, \cdots$.

Set $\mu_0 = 1.1506$, then the bifurcation of the Hopf bifurcation is -1, and $c = 1$. The cases of $\mu = 1$ and $\mu = 1.3$ are simulated as Fig. 1 and 2.

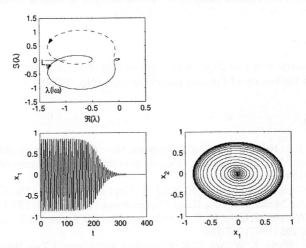

Fig. 1. $\eta = 1, K = 2, L = 1, \mu = 1$. The half-line L_1 intersects the locus $\hat{\lambda}(i\omega)$, and $k = 0 \ne 2c$, so an unstable periodic solution exists.

5 Conclusions

An inertial neuron system with discrete delays usually provides very rich dynamical behaviors. For this kind of system, by using the time delay as a bifurcation parameter,

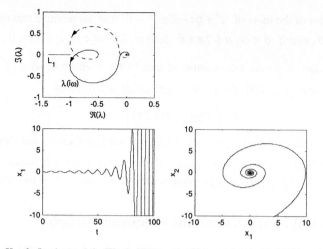

Fig. 2. $\eta = 1, K = 2, L = 1, \mu = 1.3$. The half-line L_1 does not intersect the locus $\hat{\lambda}(i\omega)$, so no periodic solution exists.

we have proved that a Hopf bifurcation occurs when this parameter passes through a critical value, showing that a family of periodic orbits bifurcates from the origin. The stability and direction of the bifurcating periodic orbits have also been analyzed by the frequency-domain methodology. It is deemed that the frequency-domain approach has great potential in bifurcation analysis of various delayed nonlinear systems.

Acknowledgements

The work described in this paper was supported by the Scientific Research Fund of Southwestern University of Finance and Economics (No. 04X16).

References

1. Hopfield, J. J.: Neurons with Graded Response Have Collective Computational Properties Like Those of Two-State Neurons. Proceedings of the National Academy of Sciences of the USA, **81** (1984) 3088–3092
2. Li, C. G., Chen, G.. R., Liao, X. F., Yu, J. B.: Hopf Bifurcation and Chaos in A Single Inertial Neuron Model with Time Delay. Physics Journal B, **41** (2004) 337-343
3. Diek, W. Wheeler, Schieve, W. C.: Stability and Chaos in An Inertial Two Neuron System. Physica D, **105** (1997) 267-284
4. Liao, X. F., Wong, K. W., Wu, Z. F.: Bifurcation Analysis on A Two-Neuron System with Distributed Delays. Physica D, **149** (2001) 123–141
5. Liao, X. F., Li, S. W., Wong, K. W.: Hopf Bifurcation On a Two-Neuron System with Distributed Delays: A Frequency Domain Approach. Nonlinear Dynamics, **31** (2003) 299–326
6. Liao, X. F., Li, S. W., Chen, G. R.: Bifurcation Analysis on A Two-Neuron System with Distributed Delays in the Frequency Domain. Neural Networks, **17** (2004) 545–561

7. Liao, X. F., Wu, Z. F., Yu, J. B.: Stability Switches and Bifurcation Analysis of A Neural Network with Continuous Delay. IEEE Transactions on Systems, Man, and Cybernetics, **29** (1999) 692–696
8. Allwright, D. J.: Harmonic Balance and the Hopf Bifurcation Theorem. Mathematical Proceedings of the Cambridge Philosophical Society, **82** (1977) 453-467
9. Mee, A. I., Chua, L. O.: The Hopf Bifurcation Theorem and Its Applications to Nonlinear Oscillations in Circuits And Systems. IEEE Transactions on Circuits and Systems, **26** (1979) 235-254
10. Moiola, J. L., Chen, G. R.: Hopf Bifurcation Analysis: A Frequency Domain Approach. World Scientific Series on Nonlinear Science Series A, Vol. 21, World Scientific, Singapore (1996)

Stability and Bifurcation of a Neuron Model with Delay-Dependent Parameters

Xu Xu[1,2] and Yanchun Liang[3]

[1] College of Mathematics, Jilin University, Key Laboratory of Symbol Computation and Knowledge Engineering of Ministry of Education 130012, Changchun, China
[2] Institute of Vibration Engineering Research, Nanjing University of Aeronautics and Astronautics, Nanjing 210016, China
[3] College of Computer Science and Technology, Jilin University, Key Laboratory of Symbol Computation and Knowledge Engineering of Ministry of Education, Changchun 130012, China
ycliang@jlu.edu.cn

Abstract. An analytical method is proposed to study the dynamics of a neuron model with delay-dependent parameters. Stability and bifurcation of this model are analyzed using stability switches and Hopf bifurcation proposition. A series of critical time delay are determined and a simple stable criterion is given according to the range of parameters. Through the analysis for the bifurcation, it is shown that a very large delay could also stabilize the system. This conclusion is quite different from that of the system with only delay-independent parameters.

1 Introduction

The dynamics of the delayed neural network are richer complicated and a great deal of papers are devoted to the stability of equilibrium, existence and stability of periodic solutions [1], and chaos of simple systems with few neurons [2]. Recently, Gopalsamy and Leung [3], Liao et al. [4] and Ruan et al. [5] studied the stability, bifurcation and chaos of the following neuron model

$$\dot{x}(t) = -x(t) + a \tanh[x(t) - bx(t-\tau) - c] \tag{1}$$

But, the aforementioned studies to Eq. (1) suppose that the parameters in this model are constant independent of time delay. However, memory intensity of the biological neuron depends usually on time history. It is easy to conceive that the parameters in neural networks will inevitably depend on time delay. The dynamics of neural network model with delay-dependent parameters is very difficult to analyze and few papers are devoted to this study. This paper proposed an analytical method to study the stable dynamics and Hopf bifurcation of this single neuron model with a parameter depending on the time delay. Special attention will be paid to the effect of the time delay on the dynamics of this system. The aims of this paper are to determine the critical time delay, to give simple stability criteria and to discuss the oscillation features.

J. Wang, X. Liao, and Z. Yi (Eds.): ISNN 2005, LNCS 3496, pp. 334–339, 2005.

2 Linear Stability Analyses and Discusses

In system (1), we suppose that $k>0$, $a>0$. Here $b(\tau)>0$ is a decreasing function of τ, which can be considered as a measure of the inhibitory influence from the past history, $\tau>0$ is the time delay and. For simplicity, $c=0$ is considered here.
Let $y(t) = x(t) - b(\tau)x(t-\tau)$, we obtain from (1) that

$$\dot{y}(t) = -ky(t) + a\,\tanh[y(t)] - ab(\tau)\tanh[y(t-\tau)] \tag{2}$$

The characteristic equation of linearized equation of system (2) at the origin is

$$D(\lambda) = \lambda + k - A + B(\tau)e^{(-\lambda\tau)} = 0 \tag{3}$$

where $A = a$ and $B(\tau) = ab(\tau)$. From Ref. [6], it is easy to see that the subsets of the parameter space defined by the equations $\lambda = 0$ and $\lambda = iw$ $(w>0)$ form the boundary of the stability region. Substituting $\lambda = iw$ into (3), we obtain

$$B(\tau)\cos(w\tau) = A - k, \quad B(\tau)\sin(w\tau) = w \tag{4}$$

Define $F(w) \equiv w^2 + (A-k)^2 - B^2(\tau)$. As analyzed in Ref. [7], if $F(w) = 0$ has no positive roots, that is $B(\tau) < |k - A|$, the stability of system does not depend on time delay. Noting that $B(\tau) < B(0)$, two regions are determined by $0 < B(0) < |k - A|$ illustrated in Figure 1. Therefore, the trivial equilibrium of system (2) is delay-independent stable when the parameters in region I, whereas it is unstable for any time delay when the parameters in region II.

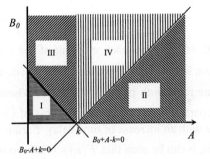

Fig. 1. Convergence in A-B_0 parameter plane

If $B(0) > |k - A|$, $F(w) = 0$ possible has positive root $w(\tau) = \sqrt{B^2(\tau) - (k-A)^2}$. Because $B(\tau)$ is a decreasing function of τ, there exists a τ_c which satisfied that $k - A + B(\tau_c) = 0$ such that $F(w) = 0$ has positive roots at $\tau \in [0, \tau_c)$ whereas it has no root when $\tau \geq \tau_c$. Substituting $w(\tau)$ into Eq. (4), it can be seen that if it has simple positive roots τ_j $(j = 0, 1, 2, \ldots)$ in the interval $(0, \tau_c)$, a pair of simple conjugate pure imaginary roots $\lambda = \pm iw(\tau)$ of Eq. (4) exists at $\tau = \tau_j$. We can determine the moving direction of the real part of characteristic roots with an increase of time delay τ. Differentiating equation (3) with respect to τ gives

$$\frac{d\lambda}{d\tau} = \frac{B\lambda e^{-\lambda\tau} - B'e^{-\lambda\tau}}{1 - B\tau e^{-\lambda\tau}} \tag{5}$$

where B' represents $dB(\tau)/d\tau$. From (3) we obtain $Be^{-\lambda\tau} = -\lambda - p$ and substitute it into Eq. (5) gives

$$\frac{d\lambda}{d\tau} = \frac{-B\lambda(\lambda+p) + B'(\lambda+p)}{B + B\tau(\lambda+p)} \tag{6}$$

where $p = k - A$. Substituting $\lambda = iw(\tau)$ into Eq (6), we obtain

$$\text{Re}(\frac{d\lambda}{d\tau}\Big|_{\lambda=iw(\tau)}) = \text{Re}(\frac{-Biw(iw+p) + B'(iw+p)}{B + B\tau(iw+p)}) = \frac{B^2 w^2 + BB'(p+\tau B^2)}{(B + B\tau p)^2 + (B\tau w)^2} \tag{7}$$

Introduce the function $S_j(\tau) = \tau - \tau_j$ which is continuous and differentiable for a given j. Noticing that $BB' = ww'$ from (3), the derivative of $S_j(\tau)$ with respect to τ is

$$\frac{dS_j(\tau)}{d\tau} = 1 - \frac{d\tau_j}{d\tau} = 1 - \frac{d\tau_j}{dw}\frac{dw}{d\tau} \tag{8}$$

From Eq. (4) we obtain

$$\frac{d\tau_j}{dw} = -\frac{(p + B^2\tau_j)w}{B^2 w^2} \tag{9}$$

Substituting Eq. (9) into (8), and noticing that $BB' = ww'$, we have

$$\frac{dS_j(\tau)}{d\tau} = 1 - \frac{d\tau_j}{d\tau} = 1 - \frac{d\tau_j}{dw}\frac{dw}{d\tau} = \frac{B^2 w^2 + BB'(p + B^2\tau_j)}{B^2 w^2} \tag{10}$$

Comparing (7) with (10), we have that $\text{sgn}(\frac{d\,\text{Re}(\lambda)}{d\tau}\Big|_{\lambda=iw(\tau_j)}) = \text{sgn}(\frac{dS(\tau)}{d\tau}\Big|_{\tau=\tau_j})$.

Therefore, if $\text{sgn}(dS_j(\tau_j)/d\tau) > 0$ each crossing of the real part of characteristic roots at τ_j must be from left to right. Thus, the characteristic equation of system adds a new pair of conjugate roots with positive real parts. Whereas $\text{sgn}(dS_j(\tau_j)/d\tau) < 0$ indicates that the real part of a pair of conjugate roots of (3) changes from positive value to negative value with an increase of time delay τ around τ_j.

From above analysis, it can be seen that $F(w) = 0$ has no real root and no stability switches occur when $\tau \geq \tau_c$. From (3) it is easy to see that $\lambda = A - k$ as $\tau \to +\infty$. If parameter A satisfies $A < k$ (the region III in Figure 1), system (2) is finally stable, whereas if $A > k$ (the region IV) it is finally unstable by the finite number of stability switches.

3 Bifurcation Analysis

Substituting $\lambda = 0$ into (3), we have that $k - A + B(\tau) = 0$. That is, steady bifurcation occurs when $\tau = \tau_c$. The derivative of $D(\lambda)$ with respect to λ is

$$dD(\lambda)/d\lambda = 1 - B(\tau)\tau e^{-\lambda\tau} \tag{11}$$

Thus, $dD(0)/d\lambda = 1 - B(\tau)\tau \neq 0$ if $\tau \neq 1/B(\tau) = 1/(A-k) \equiv \tau^*$. So if $\tau \neq \tau^*$ then there is only one root with $\text{Re}(\lambda) = 0$. Eq. (6) gives

$$\left.\frac{d\lambda}{d\tau}\right|_{\lambda=0} = \frac{-B'(\tau)}{1-B(\tau)\tau} = \frac{-B'(\tau)}{1-(A-k)\tau} > 0 \ \ if (\tau < \tau^*) \ \ and \ \ \left.\frac{d\lambda}{dA}\right|_{\lambda=0} < 0 \ \ if (\tau > \tau^*) \tag{12}$$

Therefore the number of roots λ of Eq. (3) with $\text{Re}(\lambda) > 0$ increases (decreases) if $\tau < \tau^*$ ($\tau > \tau^*$) with an increase of time delay τ around τ_c

Let $\lambda = \pm iw(\tau)$ $(w > 0)$ is a pair of pure imaginary roots of the characteristic equation (3). Suppose $dD(\lambda)/d\lambda\big|_{\lambda=iw} = 1 - \tau B(\tau)e^{-iw\tau} = 0$, we have

$$\tau B(\tau)\cos(w\tau) = 1, \ \ \tau B(\tau)\sin(w\tau) = 0 \tag{13}$$

Eqs. (4) and (13) give that $\tau = 1/(A-k)$ where $A>k$. So, the zeros solutions of $dD(\lambda = iw)/d\lambda$ occur at $\tau = 1/(A-k)$. Excluding this value, root $\lambda = iw$ is a simple root of Eq. (3).

From equations (7) and (10), we have that $dD(\lambda = iw)/d\lambda\big| = 0$ if and only if $dS_j/d\tau\big|_{\tau=\tau_j} = 0$. These imply that system (2) obeys the conditions of the Hopf bifurcation theorem [6] and undergoes a Hopf bifurcation at $\tau = \tau_j$ if τ_j is neither zero point of $dS_j/d\tau$ nor equal to $1/(A-k)$.

4   Numerical Simulations and Discussions

To verify the analysis results, an example is discussed as following. Here, we suppose that $B(\tau) = B_0 e^{-\alpha\tau}$ ($B_0 > 0, \alpha > 0$).

Choosing parameters $B_0 = 4, A = 1, k = 2$ and $\alpha = 0.12$, we plot the graph of $s_j(\tau)$ versus τ in the interval $[0, \tau_c)$, which is shown in Figure 2(a). $S_0(\tau)$ has two zeros at $\tau_1 = 0.763$ and $\tau_2 = 8.74$. And $S_0(\tau_1) > 0$ carries that a stability switch occurs toward instability at τ_1, whereas $S_0(\tau_2) < 0$ carries that the stability switch occurs towards stable at τ_2. When $\tau \geq \tau_2$ no stability switches could occur and the system remains stable. Therefore, for $\tau \in [0, \tau_1)$ and $\tau \in [\tau_2, +\infty)$, system (2) is stable, whereas it is unstable for $\tau \in [\tau_1, \tau_2)$. The bifurcation diagram is shown in Figure 2(b). With an increasing of τ, we observe that the trivial equilibrium of system (2) loses its stability when $\tau > \tau_1$ and the solutions tend to a stable limit circle which is illustrated in Figure 2(c). Whereas $\tau > \tau_2$ cause the stable limit circle disappear and the trivial equilibrium solution regain the stability shown in Figure 2(d). From this figure it can be seen that Hopf bifurcation of the trivial equilibrium occurs as τ_1 and τ_2 is crossed, respectively.

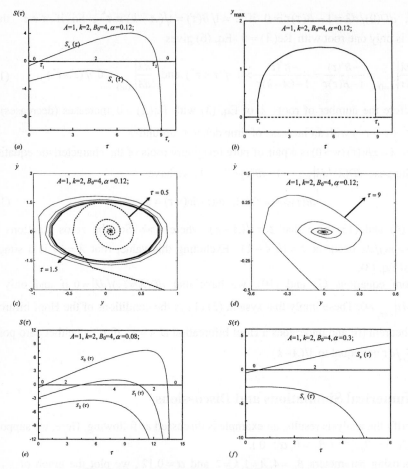

Fig. 2. Numerical simulations for different parameters

From Figure 2, it also can be seen that the stability switches can occur many times with an increasing of time delay from zero. From Figures 2(a), (e) and (f), we observe that larger the value of α, the wider the interval of time delay τ stabilized system. This shows the stability of the system is also determined by the parameter α and it can be used as a control parameter to stabilize the system.

5 Conclusions

This paper proposed an analytical method to study the dynamics of a neuron model with delay-dependent parameters. A series of critical time delay are determined and a simple stable criterion is given according to the range of parameters. Through the analysis for the bifurcation, it is shown that the trivial equilibrium may lose stability via a steady bifurcation and a Hopf bifurcation. The analytical method is not only enables one to achieve dynamic analysis for neural network model with delayed dependent parameters but also provides a scheme to control the performance of the system dynamics.

Acknowledgments

The authors are grateful to the support of the National Natural Science Foundation of China (Grant No. 60433020), the science-technology development project of Jilin Province of China (Grant No. 20030520) and the doctoral funds of the National Education Ministry of China (Grant No. 20030183060)

References

1. Zhang, Q., Wei, X., Xu, J.: Global Exponential Stability of Hopfield Neural Networks with Continuously Distributed Delays. Physics Letters A, **315** (2003) 431–436
2. Zhou, J., Liu, Z. R., Chen, G. R.: Dynamics of Periodic Delayed Neural Networks. Neural Networks, **17** (2004) 87–101
3. Gopalsamy, K., Leung, I.: Convergence under Dynamical Thresholds with Delays. IEEE Transactions on Neural Networks, **8** (1997) 341–348
4. Liao, X. F., Wong, K. W., Leung, C. S., Wu, Z. F.: Hopf Bifurcation and Chaos in a Single Delayed Neuron Equation with Nonmonotonic Activation Function. Chaos, Solitons and Fractals, **12** (2001) 1535–1547
5. Ruan, J., Li, L. and Lin, W.: Dynamics of Some Neural Network Models With Delay. Physical Review E, **63** (2001)
6. Hu, H. Y., Wang, Z. H.: Dynamics of Controlled Mechanical Systems with Delayed Feedback. Springer-Verlag, Heidelberg (2002)
7. Wang, Z. H., Hu, H. Y.: Stability Switches of Time-delayed Dynamic Systems with Unknown Parameters. Journal of Sound and Vibration, **233** (2000) 215–233

Theorem 2: If conditions in theorem 1 hold and $0 < \tau'(t) \le \tau'_M, |f(x)| \le M$, and if there is a positive number $\alpha > 0$, subjecting to

$$\alpha(\tau_M - 1) - \frac{(abM)^2}{4(1 - \alpha + |a|M)} < 0, \tag{8}$$

then the equilibrium point is global stable.

Proof: We define a Lyapunov functional as follows:

$$V = \frac{1}{2}x^2 + \alpha \int_{-\tau(t)}^{0} x^2(t+\xi)d\xi , \tag{9}$$

where α is a positive constant. Differentiate V with respect to t along the trajectory of the system in equation (3), we have

$$\dot{V} = x(t)\dot{x}(t) + \alpha(x^2(t) - x^2(t - \tau(t)) - x^2(t - \tau(t))\tau'(t))$$

$$= x(t)(-x(t) + a(f(x(t)) - bf(x(t - \tau(t)))))$$

$$+ \alpha(x^2(t) - x^2(t - \tau(t)) - x^2(t - \tau(t))\tau'(t))$$

$$\le (1 - \alpha)x^2(t) + ax(t)f(x(t)) - abx(t)f(x(t - \tau(t)))$$

$$+ \alpha(\tau'_M - 1)x^2(t - \tau(t))$$

$$\le (1 - \alpha)x^2(t) + |a|Mx^2(t) + |ab|Mx(t)x(t - \tau(t))|$$

$$+ \alpha(\tau'_M - 1)x^2(t - \tau(t))$$

$$= (1 - \alpha + |a|M)[x^2(t) + |ab|M/(1 - \alpha + |a|M)$$

$$+ |x(t)x(t - \tau(t))|] + \alpha(\tau'_M - 1)x^2(t - \tau(t))$$

$$= (1 - \alpha + |a|M)[x(t) + |ab|Mx(t - \tau(t)/(2(1 - \alpha + |a|M))]^2$$

$$+ [\alpha(\tau'_M - 1) - [(abM)^2/4(1 - \alpha + |a|M)]x^2(t - \tau(t)) < 0,$$

i.e. $\dot{V} < 0$. Thus proved the equilibrium point is global stable.

3 Chaotic Phenomenon

In this section, the activation function of the neural network is chosen as $f(x) = \tanh(x)$. We let $a = 2$, and $\tau(t) = 0.5 + \sin^2(1.3t)$.

It is theoretically difficult to analyze chaotic phenomena in a neural network with uncertain time delays. Numerical simulation is a useful method to calculate the Largest Lyapunov exponent, which can be use to verify if there is chaotic phenomenon or not.

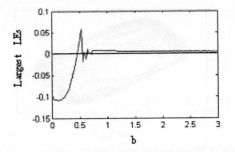

Fig. 1. Largest Lyapunov exponent of system in equation (3)

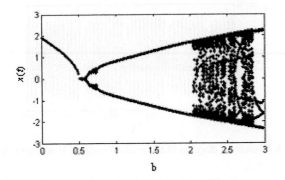

Fig. 2. Bifurcation graph of system (3) across $x(t) = x(t - \tau_M)$

Estimation of the Largest Lyapunov exponents is shown in Fig 1. The graph of the original system in equation (3) is shown in Fig 2. The bifurcation graph of the original system in equation (3) is shown in Fig 1. Both the Largest Lyapunov exponent and the bifurcation diagram describe chaotic phenomena occur with some parameter b. Phase portrait and waveform diagram of the system in equation (3) with $a=2$, $b=2.1$ are given in Fig. 3 and Fig. 4 respectively, where $\tau_M = \max(\tau(t))$, the initial value of x is 0.5, and $\tau(t) = 0.5 + \sin^2(4.3t)$. The phase portrait of the system in equation (3) with $a=2$, $b=2.1$ is given in Fig. 5. From these simulations, we can see that the chaotic phenomena existed in the proposed single neuron model.

4 Conclusions

A model for a single neuron neural network with uncertain time delays is considered. The system is stable with specific system parameters. By the use of computer simulation, the single neuron equation with uncertain time delays has been shown to exhibit rich dynamic behaviors, such as stable equilibrium, oscillations and chaos. Chaotic phenomenon in multi-uncertain time delays neural networks should be discussed in another paper.

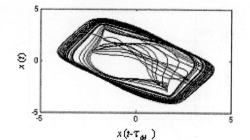

Fig. 3. Phase portrait of system (3) with a=2, b=2.1

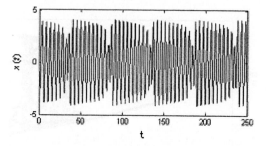

Fig. 4. Waveform diagram of the original system (3) with a=2, b=2.1

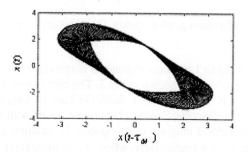

Fig. 5. Phase portrait of system (3) with a=2, b=2.1

There are many problems of dynamic system with uncertain time delays remaining unstudied. Most papers about uncertain time delays in dynamic systems are focus on stability analyses or stability control. Chaos phenomena in such a system are not deeply studied before. Our future work would be further analyzing the bifurcation for a single neuron neural network with uncertain time delay.

References

1. Lu, H., He, Z.: Chaotic Behavior in First-order Autonomous Continuous-time Systems with Delay. IEEE trans. on CAS-1, **43** (1996) 700-702
2. Wang, X.: Period-Doubling to Chaos in A Simple Neural Network. Neural Networks, IJCNN-91-Seattle Int. Joint Conference on, **2** (1991) 333-339

3. Tian, Y., Gao, F.: Simple Delay Systems with Chaotic Dynamics. Control of Oscillations and Chaos, 1997, Proceedings, 1997 1st Int. Conference, 2 (1997) 319-320
4. Lu, H., He, Y., He, Z.: A Chaos-Generator: Analyses of Complex Dynamics of a Cell Equation in Delayed Cellular Neural Networks. IEEE trans. on CAS-1, 45 (1996) 178-181
5. Zhou, S.B., Liao, X.F., Yu, J.B.: Chaotic Behavior in a Simplest Neuron Model With Discrete Time Delay. J. Electronics & Information Technology, 24 (2002) 1341-1345.
6. Liao, X.F., Wong, K.W., Leung, C.S., Wu, Z.F.: Hopf Bifurcation and Chaos in a Single Delayed Neuron Equation with Non-monotonic Activation Function. Chaos, Solitons and Fractals, 12 (2001) 1535-1547
7. Marco Gilli: Strange Attractors in Delayed Cellar Neural Networks. IEEE trans. on CAS-1, 40 (1993) 849-853
8. Gopalsamy, K., Issic K.C. Leung: Convergence under Dynamical Thresholds with Delays. IEEE trans. on neural networks, 8 (1997) 341-348
9. Liao, X.F., Wu, Z.F., Yu, J.B.: Stability Switches and Bifurcation Analysis of a Neural Network with Continuously Distributed Delay. IEEE Trans. On SMC-I, 29 (1999) 692-696
10. Gopalsmy, K., Leung, I.: Delay Induced Periodicity in a Neural Network of Excitation and Inhibition. Physica D, 89 (1996) 395-426
11. Olien, L., Belair, J.: Bifurcations Stability and Monotonicity Properties of a Delayed Neural Network Model. Physica D, 102 (1997) 349-363
12. Vladimir L. Kharitonov, Silviu-Iulian Niculescu: On the Stability of Linear SystemsWith Uncertain Delay. IEEE Trans. on Automatic Control, 48 (2003) 127-132
13. Yunping Huang, Kemin Zhou: Robust Stability of Uncertain Time-Delay Systems. IEEE Trans. on Automatic Control, 45 (2000) 2169-2173

Chaotic Synchronization of Delayed Neural Networks

Fenghua Tu, Xiaofeng Liao, and Chuandong Li

College of Computer Science and Engineering, Chongqing University 400044, China
fhtu@cqu.edu.cn

Abstract. In this paper, synchronization issue of a coupled time-delayed neural system with chaos is investigated. A sufficient condition for determining the exponential synchronization between the drive and response systems is derived via Lyapunov-Krasovskii stability theorem. Moreover, the synchronization thresholds of the coupled systems are estimated based on the proposed result. The effectiveness of our scheme is illustrated by numerical simulation.

1 Introduction

Recently, the synchronization of coupled chaotic systems has become an active research area [1-4] because of the pioneering work of Pecora and Carroll [5, 6]. Interest in the chaotic synchronization of coupled neural systems is due to its potential applications for secure communications and control. It has been studied theoretically and experimentally in many systems such as coupled circuits [5, 6], coupled laser systems [7] and neuron systems [8] with or without time delays.

This paper focuses on chaotic synchronization of coupled delayed neural networks. In this research field, some attempts have been reported in recent years. Tonnelier et al. [9] presented the analytical investigation of synchronization and desynchronization of two coupled neurons; Zhou et al. [10] studied chaos and its synchronization in two-neuron systems with discrete delays. For other results reported in the literature, we refer the readers to Refs. [13-15]. In this paper, we analyze the exponential stability of the error dynamical system associated with the interactive neuron systems. By constructing a suitable Lyapunov-Krasovskii functional, an exponential synchronization criterion is obtained. Numerical simulation shows that the proposed synchronization scheme is practical and significant.

2 Problem Formulation

Consider the following time-delayed neural system[12]:

$$\frac{dx(t)}{dt} = -x(t) + af[x(t) - bx(t-\tau) + c], t > 0, \tag{1}$$

where $f \in C^{(1)}$ is a nonlinear function, the time delay $\tau > 0$, a, b and c are constants.

J. Wang, X. Liao, and Z. Yi (Eds.): ISNN 2005, LNCS 3496, pp. 346–350, 2005.

By coupling system (1), we have

$$\frac{dx(t)}{dt} = -\alpha x(t) + af[x(t) - bx(t-\tau) + c],$$

$$\frac{dy(t)}{dt} = -\alpha y(t) + af[y(t) - by(t-\tau) + c] + K[x(t) - y(t)] \tag{2}$$

where K is the controlling parameter. Suppose that the initial conditions of system (2) are of the form

$$x(\theta) = \phi_x(\theta), \qquad y(\theta) = \phi_y(\theta), \qquad \theta \in \lfloor -\tau, 0 \rfloor \tag{3}$$

where ϕ_x, ϕ_y are real-value continuous function on $\lfloor -\tau, 0 \rfloor$.

System (2) is a typical drive-response model, where the drive system was well investigated in [11, 12]. Gopalsamy and Leung [11] analyzed the stability of this system in the case of $f(x) = \tanh(x)$ and $\alpha = 1$; Liao et al. [12] discussed the chaotic behavior of the system with

$$f(x) = \sum_{i=1}^{2} \alpha_i [\tanh(x+k_i) - \tanh(x-k_i)] \tag{4}$$

where α_i and k_i are constants. Throughout this paper, we select

$$\alpha = 1,\ a = 3,\ b = 4.5,\ a_1 = 2,\ a_2 = 1.5,\ k_1 = 1,\ k_2 = \frac{4}{3},\ \tau = 1, \tag{5}$$

which makes the considered system be chaotic.

3   Chaotic Synchronization of Coupled Delayed Neural Systems

In this section, we will investigate the synchronization of coupled time delay chaotic system (2). Moreover, we will attempt to obtain the criteria for the exponential convergence of the synchronized system. For this purpose, we denote $\Delta(t)$ the synchronization error, i.e., $\Delta(t) = x(t) - y(t)$.

Therefore,

$$\begin{aligned}\dot{\Delta}(t) &= \dot{x}(t) - \dot{y}(t) \\ &= -(\alpha + K - af'(z))\Delta(t) - abf'(z)\Delta(t-\tau) \\ &= -r(t)\Delta(t) + s(t)\Delta(t-\tau)\end{aligned} \tag{6}$$

where

$$z = (1-\lambda)[x(t) - bx(t-\tau) + c] + \lambda[y(t) - by(t-\tau) + c], \qquad 0 \le \lambda \le 1,$$

$$r(t) = \alpha + K - af'(z),\ s(t) = -abf'(z). \tag{7}$$

Definition 1. If there exist $k > 0$ and $r(k) > 1$ such that

$$\|\Delta(t)\| \le r(k)e^{-kt} \sup_{-\tau \le \theta \le 0} \|\Delta(\theta)\| \qquad \forall t > 0$$

then system (6) is said to be exponentially stable, where k is called the exponential convergence rate, and the notation $\|\cdot\|$ denotes the Euclidian norm of a vector or a square matrix.

Theorem 1. *Suppose that there exist three positive numbers p, q and k such that*

$$2r(t)p - 2kp - q - q^{-1}p^2s^2(t)e^{2k\tau} > 0 \qquad (8)$$

then system (7) is exponentially stable. Moreover

$$\|\Delta(t)\| \le \left[1 + \frac{q}{2kp}\left(1 - e^{-2k\tau}\right)\right]^{1/2} \|\varphi\|e^{-kt} \qquad \forall t > t_0 \qquad (9)$$

Proof. Construct the following Lyapunov-Krasovskii functional:

$$V(\Delta(t)) = e^{2kt} p\Delta^2(t) + \int_{-\tau}^{t} e^{2k\xi} q\Delta^2(\xi)d\xi$$

Then the time derivative of $V(\Delta(t))$ along the solution of system (6) is

$$\dot{V}(\Delta(t)) = e^{2kt}\left\{2kp\Delta^2(t) + 2p\Delta(t)[-r(t)\Delta(t) + s(t)\Delta(t-\tau)] + q\Delta^2(t)\right.$$

$$\left. - e^{-2k\tau}q\Delta^2(t-\tau)\right\}$$

$$\le e^{2kt}\left\{[2kp - 2pr(t) + q]\Delta^2(t) + q^{-1}p^2s^2(t)\Delta^2(t)e^{-2k\tau}\right\}$$

$$\equiv -\Omega\Delta^2(t)e^{2kt}$$

From (8), i.e., $\Omega = 2kp - 2pr(t) + q + q^{-1}p^2s^2(t)e^{-2k\tau} > 0$, it follows that $\dot{V}(\Delta(t)) < 0$.

To complete the proof, notice that $V(\Delta(t)) \le V(\Delta(0))$, because $\dot{V}(\Delta(t)) < 0$, and

$$V(\Delta(0)) = p\Delta^2(0) + \int_{-\tau}^{0} qe^{2k\xi}\Delta^2(\xi)d\xi \le p\|\varphi\|^2 + q\|\varphi\|^2 \int_{-\tau}^{0} e^{2k\xi}d\xi$$

$$= p\|\varphi\|^2 + q\|\varphi\|^2 \frac{1}{2k}\left(1 - e^{-2k\tau}\right) = \left[p + q\frac{1}{2k}\left(1 - e^{-2k\tau}\right)\right]\|\varphi\|^2$$

On the other hand, $V(\Delta(t)) \ge e^{2kt}p\Delta^2(t)$. We can immediately derive the following inequality:

$$e^{2kt}p\Delta^2(t) \le \left[p + q\frac{1}{2k}\left(1 - e^{-2k\tau}\right)\right]\|\varphi\|^2$$

which implies (9). The proof is thus completed.   □

Obviously, in this section, from (7), condition (8) can be transformed easily into the following inequality:

$$2[\alpha + K - af'(z)]p - 2kp - q - q^{-1}p^2a^2b^2[f'(z)]^2 e^{2k\tau} > 0 \tag{10}$$

To derive a practical and convenient criterion of exponential stability for synchronizing system (2), let

$$p = \frac{1}{2}, \quad q = \frac{1}{2}ab\,sup|f'(z)|e^{k\tau}$$

Then, from Theorem 1, the following corollary is immediate.

Corollary 1. Suppose that there exists a scalar k > 0 such that

$$K > -\alpha + k + a|sup\,f'(z)| + ab\,sup|f'(z)|e^{k\tau} \tag{11}$$

where k is the exponential convergence rate. Then system (6) with (7) is exponentially stable near $\Delta = 0$. Therefore, the coupled chaotic system (2) are exponentially synchronized with degree k.

4 Numerical Example

To support the analysis made the above section 3, an example is given in this section. Throughout this paper, we adopt the combination of the forth-order Runge-Kutta integration algorithm (time step is 0.001) with the interpolation technique for computer simulation.

Example Consider system (2) with conditions (4) and (5). In this case, $sup\,f'(z) \approx 0.18, sup|f'(z)| \approx 4.448795$. It follows from (11) that

$$K > k - 0.46 + 60.5873e^{k\tau} \tag{12}$$

According to the selected values of k and τ, we can obtain the threshold value K^* of parameter K, e.g., $K^* = 66.599$ when $k = 0.1$ and $\tau = 1$. From Corollary 1, if $K > K^*$, system (2) exponentially synchronizes, as shown in Fig. 1.

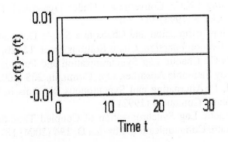

Fig. 1. Time response curve of synchronized error system with K=70.

5 Conclusions

We have investigated the exponential synchronization of coupled delayed neuron systems. By employing the Lyapunov-Krasovskii functional approach, a sufficient condition is derived and the synchronization threshold of the coupled system is estimated. The effectiveness of the proposed approach was also illustrated by computer simulation

Acknowledgements

The work described in this paper was partially supported by The National Natural Science Foundation of China (Grant no. 60271019), the Natural Science Foundation of Chongqing (Grant no. 20020611007), the Post-doctoral Science Foundation of China and the Natural Science Foundation of Chongqing

References

1. Kocarev, L., Parlitz, U.: General Approach for Chaotic Synchronization with Applications to Communication. Phys. Rev. Lett. **74** (1995) 5028
2. Cuomo, K.M., Oppenheim, A.V., Strogatz, S.H.: Synchronization of Lorenz-based Chaotic Circuits with Applications to Communications. IEEE Trans. Circuits Syst. II, **40** (1993) 626–633
3. Grassi, G., Mascolo, S.: Prentice-Hall, Englewood Cliffs, NJ (1999)
4. Yang T., Chua L.O.: Secure Communication via Chaotic Parameter Modulation. IEEE Trans. Circuits Syst. I, **43** (1996) 817–819
5. Pecora, L.M., Carroll, T.L.: Synchronization in Chaotic Systems. Phys. Rev. Lett. **64** (1990) 821–824
6. Pecora, L.M., Carroll, T.L.: Driving Systems with Chaotic Signals. Phys. Rev. A, **44** (1991) 2374–2383
7. Thornburg Jr. K.S., Moller M., Roy R., Carr T.W.: Chaos and Coherence in Coupled Lasers. Phys. Rev. E, **55** (1997) 3865
8. Pasemann F.: Synchronized Chaos and Other Coherent States for Two Coupled Neurons. Physica D, **128** (1999) 236-249
9. Tnonnelier A. et al : Synchronization and Desynchronization of Neural Oscillators. Neural Networks, **12** (1999) 1213-1228
11. Gopalsamy K., Leung I.K.C.: Convergence Under Dynamical Thresholds with Delays. IEEE Neural Networks, **8** (1994) 341–348
12. Liao X.F., et al.: Hopf Bifurcation and Chaos in a Single Delayed Neuron Equation with Non-monotonic Activation Function. Chaos Solitons Fract. **12** (2001) 1535–1547
13. Barsella A., Lepers C.: Chaotic Lag Synchronization and Pulse-induced Transient Chaos in Lasers Coupled by Saturable Absorber. Opt. Commun. **205** (2002) 397–403
14. Wang X.J., Rinzel, J.: Alternating and Synchronous Rhythms in Reciprocally Inhibitory Model Neurons. Neural Comput. **4** (1992) 84
15. Li C.D., et al.: Chaotic Lag Synchronization of Coupled Time-delayed Systems and its Applications in Secure Communication. Physica D, **194** (2004) 187–202

Chaos Synchronization for Bi-directional Coupled Two-Neuron Systems with Discrete Delays

Xiaohong Zhang[1,2] and Shangbo Zhou[2]

[1] School of Software Engineering, Chongqing University 400030, China
xhongz@yahoo.com.cn
[2] Department of Computer Science and Engineering, Chongqing University 400030, China

Abstract. In this paper, the chaos synchronization between two delayed neuron chaotic systems with linearly bi-directional coupling is investigated. Some generic criterion is developed for choosing the appropriate coupling parameters to ensure global chaos synchronization based on Krasovskii-Lyapunov theory. Finally, numerical results illustrate the effectiveness of the criterion.

1 Introduction

Recently the chaotic neural network constructed with chaotic neurons has attracted much attention [1-9]. Chaos and bifurcation behavior are investigated among the neural networks models, such as Hopfield network [9, 6], Cellular neural network [1, 3], Neural networks with time-delay [2, 8], discrete-time neural networks [4]. It is well known that neural networks are large-scale and complicated dynamical systems with time-delay, so many researchers have directed their attention to the study of simple systems. Although the analysis of simple systems with time delays is simpler than that for complex systems, it is useful since the complexity found might be applied to large networks. Among the simple delayed neural networks, several simple neuron systems with discrete and continuous time-delay are proposed and investigated [21, 2, 8, 10, 11],

$$\frac{dy}{dt} = -y(t) + a\tanh[y(t) - by(t-\tau) - c] \tag{1}$$

$$\frac{dy}{dt} = -y(t) + a\tanh[y(t) - b\int_0^\infty F(y)y(t-s)ds + c] \tag{2}$$

Where a denotes the range of the continuous variable $y(\cdot)$ while b can be considered as a measure of the inhibitory influence from the past history. In [2, 8, 10, 11], Hopf bifurcation and chaotic behavior are investigated, respectively, in the single and the two-neuron systems with time-delay.

On the other hand, because the synchronization of coupled neural networks has many applications, the control and synchronization of chaotic systems has become an active research field [12-20]. With respect to simple neuron systems, in [17], although the dynamical behaviors of two-neuron systems with discrete delays and synchroniza-

J. Wang, X. Liao, and Z. Yi (Eds.): ISNN 2005, LNCS 3496, pp. 351–356, 2005.

tion are studied, however, unidirectional coupled system is only considered and the proof given is very complex.

Motivated by the aforementioned reasons, we investigate the synchronization chaos for bi-directional coupled time-delay neural network systems with two neurons, and develop a new simple synchronization condition for bi-directional coupled in this paper.

The rest of this paper is organized as follows: section 2 describes two-neuron systems with time-delay and introduces its dynamical behaviors. In the section 3, some new simple global synchronization criterions are derived for bi-directional coupled system. In Section 4, simulation results are presented to demonstrate the correctness of the analysis on the stability of synchronization. Finally, some conclusions are made in Section 5.

2 Problem Formulation

For system (1), set $c = 0, x(t) = y(t) - by(t-\tau)$, and consider a general function $f(\cdot)$ instead of $\tanh(\cdot)$, we easily obtain the following equation:

$$\dot{x} = -x + af(x(t)) - abf(x(t-\tau)) \tag{3}$$

Actuary, it is a single delayed neuron equation, which has been carefully investigated about its dynamical behavior and bifurcation [10]. In this paper, we mainly consider the two-neuron equation with time-delay:

$$\begin{cases} \dot{x}_1 = -x_1 + a_1 f(x_2(t)) + a_1 b_1 f(x_2(t-\tau)) \\ \dot{x}_2 = -x_2 + a_2 f(x_1(t)) + a_2 b_2 f(x_1(t-\tau)) \end{cases} \tag{4}$$

With respect to dynamical behavior of (4), in [17], there are the following two results. System (4) possesses a fixed point and exists bifurcation phenomena when the parameters of system satisfy some conditions.

In the following, we consider bi-directional coupled synchronization of the identical system.

$$\begin{cases} \dot{x}_1 = -x_1 + a_1 f(x_2(t)) + a_1 b_1 f(x_2(t-\tau)) + k_1(y_1 - x_1) \\ \dot{x}_2 = -x_2 + a_2 f(x_1(t)) + a_2 b_2 f(x_1(t-\tau)) + k_2(y_2 - x_2) \\ \dot{y}_1 = -y_1 + a_1 f(y_2(t)) + a_1 b_1 f(y_2(t-\tau)) + k_1(x_1 - y_1) \\ \dot{y}_2 = -y_2 + a_2 f(y_1(t)) + a_2 b_2 f(y_1(t-\tau)) + k_2(x_2 - y_2) \end{cases} \tag{5}$$

Where, x_i, y_i $(i = 1, 2)$ are state variables, and $k_i (i = 1, 2)$ are coupled coefficients.

Define an error system with $e = \begin{bmatrix} e_1(t) \\ e_2(t) \end{bmatrix} = \begin{bmatrix} x_1(t) - y_1(t) \\ x_2(t) - y_2(t) \end{bmatrix}$, and we have

$$\begin{cases} \dot{e}_1 = -(1 + 2k_1)e_1 + a_1(f(x_2(t)) - f(y_2(t))) + a_1 b_1(f(x_2(t-\tau)) - f(y_2(t-\tau))) \\ \dot{e}_2 = -(1 + 2k_2)e_2 + a_2(f(x_1(t)) - f(y_1(t))) + a_2 b_2(f(x_1(t-\tau)) - f(y_1(t-\tau))) \end{cases} \tag{6}$$

For convenience, we set

$$F(t) = \begin{bmatrix} a_1(f(x_2(t)) - f(y_2(t))) \\ a_2(f(x_1(t)) - f(y_1(t))) \end{bmatrix}, \quad F(t-\tau) = \begin{bmatrix} a_1 b_1(f(x_2(t-\tau)) - f(y_2(t-\tau))) \\ a_2 b_2(f(x_1(t-\tau)) - f(y_1(t-\tau))) \end{bmatrix}$$

$$e = \begin{bmatrix} e_1(t) \\ e_2(t) \end{bmatrix}, \quad A = \begin{bmatrix} -(1+2k_1) & 0 \\ 0 & -(1+2k_2) \end{bmatrix}$$

by system (6), error system is given as follows:

$$\dot{e} = Ae + F(t) + F(t-\tau) \tag{7}$$

Then the coupling synchronization problem is turned into the stability of system (7).

3   Chaos Synchronization for Bi-directional Coupled System

It is very interesting that, for two identical systems (4), if the initial value $(x_1(0), x_2(0)) \neq (y_1(0), y_2(0))$, then the trajectories of two identical systems will quickly separate each other and become irrelevant [8, 10, 11, 17]. However, if the coupling coefficients satisfy certain condition, the two coupled systems will approach global synchronization for any initial value. In the following, we derive global stability condition of the error system (7) based on Krasovskii-Lyapunov theory.

Theorem 1. Assume P be a positive definite symmetric constant matrix and $f'(\cdot) < M$ ($M < \infty$), if there exits a constant $\beta > 0$, such that

$$Q_1 = PA + \beta P + \frac{1}{2}I + MP^T P < 0 \tag{8}$$

$$Q_2 = I - 2\beta P \leq 0 \tag{9}$$

Then error system (7) is global stability, i.e., the coupled system (5) synchronizes.

Proof. We construct a Lyapunov functional as followings:

$$V = \frac{1}{2}e^T Pe + \mu \int_{-\tau}^{0} e^T(t+\theta) Pe(t+\theta) d\theta$$

Differentiate V with respect to t along the trajectory of system (7), we have

$$\dot{V} = \frac{1}{2}\dot{e}^T Pe + \frac{1}{2}e^T P\dot{e} + \beta(e^T Pe - e^T(t-\tau)Pe(t-\tau))$$

$$= \frac{1}{2}e^T A^T Pe + \frac{1}{2}e^T PAe + [F(t) + F(t-\tau)]Pe + \beta e^T Pe - \beta e^T(t-\tau)Pe(t-\tau)$$

$$= e^T[PA + \beta P]e + F(t)Pe + F(t-\tau)Pe - \beta e^T(t-\tau)Pe(t-\tau)$$

$$\leq e^T[PA + \beta P]e + M\|e\|\|Pe\| + M\|e(t-\tau)\|Pe - \beta e^T(t-\tau)Pe(t-\tau)$$

$$\leq e^T[PA + \beta P]e + \frac{1}{2}(\|e\|^2 + M\|Pe\|^2) + \frac{1}{2}[\|e(t-\tau)\|^2 + M\|Pe\|^2] - \beta e^T(t-\tau)Pe(t-\tau)$$

$$= e^T[PA+\beta P]e+\frac{1}{2}e^Te+\frac{1}{2}Me^TP^TPe+\frac{1}{2}Me^TP^TPe+e^T(t-\tau)[\frac{1}{2}I-\beta P]e(t-\tau)$$

$$= e^T[PA+\beta P+\frac{1}{2}I+MP^TP]e+e^T(t-\tau)[\frac{1}{2}I-\beta P]e(t-\tau)$$

$$= e^TQ_1e+e^T(t-\tau)Q_2e(t-\tau)$$

$$< 0$$

Where, $e=\begin{bmatrix} e_1(t) \\ e_2(t) \end{bmatrix}$, $e(t-\tau)=\begin{bmatrix} e_1(t-\tau) \\ e_2(t-\tau) \end{bmatrix}$, $\|e\|^2=e^Te$

For the convenience of usage, we set $P=I$, and then corollary is given as following:

Corollary 1. If there exits a constant $\beta \geq \frac{1}{2}$ and let $f'(\cdot) < M$ ($M < \infty$), such that

$$A < -(\beta+\frac{1}{2}+M)I \tag{10}$$

Then error system (6) is global stability, i.e., the coupled system (5) synchronizes.

4 Numerical Simulation

In this section, we choose similar parameter as paper [17] with respect to chaotic system, i.e., $f(\cdot)=\sin(p\cdot)$, $p=2.81$, $a_1=1$, $b_1=1.9$, $a_2=1.71$, $b_2=0.61$, $\tau=1$.

According to Corollary 1, $M=2.81$, and if we choose $\beta=\frac{1}{2}$, then

$$A=\begin{bmatrix} -(1+2k_1) & 0 \\ 0 & -(1+2k_2) \end{bmatrix} < (\frac{1}{2}+\frac{1}{2}+2.81)I \,,$$

namely, $k_i > 1.405$ $(i=1,2)$.

The state portrait of the error system with initial conditions $x_1(t)=0.5$, $x_2(t)=-0.71$, $y_1(t)=-1.16$, $y_2(t)=-0.13$ is given in Fig. 1. State diagram of the coupled system are given in Fig. 2 and Fig. 3, respectively. The results of numerical simulation are consistent with the theoretical analyses.

5 Conclusions

Based on Krasovskii-Lyapunov theory, we investigate two delayed neuron chaotic systems with linearly bi-directional coupling, and derived some fairly simple generic synchronization criterion, which not only can be applied to the global synchronization for coupled chaotic systems via unidirectional linear error feedback, but also are more less conservation than the ones in paper [17]. The simulations demonstrate the effectiveness of the generic criterion.

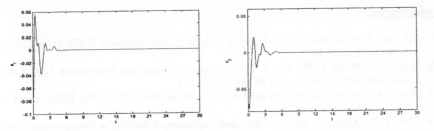

Fig. 1. Errors $e_1(t) = x_1(t) - y_1(t)$ and $e_2(t) = x_2(t) - y_2(t)$ of the coupled system.

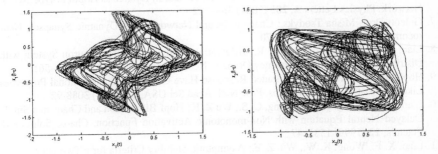

Fig. 2. State portraits for the coupled system (5).

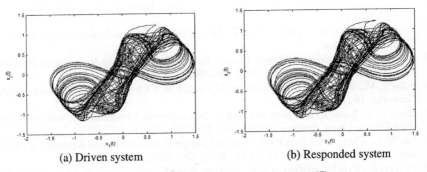

(a) Driven system   (b) Responded system

Fig. 3. State portraits for the coupled system (5).

Acknowledgements

The work described in this paper was supported by The National Natural Science Foundation of China (Grant no. 60271019), The Doctorate Foundation of the Ministry of Education of China (Grant no. 20020611007), the Post-doctoral Science Foundation of China and the Natural Science Foundation of Chongqing.

References

1. Gilli M.: Strange Attractors in Delayed Cellular Neural Networks. IEEE Trans Circuits Syst, **40** (1993) 849-53
2. Liao, X. F., Wong, K. W., Wu, Z. F.: Bifurcation Analysis on a Two-neuron System with Distributed Delays. Physica D, **149** (2001) 123–141
3. Zou, J. A., Nossek.: Bifurcation and Chaos in Cellular Neural Networks. IEEE Trans Circuits Syst, **40** (1993) 166-173
4. Chen, L. N., Kazuyuki Aihara.: Chaos and Asymptotical Stability in Discrete-time Neural Networks. Physica D, **104** (1997) 286-325
5. A. Das, A. B. Roy, Pritha Das.: Chaos in a Three Dimensional Neural Network. Applied Mathematical Modeling, **24** (2000) 511-522
6. Li, C. G., Yu, J. G., Liao, X. F.: Chaos in a Three-neuron Hysteresis Hopfield-type Neural Network. Physics Letters A, **285** (2001) 368–372
7. Gideon Dror, Misha Tsodyks.: Chaos in Neural Networks with Dynamic Synapses. Neurocomputin, **32-33** (2000) 365-370
8. Liao, X. F., Li, S. W., Chen, G. R.: Bifurcation Analysis on a Two-neuron System with Distributed Delays in the Frequency Domain. Neural Networks, **17** (2004) 545–561
9. Hopfield J. J.: Neurons with Graded Response Have Collective Computational Properties Like Those of Two-state neurons. Proc Natl. Acad Sci USA, **81** (1984) 3088-92
10. Liao, X. F., Wong, K. W., Leung, C. S., Wu Z. F.: Hopf Bifurcation and Chaos in a Single Delayed Neural Equation with Non-monotonic Activation Function. Chaos, Solitons & Fractals, **12** (1999) 1535–47
11. Liao, X. F., Wong, K. W., Wu, Z. F.: Asymptotic Stability Criteria for a Two-neuron Network with Different Time Delays. IEEE Trans NN, **14** (2003) 222–7
12. He, G. G., Cao, Z. T.: Controlling Chaos in a Chaotic Neural Network. Neural Networks, **16** (2003) 1195–1200
13. Aihara, K., Takabe, T., & Toyoda, M.: Chaotic Neural Networks. Physical Letters A, **144** (1990) 333–340
14. Degn, H., Holden, A. V., & Olsen, L. F. (Eds.): (1987). Chaos in Biological Systems. New York: Plenum Press
15. Adachi, M.: Controlling a Simple Chaotic Neural Network Using Response to Perturbation. Proceedings of NOLTA'95 1995; 989–992
16. Adachi, M., & Aihara, K.: Associative Dynamics in a Chaotic Neural Network. Neural Networks, **10** (1997) 83–98
17. Zhou, S. B., Liao, X. F.: Chaos and Its Synchronization in Two-neuron Systems with Discrete Delays. Chaos, Solitons and Fractals, **21** (2004) 133-142
18. Lu, W. L., Chen, T. P.: Synchronization of Coupled Connected Neural Networks With Delays. IEEE Trans. Circuits and Syst, **51** (2004) 2491-2503
19. Pyragas K.: Synchronization of Coupled Time-delay Systems: Analytical Estimations. Phy Rev E, **58** (1998) 3067-71
20. Frank Pasemann.: Synchronized Chaos and Other Coherent States for Two Coupled Neurons. Physica D, **128** (1999) 236–249
21. Gopalsamy K, Leung KC.: Convergence Under Dynamical Thresholds with Delays. IEEE Trans NN, **8** (1997) 341-8

Complex Dynamics
in a Simple Hopfield-Type Neural Network

Qingdu Li[1] and Xiaosong Yang[1,2]

[1] Institute for Nonlinear Systems, Chongqing University of Posts and Telecomm.,
Chongqing 400065, China
Qingdu_li@163.com
[2] Department of Mathematics, Huazhong University of Science and Technology,
Wuhan 430074, China
Yangxs@cqupt.edu.cn

Abstract. In this paper, we demonstrate complex dynamics in a classical Hopfield-type neural network with three neurons. There are no interconnections between the first one and the third one, so it may be a part with ignorable input from a complex neural network. However, the stable points, limit circles, single-scroll chaotic attractors and double-scrolls chaotic attractor have been observed as we adjust the weight from the second neuron to itself.

1 Introduction

The Hopfield neural network [1] abstracted from brain dynamics is a significant model in artificial neurocomputing. However, there exist criticisms that it is too simple because it is just a gradient descent system converging to an equilibrium point, and the brains seem more dynamical. In fact, substantial evidence has been found in biological studies for the presence of chaos in the dynamics of natural neuronal systems [2], [3], [4]). Many have suggested that this chaos plays a central role in memory storage and retrieval. From this point of view, many artificial neural networks have been proposed in order to realize more dynamical attractors as limit cycles and strange attractors in artificial neural dynamics (see the references of [5–11]).

Chaotic dynamics has been observed in high autonomous continuous time Hopfield-type neural networks [10, 11] as described by (1):

$$C_i \frac{dx_i}{dt} = -\frac{x_i}{R_i} + \sum_{j=1}^{n} w_{ij} v_j + I_i \quad i = 1, 2, ..., n \tag{1}$$

$$v_i = f_i(x_i)$$

where f_i is a monotone differentiable function which is bounded above and below and $W = (w_{ij})$ is an $n \times n$ matrix, called weight matrix or connection matrix describing the strength of connections between neurons.

So two natural questions we are interested are: Whether chaos can take place in autonomous continuous time Hopfield-type neural networks with few neurons as described by (1), and how many interconnections at least are needed?

J. Wang, X. Liao, and Z. Yi (Eds.): ISNN 2005, LNCS 3496, pp. 357–362, 2005.

Generally speaking, chaos does not occur in $n(\leq 2)$-dimensional autonomous continuous time systems [12], so we consider the questions in 3 dimension. We study three neurons that have no interconnections between the first neuron and the last one. The three neurons may be a part with ignorable input from a complex neural network. This could help to design more dynamical neural networks.

In this paper, the three neurons we will study are of a simplified form of (1), as given in Fig. 1 and (2) in next section, we will show that complex dynamics can occur in (2) for some weight matrices.

2   The 3D Hopfield-Type Neural Network

In this section, we will study a simplified Hopfield-type neural network, as shown in Fig. 1.

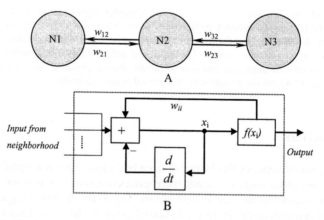

Fig. 1. A. The three-neuron Hopfield neural network; B. The detail diagram of a neuron

The dynamic of this system can be described by the following ordinary differential equations:

$$\dot{x} = -x + W \cdot f(x) \tag{2}$$

where $f(x) = [f(x_1) \quad f(x_2) \quad f(x_3)]^T$ and $W = \begin{pmatrix} w_{11} & w_{12} & \\ w_{21} & w_{22} & w_{23} \\ & w_{32} & w_{33} \end{pmatrix}$.

For simplicity, we choose the output function $f(x_i) = \frac{1}{2}(|x_i+1|-|x_i-1|)$, $i=1,2,3$, as shown in Fig. 2. Obviously, the system is symmetric with respect to the origin.

An interesting result is that (2) has a nontrivial periodic orbit for some weight matrices W, and this result was proved using monotone systems theory [13]. A more interesting question is whether (2) can exhibits chaos for some weight matrix W.

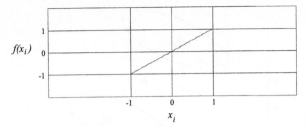

Fig. 2. The output function

An obvious statement is that (2) does not have complicated dynamics (even non-trivial periodic solutions) if the weight matrix is symmetric, therefore it is reasonable to consider the case that the weight matrix W is asymmetric.

In what follows, we will show by computer simulation that (2) does exhibit abundant dynamics for some carefully chosen weight matrices. Now take the weight matrix W to be

$$W = \begin{pmatrix} 1.2 & -1.6 & \\ 1.2 & 1+p & 0.9 \\ & 2.2 & 1.5 \end{pmatrix} \tag{3}$$

where, p is an adjustable parameter.

The number of equilibrium points of (2) and the corresponding stability are shown in the following table (Table 1), as we adjust the parameter p from $-\infty$ to $+\infty$.

Table 1. Equilibrium points of (2) as we adjust the parameter p from $-\infty$ to $+\infty$.

Region of p	Number of isolated equilibrium points	Stable	Unstable
$p \le -16.8$	9	4	5
$-16.8 < p \le -5.64$	5	2	3
$-5.64 < p < -0.2$	3	2	1
$-0.2 \le p < 0.3$	3	0	3
$0.3 \le p < -9.24$	7	2	5
$-9.24 \le p$	11	2	9

For $-0.2 \le p < 0.3$, there are only three equilibrium points, $O = \begin{pmatrix} 0 & 0 & 0 \end{pmatrix}^T$ and

$$O_u = -O_d = \frac{3}{48+5p} \left(-12 \quad \frac{3}{2} \quad \frac{207+25p}{10} \right)^T,$$ but they are all unstable. Because all

the trajectories of the system (2) are bounded, there must be some other dynamics such as limit circles or strange attractors. To illustrate this fact, we adjusted p slowly, and observed the following interesting cases.

1) Two Limit Circles
At first we observed two limit circles in the phase space as illustrated in the following figures (Fig.3) with p = -0.12.

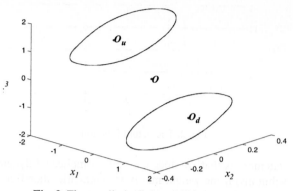

Fig. 3. The two limit circles of (2) with p=-0.12

2) Two Single-Scroll Chaotic Attractors

As we adjusted p up slowly, the two limit circles disappear, and then two strange attractors emerge. More precisely, for the parameter p=-0.03, computer simulation shows that (2) has two single-scroll strange attractors as illustrated in the following figures (Fig.4) in terms of variables $x_1x_2x_3$ and x_1x_3 , respectively and the Lyapunov exponents of (2) are 0.050, 0.000 and -0.480, respectively, and its Lyapunov dimension (or Kaplan-Yorke dimension) is *2.106*. The calculated positive Lyapunov exponent for the attractor suggests that it should be chaotic.

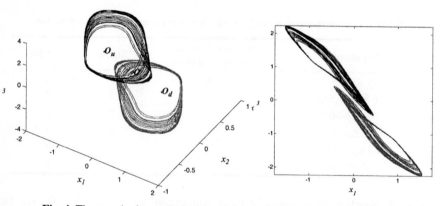

Fig. 4. The two single-scroll chaotic attractors of (2) observed with p=-0.03

3) One Double-Scroll Chaotic Attractor

As we adjust p up slowly, the two single-scroll strange attractors approach to each other and form a two-scroll attractor. For the parameter p=0, computer simulation shows (2) has an attractor as illustrated in the following figures (Fig.5). It also shows that Lyapunov exponents of (2) are *0.070, 0.000*, and *-0.638*, respectively, and its Lyapunov dimension is *2.110,* thus giving a numerical evidence of chaos in (2).

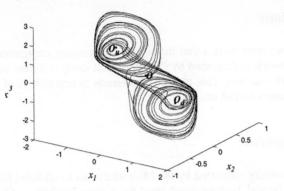

Fig. 5. The double-scrolls chaotic attractor observed with $p=0$

4) One Limit Circle

As we continue adjusting p up slowly, the two-scroll attractor become a limit circle at last. For the parameter p=0.12, a limit circle appears, as shown in Fig. 6.

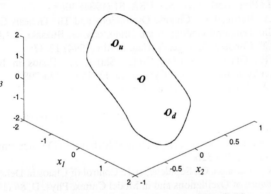

Fig. 6. The one limit circle observed with p=0.12

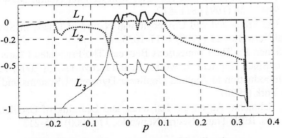
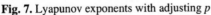

Fig. 7. Lyapunov exponents with adjusting p

In addition, we calculated the three Lyapunov exponents $[L_1\ L_2\ L_3]^T$ of this system with the initiate condition $x_0 = [0.02\ -0.1\ 0.1]^T$, as shown in Fig. 7. This figure shows the evidence that there may be other limit circles between the chaotic attractors with respect to p.

3 Conclusions

In this paper, we have shown that the simple autonomous continuous time Hopfield-type neural network as described by (2) can exhibit many different complex dynamics for some weight matrices. This may be of interests to researchers of neural networks, nonlinear dynamics and so on.

Acknowledgments

This work is partially supported by the Microelectronics and Solid Electronics Project and the Scientific and Technological Fund for Young Teachers of Chongqing University of Posts and Telecommunications.

References

1. Hopfield, J. J.: Proc. Natl. Acad. Sci. USA. **81** (1984) 3088
2. Freeman, W.J., Barrie, J.M.: Chaotic Oscillations and The Genesis Of Meaning In Cerebral cortex. In: Temporal Coding in the Brain, eds. G. Buzsaki, R.Llinas, W. Singer, A. Berthoz, and Y. Christen, Springer-Verlag, Berlin (1994) 13-37
3. Guevara, M.R., Glass, L. Mackey, M.C., Shrier, A.: Chaos in Neurobiology. IEEE Transctions on Systems, Man, and Cybernetic, **13** (1983) 790-798
4. Babloyantz, A., Lourenco, C.: Brain Chaos and Computation. International Journal of Neural Systems, **7** (1996) 461-471
5. Wheeler, D.W., Schieve, W.C.: Stability and Chaos In An Inertial Two-neuron Systems. Physica D, **105** (1997) 267-284
6. Chen, L., Aihara, K.: Chaos and Asymptotical Stability In Discrete-time Neural Networks. Phys. D, **104** (1997) 286-326
7. Babloyanz, A., Lourenco, C., Sepulche, A.J.: Control of Chaos in Delay Differential Equations in a Network of Oscillations and in Model Cortex. Phys. D, **86** (1995) 274-283
8. Das, P., Schieve, W.C., Zeng, Z.: Chaos in an Effective Four-Neuron Neural Network. Phys. Lett. A, **161** (1991) 60-66
9. Zou, F., Nossek, J. A.: A Chaotic Attractor with Cellular Neural Networks. IEEE Trans. Circuits and Syst. I, **38** (1991) 811–812
10. Bersini, H.: The Frustrated and Compositional Nature of Chaos in Small Hopfield Networks. Neural Networks, **11** (1998) 1017-1025
11. Bersini, H., Sener, P.: The Connections Between The Frustrated Chaos and the Intermittency Chaos in Small Hopfield Networks. Neural Networks, 15(2002) 197-1204
12. Wiggins,S.: Introduction to Applied Nonlinear Dynamical Systems and Chaos, Springer-Verlag, New York (1990)
13. Townley, S., Ilchmann, A., et, al.: Existence and Learning of Oscillations in Recurrent Neural Networks. IEEE Trans. Neural Networks, **11** (2000) 205-213

Adaptive Chaotic Controlling Method of a Chaotic Neural Network Model

Lidan Wang[1], Shukai Duan[1,2], and Guangyuan Liu[1]

[1] School of Electronic Information Engineering, Southwest China Normal University
Chongqing 400715, China
ldwang_swnu@163.com, liugy@swnu.edu.cn
[2] Department of Computer Science and Engineering, Chongqing University 400030, China
duansk@swnu.edu.cn

Abstract. It has been found that chaotic dynamics may exist in real brain neurons and play important roles in signal proceeding. But it is hard to set suitable parameters of system to make it be chaotic in practice. In this paper, a general adaptive controlling method of nonlinear systems with chaotic dynamics is studied. According analysis Lyapunov exponent, the effectiveness of our scheme is illustrated by a series of computer simulations.

1 Introduction

After Lorenz found the first chaos phenomena [1], chaos and chaotic neural networks (CNN) have been deeply studied by many researchers. It has been found that chaotic dynamics may exist in real brain neurons and play important roles in signal proceeding [1–4]. However, it is hard to set suitable parameters of CNN to make it show chaotic states. There have been reported a lot of controlling methods such as optimal control, adaptive control, continuous control et al [5–7]. But most of them are too complex and it is too difficult to operate in practice.

In this paper, we analyze chaotic behavior of a chaotic network and propose a new adaptive chaotic controlling method. Computer simulations show the effectiveness of the proposed method.

2 A Chaotic Neuron and Chaotic Neural Network

Here, we review a chaotic neuron and chaotic neural networks. The dynamical equation of a chaotic neuron model is described as follows [1]:

$$x(t+1) = f\left(A(t) + \alpha \sum_{d=0}^{t} k^d g(x(t-d)) - \theta\right) \tag{1}$$

where $x(t+1)$ is the output of the chaotic neuron at the $t+1$; f is a output function $f(y) = 1/(1+e^{-y/\varepsilon})$ with the steepness parameter ε ; $g(x) = x$; $A(t)$ is the external stimulation; α, k and θ are positive refractory scaling parameter, the damping factor and the threshold , respectively. If defining the internal state $y(t+1)$ as

J. Wang, X. Liao, and Z. Yi (Eds.): ISNN 2005, LNCS 3496, pp. 363–368, 2005.
© Springer-Verlag Berlin Heidelberg 2005

$$y(t+1) = A(t) + \alpha \sum_{d=0}^{t} k^d g(x(t-d)) - \theta \tag{2}$$

we can reduce equation (1) to the following equation,

$$y(t+1) = ky(t) + \alpha g\{f[y(t)]\} - a(t) \tag{3}$$

rewrite the output of the chaotic neuron as,

$$x(t+1) = f[y(t+1)] \tag{4}$$

The chaotic neural networks (CNN) consist of chaotic neurons with spatio-temporal summation of feedback inputs and externally applied inputs.

$$x_i(t+1) = f(\xi_i(t+1) + \eta_i(t+1) + \zeta_i(t+1)) \tag{5}$$

Where external inputs $\xi_i(t+1)$, feedback inputs $\eta_i(t+1)$ and refractoriness $\zeta_i(t+1)$ are defined as equation (6-8), respectively.

$$\xi_i(t+1) = \sum_{j=1}^{M} v_{ij} A_j(t) + k_e \xi_i(t) = \sum_{j=1}^{M} v_{ij} \sum_{d=0}^{t} k_e^d A_j(t-d) \tag{6}$$

$$\eta_i(t+1) = \sum_{j=1}^{N} w_{ij} x_j(t) + k_f \eta_i(t) = \sum_{j=1}^{N} w_{ij} \sum_{d=0}^{t} k_f^d x_j(t-d) \tag{7}$$

$$\zeta_i(t+1) = -\alpha g\{x_i(t)\} + k_r \zeta_i(t) - \theta_i = -\alpha \sum_{d=0}^{t} k_r^d g\{x_i(t-d)\} \tag{8}$$

In equation (7), the synaptic weights are trained by Hebbian learning, such as,

$$w_{ij} = \sum_{k=1}^{N} (x_i^k x_j^k) \quad or \quad w_{ij} = \frac{1}{N} \sum_{k=1}^{N} (2x_i^k - 1)(2x_j^k - 1) \tag{9}$$

3   Adaptive Control on Chaotic Dynamics Behavior

Based on our previous work [9–11], we propose a novel associative chaotic neural network (NACNN). We use continuously external input instead of initial input and replace spatio-temporal effect with a new exponential decreasing effect. In the traditional associative chaotic neural networks, researches considered that each historic effect working on the network as equal level, so they usually assumed decay parameters as constant values which are far from the truth in facts. Here, we explain the structure, the associative memory capacity of the proposed model.

3.1   Analysis on Chaotic Dynamics Behavior of the CNN

Fig. 1 shows a periodic and a chaotic response according to the parameter value. The firing rate ρ and the Lyapunov exponent λ are defined as follows, respectively.

$$\rho = \lim_{n \to +\infty} \frac{1}{n} \sum_{t=t_0}^{n+t_0-1} h[x(t)] \tag{10}$$

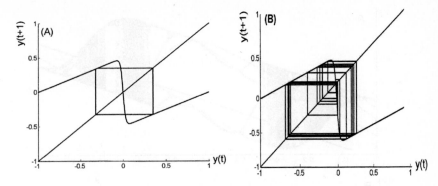

Fig. 1. Phase diagram of equation (3). (A) periodic response with k=0.5, a=0.5, α=1.0 ε =0.015. (B) chaotic response with k=0.5, a=0.3, α=1.0 ε =0.015.

$$\lambda = \lim_{n \to +\infty} \frac{1}{n} \sum_{t=t_0}^{n+t_0-1} \ln |\frac{dy(t+1)}{dy(t)}| \tag{11}$$

h is transfer function which can be assumed to $h(x)=1$ for $x \geq 0.5$ and $h(x)=0$ for $x \leq 0.5$. Fig. 2 shows the neuron has rich bifurcate and chaotic dynamical behaviors according to changing the parameter value.

From fig.2, we can find that the CNN shows rich chaotic states while parameter a is changing from 0.12 to 0.20, from 0.25 to 0.34, from 0.65 to 0.74, from 0.77 to 0.87 while k =0.7, a=0.5, α =1.0 ε =0.05, y(0)=0.5 is given. In fig. 3, the average firing rate is increasing according to emergence of the chaotic states of the CNN. We make use of this property, and adaptive chaotic controlling can be expected to realize.

3.2 Adaptive Chaotic Control of the CNN

To realize adaptive control on chaotic dynamics behavior of the CNN, we add a controlling term in equation (3, 8), which can be represented as $C(t)$. So, equation (3, 8) can be stated as followings:

$$y(t+1) = ky(t) + \alpha g\{f[y(t)]\} - a + C(t) \tag{12}$$

$$\zeta_i(t+1) = -\alpha g\{x_i(t)\} + k_r \zeta_i(t) - a_i + C(t) \tag{13}$$

Where $C(t)$ can be a random positive parameter selected from 0.12 to 0.20, or from 0.25 to 0.34, or from 0.65 to 0.74, or from 0.77 to 0.87 while in senses of section (2). In general, we need a quasi-energy function E, as used in Hopfield network.

$$E = -\frac{1}{2}\sum_{i}^{N}\sum_{j}^{N} w_{ij}x_i x_j + a_i x_i \tag{14}$$

Statement of adaptive chaotic control algorithm can be simply described as follows:

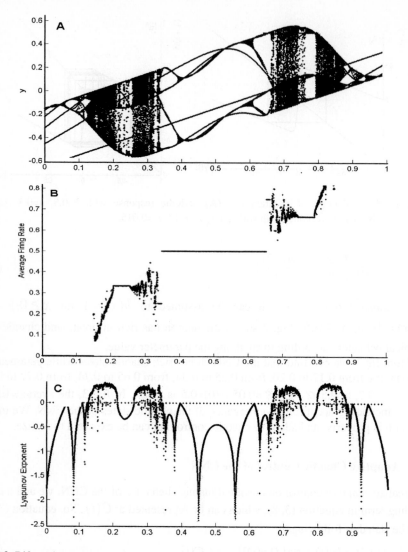

Fig. 2. Bifurcation structure diagram of equation (3) with changing the value of the parameter a from 0 to 1 when k=0.7, a=0.5, α=1.0 ε=0.05, y(0)=0.5. (A) the bifurcation structure diagram; (B) the average firing rate; (C) the Lyapunov exponent.

(1) Input initial value to the chaotic neural network.
(2) Calculate the quasi-energy function E.
(3) Produce a random positive parameter which selected from chaotic attractive region. At this step, one need compute Lyapunov exponent to confirm the state of the neural network is chaotic.
(4) Compare the quasi-energy function E and E' (which is concluding the $C(t)$) .
(5) Update state of the chaotic neural network.

In order to test the effectiveness of the proposed method, we accomplished two examples to control the state of chaotic neural network to be chaotic (Fig. 3, Fig. 4). Comparing with as it shows in fig. 2 (C), we consider that the chaotic neural network with adaptive chaotic controlling show much richer chaotic dynamical behaviors than conventional chaotic neural network. Further more, we can expect that the proposed method may improve chaotic search capacity in associations and combination optimizations.

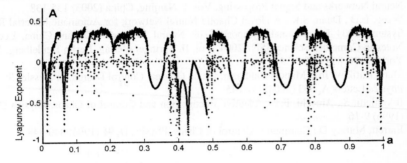

Fig. 3. The Lyapunov exponent of the CNN with adaptive chaotic controlling (1).

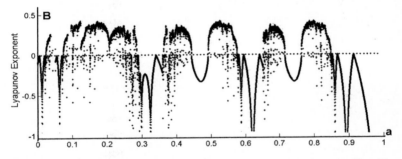

Fig. 4. The Lyapunov exponent of the CNN with adaptive chaotic controlling (2).

4 Conclusions

we analyzed chaotic behavior of a chaotic network in details. Aiming to control chaotic behavior of a chaotic neural network, we propose a new adaptive chaotic controlling method. Computer simulations show the effectiveness of the proposed method.

Acknowledgement

The work is supported by the Natural Science Foundation of Southwest China Normal University (Grant No. SWNUQ2004024).

References

1. Aihara, K., Takabe, T., Toyoda, M.: Chaotic Neural Networks. Physics Letters A, **144** (1990) 333-340
2. Yao, Y., Freeman, W.J.: Model of Biological Pattern Recognition with Spatially Chaotic Dynamics Neural Networks. Neural Networks, **3** (1990) 153-170
3. Duan, S.K., Liu, G.Y., Wang, L.D., Qiu, Y.H.: A Novel Chaotic Neural Network for Many-to-Many Associations and Successive Learning. IEEE International Conference on Neural Networks and Signal Processing, Vol. 1. Nanjing, China (2003) 135-138
4. Wang, L.D., Duan, S.K.: A Novel Chaotic Neural Network for Automatic Material Ratio System. 2004 International Symposium on Neural Networks, Dalian, China. Lecture Notes in Computer Science-ISNN2004. Vol. II. Springer-Verlag, Berlin Heidelberg, New York (2004) 813-819
5. Marat Rafikov, José Manoel Balthazar: On an Optimal Control Design for Rossler System Physics Letters A, **333** (2004)241-245
6. Boccaletti, S., Arecchi, F.T.: Adaptive Recognition and Control of Chaos. Physica D, **96** (1996) 9-16
7. Barrett, Murray D.: Continuous Control of Chaos. Physica D, **91** (1996) 340-348

Modeling Cortex Network:
A Spatio-temporal Population Approach

Wentao Huang[1,2], Licheng Jiao[1], Maoguo Gong[1], and Chuang Guo[2]

[1] Institute of Intelligent Information Processing
and Key Laboratory of Radar Signal Processing, Xidian University,
Xi'an, Shaanxi 710071, China
skyhwt@yahoo.com.cn
[2] College of Engineering, Air Force Engineering University,
Xi'an, Shaanxi 710038, China

Abstract. The cerebral cortex is composed of a large number of neurons. More and more evidences indicate the information is coded via a population approach in cerebrum, and is associated with the spatio-temporal pattern of spiking of neurons. In this paper, we present a novel model that represents the collective activity of neurons with spatio-temporal evolution. We get a density evolution equation of neuronal populations in phase space, which utilize the single neuron dynamics (integrate-and-fire neuron model). Both in theory analysis and applications, our method shows more predominance than direct simulation the large populations of neurons via single neuron.

1 Introduction

In many areas of the brain neurons are organized in populations of units with similar properties. Prominent examples are columns in the visual cortex and somatosensory, and pools of motor neurons. Given a large number of neurons within such a column or pool it is sensible to describe the mean activity of the neuronal population rather than the spiking of individual neurons. Each cubic millimeter of cortical tissue contains about 10^5 neurons. This impressive number also suggests that a description of neuronal dynamics in terms of a population activity is more appropriate than a description on the single-neuron level. Furthermore, the population approach allows us to study neural signal transmission and coding, oscillations and synchronization. Knight *et al* [1] introduce an approach to modeling and simulating of the dynamics of interacting populations of neurons. Further expositions of the present approach are given in [2–6]. But their models are restricted to homogeneous populations of neurons and absence of spatial structure information. In this paper, we present a novel spatio-temporal population approach to study the behavior of neural system, which allow us to characterize the nonhomogeneous populations of neurons.

2 The Integrate-and-Fire Neuron Model

Our implementation of the population density approach is based on an integrate-and-fire (IF) point (single compartment) neuron. Although the approach could be general-

J. Wang, X. Liao, and Z. Yi (Eds.): ISNN 2005, LNCS 3496, pp. 369–374, 2005.

ized to other neuron models, the population density based on an integrate-and-fire neuron is low-dimensional and thus can be computed efficiently [3], [4]. The basic circuit of an integrate-and-fire model consists of a capacitor C in parallel with a resistor R driven by a current $I(t)$. The driving current can be split into two components, $I(t) = I_R + I_C$. The first component is the resistive current I_R which passes through the linear resistor R. It can be calculated from Ohm's law as $I_R = (u\text{-}u_r)/R$, where $u\text{-}u_r$ is the voltage across the resistor, and u_r is the reset potential. The second component I_C charges the capacitor C. From the definition of the capacity as $C = q/(u\text{-}u_r)$ (where q is the charge and $u\text{-}u_r$ the voltage), we can get a capacitive current $I_C = C\, du/dt$. Thus

$$I(t) = \frac{u(t) - u_r}{R} + C\frac{du}{dt} .$$

(1)

We multiply (1) by R and introduce the time constant $\tau = RC$ of the `leaky integrator'. This yields the standard form

$$\frac{du}{dt} = -\frac{u(t) - u_r}{\tau} + RI(t) .$$

(2)

We refer to u as the membrane potential and to τ as the membrane time constant of the neuron. In integrate-and-fire models the form of an action potential is not described explicitly. Spikes are formal events characterized by a `firing time' $t^{(f)}$. The firing time $t^{(f)}$ is defined by a threshold criterion

$$t^{(f)} : u(t^{(f)}) = u_{th} .$$

(3)

Immediately after $t^{(f)}$, the potential is reset to a new value $u_r < u_{th}$

$$\lim_{t \to t^{(f)}, t > t^{(f)}} u(t) = u_r .$$

(4)

For $t > t^{(f)}$ the dynamics is again given by (2) until the next threshold crossing occurs. The combination of leaky integration (2) and reset (4) defines the basic integrate-and-fire model. By convention, the u is replaced by v, the rescaled potential

$$v = \frac{u - u_r}{u_{th} - u_r} .$$

(5)

We have

$$\frac{dv}{dt} = -\eta v(t) + s(t) ; \quad 0 \le v \le 1 ,$$

(6)

Where the trans-membrane potential, v, has been normalized so that $v = 0$ marks the rest state, and $v = 1$ the threshold for firing. When the latter is achieved v is reset to zero. $\eta = 1/\tau$, a frequency, is the leakage rate and s, also having the dimensions of frequency, is the normalized current due to synaptic arrivals at the neuron.

3   The Spatio-temporal Population Model

Under the statistical approach one considers a population of N neurons which distribute over a volume Ω, each following (6). At time t, $\rho(\vec{r}, t)$ gives the density of neurons and $p(v, \vec{r}, t)$ gives the ratio of state v, at location $\vec{r} = (x, y, z)$. Then

$q(v,\vec{r},t)dvd\vec{r} = \rho(\vec{r},t)p(v,\vec{r},t)dvd\vec{r}$ specifies the number of neurons, at time t, in the range of states $(v, v+dv)$ and cubage $(\vec{r},\vec{r}+d\vec{r})$, $d\vec{r} = dxdydz$. When $\vec{r} \notin \Omega$ or $v \notin [0,1]$, $q(v,\vec{r},t) \equiv 0$. The firing rate $R(\vec{r},t)$ of the population around location $\vec{r}$ is defined by the the number of spikes of neurons $n(\vec{r},\vec{r}+\Delta\vec{r};t,t+\Delta t)$ in a small time interval Δt and a small volume $\Delta\vec{r}$

$$R(\vec{r},t) = \lim_{\substack{\Delta t \to 0 \\ \Delta r \to 0}} \frac{n(\vec{r},\vec{r}+\Delta\vec{r};t,t+\Delta t)}{\Delta t \Delta \vec{r}} . \tag{7}$$

The total input intensity of per unit time, $\sigma(\vec{r},t)$, to a neuron at position $\vec{r}$ is therefore

$$\sigma(\vec{r},t) = \int_{\Omega} R(\vec{r}',t)w(\vec{r},\vec{r}')d\vec{r}' . \tag{8}$$

Here, $w(\vec{r},\vec{r}')$ is the coupling strength of two neurons as a function of their location.

Our aim is to find the evolution equation of the density function $q(\overline{X},t)$ in phase space $\overline{X} = (v,\vec{r})$ through the dynamics of individual neurons. First, we have

$$\dot{\overline{X}} = (\frac{dv}{dt},\frac{d\vec{r}}{dt}) = (-\eta v + s, \vec{0}) ; \quad 0 \le v \le 1 , \ \vec{r} \in \Omega . \tag{9}$$

From (9), we can know the evolution of phase state $\overline{X}$ with time is only affected by v. In (6), the first term of right, $-\eta v$, accounts for the continuous drift of the membrane potential v during the time when no input spike arrives, and the second term, s, which can make membrane potential v produce discrete jump, is due to excitatory and inhibitory spike arrival.

Considering a continuous medium in phase space $\overline{X}$, at time t and location $\overline{X}(t)$, it streams at a speed of $V(\overline{X},t) = \dot{\overline{X}}(t) = (-\eta v, \vec{0})$. Here, a fixed reference frame, O_x, measures the location $\overline{X}(t)$. For convenience, we introduce the second reference frame, O_z, flowing with medium at a speed of $V(\overline{X},t)$. Assume the two reference frames superposition at time $t=0$, the coordinate of the medium particle is $\overline{X}(0)$ in O_x, and $\overline{Z}(0)$ in O_z. At time t, the coordinate respectively is $\overline{X}(t)$ and $\overline{Z}(t) = \overline{Z}(0)$.

Following, we are to find out the evolution equation of the density function $q(\overline{X},t)$. Considering the small voxel D_0 around location $\overline{X}_0$ at time $t=0$, it changes to D_t at time t. Then the total quantity in voxel D_t is

$$Q(t) = \int_{D_t} q(\overline{X},t)dD_t . \tag{10}$$

The rate of variety observed by one flowing with D_t in voxel D_t is

$$\frac{d}{dt}Q(t) = \frac{d}{dt}\int_{D_t} q(\overline{X},t)dD_t = \frac{d}{dt}\int_{D_t} q(\overline{X},t)dv_t d\vec{r}_t = \frac{d}{dt}\int_{D_0} q(\overline{X},t)\xi dv_0 d\vec{r} = \int_{D_0}(\frac{dq}{dt}\xi + q\frac{d\xi}{dt})dv_0 d\vec{r} . \tag{11}$$

Here ξ is the Jacobian transformation

$$\xi = \frac{dv_t}{dv_0} . \tag{12}$$

and

$$\frac{d\xi}{dt} = \frac{d}{dt}(\frac{dv_t}{dv_0}) = \frac{d\dot{v}_t}{dv_0} = \frac{d\dot{v}_t}{dv_t}\frac{dv_t}{dv_0} = \frac{d\dot{v}_t}{dv_t}\xi . \tag{13}$$

According (13) and (11), we have

$$\frac{d}{dt}Q(t) = \int_{D_0}(\frac{dq}{dt} + q\frac{d\dot{v}_t}{dv_t})\xi dv_0 d\vec{r} = \int_{D_t}(\frac{dq}{dt} + q\frac{d\dot{v}_t}{dv_t})dv_t d\vec{r}_t = \int_{D_t}(\frac{dq}{dt} + q\frac{d\dot{v}_t}{dv_t})dD_t . \tag{14}$$

From above discussion, we can get

$$\frac{dq}{dt} = \frac{\partial q}{\partial t} + \frac{\partial q}{\partial \overline{X}}\frac{d\overline{X}}{dt} = \frac{\partial q}{\partial t} + \dot{\overline{X}}\cdot\nabla_{\overline{X}}q . \tag{15}$$

and

$$\frac{d\dot{v}_t}{dv_t} = \nabla_{\overline{X}}\cdot\dot{\overline{X}} . \tag{16}$$

Thus, we have

$$\frac{d}{dt}Q(t) = \int_{D_t}(\frac{\partial q}{\partial t} + \dot{\overline{X}}\cdot\nabla_{\overline{X}}q + q\nabla_{\overline{X}}\cdot\dot{\overline{X}})dD_t = \int_{D_t}(\frac{\partial q}{\partial t} + \nabla_{\overline{X}}\cdot(q\dot{\overline{X}}))dD_t . \tag{17}$$

If it is no external input in voxel D_t, $\frac{d}{dt}Q(t) = 0$, but, due to the synaptic input $s(t)$ from other neurons at position $\vec{r}'$, the membrane potential v of neurons at position $\vec{r}$ can jump into new state $v' = v + w(\vec{r},\vec{r}')$ at time t. Then $\frac{d}{dt}Q(t) \neq 0$,

$$\frac{d}{dt}Q(t) = \int_{D_t}dD_t \int_{\Omega}q(v-w(\vec{r},\vec{r}'),r,t)R(\vec{r}',t-\tau)d\vec{r}' - \int_{D_t}dD_t \int_{\Omega}q(v,\vec{r},t)R(\vec{r}',t-\tau)d\vec{r}' + \int_{D_t}\delta(v)R(\vec{r},t)dD_t . \tag{18}$$

where $\delta(v)$ is the Dirac function, and τ is the time delay from $\vec{r}'$ to $\vec{r}$ and is the function of location $\vec{r}'$ and $\vec{r}$, $\tau = \tau(\vec{r},\vec{r}')$. In equation (18), the first term is the per unit time number of jump into voxel D_t, and the second term is the per unit time number of escape from voxel D_t at time t, and the last term is arisen by reset. According (17) and (18), we can get

$$\frac{\partial q}{\partial t} + \nabla_{\overline{X}}\cdot(q\dot{\overline{X}}) = \int_{\Omega}q(v-w(\vec{r},\vec{r}'),\vec{r},t)R(\vec{r}',t-\tau)d\vec{r}' - \int_{\Omega}q(v,\vec{r},t)R(\vec{r}',t-\tau)d\vec{r}' + \delta(v)R(\vec{r},t) . \tag{19}$$

Formula (19) is the evolution equation of the function $q(\overline{X},t)$. From the right terms of equation (19), we have

$$-\frac{\partial}{\partial v}(\int_{v-w(\vec{r},\vec{r}')}^{v}\int_{\Omega}q(v',\vec{r},t)R(\vec{r}',t-\tau)d\vec{r}'dv' + H(-v)R(\vec{r},t)) . \tag{20}$$

Here $H(v)$ is the Heaviside step function

$$H(v) = \begin{cases} 0 & v < 0 \\ 1 & v \geq 0 \end{cases} . \tag{21}$$

Due to

$$\dot{\vec{X}} = (-\eta v, \vec{0}) \,. \tag{22}$$

we get

$$\nabla_{\vec{x}} \cdot (q \, \dot{\vec{X}}) = \frac{\partial}{\partial v}(-\eta v q) \,. \tag{23}$$

Thus, from equation (19), we have

$$\frac{\partial q}{\partial t} = -\frac{\partial}{\partial v}(-\eta v q + \int_{v-w(\vec{r},\vec{r'})}^{v} \int_{\Omega} q(v',\vec{r},t)R(\vec{r'},t-\tau)d\vec{r'}dv' + H(-v)R(\vec{r},t)), \quad 0 \le v \le 1 \,. \tag{24}$$

The flux in direction v is

$$J_v(v,\vec{r},t) = -\eta v q + \int_{v-w(\vec{r},\vec{r'})}^{v} \int_{\Omega} q(v',\vec{r},t)R(\vec{r'},t-\tau)d\vec{r'}dv' + H(-v)R(\vec{r},t), \quad 0 \le v \le 1 \,. \tag{25}$$

In directions $\vec{r} = (x,y,z)$, the fluxes are $J_x = J_y = J_z = 0$. The total flux in direction v, $J_v(v,\vec{r},t)$, at the threshold $v = 1$ yields the population activity

$$R(\vec{r},t) = J_v(1,\vec{r},t) = \int_{1-w(\vec{r},\vec{r'})}^{1} \int_{\Omega} q(v',\vec{r},t)R(\vec{r'},t-\tau)d\vec{r'}dv' \,. \tag{26}$$

Defining

$$\vec{J}(v,\vec{r},t) = (J_v, J_x, J_y, J_z) \,. \tag{27}$$

equation (24) is

$$\frac{\partial q}{\partial t} = -\nabla_{\vec{x}} \cdot \vec{J}(v,\vec{r},t), \quad 0 \le v \le 1, \ \vec{r} \in \Omega \,. \tag{28}$$

If we use following coordinate transformation:

$$\begin{cases} v' = v e^{-\eta t} \\ q'(v',\vec{r},t) = e^{-\eta t} q(v e^{-\eta t}, \vec{r}, t) \end{cases} \tag{29}$$

the partial differential equation (PDE) (19) can change to an ordinary differential equations (ODE)

$$\frac{dq'}{dt} = \int_{\Omega} q'(v' - w(\vec{r},\vec{r'})e^{-\eta t}, \vec{r}, t)R(\vec{r'}, t-\tau)d\vec{r'} - \int_{\Omega} q'(v',\vec{r},t)R(\vec{r'},t-\tau)d\vec{r'} + \delta(v')R(\vec{r},t) \,. \tag{30}$$

4 Conclusion and Future Work

We have presented a dynamics model of large populations of interacting neurons. This model represents a temporal evolution equation of the density function $q(v,\vec{r},t)$ of phase space $\vec{X} = (v,\vec{r})$. It can be extend to the condition v is a vector $\vec{v}$. We have got the different forms of the equation, namely (19), (24), (28), and (30). In order to solve these equations, we can use the numerical method. If the neurons density is homogeneous, we can deduce the equation in literature [1] from our model. When compared with direct simulation of the large populations of neurons via single neuron, our method has obvious predominance both in theory analysis and applications.

References

1. Knight, B.W., Manin, D., Sirovich, L.: Dynamical Models of Interacting Neuron Populations in Visual Cortex. In Symposium on Robotics and Cybernetics: Computational Engineering in Systems Applications. E. C. Gerf, France: Cite Scientifique (1996)
2. Knight, B.: Dynamics of Encoding in Neuron Populations: Some General Mathematical Features. Neural Computation, 12 (2000) 473-518
3. Omurtag, A., Knight, B., Sirovich, L.: On the Simulation of Large Populations of Neurons. Journal of Computational Neuroscience, 8 (2000) 51-63
4. Nykamp, D.Q., Tranchina, D.: A Population Density Approach that Facilitates Large-Scale Modeling of Neural Networks: Analysis and an Application to Orientation Tuning. Journal of Computational Neuroscience, 8 (2000) 19-50
5. Casti, A., Omurtag, A., Sornborger, A., Kaplan, E., Knight, B., Victor, J., Sirovich, L.: A Population Study of Integrate-and-Fire-or-Burst Neurons. Neural Computation, 14 (2002) 947-986
6. de Kamps, M.: A Simple and Stable Numerical Solution for the Population Density Equation. Neural Computation, 15 (2003) 2129-2146

A Special Kind of Neural Networks: Continuous Piecewise Linear Functions

Xusheng Sun and Shuning Wang

Department of Automation, Tsinghua University, Beijing 100084, China
sunxs03@mails.tsinghua.edu.cn, swang@mail.tsinghua.edu.cn

Abstract. Continuous piecewise linear functions play an import role in approximation, regression and classification, and the problem of their explicit representation is still an open problem. In this paper, we propose a new form of representation,thus it can be seen clearly that they are actually a special kind of neural networks. Then we improve a typical piecewise linear algorithm called hinging hyperplanes to solve the training problem. Simulation results show that the novel algorithm has a much better approximation effect than the previous algorithm.

1 Introduction

It is a natural idea to approximate arbitrary continuous functions using continuous piecewise linear functions. Their ability of approximation is guaranteed by the Taylor series expansion only if the space is divided into many tiny regions. The problem of their explicit representation was first proposed in[1], and much work [2],[3],[4] has been done in this field. But approximation algorithms based on these representations are not very easy to implement.

In this paper, we derive a general representation of arbitrary continuous piecewise linear functions which can be written as the linear combination of some maximal functions. Then the hinging hyperplanes algorithm[5]which was used to cope with similar optimization problem, is introduced here.

With the generalization of that algorithm, the approximation process can be very fast and effective.Finally, simulation results show the effectiveness of our training algorithm.

2 Canonical Continuous Piecewise Linear Model

Definition 1. The functions are called continuous piecewise linear functions if they satisfy the following conditions: first, the space is divided into many polyhedral regions, whose boundaries are all linear; second, the local function in each region is also linear; third, on both sides of each boundary, the values of different local functions are exactly the same.

An explicit representation called canonical representation can represent these functions in low dimensions.

J. Wang, X. Liao, and Z. Yi (Eds.): ISNN 2005, LNCS 3496, pp. 375–379, 2005.

Theorem 1. all the continuous piecewise linear functions in one dimension[1] can be represented as

$$f(\mathbf{x}) = l(\mathbf{x}, \mathbf{a}_0) + \sum_{i=1}^{M} c_i |l(\mathbf{x}, \mathbf{a}_i)| \tag{1}$$

where $\mathbf{x} = (x, 1)$,$\mathbf{a}_i \in \mathbf{R}^2$,$c_i \in \{+1, -1\}$,$i = 0, 1, ..., M$ and $l(,)$ denotes the inner product of two vectors.

Theorem 2. all the continuous piecewise linear functions in two dimensions [2]can be represented as

$$f(\mathbf{x}) = l(\mathbf{x}, \mathbf{a}_0) + \sum_{i=1}^{M} c_i |l(\mathbf{x}, \mathbf{a}_i) + d_i |l(\mathbf{x}, \mathbf{b}_i)|| \tag{2}$$

where $\mathbf{x} = (x_1, x_2, 1)$,$\mathbf{a}_i, \mathbf{b}_i \in \mathbf{R}^3$,$c_i, d_i \in \{+1, -1\}$,$i = 0, 1, ..., M$.

Another explicit representation called lattice representation can represent these functions in arbitrary dimensions, but its form seems more complicated than canonical piecewise linear representation.

Theorem 3. all the continuous piecewise linear functions [3] can be represented as

$$f(\mathbf{x}) = \min_{1 \leq i \leq M} \{\max_{j \in S_i} \{l(\mathbf{x}, \mathbf{a}_j)\}\} \tag{3}$$

where $M \in N$, S_i is a nonempty subset of the index set$\{1, 2, \cdots, m\}$ and $l(\mathbf{x}, \mathbf{a}_j)\}$has the same meaning as is defined above.

3   A New Continuous Piecewise Linear Model

In this section, we will transform the previous representations to another form that is more convenient in computational viewpoint. Here the canonical piecewise linear models in low dimensions are considered to see whether our idea makes sense.

According to the fact, for $\forall a, b \in \mathbf{R}$, $|a| = max\{0, 2a\} - a$, $|a + |b|| = max\{0, 2a + 2b, 2a - 2b\} - max\{0, 2b\} - a + b$, we can rewrite the canonical representations in one and two dimensions respectively, as

$$f(\mathbf{x}) = l(\mathbf{x}, \mathbf{a}_0) + \sum_{i=1}^{M} c_i \, max\{0, l(\mathbf{x}, \mathbf{a}_i)\} \tag{4}$$

and

$$f(\mathbf{x}) = l(\mathbf{x}, \mathbf{a}_0) + \sum_{i=1}^{M} c_i \, max\{0, l(\mathbf{x}, \mathbf{a}_i), l(\mathbf{x}, \mathbf{b}_i)\} \tag{5}$$

where $\mathbf{a}_i$ and $\mathbf{b}_i$ are vectors in appropriate dimensions.

Theorem 4. all the continuous piecewise linear functions can be represented as the following form:

$$f(\mathbf{x}) = l(\mathbf{x}, \mathbf{a}_0) + \sum_{i=1}^{M} c_i \, max\{0, l(\mathbf{x}, \mathbf{a}_{1i}), ..., l(\mathbf{x}, \mathbf{a}_{ki}))\} \qquad (6)$$

and k is equal to the dimension of the variables in functions.

We can derive Theorem 4 from Theorem 3 through a series of identical transforms. Due to the limitation of the space,we will not state the mathematical proof here. The whole process can be seen in our another paper[6].

Continuous piecewise linear functions can be treated as a special kind of neural networks with continuous piecewise linear activation functions. Since $max\{x_1, x_2, \cdots, x_{k+1}\}$ means $max\{x_1, max\{x_2, \cdots, x_{k+1}\}\}$, this kind of neural networks are actually a kind of k-hidden layers neural networks and can provide more powerful capability of approximation.

Although one-hidden layer neural networks have the universal approximation capability, but it should be based on the infinity of the neurons.Multi-hidden layer neural networks may achieve a better effect with limited parameters,but of course the training algorithm will be more complicated. So there should be a tradeoff between them.In other words,when adding neurons in one-hidden layer networks is not effective,multi-hidden layer networks should be considered.

4   Training Algorithm

In this section, a training algorithm will be design specially for the special kind of neural networks.Hinging hyperplanes algorithm[5]is proposed to approximate continuous functions with a similar model

$$f(\mathbf{x}) = l(\mathbf{x}, \mathbf{a}_0) + \sum_{i=1}^{M} \mathbf{c}_i max\{0, l(\mathbf{x}, \mathbf{a}_i)\} \qquad (7)$$

One hyperplane means one maximal function in (7). In fact, it is equivalent to the canonical representation in one dimension and can be looked upon as a kind of one-hidden layer neural networks[7]. It has the universal capability of approximation only if M tends to infinity in(7). We can generalize the Breiman's algorithm to cope with this problem. Essentially, it belongs to gradient based algorithms.

The generalized hinging hyperplanes algorithm can be implemented in the following steps:

1.Use a maximal function $\phi(\mathbf{x}) = cmax\{l(\mathbf{x}, \mathbf{a}_1), ..., l(\mathbf{x}, \mathbf{a}_k)\}$ to approximate the given function, where k is equal to the dimension of the function and $c \in \{+1, -1\}$. Start with a group of standard orthogonal basis $\mathbf{a}_1, \mathbf{a}_2, ..., \mathbf{a}_{k+1}$. Divide the data into k+1 categories in accordance with the linear function that gives the maximum namely$X_i = \{\mathbf{x}|l(\mathbf{x}, \mathbf{a}_i) = max\{l(\mathbf{x}, \mathbf{a}_1), ..., l(\mathbf{x}, \mathbf{a}_k)\}\}, Y(i) = \{y|\mathbf{x} \in X_i\}$.Fit the data X_i and Y_i with linear functions. The process will stop if the

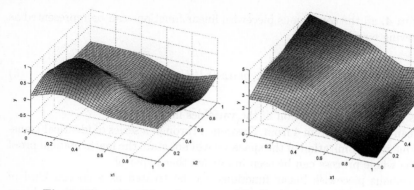

Fig. 1. The original function

Fig. 2. Approximation with ten hinging hyperplanes

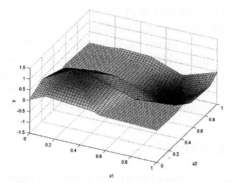

Fig. 3. Approximation with five maximal functions

approximation error does not decrease significantly. Both possible values of c will be tested, the one with the better approximation effect is chosen, and $\phi(\mathbf{x})$ is actually a minimal function when c is negative.

2.Subtract the values of all the existing maximal functions from the initial data, and add a new maximal function to approximate the error. For instance, if two maximal functions have been added, we can use $\phi_3(\mathbf{x})$ to fit $f(\mathbf{x}) - \phi_1(\mathbf{x}) - \phi_2(\mathbf{x})$. The implementation process is the same as step1 describes.

3.For every existing maximal functions, subtract all the other maximal functions from the initial data, and utilize the difference to refine the current maximal function. For instance, if three maximal functions have been added, we can use $f(\mathbf{x}) - \phi_2(\mathbf{x}) - \phi_3(\mathbf{x})$ to refit $\phi_1(\mathbf{x})$, use $f(\mathbf{x}) - \phi_1(\mathbf{x}) - \phi_3(\mathbf{x})$ to refit $\phi_2(\mathbf{x})$and then use $f(\mathbf{x}) - \phi_1(\mathbf{x}) - \phi_2(\mathbf{x})$ to refit $\phi_3(\mathbf{x})$. Thus all the maximal functions can be refitted iteratively.

5   Simulation Results

Simulation results are presented to verify the effectiveness of the new algorithm, and a two dimensional example is given here for visual inspection.

$$f(\mathbf{x}) = e^{-16\|\mathbf{x}-0.25\mathbf{e}\|^2} - e^{-16\|\mathbf{x}-0.75\mathbf{e}\|^2}, \mathbf{e} = (1,1). \tag{8}$$

1000 points uniformly distributed in the area $\mathbf{x} \in [0,1]^2$ are generated. Five maximal functions are used to fit the surface. Ten hinging hyperplanes are also used to fit the same surface to make a comparison. The parameters used by the two algorithms are equivalent. The formula $E = \sum_{i=1}^{N}(y - \hat{y})^2 / \sum_{i=1}^{N} y^2$ is used to measure the error. The error of hinging hyperplanes algorithm is 0.22, while that of the novel algorithm is 0.04. Figure1 shows the initial function, and Figure2 and Figure3 show the approximation effects of Breiman's algorithm and the generalized algorithm respectively. We can see that the new algorithm has better approximation effect than the previous one obviously.

6   Conclusions

In this paper, a novel continuous piecewise linear model is derived,and a corresponding training algorithm is designed. It improves the approximation effect of hinging hyperplanes algorithm. In fact, continuous piecewise linear functions are a special kind of neural networks with continuous piecewise linear activation functions.

References

1. Chua, L.O., Kang, S.M.: Section-wise Piecewise-linear Functions: Canonical Representation, Properties, and Applications. IEEE Trans. Circuits Systems, **30** (1977) 125-140
2. Chua, L.O., Deng, A.C.: Canonical Piecewise-linear Representation. IEEE Trans. Circuits Systems, **35** (1988) 101-111
3. Tarela, J.M., Matinez, M.V.: Region Configurations for Realizability of Lattice Piecewise-linear Models. Mathematical and Computer Modeling, **30** (1999) 17-27
4. Wang, S.: General Constructive Representations for Continuous Piecewise-linear Functions. IEEE Trans. Circuits Systems, **51** (2004) 1889-1896
5. Breiman, L.: Hinging Hyperplanes for Regression, Classification, and Function Approximation. IEEE Trans. Information Theory, **39** (1993) 999-1013
6. Wang, S., Sun, X.: Generalized Hinging Hyperplanes. Submitted to IEEE Trans. Information Theory
7. Lin, J., Unbehauen, R.: Canonical Piece-wise Linear Networks. IEEE Trans. Neural Networks, **6** (1995) 43-50
8. Pucar, P., Sjuberg. J.: On the Hinge-finding Algorithm for Hinging Hyperplanes. IEEE Trans. Information Theory, **44** (1998) 1310-1319

A Novel Dynamic Structural Neural Network with Neuron-Regeneration and Neuron-Degeneration Mechanisms

Yingtung Hsiao[1], Chenglong Chuang[1], Joeair Jiang[2],
Chiang Wang[1], and Chengchih Chien[1]

[1] Department of Electrical Engineering
Tamkang University, Taipei 251, Taiwan, China
{hsiao,clchuang,cwang,chien}@ee.tku.edu.tw
[2] Department of Bio-Industrial Mechanical Engineering
National Taiwan University, Taipei 106, Taiwan, China
jajiang@ccms.ntu.edu.tw

Abstract. This paper presents a novel neural network model with free structure style called dynamic structural neural network (DSNN). The neurons are randomly generated in a virtual three-dimensional space. The structure of the DSNN is reconfigurable for increasing the learning capacity of the network. Therefore, in this work, a structure reconfiguration algorithm is also proposed to achieve this goal. Finally, several simple pattern recognition problems are applied to the proposed DSNN to demonstrate the efficiency and performance of the network.

1 Introduction

Artificial neural networks have been successfully applied to different kind of research fields and applications due to their model-free approximation capability to complex decision making process [3]. The alignment of the neurons of the artificial neural networks is layered structure.

In present time, most of the neural networks are constructed in layer structure. The performance of this kind of structure might be degraded by the fixed structure. Compare to the real life nerve system, the structure of the conventional neural networks are too simple and form in regular alignment patterns. The performance of the neural networks may be limited by the regular alignment patterns. The only way to solve the bad performance problem is to redesign the entire neural systems. However, the costs to redesign the neural systems are too high. The automatic structure optimization algorithm of the neural network is needed to improve the performance of the neural networks.

In this article, we developed an automatic neural network generating algorithm and its learning algorithm to optimize the structure of the proposed neural network. All generated neurons shall be deployed in a virtual three-dimensional (3-D) space. Moreover, the neuron-regeneration mechanism and the neuron-regeneration mechanism, which are developed according to the neuron-regeneration technique of the real life nerve system in the territory of medical science, are proposed to automatically reconstruct the neural network for achieving higher performance.

J. Wang, X. Liao, and Z. Yi (Eds.): ISNN 2005, LNCS 3496, pp. 380–384, 2005.

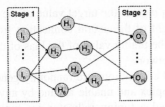

Fig. 1. Structure of the dynamic structure neural network. (DSNN)

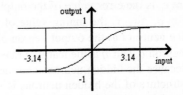

Fig. 2. Transfer function of the DSNN

This paper is organized as follows: In Section 2, the architecture of DSNN is described. And the structure optimization method of the proposed neural network is presented in Section 3. The testing results are showed in Section 4. Conclusions are given in Section 5.

2 Dynamic Structural Neural Network

The DSNN model constructs in a virtual 3-D space. Figure 1 represents the structure of the DSNN model. Each hidden neuron is labeled as H_j, where $j=1,2,...,6$. The stage 1 and the stage 2 represent the input and the output layers respectively. And n and m indicate numbers of neurons of input and output layers of the DSNN. The traditional artificial neural networks require trial-and-error processes for designing of an optimal structure for an artificial neural network under a given problem. The designer's experiences play an important role in designing of a neural network. Therefore, in this work, the structure and neural connections of DSNN are automatically generated to avoid the trial-and-error problem.

The activation function of a single neuron can be formulated as follow:

$$v_o(n) = \sum_{i \in C_o} w_{io}(n) y_i(n) + b_o(n) \tag{2}$$

$$y_o(n) = \varphi_o(v_o(n)) \tag{3}$$

where v_o is the internal activity level of the neuron o, w_{io} is weight value of neural connection from the hidden neuron i to the neuron o, b_o is the bias value of the neuron o, C_o is the neural connection set of the neuron o, φ_o represents the transfer function of the neuron o, and y_o is the actual output of the neuron o.

Because the output neurons can be trained by the target vector while we train the neural network, the weight vector of the output neuron can be tuned simply by BPN-like learning algorithm. The weight and bias tuning algorithm for output neuron of DSNN is described as follow:

$$e_o(n) = d_o(n) - y_o(n) \tag{4}$$

$$\Delta w_{io}(n) = l \cdot \eta \cdot e_o(n) | y_o(n) | \tag{5}$$

$$\Delta b_o(n) = \eta \cdot e_o(n) \tag{6}$$

$$l = \begin{cases} 1 & if \ \Delta y_o(n) > 0 \\ -1 & if \ \Delta y_o(n) < 0 \end{cases} \tag{7}$$

where e_o is the error value of the output neuron o, d_o is the target value of the output neuron o, Δw_{io} is the tuning value of the weight of the neural connection from the hidden neuron i to the output neuron o, Δb_o is the tuning value of the bias of the output neuron o, and η is the learning rate of the network.

Because the hidden neurons and their neural connections are randomly generated, the structure of the hidden neurons is very complex and unable to tune by any existing learning algorithm. Therefore, we provided a tuning indicator for tuning the weight and the bias of the hidden neurons. The tuning indicator is defined as below:

$$g_i(n) = \eta \cdot e_o(n) \,|\, y_o(n)| \tag{8}$$

where g_i is the tuning indicator of the hidden neuron i that connected to output neuron o. For the hidden neurons, the tuning of the weight and the bias are depending on the tuning indicator g_i. According to the tuning indicator g_i, we can get the average tuning indicator s_i by following equation:

$$s_i(n) = \frac{g_i(n)}{Nc} \tag{9}$$

where s_i is the average tuning indicator that use for determining the tuning range of the weight and the bias of the hidden neuron i. Nc is the total number of neurons that connected to the hidden neuron i. Since the si is obtained, the weight and bias of the output neuron of DSNN can be tuned by applying the following equations:

$$\Delta w_{ji}(n) = l \cdot \eta \cdot s_i(n) \,|\, y_i(n)| \tag{10}$$

$$l = \begin{cases} 1 & if\ \Delta y_i(n) > 0 \\ -1 & if\ \Delta y_i(n) < 0 \end{cases} \tag{11}$$

$$\Delta b_i(n) = \eta \cdot s_i(n) \tag{12}$$

where Δw_{ji} is the tuning value of the weight of connection from hidden neuron j to hidden neuron i, and Δb_i is the tuning value of the bias of the output neuron i.

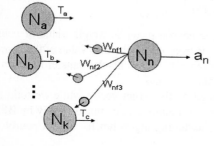

Fig. 3. The concept of the neuron-regeneration

Table 1. The Growing Rules of Free Connectors

W_L	a_n	T_R	Att/Exc
+	+	+	Att
+	+	-	Rep
+	-	+	Rep
+	-	-	Att
-	+	+	Rep
-	+	-	Att
-	-	+	Att
-	-	-	Rep

3   Automatically Structural Optimization Technique

Figure 3 shows the concept of the neuron-regeneration process; where for each neuron N_n has three free connectors. Each of them has assigned a randomly generated weighting, such as W_{nf1}, W_{nf2} and W_{nf3} in Figure 4. When the DSNN is under training process, the free connectors shall search for near neurons and try to create new connections. Table 1 shows the rules of determining the growing direction of free connectors, where a_n is the output of the neuron that belong to the corresponding free connector, WL is the weight of the free connector, TR is the tuning indicator of the near neuron, and the Att means the free connectors will be attracted to the position of the near neuron, the Rep means the free connectors will be distracted from the position of the near neuron. The attraction or repulsion process is described as below:

$$\Delta(x_{fn}, y_{fn}, z_{fn}) = D \cdot \frac{|g_j|}{L} \cdot (x_j, y_j, z_j) \tag{13}$$

$$D = \begin{cases} 1 & \text{if Attraction} \\ -1 & \text{if Repulsion} \end{cases} \tag{14}$$

where (x_{fn}, y_{fn}, z_{fn}) is the current coordinate of the free connector, g_i is the tuning indicator of the near neuron, L is the distance between the original neuron of the free connector and the near neuron, (x_j, y_j, z_j) is the coordinate of the near neuron. If the distance from free connector to any neuron of the system is less then 1, a new neural connection is established between closest neuron and the neuron that own the free connector, and the free connector is then reinitialized.

When the DSNN is under training process, some weight of the connections might be very small. If a neural connection that the weight value is very small for several iterations, then this neural connection will be defined as unnecessary connection that shall be pruned from the system to save the computational resources. The mechanism of the dynamic growth of the middle layer allows the network to increase it's learn ability by inventing new neurons. The probability P of a new neuron being created is given by:

$$P = \sum_i e_i \cdot \left(\frac{N_{max\_h} - N_h}{N_{max\_h}} \right) \tag{15}$$

where e_i is the error of the output neuron i, N_h is the current number of the hidden neurons in the middle layer. N_{max_h} is the maximum number of neurons that can be created in the virtual 3-D space. In (22), if output error of the network is high, the probability of generating a new neuron is high. After a new neuron generating, N is increased by 1. As long as N reaches the maximum number of neurons N_{max_h}, the probability P will reduce to 0 to prevent the unlimited growing of the hidden neurons.

4   Experimental Results

In this work, we have applied a problem to the proposed DSNN to verify the algorithm. The problem is a specific function as below:

$$Z = Y \cdot \exp^{-(X^2 + Y^2)} \tag{16}$$

where X and Y are input values, and Z is the output value. The ideal output graph is shown in Figure 4. The number of neurons of the initially generated DSNN is 150. The generation of the training loop is 200. Since the simulating program has completed the training process, the DSNN created 2 new neurons and 12684 connections where 12595 connections were pruned by the restructure optimal program.

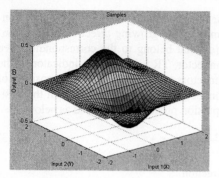 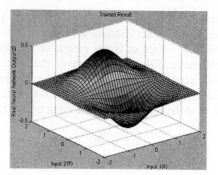

Fig. 4. The target graph of (16)   **Fig. 5.** The DSNN output graph of problem

5 Conclusions

A novel three-dimensional dynamic structure neural network architecture is proposed in this paper. The neurons and connections of DSNN are randomly generated to avoid the experience-dependence problem while the designers trying to design a properly structure for an artificial neural network under a given problem. The structure of DSNN is optimized automatically by the neuron-regeneration and the neuron-degeneration mechanisms. The DSNN also can invent new neurons into network for enhancing the performance of the DSNN. The proposed DSNN model could be applied to solve other different kind of problems in the future.

References

1. Matlab User's Guide, (1992)
2. Haykin, S: Neural Networks: A Comprehensive Foundation. Prentice Hall (1994)
3. Moon, S. W., Kong, S. G.: Block-based Neural Networks. IEEE Trans. Neural Networks, **12** (2001) 307-317
4. Ming, G., Wong, S.T., Henley, J., Yuan, X., Song, H., Spitzer, N.C., M. Poo.: Adaptation in the Chemotactic Guidance of Nerve Growth Cones. Nature, **417** (2002) 411-418
5. Pasterkamp, R.J., Peschon, J.J., Spriggs, M.K., Kolodkin, A.L.: Semaphorin 7A Promotes Axon Outgrowth Through Integrins and MAPKs. Nature **424** (2003) 398-405

A New Adaptive Ridgelet Neural Network

Shuyuan Yang, Min Wang, and Licheng Jiao

Institute of Intelligence Information Processing, Xidian University,
Xi'an 710071, China
syyang@xidian.edu.cn

Abstract. In this paper, a new kind of neural network is proposed by combining ridgelet with feed-forward neural network (FNN). The network adopts ridgelet as the activation function in hidden layer of a three-layer FNN. Ridgelet is a good basis for describing the directional information in high dimension and it proves to be optimal in representing the functions with hyperplane singularity. So the network can approximate quite a wide range of multivariate functions in a more stable and efficient way, especially those with certain kinds of spatial inhomogeneities. Both theoretical analysis and experimental results of function approximation prove its superiority to wavelet neural network.

1 Introduction

Feed-forward neural network (NN) is a good nonparametric method for function approximation which mimics the behavior of human neurons. By using the superposition of nonlinear activation functions g came from a dictionary $D=\{g(kx-b)\}$, FNN searches for the optimal solution to approximation of a function through self-learning. The solution obtained finally by the network is more tolerant and flexible than other traditional mathematical methods [1]. Many kinds of FNNs can fulfill such an approximation task such as back-propagation network (BPN), which takes a global *Sigmoid* function as its activation function and allows non-smooth fitting. However, it has to adjust all the connected weights of the network once it receives a new sample, which leads to slow convergence and low accuracy [2]. To overcome its disadvantages, people have been seeking for new dictionaries and fitting ways to get a more efficient approximation. In 1988, Daugman proposed a new way to construct NN using Gabor transform [3]; later Moody and Darken advanced radial-basis-function network (RBFN) based on regularized theory [4]; Q.Zhang presented a model of wavelet neural network (WNN) in 1992 [5].

For the good characteristic of NN, it has been widely applied in many science and engineering communities [6], [7], but the applications are mostly limited in 1-dimension (1-D). When the dimension of input increase, the scale of network increases rapidly. For example, if a map of $f\colon R \to R$ requires a network with N weights, then N^d weights are required to get a map of $g\colon R^d \to R$ with the same accuracy. That is, the increase of dimension d leads to a remarkable augment of computation complexity in space and time, and we often call it the "calamity of dimensionality"[8]. To approximate multivariate functions, finding a good activation function is the first thing to be considered to construct an efficient network. Wavelet is well

J. Wang, X. Liao, and Z. Yi (Eds.): ISNN 2005, LNCS 3496, pp. 385–390, 2005.

known for its localization property in time and frequency, but it fails in an extension to high dimension for it lacks of a directional description. Direction is an important attribute in the high dimensional space, and a good basis function should be able to give a good description of it.

As an extension of wavelet to higher dimension, ridgelet is an efficient geometrical multi-resolution analysis (GMA) tool in dealing with directional information [9-10]. It can represent any multivariate function in a more efficient and stable way than wavelet, moreover, this representation is optimal for a group of functions with hyperplane singularities. Based on it, we proposed a new adaptive network with the ridgelet being the activation function in hidden layer of FNN. Its ability and property of approximation are investigated, as well as the construction and leaning algorithm of the network. Finally some experiment simulations are demonstrated to prove its superiority.

2   Adaptive Ridgelet Neural Network

2.1   Ridgelet Approximation

In 1996, a pioneered work was done by *E.J. Candés* to represent multivariate functions with a new kind of basis function in a more stable way. Thus a new tool in harmonic analysis ridgelet is proposed [9]. Its definition is as: If $\Psi: R^d \rightarrow R$ satisfies the condition $K_\psi = \int (|\hat{\psi}(\xi)|^2 / |\xi|^d) d\xi < \infty$, then we called the functions Ψ_r as ridgelets.

Parameter $\gamma = (a, u, b)$ belongs to a space of neurons $\Gamma = \{\tau = (a,u,b), a,b \in R, a > 0,$ $u \in S^{d-1}, \|u\| = 1\}$,where a,u,b represent the scale, direction and localization respectively. Continuous ridgelet transform is defined as

$$R(f)(\gamma) = < f, \psi_\gamma > = \int_R \psi_\gamma(x) f(x) dx .$$

Denote the surface area of sphere S^{d-1} as σ_d; du is the uniform measure on S^{d-1}. Any multivariate function $f \in L^1 \cap L^2(R^d)$ can be expanded as a superposition of ridgelets:

$$f = c_\psi \int <f, \psi_\gamma > \psi_\gamma \mu(d\gamma) = c_\psi = \pi(2\pi)^{-d} K_\psi^{-1} \int <f, \psi_\gamma > \psi_\gamma \sigma_d da / a^{d+1} dudb \qquad (1)$$

Additionally it can approximate a group of functions with hypeplane singularities with a speed rapider than Fourier transform and Wavelet transform [10]. For function $Y=f(x):R^d \rightarrow R^m$, it can be divided into m mappings of $R^d \rightarrow R$. Selecting ridgelet as the basis, we get such an approximation equation:

$$\hat{y}_i = \sum_{j=1}^{N} c_{ij} \psi((u_j \cdot x - b_j)/a_j) \quad (\hat{Y} = [\hat{y}_1, \cdots, \hat{y}_m]; x, u_i \in R^d; \|u_j\|^2 = 1, i = 1,..,m) \qquad (2)$$

where c_{ij} is the superposition coefficient of ridgelets. From above we can see ridgelet is a constant on spatial lines $t=u \cdot x$, which is responsible for its great capacity in dealing with hyperplane singularities such as linear singularity and curvilinear singularity by a localized ridgelet version.

2.2 The Structure of Ridgelet Network and Its Learning Algorithm

Combination of wavelet analysis with FNN produces WNN. For the good localized property of wavelet, WNN can reconstruct a function with singularity in 1-D, but it fails in higher dimension[5]. Although the training of WNN can compensate the distortion brought by singularity, the diffuses of singularity appear for the failure of wavelet in dealing with high-dimensional singularity. Moreover, similar to other NNs, the complexity of WNN increases with d for using a tensor product form of wavelet. Accordingly a new kind of activation function, ridgelet, which has the same effect in high dimension as that of wavelet in 1-D is used in our proposed network. Replacing the *Sigmoid* function by ridgelet function in BPN, the network can approximate a wider range of functions with singularities by adaptively adjust the directions of ridgelets, as shown in Fig. 1.

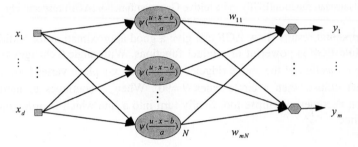

Fig. 1. Adaptive ridgelet neural network, which is a standard three-layer FNN with ridgelet function being the activation function in the hidden layer

Given a pair of sample set $S=\{(x_1, y_1),...,(x_p, y_p)\}$(where $x_i=[x_{1i}, ...,x_{di}]\in R^d$, $y_i=[y_{1i}, ...,x_{mi}]\in R^m$ and $i=1,..,P$) generated from an unknown model. Here x_i and y_i represent the i-th input sample and its corresponding output; w_{ij} is the weights connected the i-th output neuron and j-th hidden neuron, w_{i0} is the threshold of the i-th output neuron; $Z_i=[z_{i1}, ...,z_{iP}](i=1,..,N)$ is the output of hidden layer. For the p-th input sample, the output of hidden layer is shown in (3) and a linear layer follows it.

$$z_{ip} = \psi[\sum_{j=1}^{d}(\frac{u_j x_{jp} - b_{ij}}{a_{ij}})] \quad (\sum_i \|u_i\|^* = 1, \ i \in [1,N], p \in [1,P]) \tag{3}$$

$$y_{kp} = w_{k0} + \sum_{i=1}^{N} w_{ki} z_{ip} \quad (k=1,....,m) \tag{4}$$

Our goal of the training is to find some ridgelets and weights which can minimize the error between the desired and actual output. It can be reduced to an optimization problem and many kinds of optimization methods can be used to solve it.

As we all know, in FNN-based function approximation, the type of activation function determines the capability and efficiency of a network. For using ridgelet as the activation function, the proposed network can approximate functions in a many kinds of spaces as WNN, such as $L^P(R)$, *Sobolev* and *Holder*. Apart from it, it can deal with functions which WNN cannot process, such as the bounded $BV(1)$ function in T=[0,2 π] and others. Different with wavelet, the ridgelet functions in the new network are different not only in scale and localization, but also in direction. For example, when the ridgelet is of Gaussian form, the shapes of ridgelets with different directions are shown in Fig. 2.

Fig. 2. The Gaussion ridgelet function with different directions. Since the directional vector of ridgelet $\|u\|=1$, the direction $u=[u_1,.., u_d]$ can be described using $(d-1)$ angles $\theta_1,...,\theta_{d-1}$: $u_1=\cos\theta_1, u_2=\sin\theta_1\cos\theta_2,.., u_d=\sin\theta_1\sin\theta_2...\sin\theta_{d-1}$. From left to right, the three ridgelet functions are of $\theta=30°,45°,90°$. We often call them as the standard Gaussian function(SGF), elliptical Gaussian function(EGF) and additive Gaussian function(AGF) respectively

Experimental results show AGF can give a good approximation for additive functions, while EGF is powerful in product functions. With angle θ changes from $0°$ to $180°$, the projection of the above ridgelet function on x-y plane varies form a straight line to an ellipse, then a circle when $\theta=90°$. When θ continues to increase, the projection becomes an ellipse too, finally turn into a line which is orthogonal to the original line.

2.3 Learning Algorithm of the Ridgelet Neural Network

The determination of the number of hidden neurons is very important in the training of a FNN. Here we choose a way of adding neurons little by little. At each iteration, we add a hidden neuron until the approximation error is smaller than a given small number ε. Firstly define the training error J of the network as:

$$J=\sum_{i=1}^{P}e_i^2/2=\sum_{i=1}^{P}\sum_{j=1}^{m}(y_{ji}-d_{ji})^2/2 \tag{5}$$

The training of network is based on the steepest gradient descent algorithm, and the update equation of the parameters of the network is:

$$\beta(k+1)=\beta(k)+\eta\times\partial J(k)/\partial\beta \tag{6}$$

where β means w, α ,b, u and η is the learning step ($0.01<\eta<0.1$). The gradients of the network are given in the following equation:

$$\partial J(k)/\partial w_{ij}=2\eta\times\sum_{i=1}^{P}\sum_{j=1}^{m}(y_{ji}-d_{ji})z_{ip} \tag{7}$$

$$\frac{\partial J(k)}{\partial u_{jl}}=2\sum_{i=1}^{P}e_i(k)\sum_{j=1}^{m}(y_{ji}-d_{ji})w_{ji}\frac{\partial\psi(\Sigma)}{\partial(\Sigma)}(\frac{x_{jl}}{\alpha_{jl}})[\sum_{r=1,r\neq l}^{d}u_{jr}^2(k)/(\sum_{r=1}^{d}u_{jr}^2(k))^{3/2}] \tag{8}$$

$$\frac{\partial J(k)}{\partial\alpha_{jl}}=-2\eta\times\sum_{i=1}^{P}\sum_{j=1}^{m}(y_{ji}-d_{ji})w_{ji}\frac{\partial\psi(\Sigma)}{\partial(\Sigma)}[\frac{u_l x_{li}-b_{jl}}{a_{jl}(k)}]^2 \quad (l=1,..,d) \tag{9}$$

$$\frac{\partial J(k)}{\partial b_{jl}} = -2\eta \times \sum_{i=1}^{P}\sum_{j=1}^{m}(y_{ji} - d_{ji})w_{ji}\frac{\partial \psi(\Sigma)}{\partial(\Sigma)} \tag{10}$$

3 Simulations

Some simulations are taken to investigate the performance of ridgelet neural network (RNN).Considering 2-D step function and singular function $f(x_1,x_2)=4-x_1^2-x_2^2_{\{x1+4x2<1.2\}}$ shown in Fig. 3. Using 25 training samples, RNN and WNN are used to approximate them. Some comparisons of the obtained results are given, including the number of hidden neurons and errors. As shown in the figure, we can see that RNN obtain more accurate approximation with less neuron than WNN, as well as a better testing result.

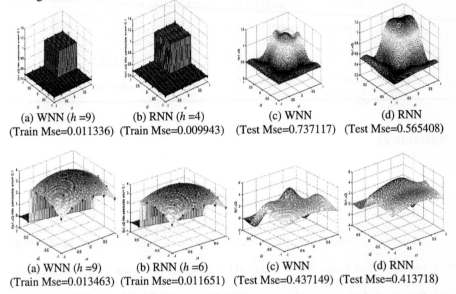

(a) WNN ($h =9$)   (b) RNN ($h =4$)   (c) WNN   (d) RNN
(Train Mse=0.011336) (Train Mse=0.009943) (Test Mse=0.737117) (Test Mse=0.565408)

(a) WNN ($h =9$)   (b) RNN ($h =6$)   (c) WNN   (d) RNN
(Train Mse=0.013463) (Train Mse=0.011651) (Test Mse=0.437149) (Test Mse=0.413718)

Fig. 3. Comparison results of WNN and RNN for two singular functions. (a) (b) are results of training, and "h" means the number of hidden neurons. (c) (d) are results of testing. The approximation error is measured by MSE and $\varepsilon =10e-3$

More functions are considered, including smooth and non-smooth functions. For singular functions, we mainly consider 2-D functions to get an intuitionistic impression. The results are shown in Table 1, where W means WNN and R means RNN.

4 Conclusions

In this paper, we proposed a ridgelet neural network based on ridgelet and FNN. It proves to be more efficient in representing multivariate functions, especially those with hyperplane singularities. Some experimental results are also demonstrated to prove its superiority to WNN.

Table 1. Approximation results of WNN and RNN for function F1-F8

	h	Train MSE	Test MSE	Expression
F1	9(W)	0.013463(W)	0.437107(W)	$f(x_1,x_2)=\begin{cases}4-x_1^2-x_2^2 & x_1+4x_2<1.2\\0 & \text{otherwise}\end{cases}$
	6(R)	0.008661(R)	0.416268(R)	
F2	9(W)	0.08831(W)	0.801428(W)	$f(x_1,x_2)=\begin{cases}1-0.2x_1^2-x_2^2 & x_1^2+2x_2^2<0.5\\0 & 3x_1^2+x_2^2>1\\\sqrt{x_1^2+(x_2^2+0.5)^2} & \text{otherwise}\end{cases}$
	7(R)	0.057806(R)	0.470874(R)	
F3	11(W)	0.020853(W)	0.662501(W)	$f(x_1,x_2)=\begin{cases}1/2 & x_1x_2\geq 0\\-1/2 & \text{otherwise}\end{cases}$
	8(R)	0.013419(R)	0.662362(R)	
F4	8(W)	0.098679(W)	0.330587(W)	$f(x_1,x_2)=\begin{cases}\sqrt{x_1^2+(x_2^2+0.5)^2} & 3x_1+x_2\leq 1\\0 & \text{otherwise}\end{cases}$
	8(R)	0.087981(R)	0.234257(R)	
F5	8(W)	0.042932(W)	0.464038(W)	$f(x_1,x_2)=\begin{cases}e^{-(x_1^2+x_2^2)} & x_2\geq x_1^2\\0 & \text{otherwise}\end{cases}$
	4(R)	0.040467	0.343489(R)	
F6	11(W)	0.056616(W)	0.213801(W)	$f(x_1,x_2)=\dfrac{\sin(\pi x_2)}{2+\sin(2\pi x_1)}$
	8(R)	0.053875(R)	0.212728(R)	
F7	15(W)	0.063444(W)	0.236392(W)	$f(x_1,x_2)=\dfrac{1+\sin(6x_1^2+6x_2^2)}{2}$
	11(R)	0.061147(R)	0.177487(R)	
F8	6(W)	0.088179(W)	0.422858(W)	$f(x_1,x_2)=e^{-x_1^2}\cos(0.75\pi(x_1+x_2))$
	6(R)	0.0893608(R)	0.424311(R)	

References

1. DeVore, R. A., Oskolkov, K. I., Petrushev, P. P.: Approximation by Feed-forward Neural Networks. Ann. Numer. Math, **4** (1997) 261-287
2. Cybenko, G.:Approximation by Superpositions of a Sigmoidal Function. Math.Control Signals System, **2** (1989) 303-314
3. Daugman, J. G.: Complete Discrete 2-d Gabor Transforms by Neural Networks for Image Analysis and Compression. IEEE Trans on Acoustics, Speech, and SP, **36 (7)** (1988) 1169-1179
4. Moody, J.: Fast Learning in Networks of Locally-tuned Processing Units. Neural Computation, **1** (1989) 281-294
5. Benveniste, A.Q.: Wavelet Networks. IEEE Trans on NN, **3** (1992) 899-898
6. Narendra, K. S., Parthasarathy, K.: Identification and Control of Dynamical Systems Using Neural Networks. IEEE Trans. Neural Networks, **1** (1990) 4–27
7. Chow, M. Y.: A Neural Network Approach to Real Time Condition Monitoring of Induction Machines". IEEE Trans. on Industrial Electronics, **38** (1991) 48-453
8. Friedman, W.: Projection Pursuit Regression. J. Amer. Statist. Assoc, **76** (1981) 817-823
9. Candes, E.J.: Ridgelet: Theory and Applications. Ph.D. Dissertation. Stanford Univ, (1998)
10. Candes, E.J.: Ridgelets: A Key to Higher-Dimensional Intermittency. Technical Report, Stanford Univ (1999)

Designing Neural Networks
Using Hybrid Particle Swarm Optimization

Bo Liu, Ling Wang, Yihui Jin, and Dexian Huang

Department of Automation, Tsinghua University, Beijing 100084, China
liub01@mails.tsinghua.edu.cn

Abstract. Evolving artificial neural network is an important issue in both evolutionary computation (EC) and neural networks (NN) fields. In this paper, a hybrid particle swarm optimization (PSO) is proposed by incorporating differential evolution (DE) and chaos into the classic PSO. By combining DE operation with PSO, the exploration and exploitation abilities can be well balanced, and the diversity of swarms can be reasonably maintained. Moreover, by hybridizing chaotic local search (CLS), DE operator and PSO operator, searching behavior can be enriched and the ability to avoid being trapped in local optima can be well enhanced. Then, the proposed hybrid PSO (named CPSODE) is applied to design multi-layer feed-forward neural network. Simulation results and comparisons demonstrate the effectiveness and efficiency of the proposed hybrid PSO.

1 Introduction

Design of neural networks is an important topic in computer science field. As we know, essentially it is an optimization problem to design a neural network. So far, many techniques have been proposed to train a network, such as gradient-descent method (i.e., BP, conjugate gradient algorithms), tabu search, simulated annealing, evolutionary computation etc. [1]. It is concluded that gradient-descent methods are very slow, easy to be trapped in local optima, short of generalization and rather sensitive to initial weights. During the past two decades, evolutionary computation (EC) techniques especially gained much attention for neural network design [2]. As a new kind of EC technique, particle swarm optimization (PSO) has been applied to many kinds of problems, including neural network design [3]. In this paper, an effective hybrid PSO is proposed for neural network design by combining differential evolution (DE) [4] and chaotic search [5]. By combining DE operation with PSO, the exploration and exploitation abilities can be well balanced, and the diversity of swarms can be reasonably maintained. Moreover, by hybridizing chaotic local search, DE operator and PSO operator, searching behavior can be enriched and the ability to avoid being trapped in local optima can be well enhanced. Simulation results and comparisons demonstrate the effectiveness and efficiency of the proposed hybrid PSO.

The remaining content of this paper is organized as follows. In Section 2, PSO, DE and CLS are briefly reviewed. Then, the hybrid PSO (CPSODE) is proposed in Section 3. Numerical simulations and comparisons are provided in Section 4. Finally, Section 5 provides some concluding remarks.

J. Wang, X. Liao, and Z. Yi (Eds.): ISNN 2005, LNCS 3496, pp. 391–397, 2005.

2 Review of PSO, DE and CLS

2.1 PSO

PSO is an EC technique through individual improvement plus population cooperation and competition, which is based on the simulation of simplified social models, such as bird flocking, fish schooling, and the swarming theory. The theoretical framework of PSO is very simple and PSO is easy to be coded and implemented with computer. Besides, it is computationally inexpensive in terms of memory requirements and CPU times. Thus, nowadays PSO has gained much attention and wide applications in various fields [3].

In PSO, it starts with the random initialization of a population (*swarm*) of individuals (*particles*) in the search space and works on the social behavior of the particles in the swarm. Therefore, it finds the global best solution by simply adjusting the trajectory of each individual towards its own best location and towards the best particle of the swarm at each time step (*generation*). However, the trajectory of each individual in the search space is adjusted by dynamically altering the velocity of each particle, according to its own flying experience and the flying experience of the other particles in the search space.

The position and the velocity of the i-th particle in the d-dimensional search space can be represented as $X_i = [x_{i,1}, x_{i,2}, ..., x_{i,d}]$ and $V_i = [v_{i,1}, v_{i,2}, ..., v_{i,d}]$ respectively. Each particle has its own best position (*pbest*) $P_i = (p_{i,1}, p_{i,2}, ..., p_{i,d})$, corresponding to the personal best objective value obtained so far at time t. The global best particle (*gbest*) is denoted by P_g, which represents the best particle found so far at time t in the entire swarm. The new velocity of each particle is calculated as follows:

$$v_{i,j}(t+1) = w v_{i,j}(t) + c_1 r_1 (p_{i,j} - x_{i,j}(t)) + c_2 r_2 (p_{g,j} - x_{i,j}(t)), j = 1,2,...,d \ . \tag{1}$$

where c_1 and c_2 are constants called acceleration coefficients, w is called the inertia factor, r_1 and r_2 are two independent random numbers uniformly distributed in the range of [0, 1].

Thus, the position of each particle is updated in each generation according to the following equation:

$$x_{i,j}(t+1) = x_{i,j}(t) + v_{i,j}(t+1), j = 1,2,...,d \ . \tag{2}$$

Due to the simple concept, easy implementation and quick convergence, nowadays PSO has gained much attention and wide applications in different fields. However, the performance of simple PSO greatly depends on its parameters, and it often suffers the problem of being trapped in local optima for the lost of diversity of swarm.

2.2 DE

Although DE [4] has the similar structure of GA, the mode to generate new candidates in DE is different from GA, and a 'greedy' selection scheme is employed in DE as well. The DE evolves a population including N individuals, and the i-th individual in the d-dimensional search space can be represented as $X_i = [x_{i,1}, x_{i,2}, ..., x_{i,d}]$. The procedure of DE can be briefly described in Fig.1.

For each individual i in the population, initialize $X_i(0)$ randomly, $g = 0$

Repeat until a stopping criterion is satisfied:

  Mutation step:

$$v_{i,j} = x_{i,j}(g) + \lambda(x_{best,j}(g) - x_{i,j}(g)) + F(x_{r1,j}(g) - x_{r2,j}(g)),$$

$$\forall i \leq N \text{ and } \forall j \leq d .$$

  Crossover step:

$$u_{i,j} = \begin{cases} v_{i,j} & \text{with probability } CR \\ x_{i,j}(g) & \text{with probability } 1 - CR \end{cases}, \quad \forall i \leq N \text{ and } \forall j \leq d .$$

  Selection step:

$$\text{if } f(U_i) < f(X_i(g)) \text{ then } x_i(g+1) = u_i \text{ else } x_i(g+1) = x_i(g), \forall i \leq N$$

$$g = g + 1$$

Fig. 1. Procedure of DE algorithm

In DE, $r1$, $r2$ are different elements of $\{1,2,...,N\}$ randomly selected for each i. λ is used to enhance the greediness of DE by incorporating the current best vector $x_{best}(g)$. F is the constant factor weighting the differential perturbation variation $(x_{r1,j} - x_{r2,j})$. And $CR \in (0,1]$ controls the proportion of perturbed elements in the new population. DE is simple fast and ease to implement, but it often suffers the problem of being trapped in local optima.

2.3 CLS

For a given cost function, by following chaotic ergodic orbits, a chaotic dynamic system may eventually reach the global optimum or its good approximation with high probability. The simple philosophy of the chaotic search includes two main steps: firstly mapping from the chaotic space to the solution space, and then searching optimal regions using chaotic dynamics instead of random search [5]. The widely used chaotic map is the following Logistic map.

$$x_{n+1} = \mu \cdot x_n (1 - x_n), 0 \leq x_0 \leq 1 . \tag{3}$$

where μ is the control parameter, x is a variable and $n = 0,1,2,..., $.

The process of the chaotic local search could be defined through the following equation:

$$cx_i^{(k+1)} = 4cx_i^{(k)}(1 - cx_i^{(k)}), i = 1,2,...,n . \tag{4}$$

where cx_i is the ith chaotic variable, and k denotes the iteration number. Obviously, $cx_i^{(k)}$ is distributed in the range $(0, 1.0)$ under the conditions that the initial $cx_i^{(0)} \in (0,1)$ and that $cx_i^{(0)} \notin \{0.25, 0.5, 0.75\}$.

The procedures of chaotic local search (CLS) can be illustrated as follows:

Step 1: Setting $k = 0$, and mapping the decision variables $x_i^{(k)}, i = 1, 2, ..., n$ among the intervals $(x_{min,i}, x_{max,i}), i = 1, 2, ..., n$ to chaotic variables $cx_i^{(k)}$ located in the interval $(0,1)$ using the following equation.

$$cx_i^{(k)} = \frac{x_i^{(k)} - x_{min,i}}{x_{max,i} - x_{min,i}}, i = 1, 2, ..., n \ . \tag{5}$$

Step 2: Determining the chaotic variables $cx_i^{(k+1)}$ for the next iteration using the logistic equation according to $cx_i^{(k)}$.

Step 3: Converting the chaotic variables $cx_i^{(k+1)}$ to decision variables $x_i^{(k+1)}$ using the following equation.

$$x_i^{(k+1)} = x_{min,i} + cx_i^{(k+1)} \left(x_{max,i} - x_{min,i} \right), i = 1, 2, ..., n \ . \tag{6}$$

Step 4: Evaluating the new solution with decision variables $x_i^{(k+1)}, i = 1, 2, ..., n$.

Step 5: If the new solution is better than $X^{(0)} = [x_1^{(0)}, \cdots, x_n^{(0)}]$ or the predefined maximum iteration is reached, output the new solution as the result of the CLS; otherwise, let $k = k + 1$ and go back to Step 2.

Due to the easy implementation and special ability to avoid being trapped in local optima, chaos has been regarded and utilized as a novel optimization technique [6]. However, simple chaotic local search (CLS) may need a large number of iterations to reach the global optimum because of the sensitivity to initial conditions.

3 Hybrid PSO (CPSODE)

By hybridizing PSO, DE and CLS, a hybrid PSO is proposed in this section, whose procedure is illustrated in Fig. 2.

It can be seen that PSO and DE operators are simultaneously used for exploration by updating particle swarm, and chaos dynamic is applied for exploitation by locally modified the best particle resulted by PSO. Besides, to maintain population diversity, several new particles are randomly generated and incorporated in the new population. Especially, the region for generating new particles is dynamically decreased so as to speed up convergence.

Step 1: Set $k = 0$. For each particle i in the population:

  Step 1.1: initialize X_i randomly.

  Step 1.2: initialize V_i randomly.

  Step 1.3: evaluate f_i.

  Step 1.4: initialize P_g with the index of the particle with the best function value among the population.

  Step 1.5: initialize P_i with a copy of X_i, $\forall i \leq N$.

Step 2: Repeat until a stopping criterion is satisfied:

  Step 2.1: find P_g such that $f[P_g] \leq f_i, \forall i \leq N$.

  Step 2.2: for each particle i, $P_i = X_i$ if $f_i < f_{best}[i], \forall i \leq N$.

  Step 2.3: perform DE operation on the P_i, and update them.

  Step 2.4: perform PSO operator for each particle i to update V_i and X_i according to equation 1 and 2.

  Step 2.5: evaluate f_i for all particles.

Step 3: Reserve the top $N/5$ particles.

Step 4: Implement the CLS for the best particle, and update the best particle using the result of CLS with variables $x_{g,i}^{(k)}, i = 1,2,...,n$.

Step 5: If a stopping criterion is satisfied, output the solution found best so far.

Step 6: Decrease the search space:

$$x_{\min i} = \max\left(x_{\min i}, x_{g,i}^{(k)} - r\left(x_{\max i} - x_{\min i}\right)\right), 0 < r < 1$$

$$x_{\max i} = \min\left(x_{\max i}, x_{g,i}^{(k)} + r\left(x_{\max i} - x_{\min i}\right)\right), 0 < r < 1$$

Step 7: Randomly generate $4N/5$ new particles within the decreased search space and evaluate them.

Step 8: Construct the new population consisting of the $4N/5$ new particles and the old top $N/5$ particles in which the best particle is replaced by the result of CLS.

Step 9: Let $k = k + 1$ and go back to step 2.

Fig. 2. Procedure of CPSODE

4   Experiments

In this section, we apply the CPSODE for designing the weight of multi-layer feed-forward networks. Three different problems with different network structures, including two classification problems [7] (Ionosphere and Cmc) and one approximation problem (Henon), are used for testing. The training objective is to minimize mean squared error (MSE). In particular, each particle in PSO denotes a set of weights of the network. The dimension of each particle is same as the number of weights of a certain network. Training a network using PSO means moving the particle among the weight space to minimize the MSE.

  We apply three training algorithms, including CPSODE, PSO and Powell-Beale's conjugate gradient algorithm for comparison. In PSO and CPSODE, population size is

10, c_1 and c_2 are set to 2.0, v_{max} is clamped to be 15% of the search space and use linearly varying inertia weight over the generations, varying from 1.2 at the beginning of the search to 0.2 at the end. The parameters of DE in CPSODE are set as follows: $CR = 0.5$, $\lambda = F = 0.8$. The Powell-Beale's method is a standard training algorithm in Matlab neural network toolbox. The total number of function evaluation is set to 20000 in each run. All experiments were conducted 20 runs. In each experiment, each data set was randomly divided into two parts: 2/3 as training set and 1/3 as test set. MSE_T and MSE_G refer to mean square error averaged over 20 runs on the training and test set, respectively. And $Error_T$ and $Error_G$ referred to the error rate of classification and generalization averaged over 20 runs for the training and test sets, respectively. The information of data sets as well as the designing results of the three algorithms are listed in Table 1.

Table 1. Information of data sets and results of the three algorithms

Case/ Architecture	Index	CPSODE	PSO	Powell-Beale
Ionosphere/34-2-2	MSE_T	1.01e-05	1.36e-01	1.82e-03
	$Error_T$	0	26.1538	0
	MSE_G	2.66e-01	9.44e-01	7.24e-01
	$Error_G$	18.6752	48.5897	28.6325
Cmc/9-5-3	MSE_T	1.62e-02	1.97e-01	1.91e-01
	$Error_T$	57.0400	76.4950	77.9600
	MSE_G	3.33e-01	2.09e-01	2.04e-01
	$Error_G$	64.0381	77.5370	80.3171
Henon/2-5-1	MSE_T	2.73e-05	2.20e-03	4.49e-05
	$Error_T$	0	0.0250	0
	MSE_G	2.79e-05	2.00e-03	6.45e-05
	$Error_G$	0	0	0

From Table 1, it can be seen that MSE_T, MSE_G, $Error_T$ and $Error_G$ of CPSODE outperform those of both PSO and Powell-Beale. So, it is concluded that CPSODE is more effective for neural network design, more robust on initial conditions and can obtain networks with better generalization property. In a word, the proposed hybrid PSO is a viable approach for neural networks design.

5   Conclusions

In this paper, an effective hybrid PSO is proposed for designing neural networks, in which PSO, differential evolution and chaotic local search are well combined. Simulation results and comparisons based on three typical problems demonstrated the effectiveness of the hybrid PSO in term of searching quality, robustness on initial conditions and generalization property. The future work is to apply such approach to design other kinds of neural networks.

Acknowledgements

This work is supported by National Natural Science Foundation of China (Grant No. 60204008 and 60374060) and National 973 Program (Grant No. 2002CB312200).

References

1. Yao, X.: Evolving Artificial Neural Networks. Proceedings of the IEEE, **87** (1999) 1423-1447
2. Wang, L.: Intelligent Optimization Algorithms with Applications. Tsinghua University & Springer Press, Beijing (2001)
3. Kennedy, J., Eberhart, R.C., Shi, Y.: Swarm Intelligence. Morgan Kaufmann Publishers, San Francisco (2001)
4. Storn, R., Price, K.: Differential Evolution: A Simple and Efficient Adaptive Scheme for Global Optimization over Continuous Spaces. Technical report, TR-95-012, International Computer Science Institute, Berkley 1995
5. Li, B., Jiang, WS.: Optimizing Complex Functions by Chaos Search. Cybernetics and Systems, **29** (1998) 409-419
6. Wang, L., Zheng, D.Z., Lin, Q.S.: Survey on Chaotic Optimization Methods. Computing Technology and Automation, **20** (2001) 1-5
7. Blake, C., Keogh, E., Merz, C.J.: UCI Repository of Machine Learning Databases, http://www.ics.uci.edu/~mlearn/MLRepository.html. University of California, Irvine, Dept. of Information and Computer Sciences, 1998

A New Strategy
for Designing Bidirectional Associative Memories

Gengsheng Zheng[1], Sidney Nascimento Givigi[2], and Weiyu Zheng[3]

[1] Gengsheng Zheng, School of Computer Science and Technology,
Wuhan University, Wuhan Hubei 430072, China
zhenggengsheng@sina.com
[2] Sidney Nascimento Givigi Junior, Department of Systems and Computer Engineering,
Carleton University, Ottawa, ON, K1S 5B6, Canada
sidney.givigi@rogers.com
[3] Weiyu Zheng, Department of Systems and Computer Engineering,
Carleton University, Ottawa, ON, K1S 5B6, Canada
wyzheng@sce.carleton.ca

Abstract. A comprehensive strategy for bidirectional associative memories (BAMs) is presented, which enhances storage capacity greatly. The design strategy combines the dummy augmentation encoding method with optimal gradient descent algorithm. The proposed method increases the storage capacity performance of BAM to its upper limit compared with original Kosko method and optimal gradient descent algorithm. Computer simulations and comparison are given based on three methods to demonstrate the performance improvement of the proposed strategy.

1 Introduction

Kosko [1] defines the bidirectional associative memory (BAM) neural network as an associative memory that uses forward and backward bidirectional search to recall an associated bipolar (or binary) vector pair $(x^{(i)}, y^{(i)})$ from an input vector pair (x, y) or just an arbitrary vector x. It bases upon correlation matrix summation. Given m bipolar row-vector pairs, these pairs are encoded in the correlation matrix of a BAM by 'sum-of-outer-product' (SOP) rule. Then the correlation matrix W is:

$$W = \sum_{i=1}^{m} (x^{(i)})^T y^{(i)}. \tag{1}$$

However, Kosko's BAM doesn't guarantee the storage of the training pairs to be the stable states of the network. There are two basic ways to overcome disadvantages. One way is to change the network architecture and the other is to modify the learning algorithm [2]-[5].

In this work, we keep the structure of the BAM unchanged and by the introduction of the dummy augmentation encoding strategy with optimal gradient descent algorithm the storage capacity is greatly enhanced. It exploits the maximum number of degrees of freedom allowed in the BAM structure and changes the thresholds automatically, while preserving the global stability.

J. Wang, X. Liao, and Z. Yi (Eds.): ISNN 2005, LNCS 3496, pp. 398–403, 2005.
© Springer-Verlag Berlin Heidelberg 2005

2 Optimal Gradient Descent Algorithm

Let $\{x^{(k)}, y^{(k)}, k = 1\cdots m\}$ denote the set of associations to be stored. $x_i^{(k)}$ denotes the ith element of vector $x^{(k)}$ and $y_j^{(k)}$ denotes the jth element of vector $y^{(k)}$, respectively. $E()$ denotes the total squared error function to be minimized by a given network. Here we make $E(w)$ the cost function. Then

$$E(w) = \frac{1}{2}\sum_{k=1}^{m}[(y_j^{(k)} - y_{dj}^{(k)})^2 + (x_i^{(k)} - x_{di}^{(k)})^2], \tag{2}$$

where $y_{dj}^{(k)}$ is the desired output for input $x_i^{(k)}$ and $x_{di}^{(k)}$ is the desired output for input $y_j^{(k)}$. And

$$x_i^{(k)} = \text{sgn}(\sum_{j=1}^{p} w_{ij} y_j^{(k)}), i = 1,\cdots,n \tag{3}$$

$$y_j^{(k)} = \text{sgn}(\sum_{i=1}^{n} w_{ij} x_i^{(k)}), j = 1,\cdots,p \tag{4}$$

where $\text{sgn}(x)$ bipolarizes x and w_{ij} is the ijth element of the matrix W. According to the idea of optimal gradient descent method, at the $(t+1)$th iteration step, the matrix $w(t+1)$ changes along the gradient direction $\dfrac{\partial E(w)}{\partial w}$ defined at $w(t)$, until the objective function $E(w(t+1))$ reaches a local minimum. So

$$w(t+1) = w(t) - \lambda\frac{\partial E(w)}{\partial w(t)}. \tag{5}$$

Function $E(w)$ is convex. This means that the only points where $\nabla E(w) = 0$ are local minima (corresponding to the feasible points). Moreover, the cost function may be a Lyapunov function if the inputs are suitably chosen. Therefore, minimization of $E(w)$ by a gradient descent rule will guarantee convergence to a stationary point, i.e. to a feasible W matrix:

$$w_{ij}(t+1) = w_{ij}(t) - \lambda\frac{\partial E(w)}{\partial w_{ij}}$$
$$= w_{ij}(t) - \lambda\sum_{k=1}^{m}[(y_j^{(k)} - y_{dj}^{(k)})x_i^{(k)} + (x_i^{(k)} - x_{di}^{(k)})y_j^{(k)}], t = 0,1,2,\cdots \tag{6}$$

In [5] it has been proved that optimal gradient descent algorithm can be better for the improvement on BAM capacity than both multiple training method and original Kosko method by removing the correlation-like form of the matrix W.

3 The Dummy Augmentation Encoding Strategy with Optimal Gradient Descent Algorithm

To improve the BAM capacity further, we combine optimal gradient descent algorithm with the dummy augmentation encoding strategy. Dummy augmentation method makes the correct recall of every training pair possible since it changes the threshold function to the input and output of the network differently. However, this is obtained at an additional cost in terms of extra weights in the modified network, i.e., more memory usage. Dummy elements are required for both the coding phase and the decoding phase of BAM.

Consider the training pairs $\{(x^{(i)}, y^{(i)}), i = 1, 2, ..., m\}$. Let us now extend each input pattern vector $x^{(i)}$ by padding it with r zero dimensions represented by the new vector $d^{(i)}$, where $d^{(i)} = (\underbrace{0, 0, ... 0}_{r})$. We denote the new input with x_{new}. Then the correlation matrix W is

$$W = \sum_{i=1}^{m} [x^{(i)} \mid \beta_1 \mid \cdots \mid \beta_r]^T y^{(i)} = \sum_{i=1}^{m} \left(x_{new}^{(i)}\right)^T y^{(i)} \tag{7}$$

For binary coding, $x_{new}^{(i)} \in \{0, +1\}^{n+r}$, where $x_{new(n+l)}^{(i)} = 0$, $(l = 1, 2 \cdots r)$. In this case, clearly $rank(x_{new}) = rank(x_{old})$, which means that no information is introduced when expanding the input patterns with 0. While, for bipolar coding, $x_{new}^{(i)} \in \{-1, +1\}^{n+r}$, where $x_{new(n+l)}^{(i)} = -1$, $(l = 1, 2 \cdots r)$, and in this case, $rank(x_{new}) \neq rank(x_{old})$, which means that new information is introduced. We note that the same analysis is valid if we pad the output patterns.

The matrix W above would be $(n + r) \times p$. The extra dimension provides a way to shift the threshold in the space it is related to, i.e., if we add an extra dimension to the input space we dynamically change the threshold of the output space and vice versa. If the thresholds are judiciously selected, the memory capacity can increase greatly with the upper bound $\min(2^n, 2^p)$ instead of $\min(n, p)$. Changing the threshold also makes it easier to separate the input patterns. If we want to implement the BAM in hardware, we may still use a $n \times p$ matrix W and just add elements to implement the threshold shift. In this way, the resulting design is not much more complex (in terms of component number) than an original BAM.

4 Computer Simulations

Computer simulations have been performed to ascertain the improvement of capacity. Three technical methods have been compared: the traditional Kosko method, optimal gradient descent algorithm, and the present comprehensive strategy. The initial conditions in the simulations are: $1 \leq m \leq 10$, $n = 7$, $p = 10$, $\lambda = 0.01$. To simplify the question, we make $threshold = 0$. We use bipolar coding of pattern pairs for it's better on average than binary coding. In Fig. 1, $x(j)$ and $y(j)$ denote the jth element of row-vector $x^{(i)}$ and $y^{(i)}$, respectively. In our test, if a pair can be recalled, it is declared a success; otherwise, it is a failure. The percentage of success against m is shown in Fig. 2.

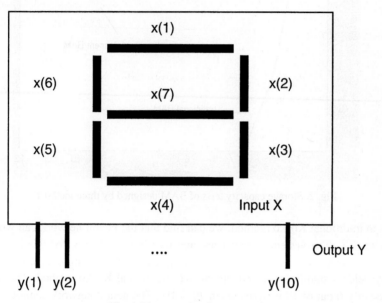

Fig. 1. Test sample of pattern pair

$$x^{(1)} = [1111110], \qquad y^{(1)} = [1000000000];$$

$$x^{(2)} = [0110000], \qquad y^{(2)} = [0100000000];$$

$$x^{(3)} = [1101101], \qquad y^{(3)} = [0010000000];$$

$$x^{(4)} = [1111001], \qquad y^{(4)} = [0001000000];$$

$$x^{(5)} = [0110011], \qquad y^{(5)} = [0000100000];$$

$$x^{(6)} = [1011011], \qquad y^{(6)} = [0000010000];$$

$$x^{(7)} = [1011111], \qquad y^{(7)} = [0000001000];$$

$$x^{(8)} = [1110000], \qquad y^{(8)} = [0000000100];$$

$$x^{(9)} = [1111111], \qquad y^{(9)} = [0000000010];$$

$$x^{(10)} = [1111011], \qquad y^{(10)} = [0000000001];$$

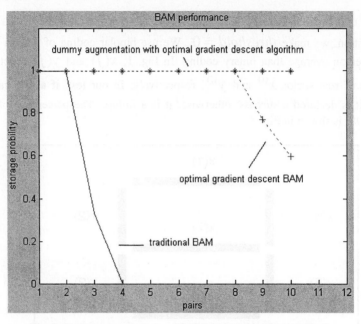

Fig. 2. Storage capacity tests of BAM designed by three methods

As to traditional Kosko method, we can find that the neural network can work only when $m \leq 3$. The system cannot store any pair if $m > 3$. So, for this case Kosko method does not work well. About optimal gradient descent algorithm, we can find that it yields a significant improvement over traditional Kosko method. The network can identify 6 out of 10 patterns when $m = 10$. The neural network cannot store and recall all patterns when $m > 8$. When we combine the dummy augmentation encoding strategy with optimal gradient descent algorithm to train the neural network, it assures recall of every pattern pair of the test set.

Table 1. Epoch numbers designed by three methods

Technical solution	Epoch numbers
Initial conditions: m=10, n=7, p=10, λ=0.01, threshold=0	
Traditional Kosko's BAM method	1
Optimal gradient descent algorithm	50000
The dummy augmentation encoding strategy with optimal gradient descent algorithm (one dimension padding)	189

Table 1 shows the number of epochs in each of the techniques when $m = 10$. For the traditional Kosko method, we do not really train the network and just some multiplications and additions are needed. Here we represent it by just one epoch, meaning we do not alter the correlation matrix once it is calculated. When optimal gradient descent algorithm is used, we need much more epochs and consequently multiplications and additions due to the fact that the cost function does not have a global mini-mum and the algorithm does not converge. As to our proposed comprehensive strat-egy, it converges in just 189 epochs. If we increase the number of extra-dimensions in the input space, the algorithm converges even faster (in terms of epoch numbers). Since the dimensions of W are also increased, according to number of multiplications and additions, the performance is almost the same.

From the simulations above, it is seen that the dummy augmentation strategy with optimal gradient descent algorithm offers significant performance advantages com-pared with traditional Kosko method and optimal gradient descent algorithm.

5   Conclusion

The concepts of optimal gradient descent and dummy augmentation in the encoding of data for a discrete BAM have been introduced to improve the storage capacity greatly. When we combine them together, we can improve the storage capacity of BAM fur-ther. This is due to the fact that adding more dimensions to the input patterns provides a mean to determine the thresholds dynamically. If we want to make the BAM behave well for noisy patterns, more dimensions must be added.

References

1. Kosko, B.: Bidirectional Associative Memories. IEEE Trans. Syst., Man, Cybern., **18** (1988) 49-60
2. Haines, K., Hecht-Nielson, R.: A BAM with Increased Information Storage Capacity. Proc. IJCNN 88, **1** (1988) 181-190
3. Wang, Y. F., Cruz, J.B., Mulligan, J.H.: Two coding Strategies for Bidirectional Associa-tive Memory. IEEE Trans. Neural Network, **1** (1990) 81-92
4. Wang, W. J., Lee, D. L.: A Modified Bidirectional Decoding Strategy Based on the BAM Structure. IEEE Trans. Neural Network, **4** (1993) 710-717
5. Perfetti, R.: Optimal Gradient Descent Learning for Bidirectional Associative Memories. Electron. Lett., **29** (1993) 1556-1557

Genetically Optimized Hybrid Fuzzy Neural Networks Based on TSK Fuzzy Rules and Polynomial Neurons

Sungkwun Oh[1], Byoungjun Park[2], and Hyunki Kim[1]

[1] Department of Electrical Engineering, The University of Suwon, San 2-2 Wau-ri,
Bongdam-eup, Hwaseong-si, Gyeonggi-do, 445-743, South Korea
ohsk@suwon.ac.kr
[2] Department of Electrical, Electronic and Information Engineering,
Wonkwang University, Korea

Abstract. We introduce an advanced architecture of genetically optimized Hybrid Fuzzy Neural Networks (gHFNN) and develop a comprehensive design methodology supporting their construction. The gHFNN architecture results from a synergistic usage of the hybrid system generated by combining Fuzzy Neural Networks (FNN) with Polynomial Neural Networks (PNN). We distinguish between two types of the linear fuzzy inference rule (TSK fuzzy rule)-based FNN structures showing how this taxonomy depends upon the type of a fuzzy partition of input variables. As to the consequence part of the gHFNN, the development of the PNN dwells on two general optimization mechanisms: the structural optimization is realized via GAs whereas in case of the parametric optimization we proceed with a standard least square method-based learning.

1 Introductory Remarks

The models should be able to take advantage of the existing domain knowledge and augment it by available numeric data to form a coherent data-knowledge modeling entity. The omnipresent modeling tendency is the one that exploits techniques of Computational Intelligence (CI) by embracing fuzzy modeling [1–6], neurocomputing [7], and genetic optimization [8].

In this study, we develop a hybrid modeling architecture, called genetically optimized Hybrid Fuzzy Neural Networks (gHFNN). In a nutshell, gHFNN is composed of two main substructures driven to genetic optimization, namely a fuzzy set-based fuzzy neural network (FNN) and a polynomial neural network (PNN). From a standpoint of rule-based architectures, one can regard the FNN as an implementation of the antecedent part of the rules while the consequent is realized with the aid of a PNN. The role of the FNN is to interact with input data, granulate the corresponding input spaces. The FNNs come with two kinds of network architectures, namely fuzzy-set based FNN and fuzzy-relation based FNN. The role of the PNN is to carry out nonlinear transformation at the level of the fuzzy sets formed at the level of FNN. The PNN that exhibits a flexible and versatile structure [9] is constructed on a basis of Group Method of Data Handling (GMDH [10]) method and genetic algorithms (GAs). The design procedure applied in the construction of each layer of the PNN

J. Wang, X. Liao, and Z. Yi (Eds.): ISNN 2005, LNCS 3496, pp. 404–409, 2005.

deals with its structural optimization involving the selection of optimal nodes with specific local characteristics and addresses specific aspects of parametric optimization. To assess the performance of the proposed model, we exploit a well-known time series data. Furthermore, the network is directly contrasted with several existing intelligent models.

2 Conventional Hybrid Fuzzy Neural Networks (HFNN)

The architectures of conventional HFNN [11], [12] result as a synergy between two other general constructs such as FNN and PNN. Based on the different PNN topologies, the HFNN distinguish between two kinds of architectures, namely basic and modified architectures. Moreover we identify also two types as the generic and advanced. The topologies of the HFNN depend on those of the PNN used for the consequence part of HFNN. The design of the PNN proceeds further and involves a generation of some additional layers. Each layer consists of nodes (PNs) for which the number of input variables could the same as in the previous layers or may differ across the network. The structure of the PNN is selected on the basis of the number of input variables and the order of the polynomial occurring in each layer.

3 Genetically Optimized HFNN (gHFNN)

The gHFNN emerges from the genetically optimized multi-layer perceptron architecture based on fuzzy set-based FNN, GAs and GMDH. These networks result as a synergy between two other general constructs such as FNN [13] and PNN [9].

3.1 Fuzzy Neural Networks Based on Genetic Optimization

We consider two kinds of FNNs (viz. FS_FNN and FR_FNN) based on linear fuzzy inference. The FNN is designed by using space partitioning realized in terms of the individual input variables or an ensemble of all variables. Table 1 represents the comparison of fuzzy rules, inference result and learning for two types of FNNs.

The learning of FNN is realized by adjusting connections of the neurons and as such it follows a standard Back-Propagation (BP) algorithm [14]. For the linear fuzzy inference-based FNN, the update formula of a connection is as shown in Table 1. η is a positive learning rate and α is a momentum coefficient constrained to the unit interval. In order to enhance the learning of the FNN and augment its performance of a FNN, we use GAs to adjust learning rate, momentum coefficient and the parameters of the membership functions of the antecedents of the rules.

3.2 Genetically Optimized PNN (gPNN)

When we construct PNs of each layer in the conventional PNN [9], such parameters as the number of input variables (nodes), the order of polynomial, and input variables available within a PN are fixed in advance by the designer. This could have frequently contributed to the difficulties in the design of the optimal network. To overcome this apparent drawback, we introduce a new genetic design approach.

Table 1. Comparison of fuzzy set with fuzzy relation-based FNNs

Structure	FS_FNN	FR_FNN
Fuzzy rules	$R^j : If\ x_i\ is\ A_{ij}\ then\ Cy_{ij} = ws_{ij} + w_{ij} \cdot x_i$	$R : If\ x_1\ is\ A_1 \cdots and\ x_k\ is\ A_k\ then\ G_i = w_{0i} + w_{1i} \cdot x_1 + w_{ki} \cdot x_k$
Inference result	$\hat{y} = f_1(x_1) + f_2(x_2) + \cdots + f_m(x_m) = \sum_{i=1}^{m} f_i(x_i)$ $$f_i(x_i) = \dfrac{\sum_{j=1}^{z}(\mu_{ij}(x_i)\cdot(ws_{ij}+x_iw_{ij}))}{\sum_{j=1}^{z}\mu_{ij}(x_i)}$$ $= \mu_{ik}(x_i)\cdot(ws_{ik}+x_iw_{ik})$ $+ \ \mu_{ik+1}(x_i)\cdot(ws_{ik+1}+x_iw_{ik+1})$	$\hat{y} = \sum_{i=1}^{n} f_i$ $= \sum_{i=1}^{n} \overline{\mu}_i \cdot (w_{0i} + w_{1i} \cdot x_1 + w_{ki} \cdot x_k)$ $= \sum_{i=1}^{n} \dfrac{\mu_i \cdot (w_{0i} + w_{1i} \cdot x_1 + w_{ki} \cdot x_k)}{\sum_{i=1}^{n} \mu_i}$
Learning	$\begin{cases} \Delta ws_{ij} = 2 \cdot \eta \cdot (y - \hat{y}) \cdot \mu_{ij} \\ \qquad + \alpha(ws_{ij}(t) - ws_{ij}(t-1)) \\ \Delta w_{ij} = 2 \cdot \eta \cdot (y - \hat{y}) \cdot \mu_{ij} \cdot x_i \\ \qquad + \alpha(w_{ij}(t) - w_{ij}(t-1)) \end{cases}$	$\begin{cases} \Delta w_{0i} = 2 \cdot \eta \cdot (y - \hat{y}) \cdot \overline{\mu}_i \\ \qquad + \alpha(w_{0i}(t) - w_{0i}(t-1)) \\ \Delta w_{ki} = 2 \cdot \eta \cdot (y - \hat{y}) \cdot \overline{\mu}_i \cdot x_k \\ \qquad + \alpha(w_{ki}(t) - w_{ki}(t-1)) \end{cases}$

4   The Algorithms and Design Procedure of gHFNN

The Premise of gHFNN: FNN
[Layer 1] Input layer.
[Layer 2] Computing activation degrees of linguistic labels.
[Layer 3] Normalization of a degree activation (firing) of the rule.
[Layer 4] Multiplying a normalized activation degree of the rule by connection. If we choose Connection point 1 for combining FNN with gPNN, a_{ij} is given as the input variable of the gPNN.

$$a_{ij} = \overline{\mu}_{ij} \times Cy_{ij} = \mu_{ij} \times Cy_{ij}\ (Here,\ Cy_{ij} = ws_{ij} + w_{ij} \cdot x_i) \tag{1}$$

[Layer 5] Fuzzy inference for the fuzzy rules. If we choose Connection point 2, f_i is the input variable of gPNN.
[Layer 6; Output layer of FNN] Computing output of a FNN.

The Consequence of gHFNN: gPNN
[Step 1] Configuration of input variables.
[Step 2] Decision of initial information for constructing the gPNN.
[Step 3] Initialization of population.
[Step 4] Decision of PNs structure using genetic design.
[Step 5] Evaluation of PNs.
[Step 6] Elitist strategy and selection of PNs with the best predictive capability.
[Step 7] Reproduction.
[Step 8] Repeating Step 4-7.

[Step 9] Construction of their corresponding layer.
[Step 10] Check the termination criterion (performance index).
[Step 11] Determining new input variables for the next layer.
  The gPNN algorithm is carried out by repeating Steps 4-11.

5 Experimental Studies

The performance of the gHFNN is illustrated with the aid of a time series of gas furnace [14]. The delayed terms of methane gas flow rate, $u(t)$ and carbon dioxide density, $y(t)$ are used as system input variables. We use two types of system input variables of FNN structure, Type I and Type II to design an optimal model from gas furnace data. Type I utilize two input variables such as $u(t-3)$ and $y(t-1)$ and Type II utilizes 3 input variables such as $u(t-2)$, $y(t-2)$, and $y(t-1)$. The output variable is $y(t)$.

Table 2. Performance index of gHFNN for the gas furnace

Structure	Premise part No. of rules (MFs)	PI	EPI	CP	Layer	Consequence part No. of inputs	Input No.				T	PI	EPI
FS_FNN	4 (2+2)	0.041	0.267	01	1	4	4	2	1	3	3	0.019	0.292
					2	4	7	12	2	10	2	0.018	0.271
					3	4	20	21	5	3	2	0.017	0.267
					4	3	22	13	29	·	2	0.016	0.263
					5	4	25	18	27	9	3	0.015	0.258
	6 (2+2+2)	0.0256	0.143	02	1	3	1	2	3	·	3	0.0232	0.130
					2	4	12	15	13	6	2	0.0196	0.120
					3	2	19	30	·	·	2	0.0194	0.115
					4	4	2	21	11	5	1	0.0188	0.113
					5	4	13	3	26	25	1	0.0184	0.110
FR_FNN	4 (2x2)	0.025	0.265		1	4	4	1	3	2	2	0.019	0.267
					2	4	7	1	13	22	3	0.026	0.251
					3	2	1	28	·	·	3	0.025	0.244
					4	2	7	6	·	·	3	0.025	0.243
					5	3	29	22	17	·	3	0.016	0.249
	8 (2x2x2)	0.033	0.119		1	4	6	5	2	8	1	0.083	0.146
					2	4	21	18	6	9	2	0.028	0.116
					3	4	4	24	5	6	2	0.022	0.110
					4	3	28	4	5	·	2	0.021	0.106
					5	3	21	18	25	·	1	0.021	0.104

Table 2 includes the results of the overall network reported according to various alternatives concerning various forms of FNN architecture, format of entire system inputs and location of the connection point (CP). Fig. 1 illustrates the optimization process by visualizing the performance index in successive cycles. Table 3 contrasts the performance of the genetically developed network with other fuzzy and fuzzy-neural networks studied in the literatures.

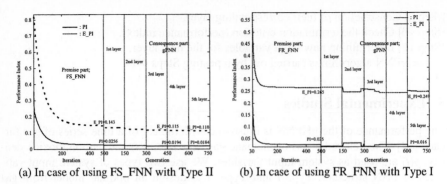

(a) In case of using FS_FNN with Type II   (b) In case of using FR_FNN with Type I

Fig. 1. Optimization procedure of gHFNN by BP learning and GAs

Table 3. Comparison of performance with other modeling methods

	Model		PI	EPI	No. of rules
	Box and Jenkin's model [14]		0.710		
	Pedrycz's model [1]		0.320		
	Xu and Zailu's model [2]		0.328		
	Sugeno and Yasukawa's model [3]		0190		
	Kim, et al.'s model [15]		0.034	0.244	2
	Lin and Cunningham's mode [16]		0.071	0.261	4
Fuzzy	Complex [4]	Simplified	0.024	0.328	4(2×2)
		Linear	0.023	0.306	4(2×2)
	Hybrid [6] (GAs+Complex)	Simplified	0.024	0.329	4(2×2)
		Linear	0.017	0.289	4(2×2)
	HCM+GAs [5]	Simplified	0.022	0.333	6(3×2)
		Linear	0.020	0.264	6(3×2)
	FNN [13]	Simplified	0.043	0.264	6(3+3)
		Linear	0.037	0.273	6(3+3)
	SOFPNN	Generic [11]	0.023	0.277	4 rules/5th layer(NA)
			0.020	0.119	6 rules/5th layer(22 nodes)
		Advanced [12]	0.019	0.264	4 rules/5th layer(NA)
			0.017	0.113	6 rules/5th layer(26 nodes)
Proposed model (gHFNN)	FS_FNN		0.017	0.267	4 rules/3rd layer(16 nodes)
			0.019	0.115	6 rules/3rd layer(10 nodes)
	FR_FNN		0.025	0.244	4 rules/3rd layer(8 nodes)
			0.022	0.110	7 rules/3rd layer(14 nodes)

6   Concluding Remarks

The genetically optimized HFNNs are constructed by combining FNNs with gPNNs. The proposed model comes with two kinds of rule-based FNNs as well as a diversity of local characteristics of PNs that are extremely useful when coping with various nonlinear characteristics of the system under consideration. In what follows, in contrast to the conventional HFNN structures and their learning, the depth (the number of layers) and the width (the number of nodes) as well as the number of entire nodes (inputs) of the proposed genetically optimized HFNN (gHFNN) can be lower. The comprehensive design methodology comes with the parametrically as well as structurally optimized network architecture.

Acknowledgements

This work has been supported by KESRI(R-2004-B-274), which is funded by MOCIE(Ministry of commerce, industry and energy)

References

1. Pedrycz, W.: An Identification Algorithm in Fuzzy Relational System. Fuzzy Sets and Systems. 13 (1984) 153-167
2. Xu, C.W., Zailu, Y.: Fuzzy Model Identification Self-learning for Dynamic system. IEEE Trans. on Syst. Man, Cybern. SMC-17 (1987) 683-689
3. Sugeno, M., Yasukawa, T.: A Fuzzy-Logic-Based Approach to Qualitative Modeling. IEEE Trans. Fuzzy Systems. 1 (1993) 7-31
4. Oh, S.K., Pedrycz, W.: Fuzzy Identification by Means of Auto-Tuning Algorithm and Its Application to Nonlinear Systems. Fuzzy Sets and Systems. 115 (2000) 205-230
5. Park, B.J., Pedrycz, W., Oh, S.K.: Identification of Fuzzy Models with the Aid of Evolutionary Data Granulation. IEE Proceedings-Control theory and application. 148 (2001) 406-418
6. Oh, S.K., Pedrycz, W., Park, B.J.: Hybrid Identification of Fuzzy Rule-Based Models. International Journal of Intelligent Systems. 17 (2002) 77-103
7. Narendra, K.S., Parthasarathy, K.: Gradient Methods for the Optimization of Dynamical Systems Containing Neural Networks. IEEE Transactions on Neural Networks, 2 (1991) 252-262
8. Michalewicz, Z.: Genetic Algorithms + Data Structures = Evolution Programs. Springer-Verlag, Berlin Heidelberg (1996)
9. Oh, S.K., Pedrycz, W., Park, B.J.: Polynomial Neural Networks Architecture: Analysis and Design. Computers and Electrical Engineering. 29 (2003) 653-725
10. Ivahnenko, A.G.: The Group Method of Data Handling: A Rival of Method of Stochastic Approximation. Soviet Automatic Control. 13 (1968) 43-55
11. Park, B.J., Oh, S.K., Jang, S.W.: The Design of Adaptive Fuzzy Polynomial Neural Networks Architectures Based on Fuzzy Neural Networks and Self-Organizing Networks. Journal of Control, Automation and Systems Engineering. 8 (2002) 126-135
12. Park, B.J., Oh, S.K.: The Analysis and Design of Advanced Neurofuzzy Polynomial Networks. Journal of the Institute of Electronics Engineers of Korea. 39-CI (3) (2002) 18-31
13. Oh, S.K., Pedrycz, W., Park, H.S.: Hybrid Identification in Fuzzy-Neural Networks. Fuzzy Sets and Systems. 138 (2003) 399-426
14. Box, D.E.P., Jenkins, G.M.: Time Series Analysis, Forecasting, and Control, 2nd edition Holden-Day, San Francisco (1976)
15. Kim, E., Lee, H., Park, M., Park, M.: A Simply Identified Sugeno-type Fuzzy Model via Double Clustering. Information Sciences. 110 (1998) 25-39
16. Lin, Y., Cunningham, G.A. III: A New Approach to Fuzzy-neural Modeling. IEEE Transaction on Fuzzy Systems. 3 (2) 190-197

Genetically Optimized Self-organizing Fuzzy Polynomial Neural Networks Based on Information Granulation

Hosung Park[1], Daehee Park[1], and Sungkwun Oh[2]

[1] School of Electrical Electronic and Information Engineering, Wonkwang University, 344-2, Shinyong-Dong, Iksan, Chon-Buk, 570-749, South Korea
[2] Department of Electrical Engineering, The University of Suwon, San 2-2 Wau-ri, Bongdam-eup, Hwaseong-si, Gyeonggi-do, 445-743, South Korea
ohsk@suwon.ac.kr

Abstract. In this study, we introduce and investigate a genetically optimized self-organizing fuzzy polynomial neural network with the aid of information granulation (IG_gSOFPNN), develop a comprehensive design methodology involving mechanisms of genetic optimization. With the aid of the information granulation, we determine the initial location (apexes) of membership functions and initial values of polynomial function being used in the premised and consequence part of the fuzzy rules respectively. The GA-based design procedure being applied at each layer of IG_gSOFPNN leads to the selection of preferred nodes with specific local characteristics (such as the number of input variables, the order of the polynomial, a collection of the specific subset of input variables, and the number of membership function) available within the network.

1 Introduction

While the theory of traditional equation-based approaches is well developed and successful in practice (particularly in linear cases) there has been a great deal of interest in applying model-free methods such as neural and fuzzy techniques for nonlinear function approximation [1]. GMDH was introduced by Ivakhnenko in the early 1970's [2]. GMDH-type algorithms have been extensively used since the mid-1970's for prediction and modeling complex nonlinear processes. While providing with a systematic design procedure, GMDH comes with some drawbacks. To alleviate the problems associated with the GMDH, Self-Organizing Neural Networks (SONN, called "SOFPNN") were introduced by Oh and Pedrycz [3], [4], [5] as a new category of neural networks or neuro-fuzzy networks. Although the SOFPNN has a flexible architecture whose potential can be fully utilized through a systematic design, it is difficult to obtain the structurally and parametrically optimized network because of the limited design of the nodes located in each layer of the SOFPNN.

In this study, in considering the above problems coming with the conventional SOFPNN, we introduce a new structure and organization of fuzzy rules as well as a new genetic design approach. The new meaning of fuzzy rules, information granules melt into the fuzzy rules. In a nutshell, each fuzzy rule describes the related information granule. The determination of the optimal values of the parameters available within an individual FPN (viz. the number of input variables, the order of the polynomial, a collection of preferred nodes, and the number of MF) leads to a structurally and parametrically optimized network through the genetic approach.

J. Wang, X. Liao, and Z. Yi (Eds.): ISNN 2005, LNCS 3496, pp. 410–415, 2005.

2 SOFPNN with Fuzzy Polynomial Neuron (FPN) and Its Topology

The FPN consists of two basic functional modules. The first one, labeled by **F**, is a collection of fuzzy sets that form an interface between the input numeric variables and the processing part realized by the neuron. The second module (denoted here by **P**) is about the function – based nonlinear (polynomial) processing. The detailed FPN involving a certain regression polynomial is shown in Table 1. The choice of the number of input variables, the polynomial order, input variables, and the number of MF available within each node itself helps select the best model with respect to the characteristics of the data, model design strategy, nonlinearity and predictive capabilities.

Table 1. Different forms of regression polynomial building a FPN

| Order of the polynomial | | No. of inputs | | |
| | | 1 | 2 | 3 |
Order	FPN			
0	Type 1	Constant	Constant	Constant
1	Type 2	Linear	Bilinear	Trilinear
2	Type 3	Quadratic	Biquadratic-1	Triquadratic-1
	Type 4		Biquadratic-2	Triquadratic-2

1: Basic type, 2: Modified type

Proceeding with the SOFPNN architecture essential design decisions have to be made with regard to the number of input variables and the order of the polynomial forming the conclusion part of the rules as well as a collection of the specific subset of input variables.

Table 2. Polynomial type according to the number of input variables in the conclusion part of fuzzy rules

Type of the consequence polynomial	Input vector	Selected input variables in the premise part	Selected input variables in the consequence part
Type T		A	A
Type T*		A	B

Where notation **A**: Vector of the selected input variables $(x_1, x_2,..., x_i)$, **B**: Vector of the entire system input variables $(x_1, x_2, ...x_i, x_j ...)$, Type T: $f(A)=f(x_1, x_2,..., x_i)$ - type of a polynomial function standing in the consequence part of the fuzzy rules, Type T*: $f(B)=f(x_1, x_2, ...x_i, x_j ...)$ - type of a polynomial function occurring in the consequence part of the fuzzy rules

3 The Structural Optimization of IG_gSOFPNN

3.1 Information Granulation by Means of Hard C-Means Clustering Method

Information granulation is defined informally as linked collections of objects (data points, in particular) drawn together by the criteria of indistinguishability, similarity or functionality [6]. Granulation of information is a procedure to extract meaningful concepts from insignificant numerical data and an inherent activity of human being carried out with intend of better understanding of the problem. We extract information for the real system with the aid of Hard C-means clustering method [7], which deals with the conventional crisp sets. Through HCM, we determine the initial location (apexes) of membership functions and initial values of polynomial function being used in the premise and consequence part of the fuzzy rules respectively.

The fuzzy rules of IG_gSOFPNN is given as follows:

$$R^j : If\ x_1\ is\ A_{ji}\ and \cdots x_k\ is\ A_{jk}\ then\ y_j - M_j = f_j\{(x_1 - v_{j1}), (x_2 - v_{j2}), \cdots, (x_k - v_{jk})\}$$

Where, A_{jk} mean the fuzzy set, the apex of which is defined as the center point of information granule (cluster). M_j and v_{jk} are the center points of new created input-output variables by information graunle.

3.2 Genetic Optimization of IG_gSOFPNN

Let us briefly recall that GAs is a stochastic search technique based on the principles of evolution, natural selection, and genetic recombination by simulating a process of "survival of the fittest" in a population of potential solutions to the given problem. The main features of genetic algorithms concern individuals viewed as strings, population-based optimization and stochastic search mechanism (selection and crossover). In order to enhance the learning of the IG_gSOFPNN and augment its performance, we use genetic algorithms to obtain the structural optimization of the network by optimally selecting such parameters as the number of input variables (nodes), the order of polynomial, input variables, and the number of MF within a IG_gSOFPNN. Here, GAs use serial method of binary type, roulette-wheel as the selection operator, one-point crossover, and an invert operation in the mutation operator [8].

4 The Algorithm and Design Procedure of IG_gSOFPNN

The framework of the design procedure of the gSOFPNN with aid of the Information granulation (IG) comprises the following steps.
[Step 1] Determine system's input variables.
[Step 2] Form training and testing data.
[Step 3] Determine apexes of MF by HCM clustering method.
[Step 4] Decide initial design information for constructing the IG_gSOFPNN structure.
[Step 5] Decide a structure of the FPN structure using genetic design.
[Step 6] Carry out fuzzy inference and coefficient parameters estimation for fuzzy identification in the selected node(FPNs). – New construction of consequence part of the fuzzy rule by HCM clustering method.
[Step 7] Select nodes (FPNs) with the best predictive capability and construct their corresponding layer.
[Step 8] Check the termination criterion.
[Step 9] Determine new input variables for the next layer.
The IG_gSOFPNN algorithm is carried out by repeating steps 3-9 of the algorithm.

5 Experimental Studies

We illustrate the performance of the network and elaborate on its development by experimenting with data coming from the NOx emission process of gas turbine power plant [9]. The input variables include AT, CS, LPTS, CDP, and TET. The output variable is NOx. We consider 260 pairs of the original input-output data. 130 out of 260 pairs of input-output data are used as learning set; the remaining part serves as a testing set. To come up with a quantitative evaluation of network, we use the standard MSE performance index.

Table 3. Computational aspects of the genetic optimization of IG_gSOFPNN

	Parameters	1st layer	2nd to 3rd layer
GA	Maximum generation	150	150
	Total population size	100	100
	Selected population size (W)	30	30
	Crossover rate	0.65	0.65
	Mutation rate	0.1	0.1
	String length	3+3+30+5	3+3+30+5
IG_gSOFPNN	Maximal no.(Max) of inputs to be selected	$1 \leq l \leq \mathrm{Max}(2\!\sim\!3)$	$1 \leq l \leq \mathrm{Max}(2\!\sim\!3)$
	Polynomial type (Type T) of the consequent part of fuzzy rules(#)	$1 \leq T \leq 4$	$1 \leq T \leq 4$
	Consequent input type to be used for Type T (##)	Type T*	Type T
	Membership Function (MF) type	Triangular Gaussian	Triangular Gaussian
	No. of MFs per input	2 or 3	2 or 3

l, T, Max: integers, # and ## : refer to Tables 1-2 respectively.

Fig. 1 shows the membership functions of one input variable (LPT) according to the partition of fuzzy input spaces by a Min-Max method and the HCM clustering method.

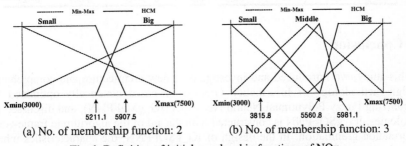

(a) No. of membership function: 2   (b) No. of membership function: 3

Fig. 1. Definition of initial membership functions of NOx

Fig. 2 illustrates the different optimization process between gSOFPNN and the proposed IG_gSOFPNN by visualizing the values of the performance index obtained in successive generations of GA when using Type T*. It also shows the optimized network architecture when using the Gaussina-like MF(the Max of inputs to be selected is set to 2 with the structure composed of 3 layers).

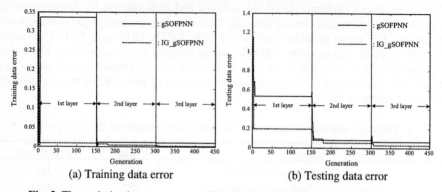

(a) Training data error   (b) Testing data error

Fig. 2. The optimization process quantified by the values of the performance index

Table 4. Comparative analysis of the performance of the network; considered are models reported in the literature

Model				PI_s	EPI_s
	Regression model			17.68	19.23
FNN model[10]	GA	Simplified		7.045	11.264
		Linear		4.038	6.028
	Hybrid	Simplified		6.205	8.868
	(GA+Complex)	Linear		3.830	5.397
Multi-FNNs[11]	Simplified			2.806	5.164
Multi-FNNs[12]	Linear			0.720	2.025
gHFPNN[13]	Max=2 (Type T*)	Triangular	3rd layer	0.008	0.082
			5th layer	0.008	0.081
		Gaussian-like	3rd layer	0.016	0.132
			5th layer	0.016	0.116
Proposed IG_gSOFPNN	Max=2 (Type T*)	Triangular	3rd layer	0.002	0.045
		Gaussian-like	3rd layer	0.001	0.027

6   Conclusion

In this study, we introduced and investigated a new architecture and comprehensive design methodology of Information Granulation-based genetically optimized Self-Organizing Fuzzy Polynomial Neural Networks (IG_gSOFPNN), and discussed their topologies. IG_gSOFPNN is constructed with the aid of the algorithmic framework of information granulation. In the design of IG_gSOFPNN, the characteristics inherent to entire experimental data being used in the construction of the gSOFPNN architecture is reflected to fuzzy rules available within a FPN. Therefore Information granulation based on HCM(Hard C-Means) clustering method was adopted.

Acknowledgements

This work has been supported by KESRI(I-2004-0-074-0-00), which is funded by MOCIE(Ministry of commerce, industry and energy)

References

1. Nie, J.H., Lee, T.H.: Rule-based Modeling: Fast construction and optimal manipulation. IEEE Trans. Syst., Man, Cybern. **26** (1996) 728-738
2. Ivakhnenko, A.G.: Polynomial Theory of Complex Systems. IEEE Trans. on Systems, Man and Cybernetics. SMC-**1** (1971) 364-378
3. Oh, S.K., Pedrycz, W.: The Design of Self-organizing Polynomial Neural Networks. Information Science. **141** (2002) 237-258
4. Oh, S.K., Pedrycz, W., Park, B.J.: Polynomial Neural Networks Architecture: Analysis and Design. Computers and Electrical Engineering. **29** (2003) 703-725
5. Oh, S.K., Pedrycz, W.: Fuzzy Polynomial Neuron-Based Self-Organizing Neural Networks. Int. J. of General Systems. **32** (2003) 237-250
6. Zadeh, L.A.: Toward a Theory of Fuzzy Information Granulation and Its Centrality in Human Reasoning and Fuzzy Logic. Fuzzy sets and Systems. **90** (1997) 111-117
7. Bezdek, J.C.: Pattern Recognition with Fuzzy Objective Function Algorithms. New York. Plenum (1981)
8. Jong, D.K.A.: Are Genetic Algorithms Function Optimizers?. Parallel Problem Solving from Nature 2, Manner, R. and Manderick, B. eds., North-Holland, Amsterdam (1992)
9. Vachtsevanos, G., Ramani, V., Hwang, T.W.: Prediction of Gas Turbine NOx Emissions Using Polynomial Neural Network. Technical Report, Georgia Institute of Technology. Atlanta. (1995)
10. Oh, S.K., Pedrycz, W., and Park, H.S.: Hybrid Identification in Fuzzy-neural Networks. Fuzzy Sets and Systems. **138** (2003) 399-426
11. Oh, S.K., Pedrycz, W., Park, H.S.: Multi-FNN Identification by Means of HCM Clustering and Genetic Algorithms. Fuzzy Sets and Systems. (2002)
12. Oh, S.K., Pedrycz, W., Park, H.S.: Multi-FNN Identification Based on HCM Clustering and Evolutionary Fuzzy Granulation. Simulation Modelling Practice and Theory. **11** (2003) 627-642
13. Oh, S.K., Pedrycz, W., Park, H.S.: Multi-layer Hybrid Fuzzy Polynomial Neural Networks: A Design in the Framework of Computational Intelligence. Neurocomputing. (2004)
14. Park, B.J., Lee, D.Y., Oh, S.K.: Rule-based Fuzzy Polynomial Neural Networks in Modeling Software Process Data. Int. J. of Control, Automations, and Systems. **1** (2003) 321-331

Identification of ANFIS-Based Fuzzy Systems with the Aid of Genetic Optimization and Information Granulation

Sungkwun Oh[1], Keonjun Park[1], and Hyungsoo Hwang[2]

[1] Department of Electrical Engineering, The University of Suwon,
San 2-2 Wau-ri, Bongdam-eup, Hwaseong-si, Gyeonggi-do, 445-743, South Korea
ohsk@suwon.ac.kr
[2] Department of Electrical, Electronic and Information Engineering, Wonkwang University,
South Korea

Abstract. In this study, we introduce a new category of ANFIS-based fuzzy inference systems with the aid of information granulation to carry out the model identification of complex and nonlinear systems. To identify the structure of fuzzy rules we use genetic algorithms (GAs). Granulation of information with the aid of Hard C-Means (HCM) clustering algorithm help determine the initial parameters of fuzzy model such as the initial apexes of the membership functions and the initial values of polynomial functions being used in the premise and consequence part of the fuzzy rules. And the initial parameters are tuned effectively with the aid of the genetic algorithms and the least square method (LSM). The proposed model is contrasted with the performance of the conventional fuzzy models in the literature.

1 Introduction

There has been a diversity of approaches to fuzzy modeling. To enumerate a few representative trends, it is essential to refer to some developments that have happened over time. In the early 1980s, linguistic modeling [2] and fuzzy relation equation-based approach [3] were proposed as primordial identification methods for fuzzy models. The general class of Sugeno-Takagi models [4] gave rise to more sophisticated rule-based systems where the rules come with conclusions forming local regression models. While appealing with respect to the basic topology (a modular fuzzy model composed of a series of rules) [5], these models still await formal solutions as far as the structure optimization of the model is concerned, say a construction of the underlying fuzzy sets - information granules being viewed as basic building blocks of any fuzzy model. Some enhancements to the model have been proposed by Oh and Pedrycz [6], yet the problem of finding "good" initial parameters of the fuzzy sets in the rules remains open.

This study concentrates on the central problem of fuzzy modeling that is a development of information granules-fuzzy sets. The design methodology emerges as a hybrid structural optimization and parametric optimization. The proposed model is evaluated with using numerical example and is contrasted with the performance of conventional models in the literature.

J. Wang, X. Liao, and Z. Yi (Eds.): ISNN 2005, LNCS 3496, pp. 416–421, 2005.

2 Information Granulation (IG)

Informally speaking, information granules [7], [8] are viewed as linked collections of objects drawn together by the criteria of proximity, similarity, or functionality. Granulation of information is an inherent and omnipresent activity of human beings carried out with intent of better understanding of the problem. In particular, granulation of information is aimed at splitting the problem into several manageable chunks. The form of information granulation themselves becomes an important design feature of the fuzzy model, which are geared toward capturing relationships between information granules.

It is worth emphasizing that the HCM clustering has been used extensively not only to organize and categorize data, but it becomes useful in data compression and model identification. For the sake of completeness of the entire discussion, let us briefly recall the essence of the HCM algorithm [9].

[Step 1] Fix the number of clusters $c(2 \leq c < n)$ and initialize the partition matrix.

[Step 2] Calculate the center vectors v_i of each cluster.

[Step 3] Update the partition matrix $\mathbf{U}^{(r)}$; these modifications are based on the standard Euclidean distance function between the data points and the prototypes.

[Step 4] Check a termination criterion.

3 ANFIS-Based Fuzzy Inference Systems (FIS) with the Aid of IG

The identification procedure for fuzzy models based on ANFIS is usually split into the identification activities dealing with the premise and consequence parts of the rules. The identification completed at the premise level consists of two main steps. First, we select the input variables $x_1, x_2, \ldots, x_k$ of the rules. Second, we form fuzzy partitions of the spaces over which these individual variables are defined. The identification of the consequence part of the rules embraces two phases, namely 1) a selection of the consequence variables of the fuzzy rules, and 2) determination of the parameters of the consequence (conclusion part).

In this study, we carry out the modeling using characteristics of input-output data set. Therefore, it is important to understand characteristics of data. To find this we use HCM clustering. By classifying data as characteristics through HCM clustering, we design the fuzzy model by means of center of classified clusters.

3.1 Premise Identification

In the premise part of the rules, we confine ourselves to a triangular type of membership function. The HCM clustering helps us organize the data into cluster, and in this way we take into account the characteristics of the experimental data. In the regions where some clusters of data have occurred, we end up with some fuzzy sets that help represent the specificity of the data set. In the sequel, the modal values of the clusters are refined (optimized) using genetic optimization (GAs).

To determine the initial membership parameters in premise part we are

[Step 1] Find the center values of each cluster from input-output data set using HCM clustering.

[Step 2] Divide the correlational fuzzy space between input variables by center values. At this time, those values become the initial values of the vertical points of the membership functions.

3.2 Consequence Identification

The characteristics of input-output data is also involved in the conclusion parts as follows:

[Step 1] Find input data set $(x_1, x_2, ..., x_k)$ that is included the respective fuzzy space because the fuzzy rules is formed by correlation between input variables.

[Step 2] Seek the corresponding output data at this time and find input-output data set $(x_1, x_2, ..., x_k ; y)$ that is contained the each fuzzy space. In this way, the center values of input-output variables in the rules of the conclusion parts are determined and these values become the initial values of the consequence polynomial functions.

4   Optimization of IG-Based FIS (ANFIS Type)

The need to solve optimization problems arises in many fields and is especially dominant in the engineering environment. There are several analytic and numerical optimization techniques, but there are still large classes of functions that are fully addressed by these techniques. This forces us to explore other optimization techniques such as genetic algorithms. To identify the fuzzy model we determine such an initial structure as the number of input variables, input variables being selected and the number of membership functions in premise part and the order of polynomial (Type) in conclusion. And then the membership parameters of the premise are optimally tuned by GAs. In what follows, we briefly review the underlying ideas of GAs, and then discuss a form of the performance index used in this identification problem.

5   Experimental Studies

This section includes comprehensive numeric studies illustrating the design of the fuzzy model. We demonstrate how IG-based FIS can be utilized to predict future values of a chaotic time series. The performance of the proposed model is also contrasted with some other models existing in the literature. The time series is generated by the chaotic Mackey–Glass differential delay equation [11] of the form:

$$\dot{x}(t) = \frac{0.2x(t-\tau)}{1+x^{10}(t-\tau)} - 0.1x(t) \tag{1}$$

The prediction of future values of this series arises is a benchmark problem that has been used and reported by a number of researchers. From the Mackey–Glass time series $x(t)$, we extracted 1000 input–output data pairs for the type from the following the type of vector format such as: [$x(t$-30), $x(t$-24), $x(t$-18), $x(t$-12), $x(t$-6), $x(t)$; $x(t$+6)] where $t = 118$–1117. The first 500 pairs were used as the training data set for IG-based FIS while the remaining 500 pairs were the testing data set for assessing the predictive performance. To come up with a quantitative evaluation of the fuzzy model, we use the standard RMSE performance index.

Table 1 summarizes the structure and parameters identification and the performance index for Max_Min-based and IG-based fuzzy model of ANFIS type with four input variables, which consist of consequence type 3.

From the table 1 it is clear that the performance of a IG-based fuzzy model of ANFIS type is better than that of a Max_Min-based fuzzy model not only after identifying the structure but also after identifying optimally the parameters.

Figure 1 depicts the values of the performance index produced in successive generation of the GAs. It is obvious that the performance of an IG-based fuzzy model is good from initial generation due to the characteristics of input-output data.

Table 1. Performance index of Max_Min-based and IG-based fuzzy model (ANFIS Type)

Model	Identification	Input variable	No. Of MFs	Type	PI	E_PI
Max/Min_FIS	Structure	$x(t$-30) $x(t$-18) $x(t$-12) $x(t)$	2x3x3x2	Type 3	0.0094	0.0091
	Parameters				0.0021	0.0020
IG_FIS	Structure	$x(t$-30) $x(t$-18) $x(t$-12) $x(t)$	2x2x2x2	Type 3	0.0007	0.0070
	Parameters				0.0005	0.0005

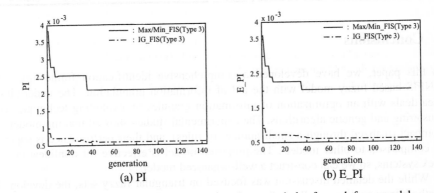

(a) PI   (b) E_PI

Fig. 1. Optimal convergence process of performance index for each fuzzy model

Model output and predicting error of training and testing data for IG-based fuzzy model is presented in figure 2.

The identification error (performance index) of the proposed model is also compared to the performance of some other models in table 2. Here the non-dimensional error index (NDEI) is defined as the root mean square errors divided by the standard deviation of the target series.

Design of Rule-Based Neurofuzzy Networks
by Means of Genetic Fuzzy Set-Based Granulation

Byoungjun Park[1] and Sungkwun Oh[2]

[1] School of Electrical, Electronic and Information Engineering, Wonkwang University,
South Korea
[2] Department of Electrical Engineering, The University of Suwon,
San 2-2 Wau-ri, Bongdam-eup, Hwaseong-si, Gyeonggi-do, 445-743, South Korea
ohsk@suwon.ac.kr

Abstract. In this paper, new architectures and design methodologies of Rule based Neurofuzzy Networks (RNFN) are introduced and the dynamic search-based GAs is introduced to lead to rapidly optimal convergence over a limited region or a boundary condition. The proposed RNFN is based on the fuzzy set based neurofuzzy networks (NFN) with the extended structure of fuzzy rules being formed within the networks. In the consequence part of the fuzzy rules, three different forms of the regression polynomials such as constant, linear and modified quadratic are taken into consideration. The dynamic search-based GAs optimizes the structure and parameters of the RNFN.

1 Introduction

Lately, CI technologies become hot issue of IT (Information technology) field and abilities of that interest. The omnipresent tendency is the one that exploits techniques of CI [1] by embracing neurocomputing imitating neural structure of a human [2], fuzzy modeling using linguistic knowledge and experiences of experts [3], [4], and genetic optimization based on the natural law [5], [6]. Especially the two of the most successful approaches have been the hybridization attempts made in the framework of CI. Neuro-fuzzy systems are one of them [7], [8]. A different approach to hybridization leads to genetic fuzzy systems [6], [9].

In this paper, new architectures and design methodologies of Rule based Neuro-fuzzy Networks (RNFN) are introduced for effective analysis and solution of nonlinear problem and complex systems. The proposed RNFN is based on the fuzzy set based neurofuzzy networks (NFN). In the consequence part of the fuzzy rules, three different forms of the regression polynomials such as constant, linear and modified quadratic are taken into consideration. Contrary to the former, we make a simplified form for the representation of a fuzzy subspace lowering of the performance of a model. This methodology can effectively reduce the number of parameters and improve the performance of a model. GAs being a global optimization technique determines optimal parameters in a vast search space. But it cannot effectively avoid a large amount of time-consuming iteration because GAs finds optimal parameters by using a given space (region). To alleviate the problems, the dynamic search-based GAs is introduced to lead to rapidly optimal convergence over a limited region or a boundary condition.

J. Wang, X. Liao, and Z. Yi (Eds.): ISNN 2005, LNCS 3496, pp. 422–427, 2005.
© Springer-Verlag Berlin Heidelberg 2005

2 Polynomial Fuzzy Inference Architecture of Neurofuzzy Networks

The overall network of conventional neurofuzzy networks (NFN [8], [9]) consists of fuzzy rules as shown in (1) and (2). The networks are classified into the two main categories according to the type of fuzzy inference, namely, the simplified and linear fuzzy inference. The learning of the NFN is realized by adjusting connection weights w_{ki} or ws_{ki} of the nodes and as such it follows a standard Back-Propagation (BP) algorithm.

$$R^i : If \ x_k \ is \ A_{ki} \ then \ Cy_{ki} = w_{ki} \tag{1}$$

$$R^i : If \ x_k \ is \ A_{ki} \ then \ Cy_{ki} = ws_{ki} + w_{ki} \cdot x_k \tag{2}$$

In this paper, we propose the polynomial fuzzy inference based NFN (pNFN). The topology of the proposed pNFN is show in Fig. 1 and consists of the aggregate of fuzzy rules such as (3). The consequence part of pNFN consists of summation of constant and input terms. The polynomial structure of the consequence part in pNFN emerges into the networks with connection weights shown in Fig. 1. This network structure of the quadratic polynomial involves simplified (Type 0), linear (Type 1), and polynomial (Type 2) fuzzy inferences. And fuzzy inference structure of pNFN can be defined by the selection of Type (order of a polynomial).

$$R^i : If \ x_k \ is \ A_{ki} \ then \ Cy_{ki} = w0_{ki} + w1_{ki} \cdot x_k + w2_{ki} \cdot x_k^2 \tag{3}$$

[Layer 1] Input layer
[Layer 2] Computing activation degrees of linguistic labels
[Layer 3] Normalization of a degree activation (firing) of the rule: An input signal x_k activates only two membership functions simultaneously and the sum of grades of these two neighboring membership functions is always equal to 1, so that can be represented in a simpler format.

$$\bar{\mu}_{ki} = \frac{\mu_{ki}}{\mu_{ki} + \mu_{ki+1}} = \mu_{ki} \tag{4}$$

[Layer 4] Multiplying a normalized activation degree of the rule by connection.

$$a_{ki} = \bar{\mu}_{ki} \times Cy_{ki} = \mu_{ki} \times Cy_{ki} \ \text{where,} \ \begin{cases} \text{Type 0} : Cy_{ki} = w0_{ki} \\ \text{Type 1} : Cy_{ki} = w0_{ki} + w1_{ki} \cdot x_k \\ \text{Type 2} : Cy_{ki} = w0_{ki} + w1_{ki} \cdot x_k + w2_{ki} \cdot x_k^2 \end{cases} \tag{5}$$

[Layer 5] Fuzzy inference for output of the rules.

$$f_k(x_k) = \sum_{i=1}^{n} a_{ki} = \sum_{i=1}^{n} \mu_{ki} \cdot Cy_{ki} = \mu_{ki} \cdot Cy_{ki} + \mu_{ki+1} \cdot Cy_{ki+1} \tag{6}$$

[Layer 6] Computing output of pNFN.

$$\hat{y} = f_1(x_1) + f_2(x_2) + \cdots + f_m(x_m) = \sum_{k=1}^{m} f_k(x_k) \tag{7}$$

The learning of the proposed pNFN is realized by adjusting connection weights w, which organize the consequence networks of pNFN in Fig. 1. The standard Back-propagation (BP) algorithm is utilized as the learning method in this study.

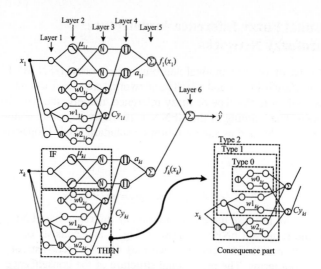

Fig. 1. Topology of the polynomial fuzzy inference based NFN

The proposed pNFN can be designed to adapt a characteristic of a given system, also, that has the faculty of making a simple structure out of a complicated model for a nonlinear system, because the pNFN comprises consequence structure with various orders (Types) for fuzzy rules.

3   Genetically Optimization for the Polynomial Based Neurofuzzy Networks

In this chapter, we introduce new architectures and design methodologies of rule based neurofuzzy networks (RNFN). For the genetically optimized architecture, the dynamic search-based GAs is proposed, and also the efficient methodology of chromosomes application of GAs for the identification of architecture and parameters of RNFN is discussed.

To search a global solution in process of optimization using GAs is stagnated by various causes. If the search range is defined as a large scale, then the number of bit increases in size for the given space. If a small-scale for the solution space is given, then the string length is sorted. Therefore, in order to improve these problems, we introduce the dynamic search-based GAs. This methodology discovers an optimal solution through adjusting search range. In order to generate the proposed RNFN, the dynamic search based GAs is used in the optimization problems of structures and parameters. From the point of fuzzy rules, these divide into the structure and parameters of the premise part, and that of consequence part. The structure issues in the premise of fuzzy rules deal with how to use of input variables (space) influencing outputs of model. The selection of input variables and the division of space are closely related to generation of fuzzy rules that determine the structure of NFN, and govern the performance of a model. Moreover, a number of input variable and a number of space divisions induce some fatal problems such as the increase of the number of fuzzy rules and the time required. Therefore, the relevant selection of input variables and the appropriate division of space are required. The structure of the consequence part of fuzzy rules is related to how represents a fuzzy subspace. Universally, the conventional methods offer uniform types to each subspace. However, it forms a complicated structure and debases the output quality of a model, because it does not consider the correlation of input variables and reflect a feature of fuzzy subspace. In this study, we apply the various forms in expressions of a fuzzy subspace. The form is selected according to an influence of a fuzzy subspace for an output criterion and provides users with the necessary information of a subspace for a system analysis.

4 Experimental Studies

In this experiment, NOx emission process is modeled using the data of gas turbine power plants. The input variables include AT, CS, LPTS, CDP, and TET. The output variable is NOx [9, 11,12]. The performance index is defined by (8).

$$E(PI \text{ or } EPI) = \frac{1}{n}\sum_{p=1}^{n}(y_p - \hat{y}_p)^2 \tag{8}$$

Table 1. Performance index of RNFN for the NOx emission process

Case	Inputs variables	Premise MFs	Premise Para.	Consequence Order	Consequence Para.	PI	E_PI
Ⓐ	**GAs** (x_1,x_2,x_3,x_4,x_5)	2+2+2+2+2	Min-Max	**GAs**	BP	7.918	10.676
	Tuned	2+2+2+2+2	GAs	Tuned	BP	7.338	9.666
Ⓑ	**GAs** (x_1,x_2,x_3,x_4,x_5)	2+2+2+2+2	GAs	**GAs**	BP	7.972	10.216
Ⓐ₃	≤3 **GAs** Tuned(x_1,x_3,x_5)	2+2+2 2+2+2	Min-Max GAs	**GAs** Tuned	BP BP	16.003 14.522	15.837 13.865
Ⓑ₃	≤3 **GAs** (x_3,x_4,x_5)	2+2+2	GAs	**GAs**	BP	9.774	10.002

Table 1 summarizes the results of the RNFN architectures. This table includes the tuning methodologies using dynamic search based GAs. Ⓐ case includes two auto-tuning processes, namely, structure and parameter tuning processes. Ⓑ case includes structure and parameter tuning processes, however, two tuning processes is not done separately but done at the same time. Ⓐ_k and Ⓑ_k add the condition of being restricted in the number of inputs of a model to Ⓐ and Ⓑ.

RNFN designed by Ⓐ method consists of 10 fuzzy rules. Ⓑ carries out the tuning process of structure and parameters at the same time unlike Ⓐ. RNFN designed by Ⓑ method consists of 10 fuzzy rules. The premise structures of the fuzzy rule are the same that of Ⓐ. The output characteristic of the architecture obtained by means of Ⓑ is better than that of Ⓐ. Therefore, RNFN generated by Ⓑ is preferred as an optimal architecture for the output performance and simplicity of a model. The preferred RNFN results from Ⓑ₃. This architecture consists of 3 inputs, x_3, x_4, and x_5, and 6 fuzzy rules. Fig. 2 shows output of RNFN topology for the NOx emission process.

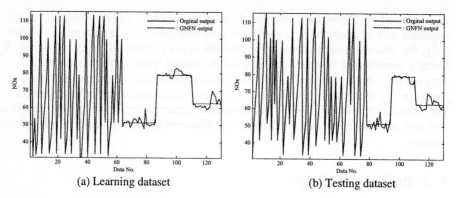

(a) Learning dataset   (b) Testing dataset

Fig. 2. Output of RNFN topology for the NOx emission process

Table 2 covers a comparative analysis including several previous models. The proposed RNFNs come with higher accuracy and improved prediction capabilities

Table 2. Comparison of performance with other modeling methods

Model		PI	E_PI	No. of rules
Regression model		17.68	19.23	
Ahn's [11]	FNN	5.835		
	AIM	8.420		
FNN [12]	Simplified	6.269	8.778	30(6+6+6+6+6)
	Linear	3.725	5.291	30(6+6+6+6+6)
Multi-FNN [9]	Simplified	2.806	5.164	30(6+6+6+6+6) ×4
	Linear	0.720	2.025	30(6+6+6+6+6)×4
Proposed model (RNFN)		7.972	10.216	10(2+2+2+2+2)
		3.453	3.974	24(6+6+6+6)
		3.287	4.067	30(6+6+6+6+6)

5   Conclusion

In this paper, new architectures and design methodologies of Rule based Neurofuzzy Networks (RNFN) has discussed for effective analysis and solution of nonlinear problem and complex systems. Also, the dynamic search-based GAs has introduced to lead to rapidly optimal convergence over a limited region or a boundary condition.

The proposed RNFN can be designed to adapt a characteristic of a given system, also, that has the faculty of making a simple structure out of a complicated model for a nonlinear system. This methodology can effectively reduce the number of parameters and improve the performance of a model. The proposed RNFN can be efficiently carried out both at the structural as well as parametric level for overall optimization by utilizing the separate (Ⓐ) or consecutive (Ⓑ) tuning technology. From the results, Ⓑ methodologies simultaneously tuning both structure and parameters, reduce parameters of consequence part, and offer the output performance better than the Ⓐ. Namely, Ⓑ method is effective in identifying a model than Ⓐ.

Acknowledgements

This work has been supported by KESRI(I-2004-0-074-0-00), which is funded by MOCIE(Ministry of commerce, industry and energy)

References

1. Pedrycz, W., Peters, J.F.: Computational Intelligence and Software Engineering. World Scientific. Singapore (1998)
2. Chan, L.W., Fallside, F.: An Adaptive Training Algorithm for Back Propagation Networks. Computer Speech and Language. 2 (1987) 205-218
3. Kang, G., Sugeno, M.: Fuzzy Modeling. Trans. of the Society of Instrument and Control Engineers. 23 (1987) 106-108
4. Park, M.Y., Choi, H.S.: Fuzzy Control System. Daeyoungsa, Korea (1990) 143-158
5. Goldberg D.E.: Genetic Algorithms in search, Optimization & Machine Learning. Addison-Wesley (1989)
6. Cordon, O., et al.: Ten years of Genetic Fuzzy Systems: Current Framework and New Trends. Fuzzy Sets and Systems. 141 (2004) 5-31
7. Horikawa, S.I., Furuhashi, T., Uchigawa, Y.: On Fuzzy Modeling Using Fuzzy Neural Networks with the Back Propagation Algorithm. IEEE Transactions on Neural Networks. 3 (1992) 801-806
8. Park, H.S., Oh, S.K.: Rule-based Fuzzy-neural Networks Using the Identification Algorithm of GA Hybrid Scheme. International Journal of Control, Automation, and Systems, 1 (2003) 101-110
9. Park, H.S., Oh, S.K.: Multi-FNN Identification Based on HCM Clustering and Evolutionary Fuzzy Granulation. International Journal of Control, Automation and Systems, 1 (2003) 194-202
10. Kondo, T.: Revised GMDH Algorithm Estimating Degree of the Complete Polynomial. Transactions of the Society of Instrument and Control Engineers, 22 (1986) 928-934
11. Ahn, T.C., Oh, S.K.: Intelligent Models Concerning the Pattern of an Air Pollutant Emission in a Thermal Power Plant. Final Report, EESRI (1997)
12. Oh, S.K., Pedrycz, W., Park, H.S.: Hybrid Identification in Fuzzy-Neural Networks. Fuzzy Sets and Systems, 138 (2003) 399-426

Design of Genetic Fuzzy Set-Based Polynomial Neural Networks with the Aid of Information Granulation

Sungkwun Oh[1], Seokbeom Roh[2], and Yongkab Kim[2]

[1] Department of Electrical Engineering, The University of Suwon, San 2-2 Wau-ri,
Bongdam-eup, Hwaseong-si, Gyeonggi-do 445-743, Korea
ohsk@suwon.ac.kr
[2] Department of Electrical Electronic and Information Engineering, Wonkwang University,
344-2, Shinyong-Dong, Iksan, Chon-Buk 570-749, Korea

Abstract. In this paper, we introduce new fuzzy-neural networks – Fuzzy Set – based Polynomial Neural Networks (FSPNN) with a new fuzzy set-based polynomial neuron (FSPN) whose fuzzy rules include the information granules obtained through Information Granulation. We investigate the proposed networks from two different aspects to improve the performance of the fuzzy-neural networks. First, We have developed genetic optimization using Genetic Algorithms to find the optimal structure for fuzzy-neural networks. Second, we have been interested in the architecture of fuzzy rules that mimic the real world, namely sub-model composing the fuzzy-neural networks. We adopt fuzzy set-based fuzzy rules as substitute for fuzzy relation-based fuzzy rules and apply the concept of Information Granulation to the proposed fuzzy set-based rules. The performance of genetically optimized FSPNN (gFSPNN) with fuzzy set-based neural neuron (FSPN) involving information granules is quantified through experimentation.

1 Introduction

A lot of researchers on system modeling have been interested in the multitude of challenging and conflicting objectives such as compactness, approximation ability, generalization capability and so on which they wish to satisfy. Fuzzy sets emphasize the aspect of linguistic transparency of models and a role of a model designer whose prior knowledge about the system may be very helpful in facilitating all identification pursuits. In addition, to build models with substantial approximation capabilities, there should be a need for advanced tools.

As one of the representative advanced design approaches comes a family of self-organizing networks with fuzzy set-based polynomial neuron (FSPN) (called "FSPNN" as a new category of neuro-fuzzy networks) [1, 4]. The design procedure of the FSPNNs exhibits some tendency to produce overly complex networks as well as comes with a repetitive computation load caused by the trial and error method being a part of the development process.

In this paper, in considering the above problems coming with the conventional FPNN [1, 4], we introduce a new structure of fuzzy rules as well as a new genetic design approach. The new structure of fuzzy rules based on the fuzzy set-based ap-

J. Wang, X. Liao, and Z. Yi (Eds.): ISNN 2005, LNCS 3496, pp. 428–433, 2005.

proach changes the viewpoint of input space division. In other hand, from a point of view of a new understanding of fuzzy rules, information granules seem to melt into the fuzzy rules respectively. The determination of the optimal values of the parameters available within an individual FSPN leads to a structurally and parametrically optimized network through the genetic approach.

2 The Architecture and Development of Fuzzy Set-Based Polynomial Neural Networks (FSPNN)

The FSPN encapsulates a family of nonlinear "if-then" rules. When put together, FSPNs results in a self-organizing Fuzzy Set-based Polynomial Neural Networks (FSPNN). Each rule reads in the form

$$\text{if } x_p \text{ is } A_k \text{ then } z \text{ is } P_{pk}(x_i, x_j, a_{pk})$$
$$\text{if } x_q \text{ is } B_k \text{ then } z \text{ is } P_{qk}(x_i, x_j, a_{qk}) \tag{1}$$

where a_{qk} is a vector of the parameters of the conclusion part of the rule while $P(x_i, x_j, a)$ denoted the regression polynomial forming the consequence part of the fuzzy rule. The activation levels of the rules contribute to the output of the FSPN being computed as a weighted average of the individual condition parts (functional transformations) P_K.

$$
\begin{aligned}
z &= \sum_{l=1}^{\text{total inputs}} \left(\sum_{k=1}^{\text{total\_rules related to input } l} \mu_{(l,k)} P_{(l,k)}(x_i, x_j, a_{(l,k)}) \middle/ \sum_{k=1}^{\text{total\_rules related to input } l} \mu_{(l,k)} \right) \\
&= \sum_{l=1}^{\text{total inputs}} \left(\sum_{k=1}^{\text{rules related to input } l} \tilde{\mu}_{(l,k)} P_{(l,k)}(x_i, x_j, a_{(l,k)}) \right)
\end{aligned}
\tag{2}
$$

Table 1. Different forms of the regression polynomials forming the consequence part of the fuzzy rules.

Order of the polynomial	No. of inputs 1	2	3
0 (Type 1)	Constant	Constant	Constant
1 (Type 2)	Linear	Bilinear	Trilinear
2 (Type 3)	Quadratic	Biquadratic-1	Triquadratic-1
2 (Type 4)	Quadratic	Biquadratic-2	Triquadratic-2

1: Basic type, 2: Modified type

3 Information Granulation Through Hard C-Means Clustering Algorithm

Information granules are defined informally as linked collections of objects (data points, in particular) drawn together by the criteria of indistinguishability, similarity or functionality [12].

Definition of the premise and consequent part of fuzzy rules using Information Granulation

The fuzzy rules of Information Granulation-based FSPN are as followings.

$$\text{if } x_p \text{ is } A^*_k \text{ then } z\text{-}m_{pk} = P_{pk}((x_i\text{-}v^i_{pk}),(x_j\text{-}v^j_{pk}),a_{pk})$$

$$\text{if } x_q \text{ is } B^*_k \text{ then } z\text{-}m_{qk} = P_{qk}((x_i\text{-}v^i_{qk}),(x_j\text{-}v^j_{qk}),a_{qk}) \tag{3}$$

Where, A^*_k and B^*_k mean the fuzzy set, the apex of which is defined as the center point of information granule (cluster) and m_{pk} is the center point related to the output variable on cluster$_{pk}$, v^i_{pk} is the center point related to the i-th input variable on cluster$_{pk}$ and a_{qk} is a vector of the parameters of the conclusion part of the rule while $P((x_i\text{-}v^i),(x_j\text{-}v^j),a)$ denoted the regression polynomial forming the consequence part of the fuzzy rule. The given inputs are $X=[x_1\ x_2 \dots x_m]$ related to a certain application and the output is $Y=[y_1\ y_2 \dots y_n]^T$.

Step 1) build the universe set

Step 2) build m reference data pairs composed of $[x_1;Y]$, $[x_2;Y]$, and $[x_m;Y]$.

Step 3) classify the universe set U into l clusters such as c_{i1}, c_{i2}, …, c_{il} (subsets) by using HCM according to the reference data pair $[x_i;Y]$.

Step 4) construct the premise part of the fuzzy rules related to the i-th input variable (x_i) using the directly obtained center points from HCM.

Step 5) construct the consequent part of the fuzzy rules related to the i-th input variable (x_i).

Sub-step1) make a matrix as (4) according to the clustered subsets

$$A^i_j = \begin{bmatrix} x_{21} & x_{22} & \cdots & x_{2m} & \vdots & y_2 \\ x_{51} & x_{52} & \cdots & x_{5m} & \vdots & y_5 \\ x_{k1} & x_{k2} & \cdots & x_{km} & \vdots & y_k \\ \vdots & \vdots & \cdots & \vdots & \vdots & \vdots \end{bmatrix} \tag{4}$$

Where, $\{x_{k1}, x_{k2}, \dots, x_{km}, y_k\} \in c_{ij}$ and A^i_j means the membership matrix of j-th subset related to the i-th input variable.

Sub-step2) take an arithmetic mean of each column on A^i_j. The mean of each column is the additional center point of subset c_{ij}. The arithmetic means of column is (5).

$$center\ points = \begin{bmatrix} v^1_{ij} & v^2_{ij} & \cdots & v^m_{ij} & \vdots & m_{ij} \end{bmatrix} \tag{5}$$

Step 6) if i is m then terminate, otherwise, set i=i+1 and return step 3.

4   Genetic Optimization of FSPNN

To retain the best individual and carry it over to the next generation, we use elitist strategy [3]. The overall genetically-driven structural optimization process of FSPNN is shown in Fig. 1.

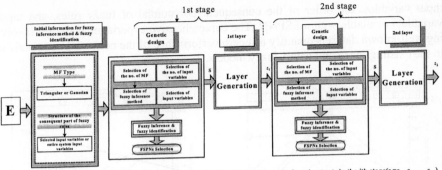

E : Entire inputs, S : Selected FSPNs, z_i : Preferred outputs in the ith stage(z_i=z_{1i}, z_{2i}, ..., z_{Wi})

Fig. 1. Overall genetically-driven structural optimization process of FSPNN

[Step 1] Determine system's input variables
[Step 2] Form training and testing data
[Step 3] specify initial design parameters
[Step 4] Decide FSPN structure using genetic design
[Step 5] Carry out coefficient parameters estimation for fuzzy identification
[Step 7] Check the termination criterion
[Step 8] Determine new input variables for the next layer

5   Experimental Studies

We illustrate the performance of the network and elaborate on its development by experimenting with data coming from the gas furnace process. The time series data (296 input-output pairs) resulting from the gas furnace process has been intensively studied in the previous literature [5], [6], [7], [8], [9], [10], [11]. Fig. 2 depicts the performance index of each layer of IG-gFSPNN with Type T* according to the increase of maximal number of inputs to be selected.

Fig. 3. illustrates the different optimization process between gFSPNN and the proposed IG-gFSPNN by visualizing the values of the performance index obtained in successive generations of GA when using Type T*. Table 2 summarizes a comparative analysis of the performance of the network with other models.

6   Concluding Remarks

In this study, we have surveyed the new structure and meaning of fuzzy rules and investigated the GA-based design procedure of Fuzzy Set-based Polynomial Neural Networks (IG-FSPNN) with information granules along with its architectural considerations. A new fuzzy rule with information granule describes a sub-system as a stand-alone system. A fuzzy system with some new fuzzy rules depicts the whole system as a combination of some stand-alone sub-system. The GA-based design procedure applied at each stage (layer) of the FSPNN leads to the selection of the preferred nodes (or FSPNs) with optimal local characteristics (such as the number of input variables, the order of the consequent polynomial of fuzzy rules, and input variables) available within FSPNN. The comprehensive experimental studies involving well-known datasets quantify a superb performance of the network in comparison to the existing fuzzy and neuro-fuzzy models.

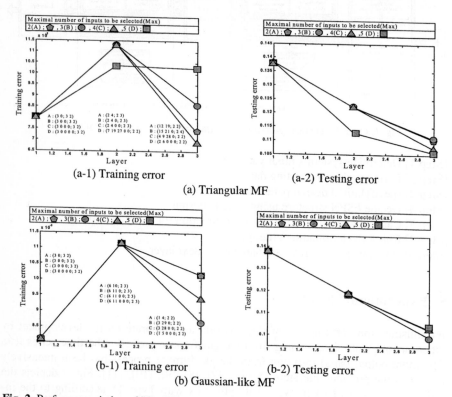

(a-1) Training error   (a-2) Testing error

(a) Triangular MF

(b-1) Training error   (b-2) Testing error

(b) Gaussian-like MF

Fig. 2. Performance index of IG-gFSPNN (with Type T*) with respect to the increase of number of layers

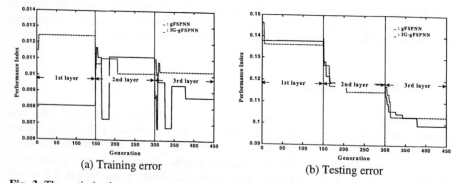

(a) Training error   (b) Testing error

Fig. 3. The optimization process quantified by the values of the performance index (in case of using Gaussian MF with Max=5 and Type T*)

Table 2. Comparative analysis of the performance of the network; considered are models reported in the literature

Model				Performance index		
				PI	PI$_s$	EPI$_s$
Box and Jenkin's model[7]				0.710		
Tong's model[8]				0.469		
Sugeno and Yasukawa's model[9]				0190		
Pedrycz's model[5]				0.320		
Oh and Pedrycz's model[6]				0.123	0.020	0.271
Kim et al.'s model[10]					0.034	0.244
Leski and Czogala's model[11]				0.047		
Proposed IG-gFSPNN	Type III (SI=6)	Triangular	3rd layer(Max=3)		0.008	0.110
		Gaussian-like	3rd layer(Max=3)		0.008	0.099

PI - performance index over the entire data set,
PI$_s$ - performance index on the training data, EPI$_s$ - performance index on the testing data.

Acknowledgements

This work has been supported by KESRI(I-2004-0-074-0-00), which is funded by MOCIE(Ministry of commerce, industry and energy)

References

1. Oh, S.K., Pedrycz, W.: Self-organizing Polynomial Neural Networks Based on PNs or FPNs: Analysis and Design. Fuzzy Sets and Systems, **142** (2003) 163-198
2. Michalewicz, Z.: Genetic Algorithms + Data Structures = Evolution Programs. Springer-Verlag, Berlin Heidelberg (1996)
3. Jong, D.K.A.: Are Genetic Algorithms Function Optimizers?. Parallel Problem Solving from Nature 2, Manner, R. and Manderick, B. eds., North-Holland, Amsterdam.
4. Oh, S.K., Pedrycz, W.: Fuzzy Polynomial Neuron-Based Self-Organizing Neural Networks. Int. J. of General Systems, **32** (2003) 237-250
5. Jang J.R.: ANFIS: Adaptive-Network-Based Fuzzy Inference System.. IEEE Trans. System, Man, and Cybern. **23** (1993) 665-685
6. Maguire, L.P., Roche, B., McGinnity, T.M., McDaid, L.J.: Predicting a Chaotic Time Series Using a Fuzzy Neural Network. Information Sciences, **112** (1998) 125-136
7. Mackey, M.C., Glass, L.: Oscillation and Chaos in Physiological Control Systems. Science, **197** (1977) 287-289
8. Oh, S.K., Pedrycz, W.: The Design of Self-organizing Polynomial Neural Networks. Information Science, **141** (2002) 237-258
9. Sugeno, M., Yasukawa, T.: A Fuzzy-Logic-Based Approach to Qualitative Modeling. IEEE Trans. Fuzzy Systems, **1** (1993) 7-31
10. Park, B.J., Pedrycz, W., Oh, S.K.: Fuzzy Polynomial Neural Networks: Hybrid Architectures of Fuzzy Modeling. IEEE Transaction on Fuzzy Systems, **10** (2002) 607-621
11. Lapedes, A.S., Farber, R.: Non-linear Signal Processing Using Neural Networks: Prediction and System Modeling. Technical Report LA-UR-87-2662, Los Alamos National Laboratory. Los Alamos New Mexico **87545** (1987)
12. Zadeh, L.A.: Toward a Theory of Fuzzy Information Granulation and Its Centrality in Human Reasoning and Fuzzy Logic. Fuzzy Sets Syst. **90** (1197) 111-117

A Novel Self-organizing Neural Fuzzy Network
for Automatic Generation
of Fuzzy Inference Systems

Meng Joo Er and Rishikesh Parthasarathi

Intelligent Systems Center (IntelliSys) 50, Nanyang Drive, Research TechnoPlaza
BorderX Block, 7th Storey, Nanyang Technological University, Singapore 637553
Tel:(65) 67906962, Fax:(65) 67906961
{emjer,rishi}@ntu.edu.sg

Abstract. This paper presents Fuzzy Multi-Agent Structure Learning (FMASL), a neural fuzzy network for unsupervised clustering and automatic structure generation of Fuzzy Inference Systems (FISs). The FMASL clustering identifies crisp clusters in an unlabeled input data and represents them by an agent, using competitive agent learning. In generating a FIS, the FMASL is used to identify the optimum number of agents (rules) of the FIS. The best action (consequent) for each agent is automatically selected using an enhanced version of Actor-Critic learning (ACL). The structure of the FIS is dynamically changed based only on experiences and no expert's knowledge is required. This is a significant feature of our approach because constructing a FIS manually for a complex task is very difficult, if not impossible. The performance of the algorithm is elucidated using the cart-pole balancing problem.

1 Introduction

Neural fuzzy networks are the realizations of the functionality of fuzzy systems using neural networks. In [1], it has been proven that 1) any rule-based fuzzy system may be approximated by a neural net and 2) any neural net (feedforward, multilayered) may be approximated by a rule-based fuzzy system. In [2], fuzzy systems have been shown to be functionally equivalent to a class of radial basis function (RBF) networks, based on the similarity between the local receptive fields of the network and the membership functions of the fuzzy system. The main issues associated with a fuzzy system are 1) Parameter estimation, which involves determining the parameters of premises and consequents, and 2) Structure identification, which concerns partitioning the input space and determining the number of fuzzy rules for a specific performance. A consistent rule base that can adapt to new inputs and can generalize any input in the operating range is necessary for knowledge representation and this is achieved using the Fuzzy Multi-Agent Structure Learning (FMASL) neural fuzzy network introduced in this paper.

The remainder of this paper is organized as follows. Section 2 introduces the FMASL-ACL architecture in details. A layer-by-layer working principle of

J. Wang, X. Liao, and Z. Yi (Eds.): ISNN 2005, LNCS 3496, pp. 434–439, 2005.

the proposed architecture is given in this section. Section 3 demonstrates the FMASL-ACL approach via the cart-pole problem and compares the results obtained with the AHC in [7]. Finally, conclusions are drawn in Section 4.

2 Overview of the FMASL-ACL Architecture

FISs are a collection of agents (rules) that attempt to generalize any input. An agent may be able to generalize the fuzzified input with a certain influence value (Ω) in the sense that real-valued descriptions of the input to the system may be matched by a degree less than one. Figure 1 shows the structure of the neural fuzzy system that is used to extract an FIS. The FMASL algorithm can be used in two different modes: 1) *Online generation mode* where agents are introduced depending on the input data. This helps to avoid the case of having fewer initial agents than there are in the input data, as cited in the RPCL algorithm of [3] and 2) *Initial generation mode* like the RPCL algorithm. A full introduction of the FMASL is provided in [4]. In this paper, we deal only with the online generation mode which is most appropriate for control applications. In our approach, the first three layers of the FIS are generated by the self-organizing FMASL and the fourth layer is estimated using ACL.

2.1 FMASL Antecedent Generation and ACL Consequent Selection

Layer 1: Input Fuzzification. Calculate the distance a_{idist} between the m-dimensional input x and all the agents (a_i) as follows:

$$a_{idist} = \gamma_i \|x - a_i\|^2 \tag{1}$$

where $\gamma_i = \frac{n_i}{\sum_{r=1}^{k} n_r}$ is the relative winning frequency of the agent a_i so far and n_i is the number of times the agent has won the input situation. We define

$$\Omega_c = arg \min_i (\sqrt{a_{idist}}) \tag{2}$$

where Ω_c is the proximity of the winning agent a_c, characterized by the shortest distance to the input. If $\Omega_c > \Omega_{th}$, a threshold which is defined as $\Omega_{th} = \sqrt{\frac{1}{\epsilon}}$, we assume that the current structure is unable to generalize the input and we introduce a new agent at the input and thus grow the FIS.

Layer 2: T-normalization. If the threshold condition is satisfied, we calculate

$$\mu_{ij}(x) = e^{\frac{-a_{idist}}{\sigma^2}} \tag{3}$$

where μ_{ij} is the jth membership value of the input x with respect to the agent a_i as the center and width σ^2. The overall membership value of the premise of agent a_i in generalizing input x is then calculated as follows:

$$\phi_{a_i}(x) = T(\mu_{i1}(x_1), \mu_{i2}(x_2)...\mu_{im}(x_m)) \tag{4}$$

where the operator T is the T-Norm. The T-Norm used here is a minimum operation. Hence, $\phi_{a_i}(x) = min(\mu_{ij}(x))$

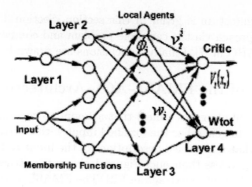

Fig. 1. FIS Structure

Layer 3: Normalization. The resulting firing strength Φ_{a_i} of each agent is calculated in this layer. This is given by:

$$\Phi_{a_i}(x) = \frac{\phi_{a_i}(x)}{\phi_A} \tag{5}$$

where ϕ_A is the cumulative sum of the membership values of all the agents. The parameters of the agents are modified as follows:

$$a_\tau^{new} = a_\tau^{old} + \Delta a_\tau \tag{6}$$

where τ includes all the agents. The Δ components are obtained as follows

$$\Delta a_c = \alpha_c(x - a_c) * \Phi_{a_c}, \quad \Delta a_r = -\alpha_c(x - a_r) * \Phi_{a_r} \tag{7}$$

where c is the winning agent and r denotes the "rest of the agents".

These three layers together help to form a compact structure with independent agents representing disjoint regions of the input space. More results of this are presented in [4]. Thus, the antecedent part of the FIS is constructed based only on experiences.

Layer 4: Conclusion. This layer of the FIS normally performs defuzzification of the fuzzy outputs into crisp values. Instead of producing the output value directly, we produce quality values for action selection (w_{tot}) and failure prediction (V) for each agent. The w_{tot} is calculated as follows:

$$W_{tot} = \sum_{i=1}^{k} \Phi_i(x)w_i \tag{8}$$

and then used to choose an action using a stochastic ϵ-greedy method. The Temporal Difference (TD) learning of [5] is used to update the weights of layer 4. The update enables the stochastic method to select the best consequent for each agent.

Actor-critic methods are TD methods that have a separate memory structure to explicitly represent the policy independent of the value function. The policy structure is known as the actor because it is used to select actions, and the estimated value function is known as the critic because it criticizes the actions made by the actor. Learning is always on-policy in the sense the critic must learn about and critique whatever policy is currently being followed. The critique is the TD error. Assuming that an action from state x has brought the system to x', the associated TD error as in [5] is given by

$$\epsilon = r_{xx'} + \gamma V_t(x') - V_t(x) \tag{9}$$

where V is the value function defined as follows:

$$V^\pi(x) = E_\pi\{\textstyle\sum_{t=0}^\infty \gamma^t r_{t+1} | x_0 = x\} \tag{10}$$

It is the expected value of the infinite-horizon sum of the discounted rewards when the system is started in state x and the policy π is followed forever. This scalar signal is the output of the critic and it drives learning in both the actor and the critic. To reduce the severity of the temporal credit-assignment problem, the reinforcement signal used by the actor learning mechanism is not the primary reinforcement but the internal reinforcement signal given by the critic (w). The learning rate of the two is also different. The updates in the critic value V and the actor value w at each time step are governed by the following equations

$$\begin{aligned} V_{t+1}(x') &= V_t(x') + \beta \epsilon_{t+1} T c_t \\ w_{t+1}(x') &= w_t(x') + \theta \epsilon_{t+1} T a_t \end{aligned} \tag{11}$$

where β and θ are the learning rates of the critic and the actor respectively and the eligibility traces Tc and Ta are given by

$$\begin{aligned} Tc_t^x &= \lambda_c Tc_{t-1}^x + (1 - \lambda_c)\Phi_i(x) \\ Ta_t^x &= \lambda_a Ta_{t-1}^x + (1 - \lambda_a)\Phi_i(x) \end{aligned} \tag{12}$$

where λ_a and λ_c in the above equation are the trace decay parameters of the actor and critic respectively. Eligibility traces are used to memorize previously visited agent-action pairs, weighted by their proximity to time step t. A more comprehensive introduction to the ACL is necessary and is available in [6].

3   Cart-Pole Balancing Problem

System Dynamics. The rigid pole in Figure 2 is hinged to the cart that is free to move on a one-dimensional (1-D) track. The pole may move only in the vertical plane of the cart and track, and the global agent can apply an impulsive force to the cart at discrete time intervals. The cart-pole model uses the following four state variables: θ-angle the pole makes with the vertical axis, $\dot\theta$-velocity of the pole, x-cart position on the track, $\dot x$-cart velocity. As in [7], the dynamics of the cart-pole system are modeled. The force F has been considered only to be a

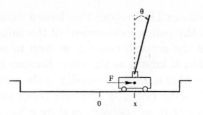

Fig. 2. Cart-Pole System

two-valued function as the pole balancing problem is well suited for bang-bang actions and also the learning becomes faster in the case of two actions. Each trial begins with all the state variables set to zero and ends either when a failure occurs or if the number of time steps exceeds 100,000. A failure depends on both the pole angle (if $|\theta| > 12°$) and the cart position (if $|x| > 2.4$m). All the trace variables Tc and Ta are set to zero and the agent is "naive" at the start of each run. The reinforcement value of -1 is given only when a failure occurs.

Selection of action is based on the stochastic ϵ-greedy method explained as follows: As the system uses only two actions, the value of W_{tot} is used to calculate

$$F_R = \frac{1}{1 + e^{(-max(-50, min(x, 50)))}} \tag{13}$$

where x takes the value of W_{tot} as the case may be. The constants 50 and -50 are used to set saturation values for right and left actions. These have no adverse effect on the learning even if ignored. The value F_R obtained gives the probability of choosing right push as the action. Initially, all agents are assigned a W value of 0, giving a probability of 0.5 for both actions. As learning progresses the W values of the agents change accordingly to choose the right action for each agent. The more positive the value is, the higher the probability of choosing the right push as the action.

Various parameters of the FMASL were initialized as follows: $\epsilon = 0.5$, $\alpha_c = 0.05$ and the width of the local agent $2\sigma^2$ is set to 1. The FMASL started with zero agents and introduced new agents as and when the situation required. The results obtained in Figure 4 show that the performance of the proposed approach is much better than that of the AHC of [7]. The agents were systematically introduced online as shown in Figure 3.

4   Conclusions

Unsupervised clustering and automatic structure generation of Fuzzy Inference Systems was accomplished in this paper. The experiment establishes three important conclusions, namely, 1) Neural fuzzy networks are very efficient in generalizing complex tasks; 2) Automatic generation and tuning of FIS is superior to fixed FIS generated based on expert's knowledge as the latter is highly subject to various factors concerning the expertise of the human being and 3) ACL is superior to supervised learning for selecting the output as the latter beats the idea of developing "intelligent agents".

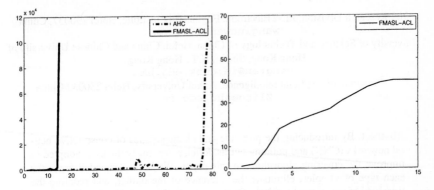

Fig. 3. Performance of the FMASL-ACL. a) Number of trials for the cart-pole system to learn to balance for 100000 trials. b) Number of fuzzy rules generated by the FMASL clustering

References

1. Hayashi, Y., Buckley, J.J.: Approximations Between Fuzzy Expert Systems and Neural Networks. Int. J. Approx. Reas., **10** (1994)
2. Jang, J.S.R., Sun, C.T.: Functional Equivalence Between Radial Basis Function Networks and Fuzzy Inference Systems. IEEE Trans. Neural Networks, **4** (1993)
3. Xu, L., Krzyzak, A., Oja, E.: Rival Penalized Competitive Learning for Clustering Analysis, RBF Net, and Curve Detection. IEEE Transaction on Neural Networks, **4** (1993)
4. Parthasarathi, R., Er, M.J.: Fuzzy Multi-Agent Structure Learning for Clustering Analysis and Intelligent Control. Accepted for Publication, International Fuzzy Systems Association (2005)
5. Sutton, R.S.: Learning to Predict by the Methods of Temporal Differences. Mach. Learn., **3** (1988) 9-44
6. Sutton, R.S., Barto, A.G.: Reinforcement Learning an Introduction. MIT Press. Cambridge, Massachusetts London
7. Barto, A.G., Sutton, R.S., Anderson, C.W.: Neuronlike Adaptive Elements That Can Solve difficult Learning Control Problems. IEEE Trans. Systems, Man, and Cybernetics, **SMC-13** (1983) 834-846

Constructive Fuzzy Neural Networks and Its Application*

Lunwen Wang[1], Ying Tan[2], and Ling Zhang[3]

[1] 702 Research Division, PLA Electronic Engineering Institute, Hefei 230037, China
wanglunwen@163.com
[2] University of Science and Technology of China, Hefei, China and Chinese University of
Hong Kong, Shatin, N.T., Hong Kong
yingtan@ie.cuhk.edu.hk
[3] Institute of Artificial intelligence, Anhui University, Hefei 230039 China
Zling@ahu.edu.cn

Abstract. By introducing the principle and characteristics of constructive neural networks (CNN) and pointing out its deficiencies, fuzzy theory is adopted to
improve the covering algorithms in this paper. We build "extended area" for
each type of samples, eliminate the inference of the outlier, and redefine the
threshold of covering algorithms. Furthermore, "sphere neighborhood" (SN) are
constructed, the membership functions of test samples are given and all of the
test samples are determined accordingly. First of all, the procedure of constructive fuzzy algorithm is given, then the model of constructive fuzzy neural networks (CFNN) is built, finally, CFNN is applied to search for communications
signals. Extensive experimental results demonstrate the efficiency and practica-
bility of the proposed algorithm.

1 Introduction

Since Hopfield published his seminal paper on *neural networks with self-feedback
connections* [1] in 1982 and Rumelhart published the monograph *parallel distributed
processing* [2] in 1986, neural networks have been widely studied again. Now people
have not only found the comprehensive application of neural networks, but also made
tremendous progress on its principle research. Regarding the comprehensive study of
neural networks, there are at least three principal methods of search-based algorithms,
programming based algorithms [3] and constructive algorithms [4] so far.

The most representative search-based algorithms are the back-propagation algorithm (BP) and simulated annealing algorithm. BP has solid theory and is widely used
in the feed-forward neural networks. But BP converges slowly, and the networks'
tolerant capacity is poor. Therefore, many improved algorithms are presented so far,
such as increasing inertial terms, using dynamic step size, combining with other
global search algorithms and simulated annealing algorithm.

Due to the deficiency of BP, programming-based algorithms use the necessary
condition as the restricted one, regard one of performances as the objective function,
and convert the networks' learning problem into a mathematical programming. With
this idea in mind, self-feedback approach, potential reduction algorithm, geometrical
iteration algorithm and support vector machines have come into being.

* This work was supported by the Natural Science Foundation of China under Grant
No.60175018, No.60135010; partially by the National Grand Fundamental Research 973
Program of China under Grant No. G1998030509.

J. Wang, X. Liao, and Z. Yi (Eds.): ISNN 2005, LNCS 3496, pp. 440–445, 2005.

In 1999, Bo Zhang and Ling Zhang presented CNN [4] based on the covering algorithms. This approach constructs a SN through a smart nonlinear transform, and converts the learning problem into a covering problem. It has little computation and is suitable for large-scale pattern classification [5]. But there is still deficiency to classify the test samples of the boundary. This paper tries to adopt the fuzzy theory to improve the performance of CNN by constructing so-called CFNN, which is also applied to search for communications signals. The experimental results show the proposed networks have improved the efficiency in the pattern classification.

2   Brief Description of CNN

In 1943, McCulloch and Pitts first presented M-P model of a neuron. Since then many artificial neural networks have been developed from this well-known model. An M-P neuron is an element with n inputs and one output. The form of its function is $y = \text{sgn}(wx - \varphi)$, where $x = (x_1, x_2, \cdots, x_n)^T$ is an input vector, $w = (w_1, w_2, \cdots w_n)$ is a weight vector, and φ is a threshold.

The node function $f(wx - \varphi)$ of the M-P model can be regarded as a function of two functions: a linear function $wx - \varphi$ and a sign (or characteristic) function $f(\cdot)$. Note that $wx - \varphi = 0$ can be interpreted as a hyperplane P in an n-dimensional space. When $wx - \varphi > 0$ input vector x falls into the positive half-space of the hyperplane P. Meanwhile, $y = \text{sgn}(wx - \varphi) = 1$. When $wx - \varphi < 0$ input vector x falls into the negative half-space of the hyperplane P, and $y = -1$. In summary, the function of an M-P neuron can geometrically be regarded as a spatial discriminator of an n-dimensional space divided by the hyperplane. People intended to use such a geometrical interpretation to analyze the behavior of neural networks. Unfortunately, as the dimension n of the space and the number m of the hyperplane P increase, the mutual intersection among these hyperplanes in n-dimensional space will become too complex to analyze. Therefore, until now, the geometrical interpretation has still rarely been used to improve the learning capacity of complex neural networks.

In order to overcome this difficulty, a new representation is presented as follows. First of all, assume that each input vector has an equal length. Thus all input vector will be restricted to an n-dimensional sphere S^n. Then $wx - \varphi > 0$ represents the positive half-space partitioned by the P and the intersection between the positive half-space and sphere S^n is called a SN as shown in Fig.1. When input x falls into the region, output $y = 1$, otherwise $y = -1$. If the weight vector w has same length as input x, then w becomes the center of the SN, and $r(\varphi)$ become its radius. If the node function is a characteristic function $sgn(v)$, then the function $\sigma(wx - \varphi)$ of a neuron is a characteristic function of the SN on the sphere. The preceding clear visual picture of an M-P neuron is great help to analysis of neural networks. In the preceding discussion, input vectors are assumed to have equal length. This is not the case in general. Assume that the domain of input vectors is a bounded set D of an n-dimensional space. S^n is an n-dimensional sphere of an $(n+1)$-dimensional space.

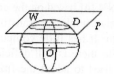
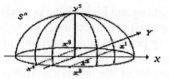

Fig. 1. A sphere neighborhood   **Fig. 2.** Input vectors and their projections

Define a transformation $T:D \rightarrow S^n$, $x \in 7D$, such that $T:D \rightarrow S_n : T(x) = (x, \sqrt{R^2 - |x|^2})$ where $R \geqslant \max(|x|, x \in D)$. Thus all points of D are projected upward on the S^n by the transform (Fig2). Obviously, a neuron (w, φ) corresponds to a characteristic function of a SN on S^n with w as its center and $r(\varphi)$ as its radius.

The size of SN is related to the partition of bounded set D of an n-dimensional space. Ref. 4 gives an example of partition of two type samples by using the covering algorithms, and the covering approach is expressed as follows.

$$\theta = \frac{1}{2}\left(d^1(k) + d^2(k)\right), \quad d^1(k) = \max_{x \in X_k}\{<a_i, x>\}, \quad d^2(k) = \min_{x \in X_k}\{<a_i, x>\}, \quad \text{where } <x,y> \text{ de-}$$

notes the inner-product of x and y, and $d^1(k)$ is minimum "distance" between a_i and x, $d^2(k)$ is maximum "distance" between a_i and x. To explain the approach of classification of two learning samples, Fig 3 takes the covering of samples as an example in the planar space. We plot a circle whose center is a_i and radius is θ, and it covers the whole samples of $\triangle$. We similarly plot another circle which covers sample $\square$. Because the two sphere fields are neighbor, the two training samples can be classified easily and inerrably.

CNN has several characters as follows. Firstly, it simplifies the complexity of multi-field cross by transforming the infinite fields into finite fields, and learning problem into covering problem. Secondly, it transforms the solution of neurons into field covering problem, and constructs needed covering fields from the given data. Thirdly, it needs little computation, can accomplish large-scale pattern classification.

The CNN take two samples as an example to discuss the construction of spherical fields in Fig 3. However, there are several types of training samples in a practice, and we will not classify the samples easily if we still adopt CNN. We take three types of samples as an example in Fig 4, where three different samples $\triangle, \square$ and $\bigcirc$ denote the 1st, 2nd, and 3rd type of sample, respectively. If we only take the classification of the 1st and 2nd type of samples into account, then the corresponding spherical fields of the real line circle whose threshold is R denote the 1st type of samples. And if we only take the classification of the 1st and 3rd type of samples into account, then the corresponding spherical fields of the broken line circle whose threshold is r denote the 1st type of samples. So the corresponding spherical fields of the 1st type of samples cannot be unique, which lead to the inconsistent result, so even is the classification of multi-samples. In addition, we cannot classify some test set if we adopt above algorithm, such as the test point ■ shown in Fig. 4. It should belong to 1st type if we classify it by 1st and 2nd type, and it should be 3rd type if we classify it by 1st and 3rd type. So we cannot classify the test point ■ exactly.

3 The Constructive Fuzzy Neural Networks

Due to above reasons, we introduce fuzzy classification mechanism to improve the covering algorithms and build a new neural networks called CFNN.

In order to eliminate the contradiction, the threshold is redefined as

$$\theta = \min_{x \in X_k}\{< a_i, x >\} \tag{1}$$

Compared with Fig 3, the threshold decreases, i.e. the element number of every type sample does not change, but the radius decreases to the least. As shown in Fig 5, in such a way, each type training sample can be projected to a unique SN.

Due to the decrease of the threshold, the areas between the different training samples become bigger. These areas are boundary areas of different type samples. In practice, although the training samples can be classified correctly, part of test samples may be located in the boundaries, not belong to any type, we call them as unidentified samples. In Fig 5, there exist larger boundary areas between the SN, if a test sample located in these boundary areas, it will not be determined which type sample it should belong to, then this sample will be regarded as unidentified sample, such as the sample ▲, ● and ■. Here we classify this sample by its fuzzy membership [6], i.e. we calculate its memberships to the different type samples according to the distances from the different type samples, and then judge its type by the calculated memberships. The membership of the sample x to the SN C can be defined as

$$\mu_C(x) = \begin{cases} 1 & x \in C \\ < w, x >/\theta & x \notin C \end{cases} \tag{2}$$

where w is the center of C, $<w, x>$ is the inner product of x and w,

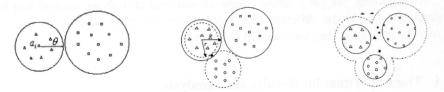

Fig. 3. Input vector and their covering   **Fig. 4.** Covering algorithms divide three types of samples   **Fig. 5.** Improved covering algorithms divide three types of samples

$\theta = \min_{x \in C}\{< w, x >\}$. So $\theta \geq < w, x >, x \notin C$.

As discussed above, there will be no unidentified samples. This is not true because there are lots of noise, interference and outliers which should not be classified into any type training samples. Here we build an "extended area" of one type sample according to $d^1(k) = \max_{x \notin X_k}\{< a_i, x >\}$, it is big enough to include its own sample, but not include the other sample. In Fig 5, there are three type samples, and there are three "extended area" can be built. The three "extended area" can be united to construct a big "extended area". All test samples which are not belong to the "extended area" will be regarded as unidentified samples, such as the test sample ⊥ in Fig. 5.

Finally, the unidentified samples are removed, the threshold of every sample is decreased to minimum, and the membership of the test sample can be calculated.

According to eq. (2), the memberships of a test sample x can be calculated as $u_1(x)$, $u_2(x)$, $\cdots$, $u_n(x)$, We can judge the sample by the principle of maximal membership, that is, if $u_k(x) = \max\{ u_1(x), u_2(x), \cdots, u_n(x) \}$, the sample is regarded as the k-th type.

Suppose training sample set be $X=\{x_1,x_2,...,x_n\}$, and X can be classified as k types, i.e. $X=\{X_1,X_2,...,X_k\}$, where, $X_j=\{x_1,x_2,...,x_c\}$, $j=1,2,...,k$; $x_i \in X$, $i=1,2,...,c_j$, $c_1+c_2+...+c_k=n$. The procedure of the CFNN algorithm can be presented as follows.

① All samples are projected to S^n, (the center is origin, and the radius is R, $R>\max(|x_i|)$, $i=1,2,\cdots,n$), and still marked;

② If X_i (at the beginning i=1) is a not empty set, a SN C_{ij} can be built, which only includes partial elements of X_i, and is labeled X_{ij} $(j=1, 2, \cdots, c_j)$;

③ If X_i is an empty set, $i = i +1$, if $i>k$, go to end, otherwise, get a element a_i from X_i which is not covered;

④ Calculate threshold θ_i according to eq. (1), then build a SN which center is a_i and radius is θ_i.

⑤ Calculate the center of gravity of all training samples covered by $C(a_i)$, then project them to the spherical surface. Suppose the projected point be $a_i{'}$, the threshold $\theta_i{'}$ can be obtained by eq. (1), then the SN $C(a_i{'})$ will be built.

⑥ If the element number of $C(a_i{'})$ is bigger than the number of $C(a_i)$, then $a_i{'} \to a_i$, $\theta_i{'} \to \theta_i$, go to ⑤, otherwise, get a SN of $C(a_i)$, suppose j=0, go to ③.

After training, we get p sphere neighborhoods, and all training samples will be classified into k-type different sets $C=\{C_1,C_2,\cdots,C_k\}$, where $C_i=C_{i1} \cup \cdots \cup C_{ij}$ $(i=1,2,\cdots,k, p \geq k)$. Then we can build neural networks.

4   The Experimental Results and Analysis

In order to testify the CFNN, we take the search of communication signals as an experiment. So far, the search of communication signals mainly make use of auto-scan method, which use a computer to control radio receivers to receive and sample signals circularly according to a given bandwidth. By comparing the samples with training samples, one can find the new appearing signals. Because of the influence of noise and channel, the power spectrum of signals fluctuates greatly and the detection performance of this method deteriorates considerably.

Although the search accuracy is highly improved by using CNN [7], there are also some deficiencies. We do the same experiment using CFNN, and make the conclusion that the proposed networks can reduce the probability of un-identification.

We acquire some data from the radio receivers as the training data classified 6 types in a day, and the frequency ranges from 15MHz to 16MHz. The test samples are the sampled data in next two days. The results are listed in Table 1, which is analyzed using general covering algorithms.

Table 1. Results of the search of communications signals by using CNN and CFNN

Methods	Number of training samples	Number of test samples	Number of unidentification	Identification rate
CNN	5572	13463	929	93.1%
CFNN	5572	13463	215	98.4%

It turns out from Table 1 that the samples unidentified by the CFNN are less than that by CNN, its identification rate is highly improved by 5.3 percent, which validates CFNN considerably.

5 Conclusion

This paper introduces principle and characteristics of CNN, analyzes the covering algorithms, adopts the fuzzy processing technique to the CNN, removes outliners from the test samples, decreases the threshold of every type sample, and calculates membership of a test sample, then separates test samples rightly. CFNN not only overcomes the difficulties of CNN, but also holds its advantages. Experimental results indicate that the improvement and application are very efficient and might be found more applications in pattern recognition and machine learning in future.

References

1. Hopfield, J.J.: Neural Networks and Physical Systems with Emergent Collective Computational Abilities. Proc. Natl. Acad. Sci, (1982) 2554–2558
2. Rumelhart, D.E., McClelland, J.L.: Parallel Distributed Processing. Cambridge, MA, MIT Press **1–2** (1986)
3. Zhang, B., Zhang, L.: Programming Based Learning Algorithm of Neural Networks with Self-feedback Connection. IEEE Trans. on Neural Networks, **6** (1995) 771–775
4. Zhang, L., Zhang, B.: A Geometrical Representation of McCulloch-Pitts Neural Model and Its Applications. IEEE Transactions on Neural Networks, **10** (1999) 925–929
5. Wu, M., Zhang, B., Zhang, L.: Neural Networks Based Classifier for Handwritten Chinese Character Recognition. Proceedings 15th International Conference on Pattern Recognition, Barcelona, **2** (2000) 561–568
6. Yager, R.R., Zadeh, L.A.: An Introduction to Fuzzy Logic Application in Intelligent Systems. Kluwer Academic Publishers (1992)
7. Wang, L., Wu, T., Zhang, L.: An Improved Neighborhood Covering Algorithm and Its Applications. Pattern Recognition & Artificial Intelligence (2003) 81–85

A Novel CNN Template Design Method Based on GIM

Jianye Zhao, Hongling Meng, and Daoheng Yu

Department of Electronics, Peking University, Beijing 100871, China
phdzjy@263.net

Abstract. In this paper, a kind of relation between CNN (cellular neural network) and GIM (Gibbs image model) is noted. Based on this relation, a new approach for CNN's template design is proposed, this approach is valid to many questions that could be processed with GIM, such as segmentation, edge detection and restoration. We also discuss the learning algorithm and hardware annealing jointed with the new approach. Simulations of some examples are shown in order to validate effectiveness of new approach.

1 Introduction

CNN [1],[2] is widely used in image processing, such as shadow detector, image thinning and image compressing [2],[3]. Though these approaches are very efficient, stochastic characteristics of the image are not utilized. GIM [4] is an image model which described the stochastic characteristic of the image. Based on GIM, many effective algorithms for image processing were proposed, such as restoration, segmentation, edge detection and texture synthesis. The origin of the GIM is in physics, so the image processing algorithm based on GIM is similar to the algorithm for ferromagnetic material based on Ising model. These algorithms for image processing are usually equal to the problem of seeking the minimum of the cost function. The definition of energy function is also useful in the area of CNN. When CNN dynamic system gets to the steady state, the energy function gets to a global or local minimum. There are some similarities between CNN's energy function and the cost function of special question based on GIM. According to these similarities, we can design CNN template and process many questions, such as segmentation and texture synthesis. In order to avoid getting local minimum of CNN's energy function (the solution correspond to local minimum is suboptimal or completely unwanted), the hardware annealing for CNN is employed.

This paper is organized in the following way: the similarities between GIM and CNN are noted in next section. Section 3 discusses the CNN template design approach that originates from GIM. We also show how to avoid getting the local minimum of energy function in this section 4. Finally concluding remarks are given in Section 5.

2 The Relationship Between GIM and CNN

We first give the definition of Gibbs distributions, the discussion is focused on 2-D random fields defined over a finite rectangular $N_1 \times N_2$ lattice of pixels, and the random field is denoted by: $L = \{(i, j) | 1 \le i \le N_1, 1 \le j \le N_2 \}$.

J. Wang, X. Liao, and Z. Yi (Eds.): ISNN 2005, LNCS 3496, pp. 446–454, 2005.

Definition (1): A *collection* of subsets of L is described as: $\eta=\{\eta_{ij}|(i,j)\in L, \eta_{ij}\subseteq L\}$

if the collection of subsets satisfy following conditions, it is a *neighborhood system*:

(a) $(i,j)\notin\eta_{ij}^{d}$ (b) If $(k,l)\in\eta_{ij}$, then $(i,j)\in\eta_{kl}$ for any $(i,j)\in L$

The symbol η^{d} is called the *dth order neighborhood system*. $\eta^{1}=\{\eta_{ij}^{1}\}$ consists of

the closet four neighbors of each pixel, $\eta^{2}=\{\eta_{ij}^{2}\}$ consists of the eight pixels

neighboring (i,j), it is shown in Fig.1.

Definition (2): A *clique* of the pair (L,η), denoted by C, is a subset of L such that:

(a) A single pixel (b) for $(i,j)\neq(k,l)$, $(i,j)\in C$ and $(k,l)\in C$ implies that $(i,j)\in\eta_{kl}^{d}$

		4	3	4	
4	2	1	2	4	
3	1	(i,j)	1	3	
4	2	1	2	4	
	4	3	4		

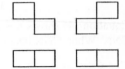

Fig. 1. (a) Neighborhood System η^{d} (d=1,2,3,4)   (b) Clique consists of 2pixels

Definition (1) and (2) present the basic idea of MRF(Markov Random Field)model. If the Gibbs distribution is utilized for characterizing MRF, GIM is gotten. To different questions, the form of Gibbs distribution is usually different. Although there are differences between forms of Gibbs distribution for special questions, such as edge detection and segmentation, the approach for template design based on Gibbs image model and CNN is similar. At first we analyze the easiest Gibbs distribution for segmentation in this paper, but our purpose is to describe the design approach. The definition of Gibbs distributions for image segmentation is described as follows:

Definition (3) Let η^{d} be a neighborhood system which is defined over the finite

lattice L, $X=\{X_{ij}\}$ defined on L is a *Gibbs Distribution*, then X satisfies:

$$P(X=x)=\frac{e^{-u(x)}}{Z}.\qquad(1)$$

$$U(x)=\sum_{c\in C}V_{c}(x)\cdot\qquad(2)$$

$$Z=\sum_{x}e^{-U(x)}.\qquad(3)$$

$V_{c}(x)$ is the *clique potential* which depends on the pixel values and clique c; Z is the partition function, and Z is a constant. Because GIM shows the relationship between the random field Y(noisy image) and an unknown random field $X=\{x_{ij}\}$ (scene), it is more convenient to process the problem based on GIM. From noisy image Y, MAP estimation of the scene random field X could be made, then the region type k of the noisy pixel is decided. Now we note the similarity between GIM and CNN.

The structure of CNN is presented in Fig.2. The cell C(i,j) is only linked with its *r-neighborhood* [1], only *1-neighborhood* of the cell C(i,j) is shown in Fig.2. Obviously, it is similar with neighborhood system η^d of GIM when d=1, 2. GIM describe the statistic characteristic of images.

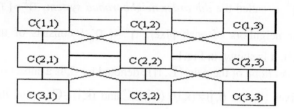

Fig. 2. The structure of CNN

Equation (4) shows the state equation of CNN, and (5) shows the energy function of CNN. When CNN become stable, (5) gets its local minimum or global minimum.

$$C\frac{dV_{Xij}(t)}{dt} = -\frac{1}{R_X}V_{Xij}(t) + \sum_{(k,l)\in N_r(i,j)}A_{ijkl}V_{Ykl}(t) + \sum_{(k,l)\in N_r(i,j)}B_{ijkl}V_{Ukl}(t) + I_{ij}. \tag{4}$$

$$E(t) = -\frac{1}{2}\sum_i\sum_j A(i,j;k,l)V_{Yij}V_{Ykl} - \sum_{i,j}IV_{Yij} + \frac{1}{2R_X}\sum_{i,j}V_{Yij}^2 - \sum_i\sum_j B(i,j;k,l)V_{Yij}V_{Ukl} \tag{5}$$

To make use of GIM to process an image question, we also need to minimize a cost function. The cost function gets to its global minimum, and then the right solution is gotten. So the similarities lie in both the structure and the processing approach. We may see how to utilize the similarities from the following example: segmentation. In order to simplify the question, we assume that the image is only corrupted by additive noise. So: $Y_{ij} = F(x_{ij}) + W_{ij}$, Y_{ij} is noisy image, $W(X_{ij})$ is Gaussian noise with mean zero and variance σ^2. In this paper, the image is processed according to its gray-level, and we map the region type q_m to the corresponding gray-level m with the function $F(X_{ij})$: $F(X_{ij}) = q_m$ if $X_{ij} = m$

We can make MAP estimation to the segmentation or restoration questions:

$$P(X = x | Y = y) = \frac{P(Y = y | X = x)P(X = x)}{P(Y = y)}. \tag{6}$$

X= x is the region type, and Y= y is the gray-level of noisy image. To make MAP estimation, (5) should be maximized. Since the noisy image is given, P(Y=y) is also definite. So maximization of the numerator is equal to maximization of (5).

$$\ln P(Y = y | X = x) = -\frac{N_1 N_2}{2}\ln(2\pi\sigma^2) - \sum_{m=1}^{M}\sum_{i,j}\frac{(y_{ij} - q_m)^2}{2\sigma^2}. \tag{7}$$

$$\ln P(X = x) = -\ln Z - \sum_{c\in C}V_c(x). \tag{8}$$

Only the clique that contains two pixels is used. The clique that contains one pixel is nonsense to the noisy image because statistic characteristic of the image can't be utilized. When the clique that contains two pixels is utilized, there is a statistic relationship between the original pixel (without noise) and its neighbor, and there is not any definite relationship between Gaussian noise added to the pixel and the neighbor's noise. So we can adopt a simple form of $V_C(x)$:

$$V_C(x) = -W(i, j; k, l) V_{ij} V_{kl}.$$ (9)

The effect of each pixel can be considered in (9) through $W(i,j;k,l)$.

From (7),(8), if we maximize function (10) we can make MAP estimation:

$$\max\left(\sum_{i,j} \sum_{k,l} W(i, j; k, l) V_{ij} V_{k,l} - \frac{1}{2\sigma^2} \left(-2\sum_{i,j} I_{ij} V_{ij} + \sum_{ij} V_{ij}^2 \right) \right).$$ (10)

In order to utilize the energy function of CNN, we write the equivalent form of (10)

$$\min\left(-\sum_{i,j} \sum_{k,l} W(i, j; k, l) V_{ij} V_{k,l} - \frac{1}{\sigma^2} \sum_{i,j} I_{ij} V_{ij} + \frac{1}{2\sigma^2} \sum_{ij} V_{ij}^2 \right).$$ (11)

CNN can be used to get minimum of (11). The same approach (minimizing cost function) is also employed in other image questions based on GIM, and many new questions can be processed with CNN. Another example is edge detection. Stochastic algorithm of edge detection includes three steps:

(1) Initial processing, processing noise image with the gradient operator
(2) Learning parameters, if the parameters are definite, the cost function is definite.
(3) Seeking minimum of the cost function.

When CNN is employed to realize the stochastic algorithm, step 1 should not be changed. To calculate the posterior probability, the conditional probability $P(y_{ij}|x_{ij} = k)$ should be given. In certain applications we are not interested in the direction of the edge elements, then k take only -1 or 1. If k= -1, then x_{ij} is not an edge element, and k=1 means that x_{ij} is an edge element. MAP estimation for edge detection of noisy image is equivalent to the maximization of next formula [5]:

$$\max(-E(x_{ij} = 1) + \ln P(y_{ij}|x_{ij} = 1)).$$ (12)

$$E(x_{ij} = 1) = -\partial_0 \sum_i \sum_j V_{i,j} - \partial_1 \sum_i \sum_j V_{i,j} V_{i,j-1} - \partial_2 \sum_i \sum_j V_{i,j} V_{i,j+1} - \cdots.$$ (13)

The maximization of (12) is also equivalent to the minimization of (14):

$$\min(-\partial_0 \sum_i \sum_j V_{i,j} - \partial_1 \sum_i \sum_j V_{i,j} V_{i,j-1} - \partial_2 \sum_i \sum_j V_{i,j} V_{i,j+1} - \cdots - \sum_i \sum_j \ln P(y_{ij}|x_{ij} = 1)).$$ (14)

Because CNN should be easily realized, we make an approximation:

$$\ln P(y_{ij}|x_{ij} = 1) \approx (-1)^0 \frac{-P(y_{ij}|x_{ij} = 0)}{1} = P(y_{ij}|x_{ij} = 1) - 1$$

Since $P(y_{ij}|x_{ij}=1) = \dfrac{1+y_{ij}}{2}$, thus $\ln P(y_{ij}|x_{ij}=1) \approx \dfrac{y_{ij}-1}{2} = \dfrac{V_{ij}-1}{2}$, so:

$$E(t) = -\frac{1}{2}\sum_{i,j}\sum_{k,l}A(i.j;k,l)V_{Yij}V_{Ykl} - \sum_{i,j}I_{ij}V_{Yij} + \frac{1}{2R_X}\sum_{i,j}V_{Yij}^2 - \sum_{i,j}\sum_{k,l}B(i.j;k,l)V_{Yij}u_{kl} \cdot \quad (15)$$

Obviously (14) is similar to CNN's energy function (15), but the parameter need to be given, if we choose ∂, the cloning template of CNN is also definite, it's difficult to find the answer directly, so we utilized Genetic Algorithm to deal with this problem. When CNN is employed to realize the stochastic algorithm, step 1 should not be changed. In next section, we will propose a general method to design cloning template for minimizing the energy function.

3 Template Design with Genetic Algorithm

Genetic algorithms(GA) are famous optimization algorithms[6]. CNN's template for edge detection is designed with GA in this section, and the design approach is universal to other similar questions. When a simple GA is applied for CNN template learning, the following steps should be carried out:

At first, we should find cost function: G.A is applied in many questions, but the most popular application is optimization. In section2, we have shown how to get the cost function for CNN: Gibbs image model is utilized.

The second step is coding. The representation of a template with a binary string is coding. To image processing question, the parameter of $V_{i,j}V_{i,j-1}$ and that of $V_{i,j}V_{i,j+1}$ is the same one. Thus the initial form of template is: $\begin{bmatrix} a_1 & a_4 & a_3 \\ a_2 & a_0 & a_2 \\ a_3 & a_4 & a_1 \end{bmatrix}$, so less

template elements need to be found. From comparison of MAP formula (14) and CNN energy function (5), some elements is found directly, and others could be sought with G.A. Obviously, template B=0, the middle elements of template A: a_0 should be greater than $\dfrac{1}{R_X}$, and CNN will become stable when time increase, but the difference is very small. Template I_{ij} corresponds to the first term: $-\partial_0\sum_i\sum_j V_{i,j}$. Now we

get the template of CNN: B=0; A= $\begin{bmatrix} a_1 & a_4 & a_3 \\ a_2 & a_0 & a_2 \\ a_3 & a_4 & a_1 \end{bmatrix}$, $a_1 = 2\partial_1$, $a_2 = 2\partial_2$, $a_3 = 2\partial_3$, $a_4 = 2\partial_4$, $a_0 > \dfrac{1}{R_X}$, and $a_0 \approx \dfrac{1}{R_X}$, $I = \partial_0$;

then the energy function is similar to the cost function (14). Once five parameters ∂_0, $\partial_1, \partial_2, \partial_3$ and ∂_4 are determined, the solution is also gotten. So we represent these five parameters as a binary string. The resolution is 0.01, so 10 bits can represent a parameter. Five parameters need to be optimized, and a binary string with 50 bits is utilized. After representing the parameters with a binary string, such operation of G.A as crossover can be performed.

The third step is initialization chromosomes group and evaluation A population of binary strings – called chromosomes – is used. These binary strings are evaluated. Some successful chromosomes will produce the next generation, and other unsuccessful chromosomes will be dead; Evolution of GA minimizes the cost function, so the optimal answer is gotten.

Fitness of the chromosome should be calculated to find good chromosome. The evaluation function takes a binary string and returns a fitness value. The fitness of chromosome C_i, which means a template, is calculated by:

$$f_i = \max \, energy - E(C_i).$$ (16)

In (16) $E(C_i)$ is the value of energy function that corresponds to chromosomes C_i, and the maxenergy is the maximum of energy function. An approximate estimation is

$$E_{max} = \frac{1}{2} \sum_{i,j} \sum_{k,l} |A(i,j;k,l)| + \sum_{i,j} \sum_{k,l} |B(i,j;k,l)| + N_1 N_2 (\frac{1}{2R_x} + |I|).$$ (17)

When initial binary strings and the evaluation method are given, the evolution of genetic algorithm could be carried out, and we will get the optimal solution of (16). Then we should carry out the fourth step: evolution Operations Procedure of genetic algorithm is described in reference [6]. New chromosomes are generated mainly through these operations:(A) Crossover,(B) Mutation,(C)Reproduction,(D) Selection. Operations of simple genetic algorithm have been described, and we find the parameter with this algorithm, and process the noisy image with CNN. Fig.3 shows the change of average fitness when iterations of genetic algorithm increase.

The probability of crossover is 0.5, the probability of mutation is 0.01. There are 200 chromosomes in initial population. After 60 generations, we get the best individual, and find the template of CNN. Normalize the parameter, then we get:

$$B=0; A = \begin{bmatrix} 0.57 & 0.87 & 0.60 \\ 0.87 & 1.05 & 0.87 \\ 0.60 & 0.87 & 0.57 \end{bmatrix}, a_0 > \frac{1}{R_x}, a_0 \approx \frac{1}{R_x}, \quad R_x = 1.0, \quad I = \partial_0 = 1.00;$$

We apply the algorithm to the image which is also processed with stochastic algorithm by Pitas[7].The image is normalized to (-1,+1), then a white Gaussian noise with mean zero and variance σ^2 is added. The initial result is gotten with gradient operator ($\nabla f = |G_x| + |G_y|$), and the threshold is 0.3.

Fig.4 is the processing result of CNN. Obviously the result is better than that of traditional. Like other stochastic algorithm for edge detection [5], CNN algorithm is to improve the quality of edge detection result. So the input of CNN is the initial edge detection result, The quality of CNN's result is much better than traditional algorithm, and the speed of algorithm is faster than other stochastic algorithm

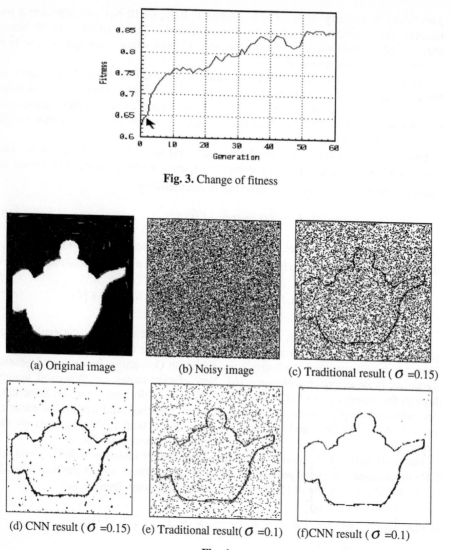

Fig. 3. Change of fitness

(a) Original image   (b) Noisy image   (c) Traditional result (σ =0.15)

(d) CNN result (σ =0.15) (e) Traditional result(σ =0.1) (f)CNN result (σ =0.1)

Fig. 4.

4 Optimal Solution

Based on GIM, many image-processing problems are changed into the corresponding combinatorial optimization questions. When these algorithms are realized with CNN, the structure of traditional CNN can't guarantee the optimal solution. There are many local minimum points in the energy function, therefore the dynamic relaxation should be employed. An efficient relation algorithm for CNN has been proposed in [8]. The algorithm, which is called *Hardware Annealing* (HA), is realized by control gain of the cell. The transfer function of CNN with annealing capability can be described as:

$$V_{Yij}(t) = \begin{cases} +1 & V_{Xij}(t) \geq 1 \\ V_{Xij}(t) & -1 \leq V_{Xij}(t) \leq 1 \\ -1 & V_{Xij}(t) \leq -1 \end{cases}.$$ (18)

$V_{Xij}(t) = g(t)V_{Xij}(t)$, g(t) is a linear function increasing from x_{ij} to g_{max} :

$$g_{max} = 1 \text{ and } 0 \leq g_{min} << 1: \quad g(t) = g_{min} + \frac{t}{T_A}.$$ (19)

After g gets its maximum value, the maximum gain is maintained during $T_A \leq t \leq T$, and the network is stabilized. Though hardware annealing CNN is more complex than traditional CNN, the result of segmentation with annealing CNN is more satisfactory. In reference [8], the reasonable application for CNN that is applied in the area of communication is shown, but the approach is not applied naturally for CNN that is utilized in the area of image processing. The reason is that the image-processing problem is not transformed into a global optimization problem. With help of GIM, the question is easily smoothed, and the effectiveness of HA is strongly confirmed. The simulation results are shown in Fig.5. Fig.5(a) is the image which is corrupted by noise, its SNR=0dB . P_{err} is the error rate of segmentation. Fig. 5(b) is the segmentation result of CNN with annealing capability, P_{err}=0.028. Fig. 5(c) is the result of CNN without HA. P_{err}=0.056.We get the template with G.A.:

$$R_X = 1, C = 1, I = 0, B = \begin{bmatrix} 0 & 0 & 0 \\ 0 & 1.00 & 0 \\ 0 & 0 & 0 \end{bmatrix}, A = \begin{bmatrix} 1.10 & 1.01 & 0.97 \\ 0.98 & 4.0 & 0.98 \\ 0.97 & 1.01 & 1.10 \end{bmatrix}.$$

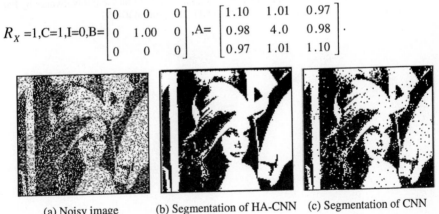

(a) Noisy image   (b) Segmentation of HA-CNN   (c) Segmentation of CNN

Fig. 5.

The simulation results confirm that the segmentation of binary image can be segmented with CNN, and the results also confirm the effectiveness of HA algorithm for CNN. The segmentation result of hardware annealing CNN is more satisfactory than that of traditional CNN

5 Conclusions

An approach for template design is proposed in this paper. Based on the similarities between GIM image model and CNN, the utilization of energy function for template design is noted. An efficient method for parameter estimation is also presented, and cloning template can be defined with G.A algorithm. To avoid getting local minimum in energy function, CNN with HA is employed. The simulation results confirm the effectiveness of this approach.

References

1. Chua, L.O., Yang, L.: Cellular Neural Network: Theory. IEEE Trans. Circuits Syst., **35** (1988) 1257-1272
2. Chua, L.O., Yang, L.: Cellular Neural Network: Applications. IEEE Trans. Circuits Syst., **35** (1988) 1257-1272
3. Gu, X., Yu, D., Zhang, L.: Image Thinning Using PCNN. Pattern Recognition Letters, **25** (2004) 1075-1084
4. Derin, H., Elliot, H.: Modeling and Segmentation of Noisy Textured Image Using GRF. IEEE Trans. on PAMI, **9** (1987) 39-55
5. Tan, L.: A Comparative Cost Function Approach to Edge Detection. IEEE Tran. on SMC, **9** (1989) 1337-1349
6. Holland, J.H.: Adaptation in Natural and Artificial Systems. The university of Michigan Press, Ann Arbor (1975)
7. Pitas, I.: Markov Image Model for Image Labeling and Edge Detection. Signal Processing, **15** (1988) 365-374
8. Bang, S.H., Sheu, B.J., Wu, H.Y.: Optimal Solutions for Cellular Neural Networks by Paralleled Hardware Annealing. IEEE Trans. on NN, **28** (1996) 440-453

A Novel Generalized Congruence Neural Networks*

Yong Chen[1,2], Guoyin Wang[1,2], Fan Jin[2], and Tianyun Yan[2]

[1] Institute of Computer Science and Technology,
Chongqing University of Posts and Telecommunications,
Chongqing 400065, China
[2] School of Computer and Communication Engineering, Southwest Jiaotong University,
Chengdu 610031, China

Abstract. All existing architectures and learning algorithms for Generalized Congruence Neural Network (GCNN) seem to have some shortages or lack rigorous theoretical foundation. In this paper, a novel GCNN architecture (BPGCNN) is proposed. A new error back-propagation learning algorithm is also developed for the BPGCNN. Experimental results on some benchmark problems show that the proposed BPGCNN performs better than standard sigmoidal BPNN and some improved versions of BPNN in convergence speed and learning capability, and can overcome the drawbacks of other existing GCNNs.

1 Introduction

Since the resurgence of artificial Neural Network (NN) in 1980s, more and more researchers come to this area and much achievement has been obtained. Most studies on learning algorithms and architectures are done while the importance of activation functions is ignored. As a matter of fact, the choice of activation functions may strongly influence the complexity and performance of neural networks. In the process of neural networks training, flexible activation functions are as important as good architectures and learning procedures [1]. Although sigmoidal functions are the most frequently used ones, there is not a priori reason why models based on sigmoidal functions should have better performance [1],[2]. Furthermore, it has some shortages in learning speed and hardware implementation. A lot of alternative activation functions were proposed, such as Rational function [3], Lorentzian function [4], etc. For a summary of the past work on activation function see [1].

Jin [5],[6] developed a Generalized Congruence Neural Network (GCNN), which employs generalized congruence function as its activation function. It is proved that GCNN has faster learning speed with slightly weaker learning ability compared with common Back-Propagation Neural Networks (BPNN) on some function approximation problems [5],[6]. In [5], the basic architecture and an error back-distribution (BD) algorithm are proposed for GCNN. Unfortunately, the BD algorithm lacks rigorous mathematical foundation and there exist some other problems, say, it can't learn all zero

* This paper is supported by the Program for New Century Excellent Talents in University, the National Natural Science Foundation of China under Grant No.60373111, the Science and Technology Research Program of the Municipal Education Committee of Chongqing of China under Grant No.040505, and the Key Lab of Computer Network and Communication Technology of Chongqing of China.

J. Wang, X. Liao, and Z. Yi (Eds.): ISNN 2005, LNCS 3496, pp. 455–460, 2005.

inputs sample [6]. Besides, it is difficult to set modulus in this architecture. Hu and Jin [7] presented an improved generalized congruence function and proposed an improved learning algorithm based on error back-distribution algorithm. Its architecture is the same as that of [5],[6] except for the improvement in activation function. This model can obtain some good results, but the problems mentioned above are still unsolved. In addition, their algorithm can only deal with the problem with single output. These disadvantages restrict its application seriously.

In this paper, a novel architecture for GCNN (BPGCNN) is proposed based on an improved generalized congruence activation function. As mentioned above, in all existing GCNN architectures, modulus setting is a complicate task because different modulus is employed in different hidden neuron [5],[6],[7]. Through our study, we find that similar results will be got by setting the same modulus for all hidden neurons. The activation function of output neurons is also changed. We use linear function instead of the generalized congruence function in output neurons for the sake of better generalization ability. Besides, an improved error Back-Propagation (BP) algorithm is also developed for BPGCNN. The BPGCNN is tested by the Two-Spirals problem and a function approximation problem.

2   Architecture and Learning Algorithm

In this section, we first introduce the improved generalized congruence function. Then, a new GCNN architecture and learning algorithm are discussed.

2.1   Improved Generalized Congruence Function

Definition 1 *Let $a, b, m \in R$. Then a is generalized congruent to b modulo m:*

$$a \equiv b \ (Gmod \ m) \tag{1}$$

if $m \mid (a - b)$.

The generalized congruence function, which is adopted as activation function by [5],[6], is defined as:

$$y = x \ (Gmod \ m) \tag{2}$$

It is shown in Fig. 1.

As mentioned in section 1, the learning ability of original GCNN model is a little weaker. It is also difficult to develop a good learning algorithm for it. Through our study, we find that this is mainly due to the discontinuity of common generalized congruence function. To avoid this problem, we adopt the following improved generalized congruence function.

$$g = \begin{cases} x(Gmod \ m) & \text{if } [|x/m|] \text{ is even;} \\ m - x(Gmod \ m) & \text{if } [|x/m|] \text{ is odd and } x \geq 0; \\ -m - x(Gmod \ m) & \text{if } [|x/m|] \text{ is odd and } x < 0. \end{cases} \tag{3}$$

where, x is an independent variable (input), g is a dependent variable (output), m is modulus, usually $m > 0$; $[\cdot]$ denotes truncate function. An example of improved generalized congruence function with $m = 0.5$ is shown Fig. 2.

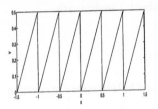
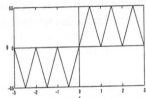
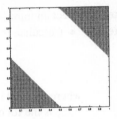

Fig. 1. Common Generalized Congruence Function (m=0.5)

Fig. 2. Improved Generalized Congruence Function (m=0.5)

Fig. 3. XOR Result

From Fig. 2, we know that the improved generalized congruence function is not differentiable on all extremum points. In order to apply BP algorithm we need make a little modification on its derivative: the derivatives of the improved generalized congruence function on these points are set to be zero. Then, its derivative should be:

$$g' = \begin{cases} -1 & \text{if } x\%m \neq 0 \text{ and } [|x/m|] \text{ is even;} \\ +1 & \text{if } x\%m \neq 0 \text{ and } [|x/m|] \text{ is odd;} \\ 0 & \text{else.} \end{cases} \quad (4)$$

Although as the most popular activation function for neural networks, sigmoidal function suffers from the saturation problem while employing BP algorithm. Its saturation region is the root of slow asymptotic convergence speed. Some improvements have been obtained with the artificial enlarging of errors for neurons operating in the saturation region [8]. It should be noted that our improved generalized congruence function can overcome the saturation problem. From Fig. 2, we can see that this function is linear in local. This make it easy for physical implementation. Besides, this function fits the XOR and N-parity problems when setting modulus as 1.0. And the two problems can be solved by just one neuron using such activation function. The result of XOR problem by an improved generalized neuron with 2 inputs and 1 output is shown in Fig 3.

2.2 BPGCNN Architecture

In most applications, we adopt three layers feedforward network: an input layer, a hidden layer and an output layer. Suppose there are n_1 inputs, n_2 hidden neurons and n_3 output neurons. In the original GCNN architecture [5],[6],[7], different hidden neurons have different moduli. It is hoped that this may enhance GCNN's learning ability [5]. But through our studies, we find that it is not as we expected. Using multiple moduli and only one modulus for the hidden neurons produces similar results. Since it is easier to set one parameter than many parameters, one modulus method is adopted in this paper. And also different from other existing GCNNs, all the moduli of output neurons in our BPGCNN architecture are set to be 0, i.e. all the output neurons don't use any activation function. This may make network's generalization ability much better.

2.3 Learning Algorithm

The improved back-propagation algorithm for BPGCNN is as follows.

Step 1 Initialize all flexible parameters, such as modulus m, learning rate η, etc. Randomize the weights w_{ij} to small random values. Set iteration step $n = 0$.

Step 2 Select an input/output pattern (x, d) from the training set sequentially.

Step 3   • Calculate the input of each hidden neuron by:

$$v_j(n) = \sum_{i=1}^{n_1} w_{ji}^{(1)}(n)x_i \qquad (5)$$

where $w_{ji}^{(1)}(n)$ is the weight between input node i and hidden node j at iteration n, x_i is the ith element of input vector x.

• Calculate the output of each hidden neuron by:

$$y_j^{(1)}(n) = g(v_j(n)). \qquad (6)$$

• Calculate the outputs of output neurons:

$$y_j^{(2)}(n) = \sum_{i=1}^{n_2} w_{ji}^{(2)}(n)y_i^{(1)}(n) \qquad (7)$$

where $w_{ji}^{(2)}(n)$ is the weight between the ith hidden neuron and the jth output neuron.

• Calculate the error signal

$$e_j(n) = d_j - y_j^{(2)}(n) \qquad (8)$$

where d_j is the jth element of the desired response vector d.

Step 4 Calculate the local gradients of the network, defined by

$$\begin{cases} \delta_j^{(2)}(n) = e_j(n) & \text{for output neuron;} \\ \delta_j^{(1)}(n) = g'(v_j(n)) \sum_k \delta_k^{(2)}(n)w_{kj}^{(2)}(n) & \text{for hidden neuron;} \end{cases} \qquad (9)$$

where g' denotes differentiation with respect to the argument.

Step 5 Adjust the weights $(l = 1, 2)$ of the network according to the generalized delta rule:

$$w_{ji}^{(l)}(n+1) = w_{ji}^{(l)}(n) + \alpha\Delta w_{ji}^{(l)}(n) + \beta\delta_j^{(l)}(n)y_j^{(l-1)}(n) \qquad (10)$$

where, $y_j^{(0)}(n) = x_j$; the learning rate α and momentum factor η can be set with different values for different layers.

Step 6 If all the training patterns have been trained at this iteration, then, go to Step 7; otherwise, go back to Step 2.

Step 7 Set $n = n + 1$. If the stop criterion is satisfied, stop; otherwise, go back to Step 2.

3   Simulation Experiments

Several experiments are done aiming to verify the BPGCNN proposed in this paper. The two Spirals problem and a function approximation problem are considered. In both cases, all initial weights were set at random in the range [-1,+1]. The metrics adopted in this paper are: the minimum number of iteration steps (min), the mean value of iteration steps ($mean$), the maximum number of iteration steps (max), the standard deviation of iteration steps ($s.d.$), and the success rate ($succ.$). The proposed BPGCNN has been implemented in C++ on an IBM PC.

3.1   Two-Spirals Problem

Two-spirals problem is an extremely difficult problem for neural networks. Baum and Lang [9] reported that it was unable to find a correct solution using a 2-50-1 common BP network when starting from randomly selected initial weights. Shang and Wah [10] used a single hidden layer network with lateral connections between each hidden neurons. With a global optimization algorithm, namely NOVEL, they obtained promising results. But an extremely long computation time of about 10 h was required even though it was executed on a multi-processor workstation machine. Cho and Chow [11] also employed a single hidden layer network with the least squares and penalized algorithm for this problem. The learning time are about 4.75h, 2.67h and 1.29 h for implementation on the different number of hidden units with 30, 40 and 50, respectively.

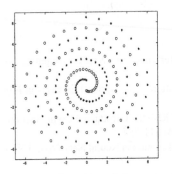

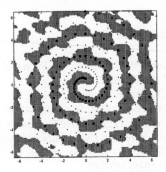

Fig. 4. The training data of the two-spirals problem

Fig. 5. Classification result for two-spirals problem

In our experiment, the training set consists of 194 patterns, as shown in Fig. 4. A 2-35-1 topology for our BPGCNN is adopted. Fig. 5 is the decision region of our BPGCNN after 2613 iteration steps. The training time is just 145 seconds.

3.2   Function Approximation

The desired function $f(x) = \sin(x)\cos(2x)$ used in [12] is used here. Twenty training patterns are sampled uniformly from the interval $[0, 2\pi]$. The BPGCNN is compared to three types of BPNN, i.e. momentum BP neural networks (MBPNN), adaptive BP neural networks (ABPNN) [13], BP neural networks with adaptive learning rate (BBPNN) [12]. The results of the three BPNNs for this problem are cited from [12]. A 1-15-1 network architecture is used in [12], and the numbers of weights and bias are 30 and 16 respectively. As BPGCNN don't use any bias, for fair comparison, a 1-20-1 network is employed. So the number of weights of our BPGCNN is 40. Table 1 shows the comparative results. We can see that BPGCNN converges much faster than the others, and its success rate is also high.

4   Conclusions

In this paper, we propose a novel improved generalized congruence neural networks (BPGCNN) and its corresponding learning algorithm. It outperforms the common BP

Table 1. Comparative results for function approximation

NN type	topology	weight	min	mean	max	s.d.	succ.(%)
MBPNN	1-15-1	30+16	561	718.25	963	131.66	16
ABPNN	1-15-1	30+16	280	577.26	970	200.01	27
BBPNN	1-15-1	30+16	63	309.82	966	230.28	74
BPGCNN	1-20-1	40+0	16	66.08	439	80.85	80

neural networks and some improved versions of BP neural networks in terms of convergence speed and learning ability in several experiments. Besides, it can overcome the shortages of other existing GCNNs. This result is encouraging and suggests that the generalized congruence neural networks may be better in some kinds of tasks.

References

1. Duch, W., Jankowski, N.: Survey of Neural Transfer Functions. Neural Computing Surveys, **2** (1999) 163-212
2. Chandra, P., Singh, Y.: A Case for the Self-adaptation of Activation Functions in FFANNs. Neurocomputing, **56** (2004) 447-454
3. Dorffner, G.: A Unified Framework for MLPs and RBFNs: Introducing Conic Section Function Networks. Cybernetics and Systems, **25** (1994) 511-554
4. Giraud, B.G., Lapedes, A., Liu, L.C., Lemm, J.C.: Lorentzian Neural Nets. Neural Networks, **8** (1995) 757-767
5. Jin, F.: Architectures and Algorithms of Generalized Gongruence Neural Networks. Journal of Southwest Jiaotong University, **6** (1998)
6. Jin, F.: Study on Principles and Algorithms of Generalized Congruence Neural Networks. International Conference on Neural Network and Brain. Publishing House of Electronics Industry, Beijing (1998) 441-444
7. Hu, F., Jin, F.: Analysis of Characteristics of Generalized Congruence Neural Networks with an Improved Algorithm. Journal of Southwest Jiaotong University, **36** (2001) 136-139
8. NG, S.C., LEUNG, S.H., LUK, A.: Fast Convergent Generalized Back-propagation Algorithm with Constant Learning Rate. Neural Processing Letters, **9** (1999) 13-23
9. Baum, E., Lang, K.: Constructing Hidden Units Using Examples and Queries. In Lippmann, R., Moody, J., Touretzky, D.(eds.): Advances in Neural Information Processing Systems. Vol. 3. Morgan Kaufmann, San Mateo, CA (1991) 904-910
10. Shang, Y., Wah, B.W.: Global optimization for Neural Network Training. IEEE Computer, **29** (1996) 45-54
11. Cho, S., Chow, T.W.S.: Training Multilayer Neural Networks Using Fast Global Learning Algorithm-least-squares and Penalized Optimization Methods. Neurocomputing, **25** (1999) 115-131
12. Plagianakos, V.P., Sotiropoulos, D.G., Vrahatis, M.N.: An Improved Backpropagation Method with Adaptive Learning Rate. Department of Mathematics, University of Patras, Technical Report No.98-02 (1998)
13. Vogl, T.P., Mangis, J.K., Rigler, J.K., Zink, W.T., Alkon, D.L.: Accelerating the Convergence of the Back-propagation Method. Biological Cybernetics, **59** (1988) 257-263

A SOM Based Model Combination Strategy

Cristofer Englund[1] and Antanas Verikas[1,2]

[1] Intelligent Systems Laboratory, Halmstad University, Box 823,
S-301 18 Halmstad, Sweden
cristofer.englund@ide.hh.se
[2] Department of Applied Electronics, Kaunas University of Technology, Studentu 50,
LT-3031, Kaunas, Lithuania
antanas.verikas@ide.hh.se

Abstract. A SOM based model combination strategy, allowing to create adaptive – data dependent – committees, is proposed. Both, models included into a committee and aggregation weights are specific for each input data point analyzed. The possibility to detect outliers is one more characteristic feature of the strategy.

1 Introduction

A variety of schemes have been proposed for combining multiple models into a committee. The approaches used most often include averaging [1], weighted averaging [1, 2], the fuzzy integral [3, 4], probabilistic aggregation [5], and aggregation by a neural network [6]. Aggregation parameters assigned to different models as well as models included into a committee can be the same in the entire data space or can be different – *data dependent* – in various regions of the space [1, 2]. The use of data-dependent schemes, usually provides a higher estimation accuracy [2, 7, 8].

In this work, we further study data-dependent committees of models. The paper is concerned with a set of neural models trained on different data sets. We call these models *specialized*. The training sets of the specialized models may overlap to varying, sometimes considerable, extent. The specialized models implement approximately the same function, however only approximately. The unknown underlying functions may slightly differ between the different specialized models. However, the functions may also be almost identical for some of the models. In addition to the set of specialized models, a *general* model, trained on the union of the training sets of the specialized models, is also available. On average, when operating in the regions of their expertise, the specialized models provide a higher estimation accuracy than the general one. However, the risk of extrapolation is much higher in the case of specialized models than when using the general model. Since training data sets of the specialized models overlap to some extent, a data point being in the extrapolation region for one specialized model may be in the interpolation region for another model. Moreover, since the underlying functions that are to be implemented by some of the

J. Wang, X. Liao, and Z. Yi (Eds.): ISNN 2005, LNCS 3496, pp. 461–466, 2005.

specialized models may be almost identical, we can expect boosting the estimation accuracy by aggregating appropriate models into a committee. It all goes to show that an adaptive – possessing data dependent structure – committee is required. Depending on a data point being analyzed, appropriate specialized models should be detected and aggregated into a committee. If for a particular data point extrapolation is encountered for all the specialized models – *outlying* data point for the specialized models – the committee should be made of only the general model. We utilize a SOM [9] for attaining such adaptive behaviour of the committee. Amongst the variety of tasks a SOM has been applied to, outlier detection and model combination are also on the list. In the context of model combination, a SOM has been used as a tool for subdividing a task into subtasks [10]. In this work, we employ a SOM for obtaining committees of an adaptive, data dependent, structure.

2   The Approach

We consider a non-linear regression problem and use a one hidden layer perceptron as a specialized model. Let $\mathbf{T}_i = \{(\mathbf{x}_i^1, \mathbf{y}_i^1), (\mathbf{x}_i^2, \mathbf{y}_i^2), ..., (\mathbf{x}_i^{N_i}, \mathbf{y}_i^{N_i})\}$, $i = 1, ..., K$ be the learning data set used to train the ith specialized model network, where $\mathbf{x} \in \Re^n$ is an input vector, $\mathbf{y} \in \Re^m$ is the desired output vector, and N_i is the number of data points used to train the ith network. The learning set of the general model – also a one hidden layer perceptron – is given by the union $\mathbf{T} = \bigcup_{i=1}^{K} \mathbf{T}_i$ of the sets $\mathbf{T}_i$. Let $\mathbf{z} \in \Re^{n+m}$ be a centered concatenated vector consisting of $\mathbf{x}$ augmented with $\mathbf{y}$. Training of the prediction committee then proceeds according to the following steps.

1. Train the specialized networks using the training data sets $\mathbf{T}_i$.
2. Train the general model using the data set $\mathbf{T}$.
3. Calculate eigenvalues λ_i ($\lambda_1 > \lambda_2 > ... > \lambda_{n+m}$) and the associated eigenvectors $\mathbf{u}_i$ of the covariance matrix $\mathbf{C} = \frac{1}{N} \sum_{j=1}^{N} \mathbf{z}_j \mathbf{z}_j^T$, where $N = \sum_{i=1}^{K} N_i$.
4. Project the $N \times (n + m)$ matrix $\mathbf{Z}$ of the concatenated vectors $\mathbf{z}$ onto the first M eigenvectors $\mathbf{u}_k$, $\mathbf{A} = \mathbf{ZU}$.
5. Train a 2–D SOM using the $N \times M$ matrix $\mathbf{A}$ of the principal components by using the following adaptation rule:

$$\mathbf{w}_j(t+1) = \mathbf{w}_j(t) + \alpha(t)h(j^*, j; t)[\mathbf{a}(t) - \mathbf{w}_j(t)] \qquad (1)$$

where $\mathbf{w}_j(t)$ is the weight vector of the jth unit at the time step t, $\alpha(t)$ is the decaying learning rate, $h(j^*, j; t)$ is the decaying Gaussian neighbourhood, and j^* stands for the index of the winning unit.
6. Map each data set $\mathbf{A}_i$ associated with $\mathbf{T}_i$ on the trained SOM and calculate the hit histograms.
7. Low-pass filter the histograms by convolving them with the following filter $h(n)$ as suggested in [11]:

$$h[n] = (M - |n|)/M \qquad (2)$$

where $2M + 1$ is the filter size. The convolution is made in two steps, first in the horizontal and then in the vertical direction. The filtered signal $y[n]$ is given by

$$y[n] = x[n] * h[n] = \sum_{m=-M}^{M} x[n-m]h[m] \qquad (3)$$

8. Calculate the discrete probability distributions from the filtered histograms:

$$P_{ij} = P(\mathbf{a} \in \mathcal{S}_j) = \frac{\text{card}\{k|\mathbf{a}_k \in \mathcal{S}_j\}}{N_i}, \quad i = 1, ..., K \qquad (4)$$

where $\text{card}\{\bullet\}$ stands for the cardinality of the set and $\mathcal{S}_j$ is the *Voronoi region* of the jth SOM unit.

9. For each specialized model $i = 1, ..., K$ determine the expertise region given by the lowest acceptable P_{ij}.

In the operation mode, processing proceeds as follows.

1. Present $\mathbf{x}$ to the specialized models and calculate outputs $\widehat{\mathbf{y}}_i$, $i = 1, ..., K$.
2. Form K centered $\mathbf{z}_i$ vectors by concatenating the $\mathbf{x}$ and $\widehat{\mathbf{y}}_i$.
3. Project the vectors onto the first M eigenvectors $\mathbf{u}_k$.
4. For each vector of the principle components $\mathbf{a}_i$, $i = 1, ..., K$ find the best matching unit ij^* on the SOM.
5. Aggregate outputs of those specialized models $i = 1, ..., K$, for which $P_{ij^*} \geq \beta_i^1$, where β_i^1 is a threshold. If

$$P_{ij^*} < \beta_i^1 \ \ \forall i \quad \text{and} \quad P_{ij^*} \geq \beta_i^2 \ \ \exists i \qquad (5)$$

where the threshold $\beta_i^2 < \beta_i^1$, use the general model to make the prediction. Otherwise use the general model and make a warning about the prediction.

We use two aggregation schemes, namely averaging and the weighted averaging. In the weighted averaging scheme, the committee output $\widehat{\mathbf{y}}$ is given by

$$\widehat{\mathbf{y}} = \frac{\sum_i v_i \widehat{\mathbf{y}}_i}{\sum_i v_i} \qquad (6)$$

where the sum runs over the selected specialized models and the aggregation weight $v_i = P_{ij^*}$.

3   Experimental Investigations

The motivation for this work comes from the printing industry. In the offset lithographic printing, four inks – cyan (C), magenta (M), yellow (Y), and black (K) – are used to create multicoloured pictures. The print is represented by C, M, Y, and K dots of varying sizes on thin metal plates. These plates are mounted on press cylinders. Since both the empty and areas to be printed are on the same plane, they are distinguished from each other by ones being water receptive and the others being ink receptive. During printing, a thin layer of water is applied to the plate followed by an application of the corresponding ink. The inked picture is transferred from a plate onto the blanket cylinder, and then onto the paper. Fig. 1 presents a schematic illustration of the ink-path.

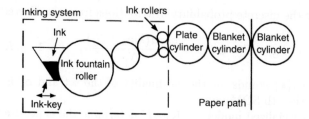

Fig. 1. A schematic illustration of the ink-path.

Ink feed control along the page is accomplished in narrow – about 4 cm wide – the so call ink zones. Thus, up to several tens of ink zones can be found along a print cylinder. The amount of ink deposited on the paper in each ink zone is determined by the opening of the corresponding ink-key – see Fig. 1. The ink feed control instrumentation is supposed to be identical in all the ink zones. However, some discrepancy is always observed. The aim of the work is to predict the initial settings of the instrumentation for each of the four inks in different ink zones depending on the print job to be run. The accurate prediction is very valuable, since the waste of paper and ink is minimized. Due to possible discrepancies between the instrumentation of different ink zones, we build a separate – specialized – neural model for each ink zone. A general model, exploiting data from all the ink zones is also built.

In this work, we consider prediction of the settings for only cyan, magenta, and yellow inks. The setting for black ink is predicted separately in the same way. Thus, there are three outputs in all the model networks. Each model has seventeen inputs characterizing the ink demand in the actual and the adjacent ink zones for C, M, and Y inks, the temperature of inks, the printing speed, the revolution speed of the C, M, and Y ink fountain rollers, and the $L^*a^*b^*$ values [12] characterizing colour in the test area. Thus, the concatenated vector $\mathbf{z}_i$ contains 20 components. The structure of all the model networks has been found by cross-validation. To test the approach, models for 12 ink zones have been built. About 400 data points were available from each ink zone. Half of the data have been used for training, 25% for validation, and 25% for testing.

There are five parameters to set for the user, namely, the number of principal components used, the size of the filtering mask, the SOM size, and the thresholds β_i^1 and β_i^2. The SOM training was conducted in the way suggested in [9]. The number of principal components used was such that 95% of the variance in the data set was accounted for. The SOM size is not a critical parameter. After some experiments, a SOM of 12×12 units and the filtering mask of 3×3 size were adopted. The value of $\beta_i^2 = 0$ has been used, meaning that a prediction result was always delivered. The value of β_i^1 was such that for 90% of the training data the specialized models were utilized.

Fig. 2 presents the distribution of the training data coming from all the specialized models on the 2–D SOM before and after the low-pass filtering of the distribution. As it can be seen from Fig. 2, clustering on the SOM surface becomes more clear after the filtering.

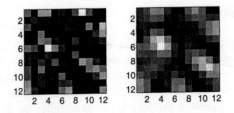

Fig. 2. The distribution of the training data on the 2–D SOM before (left) and after (right) the low-pass filtering.

Fig. 3 illustrates low-pass filtered distributions of training data coming from four different specialized models. The images placed on the left-hand side of Fig. 3 are quite similar. Thus, we can expect that functions implemented by these models are also rather similar. By contrast, the right-hand side of the figure exemplifies two quite different data distributions.

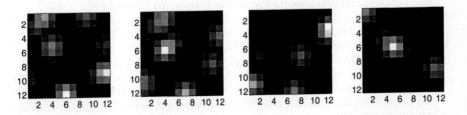

Fig. 3. The low-pass filtered distributions of the training data of four specialized models on the 2–D SOM.

Table 1 presents the average prediction error $\overline{E}$, the standard deviation of the error σ, and the maximum prediction error E^{max} for 209 data samples from the test data set. The data points chosen are "difficult" for the specialized models, since they are situated on the borders of their expertise. As it can be seen, an evident improvement is obtained from the use of the committees. The weighted committees is more accurate than the averaging one.

Table 1. Performance of the Specialized, General, and Committee models estimated on 209 unforseen test set data samples.

Model	$\overline{E}_c$ (σ_c)	$\overline{E}_m$ (σ_m)	$\overline{E}_y$ (σ_y)	E_c^{max}	E_m^{max}	E_y^{max}
Specialized	2.05 (1.52)	8.23 (4.43)	3.23 (0.75)	5.89	17.99	4.33
General	1.96 (2.13)	3.90 (2.87)	3.28 (1.19)	5.21	9.07	5.24
Committee (averaging)	0.79 (0.61)	3.63 (3.61)	1.35 (0.70)	1.72	9.49	2.12
Committee (weighted)	0.75 (0.63)	2.73 (2.46)	1.23 (0.66)	1.72	7.75	2.12

4   Conclusions

We presented an approach to building adaptive – data dependent – committees for regression analysis. The developed strategy of choosing relevant, input data point specific, committee members and using data dependent aggregation weights proved to be very useful in the modelling of the offset printing process. Based on the approach proposed, the possibility to detect outliers in the input-output space is easily implemented, if required.

References

1. Taniguchi, M., Tresp, V.: Averaging Regularized Estimators. Neural Computation, **9** (1997) 1163–1178
2. Verikas, A., Lipnickas, A., Malmqvist, K., Bacauskiene, M., Gelzinis, A.: Soft Combination of Neural Classifiers: A Comparative Study. Pattern Recognition Letters, **20** (1999) 429–444
3. Gader, P.D., Mohamed, M.A., Keller, J.M.: Fusion of Handwritten Word Classifiers. Pattern Recognition Letters, **17** (1996) 577–584
4. Verikas, A., Lipnickas, A.: Fusing Neural Networks through Space Partitioning and Fuzzy Integration. Neural Processing Letters, **16** (2002) 53–65
5. Kittler, J., Hatef, M., Duin, R.P.W., Matas, J.: On Combining Classifiers. IEEE Trans Pattern Analysis and Machine Intelligence, **20** (1998) 226–239
6. Kim, S.P., Sanchez, J.C., Erdogmus, D., Rao, Y.N., Wessberg, J., Principe, J.C., Nicolelis, M.: Divide-and-conquer Approach for Brain Machine Interfaces: Nonlinear Mixture of Competitive Linear Models. Neural Networks, **16** (2003) 865–871
7. Woods, K., Kegelmeyer, W.P., Bowyer, K.: Combination of Multiple Classifiers Using Local Accuracy Estimates. IEEE Trans Pattern Analysis Machine Intelligence, **19** (1997) 405–410
8. Verikas, A., Lipnickas, A., Malmqvist, K.: Selecting Neural Networks for a Committee Decision. International Journal of Neural Systems, **12** (2002) 351–361
9. Kohonen, T.: Self-Organizing Maps. 3 edn. Springer-Verlag, Berlin (2001)
10. Griffith, N., Partridge, D.: Self-organizing Decomposition of Functions. In Kittler, J., Roli, F., eds.: Lecture Notes in Computer Science. Vol. 1857. Springer-Verlag Heidelberg, Berlin (2000) 250–259
11. Koskela, M., Laaksonen, J., Oja, E.: Implementing Relevance Feedback as Convolutions of Local Neighborhoods on Self-organizing Maps. In Dorronsoro, J.R., ed.: Lecture Notes in Computer Science. Volume 2415. Springer-Verlag Heidelberg (2002) 981–986
12. Hunt, R.W.G.: Measuring Colour. Fountain Press (1998)

Typical Sample Selection and Redundancy Reduction for Min-Max Modular Network with GZC Function*

Jing Li[1], Baoliang Lu[1], and Michinori Ichikawa[2]

[1] Department of Computer Science and Engineering, Shanghai Jiao Tong University,
1954 Hua Shan Road, Shanghai 200030, China
jinglee@sjtu.edu.cn, blu@cs.sjtu.edu.cn
[2] Lab. for Brain-Operative Device, RIKEN Brain Science Institue, 2-1 Hirosawa,
Wako-shi, Saitama, 351-0198, Japan

Abstract. The min-max modular neural network with Gaussian zero-crossing function (M^3-GZC) has locally tuned response characteristic and emergent incremental learning ability, but it suffers from high storage requirement. This paper presents a new algorithm, called Enhanced Threshold Incremental Check (ETIC), which can select representative samples from new training data set and can prune redundant modules in an already trained M^3-GZC network. We perform experiments on an artificial problem and some real-world problems. The results show that our ETIC algorithm reduces the size of the network and the response time while maintaining the generalization performance.

1 Introduction

The min-max modular (M^3) neural network [1, 2] is an alternative modular neural network model for pattern classification. It has been used in real-world problems such as part-of-speech tagging [3] and single-trial EEG signal classification [4]. The fundamental idea of M^3 network is divide-and-conquer strategy: decomposition of a complex problem into easy subproblems; learning all the subproblems by using smaller network modules in parallel; and integration of the trained individual network modules into a M^3 network.

Using Gaussian zero-crossing (GZC) function [5] as a base network module, the M^3 network (M^3-GZC) has locally tuned response characteristic and emergent incremental learning ability. But M^3-GZC network remembers all the samples that have been presented to it. The space in memory and response time can not be satisfied when more and more samples become available, because the space complexity and time complexity of M^3-GZC network are $O(n^2)$, here n is the total number of training samples.

In this paper, we introduce a novel method called Enhanced Threshold Incremental Check (ETIC) algorithm for selecting representative samples and pruning redundant modules for M^3-GZC network. Based on ETIC, M^3 network adds

---

* This work was supported in part by the National Natural Science Foundation of China under the grants NSFC 60375022 and NSFC 60473040.

J. Wang, X. Liao, and Z. Yi (Eds.): ISNN 2005, LNCS 3496, pp. 467–472, 2005.

samples selectively. The size of the network will increase only when necessary and the corresponding response time can drop sharply.

2   ETIC Algorithm

2.1   Influence of Threshold Limits in M^3-GZC Network

Gaussian zero-crossing discriminate function is defined by

$$f_{ij}(x) = exp\left[-\left(\frac{\|x - c_i\|}{\sigma}\right)^2\right] - exp\left[-\left(\frac{\|x - c_j\|}{\sigma}\right)^2\right], \tag{1}$$

where x is the input vector, c_i and c_j are the given training inputs belonging to class C_i and class C_j ($i \neq j$), respectively, $\sigma = \lambda\|c_i - c_j\|$, and λ is a user-defined constant.

The output of M^3-GZC network is defined as following.

$$g_i(x) = \begin{cases} 1 & \text{if } y_i(x) > \theta^+ \\ Unknown & \text{if } \theta^- \leq y_i(x) \leq \theta^+ \\ -1 & \text{if } y_i(x) < \theta^- \end{cases} \tag{2}$$

where θ^+ and θ^- are the upper and lower threshold limits, and y_i denotes the transfer function of the M^3 network for class C_i, which discriminates the pattern of the M^3 network for class C_i from those of the rest of the classes.

From equation (2) we can see that the decision boundary can be easily controlled by adjusting θ^+ and θ^-. If a test sample is accepted by a M^3-GZC network with high absolute values of θ^+ and θ^-, it will be accepted by the same network with lower absolute values of θ^+ and θ^- (as depicted in Fig.1). The threshold limit can be viewed as a degree of confidence of correct classification. When the M^3-GZC net can classify a new sample with a high degree of confidence, it treats the sample as already learned, and will not change itself. While the network misclassifies or correctly classifies a new sample only in a low degree of confidence, it treats the sample as not learned or not learned well; and adds it. So the samples in accept domain and can be classified correctly will not be added to the net in future, and the size of the net can be guaranteed, will not expand if there is no new knowledge presented.

Depending on the important role of thresholds, ETIC chooses the representative samples from training data set. The algorithm can be used in two cases, one is that new training samples are available to the network; the other is that the network still has redundant samples in it, and need to be reduced. Inspired by Condensed Nearest Neighbor [6] and Reduced Nearest Neighbor [7], our algorithm in the two circumstance are called Condensed ETIC and Reduced ETIC, respectively.

2.2   Condensed ETIC Algorithm

When new training samples are available, Condensed ETIC algorithm stores samples misclassified by the current network. The network can be started from

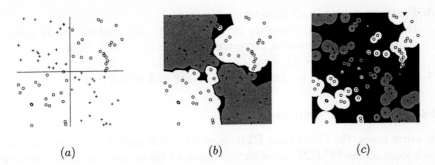

(a) (b) (c)

Fig. 1. Checkerboard problem and the corresponding decision boundary. The black denotes unknown decision regions. (a) A checkerboard problem; (b) decision boundary at $\theta^+ = 0.1$ and $\theta^- = -0.1$; (c) decision boundary at $\theta^+ = 0.8$ and $\theta^- = -0.8$.

scratch, or from a network that has been trained by previous training data. And the new training set can have only one sample or a batch of samples.

When a new training set S_{new} is presented to the network Net, the Condensed ETIC algorithm works as below:

1. Adjust θ^+ and θ^- to θ_e^+ and θ_e^-, $(|\theta_e^+| > |\theta^+|, |\theta_e^-| > |\theta^-|)$.
2. $S = S_{new}$.
3. For each sample (x, d) in S:
   If (x, d) is misclassified by current network Net:
   $S = S \backslash (x, d)$, and add (x, d) to network Net.
4. If S is not exhausted and the network Net has been changed in step 3, go to step 3 for another loop. Or else restore θ^+ and θ^- to their previous values and return Net.

2.3   Reduced ETIC Algorithm

Although Condensed ETIC algorithm can filter out many samples during learning, there are still redundant samples in the network, especially after some groups of training samples are presented to the network. Then we use the Reduced ETIC to remove these samples. The fundamental idea of Reduced ETIC algorithm is to remove samples in the M^3-GZC network if the removal does not cause any misclassification of other samples. Because M^3-GZC network will not misclassify samples in the network, the algorithm only needs to check whether the removal causes any misclassification of all the removed samples.

When an already trained M^3-GZC network has some redundant samples that need to be filtered out, the Reduced ETIC algorithm works as follows. Suppose $Net(S)$ denotes the network that has been set up based on training set S.

1. Adjust θ^+ and θ^- to θ_e^+ and θ_e^-, $(|\theta_e^+| > |\theta^+|, |\theta_e^-| > |\theta^-|)$.
2. $S_{garbage} = \Phi$, $S_{net} =$ all samples in Net.

3. For each sample (x, d) in network Net:
   If (x, d) is classified correctly by $Net\,(S_{net}\backslash (x, d))$, and all the samples in $S_{garbage}$ are also classified correctly by $Net\,(S_{net}\backslash (x, d))$, then $S_{garbage} = S_{garbage}\bigcup (x, d)$, $S_{net} = S_{net}\backslash (x, d)$.
4. Restore θ^+ and θ^- to their previous values and return Net.

2.4   Complexity Analysis

In worst cases, the Condensed ETIC algorithm will include only one sample in each loop, so the M^3-GZC network will check n^2 times, and the time complexity is $O\,(n^2)$, where n denotes the number of samples in the new training set.

Suppose there are n samples in a trained network, and m samples will be filtered out by the Reduced ETIC Algorithm. So the network will check m samples at most to decide whether to remove one sample or not. The corresponding time complexity is $O\,(m \times n)$.

3   Experimental Results

In order to verify our method, we present three experiments. The first is an artificial problem and the other two are real-world problems. All the experiments were performed on a 2.8GHz Pentium 4 PC with 1GB RAM.

3.1   Checkerboard Problem

A checkerboard problem is depicted in Fig. 1 (a). The checkerboard divides a square into four quadrants. The points labelled by circle and plus are positive and negative samples, respectively. In this experiment, we randomly generate 400 samples as training data set, and another 400 samples as test data set. We use the Condensed ETIC to build M^3-GZC network at different threshold. The results are listed in Table 1, and show that the lower the absolute value of thresholds, the smaller size and lower response time ratio, but the generalization performance can be guaranteed in a higher absolute value of thresholds.

Table 1. Results of checkerboard problem. The value in column threshold is θ_e^+, while the corresponding $\theta_e^- = \theta_e^+$. 'no check' means using the traditional algorithm to build up M^3-GZC network. The unit of 'Time' is ms.

Threshold	Accuracy	Unknown	False	Size	Time	Size Ratio	Time Ratio
no check	94.75%	1.00%	4.25%	400	13570	100%	100%
0.01	92.75%	4.50%	2.75%	77	1458	19.25%	10.74%
0.02	92.75%	4.50%	2.75%	80	1489	20.00%	10.97%
0.05	94.00%	3.00%	3.00%	85	1689	21.25%	12.45%
0.08	94.50%	2.00%	3.50%	95	2105	23.75%	15.51%
0.1	94.75%	2.25%	3.00%	97	2433	24.25%	17.93%

3.2   UCI Database

In this experiment, our algorithm is tested on five benchmark data sets from the Machine Learning Database Repository[8]. At start up, all the training samples are presented to an empty network, the Condensed ETIC algorithm and traditional algorithm are used respectively to build the network. The results are shown in Table 2. We can see that our Condensed ETIC algorithm will reduce the size of the network and speed up the response time of trained network, while the generalization ability is influenced only a little.

Table 2. Results on UCI database. Parameters of each net: $\lambda = 0.5$; $\theta^+ = 0.01$; $\theta^- = -0.01$; $\theta_e^+ = 0.1$; and $\theta_e^- = -0.1$. The unit of 'Time' is ms.

Data set	With ETIC	Accuracy	Unknown	False	Size	Time	Size Ratio	Time Ratio
balance	Y	92.0%	0.0%	8.0%	299	2510	59.8%	43.5%
	N	92.0%	0.0%	8.0%	500	5767		
car	Y	62.15%	34.14%	3.70%	436	36392	50.5%	33.7%
	N	57.87%	42.13%	0.0%	864	107878		
image	Y	82.0%	9.24%	8.76%	122	12035	58.1%	40.5%
	N	84.0%	7.33%	8.67%	210	29730		
Iris	Y	94.67%	1.33%	4.0%	36	125	48.0%	49.6%
	N	94.67%	1.33%	4.0%	75	252		
optdigits	Y	96.05%	2.62%	1.34%	1257	840613	32.9%	11.1%
	N	97.22%	1.45%	1.34%	3823	7548237		

3.3   Industry Image Classification

The database of this experiment comes from the images of glass in a product line. The images are converted into 4096 dimension vectors as training and test data. All of the data are divided into four groups; the number of data in each group is 1149, 1138, 1133 and 1197, respectively. We use the first to third data sets as the training data, and the forth as the test data. At first, the first data set was presented to an empty network, and net_1 was built. Then the second data set was presented to net_1, and net_2 was built. At last the third data set was presented to net_2, and net_3 was built. We do this experiment in two ways, one is using our Condensed ETIC algorithm, and the other is using traditional way. We also use the Reduced ETIC algorithm to reduce the size of the final network. The results are listed in Table 3, and show that the generalization performance becomes better and better when new training data sets are available. And the Reduced ETIC algorithm can prune redundant samples in the network.

4   Conclusions

In this paper we have presented a novel algorithm, called ETIC (Enhanced Threshold Incremental Check), which can select representative samples from

Table 3. Results of industry images classification. Parameters of each net: $\lambda = 1$; $\theta^+ = 0.01$; $\theta^- = -0.01$; $\theta_e^+ = 0.5$; and $\theta_e^- = -0.5$. The net_4 denotes the network built up according to Reduced ETIC algorithm from net_3. The unit of 'Time' is s.

Net	With ETIC	Accuracy	Unknown	False	Size	Time	Size Ratio	Time Ratio
net_1	Y	69.17%	0.08%	30.74%	33	20.3	2.87%	6.49%
	N	69.17%	0.00%	30.83%	1149	313.3		
net_2	Y	88.05%	0.0%	11.95%	550	119.4	24.05%	18.27%
	N	86.63%	0.33%	13.03%	2287	653.5		
net_3	Y	88.30%	0.25%	11.45%	1870	1148.6	54.68%	52.49%
	N	86.97%	0.50%	12.53%	3420	2188.2		
net_4	Y	88.30%	0.25%	11.45%	1764	1087.3	51.58%	49.69%
	N	86.97%	0.50%	12.53%	3420	2188.2		

new training data set and can prune redundant modules of an already trained M^3-GZC network. Using ETIC, the M^3-GZC network still has the locally tuned response characteristic and emergent incremental learning ability, and its size will not increase if there is no new knowledge presented. Several experimental results indicate that our ETIC algorithm can reduces the size of the network and the response time while maintaining the generalization performance.

References

1. Lu, B.L., Ito, M.: Task Decomposition Based on Class Relations: a Modular Neural Network Architecture for Pattern Classification. Lecture Notes in Computer Science. Vol. 1240. Springer (1997) 330-339
2. Lu, B.L., Ito, M.: Task Decomposition and Module Combination Based on Class Relations: A Modular Neural Network for Pattern Classification. IEEE Trans. Neural Networks, **10** (1999) 1244-1256
3. Lu, B.L., Ma, Q., Ichikawa,M., Isahara, H.: Efficient Part-of-Speech Tagging with a Min-Max Modular Neural Network Model. Applied Intelligence, **19** (2003) 65-81
4. Lu, B.L., Shin, J., Ichikawa, M.: Massively Parallel Classification of Single-Trial EEG Signals Using a Min-Max Modular Neural Network. IEEE Trans. Biomedical Engineering, **51** (2004) 551-558
5. Lu, B.L., Ichikawa, M.: Emergent On-Line Learning with a Gaussian Zero-Crossing Discriminant Function. IJCNN '02, **2** (2002) 1263-1268
6. Hart, P.E.: The Condensed Nearest Neighbor Rule. IEEE Trans. Information Theory, **14** (1968) 515-516
7. Gates, G.W.: The Reduced Nearest Neighbor Rule. IEEE Trans. Information Theory, **18** (1972) 431-433
8. Murphy, P.M., Aha, D.W.: UCI Repository of Machine Learning Database. Dept. of Information and Computer Science, Univ. of Calif., Irvine (1994)

Parallel Feedforward Process Neural Network with Time-Varying Input and Output Functions

Shisheng Zhong, Gang Ding, and Daizhong Su

School of Mechanical and Electrical Engineering, Harbin Institute of Technology,
Harbin 150001, China
dingganghit@163.com

Abstract. In reality, the inputs and outputs of many complicated systems are time-varied functions. However, conventional artificial neural networks are not suitable to solving such problems. In order to overcome this limitation, parallel feedforward process neural network (PFPNN) with time-varied input and output functions is proposed. A corresponding learning algorithm is developed. To simplify the learning algorithm, appropriate orthogonal basis functions are selected to expand the input functions, weight functions and output functions. The efficiency of PFPNN and the learning algorithm is proved by the exhaust gas temperature prediction in aircraft engine condition monitoring. The simulation results also indicate that not only the convergence speed is much faster than multilayer feedforward process neural network (MFPNN), but also the accuracy of PFPNN is higher.

1 Introduction

Artificial neuron model is extremely simple abstraction of biological neuron. Since McCulloch and Pitts proposed MP neuron model [1] in 1943, artificial neural networks have received significant attention due to their successes in applications involving pattern recognition, function approximation, time series prediction, nonlinear system modeling, etc. Unfortunately, inputs and outputs of these artificial neural networks are discrete values. However, inputs and outputs of many complicated systems are continuous time-varied functions. To solve this problem, He and Liang proposed artificial process neuron model [2] in 2000. From a point view of architecture, process neuron is similar to conventional artificial neuron. The major difference is that the input, the output and the corresponding connection weight of process neuron are not discrete values but time-varied functions.

Process neural network is comprised of densely interconnected process neurons. A particularly important element of the design of process neural network is the choice of architecture. In general, multilayer feedforward architecture is widely adopted. But neural networks of this architecture converge slowly. Moreover, the accuracy of these neural networks is low. In this paper, parallel feedforward process neural network (PFPNN) is proposed to aim to solve these problems. To simplify the time aggregation operation of PFPNN, a learning algorithm based on expansion of orthogonal basis functions is developed. The effectiveness of the PFPNN and the learning algorithm is proved by exhaust gas temperature (EGT) prediction in aircraft engine condition monitoring. Comparative simulation test results highlight the property of the PFPNN.

J. Wang, X. Liao, and Z. Yi (Eds.): ISNN 2005, LNCS 3496, pp. 473–478, 2005.

spectively   as $x_i(t) = \sum_{l=1}^{K} a_{il} b_l(t)$, $\omega_{ij}(t) = \sum_{l=1}^{K} \omega_{ij}^{(l)} b_l(t)$ and $u_{ik}(t) = \sum_{l=1}^{K} u_{ik}^{(l)} b_l(t)$,

$a_{il}, \omega_{ij}^{(l)}, u_{ik}^{(l)} \in R$. According to the orthogonal property, $\int_0^T b_p(t) b_l(t) dt = \begin{cases} 1 & l = p \\ 0 & l \neq p \end{cases}$.

Thus, Equation (7) can be substituted by

$$y(t) = \sum_{k=1}^{K} g(\sum_{j=1}^{m} v_{jk} f(\sum_{i=1}^{n} \sum_{l=1}^{K} \omega_{ij}^{(l)} a_{il} - \theta_j^{(1)}) + \sum_{i=1}^{n} \sum_{l=1}^{K} u_{ik}^{(l)} a_{il} - \theta_k^{(2)}) b_k(t) \tag{8}$$

Suppose that $d(t)$ is the desired output function, and $y(t)$ is the corresponding actual output function of PFPNN. $d(t)$ can be expanded as $d(t) = \sum_{K=1}^{K} d_k b_k(t)$, $d_k \in R$. Then, the mean square error of PFPNN can be written as

$$E = \frac{1}{2} \sum_{k=1}^{K} (g(\sum_{j=1}^{m} v_{jk} f(\sum_{i=1}^{n} \sum_{l=1}^{K} \omega_{ij}^{(l)} a_{il} - \theta_j^{(1)}) + \sum_{i=1}^{n} \sum_{l=1}^{K} u_{ik}^{(l)} a_{il} - \theta_k^{(2)}) - d_k)^2 \tag{9}$$

For analysis convenience, Q_j and Z_k are defined respectively as

$$Q_j = \sum_{i=1}^{n} \sum_{l=1}^{K} \omega_{ij}^{(l)} a_{il} - \theta_j^{(1)}, \ Z_k = \sum_{j=1}^{m} v_{jk} f(\sum_{i=1}^{n} \sum_{l=1}^{K} \omega_{ij}^{(l)} a_{il} - \theta_j^{(1)}) + \sum_{i=1}^{n} \sum_{l=1}^{K} u_{ik}^{(l)} a_{il} - \theta_k^{(2)}$$

According to the gradient descent method, the learning rules are defined as follows

$$\begin{cases} \omega_{ij}^{(l)} = \omega_{ij}^{(l)} + \lambda \Delta \omega_{ij}^{(l)} \\ v_{jk} = v_{jk} + \lambda \Delta v_{jk} \\ u_{ik}^{(l)} = u_{ik}^{(l)} + \lambda \Delta u_{ik}^{(l)} \\ \theta_j^{(1)} = \theta_j^{(1)} + \lambda \Delta \theta_j^{(1)} \\ \theta_k^{(2)} = \theta_k^{(2)} + \lambda \Delta \theta_k^{(2)} \end{cases} \tag{10}$$

Where λ is learning rate, and $\Delta \omega_{ij}^{(l)}, \Delta v_{jk}, \Delta u_{ik}^{(l)}, \Delta \theta_j^{(1)}, \Delta \theta_k^{(2)}$ can be calculated as follows

$$\begin{cases} \Delta \omega_{ij}^{(l)} = -\frac{\partial E}{\partial \omega_{ij}^{(l)}} = -\sum_{k=1}^{K} (g(Z_k) - d_k) g'(Z_k) v_{jk} f'(Q_j) a_{il} \\ \Delta v_{jk} = -\frac{\partial E}{\partial v_{jk}} = -\sum_{k=1}^{K} (g(Z_k) - d_k) g'(Z_k) f(Q_j) \\ \Delta u_{ik}^{(l)} = -\frac{\partial E}{\partial u_{ik}^{(l)}} = -\sum_{k=1}^{K} (g(Z_k) - d_k) g'(Z_k) a_{il} \\ \Delta \theta_j^{(1)} = -\frac{\partial E}{\partial \theta_j^{(1)}} = \sum_{k=1}^{K} (g(Z_k) - d_k) g'(Z_k) f'(Q_j) a_{il} \\ \Delta \theta_k^{(2)} = -\frac{\partial E}{\partial \theta_j^{(1)}} = \sum_{k=1}^{K} (g(Z_k) - d_k) g'(Z_k) \end{cases} \tag{11}$$

5 Simulation Test

Aircraft engine operates under high temperature and speed conditions. Aircraft engine condition monitoring is essential in terms of safety. EGT is an important indicator of aircraft engine. With the limitation of the temperature, EGT must be monitored to judge the deterioration of the engine. EGT is influenced by many complicated factors and varied continuously with time. In this paper, the prediction of EGT by the PFPNN is presented.

The EGT data used was taken from the aircraft of Air China. The aircraft engine type is JT9D-7R4E, its serial number is 716928. The data was measured from Jan. 4, 2000 to Dec. 12, 2000, and the sampling interval is about 7 days. We get a EGT sequence with 44 discrete values such as $\{EGT_j\}_{j=1}^{44}$. $(EGT_i, \cdots, EGT_{i+5})$ $(i=1,\cdots,33)$ can be used to generate i-th input function IF_i by nonlinear least-squares method, and the output function OF_i corresponding to IF_i can be generated by $(EGT_{i+6}, \cdots, EGT_{i+11})$. Thus, we can get 33 couples of samples such as $\{IF_i, OF_i\}_{i=1}^{33}$. $\{IF_i, OF_i\}_{i=1}^{32}$ are selected to train PFPNN. The topological structure of PFPNN is $1-10-6-1$. The error goal is set to 0.01, and the learning rate is 0.5, the max iteration number is 10000. After 1247 iterations, PFPNN has converged. $\{IF_{33}, OF_{33}\}$ is selected to test PFPNN. Test results are shown in Table 1.

Table 1. Simulation test results of PFPNN

desired value(°C)	actual value(°C)	absolute error(°C)	relative error(%)
34.5000	34.9798	0.4798	1.39
35.4000	35.0146	0.3854	1.09
35.9000	36.1383	0.2383	0.66
35.7000	35.0129	0.6871	1.92
37.1000	36.0380	1.0620	2.86
36.6000	34.5424	2.0576	5.62

To verify the property of PFPNN, multilayer feedforward process neural network (MFPNN) is trained by $\{IF_i, OF_i\}_{i=1}^{32}$. After 2351 iterations, the MFPNN has converged. $\{IF_{33}, OF_{33}\}$ is selected to test MFPNN. Test results are described in Table 2.

Table 2. Simulation test results of MFPNN

desired value(°C)	actual value(°C)	absolute error(°C)	relative error(%)
34.5000	33.9409	0.5591	1.6206
35.4000	34.7489	0.6511	1.8393
35.9000	37.3486	1.4486	4.0351
35.7000	34.5785	1.1215	3.1415
37.1000	34.3323	2.7677	7.4601
36.6000	32.8492	3.7508	10.2481

The test results as shown in Table (1) and Table (2) indicate that PFPNN has a faster convergence speed and higher accuracy than MFPNN.

6 Conclusion

A PFPNN model is proposed in this paper, which has a typical feedforward architecture with parallel links between input layer and all hidden layers. A corresponding learning algorithm is given. To simplify the learning algorithm, orthogonal basis functions are selected to expand the input functions, output functions and weight functions. The efficiency of PFPNN and the learning algorithm is proved by the EGT prediction in aircraft engine condition monitoring. The simulation test results also indicate that PFPNN has two merits. Firstly, since its inputs and outputs are time-varied functions, PFPNN can expand the application field of the artificial neural networks. Secondly, compared to the same scale MFPNN model, PFPNN model has a faster convergence speed and higher accuracy.

Acknowledgement

This study was supported by the National Natural Science Foundation of China under Grant No.60373102.

References

1. McCullon, W., Pitts, W.: A Logical Calculus of the Ideas Immanent in Nervous Activity. Bulletin of Mathematical Biophysics, 5 (1943) 115-133
2. He X.G., Liang J.Z.: Some Theoretical Issues on Process Neural Networks. Engineering Science, 2 (2000) 40-44
3. Jeffreys, H., Jeffreys, B.S.: Methods of Mathematical Physics. 3rd edn. Cambridge University Press, Cambridge (1988) 446-448

A Novel Solid Neuron-Network Chip Based on Both Biological and Artificial Neural Network Theories

Zihong Liu[1,2], Zhihua Wang[1,2], Guolin Li[1], and Zhiping Yu[2]

[1] Circuits and Systems Division, Department of Electronic Engineering,
Tsinghua University, Beijing 100084, China
Liuzihong00@mails.tsinghua.edu.cn
[2] Institute of Microelectronics, Tsinghua University, Beijing 100084, China
zhihua@tsinghua.edu.cn

Abstract. Built on the theories of biological neural network, artificial neural network methods have shown many significant advantages. However, the memory space in an artificial neural chip for storing all connection weights of the neuron-units is extremely large and it increases exponentially with the number of neuron-dentrites. Those result in high complexity for design of the algorithms and hardware. In this paper, we propose a novel solid neuron-network chip based on both biological and artificial neural network theories, combining semiconductor integrated circuits and biological neurons together on a single silicon wafer for signal processing. With a neuro-electronic interaction structure, the chip has exhibited more intelligent capabilities for fuzzy control, speech or pattern recognition as compared with conventional ways.

1 Introduction

Neural Network (NN) refers to Biological Neural Network (BNN) [1] which processes signals with groups of neurons in the realm of neurophysiology or biology. While in electronics or informatics, NN means Artificial Neural Network (ANN) that works on analog/digital circuit cells in which the mechanism for signal processing is similar to that of biological neurons [2]. In the past several decades, ANN VLSI chips have functioned in many aspects, e.g. image manipulation, pattern recognition, and smart information processing [3]-[7], with their special advantages including self-adapting, self-study, large scale parallel computation, information distributed expressing and error tolerance [7].

However, ANN chips rely on VLSI semiconductor technology while the kernel is the neural cell [7][8]. As exemplified in Fig. 1, the memory space in an ANN chip for storing all connection weights (W_i) is quite large and it increases with the number of neuron-dentrites as an exponential function. What's more, the algorithms for signal processing with ANN are often perplexing [7]. Thus, ANN has shown poorer capabilities in speech or image recognition as compared with BNN in the brain.

To improve the performances of those ANN chips for signal processing often confined by the problems mentioned above, we have proposed and built a new solid silicon-neuron interaction chip in this paper. Recorded simulation results demonstrate that utilizing both the live biological neuron-network and integrated circuits simultaneously will efficiently reduce the design complexity for algorithms and also solve the problems of huge memory space for traditional ANN chips.

J. Wang, X. Liao, and Z. Yi (Eds.): ISNN 2005, LNCS 3496, pp. 479–484, 2005.
© Springer-Verlag Berlin Heidelberg 2005

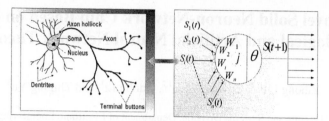

Fig. 1. Illustration of a simplified neuron-cell in ANN chips. $S_{j,i}(t)$ *(i=0,1,2,...n; j⊔N+)* are the input signals to the *n* dentrites of neuron(*j*), $S_j(t+1)$ is the corresponding output, W_i denotes the connection weight, θ represents the action threshold

2 Proposed Chip Working Principles

The basic working principles of the chip can be summarized as "(1) Signal Pre-Processing -> (2) Signal Training Operation -> (3) Signal Post Processing". Among these three blocks, both (1) and (3) comprise dedicated integrated circuits (IC) while (2) consists of cultured BNN and functioning microelectrode-array. As illustrated in Fig.2, raw input signals to be processed are transmitted to (1) that realizes signal transformation, amplitude modulation, and contains static RAM. The output signals from (1) will then stimulate the cultured neurons in (2), where the microelectrode-array for both stimulating BNN and recording the action potentials of the neurons is built on the specific silicon well substrate. The third block, (3), is designed for processing the recorded electrical activities of BNN in (2). Actually, they are fed back to the input of (2) for training the BNN. This step is often called *self-study*.

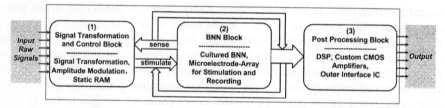

Fig. 2. Illustration of the proposed chip working principles

3 Chip Architecture and Blocks Design

Fig. 3 shows the schematic view of the chip architecture corresponding to Fig. 2.

3.1 IC Blocks: (1) and (3)

As for a certain input raw signal series, it is necessary to complete amplitude transformation in advance so as to reach the optimal stimulation effect for those neurons in Block-(2). The modulation coefficient can be set by users with trials. Fig. 4 shows the schematic circuit diagram for designation of a certain microelectrode to stimulate or sense the neurons [9]. Here we represent the iterative part as *"Unit"*. The SRAM is controlled by Reset, Address, and Stimulation/Sensing buses, while its output determines the microelectrode to stimulate or sense the right neurons. In our design, each chip contains N *"Unit"* (N is a natural number) in Block-(1), where the Address bus enables necessary units.

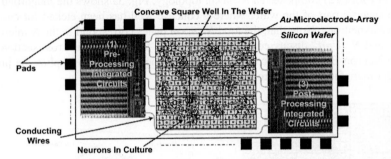

Fig. 3. Schematic representation: Top view of the chip prototype architecture

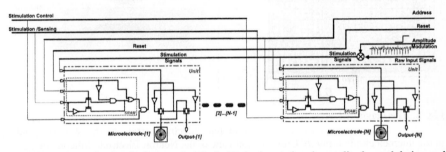

Fig. 4. The schematic circuit diagram in Block-(1) for input signals amplitude modulation and designation of a certain microelectrode to stimulate or sense the right neurons in Block-(2)

As mentioned above, the recorded electrical activities of BNN in Block-(2) will be finally sent to Block-(3) for post processing and further applications. Block-(3) mainly contains custom CMOS amplifier [9], DSP, and I/O interface IC. Some useful signal analysis methods are partly discussed in previous works [10][11].

3.2 BNN and Microelectrode-Array: Block-(2)

Before the implementation of the chip, we need to demonstrate the feasibility of culturing BNN in the specific well on the silicon wafer. As early as in 1986, Weiss S et al. [12] reported that striatal neurons of the fetal rats were successfully cultured in serum-free medium for 14 to 21 days, and synaptogenesis was seen when about day 9. In 1993, Stenger et al. [13] demonstrated that hippocampal neurons were able to grow on the substrate spread with poly-lysine and self-organize to monolayer. Moreover, rat spinal cord neurons have been cultured in defined medium on microelectrode arrays while only 1.1±0.5% astrocyte mixed [14]. Recently, new approaches to neural cell culture for long-term studies have also been explored [15]. Based on these successes, rat spinal cord neurons are being cultured without serum in a specific part

of the silicon wafer where a concave square well has been etched before implementation of Block-(1)(3) and the well was spread with poly-lysine at the bottom as shown
in Fig. 3, thus, BNN would be constructed spontaneously in about 10 days.

As seen from the top view of the chip architecture in Fig. 3, the microelectrode-
array in Block-(2) composes an $N=m×m$ topology. Fig. 5a shows the magnified view
of the specific part of the wafer where a concave well has been etched for culturing
biological neurons. Note that, the total $N=m×m$ panes correspond to the N microelec-
trodes and N "Unit" discussed in Section 3.1. Fig. 5b sketches the cross section view
of the Neuron-Electrode-Silicon interaction in Block-(2). It also implies the process
flow for manufacturing the proposed chip.

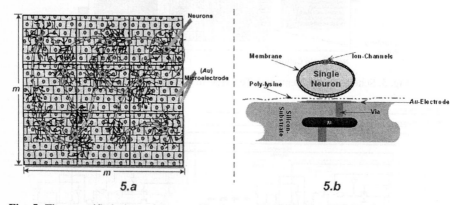

Fig. 5. The magnified view of the specific part of the wafer where a concave well has been
etched for culturing biological neurons on the microelectrode array. *a:* The $N=m×m$ topology
of the microelectrode-array. *b:* Cross section view of the Neuron-Electrode-Silicon interaction

4 Simulation Results

The purpose of the proposed chip operation is to realize easier and faster pattern recognition initially. In order to reach this, the essential thought is that under continuous
feedback of the responses of the cultured BNN, different input signals to (1) with the
same characteristic will generate the same output mode from (2). As illustrated in Fig.
5a, we apply the outputs of Block-(1) to the four electrodes in the square corners
denoted with solid circles. Thus, each microelectrode in the $(m-2)×(m-2)$ area will
record independent electrical activities. Fig. 6a marks the stimulation and recording
sites, Fig. 6b and 6c show the recorded simulation results while $m=5$ and the raw
input is a square wave for 6b and acoustic signal for 6c [16]. As seen from the input-
output comparison, the predictions above are primarily testified.

5 Conclusions

A novel solid neuron-network chip prototype based on both biological and artificial
neural network theories has been proposed and designed in detail in this paper. Combining semiconductor IC and biological neurons (BNN) together on a single silicon
wafer for signal processing, the chip has primarily exhibited more intelligent capabili-

ties for fuzzy control, speech or pattern recognition as compared with conventional ANN ways while it is with a neuro-electronic interaction architecture.

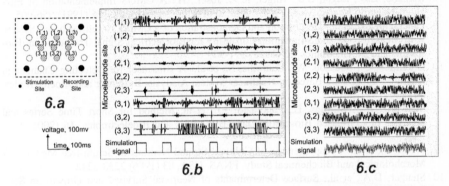

Fig. 6. Simulations for the proposed chip. *a:* Stimulation and recording electrode sites. *b:* A square wave acting as the raw input. *c:* An acoustic signal acting as the raw input

Actually, due to the BNN *self-study* and continuous feedback effect, once enough, usually two or more different signals with the same characteristic were sent to the chip, the way of synaptogenesis among the neurons cultured on the microelectrode-array in the specific square silicon well contained in the chip will vary into a fixed mode and be detected from the responses of the BNN. Hence, an easier and faster pattern recognition or fuzzy signals recognition is obtained. It simultaneously efficiently reduces the design complexity for algorithms and also resolves the problems of huge memory space for traditional ANN chips.

Acknowledgments

This work was partly supported by the National Natural Science Foundation of China (No. 60475018, No. 60372021) and National Key Basic Research and Development Program (No. G2000036508).

References

1. Nicholls, J.G., Martin, A.R., Wallace, B.G., Fuchs, P.A.: From Neuron to Brain. Sinauer Associates, Sunderland, Mass (2001)
2. Lewis, E.R.: Using Electronic Circuits to Model Simple Neuroelectric Interactions. IEEE Proceedings, **56** (1968) 931-949
3. Zhou, W.X., Jin, D.M., Li, Z.J.: Review of The Research on Realization of Analog Neural Cells. Research & Progress of Solid State Electronics, **22** (2002) 268-279
4. Bass, S.A., Bischoff, A.C., Hartnack, et al.: Neural Networks for Impact Parameter Determination. Journal of Physics G-Nuclear and Particle Physics, **20** (1994) 21-26
5. Tambouratzis, T.: A Novel Artificial Neural Network for Sorting. IEEE Transactions on Systems Man and Cybernetics Part B-Cybernetics, **29** (1999) 271-275
6. Acir, N., Guzelis, C.: Automatic Recognition of Sleep Spindles in EEG by Using Artificial Neural Networks. Expert Systems with Applications, **27** (2004) 451-458

7. Jain, L., Fanelli, A.M.: Recent Advances in Artificial Neural Networks: Design and Applications. CRC Press, Boca Raton (2000)
8. Bernabe, L.B., Edgar, S.S., Angel, R.V., JoseL, H.: A CMOS Implementation of Fitz-Hugh-Nagumo Neuron Model. IEEE Journal of Solid-State Circuits, 26 (1991) 956-965
9. Pancrazio, J.J., et al.: Description and Demonstration of a CMOS Amplifier-based System with Measurement and Stimulation Capability for Bioelectrical Signal Transduction. Biosensors and Bioelectronics, 13 (1998) 971-979
10. Gholmieh, G., Soussou, W., Courellis, S., et al.: A Biosensor for Detecting Changes in Cognitive Processing Based on Nonlinear System Analysis. Biosensor and Bioelectronics, 16 (2001) 491-501
11. Schwartz, I.B., Billings, L., Pancrazio, J.J., et al: Method for Short Time Series and Analysis of Cell-based Biosensors Data. Biosensors and Bioelectronics, 16 (2001) 503-512
12. Weiss, S., et al.: Synaptogenesis of Cultured Striatal Neurons in Serum-free Medium: A Morphological and Biochemical Study. PNAS. USA, 83 (1986) 2238-2242
13. Stenger, D.A., et al.: Surface Determinants of Neuronal Survival and Growth on Self-assembled Monolayers in Culture. Brain Research, 630 (1993) 136-147
14. Manos, P. et al.: Characterization of Rat Spinal Cord Neurons Cultured in Defined Media on Microelectrode Arrays. Neuroscience Letters, 271 (1999) 179-182
15. Potter, S.M., DeMarse, T.B.: A New Approach to Neural Cell Culture for Long-term Studies. Journal of Neuroscience Methods, 110 (2001) 17–24
16. Liu, Z.H., Wang, Z.H., Li, G.L, Yu, Z.P., Zhang, C.: Design Proposal for a Chip Jointing VLSI and Rat Spinal Cord Neurons on a Single Silicon Wafer. 2nd IEEE/EMBS International Conference on Neural Engineering Proceedings (2005) in Print

Associative Memory Using Nonlinear Line Attractor Network for Multi-valued Pattern Association

Ming-Jung Seow and Vijayan K. Asari

Department of Electrical and Computer Engineering
Old Dominion University, Norfolk, VA 23529
{mseow,vasari}@odu.edu

Abstract. A method to embed P multi-valued patterns $x^s \in \mathbb{R}^N$ into a memory of a recurrent neural network is introduced. The method represents memory as a nonlinear line of attraction as opposed to the conventional model that stores memory in attractive fixed points at discrete locations in the state space. The activation function of the network is defined by the statistical characteristics of the training data. The stability of the proposed nonlinear line attractor network is investigated by mathematical analysis and extensive computer simulation. The performance of the network is benchmarked by reconstructing noisy gray-scale images.

1 Introduction

An associative memory model can be formulated as an input – output system. The input to the system can be an N dimensional vector $x \in \mathbb{R}^N$ called the memory stimuli, and the output can be an L dimensional vector $y \in \mathbb{R}^L$ called the memory response. The relationship between the stimuli and the response is given by $y = f(x)$, where $f : \mathbb{R}^N \mapsto \mathbb{R}^L$ is the associative mapping of the memory. Each input – output pair or memory association (x, y) is said to be stored and recalled in the memory.

Designing the dynamic of the recurrent network as associative memories has been one of the major research areas in the neural networks since the pioneering work by John Hopfield [1]. This work demonstrated that a single-layer fully connected network is capable of restoring a previously learned pattern by converging from initially corrupted or partial information toward a fixed point. However, fixed-point attractor may not be suitable for patterns which exhibit similar characteristics as shown in [2-5]. As a consequence, it may be more appropriate to represent memory as a line of attraction using a nonlinear line attractor network [6].

In this paper, we propose a learning model for recurrent neural network capable of recalling memory vectors $\mathbb{R}^N$. The solution of the system is based on the concept of nonlinear line attractor, which the network encapsulate the fixed point attractor in the state space as a nonlinear line of attraction. The stability of the network is proved mathematically and by extensive computer simulation. The performance of the network is tested in reconstructing noisy gray-scale images.

J. Wang, X. Liao, and Z. Yi (Eds.): ISNN 2005, LNCS 3496, pp. 485–490, 2005.
© Springer-Verlag Berlin Heidelberg 2005

2 Learning Model

The relationship of each neuron with respect to every other neuron in a N neuron network can be expressed as a k-order polynomial function

$$y_i^s = \frac{1}{N} \sum_{j=1}^{N} \left[\sum_{v=0}^{k} w_{(v,ij)}^s \left(x_j^s \right)^v \right] \quad \text{for } 1 \leq i \leq N \tag{1}$$

for stimulus-response pair $\left(x_j^s, y_i^s \right)$ corresponding to the sth pattern. The resultant memory w^s can be expressed as

$$w_m^s = \begin{pmatrix} w_{(m,11)}^s & \cdots & w_{(m,1N)}^s \\ \vdots & \ddots & \vdots \\ w_{(m,N1)}^s & \cdots & w_{(m,NN)}^s \end{pmatrix} \quad \text{for } 0 \leq m \leq k \tag{2}$$

where w_m^s is the mth order term of the w^s. To combine w^s for P patterns to form a memory matrix w; we need to utilize statistical methods. The least squares estimation approach to this problem can determine the best fit line when the error involved is the sum of the squares of the differences between the expected outputs and the approximated outputs. Hence, the weight matrix must be found that minimizes the total least squares error

$$E\left(w_{(0,ij)}, w_{(1,ij)}, \cdots w_{(k,ij)} \right) = \sum_{s=1}^{P} \left[y_i^s - p_k \left(x_j^s \right) \right]^2 \quad \text{for } 1 \leq i, j \leq N \tag{3}$$

where

$$p_k \left(x_j^s \right) = \sum_{v=0}^{k} w_{(v,ij)} \left(x_j^s \right)^v \tag{4}$$

A necessary condition for the coefficients $w_{(0,ij)}, w_{(1,ij)},...,w_{(k,ij)}$ to minimize the total error E is that

$$\frac{dE}{dw_{(m,ij)}} = 0 \quad \text{for each } m = 0,1,\cdots,k \tag{5}$$

The coefficient $w_{(0,ij)}, w_{(1,ij)},...,w_{(k,ij)}$ can hence be obtained by solving the above equations. Fig. 1 shows the weight graph illustrating the concept of training based on the above theory. A weight graph is a graphical representation of the relationship between the ith neuron and the jth neuron for P patterns. Utilization of the weight graph can help visualize the behavior of one neuron pair.

The network architecture is a single layer fully connected recurrent neural network. The dynamics of the network can be computed as

$$x_i(t+1) = \frac{1}{N} \sum_{j=1}^{N} \left[x_i(t) + \alpha \Delta x_{ij} \right] \quad \text{for } 1 \leq i \leq N \tag{6}$$

where α is the update rate ranging in $0 < \alpha \leq 1$ and Δx_{ij} is the different between the approximated output and the actual output:

$$\Delta x_{ij} = \left\{ f \left\{ \sum_{v=0}^{k} w_{(v,ij)} x_j^v(t) \right\} - x_i(t) \right\} \tag{7}$$

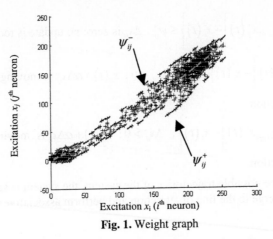

Fig. 1. Weight graph

where $f(.)$ is the activation function and it can be found by considering the distance between the approximated output and the actual output shown in the weight graph illustrated in Fig. 1. That is, in order to consider each pattern, we need to find the region where the threshold can encapsulate each pattern. Mathematically the activation function can be expressed as

$$f\left\{\sum_{v=0}^{k} w_{(v,ij)} x_j^v(t)\right\} = \begin{cases} x_i(t) & \text{if } \psi_{ij}^- \leq \left\{\left[\sum_{v=0}^{k} w_{(v,ij)} x_j^v(t)\right] - x_i(t)\right\} \leq \psi_{ij}^+ \\ \sum_{v=0}^{k} w_{(v,ij)} x_j^v(t) & \text{otherwise} \end{cases} \tag{8}$$

where

$$\psi_{ij}^- = \left\{\psi_{(1,ij)}^-, \psi_{(2,ij)}^-, \cdots, \psi_{(\Omega,ij)}^-\right\} \quad \text{for } 1 \leq i, j \leq N \tag{9}$$

$$\psi_{(l,ij)}^- = \min_{\forall s}\left[\left\{\left[\sum_{v=0}^{k} w_{(v,ij)}\left(x_j^s\right)^v\right] - x_i^s\right\}, \quad (l-1)\frac{L}{\Omega} \leq x_j^s < l\frac{L}{\Omega}\right] \quad 1 \leq l \leq \Omega \tag{10}$$

$$\psi_{ij}^+ = \left\{\psi_{(1,ij)}^+, \psi_{(2,ij)}^+, \cdots, \psi_{(\Omega,ij)}^+\right\} \quad \text{for } 1 \leq i, j \leq N \tag{11}$$

$$\psi_{(l,ij)}^+ = \max_{\forall s}\left[\left\{\left[\sum_{v=0}^{k} w_{(v,ij)}\left(x_j^s\right)^v\right] - x_i^s\right\}, \quad (l-1)\frac{L}{\Omega} \leq x_j^s < l\frac{L}{\Omega}\right] \quad 1 \leq l \leq \Omega \tag{12}$$

where L is the number of levels in x_j^s and Ω is the number of threshold regions such that $1 \leq \Omega \leq L$.

3 Stability of the Nonlinear Line Attractor Network

Let us now examine the stability and associability of the nonlinear line attractor network based on equations 6 to 8 by considering three cases:

1. If $\psi_{ij}^- \leq \left\{\left[\sum_{v=0}^{k} w_{(v,ij)} x_j^v(t)\right] - x_i(t)\right\} \leq \psi_{ij}^+$, Δx_{ij} is zero: no update is required.

2. If $\left\{\left[\sum_{v=0}^{k} w_{(v,ij)} x_j^v(t)\right] - x_i(t)\right\} \leq \psi_{ij}^-$, $\Delta x_{ij} < 0$: $x_i(t) + \alpha \Delta x_{ij}$ is moving down towards the line of attraction.

3. If $\psi_{ij}^+ \leq \left\{\left[\sum_{v=0}^{k} w_{(v,ij)} x_j^v(t)\right] - x_i(t)\right\}$, $\Delta x_{ij} > 0$: $x_i(t) + \alpha \Delta x_{ij}$ is moving up towards the line of attraction.

Based on these three possible trajectories of the system, the system is said to be stable and is able to converge to the line of attraction and perform associative recall.

4   Experimental Result

In order to confirm the effectiveness of the proposed learning algorithm, gray-scale versions of three well-known images, namely camera man, Lena, and peppers, have been used in these experiments. These training images are shown on Fig. 2.

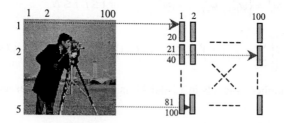

Fig. 2. Test images used in image reconstruction example

Due to computational simplicity, the original images of size 512×512 are scaled down to 100×100. In addition, each image has been divided into 500 20 dimensional vectors: we divided each image into sub-images of 1×20 (column × rows) as illustrated on Fig. 3 and trained 500 20-neurons network using the proposed method.

Fig. 3. Original image is divided into 500 20 dimensional vectors

After training the nonlinear line attractor network, distorted versions of the training image is obtained by adding 40% salt and pepper noise as shown in Fig. 4 (top). Each of these distorted images was divided as shown on Fig. 3 and was applied to their corresponding networks. After all 500 networks converges to their line of attraction, the reconstructed images was obtained and are shown on Fig. 4 (bottom). It can be seen that networks are capable of removing 40% salt and pepper noise on each image successfully. That is, none of these 500 networks diverges away from the nonlinear line of attraction.

Fig. 4. Images corrupted by 40% salt and pepper noise (top) and their reconstructed versions obtained by the nonlinear line attractor networks (bottom)

Fig. 5. 20% salt and pepper Lena image (left), reconstructed Lena image by generalized Hebb rule (middle), and reconstructed Lena image by nonlinear line attractor network (right)

The recall capability of the nonlinear line attractor is compared with the generalized Hebb rule proposed by Jankowski et. al. [8]. In this experiment, the three images shown in Fig. 2 were used for training the generalized Hebb rule similar to the fashion used for training the nonlinear line attractor network (Fig. 3). It can be seen from Fig. 5 (middle) that generalize Hebb rule is unable to reconstruct the image from its 20% distorted version (left), instead it converge to a spurious state. Our method, on the other hand, is able to recall Lena image with reasonable clarity (right).

5 Conclusion

A learning algorithm for the training of multi-valued neural network for pattern association is presented. The method represents memory as a nonlinear line of attraction.

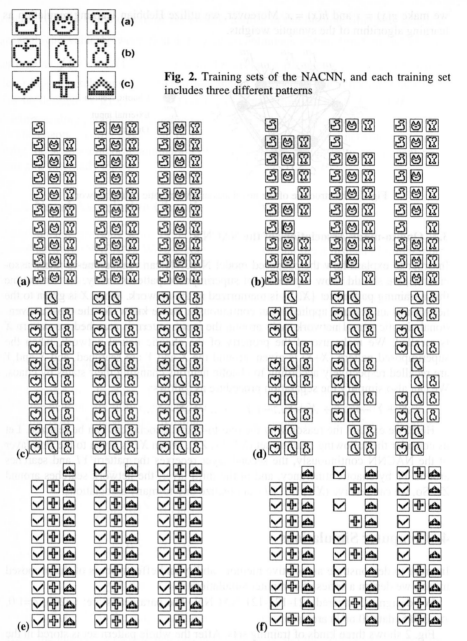

Fig. 2. Training sets of the NACNN, and each training set includes three different patterns

Fig. 3. The output of the different chaotic neural network while different inputs are given continuously: (a), (c) and (e) are output of the proposed NCANN; (b), (d) and (f) are output of the conventional chaotic network, respectively. (a) and (b) show the first pattern of training set (a) of Fig.4 are given to the first layers ; (c) and (d) show the second pattern of training set (b) of Fig.4 are given to the second layers; (e) and (f) show the third pattern of training set (c) of Fig.4 are given to the third layers.

From fig.3 we can find successful recalling times of proposed model reach 29, 28 and 28 times in 30 times outputs, successful rate reach 99.67%, 93.33% and 93.33%, respectively. However, successful recalling times of the conventional CNN are 22, 10 and 15 times, successful rate are 73.33%, 33.33%, 50.00%, respectively. Clearly, the proposed model shows high-level capacity in associative memories.

5 Conclusions

In this paper, we propose a novel associative chaotic neural network with exponential decay spatio-temporal effect which has two advantages as following: (1) the association proceeding of the proposed model NACNN is more close to real world and (2) the proposed model can realize one-to-many association perfectly. A series of Computer simulations prove its effectiveness of the proposed method.

Acknowledgements

The work is partially supported by the Natural Science Foundation of Southwest China Normal University (Grant no. SWNUQ2004024), the National Natural Science Foundation of China (Grant no. 60271019), the Natural Science Foundation of Chongqing (Grant no. 20020611007), and the Post-doctoral Science Foundation of China and the Natural Science Foundation of Chongqing.

References

1. Aihara, K., Takabe, T., Toyoda, M.: Chaotic Neural Networks. Physics Letters A, **144** (1990) 333 –340
2. Yao, Y., Freeman, W.J.: Model of Biological Pattern Recognition with Spatially Chaotic Dynamics Neural Networks. Neural Networks, USA, **3** (1990) 153-170
3. Osana, Y., Hagiwara, M.: Separation of Superimposed Pattern and Many-to-Many Associations by Chaotic Neural Networks. Proceedings IEEE International Joint Conference on Neural Networks, 1 Anchorage (1998) 514-519
4. Kaneko, K.: Clustering, Coding, Switch, Hierarchical Ordering and Control in a Network of Chaotic Elements. Physics Letters D, **41** (1990) 137-172
5. Ishii, S., Fukumizu, K., Watanabe, S.: A Network of Chaotic Elements for Information Processing. Neural Networks, **1** (1996) 25-40
6. Osana, Y., Hattori, M., Hagiwara, M.: Chaotic Bidirectional Associative Memory. International Conference on Neural Networks, **2** Houston (1997) 816 –821
7. Osana, Y., Hagiwara, M.: Successive Learning in Chaotic Neural Network. IEEE International Joint Conference on Neural Networks Proceedings, Anchorage, **2** (1998) 1510-1515
8. Kawasaki, N., Osana, Y., Hagiwara, M.: Chaotic Associative Memory for Successive Learning Using Internal Patterns. 2000 IEEE International Conference on Systems, Man, and Cybernetics, **4** (2000) 2521 – 2526
9. Osana, Y: Improved Chaotic Associative Memory Using Distributed Patterns for Image Retrieval. 2003 Proceedings of the International Joint Conference on Neural Networks, **2** (2003) 846 - 851
10. Liu, G.Y., Duan, S.K.: A Chaotic Neural Network and its Applications in Separation Superimposed Pattern and Many-to-Many Associative Memory. Computer Science, Chongqing, China, **30** (2003) 83-85

11. Duan, S.K., Liu, G.Y., Wang, L.D., Qiu, Y.H.: A Novel Chaotic Neural Network for Many-to-Many Associations and Successive Learning. IEEE International Conference on Neural Networks and Signal Processing, Vol. 1. Nanjing, China (2003) 135-138
12. Wang, L.D., Duan, S.K.: A Novel Chaotic Neural Network for Automatic Material Ratio System. 2004 International Symposium on Neural Networks, Dalian, China. Lecture Notes in Computer Science-ISNN2004, Vol. II. Springer-Verlag, Berlin Heidelberg New York (2004) 813-819

On a Chaotic Neural Network with Decaying Chaotic Noise

Tianyi Ma[1], Ling Wang[1], Yingtao Jiang[2], and Xiaozong Yang[1]

[1] Department of Computer Science and Engineering,
Harbin Institute of Technology, Harbin, Heilongjiang 15000, China
{mty,lwang}@ftcl.hit.edu.cn
[2] Department of Electrical Engineering and Computer Science,
University of Nevada, Las Vegas, NV, USA 89147
yingtao@egr.unlv.edu

Abstract. In this paper, we propose a novel chaotic Hopfield neural network (CHNN), which introduces chaotic noise to each neuron of a discrete-time Hopfield neural network (HNN), and the noise is gradually reduced to zero. The proposed CHNN has richer and more complex dynamics than HNN, and the transient chaos enables the network to escape from local energy minima and to settle down at the global optimal solution. We have applied this method to solve a few traveling salesman problems, and simulations show that the proposed CHNN can converge to the global or near global optimal solutions more efficiently than the HNN.

1 Introduction

Since the Hopfield model [1] was proposed in 1985, it has been extensively applied to solve various combinatorial optimization problems. The advantage of this approach in optimization is that it exploits the massive parallelism of a neural network. However, results obtained are not satisfactory. The reason is that Hopfield neural network (HNN) is a stable system with gradient descent mechanism. Using HNN to solve combinatorial optimization problems often has the following three shortcomings. First, an HNN can often be trapped at a local minimum on the complex energy terrain. Second, an HNN may not converge to a valid solution. Third, an HNN sometimes does not converge at all within a prescribed number of iterations.

To overcome the shortcomings of the HNN described above, chaos or nonlinear dynamics have been introduced to prevent the HNN from being trapped at local minima. There are currently two major classes of chaotic neural network (CNN).

One is internal approach, where chaotic dynamics is generated within the network controlled by some bifurcation parameters [2], [3]. Based on chaotic properties of biological neurons, Chen and Aihara [2] recently proposed chaotic simulated annealing by a neural network model with transient chaos. By adding a negative of self-coupling to a network model and gradually removing this negative self-coupling, the transient chaos have been used for searching and self-organizing, therefore reaching global optimal solutions. Wang and Smith [3] developed a method by varying the timestep of an Euler discretized HNN. This approach eliminates the need to carefully select other system parameters.

J. Wang, X. Liao, and Z. Yi (Eds.): ISNN 2005, LNCS 3496, pp. 497–502, 2005.

The other class contains CNN models employing an external approach, where an externally generated chaotic signal is added to the network as perturbation [4], [5], [6]. Zhou *et al.* [4] added a chaotic time series generated by the Henon map into the HNN. Hayakawa *et al.* [5] and Yuyao He [6] used the logistic map with different values of the bifurcation parameter as noise. Recently Wang [7] *et al.* proposed a noisy chaotic neural network by adding noise to Chen and Aihara's transiently chaotic neural network.

In this paper, a chaotic neural network is proposed by adding chaotic noise to each neuron of the discrete-time Hopfield neural network (HNN) and gradually reducing the noise to zero. The chaotic neural network can converge to discrete-time Hopfield neural network by adjusting the bifurcation parameter of chaotic noise. We apply this method to solve the four cities and ten cities traveling salesman problems, and the simulation results show that the proposed chaotic neural network can converge to the global minimum or their near global optimal solutions more efficiently than the HNN.

2   Chaotic Neural Network with Decaying Chaotic Noise

The chaotic neural network proposed in this paper is defined as follows:

$$x_{ij}(t) = \frac{1}{1 + e^{-y_{ij}(t)/\varepsilon}} . \tag{1}$$

$$y_{ij}(t+1) = ky_{ij}(t) + \alpha(\sum_{i=1}^{n}\sum_{j=1}^{n}\sum_{k=1}^{n}\sum_{il \neq j}^{n} w_{ijkl}x_{kl}(t) + I_{ij}) + \gamma z_{ij}(t) . \tag{2}$$

Where $i, j = 1, 2, \cdots n$, $x_{ij}(t)$: output of neuron ij; $y_{ij}(t)$: internal state of neuron ij; I_{ij}: input bias of neuron ij; k: damping factor of the nerve membrane; α: positive scaling parameter for the inputs; γ: positive scaling parameter for the chaotic noise; $z_{ij}(t)$: chaotic noise for neuron ij; w_{ijkl}: connection weight from neuron kl to neuron ij; ε: steepness parameter of the neuronal output function ($\varepsilon > 0$)

The chaotic noise can be generated from the following chaotic map [8]:

$$z_{ij}(t+1) = a(t)z_{ij}(t)(1 - z_{ij}(t)^2) \tag{3}$$

Fig.1 shows that the system can vary from chaotic state to the stable point by gradually reducing the bifurcation parameter $a(t)$. According to Fig.1, we know that $a(t) \in [0,1]$, $z_{ij}(t) = 0$. Therefore, we can regard this chaotic system as chaotic noise of discrete-time HNN and adjust bifurcation parameter such that chaotic neural network converges to discrete-time Hopfield neural network with decaying chaotic noise. In order to prevent the chaotic dynamics from vanishing quickly to realize sufficient chaotic searching, we use two different decaying value of parameter $a(t)$ such that $z_{ij}(t)$ searches sufficiently in chaotic space with relatively slower speed.

$$a(t+1) = (1 - \beta_1)a(t) + \beta_1 a_0 \quad \text{When } a(t) > a_1 . \tag{4}$$

$$a(t+1) = (1-\beta_2)a(t) + \beta_2 a_0 \quad \text{When } a(t) \le a_1 . \tag{5}$$

Where $a_0 \in [0,1]$, $0 < \beta_1 \le \beta_2 < 1$, a_1 is selected to make the dynamics system (3) be the process from chaotic state to stable state.

In the above model, the chaotic noise is assigned to each neuron, and the noise generators for the neurons operate with the same rule but different initial values.

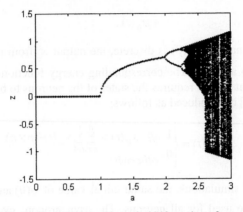

Fig. 1. Chaos and double period bifurcation of map

3 Application to the Traveling Salesman Problem

Traveling Sales Problem (TSP) is a classical combinatorial optimization problem. The goal of the TSP is to find the shortest route through n cities, that a person can visit each city once and only once, and return to the starting point. The exact solution is easily concluded for small n. As the number of possible solutions increases exponentially with n, it becomes difficult to find the best solution as the system size n increases. We adopt formulation for TSP by Hopfield and Tank [1] (1985). Namely, a solution of TSP with n cities is mapped to a network with $n \times n$ neuron. Assume $x_{ij} = 1$ to be the neuron output which represents to visit city i in visiting order j, whereas $x_{ij} = 0$ represents that city i is not visited in visiting order j. A computational energy function, minimizing the total tour length while simultaneously satisfying all constraints, has the following form:

$$E = \frac{W_1}{2}\left\{\sum_{i=1}^{n}(\sum_{j=1}^{n}x_{ij}-1)^2 + \sum_{j=1}^{n}(\sum_{i=1}^{n}x_{ij}-1)^2\right\} + \frac{W_2}{2}\sum_{i=1}^{n}\sum_{j=1}^{n}\sum_{k=1}^{n}(x_{kj+1}+x_{kj-1})x_{ij}d_{ik} \tag{6}$$

Where i) $x_{i0} = x_{in}$, ii) $x_{in+1} = x_{i1}$, iii) W_1 and W_2 are the coupling parameters corresponding to the constraints and the cost function of the tour length, respectively, and iv) d_{ij} is the distance between city i and city j.

According to the form of the energy function, we can obtain the connection weights of all neurons.

$$\sum_{k,l\neq i,j}^{n} w_{ijkl}x_{kl} + I_{ij} = -\frac{\partial E}{\partial x_{ij}} \tag{7}$$

From (2), (6) and (7), the dynamics of neuron for the TSP is derived as follows:

$$y_{ij}(t+1) = ky_{ij}(t) + \alpha\left\{-W_1\left(\sum_{l\neq j}^{n}x_{il}(t) + \sum_{k\neq i}^{n}x_{kj}(t)\right) - W_2\sum_{k\neq i}^{n}(x_{kj+1}(t) + x_{kj-1}(t))d_{ik} + W_1\right\} \tag{8}$$

$$+\gamma z_{ij}(t).$$

Although the dynamics of (8) is discrete, the output x_{ij} from (1) has a continuous value between zero and one. The corresponding energy function (6) is also continuous. Since a solution to TSP requires the states of the neurons to be either zero or one, a discrete output [2] is introduced as follows:

$$x_{ij}^{d}(t) = \begin{cases} 1 & \textit{iff } x_{ij}(t) > \sum_{k=1/l=1}^{n}\sum^{n} x_{kl}(t)/(n\times n) \\ 0 & \textit{otherwise} \end{cases}$$

In the following simulations, the same initial value of $a(0)$ and the decaying rule for $a(t)$ in (4)(5) are used for all neurons. The asynchronous cyclic updating of the neural network model is employed. One iteration corresponds to one cyclic updating of all neuron states and $a(t)$ is decreased by one step after each of iteration.

Table 1. Data on TSP with 10 cities

City	1	2	3	4	5
x-axis	0.4000	0.2439	0.1707	0.2293	0.5171
y-axis	0.4439	0.1463	0.2293	0.7610	0.9414
City	6	7	8	9	10
x-axis	0.8732	0.6878	0.8488	0.6683	0.6195
y-axis	0.6536	0.5219	0.3609	0.2536	0.2634

Table 1 shows the Hopfield –Tank original data [1] on TSP with 10 cities. The parameters of the following simulations are set as follows:

$W_1 = W_2 = 1$, $\varepsilon = 0.004$, $k = 0.9$, $\alpha = 0.015$, $\gamma = 1$, $a_0 = 0.5$, $a_1 = 2$, $a(0) = 3$
$\beta_1 = 0.001$, $\beta_2 = 0.1$

Fig.2 shows that the time evolutions of the CNN changes from chaotic behavior with large fluctuations at the early stage, through periodic behavior with small fluctuations and stable point of chaotic noise, into the later convergent stage. After about iteration 500, when $a(t)$ converges to 0.5 and $z_{ij}(t)$ converges to zero, the neural network state finally converges to a fixed point corresponding to a global shortest route 2-3-4-1-2 (tour length=1.3418). In addition, other shortest routes with the same energy value, such as 1-2-3-4-1 and 4-1-2-3-4 can be also obtained, depending on the initial values of $y_{ij}(t)$ and $z_{ij}(t)$.

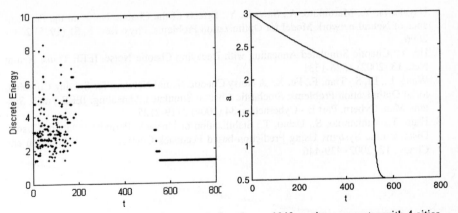

Fig. 2. Time evolutions of discrete energy function and bifurcation parameter with 4 cities

For comparison, the HNN is also applied to the same four-city and ten-city TSP to compare with the proposed method in terms of computational cost. The results with 100 different initial conditions are summarized in table 2.

Table 2. Results of 100 different initial conditions

TSP	4-city HNN	4-city CNN	10-city HNN	10-city CNN
Rate of global minima (%)	63%	96%	48%	90%
Rate of local minima (%)	21%	4%	28%	8%
Rate of infeasible solutions (%)	16%	0%	24%	2%

4 Conclusions

In this paper, we proposed a chaotic neural network with decaying chaos to solve the TSP optimization problems. By adding chaotic noise to each neuron of the discrete-time Hopfield neural network (HNN) and gradually adjusting the bifurcation parameter to reduce the noise to zero, the proposed method has stronger ability to search the global optimal solutions. Simulations of 4-city and 10-city TSP show this method can converge to the global minimum more efficiently than the HNN.

References

1. Hopfield, J. J., Tank, D. W.: Neural Computation of Decisions in Optimization Problems. Biol. Cybern, **52** (1985) 141-152
2. Chen, L., Aihara, K.: Chaotic Simulated Annealing by A Neural-Network Model with Transient Chaos. Neural Networks, 8 (1995) 915-930
3. Wang, L., Smith, K.: On Chaotic Simulated Annealing. IEEE Trans. Neural Net, **9** (1998) 716-718
4. Zhou, D., Yasuda, K., Yokoyama, R.: A Method to Combine Chaos and Neural-Network Based on the Fixed Point Theory. Trans. Inst. Elect. Eng. Japan, **117** (1997) 599-608

5. Hayakawa, Y., Marumoto, A., Sawada, Y.: Effects of the Chaotic Noise on the Performance of Neural-network Model for Optimization Problems. Phys. Rev. E, **51** (1995) 2693-2696

6. He, Y.: Chaotic Simulated Annealing with Decaying Chaotic Noise. IEEE Trans. Neural Net., **13** (2002) 1526-1531

7. Wang, L., Li, S., Tian, F., Fu, X.: A Noisy Chaotic Neural Network for Solving Combinatorial Optimization Problems: Stochastic Chaotic Simulated Annealing. IEEE Trans. System, Man, Cybern, Part B - Cybernetics, **34** (2004) 2119-2125

8. Hino, T., Yamamoto, S., Ushio, T.: Stabilization of Unstable Periodic Orbits of Chaotic Discrete-time Systems Using Prediction-based Feedback Control. Int. J. Bifurcation and Chaos., **12** (2002) 439-446

Extension Neural Network-Type 3

Manghui Wang

Institute of Information and Electrical Energy,
National Chin-Yi Institute of Technology, Taichung, Taiwan, China.
wangmh@chinyi.ncit.edu.tw

Abstract. The extension neural network types 1 and 2 have been proposed in my recent paper. In this sequel, this paper will propose a new extension neural network called ENN-3. The ENN-3 is a three layers neural network and a pattern classifier. It shows the same capability as human memory systems to keep stability and plasticity characteristics at the same time, and it can produce meaningful weights after learning. The ENN-3 can solve the linear and nonlinear separable problems. Experimental results from two benchmark data sets verify the effectiveness of the proposed ENN-3.

1 Introduction

Modern commodities have request light, thin, short and small characteristics, so it is very important to develop a neural network system to keep the system low in computation cost and memory consumption. On the other hand, there are some classification problems whose features are defined over an interval of values in our world. For example, boys can be defined as a cluster of men from age 1 to 14. Therefore, a new neural network topology, called the extension neural network (ENN) was proposed to solve these problems in our recent paper [1]. In this sequel, this paper will propose a new extension neural network that is called ENN-3. The main idea of the proposed ENN-3 is based on the operating function of brain. The brain is a highly complex, nonlinear and parallel information processing system. It has two kinds of different processing mechanisms, one is the right brain that can perform prompt response, e.g. pattern recognition, imagination; the other is the left brain that can perform certain computations, e.g. calculation, logical operation, analysis. Usually, the right and left brains are operation together to analyze outside information, for example, when the brain want to compare two pictures, it will use the right brain to recognize the image, then it also use left brain to analyze the similar degree between the two pictures. Therefore, first, the ENN-3 uses an unsupervised learning and an extension distance (ED) to measure the similarity between data and the sub-cluster center. Second, the ENN-3 uses a logical 'OR' operation in layer 3 to assemble the similar sub-clusters.

2 ENN-3 and Its Learning Algorithm

The structure of ENN-3 is a dilation of the proposed ENN that has been proposed in our recently paper [1]; the proposed ENN is a combination of the neural network and the extension theory [2]. The extension theory introduces a novel distance measure-

J. Wang, X. Liao, and Z. Yi (Eds.): ISNN 2005, LNCS 3496, pp. 503–508, 2005.

ment for classification processes, and the neural network can embed the salient features of parallel computation power and learning capability.

2.1 Architecture of the ENN-3

The schematic architecture of the ENN-3 is shown in Fig. 1. In this network, there are two connection values (weights) between input and competitive layers; one connection represents the feature's lower bound, and the other connection represents the feature's upper bound in the sub-cluster. The connections between the j-th input node and the p-th output node are w_{pj}^L and w_{pj}^U. The main function of the competitive layer is to partition the sub-clusters of input features by using the unsupervised learning. When the clustering process in the competitive layer is convergence, the output layer will use the logical 'OR' to combine the similar sub-clusters based on the trained data.

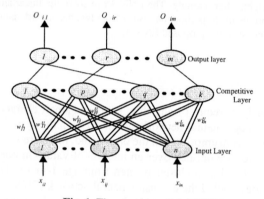

Fig. 1. The structure of the ENN-3.

2.2 Learning Algorithm of the ENN-3

The learning algorithm of the ENN-3 investigates both unsupervised and supervised learning. In first stage, the ENN-3 uses unsupervised learning to generate the sub-clusters in the competitive layer, which uses a threshold called the distance parameter (DP) λ and a Extension distance (ED) function to control the clustering process. The DP λ is used to measure the distance between the cluster center and the desired boundary. A pattern is firstly selected as the center of the first cluster, and the initial weights of the first cluster can be computed from the center with desired distance parameter λ. Then the next pattern is compared to the first cluster. It is clustered with the first if its distance is smaller than the vigilance parameter (i.e. it just equals the number of features). Otherwise it forms the center of a new cluster. This process is repeated for all patterns until a stable or desiderative cluster formation occurs. In second stage, the ENN-3 uses supervised learning to combine the similar clusters according to the target output between the competitive and output layers, the weights between competitive and output layer can directly set to 1, and their links need according to target output of training data set. Before introduce the learning algorithm, several variables have to be defined as follows: O_k^C : the output of k-th node in the competitive layer; N_p: the total number of the input pattern; n: the number of the features; k: the number of existing clusters; M_k: the number of patterns belonging to cluster k. The detailed unsupervised learning algorithm of the ENN-2 can be described as follows:

Step 1: Set the desired DP λ, produce the first pattern, and set $k=1$; $M_1=1$, then the center coordinates and weights of the first cluster are calculated as

$$Z_k = X_k \Rightarrow \{z_{k1}, z_{k2}, \cdots, z_{kn}\} = \{x_{k1}, x_{k2}, \cdots, x_{kn}\} \tag{1}$$

$$w_{kj}^L = z_{kj} - \lambda \; ; \; w_{kj}^U = z_{kj} + \lambda \; ; \; for \; j=1,2...n \tag{2}$$

Step 3: Read next input pattern $X_i = \{x_{i1}, x_{i2}, \cdots, x_{in}\}$ and calculate the Extension distance ED_p between X_i and the existing p-th cluster center as

$$ED_p = \sum_{j=1}^{n} \left[\frac{\left| x_{ij} - z_{pj} \right| - (w_{pj}^U - w_{pj}^L)/2}{\left| (w_{pj}^U - w_{pj}^L)/2 \right|} + 1 \right] \; ; \; for \; p=1,2...k \tag{3}$$

The concept of distance, the position relation between a point and an interval can be precisely expressed by means of the quantitative form [1].

Step 6: Find the s for which, $ED_s = min\{ED_p\}$. If $ED_s > n$, then create a new cluster center. When $ED_s > n$ expresses that X_i does not belong to p-th cluster. Then a new cluster center will be created,

$$k = k+1 \; ; \; M_k = 1 \; ; Z_k = X_i \Rightarrow \{z_{k1}, z_{k2}, \cdots, z_{kn}\} = \{x_{i1}, x_{i2}, \cdots, x_{in}\} \tag{4}$$

$$O_k^c = 1 \; ; \; O_p^c = 0 \; for \; all \; p \neq k \tag{5}$$

$$w_{kj}^L = z_{kj} - \lambda \; ; \; w_{kj}^U = z_{kj} + \lambda \; for \; j=1,2...n \tag{6}$$

Else, the pattern X_i belongs to the cluster s, and updates the s-th weights,

$$w_{sj}^{U(new)} = w_{sj}^{U(old)} + \frac{1}{M_s + 1}(x_{ij} - z_{sj}^{old}) \; ; \; w_{sj}^{L(new)} = w_{sj}^{L(old)} + \frac{1}{M_s + 1}(x_{ij} - z_{sj}^{old}) \tag{7}$$

$$z_{sj}^{new} = \frac{w_{sj}^{U(new)} + w_{sj}^{L(new)}}{2} \; ; \; for \; j=1,2...n \tag{8}$$

$$M_s = M_s + 1 \; ; \; O_s^c = 1 \; ; \; O_p^c = 0 \; ; \; for \; all \; p \neq S \tag{9}$$

It should be noted in this step, only one node is active to indicate a classification of the input pattern; the output of other nodes should set to 0 or non-active.

Step 7: If input pattern X_i changes from the cluster "q" (the old one) to "k"(the new one), then the weights and center of cluster "o" are modified as:

$$w_{qj}^{U(new)} = w_{qj}^{U(old)} - \frac{1}{M_q}(x_{ij} - z_{qj}^{old}) \; ; \; w_{qj}^{L(new)} = w_{qj}^{L(old)} - \frac{1}{M_q}(x_{ij} - z_{qj}^{old}) \tag{10}$$

$$z_{qj}^{new} = \frac{w_{qj}^{U(new)} + w_{qj}^{L(new)}}{2} \; ; \; for \; j=1,2...n \tag{11}$$

$$M_q = M_q - 1 \tag{12}$$

Step 8: Set $i=i+1$, repeat Step 3-Step 8 until all the patterns have been compared with the existing sub-clusters, if the clustering process has converged, go to next step; otherwise, return to Step 3.

Step 9: Using the logical 'OR' to combine the similar sub-clusters that have a similar output according to the trained data. The functions of this layer are expressed as follows:

$$NET_r = \sum W_{rp} O_p^C \tag{13}$$

$$O_r = min(1, NET_r), \ for \ r=1,2...m. \tag{14}$$

The all weights W_{rp} set to 1 in this layer, but their links need according to target output of training data set. If the clustering processing is not completely correct, then reduce DP λ to give more sub-clusters and repeat Step1-Step9.

3   Experimental Results

To verify the effectiveness of the proposed ENN-3, two benchmarks are used to illustrate the applications of the proposed ENN-3.

3.1   Simpson Data Set

The Simpson data set [3] is a perceptual grouping problem in vision, which deals with the detection of the right partition of an image into subsets as shown in Fig. 2. The proposed ENN-3 can completely partition the Simpson data set into four sub-clusters by setting the distance parameter λ of 0.05 after unsupervised learning as shown in Fig. 3. If we combine the sub-clusters 1 and 4 by logical "OR" operation, then there will appear 3 nodes in the output layer of the ENN-3, the final cluster result is shown in Fig. 3, which is an nonlinear separable problem, and the ENN-3 can easily solve this kinds of separable problems. To illustrate the stability and plasticity capability of the proposed ENN-3, if the new data is near the center of sub-cluster 4 and the $EP<n$ for a determined distance parameter (e.g. 0.1), then this new data will be included in sub-cluster 4 and ENN-3 will only tune the weights between input layer and sub-cluster 4 (i.e. the 4-th node of competitive layer). If the new data is far from all cluster centers and the $EP>n$ for a determined distance parameter (e.g. 0.05), then this new data will create a new sub-cluster and the original weights of old sub-clusters will be retained, and a new cluster will be created by adding one node in both output and competitive layers, respectively. The new clustering result is shown in Fig. 4, there are 4 clusters.

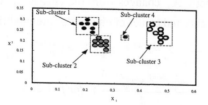

Fig. 2. The initially clustering result after unsupervised learning.

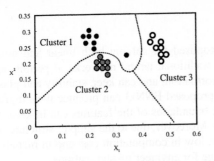
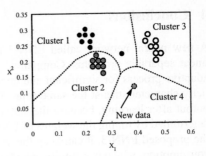

Fig. 3. The finally cluster result of the proposed ENN-3.

Fig. 4. The stability and plasticity test of the proposed ENN-3.

3.2 Iris Data Classification

The Iris data is also a benchmark-tested data set there have 150 instances [4]; it can be divided into three categories with the distinguishing variables being the length and width of sepal and petal. In this case, the structures of the proposed ENN-3 are three output nodes and four input nodes. If the system randomly chooses 75 instances from the Iris data as the training data set, and let the rest of the instances of the Iris data are the testing data set. Table 1 shows the comparison of the experimental results of the proposed ENN-3 with other typical neural networks. It can be seen from the Table 1 that the proposed ENN-3 has a shorter learning time than the traditional neural networks. As well, the accuracy rates are quite high with about 100% and 98% for training and testing patterns, respectively. If the training data set contains 150 training instances (i.e., the full Iris data) and the testing data set is equal to the training data set containing 150 training instances. Table 2 shows some typical clustering results of the proposed ENN-3 with different structures, we can see that the accuracy always proportion to the number of neurons. If the proposed ENN-3 uses 95 neurons in the competitive layer, the accuracy will reach 100%.

Table 1. Comparison of the classification performance of various neural networks.

Model	Structure	Learning Epochs	Training error	Testing Error
Perceptron	4-3	200	82.7%	78.7%
MLP	4-4-3-3	50	97.3%	96%
PNN	4-75-3	1	100%	94.7%
LVQ	4-15-3	20	92%	94.7%
CPN	4-20-3	60	89.3%	84%
ENN-3	4-15-3	2	100%	98%

Table 2. Compared results of ENN-3 with different structures.

Types	DP λ	Neuron no. Layer I	Layer II	Layer III	Learning Epochs	Accuracy rates (%)
1	0.2	4	4	3	2	94%
2	0.15	4	7	3	2	94.6%
3	0.1	4	15	3	2	98%
4	0.04	4	72	3	2	99.3%
5	0.03	4	95	3	2	100%

4 Conclusions

A new neural network called ENN-3 is proposed for solving both linear and non-linear separable problems. Compared with traditional neural networks, it permits an adaptive process for significant and new information, and can keep stability and plasticity characteristics at the same time. The proposed ENN-3 can produce meaningful output after learning, because the classified boundaries of the features can be clearly determined by tuning the weights of ENN-3. Moreover, due to the simpler structure of the proposed ENN-3, it can keep the system low in computation cost and in memory consumption, which is a significant advantage for engineering applications.

Acknowledgments

The author gratefully acknowledges the part support of the National Science Council, Taiwan, under the grant no. NSC-93-2213-E-167-006.

References

1. Wang, M.-H., Hung, C.-P.: Extension Neural Network and Its Applications. Neural Networks, **16** (2003) 779-784
2. Cai, W.: The Extension Set and Incompatibility Problem. Journal of Scientific Exploration, **1** (1983) 81-93
3. Simpson, P. K.: Fuzzy Min-Max Neural Networks-Part 2: Clustering. IEEE Trans. on Fuzzy Systems, **1** (1993) 32-45
4. Chien, Y. T.: Interactive pattern recognition. Marcel Dekker, Inc., (1978) 223-225

Pulsed Para-neural Networks (PPNN)
Based on MEXORs and Counters

Junquan Li and Yixin Yin

School of Information Engineering, University of Science and Technology Beijing,
Haidian Zone, Beijing 100083, China
Tel: 86-10-62332262; Fax: 86-10-62332873
tq097rr@163.com, yyx@ies.ustb.edu.cn

Abstract. We discuss Pulsed Para-Neural Networks (PPNN) defined as graphs consisting of processing nodes and directed edges, called axons, where axons are pure delays, whereas nodes are paraneurons processing spiketrains of constant amplitude. Pusles are received and produced in discrete moments of time. We consider PPNNs with two kinds of paraneurons: MEXOR-based paraneurons (producing a pulse at clock t if it received one and only one pulse at clock t-1) and counter-based paraneuron. We present a "brain" for a mobile robot as an example of practical PPNN.

1 Introduction

The notion of Pulsed Para-Neural Network (PPNN) appeared in 2001 in circles associated with the Artificial Brain Project conducted at the Advanced Telecommunications Research Institute (ATR), Kyoto and initially referred to a set of certain spike-train-processing elements located in 3-dimensional space at points of integer coordinates (Buller et al. 2002). The meaning of the notion was later extended to refer to a graph consisting of processing nodes and directed edges, called axons, where axons are pure delays, whereas nodes are paraneurons processing spiketrains of constant amplitude, where pulses are received and produced in discrete moments of time (Buller 2002). Nodes for PPNNs investigated to date were counter-based paraneurons (CPN) and/or MEXOR-based paraneurons (MPN). Comparing CPN and MPN with the formal neuron (FN) proposed by McCulloch and Pitts (1943), it can be noted that CPN is FN supplemented with a counter accumulating excitations, whereas MPN is FN neuron supplemented with a set of pre-synaptic inhibitions (cf. Figure 1).

As Buller et al. (2004c) wrote, the research on MEXOR Logic and pulsed paraneural networks (PPNN) aims to develop a future-generation evolvable hardware for brain-like computing. MEXOR is known in switching theory as the elementary symmetric Boolean function S^3_1 (Sasao 1999, p. 99) returning 1 if one and only one of its three arguments equals 1. Certain kind of CPNs and delayed-S^5_1-based cells were implemented by Genobyte Inc. in the second half of 90s in the CAM-Brain Machine (CBM) that was to evolve neural-like circuits in a 3-dimensional cellular automata space (Korkin at al. 2000). According to the, so called, CoDi ("collect and distribute") paradigm (Gers et al. 1997) the only role of the delayed-S^5_1-based cells was to "collect" signals, while the key processing was to be done in CPNs. As Buller at al.

J. Wang, X. Liao, and Z. Yi (Eds.): ISNN 2005, LNCS 3496, pp. 509–514, 2005.

2004c wrote, the CBM failed to evolve anything practical, nevertheless, the research on CoDi-style computation were conducted at the ATR, as well as at the Ghent University. The ATR team concentrated on spike-train processing in PPNNs (Buller 2003) and non-evolutionary synthesis of CoDi, which resulted in the development of the *NeuroMaze*[TM] *3.0 Pro* (see Liu 2002) – a CAD tool on which, among others, a module for reinforcement learning and various robotic-action drivers has been synthesized (Buller & Tuli 2002; Buller et al. 2002; Buller et al. 2004b). In Ghent, Hendrik Eeckhaut and Jan Van Campenhout (2003) discovered, that all practical spike-train processing can be done using exclusively S^3_1–based cells and provided related schemes for all sixteen 2-input Boolean functions, a circuit built on a single plane in which two signal paths can cross without interfering, and other devices. Some promising results in "MEXOR-only"-PPNN synthesis have been reported by Juan Liu & Andrzej Buller (2004).

This paper summarizes the current state of PPNN theory and proposes a "brain" for a mobile robot as a practical example of PPNN.

2 MEXOR Logic and Para-neurons

MEXOR Logic is a system of data processing that uses only one Boolean function called MEXOR and denoted "□" (Buller et al. 2004c). In this paper we deal with up-to-three-argument MEXOR working as stated in Definition 1.

Definition 1. MEXOR, denoted □, is a Boolean function that returns 1 if one and only one of its arguments equals 1.

MEXOR is not to be confused with XOR (exclusive OR) denoted "⊕". Although □(x, y) = x ⊕ y, □(x, y, z) = x ⊕ y ⊕ z if and only if xyz = 0. This property makes 3-input-MEXOR logic universally complete, i.e. covering all possible Boolean functions. Table 1 shows selected Boolean functions built exclusively from MEXORs.

Table 1. Basic Boolean functions built from MEXORs (Notations ~, &, ||, and ⊕ mean Boolean NOT, AND, and XOR, respectively).

~x	□(x, 1)
x ⊕ y	□(x, y)
x & (~y)	□(x, y, y)
x & y	□(□(x, x, y), y)
x ‖ y	□(□(x, x, y), x)
(~y) & x ‖ y & z	□(□(□(x, x, y), x), □(y, z, z))

MEXOR supplemented with a 1-clock delay becomes a MEXOR-based para-neuron (MPN) - a suitable building block for a synthesis of custom spike-train processing devices. As Buller et al. (2004c) note, a network built from CPNs meets some definitions of a neural network (eg. Hecht-Nielsen 1990, p. 2-3) and can be interpreted as a modification of the historical formal neuron proposed by McCulloch and Pitts (1943) (Fig. 1a).

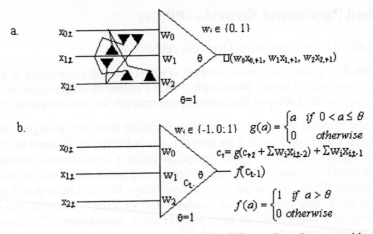

Fig. 1. Para-neurons. a. MEXOR-based para-neuron (MPN) as a formal neuron with added six pre-synaptic inhibition connections. b. Counter-based para-neuron (CPN) as a formal neuron supplemented with a counter.

Let us consider a function denoted "Δ", working according to Definition 2.

Definition 2. Assuming c_t is integer, $w_i \in \{-1, 0, 1\}$, and $x_{i,t}$ is Boolean,

$$u_t = \Delta(\, c_{t-1}, w_0 x_{0,t-1}, w_1 x_{1,t-1}, w_2 x_{2,t-1}) = f(c_{t-1}),$$

where $c_t = w_0 x_{0,t-1} + w_1 x_{1,t-1} + w_2 x_{2,t-1} + g(c_{t-2} + w_0 x_{0,t-2} + w_1 x_{1,t-2} + w_2 x_{2,t-2})$,

where:

$$f(a) = \begin{cases} 1 & if \; a > 1 \\ 0 & otherwise \end{cases} \qquad g(a) = \begin{cases} a & if \; a = 1 \\ 0 & otherwise. \end{cases}$$

Function Δ represents a special case of Counter-based para-neuron (CPN) shown in Figure 1b (here it is assumed that the threshold (θ) equals 1). Table 3 shows the possibilities of building certain Boolean functions using Δ. However, the function is the most valuable as a processor of spiketrains. The simplest example of Δ-based processing of time series is frequency divider (Table 4).

Table 2. Boolean AND and OR built using Δ (Notation $\leftarrow$ means an one-clock delay or $\Box(x)$; the asterix represents a non-essential value of the counter).

x & y	$\Delta(*, \leftarrow x, -\Box(x, y), \leftarrow y)$
x || y	$\Delta(*, \leftarrow x, \Box(x, y), \leftarrow y)$

Table 3. Function Δ working as a frequency divider.

T	0	1	2	3	4	5	6	7	8	9	...
x_t	1	1	1	1	1	1	1	1	1	1	...
c_t	0	1	0	1	0	1	0	1	0	1	...
$u_t = \Delta(c_{t-1}, x_{t-1})$	0	0	1	0	1	0	1	0	1	0	...

3 Pulsed Para-neural Networks (PPNN)

Let us define a Pulsed Para-Neural Network (PPNN)

Definition 3. A pulsed para-neural network (PPNN) is a graph consisting of processing nodes and directed edges, where each node is a pulsed para-neuron and each edge represents a specified delay of data transmission between the points it connects.

As it can be noted, the complex functions (including more than a single □ or single Δ) shown in Table 2 and Table 3 are simple cases of PPNNs represented as symbolic formulas. For more complex PPNNs a graphical notation appears more convenient.

PPNN graphical notation uses squares representing para-neurons and arrows representing delays. Thick, labeled arrows represent delays that are longer than 1 clock or unknown, thick, non-labeled arrows represent 1-clock delays, and tiny, non-labeled arrows represent connections of zero delay. Two or more points of reference may concern a given arrow: begin, end, and a possible fan-out points. A given label appears as a specified number of clocks or as a symbol of a variable and concerns only an arrow's section between two nearest points of reference (Fig. 2). Fig. 3 shows simple examples of using PPNN notation. A "brain" for a mobile robot shown in Figure 4 is an example of a practical PPNN.

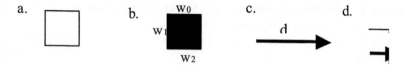

Fig. 2. PPNN graphical notation. a. MPN. B. CPN ($w_i \in \{+, -,$ c. delay (d – number of clocks), d. simplifying replacements.

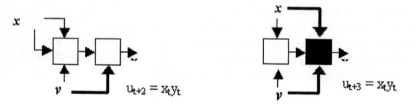

Fig. 3. PPNNs for Boolean AND ("MPN-only" and CPN-based version).

4 Concluding Remarks

PPNN provides a new quite attractive paradigm of synthesis of neural-like systems. I was shown that "MPN-only" PPNNs can be evolved using genetic algorithms. An user-friendly CAD tool for MPN/CPN-based PPNN has been developed and successfully tested via synthesizing several useful devices (timers, associative memories, "brains" for mobile robots, etc.). "MPN-only" PPNN has a good chance to be one day implemented as a new, post-FPGA-generation evolvable hardware. Nevertheless, PPNN theory still requires more lemmas and methods toward a higher-level automation of synthesis of useful devices.

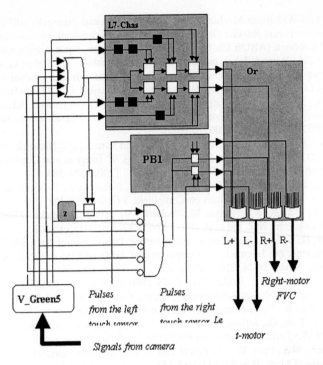

Fig. 4. PPNN-based "brain" for a mobile robot. **V_Green5** – a device processing signals from camera, recognizing object of interest and determining the direction it is seen at (one of five considered). **z** – source of "1-only" spiketrain, *FVC* – Frequency-to-Voltage Converter. L+, L- – positive and negative input to left-motor FVC; R+, R- – positive and negative input to left-motor FVC (equal-frequency spiketrains provided to L+ and R+ results in the robot's moving forward; spiketrains provided to L+ and R- results in the robot's rotation). **Or8x4** – PPNN substituting four 8-input OR-functions. **PB1** – a PPNN responsible for collision avoidance (blocking the spiketrain directed to L+ or R+ and providing a spiketrain to L- or R-, respectively, which result in robot's rotation); **L7-Chaser** – a device providing a spiketrain to both L+ and R+ (via Or8x4) when the object of interest is seen and allowing spiketrains from appropriate V_Green5's outputs to lower the frequency of the spiketrain provided to L+ or R+, which results in appropriate distortion of the robot's trajectory facilitating approaching the object-of-interest (two opional inputs have been designed for further development). For clarity, some Boolean functions are shown using traditional graphic representations.

Acknowledgement

This research was supported by NSFC, China (Grant No.60374032). The authors thank Dr. Andrzej Buller for providing materials about PPNN and valuable explanations.

References

1. Buller, A.: Pulsed Para-Neural Networks (PPNN): Basic Concepts and Definitions, Technical Report TR-HIS-0007, Advanced Telecommunications Research Institute International (ATR). Human Information Science Labs, 2002

2. Buller, A.: CAM-Brain Machines and Pulsed Para-Neural Networks (PPNN): Toward a Hardware for Future Robotic On-Board Brains, 8th International Symposium. on Artificial Life and Robotics. (AROB 8th '03), January 24-26, Beppu, Japan, (2003) 490-493

3. Buller, A., H., Hemmi, Joachimczak, M., Liu, J., Shimohara, K., Stefanski, A.: ATR Artificial Brain Project: Key Assumptions and Current State. 9th International Symposium on Artificial Life and Robotics (AROB 9th), Jan. 28-30, Japan, (2004) 166-169

4. Buller, A., Tuli, T.S.: Four-Legged Robot's Behavior Controlled by Pulsed Para-Neural Networks . Proceedings, 9th International Conference on Neural Information Processing (ICONIP'02), Singapore, 1 (2002) 239-242

5. Buller, A., Eeckhaut, H., Joachimczak, M.: Pulsed Para-Neural Network (PPNN) synthesis in a 3-D Cellular Automata Space, Proceedings. 9th International Conference on Neural Information Processing (ICONIP'02), Singapore, 1 (2002) 581-584.

6. Buller, A., Joachimczak, M., Liu, J., Stefanski A., Neuromazes: 3-Dimensional Spiketrain Processors. WSEAS Transactions on Computers, 3 (2004) 157-161

7. Buller, A., Ahson, S.I., Azim, M.: Pulsed Para-neural Networks (PPNN) Based on MEXOR Logic. In: N.R. Pal et al. (Eds.) LNCS 3316, (2004) 399-408

8. Eeckhaut, H., J. Van Campenhout: Handcrafting Pulsed Neural Networks for the CAM-Brain Machine. 8th International Symposium on Artificial Life and Robotics (AROB 8th '03). Beppu, Japan, 24-26 (2003) 494-498

9. Gers, F., de Garis, H., Korkin, M.: CoDi-1Bit: A Simplified Cellular Automata based Neuron Model. Evolution Artificiele 97, Nimes, France, (1997) 211-229

10. Hecht Nielsen R.: Neurocomputing. Addison-Wesley, (1990)

11. Korkin M., Fehr, G., Jeffrey, G.: Evolving Hardware on a Large Scale Proceedings, The Second NASA / DoD Workshop on Evolvable Hardware. Pasadena (2000) 173-81.

12. McCulloch, W.S., Pitts, W.: A Logical Calculus of the Ideas Immanent in Nervous Activity. Bulletin of Math. Bio, 5 (1943) 115-133

13. Liu, J., NeuroMaze^{TM}: User's Guide, Version 3.0. ATR Human Information Science Labs., Kyoto, Japan, (2002)

14. Liu, J., Buller, A.: Evolving Spike-train Processors, Genetic and Evolutionary Comput.. Seattle, WA, Proceedings, Part I, Springer, (2004) 408-9

15. Sasao, T.: Switching Theory for Logic Synthesis. Boston, Kluwer Academic Publ. (1999)

Using Ensemble Information in Swarming Artificial Neural Networks

Jian Tang, Zengqi Sun, and Jihong Zhu

State Key Laboratory of Intelligent Technology and Systems, Department of Computer Science and Technology, Tsinghua University, Beijing 100084, China
tjian99@mails.tsinghua.edu.cn

Abstract. Artificial neural network (ANN) ensembles are effective techniques to improve the generalization of a neural network system. This paper presents an evolutionary approach to train feedforward neural networks with Particle Swarm Optimization (PSO) algorithm, then the swarming neural networks are organized as an ensemble to give a combined output. Three real-world data sets have been used in our experimental studies, which show that the fitness-based congregate ensemble usually outperforms the best individual. The results confirm that PSO is a rapid promising evolutionary algorithm, and evolutionary learning should exploit collective information to improve generalization of learned systems.

1 Introduction

Evolutionary algorithms have been introduced into ANN's in recent years [1]. It can be used to evolve weights, architectures, and learning rules and parameters [2]. There exist many evolutionary algorithms, such as genetic algorithm, genetic programming, evolutionary programming, and evolution strategies. They not only automate the process of ANN design, but also provide an approach to study evolution and learning in the same framework [3].

Recently, a new evolutionary computation technique, the particle swarm optimization (PSO), is proposed [4]. Its development was based on observations of the social behavior of animals such as bird flocking, fish schooling, and swarm theory. Each individual in PSO is assigned with a randomized velocity according to its own and its companions' flying experiences, and the individuals, called particles, are then flown through hyperspace. Compared with GA, PSO has some attractive characteristics. It has memory, so knowledge of good solutions is retained by all particles; whereas in GA, previous knowledge of the problem is destroyed once the population changes. Besides, PSO has constructive cooperation between particles, particles in the swarm share information between them. In this paper we will apply PSO algorithm to evolve ANN classifiers.

In evolving ANNs and some other applications, most people use an evolutionary algorithm to maximize a fitness function or minimize an error function. But maximizing a fitness function is different from maximizing generalization. The evolutionary algorithm is just used as an optimization algorithm, not a population-based learning algorithm which it actually is. Since the maximum fitness is not equivalent to best generalization in evolutionary learning, the best individual with the maximum fitness

J. Wang, X. Liao, and Z. Yi (Eds.): ISNN 2005, LNCS 3496, pp. 515–519, 2005.

in a swarm may not be the one we want. Other individuals in the swarm may contain some useful information that will help to improve generalization of learned systems. Then using the whole swarming ANNs' information will be helpful, that is the so-called ensemble technique [5] which will be talked about later.

2   PSO Approach to Evolve ANNs

A Particle Swarm Optimization algorithm is defined by the evolution of a population of particles, represented as vectors in a n-dimensional real-valued search space of possible problem solutions. When applied to ANN training, each particle represents a possible ANN configuration, i.e., its weights. Therefore each vector has a dimension equal to the number of weights in the ANN.

Every particle has a position vector $\vec{x}$ encoding a candidate solution to the problem and a velocity vector $\vec{v}$. Moreover, each particle contains a small memory that stores its individual best position seen so far $\vec{p}$ and a global best position $\vec{p}_g$ obtained through communication with its neighbor particles in the same group. In this paper we have tested three kind of topology for particles' connectivity, being Gbest and Lbest, the most common ones, and 4Clusters, a new social network [6] (Figure 1). In Gbest, all particles know each other, while in Lbest, each individual has only two neighbors. 4Clusters contains four cliques of cardinality five, loosely connected. Some modest experiments indicate that Gbest and Lbest are both too extreme to get a high performance, so 4Clusters topology is selected going into the final experiment.

Fig. 1. Gbest, Lbest and 4Clusters topologies.

Information about good solutions spreads through the swarm, and thus the particles tend to move to good areas in the search space. At each time step t, the velocity is updated and the particle is moved to a new position. This new position is calculated as the sum of the previous position and the new velocity:

$$\vec{x}(t+1) = \vec{x}(t) + \vec{v}(t+1) \qquad (1)$$

The update of the velocity from the previous velocity is determined by the following equation:

$$\vec{v}(t+1) = w \times \left(\vec{v}(t) + c_1\phi_1\left(\vec{p}(t) - \vec{x}(t)\right) + c_2\phi_2\left(\vec{p}_g(t) - \vec{x}(t)\right)\right) \qquad (2)$$

where w controls the magnitude of $\vec{v}$, c_1 and c_2 are positive constants and ϕ_1 and ϕ_2 are uniformly distributed random numbers in [0, 1]. Changing velocity this way enables the particle to search around its individual best position and global best position. In our later experimental study we finally set $w = 0.729844$, $c_1 = 2.01$ and $c_2 = 2.01$.

Another important component in PSO algorithm is an appropriate fitness function which evaluates particles to guide the evolution process. In our PSO approach to evolve ANN, to improve the generalization of learned system, we split training set into two sets, being real training set and validating set. And define the fitness function as:

$$fitness = MSE_T + \lambda \cdot MSE_V \tag{3}$$

where MSE_T is the mean squared error between network outputs and desired outputs on the real training set, and MSE_V means the one on validating set. λ is a constant which balance the training accuracy and the system generalization.

3 Ensemble of Swarming ANNs

In order to integrate useful information in different particles in the swarm, we can treat each individual as a module and linearly combine them together. The reason we consider linear combination is its simplicity. It's not our purpose here to find the best combination method, we just want to show the importance of using ensemble information. Better results would be expected if we had used adaptive combination methods [7].

3.1 Majority Voting

The simplest integrating method is majority voting which works by regarding the output of the most number of ANN's as the output of the ensemble. The ensemble in this case is the whole swarm. All particles in the last iteration participate in voting and are treated equally.

3.2 Fitness-Based Linear Combination

One emergent idea in our PSO approach is to use the fitness information to compute a weight for each particle, which is an effective way to consider differences among individuals without involving much extra computational cost. In particular, we can use ANNs' ranking to generate weights for each ANN in integrating the ensemble output. That is, sort all ANNs' according to their fitness function value in decreasing order. Then the weight for the ith ANN is

$$w_i = \frac{\exp(\eta i)}{\sum_{k=1}^{n} \exp(\eta k)} \tag{4}$$

where n is the swarm size, η is a scaling factor, and it was set to 0.7 finally in our study. The ensemble output is thus the weighted voting of all particles

$$output = \sum_{i=1}^{n} w_i o_i \tag{5}$$

constants. Particle swarm optimization algorithm [7] is an effective population-based stochastic global optimization algorithm. A group of individuals (called particles) that constitute a population (called swarm) move around in defined search space. The position of each particle is a candidate solution to the problem. In every iteration every particle adjusts its position toward two points: the best solution achieved so far by itself (referred to pBest) and by whole sub-swarm (referred to gBest) at every iteration. When a particle meets a better solution than gBest during the course of moving toward gBest, this new solution will replace gBest. In the sub-swarm, gBest plays a very important role. It guides the search direction of the whole sub-swarm. Given enough iterations for search, all particles in the sub-swarm will finally converge to gBest. In the case of training component networks, when a particle evaluates its fitness, the neural network is constructed depending on its position, and the network is performed on training sets according to formulas (2), (3). The first two terms in formula (2) can be calculated with the resulting neural network. However, the last term in formula (2), correlation penalty term, requires the information of other component networks. For a sub-swarm, there is a group of candidates for the component network. Each one is represented by a particle and has different outputs. It is impossible for a particle in a sub-swarm to combine all the particles in other sub-swarms, so we select the representative network that is constructed by gBest in sub-swarm i as the component neural network i in the ensemble because all the particles in sub-swarm are nearly equal to gBest at the end of iterations as mentioned above. Thus, sub-swarms just exchange the output values of the network that is formed by gBests. Therefore, the correlation penalty term is calculated with the outputs of M gBest in M sub-swarms. When the MPPSO ends, the gBests in all sub-swarms are the component neural networks and the combination of them is the ensemble. Another issue should also be noted. The correlation penalty term of a sub-swarm is changing with time and is influenced by other sub-swarms, so the training for a component network is a dynamic optimization problem. Because pBests and gBests can be regarded as the memory of the swarm, they must be taken into the new fitness function to re-evaluate their fitness values in every iteration. The pseudocode of MPPSO algorithm is as follows:

```
BeginMPPSO;
    Initialize;
    for t =1 to the limit of iterations
        for s = to the number of sub-swarms
            Sub-swarm s receives the information from the other sub-swarms;
            for each particle in Sub-swarm s
                Evaluate fitness according to formulas (2), (3);
                Update the position of the particle as common PSO algorithm;
                re-calculate pBest and gBest;
            End
        End
    End
End MPPSO
```

Subsequently, we will prove that all the optima of error function E_i defined above in the component networks are in the same location as error function E_{ens} in the ensemble, given an appropriate value of λ.

Definition 2. *The error function of the ensemble for the n-th training pattern is*

$$E_{ens}(n) = \frac{1}{2}(F(n) - d(n))^2 = \frac{1}{2}(\frac{1}{M}\sum_{i=1}^{M}F_i(n) - d(n))^2 \qquad (4)$$

Theorem 1. *Let* $\lambda = \frac{M}{2(M-1)}$*, then* $\frac{\partial E_i(n)}{\partial F_i(n)} = \frac{1}{M} \cdot \frac{\partial E_{ens}(n)}{\partial F_i(n)}$

Proof.

$$p_i(n) = (F_i(n) - \frac{1}{M}\sum_{i=1}^{M}F_i(n))\sum_{j\neq i}(F_j(n) - \frac{1}{M}\sum_{i=1}^{M}F_i(n)) \qquad (5)$$

Hence:

$$\frac{\partial p_i(n)}{\partial F_i(n)} = (1 - \frac{1}{M})\sum_{j\neq i}(F_j(n) - \frac{1}{M}\sum_{i=1}^{M}F_i(n)) + (F_i(n) - \frac{1}{M}\sum_{i=1}^{M}F_i(n))(-\frac{M-1}{M})$$

$$= (1 - \frac{1}{M})((\sum_{i=1}^{M}F_i(n) - F_i(n)) - \frac{M-1}{M}\sum_{i=1}^{M}F_i(n))$$

$$+ (F_i(n) - \frac{1}{M}\sum_{i=1}^{M}F_i(n))(-\frac{M-1}{M})$$

$$= \frac{2(M-1)}{M}(F(n) - F_i(n)) \qquad (6)$$

The partial derivative of $E_i(n)$ with respect to the output of network i on the n-th training pattern is

$$\frac{\partial E_i(n)}{\partial F_i(n)} = F_i(n) - d(n) + \lambda\frac{\partial p_i(n)}{\partial F_i(n)} = F_i(n) - d(n) + \frac{2\lambda(M-1)}{M}(F(n) - F_i(n)) \qquad (7)$$

The partial derivative of $E_{ens}(n)$ with respect to the output of network i on the n-th training pattern is

$$\frac{\partial E_{ens}(n)}{\partial F_i(n)} = \frac{1}{M}(\frac{1}{M}\sum_{i=1}^{M}F_i(n) - d(n)) = \frac{1}{M}(F(n) - d(n)) \qquad (8)$$

Therefore, when $\frac{2\lambda(M-1)}{M} = 1$ that is $\lambda = \frac{M}{2(M-1)}$, then

$$\frac{\partial E_i(n)}{\partial F_i(n)} = \frac{1}{M} \cdot \frac{\partial E_{ens}(n)}{\partial F_i(n)} \qquad (9)$$

This shows that the gradient of the component network error is directly proportional to the gradient of the ensemble error, which means all the optima in the component network error functions are in the same location as in the ensemble error function, but the landscape is scaled by a factor of $\frac{1}{M}$. Therefore, the good solution to the component network achieved by sub-swarm is also the good solution to the whole ensemble.

3   Experiments

Australian Credit Approval data set from the UCI repository of Machine Learning databases [8] was used. This data set contains 690 patterns with 14 attributes; 6 of them are numeric and 8 discrete. The predicted class is 1 for awarding the credit and 0 for not. Three training algorithms were compared, MPPSO in this paper, PSO in [6], and EENCL in [5]. In MPPSO, the number of sub-swarms is 10, and hence λ is 0.5556. The rest of parameters in MPPSO are the same as that in PSO: weight w decreasing linearly from 0.9 to 0.4, learning rates $c_1 = c_2 = 2$, and $Vmax = 5$. Max iteration is 200. To be consistent with the literature [5], 10-fold cross-validation was used. Table 1 shows the results of MPPSO where the ensemble was constructed by the gBests in the all sub-swarms in the last generation. In Table 1, the values of predictive accuracy on test data sets were listed. The results were averaged on 10-fold cross-validation. The measures Mean, SD, Min, and Max indicate the mean value, standard deviation, minimum, and maximum value, respectively. The results of EENCL listed here were reported in [5].

Table 1. The predictive accuracy of the networks or the ensemble trained by three algorithms, on test data sets

	Mean	SD	Min	Max
MPPSO	0.866	0.030	0.822	0.913
PSO	0.858	0.039	0.797	0.910
EENCL	0.868	0.030	0.812	0.913

Table 1 shows that the ensemble trained by MPPSO achieves better performance than neural network trained by PSO. This indicates the effectiveness of negatively related ensemble. Negative correlation learning encourages different component networks to learn different aspects of the training data so that the ensemble can learn the whole training data better. MPPSO achieves comparable results to EENCL. In addition, MPPSO has no requirement that the designer need to specify the number of hidden units in advance, so it is more practical.

4   Conclusions

In this paper, multi-population particle swarm optimization algorithm was proposed to train negatively correlated neural network ensemble. The correlation penalty term was incorporated into the fitness function of the sub-swarm so that the component networks are negatively correlated. Furthermore, we proved that the minimum of the component network error function located the same position as that of the ensemble. In Liu's proof [5], there is the assumption that $F(n)$ has constant value with respect to $F_i(n)$. This assumption contradicts formula (1), so he cannot give a good suggestion about how to choose the value of parameter λ. In our proof, we don't need this assumption and give a formula to figure out the value of λ. Due to the correlation penalty term, the learning task of the ensemble is decomposed into a number of subtasks for different component networks. The goals of the component networks learning are consistent with that of the ensemble learning. Each sub-swarm is responsible for training a component network. The architecture of each component neural network in the ensemble is automatically configured. The results of classification on Australian credit card assessment problem show that MPPSO is an effective and practical method for negatively correlated ensemble.

Acknowledgements

This work was supported by the national grand fundamental research 973 program of China (No.2004CB719401) and the national high-tech research and development plan of China (No. 2003AA412020).

References

1. Hansen, L.K., Salamon, P.: Neural Network Ensembles. IEEE Transactions on Pattern Analysis and Machine Intelligence, **PAMI-12** (1990) 993–1001
2. Breiman, L.: Bagging Predictors. Machine Learning, **24** (1996) 123–140
3. Schapire, R.E.: The Strength of Weak Learnability. Machine Learning, **5** (1990) 197–227
4. Liu, Y., Yao, X.: Simultaneous Training of Negatively Correlated Neural Networks in an Ensemble. IEEE Transactions on Systems, Man and Cybernetics, Part B, (1999) 716–725
5. Liu, Y., Yao, X., Higuchi, T.: Evolutionary Ensembles with Negative Correlation Learning. IEEE Transactions on Evolutionary Computation, **4** (2000) 380–387
6. Liu, Y., Qin, Z., Shi, Z., Chen, J.: Training Radial Basis Function Networks with Particle Swarms. In: Proceeding of the IEEE International Symposium Neural Networks. Lecture Notes in Computer Science. Vol. 3173. Springer (2004) 317–322
7. Kennedy, J., Eberhart, R.: Particle Swarm Optimization. In: Proceeding of IEEE International Conference on Neural Networks (ICNN'95). Vol. 4. Perth, Western Australia (1995) 1942–1947
8. Blake, C., Merz, C.J.: UCI Repository of Machine Learning Databases, http://www.ics.uci.edu/~mlearn/MLRepository.html (1998)

Wrapper Approach
for Learning Neural Network Ensemble
by Feature Selection

Haixia Chen[1], Senmiao Yuan[1], and Kai Jiang[2]

[1] College of Computer Science and Technology, Jilin University, Changchun 130025, China
hxchen2004@sohu.com
[2] The 45th Research Institute of CETC, Beijing 101601, China
kjiang2004@sohu.com

Abstract. A new algorithm for learning neural network ensemble is introduced in this paper. The proposed algorithm, called NNEFS, exploits the synergistic power of neural network ensemble and feature subset selection to fully exploit the information encoded in the original dataset. All the neural network components in the ensemble are trained with feature subsets selected from the total number of available features by wrapper approach. Classification for a given intance is decided by weighted majority votes of all available components in the ensemble. Experiments on two UCI datasets show the superiority of the algorithm to other two state of art algorithms. In addition, the induced neural network ensemble has more consistent performance for incomplete datasets, without any assumption of the missing mechanism.

1 Introduction

Neural network ensemble (NNE) can significantly improve the generalization ability of learning system through training a finite number of neural networks and then combining their outputs [1], [2]. Bagging and boosting are the most popular methods for ensemble learning. In bagging, the training set is randomly sampled K times with replacement, producing K training sets with the same size with the original training set. K classifiers are induced with those training sets and then be used to classify new data by equal weight voting. In boosting, on the other hand, a classifier is induced by adaptively changing the distribution of the training set according to the performance of the previously induced classifiers. Classifiers are used to classify new data according to weights measured by their performance.

The methods mentioned above both require that the training and testing data set be complete. In real world applications, however, it is not unusual for the training and testing dataset to be incomplete as bad sensors, failed pixels, malfunctioning equipment, unexpected noise, data corruption, etc. are familiar scenarios. In order to get a complete dataset, the original dataset must be preprocessed. A simple approach to deal with missing values is to ignore those instances with missing values. The approach is brute as it ignores useful information in the missing data. Some other approaches rely on Bayesian techniques for extracting class probabilities from partial data, by integrating over missing portions of the features space. They are complicated in computation and are always based on the assumption that the values are missing according to a certain rule. As the data missing mechanism is difficult to interpret in

J. Wang, X. Liao, and Z. Yi (Eds.): ISNN 2005, LNCS 3496, pp. 526–531, 2005.
© Springer-Verlag Berlin Heidelberg 2005

practice, those assumptions are seldom satisfied. Another approach is to search for the optimal subset of features so that fewer features are required. However, the problem still remains if one or more of these optimal features have missing values. Furthermore, to search for the optimal subset the dataset should also be completed beforehand.

In this paper, we proposed a wrapper approach for learning NNE by feature selection, named NNEFS, to deal with this problem. The basic idea of the algorithm is to train an ensemble of neural networks, each trained with a subset of features learned by a wrapper approach. When an instance with missing values needs to be classified, only those networks with features that all presented in the given instance can vote to determine the classification of the instance. Besides its superior ability in dealing with missing values, the proposed method has following advantages in comparison with bagging and boosting methods. By reducing the dimensionality of the data, the impact of the "curse of dimensionality" is lessened. The training dataset used to train each classifier is more compact, which would fasten the induction of classifiers. Using different feature subsets to induce neural networks reduce correlation among the ensemble components. As theoretical and empirical results in [3] suggest that combining classifiers gives optimal improvements in accuracy if the classifiers are "independent", the method can give a more accurate classification. The wrapper approach used in the method can lead to further performance improvement of the ensemble.

2   Learning Neural Network Ensemble by Feature Selection

Wrapper approach is used to build NNEFS. Firstly, the original dataset is analyzed to get the compatible optimal set, a concept that will be illustrated in the following section. Then, the compatible optimal set is used to form different species for searching feature subset. For each candidate feature subset encountered in the searching process, a neural network is built to evaluate its fitness. At the end of the searching, K candidate subsets are selected and used to learn the corresponding neural networks that form the ensemble. When a particular instance with certain missing values is inputted into the system, it is classified by those neural networks that did not use the missing features. All other classifiers that use one or more of the missing features of the given instance are simply disregarded for the purpose of classifying the current instance.

2.1   Compatible Optimal Set

Assuming that we have dataset denoted by D, which have m instances with f features. Feature i is denoted by X_i, $i = 1,..., f$, and instance i by e_i, $i = 1,..., m$.

For an instance e, define the feature subset $S = \{X_i \mid$ value for X_i in e is not missing, $i = 1,..., m\}$ as a *compatible feature subset* for e. S is a compatible feature subset for D iff $\exists e \in D: S$ is a compatible feature subset for e.

A feature subset S_i is said to *dominate* a feature subset S_j in D, denoted as $S_i \underset{D}{\succ} S_j$, iff $S_i \subset S_j \wedge S_i$ and S_j are compatible feature subsets for D.

A feature subset S_i is said to be *compatible optimal* iff $\neg \exists S_j : S_j \underset{D}{\succ} S_i$.

The set S consisting of all compatible optimal feature subsets is called the *Compatible optimal set* for D.

The introduction of the concept of compatible optimal set insures that for any given instance, there is at least one component available for classification.

A pass through the dataset can find all compatible feature subsets for D and the corresponding compatible optimal sets. For each element in the compatible optimal set, species will be constructed accordingly. Since more neural networks in NNE means higher classification accuracy, we demand that at least 50 neural networks be presented in the ensemble. If the size of the compatible optimal set is less than 50, then more feature subset should be searched in the total feature space to meet the minimal demand, that is, the searching method must have the ability to find multiple solutions.

2.2 Searching for Feature Subsets

The feature selection problem can be viewed as a combinational optimization problem that can be solved using various search methods. Here a nondeterministic search algorithm, estimation of distribution algorithm (EDA), is employed, because exhaust/complete search is too time consuming and the heuristic search often trapped into local optimal. Furthermore, EDA can be readily adapted to deal with the multimodal optimization problem as demanded by this application.

EDA is an evolutionary algorithm. Instead of the crossover and mutation operators used in GA, EDA produces new populations by means of the following two steps: first, a probabilistic model of selected promising solutions is induced; second, new solutions are generated according to the induced model. EDA explicitly identifies the relationships among the variables of a problem by building models and sampling individuals from it. In this way, it can avoid the disruption of groups of related variables that might prevent GA from reaching the global optimum [4].

An individual in the searching space is corresponding to a subset of the available feature space defined by the species. Each bit of an individual indicates whether a feature is present (1) or absent (0). In each generation of the search, a 2-order Bayesian network (BN) will be used to represent the probability distribution of selected individuals for each species. The BN will be formed by d nodes, each representing a feature presented in the species. Each node has two possible values or states, 0 for absence of the feature, 1 for presence.

2.3 Individual Evaluation

The issue of individual evaluation plays an important role in this research. This paper advocates for the wrapper approach instead of the filter approach. Because filter approach only use information obtained from the dataset and disregard the bias of the particular induction algorithm to select a fine-tuned feature set. In addition, the filter approach has a tendency to select larger feature subsets since filter measures are generally monotonic. In the wrapper approach, however, the subset is selected using the induction algorithm as a black box. A search for a good subset can be done by using the performance of the neural network itself as a factor for determining the fitness of the given candidate feature subset. The other factor being used is what we call 'parsimony' that measures how many features we have managed to 'save'.

For each individual s, a multi-layer perceptron networks trained by gradient descent with momentum back propagation is constructed. The number of input nodes of the neural network is determined by the feature subset, which is obtained by function $ones(s)$ by counting 1 in s. The number of outputs c is determined by the classes in the dataset. According to the geometric pyramid rule, the number of hidden nodes is $\sqrt{ones(s) \cdot c}$. Fitness for an individual s in species with dimension d is defined as:

$$F(s) = acc\{NN(s)\}(\%) + (1 - ones(s)/d) . \tag{1}$$

Function $acc(\bullet)$ calculates the accuracy of the neural network corresponding to s in the training dataset. As the accuracy estimation has an intrinsic uncertainty, a 5-fold cross-validation is used to control this intrinsic uncertainty. The 5-fold cross-validation is repeated unless the standard deviation of the accuracy estimate is below 1%, a maximum of five times. In this way, small datasets will be cross-validated many times, while larger ones maybe only once.

It should be noted that in the training process, only instances that are complete according to the given feature subset are used to form the training dataset for the subset.

2.4 Building Neural Network Ensemble

All neural networks whose input features do not include any of the missing features in the given instance e are combined through relative majority voting to determine the classification of the instance:

$$H(e) = \underset{y}{\arg\max} \sum_{i=1}^{K} w_i I\{h_i(e) == y\} . \tag{2}$$

$I(\bullet)$ is an indicate function that evaluates to 1 if the predicate holds true. $h_i(e)$ outputs the hypothesis of e by neural network i. Error code -1 will be outputted if a missing feature in e appears in the network. w_i is the weight for neural network i. It is proportional to the number of instances in the training dataset that give correct classification. A more compact feature subset means more available instances in the training dataset and has more chance to get a larger weight. Weights defined in this way can reflect the confidence of each neural network to its hypothesis. The class that receive the highest vote among all classes is the classification for e.

3 Experiments

The algorithm was tested on two datasets form UCI machine learning repository [5]. The optical character recognition (OCR) dataset consists of 1200 training instances and 1797 test instances of digitized hand written characters. Digits 0 through 9 are mapped on a 8×8 grid, and create 64 features for 10 classes. The ionosphere radar return (ION) dataset consists of 220 training instances plus 131 test instances. 34 features in the dataset classify the data into two classes. For each dataset, four additional datasets with different percentage of missing values are produced by randomly removal of data in the original dataset.

The proposed algorithm is compared with bagging and boosting algorithms. For the later two algorithms, incomplete dataset are preprocessed by two methods. In the first method, incomplete instances are simply dismissed from consideration. In the second one, the dataset are completed with modes for the discrete variables and with means for the continuous variables. The classification performance of these three methods for the two datasets is compared in Table1 and Table2 separately.

Table 1. Performance comparison on dataset OCI, superscript means different preprocessing method dealing with missing values.

Percentage of Missing Values (%)	Classification Accuracy (%)				
	Bagging[1]	Bagging[2]	Boosting[1]	Boosting[2]	NNEFS
0	94.25	94.25	94.32	94.32	94.47
5	93.83	93.12	93.51	93.49	94.03
10	90.47	92.61	91.14	89.53	92.74
15	87.56	89.73	87.46	83.28	90.92
20	83.74	85.87	84.76	81.33	90.85

Table 2. Performance comparison on dataset ION.

Percentage of Missing Values (%)	Classification Accuracy (%)				
	Bagging[1]	Bagging[2]	Boosting[1]	Boosting[2]	NNEFS
0	94.73	94.73	94.86	94.86	95.21
5	93.44	94.08	93.36	93.52	95.03
10	89.29	91.43	91.31	90.78	94.34
15	83.37	87.75	86.32	83.61	92.57
20	81.15	82.46	81.44	77.27	91.78

NNEFS performs much better than the bagging and boosting algorithms. It gives the most accurate hypothesis for the original datasets. Moreover, relatively consistent results for different percentage of missing values are also observed. The classification accuracy reduction is no more than 4% for dataset with up to 20% missing values. This can be explained by the feature selection property employed in the system. An instance with missing values has a higher probability to be used by a more compact network. Thus a more compact network has more training data available without calling any preprocessing method that will introduce some biases.

It is also apparent in the tables that different preprocessing methods do produce different biases on ensemble learning methods. Bagging seems more compatible with the second preprocessing method, while boosting with the first, except for a few cases. Disregard of instances with missing values is amount to reduce the training dataset, and improper completion method introduces additional noises. As boosting is sensitive to noises in the dataset, performance deterioration is doomed. For bagging, however, information in the completed instances can still be exploited.

4 Conclusions

Most recent research in machine learning relies more on the wrapper approach rather than the filter approach, the proposed method can be considered as the first attempt to apply the wrapper approach feature selection to NNE. As shown in the experiments, NNEFS can achieve higher classification performance with a set of more compact neural networks. Furthermore, the specially designed mechanism in the system makes it a powerful tool to process incomplete dataset that often encountered in real application. As a preliminary study in this line, however, further researches on aspects of parameters setting, efficiency improvement, and theoretical analysis etc. are still demanded.

Acknowledgements

This paper is supported by National Natural Science Found-ation of China under grant No. 60275026.

References

1. Zhou, Z.H., Chen, S.F.: Neural Network Ensemble. Chinese Journal of Computers, **25** (2002) 1-8
2. Bryll, R., Osuna, R.G., Quek, F.: Attribute Bagging: Improving Accuracy of Classifier Ensembles by Using Random Feature Subsets. Pattern Recognition, **36** (2003) 1291-1302
3. Krogh, A., Vedelsby, J.: Neural Network Ensembles, Cross Validation, and Active Learning. In: Tesauro, G., Touretzky, D. and Leen, T. eds. Advances in Neural Information Processing Systems, Vol 7. MIT Press, Cambridge, MA (1995) 231-238
4. Inza, I., Larranaga, P., Etxeberria, Sierra, R.B.: Feature Subset Selection by Bayesian Network-based Optimization. Artificial Intelligence, **123** (2000) 157-184
5. Blake, C.L., Merz, C.J.: UCI Repository of Machine Learning Databases at http://www.ics.uci.edu/~mlearn/MLRepository.html. Irvine, CA: University of California, Dept. of Information and Computer Science (1998)

Constructive Ensemble of RBF Neural Networks and Its Application to Earthquake Prediction

Yue Liu, Yuan Li, Guozheng Li, Bofeng Zhang, and Genfeng Wu

School of Computer Engineering & Science, Shanghai University,
Shanghai 200072, China
yliu@staff.shu.edu.cn

Abstract. Neural networks ensemble is a hot topic in machine learning community, which can significantly improve the generalization ability of single neural networks. However, the design of ensemble architecture still relies on either a tedious trial-and-error process or the experts' experience. This paper proposes a novel method called CERNN (Constructive Ensemble of RBF Neural Networks), in which the number of individuals, the number of hidden nodes and training epoch of each individual are determined automatically. The generalization performance of CERNN can be improved by using different training subsets and individuals with different architectures. Experiments on UCI datasets demonstrate that CERNN is effective to release the user from the tedious trial-and-error process, so is it when applied to earthquake prediction.

1 Introduction

It is challenging to build successful artificial neural network (NN) to solve real-world problems. Neural network ensemble, a collection of a (finite) number of neural network that are trained for the same task, can be expected to perform far better than single neural network [1], [2]. In recent years, diversified ensemble methods have been proposed [3]. However, the architecture of the ensemble, e.g., the number of individual NNs, the number of hidden nodes of individual NNs, and many other parameters for training the individual NNs, are all determined manually and fixed during the training process. As well known, the ensemble architecture design involves a tedious trial-and-error process for the real world problems and depends on the users' prior knowledge and experience. An automatic approach to ensemble architecture design is clearly needed. As far as we know, Yao's group proposed a method called CNNE (Constructive Neural Network Ensemble) [4], [5] for feed forward neural networks based on negative correlation learning, in which both the number of NNs and the number of hidden nodes in individual NNs are determined using constructive algorithm. However, the number of hidden nodes of RBFNN (Radial Basis Function Neural Networks) can be selected by clustering algorithm instead of being determined incrementally from only one [6]. In this paper, we propose a novel method named CERNN (Constructive Ensemble of RBF Neural Networks) based on Bagging, in which the number of hidden nodes in RBFNNs is determined automatically with high accuracy and efficiency by using the nearest-neighbor clustering algorithm [7]. Furthermore, the CERNN employs a constructive algorithm to automatically determine

J. Wang, X. Liao, and Z. Yi (Eds.): ISNN 2005, LNCS 3496, pp. 532–537, 2005.

the number of individuals and the training epochs. The diversity among the individuals and the generalization ability of the ensemble are guaranteed using three techniques: different training subsets generated by Bootstrap algorithm; an adaptive number of hidden nodes determined by nearest-neighbor clustering algorithm; an adaptive number of training epochs.

Earthquake Prediction has been one of the difficult problems that have not been solved [8]. It is difficult to give an overall and accurate prediction objectively because many direct factors arousing the earthquake cannot be measured, and the relationship between earthquake and the observed precursors is strongly non-linear. Especially, the data are noisy, dirty and incomplete. In the previous works, Wang et al. used BP neural networks [9] and Liu et al. employed RBFNN ensemble [10] for earthquake prediction, which, however, are not convenient and need a very tedious trial-and-error process. Therefore, the CERNN method is employed.

The rest of this paper is organized as follows. In Section 2, details of CERNN method are described. In Section 3, experiments are performed on benchmark datasets and earthquake dataset. Finally, conclusions are drawn in Section 4.

2   Constructive Algorithm for RBF Neural Networks Ensemble

The performance of an ensemble has close relation with its architecture. In the previous works, the architecture of an ensemble is determined manually and fixed during the training process or produced from one using constructive algorithm. To design the architecture efficiently and automatically, we propose a novel method named CERNN (Constructive Neural Network Ensemble), which employs both nearest-neighbor clustering algorithm and constructive algorithm. CERNN is briefly described as follows and its flowchart is shown in Fig.1.

Step 1 Generate a training subset by using Bootstrap algorithm [11].

Step 2 Determine the architecture of the individual RBFNN by using nearest-neighbor clustering algorithm on the training subset. Then add it to set U, which means the RBFNN has not been trained sufficiently.

Step 3 Train the RBFNN in set U for μ times if the RBFNN hasn't been trained. Otherwise, train it for σ times. μ and σ are pre-defined by the user.

Step 4 Determine whether to continue training the individual RBFNN or not. If the RBFNN in set U doesn't satisfy the criterion for halting the RBFNN construction, it means the RBFNN is not trained sufficiently. Save the current weight matrix and go to step 3 to continue training the RBFNN. The incremental step is σ times. Otherwise, terminate training the RBFNN and add it to the set T, which means the RBFNN has been trained sufficiently.

Step 5 Check the criterion for adding a new RBFNN to the CERNN. If the criterion is satisfied, go to step 1. Otherwise, stop the modification of the CERNN architecture and go to step 6. In the optimal condition, the ensemble has only one single RBFNN when the termination criterion is satisfied.

Step 6 Finish the construction of CERNN and obtain the final CERNN.

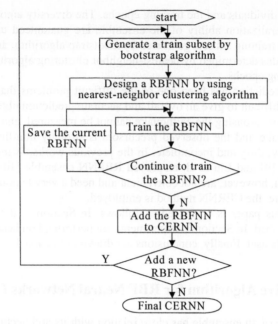

Fig. 1. Flowchart of CERNN.

2.1 Determination of the Architecture of Individual RBFNNs

The number of hidden nodes of the RBFNN is a key factor that influences its performance. In CERNN, it is determined automatically with high accuracy and efficiency by using the nearest-neighbor clustering algorithm [7]. The algorithm starts with one center, which is the first pattern in the training set. Then calculate the Euclid distances between the other patterns x_i and the already clustered centers c_j by

$$D = \min \left\| x_i - c_j \right\|, i = 2, \ldots, N, j = 1, 2, \ldots, m \tag{1}$$

Where N and m denote the number of patterns and the centers respectively. If $D \geq \delta$, add a new hidden node in the RBFNN. δ is predefined by the user. Otherwise, no more hidden node is needed.

2.2 The Criteria for Halting Individual RBFNNs and Construction of CERNN

The criterion to continue or halt training the individuals will directly influence the generalization performance of the ensemble. An individual RBFNN in set U is not worth continuing to be trained if the following formula

$$E_n^i - E_{n-1}^i < \varepsilon \tag{2}$$

is satisfied, where E_n^i and E_{n-1}^i denote the standard error rate of the ith individual RBFNN on the validation dataset, which is trained for $\mu + n\sigma$ and $\mu + (n-1)\sigma$ times respectively. In general, μ and σ equal to 100, ε is a small value predefined by the user. Otherwise, continue training the individual for σ times.

After training the individual RBFNN sufficiently, CERNN will determine whether to add a new RBFNN or not. If the error rate E on the validation dataset of the ensemble with the current RBFNN is not larger than the predefined small value e or the CERNN has reached the maximum number of individuals Max, terminate the ensemble construction. Otherwise, add a new individual RBFNN. For regression problems, E is defined by

$$E = \tfrac{1}{N}\sum_{n=1}^{N}\left(\tfrac{1}{M}\sum_{m=1}^{M}F_m(n)-d(n)\right)^2 \tag{3}$$

For classification problems, E is defined by

$$E = \tfrac{1}{N}\sum_{n=1}^{N}C(n), \;\; if \;\; O(n)=D(n), C(n)=1, \; otherwise \; C(n)=0 \tag{4}$$

where N is the number of patterns in the validation set, M is the number of individuals, $F_m(n)$ is the output of mth RBFNN on the nth pattern, and $d(n)$ represents the desired output. $\{O(n)|n=1,\ldots,N\}$ presents the outputs of CERNN on each pattern.

3   Experiments

Four data sets selected from UCI repository [12] are used to validate the performance of CERNN, which are listed in Table 1. During the experiments, the datasets are divided to three parts: training subset, validation subset, and test subset. Experiments are repeated ten times for each data set. All the final results are the average values over the ten iterations.

Table 1. A summary of the used UCI data sets.

Data set	# Train set	# Validation set	# Test set	# Attributes	#Classes
Wine	99	50	49	13	3
Breast cancer	349	175	175	9	2
Glass	107	54	53	9	6
Liver disorders	173	86	86	6	2

Experimental results shown in Table 2 indicate that CERNN achieves the same or superior performance to Bagging does and absolutely better performance than single RBFNN does on the used UCI data sets. Moreover, the major advantage of CERNN is that it is convenient to be applied to real-world problems.

From the previous earthquake cases, we can find that there exists certain relationship between the duration of the observed attributes and the magnitude of earthquake. However, this relationship is non-linear which is appropriate to be simulated by neural networks. With the help of the seismologist, fourteen attributes such as belt, gap, train release, frequency, b-time, window, Vp/Vs, s-level, slope, resistance, radon, water, stress, macro precursors, and the total number of the observed precursors are selected as input attributes. The output is the magnitude of earthquake. Forty-five famous earthquake cases erupted in China [13] are divided into three parts: 23 as training subset, 11 as validation subset, and other 11 as test subset. Table 3 shows the prediction results.

Table 2. Results of statistical accuracy (%) on the used UCI Datasets.

Data set	CERNN	Bagging	Single RBFNN
Wine	98.0	95.9	63.3
Breast cancer	98.3	96.0	51.4
Glass	54.7	54.7	45.3
Liver disorders	64.0	61.6	53.5

Table 3. Prediction results for earthquake cases using CERNN.

Earthquake Case	Actual	CERNN	Earthquake Case	Actual	CERNN
Ganzi	6.0	5.9	Qilian	5.1	5.5
Jianchuan	5.4	5.7	Liyang	5.5	5.5
Hutubi	5.4	5.7	Lingwu	5.3	6.1
Fukang	5.3	5.7	Luquan	6.3	6.4
Heze	5.9	6.0	Haicheng	7.3	6.8
Jiashi	6.0	5.6			

In the earthquake prediction field, if the difference between the actual and predicted magnitude is not larger than 0.5, we say the prediction is correct. Based on this hypothesis, the prediction accuracy of CERNN achieves 91%. In fact, besides this advantage of high accuracy, another major advantage of CERNN is that it only needs to be initialized with few parameters without any additional knowledge or experience of earthquake: $\delta = 0.8$, $\mu = 100$, $\sigma = 100$, $e = 0.1$, $\varepsilon = 0.1$, and $Max = 49$.

4 Conclusion

A novel method named CERNN (Constructive Ensemble of RBF Neural Networks) is proposed to automatically design the architecture of an ensemble. Experiments on benchmark datasets demonstrate that accuracy obtained by CERNN is the same as or better than that of other ensemble learning methods like Bagging and significantly better than that of single neural network. CERNN greatly releases the experts from the tedious trial-and-error process when it is applied to earthquake prediction.

To address the problem of automatically designing the ensemble architecture and simultaneously obtaining the same or greater accuracy, CERNN employs two key techniques. The first is to use the nearest-neighbor clustering algorithm to determine the architecture of individuals. The second is to use a constructive algorithm to determine the number of hidden nodes and the training epochs of individuals. The performance of CERNN is improved by using different training subsets, different number of hidden nodes and epochs for training individual RBFNNs. Though the problem of designing architecture has been solved by CERNN in some degree, much work has to be done in the future, in which one work is to introduce multi-objective optimization algorithm to determine when the construction of CERNN is halted.

Acknowledgements

This research is supported by the National Natural Science Foundation of China (Grant No. 60203011) and SEC E-Institute: Shanghai High Institutions Grid project.

References

1. Gutta, S., Wechsler, H.: Face Recognition Using Hybrid Classifier Systems. In: Proc. ICNN-96. IEEE Computer Society Press, Los Alamitos, CA (1996) 1017–1022
2. Hansen, L.K., Salamon, P.: Neural Network Ensembles. IEEE Trans. on Pattern Analysis and Machine Intelligence 12 (1990) 993–1001
3. Dietterich, T.G.: Ensemble Methods in Machine Learning. In: Proc. MCS 2000. Springer-Verlag, Cagliari, Italy (2000) 1–15
4. Islam, M. M., Yao, X., Murase, K.: A Constructive Algorithm for Training Cooperative Neural Network Ensembles. IEEE Trans. on Neural Networks, 14 (2003) 820–834
5. Wang, Z., Yao, X., Xu, Y.: An Improved Constructive Neural Network Ensemble Approach to Medical Diagnoses. In Proc. IDEAL'04, (2004) 572–577
6. Moody, J., Darken C.: Fast Learning in Networks of Locally-tuned Processing Units. Neural Computation, 1 (1989) 1281–294
7. Han, J., Kamber, M.: Data Mining: Concepts and Techniques. Morgan Kaufmann Publishers Inc. (2000)
8. Mei, S., Feng, D.: Introduction to Earthquake Prediction in China. Earthquake Press, Beijing (1993)
9. Wang, W., Jiang, C., Zhang, J., Zhou, S., Wang, C.: The Application of BP Neural Network to Comprehensive Earthquake Prediction. Earthquake, 19 (1999) 118–126
10. Liu, Y., Wang, Y., Li, Y., Zhang, B., Wu, G.: Earthquake Prediction by RBF Neural Network Ensemble. In Proc. ISNN'04, (2004) 962–969
11. Efron, B., Tibshirani, R.: An Introduction to the Bootstrap. Chapman & Hall, New York (1993)
12. Blake, C.L., Merz, C.J.: UCI Repository of Machine Learning Databases. University of California, Department of Information and Computer Science, Irvine, CA (1998)
13. Zhang, Z. (ed.): China Earthquake Cases (1966–1985). Earthquake Press, Beijing (1990)

The Bounds on the Rate
of Uniform Convergence for Learning Machine*

Bin Zou[1], Luoqing Li[1], and Jie Xu[2]

[1] Faculty of Mathematics and Computer Science,
Hubei University, Wuhan 430062, China
{zoubin0502, lilq}@hubu.edu.cn
[2] College of Computer Science,
Huazhong University of Science and Technology,
Wuhan 430074, China
jiexu@hust.edu.cn

Abstract. The generalization performance is the important property of learning machines. The desired learning machines should have the quality of stability with respect to the training samples. We consider the empirical risk minimization on the function sets which are eliminated noisy. By applying the Kutin's inequality we establish the bounds of the rate of uniform convergence of the empirical risks to their expected risks for learning machines and compare the bounds with known results.

1 Introduction

The key property of learning machines is generalization performance: the empirical risks (or empirical errors) must converge to their expected risks (or expected errors) when the number of examples m increases. The generalization performance of learning machines has been the topic of ongoing research in recent years [1], [2]. The important theoretical tools for studying the generalization performance of learning machines are the principle of empirical risk minimization (ERM) [3], the stability of learning machines [4], [5], [6], [7] and the leave-one out error (or the cross validation error) [8], [9], [10].

Vapnik [3] applied the Chernoff's inequality to obtain exponential bounds on the rate of uniform convergence and relative uniform convergence. These bounds are based on the capacity of a set of loss functions, the VC dimension. In [11], Cucker and Smale considered the least squares error and obtained the bounds of the empirical errors uniform converge to their expected errors over the compact subset of hypothesis space by using the Bernstein's and Hoeffding's inequalities. The bounds depend on the covering number of the hypothesis space.

The stable property of learning machines has become important recently with the work of [4], [5], [7]. According to the stable property of learning machines, the desired learning machines that we want to obtain, according to the principle of empirical risk minimization, should have the quality of stability with respect

---

* Supported in part by NSFC under grant 10371033.

J. Wang, X. Liao, and Z. Yi (Eds.): ISNN 2005, LNCS 3496, pp. 538–545, 2005.
© Springer-Verlag Berlin Heidelberg 2005

to the training samples (or the empirical datum) S. Algorithmic stability was first introduced by Devroye and Wagner [12]. An algorithm is stable at a training sample set S if any change of a single element in S yields only a small change in the output hypothesis. In the paper we discuss the generalization performance of learning machines. In order to do so, we consider the empirical risk minimization on the function sets which are eliminated noisy. By applying the Kutin's inequality we establish the bounds of the rate of uniform convergence of the empirical risks to their expected risks for learning machines on the totally bounded nonnegative function set.

The paper is organized as follows: In section 2 we introduce some notations and main tools. In section 3, we present the basic inequality. We estimate the bounds of the empirical risks uniform converge to their expected risks in section 4. Finally we compare the bounds obtained in the paper with known results in section 5.

2   Preliminaries

We introduce some notations and do some preparations. Denote by $\mathcal{X}$ and $\mathcal{Y}$ an input and an output space respectively. Let a sample set of size m

$$S = \{z_1 = (x_1, y_1), z_2 = (x_2, y_2), \cdots, z_m = (x_m, y_m)\}$$

in $\mathcal{Z} = \mathcal{X} \times \mathcal{Y}$ drawn i.i.d. from an unknown distribution function $F(z)$. The goal of machine learning is to find a function (or learning machine) f_S^α that assigns values to objects such that if new objects are given, the function f_S^α will forecast them correctly. Here α is a parameter from the set Λ. Let

$$R(f_S^\alpha) = \int \mathcal{L}(f_S^\alpha, z) dF(z), \quad \alpha \in \Lambda, \tag{1}$$

be the expected risk (or the risk functional, the error) of function f_S^α, $\alpha \in \Lambda$, where the function $\mathcal{L}(f_S^\alpha, z)$, which is integrable for any f_S^α and depends on z and f_S^α, is called loss function. Throughout the article, we require that there exists a positive constant M such that

$$0 \leq \mathcal{L}(f_S^\alpha, z) \leq M, \quad \alpha \in \Lambda. \tag{2}$$

In the case of classification, one has $0 \leq \mathcal{L}(f_S^\alpha, z) \leq 1$, $\alpha \in \Lambda$.

Let

$$Q = \{\mathcal{L}(f_S^\alpha, z) : \alpha \in \Lambda\}$$

be the set of admissible functions with respect to the sample set S, and let

$$Q_1 = \{\mathcal{L}(f_{S^{i,u}}^\alpha, z) : \alpha \in \Lambda\}$$

be the set of admissible functions with respect to the sample set

$$S^{i,u} = \{z_1, \cdots, z_{i-1}, u, z_{i+1} \cdots, z_m\}.$$

Supervised Learning on Local Tangent Space

Hongyu Li[1], Li Teng[1], Wenbin Chen[2], and I-Fan Shen[1]

[1] Department of Computer Science and Engineering
Fudan University, Shanghai, China
{hongyuli,yfshen}@fudan.edu.cn
tengli.hust@263.net
[2] Department of Mathematics
Fudan University, Shanghai, China
wbchen@fudan.edu.cn

Abstract. A novel supervised learning method is proposed in this paper. It is an extension of local tangent space alignment (LTSA) to supervised feature extraction. First LTSA has been improved to be suitable in a changing, dynamic environment, that is, now it can map new data to the embedded low-dimensional space. Next class membership information is introduced to construct local tangent space when data sets contain multiple classes. This method has been applied to a number of data sets for classification and performs well when combined with some simple classifiers.

1 Introduction

Usually raw data taken with capturing devices are multidimensional and therefore are not very suitable for accurate classification. To obtain compact representations of raw data, some techniques about dimension reduction have come forth. From the geometrical point of view, dimension reduction can be considered as discovering a low-dimensional embedding of high-dimensional data assumed to lie on a manifold.

The key of dimension reduction is to preserve the underlying local geometrical information of raw high-dimensional data while reducing insignificant dimensions. However, if the original data lie on a nonlinear manifold in nature, traditional dimension reduction methods such as Principal Component Analysis (PCA) will fail to well preserve its geometrical information in a low-dimensional space while unfolding the nonlinear manifold. That is, in the case of nonlinear manifolds, PCA often maps close points in the original space into distant points in the embedded space. Over the years, a number of techniques have been proposed to perform nonlinear mappings, such as MDS [1], SOM [2], auto-encoder neural networks [3], locally linear embedding (LLE) [4] and mixtures of linear models [5]. All of these are problematic in application in some way: multi-dimensional scaling and neural networks are hard to train and time-consuming. Mixtures of localized linear models require the user to set a number of parameters, which are highly specific to each data set and determine how well the model fits the data.

Recently, Zhang and Zha [6] proposed a fine method: local tangent space alignment (LTSA). The basic idea is that of global minimization of the reconstruction error of the

J. Wang, X. Liao, and Z. Yi (Eds.): ISNN 2005, LNCS 3496, pp. 546–551, 2005.

set of all local tangent spaces in the data set. Since LTSA does not make use of class membership, it can not usually perform well in the field of data classification. Moreover, LTSA is stationary with respect to the data and lacks generalization to new data. The purpose of this work is to explore and extend LTSA beyond the already known findings and examine the performance of LTSA and its extensions in terms of data classification. Here, two algorithmic improvements are made upon LTSA for classification. First a simple technique is proposed to map new data to the embedded low-dimensional space and make LTSA suitable in a changing, dynamic environment. Then supervised LTSA (SLTSA) is introduced to deal with data sets containing multiple classes with class membership information. This method has been applied to a number of synthetic and benchmark data for classification and performs well when combined with some simple classifiers.

The remainder of the paper is divided into the following parts. Section 2 presents a general framework of LTSA. Next, in section 3 a variation on LTSA for supervised problems will be introduced and some experiments and results will be presented in section 4. Finally, section 5 ends with some conclusions.

2 Local Tangent Space Alignment

LTSA maps a data set $X = [x^{(1)}, \ldots, x^{(N)}]$, $x^{(i)} \in R^m$ globally to a data set $Y = [y^{(1)}, \ldots, y^{(N)}]$, $y^{(i)} \in R^d$ with $d < m$. Assuming the data lies on a nonlinear manifold which locally can be approximated, the LTSA algorithm consists in two main steps: (I) locally approximating tangent spaces around each sample $x^{(i)}$, based on its k nearest neighbors combined with the process of reconstruction of the nonlinear manifold, and (II) aligning those local tangent spaces to find global lower-dimensional coordinates $y^{(i)}$ for each $x^{(i)}$.

The brief description of LTSA is presented as follows:

1. Finding k nearest neighbors $x^{(ij)}$ of $x^{(i)}$, $j = 1, \ldots, k$. Set $X_i = [x^{(ij)}, \ldots, x^{(ik)}]$.
2. Extracting local information by calculating the d largest eigenvectors $g_1, \ldots, g_d$ of the correlation matrix $(X^{(i)} - \bar{x}^{(i)}e^T)^T(X^{(i)} - \bar{x}^{(i)}e^T)$, and setting $G_i = [e/\sqrt{k}, g_1, \ldots, g_d]$. Here $\bar{x}^{(i)} = \frac{1}{k}\sum_j x^{(ij)}$.
3. Constructing the alignment matrix B by locally summing as follows:

$$B(I_i, I_i) \leftarrow B(I_i, I_i) + I - G_i G_i^T, i = 1, \ldots, N, \tag{1}$$

with initial $B = 0$.
4. computing the $d + 1$ smallest eigenvectors of B and picking up the eigenvector matrix $[u_2, \ldots, u_{d+1}]$ corresponding to the 2nd to d+1st smallest eigenvalues, and setting the global coordinates $Y = [u_2, \ldots, u_{d+1}]^T$.

LTSA is more suitable for data classification than LLE since it can potentially detect the intrinsic distribution and structure of a data set, as is illustrated in Fig. 1. It is easy from the figure to see that the 1-D global coordinates computed with LTSA (plotted in the right panel) clearly separate three bivariate Gaussians shown in the left panel each of which includes 100 sample points. LLE, however, did not perform well, mixing two of the Gaussians marked with green star and red plus respectively, which is shown in the middle panel.

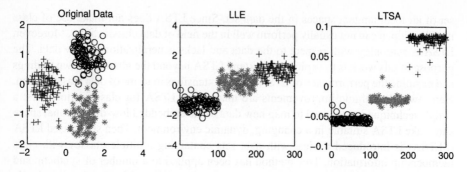

Fig. 1. Comparison LLE with LTSA. Three bivariate Gaussian (left); 1-D global coordinates with LLE (middle); 1-D global coordinates with LTSA (right). The horizontal axis in the middle and right panels represents the index of points.

3 Extensions of LTSA

In this section, our extensions of LTSA are described which, respectively, are 1) generalization of LTSA to new data, 2) supervised variant of LTSA.

3.1 Generalization of LTSA

The original LTSA is stationary with respect to the data, that is, it requires a whole set of points as an input in order to map them into the embedded space. When new data points arrive, the only way to map them is to pool both old and new points and return LTSA again for this pool. Therefore it is not suitable in a changing, dynamic environment.

Our attempt is to adapt LTSA to a situation when the data come incrementally point by point, which is similar to the generalization of LLE in [7]. We assume that the dimensionality of the embedded space does not grow after projecting a new point to it, i.e., d remains constant.

The technique proposed is based on the fact that new points are assumed to come from those parts of the high-dimensional space that have already been explicitly mapped by LTSA. It implies that when a new point arrives, a task of interpolation should be solved. Let points $x^{(i)}$, $i = 1, \ldots, N$, as an input, and compose a set X. For a new point $x^{(N+1)}$, we look for the point $x^{(j)}$ closest to it among all $x^{(i)} \in X$. If $y^{(j)}$ is the projection of $x^{(j)}$ to the embedded space, the following equation is approximately true according to LTSA:

$$y^{(j)} - \bar{y}^{(j)} = L(x^{(j)} - \bar{x}^{(j)}),$$

where L is an unknown affine transformation matrix of size $d \times m$, $\bar{x}^{(j)}$ and $\bar{y}^{(j)}$ are respectively the mean of k nearest neighbors of $x^{(j)}$ and $y^{(j)}$. The transformation matrix L can be straightforwardly determined as

$$L = (y^{(j)} - \bar{y}^{(j)})(x^{(j)} - \bar{x}^{(j)})^{+}, \tag{2}$$

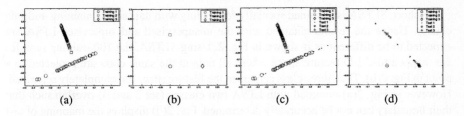

Fig. 2. Mapping **iris** data into a 2-D feature space, where there are 100 training and 50 test samples. (a) Mapping the 100 training samples with LTSA. (b)Mapping these training samples with SLTSA. (c) Mapping the 50 unknown test samples to the feature space (a). (d) Mapping these test samples to the feature space (b).

where $(\cdot)^+$ is the Moor-Penrose generalized inverse of a matrix. Because $x^{(j)}$ and $x^{(N+1)}$ lie close to each other, the transformation matrix L can be used for both points.

$$y^{(N+1)} = \bar{y}^{(j)} + L(x^{(N+1)} - \bar{x}^{(j)}). \tag{3}$$

This method is exemplified by the **iris** data set [8] which includes 150 4-D data belonging to 3 different classes. Here first 100 data points are selected as training samples and mapped from the 4-D input space to a 2-D feature space. Fig. 2(a) showed the distribution of the data in the low-dimensional feature space. Then when other data as test samples arrive, they can be appropriately interpolated in the feature space in terms of the transformation matrix L shown in Fig. 2(c).

3.2   Supervised LTSA

LTSA [6] belongs to unsupervised methods; it does not make use of the class membership of each point. Such methods are mostly intended for data mining and visualization. But they can not usually perform well in the field of classification where the membership information of training samples is known and the center of each class needs to be searched.

Consequently, in this work we propose *superevised* LTSA (SLTSA) for classification. The term implies that membership information is employed to form the neighborhood of each point. That is, nearest neighbors of a given point $x^{(i)}$ are chosen only from representatives of the same class as that of $x^{(i)}$. This can be achieved by artificially increasing the shift distances between samples belonging to different classes, but leaving them unchanged if samples are from the same class. To select the neighbors of samples, we can define a $N \times N$ distance matrix D where each entry d_{ij} represents the *Euclidean* distance between two samples $x^{(i)}$ and $x^{(j)}$. Furthermore, considering the membership information, we can get a variant D' of D,

$$D' = D + \rho\delta \tag{4}$$

where the shift distance ρ is assigned a relatively very large value in comparison with the distance between any pairs of points, $\delta_{ij} = 1$ if $x^{(i)}$ and $x^{(j)}$ are in different classes, and 0 otherwise.

In short, SLTSA is designed specially for dealing with data sets containing multiple classes. Hence the results obtained with the unsupervised and supervised LTSA are expected to be different as is shown in Fig. 2. Using SLTSA, the 100 training samples are mapped to a 2-D feature space where all data in the same class are projected to a point in Fig. 2(b). Then three class centers in the feature space are completely separated. However in Fig. 2(a) obtained with LTSA two classes (set 2 and 3) overlap such that their boundary can not be accurately determined. Fig. 2(d) displays the mapping of test samples to the feature space. As you see, test samples in different classes can be well distributed around class centers. Thus the classification of test samples can be easily implemented in the feature space when combined with simple classifiers.

4   Supervised LTSA for Classification

To examine its performance, SLTSA has been applied to a number of data sets varying in number of samples N, dimensions m and classes c. Most of the sets were obtained from the repository [8] and some are synthetic.

Here we emphasize to study the binarydigits set which consists of 20×16-pixel binary images of preprocessed handwritten digits. In our experiment only three digits: 0, 1 and 2 are dealt with and some of the data are shown in Fig. 3(a). And 90 of the 117 binary images are used as training samples and others as test samples. It is clear that the 320-dimensional binarydigits data contain many insignificant features, so removing these unimportant features will be helpful for the classification. The results after dimension reduction from 320 to 2 with LTSA and SLTSA are respectively displayed in Fig. 3(b) and 3(c) where two coordinate axes represent the two most important features of the binarydigits data. The figure shows that the feature space obtained with SLTSA provides better classification information than LTSA. Actually no cases of misclassification in the test samples occurred if using the SLTSA method, but LTSA will result in the error rate of 18.52% in the best case.

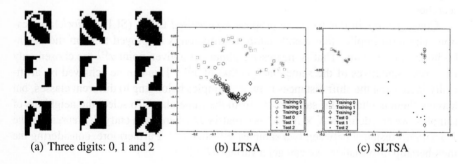

(a) Three digits: 0, 1 and 2   (b) LTSA   (c) SLTSA

Fig. 3. Recognition of three digits: 0, 1 and 2. (a) Some of digits are shown. They are represented by 20×16 binary images which can be considered as points in a 320-dimensional space. (b) Mapping these binary digits into a 2-dimensional feature space with LTSA, where these three clusters overlap and are unseparated. (c) The feature space obtained with SLTSA, which provides better clustering information.

In our experiments three classifiers were used: the minimum distance classifier; the linear discriminate analysis classifier and the k-nearest neighbor classifier. To compare the SLTSA method with other feature extraction techniques, the classifiers were also trained with PCA and (S)LLE.

Our experimental results confirm that SLTSA generally leads to better classification performance than LTSA. Besides this, classification with SLTSA is comparable with SLLE and finer than with PCA. Maybe the reason is that those points in a high-dimensional input space are more likely to lie close on nonlinear rather than linear manifolds. Thus, such linear transformation methods as PCA can not perform well.

5 Conclusions

An extension of the local tangent space alignment method to supervised mapping, SLTSA, was discussed in this paper. It takes into account class membership during selecting neighbors.

Another enhancement to LTSA is to map new data to the low-dimensional feature space embedding in the training data and make it suitable in a changing, dynamic environment. This technique for generalization to new data was applied to our experiments and has demonstrated satisfactory results.

We also compare the proposed method with other mapping techniques such as PCA, unsupervised and supervised LLE on a number of data sets, in order to gain insight into what methods are suitable for data classification. The SLTSA method has been shown to yield very promising classification results in our experiments.

Acknowledgments

This work was supported by NSFC under contract 60473104 and the Special Funds for Major State Basic Research Projects under contract G199032804.

References

1. Borg, I., Groenen, P.: Modern Multidimensional Scaling. Springer-Verlag (1997)
2. Kohonen, T.: Self-Organizing Maps. 3rd edn. Springer-Verlag (2000)
3. DeMers, D., Cottrell, G.: Non-linear Dimensionality Reduction. Advances in Neural Information Processing Systems 5 (1993) 580–587
4. Roweis, S., Saul, L.: Nonlinear Dimension Reduction by Locally Linear Embedding. Science **290** (2000) 2323–2326
5. Tipping, M.E., Bishop, C.: Mixtures of Probabilistic Principal Component Analyzers. Neural Computation **11** (1999) 443–482
6. Zhang, Z., Zha, H.: Principal Manifolds and Nonlinear Dimension Reduction via Local Tangent Space Alignment. Technical Report CSE-02-019, CSE, Penn State Univ. (2002)
7. Kouroteva, O., Okun, O., Soriano, M., Marcos, S., Pietikainen, M.: Beyond Locally Linear Embedding Algorithm. Technical Report MVG-01-2002, Machine Vision Group, University of Oulu, Finland (2002)
8. Blake, C., Merz, C.: Uci Repository of Machine Learning Databases (1998)

Study Markov Neural Network by Stochastic Graph*

Yali Zhao, Guangcheng Xi, and Jianqiang Yi

Key Laboratory of Complex System and Intelligence Science,
Institute of Automation, Academy of Sciences, Beijing 100080, China
{yali.zhao,guangcheng.xi,jianqiang.yi}@mail.ia.ac.cn

Abstract. In this paper, the progress, human learns from the outside world by image and graph, is described by stochastic graph. Markov model is given for the learning process in neural network. Then a method of computation transition matrix is presented via energy function. Based on transition matrix, probability of state is computed and implemented to cognize the objective world by stochastic graph. By applying stochastic graph to the network of two neurons, it shows that states can transform between each other. Finally, the network updates to the state of the least energy.

1 Introduction

The brain is a highly complex, nonlinear, and parallel computer (information-processing system). It has the capability to organize its structural constituents, known as neurons, so as to perform certain computations (e.g., pattern recognition [1] and perception). In its most general form, neural network is a machine that is designed to model the way in which the brain performs a particular task or function of interest.

The progress of brain work is shown in Fig. 1. The retina is where human begin to put together the relationships between the outside world represented by a visual sense, its physical image projected onto an array of receptors, and the first neural images. The retina converts an optical image into a neural image that is stored in brain. Neurons are trained by learning and relearning image in the brain. Finally, the trained neural can operate some tasks. Process of brain learning from objective world is one in which picture of concept and law formed by objective world in human brain frequently has evolved. On organization chart, receiving fields of eyes, ears and skins are painted in different part and different level of the brain with zone-corresponding form [2]. Trough changes of graph and image, this process always approximates to immediate ensemble of objective world. Learning, computing, reasoning or determining, thinking and dialectical logic are implemented by graphs and images.

From Fig. 1, it shows image is one of the most important parts in the progress of brain working. Neural network simplifies and abstracts the work of brain. It is built up by many processing units that reflect the essential characteristics of brain. Historically, the interest in neural networks is confined to the study of artificial neural network from an engineering perspective, that is to say, build machine that are capable of performing complex tasks, which don't embody the essential and character of neural network. In this paper, the character of neural network is primarily explained by stochastic graph.

---

* The research was supported by National Basic Research Program of China (973 Program) under grant No. 2003CB517106

J. Wang, X. Liao, and Z. Yi (Eds.): ISNN 2005, LNCS 3496, pp. 552–557, 2005.

2 Markov Neural Network and Transition Matrix

Neural network works in complexity environment, frequently learns from new environment. In process of learning and relearning, it implements competition and coordination between states and eventually achieves an equilibrium state [3]. It is Markov neural network (MNN) if transition among states possesses Markov property.

Consider MNN consists of n neurons (including input layer, hidden layer and output layer). Specifically, a neuron is permitted to reside in only one of two states: 1 or 0. It is either in an "on" state denoted by 1 or in an "off" state denoted by 0. $x_a \in \{0,1\}$ denotes one element of state vector $s \in S = \{s, s = (x_1, \cdots x_a, \cdots, x_n)\}$. The state of every neuron is random, so neural network system has 2^n conditions. Every neuron connects with other neuron via bi-directional links is called synapse. Weight w_{ab} is associated with neuron a and b, $w_{ab}=w_{ba}$. Input-output relation of neuron has typical sigmoid response.

Let network work asynchronously, namely, neurons may change their states only one at a time and probability of the state transition is independent of time. Thus the Markov neural network corresponds to stationary ergodic Markov chain. It is completely determined by initial probability distribution

$$\pi^{(0)} = (\pi_1(0), \cdots, \pi_i(0), \cdots, \pi_{2^n}(0)) \tag{1}$$

and transition matrix

$$P = [p_{i,j}](i, j \in \{1, 2, \cdots, 2^n\}) \tag{2}$$

Where $\pi^{(0)}$ is the initial value of state distribution vector, the ith element of $\pi^{(0)}$ is the probability that the chain is in state s_i at initial time, and

$$0 \leq \pi_i(0) \leq 1 \quad \text{for all } (i) \qquad \sum_{i=1}^{2^n} \pi_i(0) = 1 \tag{3}$$

$$0 \leq p_{i,j} \leq 1 \quad \text{for all } (i, j) \qquad \sum_{j=1}^{2^n} p_{i,j} = 1 \quad \text{for all } (i) \tag{4}$$

Corresponding to with transition matrix $P=[p_{i,j}]$, there is stochastic graph $G=(S,U)$. The ith vertex corresponds to the ith state of MNN, and arc $u_{ij}=(s_i, s_j)$ corresponds to transition probability p_{ij}. In the following text, the method of compute transition probability will be introduced in detail.

Let $s_i = (x_1, x_2, \cdots x_a, \cdots, x_n)$ be the current state of the network and $S = \{s_j, j=1,2,\cdots,2^n\}$ be the set of successors of s_i. Let $p_{i,j}^{(T)}$ denote the T-step transition probability from the current state s_i to state s_j. T is referred to simply as the temperature of the system.

From an initial state, a state always transfers to a new state that is neighbouring region $|N_i|$ of current state s_i. The state is accepted in probability manner. It's obvious that acceptance probability depends on new state, current state and temperature T. Thus the states form Markov chain. If temperature T is fixed, the changes of Markov chain wouldn't end until steady state and then decrease temperature, which is called stationary ergodic Markov chain.

Let $g_{i,j}$ denote adjacent transition probability from state s_i to adjacent state s_j:

$$g_{i,j} = \begin{cases} g(i,j)/g(i) & j \in N_i \\ 0 & j \notin N_i \end{cases} \tag{5}$$

It must satisfy condition $g(i) = \sum_{j \in N_i} g(i,j)$. Where $g_{i,j}$ is independent of temperature T. Assume that the current state transfers to adjacent states with the same probability, then $g(i,j)/g(i) = 1/|N_i|$. Where $|N_i|$ is the total number of state. Let $a_{i,j}^{(T)}$ be the acceptance probability from current state s_i to s_j, so

$$a_{i,j}^{(T)} = \min\left\{1, \exp(-(E_j - E_i)/T)\right\} \tag{6}$$

where E_i and E_j is energy of state s_i and s_j respectively under temperature T, the energy E (x_a, x_b denotes the state of neuron a and b respectively) is defined by [4]

$$E = -\frac{1}{2} \sum_{a=1}^{n} \sum_{b=1,a\neq b}^{n} w_{ab} x_a x_b + \sum_{a=1}^{n} x_a \theta_a \tag{7}$$

where θ_a is threshold. By using the adjacent transition probability $g_{i,j}$ and the acceptance probability, the desired set of transition probabilities is formulated as

$$p_{i,j}^{(T)} = \begin{cases} g_{i,j} a_{i,j}^{(T)} & j \in N_i \text{ and } i \neq j \\ 0 & j \notin N_i \text{ and } i \neq j \end{cases} \tag{8}$$

To ensure that the transition probabilities are normalized to unity, this additional condition of the probability without transition is introduced as

$$p_{ii} = 1 - \sum_{j \in N_i} p_{i,j}^{(T)} \tag{9}$$

In the case of a system with a finite number of possible states 2^n, for example, the transition probabilities constitute a 2^n-by-2^n matrix:

$$p^{(T_m)} = \begin{bmatrix} p_{1,1}^{(T_m)} & p_{1,2}^{(T_m)} & \cdots & p_{1,2^n}^{(T_m)} \\ p_{2,1}^{(T_m)} & p_{2,2}^{(T_m)} & \cdots & p_{2,2^n}^{(T_m)} \\ \vdots & \vdots & \vdots & \vdots \\ p_{1,2^n}^{(T_m)} & p_{1,2^n}^{(T_m)} & \cdots & p_{2^n,2^n}^{(T_m)} \end{bmatrix} \tag{10}$$

Where T_m is the temperature in the mth step. Let temperature decrease in every iteration using the rule $T_{m+1} = T_m q$, $T_m > 0$, $0 < q < 1$.

3 Computation of Steady Probability by Stochastic Graph

As step number is increasing, the probability of state s_i can be expressed as [5]:

$$p_i = D_i \bigg/ \sum_{j=1}^{N} D_j \tag{11}$$

where D_i is the principle minor determinant corresponding to the iith element of matrix $E-P$, E is identity matrix, P is transition matrix. Computation of probability of states by stochastic graph method is required to use topological property of minor principal D_i.

Let $G=(S,U)$ be a stochastic graph. Vertex S is the set of all states. Arc $u_{ij}=(s_i,s_j)$ is represented by transition matrix $P=[p_{i,j}]$. Let us introduce following formula

$$\Delta G = 1 - \sum_i L_i + \sum_{i,j} L_i L_j - \sum_{i,j,k} L_i L_j L_k + \cdots \tag{12}$$

where L_i is multiplication of arc forming the ith loop in G. In first summation, let i run overall possible loops. In second summation, let i,j run overall possible loops pair without common vertex. In third summation, let i,j,k run overall possible three loops without common vertex, etc. Value of ΔG is determinant of graph G

$$\Delta G = \det(E - P) \tag{13}$$

The graph, obtained by removing vertex s_i and arc incident with it from stochastic graph G, is represented by symbol ΔG.

Theorem 1 [5]. The minor D_i in the expression (11) is determined by $D_i=\Delta(G_i)$ Thus

$$p_i = \Delta(G_i) \Big/ \sum_{j=1}^{N} \Delta(G_j) \tag{14}$$

$D_i=\Delta(G_i)$ clearly shows probability of steady state of Markov neural network can be computed by using topological property of the stochastic graph.

Transition probabilities are different as temperature T is monotonically decreasing, so stochastic graph changes continually. By learning and relearning, the graph is formed. The state vector is also obtained.

4   Simulation

Consider neural network has 2 neurons, then network has four states: $s_1=(0,0)$, $s_2=(0,1)$, $s_3=(1,0)$, $s_4=(1,1)$. By training, weight matrix is got

$$w = \begin{bmatrix} 0 & 0.55 \\ 0.55 & 0 \end{bmatrix} \tag{15}$$

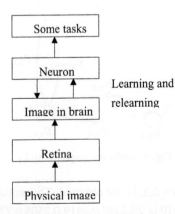

Learning and relearning

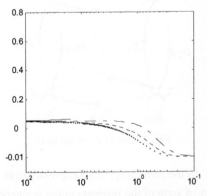

Fig. 1. Structural organization of progress of brain working.

Fig. 2. Steady probability of every state with temperature T.

Threshold matrix is $\Theta=[-0.65\ -0.3]$. The energy of each state can be obtained by Eq. (7).

Suppose initial temperature T_0 is 100 and finial temperature T_{finial} is 0.1. Fig. 2 shows the probability of every state with temperature T. It can be seen that the probability is dependent on temperature T. With the decreasing of temperature T, the probability of state s_4 becomes larger, whereas the probabilities of state s_1,s_2,s_3 become smaller. In order to express clearly, abscissa is used logarithm scale in Fig.2. From Fig. 2, it can be got some results. First, when $T\to\infty$, the state distribution of every state is equal. Second, when $T\to0$, the steady probability of s_4 is approximate to 1. Third, the steady probabilities of all states are $p_4>p_3>p_2>p_1$ with the same temperature, whereas the energy of all states is $E_4<E_3<E_2<E_1$ with the same temperature. Last, the network changes its state along the decrease of energy. and it steadies the state of the least energy finally.

Next, the process of compute steady probability will be explained in detail. Assume that T is 10, the energy of states s_1 and s_2 is 0 and -0.3 respectively. The energy between them is $E_2-E_1=-0.3$. The neighbouring states around s_1 are s_2 and s_3, then $g(i,j)/g(i)=1/3$. By (6), it is easy to obtain $a_{1,2}^{(T)}=0.5$. According to (8), $p_{1,2}^{(10)}$ is 0.5. In the same way, transition matrix is get

$$P=\begin{bmatrix} 0 & 0.5 & 0.5 & 0 \\ 0.4852 & 0.0148 & 0 & 0.5 \\ 0.4685 & 0 & 0.0315 & 0.5 \\ 0 & 0.4435 & 0.4593 & 0.0973 \end{bmatrix} \qquad (16)$$

Fig. 3 shows the stochastic graph under temperature $T=10$. Fig. 4 shows the stochastic graph of $\Delta(G_1)$. $\Delta(G_1)$ is removing vertex s_1 and arc incident with it from stochastic graph G. According to (12), the value of $\Delta(G_1)$ is obtained.

$$\Delta(G_1)=1-P_{2,2}-P_{3,3}-P_{4,4}-P_{4,2}P_{2,4}-P_{3,4}P_{4,3}+P_{2,2}P_{3,4}P_{4,3} \\ +P_{3,3}P_{4,2}P_{2,4}+P_{3,3}P_{2,2}+P_{4,4}P_{2,2}+P_{3,3}P_{4,4}-P_{2,2}P_{3,3}P_{4,4} \qquad (17)$$

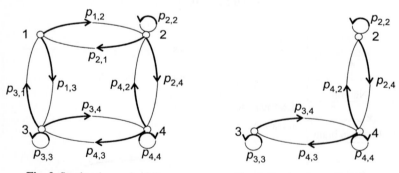

Fig. 3. Stochastic graph of G.   **Fig. 4.** Stochastic graph of $\Delta(G_1)$.

In the same way, other expressions of $\Delta(G_i)(i=2,3,4)$ are obtained. The probability of state of the progress under temperature $T=10$ is $[0.2348\ 0.2419\ 0.2505\ 0.2728]$ according to (14).

5   Conclusion

Human learns from the outside world by image and graph. Neural network simplifies and abstracts the work of brain. In this paper, Markov model for the learning process in a neural network is given and the learning progress is described by stochastic graph. Through training, the Markov neural network can implement computation by means of evolution of the graph. Hence it can implement learning, reasoning, decision making, thinking, dialectical logic, cognizing objective world by means of the graph. It can be seen that Markov neural network gets to the state of the least energy with the decreasing of temperature T, which validates the method of stochastic graph.

References

1. Kim, L.: Proposed Model of Human Consciousness System with Applications in Pattern Recognition. International Conference on Knowledge-Based Intelligent Electronic Systems, Proceedings, KES'97, (1997) 159-166
2. Ewert, J.P.: An Introduction to the Neurophysiological Fundamentals of Behavior. Springer, New York (1980)
3. Xi, G.C.: A Tentative Investigation of the Learning Process of a Neural Network System. Acta Automation Sinica, **17** (1991) 331-316
4. Burnod, Y., et al.: Cooperation Between Learning Rules in the Cerebro-cerebellar Neural Network. Neural Network World, **14** (2004) 207-220.
5. Biondi, E., Guardabassi, G., Kinaldi, S.: On the analysis of Markovian Discrete by Means of Stochastic Graphs. Automation and Remote Control, **28** (1967) 275-277

An Efficient Recursive Total Least Squares Algorithm for Training Multilayer Feedforward Neural Networks

Nakjin Choi[1], JunSeok Lim[2], and KoengMo Sung[1]

[1] School of Electrical Engineering and Computer Science, Seoul National University,
Shinlim-dong, Kwanak-gu, Seoul, Korea, 151-742
{nakjin,kmsung}@acoustics.snu.ac.kr
[2] Department of Electronics Engineering, Sejong University,
98 Kwangjin Kunja, Seoul Korea, 143-747
jslim96@chollian.net

Abstract. We present a recursive total least squares (RTLS) algorithm for multilayer feedforward neural networks. So far, recursive least squares (RLS) has been successfully applied to training multilayer feedforward neural networks. If the input data contains additive noise, the results from RLS could be biased. Such biased results can be avoided by using the RTLS algorithm. The RTLS algorithm described in this paper performs better than RLS algorithm over a wide range of SNRs and involves approximately the same computational complexity of $O(N^2)$ as the RLS algorithm.

1 Introduction

The property of an artificial neural network that is of significant importance is its ability to learn from its environment and to improve its performance through learning. The multilayer feedforward neural networks (MFNNs) have attracted a great deal of interest among artificial neural networks due to its rapid training, generality, and simplicity [1]. In the past decade, the use of the recursive least squares (RLS) algorithm applied to training MFNNs has been extensively investigated [2][3].

In RLS, the underlying assumption is that the input vector is exactly known and all the errors are confined to the observation vector. Unfortunately, in many case this assumption is not true. Because quantization errors, human errors, modeling errors, and instrument errors may preclude the possibility of knowing the input vector exactly [4]. Particularly, in the training process of MFNNs, the inputs of the hidden and output neurons of the net contains an additive noise produced as a result of the use of quantization.

To overcome this problem, the total least squares (TLS) method has been applied. This method compensates for the errors of input vector and the errors of observation vector, simultaneously. Most N-dimensional TLS solutions have been obtained by computing a singular value decomposition(SVD), generally requiring $O(N^3)$ multiplications.

J. Wang, X. Liao, and Z. Yi (Eds.): ISNN 2005, LNCS 3496, pp. 558–565, 2005.

In this paper, we present an efficient recursive total least squares (RTLS) algorithm for training multilayer feedforward neural networks. This RTLS algorithm was first presented by Nakjin Choi et al. [5][6]. This algorithm recursively calculates and tracks the eigenvector corresponding to the minimum eigenvalue, the estimated synaptic weights, from the inverse correlation matrix of the augmented sample matrix. Then, we demonstrate that this algorithm outperforms the RLS algorithm in the training of MFNNs through computer simulation. Moreover, we will show that the recursive TLS algorithm involves approximately the same computational complexity of $O(N^2)$ as RLS algorithm.

In Section II, we will explain the training of MFNNs and in Section III, derive an efficient RTLS algorithm for MFNNs training. In Section IV, the results and analysis of our experiment are present and in Section V are present our conclusions.

2   Training of Multilayer Feedforward Neural Networks

The MFNNs have attracted a great deal of interest due to its rapid training, generality, and simplicity. The architectural graph in Fig. 1 shows the layout of a MFNN for the case of a single hidden layer.

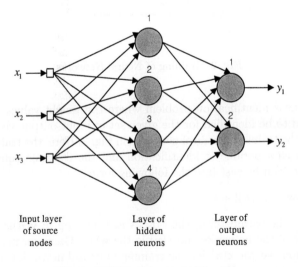

Input layer   Layer of   Layer of
of source   hidden   output
nodes   neurons   neurons

Fig. 1. MFNN with one hidden layer

An MFNN is trained by adjusting its weights according to an ongoing stream of input-output observations $\{x_k(n), y_l(n)\}$ where n denotes the time index. The objective, here, is to obtain a set of weights such that the neural networks can predict future outputs accurately. This training process is exactly the same as the parameter estimation process. Thus, training MFNN can be regarded as a

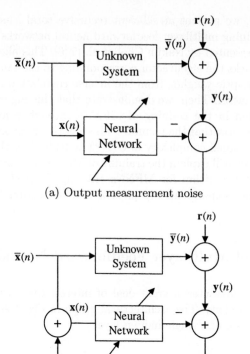

(a) Output measurement noise

(b) Input and output measurement noise

Fig. 2. Training of Neural Networks

kind of nonlinear identification problem where the weight values are unknown and thus need to be identified for the given set of input-output vectors.

Consider the training of neural networks in Fig. 2. Here, the training problem can be posed as a parameter identification and the dynamic equation for the neural network can be cast into the following form

$$\mathbf{y}(n) = \mathbf{f}[\mathbf{w}(n), \mathbf{x}(n)] + \boldsymbol{\varepsilon}(n) . \tag{1}$$

where $\mathbf{w}(n)$ is the parameter of the MFNN at time step n; $\mathbf{y}(n)$ is the observed output vector of the networks and $\epsilon(n)$ is the white Gaussian noises with zero mean. The input vector $\mathbf{x}(n)$ for the training of neural network is usually taken to be clear as shown in Fig. 2 (a). However, the fact that the unknown system input must be sampled and quantized (along with the desired signal) results in the production of a broad-band quantization noise which contaminates the neural network input. Therefore, the model in Fig. 2(b) which contains both input and output measurement noises is more practical than that in Fig. 2(a) which contains input measurement noise only.

The TLS method compensates for the errors of input vector $\mathbf{x}(n)$ and observation vector $\mathbf{y}(n)$, simultaneously.

3   Efficient RTLS Algorithm for MFNNs

The TLS problem which is a minimization problem can be described as follows

$$\text{minimize } \|[\mathbf{E}|\mathbf{r}]\|_F \quad \text{subject to} \quad (\mathbf{A} + \mathbf{E})\mathbf{x} = \mathbf{b} + \mathbf{r} . \tag{2}$$

where $\mathbf{A}$ is an $M \times N$ input matrix and $\mathbf{b}$ is an $M \times 1$ output vector. $\mathbf{E}$ and $\mathbf{r}$ are the $M \times N$ input error matrix and $M \times 1$ output error vector, respectively. Once a minimization $\mathbf{E}$ and $\mathbf{r}$ are found, then any $\mathbf{x}$ satisfying

$$(\mathbf{A} + \mathbf{E})\mathbf{x} = \mathbf{b} + \mathbf{r} . \tag{3}$$

is assumed to be used to solve the TLS problem in Eq.(2).

Let's define $M \times (N + 1)$ augmented data matrix $\overline{\mathbf{A}}$ as

$$\overline{\mathbf{A}} = [\mathbf{A}|\mathbf{b}] . \tag{4}$$

Golub and Van Loan proved that the TLS solution $\mathbf{x}_{TLS}$ involves the most right singular vector of the $\overline{\mathbf{A}}$, as follows [4]:

$$\mathbf{x}_{TLS} = -\frac{1}{\nu_{N+1,N+1}} \begin{pmatrix} \nu_{N+1,1} \\ \vdots \\ \nu_{N+1,N} \end{pmatrix} \quad \text{where} \quad \mathbf{v}_{N+1} = \begin{pmatrix} \nu_{N+1,1} \\ \vdots \\ \nu_{N+1,N+1} \end{pmatrix} . \tag{5}$$

Now, let us develop the adaptive algorithm which calculates the TLS solution recursively.

We know that the most right singular vector $\mathbf{v}_{N+1}(n)$ can be obtained from the SVD of the inverse correlation matrix $\mathbf{P}(n)$ which can be defined as

$$\mathbf{P}(n) = \mathbf{R}^{-1}(n) \quad \text{where} \quad \mathbf{R}(n) = \overline{\mathbf{A}}^H(n)\overline{\mathbf{A}}(n) \quad n = 1, 2, \ldots, M . \tag{6}$$

The minimum eigenvector $\mathbf{v}_{N+1}(n-1)$ at time index $n-1$ can be represented as a linear combination of the orthogonal eigenvectors $\mathbf{v}_1(n), \ldots, \mathbf{v}_{N+1}(n)$ at time index n.

$$\mathbf{v}_{N+1}(n - 1) = c_1(n)\mathbf{v}_1(n) + \cdots + c_{N+1}(n)\mathbf{v}_{N+1}(n) . \tag{7}$$

Because $\mathbf{P}(n)$ and $\mathbf{P}(n - 1)$ are highly correlated, the coefficient $c_{N+1}(n)$ by which the minimum eigenvector $\mathbf{v}_{N+1}(n)$ is multiplied, produces a larger value than any other coefficients. That is,

$$c_{N+1}(n) \geq c_N(n) \geq \ldots \geq c_1(n) . \tag{8}$$

Now, consider a new vector $\mathbf{v}(n)$ which can be defined as

$$\mathbf{v}(n) = \mathbf{P}(n)\mathbf{v}_{N+1}(n - 1) . \tag{9}$$

By using the singular value decomposition (SVD) $\mathbf{A}$ and Eq. (7), we can rewrite the vector $\mathbf{v}(n)$ as Eq. (10)

$$\mathbf{v}(n) = \mathbf{P}(n)\mathbf{v}_{N+1}(n - 1)$$

$$= \left(\sum_{i=1}^{N+1} \frac{1}{\sigma_i^2(n)} \mathbf{v}_i(n) \mathbf{v}_i^H(n) \right) \left(\sum_{j=1}^{N+1} c_j(n) \mathbf{v}_j(n) \right)$$

$$= \sum_{i=1}^{N+1} \frac{c_i(n)}{\sigma_i^2(n)} \mathbf{v}_i(n) . \tag{10}$$

From Eq. (8) and using the property of SVD, the order of the magnitudes of the coefficients $\mathbf{v}_i(n)$ becomes

$$\frac{c_{N+1}(n)}{\sigma_{N+1}^2(n)} \geq \frac{c_N(n)}{\sigma_N^2(n)} \geq \cdots \geq \frac{c_1(n)}{\sigma_1^2(n)} . \tag{11}$$

So, $\mathbf{v}(n)$ in Eq.(10) converges to the scaled minimum eigenvector as follows.

$$\mathbf{v}(n) \approx \frac{c_{N+1}(n)}{\sigma_{N+1}^2(n)} \mathbf{v}_{N+1}(n) . \tag{12}$$

Therefore, the minimum eigenvector $\mathbf{v}_{N+1}(n)$ can be approximately calculated from the $\mathbf{v}(n)$ as in Eq. (13).

$$\hat{\mathbf{v}}_{N+1}(n) = \frac{\mathbf{v}(n)}{\|\mathbf{v}(n)\|} . \tag{13}$$

The above procedure is summarized in Table 1.

MAD's stands for the number of multiplies, divides, and square roots. This new RTLS algorithm requires almost the same order of MAD's of $O(N^2)$. Moreover, it maintains the structure of the conventional RLS algorithm so that the complexity can be reduced by adopting various fast algorithms for RLS.

4   Simulation

Fig. 3 presents the MFNN model to identify the unknown system. It is connected with one hidden layer and one output layer. It has 1 source node, 25 hidden neurons, and 1 output neuron. The bias of hidden layer ranges from -6.0 to 6.0 which is the dynamic range of input data x. The sampling frequency is 2Hz, which satisfies the Nyquist sampling rate. In addition, the synaptic weights of the first hidden layer is fixed to 1. This model can be found in [7].

The unknown system is modeled as a nonlinear system. In this simulation, we assume that the relationship between input $\bar{x}$ and output $\bar{y}$ of the unknown system is

$$\bar{y} = \sin(3\bar{x}), \quad , -6 \leq \bar{x} \leq 6 . \tag{14}$$

In this MFNN model, the output of the first hidden layer is

$$\varphi = [\varphi_1(x - 6.0) \quad \varphi_2(x - 5.5) \quad \cdots \quad \varphi_{25}(x + 6.0)]^T . \tag{15}$$

Table 1. Summary of the new RTLS algorithm

Initialize this algorithm by setting
$\mathbf{P}(0) = \delta^{-1}\mathbf{I}, \hat{\mathbf{v}}_{N+1}(0) = \frac{1}{\|\mathbf{1}\|}$
where δ is a small positive constant, $\mathbf{I}$ is an $(N+1) \times (N+1)$ identity matrix
and $\mathbf{I}$ is an $(N+1) \times 1$ 1's vector.

Compute for each instant of time, $n = 1, 2, \ldots, L$	MAD's
1. $\bar{\mathbf{a}}(n) = \begin{pmatrix} \mathbf{a}(n) \\ b(n) \end{pmatrix}$	
2. $\mathbf{k}(n) = \frac{\lambda^{-1}\mathbf{P}(n-1)\bar{\mathbf{a}}(n)}{1+\lambda^{-1}\bar{\mathbf{a}}^H(n)\mathbf{P}(n-1)}$	2. $N^2 + 4N + 4$
3. $\mathbf{P}(n) = \lambda^{-1}\mathbf{P}(n-1) - \lambda^{-1}\mathbf{k}(n)\bar{\mathbf{a}}^H(n)\mathbf{P}(n-1)$	3. $3N^2 + 6N + 3$
4. $\mathbf{v}(n) = \mathbf{P}(n)\hat{\mathbf{v}}_{N+1}(n-1)$	4. $N^2 + 2N + 1$
5. $\hat{\mathbf{v}}_{N+1}(n) = \frac{\mathbf{v}(n)}{\|\mathbf{v}(n)\|}$	5. $N^2 + 3N + 3$
6. $\hat{\mathbf{x}}_{TLS}(n) = -\frac{1}{\hat{v}_{N+1,N+1}(n)}\begin{pmatrix} \hat{v}_{N+1,1}(n) \\ \vdots \\ \hat{v}_{N+1,N}(n) \end{pmatrix}$	6. $N+1$
where $\hat{\mathbf{v}}_{N+1}(n) = \begin{pmatrix} \hat{v}_{N+1,1}(n) \\ \vdots \\ \hat{v}_{N+1,N+1}(n) \end{pmatrix}$	

Total number of MAD's	$6N^2 + 16N + 12$

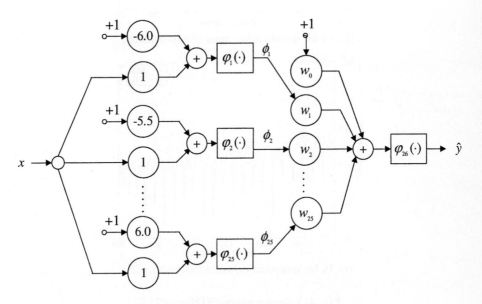

Fig. 3. MFNN model

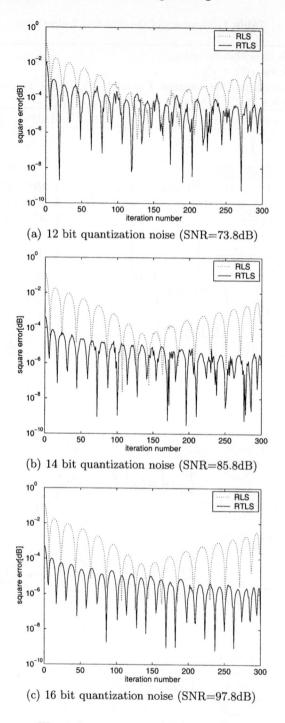

(a) 12 bit quantization noise (SNR=73.8dB)

(b) 14 bit quantization noise (SNR=85.8dB)

(c) 16 bit quantization noise (SNR=97.8dB)

Fig. 4. Learning curves(RLS vs. RTLS)

and the output of total MFNN model is

$$\hat{y} = \mathbf{w}^H \varphi \quad \text{where} \quad \mathbf{w} = [w_1 \quad w_2 \quad \cdots \quad w_{25}]^H . \tag{16}$$

The input and output measurement noises in Fig. 2 and Fig. 3 is a zero-mean white noise process that is independent of the input vector.

It is assumed that there is a 12, 14, and 16 bits μ-law pulse code modulation (PCM) quantizer which will produce SNRs of 73.8, 85.8, and 97.8 dB respectively. Fig. 4 shows the errors derived from RLS and RTLS with these SNRs. It is observed that the presented RTLS outperforms the conventional RLS and performs better,particularly, at high SNRs.

5   Conclusions

In this paper, an efficient RTLS algorithm is presented. This algorithm was found to outperform the RLS algorithm in neural network training and it involves approximately the same computational complexity as RLS algorithm. In order to validate its performance, we applied this algorithm in the training MFNN in various quantization noise conditions. In each case, the results showed that the RTLS outperforms the RLS.

References

1. Haykin, S.: Neural Networks - A Comprehensive Foundation, 2nd Ed., Prentice-Hall (1999)
2. Wong, K.W., and Leung, C.S.: On-line Successive Synthesis of Wavelet Networks. Neural Processing Lett., **7** (1998) 91-100
3. Leung, C.S., Tsoi, A.C.: Two-regularizers for Recursive Least Squared Algorithms in Feedforward Multilayered Neural Networks. IEEE Trans. Neural Networks, **12** (2001) 1314-1332
4. Golub, G.H., and Van Loan, C.F.: An Analysis of the Total Least Squares Problem. SIAM J. Numer. Anal., **17** (1980) 883-893
5. Choi, N., Lim, J.S., Song, J., Sung, K.M.: Adaptive System Identification Using an Efficient Recursive Total Least Squares Algorithm. Journal of the Acoustical Society of Korea. **22** (2003) 93-100
6. Choi, N., Lim, J.S., Song, J., Sung, K.M.: An Efficient Recursive Total Least Squares Algorithm for Fequency Estimation. ICA, **25** (2001) 6A.06.02
7. Passino, K.M.: Biomimicry for Optimization, Control and Automation. Springer-Verlag, London, UK. (2004)

A Robust Learning Algorithm for Feedforward Neural Networks with Adaptive Spline Activation Function

Lingyun Hu and Zengqi Sun

State Key Laboratory of Intelligent Technology and Systems,
Computer Science and Technology Department, Tsinghua University, Beijing 100084, China
huly02@mails.tsinghua.edu.cn, szq-dcs@mail.tsinghua.edu.cn

Abstract. This paper proposed an adaptive robust learning algorithm for spline-based neural network. Adaptive influence function was dynamically added before objective function to modify the learning gain of back-propagate learning method in neural networks with spline activation functions. Besides the nonlinear activation functions in neurons and linear interconnections between neurons, objective function also changes the shape during iteration. This employed neural network the robust ability to reject gross errors and to learn the underlying input-output mapping from training data. Simulation results also conformed that compared to common learning method, convergence rate of this algorithm is improved for: 1) more free parameters are updated simultaneously in each iteration; 2) the influence of incorrect samples is gracefully suppressed.

1 Introduction

The classical neuron model is represented as a nonlinear activation function with linear combiner. Since behavior of neural network (NN), as shown by multilayer perceptron, is mostly determined by the shape of activation function. Sigmoid functions are greatly used in NN applications.

On the other hand, representation capability of NN depends on the number of free parameters, whatever the structure of the network. To simplify the complexity of structure and computation in hardware and software implementations, adaptive activation functions is proposed to provide more free parameters for representation. Based on it, A natural solution is involving adaptive sigmoid function $\frac{2a}{1 + e^{-bx}} - c$ with gradient-based learning algorithm. Advantages of better data modeling of this method were proved by Chen and Chang [1]. Compared with it, a simpler adaptive solution is LUT (look-up-table) based on polynomial functions [2]. Complexity of this method can be controlled easily because that each LUT element represents a free parameter. But it simultaneously increases the difficult of coefficients learning and makes the activation function non-boundary.

Followed with it, Vecci [3] proposed the multilayer forward NN with adaptive spline activation functions (ASNN), whose shape can be modified through control points. Later, Guarnieri [4] presented a new structure of ASNN with gradient-based learning algorithm. Good generalization capabilities and high adaptation were achieved thorough this method with smaller structural complexity.

J. Wang, X. Liao, and Z. Yi (Eds.): ISNN 2005, LNCS 3496, pp. 566–571, 2005.

To apply ASNN into online biped gait recognition and planning, for its structural simplicity, an adaptive robust learning algorithm based on classic BP method is proposed in this paper. Object function is modified with influence factor to optimize both the connection between neurons and control points in neurons. Newly added influence function adjusts function shape according to learning errors, which guaranteed the consequently object function in form of Hampel tanh estimator.

2 The ASNN

Different to traditional sigmoid activation function neurons, spline neurons adopt piecewise polynomial spline interpolation schemes to ensure a continuous first derivative and local adaptation. For given points $\{Q_1 \cdots Q_N\}$, $Q_i = [q_{xi}, q_{yi}]^T$, a general

spline expression for that curve would be $F(u) = \bigcup\limits_{i=1}^{N-3} F_i(u)$ .Where $\bigcup$ is the con-

catenation   operator   on   local   spline   basis   functions   and

$F_i(u) = [F_{xi}(u), F_{yi}(u)]^T = \sum\limits_{j=0}^{3} C_j(u)Q_{i+j} \cdot F_i(u)$ ($0 \le u \le 1$) is the i th curve span function

controlled by four points $[Q_i \quad Q_{i+1} \quad Q_{i+2} \quad Q_{i+3}]^T$ and $C_j(u)$ are the Catmull-Rom (CR) cubic spline basis.

For input s , proper local parameters of u and i can be obtained through $s = F_{xi}(u)$ and then substitute these values into $y = F_{yi}(u)$ to generate the corresponding output. To accelerate the computation of local parameters and reduce free parameters, we uniformly sample the x-axis with fixed step length $\Delta x = q_{x,i+1} - q_{x,i}$.

This turned $F_{xi}(u)$ to a first degree polynomial of $F_{xi}(u) = u\Delta x + q_{x,i+1}$.

For a standard multilayer ASNN with M layer and N_l ($l = 1, \cdots M$) neurons per layer, quantities are defined as follows for later discussion. Structural details of ASNN and spline neuron are shown in Fig1.

For a standard multilayer ASNN with M layer and N_L ($l=1,...M$) neurons per layer, quantities are defined as follows for later discussion.

$x_k^l(t, p)$: Output of the kth neuron in the lth layer at the tth iteration related to the pth batch input of the NN. ($p=1,...P$, $x_k^0(t)$ are inputs of the network; $x_k^l(t)=1$ for bias computation. Indexes t and p are the same meaning in following definition.)

$w_{kj}^l(t, p)$: Weight of the kth neuron in the lth layer with respect to the jth neuron in the previous layer. (w_{0j}^l are bias weight)

$s_k^l(t, p)$: The linear combiner input of the kth neuron in the lth layer.

N: Number of control points in every neuron.

$i_k^l(t, p)$: Curve span index of the activation function for the kth neuron in the lth layer. ($1 \le i_k^l(t,p) \le N-3$)

$u_k^l(t, p)$: Local parameters of the $i_k^l(t)$ curve span for the kth neuron in the lth layer. ($0 \le u_k^l(t,p) \le 1$)

Four forward NN with four kinds of learning methods, as listed in table 1, are tested for ZMP curve learning. 28 control points were initialized by sigmoid function in each ASNN neuron and the value of 28 was determined empirically. More neurons and larger learning rate are designed in common NN to achieve equivalent adaptable variables and better results. Results of standard deviation as $err = \dfrac{\sqrt{\sum_{i=1}^{n}(d_i - X_i)^2}}{n}$ and object function (for NN2 and NN4 only) are also shown in table1 for comparison. Since only 1 neuron is set in output layer in NN1 and NN2, training data for these two networks is transformed to (-1, 1). To convenient comparison, error function calculation for NN3 and NN4 is also mapping to the same region.

Comparing results of the 4 networks after 300 training epochs, it can be found that NN with common learning method (NN1 and NN3) are attracted by exterior points. Learning speed of NN3 is obviously faster than that of NN1 with the same number of freedom parameters. But neither of them learned the real outliner of this curve in the first 300 steps. NN2 and NN4, especially the latter network, performed better in contour forming.

Experiment results indicated that ASNN exhibit quicker learning speed and higher approximation ability than that of common neuron network with the same BP algorithm. In the mean time, robust learning method endows the network ability to reject gross errors and to learn the underlying input-output mapping from training data. Objective function of tanh form also took a good balance between bias and variance of approximation. NN4, which binding the two advantages together, proved above discussion.

Table 1. Four kinds of NN.

	NN1	NN2	NN3	NN4
Structure of NN	Common 3-layer forward NN		ASNN	
Learning method	Common BP	Robust BP	Common BP	Robust BP
Neuron number in hidden layer	12		1	
Free Parameters	36			
Learning Rate	0.01		0.001	

5   Conclusion

This paper focused on the balance between better relationship forming ability and higher approximation characteristics for ASNN. Compared to classical BP algorithm, improvement in balance of this algorithm should own to 1): robust learning algorithm with the adaptive objective function in tanh form, which abrogates outliers for relationship learning; 2): structure of ASNN, which provides more available parameters for learning simultaneously. Its excellent performing will be brought to biped gait recognition in our future research.

References

1. Chen, C. T., Chang, W. D.: A Feedforward Neural Network with Function Shape Auto Tuning. Neural Networks, **9** (1996) 627-641
2. Piazza, F., Uncini, A. and Zenobi, M.: Neural Networks with Digital LUT Activation Function. In Proc. of the IJCNN, Beijing, China, **2** (1993) 343-349
3. Vecci, L, et.al: Learning and Approximation Capabilities of Adaptive Spline Activation Function Neural Networks. In Neural Networks, **11** (1998) 259-270
4. Stefano, G., Francesco P., Aurelio U.: Multilayer Feedforward Networks with Adaptive Spline Activation Function. IEEE Trans. on Neural Networks, **10** (1999) 672-683
5. Hu, L., Sun, Z.: A Rapid Two-Step Learning Algorithm for Spline Activation Function Neural Networks with the Application on Biped Gait Recognition. In: Fuliang, Y., Jun W., Cheng G. (eds.): Advanc4es in Neural Networks. Lecture Notes in Computer Science, **3173**. Springer-Verlag, Heidelberg Germany Dalian China (2004) 286-292
6. Hampel, F.R. et al.: Robust Statistics-The Approach Based on Influence Function. New York: Wiley (1986) 150-153

A New Modified Hybrid Learning Algorithm for Feedforward Neural Networks

Fei Han[1,2], Deshuang Huang[1] ,Yiuming Cheung[3], and Guangbin Huang[4]

[1] Intelligent Computing Lab, Hefei Institute of Intelligent Machines,
Chinese Academy of Sciences, P.O.Box 1130, Hefei, Anhui 230031, China
{hanfei1976,dshuang}@iim.ac.cn
[2] Department of Automation, University of Science and Technology of China,
Hefei, Anhui 230027, China
[3] Department of Computer Science, Hong Kong Baptist University, Hong Kong,China
ymc@comp.hkbu.edu.hk
[4] School of Electrical and Electronic Engineering, Nanyang Technological University
egbhuang@ntu.edu.sg

Abstract. In this paper, a new modified hybrid learning algorithm for feedforward neural networks is proposed to obtain better generalization performance. For the sake of penalizing both the input-to-output mapping sensitivity and the high frequency components in training data, the first additional cost term and the second one are selected based on the first-order derivatives of the neural activation at the hidden layers and the second-order derivatives of the neural activation at the output layer, respectively. Finally, theoretical justifications and simulation results are given to verify the efficiency and effectiveness of our proposed learning algorithm.

1 Introduction

Most of traditional supervised learning algorithms with feedforward neural networks (FNN) are to use the sum-of-square error criterion to derive the updated formulae. However, these learning algorithms have not considered the network structure and the involved problem properties, thus their capabilities are limited [1], [2], [3], [4].

In literatures [5],[6], two algorithms were proposed that are referred to as Hybrid-I method and Hybrid-II method, respectively. The Hybrid-I algorithm incorporates the first-order derivatives of the neural activation at hidden layers into the sum-of-square error cost function to reduce the input-to-output mapping sensitivity. On the other hand, the Hybrid-II algorithm incorporates the second-order derivatives of the neural activations at hidden layers and output layer into the sum-of-square error cost function to penalize the high frequency components in training data. All the above learning algorithms can almost improve the generalization performance to some degree, but not having the best generalization performance.

In this paper, we proposed a new error cost function which is based on the above two learning algorithms. This new modified cost function is made up of three parts. The first term is the sum-of-square errors, while the second term and the last term consider the first-order derivatives of the neural activation at the hidden layers and the second-order derivatives of the neural activation at output layer, respectively. Finally, simulation results through time series prediction are presented to substantiate the efficiency and effectiveness of our proposed learning algorithm. It can be found that

J. Wang, X. Liao, and Z. Yi (Eds.): ISNN 2005, LNCS 3496, pp. 572–577, 2005.

our proposed learning algorithm has better generalization performance than that of the above two learning algorithms. In addition, what should be stressed is that since a neural network with single nonlinear hidden layer is capable of forming an arbitrarily close approximation of any continuous nonlinear mapping [7],[8],[9], our discussion will be limited to such networks.

2 The Modified Hybrid Learning Algorithm

Before presenting new modified cost function, we first make the following mathematical notation. Assume that x_k and y_i denote the kth element of the input vector and the ith element of the output vector, respectively; $w_{j_l j_{l-1}}$ denotes the synaptic weight from the j_l th neuron at the l th layer to the j_{l-1} th neuron at $(l-1)$ th layer; $h_{j_l} = f_l(\hat{h}_{j_l})$ is the activation function of the j_l th element at the l th layer with $\hat{h}_{j_l} = \sum_{j_{l-1}} w_{j_l j_{l-1}} h_{j_{l-1}}$. The t_i and y_i denote the target and actual output values of the i th neuron at output layer, respectively.

To obtain a more efficient cost function with respect to Hybrid-I and Hybrid-II algorithms in literatures [5], [6], a new cost function based on the neural network with single nonlinear hidden layer embodying the additional hidden layer penalty term and the weights decay term at the output layer is defined as follows:

$$E_{new} = \frac{1}{N}\sum_{p=1}^{N} E_O^p + \frac{1}{N}\sum_{p=1}^{N} E_B^p + \frac{1}{N}\sum_{p=1}^{N} E_C^p.\tag{1}$$

where

$$E_O^p = \frac{1}{2}\sum_{j=1}^{J}(t_j^p - y_j^p)^2,\tag{2}$$

$$E_B^p = (\gamma_h/H)\sum_{j_2=1}^{H} f'(\hat{h}_{j_2}^p),\tag{3}$$

and

$$E_C^p = (\gamma_o/J)\sum_{j_3=1}^{J} f'(\hat{h}_{j_3}^p)(\frac{1}{2}\sum_{j_2=1}^{H}(w_{j_3 j_2})^2).\tag{4}$$

The cost terms E_O^p, E_B^p and E_C^p denote the normalized output error, the additional hidden layer penalty term and the weights decay term at the output layer for the p th training pattern, respectively. The gains γ_h and γ_o represent the relative significance among the four cost terms. N, H and J denote the number of the stored patterns, the number of the neurons at the hidden layer and the number of the neurons at the output layer, respectively. In this paper, we adopt the same activation function for all neurons at all layers but the input layer, i.e., tangent sigmoid transfer function:

$$f(x) = (1 - \exp(-2x))/(1 + \exp(-2x)).\tag{5}$$

It can be deduced that this activation function has the following property:

$$f''(x) = -2f(x)f'(x).\tag{6}$$

Assuming that the network is trained by a steepest-descent error minimization algorithm as usual, in order to make this algorithm easier to understand, we first denotes

Robust Recursive TLS (Total Least Square) Method Using Regularized UDU Decomposed for FNN (Feedforward Neural Network) Training

JunSeok Lim[1], Nakjin Choi[2], and KoengMo Sung[2]

[1] Department of Electronics Engineering, Sejong University,
98 Kwangjin Kunja 143-747, Seoul Korea
jslim@sejong.ac.kr
[2] School of Electrical Engineering and Computer Science, Seoul National University

Abstract. We present a robust recursive total least squares (RRTLS) algorithm for multilayer feed-forward neural networks. So far, recursive least squares (RLS) has been successfully applied to training multilayer feed-forward neural networks. However, if input data has additive noise, the results from RLS could be biased. Theoretically, such biased results can be avoided by using the recursive total least squares (RTLS) algorithm based on Power Method. In this approach, Power Method uses rank-1 update. and thus is apt to be in ill condition. In this paper, therefore, we propose a robust RTLS algorithm using regularized UDU factorization[1]. This method gives better performance than RLS based training over a wide range of SNRs.

1 Introduction

In an artificial neural network, the most fundamental property is the ability of the network to learn from its environment and to improve its performance through learning. The multilayer feedforward neural networks (FNNs) have attracted a great deal of interest due to its rapid training, generality and simplicity. In the past decade, the use of the recursive least squares (RLS) algorithm for training FNNs has been investigated extensively [1],[2]. In the RLS problem, the underlying assumption is that the input vector is exactly known and all the errors are confined to the observation vector. Unfortunately, this assumption,in many cases, is not true. Quantization errors, modeling errors, and instrument errors may introduce input noise and output noise [3]. Particularly, in the training of FNNs, the inputs of the hidden layer and output of the neural net may also contain an additive noise because of the use of quantization. To overcome this problem, the total least squares (TLS) method has been devised [3]. This method compensates for the errors of input vector and observation vector, simultaneously. Most N-dimensional TLS solutions have been obtained by computing singular value decomposition (SVD).

---

[1] UDU factorization is also called square root free Cholesky decomposition

J. Wang, X. Liao, and Z. Yi (Eds.): ISNN 2005, LNCS 3496, pp. 578–584, 2005.

In this paper, we present a robust recursive total least squares (RRTLS) algorithm for training multilayer feed-forward neural networks. This method is based on the algorithm proposed by the Authors [4],[5], which is called RTLS in this paper. This RTLS recursively calculates and tracks the eigenvector corresponding to the minimum eigenvalue based on the Power Method for the augmented sample matrix. This approach uses rank-1 update, and thus, is apt to be in ill condition. The proposed algorithm herein, makes the RTLS algorithm robust using regularized UDU factorization and shows that it performs better than RLS and RTLS in FNN training with input and output noise.

2   Training of Multilayer Feedforward Neural Networks

Fig. 1 shows the layout of a multiplayer feedforward neural network for a single hidden layer.

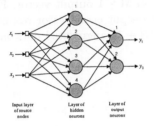
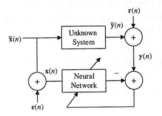

(a) FNN with one hidden layer

(b) Equivalent model with input and output noises

Fig. 1. Training of Neural Networks

An FNN is trained by adjusting its weights according to an ongoing stream of input-output observations $\{x_k(n), y_k(n) : t = 1, \ldots, N; k = 1, \ldots, p\}$ where p denotes the number of training samples sequences. The objective is to obtain a set of weights such that the neural networks will predict future outputs accurately. This training process is the same as parameter estimation process. Thus, training FNN can be considered as a nonlinear identification problem where the weight values are unknown and need to be identified for the given set of input-output vectors.

Considering the training of neural networks in Fig. 1(a), the training problem can be regarded as a parameter identification and the dynamic equation for the neural network can be expressed as follows:

$$\mathbf{y}(n) = \mathbf{f}[\mathbf{w}(n), \mathbf{x}(n)] + \varepsilon(n) .$$ (1)

where $\mathbf{w}(n)$ is the parameter of the FNN at time step n; $\mathbf{y}(n)$ is the observed output vector of the networks; and $\epsilon(n)$ is white Gaussian noises with zero mean.

In many cases, it is assumed that input is clear and output has noise. However, the measured input, in many cases, also has noise as well. This noise originates from quantizer, etc. Therefore, Fig. 2(b) model, which contains input and output measurement noise, is more practical. It is well known that the total least squares (TLS) method can be successfully applied to the system with errors or noise in input and output simultaneously [2],[3]. In Fig 1(a), we can apply TLS for noisy output of hidden layer and noisy output of neural network in Fig. 1(a).

3   TLS (Total Least Square) and an Efficient RTLS (Recursive TLS) Algorithm for FNNs

TLS problem is a minimization problem, which can be described as follows

$$\text{minimize } \|[\mathbf{E}|\mathbf{r}]\|_F \quad \text{subject to} \quad (\mathbf{A} + \mathbf{E})\mathbf{w} = \mathbf{b} + \mathbf{r}. \tag{2}$$

where $\mathbf{A}$ is an M × N input matrix, $\mathbf{b}$ is an M × 1 output vector. $\mathbf{E}$ and $\mathbf{r}$ are the M × N input error matrix and M × 1 output error vector respectively. For the TLS solution, let us define the M × (N + 1) augmented data matrix $\overline{\mathbf{A}}$ as $\overline{\mathbf{A}} = [\mathbf{A}|\mathbf{b}]$. Then Eq.(2) becomes

$$\overline{\mathbf{A}}\tilde{\mathbf{x}} = 0, \tag{3}$$

where $\tilde{\mathbf{x}} = [\mathbf{x}^T| - 1]$. Golub and Van Loan proved that the TLS solution $\mathbf{w}_{TLS}$ could be derived from the most right singular vector of the $\overline{\mathbf{A}}, \hat{\mathbf{V}}_{N+1}$, as follows [3],

$$\mathbf{x}_{TLS} = -\frac{1}{\nu_{N+1,N+1}} \begin{pmatrix} \nu_{N+1,1} \\ \vdots \\ \nu_{N+1,N} \end{pmatrix} \quad \text{where} \quad \mathbf{v}_{N+1} = \begin{pmatrix} \nu_{N+1,1} \\ \vdots \\ \nu_{N+1,N+1} \end{pmatrix}. \tag{4}$$

We have developed a recursive algorithm for the TLS solution in [4]. In the algorithm, the most right singular vector $\mathbf{v}_{N+1}(n)$ can be obtained from the inverse correlation matrix $\mathbf{P}(n)$ that is defined as

$$\mathbf{P}(n) = \mathbf{R}^{-1}(n) \quad \text{where} \quad \mathbf{R}(n) = \overline{\mathbf{A}}^H(n)\overline{\mathbf{A}}(n) \quad n = 1, 2, \dots, M. \tag{5}$$

Therefore, the minimum eigenvector $\mathbf{v}_{N+1}(n)$ can be approximately calculated from the $\mathbf{v}(n)$ as in Eq. (6).

$$\hat{\mathbf{v}}_{N+1}(n) = \frac{\mathbf{v}(n)}{\|\mathbf{v}(n)\|}. \tag{6}$$

This is a kind of Power Method. Using this algorithm, we can derive a new training algorithm for FNN based on [4],[5]. The procedure is summarized in Table 1.

Table 1. Summary of the RTLS (Recursive Total Least Square) algorithm for neural net training

---

Initialize this algorithm by setting
$\mathbf{P}(0) = \delta^{-1}\mathbf{I}, \hat{\mathbf{v}}_{N+1}(0) = \frac{1}{\|\mathbf{1}\|}$,
where δ is a small positive constant, $\mathbf{I}$ is an $(N+1) \times (N+1)$ identity matrix
and $\mathbf{I}$ is an $(N+1) \times 1$ 1's vector.

Compute for each instant of time, $n = 1, 2, \ldots, N$

---

1. $\bar{\mathbf{x}}(n) = \left(\mathbf{h}(n)^T \ y(n)\right)^T$

2. $\mathbf{k}(n) = \frac{\lambda^{-1}\mathbf{P}(n-1)\bar{\mathbf{x}}(n)}{1+\lambda^{-1}\bar{\mathbf{x}}^H(n)\mathbf{P}(n-1)}$

3. $\mathbf{P}(n) = \lambda^{-1}\mathbf{P}(n-1) - \lambda^{-1}\mathbf{k}(n)\bar{\mathbf{x}}^H(n)\mathbf{P}(n-1)$

4. $\mathbf{v}(n) = \mathbf{P}(n)\hat{\mathbf{v}}_{N+1}(n-1)$

5. $\hat{\mathbf{v}}_{N+1}(n) = \frac{\mathbf{v}(n)}{\|\mathbf{v}(n)\|}$

6. $\hat{\mathbf{w}}_{\text{TLS}}(n) = -\frac{1}{\hat{\nu}_{N+1,N+1}(n)}\left(\hat{\nu}_{N+1,1}(n)\ldots\hat{\nu}_{N+1,N}(n)\right)^T$

   where $\hat{\mathbf{v}}_{N+1}(n) = \left(\hat{\nu}_{N+1,1}(n)\ldots\hat{\nu}_{N+1,N+1}(n)\right)^T$

---

4 RRTLS (Robust Recursive TLS) Using Regularized UDU Decomposed for FNN Training

The performance of Recursive TLS is dependent mainly on the quality of P(n). However, recursive update of P(n) very often becomes ill conditioned because Power Method uses rank 1 update. In this section, we propose a method to improve the robustness against the ill condition by applying the regularized UDU factorization to the update of P(n). For the regularized UDU factorization, we introduce limits to the elements of the diagonal matrix, D by adding a loading matrix with small positive elements.

To derive the regularized UDU factorized Recursive TLS, we introduce the UDU factorization first as follows.

$$\mathbf{P}(n) = \lambda^{-1}\mathbf{P}(n-1) - \lambda^{-1}\mathbf{k}(n)\bar{\mathbf{x}}^H(n)\mathbf{P}(n-1) \tag{7}$$

$$= \lambda^{-1}\mathbf{P}(n-1) - \lambda^{-1}\mathbf{P}(n-1)\bar{\mathbf{x}}(n)\bar{\mathbf{x}}^H(n)\mathbf{P}(n-1) \tag{8}$$

Suppose $\mathbf{P}(n-1)$ has been U-D decomposed as $\mathbf{P}(n-1) = \mathbf{U}(n-1)\mathbf{D}(n-1)\mathbf{U}^T(n-1)$ at time n-1, Eq.(8) then becomes,

$$\mathbf{P}(n-1) = \mathbf{U}(n-1)\mathbf{D}(n-1)\mathbf{U}^H(n-1), \tag{9}$$

$$\mathbf{P}(n) = \lambda^{-1}\mathbf{U}(n-1)\left[\mathbf{D}(n-1) - \frac{\mathbf{gg}}{\beta}\right]\mathbf{U}(n-1), \tag{10}$$

where $\mathbf{g} = \mathbf{D}(n-1)\mathbf{e}$, $\mathbf{e} = \mathbf{U}(N-1)\bar{\mathbf{x}}(n)$ and $\beta = \mathbf{e}^T\mathbf{D}(n-1)\mathbf{e} + \lambda$. For the regularization, we introduce a loading matrix to $\mathbf{D}(n)$ as Eq.(11).

$$\mathbf{P}(n) = \mathbf{U}(n)\left[\mathbf{D}(n) + \alpha\min\left(diag\left(\mathbf{D}(n)\right)\right)\right]\mathbf{U}^T(n). \tag{11}$$

We can adopt Bierd algorithm for calculation of UDU factorization in [2]. In Table 2, we summarize the regularized recursive total least square algorithm.

Table 2. Regularized recursive total least square for training FNN

$\mathbf{P}(0) = \delta^{-1}\mathbf{I}, \quad V(0) = \left(\frac{1}{\sqrt{N+1}} \cdots \frac{1}{\sqrt{N+1}} \right)^T$, where δ is small value, and $\mathbf{I}$ is a identity matrix.
At step n, compute $\hat{\mathbf{w}}_{\mathbf{TLS}}(n)$ and update $\mathbf{U(n\text{-}1)}$ and $\mathbf{D(n\text{-}1)}$ by performing steps 1-10.

1. $\bar{\mathbf{x}}(n) = \left(\mathbf{h(n)}^{\mathbf{T}} \; y(n) \right)^{\mathbf{T}}$
   $e = \mathbf{U(n-1)}\bar{\mathbf{x}}^H(n), g = \mathbf{D(n-1)}e, \alpha_0 = \lambda$
2. For j=1,...,N_θ, go through the steps 3-5, where N_θ is the total number of the weights and thresholds.
3.
   $\alpha_j = \alpha_{j-1} + e_j g_j$
   $D_{jj}(n) = D_{jj}(n-1)\alpha_{j-1}/\alpha_j \lambda$
   $\nu_j = g_j$
   $\mu_j = -e_j/\alpha_{j-1}$
4.   For i=1,...,j -1, go through step 4. (if j=1, skip step 5).
5.
   $U_{ij}(n) = U_{ij}(n-1) + \nu_i\mu_j$
   $\nu_i = \nu_i + U_{ij}(n-1)\nu_j$
6. $\tilde{\mathbf{P}}(n) = \mathbf{U(n)D(n)U(n)}$
7. $\mathbf{P}(n) = \mathbf{U(n)} [\mathbf{D(n)} + \alpha \min(diag(\mathbf{D(n)}))] \mathbf{U}^T(n)$
8. $\mathbf{v}(n) = \mathbf{P}(n)\hat{\mathbf{v}}_{N+1}(n-1)$
9. $\hat{\mathbf{v}}_{N+1}(n) = \mathbf{v}(n)/\|\mathbf{v}(n)\|$
10. $\hat{\mathbf{w}}_{\mathbf{TLS}}(n) = -\frac{1}{\hat{v}_{N+1,N+1}(n)} \left(\hat{v}_{N+1,1} \cdots \hat{v}_{N+1,N+1} \right)^T$

5   Simulation

Figure 2(a) presents the FNN model to identify the unknown system. This is connected to one hidden layer and one output layer, and has 1 source node, 25 hidden neurons, and 1 output neuron. The bias of hidden layer ranges from -6.0 to 6.0 which is the dynamic range of input data x. In this simulation, we assume that the relationship between input x and output y of the unknown system is

$$y = \sin(3x), \quad , -6 \le x \le 6 . \tag{12}$$

In this FNN model, the output of the first hidden layer is

$$\varphi = [\varphi_1(x - 6.0) \quad \varphi_2(x - 5.5) \quad \cdots \quad \varphi_{25}(x + 6.0)]^T , \tag{13}$$

where $\varphi_i = 1/\left(1 + e^{-(b_i+x)}\right)$ and b_i is the i-th bias in Fig. 2(a). The output of total FNN model is

$$\hat{y} = \mathbf{w}^H\varphi \quad \text{where} \quad \mathbf{w} = [w_1 \quad w_2 \quad \cdots \quad w_{25}]^T . \tag{14}$$

When we compare Eq.(14) with Eq.(2), we can apply the proposed algorithm in Table 2 to Eq.(14).

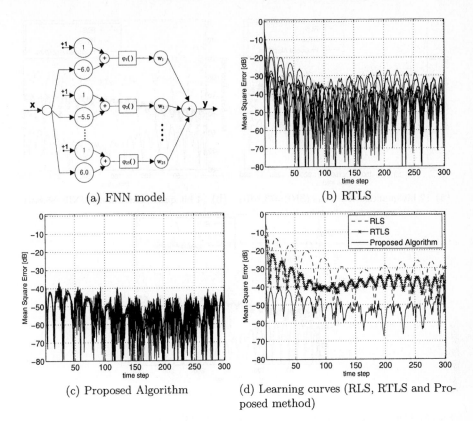

(a) FNN model

(b) RTLS

(c) Proposed Algorithm

(d) Learning curves (RLS, RTLS and Proposed method)

Fig. 2. Learning curves(RLS, RTLS and Proposed Algorithm)

In this simulation, we assume that noise exists in input and output as 10^{-4} and that α in Table 2 is set as 10. Fig. 2(b) shows the results from independent trials using RTLS in Table 1. This shows ill-conditioned effect. To the contrary, Fig. 2(c), which is the result of the proposed algorithm, shows no ill-conditioned effect. Fig. 2(d) compares the results from RLS, RTLS and the proposed method, respectively. For more explanatory result, another simulation experiment is performed. In this simulation, we assume signal to quatization error ratios of 73.8, 85.8, and 97.8 dB in input and output, which correspond to 12, 14, and 16 bit μ-law pulse code modulation (PCM) quantizers, respectively. Fig. 3 shows the learning error results from RLS, RTLS and the proposed algorithm. From the above results, it is clear that the proposed algorithm is robust enough to prevent the ill conditioned effect in RTLS algorithm.

6   Conclusions

In this paper, a robust recursive total least squares algorithm is presented to be applied in neural network training. This algorithm was found to outperform

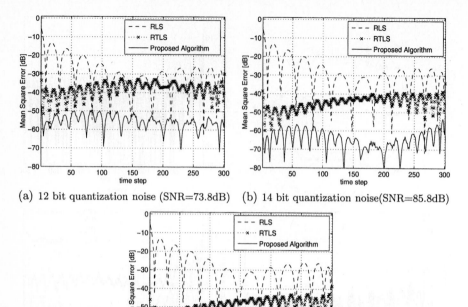

(a) 12 bit quantization noise (SNR=73.8dB)   (b) 14 bit quantization noise(SNR=85.8dB)

(c) 16 bit quantization noise (SNR=97.8dB)

Fig. 3. Comparison of learning curves in 12bit, 14bit and 16bit input and output quantizer cases (RLS, RTLS and Proposed Algorithm)

the RLS algorithm. The proposed algorithm shows no ill-conditioned effect as opposed to the RTLS as shown in Table 1 shows it. In order to validate its performance, we applied this algorithm to training multilayer feedforward neural network with noise in input and output.

References

1. Wong, K.W., Leung, C.S.: On-line Successive Synthesis of Wavelet Networks. Neural Processing Letter, **7** (1998) 91-100
2. Leung, C.S., Tsoi, A.C.: Two-regularizers for Recursive Least Squared Algorithms in Feedforward Multilayered Neural Networks. IEEE Trans. Neural Networks, **12** (2001)1314-1332
3. Golub, G.H., Van Loan, C.F.: An Analysis of the Total Least Squares Problem. SIAM J. Numer. Anal. **17** (1980) 883-893
4. Choi, N., Lim, J.S., Song, J., Sung. K.M.: Adaptive System Identi-fication Using an Efficient Recursive Total Least Squares Algorithm. Journal of the Acoustical Society of Korea, **22** (2003) 93-100
5. Choi, N., Lim, J.S., Song, J., Sung. K.M.: Sung: An Efficient Recursive Total Least Squares Algorithm For Fequency Estimation. ICA, **25** (2001) 90-92

An Improved Backpropagation Algorithm
Using Absolute Error Function*

Jiancheng Lv and Zhang Yi

Computational Intelligence Laboratory, School of Computer Science and Engineering,
University of Electronic Science and Technology of China, Chegndu 610054, China
zhangyi@uestc.edu.cn, jeson_cha@yahoo.com
http://cilab.uestc.edu.cn

Abstract. An improved backpropagation algorithm is proposed by using the Lyapunov method to minimize the absolution error function. The improved algorithm can make both error and gradient approach zero so that the local minima problem can be avoided. In addition, since the absolute error function is used, this algorithm is more robust and faster for learning than the backpropagation with the traditional square error function when target signals include some incorrect data. This paper also proposes a method of using Lyapunov stability theory to derive a learning algorithm which directly minimize the absolute error function.

1 Introduction

The backpropagation algorithm [1] for training the multi-layer feed-forward neural network is a very popular learning algorithm. The multilayer neural networks have been used in many fields such as pattern recognition, image processing, classification, diagnosis and so on. But the standard backpropagation encounters two problems in practice, i.e., slow convergence and local minima problem. Several improvements have been proposed to overcome the problems [4], [5], [6], [7], [10], [11], [12].

To accelerate the convergence and escape from the local minima, this paper proposes an improved algorithm that makes the error and gradient go to zero together. This algorithm is obtained by minimizing the absolute error function directly based on Lyapunov stability theory. Usually, the standard backpropagation algorithm is a steepest gradient descent algorithm based on the square error function[2]. The algorithm iteratively and gradually modifies parameters according to the deviation between an actual output and target signal. The square error function is used to calculate the deviation. To minimizing the deviation, the absolute error may sometimes be better than the square error. In fact, the absolute error function is less influenced by anomalous data than the square error function[3]. So, the algorithms with the absolute error function are

---

* This work was supported by National Science Foundation of China under Grant 60471055 and Specialized Research Fund for the Doctoral Program of Higher Education under Grant 20040614017.

J. Wang, X. Liao, and Z. Yi (Eds.): ISNN 2005, LNCS 3496, pp. 585–590, 2005.

more robust and learns faster than the backpropagation with the square error function when target signals include some incorrect data[3]. In [3], K. Taji et al took the differentiable approximate function for the absolute error as the objective function. However, in this paper, the absolute error will directly be used to calculate the backpropagation error. At the same time, the improved algorithm based on Lyaponuv stability theory is able to make the error and gradient go to zero together so that the evolution can escape from the local minima.

The rest part of this paper is organized as follows. In Section 2, this improved backpropagation algorithm is obtained by using Lyapunov method to minimize the absolute error function. In Section 3, the simulation are carried out to illustrate the result. Finally, conclusions is described in Section 4.

2   An Improved BP Algorithm

Consider a multiplayer feed-forward neural network with L layers and N output neurons (Fig. 1). As for the ith neuron in the l layer, let a_i^l be its output and n_i^l is its net input. It follows that

$$a_i^l = f_i^l(n_i^l),\tag{1}$$

$$n_i^l = \sum_{j=1}^{N_{l-1}} w_{ij}^l a_j^{l-1},\tag{2}$$

where N_{l-1} is the number of neurons in $l-1$ layer.

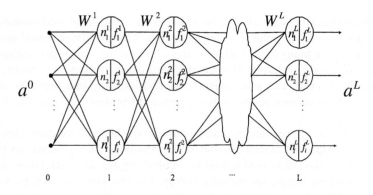

Fig. 1. An L layers BP Neural Network with N output neurons

Suppose the training data set consists of p patterns. For a specific pattern, the network outputs $\mathbf{y}_i$ and targets $\mathbf{y}_i^d$ are wrote that

$$\{\mathbf{y}^d, \mathbf{y}\} = \left\{ \begin{pmatrix} y_1^d \\ y_2^d \\ \vdots \\ y_N^d \end{pmatrix}, \begin{pmatrix} y_1 \\ y_2 \\ \vdots \\ y_N \end{pmatrix} \right\}.\tag{3}$$

The absolute error function is constructed as the cost function, it follows that

$$E = \sum_p \sum_{q=1}^{N} |y_q^d - y_q|.$$
(4)

We consider a Lyapunov function that

$$V(t) = \sum_p \sum_{q=1}^{N} |y_q^d(t) - y_q(t)|.$$
(5)

The time derivative of the Lyapunov function is given by

$$\frac{dV(t)}{dt} = \frac{\partial V(t)}{\partial w_{ij}} \cdot \frac{dw_{ij}}{dt} = \sum_p \left[\sum_{q=1}^{N} (sign(y_q^d(t) - y_q(t))) \frac{\partial(y_q^d(t) - y_q(t))}{\partial w_{ij}} \frac{dw_{ij}}{dt} \right]$$

$$= -\sum_p \left[\sum_{q=1}^{N} (sign(y_q^d(t) - y_q(t))) \frac{\partial y_q(t)}{\partial w_{ij}} \frac{dw_{ij}}{dt} \right]$$
(6)

Theorem 1. *Given any initial value $w_{ij}(0)$, if the wight is updated by*

$$w_{ij}(k) = w_{ij}(0) + \int_0^k \frac{dw_{ij}}{dt} dt,$$
(7)

where

$$\frac{dw_{ij}}{dt} = \frac{y_q^d(t) - y_q(t)}{\dfrac{\partial y_q(t)}{\partial w_{ij}}},$$
(8)

then E goes to zero along the trajectory of the weight w_{ij}.

Poof: From (6) and (8), it gives that

$$\frac{dV(t)}{dt} = -2V(t)$$
(9)

then

$$V(t) = V(0)e^{-2t}.$$
(10)

Clearly, $V(t) \to 0$ as $t \to 0$, i.e. E goes to zero with exponential rate. The proof is completed.

According to the equation (8), the weight update equation in the l layer can be wrote as follows:

$$w_{ij}^l(k+1) = w_{ij}^l(k) + \eta \frac{y^d - y}{\dfrac{\partial y}{\partial w_{ij}^l}}.$$
(11)

By chain rule, if l is the last layer L, it holds that

$$\frac{\partial y}{\partial w_{ij}^l} = \frac{\partial f_i^l(n_i^l)}{\partial n_i^l} \cdot \frac{\partial n_i^l}{\partial w_{ij}} = \dot{f}_i^l(n_i^l) \cdot a_{ij}^l. \tag{12}$$

Otherwise, we have

$$\begin{aligned}
\frac{\partial y}{\partial w_{ij}^l} &= \sum_{k=1}^{N_{l+1}} \left(\frac{\partial f_k^{l+1}(n_k^{l+1})}{\partial n_k^{l+1}} \cdot \frac{\partial n_k^{l+1}}{\partial n_k^l} \cdot \frac{\partial n_k^l}{\partial w_{ij}} \right) \\
&= \sum_{k=1}^{N_{l+1}} \dot{f}_k^{l+1}(n_k^{l+1}) \cdot \dot{f}_i^l(n_i^l) w_{ki}^{l+1} \cdot a_{ij}^l \\
&= \dot{f}_i^l(n_i^l) \cdot a_{ij}^l \sum_{k=1}^{N_{l+1}} \dot{f}_k^{l+1}(n_k^{l+1}) \cdot w_{ki}^{l+1}, \tag{13}
\end{aligned}$$

where N_{l+1} is the number of the $l+1$ layer. So, from (11), (12) and (13), the improved algorithm can be described as follows:

$$\begin{cases}
w_{ij}^l(k+1) = w_{ij}^l(k) + \dfrac{\eta}{\dot{f}_i^l(n_i) \cdot a_{ij}^l} \cdot (y^d - y), \quad l = L \\[4mm]
w_{ij}^l(k+1) = w_{ij}^l(k) + \dfrac{\eta}{\dot{f}_i^l(n_i) \cdot a_{ij}^l \cdot \displaystyle\sum_{p=1}^{N_{l+1}} \dot{f}_p^{l+1}(n_p^{l+1}) w_{pi}^{l+1}} \cdot (y^d - y), otherwise,
\end{cases} \tag{14}$$

where N_{l+1} is the number of the $l+1$ layer. In addition, to avoid the algorithm becomes unstable when the gradient goes to zero, a small constant is added to the denominator. From (14), it is clear that the evolution stops only when the error goes to zero. If the gradient goes to zero and the error is still large, a large update value makes the evolution escape from the minima.

3   Simulations and Discussions

This improved algorithm for multilayer networks makes the weight evolution escape from the local minima. The XOR problem experiment will be carried out to illustrate it. A simple architecture(2-2-1) is used to learn the problem. The bipolar sigmoid function is taken as the active function. To avoid the active function goes to the saturation[4], [5], [8], [9], the parameter η should is a relatively small constant. And, once the update value is enough large so that the evolution goes to premature saturation, the update value must be scaled down to avoid the premature saturation.

Fig. 2 and Fig. 3 show the experiment result with 500 epochs. On the left picture in Fig. 3, the 60 sample points are divide into two class ones by two decision lines. The class 0 is presented by star points and the class 1 is presented by circinal points. The Fig. 2 shows the error evolution. It is clear the evolution escapes from the local minima. Finally, the 200 random test points are provided for the network. Obviously, these points have been divided rightly on the right one in Fig. 3.

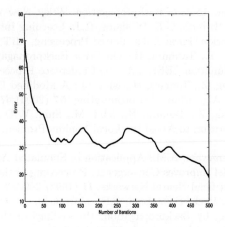

Fig. 2. The error evolution

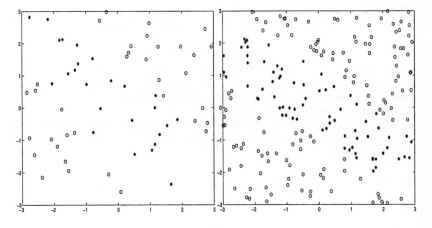

Fig. 3. The samples (left) and test result (right)

4   Conclusions

This paper gives an improved learning method for multilayer neural networks to escape local minima. This algorithm is obtained by using Lyapunov method to minimize the absolute error function. Since the algorithm makes the error and gradient go to zero together, the local minima problem could be avoided. The algorithm is more robust since it uses the absolute error function. The simulation for XOR problem shows the effectiveness of the proposed algorithm.

References

1. Werbos, P. J.: The Roots of Backpropagation. Wiley, New York (1994)
2. Rumelhart, D.E., Hinton, G.E. Williams, R.J.: Learning Internal Representations by Error Propagation. Parallel distributed Processing, MIT Press (1986) 318-362
3. Taji, K., Miyake, T., Tarmura, H.: On Error Backpropagation Algorithm Using Absolute Error Function. IEEE SMC '99 Conference Proceedings, 5 (1999) 12-15
4. Wang, X.G., Tang, Z., Tamura, H., Ishii, M.: A Modified Error Function for the Backpropagation Algorithm. Neurocomputing, 57 (2004) 477-484
5. Wang, X.G., Tang, Z., Tamura, H., Ishii, M., Sum, W.D.: An Improved Backpropagation Algorithm to Avoid the Local Minima Problem. Neurocomputing, 56 (2004) 455-460
6. Owen, C.B., Abunawass, A.M.: Application of Simulated Annealing to the Backpropagation Model Improves Convergence. Proceeding of the SPIE Conference on the Science of Artificial Neural Networks, II (1993) 269-276
7. Von Lehmen, A., Paek, E.G., Liao, P.F., Marrakchi, A., Patel, J.S.: Factors Influencing Learning by Backpropagation. Proceedings of the IEEE International Conference On Neural Networks, I (1988) 335-341
8. Fukuoka, Y., Matsuki, H., Minamitani, H., Ishida, A.: A Modified Backpropagation Method to Avoid False Local Minima. Neural Networks, 11 (1998) 1059-1072
9. Vitela, J. E., Reifman, J.: Premature Saturation in Backpropagation Networks: Mechanism and Necessary Conditions. Neural Networks, 10 (1997) 721-735
10. Wand, C., Principe, J. C.: Training Neural Networks with Additive Noise in the Desired Signal. IEEE Trans. Neural Networks, 10 (1999) 1511-1517
11. Battiti, R., Masulli, F.: BFGS Optimization for Faster and Automated Supervised Learning. Proceedings of the Internatioanl Neural Network Conference. Kluwer, paris, France (1990) 757-760
12. Kollias, S., Anastassiou, D.: An Adaptive Least Squares Algorithm for Efficient Training of Artificial Neural Networks. IEEE Trans. on Circuits and systems, 36 (1989) 1092-1101

An Improved Relative Criterion Using BP Algorithm*

Zhiyong Zhang[1,2], Jingang Liu[3], and Zhongzhi Shi[1]

[1] Institute of Computing and Technology, Chinese Academy of Sciences,
Beijing 100080, China
{zhangzy,shizz}@ics.ict.ac.cn
[2] Graduate School of the Chinese Academy of Sciences, Beijin, 100039, China
[3] Join Faculty of Computer Scientific Research, Capital Normal University,
Beijing 100037, China

Abstract. People are usually more interested in ranking a group of interrelated objects than the actual scores of them. In this paper, we presented an improved relative evaluation criterion, which is suitable for such task and can solve those problems that are difficult to deal with by common absolute criterion. We put forward the improvement algorithm and realize its function approximation by means of BP algorithm. Furthermore, we tested this criterion with the statistic data of SARS. The experimental results indicated that our improved criterion is effective in comparing a set of correlative objects objectively. Finally, we made a conclusion about the goal and efficiency of relative criterion.

1 Introduction

In general, people sort objects by summing up the absolute scores of all their properties. For example, teachers often rank students based on the sum scores of all courses. In the reality, it is possible for a mathematics department to reject a student even if he scored high in Math but relatively low in other courses, if the total score is a little lower than the admission requirement. An even worse case is that which one should be selected when there is only one opening but two candidates with same score? So the absolute evaluation criterion is not always justified or practical.

Based on the thought of treating a set of objects as a whole, Liu put forward the concept of relative criterion that can solve such puzzles easily [1]. Referring to subsection 2.2, we will know this criterion is bewildering, or even providing an opportunity to cause inequity in practice, which is named as instability in this paper. The aim of our study is to enhance its stability. The rest of the paper is organized as follows: In Section 2, we introduce the basic mathematical model of Liu's relative criterion and give some examples to explain its deficiency. In Section 3 we present our stability improvement algorithm. We discuss the key factors for applying BP algorithm in Section 4. The experimental results of verifying our improved criterion with actual figures SARS are shown in Section 5 and conclude in Section 6.

* This research was supported by NSFC (No.60435010), NSFB (No.4052025) and National Great Basic Research Program (973 project No.2003CB317000)

J. Wang, X. Liao, and Z. Yi (Eds.): ISNN 2005, LNCS 3496, pp. 591–596, 2005.

2   The Overview of Relative Criterion

2.1   Mathematical Model

Liu depict a group of objects as a matrix $U = (a_{ij})_{m \times n}$ [1]. Let U is the set of objects, m is the number of objects, n is the number of columns, A_i is the ith object in U, a_{ij} is the value of the jth property of A_i, w_j is the given weight of the jth property. Define S_j as the sum of jth column and g_{ij} denotes the relative unit scale of jth column of A_i. Let G_i denotes the relative evaluation result for A_i and D denotes a constant multiple to adjust G_i the to a readable value. We can deduce G_i as bellow.

$$G_i = D \cdot \sum_{j=1}^{n} g_{ij} = D \cdot \sum_{j=1}^{n} \left(a_{ij} \cdot w_j \middle/ \sum_{i=1}^{m} a_{ij} \right) \tag{1}$$

We depict a collection of objects with three properties in table1. Here, relative criterion1 denotes all three students will be considered, and Helen is excluded in relative criterion2. Compared with absolute criterion, Amy receives more preference in relative criterion1 because she does much better than others in Math.

Table 1. A comparison of absolute criterion and relative criterion

Students \ Topics	Math	Geog-raphy	English	Absolute Criterion Score	Ranks	Relative Criterion1 Score	Ranks	Relative Criterion2 Score	Ranks
Amy	93	65	70	74. 9	2	72.7	1	74.6	2
Boris	62	80	82	75. 2	1	70.5	2	75.4	1
Helen	20	80	85	63. 5	3	56.8	3		
P_j	30%	40%	30%			$D = 200$		$D = 150$	

2.2   Puzzled by Relative Criterion

After observing the results of last two criterions in table 1 carefully, we can get a bewildering fact. If we only consider Amy and Boris (i.e. relative criterion 2) in this case, Boris ranks higher than Amy. Once Helen is included, we will get a reverse result although the scores of Amy and Boris are not changed. What happened here? In fact, S_j is changed with Helen's entering and this influences the ultimate orders.

Suppose three participants take part in a triathlon contest, which is evaluated with relative criterion. Two players, Amy and Helen, come from the same country. Boris is good at bicycling and running, and Amy is much better than Boris in swimming. Having realized it, Helen plans to get a very bad score in swimming so that this will help Amy win the contest. This scene is similar to Tian Ji's Horse Racing, an ancient story happened in Warring States Period of China. Since any individual may have an impact on evaluation results to some extent, we can conclude that the relative evaluation is instable and can even provide an opportunity to cause inequity in practice.

3   Improved Relative Criterion

For a given task, the number of objects and properties are usually invariable and known in advance. So it is feasible to calculate the relative evaluation results in any subset of U. Here, we take A_i for example to explain the core idea of improvement.

3.1 Mathematical Deductions

Notation 1. Let $U_{i,k}^t$ denotes the kth subset of residual objects after $t_{(t\in[0,m-2])}$ objects other than A_i are removed from U. Apparently, the number of such subsets is C_{m-1}^t. Let $S_{i,k,j}^t$ denotes the sum of all elements in the jth column of $U_{i,k}^t$. Let $S_{i,j}^t$ denotes the average of all $S_{i,k,j}^t$ and let G_i^t denotes the average of relative evaluation results about A_i in all $U_{i,k}^t$ ($k \in [1, C_{m-1}^t]$).

When $t = 0$ (i.e. no object being excluded), G_i^t can be measured based on the formula (1). When $t = 1$ (i.e. only one object being removed), we have $U_{i,k}^1 = U - \{A_r\}$ ($r \in [1,m] \wedge r \neq i$). Note that A_i appears in all $U_{i,k}^1$ and A_i is excluded from a certain $U_{i,k}^1$ ($k \in [1, C_{m-1}^1]$). Thus we can deduce the equations as bellow.

$$S_{i,j}^1 = \sum_{h=1}^{m} a_{hj} - \sum_{h \neq i} a_{hj} \Big/ C_{m-1}^1 = \sum_{h=1}^{m} a_{hj} - \left(C_{m-2}^{1-1} \cdot \sum_{h \neq i} a_{hj} \Big/ C_{m-1}^1\right) \tag{2}$$

$$G_i^1 = D \cdot \sum_{j=1}^{n} \left(a_{ij} \cdot w_j \Big/ S_{i,j}^1\right) = D \cdot \sum_{j=1}^{n} \left(a_{ij} \cdot w_j \Big/ \left(\sum_{h=1}^{m} a_{hj} - C_{m-2}^{1-1} \cdot \sum_{h \neq i} a_{hj} \Big/ C_{m-1}^1\right)\right) \tag{3}$$

Where C_α^β is the number of combinations of α objects taken β at a time. As t elements being removed, we have $U_{i,k}^t = U - \{A_{r1}, \cdots, A_{rf}, \cdots A_{rt}\}$ ($rf \in [1,m] \wedge rf \neq i$). A_i appears in all $U_{i,k}^t$, but A_{rf} is excluded C_{m-2}^{t-1} times from $U_{i,k}^t$. So, we can depict G_i^t as equation (4). After all G_i^t have been calculated, they will be grouped by t. And each group contains m elements (i.e. $G_1^t, G_2^t, \cdots, G_m^t$). Let *Max* and *Stdev* represent the maximal value and the standard deviation function. Let $ratiov_t$ be a given adjustable proportion toward t. The transformation of G_i^t can be defined as equation (5).

$$G_i^t = D \cdot \sum_{j=1}^{n} \left(a_{ij} \cdot w_j \Big/ S_{i,j}^t\right) = D \cdot \sum_{j=1}^{n} \left(a_{ij} \cdot w_j \Big/ \left(\sum_{h=1}^{m} a_{hj} - C_{m-2}^{t-1} \cdot \sum_{h \neq i} a_{hj} \Big/ C_{m-1}^t\right)\right) \tag{4}$$

$$R(G_i^t) = G_i^t \Big/ (Max(G_1^t, \cdots, G_m^t) \cdot Stdev(G_1^t, \cdots, G_m^t) \cdot ratiov_t) \tag{5}$$

To A_i, there are $m-1$ transformations (i.e. $R(G_i^0), \cdots, R(G_i^{m-2})$). As shown in Fig.1, we can describe each one and build a curve to approximate these points. Two things should be made out. Firstly, what we concerned about is the place ranks of all objects and all transformations are based on the same rule, so the final relative results will not be influenced. Secondly, these points stand for the transformations about the same object, so they are interrelated and can be approximated with a curve.

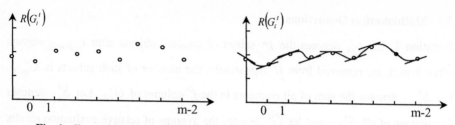

Fig. 1. all transformations of unstable relative evaluation results about the ith object

Notation 2. Let f_i'' be the first derivative of the point whose x-coordinate is equal to t, and p^t be the influence weight of σ_i when t objects being excluded. The more t is, the less p^t should be. Here we define p^t as the equation (6) and σ_i can be calculated as the equation (7). Apparently, the sum of all p^t is one.

$$P^t = (m-t-1)/(1+2+\cdots+(m-1)) = 2 \cdot (m-t-1)/((m-1) \cdot m) \tag{6}$$

$$\sigma_i = \sum_{t=0}^{m-2} (P^t \cdot f_i^{t'}) = \sum_{t=0}^{m-2} \left(2 \cdot f_i^{t'} \cdot (m-t-1)/((m-1) \cdot m)\right) \tag{7}$$

Because the coefficient σ_i implies the influence of other objects on G_i, we can calculate G_i^s, the improved stable value, as the product of G_i and $1+\sigma_i$.

$$G_i^s = G_i \cdot (1+\sigma_i) = D \cdot \sum_{j=1}^{n} (a_{ij} \cdot w_j) \bigg/ \sum_{i=1}^{m} a_{ij} \cdot \left(1 + \sum_{t=0}^{m-2} \left(2 \cdot f_i^{t'} \cdot (m-t-1)/((m-1) \cdot m)\right)\right). \tag{8}$$

3.2 Stability Improvement Algorithm

STEP1: For each element in U, supposing to A_i, do
    For each t, calculate G_i^t according to the equation (4)
    Next element in U

STEP2: For each t, calculate $Max(G_1^t, \cdots, G_m^t)$ and $Stdev(G_1^t, \cdots, G_m^t)$

STEP3: For each i, calculate $R(G_i^t)$ according to the equation (5)

STEP4: Based on function approximation method in section 4, calculate each $f_i^{t'}$
STEP5: For each i, calculate G_i^s according to the equation (8)

4   Using BP Algorithm for Function Approximation

As mentioned in section 3, we have to calculate the ratio of change for each point in Fig. 1 in order to evaluate σ_i. So, we need build a curve to approximate these points and calculate their first derivatives. Fortunately, BP algorithm is appropriate to solve problems that involve learning the relationships between a set of inputs and known outputs [3]. Apparently, inputs and outputs will correspond to x and y coordinates of all points. Trained with BP algorithm, MFNNs (Multilayer Feedforward Neural Networks) can be used as a class of universal approximators [2].

It has been proved that two-layer networks, with *sigmoid* transfer functions in the hidden layer and *linear* transfer functions in the output layer, can approximate any function to any degree of accuracy [2]. Furthermore, only one element is input or output at a time in our stability improvement algorithm. So a two-layer, $1-S^1-1$ network is competent for our job. Referring to some investigations and our experiments with MATLAB, we believe a *sigmoid/purline*, $1-6^1-1$ network would be more preferable[3,4]. Yuan compared eight *sigmoid* functions in detail [4]. We know log *sig* is popular and generates outputs between 0 and 1. Referring to section 3, we know that $R(G_i')$ are positive. Thus we apply log *sig* to the first layer transform.

Basic BP algorithm is not feasible to solve practice problems, since it will spend much time in training a network. Several variable algorithms have a variety of different computation and storage requirements, and no one is best suited to all locations. LMBP (Levenberg-Marquardt algorithm) converges in less iteration than any of other algorithms. And it appears to be the fastest training algorithm for networks of moderate size [3]. Our MFNN is a $1-6^1-1$ network, which has 12 weights and 7 biases. So we can think LMBP is competent for our task.

5 Experimental Results

In this section, we will verify our algorithm with actual figures of SARS. We picked the data from April 27, 2003 to May 17, 2003 since the cases are relatively concentrative during this period. Our task is to evaluate the criticality of each date. So these dates constitute the set of objects having three properties (i.e. diagnosed, suspected and deaths). We defined the weight of each property based on the statistic released by the Ministry of Health (a total of 5327 SARS cases and 6565 suspected cases were reported, and death toll from the disease stood at 349 [6]). So we have the definition.

$$a_1 = 5327 \quad a_2 = 6565 \quad a_3 = 349 \quad r_1 = \log(a_2)/\log(a_1) \quad r_2 = \log(a_2)/\log(a_2) \quad r_3 = \log(a_2)/\log(a_3)$$
$$w_1 = r_1/(r_1 + r_2 + r_3) = 0.2906 \quad w_2 = r_2/(r_1 + r_2 + r_3) = 0.2836 \quad w_3 = r_3/(r_1 + r_2 + r_3) = 0.4258$$

where a_1, a_2, a_3, stands for the number of three properties respectively. Because suspected cases overlay two others, we can set a_2 as the base. And we applied log function to enhance the stabilization of w_j. Note that the definition of w_j may not appropriate since our professional knowledge is limited. But this will influence results very lightly because our experiment aims to explain the improvement algorithm. In addition, the constant multiple D is set as 130. As shown in table 2, the result in April 28 is more than that in May 3 as 23 objects are removed, contrary to G^{24}.

Table 2. Two samples and their relative evaluation results as some objects are excluded

Date	Diagnosed Cases	Suspected Cases	Deaths	Absolute Criterion	G^0	G^{23}	G^{24}
4-28	203	290	7	144.216	7.085	21.879	24.069
5-3	181	251	9	127.614	7.017	21.871	24.090

Based on the formula (5), the transformations of all $G_i^{'} (t \in [0, m-2])$ can be calculated and approximated by BP algorithm. Had trained MFNN, we got approximate curve for each object. By simulating each one, we get the x and y coordinates of any point near $R(G_i^{'})(t \in [0, m-2])$. So the average changing trend of all $R(G_i^{'})$ can be calculated based on all the first derivatives of them. Finally, σ_i and G_i^{s} of each object was evaluated based on the equation (8). In Fig.2, we compare the orders of criticality among three evaluation methods. We can detect improved relative evaluation is closer to absolute criterion than original one. For the absolute results are most stable, we can conclude that the improved results are more stable than original results.

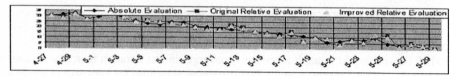

Fig. 2. The comparison of three different evaluation methods

6 Conclusions

In this paper, we put forward an improved relative criterion that can overcome the instability of original relative criterion. Based on a series of mathematical deductions, we explained stability improvement algorithm in detail. In the end, we experimented with the statistic data of SARS in order to verify this criterion. Relative criterion aims to evaluate the place ranks for a set of interrelated objects, rather than exact scores. And the more similar that all objects are, the more effective this criterion will be.

References

1. Liu, J.: The Theory Research of Relative Criterion Method in Scatter Quantity System. Proceedings of Join Faculty of Computer Scientific Research, ICT&CNU (2004) 14-18
2. Hornik, K., Stinchcombe, M., White, H.: Multilayer Feedforward Networks Are Universal Approximators, Neural Networks, 2 (1989) 359-366
3. Hagan, M.T., et al.: Neural Network Design. PWS Publishing Company (1995)
4. Yuan, C.: Artificial Neural Networks and Applications. Tsinghua University (1999)
5. http://www.mathworks.com/access/helpdesk/help/techdoc/matlab.html
6. http://big5.china.com.cn/english/jianghua/72566.htm.

Solving Hard Local Minima Problems Using Basin Cells for Multilayer Perceptron Training

Younggui Yoon[1] and Jaewook Lee[2]

[1] Department of Physics, Chung-Ang University, Seoul, 156-756, Korea
yyoon@cau.ac.kr
[2] Department of Industrial and Management Engineering, Pohang University of
Science and Technology,
Pohang, Kyungbuk 790-784, Korea
jaewookl@postech.ac.kr

Abstract. An analysis of basin cells for an error surface of an multi-layer perceptron is presented. Utilizing the topological structure of the basin cells, an escaping strategy is proposed to solve difficult local minima problems. A numerical example is given to illustrate the proposed method and is shown to have a potential to locate better local minima efficiently.

1 Introduction

During the last decades, several learning methods for training multi-layer perceptrons (MLPs) have been applied successfully to solve a number of nontrivial problems such as pattern recognition, classification, robotics and automation, financial engineering, and so on. They include the *error back-propagation* (EBP) algorithm and its variants, and second order methods such as conjugate gradient, Newton, BFGS, and Levenberg-Marguardt methods [2], [4].

When the error surface of MLPs are very rugged and have few good solutions, however, these learning methods run the risk of being trapped in a poor local minimum since they are basically hill-climbing local minimization techniques.

In this paper, we introduce a concept of a basin cell for an error surface of MLPs and present a new escaping strategy to effectively overcome such local minima problems utilizing the topological property of a basin cell.

2 Basin Cells

Our approach is based on first viewing the supervised training of a MLP as an unconstrained minimization problem [4] as follows:

$$\min_{\mathbf{w}} J(\mathbf{w}), \tag{1}$$

where the error cost function $J(\cdot)$ averaged over the training samples is a highly nonlinear function of the synaptic weight vector $\mathbf{w}$. Then we try to solve problem (1) by exploiting some geometric and topological structure of the error

J. Wang, X. Liao, and Z. Yi (Eds.): ISNN 2005, LNCS 3496, pp. 597–602, 2005.

function, J. To analyze the geometric structure of an error surface, we build a deterministic, generalized gradient system described by

$$\frac{d\mathbf{w}}{dt} = -\mathrm{grad}_R J(\mathbf{w}) \equiv -R(\mathbf{w})^{-1} \nabla J(\mathbf{w}) \tag{2}$$

where the error function J is assumed to be twice differentiable to guarantee the existence of a unique solution (or trajectory) $\mathbf{w}(\cdot) : \Re \to \Re^n$ for each initial condition $\mathbf{w}(0)$ and $R(\mathbf{w})$ is a positive definite symmetric matrix for all $\mathbf{w} \in \Re^n$. (Such an R is called a *Riemannian metric* on $\Re^n$.) It is interesting to note that many learning algorithms based on the formulation (1) can be considered as a discretized implementation of system (2) depending on the choice of $R(\mathbf{w})$. For example, if $R(\mathbf{w}) = I$, it is the naive error back-propagation algorithm; if $R(\mathbf{w}) = \nabla^2 J(\mathbf{w})$, it is the Newton method; if $R(\mathbf{w}) = [\nabla^2 J(\mathbf{w}) + \mu I]$, it is the Levenberg-Marguardt method, and so on.

A state vector $\bar{\mathbf{w}}$ satisfying the equation $\nabla J(\bar{\mathbf{w}}) = 0$ is called an *equilibrium point* (or *critical point*) of system (2). We say that an equilibrium point $\bar{\mathbf{w}}$ of (2) is *hyperbolic* if the Hessian matrix of J at $\bar{\mathbf{w}}$, denoted by $\nabla^2 J(\bar{\mathbf{w}})$, has no zero eigenvalues. Note that all the eigenvalues of $\nabla^2 J(\bar{\mathbf{w}})$ are real since it is symmetric. A hyperbolic equilibrium point is called (i) a (asymptotically) *stable* equilibrium point (or an *attractor*) if all the eigenvalues of its corresponding Hessian are positive, or (ii) an *unstable* equilibrium point (or a *repeller*) if all the eigenvalues of its corresponding Hessian are negative, or (iii) a *saddle point* otherwise. A hyperbolic equilibrium point $\bar{\mathbf{w}}$ is called an *index-k equilibrium point* if its Hessian has exactly k negative eigenvalues. A set K in $\Re^n$ is called to be an *invariant set* of (2) if every trajectory of (2) starting in K remains in K for all $t \in \Re$.

One nice property of the formulation (2) is that every local minimum of Eq. (1) corresponds to a (asymptotically) stable equilibrium point of system (2), i.e., $\bar{\mathbf{w}}$ is a stable equilibrium point of (2) if and only if $\bar{\mathbf{w}}$ is an isolated local minimum for (1) [5], [8]. Hence, the task of finding locally optimal weight vectors of (1) can be achieved when one locates corresponding stable equilibrium points of (2). Another nice property is that all the generalized gradient system have the same equilibrium points with the same index [5], i.e., if $R_1(\mathbf{w})$ and $R_2(\mathbf{w})$ are Riemannian metrics on $\Re^n$, then the locations and the indices of the equilibrium points of $\mathrm{grad}_{R_1} J(\mathbf{w})$ and $\mathrm{grad}_{R_2} J(\mathbf{w})$ are the same. This convenient property allows us to design a computationally efficient algorithm to find stable equilibrium points of (2), and thus to find local optimal solutions of (1).

For an index-k hyperbolic equilibrium point $\bar{\mathbf{w}}$, its *stable and unstable manifolds* $W^s(\bar{\mathbf{w}}), W^u(\bar{\mathbf{w}})$ are defined as follows:

$$W^s(\bar{\mathbf{w}}) = \{\mathbf{w}(0) \in \Re^n : \lim_{t \to \infty} \mathbf{w}(t) = \bar{\mathbf{w}}\},$$
$$W^u(\bar{\mathbf{w}}) = \{\mathbf{w}(0) \in \Re^n : \lim_{t \to -\infty} \mathbf{w}(t) = \bar{\mathbf{w}}\},$$

where the dimension of $W^u(\bar{\mathbf{w}})$ and $W^s(\bar{\mathbf{w}})$ is k and $n - k$, respectively.

The *basin of attraction* (or *stability region*) of a stable equilibrium point $\mathbf{w}_s$ is defined as

$$A(\mathbf{w}_s) := \{\mathbf{w}(0) \in \Re^n : \lim_{t \to \infty} \mathbf{w}(t) = \mathbf{w}_s\}.$$

One important concept used in this paper is that of basin cell. A *basin cell* of a stable equilibrium point $\mathbf{w}_s$ is defined by the closure of the basin $A(\mathbf{w}_s)$, denoted by $\overline{A(\mathbf{w}_s)}$. The boundary of the basin cell defines the *basin boundary*, denoted by $\partial \overline{A(\mathbf{w}_s)}$. Note that the basin boundary is $(n-1)$ dimensional whereas the basin cell $\overline{A(\mathbf{w}_s)}$ is n-dimensional.

The stable manifolds, basin cells, and basin boundaries are all invariant; that is, if a point is on the invariant region, then its entire trajectory lies on the region when either the process or the corresponding time reversed process is applied.

In general, the behavior of basin boundaries can be quite complicated for a nonlinear system. However, for the system of the form (2), it can be shown that, under some weak conditions, the geometrical and topological properties of the basin boundary for system (2) can be characterized as follows:

$$\partial \overline{A(\mathbf{w}_s)} = \bigcup_i^N \overline{W^s(\mathbf{w}_d^i)}. \tag{3}$$

where $\mathbf{w}_d^i$, $i = 1, ..., N$ are the index-one saddle points on the basin boundary $\partial \overline{A(\mathbf{w}_s)}$ [3], [8]. This characterization implies that the state space of system (2) is composed of the basin cells where each basin boundary is the union of the stable cells (=the closure of the stable manifolds) of all the index-one saddle points on its boundary.

3   An Escaping Strategy Using Basin Cells

In this section, we present a new escaping strategy to effectively overcome such local minima problems utilizing the concept of basin cell introduced in Section 2. The key idea of our strategy to escape from a local optimal weight vector is based on the following property: For a given stable equilibrium point $\mathbf{w}_s$ and its adjacent stable equilibrium point $\tilde{\mathbf{w}}_s$, there exists an index-one saddle point $\mathbf{w}_1$ such that the 1-D unstable manifold (or curve) $W^u(\mathbf{w}_1)$ of $\mathbf{w}_1$ converges to both $\mathbf{w}_s$ and $\tilde{\mathbf{w}}_s$ with respect to system (2) [8]. From this property, we can deduce that $\mathbf{w}_s$ and $\tilde{\mathbf{w}}_s$ take local minimum values on the 1-D curve $W^u(\mathbf{w}_1)$ whereas $\mathbf{w}_1$ takes on the local maximum value on that curve.

One possible direct escaping strategy based on this idea is to first find an index-one saddle point $\mathbf{w}_1$ starting from a given local minimum $\mathbf{w}_s$, and then to locate another adjacent local minimum $\tilde{\mathbf{w}}_s$ starting from $\mathbf{w}_1$. The latter step can be easily performed by numerically integrating (2) or using gradient descent method. The former step, although it is hard to perform directly and time-consuming, can also be accomplished by adopting the techniques suggested in [6], [7]. This direct strategy, however, inefficient since we are only interested in locating another local minimum and do not need to find an index-one saddle point, which is time intensive.

In this paper, we instead propose the following indirect strategy to escape from a local optimal weight vector $\mathbf{w}_s$:

(I) Choose an initial straight ray $\mathbf{x}(s)$, emanated from $\mathbf{x}(0) = \mathbf{w}_s$. For instance, the initial ray may be generated by joining $\mathbf{w}_s$ and any other point toward the basin boundary to form a straight line.

(II) Follow the ray $\mathbf{x}(s)$, with increasing s, until the trajectory reaches its first local minimum of $J(\mathbf{x}(\cdot))$. The parameter s is increased by a constant increment, and values of J are monitored to determine the sign of the slope of J. (The error function J would increase, pass a local maximum, and reach a local minimum along the ray. If the local minimum cannot be found, try another ray until we get the next local minimum restricted in the ray. In this case, we choose the new ray is orthogonal to the previous one.) Set the obtained weight vector as $\mathbf{x}(\bar{s})$.

(III) Obtain a new stable equilibrium point of system Eq. (2) starting from $\mathbf{x}(\bar{s})$ and set the resulting weight vector as $\tilde{\mathbf{w}}_s$. If $J(\tilde{\mathbf{w}}_s) < J(\mathbf{w}_s)$ set $\mathbf{w}_s = \tilde{\mathbf{w}}_s$ and go to step (I); Otherwise, go to step (I) to construct another ray.

Any reasonable ray is sufficient to continue the MLP training process as long as local minima are well-separated in the step (I) and one only need to evaluate the error cost function in the step (II), leading to simple and stable implementation. The algorithm fully utilize the generic topological structure of the error cost function for escaping from a local minimum trap as well as the information available from the previous training process. For these reasons, our strategy is very efficient.

4 Illustrative Example

To illustrate the proposed strategy, we conducted experiments on a 2-spirals problem [4], which is known to be a very difficult task for back-propagation networks and their relatives. 2-spirals problem is a classification problem to discriminate between two sets of training points that lie on two distinct spirals in the 2-dimensional plane. These spirals coil three times around the origin and around one another. (See Figure 1-(a).) The data set contains 2 classes of 194 instances each, where each class is discriminated by 2-output values, (1,0) denoted by "1" and (0,1) denoted by "2" in Figure 1. The architecture of a neural network is set to be a 3-layer network of size 2-20-8-1.

The proposed strategy is compared with EBP using momentum (EBPM) [4], and EBP using tunneling method (EBPT) [1]. For a performance comparison, the average number of epochs (NOE), the average time of training (TIME), the mis-classification error rate of training (Train ME), the mis-classification error rate of test (Test ME) are conducted and are given in Table 1. The results shows that the proposed algorithm has a potential to solve hard local minima problems for training MLPs. Although our proposed strategy requires separate calculation

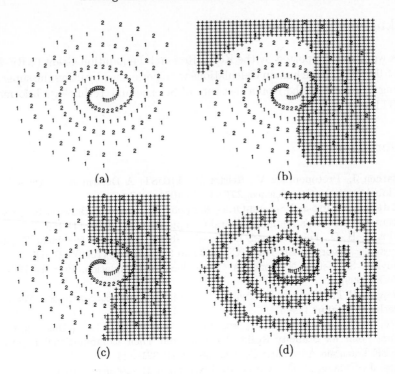

Fig. 1. Decision boundaries formed by the different training algorithms (a) Original 2-spirals problem. (b) EBPM. (c) EBPT. (d) Proposed method.

Table 1. Experimental results for the 2-spirals problem data

	EBPM	EBPT	Proposed
NOE	2000	599	399
TIME	22.703	16.657	45.625
Train ME (%)	45.36	44.33	2.0619
Test ME (%)	44.85	45.88	2.5773

for each network, it is applicable to any network of generic topological structure and improves training MLPs.

5   Conclusion

In this paper, a concept of basin cell has been introduced. The landscape of the MLP error cost function is then analyzed in terms of basin cells whose boundaries are shown to be composed of stable cells of index-one saddle points. The analysis points to a new escaping strategy to solve hard local minima problems for MLP training, where basin cell boundaries are used to escape from local minima traps. Applications of the proposed method to more large-scale training problems need to be further investigated.

Acknowledgements

This work was supported by the Korea Research Foundation Grant (KRF-2004-003-C00071), in which main Calculations were performed by using the supercomputing resources of the Korea Institute of Science and Technology Information (KISTI).

References

1. Barhen J., Protopopescu V., Reister D.: TRUST: A Deterministic Algorithm for Global Optimization. Science, **276** (1997) 1094–1097
2. Battiti, R.: First-and Second-Order Methods for Learning: Between Steepest Descent and Newton's Methods. Neural Computation, **4** (1992) 141-166
3. Chiang, H.-D., Fekih-Ahmed, L.: Quasi-stability Regions of Nonlinear Dynamical Systems: Theory. IEEE Trans. Circuits and Systems, **41** (1996) 627–635
4. Haykin, S.: Neural Networks: A Comprehensive Foundation. Prentice Hall, New York (1999)
5. Jongen, H.Th., Jonker, P., Twilt, F.: Nonlinear Optimization in $\Re^n$. Peter Lang Verlag, Frankfurt (1995)
6. Lee, J.: Dynamic Gradient Approaches to Compute the Closest Unstable Equilibrium Point for Stability Region Estimate and Their Computational Limitations. IEEE Trans. on Automatic Control, **48** (2003) 321–324
7. Lee, J., Chiang, H.-D.: A Singular Fixed-point Homotopy Method to Locate the Closest Unstable Equilibrium Point for Transient Stability Region Estimate. IEEE Trans. on Circuits and Systems- Part II, **51** (2004) 185–189
8. Lee, J., Chiang, H.-D.: A Dynamical Trajectory-Based Methodology for Systematically Computing Multiple Optimal Solutions of General Nonlinear Programming Problems. IEEE Trans. on Automatic Control, **49** (2004) 888–899

Enhanced Fuzzy Single Layer Perceptron

Kwangbaek Kim[1], Sungshin Kim[2], Younghoon Joo[3], and Am-Sok Oh[4]

[1] Dept. of Computer Engineering, Silla University, Korea
gbkim@silla.ac.kr
[2] School of Electrical and Computer Engineering, Pusan National University, Korea
sskim@pusan.ac.kr
[3] School of Electronic and Information Engineering, Kunsan National University, Korea
yhjoo@kunsan.ac.kr
[4] Dept. of Multimedia Engineering, Tongmyong Univ. of Information Technology, Korea
asoh@tit.ac.kr

Abstract. In this paper, a method of improving the learning time and convergence rate is proposed to exploit the advantages of artificial neural networks and fuzzy theory to neuron structure. This method is applied to the XOR problem, n bit parity problem, which is used as the benchmark in neural network structure, and recognition of digit image in the vehicle plate image for practical image recognition. As a result of experiments, it does not always guarantee the convergence. However, the network was improved the learning time and has the high convergence rate. The proposed network can be extended to an arbitrary layer. Though a single layer structure is considered, the proposed method has a capability of high speed during the learning process and rapid processing on huge patterns.

1 Introduction

In the conventional single layer perceptron, it is inappropriate to use when a decision boundary for classifying input pattern does not composed of hyperplane. Moreover, the conventional single layer perceptron, due to its use of unit function, was highly sensitive to change in the weights, difficult to implement and could not learn from past data [1]. Therefore, it could not find a solution of the exclusive OR problem, the benchmark. There are a lot of endeavor to implement a fuzzy theory to artificial neural network [2]. Goh et al. [3] proposed the fuzzy single layer perceptron algorithm, and advanced fuzzy perceptron based on the generalized delta rule to solve the XOR problem, and the classical problem. This algorithm guarantees some degree of stability and convergence in application using fuzzy data, however, it causes an increased amount of computation and some difficulties in application of the complicated image recognition. However, the enhanced fuzzy perceptron has shortcomings such as the possibility of falling in local minima and slow learning time [4].

In this paper, we propose an enhanced fuzzy single layer perceptron. We construct, and train, a new type of fuzzy neural net to model the linear function. Properties of this new type of fuzzy neural net include: (1) proposed linear activation function; and (2) a modified delta rule for learning. We will show that such properties can guarantee to find solutions for the problems-such as exclusive OR, 3 bits parity, 4 bits parity and digit image recognition on which simple perceptron and simple fuzzy perceptron can not.

J. Wang, X. Liao, and Z. Yi (Eds.): ISNN 2005, LNCS 3496, pp. 603–608, 2005.

2 A Fuzzy Single Layer Perceptron

A proposed learning algorithm can be simplified and divided into four steps. For each input, repeat step 1, step 2, step 3, and step 4 until error is minimized.

Step 1: Initialize weight and bias term.

Define, $W_{ij}(1 \le i \le I)$ to be the weight from input j to output i at time t, and θ_i to be a bias term in the output soma. Set $W_{ij}(0)$ to small random values, thus initializing all the weights and bias term.

Step 2: Rearrange A_i according to the ascending order of membership degree m_j and add an item m_0 at the beginning of this sequence.

$$0.0 = m_0 \le m_1 \le \cdots \le m_j \le m_J \le 1.0$$

Compute the consecutive difference between the items of the sequence.

$P_k = m_j - m_{j-1}$, where $k = 0, \cdots, n$

Step 3: Calculate a soma (O_i)'s actual output.

$$O_i = \sum_{k=0}^{J-1} P_k \times f\left(\sum_{j=k}^{J-1} W_{ij} + \theta_i \right)$$

where $f\left(\sum_{j=k}^{J-1} W_{ij} + \theta_i \right)$ is linear activation function, where $i = 1, \cdots, I$

In the sigmoid function, if the value of $\left(\dfrac{1.0}{1.0 + e^{-net}} \right)$ is between 0.0 and 0.25, $\left(\left(\dfrac{1.0}{1.0 + e^{-net}} \right) \times \left(1 - \dfrac{1.0}{1.0 + e^{-net}} \right) \right)$ is very similar to $\left(\dfrac{1.0}{1.0 + e^{-net}} \right)$. If the value of $\left(\dfrac{1.0}{1.0 + e^{-net}} \right)$ is between 0.25 and 0.75, $\left(\left(\dfrac{1.0}{1.0 + e^{-net}} \right) \times \left(1 - \dfrac{1.0}{1.0 + e^{-net}} \right) \right)$ is very similar to 0.25. If the value of $\left(\dfrac{1.0}{1.0 + e^{-net}} \right)$ is between 0.75 and 1.0, $\left(\left(\dfrac{1.0}{1.0 + e^{-net}} \right) \times \left(1 - \dfrac{1.0}{1.0 + e^{-net}} \right) \right)$ is very similar to $\left(1 - \dfrac{1.0}{1.0 + e^{-net}} \right)$.

Therefore, the proposed linear activation function expression is represented as follows:

$$f\left(\sum_{j=k}^{J-1} W_{ij} + \theta_i \right) = 1.0, \; where \; f\left(\sum_{j=k}^{J-1} W_{ij} + \theta_i \right) > 5.0$$

$$f\left(\sum_{j=k}^{J-1} W_{ij} + \theta_i \right) = \rho \times \left(\sum_{j=k}^{J-1} W_{ij} + \theta_i \right) + 0.5, where \; -5.0 \le \left(\sum_{j=k}^{J-1} W_{ij} + \theta_i \right) \le 5.0, \; \rho \in [0.1, 0.4]$$

$$f\left(\sum_{j=k}^{J-1} W_{ij} + \theta_i \right) = 0.0, \; where \; f\left(\sum_{j=k}^{J-1} W_{ij} + \theta_i \right) < -5.0$$

The formulation of the activation linear function is follow.

$$f\left(\sum_{j=k}^{J-1} W_{ij} + \theta_i\right) = \left(\frac{1}{range \times 2}\right) \times \left(\sum_{j=k}^{J-1} W_{ij} + \theta_i\right) + 0.5$$

where the range means monotonic increasing internal except for the interval between
0.0 and 1.0 of value of the $f\left(\sum_{j=k}^{J-1} W_{ij} + \theta_i\right)$.

Step 4: Applying the modified delta rule. And we derive the incremental changes for weight and bias term.

$$\Delta W_{ij}(t+1) = \eta_i \times E_i \times \sum_{k=0}^{j} P_k \times f\left(\sum_{j=k}^{J-1} W_{ij} + \theta_i\right) + \alpha_i \times \Delta W_{ij}(t)$$

$$W_{ij}(t+1) = W_{ij}(t+1) + \Delta W_{ij}(t+1)$$

$$\Delta \theta_i(t+1) = \eta_i \times E_i \times f(\theta_i) + \alpha_i \times \Delta \theta_i(t)$$

$$\theta_i(t+1) = \theta_i(t) + \Delta \theta_i(t+1)$$

where η_i is learning rate α_i is momentum.

Finally, we enhance the training speed by using the dynamical learning rate and momentum based on the division of soma.

$$if\left(Inactivation_{totalsoma} - Activation_{totalsoma} > 0\right)$$

$$then\ \Delta \eta_i(t+1) = E^2\ \eta_i(t+1) = \eta_i(t) + \Delta \eta_i(t+1)$$

$$if\left(Inactivation_{totalsoma} - Activation_{totalsoma} > 0\right)$$

$$then\ \Delta \alpha_i(t+1) = E^2\ \alpha_i(t+1) = \alpha_i(t) + \Delta \alpha_i(t+1)$$

2.1 Error Criteria Problem by Division of Soma

In the conventional learning method, learning is continued until squared sum of error is smaller than error criteria. However, this method is contradictory to the physiological neuron structure and takes place the occasion which a certain soma's output is not any longer decreased and learn no more [5]. The error criteria was divided into activation and inactivation criteria. One is an activation soma's criterion of output "1", the other is an inactivation soma's of output "0". The activation criterion is decided by soma of value "1" in the discriminate problem of actual output patterns, which means in physiological analysis that the pattern is classified by the activated soma. In this case, the network must be activated by activated somas. The criterion of activated soma can be set to the range of [0,0.1].

In this paper, however, the error criterion of activated soma was established as 0.05. On the other hand, the error criterion of inactivation soma was defined as the squared error sum, the difference between output and target value of the soma. Fig.1 shows the proposed algorithm.

3 Simulation and Result

We simulated the proposed method on IBM PC/586 with VC++ language. In order to evaluate the proposed algorithm, we applied it to the exclusive OR, 3 bits parity

```
while ((Activation_no == Target_activated_no)
 &&(Inactivation_error <= Inactivation_area))
    do {
     for (i=0; i<Pattern_no; i++)
      for(j=0; j<Out_cell_no;j++) {
        Forward Pass;
        Backward Pass;
        if(Out_cell=Activation_soma&&|error|<= Activation_area)
          Activation_number++;
      if (Out_cell=Inactivation_soma)
        Inactivation_error += error * error;
 }    }
```

Fig. 1. Learning algorithm by division of soma.

which is a benchmark in neural network and pattern recognition problems, a kind of image recognition. In the proposed algorithm, the error criterion of activation and inactivation for soma was set to 0.09.

3.1 Exclusive OR and 3 Bits Parity

Here, we set up initial learning rate and initial momentum as 0.5 and 0.75, respectively. Also we set up the range of weight [0,1]. In general, the range of weights were [-0.5,0.5] or [-1,1]. As shown in Table 1 and Table 2, the proposed method showed higher performance than the conventional fuzzy single layer perceptron in convergence epochs and convergence rates of the three tasks.

Table 1. Convergence rate in initial weight range.

Applied Problem	Fuzzy Perceptron	Proposed Algorithm	Initial Weight Range
Exclusive OR	100%	100%	[0.0, 1.0]
	89%	99%	[0.0, 5.0]
3 bit parity	83%	97%	[0.0, 1.0]
	52%	96%	[0.0, 5.0]
4 bit parity	0%	96%	[0.0, 1.0]
	0%	95%	[0.0, 5.0]

Table 2. Comparison of step number.

Epoch No	Fuzzy Perceptron	Proposed Algorithm
Exclusive OR	8 (converge)	4 (converge)
3 bit parity	13 (converge)	8 (converge)
4 bit Parity	0 (not converge)	15 (converge)

3.2 Digit Image Recognition

We extracted digit images in a vehicle plate using the method in [6]. We extracted the license plate from the image using the characteristic that green color area of the license plate is denser than other colors. The procedure of image pre-processing is presented in Fig.2 and the example images we used are shown in Fig.3.

Image Pattern	BMP File	Proposed Smoothing & Edge Detection	Proposed Learning Algorithm	Recognition

Fig. 2. Preprocessing diagram.

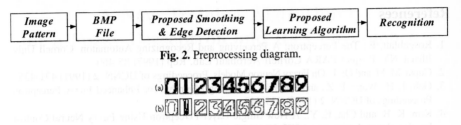

Fig. 3. (a) Digit image and (b) Training image by edge detection.

We carried out image pre-processing in order to prevent high computational load as well as loss of information. If the extracted digit images were used as training patterns, it requires expensive computation load. In contrast skeleton method causes loss of important information of images. To overcome this trade-off, we used edge information of images.

The most-frequent value method was used for image pre-processing. This method was used because blurring of boundary using common smoothing methods. Thus, it degrades both color and contour lines. The new method replaced a pixel's value with the most frequent value among specific neighboring pixels. If the difference of absolute value between neighborhoods is zero in a given area, the area was considered as background. Otherwise, it was considered as a contour. This contour was used as a training pattern. The input units were composed of 32×32 array for image patterns. In simulation, the fuzzy perceptron was not converged, but the proposed method was converged on 70 steps at image patterns. Table 3 is shown the summary of the results in training epochs between two algorithms.

Table 3. The comparison of epoch number.

Image Pattern	Epoch Number
Fuzzy perceptron	0 (not converge)
Proposed algorithm	70 (converge)

4   Conclusions

The study and application of fusion fuzzy theory with logic and inference and neural network with learning ability have been actually achieving according to expansion of automatic system and information processing, etc.

We have proposed a fuzzy supervised learning algorithm which has greater stability and functional varieties compared with the conventional fuzzy single layer perceptron. The proposed network is able to extend the arbitrary layers and has high convergence in case of two layers or more. Though, we considered only the case of single layer, the networks had the capability of high speed during the learning process and rapid processing on huge patterns. The proposed algorithm shows the possibility of the application to the image recognition besides benchmark test in neural network by single layer structure.

In future study, we will develop a novel fuzzy learning algorithm and apply it to the face recognition.

We assume that the training patterns are separable, that is, there exists a fuzzy vector $A = (a_1, \ldots, a_n)^T \in [0, 1]^n$ such that

$$\begin{cases} A \circ X^m < 0.5 & m = 1, \ldots, M \\ A \circ Y^p \geq 0.5 & p = 1, \ldots, P. \end{cases} \tag{2}$$

Beginning with $w_j^0 = 0$, $j = 1, \ldots, n$, the weights are updated according to the following formula (η is a small positive number):

$$w_j^{k+1} = \begin{cases} 0, & \text{if } \triangle_j^k < 0 \\ \triangle_j^k, & \text{if } 0 \leq \triangle_j^k \leq 1 \\ 1, & \text{if } \triangle_j^k > 1 \end{cases} \tag{3}$$

$$\triangle_j^k = w_j^k + \eta \left(O^k - \zeta^k \right) \left(\xi_j^k - 0.5 \right). \tag{4}$$

Another possible choice of the updating rule for W^k is as follows:

$$w_j^{k+1} = \begin{cases} 0, & \text{if } \tilde{\triangle}_j^k \\ \tilde{\triangle}_j^k, & \text{if } 0 \leq \tilde{\triangle}_j^k \leq 1 \\ 1, & \text{if } \tilde{\triangle}_j^k > 1. \end{cases} \tag{5}$$

$$\tilde{\triangle}_j^k = w_j^k + \eta \left(O^k - \zeta^k \right) \xi_j^k. \tag{6}$$

This is a direct analog of the usual perceptron rule ([7]). The reason for us to use formula (3)(4) instead of (5)(6) is as follows. If (5)(6) are applied, then all the components of W^k will be increased or decreased simultaneously at each step of training to get W^{k+1}. This is somehow not quite reasonable. On the other hand, let us look at (3)(4). Now assume a training vector Y^p is supplied to the network and

$$\zeta^k = f(W^k \circ Y^p) = 0 \neq O^k = 1. \tag{7}$$

Then according to (4), w_j^k will be increased if $y_j^p \geq 0.5$, which will help to increase $W^{k+1} \circ Y^p$; and w_j^k will be decreased if $y_j^p < 0.5$, which will not influence our target $f(W^{k+1} \circ Y^p) = 1$ but might help to decrease $(W^{k+1} \circ X^m)$ as desired. Similar things happen when we supply an X^m to the network for training.

2   Preliminaries

We first make an assumption on the training patterns.

Assumption I: For each $m = 1, \ldots, M$, there exists at least one m_0 such that $x_{m_0}^m \geq 0.5$.

This assumption is not restrictive. In fact, if a training pattern X^m does satisfy $x_j^m < 0.5 \ \forall \ j = 1, \ldots, n$, then we always have $W \circ X^m = \bigvee_{j=1}^{n} (w_j \wedge x_j^m) \leq \bigvee_{j=1}^{n} x_j^m < 0.5$ and its actual output is $\zeta \equiv 0$. So this X^m does not make any contribution to the updating of W, and can be dropped out from the set of the training patterns.

Because (18), (20) and $\gamma_2 > 0$, there is an integer $N_2 \geq 1$ such that

$$w_2^{k+N_2} < w_2^{k+N_2-1} = w_2^{k+N_2-2} = \cdots = w_2^k. \qquad (22)$$

Then if $\triangle_2^{k+N_2-1} \leq 0$, we have $w_2^{k+N_2} = 0 < 0.5$ and W^{k+N_2} is already the solution of (3)(4). If $0 < \triangle_2^{k+N_2-1} < 1$, we have

$$w_2^{k+N_2} = \triangle_2^{k+N_2-1} \leq w_2^{k+N_2-1} - \eta\,\gamma_2 < w_2^{k+N_2-1} = \cdots = w_2^k. \qquad (23)$$

So we will have $w_2^k < 0.5$ after w_2^k being really updated at most $\lceil \frac{w_2^{K_1}-0.5}{\gamma_2\,\eta} + 1 \rceil$ times. Hence there exists a positive integer $K_2 > K_1$ such that $w_2^k < 0.5$ if $k \geq K_2$ and (19) holds. This implies that W^k satisfies (11). $\square$

Next, we consider the case of $n > 2$. Stronger conditions are needed here to guarantee the convergence.

Theorem 2. *Suppose Assumptions I and II are satisfied, then (3) is finitely convergent under the following two conditions:*
   (a) There exits an $r_0 \leq q$, such that $y_{r_0}^p > 0.5$ for each $p = 1, \ldots, P$;
   (b) For every $m = 1, \ldots, M$ and $j = R+1, \ldots, n$, $x_j^m \geq 0.5$.

Proof. Since the number of the training patterns is finite, we have $\lambda_0 = min_{1 \leq s \leq S}$ $\{|\xi_{r_0}^{(s)} - 0.5|\} > 0$. By (a) and $x_{r_0}^m < 0.5$, we always have $\eta\,(O^k - \zeta^k)\,(\xi_{r_0}^k - 0.5) \geq 0$. Notice $w_{r_0}^0 = 0$. Then $w_{r_0}^{k+1} \geq w_{r_0}^k$. The inequality holds strictly when the training pattern ξ^k is not rightly recognized, and the least increase value is $((\eta\,\lambda_0) \wedge (1 - w_{r_0}^k))$. It can be derived that if W^k satisfies (11) before $w_{r_0}^k$ reaches to 0.5, then (3) is finitely convergent; Otherwise, if $w_{r_0}^k < 0.5$, there is an integer $N_3 \geq 1$ such that

$$w_{r_0}^{k+N_3} > w_{r_0}^{k+N_3-1} = w_{r_0}^{k+N_3-2} = \cdots = w_{r_0}^k. \qquad (24)$$

Then if $\triangle_{r_0}^{k+N_3-1} \geq 1$, we have

$$0.5 < 1 = w_{r_0}^{k+N_3} = w_{r_0}^{k+N_3+1} = \cdots .$$

If $0 < \triangle_{r_0}^{k+N_3-1} < 1$, we have

$$w_{r_0}^{k+N_3} = \triangle_{r_0}^{k+N_3-1} \geq w_{r_0}^{k+N_3-1} + \eta\,\lambda_0 > w_{r_0}^{k+N_3-1} = \cdots = w_{r_0}^k. \qquad (25)$$

So we will have $w_{r_0}^k \geq 0.5$ after $w_{r_0}^k$ being really updated at most $\lceil \frac{1}{2\eta\,\lambda_0} \rceil$ times. Hence there exists a positive integer K_3 such that $w_{r_0}^k \geq 0.5$ if $k \geq K_3$, and

$$W^k \circ Y^p = \geq w_{r_0}^k \wedge y_{r_0}^p \geq 0.5, p = 1, \ldots, P, k \geq K_3. \qquad (26)$$

Now let $k \geq K_3$. Then $w_{r_0}^k \geq 0.5$ and (26) is true. If $w_j^k < 0.5$, $j = R+1, \ldots, n$, then for each $m = 1, \ldots, M$ we have

$$W^k \circ X^m = (\bigvee_{j=1}^{R} (w_j^k \wedge x_j^m)) \vee (\bigvee_{j=R+1}^{n} (w_j^k \wedge x_j^m)) \leq (\bigvee_{j=1}^{R} x_j^m) \vee (\bigvee_{j=R+1}^{n} w_j^k) < 0.5 \qquad (27)$$

implying that W^k is already the desired solution. Otherwise, we assume $w_{j_l}^k \geq 0.5$, $l = 1, \ldots, L$, $1 \leq L \leq n - R$, $R < j_l \leq n$, and $w_j^k < 0.5$, $j \neq j_l$, $R < j \leq n$. By (b) we have

$$W^k \circ X^m \geq w_{j_l}^k \wedge x_{j_l}^m \geq 0.5, m = 1, \ldots, M \tag{28}$$

implying that W^k can not rightly recognize all $\xi^k = X^m$, even if only one $w_{j_l}^k \geq 0.5$. As a result, only X^m contribute to the updating of W^k. By Assumption II and (b) we have $\lambda_l = min_{1 \leq m \leq M \atop x_{j_l}^m \neq 0.5} \{x_{j_l}^m - 0.5\} > 0$ and $\eta \, (O^k - \zeta^k) \, (\xi_j^k - 0.5) \leq 0$, $R < j \leq n$, if $k > K_3$. Then w_j^k will decrease and $\triangle_j^k \leq 1$, $R < j \leq n$. Because (28) and $\lambda_l > 0$, there is an integer $N_{j_l} \geq 1$ such that

$$w_{j_l}^{k+N_{j_l}} < w_{j_l}^{k+N_{j_l}-1} = w_{j_l}^{k+N_{j_l}-2} = \cdots = w_{j_l}^k. \tag{29}$$

Then if $\triangle_{j_l}^{k+N_{j_l}-1} \leq 0$, we have

$$0.5 > 0 = w_{j_l}^{k+N_{j_l}} = w_{j_l}^{k+N_{j_l}+1} = \cdots.$$

Otherwise, if $0 < \triangle_{j_l}^{k+N_{j_l}-1} < 1$, we have

$$w_{j_l}^{k+N_{j_l}} = \triangle_{j_l}^{k+N_{j_l}-1} \leq w_{j_l}^{k+N_{j_l}-1} - \eta \, \lambda l < w_{j_l}^{k+N_{j_l}-1} = \cdots = w_{j_l}^k. \tag{30}$$

So there exists an integer K_{j_l} for each j_l, $l = 1, \ldots, L$, such that $w_{j_l}^k < 0.5$ if $k \geq K_{j_l}$. We set $K_4 = max_{1 \leq l \leq L}\{K_{j_l}\}$, then $w_{j_l}^k < 0.5$, $j = 1, \ldots, L$, if $k \geq K_4$, and (27) holds. So W^k satisfies (11) and (3) is finitely convergent. □

References

1. Nikov, A., Stoeva, S.: Quick Fuzzy Backpropagation Algorithm. Neural Networks. **14** (2001) 231–244
2. Castro, J.L., Delgado, M., Mantas, C.J.: A Fuzzy Rule-based Algorithm to Train Perceptrons. Fuzzy Sets and Systems. **118** (2001) 359–367
3. Pandit, M., Srivastava, L., Sharma, J.: Voltage Contingency Ranking Using Fuzzified Multilayer Perceptron. Electric Power Systems Research. **59** (2001) 65–73
4. Chowdhury, P., Shukla, K.K.: Incorporating Fuzzy Concepts along with Dynamic Tunneling for Fast and Robust Training of Multilayer Perceptrons. Neurocomputing. **50** (2003) 319–340
5. Stoeva, S., Nikov, A.: A Fuzzy Backpropagation Algorithm. Fuzzy Sets and Systems. **112** (2000) 27–39
6. Liang, Y.C., Feng, D.P., Lee, H.P., Lim, S.P., Lee, K.H.: Successive Approximation Training Algorithm for Feedforward Neural Networks. Neurocomputing. **42** (2002) 311–322
7. Wu, W., Shao, Z.: Convergence of Online Gradient Methods for Continuous perceptrons with Linearly Separable Training Patterns. Applied Mathematics Letters. **16**(2003) 999–1002

Stochastic Fuzzy Neural Network and Its Robust Parameter Learning Algorithm*

Junping Wang and Quanshi Chen

State Key Laboratory of Automobile Safety & Energy Conservation
Tsinghua University, Beijing 100084, China
wangjunping@tsinghua.org.cn

Abstract. A Stochastic Fuzzy Neural Network (SFNN) which has filtering effect on noisy input is studied. the structure of the SFNN is mended and the nodes in each layer of the SFNN are discussed. Each layer in the new structure has exact physical meaning. The number of the nodes is decreased, so is the computation amount. In the parameter learning algorithm, if noisy input data is used the LS cost function based method can cause severe biasing effects. This problem can be solved by a novel EIV cost function which contains the error variables. In this paper, the cost function is extended to multi-input single output system, and the error variables are obtained through learning algorithm to avoid repeated measurement. This method was used to train the parameters of the SFNN. The simulation results show the efficiency of this algorithm.

1 Introduction

The combination of fuzzy logic with neural network forms an integrated system named Fuzzy Neural Network (FNN) possessing the advantage of both [1]. Wang [2] has developed an adaptive fuzzy logic based on singleton fuzzy logic. But the non-singleton fuzzy logic system proposed by Mouzouris and Mendel is more useful when the input data is polluted by noise. However, they have not solved the problem of parameter and structure learning. Z L Jing and other scholars [1], [3] have developed a Stochastic Fuzzy Neural Network (SFNN) and proposed the algorithms for parameter and structure learning. But the structure of the SFNN has no physical meaning and it is just a computation network to realize the computation of stochastic fuzzy logic system. In this paper the structure of the SFNN is mended and the nodes in each layer of the SFNN are discussed. Each layer in the new structure has exact physical meaning.

While the structure is determined then the parameters of the network is learned from a set of samples. Most traditional learning algorithm are based on least squares (LS) cost function. This method provides goods results when the training data with noisy output data and known input data. While the input data contain noise, it is proved that the parameters of the network can not converge to its true value. So a new cost function named errors-in-variables (EIV) cost function was used to solve this problem [4]. But in the reference the variance of the noise was obtained through statistical estimation method with repeat measurements or given a priori value on the basis of the error

* Supported by a grant from the research fund of State Key Laboratory of Automobile Safety & Energy Conservatio (No. KF2005-006) and the China Postdoctoral Science Foundation (No.20030304145).

J. Wang, X. Liao, and Z. Yi (Eds.): ISNN 2005, LNCS 3496, pp. 615–620, 2005.
© Springer-Verlag Berlin Heidelberg 2005

variance of the measurement device. In this paper the EIV cost function is extended to Multi-input Single output (MISO) system for generality. Then the EIV cost function is used to training parameters of the SFNN and the variance of the noise is obtained through training so repeated measurement can be avoided. The simulation results show the efficiency of the algorithm.

2   The Structure of Stochastic Fuzzy Neural Network

While the input and output data contain noise the convergence of the parameters of the network will be affected while training. This can be illustrated by the theorems shown in Ref. [4]. Based on these theorems we study the SFNN with noisy input only and the noise is fitted to normal distribution. The Structure of SFNN shown in Fig.1 has five layers. The input nodes in layer-1 represent input linguistic variables. The fuzzier nodes in layer-2 perform the fuzzification of the input variable. Each node in layer-3 is a fuzzy rule. The whole nodes in layer-3 form a base of fuzzy rules and the number of fuzzy rules is optimized through structure learning. The nodes in layer-4 are fuzzy inference engine nodes. The connection of layer-3 and layer-4 acted as neural fuzzy inference engine to avoid matching processes. The layer-5 is output or defuzzifier node which maps fuzzy set to a crisp point. Fuzzy logic in SFNN possesses a non-singleton fuzzifier [2], product inference rules, center average defuzzifier and Gaussian membership function. Where, the base of the fuzzy rules is composed of L rules:

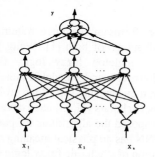

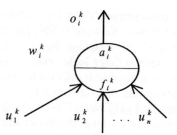

Fig. 1. Structure of The SFNN.

Fig. 2. Detail Structure of the node in the SFNN.

$$R^l : \text{IF } x_1 \text{ is } F_1^l \text{ andLand } x_i \text{ is } F_i^l, \text{ THEN } y \text{ is } G^l. \ (i=1, Ln ; l=1, LL) \tag{1}$$

Where The variables $x \in (x_1, \cdots x_n) \in U_1 \times \cdots \times U_n$ and $y \in V$ are the input and output of the system. F_i^l and G^l are fuzzy sets with the membership function $\mu_{F_i^l}(x_i)$ and $\mu_{G^l}(y)$ in input space $U_i \in R$ and output space $V \in R$ respectively. $\mu_{A_{x_i}}(x_i)$ is the membership function of the non-singleton fuzzifier which maps a crisp point to a fuzzy set. $\mu_{A_{x_i}}(x_i)$, $\mu_{F_i^l}(x_i)$ and $\mu_{G^l}(y)$ are all adopted as Gaussian membership function. m_{x_i}, σ_{x_i} are the center value and the variance of $\mu_{A_{x_i}}(x_i)$, $m_{F_i^l}, \sigma_{F_i^l}$ are for

$\mu_{F_i^l}(x_i)$ and $\bar{y}_{g^l}, \sigma_{g^l}$ for $\mu_{G^l}(y)$. These parameters are adjustable through parameters learning algorithm. L is the optimal number of the nods in layer 3, i.e., the optimal number of the fuzzy rules can be determined by structure learning algorithm seen in [1].

The detailed structure of the node in the SFNN is shown in Fig.2. In which, u_i^k and o_i^k are input and output, w_i^k and f_i^k are weight and integration functions, α_i^k is an activation function, the superscript k denotes the layer number. The nodes in each layer of the SFNN are discussed as follows.

We known from Ref.[4], according to the multiplication inference rules and $Sup-*$ combination operation, the fuzzy set has maximum value in output space V at $\hat{x}_i^l = \dfrac{\sigma_{F_i^l}^2 x_i + \sigma_{x_i}^2 m_{F_i^l}}{\sigma_{F_i^l}^2 + \sigma_{x_i}^2}$ while adopt x_i as the central value m_{x_i}. $\hat{x}_i^l$ is obtained in the input

nodes by pretreatment for x_i. So only five parameters $\sigma_{x_i}, m_{F_i^l}, \sigma_{F_i^l}, \bar{y}_{g^l}, \sigma_{g^l}$ are adjustable. As a result, the number of the learning parameters is $n + n \times L \times 2 + 2 \times L$. Compared with the number of learning parameters in the Ref.[4] which is $n \times L \times 3 + 2 \times L$, the computation of the modified SFNN reduced.

Layer-1: input nodes

$$f_i^{(1)} = u_i^{(1)} = x_i, \quad o_i^{(1)} = a_i^{(1)}(f_i^{(1)}) = \frac{\sigma_{F_i^l}^2 x_i + \sigma_{x_i}^2 m_{F_i^l}}{\sigma_{F_i^l}^2 + \sigma_{x_i}^2} = \hat{x}_i^l, \quad w_i^{(1)} = 1 \tag{2}$$

Layer-2: fuzzifier nodes

$$\begin{cases} f_i^{(2)} = -\left(\dfrac{u_i^{(2)} - m_{x_i}}{\sigma_{x_i}}\right)^2 = -\left(\dfrac{o_i^{(1)} - m_{x_i}}{\sigma_{x_i}}\right)^2 = -\left(\dfrac{\hat{x}_i^l - x_i}{\sigma_{x_i}}\right)^2 \\ o_i^{(2)} = a_i^{(2)}(f_i^{(2)}) = \exp(f_i^{(2)}), w_i^{(2)} = 1. \end{cases} \tag{3}$$

$$\begin{cases} \hat{f}_i^{(2)} = -\left(\dfrac{u_i^{(2)} - m_{F_i^l}}{\sigma_{F_i^l}}\right)^2 = -\left(\dfrac{o_i^{(1)} - m_{F_i^l}}{\sigma_{F_i^l}}\right)^2 = -\left(\dfrac{\hat{x}_i^l - m_{F_i^l}}{\sigma_{F_i^l}}\right)^2 \\ \hat{o}_i^{(2)} = \hat{a}_i^{(2)}(\hat{f}_i^{(2)}) = \exp(\hat{f}_i^{(2)}), \hat{w}_i^{(2)} = 1. \end{cases} \tag{4}$$

Layer-3: rule nodes

$$f_i^{(3)} = \prod_{i=1}^n u_i^{(3)} = \prod_{i=1}^n o_i^{(2)} \hat{o}_i^{(2)}, \quad o_i^{(3)} = a_i^{(3)}(f_i^{(3)}) = f_i^{(3)}, w_i^{(3)} = 1. \tag{5}$$

Layer-4: inference engine nodes

$$f_i^{(4)} = u_i^{(3)} = o_i^{(3)}, w_i^{(4)} = 1. \tag{6}$$

$$o_i^{(4)} = a_i^{(4)}(f_i^{(4)}) = \sup\left\{ f_i^{(3)}, \mu_{G^l}(y) \right\} = \max\left\{ o_i^{(3)} \right\} = \prod_{i=1}^n \exp\left[-\frac{(x_i - m_{F_i^l})^2}{\sigma_{F_i^l}^2 + \sigma_{x_i}^2} \right] \tag{7}$$

Layer-5: output nodes

$$u_i^{(5)} = o_i^{(4)} = \prod_{i=1}^{n} \exp\left[-\frac{(x_i - m_{F_i^l})^2}{\sigma_{F_i^l}^2 + \sigma_{x_i}^2}\right], \quad f_i^{(5)} = \sum_{l=1}^{L} \frac{\bar{y}_{g^l}}{\sigma_{g^l}} u_i^{(5)} \tag{8}$$

$$o_i^{(5)} = a_i^{(5)}(f_i^{(5)}) = \frac{f_i^{(5)}}{\sum_{l=1}^{L}(1/\sigma_{g^l})u_i^{(5)}}, \quad w_i^{(5)} = 1 \tag{9}$$

So the relationship between input and output is

$$y = f(\theta, x) = \frac{\sum_{l=1}^{L}(\bar{y}_{g^l}/\sigma_{g^l})\prod_{i=1}^{n} \exp\left[-(x_i - m_{F_i^l})^2 \big/ (\sigma_{F_i^l}^2 + \sigma_{x_i}^2)\right]}{\sum_{l=1}^{L}(1/\sigma_{g^l})\prod_{i=1}^{n} \exp\left[-(x_i - m_{F_i^l})^2 \big/ (\sigma_{F_i^l}^2 + \sigma_{x_i}^2)\right]} \tag{10}$$

3   Robust Parameter Learning Algorithm

For the given input-output data $(X_k, y_k), k = 1, 2, \cdots N$, the parameters of the SFNN can be determined by the back-propagation learning algorithm. The noisy input and output data are defined as

$$\begin{cases} X_k = X_{0,k} + n_x(k) \\ y_k = y_{0,k} + n_y(k) \end{cases} \tag{11}$$

where $X_{0,k}$, $y_{0,k}$ are the true, but unknown values of input and output data respectively. $n_x(k)$, $n_y(k)$ are the input and output noise, Its variance is σ_{x_i}, σ_y respectively. For the system with n inputs and one output, the new EIV cost function is defined as

$$E_{EIV} = \frac{1}{2}\sum_{k=1}^{N}\left[\frac{(y_k - f(\theta, \hat{X}_k))^2}{\sigma_y^2} + \sum_{i=1}^{n}\frac{(x_k^i - \hat{x}_k^i)^2}{\sigma_{x_i}^2}\right] = \frac{1}{2}\sum_{k=1}^{N}\left[\frac{e_{y_k}^2}{\sigma_y^2} + \sum_{i=1}^{n}\frac{e_{x_k}^2}{\sigma_{x_i}^2}\right] \tag{12}$$

where $\hat{X}_k$ is the estimation value of $X_{0,k}$, the subscript k denotes the k th data pair.

Theorem: If the network is trained with noisy input measurements using the EIV cost function, the estimated parameters θ of the network strongly converge to the true parameters θ^*, or

$$a.s.\lim_{N \to \infty}[\arg\min_{\theta}(E_{EIV})] = \theta^* \tag{13}$$

This theorem is easy to follow the same method to prove as for theorem in Ref.[4], so the process is omitted.

The learning algorithm requires the minimum of the EIV cost function for the parameters θ of the SFNN, the parameter update vector of the learning rule used herein is

$$\Delta\theta = -\eta \frac{\partial E_{EIV}}{\partial \theta} \tag{14}$$

where η is a small positive arbitrary value called the learning rate.

The parameters to be learned include $\theta' = [m_{F_i} \quad \sigma_{F_i} \quad \overline{y}_{g_i} \quad \sigma_{g_i}]^T$ and the noise variance σ_{x_i}, σ_y and estimation value $\hat{X}_k$. These parameters can be optimized through back-propagation learning algorithm as follow

$$\left.\begin{array}{ll}\Delta\theta'_j = -\eta \displaystyle\sum_{k=1}^{N} \frac{e_{y_k}}{\sigma_y^2} \frac{\partial f(\theta, \hat{X})}{\partial \theta'_j}, & \Delta\sigma_{x_i} = -\eta \left[\displaystyle\sum_{k=1}^{N} \frac{e_{y_k}}{\sigma_y^2} \frac{\partial f(\theta, \hat{X})}{\partial \sigma_{x_i}} + \displaystyle\sum_{k=1}^{N} \frac{e_{x_k^i}^2}{\sigma_{x_i}^3}\right], \\[4mm] \Delta\sigma_y = -\eta \displaystyle\sum_{k=1}^{N} \frac{e_{y_k}^2}{\sigma_y^3}, & \Delta\hat{x}_k^i = -\eta \left[\frac{e_{y_k}}{\sigma_y^2} \frac{\partial f(\theta, \hat{X})}{\partial \hat{x}_k^i} + \frac{e_{x_k^i}}{\sigma_{x_i}^2}\right]. \end{array}\right\} \quad (15)$$

4   A Simulation Example

The given method was simulated on the following nonlinear function.

$$z(x, y) = y\sin(-3x) + x\cos(-5y) \quad (16)$$

x and y were the inputs taken in the region $x, y \in [-1,1]$ and z was the output. The noise with the variance $\sigma^2 = 0.25$ was added and a training set with 200 simulation pairs $(x^{(k)}, y^{(k)}, z^{(k)})$, $k = 1, 2, \cdots, 200$. is given. 30 fuzzy rules were derived from the training data using subtraction clustering algorithm [5], so the initial value of the parameters of the SFNN is determined too. All the parameters were optimized through the method given in this paper and the traditional method based on the LS cost function.

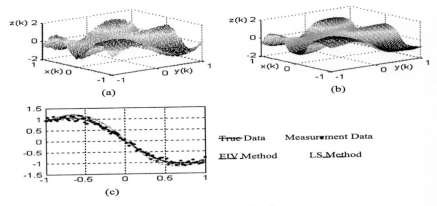

Fig. 3. Simulation Results.

The simulation results is shown in Fig.3(a)~(c). In which Fig.3(a) shows the original nonlinear function with noisy pollution in both input and output, Fig.3(b) shows approximation of the function with SFNN. From Fig.3(a) and (b) we can see that the SFNN is more effective in preventing the system from noise pollution. Fig.3(c) show the simulation results based on our method and the LS method where $x = 0.5$. The results show that the EIV cost function based method gives a better performance than the LS cost function based method.

5 Conclusion

The SFNN has filtering effect on the noisy input. In the parameter learning algorithm, if noisy input data is used the LS cost function based method can cause severe biasing effects. This problem can be solved by a novel EIV cost function which contains the error variables. This method was used to train the parameters of the SFNN. The simulation results show the efficiency of this algorithm.

References

1. Wang, J., Jing, Z.: A Stochastic Fuzzy Neural Network with Universal Approximation and Its Application. Proc. of Int. Conf. On Fuzzy Information Processing Theories and Applications. Tsinghua University Press & Springer, Beijing (2003) 497-502
2. Wang, L. X.: Adaptive Fuzzy Systems and Control. Design and Stability Analysis, Prentice-Hall (1994)
3. Wang, J., Chen, Q., Chen, Y.: RBF Kernel Based Support Vector Machine with Universal Approximation and Its Application. Proc. of ISNN2004, (2004) 512-517
4. Van Gorp, J., Soukens, J., Pintelon, R.: Learning Neural Networks with Noisy Inputs Using the Errors-in-Variables Approach. IEEE Trans. On Neural Networks, 11 (2000) 402-413
5. Yager, R.R., and Filev, D.P.: Approximate Clustering via the Mountain Method. IEEE Trans on Systems, Man and Cybernetics, (1994) 1274-1284

Applying Neural Network to Reinforcement Learning in Continuous Spaces

Dongli Wang, Yang Gao, and Pei Yang

National Laboratory for Novel Software Technology,
Nanjing University, Nanjing 210093, China
dongli@ai.nju.edu.cn

Abstract. This paper is concerned with the problem of Reinforcement Learning (RL) in large or continuous spaces. Function approximation is the main method to solve such kind of problem. We propose using neural networks as function approximators in this paper. Then we experiment with three kind of neural networks in Mountain-Car task and illustrate comparisons among them. The result shows that CMAC and Fuzzy ARTMAP perform better than BP in Reinforcement Learning with Function Approximation (RLFA).

1   Introduction

Reinforcement learning (RL) is learning what to do – how to map situations to actions – so as to maximize a numerical reward signal[1]. It is different from supervised learning and unsupervised learning. In supervised learning, teaching signals are needed. But in real world, it is always difficult for us to obtain the teaching signals in such fields as traffic control, mobile robots control, etc. Under this circumstance, supervised learning is not applicable. By Reinforcement Learning, agent can learn optimal policy through trial-and-error interactions with a dynamic environment given only an explicit object.

Most Reinforcement Learning researches are conducted within the mathematical framework of Markov decision processes (MDPs). There are many algorithms for reinforcement learning, such as TD, Q-learning, Sarsa, Dyna, etc[1]. In general problems which are in discrete state and action spaces, we estimate value functions represented as look-up tables with one entry for each state or for each state-action pair. This method has been proven to be convergent, but it is unavoidably limited. Look-up tables typically do not scale well for high-dimensional MDPs with a continuum of states and actions. First, state and action spaces must be quantized into a finite number of cells which are mapped to entries of the look-up table. It is often difficult to determine an appropriate quantization scheme to provide enough resolution (i.e., accuracy) and low quantization error. Second, the number of cells grows exponentially with the number of variables and geometrically with the number of quantization levels, which is called the curse of dimensionality.

Hence, to generalize experience with a limited subset of the state space to produce a good approximation over a much larger subset will solve this problem.

J. Wang, X. Liao, and Z. Yi (Eds.): ISNN 2005, LNCS 3496, pp. 621–626, 2005.

The kind of generalization required is often called Function Approximation (FA). It takes examples from a desired function (e.g., a value function) and attempts to generalize from them to construct an approximation of the entire function.

Neural network function approximation is a competent way for substituting look-up tables. It can avoid the curse of dimensionality, and now is a focusing field in reinforcement learning.

Although utilizing neural network to approximate value functions is a good approach, it is still difficult to find a perfect method applicable to reinforcement learning in continuous spaces. Many researchers have ever proposed various approaches to this problem, such as residual algorithms [2], wire-fitted neural network Q-learning [3] and the finite-element reinforcement learning [4].

In this paper, we present some approximation approaches which use neural networks to tackle the reinforcement learning in continuous spaces. The rest of this paper is organized as follows. In Section 2, we introduce the framework of reinforcement learning in continuous spaces; in Section 3, we discuss some neural networks in RLFA; in Section 4, we report on the results of our experiments and the comparisons of various approaches applied in the experiments on Mountain-Car task; finally in Section 5, we conclude and indicate some directions for future work.

2   Framework of RL in Continuous Spaces

Reinforcement learning is an interdisciplinary research area, which is developed from control theory, statistics, and psychology of animal learning and so on. In classical reinforcement learning, agent learns the optimal policy through trial-and-error interactions with a dynamic environment. The process can be depicted as Fig. 1.

The reinforcement learning system consists of four fundamental modules, which are input module I, reinforcement module R, policy module P and Environment module E. The input module, I, maps the environment state, s, to the internal perspective of the agent, i. Through policy module, P, the agent updates its internal knowledge and chooses an action to act on the environment according to a certain policy. And R module is to give the agent a reward, r, according to the state's transference. The environment's state,s, changes to s'under the action's effect.

At each step, the purpose of the agent is to get reward as large as possible by choosing a certain action at each step. That is to say, the fundamental of reinforcement learning is that if a certain action policy makes the agent receiving positive rewards, the trend of this action policy will be strengthened. Otherwise, the policy will be weakened.

Taking Q-learning for example, there is a Q-value, $Q(s,a)$, corresponding to each pair of state-action, (s,a), recorded in a look-up table. The traditional reinforcement learning process can be described as following.

Step 1, Observe the current state, s;

Step 2, According to s and the entries in the look-up table, choose an action which makes $Q(s,a)$ maximum, and execute it;

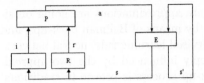

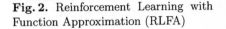

Fig. 1. Framework of Reinforcement Learning (RL)

Fig. 2. Reinforcement Learning with Function Approximation (RLFA)

Step 3, Observe next state, s', and get the current reinforcement, r, from the environment;

Step 4, Update $Q(s, a)$ according to Equation 1 (Bellman Back-up),

$$Q(s, a) = Q(s, a) + \alpha(r + \gamma \max_{a'} Q(s', a') - Q(s, a)) \tag{1}$$

Step 5, Go to Step 1, until all Q-values is stabilized, or the effect of learning is good enough.

Hereby, we can see that the Q-value function is iteratively approximated to the real function during learning.

In large or continuous spaces, we can not explore all the states in the spaces during learning. Some kind of generalization is needed. Therefore, the model of reinforcement learning under these circumstances is different from above. Here, we substitute function approximators for the lookup tables in traditional reinforcement learning. The model is depicted as Fig.2.

There is an additional module in this model, i.e. FA module. The FA module is used to approximate the Q-value function in continuous space. Taking neural networks (NN) for example, the Q-value function which is to be approximated is denoted as Q_{NN}. The inputs to the neural network in FA module are the state variables, which can be continuous, and the action variable. And the output of the neural network approximator is the Q-value corresponding to each input. Here, the back-up expression of Reinforcement Learning using Function Approximation (RLFA) is accordingly altered as Equation 2.

$$Q(s, a) = Q_{NN}(s, a) + \alpha(r + \gamma \max_{a'} Q_{NN}(s', a') - Q_{NN}(s, a)) \tag{2}$$

In RLFA, the learning process can be described as following.

Step 1, Observe the current state, s;

Step 2, According to s and the FA module, choose an action which makes $Q_{NN}(s, a)$ maximum, and execute it;

Step 3, Observe next state, s', and get the current reinforcement, r, from the environment;

Step 4, Update the value of $Q(s, a)$ according to Equation 2;

Step 5, Train the neural network in FA module incrementally using (s, a) and its Q-value, $Q(s, a)$, calculated in Step 4;

Step 6, Go to Step 1 until the neural network is stabilized or the effect of learning is good enough.

During the learning process, two separate approximations are going on simultaneously – one of the Q-value function by means of Bellman back-ups, and another one by means of some general supervised learning rule, neural network. The effect of reinforcement learning is deeply influenced by the two approximations. The convergency of Bellman back-up process is proven in traditional condition[1]. So, the performance of the latter approximation process, neural network, is critical in RLFA.

3   Neural Networks in RLFA

There are a variety of neural networks which are applicable to function approximation. In this paper, we consider three of them, BP (Back Propagation), CMAC (Cerebella Model Articulation Controller)[5, 6] and Fuzzy ARTMAP (Fuzzy Adaptive Resonance Theory Map)[7].

Backpropagation is one of the most prevailing neural algorithms in dealing with function approximation tasks at present. The errors propagate backwards from the output nodes to the inner nodes to adjust the network's weights. When it is applied to reinforcement leaning, the input of BP is the state-action pair and the output of it is the Q-value corresponding to the state-action pair.

Cerebellar Model Articulation Controllers, or CMACs, are a class of sparse coarse-coded memory that models cerebellar functionality [5]. A CMAC consists of multiple overlapping tilings of the state-action space to produce the feature representation for a final linear mapping where all the learning takes place. Each input or state-action pair activates a specific set of memory locations or features, the arithmetic sum of whose contents is the value of the stored Q-value.

Fuzzy ARTMAP is a kind of incremental supervised learning algorithm. Classical Fuzzy ARTMAP includes a pair of ART modules (ART_a and ART_b) that create stable recognition categories in response to arbitrary sequences of input patterns. These modules are linked by an inter-ART module called Mapfield whose purpose is to determine whether the correct mapping has been established from inputs to outputs or not[7].

In reinforcement learning, it is important for approximators to possess the property of the on-line learning. The on-line learning property is referred to the ability of learning new information incrementally, refining existing information quickly and without destroying old information learned previously. BP is not suitable for this kind of learning although it can approximate various functions precisely. However, CMAC and Fuzzy ARTMAP are competent in on-line problems for that they can be incrementally trained.

4   Experimental Results

In this section, we give experimental results of using the neural networks mentioned above to learn policies for Mountain-Car Task. (See Fig. 3.)

There are two continuous state variables, the position of the car, p_t, and the velocity of the car, v_t, in the problem. And at each step the car can take three possible actions denoted as a_t. In the experiment the valid ranges are $-1.2 \leq p_t \leq 0.5, -0.07 \leq v_t \leq 0.07$ and $a_t \in \{-1, 0, 1\}$. The equations describing the system are:

$$
\begin{aligned}
p_{t+1} &= bound[p_t + v_t] \\
v_{t+1} &= bound[v_t + 0.001a_t - 0.0025(\cos 3p_t)]
\end{aligned}
\tag{3}
$$

The action, a_t, can take three distinct values which represent reverse thrust, no thrust and forward thrust. The current reward is -1 everywhere except at the top of the hill. The terminal reward is 0 when the car reaches the top[6].

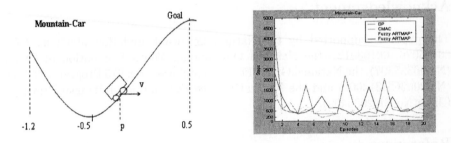

Fig. 3. Model of Mountain-Car Task   **Fig. 4.** Experiment Results Comparison

From Fig.4, we can see that CMAC achieves better performance than BP and Fuzzy ARTMAP. It can be incrementally trained and converges quite quickly. BP and Fuzzy ARTMAP show instability during learning. Although BP needs less memory space than the others, its incremental learning ability is the worst. It adjusts all the weighs in the network while learning, so it may corrupt or forget the previously learned information in the network when a new instance is fed. At the same time, it usually converges on satisfactory local minima. In order to avoid overfitting, we roll back if the performance shows overfitting. Fuzzy ARTMAP is still a good incremental learning algorithm but it does not show excellent convergence performance in reinforcement learning. What's more, the number of hidden units in Fuzzy ARTMAP increases in the learning process and is the most among the three. These dissatisfying performances might be due to the specialty of reinforcement learning, i.e., the incrementally generated instances are not accurate during learning especially at the beginning. The approximator should have excellent fault tolerant ability. In order to make up this shortcoming, we initial the Fuzzy ARTMAP with the data generated from CMAC first, and then apply Fuzzy ARTMAP to RL(see Fuzzy ARTMAP* in Fig.4). This method has improved the performance of Fuzzy ARTMAP in RL to a certain extent.

Comparatively CMAC is the best approximator among the three while applied to reinforcement learning. BP is an excellent function approximator but not good in on-line systems. Fuzzy ARTMAP is a good incremental learning algorithm and can be applicable to reinforcement learning. But it is not good enough because of its bad fault tolerant ability.

5   Conclusions and Future Work

Applying neural networks to the function approximation in reinforcement learning is a promising approach. We have tried three kinds of neural networks in this paper. The result has showed some satisfactory performance with CMAC. Meanwhile, Fuzzy ARTMAP is still a promising algorithm since many researchers have worked on the fault tolerant problem in learning algorithms. Improvements on strengthening the algorithms' fault tolerant ability and accelerating convergence should be made in future.

Acknowledgement

The paper is supported by the Natural Science Foundation of China (No. 60475026, 60103012), the National Outstanding Youth Foundation of China (No.60325207), the National Grand Fundamental Research 973 Program of China (No.2002CB312002) and the Natural Science Foundation of Jiangsu Province, China(No.BK2003409).

References

1. Richard S. Sutton and Andrew G. Barto. Reinforcement Learning: An Introduction, Bradford Books, MIT (1998)
2. Baird,L.C. Residual Algorithms: Reinforcement Learning with Function Approximation. In Armand Prieditis & Stuart Russell, eds. Machine Learning: Proceedings of Twelfth International Conference, 9-12 July, Morgan Kaufman Publishers, San Francisco, CA. (1995)
3. Chris Gaskett, David Wettergreen, and Alexander Zelinsky. Q-Learning in Continuous State and Action Spaces. In Australian Joint Conference on Artificial Intelligence (1999) 417–428
4. Remi Munos. A convergent Reinforcement Learning algorithm in the continuous case: the Finite-Element Reinforcement Learning. In Proceedings of IJCAI-97. Morgan Kaufman (1997) 826–831
5. Albus,J.S.A new approach to manipulator control:the cerebellar model articulation controller(CMAC).Journal of Dynamic Systems, Measurement, and Control, Trans. ASME, Series G, **97** (1975)
6. Richard S. Sutton. Generalization in Reinforcement learning: Successful Examples Using Sparse Coarse Coding. Advances in Neural Information Processing Systems 8,the MIT Press, Cambrige (1996) 1038-1044
7. Razvan Andonie, Lucian Sasu. A Modified Fuzzy ARTMAP Architecture for Incremental Learning Function Approximation. Neural Networks and Computational Intelligence, O. Castillo (ed.), Anaheim, California, ACTA Press, Proceedings of the IASTED International Conference on Neural Networks and Computational Intelligence (NCI 2003), Cancun, Mexico (2003) 124-129

Multiagent Reinforcement Learning Algorithm Using Temporal Difference Error

SeungGwan Lee

School of Computer Science and Information Engineering, Catholic University
43-1, Yeokgok 2-Dong, Wonmi-Gu, Bucheon-Si, Gyeonggi-Do 420-743, Korea
leesg@catholic.ac.kr

Abstract. When agent chooses some action and does state transition in present state in reinforcement learning, it is important subject to decide how will reward for conduct that agent chooses. In this paper, by new meta heuristic method to solve hard combinatorial optimization problems, we introduce Ant-Q learning method that has been proposed to solve Traveling Salesman Problem (TSP) to approach that is based for population that use positive feedback as well as greedy search, and suggest ant reinforcement learning model using TD-error(ARLM-TDE). We could know through an experiment that proposed reinforcement learning method converges faster to optimal solution than original ACS and Ant-Q.

1 Introduction

Reinforcement Learning is learning by interaction because agent achieves learning doing interaction by trial-and-error. Agent attempts action that can take in given environment while does learning, receives reinforcement value for the conduct. In this paper we introduce Ant-Q algorithm [1], [2] for combinatorial optimization has been introduced by Colorni, Dorigo and Maniezzo, and we suggest new ant reinforcement learning model using TD-error [3], [4], [5] (ARLM-TDE) to original Ant-Q learning. Proposed ARLM-TDE reinforcement learning method searches goal using TD-error, there is characteristic that converge faster to optimal solution than original ACS [6], [7] and Ant-Q. The remainder of the paper is organized as follows. In section2, we introduce original Ant-Q reinforcement learning. Section3 describes new ant reinforcement learning model using TD-error(ARLM-TDE). Section 4 presents experimental results. Finally, Section 5 concludes the paper and describes directions for future work.

2 Ant-Q Algorithm

Ant-Q learning method [1], [2] that is proposed by Coloni, Dorigo and Mauiezzo is extension of Ant System(AS) [8], [9], [10], [11], [12], it is reinforcement learning reinterpreting in view of Q-learning. In Ant-Q, An agent(k) situated in node(r)

J. Wang, X. Liao, and Z. Yi (Eds.): ISNN 2005, LNCS 3496, pp. 627–633, 2005.

Where L_{kib} is the length of the tour done by the best agent, that is the agent which did the shortest tour in the current iteration, and W is a parameter, set to 10.

4   Experimental Results

To prediction performance of ant reinforcement learning model that apply TD-error (ARLM-TDE) in Ant-Q, we measure performance through comparison with original ant model(ACS and Ant-Q). Basis environment parameter for an experiment was decided as following, and optimum value decided by an experiment usually are β=2, α=0.1, q_0=0.9, γ=0.3, λ=0.3, m=n, W=10 and AQ_0=1/(average length of edges·n). The initial position of agents assigned one agent in an each node at randomly, and the termination condition is that a fixed number of cycles or the value known as the optimum value was found. Figure 1 shows the convergence speed in case of repeated 1000 cycles to use Bayg29.TSP. The convergence speed of Ant-Q is faster in beginning. However, according as search is proceeded, we can see that ARLM-TDE converges faster.

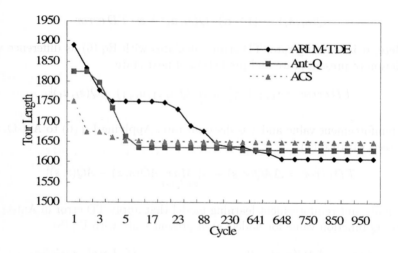

Fig. 1. Convergence Speed of each method

Figure(2,3) shows the performance by the learning rate(α) and discount rate(γ) in Ant-Q and ARLM-TDE ant model. Experiment graph used Eil51.TSP problem that have node 51 and achieved increasing the learning rate and discount rate by 0.1 step by step and the number of cycles are 2000. Figure 2 shows the results by the learning rate(α). We can see that Ant-Q and ARLM-TDE ant model display almost similar graph shape according to the learning rate(α), and the performance reduced as the learning rate(α) increases. When the learning rate(α) is 0.1 and 0.2, shown almost similar optimal tour length. But, when the

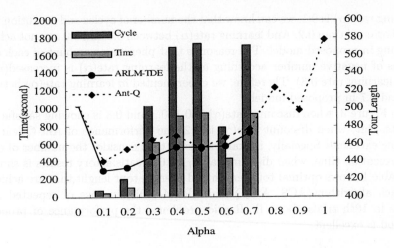

Fig. 2. Performance evaluation by the learning rate(α)

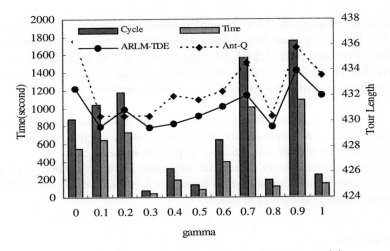

Fig. 3. Performance evaluation by the discount rate(γ)

Table 1. Performance evaluation of ARLM-TDE

Node	ACS		Ant–Q		ARLM–TDE	
	Average length	Best length	Average length	Best length	Average length	Best Length
4x4	160	160	160	160	160	160
5x5	254.14	254.14	254.14	254.14	254.14	254.14
6x6	360	360	361.66	360	360	360
7x7	510.32	506.50	509.38	502.43	494.14	494.14
8x8	660.54	654.14	653.61	648.28	640	640

learning rate(α) is 0.1, we can know that the number of cycles and execution time are shortened than 0.2. And learning rate(α) between 0.8 and 1.0 can not achieve learning in proposed model. The reason is that pheromone amount of each edge makes of negative number according as the learning rate(α) is increased(more than learning rate 0.8). Therefore, we experimented in learning rate(α) between 0.0 and 0.7 in proposed model.

In Figure 3, when discount rate(γ) is 0.1, 0.3 and 0.8 is showing satisfactory results. But, when discount rate(γ) is 0.2, the performance of Ant-Q learning is some excellent. Specially, if examine result that consider the number of cycle and execution time, when discount rate(γ) is 0.3, satisfactory results is shown.

Table 1 shows optimal tour length and average tour lenght that are achieved by each algorithms(ACS, Ant-Q and ARLM-TDE) in case of repeated 1000 cycles in 10th trials about R×R grid problems, the performance of proposed method is excellent.

5   Conclusion and Future Work

In this paper, we suggested new ant reinforcement learning model using TD-error (ARLM-TDE) to original Ant-Q learning to solve temporal-credit assignment problems. Proposed ARLM-TDE learning is method that is proposed newly to improve Ant-Q, this converged faster to optimal solution by solving temporal-credit assignment problems to use TD-error while agents accomplish tour cycle. Proposed ARLM-TDE ant model using TD-error uses difference with prediction for output of present state and prediction for output of next state at each learning step, and updated to approximate with prediction for output of present state and prediction for output of next state in present state.

Forward, we need study about reinforcement learning method that apply Eligibility factor that is measure that mean how is suitable about node selecting in present state in ARLM-TDE model.

References

1. Gambardella, L.M., Dorigo, M.: Ant-Q: A Reinforcement Learning Approach to the Traveling Salesman Problem. Proceedings of ML-95, Twelfth International Conference on Machine Learning, A. Prieditis and S. Russell (Eds.), Morgan Kaufmann, (1995) 252-260
2. Dorigo, M., Gambardella, L.M.: A Study of Some Properties of Ant-Q. Proceedings of PPSN IV-Fourth International Conference on Parallel Problem Solving From Nature, H.-M.Voigt, W. Ebeling, I. Rechenberg and H.-S. Schwefel (Eds.), Springer-Verlag, Berlin (1996) 656-665
3. Watkins, C.J.C.H.: Learning from Delayed Rewards. Ph.D. thesis, King's College, Cambridge, U.K, (1989)
4. Fiecher, C.N.: Efficient Reinforcement Learning. In Proceedings of the Seventh Annual ACM Conference On Computational Learning Theory, (1994) 88-97
5. Barnald, E.: Temporal-difference Methods and Markov Model. IEEE Transactions on Systems, Man, and Cybernetics, **23** (1993) 357-365

6. Gambardella, L.M., Dorigo, M.: Ant Colony System: A Cooperative Learning Approach to the Traveling Salesman Problem. IEEE Transactions on Evolutionary Computation, **1** (1997)
7. Stutzle, T., Dorigo, M.: ACO Algorithms for the Traveling Salesman Problem. In K. Miettinen, M. Makela, P. Neittaanmaki, J. Periaux, editors, Evolutionary Algorithms in En-gineering and Computer Science, Wiley (1999)
8. Colorni, A., Dorigo, M., Maniezzo, V.: An Investigation of Some Properties of an Ant Algorithm. Proceediings of the Parallel Parallel Problem Solving from Nature Conference(PPSn 92), R. Manner and B. Manderick (Eds.), Elsevier Publishing (1992) 509-520
9. Colorni, A., Dorigo, M., Maniezzo, V.: Distributed Optimization by Ant Colonies. Proceedings of ECAL91 - European Conference of Artificial Life, Paris, France, F.Varela and P.Bourgine(Eds.), Elsevier Publishing (1991) 134-144
10. Gambardella, L.M., Dorigo, M.: Solving Symmetric and Asymmetric TSPs by Ant Colonies. Proceedings of IEEE International Conference of Evolutionary Computation, IEEE-EC 96, IEEE Press (1996) 622-627
11. Drigo, M., Maniezzo, V., Colorni, A.: The Ant system: Optimization by a Colony of Cooperation Agents. IEEE Transactions of Systems, Man, and Cybernetics-Part B, **26** (1996) 29-41
12. Stutzle, T., Hoos, H.: The Ant System and Local Search for the Traveling Salesman Problem. Proceedings of ICEC '97 - 1997 IEEE 4th International Conference of Evolutionary, (1997)

The influence of evaluation result of FPRL again CAB is shown as Table 1.

Shown as Table 2, the table gives the executing times of FPRL-ART2 transferring to other CABs from current CAB.

Table 1. The influence of evaluation result of FPRL against CAB.

Evaluation Value	Transformation of CAB
-0.05	Increase the weights of current CAB
-0.1	Increase the weights of current CAB
-0.15	Transfer to next CAB
-0.20	Transfer to next CAB
0.05	Increase the weights of current CAB
0.1	Increase the weights of current CAB
0.15	Increase the weights of current CAB

Table 2. The execution times statistic of FPRL-ART2 when switching CAB.

The evaluation value of reinforcing current CAB	The evaluation value of transferring current CAB	The execution times of FPRL-ART2
0.1	-0.15	3
0.1	-0.2	1
0.2	-0.15	7
0.2	-0.2	3
0.3	-0.2	4
0.3	-0.25	3
0.4	-0.25	3
0.4	-0.3	3

From these two tables, we can see that the evaluation value of FPRL should be kept within limits for CAB transformation when using FPRL-ART2, if overstep the limits, the learning effect is not good as expected.

Shown as Fig.2 (a), the collision is inevitable without FPRL-ART2. With using it, R selects a CAB with successful collision avoidance (shown as Fig.2 (c)) after 6 times RL (shown as Fig.2 (b)). The deflection angle is 5^0 and acceleration is 0.

The main purpose of simulation 2 is to show the effect of FPRL-ART2 against collision avoidance. There are two Os and one R, the nodes and parameters of each layer in the neural network is the same as simulation one. The motion states of obstacles (initial position, velocity and direction) are at random in order to produce enough collision states. The simulation time is four minutes and the evaluation function of RL is shown as formula (21).

$$r = \begin{cases} 0.1 & No \quad Collision \\ -0.25 & Collision \end{cases} \tag{21}$$

Fig. 3a. Collision without FPRL-ART2.   **Fig. 3b.** Successful collision avoidance.

Collision scene between R and O without FPRL-ART2 is shown as Fig.3 (a), the scene of R avoiding O successful by using FPRL-ART2 is show as Fig.3 (b).

Table 3. Collision times between R and O with FPRL-ART2 and without FPRL-ART2.

Simulation time	Collision times (no FPRL-ART2)	Collision times (using FPRL-ART2)
1	7	3
2	14	4
3	9	5
4	2	2
5	8	6
6	8	3
7	9	3
8	6	4
9	8	5
10	9	5

It can be concluded that the collision times between R and O is effectively decreases by using FPRL-ART2 from Table 3.

5 Conclusion

The paper has given a Foremost-Policy Reinforcement Learning based ART2 neural network (FPRL-ART2) and successful used it in the research of collision avoidance problem of mobile robot. We used FPRL-ART2 to store CABs of mobile robot and took FPRL to evaluate the collision avoidance result. If the obstacles are avoided successfully, FPRL-ART2 increases the correlative weights of current CAB, or else decreases the weights. If the CAB has been selected to execute the current state, it will be more possible to be selected again when the same states are input into FPRL-ART2.The simulation experiments indicated that the collision times between R and O was effectively decreased after using FPRL-ART2.

References

1. Yang, S.X., Meng, M., Yuan, Y.: A Biological Inspired Neural Network Approach to Real-time Collision-free Motion Planning of a Nonholonomic Car-like Robot. Proceedings of the IEEE/RSJ International Conference on Intelligent Robots and Systems (2000) 239-244
2. Xiao, N.-F., Nahavandi, S.: A Reinforcement Learning Approach for Robot Control in an Unknown Environment. IEEE ICIT, Bangkok, THAILAND , **2** (2002) 1096-1099
3. Fan, J., Wu, G.F.: Reinforcement Learning and ART2 Neural Network Based Collision Avoidance System of Mobile Robot. Lecture Notes in Computer Science, **3174** (2004) 35-40
4. Chen, C.T., Quinn, R.D., Ritzmann, R.E.: A Crash Avoidance System Based Upon the Cockroach Escape Response Circuit. Proceeding of the IEEE International Conference on Robotics and Automation (1997) 2007-2012
5. Whitehead, S.D., Sutton, R.S., Ballard, D.H.: Advances in reinforcement learning and their implications for intelligent control. Intelligent Control, 1990. Proceedings., 5th IEEE International Symposium on Sept , **2** (1990) 1289-1297

A Reinforcement Learning
Based Radial-Bassis Function Network Control System*

Jianing Li, Jianqiang Yi, Dongbin Zhao, and Guangcheng Xi

Lab of Complex Systems and Intelligence Science, Institute of Automation,
Chinese Academy of Sciences, P. O. Box 2728, Beijing 100080, China
{jianing.li,jianqiang.yi,dongbin.zhao,guangcheng.xi}
@mail.ia.ac.cn

Abstract. This paper proposes a reinforcement learning based radial-basis function network control system (RL-RBFNCS) to solve non-training data based learning of radial-basis function network controllers (RBFNC). In learning process, a major contribution is by using the critic signal and the stochastic exploration method to estimate the "desired output", reinforcement learning is considered and solved from the point of view of training data based learning. Computer simulations of robot obstacle avoidance in unknown environment are conducted to show the performance of the proposed method.

1 Introduction

According to whether sufficient and consistent training data are available, learning algorithms for radial-basis function network controllers (RBFNCs) can be divided into two kinds: training data based learning such as supervised and unsupervised learning, and non-training data based learning which mainly means reinforcement learning. In general, training data based learning is a fast and effective method when training data are easy to get. But in most of real-world applications when it is expensive to obtain training data, the performance to this approach will degrade greatly. For this reason, reinforcement learning requiring no training data seems to be an attractive alternative. Enlightened by the Sutton and Barto's model presented in [1], this paper proposes a reinforcement learning based radial-basis function network control system (RL-RBFNCS), which consists of a RBFNC and a predictor, to solve non-training data based learning for RBFNCs. Based on our knowledge, there are mainly two kinds of methods for the adjustment of parameters in Sutton and Barto's model-based reinforcement learning: one such as in [2] is originated from Sutton and Barto's model; the other such as in [3] is based on gradient information. This paper is trying to solve reinforcement learning problem from another point of view. The main idea used is by employing the critic signal and the stochastic exploration method to estimate the "desired output", reinforcement learning problem is changed into training data based learning problem.

Section 2 describes the structure of a RBFNC and the RL-RBFNCS. The learning algorithm of the RL-RBFNCS is introduced in Section 3. Section 4 presents an example of robot obstacle avoidance. Finally, conclusions are given in Section 5.

---

* This work was partly supported by NSFC Projects (Grant No. 60334020, 60440420130, and 60475030), and MOST Projects (Grant No. 2003CB517106 and 2004DFB02100), China.

J. Wang, X. Liao, and Z. Yi (Eds.): ISNN 2005, LNCS 3496, pp. 640–645, 2005.

2 Structure of the RL-RBFNCS

2.1 Radial-Basis Function Network Controller (RBFNC)

This paper employs the most basic form for the construction of RBFNCs, which involves three layers, as shown in Fig. 1. The role of the input layer is for connecting the network to its environment. The hidden layer applies a nonlinear transformation from the input space to the hidden space. The output layer is comprised of linear nodes that supply the response to the network activated by the input vector. For convenience, the RBFNC is defined with two inputs and a single output. Assuming all radial basis functions for the hidden nodes use Gaussian curves, which have the same dimensions as the input vector. Next, the functions of the nodes in the hidden and the output layer shall be described by equations, in which I_i^n and O_i^n are the input and output value of the ith node in Layer n; m_{ij} and σ_{ij} are the center and width of the jth-dimensional Gaussian curve of the ith node in the hidden layer; W_i is the link weight connected with the ith hidden node in the output layer.

The hidden layer:

$$O_i^2 = \exp[-\frac{(I_1^1 - m_{i1})^2}{2\sigma_{i1}^2} - \frac{(I_2^1 - m_{i2})^2}{2\sigma_{i2}^2}] \ , i = 1,2,\cdots N \ . \tag{1}$$

The output layer:

$$I_i^3 = O_i^2 \ and \ O_1^3 = \sum_{i=1}^{N} W_i I_i^3 \ . \tag{2}$$

2.2 Structure of the RL-RBFNCS

Fig. 2 shows the structure of the RL-RBFNCS, which involves the above RBFNC and a predictor. The RBFNC can choose a proper control output according to the current input vector. The predictor performs the single or multi-step prediction of the external reinforcement signal. In this paper, we use a simple three-layer perceptron with one output node to model the predictor, which shares the same input layer with the RBFNC. The functions of the hidden and the output layer are described as follows, in which M is the number of the hidden nodes; V_{ij} and U_i are link weights of the hidden and the output layer, respectively; other symbols are defined as the same as previously.

The hidden layer:

$$I_i^2 = \sum_{j=1}^{2} V_{ij} O_j^1 \ , \ (O_j^1 = I_j^1) \ and \ O_i^2 = \frac{1 - \exp(-I_i^2)}{1 + \exp(-I_i^2)} \ , i = 1,2 \cdots M \tag{3}$$

The output layer:

$$I_1^3 = \sum_{i=1}^{M} U_i O_i^2 \ and \ O_1^3 = \frac{1 - \exp(-I_1^3)}{1 + \exp(-I_1^3)} \ . \tag{4}$$

Assuming the predictor works under single-step mode, which means a reinforcement signal is only one step behind its corresponding action. The working process of the RL-RBFNCS is described briefly as follows: at step t, the input vector $x(t)$ supplied by the environment is fed simultaneously into the RBFNC and the predictor. Based on $x(t)$, the predictor produces a signal $p(t+1)$, which is the prediction of the external reinforcement signal $r(t+1)$ but available at step t; and the RBFNC gets an output variable $y(t)$. Then using $p(t+1)$ and $y(t)$, the actual output $\hat{y}(t)$ is chosen by the stochastic exploration method introduced in the next section. Driven by $\hat{y}(t)$, the system evolves to step $t+1$ and gets $r(t+1)$ by interacting with the environment. Finally, link weights of the predictor are updated by the internal reinforcement signal $\hat{r}(t+1)$, which is the prediction error computed by $r(t+1)$ and $p(t+1)$. $\hat{r}(t+1)$, $y(t)$ and $\hat{y}(t)$ are used for the adjustment of link weights and parameters of Gaussian radial-basis functions of the RBFNC. The learning algorithms will be presented in detail in the following section.

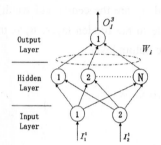

Fig. 1. Structure of the RBFNC.   **Fig. 2.** Structure of the RL-RBFNCS.

3   Learning Algorithm for RL-RBFNCS

3.1   Learning Algorithm for the Predictor

The goal of the predictor learning is to adjust link weights to minimize $\hat{r}$ at each step, which, based on the single-step mode, is computed by $\hat{r}=r-p$. Here, the gradient descent learning is used to update the link weights of the predictor. The error function is defined as $E = (r - p)^2 / 2$. According to the chain rule, the updated values of the link weights of the predictor are as follows, in which η_1 and η_2 are learning rates.

$$\Delta U_i = \eta_1(-\frac{\partial E}{\partial U_i}) = \eta_1(r-p)\left[\frac{2\exp(-I_1^3)}{(1+\exp(-I_1^3))^2}O_i^2\right]. \tag{5}$$

$$\Delta V_{ij} = \eta_2(-\frac{\partial E}{\partial V_{ij}}) = \eta_2(r-p)\left[\frac{2\exp(-I_1^3)U_i}{(1+\exp(-I_1^3))^2}\frac{2\exp(-I_i^2)O_j^1}{(1+\exp(-I_i^2))^2}\right]. \tag{6}$$

3.2   Stochastic Exploration Method

When $y(t)$ is produced by the RBFNC, the conflict between the desire to use $y(t)$ and the desire to further explore the environment to improve the performance of the

RBFNC has to be considered. This paper uses the stochastic exploration method proposed in [3] to overcome this problem, which is described as follows.

Step 1. The range of stochastic exploration is determined by the following equation, in which K and A are search-range scaling parameters.

$$\sigma(t) = \frac{K}{1 + \exp[Ap(t+1)]} \qquad (7)$$

Step 2. A Gaussian random variable $\hat{y}(t)$ is chosen by exploring the range $\sigma(t)$ around the mean point $y(t)$.

$\sigma(t)$ can be interpreted as the extent to which the output variable searches for a better action. $p(t+1)$ is the prediction of $r(t+1)$. When $p(t+1)$ is small, $\sigma(t)$ will be large, which means to broaden the search range around the $y(t)$. This can provide a higher probability to choose a $\hat{y}(t)$, which is far from $y(t)$, since $y(t)$ is regarded to be far from the best action possible for the current input vector. The similar analysis is true when $p(t+1)$ is large. By using this method, the RBFNC explores for actions possible, then a better output will be rewarded and a worse one be punished by the learning algorithm of the RBFNC introduced in the next subsection.

3.3 Learning Algorithm for the RBFNC

The goal of the RBFNC learning is to adjust link weights and centers and widths of Gaussian radial-basis functions to maximize r at each step, which means to produce an optimal action for each input vector. Basically, the difference between training data based learning and reinforcement learning is: for each input vector, the former can get the instructive signal, which is described as the desired output; and the later has only the critic signal, which represents a reward or a penalty for the output action. If the "desired output" can be produced by employing the critic signal, then a reinforcement learning problem can be changed into a training data based learning problem. In this paper, the learning of the reinforcement-based RBFNC is considered and solved just based on this idea.

Firstly, using the method presented in [4], the desired output is estimated by

$$y_d(t) \approx y(t) + \rho \frac{\partial r}{\partial y} \text{ and } \frac{\partial r}{\partial y} \approx [r - p]_{t+1} \left[\frac{\hat{y}(t) - y(t)}{\sigma(t)} \right]. \qquad (8)$$

where ρ is a real number, $\rho \in \{0,1\}$; $y_d(t)$ is the estimated output. If $r(t+1) > p(t+1)$, which means $\hat{y}(t)$ is better than $y(t)$, $\hat{y}(t)$ should be rewarded. So $y_d(t)$ is moved closer to $\hat{y}(t)$. On the other side, $y_d(t)$ is moved further away from $\hat{y}(t)$. When the desired output is produced, the reinforcement learning can be changed completely into a training data based learning. The gradient descent learning is used again to update all parameters of the RBFNC. The error function is defined as $E = (y_d - y)^2 / 2$. The updated values of the RBFNC are as follows.

$$\Delta W_i = \beta_1 (-\frac{\partial E}{\partial W_i}) = \beta_1 (y_d - y) I_i^3 \tag{9}$$

$$\Delta m_{ij} = \beta_2 (-\frac{\partial E}{\partial m_{ij}}) = \beta_2 (y_d - y) W_i I_i^3 \frac{(I_j^1 - m_{ij})}{\delta_{ij}^2} \tag{10}$$

$$\Delta \delta_{ij} = \beta_3 (-\frac{\partial E}{\partial \delta_{ij}}) = \beta_3 (y_d - y) W_i I_i^3 \frac{(I_j^1 - m_{ij})^2}{\delta_{ij}^3} \tag{11}$$

4   An Illustrative Example

The proposed RL-RBFNCS is simulated for on-line obstacle avoidance of mobile robot in unknown indoor environment that consist of walls and static obstacles, in which walls also are treated as obstacles. This paper employs the behavior-based control architecture presented in [5], which maps sensor information to control command directly. The model of the mobile robot used is a cylindrical platform and 18 infrared sensors equipped evenly in a ring, which are assumed to work in perfect mode. To reduce input dimension, the sensors around the robot are divided into six groups ($S_1 \sim S_6$), each of which consists of three neighboring sensors, as depicted in Fig. 3. The distance measured by each sensor group is defined as the smallest value. We do not consider the rotation of the robot, that means the robot coordinate system is consistent with the world coordinate all the time. Assuming the robot to move with a constant linear velocity, the control variable is defined as the angle from x-axis. r is defined as related with the distance measure supplied by sensors. As long as there are obstacles in sensor field of view, a r can be obtained at each step, so the reinforcement learning is a single-step prediction problem. Without using a normal two-valued number, the value of r is defined as a real number, $r \in \{0,1\}$, which represents a detailed and continuous degree of success or failure. When r is larger than a set safety threshold, r will keep the same changing trend with the smallest distance measure among sensor groups.

In Fig. 4, driven by the RL-RBFNCS, the robot begins to move from A and B, respectively. The region between the two dotted circles represents the detectable range of sensors. The folded lines are the trajectories of the robot center. At the start, $p(t+1)$, $y(t)$ and $\hat{y}(t)$ are set to zero; $\sigma(t)$ set to unity; all link weights are set to small nonzero values; m_{ij} are set at random in the range of detection distance; σ_{ij} are computed by $\sigma_{ij} = \sqrt{D/m}$, in which D is maximum distance of j chosen centers and m is the number of input variables. Parameters for learning are shown in Table 1. When the smallest distance measure is lower than the given threshold, the robot is backtracked two steps and $p(t+1)$, $y(t)$, $\hat{y}(t)$ and $\sigma(t)$ are set again with initial values. If there are no obstacles in sensor field of view, the robot will go along the former direction. In learning process, by adjusting the search-range scaling parameters, the search range can be broadened to speed up learning. While it should be also noticed that a too large search range will degrade learning in other situations.

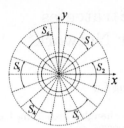

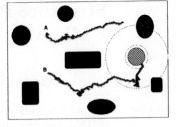

Fig. 3. Mobile robot and sensor arrangement.   **Fig. 4.** Simulation of robot obstacle avoidance.

Table 1. Parameters for learning of the RL-RBFNCS.

$\eta_1 = 0.9$	$\eta_2 = 0.9$	$K = 1$	$A = 1$	$\rho = 1$	$\beta_1 = 0.8$	$\beta_2 = 0.1$	$\beta_3 = 0.1$

5 Conclusion

This paper describes a RL-RBFNCS to solve non-training data based learning of RBFNCs, which consists of a RBFNC and a predictor. By using the critic signal and the stochastic exploration method to estimate the "desired output", the reinforcement learning is treated and realized from the point of view of training data based learning. Employing the proposed method, the RBFNC and the predictor will be gradually close to an optimal action controller and an optimal state evaluation unit, respectively. Since our method greatly lessens the quality and quantity requirements of training data for the learning of RBFNCs, the design of RBFNCs will be more practical for real-world applications. Computer simulations show the effectiveness and applicability of the proposed approach. Future work will focus on adaptive adjustment for the structure of RBFNCs and multi-step prediction problem.

References

1. Barto, A.G., Sutton R.S., Anderson C.W.: Neuronlike Adaptive Elements That Can Solve Difficult Learning Control Problems. IEEE Trans. Syst, Man, Cybern, **13** (1983) 834-846.
2. Ye, C., Yung, N.H.C., Wang D.W.: A Fuzzy Controller with Supervised Learning Assisted Reinforcement Learning Algorithm for Obstacle Avoidance. IEEE Trans. Syst, Man, Cybern, **33** (2003) 17-27
3. Lin, C.T., Lee, G.S.G.: Reinforcement Structure/Parameter Learning for Neural-Network-Based Fuzzy Logic Control Systems. IEEE Trans. Fuzzy Syst, **2** (1994) 46-63
4. Lin, C.T., Lin, C.T.: Reinforcement Learning for an ART-Based Fuzzy Adaptive Learning Control Network. IEEE Trans. Neural Network, **7** (1996) 709-731
5. Brooks, R.A.: Robust Layered Control Systems for a Mobile Robot. IEEE Trans. Robot. Automat., **2** (1986) 14-23

Structure Pruning Strategies for Min-Max Modular Network[*]

Yang Yang and Baoliang Lu

Department of Computer Science and Engineering, Shanghai Jiao Tong University,
1954 Hua Shan Rd., Shanghai 200030, China
alayman@sjtu.edu.cn, blu@cs.sjtu.edu.cn

Abstract. The min-max modular network has been shown to be an efficient classifier, especially in solving large-scale and complex pattern classification problems. Despite its high modularity and parallelism, it suffers from quadratic complexity in space when a multiple-class problem is decomposed into a number of linearly separable problems. This paper proposes two new pruning methods and an integrated process to reduce the redundancy of the network and optimize the network structure. We show that our methods can prune a lot of redundant modules in comparison with the original structure while maintaining the generalization accuracy.

1 Introduction

The min-max modular (M^3) network is an efficient modular neural network model for pattern classification[1][2], especially for large-scale and complex multi-class problems[3]. It divides a large-scale, complex problem into a series of smaller two-class problems, each of which is solved by an independent module. We can conduct the learning tasks of every module in parallel and integrate them to get a final solution to the original problem according to two module combination principles. These combination principles also successfully guide the emergent incremental learning[4]. However, we need to learn too many modules when the size of the training data is large while subproblems are small. Considering the situation of incremental learning, since the training data are presented to the network continually and more and more modules are built, the classifier will be inefficient to respond to novel inputs.

To improve the response performance of min-max modular network, we consider reducing its redundancy at two phases. First is the recognition phase. In this phase, we can dynamically decide which modules have influence on the final result and need computing. The other is the training phase. We can hold a training process to the network to prune the redundant modules for the training data. The pruned modules are supposed to be redundant for the whole input space.

---

[*] This work was supported in part by the National Natural Science Foundation of China via the grants NSFC 60375022 and NSFC 60473040.

J. Wang, X. Liao, and Z. Yi (Eds.): ISNN 2005, LNCS 3496, pp. 646–651, 2005.

In addition, other two lines of research for training optimization can be considered. The idea of the first line arises from instance reduction techniques[5]. Since the nearest-neighbor(NN) algorithm and its derivatives suffer from high computational costs and storage requirement, the instance filtering and abstraction approaches are developed to get a small and representative prototype set. We can firstly build such a condensed prototype set, and then use it to construct the modules of the M^3 network. This way is instance pruning. The other line of research is structure pruning. Lian and Lu first developed a back searching (BS) algorithm to prune redundant modules[6]. We extend their work and propose an integrated process to gain a larger module reduction rate and maintain classification accuracy.

2   Redundancy Analysis

According to the task decomposition principle of the M^3 network, a K-class problem is divided into $K \times (K-1)/2$ two-class problems, each of which can be further decomposed into a number of subproblems. A M^3 network with MIN and MAX integrating units can solve a two-class subproblem, and each network module learns a subproblem. Let's consider the situation of incremental learning [4] mentioned before. Suppose the training set of each subproblem has only two different elements. Let $\mathcal{T}$ be the total training set, X_l be the input vector, where $1 \le l \le L$, and L is the number of training data. The desired output y is defined by

$$y = \begin{cases} 1 - \epsilon, & \text{if } X_l \in \text{class } \mathcal{C}_i \\ \epsilon, & \text{if } X_l \in \text{class } \overline{\mathcal{C}}_i \end{cases} \tag{1}$$

where ϵ is a small positive real number, $\overline{\mathcal{C}}_i$ denotes all the classes except $\mathcal{C}_i$. Accordingly, the training set of a subproblem has the following form:

$$\mathcal{T}_{ij}^{(u,v)} = \{(X_l^{(iu)}, 1 - \epsilon) \cup (X_l^{(jv)}, \epsilon)\}$$
$$\text{for } u = 1, \cdots, L_i, \; v = 1, \cdots, L_j, \; i, j = 1, \cdots, K, \text{ and } j \ne i \tag{2}$$

where L_i and L_j are the numbers of training data belonging to class $\mathcal{C}_i$ and class $\mathcal{C}_j$, respectively. Hence the two-class problem has $L_i \times L_j$ subproblems. We use first minimization rule then maximization rule to construct the network. Fig. 1 shows the network structure. Since $\mathcal{T}_{ij}^{(u,v)}$ has only two instances, it is obviously a linearly separable problem and can be discriminated by a hyperplane. The optimal hyperplane is the perpendicular bisector of the line joining the two instances. In this way, the decision boundary established by the network can be written as a piecewise linear discriminant function:

$$\bigcup_{1 \le u \le L_i} \left(\bigcap_{1 \le v \le L_j} L_{i,j}^{u,v} \right) \tag{3}$$

where $L_{i,j}^{u,v}$ is the hyperplane determined by $M_{i,j}^{u,v}$ trained on $\mathcal{T}_{ij}^{(u,v)}$. And it is trivial to prove that the decision boundary is the same as that of the nearest

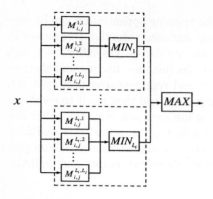

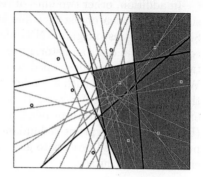

Fig. 1. M^3 network for the two-class problem of C_i and C_j

Fig. 2. An example of redundant modules, loops denote $class_1$, and squares denote $class_2$

neighbor classifier. In fact there are a lot of redundancies in the expression above. For example, we can see that only black lines contribute to the boundary, and cyan ones are redundant in Fig. 2, where a 2-D classification problem is depicted.

3   Pruning Algorithms

A backward searching(BS) algorithm for pruning redundant modules has been developed[6], which searches the useful modules based on outputs of every MIN unit. Take $\mathrm{MIN}_k (1 \leq k \leq L_i)$ as an example, set $\mathcal{T}_k = \{a_k, b_1, b_2, \cdots, b_{L_j}\}$, where $a_i \in C_i$, $b_1, b_2, \cdots, b_{L_j} \in C_j$. For each instance in $\mathcal{T}_k$, calculate the output of MIN_k and mark the module which gives the minimal value. After all the instances in $\mathcal{T}_k$ are processed, delete those modules without mark. BS algorithm can take off a lot of units while maintaining the original decision boundary. However, it fails to consider removing related redundant MIN units. Here we extend BS algorithm to get an integrated pruning process.

3.1   Our Methods

Because a multi-class problem can be decomposed into a set of two-class problems, here we discuss the redundancy reduction problem of the two-class network for simplicity. Suppose two classes are $class_1$ and $class_2$, which include M and N instances respectively. So the M^3 network contains M MIN units and a MAX unit, each MIN unit contains N modules. We propose two kinds of M^3 Structure Pruning Strategy, called M^3SPS1 and M^3SPS2, respectively. They have different criterions of redundancy.

M^3SPS1. The idea of M^3SPS1 is straightforward, which regards the unit used by no instance in the training set as redundant. Let $\mathcal{T}$ be the training set. The algorithm is as follows:

1. Set Flag(MIN_i) = FALSE ($i = 1, 2, \cdots, M$).
2. For each instance I in $\mathcal{T}$:
   (a) Calculate y(I).
   (b) Find $\text{MIN}_i (1 \leq i \leq M)$ which has the same value as y(I), and set Flag(MIN_i) = TRUE.
3. Delete $\text{MIN}_j (1 \leq j \leq M)$ that Flag(MIN_j)=FALSE.
4. For each MIN_i retained:

   Execute BS algorithm and prune redundant modules.

5. End.

This approach is consistent with the training data because it just deletes those units which are useless for the classification of the training data.

M^3SPS2. The idea of M^3SPS2 is similar to Reduced Nearest Neighbor[7] and DROP1[5]. All of them try to conduct reduction without hurting the classification accuracy of the training data set or the subset retained. However, M^3SPS2 prune modules other than instances. Moreover, all the instances are guaranteed to be classified correctly no matter whether their related modules have been removed or not. The algorithm is described as follows:

1. Set Flag(MIN_i) = FALSE , and create empty list(MIN_i), $i = 1, 2, \cdots, M$
2. For each instance I in $\mathcal{T}$:
   (a) Calculate y(I).
   (b) Find $\text{MIN}_i (1 \leq i \leq M)$ which has the same value as y(I), set Flag(MIN_i) = TRUE, and insert I to list(MIN_i).
3. For each MIN_i that Flag(MIN_i) = TRUE

   if $\forall I \in$ list(MIN_i) can be classified correctly without MIN_i :

   Set Flag(MIN_i) = FALSE.

   For each I $\in$ list(MIN_i):

   insert I to list(MIN_j), where $1 \leq j \leq M$ and MIN_j gives new y(I) instead of MIN_i.

4. Delete $\text{MIN}_i (1 \leq i \leq M)$ that Flag(MIN_i) = FALSE.
5. For each $\text{MIN}_k (1 \leq k \leq M)$ retained:

   Execute BS algorithm and prune redundant modules.

6. End.

This approach is also consistent with the training data. Note that it is sensitive to the presentation order of the MIN units. We can make a search to determine the removing sequence. The basic assumption here is that a module composed by an inner instance is likely to be useless. Since y(I) indicates the perpendicular distance between the instance I and decision boundary, so memorize y($Instance^{(1i)}$) at step 2 and examine the MIN_i in the descending order of y($Instance^{(1i)}$) at step 3 , where $1 \leq i \leq$ M, $Instance^{(1i)} \in class_1$ and it is the corresponding instance of MIN_i.

4   Experimental Results

We present five experiments to verify our methods, including the two-spirals problem and 4 real world problems. Three of them were conducted on the benchmark data sets from the Machine learning Database Repository[8]: Iris Plants, Image Segmentation and Vehicle Silhouettes. The last one was carried on a data set of glass-board images from an industrial product line, which was used to discriminate the eligible glass-boards. Table 1 shows all the data sets and the numbers of classes, dimensions, training and test samples. Table 2 shows the classification accuracy, response time, and the retention rates of modules and MIN units, where

$$retention\ rate = \frac{No.\ of\ modules/MINs\ after\ pruning}{No.\ of\ modules/MINs\ of\ the\ original\ structure}. \quad (4)$$

Since a multi-class problem can be decomposed into a number of two-class problems, each of which is solved by a M^3 network, for iris, segment and vehicle problems, we record the response time both in parallel and in series. We also list the accuracy and response time of the nearest-neighbor classifier for comparison. All the experiments were performed on a 3GHz Pentium 4 PC with 1GB RAM.

Table 1. Numbers of classes, dimensions, training and test data

	Two-spirals	Iris	Segment	Vehicle	Glass image
Class	2	3	7	4	2
Dimension	2	4	19	18	160
Training	95	135	1540	600	174
Test	96	15	770	246	78

Table 2. Experimental results. For the last three problems, the left sub-column of "time" denotes sequential time and right one denotes parallel time(CPU time, in millisecond)

Problems	1-NN		M^3SPS1					M^3SPS2				
	acc	time	acc	time		MINs	modules	acc	time		MINs	modules
Two-spirals	1.00	31	1.00	32		1.0	0.149	0.99	30		0.432	0.044
Glass image	0.897	47	0.895	63		0.477	0.282	0.885	47		0.075	0.044
Iris	0.982	15	0.982	16	15	0.289	0.042	0.982	15	15	0.067	0.013
Segment	0.958	328	0.956	766	177	0.370	0.036	0.904	453	164	0.053	0.005
Vehicle	0.638	47	0.638	78	41	0.75	0.104	0.630	63	37	0.298	0.047

From the experimental results in Table 2, several observations can be made. M^3SPS1 almost maintains the decision boundary of the original structure, but the pruning ability is limited. M^3SPS2 gains a significant module reduction rate, but the classification accuracy dropped by an average of 1.3%. This is likely due to that the decision boundaries sometimes change dramatically and the classification of the positive points may be affected since too many MIN units have been removed. When handling complex multi-class problems, we can make use of the simple decomposition rules of M^3 and learn the two-class problems in parallel. Then the time can cut down greatly. Take Segment as an example, parallel M³SPS2 costs only half the response time as nearest neighbor classifier. And the superiority will be more distinct in solving more complex multi-class problems.

5   Conclusions and Future Work

We have presented two new pruning algorithms for dealing with redundant MIN units and developed a process integrating Back Searching algorithm to reduce the redundancy of the network and optimize the M^3 network structure. The process pays attention to not only massive units but also little modules, and improves the structure significantly. The experiments verified the validity of the structure pruning strategies. Compared with instance reduction methods, they directly concern with decision boundaries, because a module in fact presents a class boundary. So it may express complex concepts more delicately. However, more investigations on the criterion of redundancy are still needed. And it should be noted that this study has examined only hyperplane base classifier. Since the M^3 network is a flexible framework, the user can choose proper base classifier and module size according to different requirements. One future work will investigate methods for pruning the min-max modular networks with other base classifiers, such as the Gaussian zero-crossing function[9], which has an adaptive locally tuned response characteristic.

References

1. Lu, B. L., Ito, M.: Task Decomposition Based on Class Relations: a Modular Neural Network Architecture for Pattern Classification. In: Biological and Artificial Computation: From Neuroscience to Technology. Lecture Notes in Computer Science. Vol. 1240. Springer-Verlag (1997) 330-339
2. Lu, B. L., Ito, M.: Task Decomposition and Module Combination Based on Class Relations: a Modular Neural Network for Pattern Classification. IEEE Trans. Neural Networks, 10 (1999) 1244-1256
3. Lu, B. L., Shin, J., Ichikawa, M.: Massively Parallel Classification of Single-trial EEG Signals Using a Min-Max Modular Neural Network. IEEE Trans. Biomedical Engineering, 51 (2004) 551-558
4. Lu, B. L., Ichikawa, M.: Emergent On-line Learning in Min-max Modular Neural Networks. In: Proc. of IEEE/INNS Int. Conf. on Neural Networks, Washington DC, USA (2001) 2650-2655
5. Wilson, D.R., Martinez, T.R.: Reduction Techniques for Instance-based Learning Algorithms. Machine Learning, 38 (2000) 257-286
6. Lian, H. C., Lu, B. L.: An Algorithm for Pruning Redundant Modules in Min-Max Modular Network (in Chinese). In: Proc. 14th National Conference on Neural Network, Hefei University of Technology Press (2004) 37-42
7. Gates, G. W.: The Reduced Nearest Neighbor Rule. IEEE Transaction on Information Theory, IT-18-3 (1972) 431-433
8. Blake, C. L., Merz, C. J.: UCI. In: ftp://ftp.ics.uci.edu/pub/machine-learning-databases (1998)
9. Lu, B. L., Ichikawa, M.: Emergent Online Learning with a Gaussian Zero-crossing Discriminant Function. In: Proc. IJCNN'02 (2002) 1263-1268

Sequential Bayesian Learning
for Modular Neural Networks

Pan Wang[1], Zhun Fan[2], Youfeng Li[1], and Shan Feng[3]

[1] Wuhan University of Technology, Wuhan 430070, China
{Wang,Li,jfpwang}@tom.com
[2] Technical University of Denmark
zf@mek.dtu.dk
[3] Huazhong University of Science and Technology, Wuhan 430074, China

Abstract. In this paper, we present a distributed computing method, namely Sequential Bayesian Learning for modular neural networks. The method is based on the idea of sequential Bayesian decision analysis to gradually improving the decision accuracy by collecting more information derived from a series of experiments and determine the combination weights of each sub-network. One of the advantages of this method is it emulates humans' problems processing mode effectively and makes uses of old information while new data information is acquired at each stage. The results of experiments on eight regression problems show that the method is superior to simple averaging on those hard-to-learn problems.

1 Introduction

Modular neural network is an effective kind of connectionism models that consists of a group of sub-networks combined to solve complex problems. It often produces superior results than single well-trained neural network does. It has been a hot topic in many areas such as pattern recognition and classification, image processing, system identification, language/speech processing, control, modeling, target detection/recognition, fault diagnosis, etc.

Here, we shall use the term modularity in the widest meaning, that is, modular neural network is a system composed of a group of neural networks, which are independent, inter-connected, co-operative in structure level or in function level. The basic unit in this system is a module. Therefore, in the literature the paradigms such as multiple neural networks, hybrid neural networks, distributed neural networks and committee machine could be unified under the aforementioned framework. In this meaning, we give corresponding architecture and description:

$$\text{MNN}=<X, C, SN, IU, Y> \tag{1}$$

Where, $X \in D \subseteq R^n$, is the input vector; C represents a classifier whose function is to decompose input space or I/O space in the system; SN represents a set of subnets $\{Net_i\}_{i=1}^K$; IU represents the integrating unit which performs adaptive combination of modules; $Y \in E \subseteq R^m$, is the output vector. The corresponding network architecture is illustrated in Fig.1

In addition, if the classifier doesn't work (i.e. the input or I/O space is classified into one class), the corresponding modular neural network is named as "neural net-

J. Wang, X. Liao, and Z. Yi (Eds.): ISNN 2005, LNCS 3496, pp. 652–659, 2005.
© Springer-Verlag Berlin Heidelberg 2005

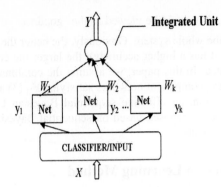

Fig. 1. Network Architecture.

work ensemble"; If any *Net_i* is composed of some sub-nets, the corresponding modular neural network should be named as "hierarchical modular neural networks". In this paper, we focus on the "neural network ensemble".

In the past few years, one of the research directions on modular neural network is how to combine the outputs of the component networks to form the output of the entire system that has the best performance. In the context of regression problems, diverse linear combination methods, such as simple averaging [1], MLS-OLCs [2], fuzzy integral [3], etc, are presented to integrate the component networks. Whatever any specific linear combination method is used, if K neural networks arc selected to form an entire system, the outputs of the component networks are then combined through weighted sum where a combination weight $w_i (i = 1, 2, ..., K)$ is assigned to the i-th component network. Consequently, the output vector $\overline{y}$ of the entire system is determined according to Eq. (1) where y_i is the output vector of the i-th component network.

$$\overline{y} = \sum_{i=1}^{K} w_i y_i \tag{2}$$

For simplicity, here we assume that each component network has only one output variable, i.e. the function to be approximated is $f : R^m \rightarrow R$. It can be easily generalized to the situations where each component network has more than one output variable.

Note that the combination weights are different from the connection weights belonging to a specific component network. The former are inter-network connection coefficients in the whole system while the latter are intra-network connection coefficients in the corresponding component network. Particularly, the sum of the combination weights are constrained to unity and each combination weight is set to be positive, that is,

$$\sum_{i=1}^{K} w_i = 1 \tag{3}$$

$$w_i \geq 0 \tag{4}$$

Here w_i ($i = 1,2,..., K$) could be viewed as the "goodness" of the i-th component network's behavior in the whole system. Obviously, the better the component network behaves (for example, it has a higher accuracy), the larger the corresponding combination weight is assigned. In this paper, we assume the combination weights associated to the corresponding component networks satisfying Eq.(3) and Eq.(4).

In the following section, an effective sequential Bayesian Learning method for modular neural network is proposed based on sequential Bayesian decision analysis with 8 regression experiments.

2   Sequential Bayesian Learning Method

It's thought that Buntine and Weigend first researched Bayesian approach for neural networks [4, 5]. From then on, lots of valuable work has been made in both theories and applications [5]. But the Bayesian approach on how to combine the output(s) of component networks (sub-nets) of a modular neural network is very rare. In this section, a novel Sequential Bayesian Learning algorithm is proposed.

If the unknown vector has prior density, the posterior density $p(\theta \mid x)$ determined by a certain observed vector x is:

$$p(\theta \mid x) = \frac{\pi(\theta) * p(x \mid \theta)}{p(x)} \tag{5}$$

When θ is continuous,

$$p(x) = \int_{\vartheta \in \Theta} p(x \mid \theta)\pi(\vartheta) \tag{6}$$

When θ is discrete,

$$p(x) = \sum_{\vartheta \in \Theta} p(x \mid \theta)\pi(\vartheta) \tag{7}$$

Where $p(x \mid \theta)$ likelihood function (LF), $p(x)$ is marginal density (or predicted function).

The proper determination of the combination weights is a key in terms of linear combination methods. Generally speaking, if we consider the process of determining the combination weights as a decision making problem, the way can be basically categorized into three types [6]: the first is determining the weights according to sufficient prior information before training the component networks; the second is determining the weights after finishing (independently) training all the component networks (sub-nets); the last is determining the weights during the process of the training. They are respectively analogous to before-the-event decision, after-the-event decision, during -the- event decision in terms of decision analysis. In practice, methods of the last two types are more frequently used due to the fact that in practice only partial prior knowledge is available in most cases. Most methods of the second type use some kind of optimization technology, such as RLS, GA[1], etc., to search the optimal combination weights. As for the third type, many schemes are presented, one of which is Bayesian learning method for combined multiple neural networks [5]. The method considers the determination of the weights as a statistical decision problem, where each individual network is a decision-maker, and the corresponding combination weight is its reliability.

One of the advantages of Bayesian decision analysis is that it can collect samples by means of experiments [4]. The new information derived from those samples can help the analyst gain more insights of the data set and thus enable him or her to take better actions to design experiments and select data in terms of reducing the expectation loss.

Based on the basic idea of sequential Bayesian decision analysis – gradually improving the decision accuracy by collecting more information derived from a series of experiments, we present a method - sequential Bayesian learning method for modular neural network. This method mainly consists of two steps: the first step is to train the component networks until either they all arrive at a preset training accuracy or the training stage comes to the maximum stage; the second step is to learn the combination weights of the individual networks in a sequential fashion (in other words, it is an iterative process.). At each stage an evaluation set is firstly constructed in a way that it consists of the previous set and the data set newly acquired at current stage. The next is to compute the generalized errors of each component network on the corresponding evaluation set and the LF value of each component network. Finally the combination weights are adjusted in light of Bayesian Theorem. Initial prior weights are set to be equal according to Maximum Entropy Principle in terms of Bayesian analysis.

The method is summarized in Table 1. From the step (4.a), we can easily find that the construction of evaluation set is of inheritable. On the other hand, in the process of learning the combination weights, the (intra-network) connection weights of each component network does not change and only the combination weights change. So the generalization error of each network over the evaluation set at each stage is also inheritable, namely, the generalization error of each network at each stage is the sum of the one at previous stage and the one over the data set newly acquired at current stage.

Considering that the parameters (i.e., the combination weights) represent the "goodness" of each component network in the whole system, we take the generalization performance of each network over an evaluation set into account when we construct LF of each network. In other words, the "goodness" of each network is measured by its generalization performance. The better the generalization performance is, the higher the goodness is. Based on this idea, two different LFs as Eq.(8) and Eq. (9) are constructed as follows:

$$\omega = P/L \tag{8}$$

$$\omega_j = \frac{1/sse_j}{\sum_{k=1}^{K} 1/sse_k} \tag{9}$$

For Eq.(8) [5], it is called ε-level approximation correctness rate with the following definition:

Assume that there are L instances in an evaluation set. If P(P<=L) instances can be approximated with an squared error less than ε by an individual network, then we get ε-level approximation correctness rate of this network as Eq.(8). For Eq.(9) we presented, sse_j represents the sum of squared error of the j-th network.

Table 1. Algorithm Procedure.

Input: training set, evaluation set, training parameters (i.e., the learning rate, the error goal, the maximum stage Procedure:

  1.set the initial combination weights to be equal (if no prior information can be acquired), that is $w_j^0 = 1 / K \, (j = 1, 2, ..., K)$;

  2.train all the component networks until the error goal or the maximum stage is arrived at;

  3.split the evaluation set into S parts, each one denoted as $\{ES_i\}_{i=1}^S$;

  4.for i=1 to S : a. compute the LF value $\omega_j^i \, (j = 1, 2, ..., K)$ according to Eq. (8) or Eq.

    (9) in $\{ES_1 \cup ES_2 \cdots \cup ES_i\}$;

    b. update the combination weights by using Bayesian theorem:

      if $\displaystyle\sum_{j=1}^K w_j^{i-1} \omega_j^i = 0$

$$w_j^i = w_j^{i-1} \tag{10}$$

    Else

$$w_j^i = \frac{w_j^{i-1} \omega_j^i}{\displaystyle\sum_{j=1}^K w_j^{i-1} \omega_j^i} \tag{11}$$

    for j=1 to K

      if $w_j^i < 0.1 / K$

      $w_j^i = 0$ and normalize the combination weights.

Output: modular neural network, including the connection weights of each component network and the combination weights.

---

3 Empirical Studies

We applied the method in eight regression problems. Because the result derived from Eq. (8) may be different from that from Eq. (9), we respectively call them Algorithm 1 and Algorithm 2 for convenience. In addition, we compared our results with those using the method of simple average.

3.1 Representation of Problems

The eight regressions problems are taken from [8], specified in Table 2. The first column lists the name of each problem. The second column lists the function expression corresponding to each problem. The third column lists the defined interval on the variables in each function. For simplicity, we (orderly) give each problem a short name: MH2, MH3. FI, F2, G, M, PL, PO.

3.2 Experimental Setup and Results

All the experiments are done in MATLAB. We use the neural network toolbox to create and train and test each network. For each problem, we do five experiments in order to reduce the randomness.

Table 2. Regression Problems to be Tested.

Name	Function	Variable(s)
2-D Mexican Hat	$y = \dfrac{\sin\lvert x\rvert}{x}$	$x \sim U[-2\pi, 2\pi]$
3-D Mexican Hat	$y = \dfrac{\sin\sqrt{x_1^2 + x_2^2}}{\sqrt{x_1^2 + x_2^2}}$	$x_i \sim U[-4\pi, 4\pi]$
Friedman #1	$y = 10\sin(\pi x_1 x_2) + 20(x_3 - 0.5)^2 + 10x_4 + 5x_5$	$x_i \sim U[0,1]$
Friedman #2	$y = \sqrt{x_1^2 + \left(x_2 x_3 - \left(\dfrac{1}{x_2 x_4}\right)\right)^2}$	$x_1 \sim U[0,100]$ $x_2 \sim U[40\pi, 560\pi]$ $x_3 \sim U[0,1]$ $x_4 \sim U[1,11]$
Gabor	$y = \dfrac{\pi}{2}\exp\!\left[-2(x_1^2 + x_2^2)\right]\cos[2\pi(x_1 + x_2)]$	$x_i \sim U[0,1]$
Multi	$y = 0.79 + 1.27x_1x_2 + 1.56x_1x_4 + 3.42x_2x_3 + 2.06x_3x_4x_5$	$x_i \sim U[0,1]$
Plane	$y = 0.6x_1 + 0.3x_2$	$x_i \sim U[0,1]$
Polynomial	$y = 1 + 2x + 3x^2 + 4x^3 + 5x^4$	$x \sim U[0,1]$

First step is to generate a data set. Then the data set is orderly partitioned into three parts: training set used to learn the connection weights of each component network, evaluation set used to learn the inter-network combination weights, test set used to check the generalization performance of the whole system. The size of each data set and variables' distribution interval are showed in Table 3, where x-y-z represents the size of these sets.

The second step is to construct the architecture of each individual network (Here we adopt BP neural networks with different architecture. Table 4 shows the architecture of each modular neural network, where x-y-z (or x-y_1-y_2-z) means the number of units in input, hidden and output layer are x, y (or y_1 and y_2) and z. The activation function of the neurons in the hidden layer(s) is sigmoid function and the one in output layer is linear.

The third step is to train all the component neural networks. Firstly, we orderly split the training set into K (the number of the component networks) subsets. Then on each subset a component network is trained. The training parameters are set to the default values of MATLAB except for the error goal, the maximum stage and the learning rate. The learning rate is 0.02. More parameters are showed in Table 3 and Table 4.

Finally, we use the method proposed in this paper to determine the combination weights. As far as evaluation set is concerned, it is randomly split into five subsets. At the first stage (we mentioned above that the process of determining the combination weights is an iterative one) the first subset is chosen to be the current evaluation set, and at the second stage the first two subsets are chosen to be the current evaluation set, ..., and so on. At the final stage, i.e., the fifth stage, all the subsets are chosen to be the current evaluation set.

Table 3. Parameters in experiments.

Prob-lem	Size of data-set	Partition of dataset	Number of component networks	Training error limit	Parameters
MH2	3000	1200-1000-800	4	le-6	5000
MH3	4000	2000-1000-1000	5	le-5	3000
Fl	5000	3200-1000-800	8	le-6	5000
F2	21000	9000-10000-2000	6	1	3000
G	3000	1200-1000-800	6	le-6	5000
M	3000	1200-1000-800	8	1 e-4	5000
PL	3000	1200-1000-800	5	le-6	5000
PO	3000	1200-1000-800	4	1 e-6	5000

Table 4. Architecture of Component Networks

Problem	The architecture of each component network
MH2	1-10-1, 1-10-1,1-12-1,1-12-1
MH3	2-8-8-1, 2-8-8-1, 2-8-10-1, 2-8-10-1, 2-10-10-1
F1	5-18-1,5-18-1,5-20-1,5-20-1,5-22-1,5-22-1,5-24-1,5-24-1
F2	4-25-1,4-25-1,4-30-1,4-30-1,4-35-1,4-35-1
G	2-12-1,2-12-1,2-14-1,2-6-8-1,2-8-8-1,2-8-8-1
M	5-10-1, 5-10-1, 5-12-1, 5-12-1, 5-14-1, 5-14-1, 5-16-1, 5-16-1
PL	2-8-1,2-8-1,2-10-1,2-5-5-1,2-5-5-1
PO	1-10-1,1-10-1, 1-12-1,1-12-1

Table 5 demonstrates the average generalization errors, from which we can see that the method proposed in this paper and the method of simple averaging show little difference in performance in problem MH2, G, M, PL, PO and show great difference in problem MH3, FI, F2. In terms of the training complexity, the first five problems are easier than the last three. Probably this is the main reason for the difference. The performances of algorithm 1 and algorithm 2 are close on average.

Table 5. Average MSE.

	Algorithm1	Algorithm2	Algorithm3
MH2	3.227e-7	3.282e-7	6.734e-6
MH3	3.510e-5	1.717e-5	0.001193
F1	2.086e-5	3.322e-5	0.357947
F2	3.90322	3.99178	1225.76
G	1.416e-5	5.992e-6	1.695e-5
M	0.000734	0.000474	0.00053
PL	5.905e-7	3.383e-7	6.112e-7
PO	5.547e-7	5.987e-7	5.559e-7

4 Conclusions

Gradually improving the decision accuracy by collecting more information derived from a series of experiments is the basic idea of sequential Bayesian decision analysis. Based on this idea we present sequential Bayesian learning method for modular neural network: after finishing independently training all the component networks, the combination weight of each network is determined in a sequential (iterative) way. At each stage the likelihood function (LF) value of each component network is firstly computed according to its generalization error over the corresponding evaluation set (an evaluation set is constructed like this: each evaluation set is the union of the one at previous stage and the data set newly acquired at current stage). And then the combination weights are adjusted in light of Bayesian Theorem where the current weights are posterior and the previous weights are prior with initial prior weights are set to be equal. The results of experiments on the eight problems show that the method is superior to simple averaging on those hard-to-learn problems.

References

1. Perrone, M. D., Cooper, L. N.: When Network Disagree: Ensemble Methods for Hybrid Neural Networks. In: Mammone R. J. Eds.: Artificial Neural Networks for Speech and Vision. Chapman & Hall, London, UK (1994) 126-142
2. Sherif, Hasbem.: Optimal Linear Combinations of Neural Networks. Neural Networks, **5** (1994) 1-32
3. Cho, S.-B, Kim, Jin H.: Combining Multiple Neural Networks by Fuzzy Integral for Robust Classification. IEEE Trans. on System, Man. and Cybernetics, **2** (1995) 380-384
4. Buntine, W. L., Weigend, A.S.: Bayesian Back-propagation. Complex Systems, **6** (1991) 603-643
5. Lampinen, J., Vehtari, A.: Bayesian Approach for Neural Networks—Review and Case Studies. Neural Networks, **14** (2001) 257-274
6. Chen, T.: Decision Analysis. Science Press, Beijing (1987)
7. Cai, J.: A Study on the Computational Intelligence Applied for Decision Support. Ph D Dissertation. Huazhong University of Science and Technology, Wuhan, China (1996)
8. Zhou, Z. H, Wu, J. X, Tang, W.: Ensembling Neural Networks: Many Could Be Better Than All. Artificial Intelligence, **1-2** (2002) 239-263

A Modified Genetic Algorithm for Fast Training Neural Networks

Dongsun Kim[1], Hyunsik Kim[1], and Duckjin Chung[2]

[1] DMB Project Office, Korea Electronics Technology Institute,
455-6 MaSanri, JinWimyon, PyungTaeksi, KyungGido 451-865, Korea
{dskim,hskim}@keti.re.kr
[2] Information Technology and Telecommunications, INHA University,
253 Younghyun-Dong, Nam-Gu, Incheon 402-751, Korea
djchung@inha.ac.kr

Abstract. The training of feed-forward Neural Networks (NNs) by back-propagation (BP) is much time-consuming and complex task of great importance. To overcome this problem, we apply Genetic Algorithm (GA) to determine parameters of NN automatically and propose a efficient GA which reduces its iterative computation time for enhancing the training capacity of NN. Proposed GA is based on steady-state model among continuous generation model and used the modified tournament selection, as well as special survival condition. To show the validity of the proposed method, we compare with conventional and the survival-based GA using mathematical optimization problems and set covering problem. In addition, we estimate the performance of training the layered feedforward NN with GA and BP.

1 Introduction

One of the most popular training algorithms for feed forward Neural Networks (NNs) is backpropagation (BP) algorithm. However, it has some shortcomings that can hardly be overcome because it is a kind of learning supervised algorithm based on gradient descent method, which leads to falling into local minimum and is very inefficient in searching for global minimum of the search space [1,2]. Recently, interest in training recurrent Neural Network (NN) based on Genetic Algorithm (GA) has been growing rapidly to overcome this problem, which includes several key issues in designing GA such as choice of running parameters and diagram of training method [3,4,5]. The interest of this paper is to explore possible benefits arising from the interactions between NNs and GA one of evolutionary search procedures due to its fast training capacity of NN. GA is powerful search and optimization algorithm, which are computational model based on Darwin's biological evolution theory of genetic selection and natural elimination. The GA, however, takes a long computation time in some specific problems because of its iteratively adaptive process for evolution [4,5,6]. Thereby, it is indispensable to improve GA for reducing the computation time and preventing from local minima efficiently. In this paper, we propose a robust GA based on

J. Wang, X. Liao, and Z. Yi (Eds.): ISNN 2005, LNCS 3496, pp. 660–665, 2005.

steady-state model among continuous generation model for training NNs. We introduce the modified tournament selection as well as special survival condition with replaced whenever the offsprings fitness is better than worse-fit parents for GA. In order to show the validity of the proposed algorithm, we compared a conventional and the survival-based GA with high convergence speed on mathematical optimization problems and set covering problem (SCP). In addition, we apply GAs to layered feedforward NN and analyze the performance using MATLAB simulation.

2   Genetic Algorithm for Neural Networks

GA has several advantages over other optimization algorithms because it is a derivative-free stochastic optimization method based on the features of natural selection and biological evolution. One of the most significant advantages is its robustness of getting trapped in local minima and the flexibility of facilitating parameter optimization in complex models such as NNs. In this reason, we focus on the learning network parameters and optimizing the connection weights using GA for designing an NN architecture. GA considers a solving model of problem as a gene and the collection of genes is called the population. Actually the goal of the GA is to come up with a best value, but not necessarily optimal solution to the problem. The general GA uses several simple operations in order to simulate evolution shown as Figure 1 [2]. After initial population generated randomly, the fitness for each individuals in the population is calculated. Selection operator reproduces the individuals selected to form a new population according to each individuals fitness. Thereafter crossover and mutation on the population are performed and such operations are repeated until some condition is satisfied. Crossover operation swaps some part of genetic bit string within parents. It means as crossover of genes in real world that descendants are inherited characteristics from both parents. Mutation operation is a inversion of some bits from whole bit string at very low rate [7]. These factors increase the diversity of genes and influence on each individual in the population evolves to get higher fitness.

2.1   Chromosome Representation

A feed-forward NN can be thought of as a weighted digraph with no closed paths and described by an upper or lower diagonal adjacency matrix with real valued elements. The nodes should be in a fixed order according to layers. An adjacency matrix is an N × N array in which elements [1].

$$\eta_{ij} = 0 \qquad \text{if} < i,j > \notin \mathbf{E} \qquad \text{for all} \quad i \leq j \qquad (1)$$

$$\eta_{ij} \neq 0 \qquad \text{if} < i,j > \in \mathbf{E} \qquad \text{for all} \quad i \leq j \qquad (2)$$

where i, j = 1,2,. . . , N and $< i,j >$ is an ordered pair and represents an edge or link between neurons i and j, $\mathbf{E}$ is the set of all edges of the graph

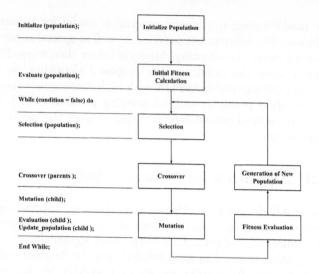

Fig. 1. The Genetic Algorithm

and N is the total number of neurons in the network. Here, the biases of the network, η_{ij} is not 0 if i equals j for all $< i, j >$. Thus, an adjacency matrix of a digraph can contain all information about the connectivity, weights and biases of a network. A layered feed-forward network is one such that a path from input node to output node will have the same path length. Thus, an n-layered NN has the path length of n. The adjacency matrix of the corresponding feed-forward NN will be an upper or lower diagonal matrix [1]. For example, a three-layered feed-forward NN is shown in Figure 2.

2.2   Fitness Function

In order to estimate the performance of the proposed algorithm, we used three fitness functions: the mathematical optimization function, the set covering problem (SCP) [8] and neural networks problem which is the sum squared error.

Mathematical Optimization Problem (MOP): The used optimization function is

$$F(x, y) = 21.5 + x \sin(4\pi x) + y \sin(20\pi y) \quad 0 \leq x \leq 12.1, \quad 4.1 \leq y \leq 5.8 \quad (3)$$

It is the problem that finds the maximum value in given range. Also, it is difficult to find the optimal solution from such a cost surface with local minima [6]. The proposed algorithm needed only 1603 crossovers on average to find the optimal solution, while the survival-based algorithm required 2363 crossovers on average for the same solution on our simulation.

Set Covering Problem (SCP): The set covering problem can be characterized by a well-known combinatorial optimization problem and non-deterministic polynomial-time hard (NP-hard). The SCP can be defined the problem of covering the rows of zero-one matrix by a subset of the columns at minimal cost within m-row, n-column. The SCP is important to the minimization of logic circuit or the optimization of the resource selection problem. In our study, we chose fitness function that obtains minimal expression of Boolean function that composed of 19 rows 63 columns with non-unicost. The objective was to find a minimum-sized set of rows whose elements covered all of the 63 columns.

Neural Networks Problem (NNP): The fitness function for NNP is the Sum Squared Error (SSE) for neural networks. If y_d is the target output and y is the actual output, it can be defined as

$$J(\omega, \theta) = \sum_{d \in \text{outputs}} e^2 \qquad where \quad e = y_d - y \tag{4}$$

The ω and θ denote the weights and biases linking the neuron unit to the previous neuron layer. Obviously the objective is to minimize J subject to weights and biases ω_{ji}, ω_{kj}, θ_j and θ_k

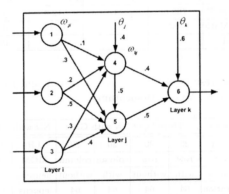

Fig. 2. Three layer Neural Network

2.3   Weight Crossover

Two types of crossover are verified, firstly row wise crossover and secondly column wise crossover. In row wise crossover, an offspring is generated by choosing alternative rows from parent chromosome matrices as shown in Figure 3 (a). Column wise crossover is similarly done like column wise and it is described in Figure 3 (b). We use an *N-I* point crossover, where *N* is the number of rows or columns.

3   Experimental Results

The relative comparison of computation time between survival-based GA and proposed GA on software simulation as shown in Table 1. In Table 1, the total

Immunity Clonal Synergetic Learning of Unbalanced Attention Parameters in Synergetic Network

Xiuli Ma and Licheng Jiao

Institute of Intelligent Information Processing, Xidian University, 710071 Xi'an, China
xlma@mail.xidian.edu.cn

Abstract. In this paper, we discuss the dynamical behavior of synergetic neural network and propose a new unbalanced attention parameters optimization algorithm based on Immunity Clonal algorithm (ICA). In comparison with the algorithms with balanced attention parameters and that with unbalanced attention parameters on GA, the new method has automatic balance ability between exploration and exploitation and is not easy to get into local optimum. In addition, iterative step is adjusted adaptively. Experiments on textural images and remote sensing images show that the presented algorithm has not only faster convergent rate but also better recognition rate.

1 Introduction

Since Synergetic Neural Network (SNN) [1] was proposed by Haken in the 1990s, its learning algorithm has been widely studied especially on the selection of prototype pattern vectors and the setting of attention parameters and so on.

Under balanced attention parameters, the output of SNN can be deduced directly from the initial order-parameter set and this makes SNN lose the self-learning ability on condition that the prototype pattern vectors are given. That is, once the target pattern is recognized incorrectly, SNN cannot recognize it correctly by further learning. For the self-learning ability, the main characteristic of SNN, is restrained and as a result the recognition ability of SNN is weaken, which affects its application. So the study on the learning algorithm and dynamical behavior of SNN under unbalanced attention parameters is very important and significant.

Wang et al [2] studied the properties of attention parameters and proposed a simple scheme to train them based on award-penalty learning mechanism. But this learning mechanism has torpid convergent rate and easily gets into local optimum, which affects SNN to achieve the optimal performance. Then Hu [3] studied the dynamical behavior of dynamical equation in SNN under unbalanced attention parameters and got the relationship between attention parameters and the stable fixed points of SNN's solutions but he did not propose an effective learning algorithm to train attention parameters. Subsequently, Wang et al [4] proposed an attention parameters optimization algorithm base on GA, which is easy to get into local optimum. In addition, the search space limited to [0,6] and iterative step equal to 0.1 are unreasonable.

Considering that, we propose to optimize attention parameters by Immunity Clonal algorithm (ICA) [5], which has global search capacity and can avoid torpid convergent rate and local optimum of GA. In our algorithm, the iterative step is adjusted adaptively [6] for it determines the stability of SNN.

J. Wang, X. Liao, and Z. Yi (Eds.): ISNN 2005, LNCS 3496, pp. 666–671, 2005.

2 Synergetic Neural Network

Haken's synergetic approach [1] to pattern recognition is based on a parallel between pattern recognition and pattern formation. Suppose that the M prototype pattern vectors V_k and the status vectors q are N-dimensional column vectors with zero mean and unit length, M is less than N for the sake of linear independence of prototype pattern vectors. Haken proposed that the synergetic recognition method can be described as a dynamical procedure with potential function:

$$V = -\frac{1}{2}\sum_{k=1}^{M}\lambda_k(v_k^+q)^2 + \frac{1}{4}B\sum_{k\neq k'}(v_k^+q)^2(v_{k'}^+q)^2 + \frac{1}{4}Cq(q^+q)^2 \ , \tag{1}$$

where λ_k is attention parameter, B and C are constants. Potential function V is made of three terms. The first term defines the minima on the potential surface at the prototypes. The depth of each minimum is controlled by the attention parameters λ_k. The second term defines the competition among prototypes and controls the location of ridges in the potential surface. This is parameterized by B. The third term is required to limit the first term and is parameterized by C.

In order to reduce the dimension of the system, we re-express the potential in terms of order parameters,

$$\xi_k = (v_k^+, q) = v_k^+q \ . \tag{2}$$

Corresponding dynamical equation of order parameters is

$$\dot{\xi} = \lambda_k\xi_k - B\sum_{k\neq k'}\xi_{k'}^2\xi_k - C(\sum_{k'=1}^{M}\xi_{k'}^2)\xi_k \ . \tag{3}$$

In synergetic pattern recognition system, attention parameters provide selectivity mechanism. Study on Haken model proves that the survival of order parameters can be determined by the observation on the initial order-parameter set and attention parameters. For convenience, equation (3) can be expressed as

$$\dot{\xi}_k = \xi_k(\lambda_k - D + B\xi_k^2), \ D = (B+C)\sum_{j=1}^{M}\xi_j^2 \ . \tag{4}$$

The dynamic equation (4) can be denoted as three-layer network and its output is projected onto the third layer by $q(t) = \sum_{k=1}^{M}\xi_k v_k$ which realizes the association function and last the recognition process of SNN is accomplished.

3 Unbalanced Attention Parameters Optimization of SNN

3.1 Stable Fixed Points of V

When $B = C = \lambda_k > 0$, Haken [1] has proved that the equation (3) has the following properties:

  1.All the prototypes are possible final states.
  2.There are no other possible final states.
  3.Order parameter with largest absolute value will grow while others will decay.

The first result assures us that all memories are accessible. The second result proves that the system will not learn any spurious memories. The third property implies that the final solution can be predicted from the initial conditions, so that the integrating system loses the ability of recognizing test pattern correctly by further learning.

When $B > 0, C > 0$ and $\lambda_k > 0$, Hu [3] has proved that the equation (3) has the following theorem.

Theorem 1 Suppose $(B + C)\lambda_i - C\lambda_j > 0, 1 \le i \ne j \le M$. The stationary points in corresponding SNN can be divided into three classifications: an unstable fixed point, $2M$ stable points and $3^M - 2M - 1$ saddle points.

Theorem 1 represents that stable points of V correspond to M prototype patterns and distribute as original symmetry. Under the condition of unbalanced attention parameters, they do not distribute symmetrically and the depths of their corresponding attractive domains differ from each other and are determined by attention parameters λ_k. The ranges of their corresponding attractive domains alter with their depths, which affect the competition among prototype patterns.

Considering that, we propose to optimize attention parameters by Immunity Clonal algorithm (ICA), which has automatic balance ability between exploration and exploitation and can avoid torpid convergence and local optimum of GA. In addition, the iterative step is adjusted adaptively.

3.2 Unbalanced Attention Parameters Optimization of SNN on ICA

In order to enhance the diversity of the population in GA and avoid prematurity, Immunity Clonal algorithm (ICA) [5] was proposed by Du *et al.* The clonal operator is an antibody random map induced by the affinity including: clone, clonal mutation and clonal selection. The state transfer of antibody population is denoted as

$$C_{MA} : A(k) \xrightarrow{\ clone\ } A'(k) \xrightarrow{\ mutation\ } A''(k) \xrightarrow{\ selection\ } A(k+1).$$

The essential of clonal operator is producing a variation population around the parents according to their affinity. Then the searching area is enlarged.

In this paper, the optimization of unbalanced attention parameters is to search the optimal solutions in attention-parameter space using ICA. Here, the iterative step is adjusted adaptively. The detailed algorithm is described as follows.

Step 1: Initialize the antibody population.
Initialize the antibody population $A(0)$ size of N randomly. Every individual represents a set of attention parameters of SNN.

Step 2: Calculate the affinity.
The affinity is donated by the classification accuracy rate of training samples.

Step 3: Clone.
Clone every individual in the kth parent $A(k)$ to produce $A'(k)$. The clonal scale is a constant or determined by the affinity.

Step 4: Clonal mutation.
Mutate $A'(k)$ to get $A''(k)$ in probability of p_m.

Step 5: Calculate the affinity.
Calculate the affinity of every individual in new population $A''(k)$.

Step 6: Clonal selection.
Select the best individual, which has better affinity than its parent, into the new parent population $A(k+1)$.

Step 7: Calculate the affinity.
Calculate the affinity of every individual in new population $A(k+1)$.

Step 8: Judge the halt conditions reached or not.
The halt condition is the specified iterative number or the classification rate keeping unchanged for several times. If the searching algorithm reaches the halt condition, terminate the iteration and make the preserved individual as the optimal attention parameters, or else preserve the best individual in current iteration and then turn its steps to Step 3.

4   Experiments

4.1   Classification of Remote Sensing Images

In order to verify the proposed method, we compare it with the algorithm with balanced attention parameters and that with unbalanced attention parameters on GA. The remote sensing image set is of size 1064 and includes boats and planes 456 and 608 respectively. The images are integrated and misshapen binary images and have different rotary angles. The image set is divided into two parts: training set and testing set. In the former, boats and planes are 120 and 160, and in the latter, 336 and 448 respectively. Additionally, relative moments [7] are used to feature extraction.

In this experiment, SCAP algorithm [8] is used to get prototype pattern vectors. The iterative step is adjusted adaptively [6] with $0.001/D$ and $B=C=1$. Iterative number is 80. The parameters in ICA are defined empirically as follows. The size of initial population is 5 and mutation probability is 0.2. The clonal scale is determined by the affinity. In GA, the size of initial population is 20, crossover probability 0.9 and mutation probability 0.1. GA operators are fitness proportional model selection operator, arithmetic crossover operator and un-uniform mutation operator. The halt condition is the specified iterative number or the classification accuracy rate (CAR) keeping unchanged for several times. The statistical average results of 10 times are shown in Table 1. The optimal attention parameters got by the presented method are $\lambda_1 = 1.2838$ and $\lambda_2 = 0.7877$, and simultaneously the optimal CAR is up to 98.822%.

Table 1. Comparison among different methods.

		$\lambda_k = B = C = 1$	Method on GA	The proposed method
Training time (s)		0	144.9672	105.0580
Testing time (s)		0	145.1202	105.2156
CAR of training set (%)		96.333	98.000	98.333
CAR of testing set (%)	Boat	91.670	96.102	96.428
	Plane	100.00	99.020	99.369
Average CAR (%)		96.335	97.769	98.109

From Table 1, it can be seen that the proposed method has advantageous over others no matter on training and testing time or on CAR. During the training process, we have observed that our algorithm has faster convergent rate and is not easy to get into local optimum. The application of SCAP affects the classification accuracy rate for it gets prototype pattern vectors by averaging the training samples simply.

4.2 Classification of Brodatz Texture by Brushlets Features

For comparing the proposed algorithm with others, 16 similar textural images are chosen from Brodatz, namely, D006, D009, D019, D020, D021, D024, D029, D053, D055, D057, D078, D080, D083, D084, D085, and D092. Every image of size 640×640 is segmented into 25 nonoverlapping images as a class. Then the samples are 16 kinds in all and each class has 25 samples. We randomly select 8 samples from each class as training data and others as testing data. Simultaneously, Brushlets [9] is used to feature extraction.

In this experiment, except that the iterative step is $0.1/D$, other parameters' setting is the same as the above one. The statistical average values of 10 times are shown in Table 2. We give the optimal attention parameters got by the presented method in Table 3 and at that time the optimal recognition rate is up to 98.162%.

Table 2. Comparison among different methods.

	$\lambda_k = B = C = 1$	Method on GA	The proposed method
Training time (s)	0	191.2607	36.3031
Testing time (s)	0	191.4811	36.4373
CAR of training set (%)	89.844	96.172	98.438
CAR of testing set (%)	93.750	94.853	96.985

From Table 2, it can be drawn the same conclusion that the proposed method has not only faster convergent rate but also higher classification accuracy rate.

Table 3. The optimal attention parameters got by the proposed algorithm.

λ_1	λ_2	λ_3	λ_4	λ_5	λ_6	λ_7	λ_8
1.8201	1.9234	2.0216	2.3017	2.3106	2.2271	1.6912	2.2429
λ_9	λ_{10}	λ_{11}	λ_{12}	λ_{13}	λ_{14}	λ_{15}	λ_{16}
2.4425	2.0336	2.3439	2.1590	1.9920	1.9174	2.3907	1.9163

5 Conclusions

An unbalanced attention parameters optimization algorithm based on Immunity Clonal algorithm (ICA) for Synergetic Neural Network (SNN) is proposed in this paper, which searches the optimal solution in attention-parameter space using the global and local searching ability of ICA. Compared with the algorithm with balanced attention parameters and that with unbalanced attention parameters on GA, the new method has automatic balance ability between exploration and exploitation and is not easy to get into local optimum. In addition, iterative step is adjusted adaptively. Experiments on textural images and remote sensing images show that the presented method has not only faster convergent rate but also higher classification accuracy rate. For the emphasis is put on optimizing attention parameters, SCAP is applied to get prototype pattern vectors. SCAP is rapid and simple for it gets prototype pattern vectors by averaging the training samples simply, which has certain effect on SNN's performance and then on its classification accuracy rate.

References

1. Haken, H.: Synergetic Computers and Cognition–A Top-Down Approach to Neural Nets. Springer-Verlag, Berlin (1991)
2. Wang, F.Y., Level, P.J.A., Pu, B.: A Robotic Vision System for Object Identification and Manipulation Using Synergetic Pattern Recognition. Robot. Comput. Integrated Manufacturing. 10 (1993) 445–459
3. Hu, D.L., Qi, F.H.: Study on Unbalanced Attention Parameters in Synergetic Approach on Pattern Recognition. Acta Electronica Sinica. 27 (1999) 15–17
4. Wang, H.L., Qi, F.H., Zhan, J.F.: A Genetic-Synergetic Learning Algorithm Under Unbalanced Attention Parameters. Acta Electronica Sinica. 28 (2000) 25–28
5. Jiao, L.C., Du, H.F.: Development and Prospect of The Artificial Immune System. Acta Electronica Sinica. 31 (2003) 1540–1548
6. Zhao, T., Qi, F.H., Feng, J.: Analysis of The Recognition Performance of Synergetic Neural Network. Acta Electronica Sinica. 28 (2000) 74–77
7. Wang, B.T., Sun, J.A., Cai, A.N.: Relative Moments and Their Applications to Geometric Shape Recognition. Journal of Image and Graphics. 6 (2001) 296–300
8. Wagner, T., Boebel, F.G.: Testing Synergetic Algorithms with Industrial Classification Problems. Neural Networks. 7 (1994) 1313–1321
9. Meyer, F.G., Coifman, R.R.: Brushlets: A Tool for Directional Image Analysis and Image Compression. Applied and Computational Harmonic Analysis. 4 (1997) 147–187

Optimizing Weights of Neural Network
Using an Adaptive Tabu Search Approach

Yi He[1,3], Yuhui Qiu[1], Guangyuan Liu[2,*], and Kaiyou Lei[1]

[1] Faculty of Computer & Information Science, Southwest-China Normal University,
Chongqing 400715, China
cqnuheyi@163.com
[2] School of Electronic & Information Engineering, Southwest-China Normal University,
Chongqing 400715, China
[3] School of Management, Chongqing Normal University, Chongqing 400047, China

Abstract. Feed forward Neural Network (FNN) has been widely applied to many fields because of its ability to closely approximate unknown function to any degree of desired accuracy. Gradient techniques, for instance, Back Propagation (BP) algorithm, are the most general learning algorithms. Since these techniques are essentially local optimization algorithms, they are subject to converging at the local optimal solutions and thus perform poorly even on simple problems when forecasting out of samples. Consequently, we presented an adaptive Tabu Search (TS) approach as a possible alternative to the problematical BP algorithm, which included a novel adaptive search strategy of intensification and diversification that was used to improve the efficiency of the general TS. Taking the classical XOR problem and function approximation as examples, a compare investigation was implemented. The experiment results show that TS algorithm has obviously superior convergence rate and convergence precision compared with other BP algorithms.

1 Introduction

The multi-layer Feed forward Neural Network (FNN) is the most popular and widely applied Neural Network due to its superior ability of non-linearity to approximate unknown function to any degree of desired accuracy which has been widely applied to many fields, such as pattern recognition, image processing, finance prediction and signal and information processing, etc. Most applications of FNNs use some variation of the gradient technique, Back Propagation (BP) algorithm, to optimize neural networks [1-3], but it is essentially a local optimizing method and thus has some inevitable drawbacks, such as easily trapping into the local optimal, and dissatisfying generalization capability, etc. Consequently, many researchers began to try to use some meta-heuristic methods, especially Genetic Algorithm (GA), as an alternative approach of the BP algorithm for optimizing Neural Network and achieved a certain extent of success [4]. In recent years, the popularity of Tabu Search (TS) has grown significantly as a global search technique and it has achieved widespread success in

---

* Corresponding author, Prof. Guangyuan Liu, E-mail: liugy@swnu.edu.cn. This work was partly supported by key Project of Chinese Ministry of Education (104262) and fund project of Chongqing Science and Technology Commission (2003–7881), China.

J. Wang, X. Liao, and Z. Yi (Eds.): ISNN 2005, LNCS 3496, pp. 672–676, 2005.

solving practical optimization problems in different domains, such as combinatorial optimization, quadratic assignment problem, production scheduling and Neural Network, etc. [5-8]. This paper extended the line of the research by applying TS to the difficult problem of neural network optimization. Taking the classical XOR problem and *sinc* function approximation as examples, many experiments were performed. The results show that the TS algorithm has obviously superior convergence rate, convergence precision and better generalization capability compared with the traditional BP algorithms.

2   Tabu Search

Tabu search, proposed by Glover [9], is a meta-heuristic method that has received widespread attention recently because of its flexible framework and several significant successes in solving NP-hard problems [7-8]. The method of neighborhood exploration and the use of short-term and long-term memories distinguish Tabu Search from local search and other heuristic search methods, and result in lower computational cost and better space exploration [10]. TS involves a lot of techniques and strategies, such as short-term memories (tabu list), long-term memories and other prior information about the solution can used to improve the intensification and diversification of the search. It can be confirmed that the strategy of intensification and diversification is very important at most time, therefore, a novel adaptive search strategy of intensification and diversification, proposed in literature [8] by us, was employed to improve the efficiency of TS for neural network optimization.

3   An Adaptive TS Approach for FNNs

In this section, we presented our TS technique for optimizing the FNN based on an adaptive strategy of intensification and diversification. It is detailed as follows:

3.1   Tabu Objective

The error function, *mse* (mean squared error), was selected as the tabu objective. But *mse* is a real value and changes continuously. As a result, a certain *mse* is hard to be considered tabu, and hence, in order for a *TL* (tabu list) to be used, a proximity criterion, *tabu criterion* (*TC*), was given. In other words, by implementing this *TC,* an *mse* would be considered tabu if it falls in a given range around the *mse* in the list. For instance, if the *TC* was set to 0.01, the new *mse* would be compared to the *mse* in the *TL* within ±0.01% .

3.2   Neighborhood

The neighborhood was defined that each of the weights was added to an incremental value respectively. This incremental value is a real value, which was picked out from a given range at random. We call this range *weights offset region* (*WOR*). For instance, if the *WOR* was set to ±0.2, the incremental value would be a real value picked out from the range [–0.2 +0.2] at random.

3.3 A Novel Adaptive Search Strategy of Intensification and Diversification

In TS, intensification strategies are used to encourage TS to search more thoroughly the neighborhood of elite solutions in order to obtain the global optimal, and diversification strategies are used to encourage TS to search solution space as widely as possible, especially the unvisited regions before. They are very important and the coordinative important, unfortunately, they are often conflicting in many cases, because it is usually hard to determine what time for intensification search and what time for diversification search. In order to eliminate the conflicting, an adaptive search strategy of intensification and diversification was proposed to improve the efficiency of the general TS, which achieved the goal by adjusting dynamically the number of intensification elements and diversification elements in candidate set, for TSPs in the literature [8]. Since this strategy needs not the special information of problems, it can be applied to other problems such as Neural Network optimization. Demonstrated with Fig.1 in the literature [8], the elements in neighborhood as well as candidate set are divided into two equal parts, namely intensification elements and diversification elements. The difference between them is that their WOR are different. The WOR_1 for intensification elements is very small, e.g., it was set to $\pm0.1\sim\pm1.0$. As a result, the solution space exploration is restricted to a very small region. On the contrary, the WOR_2 for diversification elements is very large, e.g., it was set to $\pm1.0\sim\pm10.0$, and consequently the solution space exploration is expanded to a bigger region. Except for this, other operations or control strategies are as same as its in literature [8]. Owing to the adaptivity of this strategy, we called the strategy-based TS as an adaptive Tabu Search.

4 Simulation Experiments

In order to examine the feasibility and validity of this TS technique, XOR problem was taken as a test sample. The FNN architecture was set to 2-2-1. The former two layers used the sigmoid transfer function *logsig* and the output layer used the linear transfer function *purelin*. Error function was set to *mse* (mean squared error). Error goal precision was set to 1.0×10^{-6}. The initial weights were set to the real values randomly picked out from the range [-1 +1]. Some key parameters in TS was set to several different combinations, detailed in table 1.

Table 1. The combination of some key parameters in TS.

Group	TL	TC	NL
1	50	0.01%	100
2	100	0.01%	200
3	200	0.01%	300

In the Tab.1, TL denotes the size of tabu list; TC denotes tabu criterion; NL denotes the size of neighborhood. For each group, the size of candidate set was set to 10; the max iteration steps were set to 2000, and it was regarded as the termination condition. WOR_1 and WOR_2 were set to a real value randomly from the range $\pm0.1\sim\pm1.0$ and the range $\pm1.0\sim\pm10.0$ respectively.

For each group, the experiment has been performed for 100 runs. The detailed results were presented in Tab. 2.

In order to further examine the generalization capability of the TS technique, a compare experiments with BP algorithm with momentum was implemented for *sinc* function approximation ($f(x) = \sin(x)/x$, $x \in [-4\pi, 4\pi]$). In the experiments, all the controllable factors, including network architecture, transfer function, initial weights, error goal function and sample points, etc. are set complete same. The approximation comparison is demonstrated with Fig.1. It's evident that the approximation effect of TS is better than that of BP algorithm and TS has hardly any overfitting.

Table 2. The statistics of convergence rate and convergence precision.

Group	Convergence Rate	Error goal precision	Actual convergence error precision (*mse*)			
			min	max	mean	Standard deviation
1	94%	1.0×10^{-6}	1.2785×10^{-9}	0.1258	0.0075	0.0299
2	95%	1.0×10^{-6}	3.0759×10^{-9}	0.1264	0.0063	0.0276
3	95%	1.0×10^{-6}	9.7382×10^{-10}	0.1256	0.0063	0.0274

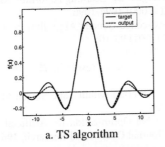
a. TS algorithm

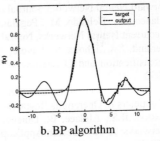
b. BP algorithm

Fig. 1. The comparison of *sinc* function approximation.

Table 3. Comparison of learning results for XOR between TS and other BP algorithms.

Learning algorithm	Maximum learning steps	Error precision	Convergence rate
Simple BP algorithm	8000	1.0×10^{-3}	89%
Steepest algorithm	8000	1.0×10^{-3}	86%
Super-Linear-BP algorithm	8000	1.0×10^{-3}	82%
TS algorithm (Group 1)	2000	1.0×10^{-6}	94%

5 Discussion

5.1 Comparison with Other BP Algorithms

In the literature [3], a super-linearly convergent BP learning algorithm for FNN was proposed, and used to solve the XOR problem. In addition, the results compared to other traditional BP learning algorithms were given. Now cited these results to compare to the results of the first group experiments in this paper, detailed in Tab.3.

5.2 Conclusion

After the observation and analysis of Tab.3 and Fig.1, it can be concluded that this TS approach has obviously superior convergence rate, convergence precision and better generalization capability compared with other traditional or modified BP algorithms. The utilization of the adaptive search strategy of intensification and diversification enhanced significantly the efficiency of TS. This work demonstrates that the importance of using global search techniques for optimizing neural networks, and it also demonstrates that the extended TS algorithm should be considered as a possible solution to the hard optimization problems.

References

1. Engoziner, S, Tomsen, E: An Accelerated Learning Algorithm for Multiplayer Perception: Optimization Layer by Layer. IEEE Transactions on Neural Network, 6 (1995) 31-42
2. Wang, X.G., Tang, Z., Tamura, H., Ishii, M., and Sun, W.D.: An Improved Backpropagation Algorithm to Avoid the Local Minima Problem. Neurocomputing, 56 (2004) 455-460
3. Liang, J., He, X., Huang, D.: Super-Linearly Convergent BP Learning Algorithm for Feedforward Neural Networks. Journal of Software, 11 (2000) 1094-1096
4. Blanco, A., Delgado, M.,. Pegalajar, M.C.: A Real-coded Genetic Algorithm for Training Recurrent Neural Networks. Neural Networks, 14 (2001) 93-105
5. Ferland, J.A., Ichoua, S., Lavoie, A., and Gagné, E.: Scheduling Using Tabu Search with Intensification and Diversification. Computer & Operations Research, 28 (2001) 1075-1092
6. Salhi, S.: Defining Tabu List Size and Aspiration Criterion within Tabu Search Methods. Computers & Operations Research, 29 (2002) 67-86
7. Sexton, R.S., Alidaee, B., Dorsey, R.E. et al.: Global Optimization for Artificial Neural Networks: A Tabu Search Application. European Journal Operational Research, 106 (1998) 570-584
8. He, Y., Liu, G., and Qiu, Y.: A Novel Adaptive Search Strategy of Intensification and Diversification in Tabu Search. Journal of Computer Research & Development, 41 (2004) 162 ~166
9. Glover, F. and Laguna, M.: Tabu Search, Boston, Kluwer Academic Publishers (1997)
10. Sadiq, H.Y., Sait, M., and Adiche, H.: Evolutionary Algorithms, Simulated Annealing and Tabu Search: a Comparative Study. Engineering Applications of Artificial Intelligence, 14 (2001) 167-181

Semi-supervised Learning for Image Retrieval Using Support Vector Machines

Ke Lu, Jidong Zhao, Mengqin Xia, and Jiazhi Zeng

School of Computer Science and Engineering,
University of Electronic Science & Technology of China, Chengdu, Sichuan 610054, China
kel@uestc.edu.cn

Abstract. We study the problem of image retrieval based on semi-supervised learning. Semi-supervised learning has attracted a lot of attention in recent years. Different from traditional supervised learning. Semi-supervised learning makes use of both labeled and unlabeled data. In image retrieval, collecting labeled examples costs human efforts, while vast amounts of unlabelled data are often readily available and offer some additional information. In this paper, based on Support Vector Machine (SVM), we introduce a semi-supervised learning method for image retrieval. The basic consideration of the method is that, if two data points are close to each, they should share the same label. Therefore, it is reasonable to search a projection with maximal margin and locality preserving property. We compare our method to standard SVM and transductive SVM. Experimental results show efficiency and effectiveness of our method.

1 Introduction

Due to the rapid growth in the volume of digit images, there is an increasing demand for effective image management tools. Consequently, content-based image retrieval (CBIR) is receiving widespread research interest [2].

In recent years, much research has been done to semi-supervised learning. Different from traditional supervised learning which only makes use of labeled data; semi-supervised learning makes use of both labeled and unlabeled data. Generally, the unlabeled data can be used to better describe the intrinsic geometrical structure of the data space, and hence improve the classification performance. Also, most previous learning algorithms only consider the Euclidean structure of the data space. However, in many cases, the objects of interest might reside on a low-dimensional manifold which is nonlinearly embedded in the ambient space (data space). In such cases, the nonlinear manifold structure is much more important than the Euclidean structure. Specifically, the similarity between objects should be described by the geodesic distance rather than the Euclidean distance.

In this paper, we introduce a new algorithm for image retrieval. Our algorithm is intrinsically based on Support Vector Machines and Locality Preserving Projections (LPP). LPP is a recently proposed algorithm for linear dimensionality reduction. We first build a nearest neighbor graph over all the data points (labeled and unlabelled) which models the local geometrical structure of the image space. By combining SVM and LPP, we can obtain a classifier which maximizes the margin and simultaneously preserves the local information. In many information retrieval tasks, the local information is much more reliable than the global structure. Also, our method explicitly

J. Wang, X. Liao, and Z. Yi (Eds.): ISNN 2005, LNCS 3496, pp. 677–681, 2005.

considers the manifold structure of the data space. It would be important to note that Euclidean space is just a special manifold. As a result of these properties, our method can produce better results.

The rest of this paper is organized as follows. Section 2 gives a brief description of Support Vector Machines and Locality Preserving Projections. In Section 3, we present a new semi-supervised method for image retrieval. The experimental results are shown in section 4. Finally, we give conclusions in section 5.

2   A Brief Review of SVM and LPP

In this section, we give a brief review of Support Vector Machine [3] and Locality Preserving Projections [1].

2.1   Support Vector Machines

Support Vector Machines are a family of pattern classification algorithms developed by Vapnik and collaborators. SVM training algorithms are based on the idea of structured risk minimization rather than empirical risk minimization, and give rise to new ways of training polynomial, neural network, and radial basis function (RBF) classifiers. SVMs make no assumptions on the distribution of the data and can, therefore, be applied even when we do not have enough knowledge to estimate the distribution that produced the input data [3].

We shall consider SVM in the binary classification setting. We assume that we have a data set $D = \{\mathbf{x}_i, y_i\}_{i=1}^{t}$ of labeled examples, where $y_i \in \{-1, 1\}$, and we wish to select, among the infinite number of linear classifiers that separate the data, one that minimizes the generalization error, or at least minimizes an upper bound on it. It is shown that the hyperplane with this property is the one that leaves the maximum margin between the two classes. Given a new data point $\mathbf{x}$ to classify, a label is assigned according to its relationship to the decision boundary, and the corresponding decision function is

$$f(\mathbf{x}) = sign(\sum_{i=1}^{t} \alpha_i y_i \langle \mathbf{x}_i, \mathbf{x} \rangle - b)$$

2.2   Locality Preserving Projections

LPP is a recently proposed linear dimensionality reduction algorithm [1]. For a data point $\mathbf{x}$ in the original space, we consider the projection $\mathbf{w} \in R^n$ such that $y = \mathbf{w}^T \mathbf{x}$. LPP aims to discover the intrinsic manifold structure by Euclidean embedding. Give a local similarity matrix S, the optimal projections can be obtained by solving the following minimization problem:

$$\mathbf{w}_{opt} = \arg\min_{\mathbf{w}} \sum_{i=1}^{m} \left(\mathbf{w}^T \mathbf{x}_i - \mathbf{w}^T \mathbf{x}_j \right)^2 S_{ij}$$

$$= \arg\min_{\mathbf{w}} \mathbf{w}^T X L X^T \mathbf{w}$$

with the constraint

$$\mathbf{w}^T XDX^T \mathbf{w} = 1$$

where $L = D - S$ is the graph Laplacian [1] and $D_{ii} = \sum_j S_{ij}$. D_{ii} measures the local density around $\mathbf{x}_i$. The bigger D_{ii} is, the more important $\mathbf{x}_i$ is. With simple algebra steps, we can finally get the following generalized eigenvector problem:

$$XLX^T \mathbf{w} = \lambda XDX^T \mathbf{w}$$

For the detailed derivation of LPP, please see [1].

3 Semi-supervised Induction

Semi-supervised learning has received a lot of attentions in recent years. So far most of the efforts have been invested in a transductive setting that predicts only for observed inputs. Yet, in many applications there is a clear need for inductive learning, for example, in image retrieval or document classification. Unfortunately, most existing semi-supervised learners do not readily generalize to new test data. A brute force approach is to incorporate the new test points and re-estimate the function using semi-supervised learning, but this is very inefficient. Another problem of semi-supervised transduction is the computational complexity. In this section, we introduce a new semi-supervised learning algorithm which efficiently combines the characters of both SVMs and LPP.

Let $X = [\mathbf{x}_1, \cdots, \mathbf{x}_m]$ denote the set of data points, belonging to c classes. Suppose the first t data points are labeled, and the rest $m - t$ are unlabelled. We construct a graph as follows: for any pair of data points, we put an edge between them if they are sufficiently close to each other. Correspondingly, the weights are defined as follows:

$$S_{ij} = \begin{cases} e^{-\|\mathbf{x}_i - \mathbf{x}_j\|^2} & \text{if } \mathbf{x}_i \text{ is among the } k \text{ nearest neighbors of } \mathbf{x}_j \\ & \text{or } \mathbf{x}_j \text{ is among the } k \text{ nearest neighbors of } \mathbf{x}_i \\ 0 & \text{otherwise} \end{cases}$$

Note that, the above definition reflects the intrinsic manifold structure of the data space.

Recall that SVM aims to maximize the margin between two classes. The loss function of SVM can be described as follows:

$$L(\mathbf{x}_i, y_i, \mathbf{w}) = \frac{2}{\mathbf{w}},$$

$$y_i(\mathbf{x}_i \cdot \mathbf{w} + b) - 1 \geq 0, \forall i$$

One disadvantage of the above loss function is that it only takes into account the labeled data, while the unlabeled data is ignored. A natural extension is to incorporate the unlabeled data and preserve the graph structure. Specifically, we expect that if two data points are close to each other, they tend to be classified into the same class.

Let f be the classifier such that $f(\mathbf{x}_i)$ is the estimated label for $\mathbf{x}_i$. Thus, we expect to minimize the following loss function:

$$G(\{\mathbf{x}\}_i, f) = \sum_{ij} \|f(\mathbf{x}_i) - f(\mathbf{x}_j)\|^2 S_{ij}$$

Suppose f is linear, we have $f(\mathbf{x}) = \mathbf{w}^T \mathbf{x}$. Thus, by simple algebra formulation, we have:

$$G(\{\mathbf{x}\}_i, f) = \mathbf{w}^T XLX^T \mathbf{w}$$

where L is the graph Laplacian.

Now the optimal classifier can be obtained as follows:

$$\mathbf{w}^{opt} = \arg\min_{\mathbf{w}} \|\mathbf{w}\|^2 - \alpha \mathbf{w}^T XLX^T \mathbf{w}$$

We can also extend the algorithm to nonlinear case by kernel method [3]. We consider the classifier in a reproducing kernel Hilbert space H_K. Given a kernel function K, the RKHS H_K is defined as follows:

$$H_K = \{f \mid f(\mathbf{x}) = \sum_{i=1}^{m} \alpha_i K(\mathbf{x}_i, \mathbf{x}), \ \alpha_i \in R\}$$

Thus, we can solve the above optimization problem in H_K, which gives us a nonlinear classifier.

4 Experimental Results

4.1 Experimental Design

We performed several experiments to evaluate the effectiveness of the proposed approaches over a large image database. The database we use consists of 10,000 images of 79 semantic categories selected from the Corel Image Gallery. Three types of color features and three types of texture features are used in our system. The dimension of the feature space is 435. We designed an automatic feedback scheme to simulate the retrieval process conducted by real users. At each iteration, the system marks the first three incorrect images from the top 100 matches as irrelevant examples, and also selects at most 3 correct images as relevant examples (relevant examples in the previous iterations are excluded from the selection). These automatic generated feedbacks are added into the query example set to refine the retrieval results. To evaluate the performance of the algorithms, we define the retrieval accuracy as follows:

$$Accuray = \frac{\text{relevant images retrieved in top N returns}}{N}$$

4.2 Image Retrieval Using Semi-supervised Induction

Image retrieval is essentially a classification problem. Given a query image, we need to find an optimal classifier which separates the relevant images and irrelevant images.

We compare the semi-supervised induction to the standard SVM and the transductive SVM [3]. In figure 1, we show the retrieval accuracy with top 20 returns. In figure 2, we show the retrieval accuracy with top 50 returns. As can be seen, semi-supervised induction outperforms both the standard SVM and transductive SVM. The standard SVM only achieved 51% accuracy after 8 iterations in the top 20 returns; transductive SVM achieved 54% accuracy; and semi-supervised induction achieved 63% accuracy, respectively. For the top 50 returns, semi-supervised induction also achieved 6% improvement over transductive SVM and 8% improvement over the standard SVM.

Our experimental results show that semi-supervised induction can make efficient use of unlabelled data.

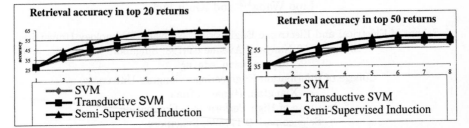

Fig. 1. Retrieval accuracy with top 20 returns.   **Fig. 2.** Retrieval accuracy with top 50 returns.

5   Conclusions

In this paper, we introduce a new semi-supervised learning algorithm which combines the Support Vector Machines and Locality Preserving Projections. The new algorithm maximizes the margin between two classes and simultaneously preserves the local information. The new method explicitly considers the manifold structure. We have applied the new method to image retrieval. Experimental results show that the unlabelled can be used to enhance the retrieval performance.

There are still several questions that remain unclear. For example, it is unclear under those conditions, the local information is especially important. Also, it remains unclear how to define the locality. Specifically, it remains unclear how to determine the parameter k in the k-nearest neighbor search. We are currently working on these problems.

References

1. He, X., Niyogi, P.: Locality Preserving Projections. Advances in Neural Information Processing Systems Vancouver, Canada **16** (2003)
2. He, X., King, O., Ma, W., Li, M., Zhang, H.: Learning a Semantic Space from User's Relevance Feedback for Image Retrieval. IEEE Trans. On Circuit and Systems for Video Technology, **13** (2003)
3. Vapnik, V.: The Nature of Statistical Learning Theory. New York, Springer Verlag (1995)
4. Joachims, T.: Transductive Inference for Text Classification Using Support Vector Machines. Proc. 16th International Conference on Machine Learning, San Francisco, USA (1999) 200-209

A Simple Rule Extraction Method
Using a Compact RBF Neural Network

Lipo Wang[1,2] and Xiuju Fu[3]

[1] School of Electrical and Electronic Engineering, Nanyang Technology University,
Block S1, 50 Nanyang Avenue 639798, Singapore
elpwang@ntu.edu.sg
[2] College of Information Engineering, Xiangtan University,
Xiangtan, Hunan, China
[3] Institute of High Performance Computing
Science park 2 117528, Singapore
fuxj@ihpc.a-star.edu.sg

Abstract. We propose a simple but efficient method to extract rules from the radial basis function (RBF) neural network. Firstly, the data are classified by an RBF classifier. During training the RBF network, we allow for large overlaps between clusters corresponding to the same class to reduce the number of hidden neurons while maintaining classification accuracy. Secondly, centers of the kernel functions are used as initial conditions when searching for rule premises by gradient descent. Thirdly, redundant rules and unimportant features are removed based on the rule tuning results. Simulations show that our approach results in accurate and concise rules.

1 Introduction

The objective of rule extraction is to obtain a comprehensible description of data. As an important task of data mining, rule extraction has attracted much attention [14][16][4][6][7][17].

Usually, a rule consists of an IF part and a THEN part. The premise parts of rules are composed of combinations of attributes. There are three kinds of rule decision boundaries, i.e., hyper-rectangular, hyper-plane, and hyper-ellipse. Due to its explicit form and perceptibility, hyper-rectangular decision boundary is often prefered in rule extraction, such as rules extracted from the MLPs [2][12] and from RBF neural networks [14][4][6]. In order to obtain symbolic rules with hyper-rectangular decision boundaries, a special *interpretable* MLP (IMLP) was constructed in [2]. In an IMLP network, each hidden neuron receives a connection from only one input unit, and the activation function used for the first hidden layer neurons is the threshold function. In [12], the range of each input attribute was divided into intervals.

Due to the local nature of their kernel functions and the global approximation capability, RBF neural networks have been used as tools for rule extraction by many researchers. In [9], before attributes were input into the rule extraction system based on an RBF neural network, redundant inputs were removed;

J. Wang, X. Liao, and Z. Yi (Eds.): ISNN 2005, LNCS 3496, pp. 682–687, 2005.

however, some information was lost with the removal, i.e., the accuracy may be lower. Huber [10] selected rules according to importance; however, the accuracy was reduced with pruning. McGarry [14] extracted rules from the parameters of Gaussian kernel functions and weights in RBF neural networks. When the number of rules was small, the accuracy was low. When the accuracy was acceptable, the number of rules became large.

We propose a simple and yet efficient algorithm to extract rules from the RBF neural network in this paper. The paper is organized as follows. A modification is described for reducing the number of hidden neurons when constructing the RBF classifier (Section 2). In Section 3, the proposed rule extraction algorithm is introduced. We show that our technique leads to a compact rule set with desirable accuracy in experimental simulations (Section 4). Finally, Section 5 presents the conclusions of this paper.

2   An Efficient RBF Classifier

The three-layer RBF neural network [1][18][19] can be represented by the following equations (assume M classes are in the data set):

$$y_m(\mathbf{X}) = \sum_{j=1}^{K} w_{mj}\o_j(\mathbf{X}) + w_{m0}b_m \quad . \tag{1}$$

Here y_m is the m-th output of the network. $\mathbf{X}$ is the input pattern vector (n-dimension). $m = 1, 2, ..., M$. K is the number of hidden neurons, M is the number of output neurons. w_{mj} is the weight connecting the j-th hidden neuron to the m-th output node. b_m is the bias, w_{m0} is the weight connecting the bias and the m-th output node. $\o_j(\mathbf{X})$ is the activation function of the j-th hidden neuron:

$$\o_j(\mathbf{X}) = e^{\frac{-||\mathbf{X}-\mathbf{C_j}||^2}{2\sigma_j{}^2}} \quad , \tag{2}$$

where $\mathbf{C_j}$ and σ_j are the center and the width for the j-th hidden neuron, respectively, and are adjusted during learning.

Two kinds of overlaps are involved in classification with RBF neural networks. One is the overlaps between clusters of different classes, which can improve the performance of the RBF classifier in rejecting noise when tackling noisy data [13]. In [11] and [15], overlapping Gaussian kernel functions are created to map out the territory of each cluster with a less number of Gaussians. Different with previous methods, we propose to allow overlaps between clusters of the same class. Let us consider a simple clustering example shown in Fig.1(a). suppose cluster A has been formed and its members "removed" from the data set V. Suppose pattern 2 is subsequently selected as the initial center of a new cluster and cluster B is thus formed with a pre-defined θ. Clusters C through G are then formed similarly in sequence. We see that clusters B, C, and D are quite small and therefore the effectiveness of the above clustering algorithm needs to be improved.

For implementing the proposed modification, a copy V_c of the original data set V is generated first. When a cluster that satisfies θ-criterion (a qualified cluster), e.g., cluster A in Fig.1(b) (same as cluster A in Fig.1(a)), is generated, the members in this cluster are "removed" from the copy data set V_c. The patterns in the original data set V remain unchanged. Subsequently, the initial center of the next cluster is selected from the *copy data set* V_c; however, the candidate members of this cluster are patterns in the *original data set* V, thus include the patterns in cluster A. Subsequently, when pattern 2 is selected as an initial cluster center, a much larger cluster B, which combines clusters B, C, and D in Fig.1(a) can still satisfy the θ-criterion and can therefore be created.

By allowing for large overlaps between clusters for the same class, we obtain more efficient RBF networks while maintaining classification accuracy, i.e., the number of hidden neurons can be reduced without sacrificing classification accuracy.

3   Rule Extraction Based on RBF Classifiers Using Gradient Descent

The jth premise of Rule i (corresponding to hidden neuron i) is: *if input j is within the interval* (L_{ji}, U_{ji}). If an input pattern satisfies n premises, then the class label of the input pattern is k_i.

Before starting the tuning process, all of the premises of the rules must be initialized first. Let us assume that the number of attributes is n. The number of initial rules equals to the number of hidden neurons in the trained RBF network. The number of the premises of rules equals to n. The upper limit $U(j, i)$ and the lower limit $L(j, i)$ of the jth premise in the ith rule are initialized according to the trained RBF classifier as $U_0(j, i) = \mu_{ji} + \sigma_i$ and $L_0(j, i) = \mu_{ji} - \sigma_i$, respectively. μ_{ji} is the jth item of the center of the ith kernel function. σ_i is the width of the ith kernel function.

The upper limit $U(j, i)$ and the lower limit $L(j, i)$ of the jth premise in the ith rule are

$$U_{t+1}(j, i) = U_t(j, i) + \Delta U_t(j, i) \quad , \tag{3}$$

$$L_{t+1}(j, i) = L_t(j, i) + \Delta L_t(j, i) \quad , \tag{4}$$

where $\Delta U_t(j, i) = \eta f_t(\frac{\partial E}{\partial U(j,i)})$ and $\Delta L_t(j, i) = \eta f_t(\frac{\partial E}{\partial L(j,i)})$. η is the tuning rate. Initially $\eta = 1/N_I$, where N_I is the number of iteration steps for adjusting a premise. N_I is set to be 20 in our experiments. E is the rule error rate. If the sign of $f_{t-1}(\frac{\partial E}{\partial U(j,i)})$ is different with the sign of $f_t(\frac{\partial E}{\partial U(j,i)})$, $\eta = 1.1\eta$.

$$f_t\left(\frac{\partial E}{\partial U(j,i)}\right) = \begin{cases} 1 & , & \frac{\partial E}{\partial U(j,i)}|_t < 0 \\ -1 & , & \frac{\partial E}{\partial U(j,i)}|_t > 0 \\ f_{t-1}\left(\frac{\partial E}{\partial U(j,i)}\right) & , & \text{if } \frac{\partial E}{\partial U(j,i)}|_t = 0 \\ -f_{t-1}\left(\frac{\partial E}{\partial U(j,i)}\right) & , & \text{if } \frac{\partial E}{\partial U(j,i)}|_t = 0 \text{ for } \frac{1}{3}N_I \text{ consecutive iterations.} \end{cases} \tag{5}$$

$$f_t\left(\frac{\partial E}{\partial L(j,i)}\right) = \begin{cases} -1 & , \quad \frac{\partial E}{\partial L(j,i)}\big|_t < 0 \\ 1 & , \quad \frac{\partial E}{\partial L(j,i)}\big|_t > 0 \\ f_{t-1}\left(\frac{\partial E}{\partial L(j,i)}\right) & , \quad \text{if } \frac{\partial E}{\partial L(j,i)}\big|_t = 0 \\ -f_{t-1}\left(\frac{\partial E}{\partial L(j,i)}\right) & , \quad \text{if } \frac{\partial E}{\partial L(j,i)}\big|_t = 0 \text{ for } \frac{1}{3}N_I \text{ consecutive iterations.} \end{cases} \tag{6}$$

Two rule tuning stages are used in our method. In the first tuning stage, the premises of m rules (m is the number of hidden neurons of the trained RBF network) are adjusted using gradient descent for minimizing the rule error rate. Redundant rules which do not make contribution to the improvement of the rule accuracy will be removed. Before implementing the second tuning stage, unimportant features will also be removed. With fewer rules and premises, remained rules will be adjusted using gradient descent again. High rule accuracy can be obtained.

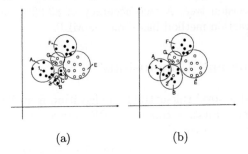

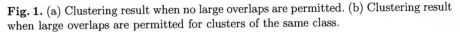

(a)   (b)

Fig. 1. (a) Clustering result when no large overlaps are permitted. (b) Clustering result when large overlaps are permitted for clusters of the same class.

4   Experimental Results

We use the Breast cancer and Glass data sets available at UCI database to test our method. To avoid missing attributes and for comparison, only attributes 2, 3, 4, 5, 6, 7 and 8 were used in the Glass data set. There are and 9 attributes, and 699 patterns in the Breast cancer data set. 16 patterns with missing attribute are removed. Of the 683 patterns left, 444 were benign, and the rest were malign. In these 683 patterns, 274 patterns are used for training, 204 for validation, and 205 for testing.

We set the ratio between the number of in-class patterns and the total patterns in a cluster as $\theta = 85\%$ (the θ-criterion) in our experiments. Table 1 shows that when large overlaps among clusters of the same class are permitted, both the number of hidden neurons and the classification error rate are decreased. The classification results shown in Table 1 are the average values of 5 independent experiments with randomly selected initial cluster centers.

We obtain 4 symbolic rules for Breast cancer data set. The average number of premises in each rule is 2. The accuracy of the symbolic rules that we obtain through the proposed method is 96.35% for the testing data set. There are 6

Table 1. Reduction in the number of hidden units in the RBF network when large overlaps are allowed between clusters for the same class.

Results		Breast	Glass
classification accuracy	small overlap	97.08%	78.41%
	large overlap	98.54%	85.09%
number of hidden units	small overlap	34	13
	large overlap	11	10

rules for the Glass data set. The average number of premises in each rule is 3.33, and the accuracy of extracted rules is 86.21%.

Setiono [16] extracted 2.9 rules and obtained 94.04% accuracy for the Breast cancer data set based on the pruned MLP. In [8], 2 rule extraction results are shown for the same the Glass data set. A rule accuracy of 83.88% was obtained based on C4.5 decision tree. A rule accuracy of 83.33% was obtained by the GLARE rule extraction method based on the MLP.

5   Discussions and Conclusions

We have proposed a novel rule-extraction algorithm from RBF networks based on gradient descent. First, a compact RBF network is obtained by allowing for large overlaps among the clusters belonging to the same class. Second, the rules are initialized according to the training result. Next, premises of each rule are tuned using gradient descent. The unimportant rules and features which do not affect the rule accuracy are removed from the rule set. Experimental results show that our rule extraction technique is simple to implement, and concise rules with high accuracy are obtained. In addition, rules extracted by our algorithm have hyper-rectangular decision boundaries, which is desirable due to its explicit perceptibility.

References

1. Bishop, C.M.: Neural Network for Pattern Recognition. Oxford University Press, New York (1995)
2. Bologna, G., Pellegrini, C.: Constraining the MLP Power of Expression to Facilitate Symbolic Rule Extraction. IEEE World Congress on Computational Intelligence, **1** (1998) 146-151
3. Fu, X.J., Wang, L., Chua, K.S., Chu, F.: Training RBF Neural Networks on Unbalanced Data. Proc. 9th International Conference on Neural Information Processing (ICONIP 2002), **2** (2002) 1016-1020
4. Fu, X.J., Wang L.: Rule Extraction from an RBF Classifier Based on Class-Dependent Features. Proc. 2002 IEEE Congress on Evolutionary Computation (CEC 2002), **2** (2002) 1916 - 1921
5. Fu, X.J., Wang L.: A GA-Based Novel RBF Classifier with Class-dependent Features. Proc. 2002 IEEE Congress on Evolutionary Computation (CEC 2002), **2** (2002) 1890 - 1894

6. Fu, X.J., Wang L.: Rule Extraction Based on Data Dimensionality Reduction. Proc. 2002 IEEE International Joint Conference on Neural Networks (IJCNN 2002), **2** (2002) 1275 - 1280
7. Fu, X.J., Wang L.: Rule Extraction by Genetic Algorithms Based on a Simplified RBF Neural Network. Proc. 2001 IEEE Congress on Evolutionary Computation (CEC 2001), **2** (2001) 753-758
8. Gupta, A., Sang, P., Lam, S.M.: Generalized Analytic Rule Extraction for Feedforward Neural Networks IEEE Transactions on Knowledge and Data Engineering, **11** (1999) 985 -991
9. Halgamuge, S.K., Poechmueller, W. , Pfeffermann, A., Schweikert, P., Glesner, M.: A New Method for Generating Fuzzy Classification Systems Using RBF Neurons with Extended RCE Learning Neural Networks. Proc. IEEE World Congress on Computational Intelligence, **3** (1994) 1589-1594
10. Huber, K. -P., Berthold, M. R.: Building Precise Classifiers with Automatic Rule Extraction. IEEE International Conference on Neural Networks, **3** (1995) 1263-1268
11. Kaylani, T., Dasgupta, S.: A New Method for Initializing Radial Basis Function Classifiers Systems. Proc. IEEE International Conference on Man, and Cybernetics, **3** (1994) 2584-2587
12. Lu, H.J., Setiono, R., Liu, H.: Effective Data Mining Using Neural Networks. IEEE Transactions on Knowledge and Data Engineering, **8** (1996) 957-961
13. Maffezzoni, P., Gubian, P.: Approximate Radial Basis Function Neural Networks (RBFNN) to Learn Smooth Relations from Noisy Data. Proceedings of the 37th Midwest, Symposium on Circuits and Systems **1** (1994) 553-556
14. McGarry, K.J., MacIntyre, J.: Knowledge Extraction and Insertion from Radial Basis Function Networks. IEE Colloquium on Applied Statistical Pattern Recognition (Ref. no. 1999/063) (1999) 15/1-15/6
15. Roy, A., Govil S., Miranda, R.: An Algorithm to Generate Radial Basis Function (RBF)-like Nets for Classification Problems. Neural networks, **8** (1995) 179-201
16. Setiono, R.: Extracting M-of-N Rules from Trained Neural Networks. IEEE Transactions on Neural Networks, **11** (2000) 512 -519
17. Wang, L., Fu, X.J: Data Mining with Computational Intelligence. Springer, Berlin (2005)
18. Fu, X. J., Wang, L.: Data Dimensionality Reduction with Application to Simplifying RBF Network Structure and Improving Classification Performance. IEEE Trans. System, Man, Cybern, Part B - Cybernetics, **33** (2003) 399-409
19. Fu, X.J, Wang, L.: Linguistic Rule Extraction from a Simplified RBF Neural Network. Computational Statistics, **16** (2001) 361-372

Automatic Fuzzy Rule Extraction
Based on Fuzzy Neural Network*

Li Xiao[1] and Guangyuan Liu[2,**]

[1] Faculty of Computer and Information Science,
Southwest China Normal University, Chongqing 400715, China
xsxiaoli@swnu.edu.cn
[2] School of Electronic and Information Engineering,
Southwest China Normal University, Chongqing 400715, China
liugy@swnu.edu.cn

Abstract. In this paper, a hybrid algorithm based on tabu search (TS) algorithm and least squares (LS) algorithm, is proposed to generate an appropriate fuzzy rule set automatically by structure and parameters optimization of fuzzy neural network. TS is used to tune the structure and membership functions simultaneously, after which LS is used for the consequent parameters of the fuzzy rules. A simulation for a nonlinear function approximation is presented and the experimental results show that the proposed algorithm can generate fewer rules with a lower average percentage error.

1 Introduction

During the past couple of years, the fuzzy neural network (FNN) has emerged as one of the most active and fruitful way for obtaining fuzzy rules. Much previous work has concentrated on quantitative approximation, while paying little attention to the qualitative properties of the resulting rules. But FNN with a large rule base is often not much different from a black-box neural network because of the unnecessarily complex and less transparent linguistic description of the system. So the problems we are often confronted in FNN modeling are how to appropriately decide the number of fuzzy rules and precisely define the parameters of each fuzzy rule. There have been a lot of approaches to resolve this problem. A typical method is firstly to decide the number of rules by resorting to similarity analysis, genetic algorithm (GA) or expert knowledge [1]-[2], [5]. Then, the Gradient techniques [1]-[2], [5] are currently the most widely used method to adjust the parameters. GA is also applied for learning both the structure and parameters in FNN [3]. Since Gradient techniques may get trapped in a local minimum and GA is easy to premature, a hybrid algorithm based on tabu search (TS) and least squares (LS), which is called FNN-based hybrid algorithm (FNN-HA), was investigated for automatic generation of fuzzy rules by optimizing the structure and parameters of FNN simultaneously. TS [4] is a meta-heuristic meth-

---

* Supported by key project of Chinese Ministry of Education (104262) and fund project of Chongqing Science and Technology Commission (2003–7881), China.
** Corresponding author.

J. Wang, X. Liao, and Z. Yi (Eds.): ISNN 2005, LNCS 3496, pp. 688–693, 2005.

odology which has been successfully applied to many fields such as combinational optimization problems, and especially it includes memory structure for escaping local minima. Additionally strategy of diversification and intensification was used to improve performance of TS in this paper.

This paper is organized as follows: The structure of the used FNN is presented in section 2.Section 3 investigates the techniques of the hybrid algorithm. Section 4 and section 5 give the experiment data and conclusion, respectively.

2 Fuzzy Neural Network

The FNN (Fig.1) used in this paper is based on Takagi-Sugeno-Kang (TSK) fuzzy models [11], and the fuzzy inference rule is as follows:

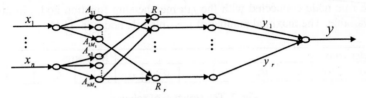

Fig. 1. Structure of FNN.

R_k: if x_1 is A_{1m_1} and x_2 is A_{2m_2},..., and x_n is A_{nm_n}, then $y_k = w_{0k} + w_{1k}x_1 + w_{2k}x_2 + \cdots w_{nk}x_n$. where R_k is the kth rule and n is the number of input invariables. A_{im_i} ($m_i = 1,2,\cdots M_i$) is the fuzzy set of input variable $x_i (i = 1,2,\cdots n)$, where M_i is the number of fuzzy sets defined on x_i . For the TSK model, the consequence y_k is polynomials in the input variables, and w_{ik} is the weight of the ith input variable in kth rule.

The five layers in FNN can be characterized as follows:

Layer 1: The nodes in this layer simply transmit the input values to the next layer.

Layer 2: Each node in this layer implements a particular membership function, which is in the form of Gaussian function used in this paper (1).

$$u_{ij}(x_i) = \exp\left[-(x_i - c_{ij})^2 / \sigma_{ij}^2\right] \quad i = 1,2,\cdots n \quad j = 1,2,\cdots M_i \tag{1}$$

where c_{ij} and σ_{ij} are the center and width parameters corresponding to the $u_{ij}(x_i)$.

Layer 3: Each node called rule node in this layer is used to compute each rule's firing strength (2), where r is the number of rule nodes.

$$f_k = \prod_{i=1}^{n} u_{ij}(x_i) \quad i = 1,2,\cdots n \quad j = 1,2,\cdots M_i \ k = 1,2,\cdots r \ , \tag{2}$$

$$\overline{f_k} = f_k \Big/ \sum_{k=1}^{r} f_k \quad k = 1,2,\cdots r \ . \tag{3}$$

$$y = \sum_{k=1}^{r} \overline{f_k} y_k \ . \tag{4}$$

Layer 4: Nodes in layer 4 are used for normalization (3).

Layer 5: The system output is computed as the sum of all incoming signals (4).

3 The FNN-Based Hybrid Algorithm: FNN-HA

In FNN-HA TS is used to optimize the structure and membership functions simultaneously after which LS is used for the consequent parameters of the rules.

3.1 Membership Functions and Structure Optimization Using TS Algorithm

The main techniques of TS algorithm used in FNN-HA is as follows:

Solution Structure. The solution structure used in this paper includes two parts (Fig.2). The former part is an array of the parameters c_{ij} and σ_{ij} of all the membership functions. The total number of parameters is $2 * \sum_{i=1}^{n} M_i$, and each parameter is represented by real number of itself. The later part is a binary array indicating the rule nodes $R_{j_1 j_2 \cdots j_n}$ ($j_i = 1,2,\cdots M_i$) in layer 3 of the FNN, where 1 and 0 are denoted to the existence and non-existence of the corresponding rule node respectively. $R_{j_1 j_2 \cdots j_n}$ defines the rule node connected with the j_ith membership function node in layer 2 of the ith variable. The maximal number of possible rule nodes is $\prod_{i=1}^{n} M_i$.

Parameter array								Rule array			
c_{11}	σ_{11}	...	c_{21}	σ_{21}	...	c_{n1}	σ_{n1}	...	$R_{11\cdots1}$	$R_{11\cdots2}$	...	$R_{M_1M_2\cdots M_n}$

Fig. 2. The structure of solution.

Neighborhood and Candidate Set. The whole neighborhood is divided into two parts, where the solutions in the former part are called intensification solutions, and the solutions in the later part are called diversification solutions. Both of the solutions are generated by the following operations:

We begin with generating a random value $\delta(i)$ ($0 < \delta(i) < 1$) for each element in the solution. If $\delta(i)$ is bigger than a given threshold P, we add an offset $\lambda(i)$ to the ith element if it is in the parameter array or have the negation operation for the ith element if it is in the rule array. The offset $\lambda(i)$ is generated at random within a given region called *POR (Parameters Offset Region)*. The threshold P of intensification solutions and diversification solutions are P1 and P2 respectively, where P1 is bigger than P2.

Candidate set is also composed of intensification solutions and diversification solutions, both of which are the best solutions selected from the intensification part and diversification part of the neighborhood respectively. The number of intensification and diversification solutions is changed by adaptive search strategy [7] during the whole search process.

Goal Function. As the hybrid algorithm is to make the FNN transparent and interpretable, the goal function G considers two factors: root mean-squared error (*RMSE*) between FNN outputs and desired outputs, and the complexity of the FNN. So the goal function is chosen as,

$$G = RMSE + \alpha * N$$

where α is the a constant which balances the impact of *RMSE* and N, N is the number of rules involved in FNN, and *RMSE* is as follows:

$$RMSE = \sqrt{\sum_{i=1}^{d} (T(i) - O(i))^2 / d}$$

where d is the number of total training pattern, $T(i)$ and $O(i)$ are the desired output and network output for the ith training pattern respectively.

Tabu Objective, Tabu List and Tabu Criterion. The tabu objective used in this paper is the goal function value G, and *tabu list (TL)* is generated by adding the last G to the beginning of the list and discarding the oldest G from the list if TL reached a give size. Since G is a real value, it's inefficient to tabu it directly, so in order for a TL to be used, the *tabu criterion (TC)* showed in [10] is used here. By implementing this TC a G would be considered tabu if it falls in a given range of the values in the TL. For instance, if the TC set to 0.01, the new G would be compared to TL within ±0.01% of the corresponding G in the list.

TS Algorithm Description. The algorithm description is summarized as follows:
Step1. Originate an initial solution x^{now} randomly, and let $TL= \phi$.
Step2. Generate neighborhood $N(x^{now})$ according to the method shown in 3.1
Step3. If the best solution x^{N-best} in the $N(x^{now})$ satisfies the aspiration criteria, then
   $x^{now} = x^{N-best}$, go Step5.
Step4. Select solutions from $N(x^{now})$ to constitute the candidate set $C(x^{now})$ using the method shown in 3.1, and find out the best solution x^{C-best} from $C(x^{now})$, the state of which isn't tabu. Then let $x^{now} = x^{C-best}$.
Step5. Update the TL and the number of intensification solutions and diversification solutions in the candidate set according to the adaptive search strategy.
Step6. Stop when the termination condition is satisfied. Otherwise go back to Step2.

3.2 Determination of Consequent Parameters Using Least Squared Algorithm

Because of the TSK functional rule consequent, least squared (LS) algorithm is used as the consequent parameters learning method, the speed of which is faster than Gradient learning method.

The FNN output of the ith pattern can be calculated by the following equation:

$$O(i) = \sum_{k=1}^{r} f_k * y_k = \sum_{k=1}^{r} f_k * (w_{0k} + w_{1k}x_1(i) + w_{2k}x_2(i) + \cdots w_{nk}x_n(i)) = f(i)w ,$$

where $f(i) = [f_1x(i) \quad f_2x(i) \quad \cdots \quad f_rx(i)]$, $x(i) = [1 \quad x_1(i) \quad \cdots \quad x_n(i)]$, $w = [w(1) \quad w(2) \quad \cdots \quad w(r)]^T$, $w(k) = [w_{0k} \quad w_{1k} \quad \cdots \quad w_{nk}]$, $k = 1,2,\cdots r$.
Let $T(i) = f(i)w + e(i)$, where $e(i)$ is the estimation error of the ith pattern .We have
$T = [T(1) \quad T(2) \quad \cdots \quad T(d)]^T$, $\phi = [f(1)^T \quad f(2)^T \quad \cdots \quad f(d)^T]^T$, $E = [e(1) \quad e(2) \quad \cdots \quad e(d)]^T$.
then $T = \phi w + E$ Based on LS, the w matrix which minimizes E can be estimated using $w = T(\phi^T \phi)^{-1}\phi^T$

4 Simulation

The simulation example is a three-input nonlinear function approximation, which is widely used to verify the approaches adopted in [8]-[9]:
$$t = (1 + x^{0.5} + y^{-1} + z^{-1.5})^2 .$$

A total of 216 training data are randomly chosen from the input ranges $[1,6]\times[1,6]\times[1,6]$.
3 membership functions are initially defined in each input variable, so the number of
possible rules is 27. Using FNN-HA the number of generated rules is reduced to 7,
and the invariables x, y, and z have 2, 3, and 3 membership functions, respectively.
The search processes are shown in Fig.3 and Fig.4, and 7 rules are shown in Table 1.

To compare the performance, we adopt the same performance index in [8]-[9]:

$$APE = \frac{1}{d}\sum_{i=1}^{d}\left(|T(i)-O(i)|/|T(i)|\right)\times100\% \ .$$

Another 125 data are randomly selected from the same operating range to check
the generalization of the FNN optimized by FNN-HA. Comparisons with ANFIS [8],
orthogonal least squares (OLS) [9] and generalized dynamic FNN (GD-FNN) [9] are
shown in Table 2, where the APE_{trn} and APE_{chk} indicate the training APE and checking
APE respectively. We see from Table 2 that FNN-HA has the smallest checking error
and the least parameters although its training error is larger than the ANFIS.

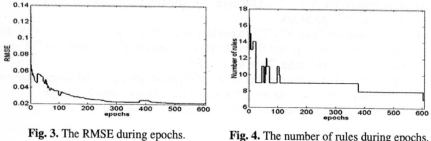

Fig. 3. The RMSE during epochs.   **Fig. 4.** The number of rules during epochs.

Table 1. Fuzzy rules generated with FNN-HA.

Number of fuzzy rules	Premise parameters			Consequent parameters
	x	y	z	
1	A (1.533 3.031) *	A (2.353 4.803)	A (3.044 3.526)	t=2.458x+3.764y−1.609z+ 67.36
2	A (1.533 3.031)	A (2.353 4.803)	A (2.484 4.395)	t=0.5675x −0.2217y+1.167z+2.978
3	A (1.533 3.031)	A (1.966 2.991)	A (3.044 3.526)	t=1.463x−9.931y−5.739z+107.4
4	A (1.533 3.031)	A (1.966 2.991)	A (2.484 4.395)	t=1.264x−0.2115y+4.339z−2.936
5	A (1.533 3.031)	A (3.795 -0.178)	A (1.077 -1.460)	t=9.824x+38.62y−16.85z+64.31
6	A (1.533 3.031)	A (3.795 -0.178)	A (3.044 3.526)	t=2.802x−49.51y−0.1041z+1.569
7	A (2.240 0.436)	A (1.966 2.991)	A (2.484 4.395)	t=1.364x−0.6705y−0.09399z+4.913

$*A(\sigma,c)$ indicates a Gaussian membership function with the width σ and the center c .

Table 2. Comparisons of the FNN-HA with other methods.

Model	Training set size	Checking set size	$APE_{trn}(\%)$	$APE_{chk}(\%)$	Parameter number
ANFIS	216	125	0.043	1.066	50
OLS	216	125	2.430	2.56(e_m=1.3949)*	66
GD-FNN	216	125	2.110	1.54(e_m=0.8781)	64
FNN-HA	216	125	0.140	0.21(e_m=0.1655)	44

* e_m indicates the maximum difference between the desired output and the calculated output.

5   Conclusion

In this paper, a novel hybrid algorithm based on TS and LS is proposed to generate an
appropriate fuzzy rule set automatically through structure and parameters optimiza-

tion of fuzzy neural network. The simulation results have shown that using the hybrid algorithm FNN-HA, the redundant fuzzy rules can be pruned efficiently so as to obtain a more compact FNN structure with the higher accuracy.

References

1. Setnes, M., Koene, A., Babuska, R., Bruijn, P.M.: Data-Driven Initialization and Structure Learning in Fuzzy Neural Networks. IEEE World Congress on Computational Intelligence, Fuzzy Systems Proceedings, Vol.2 (1998) 1147–1152
2. Ma, M., Xu, Y. and Zhang, L.: Automatic Fuzzy Rule Extraction Based Weighted Fuzzy Neural Networks. Computer Application, Vol.23 (2003) 15–17 (in Chinese with English Abstract)
3. Zhou, Z.J., Mao, Z.Y.: On Designing an Optimal Fuzzy Neural Network Controller Using Genetic Algorithms. Proceedings of the 3rd World Congress on Intelligent Control and Automation, Hefei, China (2000) 391-395
4. Glover F. and Laguna M.: Tabu Search. Boston, Kluwer Academic Publishers (1997)
5. Y. H. Lin and G. A. Cunningham III: A New Approach to Fuzzy-neural System Modeling. IEEE Trans. on Fuzzy Syst., Vol.3 (1995) 190–198
6. Guangyuan Liu, Yonghui Fang, Xufei Zheng, et al.: Tuning Neuro-fuzzy Function Approximator by Tabu Search. International Symposium on Neural Networks, Dalian, China (2004) Part 1, 276–281
7. Yi He, Guangyuan Liu, and Yuhui Qiu: A Novel Adaptive Search Strategy of Intensification and Diversification in Tabu Search. Journal of Computer Research and Development, Vol.41 (2004) 162–166 (in Chinese with English Abstract)
8. J.-S. R. Jang: ANFIS: Adaptive-Network-Based Fuzzy Inference System. IEEE Trans. on Syst., Man and Cybernetics, Vol. 23 (1993) 665–684
9. S.Wu and M. J. Er: A Fast Approach for Automatic Generation of Fuzzy Rules by Generalized Dynamic Fuzzy Neural Networks. IEEE Trans. on Fuzzy Syst., Vol.9 (2001) 578–594
10. R.S. Sexton, B. Alidaee, R.E. Dorsey, et al.: Global Optimization for Artificial Neural Networks: A Tabu Search Application. European Journal of Operational Research 106 (1998) 570–584
11. T. Takagi, and M. Sugeno: Fuzzy Identification of Systems and Its Applications to Modeling and Control. IEEE Trans. on Syst., Man and Cybernetics, Vol.15 (1985) 116–132

Neural Networks for Nonconvex Nonlinear Programming Problems: A Switching Control Approach

Changyin Sun[1,2] and Chunbo Feng[2]

[1] College of Electrical Engineering, Hohai University, Nanjing 210098, China
[2] Research Institute of Automation, Southeast University, Nanjing 210096, China
cysun@ieee.org

Abstract. In this paper, neural networks based on switching control approach are proposed, which is aimed at solving in real time a much wider class of nonconvex nonlinear programming problems where the objective function is assumed to satisfy only the weak condition of being regular functions. By using the gradient of the involved functions, the switching control approach proposed is shown to obey a gradient system of differential equation, and its dynamical behavior, trajectory convergence in finite time, and optimization capabilities, for nonconvex problems, are rigorously analyzed in the framework of the theory of differential equations, which are expected to enable to gain further insight on the geometrical structure of the energy landscape (objective function) of each specific class of nonlinear programming problems which is dealt with.

1 Introduction

Neural Networks have been shown to be effective for finding good solutions to difficult optimization problems. A Lyapunov function, determined by the network's interconnection weight matrix, bias current vector, and output functions has been found for the neural network model we consider [1-7], and techniques have been developed for mapping a number of optimization problems onto neural networks such that the problem's optimal solution is represented by the lowest point on this Lyapunov energy surface [4],[5]. However, for nonconvex problems, these mappings also introduce numerous local minima representing infeasible state or poor feasible solutions. These local minima present a difficult challenge for the network in finding an optimal, or a nearly optimal solution.

Two methods that attempt to deal with the problem presented by local minima are mean field annealing and tabu searching [5]. In mean field annealing method, sigmoid output functions are used whose slopes are gradually increased until the state converges to a local minimum of the final energy surface that results from threshold output functions. An improvement to mean field annealing is the convex relaxation method [5]. In convex relaxation, the neural currents are then gradually modified until the energy surface become nonconvex where the global minimum solves the problem. In contrast to the above methods that use gradual cooling to arrive at a single solution, tabu searching instead employs searching to travel through many solutions on the nonconvex energy surface.

In this paper, a two-phase optimization neural network model based a switching control approach is investigated. The model is composed of two phases, a function minimization phase and a phase seeking initial point. The algorithm starts from any

J. Wang, X. Liao, and Z. Yi (Eds.): ISNN 2005, LNCS 3496, pp. 694–699, 2005.
© Springer-Verlag Berlin Heidelberg 2005

initial point, then a function minimization phase run, so a local minimum is obtained. If the local minimum is not global minimum, the algorithm switches to the second phase, then the new initial point will be obtained. Repeating this procedure, the algorithm may obtain the minima and initial point sequentially. In other words the algorithm always searches the minimum at which the value of the function is less than or equals to that at the latest found minimum. The algorithm stops till the global minimum obtains. The performance of the model is analyzed rigorously. The conclusion is drawn that the global minimum can obtain using this model, no matter what the function has finite local minimum point or infinite local minimum point. Illustrative result shows the effectiveness of the model proposed.

2 Neural Networks for Nonconvex Programming Problems

Let us consider the nonlinear programming problem

$$\text{minimize } f(x), \quad x \in R^n \tag{1}$$

For simplicity, let $z = f(x)$. Suppose that there exists $M \in R$ such that $f(x) > M$. In many literatures [4], the function $f(x)$ is assumed to be convex and twice continuously differentiable. Here, the function $f(x)$ is only assumed to be twice continuously differentiable. Therefore the function $f(x)$ may be convex or nonconvex.

To solve (1), we consider a recurrent neural network model below

$$\begin{cases} \dot{x} = -W\nabla f(x), \\ x_0 = \overline{x}, \forall \overline{x} \in R^n \end{cases} \tag{2}$$

where $\overline{x}$ is any initial state and $x \in R^n$; W is a certain symmetry positive definite matrix. According to the function $f(x)$ is assumed to be twice continuously differentiable, there exists one solution of (2).

Let $L(x_0) = \{x | f(x) \le f(x_0)\}$, then $L(x_0)$ is bounded set. Similar to the results in [6], we further give the following results.

Theorem 1. When $t \to \infty$, trajectory $x(t, x_0)$ of (2) tend to an equilibrium point x^* of (2). That is,

$$\lim_{t \to \infty} x(t, x_0) = x^*, \quad \forall x_0 \in R^n$$

and $\nabla f(x^*) = 0$.

Proof. $\forall x_0 \in R^n$, there exists unique trajectory $x(t, x_0)$ of (2). Along this trajectory, it must follows,

$$\frac{df}{dt} = \nabla f^T \frac{dx}{dt} = -\nabla f(x)^T W \nabla f(x) \le 0. \tag{3}$$

Clearly to see along $x(t, x_0)$, $f(x(t, x_0))$ is monotonic decreasing. So, $\gamma^+(x_0) = \{x(t, x_0) | t \ge 0\} \subseteq L(x_0)$. Note $L(x_0)$ is bounded, $\gamma^+(x_0)$ is bounded.

Select strict monotonic increasing number sequences $\{\overline{t}_n\}: 0 \leq \overline{t}_1 < \cdots < \overline{t}_n < \cdots$,
$\overline{t}_n \to +\infty$, then $\{x(\overline{t}_n, x_0)\}$ is bounded point set, therefore there exists x^*, such that

$$\lim_{n \to \infty} x(\overline{t}_n, x_0) = x^*, \quad \forall x_0 \in R^n$$

From (3), it follows that

$$\frac{df}{dt} = 0 \Leftrightarrow \nabla f(x) = 0 .$$

If let S be invariant set of equilibrium points, by the Lasalle Principle, when $t \to \infty$, there must be $x(t, x_0) \to x^* \in S$, $\nabla f(x^*) = 0$. Therefore, from any initial state x_0, the trajectory $x(t, x_0)$ of (2) tend to x^*.

From Theorem 1, if x^* is global minima of (1), the neural network model run best. But unfortunately, it is often local minima. Therefore, we must seek a method to climb out of local minima. In the following, a two-phase optimization neural network model based a switching control approach is investigated to meet this gap.

Supposed x^* is one of equilibrium points of $z = f(x)$. The tangent plane equation through x^* is $z = f(x^*)$. In order to escape from local minima, we cut the energy landscape (object function) $z = f(x)$ by the tangent plane. Let

$$h(x) = f(x) - f(x^*) + \varepsilon \geq 0 \tag{4}$$

where $\varepsilon > 0$. If for some ε, there exists zero solution of (4), similar to the result in [6], we give the following Theorem.

Theorem 2. x^* is global minima of $f(x)$, if and only if for any $\varepsilon > 0$, there doesn't exist solution of (4).

In order to solve (4), we consider a recurrent neural network model below

$$\begin{cases} \dot{\hat{x}} = -Dh(\hat{x}), \\ \hat{x}_0 = x^* + \varepsilon_1, \forall \varepsilon_1 \in R^n. \end{cases} \tag{5}$$

where $\hat{x}$ is any initial state and $\hat{x} \in R^n$; D is a certain symmetry positive definite matrix.

3 Trajectory Convergence in Finite Time

In this section, we give a proof of trajectory convergence in finite time. Suppose that there exist two strict monotonic increasing number sequences $\{t_{in}\}$ and $\{t_{jn}\}$, $0 = t_{i0} < t_{j0} < t_{i1} < t_{j1} \cdots < t_{in} < t_{jn} < \infty$, $t_{in} \to +\infty$, $t_{jn} \to +\infty$, such that $x(t_{in})$ is initial state of the n th two-phase optimization process, and $\nabla f(x(t_{jn})) = 0$, $f(x(t_{j(n+1)})) < f(x(t_{jn}))$, $n = 1, 2, 3, \cdots$.

Lemma 1. Suppose trajectory of (2) is $x_n(t) = x(t; t_{in}, x_{in})$; $t \in [t_{in}, t_{jn}]$, $n = 0, 1, 2, 3, \cdots$. Let

$$l = \sum_{n=0}^{\infty} \int_{t_{in}}^{t_{jn}} \|\dot{x}_n(t)\|^2 \, dt$$

then

$$l \le \frac{1}{\lambda_{\min}(W^{-1})} (f(x_0(t_{i0})) - M)$$

Proof. From (2), we can obtain

$$\frac{df}{dt} = \nabla f^T \frac{dx}{dt} = -(-W\nabla f)^T W^{-1} \frac{dx}{dt} = -\left(\frac{dx}{dt}\right)^T W^{-1} \frac{dx}{dt}.$$

So,

$$\dot{f} \le -\lambda_{\min}(W^{-1})\|\dot{x}\|^2,$$

Therefore

$$\int_{t_{in}}^{t_{jn}} \|\dot{x}_n(t)\|^2 \, dt \le -\frac{1}{\lambda_{\min}(W^{-1})} \int_{t_{in}}^{t_{jn}} \dot{f}(x_n(t)) dt = \frac{1}{\lambda_{\min}(W^{-1})} (f(x_n(t_{in})) - f(x_n(t_{jn}))),$$

then

$$l = \sum_{n=0}^{\infty} \int_{t_{in}}^{t_{jn}} \|\dot{x}_n(t)\|^2 \, dt = \frac{1}{\lambda_{\min}(W^{-1})} \Big[f(x_0(t_{i0})) + (f(x_1(t_{i1})) - f(x_0(t_{j0}))) + \cdots$$

$$+ (f(x_{n+1}(t_{i(n+1)})) - f(x_n(t_{jn}))) + \cdots - f(x_\infty(t_{j\infty})) \Big]$$

$$\le \frac{1}{\lambda_{\min}(W^{-1})} (f(x_0(t_{i0})) - f(x_\infty(t_{j\infty})))$$

$$\le \frac{1}{\lambda_{\min}(W^{-1})} (f(x_0(t_{i0})) - M).$$

Thus the proof of the Lemma 1 is completed.

Theorem 3. Suppose trajectory of (2) is $x_n(t) = x(t; t_{in}, x_{in})$; $t \in [t_{in}, t_{jn}]$, $n = 0, 1, 2, 3, \cdots$. The switching control approach for nonconvex programming problem proposed in Section 2 can run till global minima of (1) is obtained.

Proof. By Lemma 1, we know that $l = \sum_{n=0}^{\infty} \int_{t_{in}}^{t_{jn}} \|\dot{x}_n(t)\|^2 \, dt$ is bounded. Note that $\int_{t_{in}}^{t_{jn}} \|\dot{x}_n(t)\|^2 \, dt \ge 0$. So when $n \to \infty$, $\int_{t_{in}}^{t_{jn}} \|\dot{x}_n(t)\|^2 \, dt \to 0$. By the Hölder Inequation, it is easily obtained $\int_{t_{in}}^{t_{jn}} \|\dot{x}_n(t)\| dt \to 0$. Therefore, by the fundamental Cauchy Limit Theorem, for $\forall \varepsilon > 0$, there exists N such that for $n > N$, $\int_{t_{in}}^{t_{jn}} \|\dot{x}_n(t)\| dt < \varepsilon$. Therefore, $\|x_n(t_{jn}) - x(t_{in})\| = \left\| \int_{t_{in}}^{t_{jn}} \dot{x}_n(t) dt \right\| < \int_{t_{in}}^{t_{jn}} \|\dot{x}_n(t)\| dt < \varepsilon$. Thus by fundamental

Cauchy Limit Theorem, $\lim_{n\to\infty} x_n(t_{in})$ exists. Let $\lim_{n\to\infty} x_n(t_{in}) = x_{g\min}$. By the fundamen-

tal Heine Limit Theorem, we obtain $\lim_{n\to\infty} x_n(t_{in}) = \lim_{t\to\infty} x_n(t) = x_{g\min}$. With the

continuity of $f(x)$, we obtain

$$\lim_{t\to\infty} f(x_n(t)) = f\left(\lim_{t\to\infty}(x_n(t))\right) = f(x_{g\min}).$$

On the other hand, by $f(x_n(t_{j(n+1)})) < f(x_n(t_{jn}))$ and $f(x) > M$,

$$\lim_{t\to\infty} f(x_n(t)) = \inf_{x\in R^n} f(x).$$

Therefore,

$$f(x_{g\min}) = \inf_{x\in R^n} f(x),$$

That is, $x_{g\min}$ is one of global minima point of $f(x)$.

4 Experimental Result

The following example is illustrated to verify the effectiveness of the method proposed. We consider the following global optimization problem.

$$\min_x \frac{1}{2}\sum_{i=1}^{2}\left(x_i^4 - 16x_i^2 + 5x_i\right)$$

It is easily to verify that there exist at least four local minima points, and $x^* = [-2.912682, -2.912682]$ is global minima point.

Let $f(x) = \frac{1}{2}\sum_{i=1}^{2}\left(x_i^4 - 16x_i^2 + 5x_i\right)$, $h(x) = f(x) - f(x^*) + \varepsilon$, $W = D = I$, $\varepsilon = 0.01$.

Firstly,

Select the initial state $x_0 = [4, -6]$, then $x_1^* = [2.7468. -2.9036]$, at this point $f(x) = -64.1956$.

Secondly.

Select $\varepsilon = 0.01$, then $\hat{x} = [-3.4552, -2.1297]$.

Thirdly,

Select new initial state $x_0 = [-3.4552, -2.1297]$, then $x_2^* = [-2.9035. -2.9035]$, at this point $f(x_2^*) = -78.3323$.

Repeat the process above, the function value of $f(x)$ doesn't descend. We can obtain the global minima of $f(x)$ is $f(x_2^*) = -78.3323$.

5 Conclusions

In this paper, neural networks based on switching control approach are introduced, which is aimed at solving in real time a much wider class of nonconvex nonlinear programming problems by using the gradient of the involved functions, the switching control approach proposed is shown to obey a gradient system of differential equation, and its dynamical behavior, trajectory convergence in finite time, and optimization capabilities, for nonconvex problems, are rigorously analyzed in the framework of the theory of differential equations, which are expected to enable to gain further insight on the geometrical structure of the energy surface (objective function) of each specific class of nonlinear programming problems which is dealt with.

Acknowledgements

This work was supported by the China Postdoctoral Science Foundation, Jiangsu Planned Projects for Postdoctoral Research Funds and a start-up grant from Hohai University of China.

References

1. Forti, M., Nistri, P., Quincampoix, M.: Generalized Neural Networks for Nonsmooth Nonlinear Programming Problems. IEEE Trans. Circuits and Systems I, **51** (2004) 1741-1754
2. Sudharsanan, S., Sundareshan, M.: Exponential Stability and a Systematic Synthesis of a Neural Network for Quadratic Minimization. Neural Networks, **4** (1991) 599-613
3. Kennedy, M., Chua, L.: Neural Networks for Linear and Nonlinear Programming. IEEE Trans. Circuits & Systems, **35** (1988) 554-562
4. Xia, Y., Wang, J.: Global Exponential Stability of Recurrent Neural Networks for Solving Optimization and Related Problems. IEEE Transaction on Neural Networks, **11** (2000) 1017-1022
5. Beyer, D., Ogier, R.: Tabu Learning: A Neural Networks Search Method for Solving Nonconvex Optimization Problems. In: Proceedings of IEEE (2000) 953-961
6. Zhao, H., Chen, K.: Neural Networks for Global Optimization. Control Theory and Applications, **19** (2002) 824-828
7. Forti, M., Tesi, A.: A New Method to Analyze Complete Stability of PWL Cellular Neural Networks. Int. J. Bifurcation Chaos, **11** (2001) 655-676

From the instinctive view, the point X_i^{new} is in a flat basin or a narrow valley where X_i^{opt} is inside. The objective function has the same values at the basin or valley bottom, which compose a local optimum set. X_i^{new} is also said to be within the neighbourhood of X_i^{opt}, if

$$Opt(X_i^{new}) = X_i^{opt}. \tag{6}$$

The WFP algorithm simulates the Huygens theory, which stated that an expanding sphere of light behaves as if each point on the wave front were a new source of radiation of the same frequency and phase. The wave front repeats propagation from the new point source X_i^{new} at the currently known boundary until the in-neighbourhood condition in Definition 2 is not satisfied in a certain dimension. Then the neighbourhood boundary point is determined and no more searching will be continued in this dimension. When the neighbourhood points are determined in all dimensions, the whole local optimum X_i^{opt} neighbourhood boundary is determined as the envelopment surface of all the wave fronts. After an iterative neighbourhood boundary search, the determined neighbourhood D_i is found and eliminated from definition space, which prevents redundant local optimization and cuts down the computation time. But the irregularity of the undivided solution domain leads to a more difficult situation for further optimization. Neural networks provide a powerful tool for solving complex non-linear optimization problems.

2.2 Levenberg-Marquardt Neural Network Predictions

Artificial neural networks have very close ties to optimization. Neural Networks (NNs) have been studied since 1943 [8],[9]. The original description of the Levenberg-Marquardt algorithm was given in [10] and its application to neural network training was described in [11]-[13]. In the present study, a Levenberg-Marquardt neural network-based predictor was implemented to provide a mapping tool from the local optimum to the neighbourhood boundary in irregular-shaped nonconvex domains. The outstanding advantage of NN prediction in this problem is that it eliminates the unnecessary costly WFP searching iterations in known solution space.

The neural network-based neighbourhood boundary algorithm consisted of three stages. First, the NN was trained by the relationship between neighbourhood points and the corresponding local optimums, which generated the multi-layer neural network weights. Secondly, the NN model was used to replace the wave front propagation and provided the mapping between boundary points and local optimums. Then the predicted neighbourhood was deleted from the solution space and more boundary information became known as time passes. As a result, the NN model was dynamically retrained and corrected so as to match all the complex local characteristics of the objective function. The neighbourhood boundary point coordinates were passed forward from the input layer to the output layer through a hidden layer with six nodes (Fig. 2). The number of the output nodes was two for the local optimum point coordinates for 2D problems. A hyperbolic tangent function was used as transfer function at each node of the network. An explanation of the training and simulating process is provided in the following section.

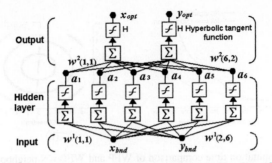

Fig. 2. The Levenberg-Marquardt neural network with one hidden layer for 2D problems.

3   Numerical Examples

A nonlinear function with a complex surface was used to test the new algorithm. Although a much larger real data example could be used to test the limits of the algorithm, it is more important to illustrate code accuracy and efficiency by a relatively small typical example. The example was run on a PC with an Intel Pentium CPU (650MHz) and 128 MB memory. The program was implemented in C++ language.

Test function:

$$F(x,y) = 0.5 - \frac{(\sin^2(\sqrt{x^2+y^2})-0.5)}{\left[1+0.001(x^2+y^2)\right]^2}.$$

(7)

The definition domain was $x \in [-100,100]$, $y \in [-100,100]$. The global maximum value 1 was located at (0,0). There were infinite secondary global maximums around the actual one (Fig. 3a). Because of the nature of oscillation, it is very difficult to find the global maximum by conventional optimization methods. In this example, a subdomain $x \in [-8,8]$, $y \in [-8,8]$ containing the global maximum was chosen as the searching space. The minimum searching step was set to 0.05. By using the WFT algorithm alone, 317 local optimums (sets) were found, as shown in Fig. 3b. The global maximum 1 was determined at (-8.375587e-009, -3.517774e-009) in 12.45 seconds. As more neighbourhoods were deleted from the searching space, the computation time increased greatly in the undivided irregular-shaped space. Among the total 317 local optimums (sets), the first 150 local optimums (sets) took only 17% of the total computation time. The last 167 local optimums (sets) took nearly 83% (Fig. 4a).

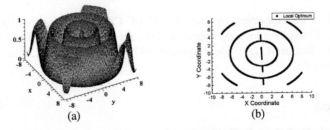

(a)   (b)

Fig. 3. (a) The objective function surface; (b) The local optimums by WFP algorithm alone.

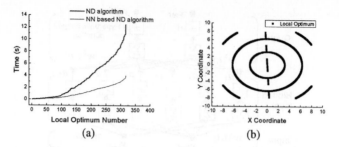

Fig. 4. (a) The computation time comparison of WFP and WFP-NN neighbourhood determination algorithm; (b) The local optimums found by WFP-NN neighbourhood determination.

The NN prediction was used to minimize the final searching time in irregular undivided space. The neural network was trained by the known mapping relationships between boundary points and the local optimums. Since an optimum was surrounded by more than $2n$ (n is the definition space dimension number) boundary points, the input points with the same output optimum composed the new boundary. With the help of the neural network, the final searching in the irregular undivided space became more efficient. The global maximum 1 was determined at (-8.691454e-009, 0.000000e+000) in 4.07 seconds. The local optimum distribution found by the WFP-NN hybrid algorithm was nearly the same as the one by the WFP algorithm. Fig. 3 (b) and Fig. 4 (b) show a good match of the final local optimums (sets) between the NN-based and non-NN-based WFP algorithms.

4   Conclusions

A neural network-based WFP algorithm was proposed in neighbourhood determination global optimization. The mapping relationships between known neighbourhood boundary and local optimums (sets) were simulated by neural network. The neural network prediction was evoked when the iteration of local optimum determination became costly. By predicting a new local optimum and its neighbourhood, unnecessary optimizations were avoided and high efficiency could be achieved, especially in the irregular-shaped undivided space. A numerical example showed an obvious improvement in global optimization by using the neural network-based WFP method.

Acknowledgements

This work was supported by the NNSF of China (Grants No.10402015 and 60473101), China Postdoctoral Foundation (Grant No.2004035309) and the National Key Basic Research and Development (973) Program of China (Grant No. 2004CB318205).

References

1. Backus, G. E., Gilbert, F: The Resolving Power of Gross Earth Data. Geophysical Journal of the Royal Astronomical Society, **16** (1968) 169-205

2. Liu, P., Ji, C., et al. : An Improved Simulated Annealing-Downhill Simplex Hybrid Global Inverse Algorithm. Chinese Journal of Geophysics, **38** (1995) 199-205
3. Ji, C. and Yao, Z. : The Uniform Design Optimized Method for Geophysics Inversion Problem. Chinese Journal of Geophysics, **39** (1996) 233-242
4. Ai, Y., Liu, P. and Zheng, T: Adaptive Global Hybrid Inversion. Science in China (series D), **28** (1991) 105-110
5. Hibbert, D. B.: A Hybrid Genetic Algorithm for the Estimation of Kinetic Parameters. Chemometrics and Intelligent Laboratory Systems, **19** (1993) 319-329
6. Chunduru, R., Sen, M. K., et al.: Hybrid Optimization Methods for Geophysical Inversion. Geophysics, **62** (1997) 1196-1207
7. Macias, C. C., Sen, M. K. et al.: Artificial Neural Networks for Parameter Estimation in Geophysics. Geophysical Prospecting, 48 (2000) 21-47
8. McCulloch, W.S., Pitts, W.H.: A Logical Calculus of Ideas Imminent in Nervous Activity. Bullet. Math. Biophys., **5** (1943) 15-33
9. Hopfield, J., Tank, D.W.: Neural Computation of Decision in Optimization Problems. Biol. Cybernet, **52** (1985) 41-52
10. Marquardt, D.: An Algorithm for Least-Squares Estimation of Nonlinear Parameters, SIAM Journal Applied Math, **11** (1963) 431-441
11. Hagan, M. T., Menhaj, M.: Training Feedforward Networks with the Marquardt Algorithm, IEEE Trans. on Neural Networks, 5 (1994) 989-993
12. Ampazis, N., Perantonis, S. J.: Two Highly Efficient Second-Order Algorithms for Training Feedforward Networks. IEEE Trans. on Neural Networks, 13 (2002) 1064-1074
13. Sun, W.T., Shu, J.W., et al.: Dynamic File Allocation in Storage Area Networks with Neural Network Prediction. Lecture Notes in Computer Science, Vol. **3174**, Springer-Verlag, Berlin Heidelberg New York (2004) 719-724

A Neural Network Methodology of Quadratic Optimization with Quadratic Equality Constraints[*]

Yongqing Yang[1,2], Jinde Cao[2,**], and Daqi Zhu[1]

[1] School of Science, Southern Yangtze University, Wuxi 214063, China
[2] Department of Mathematics, Southeast University, Nanjing 210096, China
{yongqingyang,jdcao}@seu.edu.cn

Abstract. This paper presents a feedback recurrent neural network for solving the quadratic programming with quadratic equality constraint (QPQEC) problems based on project theory and energy function. In the theoretical aspect, we prove that the proposed neural network has one unique continuous solution trajectory and the equilibrium point of neural network is stable and convergent when the initial point is given. Employing the idea of successive approximation and convergence theorem from [6], the optimal solution of QPQEC problem can be obtained. The simulation result also shows that the proposed feedback recurrent neural network is feasible and efficient.

1 Introduction

Many engineering problems can be solved by transforming the original problems into constrained optimization problems, which need to be solved in real-time. However classical methods cannot render real-time solution to these problems, especially in large-scale optimization problems. Neural networks, on the other hand, are capable of producing a general solution method.

Tank and Hopfield in 1986 first proposed a neural network for linear programming that was mapped onto a closed-loop circuit [1]. Since then, many researchers have been attracted to study the ability of the optimization computation based on the neural network methodology, and various optimization network models and computational techniques have been reported. Special, Kennedy and Chua developed a neural network with both penalty function and the gradient method for solving nonlinear programming problem [2]. To avoid using penalty function, prime-dual neural networks and projection neural networks are proposed for solving linear programming, quadratic programming and convex programming [3–5,7–9], with the linear equality constraint. Nevertheless, there are

---

[*] This work was jointly supported by the National Natural Science Foundation of China under Grant 60373067, the Natural Science Foundation of Jiangsu Province, China under Grants BK2003053 and BK2004021, and the Foundation of Southern Yangtze University.
[**] Corresponding author

J. Wang, X. Liao, and Z. Yi (Eds.): ISNN 2005, LNCS 3496, pp. 706–711, 2005.

fewer neural network models for the quadratic programming with quadratic equality constraint problems(QPQEC), which frequently occur in many applications, such as adaptive filtering, FIR filter design with time-frequency constraints, switched nonlinear dynamic system, etc. All of these optimization computation problems mentioned above can be simplified and then expressed in a quadratic programming with quadratic constraint as

$$\min f(x) = X^T A_0 X + C_0^T X$$
$$s.t. \begin{cases} X^T A_i X = c_i, i = 1, 2, ..., m, \\ l \leq X \leq h, \end{cases} \tag{1}$$

where $A_0, A_1, ..., A_m$ are positive definite and $A_1, A_2, ..., A_m$ are linearly independent. Obviously, the feasible set S will certainly not be convex, and in general will have a number of separate connected components.

In [9], Y. Tan and C. Deng proposed a neural network for solving quadratic objection function and a quadratic constraint based on penalty function and gradient method. Its energy function can be viewed as an inexact penalty function, and the true optimal solutions can only be obtained when the penalty parameter is infinite. In this paper, we assume the feasible set $S = \{X | X \in R^n, X^T A_i X = c_i, i = 1, 2, ..., m.\}$ is not-empty and the optimization solution of Eq.(1) exists. Based on the idea of project theory and energy function, a feedback recurrent neural network is constructed and the exact optimal solutions is obtained by convergence theorem.

This paper is divided into five sections. In Section 2, a recurrent neural network model for solving QPQEC problems is constructed. In Section 3, the stability and convergence of the proposing network is analyzed and the feedback subnetworks are obtained based on convergency. In Section 4, a simulation example is provided. Conclusions are given to conclude the paper in Section 5.

2   The Neural Network of Quadratic Programming with Quadratic Equality Constraints

A neural network is a large dimensional nonlinear dynamic system composed of neurons. From the viewpoint of a dynamic system, the final behavior of the system is fully determined by its attraction points, if they exist. In order to construct a neural network of QPQEC problem, the key step is to define energy function $E(X, M)$ such that the lowest energy state corresponds to the optimal solution X^* of (1). Based on the energy function, we propose the feedback recurrent neural subnetworks for solving (1).

Take M_1 as a lower bound of the optimal value of QPQEC, i.e. $M_1 \leq f(X^*)$. Let $d(X, M_1) = X^T A_0 X + C_0^T X - M_1$ and $F(X, M_1) = d(X, M_1)[d(X, M_1) + |d(X, M_1)|]/2$, we have $E(X, M_1) = F(X, M_1) + \frac{1}{2} \sum_{i=1}^{m} (X^T A_i X - c_i)^2$.

First, consider the following optimization problem with a bound constraint

$$\text{minimize } E(X, M_1)$$
$$\text{s.t. } X \in \Omega \tag{2}$$

where $\Omega = \{X \in R^n | l_i \le x_i \le h_i, i = 1, 2, ..., n\}$. The corresponding neural network for solving (2) is described by

$$\frac{dX}{dt} = -\{X - P_\Omega(X - \nabla E(X, M_1))\}, \tag{3}$$

By the result given in [6,7], the following theorem is immediately obtained.

Theorem 1. (i) $E(X, M_1)$ is a nonnegative, continuously differential convex function.

(ii) The gradient $\nabla E(X, M_1)$ of $E(X, M_1)$ is monotone and locally Lipschitz continuous. The Hessian matrix $\nabla^2 E(X, M_1)$ is symmetrical and positive semi-define.

Theorem 2. (i) For any initial point $X(t_0) = X_0$, there exists a unique continuous solution trajectory $X(t)$ for (3). Moreover, the solution $X(t)$ of (3) will approach exponentially the feasible set when initial point $X_0 \notin \Omega$ and $X(t) \in \Omega$ provided that $X_0 \in \Omega$.

(ii) If X_1 is an equilibrium point of (3), then X_1 is an optimal solution of (2) with $X_0 \in \Omega$.

Proof: (i) Since $-X + P_\Omega(X - \nabla E(X, M_1))$ is locally Lipschitz continuous, there exists an unique continuous solution $X(t)$ of (3).

Next, let $X(t_0) = X_0$. Since

$$\frac{dX}{dt} + X = P_\Omega(X - \nabla E(X, M_1)),$$

$$\int_{t_0}^t (\frac{dX}{dt} + X)e^s ds = \int_{t_0}^t e^s P_\Omega(X - \nabla E(X, M_1)) ds,$$

then $X(t) = e^{-(t-t_0)} X(t_0) + e^{-t} \int_{t_0}^t e^s P_\Omega(X - \nabla E(X, M_1)) ds$.

By the integration mean value theorem, we have

$$X(t) = e^{-(t-t_0)} X(t_0) + e^{-t} P_\Omega(\widehat{X} - \nabla E(\widehat{X}, M_1)) \int_{t_0}^t e^s ds$$

$$= e^{-(t-t_0)} X(t_0) + P_\Omega(\widehat{X} - \nabla E(\widehat{X}, M_1))(1 - e^{-(t-t_0)}),$$

then, when initial point $X_0 \notin \Omega$, the solution $X(t)$ of (3) will approach exponentially the feasible set. Provided that $X_0 \in \Omega$ and since $P_\Omega(\widehat{X} - \nabla E(\widehat{X}, M_1)) \in \Omega, X(t) \in \Omega$.

(ii) Clearly, the equilibrium point of (3) satisfies $\nabla E(X, M_1) = 0$. Thus, X_1 is an optimal solution of (2).

3   Stability and Convergence Analysis

In order to analyze the stability and convergency of the neural network corresponding to (2), we need the following lemma.

Lemma 1. Assume that the set $\Omega \subseteq R^n$ is a closed convex set. Then,

$$(v - P_\Omega(v))^T(P_\Omega(v) - u) \geq 0, \ v \in R^n, u \in \Omega, \quad and$$

$$\|P_\Omega(u) - P_\Omega(v)\| \leq \|u - v\|.$$

We present and prove the following theorems.

Theorem 3. The equilibrium point of neural network (3) is stable and convergence in the Lyapunov sense.

Proof. Let X_1^* be any minimum point of $E(X, M_1)$ and

$$V(X) = \int_0^1 (X - X_1^*)^T \nabla E(X_1^* + s(X - X_1^*), M_1)ds.$$

From Theorem 1, $E(X, M_1)$ is a nonnegative, continuously differential convex function, and $\nabla E(X, M_1)$ is monotone and $\nabla^2 E(X, M_1)$ is symmetric and positive semi-define. Then

$$\frac{dV(X)}{dt} = \nabla E(X, M_1)^T \frac{dX}{dt} = \nabla E(X, M_1)^T[P_\Omega(X - \nabla E(X, M_1)) - X].$$

Let $v = X - \nabla E(X, M_1), u = X$, since $\nabla E(X, M_1)$ is monotone, by Lemma 1, we have

$$[X - \nabla E(X, M_1) - P_\Omega(X - \nabla E(X, M_1))]^T[P_\Omega(X - \nabla E(X, M_1)) - X] \geq 0$$

i.e.

$$\nabla E(X, M_1)^T P_\Omega(X - \nabla E(X, M_1)) - X \leq -\|P_\Omega(X - \nabla E(X, M_1)) - X\|^2 .$$

Then,

$$\frac{dV(X)}{dt} \leq -\| P_\Omega(X - \nabla E(X, M_1)) - X \|^2 \leq 0.$$

Thus, the equilibrium point is stable in the Lyapunov sense.

Note that, $\frac{dV(X)}{dt} = 0$, if and only if $P_\Omega(X - \nabla E(X, M_1)) - X = 0$, i.e. $\frac{dX}{dt} = 0$. It follows from the invariance principle of LaSalle that the trajectories of the neural network in (3) is convergent.

Theorem 3 shows that neural network (3) is Lyapunov stable and converges to the solution of (2). However, it cannot guarantee to converge to a solution of (1). By using the network (3) as a subnetwork, some feedback recurrent networks are given for solving the QPQEC. Similar to the proof of [6], we have the following convergence theorem.

Theorem 4. Suppose that X^* is an optimal solution of (1) and M_1 is an estimated value of the lower bound of its optimal value, i.e., $M_1 \leq f(X^*)$, the energy function sequence is

$$E(X, M_k) = F(X, M_k) + \frac{1}{2}\sum_{i=1}^m (X^T A_i X - c_i)^2, k = 1, 2, ...,$$

where $F(X, M_k) = d(X, M_k)[d(X, M_k) + |d(X, M_k)|]/2$, $f(x) = X^T A_0 X + C_0^T X$, $d(X, M_k) = f(X) - M_k$, $M_{k+1} = M_k + \sqrt{E(X_k^*, M_k)}$ and X_k^* is the optimal solution of (2), then

(i) $M_k \leq M_{k+1}$, $M_k \leq f(X^*)$ and $f(X_k^*) \leq f(X^*)$,

(ii) $\{M_k\}$ has limiting point M^*, that is $\lim_{k \to +\infty} = M^*$,

(iii) the sequence $\{X_k^*\}$ has limiting point $\overline{X}$, any limiting point $\overline{X}$ of $\{X_k^*\}$ satisfies $f(\overline{X}) = f(X^*) = M^*$, that is, every limiting point of $\{X_k^*\}$ is the optimal solution of (1).

By Theorem 4, the feedback recurrent neural subnetworks for solving (1) are given below.

$$\begin{cases} \dfrac{dX}{dt} = -X + P_\Omega(X - \nabla E(X, M_k)), \\ X(0) = X_{k-1}^*, \end{cases} \quad (k = 1, 2, ...). \quad (4)$$

4   Simulation Example

In this section, we give an example to illustrate the effectiveness of the proposed feedback recurrent neural networks for solving QPQEC problem. The simulation is conducted in MATLAB, and the ordinary differential equation is solved by Runge-Kutta method.

Example 1. [8] Considering the QPQEC problem

$$\min f(x) = X^T Q X$$
$$\text{s.t. } X^T A X = 1 \quad (5)$$

where

$$Q = \begin{pmatrix} 89.0434 & 38.2002 & 10.6773 \\ 4.7274 & 79.0866 & 4.0530 \\ 4.0860 & 10.6929 & 67.0491 \end{pmatrix}, \quad A = \begin{pmatrix} 49.5399 & 27.6868 & 31.3546 \\ 15.5941 & 22.9994 & 16.8984 \\ 10.7686 & 31.0746 & 22.9243 \end{pmatrix}$$

The optimal solution to this problem is $x^* = [0.0796855, 0.0498699, 0.0633523]^T$ and the optimal value is $f(x^*) = 1.3229$, the error of the constrained condition is 1.65761e-05.

Example 2. Let us consider the following QPQEC problem with bound constraint. A quadratic object function with quadratic constraint as

$$\min f(x) = X^T A_0 X + C_0^T X$$
$$\text{s.t.} \begin{cases} X^T A_1 X = c_1, \\ X^T A_2 X = c_2, \\ l \leq X \leq h \end{cases} \quad (6)$$

where

$$A_0 = \begin{pmatrix} 1 & -1 & 0 \\ -1 & 2 & -1/2 \\ 0 & -1/2 & 1 \end{pmatrix}, \quad A_1 = \begin{pmatrix} 2 & 1 & 0 \\ 1 & 1 & 0 \\ 0 & 0 & 1 \end{pmatrix}, \quad A_2 = \begin{pmatrix} 1 & 0 & -1 \\ 0 & 2 & 0 \\ -1 & 0 & 3 \end{pmatrix},$$

$$C_0 = \begin{bmatrix} -1 & -2 & 5 \end{bmatrix}^T, \ l = \begin{bmatrix} 0 & 2 & 1 \end{bmatrix}^T, \ h = \begin{bmatrix} 5 & 8 & 10 \end{bmatrix}^T, \ c_1 = 6, \ c_2 = 8.$$

Using the feedback recurrent neural subnetworks to solve (6), the optimal solution is $x^* = [0.4344, \ 1.9763, \ 0]^T$, and the optimal value is $f(x^*) = 1.8960$.

5 Conclusions

We have proposed a feedback recurrent neural network for solving QPQEC problem in this paper, and constructed nonnegative, continuously differential convex energy function $E(X, M_k)$. Based on the project theory, a recurrent neural subnetwork is established. Moreover, the minimizer point of $E(X, M_k)$ is obtained by the equilibrium point of neural subnetwork. For an initial point and the lower bound of objective function, we can find the optimal solution of QPQEC problem by successive approximation.

References

1. Tank, D. W., Hopfield, J. J.: Simple 'Neural' Optimization Network:an A/D Converter, Signal Decision Circuit and a Linear Programming Circuit, IEEE Trans. Circuits and Systems, **33** (1986) 533-541
2. Kennedy, M. P., Chua, L. O.: Neural Networks for Nonlinear Programming, IEEE Trans. Circuits and Systems, **35** (1988) 554-562
3. Xia, Y., Wang, J.: Global Exponential Stability of Recurrent Neural Networks for Solving Optimization and Related Problems, IEEE Trans. Neural Networks, **11** (2000) 1017-1022
4. Xia, Y., Wang, J.: A Recurrent Neural Networks for Nonlinear Convex Optimization Subject to Nonlinear Inequality Constraints. IEEE Trans. Circuits and Systems-I, **51** (2004) 1385-1394
5. Xia, Y., Wang, J.: A General Projection Neural Networks for Solving Monotone Variational Inequalities and Related Optimization Problems, IEEE Trans. Neural Networks, **15** (2004) 318-328
6. Leung, Y., Gao, X. B.: A Hogh-performance Feedback Neural Network for Solving Convex Nonlinear Programming Problems, IEEE Trans. Neural Networks, **14** (2003) 1469-1477
7. Leung, Y., Chen, K. Z., Gao, X. B., Leung, K. S.: A new Gradient-based Neural Network for Solving Linear and Quadratic Programming Problems, IEEE Trans. Neural Networks, **12** (2001) 1074-1083
8. Tan, Y., Deng, C.: Solving for a Quadratic Programming with a Quadratic Constraint Based on a Neural Network Frame, Neurocomputing, **30** (2000) 117-127
9. Tao, Q., Cao, J., Sun, D.: A Simple and High Performance Neural Network for Quadratic Programming Problems, Applied Mathematics and Computation, **124** (2001) 251-260

The new neural network of size n is a fully connected network with n continuous valued units. Let ω_{ij} be the weight of the connection from neuron i to neuron j. Since $E(\mathbf{x})$ are twice continuous differentiable, we can define the connection coefficients as follows:

$$\omega_{ij} = \frac{\partial E^2(\mathbf{x})}{\partial x_1 \partial x_2}, \qquad i,j = 1,2,\cdots,n. \tag{4}$$

The basis structural of the neural network is as follows.

We define $\mathbf{x} = (x_1, x_2, \cdots, x_n)$ as the input vector of the neural network, $\mathbf{y} = (y_1, y_2, \cdots, y_n)$ as the output vector, and $V(t) = (v_1(t), v_2(t), \cdots, v_n(t))$ as the state vector of neurons. $v_i(t)$ is the state of neuron i at the time t. And this new neural network is a type of Hopfield-like neural network. These kinds of neural networks have two kinds of prominent properties: the state's difference of network corresponds to the negative gradient of the energy function; the connection weights of neurons correspond to the second partial derivatives of the energy function (Hessian matrix).

2.2   Stability Analysis

Consider the following problems:

(PII$_\rho$):   min $H(\mathbf{x}, \rho, \epsilon)$ s.t. $\mathbf{x} \in X$

For the stability analysis of (2), we have the following theorems of the Lyapunov stability theory.

Theorem 2.2.1 Let $\mathbf{x}^*$ be an equilibrium point of dynamics system (3) under the parameter (ρ, ε). If $\mathbf{x} \neq 0$ and $E(\mathbf{x}) \neq 0$, then $\mathbf{x}^*$ is the stable point of dynamics system (3). And if weight coefficient matrix $(\omega_{ij})_{n \times n}$ is positive semi-definite, then $\mathbf{x}^*$ is a locally optimal solution to the problem (PIIρ).

Proof. From (1),(2) and (3), we have

$$\frac{dE(\mathbf{x})}{dt} = \sum_{k=1}^{n} \frac{\partial E}{\partial x_k} \frac{dx_k}{dt} = \sum_{k=1}^{n} \frac{\partial E}{\partial x_k} (-\beta_k) \frac{\partial E}{\partial x_k} \leq 0.$$

According to the Lyapunov stable theorem, $\mathbf{x}^*$ is the stable point of dynamics system (3). The second conclusion is obvious.   □

Let $F(\mathbf{x}, \rho) = f(\mathbf{x}) + \rho \sum_{k=1}^{m} \sqrt{\max\{g_i(\mathbf{x}), 0\}}$. Consider an optimal problem:

(Pρ)   min $F(\mathbf{x}, \rho)$ s.t. $\mathbf{x} \in X$.

Since $\lim_{\varepsilon \to 0} H(\mathbf{x}, \rho, \varepsilon) = F(\mathbf{x}, \rho)$, we will first study some relationship between (Pρ) and (PIIρ).

Lemma 2.2.1 For any $\mathbf{x} \in X$ and $\epsilon > 0$, we have

$$0 \leq F(\mathbf{x}, \rho) - H(\mathbf{x}, \rho, \epsilon) \leq \frac{12}{7} m \rho \epsilon^{\frac{1}{2}}, \tag{5}$$

where $\rho > 0$.

Proof. By the definition of $q_\epsilon(t)$, we have

$$0 \leq p(t) - q_\epsilon(t) \leq \frac{12}{7}\epsilon^{\frac{1}{2}}.$$

As a result,

$$0 \leq p(g_i(\mathbf{x})) - q_\epsilon(g_i(\mathbf{x})) \leq \frac{12}{7}\epsilon^{\frac{1}{2}} \qquad \forall \mathbf{x} \in X, i = 1, 2, ..., m.$$

Adding up for above all i, we obtain

$$0 \leq \sum_{i \in I} p(g_i(\mathbf{x})) - \sum_{i \in I} q_\epsilon(g_i(\mathbf{x})) \leq \frac{12}{7}m\epsilon^{\frac{1}{2}}.$$

Hence,

$$0 \leq F(\mathbf{x}, \rho) - H(\mathbf{x}, \rho, \epsilon) \leq \frac{12}{7}m\rho\epsilon^{\frac{1}{2}}. \qquad \square$$

From Lemma 2.2.1, we easily obtain the following some theorems.

Theorem 2.2.2 Let $\{\varepsilon_j\} \to 0$ be a sequence of positive numbers. Assume that x_j is a solution to $\min_{\mathbf{x} \in X} H(\mathbf{x}, \rho, \varepsilon_j)$ for some $\rho > 0$. Let $\bar{\mathbf{x}}$ be an accumulation point of the sequence $\{\mathbf{x}_j\}$. Then $\bar{\mathbf{x}}$ is an optimal solution to $\min_{\mathbf{x} \in R^n} F(\mathbf{x}, \rho)$.

Definition 2.2.1 A vector $\mathbf{x}_\varepsilon \in X$ is ε-feasible or ε-solution if $g_i(\mathbf{x}_\varepsilon) \leq \varepsilon$, for all $i \in I$.

Theorem 2.2.3 Let $\mathbf{x}^*$ be an optimal solution to (P_ρ) and $\bar{\mathbf{x}} \in X$ an optimal solution to (PII_ρ). Then

$$0 \leq F(\mathbf{x}^*, \rho) - H(\bar{\mathbf{x}}, \rho, \epsilon) \leq \frac{12}{7}m\rho\epsilon^{\frac{1}{2}} \tag{6}$$

Theorem 2.2.4 Let $\mathbf{x}^*$ be an optimal solution to (P_ρ) and $\bar{\mathbf{x}} \in X$ an optimal solution to (PII_ρ). Furthermore, let $\mathbf{x}^*$ be feasible to (P) and $\bar{\mathbf{x}}$ be ϵ-feasible to (P). Then

$$0 \leq f(\mathbf{x}^*) - f(\bar{\mathbf{x}}) \leq \frac{24}{7}m\rho\epsilon^{\frac{1}{2}}. \tag{7}$$

Theorem 2.2.5 If $\mathbf{x}^*$ is an optimal solution to the problem $(PII \rho)$, then $\mathbf{x}^*$ is an equilibrium point of dynamics system (3) under the parameter (ρ, ε).

Theorems 2.2.1 and 2.2.5 show that an equilibrium point of the dynamic system yields an approximate optimal solution to the optimization problem $(PII\rho)$. Theorems 2.2.2 and 2.2.3 mean that an approximate solution to $(PII\rho)$ is also an approximate solution to $(P\rho)$ when ε is sufficiently small. Moreover, an approximate solution to $(PII\rho)$ also becomes an approximate optimal solution to (P) by Theorem 2.2.4 if the approximate solution is ε-feasible. Therefore, we may obtain an approximate optimal solution to (P) by finding an approximate solution to $(PII\rho)$ or an equilibrium point of the dynamic system (3).

3   Applications to Nonlinear Optimization Problems

In order to get an approximate optimal solution to (P) and an equilibrium point of the new neural network system, we propose the following Algorithm I. By Algorithm I, we get an approximate optimal solution to (P) by Theorem 2.2.4, and an equilibrium point of the dynamic system (3) of the neural network.

Algorithm I

   Step 1: Given $\mathbf{x}^0, \epsilon > 0, \epsilon_0 > 0, \rho_0 > 0, 0 < \eta < 1$ and $N > 1$.
   Let $j = 0$. To construct energy function (2)
   and dynamical differentiable system (3).

   Step 2: Using the violation $\mathbf{x}^j$ as the starting point for evaluating the
   following penalty function solve the problem: $\min_{\mathbf{x} \in X} H(\mathbf{x}, \rho_j, \epsilon_j)$.

   Let $\mathbf{x}^j$ be the optimal solution.

   Step 3: If $\mathbf{x}^j$ is ϵ-feasible to (P),
   then stop and get an approximate solution $\mathbf{x}^j$ of (P)
   and an equilibrium point of dynamics system (3),
   otherwise, let $\rho_{j+1} = N\rho_j$ and $\epsilon_{j+1} = \eta\epsilon_j$
   and set $j := j + 1$ and go to Step 2.

We give the following numerical results.

Example 3.1 Consider the Rosen-Suzki problem{[9]}:

$$(P3.1) \ \min f(\mathbf{x}) = x_1^2 + x_2^2 + 2x_3^2 + x_4^2 - 5x_1 - 5x_2 - 21x_3 + 7x_4$$
$$\text{s.t. } g_1(\mathbf{x}) = 2x_1^2 + x_2^2 + x_3^2 + 2x_1 + x_2 + x_4 - 5 \le 0$$
$$g_2(\mathbf{x}) = x_1^2 + x_2^2 + x_3^2 + x_4^2 + x_1 - x_2 + x_3 - x_4 - 8 \le 0$$
$$g_3(\mathbf{x}) = x_1^2 + 2x_2^2 + x_3^2 + 2x_4^2 - x_1 - x_4 - 10 \le 0.$$

Let starting point $\mathbf{x}^0 = (0,0,0,0)$, $\varepsilon = 10^{-6}$, $\varepsilon_0 = 1$, $\rho_0 = 10, \eta = 0.5$ and N=2. We use Algorithm I to solve (P3.1) under parameter $\rho_{j+1} = 10\rho_j$ and error $\varepsilon_{j+1} = 0.5\varepsilon_j$ in the following results Table 1.

Table 1. Results of (P3.1) are obtained by Algorithm I.

No. iter.	ρ_k	Cons. error $e(k)$	Objective value	e-Solution (x_1, x_2, x_3, x_4)
1	10	1.624681	-46.624520	(0.207426,0.861010,2.133251,-1.075119)
2	20	0.421837	-44.881344	(0.179257,0.841571,2.041896,-0.995292)
3	40	0.106477	-44.399382	(0.171953,0.836970,2.017124,-0.972714)
4	80	0.026691	-44.275467	(0.170095,0.835921,2.010785,-0.966842)
12	20480	0.000000	-44.233834	(0.169568,0.834758,2.009021,-0.964387)

It is easy to check a point $\mathbf{x}^{12} = (0.169568,0.834758,2.009021,-0.964387)$ at the 12'th iteration is feasible solution, which its objective value $f(\mathbf{x}^{12}) = -44.233837$ is better than the objective value $f(\mathbf{x}') = -44$ at the best solution in [9] $\mathbf{x}' = (0,1,2,-1)$ to (P3.1).

4   Conclusions

In this paper we have studied a Hopfield-like network when applied to nonlinear optimization problems. An energy function of the neural network with its neural dynamics is constructed based on the method of penalty function with two order differential. The system of the neural networks has been shown to be stable and its equilibrium point of the neural dynamics also yields an approximate optimal solution for nonlinear constrained optimization problems. An algorithm is given to find out an approximate optimal solution to its optimization problem, which is also an equilibrium point of the system. The numerical example shows that the algorithm is efficient.

References

1. Hopfield, J. J., Tank, D.W.: Neural Computation of Decision in Optimization Problems. Biological Cybernetics, **58** (1985) 67-70
2. Joya, G., Atencia, M.A., Sandoval, F.: Hopfield Neural Networks for Optimizatiom: Study of the Different Dynamics. Neurocomputing, **43** (2002) 219-237
3. Chen, Y.H., Fang, S.C.: Solving Convex Programming Problems with Equality Constraints by Neural Networks. Computers Math. Applic. **36** (1998) 41-68
4. Staoshi M.: Optimal Hopfield Network for Combinatorial Optimization with Linear Cost Function, IEEE Tans. On Neural Networks, **9** (1998) 1319-1329
5. Xia Y.S., Wang, J.: A General Methodology for Designing Globally Convergent Optimization Neural Networks, IEEE Trans. On Neural Networks, **9** (1998) 1331-1444
6. Zenios, S.A., Pinar, M.C., Dembo, R.S.: A Smooth Penalty Function Algorithm for Network-structured Problems. European J. of Oper. Res. **64** (1993) 258-277
7. Meng,Z.Q., Dang,C.Y., Zhou G.,Zhu Y., Jiang M.: A New Neural Network for Nonlinear Constrained Optimization Problems, Lecture Notes in Computer Science, Springer. **3173** (2004) 406-411
8. Yang, X.Q., Meng, Z.Q., Huang, X.X., Pong, G.T.Y.: Smoothing Nonlinear Penalty Functions for Constrained Optimization. Numerical Functional Analysis Optimization, **24** (2003) 351-364
9. Lasserre, J.B.: A Globally Convergent Algorithm for Exact Penalty Functions. European Journal of Opterational Research, **7** (1981) 389-395
10. Fang, S.C., Rajasekera, J.R., Tsao, H.S.J.: Entropy Optimization and Mathematical Proggramming. Kluwer (1997)

A Neural Network Algorithm
for Second-Order Conic Programming

Xuewen Mu[1], Sanyang Liu[1], and Yaling Zhang[2]

[1] Department of Applied Mathematics, Xidian University,
Xi'an 710071, China
xdmuxuewen@hotmail.com
[2] Department of Computer Science, Xi'an Science and Technology University,
Xi'an 710071, China
zyldella@xust.edu.cn

Abstract. A neural network algorithm for second-order conic programming is proposed. By the Smooth technique, a smooth and convex energy function is constructed. We have proved that for any initial point, every trajectory of the neural network converges to an optimal solution of the second-order conic programming. The simulation results show the proposed neural network is feasible and efficient.

1 Introduction

In a second-order conic programming ($SOCP$) a linear functions is minimized over the intersection of an affine set and the product of second-order cones. SOCP are nonlinear convex problems, and the linear and convex quadratic programs are special cases. the primal-dual problems of SOCP is given as [1]

$$(P) \quad min\{c^T x : Ax = b, x \in K\}$$
$$(DP) \, max\{b^T y : A^T y + s = c, s \in K\}$$

where $x = (x_1^T, x_2^T, \cdots, x_N^T)^T, s = (s_1^T, s_2^T, \cdots, s_N^T)^T \in R^n, y \in R^m$, are the variables, and the parameters are $A = (A_1, A_2, \cdots, A_N) \in R^{m \times n}, A_i \in R^{m \times n_i}$, $c = (c_1^T, c_2^T, \cdots, c_N^T)^T \in R^n, s_i, x_i, c_i \in R_i^n, i = 1, 2, \cdots, N, n_1 + n_2 + \cdots n_N = n$. K_i is the standard second-order cone of dimension n_i, which is defined as

$$K = K_1 \times K_2 \times \cdots K_N, K_i = \left\{ x_i = \begin{bmatrix} x_{i1} \\ x_{i0} \end{bmatrix} : x_{i1} \in R^{(n_i-1)}, x_{i0} \in R, \|x_{i1}\| \le x_{i0} \right\}$$

where the norm is the standard Euclidean norm, $i.e. \|u\| = (u^T u)^{1/2}$, when $u \in R^n$. When $u \in R^{n \times n}$, the norm is the Frobenius norm of the matrix.

There are many applications in the engineering for the SOCP, such as filter design, antenna array weight design, truss design, see [2, 3]. Issues involving the solution of large scale or very large scale SOCP are ubiquitous in the engineering problems. They are generally intended to be solved in real time. Most of the traditional algorithms are iterative method. They require much more computation time and cannot satisfy the real-time requirement. Neural networks

J. Wang, X. Liao, and Z. Yi (Eds.): ISNN 2005, LNCS 3496, pp. 718–724, 2005.

are a kind of self-adaptive, self-organizing, and self-learning nonlinear networks which are massively parallel, distributed, and of high error-correction capability. Their algorithms have very rapid convergence and very good stability. They are considered as an efficient approach to solve large-scale or very large-scale optimization problems in various areas of applications. Neural network for solving optimization problems have been rather extensively studied over the years and some important results have also been obtained[4–6]. Although there are some algorithms for convex programming to been studied, but these algorithms are not efficient for SOCP. In this paper, based on the gradient, a neural network algorithm for SOCP is proposed. By the Smooth technique, a smooth and convex energy function is constructed. We have proved that for any initial point, every trajectory of the neural network converges to an optimal solution of SOCP. The simulation results show the proposed neural network is feasible and efficient.

2   A Neural Network Algorithm for SOCP

Suppose that the strictly feasible primal and dual starting points of SOCP exist, based on the duality theorem, solving the primal and dual program of SOCP is equivalent to solve the following system [1].

$$\begin{cases} c^T x - b^T y = 0, Ax - b = 0, x \in K \\ A^T y + s = c, s \in K \end{cases} \tag{1}$$

(1) is the optimal condition for the SOCP.

Because the standard second-order cone constraints are nonsmooth, and the nonsmoothness can cause some problems. We use the smooth technique to obtain the alternate formulation [7].

$$K_i = \left\{ x_i = \begin{bmatrix} x_{i1} \\ x_{i0} \end{bmatrix} : x_{i1} \in R^{(n_i-1)}, x_{i0} \in R, e^{\frac{1}{2}(\|x_{i1}\|^2 - x_{i0}^2)} - 1 \le 0, x_{i0} \ge 0 \right\}$$

We definite some functions as follows.

$$g_i(x) = e^{\frac{1}{2}(\|x_{i1}\|^2 - x_{i0}^2)} - 1, x_{i0} \ge 0, h_i(s) = e^{\frac{1}{2}(\|s_{i1}\|^2 - s_{i0}^2)} - 1, s_{i0} \ge 0, i = 1, 2, \cdots, N$$

By paper [7], we know the functions g_i, h_i is differentiable convex functions. We definite the following functions.

$$F_i(x) = \frac{1}{2}g_i(x)[g_i(x) + |g_i(x)|], H_i(s) = \frac{1}{2}h_i(s)[h_i(s) + |h_i(s)|], i = 1, 2, \cdots, N$$

Theorem 1 $F_i(x), H_i(s)$ are differentiable convex functions and

$$F_i(x) = 0 \Longleftrightarrow g_i(x) \le 0, H_i(s) = 0 \Longleftrightarrow h_i(s) \le 0, i = 1, 2, \cdots, N$$

Furthermore,

$$\nabla_x F_i(x) = \begin{cases} 2g_i(x)\nabla_x g_i(x), & g_i(x) \ge 0 \\ 0, & g_i(x) < 0 \end{cases} \text{ and } \nabla_s H_i(s) = \begin{cases} 2h_i(s)\nabla_s h_i(s), & h_i(s) \ge 0 \\ 0, & h_i(s) < 0 \end{cases}$$

is local Lipschitz continuous, where $i = 1, 2, \cdots, N$.

Proof Obviously

$$F_i(x) = \begin{cases} g_i^2(x), g_i(x) \geq 0 \\ 0, \qquad g_i(x) < 0 \end{cases}$$

Let $w = g_i(x)$, then $F_i(x)$ can be viewed as the compound function as follows.

$$F_i(x) = \varphi(w) = \begin{cases} w^2, w \geq 0 \\ 0, \quad w < 0 \end{cases}$$

Apparently $F_i(x) = \varphi(w)$ is a nondecreasing differentiable convex function, and $w = g_i(x)$ is a differentiable convex function, hence their compound function $F_i(x)$ is a differentiable convex function[8]. We can similarly prove that $H_i(s)$ is a differentiable convex function. By the proof above, we obviously have

$$F_i(x) = 0 \Longleftrightarrow g_i(x) \leq 0, H_i(s) = 0 \Longleftrightarrow h_i(s) \leq 0, i = 1, 2, \cdots, N$$

Now we prove $\nabla_s H_i(s)$ is local Lipschitz continuous. Given $D \in R^n$ is a bounded and closed convex region, for two arbitrary $s_1, s_2 \in D$, then there is

$$\begin{aligned}
&\|\nabla_s H_i(s_2) - \nabla_s H_i(s_1)\| \\
&= \|[h_i(s_2) + |h_i(s_2)|]\nabla_s h_i(s_2) - [h_i(s_1) + |h_i(s_1)|]\nabla_s h_i(s_1)\| \\
&= \|[h_i(s_2) + |h_i(s_2)| - h_i(s_1) - |h_i(s_1)|]\nabla_s h_i(s_2) \\
&\quad + [h_i(s_1) + |h_i(s_1)|](\nabla_s h_i(s_2) - \nabla_s h_i(s_1))\| \\
&\leq \|\nabla_s h_i(s_2)\|\,|[h_i(s_2) - h_i(s_1) + |h_i(s_2)| - |h_i(s_1)|]| \\
&\quad + 2|h_i(s_1)|\|\nabla_s h_i(s_2) - \nabla_s h_i(s_1)\| \\
&\leq 2\|\nabla_s h_i(s_2)\|\,|h_i(s_2) - h_i(s_1)| + 2|h_i(s_1)|\|\nabla_s h_i(s_2) - \nabla_s h_i(s_1)\|
\end{aligned}$$

By the continuous differentiability of $h_i(s), \nabla_s h_i(s)$ on D, there is an M, for arbitrary $s \in D$, which satisfy $|h_i(s)| \leq M, \|\nabla_s h_i(s)\| \leq M, \|\nabla_s^2 h_i(s)\| \leq M$. Since, $\|h_i(s_2) - h_i(s_1)\| = \|\nabla_s h_i(\bar{s})(s_2 - s_1)\| \leq \|\nabla_s h_i(\bar{s})\|\|s_2 - s_1\| \leq M\|s_2 - s_1\|$, $\|\nabla_s h_i(s_2) - \nabla_s h_i(s_1)\| = \|\nabla_s^2 h_i(\tilde{s})(s_2 - s_1)\| \leq \|\nabla_s^2 h_i(\tilde{s})\|\|s_2 - s_1\| \leq M\|s_2 - s_1\|$, where $\bar{s} = \theta_1 s_1 + (1 - \theta_1)s_2, \tilde{s} = \theta_2 s_1 + (1 - \theta_2)s_2 \in D, 0 \leq \theta_1, \theta_2 \leq 1$. Let $L = 4M^2$, then

$$\|\nabla_s H_i(s_2) - \nabla_s H_i(s_1)\| \leq 4M^2\|s_2 - s_1\| \leq L\|s_2 - s_1\|$$

That is to say $\nabla_s H_i(s)$ is Lipschitz continuous on D. We can similarly prove that $\nabla_x F_i(x)$ is local Lipschitz continuous. This completes the proof of the theorem.

Formula (1) is equivalent to the following formula

$$\begin{cases} c^T x - b^T y = 0, Ax - b = 0, g_i(x) \leq 0, x_{i0} \geq 0, i = 1, 2, \cdots, N \\ A^T y + s = c, h_i(s) \leq 0, h_{i0} \geq 0, i = 1, 2, \cdots, N \end{cases} \tag{2}$$

Let $z^- = (1/2)(z - |z|), z \in R^n$, it is easy to prove the following formula [6].

$$z \geq 0 \Longleftrightarrow \|z^-\| = 0 \Longleftrightarrow (1/2)z^T(z - |z|) = 0$$

So we have

$$x_{i0} \geq 0 \Longleftrightarrow \|x_{i0}^-\|^2 = 0, s_{i0} \geq 0 \Longleftrightarrow \|s_{i0}^-\|^2 = 0, i = 1, 2, \cdots, N \qquad (3)$$

Now we construct an appropriate energy function which captures the duality and optimality of the SOCP problem as follows:

$$E(x, y, s) = \frac{1}{2}\|c^T x - b^T y\|^2 + \frac{1}{2}\|Ax - b\|^2 + \frac{1}{2}\|A^T y + s - c\|^2$$

$$+ \sum_{i=1}^{N}\{F_i(x) + H_i(s) + \frac{1}{2}(\|x_{i0}^-\|^2 + \|s_{i0}^-\|^2)\}$$

Theorem 2 $E(x, y, s) = 0 \Longleftrightarrow x$ and $(y^T, s^T)^T$ are the optimal solutions of (P) and (DP), that is, all equalities and inequalities are satisfied in (1).

Proof By Theorem 1 and formula (3), we can easily prove the theorem 2.

Theorem 3 $E(x, y, s)$ is a differentiable convex function.

Proof Obviously, $\frac{1}{2}\|c^T x - b^T y\|^2$, $\frac{1}{2}\|Ax - b\|^2$, and $\frac{1}{2}\|A^T y + s - c\|^2$ are differentiable convex functions. By Theorem 1, we have $H_i(s), F_i(x), \|x_{i0}^-\|^2, \|s_{i0}^-\|^2$ are differentiable convex functions. So $E(x, y, s)$ is a differentiable convex function.

It is easy to compute the following gradient functions.

$$\nabla_x[\frac{1}{2}\|c^T x - b^T y\|^2] = (c^T x - b^T y)c, \nabla_y[\frac{1}{2}\|c^T x - b^T y\|^2] = (c^T x - b^T y)b$$
$$\nabla_y[\frac{1}{2}\|A^T y + s - c\|^2] = A(A^T y + s - c), \nabla_s[\frac{1}{2}\|A^T y + s - c\|^2] = A^T y + s - c$$
$$\nabla_x[\frac{1}{2}\|Ax - b\|^2] = A^T(Ax - b)$$
$$\nabla_x[\frac{1}{2}\|x_{i0}^-\|^2] = (x_{i0}^-)e_k, \nabla_s[\frac{1}{2}\|s_{i0}^-\|^2] = (s_{i0}^-)e_k, k = \sum_{j=1}^{i} n_j, i = 1, 2, \cdots, N.$$

Where $e_k \in R^n$ denote that real n-dimensional unit column vectors. Thus, the neural network for solving (P) and (DP) can be formulated as follows:

$$\frac{dz}{dt} = -\nabla E(z), z(0) = z^0 \qquad (4)$$

where $z = (x^T, y^T, s^T)^T \in R^{2n+m}$, that is to say that

$$\left\{ \begin{array}{l} \frac{dx}{dt} = -(c^T - b^T y)c - A^T(Ax - b) - \sum_{i=1}^{N}(x_{i0}^-)e_k - \sum_{i=1}^{N}\nabla_x F_i(x) \\ \frac{dy}{dt} = (c^T - b^T y)b - A(A^T y + s - c) - \sum_{i=1}^{N}\nabla_s H_i(s) \\ \frac{ds}{dt} = -(A^T y + s - c) - \sum_{i=1}^{N}(s_{i0}^-)e_k \\ x(0) = x^0, y(0) = y^0, s(0) = s^0 \end{array} \right\}$$

3 Stability Analysis

In order to discuss the stability of the neural network in (4), we first prove the following theorems.

Theorem 4. The initial value problem of the system of differential equations in (4) has unique solutions.

Proof. By Theorem 1, $\nabla_x F_i(x), \nabla_s F_i(s)$ is local Lipschitz continuous, so $\sum_{i=1}^{N} \nabla_x F_i(x)$ and $\sum_{i=1}^{N} \nabla_s H_i(s)$ are local Lipschitz continuous. Because $(x_{i0})^-$, $(s_{i0})^-$ are Lipschitz continuous [6], then $-\nabla E(z)$ is local Lipschitz continuous. Hence the system of (4) has unique solution by the existence and uniqueness theorem of the initial value problem of a system of differential equations[8].

Theorem 5. Let $M = \{z = (x^T, y^T, s^T)^T \in R^{2n+m} | \nabla E(z) = 0\}$ be the set of equilibrium points of (4), and $\Omega = \{z = (x^T, y^T, s^T)^T \in R^{2n+m} | x \text{ and } (y^T, s^T)^T$ are the optimal solutions of (P) and (DP), respectively$\}$ be the set of optimal solutions of (P) and (DP), then $M = \Omega$.

Proof. Suppose that $z \in M, z^* \in \Omega$, then $E(z^*) = 0$ by Theorem 2. Since $E(z)$ is a differentiable convex function by Theorem 3, then from the necessary and sufficient conditions of convex functions, we have

$$0 = E(z^*) \geq E(z) + (z^* - z)^T \nabla E(z).$$

hence, $(z - z^*)^T \nabla E(z) \geq E(z)$. We have $E(z) \leq 0$ by $z \in M$. Nevertheless $E(z) \geq 0$, so we obtain $E(z) = 0$. Thus $z \in \Omega$ by Theorem 2.Therefore,$M \subseteq \Omega$.
Let $z \in \Omega$,then $E(z) = 0$ by Theorem 2.Hence by Theorem 1 and (3),we have

$$\sum_{i=1}^{N} \nabla_x F_i(x) = 0, \sum_{i=1}^{N} \nabla_s H_i(s) = 0, \sum_{i=1}^{N} (x_{i0}^-)e_k = 0, \sum_{i=1}^{N} (s_{i0}^-)e_k = 0.$$

So we have $\nabla E(z) = 0$ by the definition of $\nabla E(z)$,and we have $z \in M$.Therefore $\Omega \subseteq M$. Based on the above analysis,we have $M = \Omega$.This completes the proof.

Theorem 6. Suppose (P) and (DP) have unique optimal solution $z^* = ((x^*)^T, (y^*)^T, (s^*)^T)^T$, then z^* is globally, uniformly,and asymptotically stable.

Proof. Suppose that the initial point $z^0 = ((x^0)^T, (y^0)^T, (s^0)^T)^T$ is arbitrarily given and that $z(t) = z(t; t_0; z_0)$ is the solution of the initial value problem of the system of differential equations in (4). Let $L(z(t)) = (1/2)\|z(t) - z^*\|_2^2$, obviously, $L(z) \geq 0, L(z^*) = 0$, $L(z)$ is a positive unbounded function. Since $E(z)$ is a differentiable convex function by Theorem 3, then we have $0 = E(z^*) \geq E(z) + (z^* - z)^T \nabla E(z)$. hence, $(z^* - z)^T \nabla E(z) \leq -E(z)$. Therefore, for $z \neq z^*$,

$$\frac{dL(z(t))}{dt} = \nabla L(z)^T \frac{dz}{dt} = (z - z^*)^T [-\nabla E(z)] = (z^* - z)^T \nabla E(z) \leq -E(z) < 0$$

That is, when $z \neq z^*$, along the trajectory $z = z(t)$, $L(z(t))$ is negative definite. Therefore, z^* is globally, uniformly, asymptotically stable by the Lyapunov theorem [9].

Similar to the proof of Theorem 6, we have the following Theorem.

Theorem 7 Any $z^* \in \Omega$ is stable.

Theorem 8 Suppose that $z = z(t, z^0)$ is a trajectory of (4) in which the initial point is $z^0 = z(0, z^0)$, then

a) $r(z^0) = \{z(t, z^0) | t \geq 0\}$is bounded;
b) there exists $\bar{z}$ such that $\lim_{t \longrightarrow +\infty} z(t, z^0) = \bar{z}$
c)$\nabla E(\bar{z}) = 0$

Proof. See paper [6].

By Theorems 7 and 8, we have the following theorem.

Theorem 9. Suppose that (P) and (DP) have infinitely many solutions. Then for any initial point $z^0 \in R^{2n+m}$, the trajectory corresponding to the neural network in (4) convergent to an optimal solution z^* of (P) and (DP).

4   Simulation Experiments

In order to verify the feasibility and efficiency of the proposed network, we use the Euler's method to solve a simple problem.

Example. For (P) and (DP), let $A = [1\ 0\ 0\ 0; 0\ 1\ 0\ 0; 0\ 0\ 4\ 0; 0\ 0\ 0\ 1]; b = (2, 2, 1, 3)^T; c = (1, 2, 3, 5)^T, n_1 = n_2 = 2$ when the initial point are taken as $x_0 = (0, 1, 0, 0)^T; y_0 = (0, 1, 0, 0)^T; s_0 = (0, 1, 0, 0)^T$, we obtain the optimal solution is $x = (2.00, 2.00, 0.25, 3.00)^T$, $y = (1.4769, 1.5231, 0.7532, 4.9989)^T$, $s = (-0.4769, 0.4769, -0.0128, 0.0011)^T$ the optimal value is 21.7500. When the initial point are taken as $x_0 = (0, 0, 0, 0)^T; y_0 = (0, 0, 0, 0)^T; s_0 = (0, 0, 0, 0)^T$, we obtain the similar optimal solution and value.

In a word, whether or not an initial point is taken inside or outside the feasible region, the proposed network always converges to the theoretical optimal solution.

5   Conclusion

We have proposed in this paper a neural network for solving SOCP problems, and have also given a complete proof of the stability and convergence of the network. The simulation results show the proposed neural network is feasible and efficient.

References

1. Lobo, M. S., Vandenberghe, L., Boyd, S., Lebret, H.: Application of Second Order Cone Programming. Linear Algebra and its Applications, **284** (1998) 193-228
2. Lebret, H., Boyd, S.: Antenna Array Pattern Synthesis via Convex Optimization. IEEE Transactions on Signal Processing, **45** (1997) 526-532
3. Lu, W.S., Hinamoto, T.: Optimal Design of IIR Digital Filters with Robust Stability Using Conic-Quadratic-Programming Updates. IEEE Transactions on Signal Processing, **51** (2003) 1581-1592
4. Wang, J.: A Deterministic Annealing Neural Network for Conex Programming. Neural networks, **7** (1994) 629-641
5. Danchi Jiang, Jun Wang: A Recurrent Neural Network for Real-time Semidefinite Programming. IEEE Transaction on Neural Networks, **10** (1999) 81-93

6. Leung, Y., Chen, K., Jiao, Y., Gao, X., Leung, K.S.: A New Gradient-Based Neural Network for Solving Linear and Quadratic Programming Problems. IEEE Transactions on Neural Networks, **12** (2001) 1074-1083

7. Benson, H.Y., Vanderbei, R.J.: Solving Problems with Semidefinite and Related Constraints Using Interior-Point Methods for Nonlinear Programming. Math. Program, **95** (2003) 279-302

8. Avriel, M.: Nonlinear Programming: Analysis and Methods. Prentice-Hall, Englewood Cliffs, NJ (1976)

9. Scalle, J.L., Lefschetz, S.: Stability by Lyapunov's Direct Method with Applications. Academic, New York (1961)

Application of Neural Network to Interactive Physical Programming

Hongzhong Huang[1] and Zhigang Tian[2]

[1] School of Mechatronics Engn.,
University of Electronic Science and Technology of China,
Chengdu, Sichuan 610054, China
hzhhuang@dlut.edu.cn
[2] Department of Mechanical Engineering, University of Alberta,
Edmonton, Alberta, T6G 2G8, Canada,
ztian@ualberta.ca

Abstract. A neural network based interactive physical programming approach is proposed in this paper. The approximate model of Pareto surface at a given Pareto design is developed based on neural networks, and a map from Pareto designs to their corresponding evaluation values is built. Genetic algorithms is used to find the Pareto design that best satisfies the designer's local preferences. An example is given to illustrate the proposed method.

1 Introduction

Physical programming developed by Messac [1], has been successfully applied to high-speed-civil-transport plane design [1], control, structure design [2], interactive design [3], [4] and robust design [5]. Interactive physical programming is based on physical programming. It takes into account the designer's preferences during the optimization process, and allows for design exploration at a given Pareto design.

Based on the Tappeta, Renaud, and Messac's work [4], this paper mainly obtains the following achievements: (1) The approximation to the Pareto surface around a given Pareto design is developed using neural network for design exploration. (2) A map from Pareto designs to their corresponding evaluation values, called the designer's local preferences model, is built using neural networks. (3) Genetic algorithms is used in a optimization process with the designer's local preferences model as objective function to search for the Pareto design that best satisfies the designer's local preferences. The obtained Pareto design is further used as the aspiration point in a compromise programming problem [4] to obtain the final optimal design.

2 Interactive Physical Programming Based on Neural Networks

Interactive physical programming takes into account the designer's preferences during the optimization process, which enables the designer to partly control the optimization process. The flow chart of interactive physical programming is shown in Figure 1, with detailed explanations given as follows.

J. Wang, X. Liao, and Z. Yi (Eds.): ISNN 2005, LNCS 3496, pp. 725–730, 2005.
© Springer-Verlag Berlin Heidelberg 2005

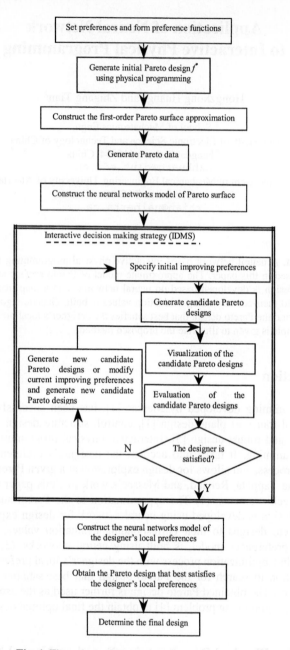

Fig. 1. Flow chart of interactive physical programming.

2.1 First-Order Pareto Surface Approximation around a Given Pareto Design

The initial Pareto design that best satisfies the designer's initial preferences, f^*, is generated by solving a physical programming problem. The next step is trying to achieve the approximation to the Pareto surface around f^* as accurately as possible, so that the designer can explore other Pareto designs in the following interactive decision making process. We make sensitivity analysis at Pareto design f^*, and generate the first-order Pareto surface approximation [4]. The first-order Pareto surface approximation is represented by a linear equation describing the relationship among all the objective functions.

2.2 Pareto Designs Generation and Neural Network Model of Pareto Surface

To represent the Pareto surface more accurately, more information about other Pareto designs around f^* is required. Usually, to a multiobjective optimization problem with m objectives, at least $m(m-1)/2$ Pareto designs are required. A projection method including two steps, the predictor step and the corrector step, is used to generate these Pareto designs [4].

Pareto surface is highly nonlinear, nonsmooth, and has discontinuity. Tappeta et al. used second-order polynomial to represent the Pareto surface [4], however, it can't describe the attributes of the Pareto surface mentioned above. Neural networks [6], [7] is good at representing complex nonlinear model, and it can describe the Pareto surface more accurately. There are interior relationships among the objective functions through design variables, and these relationships should be embodied as much as possible between inputs and outputs. Therefore, the neural network model of the Pareto surface is built with $(f_1, f_2, ..., f_{m-1})$ as input and $\sqrt[m+1]{f_1 \cdot f_2 \cdot .... \cdot f_{m-1} \cdot f_m^2}$ as output.

2.3 Interactive Decision Making Strategy

The flow chart of the interactive decision making strategy is shown in Figure 1. The Pareto design f^* is Pareto optimal, and there's no other feasible design that can improves all the objective functions. But the designer may want to improve some objectives at the expense of some other objectives, this is called the designer's improving preferences. The improving preferences can be specified by qualitative sentences, e.g., improve f_i, f_j, and sacrifice f_k.

After specifying the improving preferences, a set of candidate Pareto designs that satisfy the improving preferences are generated around the current Pareto design f^* using the neural network model of Pareto surface. They are presented to the designer using the Pareto visualization tool, which will help the designer to evaluate these candidate Pareto designs.

These candidate Pareto designs are evaluated using qualitative-quantitative analysis [8]. The qualitative part is evaluating the candidate Pareto designs with Analytic Hierarchy Process (AHP). The quantitative part is evaluating the candidate Pareto designs with quantitative criteria based on the preference functions of all the objectives. We combine four proposed quantitative criteria with the AHP approach to evaluate the candidate Pareto designs, and determine an evaluation value with respect to each candidate Pareto design to represent the designer's preference on the Pareto design.

After examining the candidate Pareto designs presented, the designer can generated a different set of approximate Pareto designs with current improving preferences, or modify current improving preferences and generate new approximate Pareto designs, or select one of the candidate designs presented if he is satisfied with it and then turn to the next step.

Finally, with the objective functions of the Pareto designs as inputs, the corresponding evaluation values as outputs to train a neural network, the neural network model of the designer's local preferences can be built. The neural network model of the map is called the neural network model of the designer's local preferences. The method developed in reference [4] to determine the Pareto design that best satisfies the designer's local preferences is just selecting the Pareto design that best satisfies the designer's local preferences from the candidates Pareto designs already generated. There're infinite Pareto designs around f^*, and it's obvious that the designer can't inspect all of them. Thus, the neural network based model gives us a continuous and more accurate model of the designer's local preference.

2.4 Determine the Final Design

With $(f_1, f_2, ..., f_{m-1})$ as design variables, the evaluation value corresponding to the Pareto design as objective function, genetic algorithms [9] is used to search for the Pareto design f_{local} that minimizes the evaluation value. f_{local} is thus the Pareto design that best satisfies the designer's local preferences. f_m is calculated via the design variables $(f_1, f_2, ..., f_{m-1})$ using the neural network model of the Pareto surface, and then the corresponding evaluation value can be calculated using the neural network model of the designer's local preferences.

The obtained Pareto design that best satisfies the designer's local preferences, f_{local}, is on the neural network model of the Pareto surface, and not on the real Pareto surface (although there's minor difference between them). With f_{local} as the aspiration point, a compromise programming problem is solved [4], and the final design x_{final} can be obtained. The objective functions vector corresponding to x_{final}, f_{final}, is on the real Pareto surface.

3 Example

A symmetrical pinned-pinned sandwich beam that supports a motor is considered [2]. A vibratory disturbance (at 10Hz) is imparted form the motor onto the beam. The mass of the motor is ignored in the following analysis. The objectives of this problem are fundamental frequency, cost, width, length, height and mass. The design variables are $x = \{d_1, d_2, d_3, b, L\}$, where L denotes the length of the beam, b is the width, and d_1, d_2 and d_3 represent the heights of the three pieces of the beam [2].

The region limits of the design objectives' preference functions are shown in Table 1. The steps are shown as follows.

(1) A physical programming problem is solved to obtained the initial Pareto design f^* and the corresponding design variables

$$f^* = (162.94, 346.27, 0.8747, 3.9967, 1867.6, 0.3599). \tag{1}$$

$$x^* = (0.3233, 0.3333, 0.3599, 0.8747, 3.9967). \tag{2}$$

(2) Through the sensitivity analysis of the Pareto surface at f^*, the first-order Pareto surface approximation around f^* is obtained. 30 Pareto designs around f^* are obtained. The neural network model of the Pareto surface is built.

(3) Through the interactive decision making process, the candidate Pareto designs are generated, visualized and evaluated. The neural network model of the designer's local preferences is built.

(4) Genetic algorithms is used in the optimization process to obtain the Pareto design that best satisfies the designer's local preferences. Then, a corresponding compromise programming problem is solved, and the final design can be obtained

$$f_{\text{final}} = (160.3097, 325.7527, 0.7799, 3.9479, 1834.2, 0.3477). \tag{3}$$

$$x_{\text{final}} = (0.2954, 0.3230, 0.3477, 0.7799, 3.9479). \tag{4}$$

Table 1. Physical programming region limits table.

Design objectives	Class type	g_{i5}	g_{i4}	g_{i3}	g_{i2}	g_{i1}
Fundamental frequency f/Hz	2-S	100	110	120	150	200
Cost c/\$·m^{-3}	1-S	2000	1950	1900	1800	1000
Width b/m	2-S	0.30	0.35	0.40	0.45	0.55
Length L/m	2-S	3.0	3.3	3.8	4.0	6.0
Mass m/kg	1-S	2800	2700	2600	2500	2000
Height h/m	1-S	0.60	0.55	0.50	0.40	0.30

4 Conclusions

The interactive nature of the proposed interactive physical programming approach enables the designer to partly control the optimization process, which can improve the design efficiency and design result, and avoid wasting lots of time in the wrong directions during the design process. Neural networks, a powerful nonlinear modeling tool, is used to construct the Pareto surface model and the designer's local preferences model, which makes them more accurate and reasonable. The continuous model of the designer's local preferences can be obtained in this way, and thus the continuous optimization can be implemented. From the view of continuous optimization, the design that best satisfies the designer's preferences can be obtained.

Acknowledgements

This research was partially supported by the National Excellent Doctoral Dissertation Special Foundation of China under the contract number 200232.

References

1. Messac, A., Hattis, P.D.: High Speed Civil Transport (HSCT) Plane Design Using Physical Programming. AIAA/ASME/ASCE/AHS Structures, Structural Dynamics & Materials Conference-collection of Technical Papers, **3** (1995) 10-13
2. Messac, A.: Physical Programming: Effective Optimization for Computational Design. AIAA Journal, **34** (1996) 149-158
3. Messac, A., Chen, X.: Visualizing the Optimization Process in Real-Time Using Physical Programming. Engineering Optimization, **32** (2000) 721-747
4. Tappeta, R.V., Renaud, J.E., Messac, A.: Interactive Physical Programming: Tradeoff Analysis and Decision Making in Multiobjective Optimization. AIAA Journal, **38** (2000) 917-926
5. Chen, W., Sahai, A., Messac, A.: Exploration of the Effectiveness of Physical Programming in Robust Design. Journal of Mechanical Design, Transactions of the ASME, **122** (2000) 155-163
6. Yan, P.F., Zhang, C.S.: Artificial Neural Network and Simulated Evolution Computation. Tsinghua University Press, Beijing (2000)
7. Huang, H.Z., Huang, W.P., Wang, J.N.: Neural Network and Application to Mechanical Engineering. Mechanical Science and Technology, **14** (1995) 97-103
8. Wang, Y.L.: System Engineering: Theory, Method and Application. Higher Education Press, Beijing (1998)
9. Huang, H.Z., Zhao, Z.J.: Genetic Algorithm Principle, Realization and Their Application Research, Prospect in Mechanical Engineering. Journal of Machine Design. **17** (2000) 1-6

Application of the "Winner Takes All" Principle in Wang's Recurrent Neural Network for the Assignment Problem

Paulo Henrique Siqueira[1], Sergio Scheer[2], and Maria Teresinha Arns Steiner[3]

[1] Federal University of Paraná, Department of Drawing, Postfach 19081
81531-990 Curitiba, Brazil
paulohs@ufpr.br
[2] Federal University of Paraná, Department of Civil Construction, Postfach 19011
81531-980 Curitiba, Brazil
scheer@ufpr.br
[3] Federal University of Paraná, Department of Mathematics, Postfach 19081
81531-990 Curitiba, Brazil
tere@mat.ufpr.br

Abstract. One technique that uses Wang's Recurrent Neural Networks with the "Winner Takes All" principle is presented to solve the Assignment problem. With proper choices for the parameters of the Recurrent Neural Network, this technique reveals to be efficient solving the Assignment problem in real time. In cases of multiple optimal solutions or very closer optimal solutions, the Wang's Neural Network does not converge. The proposed technique solves these types of problem. Comparisons between some traditional ways to adjust the RNN's parameters are made, and some proposals concerning to parameters with dispersion measures of the problem's cost matrix' coefficients are show.

1 Introduction

The Assignment problem (AP) is a classical combinatorial optimization problem of the Operational Research. The objective of this problem is assigning a number of elements to the same number of positions, and minimizing the linear cost function. This problem is known in literature as Linear Assignment problem or problem of Matching with Costs [1]-[3]. Beyond traditional techniques, as the Hungarian method and the Simplex method, diverse ways of solving this problem has been presented in the last years.

In problems of great scale, i.e., when the problem's cost matrix is very large, the traditional techniques do not reveal efficiency, because the number of restrictions and the computational time are increased. Since the Hopfield and Tank's publication [4], lots of works about the use of Neural Networks to solving optimization problems had been developed [5]-[8]. The Hopfield's Neural Network [5], converges to the optimal solution of any Linear Programming problem, in particular for the AP.

Wang considered a Recurrent Neural Network to solve the Assignment problem [6], however, the necessary number of iterations to achieve an optimal solution is increased for problems of great scale. Moreover, in problems with multiple optimal solutions or very closer optimal solutions, such network does not converge.

J. Wang, X. Liao, and Z. Yi (Eds.): ISNN 2005, LNCS 3496, pp. 731–738, 2005.

In this work, one technique based on the "Winner Takes All" (WTA) principle is presented, revealing efficiency solving the problems found in the use of Wang's Recurrent Neural Network (WRNN). Some criteria to adjust the parameters of the WRNN are presented: some traditional ways and others that use dispersion measures between the cost matrix' coefficients.

This work is divided in 6 sections, including this introduction. In section 2, the AP is defined. In section 3, the WRNN is presented and an example of a multiple optimal solutions' problem is show. In section 4, a technique based on the WTA principle is presented and an example of its application is show. In section 5, some alternatives for parameters' construction of the WRNN are presented, and in section 6 some results are presented and some conclusions are made.

2   Problem Formulation

The Linear Assignment problem can be formulated as follows:

$$\text{Minimize} \quad C = \sum_{i=1}^{n}\sum_{j=1}^{n} c_{ij}x_{ij} \tag{1}$$

$$\text{Subject to} \quad \sum_{i=1}^{n} x_{ij} = 1, \quad j = 1, 2, ..., n \tag{2}$$

$$\sum_{j=1}^{n} x_{ij} = 1, \quad i = 1, 2, ..., n \tag{3}$$

$$x_{ij} \in \{0, 1\}, \quad i, j = 1, 2, ..., n, \tag{4}$$

where c_{ij} and x_{ij} are, respectively, the cost and the decision variable associated to the assignment of element i to position j. The usual representation form of c in the Hungarian method is the matrix form. When $x_{ij} = 1$, element i is assigned to position j.

The objective function (1) represents the total cost to be minimized. The set of constraints (2) and (3) guarantees that each element i will be assigned for exactly one position j. The set (4) represents the zero-one integrality constraints of the decision variables x_{ij}. The set of constraints (4) can be replaced by:

$$x_{ij} \geq 0, \quad i, j = 1, 2, ..., n. \tag{5}$$

Consider the $n^2 \times 1$ vectors c^T, that contains all the rows of matrix c; x, that contains the decision elements x_{ij}, and b, that contains the number 1 in all positions. The matrix form of the problem described in (1)-(4) is given to [9]:

Minimize $C = c^T x$ $\qquad\qquad\qquad\qquad\qquad\qquad\qquad$ (6)
Subject to $Ax = b$ $\qquad\qquad\qquad\qquad\qquad\qquad\qquad$ (7)
$\qquad\qquad x_{ij} \geq 0, \quad i, j = 1, 2, ..., n,$

where matrix A has the following form:

$$A = \begin{bmatrix} I & I & \cdots & I \\ B_1 & B_2 & \cdots & B_n \end{bmatrix} \in \Re^{2n \times n^2}$$

where I is an $n \times n$ identity matrix, and each B_i matrix, for $i = 1, 2..., n$, contains zeros, with exception of ith row, that contains the number 1 in all positions.

3   Wang's Recurrent Neural Network for the Assignment Problem

The Recurrent Neural Network proposed by Wang, published in [6], [7] and [9], is characterized by the following differential equation:

$$\frac{du_{ij}(t)}{dt} = -\eta \sum_{k=1}^{n} x_{ik}(t) - \eta \sum_{l=1}^{n} x_{lj}(t) + \eta \theta_{ij} - \lambda c_{ij} e^{-\frac{t}{\tau}}, \tag{8}$$

where $x_{ij} = g(u_{ij}(t))$ and the steady state of this Neural Network is a solution for the AP. The threshold is defined as the $n^2 \times 1$ vector $\theta = A^T b = (2, 2, ..., 2)$. Parameters η, λ and τ are constants, and empirically chosen [9], affecting the convergence of the network. Parameter η serves to penalize violations in the problem's constraints' set, defined by (1)-(4). Parameters λ and τ control the objective function's minimization of the AP (1). The Neural Network matrix form can be written as:

$$\frac{du(t)}{dt} = -\eta(Wx(t) - \theta) - \lambda c e^{-\frac{t}{\tau}}, \tag{9}$$

where $x = g(u(t))$ and $W = A^T A$. The WRNN's convergence properties are demonstrated in [8]-[12].

3.1   Multiple Optimal Solutions and Closer Optimal Solutions

In some cost matrices, the optimal solutions are very closer to each other, or in a different way, some optimal solutions are admissible. The cost matrix c given below:

$$c = \begin{pmatrix}
1.4 & 6.1 & 3.1 & 0.4 & 2.2 & 4.4 & 0.1 & 0 \\
1.3 & 0.2 & 3.3 & 0.2 & 1.2 & 0.4 & 1.5 & 8.2 \\
1.7 & 2.9 & 2.8 & 0.1 & 9.8 & 0.4 & 4.2 & 6.8 \\
0.5 & 0 & 0 & 1.0 & 1.1 & 0.9 & 2.4 & 0.9 \\
2.8 & 7.7 & 1.2 & 0.5 & 0.9 & 6 & 4.6 & 5.9 \\
0.5 & 0.2 & 0 & 1.8 & 8.5 & 4.9 & 4.4 & 0.9 \\
0.6 & 0.7 & 0.7 & 6.9 & 0.1 & 0 & 3.2 & 3.8 \\
1.4 & 5.4 & 3.7 & 1.1 & 3 & 0 & 1.3 & 0.7
\end{pmatrix}, \tag{10}$$

has the solutions x^* and $\hat{x}$ given below:

$$x^* = \begin{pmatrix}
0 & 0 & 0 & 0 & 0 & 0 & 1 & 0 \\
0 & 1 & 0 & 0 & 0 & 0 & 0 & 0 \\
0 & 0 & 0 & 0.5 & 0 & 0.5 & 0 & 0 \\
0.5 & 0 & 0.5 & 0 & 0 & 0 & 0 & 0 \\
0 & 0 & 0 & 0.5 & 0.5 & 0 & 0 & 0 \\
0.5 & 0 & 0.5 & 0 & 0 & 0 & 0 & 0 \\
0 & 0 & 0 & 0 & 0.5 & 0.5 & 0 & 0 \\
0 & 0 & 0 & 0 & 0 & 0 & 0 & 1
\end{pmatrix} \text{ and } \hat{x} = \begin{pmatrix}
0 & 0 & 0 & 0 & 0 & 0 & 1 & 0 \\
0 & 1 & 0 & 0 & 0 & 0 & 0 & 0 \\
0 & 0 & 0 & 0 & 0 & 1 & 0 & 0 \\
0 & 0 & 1 & 0 & 0 & 0 & 0 & 0 \\
0 & 0 & 0 & 1 & 0 & 0 & 0 & 0 \\
1 & 0 & 0 & 0 & 0 & 0 & 0 & 0 \\
0 & 0 & 0 & 0 & 1 & 0 & 0 & 0 \\
0 & 0 & 0 & 0 & 0 & 0 & 0 & 1
\end{pmatrix},$$

where x^* is founded after 6,300 iterations using the WRNN, and $\hat{x}$ is an optimal solution.

The solution x^* isn't feasible, therefore, some elements x_{ij}^* violate the set of restrictions (4), showing that the WRNN needs adjustments for these cases. The simple decision to place unitary value for any one of the elements x_{ij}^* that possess value 0.5 in solution x^* can become unfeasible or determine a local optimal solution. Another adjustment that can be made is the modification of the costs' matrix' coefficients, eliminating ties in the corresponding costs of the variable x_{ij}^* that possess value different from 0 and 1. In this manner, it can be found a local optimal solution when the modifications are not made in the adequate form. Hence, these decisions can cause unsatisfactory results.

4   Use of a Technique Based on the WTA Principle for the AP

The method considered in this work uses one technique based on the "Winner Takes All" principle, speeding up the convergence of the WRNN, besides correcting eventual problems that can appear due the multiple optimal solutions or very closer optimal solutions.

The second term of equation (9), $Wx(t) - \theta$, measures the violation of the AP's constraints. After a certain number of iterations, this term does not suffer substantial changes in its value, evidencing the fact that problem's restrictions are almost satisfied. At this moment, the method considered in this section can be applied.

When all elements of x satisfy the condition $Wx(t) - \theta \leq \delta$, where $\delta \in [0, 2]$, the proposed technique can be used in all iterations of the WRNN, until a good approach of the AP be found. An algorithm of this technique is presented as follows:

Step 1: Find a solution x of the AP, using the WRNN. If $Wx(t) - \theta \leq \delta$, then go to Step 2. Else, find another solution x.

Step 2: Given the matrix of decision x, after a certain number of iterations of the WRNN. Let the matrix $\bar{x}$, where $\bar{x} = x$, $m = 1$, and go to step 3.

Step 3: Find the mth biggest array element of decision, $\bar{x}_{kl}$. The value of this element is replaced by the half of all elements sum of row k and column l of matrix x, or either,

$$\bar{x}_{kl} = \frac{1}{2} \left(\sum_{i=1}^{n} x_{il} + \sum_{j=1}^{n} x_{kj} \right). \tag{11}$$

The other elements of row k and column l become nulls. Go to step 4.

Step 4: If $m \leq n$, makes $m = m + 1$, and go to step 3. Else, go to step 5.

Step 5: If a good approach to an AP solution is found, stop. Else, make $x = \bar{x}$, execute the WRNN again and go to Step 2.

4.1   Illustrative Example

Consider the matrix below, which it is a partial solution of the AP defined by matrix c, in (10), after 14 iterations of the WRNN. The biggest array element of $\bar{x}$ is in row 1, column 7.

$$\bar{x} = \begin{pmatrix} 0.0808 & 0.0011 & 0.0168 & 0.1083 & 0.0514 & 0.0033 & *0.422 & 0.3551 \\ 0.0056 & 0.2827 & 0.0168 & 0.1525 & 0.1484 & 0.1648 & 0.1866 & 0 \\ 0.1754 & 0.0709 & 0.0688 & 0.3449 & 0 & 0.3438 & 0.0425 & 0.0024 \\ 0.1456 & 0.2412 & 0.2184 & 0.0521 & 0.1131 & 0.0747 & 0.0598 & 0.1571 \\ 0.0711 & 0 & 0.2674 & 0.272 & 0.3931 & 0.0024 & 0.0306 & 0.0061 \\ 0.2037 & 0.2823 & 0.2956 & 0.0366 & 0 & 0.0025 & 0.0136 & 0.2186 \\ 0.1681 & 0.174 & 0.1562 & 0 & 0.3053 & 0.2016 & 0.0369 & 0.0144 \\ 0.1142 & 0.0031 & 0.0138 & 0.0829 & 0.0353 & 0.2592 & 0.251 & 0.2907 \end{pmatrix}$$

After the update of this element through equation (11), meet the result given below. The second biggest element of $\bar{x}$ is in row 5, column 5.

$$\bar{x} = \begin{pmatrix} 0 & 0 & 0 & 0 & 0 & 0 & 1.0412 & 0 \\ 0.0056 & 0.2827 & 0.0168 & 0.1525 & 0.1484 & 0.1648 & 0 & 0 \\ 0.1754 & 0.0709 & 0.0688 & 0.3449 & 0 & 0.3438 & 0 & 0.0024 \\ 0.1456 & 0.2412 & 0.2184 & 0.0521 & 0.1131 & 0.0747 & 0 & 0.1571 \\ 0.0711 & 0 & 0.2674 & 0.272 & *0.393 & 0.0024 & 0 & 0.0061 \\ 0.2037 & 0.2823 & 0.2956 & 0.0366 & 0 & 0.0025 & 0. & 0.2186 \\ 0.1681 & 0.174 & 0.1562 & 0 & 0.3053 & 0.2016 & 0 & 0.0144 \\ 0.1142 & 0.0031 & 0.0138 & 0.0829 & 0.0353 & 0.2592 & 0 & 0.2907 \end{pmatrix}$$

After the update of all elements of $\bar{x}$, get the following solution:

$$\bar{x} = \begin{pmatrix} 0 & 0 & 0 & 0 & 0 & 0 & 1.0412 & 0 \\ 0 & 1.0564 & 0 & 0 & 0 & 0 & 0 & 0 \\ 0 & 0 & 0 & 1.0491 & 0 & 0 & 0 & 0 \\ 1.0632 & 0 & 0 & 0 & 0 & 0 & 0 & 0 \\ 0 & 0 & 0 & 0 & 1.0446 & 0 & 0 & 0 \\ 0 & 0 & 1.0533 & 0 & 0 & 0 & 0. & 0 \\ 0 & 0 & 0 & 0 & 0 & 1.0544 & 0 & 0 \\ 0 & 0 & 0 & 0 & 0 & 0 & 0 & 1.0473 \end{pmatrix}.$$

This solution is presented to the WRNN, and after finding another x solution, a new $\bar{x}$ solution is calculated through the WTA principle.

This procedure is made until a good approach to feasible solution is founded. In this example, after more 5 iterations, the matrix $\bar{x}$ presents one approach of the optimal solutions:

$$\bar{x} = \begin{pmatrix} 0 & 0 & 0 & 0 & 0 & 0 & 0.9992 & 0 \\ 0 & 0.9996 & 0 & 0 & 0 & 0 & 0 & 0 \\ 0 & 0 & 0 & 0.9994 & 0 & 0 & 0 & 0 \\ 0.999 & 0 & 0 & 0 & 0 & 0 & 0 & 0 \\ 0 & 0 & 0 & 0 & 0.9985 & 0 & 0 & 0 \\ 0 & 0 & 1 & 0 & 0 & 0 & 0. & 0 \\ 0 & 0 & 0 & 0 & 0 & 1.0003 & 0 & 0 \\ 0 & 0 & 0 & 0 & 0 & 0 & 0 & 0.9991 \end{pmatrix}.$$

Two important aspects of this technique that must be taken in consideration are the following: the reduced number of iterations necessary to find a solution feasible, and the absence of problems related to the matrices with multiple optimal solutions. The adjustments of the WRNN's parameters are essential to guarantee the convergence of this technique, and some forms of adjusting are presented on the next section.

5   Some Methodologies for Adjusting the WRNN's Parameters

In this work, the used parameters play basic roles for the convergence of the WRNN. In all the tested matrices, $\eta = 1$ had been considered, and parameters τ and λ had been calculated in many ways, described as follows.

One of the most usual forms to calculate parameter λ for the AP can be found in [6], where λ is given by:

$$\lambda = \eta / C_{max},$$

(12)

where $C_{max} = \max\{c_{ij}; i, j = 1, 2, ..., n\}$.

The use of dispersion measures between the c matrix coefficients had revealed to be efficient adjusting parameters τ and λ. Considering D as the standard deviation between the c cost matrix' coefficients, the parameter λ can be given as:

$$\lambda = \eta / D.$$

(13)

Another way to adjust λ is to consider it a vector, defined by:

$$\bar{\lambda} = \eta \left(\frac{1}{d_1}, \frac{1}{d_2}, ..., \frac{1}{d_n} \right),$$

(14)

where d_i, for $i = 1, 2..., n$, represents the standard deviation of each row of the matrix c. Each element of the vector $\bar{\lambda}$ is used to update the corresponding row of the x decision matrix. This form to calculate λ revealed more to be efficient in cost matrices with great dispersion between its values, as shown by the results presented in the next section.

A variation of the expression (12), that uses the same principle of the expression (14), is to define λ by the vector:

$$\bar{\lambda} = \eta \left(\frac{1}{c_{1\,max}}, \frac{1}{c_{2\,max}}, ..., \frac{1}{c_{n\,max}} \right),$$

(15)

where $c_{i\,max} = \max\{c_{ij}; j = 1, 2, ..., n\}$. This definition to λ also produces good results in matrices with great dispersion between its coefficients.

The parameter τ depends on the necessary number of iterations for the convergence of the WRNN. When the presented correction WTA technique isn't used, the necessary number of iterations for the convergence of the WRNN varies between 1,000 and 15,000 iterations. In this case, τ is a constant, such that:

$$1,000 \leq \tau \leq 15,000.$$

(16)

When the WTA correction is used, the necessary number of iterations varies between 5 and 300. Hence, the value of τ is such that:

$$5 \leq \tau \leq 300.$$

(17)

In this work, two other forms of τ parameter adjustment had been used, besides considering it constant, in the intervals showed in expressions (16) and (17). In one of the techniques, τ is given by:

$$\bar{\tau} = \left(\frac{m_1}{M} d_1, \frac{m_2}{M} d_2, ..., \frac{m_n}{M} d_n \right),$$

(18)

where m_i is the coefficients average of ith row of matrix c, d_i is the standard deviation of ith row of matrix c, and M is the average between the values of all the coefficients of c.

The second proposal of adjustment for τ uses the third term of the WRNN definition (8). When $c_{ij} = c_{max}$, the term $\lambda c \exp(-k/\tau) = g$ must be great, so that x_{ij} has minor value, minimizing the final cost of the AP. Isolating τ, and considering $c_{ij} = c_{max}$ and $\lambda_i = 1/d_i$, where $i = 1, 2..., n$, τ is got, as follows:

$$\tau_i = \frac{k}{\ln\left(\dfrac{c_{i\,max}}{g.d_i} \right)},$$

(19)

The parameters' application results given by (12)-(19) are presented in the next section.

6   Results and Conclusions

In this work, 63 matrices (with dimensions varying of 3×3 until 20×20) had been used to test the parameters' adjustments' techniques presented in the previous section, beyond the proposed   WTA correction applied to the WRNN. These matrices had been generated randomly, with some cases of multiple optimal solutions and very closer optimal solutions.

The results for the 26 tested matrices with only one optimal global and the 37 matrices with multiple optimal solutions and/or very closer optimal solutions appear in Table 1. To adjust λ, the following expressions had been used in Table 1: (12) in the first and last column; (13) in the second column; (15) in the third column; and (14) in fourth and fifth columns. To calculate τ, the following expressions they had been used: (17) in the three firsts columns; (18) in the fourth column; (19) in the fifth column; and (16) in the last column. The results of the WRNN application, without the use of the proposed correction in this work, are meet in the last column of Table 1.

In the last row of the Table 1, the numbers of iterations of each technique is given by the average between the numbers of iterations found for all tested matrices. The results had been considered satisfactory, and the adjustments of the parameters that result in better results for the WTA correction are the ones that use the standard deviation and the average between the cost matrix' coefficients, and the use of parameters in vector form revealed to be more efficient for these matrices.

The results shown in Table 1 reveal that the dispersion techniques between the coefficients of matrix c are more efficient for the use of the correction WTA in matrices with multiple optimal solutions. The pure WRNN has slower convergence when the adjustments described by (13)-(15) and (17)-(19) are applied for the parameters λ and τ, respectively.

4.2   Calculating the Second Eigenvalue

Suppose $\mathbf{v}$ is a valid solution, then exists a vector $\mathbf{v}'$ such that $\mathbf{v}' = \mathbf{v} - \hat{\mathbf{e}}^1$. Thus,

$$\mathbf{u}' = \mathbf{T}\mathbf{v}' = \mathbf{T}(\mathbf{v} - \hat{\mathbf{e}}^1) = \mathbf{T}\mathbf{v} - \mathbf{T}\hat{\mathbf{e}}^1 = \mathbf{T}\mathbf{v} - \lambda_1\hat{\mathbf{e}}^1 = \mathbf{u} - \lambda_1\hat{\mathbf{e}}^1. \quad (12)$$

From (10), it follows that

$$u_{xi} = -A\sum_y\sum_j \delta_{xy}v_{yj} - A\sum_y\sum_j \delta_{ij}v_{yj} + C\sum_y\sum_j \delta_{xy}\delta_{ij}v_{yj}$$

$$= -A\sum_j v_{xj} - A\sum_y v_{yi} + Cv_{xi}.$$

Since v is a valid solution, then $\sum_j v_{xj} = 1$ and $\sum_y v_{y,i} = 1$. Then,

$$u_{xi} = -2A + Cv_{xi} \quad and \quad u = -2Ae^1 + Cv. \quad (13)$$

Applying (13) and $\lambda_1 = -2NA + C$ to (12), it given that

$$\mathbf{u}' = \mathbf{T}\mathbf{v}' = -2Ae^1 + Cv - (-2NA + C)\hat{\mathbf{e}}^1 = C(v - \hat{\mathbf{e}}^1) = Cv' = \lambda_2v'. (14)$$

This shows that $\lambda_2 = C$ is a eigenvalue of the matrix T.

4.3   Calculating the Third Eigenvalue

We will use the same method as that in [4] for the calculation. Let $v^{(a)}$ be a valid form of v, and denoted

$$S = \sum_a v^{(a)}v^{(a)t} \quad and \quad S_{xi,yj} = \sum_a v_{xi}^{(a)}v_{yj}^{(a)}. \quad (15)$$

Consider the following cases:

1. $i = j$ and $x \neq y$. Then, v is a valid tour, it is not possible for $v_{xi} = 1$ and $v_{yj} = 1$ if $x \neq y$, therefore all the elements of S where $i = j$ and $x \neq y$ are zero.
2. $i = j$ and $x = y$. Then, $(N - 1)!$ permutation exit for a fixed position of one city. therefore $S_{xi,xi} = (N - 1)!$.
3. $i \neq j$ and $x = y$. It is not possible for $v_{xi} = 1$ and $v_{xj} = 1$ if $i \neq j$. So, all the elements of S where $x = y$ and $i \neq j$ are zero.
4. $i \neq j$ and $x \neq y$. In this case, the position of two cities in the tour are fixed and the number of possible permutation of the others is $(N - 2)!$. $S_{xi,yj} = (N - 2)!$ when $i \neq j$ and $x \neq y$.

From above, the matrix T can be expressed in terms of S by:

$$T = \frac{A}{(N - 2)!}S + (C - NA)I - Ae^1e^{1t}. \quad (16)$$

It clear that the eigenvalues of S in the valid subspace must be zero by the definition of S. The first term of (16) just scales the eigenvalue of S, and the second term only affects the eigenvalue in the direction of e^1. Thus, the subtraction of $(C - NA)I$ will give an eigenvalue of $(C - NA)$ for the invalid subspace of T, i.e., $\lambda_3 = C - NA$.

5   Dynamics of TSP Network and the Parameters Setting

Aiyer el at [4] has pointed out that the term $-(\frac{u}{\tau})$ in equation (1) do not affect the essential behavior of the network. To simplify the analysis, we assume that

$$\frac{d\mathbf{u}}{dt} = \mathbf{Tv} + \mathbf{i}^b. \tag{17}$$

The output $\mathbf{u}$ can be decomposed into its valid component $\mathbf{u}^{val}$, invalid subspace component $\mathbf{u}^{inv}$ plus a component $\mathbf{u}^1 = (\mathbf{u}.\hat{e}^1)\hat{e}^1$ in the direction of e^1, i.e.,

$$\mathbf{u} = \mathbf{u}^1 + \mathbf{u}^{val} + \mathbf{u}^{inv}. \tag{18}$$

The bias can be written as

$$\mathbf{i}^b = (A + B - C/2)\mathbf{e}^1 = (2A - C/2)\mathbf{e}^1. \tag{19}$$

The corresponding Liapunov function is

$$E = -0.5 \cdot \mathbf{u}^\top \mathbf{Tu} - (\mathbf{i}^b)^\top \mathbf{u}. \tag{20}$$

Then

$$
\begin{aligned}
2E &= -\mathbf{u}^\top \mathbf{Tu} - 2(\mathbf{i}^b)^\top \mathbf{u} \\
&= -\lambda_1 |\mathbf{u}^1|^2 - \lambda_2 |\mathbf{u}^{val}|^2 - \lambda_3 |\mathbf{u}^{inv}|^2 - 2(2A - C/2)N(\hat{e}^1.\mathbf{u}) \\
&= -\lambda_1 |\mathbf{u}^1|^2 - 2(2A - C/2)N|\mathbf{u}^1| - C|\mathbf{u}^{val}|^2 - (C - NA)|\mathbf{u}^{inv}|^2 \\
&= -\lambda_1 \left(|\mathbf{u}^1| + (2A - C/2)N/\lambda_1 \right)^2 - C|\mathbf{u}^{val}|^2 \\
&\quad -(C - NA)|\mathbf{u}^{inv}|^2 + (2A - C/2)^2 N^2/\lambda_1 \\
&= -(C - 2NA) \left(|\mathbf{u}^1| + (2A - C/2)N/\lambda_1 \right)^2 - C|\mathbf{u}^{val}|^2 \\
&\quad -(C - NA)|\mathbf{u}^{inv}|^2 + (2A - C/2)^2 N^2/\lambda_1. \tag{21}
\end{aligned}
$$

When $C - NA < 0$ and $C - 2NA < 0$, since A, C and N are positive, to minimize E what is required is that

$$|\mathbf{u}^1| = -(2A - C/2)N/\lambda_1 > 0, \quad |\mathbf{u}^{inv}| \longrightarrow 0, \quad |\mathbf{u}^{val}| \longrightarrow \infty. \tag{22}$$

So to minimize E the following condition should be meet

$$\begin{cases} C - NA < 0 \\ 2A - \frac{C}{2} < 0 \end{cases} \implies 4A < C < NA. \tag{23}$$

6   Simulation

In this section, a 30-city's example is employed to verify the theoretical results in this paper. 30 city's site are distributed in an unit square randomly, each site are varied from $[0.0000, 0.0000]$ to $[1.0000, 1.0000]$. The simulation result showed that the value of parameter D is effected slightly to the convergence of the network when it be limited in a certainly range. In our simulating, we set $A = B = 10$, $C = 80$, and $D = 0.1$. 100 different initialize value of city site are performed and 97 of that can be converged to the valid state in the above parameter setting. We select a good result as following fig.

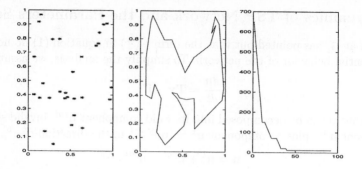

Fig. 1. The left figure is initialized city's site that random distributed in unit square; The middle one is the finally city sequence; The right one represent the energys changing.

7 Conclusions

By analyzing the eigenvalues of the connection matrix of a enhanced formulation of energy function, we shown the relation between the dynamics of network and the constants in the energy function, and given a group of criterion for parameter setting. With this enhanced formulation, a 30 city TSP problem's simulation have provided in this paper. The simulation's result shown that the new criterion of parameter setting is performed well.

References

1. Hopfield, J.J., Tank, D.W.: 'Neural' Computation of Decisions in Optimization Problemss. Biological Cybern, **52** (1985) 141-152
2. Wilson, G.W. , Pawley, G.S.: On the Stability of the Travelling salesman Problem Algorithm of Hopfield and Tank. Biol. Cybern., **58** (1988) 63-70
3. Kanmgar-Parsi, B., Kamgar-Parsi, B.: Dynamical Stability and Parameter Selection in Neural Optimization. Proceedings Int. Joint Conference On Neural Networks, **4** (1992) 566-771
4. Aiyer, S.V.B., Niranjan, M., Fallside, F.: A Theoretical Investigation into the Performance of the Hopfield Model. IEEE Trans. Neural Networks, **1** (1990) 204-215
5. Talavan, P.M., Yanez: Parameter Setting of the Hopfield Network Applied to TSP. Neural Networks, **15** (2002) 363-373
6. Brandt, R.D. , Wang, Y. Laub, A.J., Mitra, S.K.: Alternative Networks for Solving the Traveling Salesman Problem and the List-matching Problem. Proceedings Int. Joint Conference on Neural Networks, **2** (1988) 333-340
7. Matsuda, S.: Optimal Hopfield Network for Combinatorial Optimization with Linear Cost Function. IEEE Trans. Neural Networks, **9** (1999) 1319-1330

Solving Optimization Problems Based on Chaotic Neural Network with Hysteretic Activation Function

Xiuhong Wang[1], Qingli Qiao[2], and Zhengqu Wang[1]

[1] Institute of Systems Engineering, Tianjin University, Tianjin 300072, China
Wangxh1965@Eyou.com
[2] School of life Science & Technology, Shanghai Jiaotong University, Shanghai 200030, China
Qlqiao@Sjtu.edu.cn

Abstract. We present a new chaotic neural network whose neuron activation function is hysteretic, called hysteretic transiently chaotic neural network (HTCNN) and with this network, a combinatorial optimization problem is solved. By using hysteretic activation function which is multi-valued, has memory, is adaptive, HTCNN has higher ability of overcoming drawbacks that suffered from the local minimum. We proceed to prove Lyapunov stability for this new model, and then solve a combinatorial optimization problem- Assignment problems. Numerical simulations show that HTCNN has higher ability to search for globally optimal and has higher searching efficiency.

1 Introduction

Hopfield neural networks (HNN) [1] have been recognized as powerful tools for op-timization. In order to overcome the shortcoming of the HNN that they often suffer from the local minima, many researchers have investigated the characteristics of the chaotic neural networks (CNN) and attempted to apply them to solve optimization problems [2]. In fact, most of these chaotic neural networks have been proposed with monotonous activation function such as sigmoid function. Recently, the neuro-dynamics with a non-monotonous have been reported [3], they have found that the non-monotonous mapping in a neuron dynamics posses an advantage of the memory capacity superior to the neural network models with a monotonous mapping.

We'll propose a chaotic neural network model with a non-monotonous mapping-hysteresis. Hysteretic neuron models have been proposed [4] for association memory. And have been demonstrated performing better than non-hysteretic neuron models, in terms of capacity, signal-to-noise ratio, recall ability, etc. It is likely that the hysteresis in our neural network is responsible for the network overcoming local minima in its trajectory towards a solution.

2 Hysteretic Neuron and Hysteretic Hopfield Neural Network

Our hysteretic neuron activation function is depicted in Fig.1. Mathematically, our hysteretic neuron activation function is described as:

$$y(x / \dot{x}) = \phi(x - \lambda(\dot{x}) = \tanh(\gamma(\dot{x})(x - \lambda(\dot{x}))) \tag{1}$$

Where $\gamma(\dot{x}) = \begin{cases} \gamma_\alpha, \dot{x} \geq 0 \\ \gamma_\beta, \dot{x} < 0 \end{cases}$, $\lambda(\dot{x}) = \begin{cases} -\alpha, \dot{x} \geq 0 \\ \beta, \dot{x} < 0 \end{cases}$ and $\beta > -\alpha$, $(\gamma_\alpha, \gamma_\beta) > 0$.

J. Wang, X. Liao, and Z. Yi (Eds.): ISNN 2005, LNCS 3496, pp. 745–749, 2005.

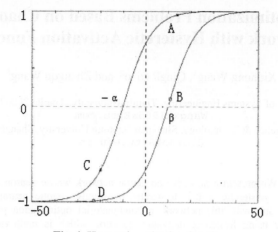

Fig. 1. Hysteretic activation function.

Observed that, in the special case when $\beta = \alpha$ and $\gamma_\alpha = \gamma_\beta$, the activation function becomes the conventional sigmoid function. If x is positive at one time point and increases in value at the next time point, that is $\dot{x} > 0$, the activation function remains along segment C-A. On the other hand, if x is positive at one time point and decreases at the next time point, that is $\dot{x} < 0$, then the activation function jumps from hysteretic segment C-A to segment B-D. Observed that this neuron's output y not only depends on its input x, but also on derivative information, namely $\dot{x}$. It is the latter information that provides the neuron with memory and distinguishes it from other neurons. We also note that the hysteretic neuron's activation function has four parameters associated with it, namely, $\alpha, \beta, \gamma_\varepsilon, \gamma_\beta$. And we can think about tuning all of its parameters in order to maximize its performance. So it seems that the hysteretic neuron has much more flexibility than the usual neuron.

By combining N neurons, each having the hysteretic activation function given in (1), we form the hysteretic Hopfield neural network (HHNN). The dynamics of HHNN is defined by

$$\frac{dx_i}{dt} = -\frac{x_i}{\tau} + \sum_j w_{ij} y_j + I_i \tag{2}$$

$$y(x/\dot{x}) = \phi(x - \lambda(\dot{x}) = \tanh(\gamma(\dot{x})(x - \lambda(\dot{x}))) \tag{3}$$

Where y_i, x_i, and I_i are output value, internal input value, and bias of a neuron i, respectively, and w_{ij} is a synaptic weight from neuron j to i. τ is the time constant, and $\tau = R_i C_i$, where, R_i and C_i are input resistance and capacitor of neuron i. Usually, set $\tau = 1$.

HHNN includes memory because of using hysteretic function as neuron's activation function. And due to a change in the direction of the input, a system can pull itself out of a saturated region by jumping from one segment of the hysteretic activation function to the other segment. This make the HHNN has a tendency to overcome local minima.

3 Stability Analysis of the HHNN

We demonstrate stability of the HHNN by Lyapunov direct method.

For a function $E(y)$ to be a Lyapunov function, the following three properties must be satisfied: 1) $E(y_0)=0$; 2) $E(y)>0$, $y \neq y_0$; And, 3) $E(y)$ should have partial derivatives with respect to all y. Given that $E(y)$ is a Lyapunov function, and if $dE(y)/dt \leq 0$, then y_0 is stable in the sense of Lyapunov.

For the system HHNN, we propose the following energy function:

$$E = -\frac{1}{2}\sum_{i=1}^{n}\sum_{j=1}^{n} w_{ij} y_i y_j \; - \sum_{i=1}^{n} y_i I_i + \sum_{i=1}^{n}\frac{1}{R_i}\int_0^{y_i}\phi^{-1}(y)dy \tag{4}$$

E in (4) is indeed a Lyapunov function, here we focus on demonstrating the truth of $dE(y)/dt \leq 0$. Assuming $w_{ij}=w_{ji}$, and $w_{ii}=0$, differentiating (4), we find that

$$\frac{dE}{dt} = -\sum_i \frac{dy_i}{dt}(\sum_j w_{ij} y_j - \frac{x_i}{R_i} + I_i) = -\sum_i C_i(\frac{dy_i}{dt})(\frac{dx_i}{dt})$$

$$= -\sum_i C_i\frac{dy_i}{dx_i}(\frac{dx_i}{dt})^2 = -\sum_i C_i\frac{\Delta y_i}{\Delta x_i}(\frac{dx_i}{dt})^2 \tag{5}$$

We analyze (5) by considering the following transitions. (See Fig.1)

1) Transition 1 (A→B) for which $\Delta x_i <0 \Rightarrow \Delta y_i <0 \Rightarrow dE/_{dt} <0$

2) Transition 2 (B→A) for which $\Delta x_i >0 \Rightarrow \Delta y_i >0 \Rightarrow dE/_{dt} <0$

3) Transition 3 (C→A, B→D) for which transition occur along the sigmoid, which is a no decreasing function i.e., for which $dE(y)/dt <0$.

Because $dE(y)/dt <0$ for all possible transitions, we have show that the equilibrium point for the HHNN is stable in the sense of Lyapunov.

4 Hysteretic Chaotic Neural Network

In order to make use of the advantages of both the chaotic neurodynamics and the hysteretic neural networks, we introduce chaotic dynamics into HHNN to create hysteretic chaotic neural network so as to solve optimal problems more efficiently.

We propose a hysteretic transiently chaotic neural network (HTCNN), and the continuous dynamics of the HTCNN is

$$y(x / \dot{x}) = \phi(x - \lambda(\dot{x}) = \tanh(\gamma(\dot{x})(x - \lambda(\dot{x}))) \tag{6}$$

$$\frac{dx_i}{dt} = -\frac{x_i}{\tau} + \sum_j w_{ij} y_j + I_i - z(y_i - I_0) \tag{7}$$

$$dz/_{dt} = -\beta_0 z \tag{8}$$

Where $z(t)$ is the self-feedback connection weight, β_0 $(0<\beta_0<1)$ is damping factor, and I_0 is a positive parameter. We set z to decay exponentially as eqn. (8) and make the neural network actually has transiently chaotic dynamics. Damping factor β_0 control the damping speed of the self-feedback strength, therefore, it control the chaotic dynamics and convergence speed of the network.

Using the Eular discretization, the difference equation is written in the form:

$$x_i(t+1) = kx_i(t) + \alpha_0 \left(\sum_j w_{ij} y_j(t) + I_i \right) - z(t)(y_i - I_0) \tag{9}$$

$$z(t+1) = z(t)(1 - \beta_0) \tag{10}$$

Where $k=(1-\Delta t/2)$, and $\alpha_0 = \Delta t$.

5 Hysteretic Transiently Chaotic Neural Network (HTCNN) for the Assignment Problems

Assignment problems (AP) are typical combinational optimization problems that can be used as models for many applications such as manufacturing, planning and flexible manufacturing systems. The simplest case of these problems can be illustrated as: Let a number n of jobs be given that have to be performed by n machines, where the cost depend on the specific assignments. Each job has to be assigned to one and only one machine, and each machine has to perform one and only job. The problem is to find such an assignment that the total cost of the assignments became minimum.

Give two lists of elements and a cost value for the paring of any two elements from these lists; the problem is to find the particular one-to-one assignment or match between the elements of the two lists that results in an overall minimum cost. We use capital letters to describe the elements of one list (i.e. X=A, B, C, etc.) and enumerate the elements of the other list (i.e. i=1,2,3,etc.). Additionally, we assume that the two lists contain the same number of elements n. A one-to-one assignment means that each element of X has to be assigned to exactly one element of i. The cost P_{Xi} for every possible assignment between X and i is given for each.

The assignment problem is represented by a two dimensional quadratic matrix of units, whose outputs are denoted by v_{Xi} which is a "decision" variable, with $v_{Xi}=1$ meaning that the element X should be assigned to the element i, and $v_{Xi}=0$ meaning that the pairing between X and i should not be made. This way, a solution to the AP can be uniquely encoded by the two dimensional matrix of the outputs v_{Xi} after all units converge to 0 or 1. The constraints of the one-to-one assignment require that the outputs of the network after convergence should produce a permutation matrix with exactly one unit "on" in each row and column, Such as Fig.2. In this example, the output-matrix determines the assignment of elements A to 7, B to 1, C to 6 etc. The solution encoded by the output-matrix is optimal with an overall cost of 165.

The energy function is

$$E_{AP} = \frac{A}{2} \left(\sum_X \left(\sum_i v_{Xi} - 1 \right)^2 + \sum_i \left(\sum_X v_{Xi} - 1 \right)^2 + D \sum_X \sum_i P_{Xi} v_{Xi} \right) \tag{11}$$

A and D are constant parameters, the weight values $w_{Xi,Yj}$ and external bias I_{Xi} of the neural network neurons are

$$w_{Xi,Yj} = -A(\delta_{XY}(1 - \delta_{ij}) + \delta_{ij}(1 - \delta_{XY})) \tag{12}$$

$$I_{Xi} = 2A - Dp_{Xi} \tag{13}$$

Where $\delta_{ij} = 1$, if i=j, otherwise $\delta_{ij} = 0$. Introduce (12), (13) into (9) and then derive the HTCNN for solving the assignment problem, where, the neuron activation function is given by the hysteretic function as follows:

$$
\begin{array}{c}
\rightarrow i \\
\downarrow \\
X
\end{array}
\begin{pmatrix}
68 & 68 & 93 & 38 & 52 & 83 & 4 \\
6 & 53 & 67 & 1 & 38 & 7 & 42 \\
68 & 59 & 93 & 84 & 53 & 10 & 65 \\
42 & 70 & 91 & 76 & 26 & 5 & 73 \\
33 & 65 & 75 & 99 & 37 & 25 & 98 \\
72 & 75 & 65 & 8 & 63 & 88 & 27 \\
44 & 76 & 48 & 24 & 28 & 36 & 17
\end{pmatrix}
\rightarrow
\begin{pmatrix}
0 & 0 & 0 & 0 & 0 & 0 & 1 \\
1 & 0 & 0 & 0 & 0 & 0 & 0 \\
0 & 0 & 0 & 0 & 0 & 1 & 0 \\
0 & 0 & 0 & 0 & 1 & 0 & 0 \\
0 & 1 & 0 & 0 & 0 & 0 & 0 \\
0 & 0 & 0 & 1 & 0 & 0 & 0 \\
0 & 0 & 1 & 0 & 0 & 0 & 0
\end{pmatrix}
$$

Fig. 2. Cost matrix and output matrix of the neural network of a 7×7 AP.

$$
v_{xi} = \begin{cases}
0.5\tanh(\gamma_{Xi}^{\alpha}(u_{Xi} + \alpha_{Xi})) + 0.5; \dot{u}_{Xi} \geq 0 \\
0.5\tanh(\gamma_{Xi}^{\beta}(u_{Xi} - \beta_{Xi})) + 0.5; \dot{u}_{Xi} < 0
\end{cases}
\tag{14}
$$

Where, v_{Xi} and u_{Xi} are output value and internal input value of neuron i.

We use HTCNN to solve the above specified 7×7 assignment problem. The parameter are chosen as A=1, D=1, I_0=0.4, k=0.975, α_0=0.015, z(0)=0.45, β_0=0.02, $\gamma_{Xi}^{\alpha} = \gamma_{Xi}^{\beta} = 50$, $\alpha_{Xi} = \beta_{Xi} = 0.02$, and, the results with 100 different initial conditions in HNN, HHNN, TCNN and HTCNN are summarized in Table.1.

It is shown that HTCNN has higher ability to search for globally optimal solution.

Table 1. Result of HTCNN, TCNN, HHNN, and HNN for AP.

Neural network	HTCNN	TCNN	HHNN	HNN
Average of valid solutions	171.33	182.3	320	402.3
Average iterations for convergence	302	435	295	324

6   Conclusions

We propose a chaotic neural network model with a non-monotonous mapping-hysteresis as the neuron activation function. Multivalent, having memory, and being adaptive characterize the hysteretic activation function, and this is due to a change in the direction of the input, a system can pull itself out of a saturated region by jumping from one segment of the hysteretic activation function to the other segment. We have proved Lyapunov stability for the HTCNN, and used the HTCNN to solve the assignment problem to show the HTCNN's ability for optimization problems.

References

1. Hopfield, J.J, Tank, D.W.: "Neural" Computation of Decisions in Optimization Problems. Biolog. Cybern, **52** (1985) 141-152
2. Wang, X., Qiao, Q.: Solving Assignment Problems with Chaotic Neural Network. Journal of Systems Engineering, **16** (2001) 46-49
3. Nakagawa, M.: An Artificial Neuron Model with a Periodic Activation Function. Journal of the Physical Society of Japan, **64** (1995) 1023
4. Yanfai, H., Sawada, Y.: Associative Memory Network Composed of Neurons with Hysteretic Property. Neural Networks, **3** (1990) 223

An Improved Transiently Chaotic Neural Network for Solving the K-Coloring Problem

Shenshen Gu

School of Computer Engineering and Science, Shanghai University,
200072 Shanghai, P.R. China
gushenshen@163.com

Abstract. This paper applies a new version of the transiently chaotic neural network (TCNN), the speedy convergent chaotic neural network (SCCNN), to solve the k-coloring problem, a classic NP-complete graph optimization problem, which has many real-world applications. From analyzing the chaotic states of its computational energy, we reach the conclusion that, like the TCNN, the SCCNN can avoid getting stuck in local minima and thus yield excellent solutions, which overcome the disadvantage of the Hopfield neural network (HNN). In addition, the experimental results verify that the SCCNN converges more quickly than the TCNN does in solving the k-coloring problem, which leads it to be a practical algorithm like the HNN. Therefore, the SCCNN not only adopts the advantages of the HNN as well as the TCNN but also avoids their drawbacks, thus provides an effective and efficient approach to solve the k-coloring problem.

1 Introduction

The k-coloring problem is the problem of coloring the vertices of a grpah G with k colors such that adjacent vertices have different colors. This problem is widely applied in various fields such as frequency assignment, VLSI layout design, etc. Nevertheless, the k-coloring problem is proved to be NP-complete [1].

To approximate the k-coloring problem more effectively and efficiently, many heuristic algorithms have been proposed. Among these algorithms, the algorithm based on the Hopfield neural network (HNN) is an effective approach. Introduced by Hopfield and Tank in 1985, the HNN is widely utilized to solve a large number of classic combinatorial optimization problems such as the travelling salesman problem (TSP) [2]. As far as the k-coloring problem is concerned, Takefuji et al. and Berger have proposed several HNN-based algorithms that were proved to be effective to some extent [3][4].

However, the HNN has some obvious defects, and the most serious of which is that the HNN is apt to get stuck in local minima owing to the steepest descent dynamics of the method, which degrades its effectiveness. Fortunately, by probing into the chaotic behavior of nonlinear dynamics, Chen and Aihara proposed the transiently chaotic neural network (TCNN) model, which can overcome this disadvantage [5]. Up to now, the TCNN has been applied to solve the TSP[5],

J. Wang, X. Liao, and Z. Yi (Eds.): ISNN 2005, LNCS 3496, pp. 750–755, 2005.

the maximum clique problem (MCP)[6], etc., and the experimental results reveal that the performance of the TCNN is considerably better than that of those conventional Hopfield neural networks. However, it is verified that the efficiency of the TCNN is worse than that of the HNN. As a result, in some real-time applications, the TCNN is not very practical.

To overcome this disadvantage, in this paper, we propose an improved TCNN, the speedy convergent chaotic neural network (SCCNN), by adjusting the original simulated annealing mechanics. From analyzing the chaotic states of its computational energy, it is justified that, like TCNN, this model possesses rich chaotic behavior in solving the k-coloring problem, which prevents it from getting stuck in local minima. In addition, experimental results also verify that this model converges much faster than the TCNN in solving the k-coloring problem.

This paper is organized as follows: In the next section, a fundamental introduction to the HNN-based algorithm for the k-coloring problem is presented to give a picture of how the HNN can be used to solve this problem. In Section 3, the TCNN model for the k-coloring problem is given, and its disadvantage in convergent speed is justified. In Section 4, the improved TCNN model, SCCNN, is proposed, and the feasibility, efficiency, robustness and scalability of the new algorithm are analyzed in detail, and the experimental results on k-colorable graph instances will be given for the HNN, TCNN and SCCNN. Finally, the conclusions of this paper are presented in Section 5.

2   HNN for the K-Coloring Problem

The Hopfield neural network is a recurrent network with N neurons connected with a symmetric matrix $W \equiv (\omega_{ij})_{N \times N}$, in which $\omega_{ii} = 0, (i = 1, 2, \ldots, N)$. And the evolution of the neurons is defined by Eqs. 1.

$$\begin{cases} \frac{du_i}{dt} = -u_i + \sum_{j=1}^{N} \omega_{ij} v_j + I_i \\ v_i = g_i(u_i(t)) \end{cases} \tag{1}$$

$(i = 1, 2, \ldots, N)$, where $g(u_i(t)) \equiv 1/(1 + e^{-u_i(t)/\varepsilon})$ is a sigmoid and ε its gain, I_i is the firing threshold of neuron i, u_i is the internal state of neuron i and v_i is the output. This update strategy minimizes the following Liapunov energy function:

$$E = -\frac{1}{2} \sum_{i=1}^{N} \sum_{j=1}^{N} \omega_{ij} v_i v_j - \sum_{i=1}^{N} v_i I_i \tag{2}$$

Before introducing the HNN for the k-coloring problem, some preliminaries should be given. A graph G with n vertices can be presented by an $n \times n$ adjacent matrix $A_{n \times n}$, in which $a_{ij} = 1$ if vertex i and vertex j are connected; and $a_{ij} = 0$, otherwise. When the HNN is applied to color the n-vertex graph with k colors, it has n groups of neurons, and each group has k neurons. So it has $n \times k$ neurons totally. When the HNN reaches a valid solution, each group has one and only

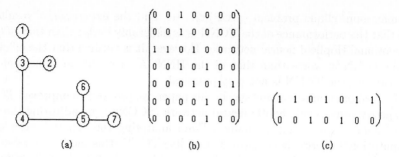

Fig. 1. (a) A graph G with 7 vertices; (b) Adjacent matrix of graph G; (c) Neural representation.

one neuron fired. Consider Fig.1(a), it is a graph with 7 vertices, and its adjacent matrix is given in Fig.1(b). Since this graph is proved to 2-colorable; therefore, if we color this graph with 2 colors, the valid solution can be expressed in the form of Fig.1(c), which means that, in this graph, vertex 1,2,4,6 and 7 can be colored with color 1, and vertex 3 and 5 can be colored with color 2.

Therefore, it is obvious that the k-coloring problem can be express by the energy function as follows:

$$E = \frac{A}{2}\sum_{x=1}^{n}\sum_{i=1}^{k}\sum_{j=1}^{k}v_{xi}v_{xj} + \frac{B}{2}\left[\sum_{x=1}^{n}\sum_{i=1}^{k}v_{xi} - n\right]^2 + C\sum_{x=1}^{n}\sum_{y=1}^{n}\sum_{i=1}^{k}a_{xy}v_{xi}v_{yi} \quad (3)$$

where the first term restricts that each group of neurons has only one neuron fired, the second term ensures that n neurons will be fired totally and the last term will be minimized when all the adjacent vertices have different colors.

Compared with the standard energy function of the HNN in Eqs.2, the energy function of the HNN for the k-coloring problem can be expressed as follows:

$$E = \frac{1}{2}\sum_{x=1}^{n}\sum_{i=1}^{k}\sum_{y=1}^{n}\sum_{j=1}^{k}\omega_{xi,yi}v_{xi}v_{yj} - \sum_{x}^{n}\sum_{i}^{k}v_{xi}I_{xi} \quad (4)$$

where $\omega_{xi,yj} = -A\delta_{xy}(1-\delta_{ij}) - B - Ca_{xy}\delta_{ij}$, ($\delta_{ij} = 1$ if $i = j$; and 0, otherwise).

However, it is uncovered that the HNN may get stuck in local minima owing to the steepest descent dynamics of the method. In the case of the k-coloring problem, the algorithms based on the HNN may reach an invalid solution, thus degrades its performance. Recall the graph in Fig.1(a), we apply the HNN to it. A round consists of 100 runs of this algorithm, each with different set initial states. We find that 30 percent results of the HNN are invalid solutions.

3   TCNN for the K-Coloring Problem

From the above section, we can reach the conclusion that the conventional HNN is apt to be trapped in local minima, thus providing invalid solutions. To overcome this disadvantage, Chen and Aihara proposed the TCNN that can avoid getting stuck in local minima. As far as the k-coloring problem is concerned, by adding an additional term that modifies the energy landscape to the energy function in Eqs.4 and by using Euler discretization, the framework of the TCNN for the k-coloring problem can be acquired as follows:

$$
\begin{cases}
v_{xi}\left(t\right) = 1/(1 + e^{-u_{xi}(t)/\varepsilon}) \\
u_{xi}\left(t+1\right) = \kappa u_{xi}\left(t\right) + \alpha \left(\sum\limits_{y=1}^{n} \sum\limits_{j=1}^{k} \omega_{xi,yj} v_{yj}\left(t\right) + I_{xi} \right) - z\left(t\right)\left(v_{xi}\left(t\right) - I_0\right) \\
z\left(t+1\right) = \left(1-\beta\right)z\left(t\right)
\end{cases}
\tag{5}
$$

$(x, y = 1, 2, \cdots, n;\ i, j = 1, 2, \cdots, k)$, where
α = positive scaling parameter for inputs,
β = damping factor of the time dependent $z\left(t\right)$, $(0 \leq \beta \leq 1)$,
κ = damping factor of nerve membrane, $(0 \leq \kappa \leq 1)$,
$z\left(t\right)$ = self feedback connection weight, $z\left(t\right) \geq 0$.

We also apply the TCNN to the graph in Fig.1(a). A round consists of 100 runs of the TCNN, each with a different set of initial states. We surprisingly find that all the results are valid solutions. It seems that TCNN is considerably effective than the HNN.

In Fig.2(a), which shows the time evolution of the computational energy E, we can observe the overall neuron dynamics. The rich chaotic behavior is obvious when $t < 900$, and then the value of energy deduced gradually and finally reached a global minimum. It suggests that, unlike the conventional HNN, the TCNN starts from deterministically chaotic dynamics, through a reversed period-doubling route with decreasing the value of z, which corresponds to the temperature of usual annealing, finally reaches a stable equilibrium solution [5]. This distinctive character of the TCNN may enhance its performance to solve the k-coloring problem.

However, it is shown from this figure that the TCNN spends about 1300 iterations to converge. It is obvious that the chaotic dynamics lasts too long. As a result, in some real-time situations, TCNN is not very practical.

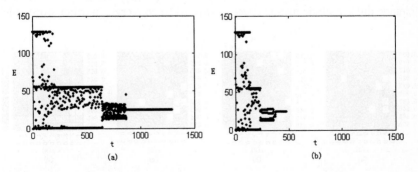

Fig. 2. Time evolution of the computational energy E. (a) TCNN; (b) SCCNN.

4   SCCNN for the K-Coloring Problem

To improve the efficiency of the TCNN, some new algorithms have been proposed by probing new simulated annealing mechanics to it [7][8]. And in this paper, we also proposed a new version of TCNN, the speedy convergent chaotic neural network (SCCNN), by adjusting its simulated annealing mechanics. And the framework of the SCCNN is given as below:

$$
\begin{cases}
v_{xi}(t) = 1/(1 + e^{-u_{xi}(t)/\varepsilon}) \\
u_{xi}(t+1) = \kappa u_{xi}(t) + \alpha \left(\sum\limits_{y=1}^{n} \sum\limits_{j=1}^{k} \omega_{xi,yj} v_{yj}(t) + I_{xi} \right) - z(t)(v_{xi}(t) - I_0) \\
\begin{cases}
z(t+1) = (1 - \beta_1) z(t), & when \ z(t) > z(0)/2 \\
z(t+1) = (1 - \beta_2) z(t), & when \ z(t) \le z(0)/2
\end{cases}
\end{cases}
\tag{6}
$$

Starts from a smaller β, β_1, this model is consistent with the TCNN. However, after the reversed period-doubling process, β is replaced by a larger value β_2, which accelerates the speed of convergence. The SCCNN is also applied to the graph in Fig.1(a). All the results are valid solutions in 100 runs, and we find that the number of iteration is greatly reduced compared with the TCNN. In Fig.2(b), we can observe that the rich chaotic behavior is obvious when $t < 400$, and then the value of energy deduces gradually and finally reaches a global minimum after about 500 iterations. The optimal performance of the SCCNN is studied in terms of feasibility, efficiency, robustness and scalability in a two-parameter space of α and $z(0)$, two important parameters that govern the performance of the algorithm. In the case of the k-coloring problem, feasibility is defined as whether the valid solution is obtained in a test; efficiency is defined as the number of iterations for convergence; robustness refers to the distribution of feasible regions; and scalability is concerned with the dependency of these measures on problem size. Fig.3(a)-(c) show the feasibility obtained for 3 graphs that are 3-colorable with $|V| = 10, 20$ and 30 respectively. And Fig.3(d) shows the number of iterations for $|V| = 20$. In Fig.3(a)-(c), a black region means the valid solution is obtained. It is obvious that the feasible region account for a considerably large portion of the parameter space, suggesting good robustness in choosing parameters. In addition, the feasible region does not shrink obviously when the size of problem increases, thus revealing good scalability. In Fig.3(d), the grayscale measures efficiency, darker shade means fewer iterations to converge, in other words higher

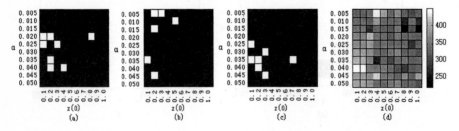

Fig. 3. Feasibility and efficiency graphs of the SCCNN.

Table 1. Iterations and optimal rates of three algorithms

|V|	Iterations			Optimal rates		
	HNN	TCNN	SCCNN	HNN	TCNN	SCCNN
10	167	1371	278	93%	98%	96%
20	258	1499	349	91%	94%	94%
30	356	1595	437	80%	91%	92%

efficiency. And it suggests that the SCCNN is an efficient algorithm since the dark shade region account for a relatively large portion of the parameter space. The iterations and optimal rates of the HNN, TCNN and SCCNN on these graph instances are listed in Table 1. It is obvious that the SCCNN can realize excellent optimal performances with much fewer iterations compared with the TCNN.

5 Conclusions

In this paper, the SCCNN is proposed by adjusting the simulated annealing mechanics of the TCNN, attempting to solve the k-coloring problem more efficiently. Experimental results reveal that the optimal performance of the SCCNN is equivalent to the TCNN. And, the SCCNN converges much more quickly than the TCNN, thus providing a practical approach in real-time situations.

References

1. Garey, M.R., Johnson, D.S.: Computers and Intractability: A Guide to the Theory of Np-Completeness. W. H. Freeman, New York (1979)
2. Hopfield, J., Tank, D.: Neural Computation of Decisions in Optimization Problems. Biological Cybernetics, Vol. 52 (1985) 141-152
3. Takefuji, Y., Lee, K.C.: Artificial Neural Networks for Four-Coloring Map Problem and for K-Colorability Problems. IEEE Transactions on Circuits and Systems, **38** (1991) 326-333
4. Berger, M.O.: k-Coloring Vertices Using a Neural Network with Convergence to Valid Solutions. Proceedings of IEEE World Congress on Computational Intelligence, **7** (1994) 4514-4517
5. Chen, L., Aihara, K.: Chaotic Simulated Annealing by a Neural Network Model with Transient Chaos. Neural Networks, **8** (1995) 915-930
6. Gu, S.S., Yu, S.N.: A Chaotic Neural Network for the Maximum Clique Problem. Advances in Artificial Intelligence, Vol. 3060. Springer-Verlag, Berlin Heidelberg New York (2004) 391-405
7. Xie, C.Q., He, C.: Simulated Annealing Mechanics in Chaotic Neural Networks. Journal of Shanghai Jiaotong University, Vol. 37. SJTU Press (2003) 36-39
8. Kang, B., Li X.Y., Lu B.C.: Improved Simulated Annealing Mechanics in Transiently Chaotic Neural Network. Proceedings of 2004 IEEE International Conference on Communications, Circuits and Systems, **2** (2004) 1057-1060

Since Hopfield and Tank applied Hopfield Neural Network (HNN) to optimization problems [5], it has been recognized as a powerful tool for optimization. However, HNN suffer from the local minimum problems whenever applied to optimization problems [6]. Compared with HNN, transiently chaotic neural network (TCNN) [7] can overcome the shortcomings by introducing chaos which is generated by negative self-feedback into HNN. With a time-variant parameter to control the chaos, TCNN goes through an inverse bifurcation process and gradually approaches to HNN with converging to a stable equilibrium point.

To apply TCNN model to the optimization problem of vehicle routing, the most important step is mapping the objectives and constraints of the problem onto the energy function of the network. The Lyapunov energy function is defined as follows:

$$E = \frac{A}{2}\sum_{i=0}^{n}\sum_{\substack{j=0\\j\neq i}}^{n}C_{ij}\cdot x_{ij} + \frac{B}{2}\sum_{i=0}^{n}(\sum_{\substack{j=0\\j\neq i}}^{n}x_{ij}-1)^2 + \frac{C}{2}\sum_{j=0}^{n}(\sum_{\substack{i=0\\j\neq i}}^{n}x_{ij}-1)^2.\tag{7}$$

In Eq. (7), A, B, C are all arbitrary and positive constants. The first term minimizes the total link cost of a routing path by taking into account the cost of all existing links; the second and the third term derives the neurons towards convergence to a valid route consisting of connected nodes.

The differential equations describing the network dynamics of TCNN for the SVRP can be written as follows:

$$x_{ij}(t) = \frac{1}{1+e^{-y_{ij}(t)/\varepsilon}}, \qquad i,j=0,1,\cdots,n\ .\tag{8}$$

$$y_{ij}(t+1) = ky_{ij}(t) - \alpha[\frac{A}{2}C_{ij} + \frac{B}{2}(2\sum_{a\neq j}^{n}x_{ia}-1) + \frac{C}{2}(2\sum_{l\neq i}^{n}x_{lj}-1) - z(t)[x_{ij}(t)-I_0].\tag{9}$$

$$z(t+1) = (1-\beta)z(t)\ .\tag{10}$$

where x_{ij} is assumed to be the neuron output which represents to visit city i in visiting order j; $y_i(t)$ is the internal state of neuron; α is positive scaling parameter for neuronal inputs; k is damping factor of nerve membrane($0<k<1$); β is damping factor of $z(t)$ ($0<\beta<1$); I_0 is positive parameter; ε is steepness parameter of the output function ($\varepsilon>0$); $z(t)$ is self-feedback connection weight or refractory strength ($z(t)\geq0$), it corresponds to the temperature in the simulated annealing and controls the speed of convergence and inverse divaricating. If $z(t)$ is a positive constant, TCNN will become CNN [8]. When $z(t)=0$, TCNN will degenerate into HNN.

To improve the computation performance, we use the method of CHEN as follows.

$$x_{ij}^D(t) = \begin{cases} 1 & \text{iff } x_{ij}(t) > \sum_k\sum_l x_{kl}(t)\Big/n\times n \\ 0 & \text{otherwise} \end{cases}.\tag{11}$$

3.3 STCNN Algorithm

The proposed algorithm generates all possible routes that can be visited by several vehicles and selects the optimal routes that have the minimum total cost. In detail, the algorithm is composed of the following steps:

Step 1. Data initialization. n, H_i, q_i, C_{ijk}, Q are given. Let $l = 0$.

Step 2. Route partition. Set $l = l+1$. If $l > n$, the generation process is finished, go to Step 5. Otherwise use the improved sweeping procedure in "GR" to assign the demand points to the vehicle and form k routes.

Step 3. Route Scheduling. Set a virtual demand $q = 0$ for the depot and add it to the vertex sets, then schedule the virtual k TSPs by applying TCNN procedure in "RS". If the process is finished, store the cost $C_{total,l} = \sum C_k$. If $l = 1$, then $C_{total} = C_{total,l}$. Go to Step 2. Otherwise angle the last axis in "GR" as the new axis of the procedure. Go to Step 2.

Step 4. Result testing. If $C_{total,l} < C_{total}$, $C_{total} = C_{total,l}$ and the closed routes R_l are written. Go to Step 2. If $C_{total,l} = C_{total}$ and the fleet size k and the point set of each route are the same as those of the last minimum. Otherwise, go to step 2.

Step 5. Rewriting of the optimal solution. Adjust the orders of the routes R_l to use the depot as the start point, which cannot change the order between other points.

4 Examples for CVRP

To demonstrate the applicability of the proposed method for CVRP, an example with 30 demand nodes is solved. When the whole process is repeated seven times, the problem is divided into 5 fleets which have 9, 8, 6, 4 and 3 nodes. The global minimum is 9.8302 and the computation time is 357 second, which are quite better and smaller than those of Rodriguez et al. The parameters of the route including 9 demand points are as follows: $A = B = C = 1$, $k = 0.995$, $I_0 = 0.5$, $\alpha = 0.015$, $\varepsilon = 0.004$. Figure 1 shows the time evolutions of the energy function when $\beta = 0.003$ and $\beta = 0.03$.

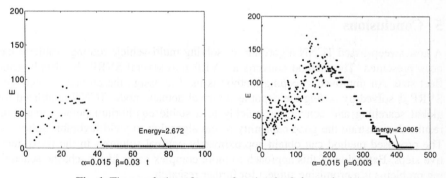

Fig. 1. Time evolution of energy function for 9 demand points.

It is shown that when β is extremely large, the chaotic dynamics vanishes very quickly. In contrast, when β is very small, the chaotic dynamics lasts very long to converges to the stable equilibrium point.

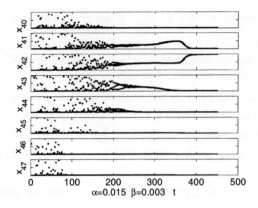

Fig. 2. Time evolutions of part neuron output x_{ij} for 9 demand points.

Figure 2 shows the time evolution of some neuron outputs with $\alpha = 0.015$, $\beta = 0.003$. TCNN behaves erratically and unpredictably during the first 250 iterations and eventually converges to a stable fixed point around iteration 380 through the reversed period-doubling bifurcation. The rate of global optimization reaches 100% and the optimal routes are listed in Table1.

Table 1. Optimal routes for 30 demand points using STCNN algorithm.

Routes	Optimal routes	Written routes
Route 1	18-0-25-14-18	0-25-14-18-0
Route 2	6-17-0-22-8-6	0-22-8-6-17-0
Route 3	29-21-12-7-0-26-19-29	0-26-19-29-21-12-7-0
Route 4	0-2-16-20-23-13-11-10-28-0	0-2-16-20-23-13-11-10-28-0
Route 5	15-3-4-9-1-27-24-5-30-0-15	0-15-3-4-9-1-27-24-5-30-0

5   Conclusions

A new sweep-based TCNN algorithm for solving multi-vehicle routing problems has been presented. The method converts a CVRP into several SVRPs by deciding the fleet size. An improved sweeping method is used to assign the problem effectively. SVRP is solved by setting the depot as a virtual demand node. TCNN is utilized for global searching and converging quickly to a stable equilibrium point. Simulation results demonstrate the good capability of the algorithm to yield favorable solutions. The proposed method can obtain approximately optimal solutions in the fixed itera-tive steps. Extension of this approach to more complex cases like stochastic schedul-ing problems is a promising subject for further research.

Acknowledgements

This work was supported by Nature Science Foundation of China (No. 60375039).

References

1. Lenstra, J.K., Rinnooy Kan, A.H.G.: Complexity of Vehicle Routing and Scheduling Problems. Neural Networks, **11** (1981) 221-227
2. Rodriguez, P., Nussbaum, M., Baeza, R., Leon, G.: Using Global Search Heuristics for the Capacity Vehicle Routing Problem. Comp. Oper. Res., **25** (1998) 407-417
3. Achuthan, N.R., Caccetta, L., Hill, S.P.: A New Subtour Elimination Constraint for the Vehicle Routing Problem. European J. of Operation Research, **91** (1996) 573-586
4. Gillett, B., Miller, L.: A Heuristic Algorithm for the Vehicle Dispatch Problem. Operations Research, **22** (1974) 340-349
5. Hopfield, J.J., Tank, D.W.: Neural Computations of Decisions in Optimization Problems. Biological Cybernetics, **52** (1985) 141-152
6. Wilson G.V., Pawley, G.S.:On the Stability of the Traveling Salesman Problem Algorithm of Hopfield and Tank. Biological Cybernetics, **58** (1988) 63-70
7. Chen, L., Aihara, K.: Chaos Simulated Annealing by a Neural Network Model with Transient Chaos. Neural Networks, **8** (1995) 915-930
8. Aihara,K., Takabe, T., Toyoda, M.: Chaotic Neural Networks. Physical Letters A, **144** (1990) 333-340

Transient Chaotic Discrete Neural Network
for Flexible Job-Shop Scheduling*

Xinli Xu, Qiu Guan, Wanliang Wang , and Shengyong Chen

Information Engineering Institute, Zhejiang University of Technology,
Hangzhou, Zhejiang 310014, China
{xxl,gq,wwl,csy}@zjut.edu.cn

Abstract. As an extension of the classical job-shop scheduling problem, the flexible job-shop scheduling problem (FJSP) allows an operation to be performed by one machine out of a set of machines. To solve the problem in real job shops, this paper presents a method of the discrete neural network with transient chaos (TDNN). The method considers various constraints in a FJSP. Furthermore, a new computational energy function for FJSP is proposed. A production scheduling program is developed in this research for validation and implementation of the proposed method in practical engineering situations. The experimental results show that the method can converge to the global optimum or near to the global optimum in reasonable and finite time.

1 Introduction

The flexible job-shop scheduling problem (FJSP) is an extension of the classic one which allows an operation to be performed by one machine out of a set of machines [1]. It has the characteristics: there are more than one kinds of job, some operations can be processed on the same machine, some operations must wait for assistant time before starting on some machine, working abilities of various machines are different, and all tasks must be completely solved in the time limit of delivery. Subjecting to the time limit of delivery and the assistant time, the objective of scheduling is to minimize the maximum among completion-dates (the time at which the last operation of a job is completed). Obviously, it is more complicated than the classical job-shop scheduling problem (JSP).

Recently, a few novel and intelligent procedures for solving FJSP have been proposed by researchers, typically such as simulated annealing algorithm [2] and genetic algorithm [3]. Many studies show that existing neural network methods based on chaos [4],[5],[6] are very effective for solving combinatorial optimization problems, especially for JSP [7], but they are few suitable for solving FJSP.

Therefore, this paper investigates the requirements of practical engineering and introduces all possible constraints of FJSP. A new computational energy function of Hopfield neural networks (HNN) is proposed and a discrete neural network with transient chaos is presented. A production scheduling program is also developed for solving practical job-shop scheduling problems (JSP).

* Foundation item: this research is supported by the National Natural Science Foundation of China (NSFC Grant No.60374056 and 60405009)

J. Wang, X. Liao, and Z. Yi (Eds.): ISNN 2005, LNCS 3496, pp. 762–769, 2005.

2 Scheduling Strategies for FJSP

In order to reduce the complexity of the problem and find the optimal solution, we decompose FJSP into some steps as follow:

Step 1. When the tasks are delivered to the job shop, first of all, jobs to be processed and machines available (except for the mistaken machines) are determined and they are formed into a set of jobs and a set of machines, respectively. Compared with different batches, the optimal scheme can be determined. The size of the batch describes the ratio of the quantity of jobs to be processed once to the number of total jobs.

Step 2. Based on the processing database, the machine and the processing time corresponding the operation in the set of jobs can be determined. Since available processing times of machines are different every day. For simplicity, the actual processing time of the machines is transformed into the standard one, i.e., the efficiency of the machine with long time is increased. In fact, the machine is used according to the actual one. However, if the processing time of the operation is less than the standard one of the according machine, the working efficiency will not be improved.

Step 3. The intervals of any two neighboring operations for the same job are calculated. If there are no delays among the operations of the job, the intervals are regarded as the processing time of the operation in precedence. For starting in time, the interval T_{ik} is calculated according to Eq. (1).

$$\begin{cases} T_{ik} = t_{ik} & t_{ik} \le t_{i,k+1} \\ T_{ik} = t_{ik} + (n_i - 1)(t_{ik} - t_{i,k+1}) & t_{ik} > t_{i,k+1} \end{cases} \tag{1}$$

where t_{ik} and $t_{i,k+1}$ is the processing time of the k-th and $(k+1)$-th operation for the i-th job, respectively. In addition, there is also efficiency conversion in this step.

Step 4. Execute the algorithm with transient chaotic neural network and obtain the solution of scheduling. The maximum completion-dates of all jobs can be found in the Gantt chart, which is a graph of resource allocation over time.

Step 5. Save the scheduling results into the database, and draw the Gantt charts of schedule for decision-making and analyzing.

3 Discrete Neural Network with Transient Chaos

Using HNN to solve JSP, a permutation matrix is generally applied to represent it [8],[9]. For a n-job m-machine FJSP, if the number of operations is less than that of machines, blank operations will be appended simply so that the number of operations is equal to that of machines. Considering otherwise the batch of jobs, so the total numbers of machines and jobs are transformed into n' and m', respectively. There are $m'n'$ rows and $(m'n'+1)$ columns in the matrix.

3.1 Energy Function of FJSP

For simplicity, we give some assumptions here. One job is not suspended by another job in the course of being processed until it is completed. For the same job, if one operation is completed, the next operation will be processed on another machine as long as the interval of the two operations satisfies with a given tolerance. All jobs are delivered in non-delay time from one machine to another.

Based on the assumptions, the characteristics above-mentioned, and the classical constraints [7], we firstly consider the classical inhibitions of FJSP as follow

$$E_1 = \frac{A}{2}\sum_{x=1}^{m'n'}\sum_{i=1}^{m'n'}\sum_{\substack{j=1\\j\neq i}}^{m'n'+1} v_{xi}v_{xj} + \frac{B}{2}(\sum_{x=1}^{m'n'}\sum_{i=1}^{m'n'+1} v_{xi} - N_o)^2 + \frac{C}{2}\sum_{\substack{x=1\\x\neq i-1}}^{m'n'}\sum_{\substack{i=2\\i\neq x+1}}^{m'n'+1} v_{xi}v_{(i-1)(x+1)}$$

$$+\frac{D_1}{2}(\sum_{x=1}^{m'n'} v_{x1} - N_m)^2 + \frac{D_2}{2}\sum_{i=1}^{m'n'+1}\sum_{k_1=1}^{n'}\sum_{k_2=1}^{m'}\sum_{\substack{k_3=1\\k_3\neq k_2}}^{m'} v_{[k_2+(k_1-1)m']i}v_{[k_3+(k_1-1)m']i} \quad (2)$$

$$+\frac{D_3}{2}\sum_{i=1}^{m'n'+1}\sum_{k_1=1}^{m'}\sum_{k_2=1}^{n'}\sum_{\substack{k_3=1\\k_3\neq k_2}}^{n'} v_{[k_1+(k_2-1)m']i}v_{[k_1+(k_3-1)m']i} \; .$$

where A, B, C, D_1, D_2, and D_3 are all random positive constants. v_{xi} is the output of the neuron in the position (x,i) of the matrix. N_o is the actual number of operations and N_m is the number of machines used to start at time 0. The first two terms in Eq.(2) are row inhibition and global inhibition in the matrix, respectively, the third term is unsymmetrical inhibition, and the rest are the column inhibitions.

Secondly the term of parallel processing inhibition is introduced as follow:

$$E_2 = \frac{D_4}{2}\sum_{k=1}^{l}\sum_{i=1}^{m'n'+1}\sum_{k_1=1}^{n'}\sum_{k_2=1}^{m'}\sum_{\substack{k_3=1\\k_3\neq k_1}}^{n'}\sum_{\substack{k_4=1\\k_4\neq k_2}}^{m'} v_{[k_2+(k1-1)m']i}v_{[k_4+(k_3-1)m']i}Flag_{kk_1k_2}Flag_{kk_3k_4} \; . \quad (3)$$

where l is the number of machines with the same assignment, $Flag$ is the tag of operations processed synchronously, and D_4 is a random positive constant.

If the batch $p < 1$, the batch-processing inhibition will be also considered as

$$E_3 = \frac{D_5}{2}\sum_{i=1}^{m'n'+1}\sum_{k=1}^{n}\sum_{k_1=1}^{1/p}\sum_{k_2=1}^{m'}\sum_{\substack{k_3=1\\k_3\neq k_1}}^{1/p}\sum_{\substack{k_4=1\\k_4\neq k_2}}^{m'} v_{[k_2+((k1-1)n+k-1)m']i}v_{[k_4+((k_3-1)n+k-1)m']i} \; . \quad (4)$$

where D_5 is a random positive constant.

In addition, the time limit F_{time} and performance (the maximum completion-dates F_{max}) inhibition must be taken into account as

$$E_4 = \lambda_1 G(F_{time} - F_{max}) \cdot v_{xi} + \lambda_2 G(F_{max\,min} - F_{max}) \cdot v_{xi} \; , \quad (5)$$

where $G(z) = \begin{cases} 0 & z \geq 0 \\ \dfrac{z^2}{2} & z < 0 \end{cases}$. \quad (6)

λ_1 and λ_2 are random positive constants. $F_{max\,min}$ is the optimal solution.

Finally, the computation energy function is $E_{sum} = E_1 + E_2 + E_3 + E_4$.

3.2 Discrete Neural Network with Transient Chaos (TDNN) for FJSP

Step1 (Start). Set the initial values of the self-feedback weight $z(1) = z_0$ and the maximum among completion-dates $F_{max}(1) = M$, where z_0 and M are all constants.

Step2 (Chaotic Search). Firstly, calculate the maximum $F_{\max}$ among completion-dates.

Set $E = E_1 + E_2 + E_3$. Equating the terms between E and the energy function of discrete Hopfield neural network (DHNN) [7], the connection weight and the bias current are decided. Putting them into the linear differential equation of DHNN, the equation can be obtained:

$$u_{xi}(t) = -A \sum_{\substack{j=1 \\ j \neq i}}^{m'n'+1} v_{xj}(t) - B \sum_{y=1}^{m'n'} \sum_{j=1}^{m'n'+1} v_{yj}(t) - C(1-\delta_{i1})(1-\delta_{x(i-1)})(1-\delta_{i(x+1)})v_{(i-1)(x+1)}(t)$$

$$- D_1 \sum_{y=1}^{m'n'} \delta_{i1} v_{y1}(t) - D_2 \sum_{k_1=1}^{n'} \sum_{k_2=1}^{m'} \sum_{\substack{k_3=1 \\ k_3 \neq k_2}}^{m'} \delta_{x[k_2+(k_1-1)m']} v_{[k_3+(k_1-1)m']i}(t)$$

$$- D_3 \sum_{k_1=1}^{m'} \sum_{k_2=1}^{n'} \sum_{\substack{k_3=1 \\ k_3 \neq k_2}}^{n'} \delta_{x[k_1+(k_2-1)m']} v_{[k_1+(k_3-1)m']i}(t) \tag{7}$$

$$-D_4 \sum_{k=1}^{l} \sum_{k_1=1}^{n'} \sum_{k_2=1}^{m'} \sum_{\substack{k_3=1 \\ k_3 \neq k_1}}^{n'} \sum_{\substack{k_4=1 \\ k_4 \neq k_2}}^{m'} \delta_{x[k_2+(k_1-1)m']} v_{[k_4+(k_3-1)m']i}(t) Flag_{kk_1k_2} Flag_{kk_3k_4}$$

$$-D_5 \sum_{k=1}^{n} \sum_{k_1=1}^{1/p} \sum_{k_2=1}^{m'} \sum_{\substack{k_3=1 \\ k_3 \neq k_1}}^{1/p} \sum_{\substack{k_4=1 \\ k_4 \neq k_2}}^{m'} \delta_{x[k_2+((k_1-1)n+k-1)m']} v_{[k_4+((k_3-1)n+k-1)m']i}(t) + BN_o + D_1 \delta_{i1} N_m \,,$$

$$v_{xi}(t) = \begin{cases} 1 & u_{xi}(t) \geq 0 \\ 0 & u_{xi}(t) < 0 \end{cases}, \tag{8}$$

where $\delta_{ij} = \begin{cases} 1 & i = j \\ 0 & i \neq j \end{cases}.$ \hfill (9)

In addition, u_{xi} is the input of the neuron in the position (x,i) of the matrix. Solve the Eq. (7), (8) and (9). When the neural network is stable and $E = 0$, the Gantt charts can be constructed in terms of the outputs of neurons that are "1". However, the assistant time of each operation must be considered for calculating the Gantt charts. The maximum completion-dates $F_{\max}$ can be found in Gantt charts. Or else set $F_{\max} = M$.

Secondly, solve the equation of TDNN.

Based on E_{sum}, the equation of TDNN is also decided as follow in the same way

$$u'_{xi}(t) = -A \sum_{\substack{j=1 \\ j \neq i}}^{m'n'+1} v_{xj}(t) - B \sum_{y=1}^{m'n'} \sum_{j=1}^{m'n'+1} v_{yj}(t) - C(1-\delta_{i1})(1-\delta_{x(i-1)})(1-\delta_{i(x+1)})v_{(i-1)(x+1)}(t)$$

$$- D_1 \sum_{y=1}^{m'n'} \delta_{i1} v_{y1}(t) - D_2 \sum_{k_1=1}^{n'} \sum_{k_2=1}^{m'} \sum_{\substack{k_3=1 \\ k_3 \neq k_2}}^{m'} \delta_{x[k_2+(k_1-1)m']} v_{[k_3+(k_1-1)m']i}(t) \tag{10}$$

$$- D_3 \sum_{k_1=1}^{m'} \sum_{k_2=1}^{n'} \sum_{\substack{k_3=1 \\ k_3 \neq k_2}}^{n'} \delta_{x[k_1+(k_2-1)m']} v_{[k_1+(k_3-1)m']i}(t)$$

$$- D_4 \sum_{k=1}^{l} \sum_{k_1=1}^{n'} \sum_{k_2=1k_3=1}^{m'} \sum_{k_4=1}^{n'} \sum_{\substack{k_4=1 \\ k3 \neq k_1\, k_4 \neq k_2}}^{m'} \delta_{x[k_2+(k1-1)m']} v_{[k_4+(k_3-1)m']i}(t) Flag_{kk_1k_2} Flag_{kk_3k_4}$$

$$- D_5 \sum_{k=1}^{n} \sum_{k_1=1}^{1/p} \sum_{k_2=1k_3=1}^{m'} \sum_{k_4=1}^{1/p} \sum_{\substack{k_4=1 \\ k3 \neq k_1\, k_4 \neq k_2}}^{m'} \delta_{x[k_2+((k1-1)n+k-1)m']} v_{[k_4+((k_3-1)n+k-1)m']i}(t)$$

$$+ BN_o + D_1\delta_{i1}N_m - \lambda_1 H(F_{time} - F_{max}) - \lambda_2 H(F_{max\,min} - F_{max}) - z(t)(v_{xi}(t) - I_0),$$
$$z(t+1) = z(t)(1-\beta), \tag{11}$$

The output equation is identical with Eq. (10), where

$$H(z) = \begin{cases} 0 & z \geq 0 \\ z & z < 0 \end{cases}. \tag{12}$$

I_0 is a positive constant and $\beta(0 < \beta < 1)$ is the attenuation gene. The variant $z(t)$ corresponds to the temperature in the simulated annealing, changes according to the exponent and thus controls the speed of convergence and inverse divaricating.

Step3 (Convergence Test). If $t > T$, the number of continuous appearances of $v_{xi}(t) = v_{xi}(t-1)$ or $F_{max}(t) = F_{max\,min}$ (if $F_{max\,min}$ is   unknown,   then   set $F_{max\,min} = \min_{k \in [1,t-1]} \{F_{max}(k)\}$) arrives to N_1, then stop; else go to step 2. Where, T is the max number of steps and $N_1 > 1$.

4 An Example for FJSP

This section demonstrates a FJSP example from a practical mechanical factory. The basic task processed or mended in the fitting shop every month is called job $P_i (i = 1,2,3,4)$, and the number of jobs are 40, 40, 70 and 50, respectively. Table 1 shows the processing time (hour) of each machine every day and Table 2 shows the machine assignment and the processing time (minute) of each operation.

There are three basic constraints involved in the problem: there is nondelay between the fourth operation and the fifth operation of job P_1, two jobs can be processed at the same time on machine M_1 for job P_2 or P_3, and it takes two hours to exchange jobs once on machine M_7 or M_8. The objective is to find the optimal scheduling solution in order that the completion-date of all jobs is minimum.

Table 1. The number is described as the processing time (hour) of each machine $M_i (i = 1,2,...,8)$ every day. For machine M_1, there two machines M_{11} and M_{12}.

M_{11}	M_{12}	M_2	M_3	M_4	M_5	M_6	M_7	M_8
16	8	8	16	8	8	16	16	16

At first, some strategies are adopted in the software of production scheduling:

Step 1. Comparing with the scheduling of different batches (1/1, 1/2, 1/3, and 1/4), we find that it is optimal if the batch is $p = 1$. Since it takes two hours to exchange jobs once on machine M_7 or M_8 and machine M_7 and M_8 are all block-net machines.

Table 2. The two terms are described as the machine assignment and the processing time (minute) of each operation $j(j=1,2,...,9)$, respectively.

Job	1	2	3	4	5	6	7	8	9
P_1	M_{11}	M_2	M_{11}	M_3	M_4	M_{12}	M_5	M_7	M_{11}
	120	95	60	320	70	95	160	60	60
P_2	M_{11}	M_8	M_6	M_{11}	M_7	M_{11}			
	80	140	480	70	85	5			
P_3	M_{11}	M_8	M_{11}	M_3	M_{11}	M_7			
	20	95	20	60	55	75			
P_4	M_{12}	M_8	M_6						
	95	160	70						

Step 2. The standard available time of each machine every day is regarded as 8 hour. Based on the method of efficiency conversion, the processing time (minute) of each operation is obtained and it is shown as the upper number in Table 3.

Step 3. Then the intervals (minute) between any two neighboring operations are calculated and also shown as the under number in Table3.

Table 3. The two terms are described as the processing time (minute) and the intervals (minute) of each operation $j(j=1,2,...,9)$, respectively.

Job	1	2	3	4	5	6	7	8	9
P_1	2400	3800	1200	6400	2800	3800	6400	1200	1200
	120	2630	60	320	70	95	5230	60	60
P_2	800	2800	9600	700	1700	100			
	80	140	8917.5	70	1651.25	100			
P_3	3500	3325	350	2100	962	2625			
	20	2980	20	1151.25	55	2625			
P_4	4750	4000	1750						
	830	2285	1750						

Step 4. Construct a matrix representation of FJSP. As observed in Table 4, because the batch of the problem is $p=1$, the new number of total jobs and machines are $n'=4$ and $m'=11$, respectively. The number of all neurons for the matrix is 1980.

Secondly, set parameters $A=0.5$, $B=0.1$, $C=0.2$, $D_1=0.5$, $D_2=0.3$, $D_3=0.3$, $D_4=0.3$, $D_5=0.3$, $\lambda_1=0.00001$, $\lambda_2=0.00001$, $z_0=0.001$, $\beta=0.008$, $I_0=0.65$, $N=10$, and $N_1=10$. Execute the algorithm with TDNN independently for 100 times, where $l=2$, $N_o=24$, and $N_s=2$. The average counts of calculation when the neural network becomes stable arrive to 49 and the outputs as the permutation matrix are shown in Figure 1.

Finally, the Gantt charts of machines shown in Figure 2 are obtained based on the permutation matrix in Figure 1. The complete time of all jobs is 12460 minute.

5 Conclusions

FJSP is very important for the modern manufacturing industry and this problem is more complicated than the classical one. In this paper, parallel processing and batch processing inhibitions are combined in the energy function and the time limit of

$$\sum_{i=1}^{n}\sum_{t=1}^{d}\{x_{ie}(y_{ik}(t)-z_{ik}(t))-d_{i}-t_{ie}\}\geq 0 \quad i\in[1,\cdots,n] \tag{5}$$

Minimizing the total penalty for early and tardy jobs,

$$Min\ Z = \sum_{i=1}^{n}\sum_{t=1}^{d_i}\sum_{k=1}^{n_i}[z_{ik}(t)\times \max(0,d_i - x_{ie})+y_{ik}(t)\times \max(0,x_{ie}-d_i)] \tag{6}$$

3   CNN Model

Neurons are basic elements of NNs. A common neural cell or neuron is defined by the linearly weighted summation of its input signals, and serially connected non-linear activity function $F(T_i)$.

$$T_i = \sum_{j=1}^{n} W_{ij} X_j \tag{7}$$

$$Y_i = F(T_i) \tag{8}$$

Links among neurons are through their weights. They represent the scheduling restrictions, also reflect the adaptation or adjustment to resolve constraint conflicts through proper feedback links. JC block is designed to check constraints (3) and to adjust the related job's start time if the constraints are not met. It is constructed with one neuron of I type, two neurons of C type, and two neurons of S type.

4   Neural-Based Approach to Production Scheduling

To improve the CNN's optimization ability and its performance, a neural-based approach is proposed with the capability of optimizing the start time of expanded job-shop scheduling.

4.1   Gradient Search Algorithm for Scheduling Optimization

Define an energy function.

$$E = Z \tag{9}$$

The evolution of neuron S_{ik} of type S follows,

$$S_{ik}(t+1) = S_{ik}(t) - \lambda \frac{\partial E}{\partial S_{ik}} \tag{10}$$

where λ is a positive coefficient and $i(1,2,\cdots,n)$. for

$$Z = \max_{i}\{\sum_{k=1}^{n_i}\sum_{t=1}^{d_i}\{z_{ik}(t)(x_{ie}+t_{ie})\},\sum_{k=1}^{n_i}\sum_{t=1}^{d_i}\{y_{ik}(t)(x_{ie}-d_i)\}\} \tag{11}$$

4.2 Gradient CNN

The two components in the gradient CNN cooperate as follows:

1. Generate the start time for each operation randomly, and set these as the initial outputs of corresponding neurons of group S at the bottom of the CNN.
2. Run SC blocks to force process sequences to be right and then run RC and JC blocks iteratively, to get a feasible scheduling solution.
3. Repeat Step 2 until all neurons of the S group are in steady states without changes. Then a feasible solution is found.
4. Check whether the feasible solution is satisfied. If not, go to Step 5, otherwise stop the search process.
5. Go back to Step 2.

5 Hybrid Approach to Expanded Job-Shop Scheduling

GA is a new type of search algorithm which is analogous to the natural selection and genetic mechanism, which was put forward by Professor John Holland and his colleague[11].

Unification of GA and gradient CNN. The hybrid scheduling method is as follows:

Step 1. Initialization - assign a starting time to each operation.
Step 2. Prescribe the sequence of operations by chromosomes.
Step 3. Run the CNN in hybrid mode.
Step 4. Return to Step 3 if conflicts exist; otherwise, go to Step 5.
Step 5. Apply the gradient search algorithm to neurons of S type if starting time optimization is needed, then return to Step 3; otherwise, go to Step 6.
Step 6. Apply the GA's crossover if an adjustment of the sequence of operations is needed, then return to Step 2; otherwise, end.

6 Experiments Analysis

The well-known 6×6 benchmark problem is taken as the test example. It is known that the minimal completion time of the last job is 55.

(1) The CNN and gradient CNN are applied to the benchmark problem. The results are shown in Table 1 where the completion time of the last job comes from experiments performed 1000 times. Compared with results obtained using CNN, those obtained using gradient CNN are better.
(2) The hybrid method is applied to the benchmark problem also. It is seen from Table 1 that it is almost impossible to obtain optimal scheduling for the 6×6 benchmark problem using gradient CNN.

The results are shown in Fig.1. Note that the solutions are different from that given by Muth and Thompson [12]. The average computing time is 5.1s using gradient CNN. And the average computing time is 7.7s using the hybrid method. Compared with CNN, we can see that the performance of the hybrid method exceeds that of the gradient CNN with a slight computing time cost.

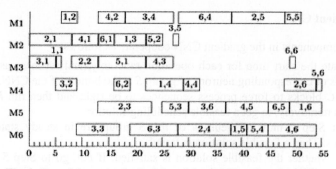

Fig. 1. One of the optimal schedules of the 6×6 benchmark problem.

Table 1. Results of 6×6 benchmark using CNN and gradient CNN.

Scheduling method	Due date setting	Initial time	Completion time
CNN	+∞	0	57.27
		Random	91.35/107.31/47.23
Gradient CNN	+∞	Random	52.4/67.55/49.33
CNN	100	0	51.28
		Random	48.6/51.44/43.23
Gradient CNN	100	Random	41.6/49.28/39
CNN	80	0	53.55
		Random	45.16/50.32/41.6
Gradient CNN	80	Random	41.2/50.36/39.63

7 Conclusions

In this paper, we proposed a new hybrid approach combining ANN and GA for job-shop scheduling. CNN is easy to implement owing to its object-oriented character. The computational ability of the hybrid approach is strong enough to deal with complex scheduling problems, thanks to NN's parallel computability and GA's searching efficiency. Many experiments have shown that this hybrid approach is extremely efficiently.

Acknowledgements

This research is supported by Natural Science foundation of GANSU province(grant NO ZS032-B25-008) and 〔grant NO ZS031-A25-015-G〕

References

1. Solimanpur, M., Vrat, P., Shankar, R. :A Neuro-Tabu Search Heuristic for the Flow Shop Scheduling Problem. Computers and Operations Research, **13** (2004) 2151-2164
2. Geyik, Faruk, Cedimoglu, Ismail Hakki.:The Strategies and Parameters of Tabu Search for Job-Shop Scheduling. **4** (2004) 439-448

3. Tang, L., Liu, J., Rong, A., Yang, Z.: A Review of Planning and Scheduling Systems and Methods for Integrated Steel Production. European Journal of Operational Research. **1** (2001) 1-20

4. Feng, Yuanjing, Feng, Zuren, Peng, Qinke.: Intelligent Hybrid Optimization Strategy and its Application to Flow-Shop Scheduling. Journal of Xi'an Jiaotong University **8** (2004) 779-782

5. Fonseca, Daniel J., Navaresse, Daniel.: Artificial Neural Networks for Job Shop Simulation. Advanced Engineering Informatics. **4** (2002) 241-246

6. Ping Chen, Junqin Liu, Jing Luo: Study on Pattern Recognition of the Quality Control Chart Based on Neural Network. Proceedings of the World Congress on Intelligent Control and Automation (WCICA) (2002) 790-793

7. Dominic, P.D.D., Kaliyamoorthy, S., Kumar, M. Saravana.: Efficient Dispatching Rules for Dynamic Job Shop Scheduling. International Journal of Advanced Manufacturing Technology, **2** (2004) 70–75

8. Wang, L., Zheng, D.Z.: A Modified Genetic Algorithm for Job Shop Scheduling. International Journal of Advanced Manufacturing Technology, **1** (2002) 72-76

9. Wang, Zhongbin, Chen, Yuliu, Wang, Ningsheng.: Research on Dynamic Process Planning System Considering Decision about Machines. Proceedings of the World Congress on Intelligent Control and Automation (WCICA) (2004) 2758-2762

10. Haq, A. Noorul, Ravindran, D., Aruna, V.etal.: A Hybridisation of Metaheuristics for Flow Shop Scheduling. International Journal of Advanced Manufacturing Technology, **5** (2004) 376-380

11. Goldberg, D. E. Genetic Algorithms in Search, Optimization and Machine Learning, Reading, MA: Addison-Wesley (1989)

12. Muth, & Thompson, G. L.:Industrial Scheduling, Englewood Cliffs, NJ: Prentice Hall (1963)

An Effective Algorithm
Based on GENET Neural Network Model for Job Shop
Scheduling with Release Dates and Due Dates

Xin Feng[1], Hofung Leung[2], and Lixin Tang[1]

[1] Department of Systems Engineering, Northeastern University, Shenyang, China
fengxinemail@sohu.com
[2] Department of Computer Science and Engineering, The Chinese University of Hong Kong,
Shatin, N.T., Hong Kong, China
lhf@cse.cuhk.edu.hk

Abstract. The job shop scheduling with release dates and due dates is considered. An effective algorithm based on GENET neural network model is presented for solving this type of problem. Two heuristics are embedded into the progressive stochastic search scheme in GENET network for the objective of minimizing the weighted number of tardy jobs. Experimental results indicate that the presented algorithm is competent to attain significant reductions in number of tardy jobs in relation to priority heuristic algorithms.

1 Introduction

Scheduling problems with release dates and due date constraints have been of practical interest to production. Failure to meet due dates can result in delay penalties, loss of customer goodwill and can have a significant negative impact in a competitive market. This type of problem seems to be more difficult than the makespan problem because there is no single, universally accepted criterion for evaluating the quality of a schedule [1]. Most of relational papers restrict within one machine and parallel machine [2][3] problem with common due date. And main efforts still concentrated on developing various dispatching rules [4]. To job shop involving release dates and due dates, the literature is quite scarce. Several local search approaches have been reported, including simulated annealing [5], tabu search [6] and genetic algorithms [7]. The drawback of this category of solving methods is that the execution can easily be trapped in local optima, and so far it is still unknown how to derive efficient neighborhood definitions for tardiness. In fact, because of their lack of flexibility to complicate constraints, these methods are not directly applicable to real world. Davenport et al. proposed GENET [8], a genetic neural network model which has the property of adapting its connection weights to modify the landscape of the search surface for escaping from local optima until a feasible solution can be obtained. In [9], Lam and Leung proposed an effective search scheme for GENET, the progressive stochastic search (PSS). In this paper, we present an effective algorithm based on GENET for the job shop scheduling problem with release dates and due dates. Two heuristics are embedded into the PSS schema in GENET network for the objective of minimizing the weighted number of tardy jobs.

J. Wang, X. Liao, and Z. Yi (Eds.): ISNN 2005, LNCS 3496, pp. 776–781, 2005.

2 Formulation of The Problem

The job shop scheduling problem with release dates and due dates can be described as follows: a set J of n jobs which each has different release date r_i and due date d_i, $i \in \{1,...,n\}$, must be performed on a set M of m machines. Each job $J_i \in J$ requires the scheduling of a set of m operations according to a specified ordering. Each operation needs to be processed during an uninterrupted time period of a given length on a given machine. Each machine can handle at most one operation at a time.

Let O_{ik}^p represents operation k of job i to be processed on machine p, S_{ik}^p and P_{ik}^p represent the starting time and processing time of O_{ik}^p, respectively. All the r_i, d_i, S_{ik}^p and P_{ik}^p are assumed to be integers. The following set of constraints will be satisfied:

- Precedence constraints:
  If O_{il}^q is the immediately successors of O_{ik}^p in J_i

$$S_{ik}^p + P_{ik}^p \leq S_{il}^q \quad i \in \{1,...,n\} \quad k,l \in \{1,...,m\} \quad p,q \in M \tag{1}$$

- Capacity constraints:

$$S_{ik}^p + P_{ik}^p \leq S_{jl}^p \vee S_{jl}^p + P_{jl}^p \leq S_{ik}^p \quad i,j \in \{1,...,n\} \quad k,l \in \{1,...,m\} \quad p \in M \tag{2}$$

- Release date constraints:

$$r_i \leq S_{i1}^p, \quad i \in \{1,...,n\} \quad p \in M \tag{3}$$

- Due date constraints:

$$S_{im}^p + P_{im}^p \leq d_i, \quad i \in \{1,...,n\} \quad p \in M \tag{4}$$

We focus on the schedule objective of minimizing the weighted number of tardy jobs T_n, i.e., the number of jobs exceeds its due date. Let $T(i)$ indicate whether job i is tardy or not, i.e., $T(i)=1$ means job i is tardy $(T_n>0)$ while $T(i)=0$ means it is completed in time. The weighted number of tardy jobs is given by

$$T_n = \sum_{i=1}^{n} \omega_i T(i) \tag{5}$$

3 Model of GENET

Generally, as a genetic neural network model, GENET is used to solve constraint satisfaction problems with binary constraints which based on the iterative repair approach.

3.1 Connections Architecture

Consider a set U of variables, D_u is the domain of u ($u \in U$), C is a set of constraints. Each variable u is represented in GENET N by a cluster of label nodes $\langle u,v \rangle$, one for each value $v \in D_u$. Each label node $\langle u,v \rangle$ is associated with an output $V_{\langle u,v \rangle}$, which is 1 if value v is assigned to variable u, and 0 otherwise. A label node is *on* if its output is 1; otherwise, it is *off*. In a valid state, only one label node in a cluster may be *on* at any one time. A constraint $c \in C$ is represented by weighted connections between incompatible label nodes. For instance, two label nodes $\langle u_1, v_1 \rangle$ and $\langle u_2, v_2 \rangle$ are connected if $u_1 = v_1 \wedge u_2 = v_2$ violates c with connection weights $W_{\langle u1,v1 \rangle \langle u2,v2 \rangle}$. The input $I_{\langle u,v \rangle}$ to a label node $\langle u,v \rangle$ is the weighted sum of all its connected nodes' outputs.

3.2 Running Mechanisms and the PSS Scheme

GENET performs incremental repair process to minimize the number of constraint violations by changing the label node which is *on* in a cluster. In PSS [9], a list F of variables which dictates the sequence of variables to repair must be kept. When a variable is designated to be repaired, it always has to choose a new value even if its original value should give the best cost value. At the mean time, the search paths are slightly "marked" by updating the connection weights according to the following rule:

$$W_{\langle u,v \rangle \langle x,y \rangle}^{new} = W_{\langle u,v \rangle \langle x,y \rangle}^{old} + \varepsilon \tag{6}$$

ε is a parameter which can be turned according to the need of problems. The repair process is repeated until the list F becomes empty, and a feasible solution can be obtained.

4 Description of Approach

4.1 Construct Network for Proposed Problem

Calculate the number of variables by $m \times n$. The domain of each variable S_{ik}^p (starting time of operation O_{ik}^p) is represented as a interval $[r_{ik}^p, d_{ik}^p - P_{ik}^p]$. Let $pred\ (O_{ik}^p)$ is the set of predecessors of O_{ik}^p in J_i and $succ(O_{ik}^p)$ is the set of successors of O_{ik}^p in J_i, we have

$$r_{ik}^p = r_i + \sum_{O_{il}^q \in pred(O_{ik}^p)} P_{il}^q \tag{7}$$

$$d_{ik}^p = d_i - P_{ik}^p - \sum_{O_{il}^q \in succ(O_{ik}^p)} pt_{il}^q \tag{8}$$

Build up clusters and their corresponding label nodes. All label nodes are connected according to two different constraint relationships: inequality constraints and disjunctive constraints, responding to precedence constraint and capacity constraint. For example, consider two starting time variables u_1 and u_2, $D_{u_1} = D_{u_2} = \{1,2,3\}$, and a disjunctive constraint $u_2 - u_1 \geq 1 \vee u_1 - u_2 \geq 1$, the incompatible connections in GENET network can be construct in Fig. 1.

4.2 Heuristics and Embedded Approach

Immediately after the network initialization by using min-conflict heuristic, all clusters are added to a list F in an arbitrary order. We select the cluster O_*^{curr} by two priority heuristic rules. Let F^{curr} denotes list F current state. First, we use the earliest due date first (EDF) rule to obtain the set O_1^{curr} of clusters satisfied EDF in F^{curr}

$$O_1^{curr} = \min\{d_{ik}^p \mid O_{ik}^p \in F^{curr}\} \tag{9}$$

Then we use the shortest processing time (SPT) rule to obtain the set O_2^{curr} of clusters satisfied SPT in O_1^{curr}

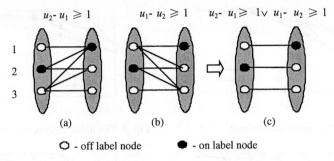

$u_2 - u_1 \geqslant 1$ $u_1 - u_2 \geqslant 1$ $u_2 - u_1 \geqslant 1 \vee u_1 - u_2 \geqslant 1$

(a)   (b)   (c)

$\bigcirc$ - off label node $\bullet$ - on label node

Fig. 1. Example capacity constraint in GENET network.

$$O_2^{curr} = \min\{P_{ik}^p \mid O_{ik}^p \in O_1^{curr}\} \tag{10}$$

At last, the cluster O_*^{curr} which will be removed from list F can be selected from the set of O_2^{curr} randomly.

5 Simulation Study

Refer to the literature [7] [10], the computational tests can be conducted with parameters and their uniform distribution ranges: Processing time - [1,9]; Used machine - [1,m]; Release date - [1,2×m]. Due date come from formula $d_i = r_i + AP_i$, P_i denotes the total processing time of job J_i. The values of A increase from 1.4 to 3.0 in step of 0.2 according to the sizes and parameters of problems from tight to loose. In general, A up to 1.5 indicates a rather tight due date; Designated running time is regarded as stopping criteria. Here the running time is no more than 30 seconds; All connection weighted in network is 1 initially, and ε is 1; The tardy weighted is 1.

The combination of parameter levels gives 6 problem scenarios, for each scenario, 5 instances are randomly generated by Taillard's random number generator [11]. Thus totally 30 problems are used in 6 different tightness which yield a total of 180 instances to measure the performance of the proposed algorithm. All the algorithms are coded in the VC++ programming language. All the timings are measured in seconds. All the timing results are the search time only.

We make a comparison with our algorithm (GPEFS) to dispatch rules of shortest processing time (SPT), earliest finish time (EFT) and earliest due date first (EDF). The performance is measured by success rate which is the ratio of the number of jobs feasibly scheduled to the number of all given jobs. This accords with the schedule objective that the number of tardy jobs is minimized. Fig. 2 - 7 illustrates mean success rate over the 5 instances versus difference A for different sizes problems. From the results, we can find that GPEFS algorithm has better performance than any dispatch rules in the case.

6 Conclusions

Experimental results, generated over a range of shop sizes with different due date tightness, indicate that the presented algorithm in this paper is an efficient scheduling alternative for the job shop scheduling problem with release dates and due dates.

780   Xin Feng, Hofung Leung, and Lixin Tang

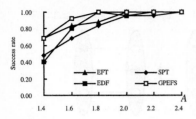

Fig. 2. The mean success rate of 5×5.

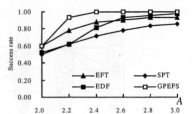

Fig. 3. The mean success rate of 10×5.

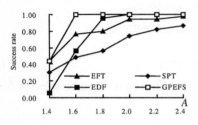

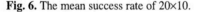

Fig. 4. The mean success rate of 10×10.

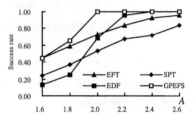

Fig. 5. The mean success rate of 15×10.

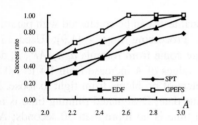

Fig. 6. The mean success rate of 20×10.

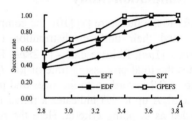

Fig. 7. The mean success rate of 30×10.

Acknowledgements

This research is partly supported by National Natural Science Foundation for Distinguished Young Scholars of China (Grant No. 70425003), National Natural Science Foundation of China (Grant No. 60274049 and Grant No. 70171030), Fok Ying Tung Education Foundation and the Excellent Young Faculty Program and the Ministry of Education, China. It is also partly supported by Integrated Automation Lab in Process Industry.

References

1. Baker, K.R.: Sequencing Rules and Due-Date Assignments in a Job Shop. Management Science, **30** (1984) 1093-1104
2. Gordon, V., Proth, J.M. and Chu, C.: A Survey of the State-of-the-art of Common Due Date Assignment and Scheduling Research. European Journal of Operational Research, **139** (2002) 1–25
3. Chen, Z.L. and Lee, C.Y.: Parallel Machine Scheduling with a Common Due Window. European Journal of Operational Research, **136** (2002) 512-527
4. Mosheiov, G., Oron, D.: A Note on the SPT Heuristic for Solving Scheduling Problems with Generalized Due Dates. Computers & Operations Research, **31** (2004) 645-655

5. He, Z., Yang, T. and Tiger, A.: An Exchange Heuristic Imbedded with Simulated Annealing for Due-Dates Job-Shop Scheduling. European Journal of Operational Research, **91** (1996) 99-117
6. Armentano, V.A., Scrich, C.R.: Tabu Search for Minimizing Total Tardiness in a Job Shop. Int. J. Production Economics, **63** (2000) 131-140
7. Mattfeld, D.C., Bierwirth, C.: An Efficient Genetic Algorithm for Job Shop Scheduling with Tardiness Objectives. European Journal of Operational Research, **155** (2004) 616–630
8. Davenport, A., Tsang, E., Wang, C.J., Zhu, K.: Genet: A Connectionist Architecture for Solving Constraint Satisfaction Problems by Iterative Improvement. Proc. 12th National Conference on Artificial Intelligence, (1994) 325-330
9. Lam, B.C.H. and Leung, H.F.: Progressive Stochastic Search for Solving Constraint Satisfaction Problems. Proc. 15th IEEE International Conference on Tools with Artificial Intelligence, (2003) 487-491
10. Hall, N.G., Posner, M.E.: Generation Experimental Data For Computational Testing with Machine Scheduling Applications. Operations Research, **49** (2001) 854-865
11. Taillard, E.: Benchmarks for Basic Scheduling Problems. European Journal of Operational Research, **64** (1993) 278-285

Fuzzy Due Dates Job Shop Scheduling Problem Based on Neural Network

Yuan Xie, Jianying Xie, and Jie Li

Department of Automation, Shanghai Jiaotong University, Postbox 280,
Shanghai 200030, China
{Hei_xieyuan,lijieDLP}@sjtu.edu.cn

Abstract. A job shop scheduling problem with fuzzy due dates is discussed. The membership function of a fuzzy due date assigned to each job denotes the degree of satisfaction of a decision maker for the completion time of this job. The performance criterion of proposed problem is to maximize the minimum degree of satisfaction over given jobs, and it is an NP-complete problem. Thus artificial neural network is considered to search optimal jobs schedule.

1 Introduction

In prior literature on scheduling problems, due dates are often assumed to be crisp numbers. However, in many practical situations, due dates may be vague and it is too difficult to decide the due dates accurately. In some cases, due dates are certain intervals and it may be proper to deal with due dates as fuzzy values. Since Ishii et al. [1] introduced the concept of fuzzy due dates to scheduling problems, fuzzy due dates scheduling problems have been investigated by many researchers [2]. In this paper we analyze a kind of job shop scheduling problem with fuzzy due dates which is also an NP-complete problem. In order to get the optimal jobs schedule, artificial neural networks are introduced who have been successful developed to solve a wide rage of deterministic (non-fuzzy) scheduling problem [3], [5].

2 Formulation of Fuzzy Due Dates Job Shop Scheduling Problem

2.1 Job Shop Problem

Job shop scheduling problem is one of the most well-know problems in the area of scheduling and general assumption of job shop scheduling problem can be written as follows. There are given n jobs $i = 1, 2, \dots, n$ and m machines $M_1, M_2, \dots, M_m$. Job i consists of a sequence of n_i operations $O_{i1}, O_{i2}, \dots, O_{in_i}$ which must be processed in this order, i.e. we have precedence constraints of the form $O_{ij} \rightarrow O_{i(j+1)}$ ($j = 1, 2, \dots, n_i -1$). There is a machine $\mu_{ij} \in \{M_1, M_2, \dots, M_m\}$ and processing time p_{ij} associated with each operation O_{ij}. O_{ij} must be processed for p_{ij} time units on machine μ_{ij}. Let C_{ij} be the completion time of operation O_{ij}. The problem is to find a feasible schedule which minimizes some objective functions depending on the finishing time C_i of the last operations O_{in_i} of jobs.

J. Wang, X. Liao, and Z. Yi (Eds.): ISNN 2005, LNCS 3496, pp. 782–787, 2005.

2.2 Fuzzy Due Dates

A fuzzy due date is associated with each job. It means the due date is not so strict and can be an interval on the positive part of real line. The membership function of a fuzzy due date assigned to each job represents the degree of satisfaction of a decision maker for the completion time of that job. This membership is shown in Fig. 1, we can see that the membership function is specified by d_i^L and d_i^U ($i = 1, 2, \ldots, n$). d_i^L can be viewed as the earliest due date for job i, and d_i^U is the latest due date.

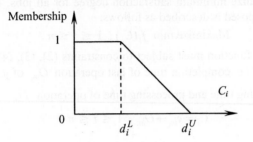

Fig. 1. Membership of fuzzy due date for job i.

In mathematics, membership function $f_i(C_i)$ of job i showed in Fig. 1 can be denoted by following formula:

$$f_i(C_i) = \begin{cases} 1 & C_i \le d_i^L \\ \dfrac{d_i^U - C_i}{d_i^U - d_i^L} & d_i^L < C_i < d_i^U \\ 0 & C_i \ge d_i^L \end{cases} . \tag{1}$$

where C_i is the completion time of job i.

2.3 Fuzzy Job Shop Scheduling Problem

In contrast to the deterministic $n \times m$ job shop scheduling problem, in this paper, a job shop scheduling problem incorporating fuzzy due dates is formulated as a fuzzy job shop scheduling problem (FJSSP). Let s_{ij} denote the starting time of operation j of job i, i.e. operation O_{ij}, whose processing time is p_{ij}. The last operation n_i of job i must begins at time s_{in_i}. For the first operation O_{i1} of job i, the starting time must be not less than zero. The constraint is

$$s_{i1} \ge 0, \; 1 \le i \le n . \tag{2}$$

Assuming operation $O_{i(j+1)}$ succeeds O_{ij}, then the inequalities denoting precedence constraints in order for a set of s_{ij} to be feasible are:

$$s_{ij} - s_{i(j+1)} + p_{ij} \le 0, \; 1 \le i \le n , 1 \le j \le n_i - 1 . \tag{3}$$

Let two operations O_{ij} and O_{kl} be assigned to be processed on same machine in their respective operation orders. It means the machine μ_{ij} which O_{ij} must be processed on is exactly the same one for O_{kl}. It is also necessary to ensure that O_{ij} and O_{kl} are not processed simultaneously on the same machine, then

$$s_{ij} + p_{ij} - s_{kl} \leq 0 \text{ or } s_{kl} + p_{kl} - s_{ij} \leq 0, 1 \leq i, j \leq n, 1 \leq j \leq n_i, 1 \leq l \leq n_k. \tag{4}$$

Above constraints enforce that each machine does not process more than one job at any time.

In this study, there is a fuzzy due date with each job whose membership is showed in Fig. 1. Since the fuzzy date represents the satisfaction degree of decision maker about completion time of this job. The objective of scheduling is to get an optimal schedule to maximize minimum satisfaction degree for all jobs. The objective function of FJSSP proposed is described as follows:

$$\text{Maximize min } f_i(C_i), \ 1 \leq i \leq n. \tag{5}$$

Here objective function must subject to constraints (2), (3), (4). C_i is the completion time of job i, i.e. completion time of last operation O_{in_i} of job i. Thus C_i equals to the sum of starting time and processing time of operation O_{in_i}:

$$C_i = s_{in_i} + O_{in_i}, \ 1 \leq i \leq n. \tag{6}$$

3   Modified Hopfield Neural Network

From description of FJSSP above, we can conclude that the problem is NP-complete problem like deterministic (non-fuzzy) job shop scheduling problem. The artificial neural networks have been successful applied in mapping and solving combinatorial optimization problem [4]. Foo and Takefuji [3] first used Hopfield neural network to solve job shop problem scheduling problem. And it is also useful to deal with FJSSP. In order to construct the energy function of neural network, original objective function of FJSSP needs to be modified as part of energy function. Because the satisfaction degree of fuzzy due dates of jobs is always not great than one, original objective function (5) to maximize can be converted into a minimizing objective function by subtracting original objective function from one:

$$F_i(C_i) = 1 - f_i(C_i). \tag{7}$$

$F_i(C_i)$ denotes dissatisfaction degree of fuzzy due dates of jobs. So new objective function is to minimize the maximum dissatisfaction degree:

$$\text{Minimize max } F_i(C_i) = \text{Minimize max } (1 - f_i(C_i)). \tag{8}$$

From formula (8), it can be concluded that new objective function is equivalent to original one.

According to new objective function (8) and constraints (2), (3), (4), the energy function of modified Hopfield neural network can be induced as follows:

$$E = \max F_i(C_i) + \sum_{i=1}^{n} A_1 H(-s_{i1}) + \sum_{i}^{n} \sum_{j=1}^{n_i-1} A_2 H(s_{ij} - s_{i(j+1)} + p_{ij})$$

$$+ \sum_{\substack{i=1 \\ i \neq k}}^{n} \sum_{\substack{k=1 \\ \mu_{ij} = \mu_{kl}}}^{n} A_3 \{\min[H(s_{kl} + p_{kl} - s_{ij}), H(s_{ij} + p_{ij} - s_{kl})]\} \tag{9}$$

where A_i ($i = 1, 2, 3$) represents arbitrary positive constants. The first item of energy function makes new objective function of FJSSP minimal. The second and third items correspond to constraint (2) and (3) respectively. The function $H(\bullet)$ is penalty function:

$$H(x) = \begin{cases} e^x & x > 0 \\ 0 & x \leq 0 \end{cases}. \tag{10}$$

The last item in formula (9) describes disjunctive constraints (4). The item relies on a zero-one variable w_{ik} to specify operation sequence, i.e.,

$$w_{ik} = \begin{cases} 1, & \text{if } job \ i \ precedes \ job \ k \\ 0, & others \end{cases}. \tag{11}$$

Then the last item can be replaced by two items:

$$\sum_{\substack{i=1 \\ i \neq k}}^{n} \sum_{\substack{k=1 \\ \mu_{ij}=\mu_{kl}}}^{n} A_3 H[s_{kl} + p_{kl} - B(1 - w_{ik}) - s_{ij}] + \sum_{\substack{i=1 \\ i \neq k}}^{n} \sum_{\substack{k=1 \\ \mu_{ij}=\mu_{kl}}}^{n} A_3 H[s_{ij} + p_{ij} - Bw_{ik} - s_{kl}] \tag{12}$$

where B is an arbitrary positive number greater than the maximum value among all processing times. Hence, the entire energy function for solving FJSSP can be rewritten:

$$E = \max F_i(C_i) + \sum_{i=1}^{n} A_1 H(-s_{i1}) + \sum_{i}^{n} \sum_{j=1}^{n_i - 1} A_2 H(s_{ij} - s_{i(j+1)} + p_{ij})$$

$$+ \sum_{\substack{i=1 \\ i \neq k}}^{n} \sum_{\substack{k=1 \\ \mu_{ij}=\mu_{kl}}}^{n} A_3 H[s_{kl} + p_{kl} - B(1 - w_{ik}) - s_{ij}] + \sum_{\substack{i=1 \\ i \neq k}}^{n} \sum_{\substack{k=1 \\ \mu_{ij}=\mu_{kl}}}^{n} A_3 H[s_{ij} + p_{ij} - Bw_{ik} - s_{kl}]. \tag{13}$$

For the neural network defined by formula (13), neuron s_{ij} evolves as follows (14):

$$s_{ij}(t+1) = s_{ij}(t) - \frac{\partial E}{\partial t} \times \Delta \tag{14}$$

Where E is energy function in formula (13). Δ is positive constant.

4   Software Simulations

For evaluating the performance of modified Hopfield neural network described in section 3, the classical 6-job 6-machine (FT6) job shop problem [6] is selected as simulation example. The sequence constraints of all jobs are the same: in order from operations 1 to 6. Table 1 presents the original data of this $6|6|J|F(C_i)$ FJSSP. In the table 1, (μ_{ij}, p_{ij}) means that the corresponding operation of some job will be processed on machine μ_{ij} with its processing time being p_{ij}. d_i^L and d_i^U are low bound and upper bound of fuzzy due date respectively. Here fuzzy due dates assumed to be same for all jobs.

In this example, let A1 = 500, A2 = 200, A3 = 250, B=12. So the optimal result can be obtained after iteration of proposed neural network. Fig. 2 shows the optimal solution in the mode of Gantt chart. In Fig. 2, a block means an operation with the length of the block equivalent to its processing time, the number pairs (i, j), inside or above the block, means that the relative operation is the jth operation of job i. The maximum completion time of jobs is 55. Thus the minimum satisfaction degree for fuzzy due dates of all jobs is 0.75.

Table 1. Original data for FT6 job shop problem with fuzzy due dates.

Job	Operation (μ_{ij}, p_{ij})						d_i^L	d_i^U
	1	2	3	4	5	6		
1	3,1	1,3	2,6	4,7	6,3	5,6	50	70
2	2,8	3,5	5,10	6,10	1,10	4,4	50	70
3	3,5	4,4	6,8	1,9	2,1	5,7	50	70
4	2,5	1,5	3,5	4,3	5,8	6,9	50	70
5	3,9	2,3	5,5	6,4	1,3	4,1	50	70
6	2,3	4,3	6,9	1,10	5,4	3,1	50	70

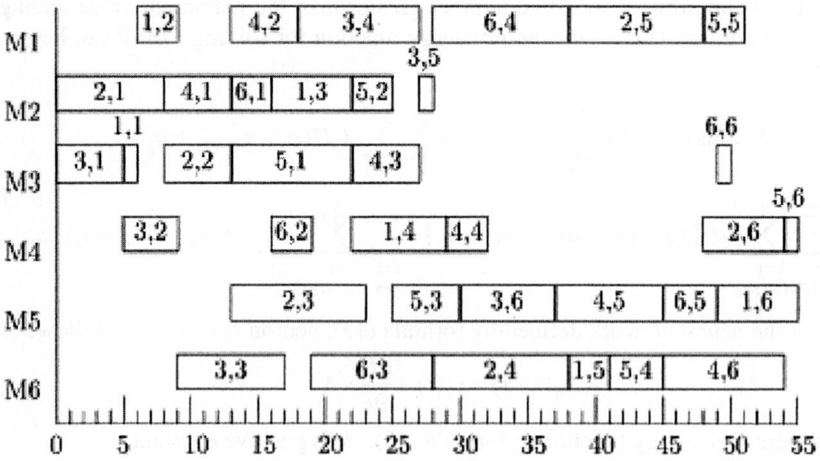

Fig. 2. An optimal schedule.

5 Conclusions

A job shop scheduling problem with fuzzy due date is discussed and fuzzy due date represents the satisfaction degree of completion time of job. The objective is to maximize the minimum satisfaction degree for all jobs. For this NP-completion problem, a modified Hopfield neural network is proposed to search the optimal scheduling. The objective function companying with constraint conditions is transformed into energy function of neural network. According iteration of neural network, the optimal or near optimal schedules can be reached.

References

1. Ishii, H., Tada, M., Masuda, T.: Two Scheduling Problems with Fuzzy Due-dates. Fuzzy Sets and Systems, **46** (1992) 339-347
2. Han, S., Ishii, H., Fujii, S.: One Machine Scheduling System with Fuzzy Duedates. European Journal Operational Research, **79** (1994) 1-12
3. Foo, Y. S., Takefuji, Y.: Integer Linear Programming Neural Networks for Job Shop Scheduling. Proc. IEEEIJCNN, IEEE, San Diego, **2** (1988) 341-348
4. Looi, C.: Neural Network Methods in Combinatorial Optimization. Computers and Operations Research, **19** (1992) 191-208
5. Zhang, C. S., Yan, P. F.: Neural Network Method of Solving Job Shop Scheduling Problem. ACTA Automation Sinica, **21** (1995) 706-712
6. Fisher, H., Thompson, G. L.: Probabilistic Learning Combinations of Local Job shop Scheduling Rules. In Industrial Scheduling, Muth, J., Thompson, G., Eds. Englewood Cliffs, NJ: Prentice Hall, (1963) 225-251

Heuristic Combined Artificial Neural Networks to Schedule Hybrid Flow Shop with Sequence Dependent Setup Times

Lixin Tang and Yanyan Zhang

Department of Systems Engineering, Northeastern University,
Shenyang, Liaoning 110000, China

Abstract. This paper addresses the problem of arranging jobs to machines in hybrid flow shop in which the setup times are dependent on job sequence. A new heuristic combined artificial neural network approach is proposed. The traditional Hopfield network formulation is modified upon theoretical analysis. Compared with the common used permutation matrix, the new construction needs fewer neurons, which makes it possible to solve large scale problems. The traditional Hopfield network running manner is also modified to make it more competitive with the proposed heuristic algorithm. The performance of the proposed algorithm is verified by randomly generated instances. Computational results of different size of data show that the proposed approach works better when compared to the individual heuristic with random initialization.

1 Introduction

Hybrid flow shop scheduling problem aims at arranging n jobs on total M machines that belong to S stages. This paper adopts the minimal sum of setup times as the measurement of optimal schedule results. Setup times are denoted to be the work to prepare the resources or tools for tasks. Scheduling problems involving setup times can be broadly found in many areas of industry, such as steel sheet color coating operation, steelmaking & continuous casting production and steel tube production of iron and steel plant, where even small unnecessary time consumption may lead to enormous corresponding cost. Okano et. al.[1] address the finishing line scheduling in the steel industry, considering all property constraints and setup times. The generated production runs outperform the original ones in quality, speed and quantity.

Hopfield neural networks have been used to combinatorial optimization problems[2], [3], [4]. However, some limitations of the approach were discovered[5] such as undesired local minima and erratic penalty parameter selection. To overcome these limitations, many modifications have been tried in this field. Recent researches[6], [7] showed that proper modifications to the original Hopfield network are effective and efficient with respect to solution quality and computational time.

2 Problem Representation and Formulation

2.1 Problem Representation

The features of hybrid flow shop scheduling can be characterized as follows.

1) The number of jobs to be scheduled has been determined.

J. Wang, X. Liao, and Z. Yi (Eds.): ISNN 2005, LNCS 3496, pp. 788–793, 2005.

2) Each job to be processed goes through s independent stages, which can not be treated at the same time. All jobs follow the same route in the shop floor.

3) Each machine can only process one job at the same time. The processing at each stage can not be interrupted.

4) Machine setup time is dependent on job sequence.

5) There is at least one identical machine at each stage, at least one stage having more than two machines, at the beginning of scheduling, all machines are available.

2.2 Compact Network Formulation for Scheduling Problems

Assume there are n jobs, S stages, M machines in the shop. Let i, i_1 be the identifier of stages, $i, i_1 = 1,2,...S$, M_i be the number of machines at stage i, $M = M_1 + M_2 + ... + M_s$, j, j_1 be the machine number, $j, j_1 = 1,..., M_i$, $i = 1,2,...,S$, k, k_1 be the processing position on a machine, $k, k_1 = 1,2,...,S \times n$. we denote the output of the neuron with V_{ijk}, the immediate succeeding processing position of k on the same machine with I_k, $I_k \in \{k + S, k + 2 \times S,..., S \times n\}$. $S_{ijk I_k}$ is the setup time between two adjacent jobs on machine j at stage i, the former job is processed on position k and the later job is processed on position I_k.

The total number of neurons is $M \times S \times n$, by taking 3 stages as example (the number of machines at each stage is 2 uniformly), the desirable permutation matrix is as follows:

Table 1. The permutation matrix of the feasible result.

	1	2	3	4	5	6	7	...	3n
1	1								
2			1						
3		1							
4					1				
5			1			1			
6									

Where the horizontal numbers denote the processing position, and the vertical numbers represent a certain machine. Position 1 denotes where the first operation of job 1 is arranged, and position 2 the second operation of job 1, position 3 the last operation of job 1. While position 4 tells the processing place of the first operation of job 2, and position 5 the second operation of job 2, position 6 the last operation of job 2. Then we get the following processing routes:

The processing route of job 1 is : machine 1– machine 3– machine 5

The processing route of job 2 is : machine 2– machine 4– machine 5

The rest routes may be deduced by analogy.

Objective function:

$$\sum_{i=1}^{S}\sum_{j=1}^{M_i}\sum_{k=1}^{Sn}\sum_{I_k} S_{ijk I_k} V_{ijk} V_{ij I_k} \cdot \tag{1}$$

subject to:

$$\left(\sum_{i=1}^{S}\sum_{j=1}^{Mi}\sum_{k=1}^{Sn} v_{ijk} - Sn\right)^2 = 0 \cdot \tag{2}$$

$$\sum_{i=1}^{S}\sum_{i_1=1}^{S}\sum_{j=1}^{Mi}\sum_{j_1=1,j_1\neq j}^{Mi}\sum_{k=1}^{Sn} v_{ijk}v_{i_1 j_1 k} = 0 \tag{3}$$

$$\sum_{i=1,2,...S}\sum_{k\in Oi}\left(\sum_{j-1}^{Mi} v_{ijk} - 1\right)^2 = 0 \tag{4}$$

Thus the energy function is:

$$E = \frac{C_1}{2}\left(\sum_{i=1}^{S}\sum_{j=1}^{Mi}\sum_{k=1}^{Sn} v_{ijk} - Sn\right)^2$$

$$+\frac{C_2}{2}\sum_{i=1}^{S}\sum_{i_1=1}^{S}\sum_{j=1}^{Mi}\sum_{j_1=1,j_1\neq j}^{Mi}\sum_{k=1}^{Sn} v_{ijk}v_{i_1 j_1 k} +\frac{C_3}{2}\sum_{i=1,2,...S}\sum_{k\in Oi}\left(\sum_{j-1}^{Mi} v_{ijk} - 1\right)^2 \tag{5}$$

$$+\frac{C_4}{2}\sum_{i=1}^{S}\sum_{j=1}^{Mi}\sum_{k=1}^{Sn}\sum_{lk} S_{ijklk}v_{ijk}v_{ijlk}$$

Constraints (2) require that all operations of all jobs must be scheduled. Constraints (3) ensure that each operation can only be arranged on one position. Constraints (4) mean that each job must select one machine from each stage. C_1, C_2, C_3, C_4 are positive coefficients. The definition of set O_i is as follows: in the permutation matrix, not all the visiting positions on each machine are valid, for example, on machine 1, position 2 and 3 are invalid. All the valid positions compose the elements in set O_i of stage i.

Smith et. al.[7] demonstrated that the performance of the Hopfield neural networks depend on the formulation used. Since proper penalty parameters selection has long been the hardest task the network confronts, the simpler the parameters are, the easier the selection will be. After carefully examining constraints terms equations (2) – (4) it is not difficult to find that if constraints (4) are satisfied, other constraints are naturally satisfied, when constraints (4) are violated, other constraints can not be satisfied. So some constraints are unnecessary in nature. Based on the above observation, we get the simplified formulation as follows.

$$E = \frac{C_1}{2}\sum_{i=1,2,...S}\sum_{k\in Oi}\left(\sum_{j-1}^{Mi} v_{ijk} - 1\right)^2 +\frac{C_2}{2}\sum_{i=1}^{S}\sum_{j=1}^{Mi}\sum_{k=1}^{Sn}\sum_{lk} S_{ijklk}v_{ijk}v_{ijlk} \tag{6}$$

We deduce the connection weights $W_{ijk, i_1 j_1 k_1}$ and the threshold value θ_{ijk} of the networks as follows:

$$W_{ijk i_1 j_1 k_1} = -\frac{C_1}{2}\delta_{ii_1}\delta_{kk_1} -\frac{C_2}{2}\delta_{ii_1}\delta_{j_1}S_{ijklk} \quad i, j = 1,2,...,S, \quad k, k_1 \in O_i \tag{7}$$

$$\theta_{ijk} = C_1 \times \delta_{k, k_1} \qquad k, k_1 \in O_i \qquad j = 1,2,...,M_i \qquad i = 1,2,...,S \tag{8}$$

$$\text{Where } \delta_{ij} = \begin{cases} 1 & i = j \\ 0 & i \neq j \end{cases} \tag{9}$$

3 Modifications to the Hopfield Network and the Improving Strategy

Smith et. al[7] observed that the Hopfield neural networks work hardest and most productively during the first a few iterations, and that the improvements of energy function slow thereafter, until a local minima is encountered. In their case, they terminated the updating after a small number of iterations and introduce stochasticity into the neurons states, based on which the network begins to run again.

This paper proposes an improving strategy. The neurons update in a sequential asynchronous manner. The basic steps can be summarized as follows.

Step 1: Initialization. Set the initial states of all neurons at zero.

Step 2: Calculate the connection weights and the threshold values of the networks.

Step 3: Randomly select a neuron and calculate its state as the traditional Hopfield networks do, and update the states of all other neurons.

Step 4: Decide whether all the neurons have been selected, if yes, go to *step 5*; otherwise, return to *step 3*.

Step 5: Check the feasibility of current networks state, if yes, go to *step 6*; otherwise, return to *step 3*.

Step 6: Begin the updating process from the steady state that the network currently has. Calculate current value of the objective function.

Step 7: Select a neuron at random and change its state, the states of other neurons are computed as before, keeping the feasibility of the solution.

Step 8: If the objective value is decreased, accept the updated state, the current objective value is updated accordingly; otherwise reject the update and choose another neuron to continue such process.

Step 9: Check whether all possible neurons have been searched, if yes, algorithm ends; otherwise return to *step 6*.

To avoid unnecessary computation and time cost, each states searching should be guaranteed in the feasible region, which is accomplished by randomly selecting a neuron with value 1 and another neuron with value 0 belonging to the machines at the same stage.

To further improve the quality of the obtained solution, we design the heuristic algorithm based on the concept of nearest neighborhood searching. That is, of all the jobs assigned on one machine, select one job except the current one as the first job, choose the one with the smallest setup time as its immediate succeeding job. This process continues until all the jobs on this machine have been sequenced. Record the sum of setup times beginning with each job and choose the sequence with the smallest one as the final processing sequence on this machine. The improvement process ends when all the machines' job sequences have been optimized.

4 Computational Experiments

To demonstrate the performance of our algorithm, we randomly generate the setup time in [1, 100] as the data sets. For each size of instances, we run the algorithm 100 times and choose the one with the best results. Table 2 gives the average optimality of 10 independent results.

Table 2. Performance comparison between the proposed approach and the individual Heuristic.

size		50		100		150		200	
stages	structure	NN+H	H	NN+H	H	NN+H	H	NN+H	H
2 stages	2×2	1.0119	1.0259	1.0040	1.0240	1.0086	1.0126	1.0062	1.0211
	2×3	1.0148	1.0338	1.0142	1.0415	1.0114	1.0300	1.0001	1.0176
	4×2	1.0000	1.1527	1.0000	1.0980	1.0000	1.1085	1.0000	1.0661
3 stages	2×2×2	1.0154	1.0276	1.0119	1.0176	1.0059	1.0111	1.0096	1.0121
	1×2×3	1.0230	1.0253	1.0094	1.0300	1.0018	1.0337	1.0118	1.0152
	3×2×1	1.0115	1.0538	1.0095	1.0219	1.0146	1.0165	1.0089	1.0202
4 stages	2×2×2×2	1.0209	1.0336	1.0079	1.0132	1.0098	1.0191	1.0020	1.0136
	1×2×3×4	1.0021	1.0946	1.0000	1.1078	1.0000	1.0884	1.0000	1.1050
	4×3×2×1	1.0000	1.0634	1.0050	1.0477	1.0000	1.0578	1.0000	1.0363
average		**1.0111**	**1.0567**	**1.0069**	**1.0446**	**1.0058**	**1.0420**	**1.0043**	**1.0341**

Where the first row gives the number of jobs, the first column the number of stages in the shop, and the second column the allocation of machines; H - the Heuristic algorithm with random initialization, NN+H - the Heuristic combined Neural Networks algorithm.

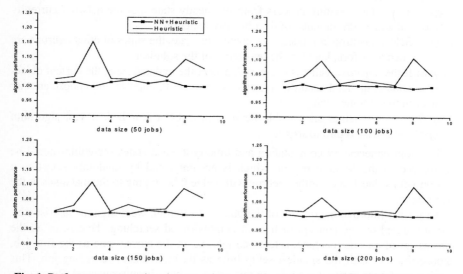

Fig. 1. Performance comparison between the proposed approach and the individual Heuristic.

For different scale of jobs, the illustration of the two algorithms is given in Fig. 1., which indicates that, the hybrid algorithm of Heuristic and Neural Networks outperforms the individual Heuristic with random initialization with respect to the optimality, no matter in uniform (the number of machines at each stage is the same), convergent or breakdown shop floor. The average improvements of four scales of jobs are 4.32%, 4.08%, 3.47% and 2.88%.

5 Conclusions

In this paper we have developed heuristic combined discrete Hopfield neural networks for solving hybrid flow shop scheduling problems. We gave the problem representation, the modified Hopfield network formulation and running strategy. Different size of data has been used to evaluate the performance of the modification to the model and to the running strategy. Simulation results indicate that the modified model is superior to the original one. Also the results imply that more efficient formulation will exhibit better scaling properties. Such promising results show us the prospect that further extension and improvement will make it possible for more general or complex scheduling problems.

Acknowledgements

This research is partly supported by National Natural Science Foundation for Distinguished Young Scholars of China (Grant No. 70425003), National Natural Science Foundation of China (Grant No. 60274049) and (Grant No. 70171030), Fok Ying Tung Education Foundation and the Excellent Young Faculty Program and the Ministry of Education, China. It is also partly supported by Integrated Automation Lab in Process Industry.

References

1. Okano, H., Davenport, A.J., Trumbo, M., Reddy, C., Yoda, K., and Amano, M.: Finishing Line Scheduling in the Steel Industry. Journal of Research & Development, **48** (2004) 811-830
2. Smith, K. A.: Neural Networks for Combinatorial Optimization: A Review of More Than a Decade of Research. INFORMS Journal on Computing, **11** (1999)
3. Mendes, A., Aguilera, L.: A Hopfield Neural Network Approach to the Single Machine Scheduling Problem. Pre-prints of IFAC/Incom98/Information Control In Manufacturing (1998)
4. Hopfield, J.J., Tank, D.W.: Neural Computation of Decisions in Optimization Problems. Biological Cybernetics, **52** (1985) 141-152
5. Wilson, G.V., Pawley, G.S.: On the Stability of the TSP Algorithm of Hopfield and Tank. Biological Cybernetics, **58** (1988) 63–70
6. Smith, K.A., Palaniswami, M., Krishnamoorthy, M.: Neural Techniques for Combinatorial Optimization with Applications. IEEE Transactions on Neural Networks, **9** (1998) 1301–1318
7. Smith, K.A., Abramson, D., Duke, D.: Hopfield Neural Networks for Timetabling: Formulations, Methods, and Comparative Results. Computers & Industrial Engineering, **44** (2003) 283-305

A Neural Network Based Heuristic
for Resource-Constrained Project Scheduling*

Yongyi Shou

School of Management, Zhejiang University, Hangzhou, Zhejiang 310027, China
yshou@zju.edu.cn

Abstract. Resource-constrained project scheduling allocates scarce resources over time to perform a set of activities. Priority rule-based heuristics are the most widely used scheduling methods though their performance depends on the characteristics of the projects. To overcome this deficiency, a feed-forward neural network is designed and integrated into the scheduling scheme so as to automatically select the suitable priority rules for each stage of project scheduling. Testing on Patterson's classic test problems and comparison with other heuristics show that the proposed neural network based heuristic is able to improve the performance of project scheduling.

1 Introduction

The resource-constrained project scheduling problem (RCPSP) is a typical scheduling problem which allocates scarce resource over time to perform a set of activities in order to minimize the project makespan. The problem is known to be strongly NP-hard. Many optimization methods have been proposed for the problem, including implicit enumeration methods, zero-one programming, dynamic programming, and etc. However, due to the inherent complexity of the RCPSP, it is no surprise that the majority of current algorithms are heuristic in nature. Comprehensive reviews of the state-of-the-art of the RCPSP could be found in the literature [1] and [2].

A number of heuristics have been proposed to improve the efficiency of resource allocation of resource-constrained project scheduling. Of the heuristics in the current literature, priority rule-based methods constitute the most important class of scheduling methods. Several reasons for their popularity are named by Kolisch [3]. First, they are straightforward and easy to implement, and that is why most commercial scheduling software prefers simple priority rules. Second, they are inexpensive in terms of computer time and memory required which makes them amenable for large instances of hard problems. Third, they are easy to be integrated into more complex metaheuristics, such as genetic algorithm and etc. This paper attempts to integrate artificial neural network (ANN) into priority rule-based scheduling methods.

2 Problem Description

A classic resource-constrained project scheduling problem can be represented by an oriented and acyclic activity-on-node (AON) network: (1) A finite set of J activities

---

* Project 70401017 is supported by National Natural Science Foundation of China.

J. Wang, X. Liao, and Z. Yi (Eds.): ISNN 2005, LNCS 3496, pp. 794–799, 2005.

or jobs, and each activity is indexed so that the preceding activity must have a smaller index and the J-th activity must be the only end activity; the processing time of the j-th activity is p_j, its start time ST_j, its completion time CT_j, $CT_j = ST_j + p_j$. (2) A finite set of precedence constraints: $S = \{S_j: j = 1,...,J\}$ where S_j is the set of activities immediately succeeding activity j. (3) A finite set of renewable resources with limited capacities, denoted by R. The capacity of resource r is K_r, and k_{jr} units of resource r is required by activity j for each period that activity j is in process.

For a classic RCPSP, an optimal project schedule is one that conforms the above-mentioned constraints and minimizes the project duration. Let I_t be the set of in-process activities at time t for a give schedule, the problem can be formulated as:

$$\min CT_J \tag{1}$$

$$\text{s.t. } ST_h \le CT_j, \qquad \forall h \in S_j \tag{2}$$

$$\sum_{j \in I_t} k_{jr} \le K_r, \qquad \forall r,t \tag{3}$$

3 Priority Rule-Based Heuristics

Priority rule-based heuristics consist of at least two components, including a schedule generation scheme (SGS) and priority rules. An SGS determines how a schedule is constructed gradually, building a feasible full schedule for all activities by augmenting a partial schedule covering only a subset of activities in a stage-wise manner. Two schemes are usually distinguished. In the serial SGS, a schedule is built by selecting the eligible activities in order and scheduling them one at a stage as soon as possible without violating the constraints. In the parallel SGS, a schedule proceeds by considering the time periods in chronological order and in each period all eligible activities are attempted to start at that time if resource availability allows. For each feasible RCPSP instance, a serial SGS searches among the set of active schedules which always contains at least one optimal schedule for makespan minimization [3]. Therefore, the serial SGS is adopted in this paper.

The serial SGS divides the set of activities into three disjoint subsets: scheduled, eligible, and ineligible. An activity that is already in the partial schedule is considered as scheduled. Otherwise, an activity is called eligible if all its predecessors are scheduled, and ineligible otherwise. The subsets of eligible and ineligible activities form the subset of unscheduled activities. The scheme proceeds in $N = J$ stages, indexed by n. On the n-th stage, the subset of scheduled activities is denoted as S_n and to the subset of eligible activities as decision set D_n. On each stage, if more than one activity is eligible, one activity j from D_n is selected using a priority rule and scheduled to begin at its earliest feasible start time. Then activity j is moved from D_n to S_n which may render some ineligible activities eligible if now all their predecessors are scheduled. The scheme terminates on stage N when all activities are scheduled.

Priority rules serve to resolve conflicts between activities competing for the allocation of scarce resources. In situations where the decision set contains more than one candidate, priority values are calculated from numerical measures which are related to properties of the activities, the complete project, or the incumbent partial schedule.

Some well-known priority rules are listed in Table 1, in which ES_j and LS_j denote the earliest start time and latest start time for activity j according to the critical path method (CPM). Shortest processing time (SPT), most total successors (MTS), latest start or finish time (LST, LFT), slack (SLK), and greatest rank positional weight (GRPW) are network and time-based priority rules; total resource demand (TRD) and total resource scarcity (TRS) are resource-based priority rules; and weighted resource utilization ratio and precedence (WRUP) is a composite priority rule which is a combination of MAX-MIS (most immediate successors) and MAX-TRS [4,5].

Table 1. Priority rules for RCPSP heuristics.

Rule	Extremum	Definition
Shortest processing time (SPT)	MIN	p_j
Most total successors (MTS)	MAX	$|\{h \mid h \text{ is } j\text{'s successor}\}|$
Latest start time (LST)	MIN	LS_j
Latest finish time (LFT)	MIN	$LS_j + p_j$
Minimal slack (SLK)	MIN	$LS_j - ES_j$
Greatest rank positional weight (GRPW)	MAX	$p_j + \sum_{h \in S_j} p_h$
Total resource demand (TRD)	MIN	$\sum_r k_{jr}$
Total resource scarcity (TRS)	MIN	$\sum_r k_{jr} / K_r$
Weighted resource utilization ration and precedence (WRUP*)	MAX	$w|S_j| + (1-w) TRS_j$

* In this paper $w = 0.7$ since the authors reported this setting as producing the best results [5].

4 Application of Neural Network

Previous studies confirmed that different priority rules are suitable for different kinds of projects. Hence, it is possible to utilize a combination of priority rules to improve the performance of scheduling. For example, Ash [6] attempted to choose heuristics through simulations, and Schirmer [7] proposed to improve the scheduling quality using parameterized composite priority rules. In these methods, one or more priority rules are employed for scheduling, yet the same set of priority rules are used throughout all the stages of one scheduling scheme. They lack the ability to forecast which priority rules are suitable for a specific situation. When the partial schedule develops, the scenario of the unscheduled activities along with the remaining resources keeps changing. The best priority rules in one stage might not be the best in the next stage since the scheduling configuration has been changed. An experienced project manager may be able to look ahead and know which priority rule is suitable for the current stage. However, selecting the rules manually is a cumbersome task even for small size problems, and it is impossible when the problem size increases. An artificial neural network is a suitable choice for this purpose.

In the n-th stage of a serial SGS, a partial schedule is known, i.e., n activities have been scheduled and their demands on resources are fulfilled. Thus, in this stage, the characteristics of the set of unscheduled activities and available resources shall indicate which priority rule is the suitable one. For the RCPSP, the resource factor (RF) and the resource strength (RS) are criteria widely used: RF determines the number of resources requested per activity, RS expresses the scarcity of the resources [8]. However, these indices are used to measure the factors of a complete project. Therefore, they require modification to suit the scenarios of partial schedules.

Partial network complexity (PNC) is defined as the average number of immediate successors per unscheduled activity. Partial resource factor (PRF) is defined as

$$PRF = \frac{1}{|U|} \frac{1}{|R|} \sum_{j \in U} \sum_{r \in R} \begin{cases} 1 & \text{if } k_{jr} > 0, \\ 0 & \text{otherwise.} \end{cases} \tag{4}$$

where U is the set of unscheduled activities. And a normalized index for partial resource strength (PRS) could be measured by

$$PRS = \frac{1}{|R|} \sum_{r \in R} \frac{K_r - K_r^{\max}}{K_r^{\max} - K_r^{\min}} \tag{5}$$

where $K_r^{\min}$ is the minimal amount of resource r required for executing the unscheduled activities, and $K_r^{\max}$ is the maximal amount of resource r required for implementing the unscheduled activities at their CPM times, i.e., any more units of resource r does not help to shorten the project makespan.

A feed-forward neural network is then designed to take these three characteristics as the input for describing the current scheduling configuration. The output is the priority rules as listed in Table 1. A certain number of neurons constitute the hidden layer. The diagram of the proposed ANN system is depicted in Fig. 1.

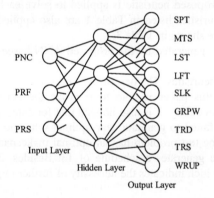

Fig. 1. Neural network for selecting priority rules in each stage of scheduling scheme.

The sigmoid function is used in selecting the priority rules. An output value less than 0.5 means that the corresponding priority rule is discarded, while an output value approximating 1.0 means that the corresponding priority rule is highly recommended. If there are more than one suitable priority rules, one of these rules is selected randomly. And if there is no suitable priority rule, the pure random sampling technique will be adopted, i.e., an activity will be selected from the decision set randomly. The artificial neural network is trained by some known optimal solutions to RCPSP instances.

Thus, the artificial neural network could be integrated in a serial SGS to form the following project scheduling heuristic as shown in Table 2.

Table 2. ANN-based heuristic for resource-constrained project scheduling.

Initialization

$\quad n \leftarrow 1$

$\quad ST_1 \leftarrow 0$

$\quad S_n \leftarrow \{1\}$

Execution

$\quad$ **for** $n \leftarrow 2$ **to** J

$\quad\quad$ **calculate** *PNC, PRF, PRS*

$\quad\quad$ **select** a priority rule according to the ANN in Fig. 1

$\quad\quad$ **calculate** D_n

$\quad\quad$ **select** j^* from D_n according to the priority rule selected

$\quad\quad$ $ST_{j*} \leftarrow$ the earliest resource-feasible time

$\quad\quad$ $S_n \leftarrow S_n \cup \{j^*\}$

$\quad$ **endfor**

Result

$\quad$ ST_j is the start time for activity j.

---

5 Testing and Comparison

The Patterson's test problems are adopted to test the proposed heuristic [9]. The problem set includes 110 scheduling problems, each with 7-50 activities and 1-3 renewable resources. The proposed heuristic is applied to solve each test problem and the priority rule-based heuristics listed in Table 1 are also applied for comparison. The computation results are shown in Table 3.

74.5% of schedules generated by the proposed ANN-based heuristic are optimal. And the average error to optimal solutions is 1.1%, which is much smaller than the errors of other tested heuristics. The descriptive statistics verified sufficiently that the method proposed in this paper is able to improve the performance of resource-constrained project scheduling. It is also noticed that for some instances the proposed ANN-based heuristic failed to generate near optimal schedules. The maximum error is 18.2%, for which the test problem has an optimal makespan of 11 time units and the proposed heuristic generates a schedule of 13. Besides, 23.6% schedules have errors within 10.0%, which indicates the necessity of further improvement.

Table 3. Descriptive statistics of the results of tested heuristics.

Heuristic	Minimum error (%)	Maximum error (%)	Average error (%)	Standard deviation (%)
ANN heuristic	0.0	18.2	1.1	2.5
MIN SPT	0.0	35.0	12.6	9.0
MAX MTS	0.0	75.0	17.7	12.7
MIN LST	0.0	75.0	27.2	13.3
MIN LFT	0.0	75.0	24.6	13.1
MIN SLK	0.0	35.0	12.6	9.0
MAX GRPW	0.0	75.0	23.4	12.2
MIN TRD	0.0	42.9	13.7	9.6
MIN TRS	0.0	42.9	10.4	8.1
MAX WRUP	0.0	79.0	19.7	13.8

6 Conclusions

An artificial neural network is designed to select the suitable priority rules in each stage of project scheduling. Three characteristics are developed to describe the progressive scheduling configuration, and used as the input data for the neural network which is integrated into the serial schedule generation scheme. Testing on Patterson's classic scheduling problems shows that the proposed method is capable to generate better solutions than other priority rule-based heuristics.

References

1. Herroelen, W., de Reyck, B., Demeulemeester, E.: Resource-Constrained Project Scheduling: A Survey of Recent Developments. Comput. Oper. Res., 25 (1998) 279-302
2. Kolisch, R. and Padman, R.: An Integrated Survey of Deterministic Project Scheduling. Omega-Int. J. Manage. Sci. 29 (2001) 249-272
3. Kolisch, R.: Serial and Parallel Resource-Constrained Project Scheduling Methods Revisited: Theory and Computation. Eur. J. Oper. Res., 90 (1996) 320-333
4. Alvarez-Valdes, R., Tamarit, J.M.: Heuristic Algorithms for Resource-Constrained Project Scheduling: A Review and an Empirical Analysis. In: Slowinski, R., Weglarz, J. (eds.): Advances in Project Scheduling. Elsevier, Amsterdam (1989) 113-134
5. Ulusoy, G., Ozdamar, L.: Heuristic Performance and Network/Resource Characteristics in Resource Constrained Project Scheduling. J. Oper. Res. Soc., 40 (1989) 1145-1152
6. Ash, R.C.: Activity Scheduling in the Dynamic, Multi-Project Setting: Choosing Heuristics through Deterministic Simulation. In: Farrington, P.A. et al. (eds.): Proceedings of the 1999 Winter Simulation Conference. ACM Press, New York (1999) 937-941
7. Schirmer, A.: Resource-constrained Project Scheduling: An Evaluation of Adaptive Control Schemes for Parameterized Sampling Heuristics. Int. J. Prod. Res., 39 (2001) 1343-1365
8. Kolisch, R., Sprecher, A., Drexl, A.: Characterization and Generation of a General Class of Resource-Constrained Project Scheduling Problems. Manage. Sci., 41 (1995) 1693-1703
9. Patterson, J.H.: A Comparison of Exact Approaches for Solving the Multiple Constrained Resource, Project Scheduling Problem. Manage. Sci., 30 (1984) 854-867

Functional-Link Net
Based Multiobjective Fuzzy Optimization

Ping Wang[1], Hongzhong Huang[2], Ming J. Zuo[3], Weidong Wu[4], and Chunsheng Liu[4]

[1] School of Automation, Chongqing University of Posts and Telecommunications,
Chongqing, 400065, China
wangping@cqupt.edu.cn
[2] School of Mechatronics Engineering,
University of Electronic Science and Technology of China, Chengdu, 610054, China
hzhhuang@dlut.edu.cn
[3] Department of Mechanical Engineering, University of Alberta,
Edmonton, Alberta, T6G 2G8, Canada
ming.zuo@ualberta.ca
[4] Department of Mechanical Engineering, Heilongjiang Institute of Science and Technology,
Harbin, 150027, China
{wu-weidong,liu_chunsheng}@163.com

Abstract. The principle of solving multiobjective optimization problems with fuzzy sets theory is studied. Membership function is the key to introduce the fuzzy sets theory to multiobjective optimization. However, it is difficult to determine membership functions in engineering applications. On the basis of rapid quadratic optimization in the learning of weights, simplification in hardware as well as in computational procedures of functional-link net, discrete membership functions are used as sample training data. When the network converges, the continuous membership functions implemented with the network. Membership functions based on functional-link net have been used in multiobjective optimization. An example is given to illustrate the method.

1 Introduction

More than thirty years have passed since the introduction of the concept of fuzzy optimization by Bellman and Zadeh [1]. Many new techniques in optimization such as fuzzy mathematical programming, fuzzy dynamic programming, and fuzzy multiobjective optimization have been developed which have enabled a higher qualitative level decision making under uncertainty. Though hundreds of papers have been published in fuzzy optimization, many problems of theoretical as well as empirical types remain to be solved. It is expected that this field of research will remain fertile - in methodology as well as in applications – in the decades to come.

The latest development in the area of intelligent techniques is a combination of fuzzy sets theory with neural networks and genetic algorithms [2].

Development of the membership function is the key in using the fuzzy sets theory for solving multiobjective optimization problems. In engineering applications, it remains to be difficult to determine the membership functions of many fuzzy entities. In the present work, we utilize rapid quadratic optimization in the learning of weights, simplification in hardware as well as in computational procedures of functional-link

J. Wang, X. Liao, and Z. Yi (Eds.): ISNN 2005, LNCS 3496, pp. 800–804, 2005.

net [3, 4] and use discrete membership functions as sample training data to train the neural network. The obtained membership functions obtained from the functional-link net are then used in fuzzy multiobjective optimization.

2 Fuzzy Model of Multiobjective Optimization

Consider the following multiobjective optimization problem

$$
\left.
\begin{aligned}
&\text{Find } X = [x_1, x_2, \cdots, x_n]^{\mathrm{T}} \\
&\text{which minimizes } F(X) = [F_1(X), F_2(X), \cdots, F_i(X), \cdots, F_m(X)]^{\mathrm{T}} \\
&\text{subject to } g_j(X) \leq 0; \quad j = 1, 2, \cdots, p \\
&\qquad\qquad\; h_k(X) = 0; \quad k = 1, 2, \cdots, q
\end{aligned}
\right\}
\tag{1}
$$

Because the m objectives of the problem may compete with one another, it is often impossible to minimize all these objectives simultaneously. As a result, a membership function is needed to indicate the degree of meeting the requirement of each objective. We propose the following computational procedure to find the optimal solution of the multiobjective optimization problem using fuzzy sets theory:

(1) Find the minimum and maximum possible values of each individual objective function.

(2) From the extreme values of the ith objective function $F_i(X)$, construct the membership function of this fuzzy objective function.

(3) The measure of the optimization of all m objectives is expresses as

$$
\tilde{D} = \bigcap_{i=1}^{m} \tilde{F}_i
\tag{2}
$$

The membership function of the fuzzy entity $\tilde{D}$ is given by

$$
\mu_{\tilde{D}}(X) = \bigwedge_{i=1}^{m} \mu_{\tilde{F}_i}(X)
\tag{3}
$$

(4) Find the optimal solution X of the following multiobjective optimization problem

$$
\mu_{\tilde{D}}(X^*) = \max \mu_{\tilde{D}}(X) = \max \bigwedge_{i=1}^{m} \mu_{\tilde{F}_i}(X)
\tag{4}
$$

3 Functional-Link Net Architecture and Learning Algorithm

The topology of a functional-link net consists of a functional link and a single-layered flat network (Fig. 1). The functional link alone acts on each input data point and expands the input pattern with the same function.

Similar to a multi-layer perceptron, the learning process of the functional-link net involves two phases: the feed-forward process and the error back-propagation process.

Let $\hat{y}_v$ and y_v denote the desired (or target) output and the actual output of the vth output neuron, respectively. Then, the error function can be defined as

$$
\varepsilon = \sum_{v=1}^{h} (\hat{y}_v - y_v)^2 / 2
\tag{5}
$$

The two-objective optimization problem can be stated as follows:

$$\text{Find } X = [x_1, x_2]^T$$

$$\text{which minimizes } F_1(X) = 2\sqrt{2}x_1 + x_2$$

$$F_2(X) = 20/(x_1 + 2\sqrt{2}x_2)$$

$$\text{subject to } 20\left(\sqrt{2}x_1 + x_2\right)\Big/\left(\sqrt{2}x_1^2 + 2x_1x_2\right) - 20 \leq 0$$

$$20\sqrt{2}x_1\Big/\left(\sqrt{2}x_1^2 + 2x_1x_2\right) - 20 \leq 0$$

$$20x_2\Big/\left(\sqrt{2}x_1^2 + 2x_1x_2\right) - 15 \leq 0$$

$$1\times10^{-5} \leq x_1 \leq 5\times10^{-4}$$

$$1\times10^{-5} \leq x_2 \leq 5\times10^{-4}$$

where x_1 and x_2 are the cross sectional areas of the bars, as indicated in Fig. 2.

We follow the proposed method outlined in the previous section to solve this multiobjective optimization problem and obtain the following optimal solution:

$$X_n^* = \left[1.32\times10^{-4}, 1.01\times10^{-4}\right]^T, \ F_1\left(X_n^*\right) = 3.73, \ F_2\left(X_n^*\right) = 4.79.$$

6 Conclusions

The key to solving multiobjective optimization problems with fuzzy sets theory is to determine the membership functions. In this paper, we have proposed a method to use the functional-link nets to represent the membership functions of objectives. These membership functions are then used in solving the multiobjective optimization problem. Because a functional-link net enhances the representation of input data by extending input vector, the network architecture is simplified. Moreover, a single-layer functional-link net is competent to perform what the multi-layer perceptron can do.

Acknowledgements

Financial supports from the National Excellent Doctoral Dissertation Special Foundation of China under contract number 200232, and the Natural Sciences and Engineering Research Council of Canada are gratefully acknowledged.

References

1. Bellman, R.E., Zadeh, L.A.: Decision Making in a Fuzzy Environment. Management Science, **17** (1970) 141-162
2. Jimenez, F., Cadenas, J.M., Verdegay, J.L. and Sonchez, G.: Solving Fuzzy Optimization Problems by Evolutionary Algorithms. Information Sciences. **152** (2003) 303-311
3. Pao, Y.H.: Adaptive Pattern Recognition and Neural Networks. Addison-Wesley, Reading, Massachusettes (1988)
4. Pao, Y.H., Takefuji, Y.: Functional-link Net Computing: Theory. System Architecture, and Functionalities. Computer, **25** (1992) 76-79

Optimizing the Distributed Network Monitoring Model with Bounded Bandwidth and Delay Constraints by Neural Networks

Xianghui Liu[1], Jianping Yin[1], Zhiping Cai[1], Xicheng Lu[1], and Shiming Chen[2]

[1] School of Computer Science, National University of Defense Technology,
Changsha , Hunan 410073 , China
LiuXH@tom.com
[2] Ende Technology, Changsha, Hunan 410073 , China

Abstract. Designing optimal measurement infrastructure is a key step for network management. In this work the goal of the optimization is to identify a minimum aggregating nodes set subject to bandwidth and delay constraints on the aggregating procedure. The problem is NP-hard. In this paper, we describe the way of using Hopfield networks for finding aggregating nodes set. The simulation indicates that Hopfield networks can produce much better result than the current method of randomly picking aggregating nodes.

1 Introduction

The explosive growth of Internet has emerged a massive need for monitoring technology that will support this growth by providing IP network managers with effective tools for monitoring network utilization and performance [1], [2]. Monitoring of the network-wide state is usually achieved through the use of the Simple Network Management Protocol (SNMP) with two kinds of entities: one management center and some monitoring nodes. The management center sends SNMP commands to the monitoring nodes to obtain information about the network and this function is performed by a centralized component responsible for aggregating all monitoring nodes [3]. Yet such processing queries have some inherent weaknesses. Firstly it can adversely impact router performance and result in significant volumes of additional network traffic. Secondly aggregating procedure is its time dependency. The support of knowledge of the up-to-date performance information requires the establishment of reliable, low delay and low cost aggregating routes [4], [5].

In above traditional centralized monitoring system, although the center provides a network-wide view but has some inherent weaknesses as being pointed out and not suitable for large scale network. Taking into account the issues of scalability and network-wide view for large service provider networks, an ideal monitoring architecture is a hierarchical system which implied that there is a management center but the resource intensive tasks such as polling are distributed. Between the management center and the monitoring nodes, there exists a set of aggregating nodes. The aggregating nodes are distributed and each node is responsible for an aggregating domain consisting of a subset of the network nodes. Information gathered from the individual monitoring nodes is then aggregated. Such a hierarchical architecture overcomes the weaknesses while still maintaining a network-wide view [4], [5]. In particular, the

J. Wang, X. Liao, and Z. Yi (Eds.): ISNN 2005, LNCS 3496, pp. 805–810, 2005.

most recently works addresses the problem of minimizing the number of aggregating nodes while keeping the aggregating bandwidth or delay within predefined limits individually. And all these problems are NP-Hard with solutions to this problem by using heuristics based on the aggregating load and the maximum assignment of monitoring nodes. The difficulties of using heuristics for optimal distributed network monitoring model is that after a possible aggregating node is picked, the algorithm tries to assign the maximum number of un-assigned monitoring nodes to the it without violating bandwidth and delay constraints. Unfortunately the general problem that assigns the maximum number of un-assigned monitoring nodes without violating constraints is also NP-Hard and all the heuristics only consider some special situation now [4], [5].

As the idea of using neural networks to provide solutions to difficult NP-Hard optimization problems has been pursued for over a decade and have some significant results. Certain neural networks embody a class of heuristic methods that lend themselves to combinatorial optimization problems. The Hopfield neural network model-- a recurrent neural network--is a prototypical example. The network evolves so as to minimize an energy function parameterized by the network's weights and biases. Thus, if the objective function of an optimization problem is mapped to the energy function of the Hopfield network, the former is also minimized during the network's operation.

The Hopfield model embodies a class of energy-minimization heuristics that are (1) problem independent, and (2) inherently parallel. And it is a single-layered feedback neural-network model which can be used to solve optimization problems if problem constraints and the optimization goal can be represented as a Lyapunov function. If we incorporate the objective function of the application problem into the energy function in an appropriate way, the objective function will also reach its minimum at equilibrium state. In this paper, we consider mapping the optimizing distributed monitoring modal with bounded bandwidth and delay constraints problem onto the Hopfield model of analog neurons [6], [7], [8].

2 Problem Formulation

We represent the whole monitoring domain of our model as an undirected graph $G(V,E)$, where $V = \{v_1, v_2, \cdots v_n\}$ is the set of all nodes or routers that are in the monitoring domain and. $E \subseteq V \times V$ represents the set of edges. The node set $S_m(S_m \subseteq V \wedge S_m \neq \Phi)$ represents the monitoring nodes in the monitoring domain. Each node $v(v \in S_m)$ ψgenerates an aggregating traffic of w_i ψbps. This aggregating traffic is destined to the relative aggregating node which has been assigned to. We define function $L : E \rightarrow R^+$ and $B : E \rightarrow R^+$ which assign a non-negative weight to each link in the network and represent the actual aggregating bandwidth used and the amount of link bandwidth allocated for aggregating traffic for each of the edges. And we also define edge-delay function $D : E \rightarrow R^+$ which assigns a non-negative weight to each of the edges. The value $D(e)$ associated with edge $e \in E$ is a measure (estimate) of total delay that packets experience on the link. Let the set $E(Path(u,v)) = \{e_1, e_2, \cdots, e_m\}$ represents the links in the path between node u and v.

The optimal aggregating node location and monitoring node assignment problem can therefore be stated as follows: Given a network $G(V, E)$, determine (1) a minimum subset of nodes $S_a(S_a \subseteq V)$ on which to place aggregating node such that the bandwidth constraint on each and every link $L(e) \le B(e)$ is satisfied (where $B(e)$ is the maximum bandwidth that can be used for aggregating on link e) and the delay constraint on every node $v(v \in S_m)$ satisfy $Delay(Path(v, w)) \le \delta$ (where δ is the maximum delay that can be used for aggregating by a node defined as w). (2) A mapping λ which maps a monitoring node to its aggregating node. That is, for any node $v(v \in S_m)$, if $\lambda(v) = w$, then node v is assigned to the aggregating node w. Note in some situation, we can use additional constraints to decide whether the monitoring node v can be aggregated by itself.

Now we define some variable to describe the integer program formulation about the problem. The binary variable x_{ij} indicates whether monitoring node v_i is aggregated by node v_j, where $v_i \in S_m$ and $v_j \in V$. The binary variable b_e^{ij} indicates whether edge e belongs to the $Path(v_i, v_j)$ between node v_i and v_j. The binary variable y_j indicates whether node v_j is an aggregating node or not .The problem can naturally expressed as an integer programming formulation:

The objective is: $Minimize \sum_{j=1}^{|V|} y_j$ and the constraints are below:

$$\sum_{j=1}^{|V|} x_{ij} = 1 (\forall v_i \in S_m) \tag{1}$$

$$x_{ij} \le y_j \left(\forall v_i \in S_m, \forall v_j \in V\right) \tag{2}$$

$$\sum_i \sum_j b_e^{ij} L(v_i) x_{ij} \le B(e)\left(\forall v_i \in S_m, \forall v_j \in V, e \in E\right) \tag{3}$$

$$\sum_{e \in E} b_e^{ij} D(e) x_{ij} \le \delta\left(\forall v_i \in S_m, \forall v_j \in V\right) \tag{4}$$

$$x_{ij} \in \{0,1\}\left(\forall v_i \in S_m, \forall v_j \in V\right) \tag{5}$$

$$y_j \in \{0,1\}\left(\forall v_j \in V\right) \tag{6}$$

The first constraint makes sure that each monitoring node v_i $\psi$is aggregated by exactly one aggregating node. The second constraint guarantees that a node v_j ψmust be an aggregating node if some other monitoring node v_i $\psi$is assigned to (aggregated by) it. The third constraint ensures the aggregating traffic on every link e not exceed the predefined bandwidth limits $B(e)$. The fourth constraint ensures that delay during aggregating procedure not exceeds the delay constraint on the path between each monitoring node and its aggregating node.

It is well-known that the integer programming formulation has an exponential running time in the worst case. In the previous work the greedy algorithm normally consists of two steps. In the first step algorithm greedily repeatedly picks a candidate aggregating node (based on the greedy selection criteria) if there are any monitoring nodes still present in the network that does not have an aggregating node assigned to it. After a candidate aggregating node is picked, the algorithm assigns candidate monitoring set to it without violating bandwidth or delay constraint. The repeat will interrupt when all monitoring nodes have been assigned, and the approximate aggregating node set includes all pickup additional aggregating nodes. Unfortunately the general problem that assigns the maximum number of un-assigned monitoring nodes without violating constraints is also NP-Hard and all the heuristic algorithm only consider some special situation. And obtaining the globally optimal solution is not as imperative as arriving at a near-optimal solution quickly. Certainly, one of the principal advantages of neural techniques is the rapid computation power and speed which can be obtained and this consideration is even more valuable in industrial situations. So the below we consider using Hopfield network to solve the problem.

3 Mapping Optimization Problems to Hopfield Network

The Hopfield network comprises a fully interconnected system of neurons. Neuron has internal state and output level. The internal state incorporates a bias current and the weighted sums of outputs from all other neurons which determine the strength of the connections between neurons. The relationship between the internal state of a neuron and its output level is determined by an activation function which is bounded below by zero and above.

Provided the weights are symmetric, the Hopfield networks can be used as an approximate method for solving 0–1 optimization problems because the network converges to a minimum of the energy function. The proof of stability of such Hopfield networks relies upon the fact that the energy function is a Lyapunov function. Hopfield and Tank showed that if a combinatorial optimization problem can be expressed in terms of a quadratic energy function of the general form, a Hopfield network can be used to find locally optimal solutions of the energy function, which may translate to local minimum solutions of the optimization problem. Typically, the network energy function is made equivalent to the objective function which is to be minimized, while each of the constraints of the optimization problem is included in the energy function as penalty terms.

A combinatorial optimization problem is a discrete problem with large but finite set of possible solutions. Typically if the problem is of size N, there are maybe $N!$ possible solutions of which we want one that minimizes the cost function that is constructed of an objective and constraints.

Approximation algorithms are often used to deal with NP-hard problems. Such algorithms perform in polynomial time but they do not assure finding an optimal solution. Hopfield networks can be considered approximation algorithms when used to solve optimization problems as being described. The following steps are taken when problem is mapped into Hopfield Network.

With specifying state variables $S_i \in \{0,1\}$, so that the vector $\vec{S} = (S_i)$ would represent the possible solution to a problem. Formulating the objective function $obj(S)$ and constraints $cons(S)$. Then the cost function can be rewritten as $H = \alpha\ obj(S) + \beta\ cns(S)$. Matching H with the energy function E and obtaining the weights and biases that allow us to construct a Hopfield style network.

Applying update rules to get a stable state that will represent the local minimum of the cost function. Hopefully we get a feasible solution.

4   Simulation

In this section, we evaluate the performance of proposed algorithm with the heuristic algorithm on several different topologies and parameter settings. For simplicity, we make the reasonable assumption of shortest path routing.

The network graph used in this study is the Waxman model. We generate different network topologies by varying the Waxman parameter β for a fixed parameter $\alpha = 0.2$. The varying β gives topologies with degrees of connectivity ranging from 4 to 8 for a given number of nodes in the network. Each link allocates a certain fraction of their bandwidth for aggregating traffic based on the capacity constraint imposed for the link. In our simulation, the fraction is the same for all links and the value is 5% and 10% respectively. The delay for every link changes from 0.1 to 2.0 with a fix delay tolerance parameter $\delta = 10$. Simulation results presented here are averaged over 4 different topologies.

Random Picking: the heuristic is to pick a possible aggregating node from V randomly. This heuristic serves as a base-line comparison for the neural network algorithm proposed in the paper. Once we select an aggregating node, we need to determine the set of monitoring nodes to be assigned to that it. Ideally, we would like to assign maximum number of unassigned monitoring nodes to a new aggregating node. The algorithm is present in paper [4].

Simulation results has been omitted due to paper size limitations and the simulation shows the Hopfield network have more better result than the randomly method.

5   Conclusions

The primary motivation for work to design good measurement infrastructure it is necessary to have a scalable system at a reduced cost of deployment. As the model is NP-Hard and the current heuristics algorithm is that after a possible aggregating node is picked, the algorithm tries to assign the maximum number of un-assigned monitoring nodes to it without violating bandwidth and delay constraints. Unfortunately the general problem that assigns the maximum number of un-assigned monitoring nodes without violating constraints is also NP-Hard.

In this paper, we have demonstrated that neural-network techniques can compete effectively with more traditional heuristic solutions to practical combinatorial optimization problems .Combining this knowledge with the fact that neural networks have the potential for rapid computational power and speed through hardware implementation, it is clear that neural-network techniques are immensely useful for solving optimization problems of practical significance.

References

1. Asgari, A., Trimintzios, P., Irons, M., Pavlou, G., and den Berghe, S.V., Egan, R.: A Scalable Real-Time Monitoring System for Supporting Traffic Engineering. In Proc. IEEE Workshop on IP Operations and Management (2002)
2. Breitbart, Y., Chan, C.Y., Garofalakis, M., Rastogi, R., and Silberschatz, A.: Efficiently Monitoring Bandwidth and Latency in IP Networks. In Proc. IEEE Infocom (2002)

3. Breitgand, D., Raz, D., and Shavitt, Y.: SNMP GetPrev: An Efficient Way to Access Data in Large MIB Tables. IEEE Journal of Selected Areas in Communication, **20** (2002) 656–667

4. Li, L., Thottan, M., Yao, B., Paul, S.: Distributed Network Monitoring with Bounded Link Utilization in IP Networks. In Proc. of IEEE Infocom (2003)

5. Liu, X.H., Yin, J.P., Lu, X.-C., Cai, Z.-P., Zhao, J.-M.: Distributed Network Monitoring Model with Bounded Delay Constraints. Wuhan University Journal of Natural Sciences

6. Miettinen, K., Mäkelä, M., and Neittaanmäki, P., and Périaux, J. (eds.), Evolutionary Algorithms in Engineering and Computer Science, John Wiley & Sons (1999)

7. Chellapilla, K., and Fogel, D.: Evolving Neural Networks to Play Checkers without Relying on Expert Knowledge. IEEE Trans. on Neural Networks, **10** (1999) 1382-1391

8. Del, R.: A Dynamic Channel Allocation Technique Based on Hopfield Neural Networks. IEEE Transactions On Vehicular Technology

Stochastic Nash Equilibrium
with a Numerical Solution Method

Jinwu Gao[1] and Yankui Liu[2]

[1] Uncertain Systems Laboratory, School of Information, Renmin University of China
Beijing 100872, China
[2] College of Mathematics and Computer Science, Hebei University
Baoding 071002, China

Abstract. Recent decades viewed increasing interests in the subject of decentralized decision-making. In this paper, three definitions of stochastic Nash equilibrium, which casts different decision criteria, are proposed for a stochastic decentralized decision system. Then the problem of how to find the stochastic Nash equilibrium is converted to an optimization problem. Lastly, a solution method combined with neural network and genetic algorithm is provided.

Keywords: decentralized decision system; Nash equilibrium; stochastic programming; neural network; genetic algorithm

1 Introduction

In modern organizations and engineering systems, there are often multiple decision makers. Each of them seeks one's own interest by making decisions. Very often, these decision makers are noncooperative and their decisions intervene with each other, which calls forth the problem of finding Nash equilibrium. Such decentralized decision-making problem has intrigued more and more economists, operations researchers, and engineers for its widely applications to the area of economics [2][12] and engineering [7][10].

In practice, uncertainty is often involved in decentralized decision systems. For instance, in economic systems, costs, demands and many other elements are often subject to fluctuations and difficult to measure. These situations emphasize the need for models which are able to tackle uncertainty inherent in decision systems. In literature, Anupindi *et al.* [1] investigated a decentralized distribution system with stochastic demands. Wang et al. [13] investigated a two-echelon decentralized supply chain. Gao *et al.* [5] solved a stochastic bilevel decentralized decision system by multilevel programming. In this paper, we will focus on the Nash equilibrium of the decision makers in stochastic environments. According to different decision criteria, three definitions of stochastic Nash equilibrium are discussed. In order to find the stochastic Nash equilibrium, we established an optimization problem, whose optimal solution is just the Nash equilibrium provided the optimum equals to zero. Lastly, we provide a solution method by integrating stochastic simulation, neural networks, and genetic algorithm.

J. Wang, X. Liao, and Z. Yi (Eds.): ISNN 2005, LNCS 3496, pp. 811–816, 2005.

2   Stochastic Nash Equilibriums

In literature, decentralized decision-making problem often involves the problem of Nash equilibrium in the decision makers. Moreover, resources, costs, demands and many other elements are often subject to fluctuations and difficult to measure. Probability theory has been widely recognized as the means for dealing with such uncertainty. Assume that the uncertain system parameters are independent stochastic variables, a general stochastic decentralized decision system is described by the following notations:

$i = 1, 2, \cdots, m$ index of the decision makers

y_i control vector of the ith decision maker

$f_i(y_1, y_2, \cdots, y_m, \xi_i)$ objective function of the ith decision maker, where ξ_i is a random vector into which problem parameters are arranged

$g_i(y_1, y_2, \cdots, y_m, \xi_i)$ constraint function of the ith decision maker, where ξ_i is a random vector into which problem parameters are arranged

Due to the randomness of system parameters, the objective functions and constraints are also stochastic. With the requirement of taking the randomness into account, appropriate formulations of stochastic Nash equilibrium is to be developed to suit the different purpose of management.

Generally, the decision makers want to optimize the expected value of their stochastic objective functions subject to some expected constraints. Then the $i^{\text{th}}(i = 1, 2, \cdots, m)$ decision maker can get his optimal solution via the following stochastic expected value model:

$$\begin{cases} \max_{y_i} \mathrm{E}[f_i(y_1, y_2, \cdots, y_m, \xi_i)] \\ \text{subject to:} \\ \quad \mathrm{E}[g_i(y_1, y_2, \cdots, y_m, \xi_i)] \leq 0. \end{cases} \qquad (1)$$

It is obvious that the decision of the ith decision maker depends on the decisions of the others denoted by

$$y_{-i} = (y_1, \cdots, y_{i-1}, y_{i+1}, \cdots, y_m).$$

For this case, the Nash equilibrium of the decision makers is defined as follows.

Definition 1. (Gao *et al.* [5]) *A feasible array* $(y_1^*, y_2^*, \cdots, y_m^*)$ *is called a Nash equilibrium, if it satisfies that*

$$\mathrm{E}\left[f_i(y_1^*, y_2^*, \cdots, y_{i-1}^*, y_i, y_{i+1}^*, \cdots, y_m^*, \xi_i)\right] \leq \mathrm{E}\left[f_i(y_1^*, y_2^*, \cdots, y_m^*, \xi_i)\right] \qquad (2)$$

for any feasible $(y_1^*, y_2^*, \cdots, y_{i-1}^*, y_i, y_{i+1}^*, \cdots, y_m^*)$ *and* $i = 1, 2, \cdots, m$.

Sometimes, the decision makers may want to optimize some critical values of his stochastic objective functions with given confidence levels subject to some chance constraints. Then the $i^{\text{th}}(i = 1, 2, \cdots, m)$ decision maker can get his optimal solution via the stochastic chance-constrained programming model [3]:

$$\begin{cases} \max_{\boldsymbol{y}_i} \bar{f}_i \\ \text{subject to:} \\ \quad \Pr\{f_i(\boldsymbol{y}_1, \boldsymbol{y}_2, \cdots, \boldsymbol{y}_m, \boldsymbol{\xi}_i) \geq \bar{f}_i\} \geq \alpha_i \\ \quad \Pr\{g_i(\boldsymbol{y}_1, \boldsymbol{y}_2, \cdots, \boldsymbol{y}_m, \boldsymbol{\xi}_i) \leq 0\} \geq \beta_i. \end{cases} \quad (3)$$

And the Nash equilibrium is defined as follows.

Definition 2. *(Gao et al. [5]) A feasible array $(\boldsymbol{y}_1^*, \boldsymbol{y}_2^*, \cdots, \boldsymbol{y}_m^*)$ is called a Nash equilibrium, if it satisfies that*

$$\max\left\{\bar{f}_i \mid \Pr\left\{f_i(\boldsymbol{y}_1^*, \cdots, \boldsymbol{y}_{i-1}^*, \boldsymbol{y}_i, \cdots, \boldsymbol{y}_m^*, \boldsymbol{\xi}_i) \geq \bar{f}_i\right\} \geq \alpha_i\right\}$$
$$\leq \max\left\{\bar{f}_i \mid \Pr\left\{f_i(\boldsymbol{y}_1^*, \boldsymbol{y}_2^*, \cdots, \boldsymbol{y}_m^*, \boldsymbol{\xi}_i) \geq \bar{f}_i\right\} \geq \alpha_i\right\} \quad (4)$$

for any feasible $(\boldsymbol{y}_1^*, \boldsymbol{y}_2^*, \cdots, \boldsymbol{y}_{i-1}^*, \boldsymbol{y}_i, \boldsymbol{y}_{i+1}^*, \cdots, \boldsymbol{y}_m^*)$ and $i = 1, 2, \cdots, m$.

In many cases, the decision makers may concern some events (e.g., objective function's being greater than a prospective value), and wish to maximize the chance functions of satisfying these events. Then the $i^{\text{th}}(i = 1, 2, \cdots, m)$ decision maker can get his optimal solution via the stochastic dependent-chance programming model [11]:

$$\begin{cases} \max_{\boldsymbol{y}_i} \Pr\{f_i(\boldsymbol{y}_1, \boldsymbol{y}_2, \cdots, \boldsymbol{y}_m, \boldsymbol{\xi}_i) \geq \bar{f}_i\} \\ \text{subject to:} \\ \quad g_i(\boldsymbol{y}_1, \boldsymbol{y}_2, \cdots, \boldsymbol{y}_m, \boldsymbol{\xi}_i) \leq 0. \end{cases} \quad (5)$$

And the Nash equilibrium in this situation is defined as follows.

Definition 3. *A feasible array $(\boldsymbol{y}_1^*, \boldsymbol{y}_2^*, \cdots, \boldsymbol{y}_m^*)$ is called a Nash equilibrium, if it satisfies that*

$$\Pr\left\{\begin{array}{l} f_i(\boldsymbol{y}_1^*, \cdots, \boldsymbol{y}_{i-1}^*, \boldsymbol{y}_i, \cdots, \boldsymbol{y}_m^*, \boldsymbol{\xi}_i) \geq \bar{f}_i \\ g_i(\boldsymbol{y}_1^*, \cdots, \boldsymbol{y}_{i-1}^*, \boldsymbol{y}_i, \cdots, \boldsymbol{y}_m^*, \boldsymbol{\xi}_i) \leq 0 \end{array}\right\}$$
$$\leq \Pr\left\{\begin{array}{l} f_i(\boldsymbol{y}_1^*, \boldsymbol{y}_2^*, \cdots, \boldsymbol{y}_m^*, \boldsymbol{\xi}_i) \geq \bar{f}_i \\ g_i(\boldsymbol{y}_1^*, \boldsymbol{y}_2^*, \cdots, \boldsymbol{y}_m^*, \boldsymbol{\xi}_i) \leq 0 \end{array}\right\} \quad (6)$$

for any feasible $(\boldsymbol{y}_1^*, \boldsymbol{y}_2^*, \cdots, \boldsymbol{y}_{i-1}^*, \boldsymbol{y}_i, \boldsymbol{y}_{i+1}^*, \cdots, \boldsymbol{y}_m^*)$ and $i = 1, 2, \cdots, m$.

3 Discussion

It is well-known that computing Nash equilibrium is a very tough task. In this section, we construct an optimization problem whose optimal solution is the

Nash equilibrium provided that the optimum is zero. Then we can find the Nash equilibrium by solving the optimization problem.

Suppose that the ith decision maker knows the strategies $\boldsymbol{y}_{-i}$ of other decision makers, then the optimal reaction of the ith decision maker is represented by a mapping $\boldsymbol{y}_i = r_i(\boldsymbol{y}_{-i})$ that solves the subproblem (1) or (3) or (5).

It is clear that the Nash equilibrium of the m decision makers will be the solution of the system of equations

$$\boldsymbol{y}_i = r_i(\boldsymbol{y}_{-i}), \quad i = 1, 2, \cdots, m. \tag{7}$$

In other words, the Nash equilibrium is a fixed point of the vector-valued function

$$r(\boldsymbol{y}_1, \boldsymbol{y}_2, \cdots, \boldsymbol{y}_m) = (r_1, r_2, \cdots, r_m).$$

Let

$$R(\boldsymbol{y}_1, \boldsymbol{y}_2, \cdots, \boldsymbol{y}_m) = \sum_{i=1}^{m} \|\boldsymbol{y}_i - r_i(\boldsymbol{y}_{-i})\|.$$

Consider the following minimization problem:

$$\begin{cases} \min\limits_{\boldsymbol{y}_1, \boldsymbol{y}_2, \cdots, \boldsymbol{y}_m} R(\boldsymbol{y}_1, \boldsymbol{y}_2, \cdots, \boldsymbol{y}_m) \\ \text{subject to:} \\ \quad (\boldsymbol{y}_1, \boldsymbol{y}_2, \cdots, \boldsymbol{y}_m) \in \bigcap\limits_{i=1}^{m} D_i, \end{cases} \tag{8}$$

where D_i is the feasible set of model (1) or (3) or (5). If the optimal solution $(\boldsymbol{y}_1^*, \boldsymbol{y}_2^*, \cdots, \boldsymbol{y}_m^*)$ satisfies that

$$R(\boldsymbol{y}_1^*, \boldsymbol{y}_2^*, \cdots, \boldsymbol{y}_m^*) = 0, \tag{9}$$

then $\boldsymbol{y}_i^* = r_i(\boldsymbol{y}_{-i}^*)$ for $i = 1, 2, \cdots, m$. Hence, $(\boldsymbol{y}_1^*, \boldsymbol{y}_2^*, \cdots, \boldsymbol{y}_m^*)$ must be a Nash equilibrium.

In a numerical solution process, if the optimal solution $(\boldsymbol{y}_1^*, \boldsymbol{y}_2^*, \cdots, \boldsymbol{y}_m^*)$ satisfies that

$$R(\boldsymbol{y}_1^*, \boldsymbol{y}_2^*, \cdots, \boldsymbol{y}_m^*) \leq \varepsilon, \tag{10}$$

where ε is a small positive number, then it can be regarded as a Nash equilibrium. Otherwise, the system of equations (7) might be considered inconsistent. That is, there is no Nash equilibrium.

4   A Numerical Solution Method

For a particular optimization problem (8), there may be effective numerical solution methods. Here, we only give a feasible method, which is integrated by neural network and genetic algorithm, for general cases.

4.1   Neural Networks

A functions involved stochastic parameters is called an uncertain function. In order to solve model (8), we should first cope with uncertain functions. Certainly, stochastic simulation is able to compute uncertain functions. However, it is a time-consuming process because we need many times of stochastic simulation to compute uncertain functions in the solution process.

As we know, a neural network with an arbitrary number of hidden neurons is a universal approximator for continuous functions [4][8] and has high speed of operation after it is well-trained on a set of input-output data. In order to speed up the solution process, we may first use stochastic simulation to generate a set of input-output data. Then, we train a neural network on the set of input-output data by using the popular backpropagation algorithm. Finally, the trained network can be used to evaluate the uncertain function. As a result, much computing time is saved.

4.2   Genetic Algorithm

Genetic algorithm [6][9] is a stochastic search and optimization procedure motivated by natural principles and selection. Genetic algorithm starts from a population of candidate solutions rather than a single one, which increases the probability of finding a global optimal solution. Then genetic algorithm improves the population step by step by biological evolutionary processes such as crossover and mutation, in which the search uses probabilistic transition rules, and can be guided towards regions of the search space with likely improvement, thus also increasing the probability of finding a global optimal solution. In brief, the advantage of genetic algorithm is just able to obtain the global optimal solution fairly. In addition, genetic algorithm does not require any key properties of the problem, thus can be easily used to solve complex decision systems. Herein, we give a genetic algorithm procedure to search for the Nash equilibrium.

Genetic Algorithm for Nash Equilibrium:
Step 1. Generate a population of chromosomes

$$(y_1^{(j)}, y_2^{(j)}, \cdots, y_m^{(j)}), \ j = 1, 2, \cdots, pop\_size$$

at random from the feasible set.
Step 2. Calculate the the objective values of chromosomes.
Step 3. If the best chromosome satisfies inequality (10), goto Step 8.
Step 4. Compute the fitness of each chromosome according to the objective values.
Step 5. Select the chromosomes by spinning the roulette wheel.
Step 6. Update the chromosomes by crossover and mutation operations.
Step 7. Repeat Steps 3–6 for a given number of cycles.
Step 8. Return the best solution $(y_1^*, y_2^*, \cdots, y_m^*)$ with the objective value

$$\sum_{i=1}^{m} \|y_i^* - r_i(y_{-i}^*)\|.$$

5 Concluding Remarks

In this paper, three definitions of stochastic Nash equilibriums were discussed. Then the problem of finding Nash equilibriums was converted to an optimization problem. A numerical solution method that is integrated by neural network and genetic algorithm was also given.

Acknowledgments

This work was supported by National Natural Science Foundation of China (No.60174049), and Specialized Research Fund for the Doctoral Program of Higher Education (No.20020003009).

References

1. Anupindi, R., Bassok, Y., and Zemel, E.: A General Framework for the Study of Decentralized Distribution Systems. Manufacturing & Service Operations Management, **3** (2001) 349–68
2. Bylka, S.: Competitive and Cooperative Policies for the Vendor-buyer System. International Journal of Production Economics, **81-82** (2003) 533–44
3. Charnes, A. and Cooper, W.W.: Chance-constrained Programming. Management Science, **6** (1959) 73–79
4. Cybenko, G.: Approximations by Superpositions of a Sigmoidal Function. Mathematics of Control, Signals and Systems, **2** (1989) 183–192
5. Gao, J., Liu, B., and Gen, M.: A Hybrid Intelligent Algorithm for Stochastic Multi-level Programming. Transactions of the Institute of Electrical Engineers of Japan, **124-C** (2004) 1991-1998
6. Goldberg, D.E.: Genetic Algorithms in Search. Optimization and Machine Learning, Addison-Wesley, MA, (1989)
7. Hirasawa, K., Yamamoto, Y., Hu, J., Murata, J., and Jin, C.: Optimization of Decentralized Control System Using Nash Equilibrium Concept. Transactions of the Institute of Electrical Engineers of Japan, **119-C** (1999) 467-73
8. Hornik, K., Stinchcombe, M., and White, H.: Multilayer Feedforward Networks are Universal Approximators. Neural Networks, **2** (1989) 359–366
9. Holland, J.H.: Adaptation in Natural and Artificial Systems. University of Michigan Press, Ann Arbor, (1975)
10. Ji, X., Wang, C.: Nash Equilibrium in the Game of Transmission Expansion. Proceedings of 2002 IEEE Region 10 Conference on Computer, Communications, Control and Power Engineering, **3** (2002) 1768-1771
11. Liu, B.: Dependent-chance Programming: A Class of Stochastic Programming. Comput. Math. Appl., **34** (1997) 89–104
12. Vriend, N.J.: A Model of Market-making. European Journal of Economic and Social Systems, **15** (2002) 185-202
13. Wang, H., Guo, M., and Efstathiou, J.: A Game-theoretical Cooperative Mechanism Design for a Two-echelon Decentralized Supply Chain. European Journal of Operational Research, **157** (2004) 372-388

Generalized Foley-Sammon Transform with Kernels

Zhenzhou Chen[1,2] and Lei Li[1]

[1] Institute of Software, Zhongshan University, Guangzhou 510275, China
[2] School of Mathematics and Computational Science, Zhongshan University,
Guangzhou 510275, China
chenzhenzhou2324@sina.com

Abstract. Fisher discriminant based Foley-Sammon Transform (FST) has great influence in the area of pattern recognition. On the basis of FST, the Generalized Foley-Sammon Transform (GFST) is presented. The main difference between the GFST and the FST is that the transformed sample set by GFST has the best discriminant ability in global sense while FST has this property only in part sense. Linear discriminants are not always optimal, so a new nonlinear feature extraction method GFST with Kernels (KGFST) based on kernel trick is proposed in this paper. Linear feature extraction in feature space corresponds to non-linear feature extraction in input space. Then, KGFST is proved to correspond to a generalized eigenvalue problem. Lastly, our method is applied to digits and images recognition problems, and the experimental results show that present method is superior to the existing methods in term of space distribution and correct classification rate.

1 Introduction

Fisher discriminant based Foley-Sammon Transform (FST)[1] has great influence in the area of pattern recognition. Guo et al. [2] proposed a generalized Foley-Sammon transform (GFST) based on FST. The main difference between the discriminant vectors constituting GFST and the discriminant vectors calculated with FST is that the projected set on the discriminant vectors constituting GFST has the best separable ability in global sense. It is clear that the properties of the scatter matrices of the sample set in the subspace spanned by the vectors of GFST are good in terms of separable ability.

GFST is a linear feature extraction method, but the linear discriminant is not always optimal. Worse, even rather simple problem like the classical XOR-problem can not be solved using linear functions. Hence, the question arises if one can modify the linear discriminants such that one remains their good properties but gets the flexibility to solve problems which need non-linear decision functions. We know that the kernel trick amounts to performing algorithm as before, but implicitly in the Kernel Hilbert space (feature space) connected to the kernel function used. So the generalized Foley-Sammon transform with kernels (KGFST) is proposed in this paper.

J. Wang, X. Liao, and Z. Yi (Eds.): ISNN 2005, LNCS 3496, pp. 817–823, 2005.

The remainder of the paper is organized as follows: Section 2 gives a brief review of GFST. Section 3 shows how to express GFST method as a linear algebraic formula in feature space and proves that the KGFST can be converted to a generalized eigenvalue problem that can be calculated easily. Section 4 provides some experiments of KGFST and other methods. Finally, section 5 gives a brief summary of the present method.

2   Generalized Foley-Sammon Transform (GFST)

Let $Z = \{(x_1, y_1), ..., (x_n, y_n)\} \subseteq R^l \times \{\omega_1, ..., \omega_C\}$. The number of samples in each class ω_i is n_i. Suppose the mean vector, the covariance matrix and a priori probability of each class ω_i are m_i, S_i, P_i, respectively. The global mean vector is m_0. Then the between-class scatter matrix S_B and the within-class scatter matrix S_W are determined by the following formulae:

$$S_B = \sum_{i=1}^{C} P_i(m_i - m_0)(m_i - m_0)^T, \quad S_W = \sum_{i=1}^{C} P_i S_i.$$

The Fisher criterion and the Generalized Fisher criterion can be defined as:

$$J_F(w) = \frac{w^T S_B w}{w^T S_W w}, \quad J_G(w_i) = \frac{\sum_{k=1}^{i} w_k^T S_B w_k}{\sum_{k=1}^{i} w_k^T S_W w_k}.$$

Let w_1 be the unit vector which maximizes $J_F(w)$, then w_1 is the first vector of Generalized Foley-Sammon optimal set of discriminant vectors, the ith vector (w_i) of Generalized Foley-Sammon optimal discriminant set can be calculated by optimizing the following problem(GFST) [2]:

$$\begin{cases} \max[J_G(w_i)] \\ s.t. \quad w_i^T w_j = 0, \quad j = 1, \cdots, i-1 \ . \\ ||w_i|| = 1 \end{cases} \tag{1}$$

3   Generalized Foley-Sammon Transform with Kernels

GFST is a linear feature extraction method, but the linear discriminant is not always optimal. Worse, even rather simple problem like the classical XOR-problem can not be solved using linear functions. Hence, the question arises if one can modify the linear discriminants such that one remains their good properties but get the flexibility to solve problems which need non-linear decision functions.

The answer is positive and we can formulate GFST using kernel functions which are widely used in SVM [3], Kernel PCA [4] and other kernel algorithms [5],[6],[7]. Kernel trick will allow for powerful, non-linear decision functions while still leaving us with a relatively simple optimization problem. The kernel trick amounts to performing algorithm as before, but implicitly in the Kernel Hilbert space H. Since for each kernel function $k()$ there exists a mapping $\Phi : X \to H$, such that $k(x, z) = (\Phi(x) \cdot \Phi(z))$, one is looking for a discriminant of the form: $f(x) = w^T \Phi(x)$, where $w \in H$ is a vector in feature space H. As the only way for us to work in feature space H is by using the kernel functions, we need a formulation of GFST that only use Φ in such dot-products.

3.1   From Input Space to Feature Space: The Derivation of KGFST

Let Φ be a mapping from X to H, to find the linear Fisher discriminant in H we need maximize:

$$J_F(w) = \frac{w^T S_B^\Phi w}{w^T S_W^\Phi w}. \tag{2}$$

In feature space, suppose the mean vector, the covariance matrix and a priori probability of each class ω_i are m_i^Φ, S_i^Φ, P_i, respectively. The global mean vector is m_0^Φ. The between-class scatter matrix and within-class scatter matrix in feature space are S_B^Φ and S_W^Φ, which are determined by the following formulae:

$$S_B^\Phi = \sum_{i=1}^{C} P_i(m_i^\Phi - m_0^\Phi)(m_i^\Phi - m_0^\Phi)^T, \quad S_W^\Phi = \sum_{i=1}^{C} P_i S_i^\Phi.$$

According to reproduced kernel theory, every solution $w \in H$ can be written as: $w = \sum_{i=1}^{n} a_i \Phi(x_i)$. Then we can get

$$w^T m_i^\Phi = \frac{1}{n_i} \sum_{j=1}^{n} \sum_{k=1}^{n_i} a_j \Phi(x_j) \cdot \Phi(x_{ik}) = \mathbf{a}^T M_i, (M_i)_j = \frac{1}{n_i} \sum_{k=1}^{n_i} k(x_j, x_{ik}). \tag{3}$$

According to S_B^Φ and (3), the numerator of (2) can be rewritten as follow:

$$w^T S_B^\Phi w = \mathbf{a}^T M \mathbf{a}, \tag{4}$$

where $M = \sum_{i=1}^{C} P_i(M_i - M_0)(M_i - M_0)^T$, $(M_0)_j = \frac{1}{n} \sum_{k=1}^{N} k(x_j, x_k)$.

For the same reason, the denominator of (2) can be rewritten as follow:

$$w^T S_W^\Phi w = w^T [\frac{1}{n} \sum_{i=1}^{C} \sum_{k=1}^{n_i} (\Phi(x_{ik}) - m_i^\Phi)(\Phi(x_{ik}) - m_i^\Phi)^T]w = \mathbf{a}^T N \mathbf{a}, \tag{5}$$

where $N = \frac{1}{n} \sum_{i=1}^{C} K_i(I - 1_{n_i})K_i^T$, K_i is a $n \times n_i$ matrix with $(K_i)_{lm} = k(x_l, x_{im})$, I is a identity and 1_{n_i} is a matrix with all entries equalling $1/n_i$.

Combining (4) and (5), the Fisher's linear discriminant in H is given as:

$$J(\mathbf{a}) = \frac{\mathbf{a}^T M \mathbf{a}}{\mathbf{a}^T N \mathbf{a}}. \tag{6}$$

The unit vector condition $w_i \cdot w_i = 1$ and the orthogonal conditions $w_i \cdot w_j = 0$ in input space can be rewritten in feature space as follows:

$$w_i \cdot w_i = \mathbf{a}_i^T K \mathbf{a}_i = 1, w_i \cdot w_j = \mathbf{a}_i^T K \mathbf{a}_j = 0, (j = 1, \ldots, i-1). \tag{7}$$

According to the results of (1), (6) and (7), one can easily get the set of discriminant vectors constituting KGFST in H. Let $\mathbf{a}_1$ be the vector which maximizes $J(\mathbf{a})$ and $\mathbf{a}_1^T K \mathbf{a}_1 = 1$, then $\mathbf{a}_1$ is the first vector of KGFST optimal set of discriminant vectors, the ith vector $(\mathbf{a}_i)$ of KGFST optimal discriminant set can be calculated by optimizing the following problem:

$$\begin{cases} \max[J(\mathbf{a}_i)], & see \ (9) \\ s.t. \quad \mathbf{a}_i^T K \mathbf{a}_j = 0, \quad j = 1, \cdots, i-1 \ . \\ \mathbf{a}_i^T K \mathbf{a}_i = 1 \end{cases} \tag{8}$$

3.2   Converting KGFST to a Generalized Eigenvalue Problem

First let's rewrite the discriminant criterion of KGFST:

$$J(\mathbf{a}_i) = \frac{\sum\limits_{j=1}^{i-1} \mathbf{a}_j^T M \mathbf{a}_j + \mathbf{a}_i^T M \mathbf{a}_i / \mathbf{a}_i^T K \mathbf{a}_i}{\sum\limits_{j=1}^{i-1} \mathbf{a}_j^T N \mathbf{a}_j + \mathbf{a}_i^T N \mathbf{a}_i / \mathbf{a}_i^T K \mathbf{a}_i} = \frac{\mathbf{a}_i^T \tilde{M}_i \mathbf{a}_i}{\mathbf{a}_i^T \tilde{N}_i \mathbf{a}_i}, \tag{9}$$

where $\tilde{M}_i = (\sum\limits_{j=1}^{i-1} \mathbf{a}_j^T M \mathbf{a}_j) K + M, \tilde{N}_i = (\sum\limits_{j=1}^{i-1} \mathbf{a}_j^T N \mathbf{a}_j) K + N, (\tilde{M}_1 = M, \tilde{N}_1 = N)$.

One can maximize (9) by optimizing the following equivalent problem:

$$\begin{cases} \max[J_E(\mathbf{a}_i) = \mathbf{a}_i^T \tilde{M}_i \mathbf{a}_i] \\ s.t. \quad \mathbf{a}_i^T \tilde{N}_i \mathbf{a}_i = 1 \end{cases} \tag{10}$$

The Lagrangian of (10) for the first discriminant vector $\mathbf{a}_1$ is:

$$L(\mathbf{a}_1, \lambda) = \mathbf{a}_1^T \tilde{M}_1 \mathbf{a}_1 - \lambda(\mathbf{a}_1^T \tilde{N}_1 \mathbf{a}_1 - 1).$$

On the saddle point, the following condition must be satisfied:

$$\frac{\partial L(\mathbf{a}_1, \lambda)}{\partial \mathbf{a}_1} = 2\tilde{M}_1 \mathbf{a}_1 - 2\lambda \tilde{N}_1 \mathbf{a}_1 = 0,$$

i.e. discriminant vector $\mathbf{a}_1$ is the eigenvector corresponding to the largest eigenvalue of the generalized eigenvalue problem: $\tilde{M}_1 \mathbf{a}_1 = \lambda \tilde{N}_1 \mathbf{a}_1$.

According to (8) and (10), the Lagrangian for the discriminant vector $\mathbf{a}_i$ is:

$$L(\mathbf{a}_i, \lambda) = \mathbf{a}_i^T \tilde{M}_i \mathbf{a}_i - \lambda(\mathbf{a}_i^T \tilde{N}_i \mathbf{a}_i - 1) - \sum\limits_{j=1}^{i-1} \mu_j \mathbf{a}_i^T K \mathbf{a}_j.$$

Just like above, on the saddle point, the following condition must be satisfied:

$$\frac{\partial L(\mathbf{a}_i, \lambda)}{\partial \mathbf{a}_i} = 2\tilde{M}_i \mathbf{a}_i - 2\lambda \tilde{N}_i \mathbf{a}_i - \sum_{j=1}^{i-1} \mu_j K \mathbf{a}_j = 0. \tag{11}$$

If both sides of (11) multiply $\mathbf{a}_k^T K \tilde{N}_i^{-1} (k < i)$, one can get:

$$2\mathbf{a}_k^T K \tilde{N}_i^{-1} \tilde{M}_i \mathbf{a}_i - \sum_{j=1}^{i-1} \mu_j \mathbf{a}_k^T K \tilde{N}_i^{-1} K \mathbf{a}_j = 0, \ k = 1, \cdots i-1. \tag{12}$$

Let $\boldsymbol{u} = [\mu_1, \cdots, \mu_{i-1}]^T$, $D = [\mathbf{a}_1, \cdots, \mathbf{a}_{i-1}]^T$, then (12) can be rewritten as $2DK\tilde{N}_i^{-1}\tilde{M}_i\mathbf{a}_i = DK\tilde{N}_i^{-1}KD^T\boldsymbol{u}$, i.e.

$$\boldsymbol{u} = 2(DK\tilde{N}_i^{-1}KD^T)^{-1}DK\tilde{N}_i^{-1}\tilde{M}_i\mathbf{a}_i. \tag{13}$$

We know that in (11): $\sum_{j=1}^{i-1} \mu_j K \mathbf{a}_j = KD^T\boldsymbol{u}$. Substituting $\boldsymbol{u}$ of (11) with (13), then the following formula is obtained:

$$P\tilde{M}_i\mathbf{a}_i = \lambda \tilde{N}_i\mathbf{a}_i, \tag{14}$$

where $P = I - KD^T(DK\tilde{N}_i^{-1}KD^T)^{-1}DK\tilde{N}_i^{-1}$. So $\mathbf{a}_i$ is the eigenvector corresponding to the largest eigenvalue of the generalized eigenvalue problem (14).

After $\mathbf{a}_i$ has been obtained, one should normalize $\mathbf{a}_i$ with $\mathbf{a}_i^T K \mathbf{a}_i = 1$.

4   Computational Comparison

In this section, we compare the performance of KFGST against GFST and KPCA. We implemented all these methods in Matlab 6.5 and ran them on a 2.0G MHz Celeron machine.

4.1   The Datasets and Algorithms

The following datasets are used in our experiments:

Dataset A: The "Pendigits" database from the UCI repository. Pendigits is a Pen-based recognition problem of handwritten digits. The digits written by 30 writers are used for training and the digits written by other 14 writers are used for testing. Each pattern contains one class attribute and 16 input features. We produce a subset of Pendigits A3 which is a 3-class problem.

Dataset B: The database of US Postal Service (USPS). USPS is also a recognition problem of handwritten digits which are collected from postal code in real life. All patterns are 0-9 digit representations and all input features are normalized. We also produce a subset of USPS B3 which is a 3-class problem.

Dataset C: The dataset of Chair Images (ChairImg). There are 25 models of chairs in ChairImg and each pattern of ChairImg is 16×16 images $r_0 \cdots r_4$ (grey-level images and edge data). ChairImg consists of three training sets (Train25, Train89 and Train100) and one testing set (Test100).

To compare the methods above, we use linear support vector machines (SVM) and K-nearest neighbors (KNN) algorithm as classifiers.

4.2 Results and Analysis

Figure 1 describes the space distribution of A3 on the features extracted by GFST and KGFST. We can see that classification ability of the features extracted by KGFST is more powerful than that of the features extracted by GFST: the between-class scatter of A3 on the features extracted by KGFST is larger than that by GFST while the within-class scatter of A3 on the features extracted by KGFST is smaller than that by GFST.

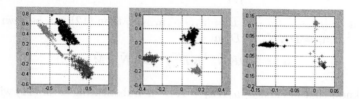

Fig. 1. (left) space distribution of A3 on the features extracted by GFST; (middle) space distribution of A3 on the features extracted by KGFST with polynomial kernel (rank=3); (right) space distribution of A3 on the features extracted by KGFST with RBF kernel (width=2);

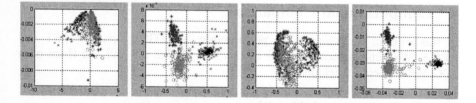

Fig. 2. (first) space distribution of B3 on the features extracted by KPCA with polynomial kernel (rank=2); (second) space distribution of B3 on the features extracted by KGFST with polynomial kernel (rank=2); (third) space distribution of B3 on the features extracted by KPCA with RBF kernel (width=300); (fourth) space distribution of B3 on the features extracted by KGFST with RBF kernel (width=300);

Figure 2 describes the space distribution of B3 on the features extracted by KPCA and KGFST. From figure 2, we can see that the overlap of features extracted by KGFST is smaller than that of features extracted by KPCA. The separable ability of the features extracted by KGFST is stronger than that of the features extracted by KPCA.

We also can compare the performance of them using classification errors (Table 1). From table 1, it is easy to conclude that KGFST is better than GFST and KPCA: the error classification number of two classifiers (SVM and KNN) on the features extracted by KGFST on dataset C is smaller than on the features extracted by the other two methods.

Table 1. classification ability of feature set extracted by methods above on dataset C

input vector	Training set	GFST error number		KPCA error number		KGFST error number	
		KNN	SVM	KNN	SVM	KNN	SVM
$r_0 \cdots r_4$	Train25	752	744	1314	837	248	248
	Train89	722	708	482	644	34	34
	Train100	716	707	578	726	58	58
r_0	Train25	1814	1608	1429	928	393	392
	Train89	1691	1541	557	883	54	56
	Train100	1680	1547	683	1004	100	100

5 Conclusion

In this paper, a new nonlinear feature extraction method KGFST based on kernel trick and GFST is proposed. Linear feature extraction in feature space corresponds to non-linear feature extraction in input space. KGFST can provide non-linear decision functions while remaining the good properties of GFST. KGFST is proved to correspond to a generalized eigenvalue problem. Lastly, our method is applied to digits and images recognition problems, and the experimental results show that present method is superior to the existing methods in term of space distribution and correct classification rate.

There are also some questions of KGFST in practice: speed of KGFST and selection of kernel model [8].

References

1. Foley, D.H., Sammon, J.W.: An Optimal Set of Discriminant Vectors. IEEE Trans on Computers, **24** (1975) 281-289
2. Guo, Y.F., Li, S.J., et al.: A Generalized Foley-Sammon Transform Based on Generalized Fisher Discriminant Criterion and Its Application to Face Recognition. Pattern Recognition Letters, **24** (2003) 147-158
3. Mika, S., Schölkopf, B., et al.: Kernel PCA and De-noising in Feature Spaces. In M. S. Kearns, S. A. Solla, and D. A. Cohn, editors, Advances in Neural Information Processing Systems. Vol. 11. MIT Press, Cambridge, MA (1999) 536 – 542
4. Smola, A.J., Schölkopf, B.: A Tutorial on Support Vector Regression. Technical Report (1998)
5. Bach, F.R., Jordan, M.I.: Kernel Independent Component Analysis. (Kernel Machines Section) **3** (2002) 1-48
6. Suykens, J.A.K., Gestel, T.V., et al.: Least Squares Support Vector Machines, World Scientific, Singapore,ISBN 981-238-151-1 (2002)
7. Weinberger, K., Sha, F., Saul, L.: Learning a Kernel Matrix for Nonlinear Dimensionality Reduction. Appearing in Proceedings of the 21st International Conference on Machine Learning, Banff, Canada (2004)
8. Glenn, F., Murat, D., et al.: A Fast Iterative Algorithm for Fisher Discriminant using Heterogeneous Kernels. Appearing in Proceedings of the 21st International Conference on Machine Learning, Banff, Canada (2004)

Sparse Kernel Fisher Discriminant Analysis*

Hongjie Xing[1,2,3], Yujiu Yang[1,2], Yong Wang[1,2], and Baogang Hu[1,2]

[1] National Laboratory of Pattern Recognition
Institute of Automation, Chinese Academy of Sciences
P.O. Box 2728, Beijing 100080, China
{hjxing,yjyang,wangyong,hubg}@nlpr.ia.ac.cn
[2] Beijing Graduate School, Chinese Academy of Sciences
P.O. Box 2728, Beijing 100080, China
[3] Faculty of Mathematics and Computer Science, Hebei University
Baoding, Hebei 071002, China

Abstract. This paper presents a method of Sparse Kernel Fisher Discriminant Analysis (**SKFDA**) through approximating the implicit within-class scatter matrix in feature space. Traditional Kernel Fisher Discriminant Analysis (**KFDA**) has to use all the training samples to construct the implicit within-class scatter matrix while SKFDA needs only small part of them. Based on this idea, the aim of sparseness can be obtained. Experiments show that SKFDA can dramatically reduce the number of training samples used for constructing the implicit within-class scatter matrix. Numerical simulations on "Banana Shaped" and "Ripley and Ionosphere" data sets confirm that SKFDA has the merit of decreasing the training complexity of KFDA.

1 Introduction

Kernel Fisher Discriminant Analysis (or KFDA)[1] is a nonlinear discriminant method based on Fisher linear discriminant[2]. The main idea of it can be described as follows. The training samples are mapped into a high dimensional feature space by a nonlinear function, then Fisher linear discriminant is performed toward these mapped samples, so a discriminant hyperplane can be obtained in the feature space. When the hyperplane mapped back into the sample space, it will become a nonlinear curved surface. So we can take KFDA as an nonlinear generalization of Fisher Discriminant Analysis (or **FDA**).

The within-class scatter matrix needs to be determined before the training of KFDA. Although the construction of within-class scatter matrix can be performed implicitly by kernel function matrices, all training samples have to be used. The problem can be overcome by Sparse Kernel Fisher Discriminant Analysis (or SKFDA). In order to achieve SKFDA, the mapped samples in the feature space are assumed to obey multivariate normal distributions. The sparseness of within-class scatter matrix can be obtained by approximating the sum of covariance matrices for training samples of two classes in the feature space.

---

* This work is supported by Natural Science of Foundation of China (#60275025, #60121302).

J. Wang, X. Liao, and Z. Yi (Eds.): ISNN 2005, LNCS 3496, pp. 824–830, 2005.
© Springer-Verlag Berlin Heidelberg 2005

This paper is organized as follows, Section 2 describes KFDA and Section 3 discusses SKFDA method in detail. The result concerning experiments are reported based on one synthetic data set and two real data sets in Section 4. Finally, some concluding remarks are given in Section 5.

2   KFDA

Given a two-class training set $\mathbf{X} = (\mathbf{x}_i, y_i)_{i=1}^n$, where $\mathbf{x}_i \in \mathbf{X} \subseteq \mathbf{R}^n (i = 1, 2, \ldots, n)$ and $y_i \in \{-1, 1\}$ is the class label of the training sample $\mathbf{x}_i$. There are n_1 training samples with positive labels denoted by $\mathbf{X}_1 = \{\mathbf{x}_1^1, \mathbf{x}_2^1, \ldots, \mathbf{x}_{n_1}^1\}$ while $\mathbf{X}_2 = \{\mathbf{x}_1^2, \mathbf{x}_2^2, \ldots, \mathbf{x}_{n_2}^2\}$ for n_2 training samples with negative labels.

KFDA is the nonlinear extension of the linear FDA. The training samples are mapped into a feature space $\mathbf{Z}$ by a nonlinear mapping function $\phi : \mathbf{X} \to \mathbf{Z}$. FDA then is performed in the feature space $\mathbf{Z}$ towards the mapped training samples $\{(\phi(\mathbf{x}_i), y_i)\}_{i=1}^n$. Finally the nonlinear discriminant in the sample space can be performed implicitly. To find the Kernel Fisher Discriminant (or KFD), we need to maximize the criterion:

$$J(\mathbf{v}) = \frac{\mathbf{v}^T \mathbf{S}_B^\phi \mathbf{v}}{\mathbf{v}^T \mathbf{S}_W^\phi \mathbf{v}}, \tag{1}$$

where $\mathbf{v} \in \mathbf{Z}$, $\mathbf{S}_B^\phi$ and $\mathbf{S}_W^\phi$ are between-class scatter matrix and within-class scatter matrix, respectively. The definitions of them are as follows[1]:

$$\mathbf{S}_B^\phi = (\mathbf{m}_1^\phi - \mathbf{m}_2^\phi)(\mathbf{m}_1^\phi - \mathbf{m}_2^\phi)^T \tag{2}$$

$$\mathbf{S}_W^\phi = \mathbf{S}_{W1}^\phi + \mathbf{S}_{W2}^\phi = \sum_{i=1,2} \sum_{\mathbf{x} \in \mathbf{X}_i} (\phi(\mathbf{x}) - \mathbf{m}_i^\phi)(\phi(\mathbf{x}) - \mathbf{m}_i^\phi)^T$$
$$= \sum_{i=1,2} \sum_{k=1}^{n_i} \phi(\mathbf{x}_k^i)(\phi(\mathbf{x}_k^i))^T \tag{3}$$

and $\mathbf{m}_i^\phi = \frac{1}{n_i} \sum_{\mathbf{x} \in \mathbf{X}_i} \phi(\mathbf{x}) = \frac{1}{n_i} \sum_{k=1}^{n_i} \phi(\mathbf{x}_k^i)$.

Nevertheless, if the dimension of the feature space $\mathbf{Z}$ is too high or close to infinite, it is impossible to solve the problem above. Kernel trick can deal with this problem easily because it can convert the problem in the feature space into the sample space. Since the KFD utilizes the inner product in the feature space, according to the theory of functional analysis, the inner product can be implied in the sample space by Mercer kernel[3], i.e. $k(\mathbf{x}, \mathbf{x}') = <\phi(\mathbf{x}) \cdot \phi(\mathbf{x}') >$, where $< \cdot >$ denotes inner product. There are three kinds of kernel functions in common use[3]:

1. **Polynomial kernel function:** $k(\mathbf{x}, \mathbf{y}) = (< \mathbf{x} \cdot \mathbf{y} > +1)^d (d = 1, 2, \ldots)$;
2. **RBF kernel function:** $k(\mathbf{x}, \mathbf{y}) = \exp(\frac{-\|\mathbf{x} - \mathbf{y}\|^2}{2r^2})$;
3. **Sigmoid kernel function:** $k(\mathbf{x}, \mathbf{y}) = \tanh(a < \mathbf{x} \cdot \mathbf{y} > +b)$.

To perform FDA in the feature space $\mathbf{Z}$, according to the theory of Reproducing Kernel[4], vector $\mathbf{v} \in \mathbf{Z}$ can be expanded into the linear summation of all the training samples, i.e. :

$$\mathbf{v} = \sum_{i=1}^n \alpha_i \phi(\mathbf{x}_i). \tag{4}$$

Then the equation (1) can be changed into:

$$\mathbf{J}(\alpha) = \frac{\alpha^T \mathbf{M} \alpha}{\alpha^T \mathbf{N} \alpha}. \tag{5}$$

and $\alpha = [\alpha_1, \alpha_2, \ldots, \alpha_n]^T$. $\mathbf{M} = (\mathbf{M}_1 - \mathbf{M}_2)(\mathbf{M}_1 - \mathbf{M}_2)^T$ while $(\mathbf{M}_i)_j = \frac{1}{n_i} \sum_{k=1}^{n_i} k(\mathbf{x}_j, \mathbf{x}_k^i)$, where $i = 1, 2$ and $j = 1, 2, \ldots, n$. $\mathbf{N} = \sum_{i=1,2} \mathbf{K}_i (\mathbf{I} - \mathbf{1}_{n_i}) \mathbf{K}_i^T$, where $(\mathbf{K}_i)_{lk} = k(\mathbf{x}_l, \mathbf{x}_k^i)$, $l = 1, 2, \ldots, n, k = 1, 2, \ldots, n_i$, $\mathbf{I}$ is the identity and $\mathbf{1}_{n_i}$ the matrix with all entries $1/n_i$. To obtain the solution of vector α, the eigenvalue and eigenvector of matrix $\mathbf{N}^{-1}\mathbf{M}$ need to be calculated. Whereas vector α can be calculated directly by $\alpha = \mathbf{N}^{-1}(\mathbf{M}_1 - \mathbf{M}_2)$[5].

When the number of the training samples is more than the dimension of feature space, the regularization method[6] can be used, that is, adding a multiple of identity matrix, then the matrix $\mathbf{N}$ can be substituted by $\mathbf{N}_\lambda$, where $\mathbf{N}_\lambda = \mathbf{N} + \lambda \mathbf{I}$ and $0 \le \lambda \le 1$.

Finally, the Fisher linear discriminant function in the feature space $\mathbf{Z}$ can be expressed as:

$$f(\mathbf{x}) = sgn(\mathbf{v}^T \phi(\mathbf{x}) + b) = sgn(\sum_{j=1}^{n} \alpha_j k(\mathbf{x}_j, \mathbf{x}) + b). \tag{6}$$

where the symbolic function $sgn(t) = \begin{cases} 1 & t > 0 \\ 0 & t \le 0 \end{cases}$, and

$$b = -\frac{\mathbf{v}^T (\mathbf{m}_1^\phi + \mathbf{m}_2^\phi)}{2} = -\frac{\alpha^T (\mathbf{K}_1 + \mathbf{K}_2)}{2}. \tag{7}$$

3   SKFDA

SKFDA assumes that the mapped samples in the feature space $\mathbf{Z}$ obey gaussian distributions which is similar to sparse kernel principal component analysis[7]. After a zero-mean operation, the mapped samples satisfy $\overline{\phi(\mathbf{x}_k^i)} \sim \mathbf{N}(0, \mathbf{C}^i)$, where the covariance matrix $\mathbf{C}^i$ is specified by:

$$\mathbf{C}^i = \sigma_i^2 \mathbf{I} + \sum_{j=1}^{n_i} w_j^i \overline{\phi(\mathbf{x}_j^i)}(\overline{\phi(\mathbf{x}_j^i)})^T = \sigma_i^2 \mathbf{I} + \sum_{j=1}^{n_i} w_j^i \overline{\phi_j^i}(\overline{\phi_j^i})^T = \sigma_i^2 \mathbf{I} + (\overline{\boldsymbol{\Phi}^i})^T \mathbf{W}^i \overline{\boldsymbol{\Phi}^i} \tag{8}$$

with diagonal matrix $\mathbf{W}^i = diag(w_1^i, \ldots, w_{n_i}^i)$. Then the within-class scatter matrix $\mathbf{S}_W^\phi$ can be approximated by matrix $\sum_{i=1,2} (\overline{\boldsymbol{\Phi}^i})^T \mathbf{W}^i \overline{\boldsymbol{\Phi}^i}$.

To sparsify the within-class scatter matrix $\mathbf{S}_W^\phi$, the covariance σ_i^2 needs to be fixed and the wights $w_k^i(k = 1, 2, \ldots, n_i)$ need to be optimized. Through maximum likelihood estimation, the most weights w_k^i are zero. Then the aim of sparsifying the matrix $\mathbf{S}_W^\phi$ can be obtained.

3.1 Optimizing the Weights

$\overline{\phi^i} \sim \prod_{k=1}^{n_i} \mathbf{N}(\mathbf{0}, \mathbf{C}^i)$ can be deduced by $\overline{\phi_k^i} = \overline{\phi(\mathbf{x}_k^i)} \sim \mathbf{N}(\mathbf{0}, \mathbf{C}^i)$. Ignoring the independent terms of the weights, its log likelihood is given by:

$$\mathcal{L} = -\frac{1}{2}[n_i \log |\mathbf{C}^i| + tr((\mathbf{C}^i)^{-1}(\overline{\mathbf{\Phi}^i})^T(\overline{\mathbf{\Phi}^i}))]. \tag{9}$$

Differentiating (9) with respect to w_k^i, we can get:

$$\begin{aligned}\frac{\partial \mathcal{L}}{\partial w_k^i} &= \frac{1}{2}[(\overline{\phi_k^i})^T(\mathbf{C}^i)^{-1}(\overline{\mathbf{\Phi}^i})^T\overline{\mathbf{\Phi}^i}(\mathbf{C}^i)^{-1}\overline{\phi_k^i} - n_i(\overline{\phi_k^i})^T(\mathbf{C}^i)^{-1}\overline{\phi_k^i}]\\&= \frac{1}{2(w_k^i)^2}[\sum_{j=1}^{n_i}(\mu_{jk}^i)^2 + n_i\Sigma_{kk}^i - n_i w_k^i]\end{aligned} \tag{10}$$

with $\mathbf{\Sigma}^i = ((\mathbf{W}^i)^{-1} + \sigma_i^{-2}\mathbf{K}^i)^{-1}$ and $\mu_j^i = \sigma_i^{-2}\mathbf{\Sigma}^i \mathbf{k}_j^i$, where $\mathbf{K}^i = \overline{\mathbf{\Phi}^i}(\overline{\mathbf{\Phi}^i})^T$ and $\mathbf{k}_j^i = [(\overline{\phi(\mathbf{x}_j^i)})^T(\overline{\mathbf{\Phi}^i})^T]^T$. The detail description please refer to[7][8]. However, note the Woodbury inversion identity[8] $(\beta^{-1}\mathbf{I} + \mathbf{\Phi}\mathbf{A}^{-1}\mathbf{\Phi}^T)^{-1} = \beta\mathbf{I} - \beta\mathbf{\Phi}(\mathbf{A} + \beta\mathbf{\Phi}^T\mathbf{\Phi})^{-1}\mathbf{\Phi}^T\beta$ and the determinant identity[9] $|\mathbf{A}||\beta^{-1}\mathbf{I} + \mathbf{\Phi}\mathbf{A}^{-1}\mathbf{\Phi}^T| = |\beta^{-1}\mathbf{I}| + |\mathbf{A} + \beta\mathbf{\Phi}^T\mathbf{\Phi}|$ during the course of consequence. Setting (10) to zero provides re-estimation equations for the weights[10]:

$$(w_k^i)^{new} = \frac{\sum_{j=1}^N(\mu_{jk}^i)}{n_i(1 - \Sigma_{kk}^i/w_k^i)} \tag{11}$$

3.2 FDA in the Feature Space

Let $\widetilde{\mathbf{\Phi}^i} = \mathbf{W}^{\frac{1}{2}}\overline{\mathbf{\Phi}^i}$, then $(\overline{\mathbf{\Phi}^i})^T\mathbf{W}^i\overline{\mathbf{\Phi}^i} = (\mathbf{W}^{\frac{1}{2}}\overline{\mathbf{\Phi}^i})^T(\mathbf{W}^{\frac{1}{2}}\overline{\mathbf{\Phi}^i}) = (\widetilde{\mathbf{\Phi}^i})^T\widetilde{\mathbf{\Phi}^i}$. According to the discriminant method in section 2, $\mathbf{M} = (\mathbf{M}_1 - \mathbf{M}_2)(\mathbf{M}_1 - \mathbf{M}_2)^T$, where $(\mathbf{M}_i)_j = \frac{1}{n_i}\sum_{k=1}^{n_i} k(\mathbf{x}_j, \mathbf{x}_k^i)$. It is given $\mathbf{N} = \sum_{i=1,2} \mathbf{\Phi}(\widetilde{\mathbf{\Phi}^i})^T\widetilde{\mathbf{\Phi}^i}\mathbf{\Phi}^T = \sum_{i=1,2} \mathbf{K}_i(\mathbf{I} - \mathbf{1}_{n_i})\mathbf{W}(\mathbf{I} - \mathbf{1}_{n_i})\mathbf{K}_i^T$ which can imply the sparseness of the within-class scatter matrix implicitly, where $\mathbf{\Phi} = [\phi(\mathbf{x}_1), \phi(\mathbf{x}_2), \dots, \phi(\mathbf{x}_n)]^T$.

Replacing the matrix $\mathbf{N}$ by $\mathbf{N}_\lambda$, so $\alpha = \mathbf{N}_\lambda^{-1}(\mathbf{M}_1 - \mathbf{M}_2)$, finally we can find the discriminant function:

$$f(\mathbf{x}) = sgn(\mathbf{v}^T\phi(\mathbf{x}) + b) = sgn(\sum_{j=1}^n \alpha_j k(\mathbf{x}_j, \mathbf{x}) + b), \tag{12}$$

where the scalar b can be obtained by (7).

4 Numerical Experiments

To observe the experiment results of SKFDA, the noise variances $\sigma_i^2(i = 1, 2)$ need to be specified firstly. But measuring it can only be performed in the feature space, thus the choice of it have no criteria so that we can only take the different values of it by experiments. The kernel functions in this paper are all RBF kernel functions $k(\mathbf{x}, \mathbf{x}') = \exp(-\frac{\|\mathbf{x} - \mathbf{x}'\|}{2r^2})$. The other kinds of kernel function can also be chosen. The paper only analysises the RBF situation.

In the present study we present a new way of scaling the kernel function. The new approach will enlarge the kernel by acting directly on the distance measure to the boundary, instead of the positions of SVs as used before. Experimental study shows that the new algorithm works robustly, and overcomes the susceptibility of the original method.

2 Scaling the Kernel Function

The SVM solution to a binary classification problem is given by a discriminant function of the form [1, 2]

$$f(\mathbf{x}) = \sum_{s \in SV} \alpha_s y_s K(\mathbf{x}_s, \mathbf{x}) + b \tag{1}$$

A new out-of-sample case is classified according to the sign of $f(\mathbf{x})$. The support vectors are, by definition, those $\mathbf{x}_i$ for which $\alpha_i > 0$. For separable problems each support vector $\mathbf{x}_s$ satisfies

$$f(\mathbf{x}_s) = y_s = \pm 1 .$$

In general, when the problem is not separable or is judged too costly to separate, a solution can always be found by bounding the multipliers α_i by the condition $\alpha_i \leq C$, for some (usually large) positive constant C.

2.1 Kernel Geometry

It has been observed that the kernel $K(\mathbf{x}, \mathbf{x}')$ induces a Riemannian metric in the input space S [3, 4]. The metric tensor induced by K at $\mathbf{x} \in S$ is

$$g_{ij}(\mathbf{x}) = \frac{\partial}{\partial x_i} \frac{\partial}{\partial x'_j} K(\mathbf{x}, \mathbf{x}') \bigg|_{\mathbf{x}'=\mathbf{x}} . \tag{2}$$

This arises by considering K to correspond to the inner product

$$K(\mathbf{x}, \mathbf{x}') = \phi(\mathbf{x}) \cdot \phi(\mathbf{x}') \tag{3}$$

in some higher dimensional feature space H, where ϕ is a mapping of S into H. The inner product metric in H then induces the Riemannian metric (2) in S via the mapping ϕ.

The volume element in S with respect to this metric is given by

$$dV = \sqrt{g(\mathbf{x})} \, dx_1 \cdots dx_n \tag{4}$$

where $g(\mathbf{x})$ is the determinant of the matrix whose (i, j)th element is $g_{ij}(\mathbf{x})$. The factor $\sqrt{g(\mathbf{x})}$, which we call the *magnification* factor, expresses how a local volume is expanded or contracted under the mapping ϕ. Amari and Wu [4] suggest that it may be beneficial to increase the separation between sample points in S which are close to the separating boundary, by using a kernel $\tilde{K}$, whose corresponding mapping $\tilde{\phi}$ provides increased separation in H between such samples.

The problem is that the location of the boundary is initially unknown. Amari and Wu therefore suggest that the problem should first be solved in a standard way using some initial kernel K. It should then be solved a second time using a conformal transformation $\tilde{K}$ of the original kernel given by

$$\tilde{K}(\mathbf{x}, \mathbf{x}') = D(\mathbf{x})K(\mathbf{x}, \mathbf{x}')D(\mathbf{x}') \tag{5}$$

for a suitably chosen positive function $D(\mathbf{x})$. It is easy to check that $\tilde{K}$ satisfies the Mercer positivity condition. It follows from (2) and (5) that the metric $\tilde{g}_{ij}(\mathbf{x})$ induced by $\tilde{K}$ is related to the original $g_{ij}(\mathbf{x})$ by

$$\tilde{g}_{ij}(\mathbf{x}) = D(\mathbf{x})^2 g_{ij}(\mathbf{x}) + D_i(\mathbf{x})K(\mathbf{x}, \mathbf{x})D_j(\mathbf{x})$$
$$+ D(\mathbf{x})\Big\{ K_i(\mathbf{x}, \mathbf{x})D_j(\mathbf{x}) + K_j(\mathbf{x}, \mathbf{x})D_i(\mathbf{x}) \Big\} \tag{6}$$

where $D_i(\mathbf{x}) = \partial D(\mathbf{x})/\partial x_i$ and $K_i(\mathbf{x}, \mathbf{x}) = \partial K(\mathbf{x}, \mathbf{x}')/\partial x_i |_{\mathbf{x}'=\mathbf{x}}$. If $g_{ij}(\mathbf{x})$ is to be enlarged in the region of the initial class boundary, $D(\mathbf{x})$ needs to be largest in that vicinity, and its gradient needs to be small far away. Amari and Wu consider the function

$$D(\mathbf{x}) = \sum_{i \in SV} e^{-\kappa \|\mathbf{x} - \mathbf{x}_i\|^2} \tag{7}$$

where κ is a positive constant. The idea is that support vectors should normally be found close to the boundary, so that a magnification in the vicinity of support vectors should implement a magnification around the boundary. A possible difficulty of (7) is that $D(\mathbf{x})$ can be rather sensitive to the distribution of SVs, consider magnification will tend to be larger at the high density region of SVs and lower otherwise. A modified version was proposed in [5] which consider different κ_i for different SVs. κ_i is chosen in a way to accommodate the local density of SVs, so that the sensitivity with respect to the distribution of SVs is diminished. By this the modified algorithm achieves some improvement, however, the cost it brings associated with fixing κ_i is huge. Also its performance in high dimensional cases is uncertain. Here, rather than attempt further refinement of the method embodied in (7), we shall describe a more direct way of achieving the desired magnification.

2.2   New Approach

The idea here is to choose D so that it decays directly with distance, suitably measured, from the boundary determined by the first-pass solution using K. Specifically we consider

$$D(\mathbf{x}) = e^{-\kappa f(\mathbf{x})^2} \tag{8}$$

where f is given by (1) and κ is a positive constant. This takes its maximum value on the separating surface where $f(\mathbf{x}) = 0$, and decays to $e^{-\kappa}$ at the margins of the separating region where $f(\mathbf{x}) = \pm 1$.

3   Geometry and Magnification

3.1   RBF Kernels

To proceed, we need to consider specific forms for the kernel K. Here, we consider the Gaussian radial basis function kernel

$$K(\mathbf{x}, \mathbf{x}') = e^{-\|\mathbf{x}-\mathbf{x}'\|^2/2\sigma^2} .$$ (9)

It is straightforward to show that the induced metric is Euclidean with

$$g_{ij}(\mathbf{x}) = \frac{1}{\sigma^2} \delta_{ij}$$ (10)

and the volume magnification is the constant

$$\sqrt{g(\mathbf{x})} = \frac{1}{\sigma^n} .$$ (11)

3.2   Conformal Kernel Transformations

For illustration, we consider a simple toy problem as shown in Fig.1(a), where 100 points have been selected at random in the square as a training set, and classified according to whether they fall above or below the curved boundary, which has been chosen as e^{-4x^2} up to a linear transform. Our approach requires a first-pass solution using conventional methods. Using a Gaussian radial basis kernel with width 0.5 and soft-margin parameter $C = 10$, we obtain the solution shown in Fig.1(b). This plots contours of the discriminant function f, which is of the form (1). For sufficiently large samples, the zero contour in Fig.1(a) should coincide with the curve in Fig.1(b).

To proceed with the second-pass we need to use the modified kernel given by (5) where K is given by (9) and D is given by (8). It is interesting first to calculate the general metric tensor $\tilde{g}_{ij}(\mathbf{x})$ when K is the Gaussian RBF kernel (9) and $\tilde{K}$ is derived from K by (5). Substituting in (6), and observing that in this case $K(\mathbf{x}, \mathbf{x}) = 1$ while $K_i(\mathbf{x}, \mathbf{x}) = K_j(\mathbf{x}, \mathbf{x}) = 0$, we obtain

$$\tilde{g}_{ij}(\mathbf{x}) = \frac{D(\mathbf{x})^2}{\sigma^2} \delta_{ij} + D_i(\mathbf{x})D_j(\mathbf{x}) .$$ (12)

Observing that $D_i(\mathbf{x})$ are the components of $\nabla D(\mathbf{x}) = D(\mathbf{x})\nabla \log D(\mathbf{x})$, it follows that the ratio of the new to the old magnification factors is given by

$$\sqrt{\frac{\tilde{g}(\mathbf{x})}{g(\mathbf{x})}} = D(\mathbf{x})^n \sqrt{1 + \sigma^2 \|\nabla \log D(\mathbf{x})\|^2} .$$ (13)

This is true for any positive scalar function $D(\mathbf{x})$. Let us now use the function given by (8) for which

$$\log D(\mathbf{x}) = -\kappa f(\mathbf{x})^2$$ (14)

where f is the first-pass solution given by (1) and shown, for example, in Fig.1(b). This gives

$$\sqrt{\frac{\tilde{g}(\mathbf{x})}{g(\mathbf{x})}} = \exp\left\{-n\kappa f(\mathbf{x})^2\right\} \sqrt{1 + 4\kappa^2\sigma^2 f(\mathbf{x})^2 \|\nabla f(\mathbf{x})\|^2} . \tag{15}$$

This means that

1. *the magnification is constant on the separating surface* $f(\mathbf{x}) = 0$;
2. *along contours of constant* $f(\mathbf{x})$, *the magnification is greatest where the contours are closest.*

These two properties are illustrated in Fig.1(c).

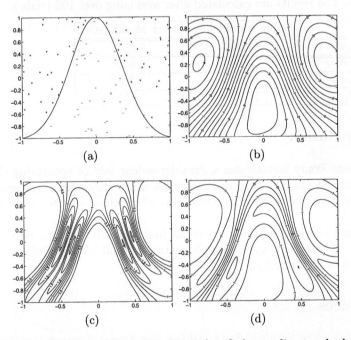

Fig. 1. (a) A training set of 100 random points classified according to whether they lie above (+) or below (−) the Gaussian boundary shown. (b) First-pass SVM solution to the problem in (a) using a Gaussian kernel. The contours show the level sets of the discriminant function f defined by (1). (c) Contours of the magnification factor (15) for the modified kernel using $D(\mathbf{x}) = \exp\{-\kappa f(\mathbf{x})^2\}$ with f defined by the solution of (b). (d) Second-pass solution using the modified kernel.

4   Simulation Results

The only free parameter in the new approach is κ. It is clear that κ is scale-invariant and independent of the input dimension. Through experimental study, we find that in most cases a suitable κ is approximately the reciprocal of $|f|_{max}$, the maximum of the absolute value of $f(\mathbf{x})$ in the first pass solution.

After applying the modified kernel $\tilde{K}$, we solve the classification problem in Fig.1(a) again, and obtain the solution in Fig.1(d). Comparing this with the first-pass solution of Fig.1(b), notice the steeper gradient in the vicinity of the boundary, and the relatively flat areas remote from the boundary. We have repeated the experiment 10000 times, with a different choice of 100 training sites and 1000 test sites on each occasion, and have found an average of 14.5% improvement in classification performance.

We also apply the new method for some real-world problems and obtain encouraging results. For instance, for the Mushroom dataset in the UCI Machine Learning Repository, we observe the improvement as shown in Table.1 (The misclassification rates are illustrated. The number of training and testing examples are 100 and 1000, respectively, which are both randomly chosen from the database. The results are calculated after averaging over 100 trials.).

	Before Modification	After Modification
$C = 10, \sigma = 0.6$	11.20%	7.05%
$C = 10, \sigma = 1.0$	4.02%	2.95%
$C = 50, \sigma = 0.6$	10.86%	7.46%
$C = 100, \sigma = 0.6$	11.97%	7.75%

5   Conclusion

The present study investigates a data-dependent way of optimizing the kernel functions in SVMs. The proposed algorithm is a modification of the one in [4, 5]. Compared with the original, the new algorithm achieves better performance in term of that it is more robust with respect to the data distribution. The new algorithm is also simple and has only one free parameter. It is therefore valuable as a general methodology for supplementing normal SVM training to enhance classification performance.

References

1. Vapnik, V.: The Nature of Statistical Learning Theory. Springer, New York (1995)
2. Scholkopf, B., Smola, A.: Learning with Kernels. MIT Press (2002)
3. Burges, C.: Geometry and Invariance in Kernel Based Method. In: Scholkopf, B., Burges, C., Smola, A. (eds.) Advances in Kernel Methods, MIT Press (1999) 89–116
4. Amari, S., Wu, S.: Improving Support Vector Machine Classifiers by Modifying Kernel Functions. Neural Networks, **12** (1999) 783–789
5. Wu, S., Amari, S.: Conformal Transformation of Kernel Functions: A Data-Dependent Way to Improve Support Vector Machine Classifiers. Neural Processing Letters, **15** (2001) 59–67

Online Support Vector Machines with Vectors Sieving Method

Liangzhi Gan[1,2], Zonghai Sun[1], and Youxian Sun[1]

[1] National Laboratory of Industrial Control Technology, Zhejiang University,
Hangzhou 310027, China
{lzgan,zhsun}@iipc.zju.edu.cn
[2] Electric Engineering Department of Xuzhou Normal University, Xuzhou 221011, China
lzh_box@163.com

Abstract. Support Vector Machines are finding application in pattern recognition, regression estimation, and operator inversion. To extend the using range, people have always been trying their best in finding online algorithms. But the Support Vector Machines are sensitive only to the extreme values and not to the distribution of the whole data. Ordinary algorithm can not predict which value will be sensitive and has to deal with all the data once. This paper introduces an algorithm that selects promising vectors from given vectors. Whenever a new vector is added to the training data set, unnecessary vectors are found and deleted. So we could easily get an online algorithm. We give the reason we delete unnecessary vectors, provide the computing method to find them. At last, we provide an example to illustrate the validity of algorithm.

1 Introduction

Support Vector Machines (SVMs) are known to give good results on pattern recognition, regression estimation problems despite the fact that they do not incorporate problem domain knowledge. However, they exhibit classification speeds which are substantially slower than those of neural networks (LeCun et al., 1995). Furthermore, the SVMs optimize a geometrical criterion that is the margin and is sensitive only to the extreme values and not to the distribution of the whole data. Ordinary algorithm can not predict which values will be sensitive and has to deal with all the data at a time. This not only expends huge memory for storage, but also restricts the applying range of SVMs. This paper provides a method to select promising vectors from given vectors(we call it vectors sieve method) on separable cases so that the SVMs can be used when an online algorithm is necessary.

This paper is organized as follows. In Section 2 we summarize the properties of SVMs. We provide the theoretical basics of the algorithm and prove them in Section 3. The online algorithm is given in Section 4. In Section 5, we illustrate the result of the algorithm. At last, the conclusion is provided in Section 6.

2 Support Vector Machines

For details on the basic SVMs approach, the reader is referred to (Boser, Guyon & Vapnik, 1992; Cortes & Vapnik, 1995; Vapnik, 1995). We summarize conclusions of SVMs in this section.

J. Wang, X. Liao, and Z. Yi (Eds.): ISNN 2005, LNCS 3496, pp. 837–842, 2005.

Given separable vectors $x_i (i = 1, \cdots, l)$, $x_i \in R^n$ with labels $y_i = \pm 1$, we are required to find a hyperplane $w * x + b = 0$ which separates the vectors according to corresponding y_i with maximal margin. The problem is equivalent to:

$$Min \qquad \qquad g(w) = \frac{1}{2}\|w\|_2^2$$
$$s.t \qquad \qquad y_i(w * x_i + b) \geq 1, i = 1 \cdots l$$
(1)

In order to solve it, one has to find the saddle point of Lagrange function

$$L(w,b,\alpha) = \frac{1}{2}(w * w) - \sum_{i=1}^{l} \alpha_i (y_i[x_i * w + b] - 1)$$
(2)

Where $\alpha_i \geq 0$ are the Lagrange multipliers. The Wolfe dual problem is

$$Max \ W(\alpha) = \sum_{i=1}^{l} \alpha_i - \frac{1}{2} \sum_{i,j=1}^{l} \alpha_i \alpha_j y_i y_j (x_i * x_j)$$
$$s.t \qquad \sum_{i=1}^{l} y_i \alpha_i = 0, \alpha_i \geq 0$$
(3)

We find that the optimal hyperplane has the form

$$f(x,\alpha) = \sum_{i=1}^{l} y_i \alpha_i (x_i * x) + b_0$$
(4)

Notice that the vectors $x_i, i = 1 \cdots l$ only appear inside an inner product. To get a potentially better representation of the data, we map the vectors into an alternative space, generally called *feature space* (an inner product space) through a replacement:

$$x_i * x_j \rightarrow \phi(x_i) * \phi(x_j)$$

The functional form of the mapping $\phi(x_i)$ does not need to be known since it is implicitly defined by the choice of kernel $K(x_i, x_j) = \phi(x_i) * \phi(x_j)$ or other inner product in Hilbert space. Then (4) has the form of:

$$f(x,\alpha) = \sum_{i=1}^{l} y_i \alpha_i K(x_i, x) + b_0$$
(5)

3 Theoretical Basics of the Algorithm

Some properties of the optimal hyperplane are given.

Theorem 1. (Uniqueness Theorem)
The optimal hyperplane is unique, but the expansion of vector w on support vectors is not unique (Vapnik, 1998)

Theorem 2. (Irrelevant Vector Theorem)

Given linear separable data (x_i, y_i), $x_i \in R^n$, $y_i = \pm 1, i = 1 \cdots l$ and required to find the optimal hyperplane, we define two subsets:

$$I = \{x_i : y_i = 1\}, \quad \Pi = \{x_i : y_i = -1\}$$

Without losing generality, we suppose that $x' \in I$. Changing the label of x' (*i.e.* $y' = -1$), we construct two new subsets:

$$I' = I \setminus \{x'\}, \quad \Pi' = \Pi \cup \{x'\}$$

If I' and Π' are not separable by a hyperplane (such an x' is called *irrelevant vector*), the following propositions are true:

1. Dataset $I \cup \Pi$ and $I' \cup \Pi$ have the same optimal hyperplane;

2. $\forall x \in R^n$ whose label is $y \in \{+1, -1\}$, if dataset $I \cup \Pi \cup \{x\}$ is separable according to labels, then datasets $I \cup \Pi \cup \{x\}$ and $(I \cup \Pi \setminus \{x'\}) \cup \{x\}$ have the same optimal hyperplane, where x' is an irrelevant vector.

Proof.

1. Without losing generality, suppose $w_0 * x + b_0 = 0$ is the optimal hyperplane of dataset $I' \cup \Pi$, $\gamma = 2 / \|w\|_2$ is the margin, so we have

$$w_0 * x_i + b_0 - 1 \geq 0, \forall x_i \in I'$$
$$w_0 * x_i + b_0 + 1 \leq 0, \forall x_i \in \Pi$$

Since I' and Π' are not separable, $w_0 * x' + b_0 - 1 \geq 0$. For if $w_0 * x' + b_0 - 1 < 0$, we have

$$\left. \begin{array}{l} w_0 * x' + b_0 - 1 < 0 \\ w_0 * x_i + b_0 + 1 \leq 0, \forall x_i \in \Pi \end{array} \right\} \Rightarrow w_0 * x_i + b_0 - 1 < 0, \forall x_i \in \Pi'$$

Integrating $w_0 * x_i + b_0 - 1 \geq 0, \forall x_i \in I'$, I' and Π' are separable. This conflicts with the supposition that I' and Π' are not separable. So we know $w_0 * x' + b_0 - 1 \geq 0$. From *theorem 1*, we know that Dataset $I \cup \Pi$ and $I' \cup \Pi$ have the same optimal hyperplane.

2. Because dataset $I \cup \Pi \cup \{x\}$ is separable according to labels, we can suppose $w_0 * x + b_0 = 0$ is the optimal hyperplane of dataset $(I \cup \Pi \setminus \{x'\}) \cup \{x\}$ and $\gamma = 2 / \|w\|_2$ is the margin, without loss of generality, suppose $y = -1$, since I' and Π' are not separable, $I' = I \setminus \{x'\}$ and $\Pi' = \Pi \cup \{x'\} \cup \{x\}$ are not separable. According to proposition 1, proposition 2 is true. Proof ends.

According to Irrelevant Vectors Theorem, we can delete *irrelevant vectors* from training datasets without effect on the training result.

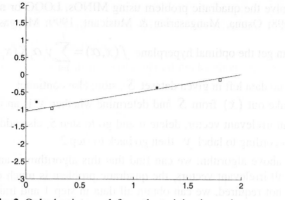

Fig. 2. Only 4 points are left, so the training is much easy.

6 Conclusion

This paper introduced a method to sieve vectors. We gave the theorem and proved it in section 3 (Irrelevant Vector Theorem). All the irrelevant vectors have no effects on the hyperplane even when new data are added into the set, so we could find and delete them one by one. Fortunately we have the Gordan Theorem, which helps us find irrelevant vectors. Having the theoretical basics and computing method, we gave an online algorithm to train SVMs. In fact, this algorithm could be used as online method or batch method.

References

1. Boser, B.E., Guyon, I.M., Vapnik, V.: A Training Algorithm for Optimal Margin Classifiers. Fifth Annual Workshop on Computational Learning Theory, Pittsburgh, ACM (1992) 144-152
2. Cortes, C., Vapnik, V.: Support Vector Networks. Machine Learning, **20** (1995) 273–297
3. Vapnik, V.: The Nature of Statistical Learning Theory, Springer, New York (1995)
4. Scholkopf, B., Mika, S., Burges, C.J.C., Knirsch, P.: Input Space Versus Feature Space in Kernel-Based Methods. IEEE Trans. on Neural Networks, (1999) 1000-1016
5. Aronszajn, N.: Theory of Reproducing Kernels. Trans. Amer. Math. Soc, (1950) 337–404
6. LeCun, Y., Jackel, L., Bottou, L., Brunot, A., Cortes, C., Dener, J., Drucker, H., Guyon, I., Muller, U., Sackinger, E., Simard, P., and Vapnik, V.: Comparison of Learning Algorithms for Handwritten Digital Recognition. International Conference on Artificial Neural Networks, Ed. F. Fogelman, P. Gallinari (1995) 53-60
7. Vapnik, V.: Statistical Learning Theory, Wiley, New York (1998) 401-408.
8. Platt, J.: Sequential Minimal Optimization: A Fast Algorithm for Training Support Vector Machine. Technical Report MSR-TR-98-14. Microsoft Research (1998)
9. Osuna, E., Freund, R., Girosi, F.: An Improved Algorithm for Support Vector Machines. Proc. Of NNSP'97 (1997)
10. Mangasarian, O.L., Musicant, D.R.: Successive Overrelaxation for Support Vector Machine. IEEE Transactions on Neural Networks, **10** (1999) 1032-1037
11. Mangasarian, O.L., Musicant, D.R.: Active Support Vector Machine Classification. Advances in Neural Information Processing Systems (NIPS 2000) (2000)
12. Mangasarian, O.L., Musicant, D.R., Lagrangian Support Vector Machines. Journal of Machine Learning Research, **1** (2001) 161-177

Least Squares Support Vector Machine Based on Continuous Wavelet Kernel*

Xiangjun Wen, Yunze Cai, and Xiaoming Xu

Automation Department, Shanghai Jiaotong University, Shanghai 200030, China
{Wenxiangjun,yzcai,xmxu}@sjtu.edu.cn

Abstract. Based on the continuous wavelet transform theory and conditions of the admissible support vector kernel, a novel notion of multidimensional wavelet kernels is proposed for Least Squares Support Vector Machine (LS-WSVM) for pattern recognition and function estimation. Theoretic analysis of the wavelet kernel is discussed in detail. The good approximation property of wavelet kernel function enhances the generalization ability of LS-WSVM method and some experimental results are presented to illustrate the effectiveness and feasibility of the proposed method.

1 Introduction

As a novel breakthrough to neural network, Support Vector Machine (SVM), originally introduced by Vapnik [2] within the frame of the statistical learning theory, have been frequently used in a wide range of fields, including pattern recognition [3], regression [4], [5] and others. In this kernel-based method, one maps the input data into a higher dimensional space and constructs an optimal separating hyper plane in this feature space. Kernel functions and regularization parameters are chosen such that a regularized empirical risk is minimized.

Presently, it has been shown that the generalization ability of learning algorithms mostly depend on the tricks of kernel, and it is vitally important to select a kernel with good reproducing property in Reproducing Kernel Hilbert Space (RKHS) [2], [13]. In functional analysis, kernels associated with Hilbert functions space rely on certain basis functions and they act as the their representer of evaluation [11]. It is very similar with the wavelet function in wavelet analysis. According to wavelet theory, wavelet function is a set of bases that can approximate arbitrary function and has been proved to be a perfect tool in this field. Noting that the wavelet networks [6] can greatly remedy the weakness of both wavelets and neural networks and has been widely used in the fields of classification and approximation with great success. It is a valuable issue whether a better performance could be obtained if we combine the wavelet decompositions with kernel method. Actually it has caused great interest of many researchers in the last few years [8], [9], [10]. Here, we proposed a least squares version of Support Vector Machine based on a continuous wavelet kernel and develop a framework for pattern recognition and regression estimator.

---

* This work was supported by the national 973 key fundamental research project of China under grant 2002CB312200 and national 863 high technology projects foundation of China under grant 2002AA412010.

J. Wang, X. Liao, and Z. Yi (Eds.): ISNN 2005, LNCS 3496, pp. 843–850, 2005.
© Springer-Verlag Berlin Heidelberg 2005

This paper is organized as follows. In the next section we first give a brief review on LS-SVM method for regression and classification problem, then in section 3 we focus on a practical approach to construct an admissible wavelet kernels based on continuous wavelet decomposition. In section 4, numerical experiments are presented to assess the applicability and the feasibility of the proposed method. Finally, Section 5 concludes the work done.

2 Least Squares Support Vector Machines

Given a training data set D of l samples independent and identically drawn (i.i.d.) from an unknown probability distribution $\mu(X, Y)$ on the product space $Z = X \times Y$:

$$D = \{z_1 = (x_1, y_1), \ldots z_n = (x_l, y_l)\} \tag{1}$$

where the input data X is assumed to be a compact domain in a Euclidean space R^d and the output data Y is assumed to be a closed subset of R. Learning from the training data can be viewed as a multivariate function f approximation (or arbitrary hyper-plane for classification) that represents the relation between the input X and output Y [1], [2]. By some nonlinear mapping $\Phi(\cdot)$, input X is mapped into a hypothesis space R^X (Feature Space) in which the learning machine (algorithm) selects a certain function f (or separating hyper-plane).

A. LS-SVM for Function Estimation [14]
In the case of Least Squares Support Vector Machine (LS-SVM), function estimation in RKHS is defined:

$$f(x) = w^T \Phi(x) + b \tag{2}$$

One defines the optimization problem.

$$\min_{w,b,e} J(w,e) = \frac{1}{2} w^T w + \gamma \frac{1}{2} e^T e \tag{3}$$

s.j.

$$y_k = w^T \Phi(x_k) + b + e_k, k = 1, \ldots, l \tag{4}$$

where $e \in R^{l \times l}$ denotes the error vector, regularization parameter γ denotes an arbitrary positive real constant.

The conditions for optimality lead to a set of linear equations:

$$\begin{bmatrix} 0 & \vec{1}^T \\ \vec{1} & \Omega + \gamma^{-1}I \end{bmatrix} \begin{bmatrix} b \\ \alpha \end{bmatrix} = \begin{bmatrix} 0 \\ y \end{bmatrix} \tag{5}$$

where $y = [y_1, y_2, \ldots, y_l]^T$, $\vec{1} = [1, \ldots, 1]_{l \times l}^T$ $\alpha = [\alpha_1, \ldots, \alpha_l]^T$.

The resulting LS-SVM model for function estimation becomes:

$$f(x) = \sum_{k=1}^{l} \alpha_k K(x, x_k) + b \tag{6}$$

where α_k, b are the solution to the linear system (5).

B. LS-SVM for Classification [15]

Similar to function estimation, the solutions for classification lead to a set of linear equations:

$$\begin{bmatrix} 0 & -y^T \\ y & \Omega + \gamma^{-1}I \end{bmatrix} \begin{bmatrix} b \\ \alpha \end{bmatrix} = \begin{bmatrix} 0 \\ 1 \end{bmatrix} \tag{7}$$

where $\quad y = [y_1, y_2, \ldots, y_l]^T, \qquad \overline{1} = [1, \ldots, 1]_{1 \times l}^T \quad, \qquad \alpha = [\alpha_1, \ldots, \alpha_l]^T$.

$Z = [y_1 \Phi(x_1), \ldots, y_l \Phi(x_l)]^T$, $\Omega = ZZ^T$.

The resulting LS-SVM model for classification is:

$$f(x) = \text{sgn}(\sum_{k=1}^{l} \alpha_k K(x, x_k) + b) \tag{8}$$

C. RKHS and Conditions of Support Vector Kernels

RKHS is a Hilbert space of functions with special properties. For simplify, only a wavelet RKHS and its corresponding reproducing kernel which satisfy the Mercer condition as follows is considered in this paper.

Lemma 1[16]: Supposed any continuous symmetry function $K(x, y) \in L^2 \otimes L^2$ is positive define kernel $\Longleftrightarrow$

$$\iint_{L^2 \otimes L^2} K(x, y) g(x) g(y) dx dy \geq 0, \ \forall g \in L^2, g \neq 0, \int g^2(u) du < \infty \tag{9}$$

The kernel that satisfies this Mercer condition is called as an admissible Support Vector (SV) kernel.

Lemma 2[5], [12]: A translation invariant kernel $K(x, y) = K(x - y)$ is an admissible SV kernel if and only if the Fourier transform

$$F[K](w) = (2\pi)^{\frac{d}{2}} \int_{R^d} \exp(-jwx) K(x) dx \geq 0 \tag{10}$$

3 Least Squares Wavelet Support Vector Machines

In this section, we propose a practical way to construct admissible SV kernels based on continuous wavelet transform. We assume that the reader is familiar with the relevant theory of wavelet analysis and briefly recall some important concept about continuous wavelet decomposition.

A. Translation Invariant Wavelet Kernel

The wavelet analysis provides a frame of time-frequency in which any function $f(x) \in L^2(R)$ can be reconstructed from its wavelet transform.

Let $\varphi(x)$ be a mother wavelet, and let a and b denote the dilation and translation factor, respectively, $a, c \in R$, $a \neq 0$, then according to wavelet theory

$$\varphi_{a,c}(x) = |a|^{-1/2} \varphi(\frac{x - c}{a}) \tag{11}$$

If $\varphi(x) \in L^2(R)$ satisfies admissibility conditions of a mother wavelet as follows:

$$C_\varphi = \int_{-\infty}^{+\infty} |\omega|^{-1} |\hat{\varphi}(\omega)|^2 \, d\omega \tag{12}$$

Then, wavelet transform of f(x) is defined as follows.

$$W_{a,c}(f) = < f(x), \varphi_{ac}(x) >_{L^2} \tag{13}$$

We can reconstructed f (x) as follows:

$$f(x) = \frac{1}{C_\varphi} \int_{-\infty}^{+\infty} \int_0^{+\infty} W_{a,c}(f) \varphi_{a,c}(x) \frac{da}{a^2} dc \tag{14}$$

For a common multidimensional wavelet function, we can write it as the product of one-dimensional (1-D) wavelet function according to tensor products theory proposed by Aronszajn [17].

$$\Phi(x) = \prod_{i=1}^{d} \varphi(x_i) \tag{15}$$

where $\{X=(x_1, ..., x_d) \in R^d\}$.

For a common case, let choose Mexican hat wavelet function

$$\varphi(x) = (1 - x^2) \exp(-\frac{x^2}{2}) \tag{16}$$

to construct wavelet kernel in this paper. This function is very popular in vision analysis and it is known as a mother wavelet, which can span the frame of $L^2(R)$ [7].

Theorem 1: Given the mother wavelet (16) and let a and c denote the dilation and translation factor, respectively, $a_i, c_i \in R$, $a_i \neq 0$, for $x_i, x_i' \in R^d$, if the inner-product of wavelet kernel is defined as follows:

$$K(x, x') = \prod_{i=1}^{d} \varphi(\frac{x_i - c_i}{a_i}) \varphi(\frac{x_i' - c_i'}{a_i}) \tag{17}$$

Then the translation invariant wavelet kernel of (17) is an admissible SV kernel as follows:

$$K(x, x') = K(x - x') = \prod_{i} (1 - \frac{\| x_i - x_i' \|^2}{a_i^2}) \exp(-\frac{\| x_i - x_i' \|^2}{2a_i^2}) \tag{18}$$

Due to page limit, the proof of theorem 1 is omitted.

Remark 1. Rewritten (11) with multiscale form:

$$\phi_i(x) = \varphi_{j,n}(x) = a_0^{-j/2} \varphi(a_0^{-j} x - nb_0) \tag{19}$$

where $a_0, b_0 \in R$, $j, n \in Z$, i denote a multi index. It is shown in [7], [9] that the denumerable family functions $\phi_i(x)$ constitute a frame of $L^2(R)$. For example, it is known that Morlet wavelet and Mexican hat wavelet yield a tight frame (very close to 1) for $a_0 = 2, b_0 = 1$. Moreover, the family functions (19) can lead to an orthonormal basis of $L^2(R)$ while selecting an orthonormal mother wavelet with $a_0 = 2, b_0 = 1$

according to Mallat & Meyer Theorem [19]. Furthermore, we can use wavelet decomposition to design universal kernels with the inner product type, and this will be extremely useful for multiscale wavelet kernels with multiresolution structure and need deeper theoretical investigation[1].

B. LS-WSVM

The goal of our LS-WSVM is to find the optimal wavelet coefficients in the space spanned by the multidimensional wavelet basis. Thereby, we can obtain the optimal estimate function or decision function.

The resulting LS-WSVM model for function estimation becomes:

$$f(x) = \sum_{k=1}^{l} \alpha_k \prod_{i=1}^{d} \varphi(\frac{x_i^k - x_i^{k'}}{a_i^k}) + b \tag{20}$$

And the decision function for classification is:

$$f(x) = \text{sgn}(\sum_{k=1}^{l} \alpha_k \prod_{i=1}^{d} \varphi(\frac{x_i^k - x_i^{k'}}{a_i^k}) + b) \tag{21}$$

4   Simulation Results

In this section, we validate the performance of wavelet kernel with three numerical experiments, the classification of Two-spiral benchmark problem, approximation of a single-variable function and two-variable function.

For comparison, we show the results obtained by Gaussian kernel and wavelet kernel, respectively. Since SVM cannot optimize the parameters of kernels, it is difficult to determine $d \times l$ parameters $a_i^k, i = 1,...l; k = 1,...,d$. For simplicity, let $a_i^k = a$ such that the number of parameters becomes only one. Note it plays the similar role as the kernel width parameter δ of Gaussian kernel $K(x, x') = \exp(-\frac{\| x - x' \|^2}{2\delta^2})$, we can adopt a cross-validation method to optimize these parameters.

A. Classification on Two-Spiral Benchmark Problem

The Two-spiral problem is a famous benchmark problem to test the generalize ability of a learning algorithms for it is extremely hard to solve. The train data are shown on Fig.1 with two classes indicated by '+' and 'o' (with 126 points for each class) in a two-dimensional input space. Points in between the training data located on the two spirals are often considered as test data for this problem but are not shown on the figure.

Note that the parameters a of wavelet kernel play a similar role to that of the kernel bandwidth of a Gaussian kernel, we choose the same dilation parameters as the δ in this experiment for comparing our wavelet kernel. In this experiment, although we give these parameters according to empirical knowledge without any optimization such as cross-validation. The excellent generalization performance for least squares SVM machine with wavelet kernels is clear from the decision boundaries shown on the figures. The two-spiral classification results is illustrated in Fig. 1 (a) and (b)

---

[1] Some experiments of multiscale wavelet kernels have been carried out in our work, but the latter point is deferred to our future work

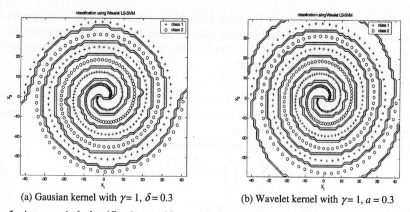

(a) Gausian kernel with $\gamma = 1$, $\delta = 0.3$ (b) Wavelet kernel with $\gamma = 1$, $a = 0.3$

Fig. 1. A two-spiral classification problem with the two classes indicated by '+' and 'o' and 126 training data for each class. The dash-line shows the expected decision boundaries.

B. Approximation of a Single-Variable Function.

In this experiment, let choose a univariate *sinc* target function [18]

$$f(x) = \frac{\sin x}{x} \tag{22}$$

over the domain [-10 10].

The train data consists of 100 points sampled randomly over [-10 10], and the y-values $y = \frac{\sin(x)}{x} + \delta$ are corrupted by additive Gaussian noise δ with zero mean and standard derivation σ. We have uniformly sampled examples of 200 points for test data. The results with two criteria, the normalized root of mean-square-error (NRMSE) and maximal-absolute-error (MAXE). We preferred to choose these tow criterions for assessing the extrapolation ability of our procedure.

For comparison, we show the result with Gaussian kernel, with wavelet kernel and with the wavelet network [6], and the parameter of Gaussian kernel and wavelet kernel is chosen as the same. Table 1 and 2 gives the approximation errors with or without noise.

Table 1. Approximation results of Sinc function(without noise).

Method	NRMSE (train)	MAXE (train)	NRMSE (test)	MAXE (test)
Wavenet	0.1004	0.0686	0.1464	0.0800
Gaussian	0.0018	0.0015	0.0028	0.0030
Mexican	0.0012	0.0014	0.0016	0.0012

Table 2. Approximation results of Sinc function(noise level $\sigma = 0.1$).

Method	NRMSE (train)	MAXE (train)	NRMSE (test)	MAXE (test)
Wavenet	0.1601	0.2364	0.2294	0.1577
Gaussian	0.0367	0.0219	0.0544	0.0419
Mexican	0.0310	0.0141	0.0269	0.0156

C. Approximation of a Two-Variable Function.

In this experiment, let choose a two-dimensional *sinc* target function [18]:

$$f(x) = \frac{\sin\sqrt{x_1^2 + x_2^2}}{\sqrt{x_1^2 + x_2^2}} \tag{23}$$

over the domain $[-5\ 5] \times [-5\ 5]$.

We have uniformly sampled examples of 100 points for the train data, and 2500 points as the testing examples. Table 3 shows the approximation results obtained by Gaussian kernel and wavelet kernel, respectively.

Table 3. Approximation results of two-dimensional Sinc function.

Method	NRMSE (train)	MAXE (train)	NRMSE (test)	MAXE (test)
Wavenet	0.46841	0.10771	0.5493	0.13093
Gaussian	0.0055	0.0044	0.2364	0.1210
Mexican	0.0026	0.0019	0.2187	0.1305

We have compared the approximation results obtained by Wavelet network, Gaussian kernel and wavelet kernel, respectively. To summarize, two kernel-based methods have greatly outperformed the wavelet networks in these two simulations, and our wavelet kernel has better performance than Gaussian kernel.

5   Conclusions and Discussion

In this paper, we discussed a practical way to construct an admissible wavelet kernel. This work provides a new approach for function estimation and pattern recognition, and some simulations show that the proposed method is feasible. Noting that several drawbacks still exist on wavelet networks such as the wavelons requested drastically with the model order *l*, the approximation result mostly depended on initialized parameters and the existence of multiple local minima [6], however, our LS-WSVM based on wavelet kernel take full advantage of the wavelet decomposition while overcome these problems. Most of all, for wavelet kernel based on wavelet decomposition is not only orthonormal (or approximately orthonormal, whereas the Gaussian kernel is correlative or even redundancy), but also suitable for local signal analysis and signal-noise separation, it is not surprising that LS-WSVM give better approximation on function estimation and show good generalization ability on classification problem. In general, the LS-WSVM methodology inspired by wavelet networks and kernel method might offer a better alternative to wavelet networks.

References

1. Poggio, T., Rifkin, R., Mukherjee, S., Niyogi, P.: General Conditions for Predictivity in Learning Theory. Nature, **428** (2004) 419-422.
2. Vapnik, V.: The Nature of Statistical Learning Theory (the second edition). New York: Springer-Verlag (1998)

3. Burges, C. J. C.: A Tutorial on Support Vector Machines for Pattern Recognition. Data Mining Knowl. Disc, **2** (1998) 1–47
4. Drucker, H., Burges, C.J.C., Kaufman, L. (ed.): Support Vector Regression Machines. In : Mozer, M., Jordan, M. , Petsche, T.(eds.): Advances in Neural Information Processing Systems,Vol. 9. Cambridge, MA, MIT Press. (1997) 155-161
5. Smola, A.J., Schölkopf, B.: A Tutorial on Support Vector Regression. Neuro COLT Technical Report NC-TR-98-030, Royal Holloway College, University of London, UK (1998)
6. Qinghua, Z., Benvenisite: Wavelet Networks. IEEE Transactions on Neural Networks, **3** (1992)889-898
7. Daubechies, I.: Ten Lectures on Wavelets. CBMS-NSF Conference Series in Applied Mathematics, **137** (1992) 117-119
8. Onural, L., Kocatepe, M.(ed.): A Class of Wavelet Kernels Associated with Wave Propagation. Acoustics, Speech, and Signal Processing, ICASSP-94. **3** (1994) 9-12
9. Rakotomamonjy, A. and Canu, S. Frame Kernel for Learning. ICANN, LNCS **2415**. (2002) 707-712
10. Li, Z., Weida, Z., Licheng, J.: Wavelet Support Vector Machine. IEEE Transactions on Systems, Man, and Cybernetics—Part B: Cybernetics, **34** (2004) 34-39
11. Scholkopf, B.: A Generalized Representer Theorem. Technical Report 2000-81, Neuro-Colt2 Technical Report Series (2000)
12. Evgeniou,.T., Pontil,M.and Poggio,T.: Regularization Networks and Support Vector Machines. Advances in Computational mathematics, **13** (2000) 1-50
13. Suykens, J., Horvath,G., Basu, S., Micchelli, C., Vandewalle, J. (Eds.): Advances in Learning Theory: Methods, Models and Applications. NATO Science Series III: Computer and Systems Sciences, **190** IOS Press, Amsterdam (2003) 89-110
14. Saunders C., Gammerman A., Vovk V.: Ridge Regression Learning Algorithm in Dual Variables. Proceedings of the 15th International Conference on Machine Learning ICML-98, Madison-Wisconsin(1998)
15. Suykens J.A.K., Vandewalle J.: Least Squares Support Vector Machine Classifiers. Neural Processing Letters, **9** (1999) 293-300.
16. Mercer, J.: Functions of Positive and Negative Type and Their Connection with the Theory of Integral Equations. Transactions of the London Philosophical Society A, **209** (1909) 415–446.
17. Aronszajn, N.: Theory of Reproducing Kernels. Transactions of the American Society. **68** (1950) 337–404.
18. Cherkassky V. and Yunqian Ma: Practical selection of SVM Parameters and the Noise Estimation for SVM regression. Neural Network (2004) 113-126
19. Mallat, S.: A Wavelet Tour of Signal Processing. 2nd edn. Academic Press (2003)

Multiple Parameter Selection for LS-SVM Using Smooth Leave-One-Out Error

Liefeng Bo, Ling Wang, and Licheng Jiao

Institute of Intelligent Information Processing and National Key Laboratory
for Radar Signal Processing, Xidian University, Xi'an 710071, China
{blf0218,wliiip}@163.com

Abstract. In least squares support vector (LS-SVM), the key challenge lies in the selection of free parameters such as kernel parameters and tradeoff parameter. However, when a large number of free parameters are involved in LS-SVM, the commonly used grid search method for model selection is intractable. In this paper, SLOO-MPS is proposed for tuning multiple parameters for LS-SVM to overcome this problem. This method is based on optimizing the smooth leave-one-out error via a gradient descent algorithm and feasible to compute. Extensive empirical comparisons confirm the feasibility and validation of the SLOO-MPS.

1 Introduction

In classification learning, we are given a set of samples of input vector along with corresponding output, and the task is to find a deterministic function that best represents the relation between the input-output pairs. The presence of noise (including input noise and output noise) implies that the key challenge is to avoid over-fitting on the training samples.

A very successful approach for classification is Support Vector Machines (SVMs) [1-2] that attempt to minimize empirical risk while simultaneously maximize the margin between two classes. This is a highly effective mechanism for avoiding over-fitting, which leads to good generalization ability. At present, SVMs have been widely used in pattern recognition, regression estimation, probabilistic density estimation and time series prediction. In this paper, we focus on least squares support vector machine (LS-SVM) [3-4], where one uses equality constraints instead of inequality constraints and a least squares error term in order to obtain a linear set of equations in the dual space. This expression is close related to regularization networks.

In LS-SVM, the key challenge lies in the selection of free parameters, i.e. kernel parameters and tradeoff parameter. A popular approach to solve this problem is grid search [5] where free parameters are firstly discretized in an appropriate interval, and then model selection criterion is performed on every parameters vector. The computational complexity of this approach increases exponentially with the number of free parameters. As a result it becomes intractable when a large number of free parameters are involved.

Motivated from that leave-one-out error of LS-SVM can be expressed as closed form, we propose an algorithm, named SLOO-MPS for tuning multiple parameters for LS-SVM. SLOO-MPS is constructed by two steps, i.e. replacing step function in leave-one-out error with sigmoid function and optimizing the resulting smooth

J. Wang, X. Liao, and Z. Yi (Eds.): ISNN 2005, LNCS 3496, pp. 851–856, 2005.

leave-one-out error via a gradient descent algorithm. Extensive empirical comparisons confirm the feasibility and validation of the SLOO-MPS.

2   Least Squares Support Vector Machine

In this section, we briefly introduce least squares support vector machine. For more details, the interested reader can refer to [6]. In the feature space, LS-SVM models take the form

$$y = \mathbf{w}^T \varphi(\mathbf{x}) \tag{1}$$

where the nonlinear mapping $\varphi(\mathbf{x})$ maps the input data into a higher dimensional feature space whose dimensionality can be infinite. Note that the bias is ignored in our formulation. In LS-SVM, the following optimization problem is formulated

$$\min\left(\frac{1}{2}\mathbf{w}^T\mathbf{w} + \frac{C}{2}e_i^2\right) \tag{2}$$

$$s.t. \quad y_i = \mathbf{w}^T\varphi(\mathbf{x}_i) + e_i \quad i = 1,\cdots l.$$

Wolfe dual of (2) is

$$\min\left(\frac{1}{2}\sum_{i,j=1}^{l}\alpha_i\alpha_j\varphi(\mathbf{x}_i)^T\varphi(\mathbf{x}_j) + \frac{1}{2}\sum_{i=1}^{l}\frac{\alpha_i^2}{C} - \sum_{i=1}^{l}\alpha_i y_i\right). \tag{3}$$

For computational convenience, the form $\varphi(\mathbf{x}_i)^T\varphi(\mathbf{x}_j)$ in (3) is often replaced with a so-called kernel function $K(\mathbf{x}_i,\mathbf{x}_j) = \varphi(\mathbf{x}_i)^T\varphi(\mathbf{x}_j)$. Then (3) is translated into (4)

$$\min\left(\frac{1}{2}\alpha^T\left(K + \frac{1}{C}I\right)\alpha - \alpha^T\mathbf{y}\right). \tag{4}$$

where I denotes a unit matrix. According to KKT condition, the equality

$$\left(K + \frac{1}{C}I\right)\alpha = \mathbf{y} \tag{5}$$

holds true.

Any kernel function that satisfies the Mercer's theorem can be expressed as the inner product of two vectors in some feature space and therefore can be used in LS-SVM.

3   Smooth Leave-One-Error for LS-SVM

Cross-validation is a method for estimating generalization error based on re-sampling. The resulting estimate of generalization error is often used for choosing free parameters. In k-fold cross-validation, the available data are divided into k subsets of (approximately) equal size. Models are trained k times, each time leaving out one of the subsets from training. The k-fold cross-validation estimate of generalization error is mean of the testing errors of k models on the removed subset. If k equals the samples size, it is called leave-one-out cross-validation that has been widely studied due to its mathematical simplicity. Let f_k be the residual error for the kth training samples during the kth iteration of the leave-one-out cross validation procedure. Then the following theorem holds true.

Theorem 1. [7]: $f_k = \dfrac{\left(\mathbf{H}^{-1}\mathbf{y}\right)_k}{\mathbf{H}_{kk}^{-1}}$, where $\mathbf{H} = \left(\mathbf{K} + \dfrac{1}{C}\mathbf{I}\right)$, $\mathbf{H}_{kk}^{-1}$ denotes the kth diagonal element of $\mathbf{H}^{-1}$. Define

$$\mathbf{f} = \left(\mathbf{H}^{-1}\mathbf{y}\right) \odot D\left(\mathbf{H}^{-1}\right), \tag{6}$$

where $\odot$ denotes elementwise division. According to Theorem 1, leave-one-out error of LS-SVM is given by

$$loo(\theta) = \frac{1}{l}\sum_{k=1}^{l}\left(\frac{1 - y_k \operatorname{sgn}(y_k - f_k)}{2}\right), \tag{7}$$

where θ denotes free parameters of kernel function and $\operatorname{sgn}(x)$ is 1, if $x \geq 0$, otherwise $\operatorname{sgn}(x)$ is -1. There exists a step function $\operatorname{sgn}(\cdot)$ in leave-one-out error $loo(\theta)$; thereby, it is not differentiable. In order to use a gradient descent approach to minimize this estimate, we approximate the step function by a sigmoid function

$$\tanh(\gamma t) = \frac{\exp(\gamma t) - \exp(-\gamma t)}{\exp(\gamma t) + \exp(-\gamma t)}, \tag{8}$$

where we set $\gamma = 10$. Then smooth leave-one-out error can be expressed as

$$loo(\theta) = \frac{1}{l}\sum_{i=1}^{l}\left(\frac{1 - y_i \tanh\left(\gamma(y_i - f_i)\right)}{2}\right). \tag{9}$$

According to the chain rule, the derivative of $loo(\theta)$ is formulated as

$$\frac{\partial\left(loo(\theta)\right)}{\partial\theta_k} = \frac{1}{l}\sum_{i=1}^{l}\left(\frac{\partial\left(loo(\theta)\right)}{\partial f_i}\frac{\partial f_i}{\partial\theta_k}\right). \tag{10}$$

Thus we need to calculate $\dfrac{\partial\left(loo(\theta)\right)}{\partial f_i}$ and $\dfrac{\partial f_i}{\partial\theta_k}$, respectively. Together with $\dfrac{\partial\left(\tanh(t)\right)}{\partial t} = \operatorname{sech}^2(t)$, we have

$$\frac{\partial\left(loo(\theta)\right)}{\partial f_i} = \frac{\gamma y_i \operatorname{sech}^2\left(\gamma(y_i - f_i)\right)}{2}. \tag{11}$$

In terms of (11), (10) is translated into

$$\frac{\partial\left(loo(\theta)\right)}{\partial\theta_k} = \frac{1}{l}\left(\frac{\gamma\mathbf{y} \otimes \operatorname{sech}^2\left(\gamma(\mathbf{y} - \mathbf{f})\right)}{2}\right)^T\left(\frac{\partial\mathbf{f}}{\partial\theta_k}\right), \tag{12}$$

where $\otimes$ denotes elementwise multiplication The major difficulty to calculate $\left(\dfrac{\partial \mathbf{f}}{\partial \theta_k}\right)$ lies in obtaining the derivative of $\mathbf{H}^{-1}$. A good solution is based on the equality: $\mathbf{H}^{-1}\mathbf{H} = \mathbf{I}$. Differentiating that with respect to θ_k, we have

$$\frac{\partial \mathbf{H}^{-1}}{\partial \theta_k} = -\mathbf{H}^{-1}\frac{\partial \mathbf{H}}{\partial \theta_k}\mathbf{H}^{-1}. \tag{13}$$

Thus $\dfrac{\partial \mathbf{f}}{\partial \theta_k}$ is given by

$$\frac{\partial \mathbf{f}}{\partial \theta_k} = \left(\frac{\partial\left(\mathbf{H}^{-1}\mathbf{y}\right)}{\partial \theta_k}\right)\odot D\left(\mathbf{H}^{-1}\right) - \left(\mathbf{H}^{-1}\mathbf{y}\right)\odot\left(D\left(\mathbf{H}^{-1}\right)\right)\odot\left(D\left(\mathbf{H}^{-1}\right)\right)\otimes\left(\frac{\partial\left(D\left(\mathbf{H}^{-1}\right)\right)}{\partial \theta_k}\right)$$

$$= -\left(\mathbf{H}^{-1}\frac{\partial \mathbf{H}}{\partial \theta_k}\mathbf{H}^{-1}\mathbf{y}\right)\odot D\left(\mathbf{H}^{-1}\right) + \left(\mathbf{H}^{-1}\mathbf{y}\right)\odot\left(D\left(\mathbf{H}^{-1}\right)\right)\odot\left(D\left(\mathbf{H}^{-1}\right)\right)\otimes D\left(\mathbf{H}^{-1}\frac{\partial \mathbf{H}}{\partial \theta_k}\mathbf{H}^{-1}\right) \tag{14}$$

where $D(\mathbf{A})$ denotes diagonal elements of matrix $\mathbf{A}$. Combining (12) and (14), we can compute the derivative of smooth leave-one-out error with respect to θ_k.

4   Empirical Study

In this section, we will employ SLOO-MPS to tune the weights of the linear mixture kernel

$$\mathbf{H} = \sum_{i=1}^{m}\theta_i^2\mathbf{K}_i + \mathbf{I}, \tag{15}$$

where the mixing weights are positive to assure the positive semidefiniteness of $\mathbf{H}$. Since the weights of mixing kernel can be adjusted, it is reasonable to fix C to 1. Conjugate gradient algorithm is used to optimize the smooth leave-one-out error (10).

Kernel matrices are constructed by Gaussian and Laplace kernel

$$K\left(\mathbf{x}_i,\mathbf{x}_j\right) = \exp\left(-\beta\sum_{m=1}^{d}\left(\mathbf{x}_i^m - \mathbf{x}_j^m\right)^2\right), \tag{16}$$

$$K\left(\mathbf{x}_i,\mathbf{x}_j\right) = \exp\left(-\beta\sum_{m=1}^{d}\left|\mathbf{x}_i^m - \mathbf{x}_j^m\right|\right). \tag{17}$$

Three kinds of mixing schemes are evaluated. The first scheme is to mix 5 Gaussian kernels with $\beta \in \{0.01, 0.1, 1, 10, 100\}$. The second scheme is to mix 5 Laplace kernels with $\beta \in \{0.01, 0.1, 1, 10, 100\}$. The third scheme is to mix 5 Gaussian kernels and 5 Laplace kernel with $\beta \in \{0.01, 0.1, 1, 10, 100\}$. Thus all the free parameters of LS-SVM can be selected by SLOO-MPS, hence our algorithm is very automatic.

In order to how well SLOO-MPS woks, we test it on the ten benchmark data sets from UCI Machine learning Repository [8]. These data sets have been extensively used in testing the performance of diversified kinds of learning algorithms. One-against-one method is used to extend LS-SVM to multi-class classifiers. Ten-fold cross validation errors on benchmark data sets are summary in Table 1.

Table 1. Ten-fold cross validation errors on benchmark data sets.

Problems	Size	Dim	Class	Gaussian	Laplace	Mix
Australian Credit	690	15	2	15.22	14.20	14.15
Breast Cancer	277	9	2	23.74	26.24	23.89
German	1000	20	2	23.50	23.50	23.50
Glass	214	9	6	29.87	21.00	21.23
Heart	270	13	2	17.41	14.07	14.41
Ionosphere	351	34	2	4.86	5.99	4.14
Liver disorders	345	6	2	29.89	25.82	25.97
Vehicle	846	18	4	17.38	20.45	17.50
Vowel	528	10	11	0.95	1.57	0.76
Wine	178	13	3	1.11	1.70	1.11
Mean	/	/	/	16.39	15.45	**14.67**

From Table 1, we can see that mix kernel is better than either of Gaussian or Laplace kernel. This suggests that Gaussian and Laplace kernels indeed provide complementary information for the classification decision and SLOO-MPS approach is able to find a combination that exploits this complementarity.

We also test SLOO-MPS on the Olivetti Research Lab (ORL) face data set in Cambridge (http://www.cam-orl.co.uk/facedatabase.html). The ORL data set contains 40 distinct subjects, with each containing 10 different images taken at different time, with the lighting varying slightly. The experiment is similar to that done by Yang [9]. The leave-one-out errors for different method are summarized in Table 2. We can see that our method obtain the best performance on this data set.

Table 2. Performance on benchmark data sets.

Method	Reduced Space	Misclassification Rate
Eigenface	40	2.50
Fisherface	39	1.50
ICA	80	6.25
SVM, d=4	N/A	3.00
LLE # neighbor=70	70	2.25
ISOMAP, $\varepsilon = 10$	30	1.75
Kernel Eigenface, d=2	40	2.50
Kernel Eigenface, d=3	40	2.00
Kernel Fisherface (P)	39	1.25
Kernel Fisherface (G)	39	1.25
SLOO-MPS(Mix)	N/A	**0.75**

5 Conclusion

SLOO-MPS is presented for tuning the multiple parameters for LS-SVM. Empirical comparisons show that SLOO-MPS works well for the various data sets.

References

1. Boser, B., Guyon, I. and Vapnik, V.: A Training Algorithm for Optimal Margin Classifiers. In Haussler, D. Proceedings of the 5th Annual ACM Workshop on Computational Learning Theory (1992) 144-152
2. Cotes, C. and Vapnik, V.: Support Vector Networks. Machine Learning, **20** (1995) 273-279
3. Suykens, J.A.K. and Vandewalle, J.: Least Squares Support Vector Machine Classifiers. Neural Processing Letters, **9** (1999) 293-300
4. Van Gestel, T., Suykens, J., Lanckriet, G., Lambrechts, A., De Moor, B. and Vandewalle, J.: Bayesian Framework for Least Squares Support Vector Machine Classifiers. Gaussian Processes and Kernel Fisher Discriminant Analysis. Neural Computation, **15** (2002) 1115-1148
5. Hsu, C.W. and Lin, C.J.: A Comparison of Methods for Multiclass Support Vector Machines. IEEE Transactions on Neural Networks, **13** (2002) 415-425
6. Vapnik, V.: Statistical Learning Theory. Wiley-Interscience Publication. New York (1998)
7. Sundararajan, S. and Keerthi, S.S.: Predictive Approaches for Choosing Hyper- parameters in Gaussian Processes. Neural Computation, **13** (2001) 1103-1118
8. Blake, C.L. and Merz, C.J.: UCI Repository of Machine Learning Databases (1998). Available at: http://www.ics.uci.edu/~mlearn/MLRepository.html
9. Yang, M.H. Kernel: Eigenfaces vs. Kernel Fisherfaces: Face Recognition Using Kernel Method. Proceedings of the Fifth International Conference on Automatic Face and Gesture Recognition (2002)

Trajectory-Based Support Vector Multicategory Classifier

Daewon Lee and Jaewook Lee

Department of Industrial and Management Engineering,
Pohang University of Science and Technology,
Pohang, Kyungbuk 790-784, Korea
{woosuhan,jaewookl}@postech.ac.kr

Abstract. Support vector machines are primarily designed for binary-class classification. Multicategory classification problems are typically solved by combining several binary machines. In this paper, we propose a novel classifier with only one machine for even multiclass data sets. The proposed method consists of two phases. The first phase builds a trained kernel radius function via the support vector domain decomposition. The second phase constructs a dynamical system corresponding to the trained kernel radius function to decompose data domain and to assign class label to each decomposed domain. Numerical results show that our method is robust and efficient for multicategory classification.

1 Introduction

The support vector machine (SVM), rooted in the statistical learning theory, has been successfully applied to diverse pattern recognition problems [1], [6]. The main idea of a conventional SVM is to construct a optimal hyperplane to separate 'binary class' data so that the margin is maximal. For multiclass problems, several approaches for multiclass SVMs [4] have been proposed. Most of the previous approaches try to reduce a multiclass problem to a set of multiple binary classification problems where a conventional SVM can be applied. Those approaches, however, have some drawbacks in that they not only generate inaccurate decision boundaries in some region due to the unbalanced data size for each class, but also suffer from a masking problem: some class of data is overwhelmed by others, resulting in being ignored in the decision step.

To overcome such difficulties, in this paper, we propose a novel efficient and robust classifier for multicategory classifications. The proposed method consists of two phases. In the first phase, we build a trained kernel radius function via support vector domain decomposition [2], [3]. In the second phase, we construct a dynamical system corresponding to the trained kernel radius function and decomposes the data domain into a small number of disjoint regions where each region, which is a basis of attraction itself for the constructed system, is classified by the class label of the corresponding stable equilibrium point. As a result, we can classify an unknown test data by using the dynamical system and the class information for each decomposed region.

J. Wang, X. Liao, and Z. Yi (Eds.): ISNN 2005, LNCS 3496, pp. 857–862, 2005.

2 The Proposed Method

2.1 Phase I: Building a Trained Kernel Radius Function via Support Vector Domain Decomposition

The basic idea of support vector domain decomposition is to map data points by means of a inner-product kernel to a high dimensional feature space and to find, not the optimal separating hyperplane, but the smallest sphere that contains most of the mapped data points in the feature space. This sphere, when mapped back to the data space, can decompose a data domain into several regions. The support vector domain decomposition builds a trained kernel radius function as follows [2], [3], [9]: let $\{x_i\} \subset \mathcal{X}$ be a given data set of N points, with $\mathcal{X} \subset \Re^n$, the data space. Using a nonlinear transformation Φ from $\mathcal{X}$ to some high dimensional feature-space, we look for the smallest enclosing sphere of radius R described by the following model:

$$\min R^2$$
$$\text{s.t.} \quad \|\Phi(x_j) - a\|^2 \le R^2 + \xi_j,$$
$$\xi_j \ge 0, \quad \text{for } j = 1, \ldots, N \tag{1}$$

where a is the center and ξ_j are slack variables allowing for soft boundaries. To solve this problem, we introduce the Lagrangian

$$L = R^2 - \sum_j (R^2 + \xi_j - \|\Phi(x_j) - a\|^2)\beta_j - \sum_j \xi_j \mu_j + C \sum_j \xi_j,$$

the solution of the primal problem (1) can be obtained by solving its dual problem:

$$\max \quad W = \sum_j \Phi(x_j)^2 \beta_j - \sum_{i,j} \beta_i \beta_j \Phi(x_i) \cdot \Phi(x_j)$$
$$\text{subject to } 0 \le \beta_j \le C, \ \sum_j \beta_j = 1, j = 1, ..., N \tag{2}$$

Only those points with $0 < \beta_j < C$ lie on the boundary of the sphere and are called support vectors (SVs).

The trained kernel radius function, defined by the squared radial distance of the image of x from the sphere center, is then given by

$$f(x) := R^2(x) = \|\Phi(x) - a\|^2 \tag{3}$$
$$= K(x, x) - 2 \sum_j \beta_j K(x_j, x) + \sum_{i,j} \beta_i \beta_j K(x_i, x_j)$$

where the inner products of $\Phi(x_i) \cdot \Phi(x_j)$ are replaced by a kernel function $K(x_i, x_j)$. One widely used kernel function is the Gaussian kernel which has the form of $K(x_i, x_j) = \exp(-q\|x_i - x_j\|^2)$ with width parameter q.

2.2 Phase II: The Class Label Assignments of the Decomposed Regions

In the second phase, we first construct the following generalized gradient descent process corresponding to the trained kernel radius function (3)

$$\frac{dx}{dt} = -\text{grad}_G f(x) \equiv -G(x)^{-1}\nabla f(x) \tag{4}$$

where $G(x)$ is a positive definite symmetric matrix for all $x \in \Re^n$. (Such an G is called a *Riemannian metric* on $\Re^n$.) A state vector $\bar{x}$ satisfying the equation $\nabla f(\bar{x}) = 0$ is called an *equilibrium point* of (4) and called a (asymptotically) *stable equilibrium point* if all the eigenvalues of its corresponding Jacobian matrix, $J_f(\bar{x}) \equiv \nabla^2 f(\bar{x})$, are positive. The *basin of attraction* (or *stability region*) of a stable equilibrium point $\bar{x}$ is defined as the set of all the points converging to $\bar{x}$ when the process (4) is applied, i.e.,

$$A(\bar{x}) := \{\mathbf{x}(0) \in \Re^n : \lim_{t \to \infty} \mathbf{x}(t) = \bar{x}\}.$$

One nice property of system (4) is that under fairly mild condition, it can be shown that the whole data space is composed of the closure of the basins, that is to say,

$$\Re^n = \bigcup_i^N \text{cl}(A(\bar{\mathbf{x}}_i))$$

where $\{\bar{\mathbf{x}}_i; i = 1, ..., N\}$ is the set of all the stable equilibrium points ([7], [8]). (See Fig. 1.)

This property of the constructed system (4) enables us to decompose the data domain into a small number of disjoint regions (i.e., basins of attraction) where each region is represented by the corresponding stable equilibrium point.

Next we define the set of the training data points converging to a stable equilibrium point $\bar{x}_k$ by $\langle \bar{x}_k \rangle$ and apply a majority vote on the set of $\langle \bar{x}_k \rangle$ to determine the class label of the corresponding stable equilibrium point $\bar{x}_k$. Then each point of a decomposed region, say $A(\bar{x}_k)$, is assigned to the same class label as that of the corresponding stable equilibrium point $\bar{x}_k$. As a result, if we want to classify an unknown data, by applying the process (4) to the test point, we assign the class label of the corresponding stable equilibrium point to which the test data point converges.

3 Simulation Results and Discussion

The proposed algorithm for the multicategory classification has been simulated on five benchmark data sets. Description of the data sets is given in Table 1.

The performance of the proposed method is compared with two widely used multiclass SVM methods; one-against-rest and one-against-one. The criteria for

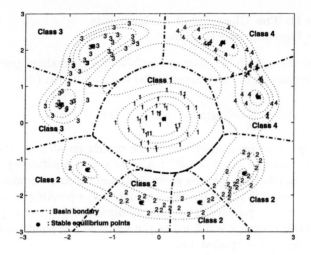

Fig. 1. Topographic map of the trained kernel radius function

Table 1. Benchmark data description

	No. of classes	No. of training data	No. of test data
Iris	3	100	50
Wine	3	118	60
Sonar	2	104	104
Ring	4	160	40
Orange	8	170	30

comparison are the training and the test mis-classification error rate. Simulation results are shown in Table 2. Experimental results demonstrate that the proposed method achieves a much better accuracy in the large-categorical classification problem, whereas, in the three- or four-category classification problems, it has a slightly better performance to previously reported methods.

In addition to this experimental result, the proposed method has several nice features: Firstly, the process (4) can be implemented as various discretized learning algorithms depending on the choice of $G(x)$ [7]. For example, if $G(x) = I$, it is the steepest descent algorithm; if $G(x) = B_f(x)$ where B_f is a positive definite matrix approximating the Hessian, $\nabla^2 f(x)$, then it is the Quasi-Newton method; if $G(x) = [\nabla^2 f(x) + \mu I]$, it is the Levenberg-Marquardt method, and so on. This convenient property allows us to design a computationally efficient algorithm to find a stable equilibrium point of (4). Secondly, our proposed method is robust to the change of a model parameter, q. In the original SVMs, a model selection is very difficult because a decision boundary is determined by only support vectors that are very sensitive to the change of a model parameter. On the contrary, the proposed method that focuses on the stable equilibrium points is reported, through numerous experiments, to be relatively less sensitive to the change of a

Multi-category Classification
by Least Squares Support Vector Regression*

Jingqing Jiang[1,2], Chunguo Wu[1], and Yanchun Liang[1,**]

[1] College of Computer Science and Technology, Jilin University,
Key Laboratory of Symbol Computation and
Knowledge Engineering of the Ministry of Education, Changchun 130012, China
[2] College of Mathematics and Computer Science,
Inner Mongolia University for Nationalities, Tongliao 028043, China

Abstract. Multi-category classification is a most interesting problem in the
fields of pattern recognition. A one-step method is presented to deal with the
multi-category problem. The proposed method converts the problem of classification to the function regression and is applied to solve the converted problem
by least squares support vector machines. The novel method classifies the samples in all categories simultaneously only by solving a set of linear equations.
Demonstrations of computer experiments are given and good performance is
achieved in the simulations.

1 Introduction

Support vector machines (SVMs) are based on the statistical learning theory developed by Vipnik. SVMs have been successful for solving classification and function
regression problems. Multi-category classification, the number of classes to be considered is more than two, is a most interesting problem in the fields of pattern recognition. This problem is conventionally solved by a decomposing and reconstruction
procedure. During the decomposing phase, training data are partitioned into two
classes in several manners and two-class learning machines are trained. Then, in the
reconstruction phase, many binary classifications are combined using voting scheme,
and one or none of the classes is selected. In this architecture, the datum is unclassifiable if it belongs to more than one class or it belongs to none. Shigeo Abe and Takuya
Inoue [1] proposed fuzzy support vector machines to overcome the unclassifiable
problem. Erin J. Bredensteiner [2] showed how the linear programming (LP) approaches based on the work of Mangasarian and quadratic programming (QP) approaches based on Vapnik's Support Vector Machine (SVM) can be combined to
yield two new approaches to the multi-category problem. J. Weston and C. Watkins
[3] introduced two methods to solve multi-category problem in one step. One of the
methods needs to solve a quadratic program problem, and the other needs to solve a
linear program. Cecilio Angulo et al. [4] proposed K-SVMR (Support Vector Classi-
fication-Regression) for multi-category classification problem. This machine evaluates all the training data into a 1-versus-1-versus-rest structure during the decompos-

---

* Supported by the National Natural Science Foundation of China (60433020), the science-
technology development project of Jilin Province of China (20030520), and the doctoral
funds of the National Education Ministry of China (20030183060).
** To whom correspondence should be addressed. E-mail: ycliang@jlu.edu.cn

J. Wang, X. Liao, and Z. Yi (Eds.): ISNN 2005, LNCS 3496, pp. 863–868, 2005.

ing phase by using a mixed classification and regression support vector machine formulation. For the reconstruction, a specific pulling scheme considering positive and negative votes has been designed.

Based on the equality constrains instead of the inequality ones, Suykens and Vandewalle [5] designed the least squares support vector machine (LS-SVM) classifiers. Suykens et al [6] further investigated least squares support vector machines for function regression. For binary classification problem, Massimiliano Pontil et al [7] studied the relation between support vector machines for regression (SVMR) and support vector machines for classification (SVMC). They showed that for a certain choice of the parameters there exists a SVMR solution equaling to a given SVMC solution. Based on the function regression, we propose a one-step method to deal with multi-category classification problem. This method solves the function regression by least squares support vector machines. It only needs to solve a set of linear equations

2 Multi-category Classification Problem

Suppose that the independent and identically distributed samples according to an unknown distribution function are given as follows: $(x_1, y_1), ..., (x_N, y_N)$. x_i $(i=1, ..., N)$ is a d dimension vector and y_i is a class-label representing the class of the sample x_i.

The multi-category classification problem is to construct a decision function $f(x, w)$ which classifies a sample x. It is selected from a set of functions defined by parameter w. The aim is to choose parameter w such that for any x the decision function could provide a classification $f(x, w)$ which is close to the class-label of sample x. Therefore, we consider this problem as a function regression problem: selecting a function $f(x, w)$ according to the given samples $(x_1, y_1), ..., (x_N, y_N)$ such that the class-label of x is estimated by using the information from the value of function $f(x, w)$.

3 Least Squares Support Vector Machines for Regression

According to [6], let us consider a given training set of N samples $\{x_i, y_i\}_{i=1}^N$ with the ith input datum $x_i \in R^n$ and the ith output datum $y_i \in R$. The aim of support vector machines model is to construct the decision function takes the form:

$$f(x, w) = w^T \varphi(x) + b \qquad (1)$$

where the nonlinear mapping $\varphi(\cdot)$ maps the input data into a higher dimensional feature space. In least squares support machines for function regression the following optimization problem is formulated

$$\min_{w, e} \ J(w, e) = \frac{1}{2} w^T w + \gamma \sum_{i=1}^{N} e_i^2 \qquad (2)$$

subject to the equality constraints

$$y_i = w^T \varphi(x_i) + b + e_i, \quad i = 1, ..., N \qquad (3)$$

This corresponds to a form of ridge regression. The Lagrangian is given by

$$L(w,b,e,\alpha) = J(w,e) - \sum_{i=1}^{N} \alpha_i \{ w^T \varphi(x_i) + b + e_i - y_i \} \tag{4}$$

with Lagrange multipliers α_k. The conditions for the optimality are

$$\begin{cases} \dfrac{\partial L}{\partial W} = 0 \rightarrow w = \sum_{i=1}^{N} \alpha_i \varphi(x_i) \\ \dfrac{\partial L}{\partial b} = 0 \rightarrow \sum_{i=1}^{N} \alpha_i = 0 \\ \dfrac{\partial L}{\partial e_i} = 0 \rightarrow \alpha_i = \gamma e_i \\ \dfrac{\partial L}{\partial \alpha_i} = 0 \rightarrow w^T \varphi(x_i) + b + e_i = 0 \end{cases} \tag{5}$$

for $i=1,...,N$. After elimination of e_i and w, the solution is given by the following set of linear equations

$$\begin{bmatrix} 0 & \vec{1}^T \\ \vec{1} & \Omega + \gamma^{-1}I \end{bmatrix} \begin{bmatrix} b \\ a \end{bmatrix} = \begin{bmatrix} 0 \\ y \end{bmatrix} \tag{6}$$

where $y = [y_1,..., y_N]^T, \vec{1} = [1,...,1]^T, \alpha = [\alpha_1,..., \alpha_N]^T$ and the Mercer condition

$$\Omega_{kl} = \varphi(x_k)^T \varphi(x_l) = \psi(x_k, x_l) \qquad k,l = 1,..., N \tag{7}$$

has been applied.

Set $A = \Omega + \gamma^{-1}I$. For A is a symmetric and positive-definite matrix, A^{-1} exists. Solve the set of linear Eqs. (6), the solution can be obtained as

$$\alpha = A^{-1}(y - b\vec{1}) \qquad b = \frac{\vec{1}^T A^{-1} y}{\vec{1}^T A^{-1} \vec{1}} \tag{8}$$

Substituting w in Eq. (1) with the first equation of Eqs. (5) and using Eq. (7) we have

$$f(x,w) = y(x) = \sum_{i=1}^{N} \alpha_i \psi(x, x_i) + b \tag{9}$$

where α_i, b are the solution to Eqs. (6). The kernel function $\psi(\cdot)$ can be chosen as linear function $\psi(x, x_i) = x_i^T x$, polynomial function $\psi(x, x_i) = (x_i^T x + 1)^d$ or radial basis function $\psi(x, x_i) = \exp\{ -\|x - x_i\|_2^2 / \sigma^2 \}$.

4 Multi-category Classification by Least Squares Support Vector Machines for Regression (LSSVRC)

In this paper, we solve the multi-category classification problem by function regression. Consider the class-label as the value of the regression function, and for each x_i the value of the regression function on x_i is as close as to its class-label. The goal is to choose a regression function such that it reflects the relation between x_i and its class-label. The steps are as follows:

Step 1. Set class-label for each class. The class-label is usually set as decimal integer, such as 1, 2, ..., n.

Step 2. Solve the set of linear Eqs. (6), and find the solution α_i, b.

Step 3. Put the solution α_i, b into Eq. (9), and obtain the regression function $y(x)$.

We choose the radial basis function as a kernel function. For multi-category classification the regression function $y(x)$ is used as classifier. When the value of the regression function $y(x)$ is located in the specified region of class-label for a given sample x, sample x is classified by the regression function $y(x)$ correctly. Note that the specified region of each class-label should not be overlapped. Otherwise, some samples would be classified into more than one class.

5   Numerical Experiments

The experiments are implemented on a DELL PC, which utilizes a 2.8GHz Pentium IV processor with 512MB memory. The OS is Microsoft Windows XP operating system. All the programs are compiled under Microsoft's Visual C++ 6.0.

We use two types of experiments to demonstrate the performance of the new method: the experiments in the plane with artificial data sets which can be visualized, and the experiments with benchmark data sets. To run the proposed method, two parameters must be predetermined for each data set. They are smoothing factor γ and band width σ (special for Gaussian kernel function) and listed in Table 1 and Table 2, respectively.

To demonstrate the proposed method we give four artificial data sets. "Strip" data set contains ten classes and it has 1000 samples, among which 500 samples are for training and the rest 500 samples are for testing. Partition 10 strip regions along y axis on a two-dimensional plane. The value of x is in the interval [0, 40]. The width of each strip region is 2 and the space between two strip regions is 2. The class-labels of "Strip_1" data set are assigned with 1,2,...,10. The class-labels of "Strip_2" data set are assigned with x coordinate of the middle of the corresponding strip region. "Checkboard_1" data set has 4000 samples which are partitioned into 2 classes. 2000 samples are for training and the rest 2000 samples are for testing. The region [0, 2]$\times$[0.2] is partitioned into 16 regions. The class-label of each region is assigned with 1 or -1, and the labels of the neighbor regions are different. "Checkboard_2" data set is similar to "Checkboard_1". The samples are partitioned into 3 classes. Fig. 1 shows the "Strip_1" data set and "Checkboard_2" data set. Table 1 shows the correct rate of classification on the artificial data sets.

Table 1. Correct rate on artificial data sets.

Name	Points	Classes	γ	σ	Training correct rate	Testing correct rate
Strip_1	1000	10	800	3	1.00000	0.99600
Strip_2	1000	10	100	2	1.00000	0.99600
Checkboard_1	4000	2	8	0.1	0.99200	0.97500
Checkboard_2	4000	3	15	0.1	0.98850	0.96350

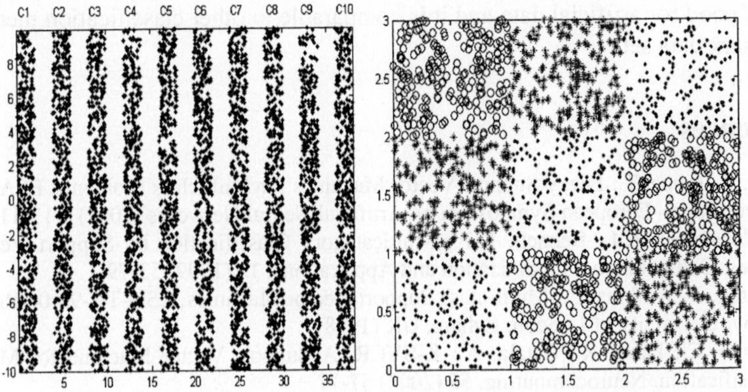

Fig. 1. "Strip_1" data set (left) and "Checkboard_2" data set (right).

The proposed method is applied to three UCI data sets (iris, wine and glass) available at the UCI Machine Learning Repository (http://www.ics.uci.edu/mlearn/MLRepository.html). The selection on the samples is benefit for the comparison of the obtained results with those presented in [3] and [4]. In the simulation, ten percent of the samples are used as a test set for "Iris" and "Glass", and twenty percent of the samples are used as a test set for "Wine".

Table 2 shows the testing correct rate on benchmark data sets, where qp-mc-sv and KSVCR represent the results from references [3] and [4], respectively, LSSVRC represents the result of the proposed method, Attr and Cla represent the numbers of attributes and categories, respectively. From Table 2 it can be seen that the multi-class classification problem can be dealt with by solving the regression problem and the correct rate of the proposed method is comparable to other multi-class classification methods. For the "Glass" data set, all of the correct rates in the three methods are small because the data are unbalanced, however, the result from LSSVRC is the best one. For the "Wine" data set, the correct rate in LSSVRC is smaller than that in KSVCR but it is larger than that in qp-mc-sv. For the "Iris" data set, the correct rate using the proposed method reaches 100%.

Table 2. Testing correct rate on benchmark data sets.

Name	Points	Attr	Cla	γ	σ	qp-mc-sv	KSVCR	LSSVRC
Iris	150	4	3	3000	5	0.986700	0.980700	1.000000
Wine	178	13	3	300	180	0.964000	0.977100	0.972973
Glass	214	9	7	2000	2	0.644000	0.695300	0.722222

6 Conclusions and Discussions

A novel one-step method for multi-category classification problem is proposed in this paper. It converts the classification problem to a function regression problem and solves the function regression by least squares support vector machines. This method only needs to solve a set of linear equations. The performance of the proposed algorithm is good for artificial data and it is comparable to other classification methods for benchmark data set.

References

1. Abe, S., Inoue, T.: Fuzzy Support Vector Machines for Multiclass Problems. ESANN'2002 Proceedings - European Symposium on Artificial Neural Networks (2002) 113-118
2. Bredensteiner, E.J., Bennett, K.P.: Multicategory Classification by Support Vector Machines. Computational Optimization and Applications, 12 (1999) 53-79
3. Weston, J., Watkins, C.: Multi-class Support Vector Machines. CSD-TR-98-04 Royal Holloway, University of London, Egham, UK (1998)
4. Angulo, C., Parra, X., Catala, A.: K-SVCR. A Support Vector Machine for Multi-class Classification. Neurocomputing, 55 (2003) 57-77
5. Suykens, J. A.K., Vandewalle, J.: Least Squares Support Vector Machine Classifiers. Neural Process Letter, 9 (1999) 293-300
6. Suykens, J.A.K., Lukas, L., Vandewalle, J.: Sparse Approximation Using Least Squares Support Vector Machines. IEEE International Symposium on Circuits and Systems (2000) 757-760
7. Pontil, M., Rifkin, R., Evgeniou, T.: From Regression to Classification in Support Vector Machines. ESANN'99 Proceedings - European Symposium on Artificial Neural Networks (1999) 225-230

Twi-Map Support Vector Machine
for Multi-classification Problems

Zhifeng Hao[1], Bo Liu[2], Xiaowei Yang[1], Yanchun Liang[3], and Feng Zhao[1]

[1] Department of Applied Mathematics, South China University of Technology,
Guangzhou 510640, China
xwyang@scut.edu.cn
[2] College of Computer Science and Engineering, South China University of Technology,
Guangzhou 510640, China
[3] College of Computer Science and Technology, Jilin University, Changchun 130012, China

Abstract. In this paper, a novel method called Twi-Map Support Vector Machines (TMSVM) for multi-classification problems is presented. Our ideas are as follows: Firstly, the training data set is mapped into a high-dimensional feature space. Secondly, we calculate the distances between the training data points and hyperplanes. Thirdly, we view the new vector consisting of the distances as new training data point. Finally, we map the new training data points into another high-dimensional feature space with the same kernel function and construct the optimal hyperplanes. In order to examine the training accuracy and the generalization performance of the proposed algorithm, One-against-One algorithm, Fuzzy Least Square Support Vector Machine (FLS-SVM) and the proposed algorithm are applied to five UCI data sets. Comparison results obtained by using three algorithms are given. The results show that the training accuracy and the testing one of the proposed algorithm are higher than those of One-against-One and FLS-SVM.

1 Introduction

There exist many multi-classification problems in practice engineering, which are usually converted into binary classification problems. Support vector machine [1] is originally designed for binary classification problems. At present, it has been widely applied to multi-classification problems.

In the previous work, Vapnik proposed one-against-all algorithm [2], in which a $C-$class problem is transformed into C two-class problems. However, there exist some unclassifiable regions and the accuracy of this algorithm is very low. Later, Vapnik [1] introduced the continuous decision functions instead of the discrete decision functions. KreBel [3] converted the $C-$class problem into $\frac{C(C-1)}{2}$ two-class problems, which is called pairwise classification (One-against-One). In this algorithm, the unclassifiable region still remains. To overcome this drawback, Platt et al. [4] put forward to a decision-tree-based pairwise classification, which was called Decision Directed Acyclic Graph (DDAG). Inoue and Abe [5] presented a fuzzy support vector machine for one-against-all classification, in which the fuzzy membership function is defined instead of continuous decision function. Recently, they proposed a fuzzy support vector machine based on pairwise classification [6]. Tsujinishi and Abe [7] proposed a new method depending on pairwise classification, in which the average

J. Wang, X. Liao, and Z. Yi (Eds.): ISNN 2005, LNCS 3496, pp. 869–874, 2005.

membership function is defined instead of the minimum membership function. In the work of Cheong et al. [8], a kernel-based self-organizing map (KSOM) is introduced to convert a multi-classification problem into a binary tree, in which the binary decisions are made by SVM.

In the work of Hsu and Lin [9], comparisons of various methods mentioned above are given. If the training data points are linearly separable in the high-dimensional feature space, the methods can work well. Otherwise, the accuracy would be reduced. It is well known that the linearly inseparable case in high-dimension space is often caused by the inappropriate selection of kernel function and the selection of the appropriate kernel function and hyperparameters is the key to solve the mentioned problem. However, up to now, the selection of the proper kernel function and parameters is still an open problem. To improve the accuracy of the first classification in the case of the same kernel function and hyperparameters, a novel algorithm is proposed for solving multi-classification problems in this paper.

The rest of this paper is organized as follows. In Section 2, One-against-One algorithm is briefly reviewed and FLS-SVM is given in Section 3. Section 4 shows the proposed algorithm. Simulation experiments are given in Section 5. In Section 6, some conclusions are presented.

2   One-Against-One Algorithm

According to the conventional pairwise classification [3], one needs to determine $\frac{C(C-1)}{2}$ decision functions for a $C-$class classification problem. The optimal hyperplane between class i and j is defined in the following

$$D_{ij}(\mathbf{x}) = w_{ij}^T \mathbf{x} + b_{ij} = 0. \tag{1}$$

where w_{ij}^T is an $m-$dimensional vector, b_{ij} is a scalar.

For the input vector $\mathbf{x}$, one can compute

$$D_i(\mathbf{x}) = \sum_{j \neq i, j=1}^{C} sign(D_{ij}(\mathbf{x})). \tag{2}$$

and classify $\mathbf{x}$ into the class

$$\arg \max_{i=1,\ldots,C} (D_i(\mathbf{x})). \tag{3}$$

3   FLS-SVM

In order to overcome the drawback of One-against-One algorithm, Tsujinishi and Abe [7] introduced the fuzzy membership functions based on the One-against-One classification. Constructing the hyperplanes in the feature space, they defined fuzzy membership functions $m_{ij}(x)$ as follows:

$$m_{ij}(\mathbf{x}) = \begin{cases} 1 & D_{ij}(\mathbf{x}) \geq 1, \\ D_{ij}(\mathbf{x}) & otherwise . \end{cases} \tag{4}$$

In their work, using the minimum operator or the average operator of the membership functions $m_{ij}(x)$ for class i, a new membership function $m_i(x)$ is given by

$$m_i(\mathbf{x}) = \min_{j=1,\ldots,C}(m_{ij}(\mathbf{x})).$$

(5)

or

$$m_i(\mathbf{x}) = \frac{1}{C-1}\sum_{j\neq i, j=1}^{C} m_{ij}(\mathbf{x}).$$

(6)

And the input data point $\mathbf{x}$ is classified into the class

$$\arg\max_{i=1,\ldots,C}(m_i(\mathbf{x})).$$

(7)

4   The Proposed Algorithm

Let $S = \{(\mathbf{x}_1, y_1), (\mathbf{x}_2, y_2), \ldots, (\mathbf{x}_l, y_l)\}$ be a training set, where $\mathbf{x}_i \in R^n$ and $y_i \in \{1, 2, \ldots, C\}$. In order to improve the accuracy of the algorithms mentioned above, we propose a novel algorithm in this section, which is called Twi-Map Support Vector Machine (TMSVM).

Firstly, transform the multi-classification into the binary classification based on One-against-One algorithm. Secondly, construct $\frac{C(C-1)}{2}$ hyperplanes in the feature space and calculate the distances from each data point $\mathbf{x}_i$ in the training set to every hyperplane. Finally, expand $\mathbf{x}_i$ into a $\frac{C(C-1)}{2}$ dimensional vector

$$\mathbf{x}_k^{new} = (D_{12}(\mathbf{x}_i), \cdots, D_{1C}(\mathbf{x}_i), D_{23}(\mathbf{x}_i), \cdots, D_{2C}(\mathbf{x}_i), \cdots, D_{(C-1)(C)}(\mathbf{x}_i)).$$

Using the same kernel function, we map the new data points $\mathbf{x}_k^{new}$ into another high-dimensional space, in which one can construct the optimal hyperplanes based on One-against-One algorithm. The decision rule is a combination of One-against-One and FLS-SVM. For the input vector $\mathbf{x}_k$, we can compute

$$D^{new}{}_i(\mathbf{x}_k^{new}) = \sum_{j\neq i, j=1}^{C} sign(D^{new}{}_{ij}(\mathbf{x}_k^{new})).$$

(8)

and then classify $\mathbf{x}_k$ into the $j-th$ class, where

$$j = \arg\max_{i=1,2,\ldots,C}(D^{new}{}_i(\mathbf{x}_k^{new})).$$

(9)

If the data point $\mathbf{x}_k$ cannot be classified, the rules of FLS-SVM are adopted.

After one map from the first feature space to the second one, in general, the distances between the training data points and hyperplanes could be enlarged and the accuracy would be improved.

From the algorithm procedures, one can conclude that the computational complexity of the presented method is high. However, the same method can be applied to One-against-All algorithm, which can reduce the complexity of the proposed approach.

5 Simulation Experiments

The simulated experiments are run on a PC with a 2.8GHz Pentium IV processor and a maximum of 512MB memory. All the programs are written in C++, using Microsoft's Visual C++ 6.0 compiler. In order to evaluate the performance of the proposed algorithm, One-against-One, FLS-SVM and TMSVM are applied to five UCI data sets available from the UCI Machine Learning Repository [10]. Data preprocessing is in the following:

- Iris dataset: This dataset involves 150 data with 4 features and 3 classes. We choose 100 samples of them randomly for training, and the rest for testing.
- Glass dataset: This dataset consists of 214 data points with 9 features and 6 classes. We select 124 randomly for training, and the rest for testing.
- Abalone dataset: This dataset consists of 4177 data with 6 features and 29 classes and hasn't class 2. We reconstructed the data by combining classes 1, 3, 4 into one class, class 5 into one class, class 6 into one class, class 7 into one class, classes 8, 9 into one class, classes 10, 11 into one class, classes 12, 13 into one class, the rest classes into one class. We select 2177 samples randomly for training, and the rest for testing.
- King-Rook-VS-King (KRK) dataset: This dataset includes 28056 data points with 6 features and 18 classes. We select several classes of them as follows: class "two" belongs to class 1, class "three" belongs to class 2, class "four" belongs to 3, class "five" belongs to class 4, class "six" belongs to class 5, class "seven" belongs to class 6 and then choose 4089 samples randomly for training, and the rest 1000 samples for testing.
- Thyroid dataset: This dataset involves 3772 data points for training and 3428 data points for testing with 21 features and 3 classes.

An RBF kernel function is employed, the parameter values and the results are shown in Table 1.

Table 1. Comparison of the Results Obtained by One-Against-One, FLS-SVM and TMSVM.

Dataset	Parameter (γ,σ)	One-against-One (LS-SVM) (%)		FLS-SVM (%)		TMSVM (%)	
		Testing	Training	Testing	Training	Testing	Training
Iris	(2,0.4)	94	99	94	99	96	100
Glass	(1,1)	68.888	83.870	68.888	83.870	70	85.484
Abalone	(3,25)	45.35	48.002	45.6	48.277	46.7	51.539
KRK	(2,1)	86.1	100	86.9	100	87.3	100
Thyroid	(1,1)	94.136	94.936	94.340	95.201	95.770	97.853

From Table 1, we can easily conclude that after one map from the first feature space to the second one, the training and testing accuracy of TMSVM is higher than those of One-against-One and FLS-SVM. Of course, the spending time of TMSVM is longer, which can be seen from the algorithm procedures.

6 Conclusions

Support vector machines are originally designed for binary classifications. As for multi-class classifications, they are usually converted into binary classifications. Up to now, several methods have been proposed to decompose and reconstruct multi-classification problems. In the high-dimensional feature space, if the training data points were still linearly inseparable, the accuracy of these methods would be reduced. In order to solve this problem, a novel method is presented in this paper. One-against-One, FLS-SVM and TMSVM are applied to five UCI data sets to examine the training accuracy and the generalization performance of TMSVM. Comparison results obtained by using three algorithms are given. The results show that the training accuracy of TMSVM is higher than those of One-against-One and FLS-SVM, and its generalization performance is also comparable with them. In the future, the proposed algorithm will be applied to the supervised feature extraction.

Acknowledgements

This work has been supported by the National Natural Science Foundation of China (19901009, 60433020), the Doctoral Foundation of the National Education Ministry of China (20030183060), Natural Science Foundation of Guangdong Province (970472, 000463), Excellent Young Teachers Program of Ministry of Education of China, Fok Ying Tong Education Foundation (91005) and Natural Science Foundation of South China University of Technology (E512199, D76010).

References

1. Vapnik, V.N.: Statistical Learning Theory. John Wiley & Sons (1998)
2. Vapnik, V.N.: The Nature of Statistical Learning Theory. Springer-Verlag, London, UK (1995)
3. KreBel, U.H.G.: Pairwise Classification and Support Vector Machines. Schölkopf, B., Burges, C.J. and Smola, A.J. Editors, Advances in Kernel Methods: Support Vector Learning. MIT Press Cambridge (1999) 255-268
4. Platt, J.C., Cristianini, N., Shawe.-Taylor, J.: Large Margin DAGs for Multiclass Classification. S.A. Solla, T.K. Leen, and K.R. Müller, Editors, Advances in Neural Information Processing Systems, MIT Press Cambridge, 12 (2000) 547-553
5. Inoue, T., Abe, S.: Fuzzy Support Vector Machines for Pattern Classification. Proceedings of International Joint Conference on Neural Networks (IJCNN'01), 2 (2001) 1449-1454
6. Abe, S., Inoue, T.: Fuzzy Support Vector Machines for Multiclass Problem. Proceedings of 10th European Symposium on Artificial Neural Networks (ESANN'2002), Bruges, Belgium (2002) 113-118

7. Tsujinishi, D., Abe, S.: Fuzzy Least Squares Support Vector Machines for Multiclass Problems. Neural Network, **16** (2003) 785-792

8. Cheong, S., Oh, S.H., Lee, S.Y.: Support Vector Machines with Binary Tree Architecture for Multi-class Classification. Neural Information Processing-Letters and Reviews, **2** (2004)

9. Hsu, C.W., Lin, C.J.: A Comparison of Methods for Multiclass Support Vector Machines. IEEE Transaction on Networks, **13** (2002) 415-425

10. Murphy, P.M., Aha, D.W.: UCI Repository of Machine Learning Database (1992)

Fuzzy Multi-class SVM Classifier
Based on Optimal Directed Acyclic Graph Using
in Similar Handwritten Chinese Characters Recognition

Jun Feng[1,2], Yang Yang[2], and Jinsheng Fan[1]

[1] Department of Computer Science,
Shijiazhuang Railway Institute, Shijiazhuang 050043, China
fengjun71@sohu.com, fanjsh@sjzri.edu.cn
[2] Information Engineering School,
University of Science and Technology Beijing, Beijing 100083, China
yyang@ustb.edu.cn

Abstract. This paper proposes a method to improve generalization perform-ance of multi-class support vector machines (SVM) based on directed acyclic graph (DAG). At first the structure of DAG is optimized according to training data and Jaakkola-Haussler bound, and then we define fuzzy membership func-tion for each class which is obtained by using average operator in the testing stage and the final recognition result is the class with maximum membership. As a result of our experiment for similar handwritten Chinese characters recog-nition, the generalization ability of the novel fuzzy multi-class DAG-based SVM classifier is better than that of pair-wise SVM classifier with other com-bination strategies and its execution time is almost the same as the original DAG.

1   Introduction

Support vector machines (SVM) are originally formulated for two-class classification problem [1] and extension to multi-class problems is not unique. In one-against-all classification, one class is separated from the remaining classes. However, unclassifi-able regions exist. In pair-wise classification, the k-class problem is converted into $k(k-1)/2$ two-class problems and max voting strategy is often used to determine the final result. Unclassifiable regions reduce, but still they remain. Directed acyclic graph (DAG) based pair-wise classifier was proposed to resolve unclassifiable re-gions [2]. But the generalization ability depends on the tree structure. In this paper, we propose an approach based on Jaakkola-Haussler error bound to optimize the DAG structure. To further improve the generalization ability, we introduce fuzzy membership function into DAG SVM, in which the membership function for each class is defined using the average operator.

Off-line handwritten Chinese characters recognition widely used in many areas is known as a challenging pattern recognition task. Investigation shows that the exis-tence of lots of similar characters is a key factor affecting recognition rate [3]. The technique for similar handwritten Chinese characters recognition is still immature and much deeper researches should be made to solve the difficult problem. This paper applies the novel fuzzy multi-class DAG-based SVM classifier to similar handwritten

J. Wang, X. Liao, and Z. Yi (Eds.): ISNN 2005, LNCS 3496, pp. 875–880, 2005.

Chinese characters recognition and experimental results show superior performance of our method to some other pair-wise classification strategies.

The paper is organized as follows. The next section briefly explains two-class L^2-Norm SVM. Section 3 describes the novel fuzzy multi-class SVM classifier based on optimal DAG in detail. Then we introduce the method of feature extraction for similar handwritten Chinese characters adopted in our recognition scheme in section 4. Section 5 presents experimental results indicating the effectiveness and practicality of our method. Finally, we conclude this paper in section 6.

2   Two–Class L^2-Norm SVM Classification

Here we emphatically discuss L^2-Norm SVM classifier with nonlinear soft margin. Let training data set be $T = \{(\mathbf{x}_1, y_1),..., (\mathbf{x}_n, y_n)\} \in (\mathbf{X} \times \mathbf{Y})^n$, where $\mathbf{x}_i \in \mathbf{X} = \mathbf{R}^m (i = 1,..., n)$ and $y_i \in \mathbf{Y} = \{1, -1\}$. The problem of constructing L^2-Norm SVM can be described as formula (1) and (2).

$$\min_{\mathbf{w} \in H, \xi \in R^n} \frac{1}{2} \|\mathbf{w}\|^2 + C \sum_{i=1}^{n} \xi_i^2 \cdot \tag{1}$$

$$\text{s.t. } y_i(\mathbf{w} \cdot \varphi(\mathbf{x}_i)) \geq 1 - \xi_i, i = 1,..., n \cdot \tag{2}$$

By applying Lagrangian function and Wolfe dual principle, we obtain the dual formulation of the above optimal problem as follows:

$$\min_{\alpha} \frac{1}{2} \sum_{i=1}^{n} \sum_{j=1}^{n} y_i y_j \alpha_i \alpha_j \left(K(x_i, x_j) + \frac{1}{2C} \delta_{ij} \right) - \sum_{i=1}^{n} \alpha_i \cdot \tag{3}$$

$$\text{s.t. } \alpha_i \geq 0 \ (i = 1, 2,..., n) \cdot \tag{4}$$

Here, $\delta_{ij} = \begin{cases} 1, i = j \\ 0, i \neq j \end{cases}$. In this study we use RBF kernel function as follows:

$$K(x, x_i) = \exp(-\frac{1}{2\sigma^2} \|x - x_i\|^2) \cdot \tag{5}$$

Thus we can determine the following homogeneous decision function where Lagrange multipliers $\alpha^* = (\alpha_1^*,..., \alpha_n^*)^T$ are the optimal solution to the problem (3)~(4).

$$D(\mathbf{x}) = \text{sgn}(\sum_{i=1}^{n} \alpha_i^* y_i K(x, x_i)) \cdot \tag{6}$$

3   Methodology

3.1   DAG-Based SVM

As one of classifier combination strategies for pair-wise SVM, DAG-based SVM also constructs $k(k-1)/2$ binary classifiers for k-class problem in the training stage. Let D_{ij} denote the decision function for class i against class j . Fig. 1(a) shows the decision functions of pair-wise classification for three-class problem and the shaded region is unclassifiable while using max voting strategy.

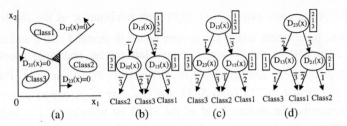

Fig. 1. Explanation of DAG for three-class problem: (a) the decision functions, (b) the first structure, (c) the second structure, (d) the third structure.

In the DAG classification stage, there are different tree structures which can be selected such as shown in Fig. 1(b)~(d). In the figure, $\bar{i}$ shows that x does not belong to class i. The DAG is equivalent to operating on a list [2]. Now consider a data point 'x' located in the shaded region of Fig. 1(a). In case of a list 1-3-2 (shown in Fig. 1(b)), firstly class 1 would be eliminated because of $D_{12}(x) < 0$ and the list becomes 3-2. Then class 2 would be removed because of $D_{32}(x) > 0$ and DAG terminates because only class 3 remains in the list. As a result, the data point 'x' is classified as class 3. Similarly, we classify 'x' into class 2 in case of a list 1-2-3 (shown in Fig. 1(c)) and into class 1 in case of a list 2-1-3 (shown in Fig. 1(d)). Obviously, unclassifiable regions are resolved but the output of the DAG depends on the sequence of binary classifiers in nodes.

3.2 Optimizing Decision Trees

In DAG-based SVM the unclassifiable regions are assigned to the classes associated with the leaf nodes. Moreover, there exists cumulative error because the DAG belongs to serial classification. Therefore, there are different generalization performances while adopting different DAG structures. We wish select the optimal structure whose performance is better. Here we use Jaakkola-Haussler error bound [4] (a kind of estimate of LOO error bound) to determine the optimal structure in the training stage. Jaakkola-Haussler bound E_{ij} for class i and j (with RBF kernel) is defined as

$$E_{ij} = \frac{1}{n}\left|\sum_{t=1}^{n} \text{sgn}(-y_t D_{ij}(x_t) + \alpha_t^*)\right| \cdot \qquad (7)$$

We optimize the structure so that the class pairs with higher generalization abilities are put in the upper nodes of the tree. The algorithm to determine the node sequence of the list is described as follows:

Step 1. Initialize class label set: $SET\ 1 = \{1, 2, ..., k\}$ and generate the initial list: $List = [12...k]$. Moreover, let left pointer point at the first element and right pointer at the kth element of $List$;

Step 2. Calculate $k(k-1)/2$ Jaakkola-Haussler error bounds, and sort E_{ij} in ascending order;

Step 3. Select the class pair (i, j) in $SET1$ with minimum E_{ij} as the top level of the DAG, i.e., the element where left pointer points is changed into i and the element where right pointer points into j. Move left pointer and right pointer respectively: $LeftPointer \leftarrow LeftPointer + 1, RightPointer \leftarrow RightPointer - 1$. Produce two subset $SET21$ and $SET22$ from $SET1$: $SET 21 \leftarrow SET 1 - \{i\}, SET 22 \leftarrow SET 1 - \{j\}$;

Step 4. Determine the second level of the DAG by utilizing $SET21$ and $SET22$ as Step 3. Repeat the above step until left pointer is the same as right pointer.

3.3 Fuzzy Multi-class SVM Based on DAG

To further improve performance, we introduce fuzzy membership function into DAG SVM in the classification stage. First, we define the one-dimensional membership function $m_{ij}(x)$ in the direction perpendicular to the optimal separating hype-plane $D_{ij}(x)$ as formula (8). Then, let $M_i(x)$ denote the membership function for class i.

$$m_{ij}(x) = \begin{cases} 1, D_{ij}(x) \geq 1 \\ 0, D_{ij}(x) \leq -1 \\ (D_{ij}(x)+1)/2, |D_{ij}(x)| < 1 \end{cases} \quad (8)$$

The fuzzy DAG-based SVM classification algorithm is described as follows:

Step 1. Initialize $M_i(x)$ as unknown state: $M_i(x) = -1$. Let left pointer point at the first element and right pointer at the kth element of $List$ which is obtained by 3.2;

Step 2. Construct the class pair (i, j) from the corresponding elements where left pointer and right pointer point separately. Calculate $m_{ij}(x)$ and $m_{ji}(x) = 1 - m_{ij}(x)$;

Step 3. Renew $M_i(x)$ by utilizing $m_{ij}(x)$. If $M_i(x) = -1$, then $M_i(x) \leftarrow m_{ij}(x)$; If $M_i(x) \neq -1$, then $M_i(x) \leftarrow [M_i(x) + m_{ij}(x)]/2$, i.e., use the average operator to renew $M_i(x)$. Similarly, we can renew $M_j(x)$ by utilizing $m_{ji}(x)$;

Step 4. If $m_{ij}(x) > m_{ji}(x)$, then $RightPointer \leftarrow RightPointer - 1$, else $LeftPointer \leftarrow LeftPointer + 1$;

Step 5. Repeat step 2, 3, and 4 until left pointer is the same as right pointer. Finally, the data point 'x' is classified into the class $l = \underset{i=1,2,...,k}{\arg\max} M_i(x)$.

4 Feature Extraction

We adopt the technique of combining wavelet transform with elastic meshing to extract the features of handwritten Chinese characters. First, a two-dimensional Chinese character image (normalized as 64×64) is decomposed into four sub-images based on the analysis of its one order two-dimensional wavelet transform. The different four sub-images, i.e., LL, LH, HL and HH sub-image, indicate fuzzy sub-image, horizontal stroke, vertical stroke and slant stroke respectively. Next, a set of elastic meshes are constructed and applied to each sub-image. The probability distribution of pixels within each mesh is then calculated to form the feature vector of this character. Fig. 2 illustrates the feature extraction process of a character "申". In our experiment we find that the peak recognition rate of the similar set is obtained when the number of meshes in each sub-image is 8×8.

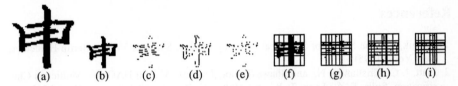

(a)   (b)   (c)   (d)   (e)   (f)   (g)   (h)   (i)

Fig. 2. Illustration of feature extraction process: (a) normalized character, (b) LL sub-image, (c) LH sub-image, (d) HL sub-image, (e) HH sub-image, (f) meshed LL sub-image, (g) meshed LH sub-image, (h) meshed HL sub-image, (i) meshed HH sub-image.

5   Performance Evaluation

The similar Chinese character set "田甲由申中" is used to evaluate the recognition performance. Each character is written by 850 writers with 400 samples used for training and the remaining 450 ones for testing. Daubechies orthonormal bases of compactly supported wavelets are adopted in the experiment. We set the model parameters $C = 32768$ and $\sigma = 2$. The experiments ran on a Pentium Ⅳ (1.6 GHz) PC with 128 MB RAM. Table 1 shows the recognition rates of the testing data and average classification time of each sample for the Max Voting-based pair-wise SVM, pair-wise fuzzy SVM [5], DAG SVM, optimal DAG and fuzzy DAG presented in this paper. The experimental results show that the recognition performance of fuzzy DAG-based SVM is superior to that of other pair-wise multi-class SVM. Moreover, classification by fuzzy DAG is much faster than by Max Voting and fuzzy SVM and is almost the same as the original DAG.

Table 1. Comparison of recognition correct rates and classification time of each sample with other methods for similar handwritten Chinese character set '田甲由申中'.

Method	Data	Training set RR(%)	Time(s)	Testing set RR(%)	Time(s)
Max Voting		99.75	0.016	86.67	0.015
Fuzzy SVM	Minimum operator	99.80	0.015	86.76	0.015
	Average operator	99.70	0.014	87.02	0.015
DAG	Max.	99.80	0.0065	86.67	0.0081
	Min.	99.75	0.0055	86.01	0.0064
	Ave.	99.77		86.36	
Optimal DAG		99.75	0.009	86.58	0.008
Fuzzy DAG		99.80	0.008	88.35	0.008

6 Conclusions

This paper proposes a method to improve the performance of DAG-based SVM by optimizing the tree structure in the training stage and then using fuzzy class membership function in the testing stage. The experimental results show that the classification performance is improved remarkably and so the proposed method is very promising.

References

1. Vapnik, V.: The Nature of Statistical Learning Theory. Springer-Verlag, Berlin Heidelberg New York (1995)
2. Platt, J. C., Cristianini, N., and Shawe-Taylor, J.: Large Margin DAGs for Multiclass Classification. Solla, S. A., Leen, T. K., and Müller, K.-R. (eds.): Advances in Neural Information Processing Systems. MIT Press 12 (2000) 547-553
3. Lin, Z.Q. and Guo, J.: An Algorithm for the Recognition of Similar Chinese Characters. Journal of Chinese Information Processing, 16 (2002) 44-48
4. Vapnik, V. and Chapelle, O.: Bounds on Error Expectation for Support Vector Machines. Neural Computation, 12 (2000) 2013-2036
5. Tsujinishi, D. and Abe, S.: Fuzzy Least Squares Support Vector Machines for Multiclass Problems. Neural Networks, 16 (2003) 785-792

A Hierarchical and Parallel Method for Training Support Vector Machines[*]

Yimin Wen[1,2] and Baoliang Lu[1]

[1] Department of Computer Science and Engineering, Shanghai Jiao Tong University,
1954 Hua Shan Rd., Shanghai 200030, China
{wenyimin,blu}@cs.sjtu.edu.cn
[2] Hunan Industry Polytechnic, Changsha 410007, China

Abstract. In order to handle large-scale pattern classification problems, various sequential and parallel classification methods have been developed according to the divide-and-conquer principle. However, existing sequential methods need long training time, and some of parallel methods lead to generalization accuracy decreasing and the number of support vectors increasing. In this paper, we propose a novel hierarchical and parallel method for training support vector machines. The simulation results indicate that our method can not only speed up training but also reduce the number of support vectors while maintaining the generalization accuracy.

1 Introduction

In the last decade, there are many surges of massive data sets. It is necessary to develop efficient methods to deal with these large-scale problems. Support vector machine (SVM) [1] has been widely used in a wide variety of problems and is a candidate tool for solving large-scale classification problems. However, the essence of training SVMs is solving a quadratic convex optimization problem whose time complexity is $O(N^3)$, here N is the number of training samples.

In the literature of machine learning, the divide-and-conquer principle is always used to handle large-scale problems and can be implemented in series or in parallel. In sequential learning approach, a large-scale problem is divided into many smaller subproblems that are learned sequently. These approaches include the advanced working set algorithms, which use only a subset of the variables as a working set while freezing the others [2], [3], such as Chunking, SMO, SVM^{light}, and LibSVM. The shortcoming of these approaches is that a large number of iterations are needed and this will lead to memory thrashing when training data set is large. In parallel learning approach, a large-scale problem is divided into many smaller subproblems that are parallelly handled by many modules. After training, all the trained modules are integrated into a modular system [4], [5], [6].

---

[*] This work was supported by the National Natural Science Foundation of China under the grants NSFC 60375022 and NSFC 60473040. This work was also supported in part by Open Fund of Grid Computing Center, Shanghai Jiao Tong University.

J. Wang, X. Liao, and Z. Yi (Eds.): ISNN 2005, LNCS 3496, pp. 881–886, 2005.

This kind of method has two main advantages over traditional SVMs approaches. 1) It can dramatically reduce training time. 2) It has good scalability. However, these methods will lead to increasing of the number of support vectors and often decrease generalization accuracy slightly.

Schölkopf [7] and Syed[8] have pointed out that support vectors (SVs) summarize classification information of training data. Based on their work and our previous work [9], we propose a novel hierarchical and parallel method for training SVMs. In our method, we first divide the training data of each class into $K(> 1)$ subsets, and construct K^2 classification subproblems. We train SVMs parallelly on these K^2 classification subproblems and union their support vectors to construct another K classification subproblems. After that, we train SVMs parallelly on these K subproblems and take union of all their support vectors to construct the last subproblem. At last, we train a SVM as the final classifier. All the experiments illustrate that our method can speed up training and reduce the number of support vectors while maintaining the generalization accuracy. The paper is organized as follows: Section 2 will introduce our hierarchical and parallel training method. The experiments and discussions are presented in Section 3. Finally, Section 4 is conclusions.

2 Hierarchical and Parallel Training Method

Given positive training data set $\mathcal{X}^+ = \{(X_i, +1)\}_{i=1}^{N^+}$ and negative training data set $\mathcal{X}^- = \{(X_i, -1)\}_{i=1}^{N^-}$ for a two-class classification problem, where X_i denotes the ith training sample, and N^+ and N^- denote the number of positive training samples and negative training samples, respectively. The entire training data set can be defined as $\mathcal{S} = \mathcal{X}^+ \bigcup \mathcal{X}^-$. In order to train a SVM on S to classify future samples, our method includes three steps:

In the first step, we divide $\mathcal{X}^+$ and $\mathcal{X}^-$ into K roughly equal subsets respectively, according to a given partition value K,

$$\mathcal{X}^+ = \bigcup_{i=1}^{K} \mathcal{X}_i^+, \quad \mathcal{X}_i^+ = \{(X_j, +1)\}_{j=1}^{N_i^+}, \quad i = 1, 2, ..., K$$

$$\mathcal{X}^- = \bigcup_{i=1}^{K} \mathcal{X}_i^-, \quad \mathcal{X}_i^- = \{(X_j, -1)\}_{j=1}^{N_i^-}, \quad i = 1, 2, ..., K \tag{1}$$

where: $N_i^+ = \lfloor N^+/K \rfloor, i = 1, 2, ..., K - 1; N_K^+ = N^+ - \sum_{i=1}^{K-1} N_i^+; N_i^- = \lfloor N^-/K \rfloor, i = 1, 2, ..., K - 1; and N_K^- = N^- - \sum_{i=1}^{K-1} N_i^-$.

According to (1), the original classification problem is divided into K^2 smaller classification subproblems as follows,

$$\mathcal{S}_{i,j}^1 = \mathcal{X}_i^+ \bigcup \mathcal{X}_j^-, \quad 1 \le i, j \le K \tag{2}$$

Because these K^2 smaller subproblems need not to communicate with each other in learning phase, they can be handled simultaneously by traditional method like SVM^{light} and K^2 sets of support vectors, $\mathcal{SV}_{i,j}^1 (1 \le i, j \le K)$, are obtained.

In the second step, taking union of each of the K sets of support vectors to construct the following K classification problems,

$$S_i^2 = \bigcup_{j=1}^{K} SV_{j,permu(j+i-1)}^1, \quad 1 \le i \le K \tag{3}$$

where $permu(n)$ is defined as:

$$permu(n) = \begin{cases} n - K, & \text{for } n > K \\ n, & \text{otherwise} \end{cases} \tag{4}$$

The approach that takes K unions from $SV_{i,j}^1 (1 \le i, j \le K)$ can be defined as "cross-combination", which can ensure all classification information are maintained. All the subproblems $S_i^2 (1 \le i \le K)$ are handled parallelly and K support vector sets $SV_i^2 (1 \le i \le K)$ will be gotten.

At the last step, we take a union of $SV_i^2 (1 \le i \le K)$, i.e. $S_{final} = \bigcup_{i=1}^{K} SV_i^2$. We train a SVM on S_{final} to get the final classifier. The procedure of our method for constructing a full-K-tree is depicted in Fig. 1.

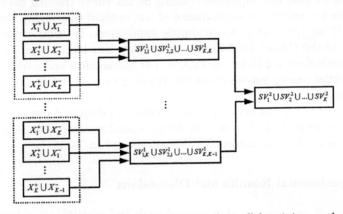

Fig. 1. Illustration of our hierarchical and parallel training method

3   Experiments

3.1   Data Sets and Simulation Environment

In order to validate our method systematically, we perform two experiments on UCI data sets [10]. The first data set is Forest coverType and the second one is Space shuttle. In this paper, multi-class classification problems are handled by one-against-one technique, i.e. a M-class classification problems is divided into $M(M-1)/2$ two-class classification subproblems by combining every two classes. The total training time is the sum of all the training time of two-class classification subproblems. Given a test instance, all the trained $M(M-1)/2$ sub-classifiers vote for the final classification. We take SVM^{light} [3] as sequential

Table 1. The problem statistics and the selections of parameters of SVMs

Problems	#attributes	#class	#training data	#test data	c	σ
Forest coverType	54	7	290504	290508	128	0.25
Space shuttle	9	5	43483	14494	1000	50

training method for its friendly interface and integrate it into our hierarchical and parallel training method. The kernel used is the radial-basis function: $\exp(-\frac{1}{2\sigma^2}\|X - X_i\|^2)$. For Forest coverType data, we take one half of it as training data and the rest one half as test data. These training and test data are normalized in the range $[0, 1]$. In this experiment, The sequential training is performed on a PC that has 3.0GHz CPU with 1GB RAM, while the hierarchical and parallel training is performed on a cluster IBM e1350, which has one management node and eight computation nodes. The management node has two 2.0GHz CPUs with 2GB RAM and the computation node has two 2.0GHz CPUs with 1.5GB RAM. So, the sequential training has an advantage on speed over our method. We implement our method by using MPI. Because the limitation of computing resource, the partition K is only took from 1 to 4. Here, $K = 1$ means that we take the sequential training on the entire training data.

In order to explore the performance of our method thoroughly, we take the values of K as 1,2,...,30 in Space shuttle experiment. Because of the resource limitation of the cluster IBM e1350, we simulate our hierarchical and parallel training method on a PC, i.e. we execute our hierarchical and parallel training in a sequential mode, but we count the training time in parallel mode. In this experiment, the original training and test data are used, but we exclude the samples of the sixth and seventh classes in training data and test data, because they can not be parted when $K > 6$. All the classification problems statics and the selection of the parameters of SVMs are showed in Table. 1.

3.2   Experimental Results and Discussions

From Table. 2, we can see that even though the partition K takes different values, the accuracy of our method is almost the same as the accuracy of the sequential method. Furthermore, our method can significantly reduce not only the training time but also the number of support vectors. From Table. 2, we see that the largest speedup is 5.05 when $K = 4$, and the number of support vectors is reduced 5.7% at $K = 3$. Fig. 2 shows that both the training time and the number of support vectors of all the two-class subproblems consistently decrease by using our hierarchical and parallel training method. From Table. 3, we can see that the accuracy of the classifiers with $K > 1$ is almost the same as the accuracy of the classifier with $K = 1$. From Fig. 3, we can also see that the largest speedup of our method is larger than 10, and the largest reduction of the number support vectors is 3%.

The reason of reducing training time in our method lies in the fact that a large number of non-support vectors are filtered out in the first two steps but a training instance maybe used again and again in the sequential method such as SVM^{light}. The reason of reducing the number of support vectors is that the training data partition lead to simpler classifiers for subproblems. As a result, the split of the training data takes an affect like editing training samples [11]. When K increases, it can be expected that the number of support vectors will increase. Fig. 3 shows an increment trend of support vectors. In the worst case, no training instances are filtered out in the first step and $S_i^2 (1 \leq i \leq K)$ will be the same as S, so the number of support vectors in our method will be equal to the number of support vector in sequential method. For many large-scale classification problems, the number of support vectors always take a small proportion of the entire training data. Consequently, our method will take higher performance than sequential training method.

Table 2. The number of SVs and training time in Forest coverType problem with different values of K

Partitions	Accuracy	#SVs	Training time(s)	Speedup
$K = 1$	0.936721	47146	13037	1
$K = 2$	0.936153	44455	5281	2.47
$K = 3$	0.936105	44300	3275	3.98
$K = 4$	0.936752	44365	2582	5.05

Table 3. The accuracy variation in Space shuttle problem with different partition K

$K = 1$	$K = 2, 3, 4, ..., 30$	
	Mean	Variance
0.99897	0.99894	0.00036

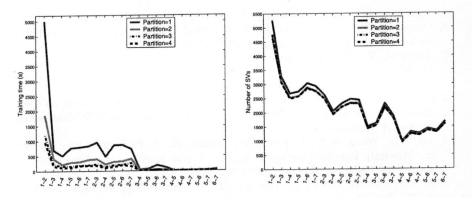

Fig. 2. Training time and the number of SVs for every two-class classification subproblems in Forest coverType classification. The digits from 1 to 7 below the horizontal axes denote the seven classes in the Forest coverType data, one pair of two digits means a two-class classification subproblem

4   Conclusions

In this paper we have presented a hierarchical and parallel methods for training SVMs. Several experimental results indicate that the proposed method has two attractive features. The first one is that it can reduce training time while keeping the generalization accuracy of the classifier. The second one is that the number of support vectors generated by our method is smaller than that of the SVMs

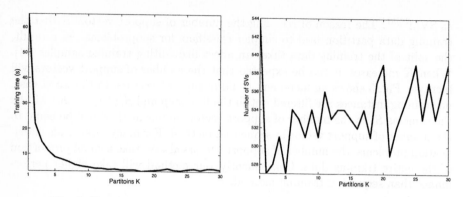

Fig. 3. Training time and the number of SVs in Space shuttle problem

trained by the traditional method. This advantage will reduce response time of the classifier and simplify implementation of the classifier in both software and hardware. We believe that our method might provide us with a promising approach to deal with large-scale pattern classification problems.

References

1. Vapnik, V.N.: Statistical Learning Theory. Wiley Interscience (1998)
2. Osuna, E., Freund, R., Girosi, F.: An Improved Training Algorithm for Support Vector Machines. In: Proceedings of IEEE NNSP'97 (1997) 276-285
3. Joachims, T.: Making Large-scale Support Vector Machine Learning Pratical. In: Schölkopf, B., Burges, C.J., Smola, A.J.(eds.), Advances in Kernel Methods-Support Vector Learning. MIT Press (2000) 169-184
4. Lu, B.L., Ito, M.: Task Decomposition and Module Combination Based on Class Relations: A Modular Neural Network for Pattern Classification. IEEE Transaction on Neural Networks, **10** (1999) 1244-1256
5. Lu, B.L., Wang, K.A., Utiyama, M., Isahara, H.: A Part-versus-part Method for Massively Parallel Training of Support Vector Machines. In: Proceedings of IJCNN'04. Budapest, Hungary (2004) 735-740
6. Schwaighofer, A., Tresp, V.: The Bayesian Committee Support Vector Machine. In: Dorffner, G., Bischof, H., and Hornik, K.(eds.), Proceedings of ICANN 2001. Lecture Notes in Computer Science, Springer Verlag, **2130** (2001) 411-417
7. Schölkopf, B., Burges, C., and Vapnik, V.N.: Extracting Support Data for a Given Task. In: Proceedings of the First International Conference on Knowledge Discovery & Data Mining. Menlo Park, CA (1995) 252-257
8. Syed, N.A., Liu, H., Sung, K.K.: Incremental Learning with Support Vector Machines. In: Proceedings of the Workshop on Support Vector Machines at the International Joint Conference on Artificial Intelligence. Stockholm, Sweden (1999)
9. Wen, Y.M. and Lu, B.L.: A Cascade Method for Reducing Training Time and the Number of Support Vectors. In: Proceedings of ISNN2004. Dalian, China. Lecture Notes in Computer Science, Springer Verlag, **3173** (2004) 480-485
10. Blake, C.L., and Merz, C. J.: UCI. In: ftp://ftp.ics.uci.edu/pub/machine-learning-databases (1998)
11. Ke, H.X. and Zhang, X.G.: Editing Support Vector Machines. In: Proceedings of IJCNN'01. Washington, USA (2001) 1464-1467

Task Decomposition Using Geometric Relation for Min-Max Modular SVMs*

Kaian Wang, Hai Zhao, and Baoliang Lu

Department of Computer Science and Engineering, Shanghai Jiao Tong University,
1954 Hua Shan Rd., Shanghai 200030, China
blu@cs.sjtu.edu.cn

Abstract. The min-max modular support vector machine (M^3-SVM) was proposed for dealing with large-scale pattern classification problems. M^3-SVM divides training data to several sub-sets, and combine them to a series of independent sub-problems, which can be learned in a parallel way. In this paper, we explore the use of the geometric relation among training data in task decomposition. The experimental results show that the proposed task decomposition method leads to faster training and better generalization accuracy than random task decomposition and traditional SVMs.

1 Introduction

Support vector machines (SVMs)[1] have been successfully applied to various pattern classification problems, such as handwritten digit recognition, text categorization and face detection, due to their powerful learning ability and good generalization performance. However, SVMs require to solve a quadratic optimization problem and cost training time that are at least quadratic to the number of training samples. Therefore, to learn a large-scale problem by using traditional SVMs is a hard task.

In our previous work, we have proposed a part-versus-part task decomposition method[2, 3] and developed a new modular SVM for solving large-scale pattern classification problems[4], which called min-max modular support vector machines (M^3-SVMs). Two main advantage of M^3-SVMs over traditional SVM is that massively parallel training of SVMs can be easily implemented in cluster systems or grid computing systems, and large-scale pattern classification problems can be solved efficiently.

In this paper, we explore the use of the geometric relation among training data in task decomposition, and try to investigate the influence of different task decomposition methods to the generalization performance and training time.

* This work was supported in part by the National Natural Science Foundation of China via the grants NSFC 60375022 and NSFC 60473040, as well as Open Fund of Grid Computing Center, Shanghai Jiao Tong University.

J. Wang, X. Liao, and Z. Yi (Eds.): ISNN 2005, LNCS 3496, pp. 887–892, 2005.

2   Min-Max Modular Support Vector Machine

Let $\mathcal{X}^+$ and $\mathcal{X}^-$ be the given positive and negative training data set for a two-class problem $\mathcal{T}$,

$$\mathcal{X}^+ = \{(x_i^+, +1)\}_{i=1}^{l^+}, \ \ \mathcal{X}^- = \{(x_i^-, -1)\}_{i=1}^{l^-} \tag{1}$$

where $x_i \in \mathbf{R}^n$ is the input vector, and l^+ and l^- are the total number of positive training data and negative training data of the two-class problem, respectively.

According to [4], $\mathcal{X}^+$ and $\mathcal{X}^-$ can be partitioned into N^+ and N^- subsets respectively,

$$\mathcal{X}_j^+ = \{(x_i^{+j}, +1)\}_{i=1}^{l_j^+}, \ \ for \ j = 1, \ldots, N^+ \tag{2}$$

$$\mathcal{X}_j^- = \{(x_i^{-j}, -1)\}_{i=1}^{l_j^-}, \ \ for \ j = 1, \ldots, N^- \tag{3}$$

where $\cup_{j=1}^{N^+}\mathcal{X}_j^+ = \mathcal{X}^+$, $1 \le N^+ \le l^+$, and $\cup_{j=1}^{N^-}\mathcal{X}_j^- = \mathcal{X}^-$, $1 \le N^- \le l^-$.

After decomposing the training data sets $\mathcal{X}^+$ and $\mathcal{X}^-$, the original two-class problem $\mathcal{T}$ is divided into $N^+ \times N^-$ relatively smaller and more balanced two-class sub-problems $\mathcal{T}^{(i,j)}$ as follows:

$$(\mathcal{T}^{(i,j)})^+ = \mathcal{X}_i^+, \ \ (\mathcal{T}^{(i,j)})^- = \mathcal{X}_j^- \tag{4}$$

where $(\mathcal{T}^{(i,j)})^+$ and $(\mathcal{T}^{(i,j)})^-$ denote the positive training data set and the negative training data set of subproblem $\mathcal{T}^{(i,j)}$ respectively.

In the learning phase, all the two-class sub-problems are independent from each other and can be efficiently learned in a massively parallel way.

After training, the $N^+ \times N^-$ smaller SVMs are integrated into a M^3-SVM with N^+ MIN units and one MAX unit according to two combination principles [3, 4] as follows,

$$T^i(x) = \min_{j=1}^{N^-} T^{(i,j)}(x) \ \text{ for } i = 1, \ldots, N^+ \text{ and } T(x) = \max_{i=1}^{N^+} T^i(x) \tag{5}$$

where $T^{(i,j)}(x)$ denotes the transfer function of the trained SVM corresponding to the two-class subproblem $\mathcal{T}^{(i,j)}$, and $T^i(x)$ denotes the transfer function of a combination of N^- SVMs integrated by the MIN unit.

3   Task Decomposition Using Geometric Relation

M^3-SVM needs to divide training data set into several sub-sets in the first step. How to divide training data set effectively is an issue which needs to investigate furthermore. Although dividing training data set randomly is a simple and straightforward approach, the geometric relation among the original training data may be damaged[3]. The data belonging to a cluster may be separated into different sub-sets, and the data in center of a cluster may be moved to the

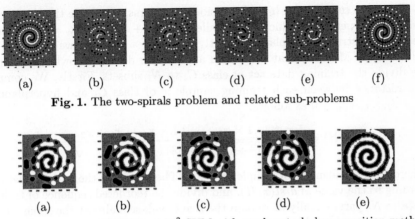

(a)   (b)   (c)   (d)   (e)   (f)

Fig. 1. The two-spirals problem and related sub-problems

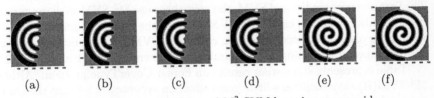

(a)   (b)   (c)   (d)   (e)

Fig. 2. The response of SVMs and M³-SVM with random task decomposition method

boundary for the new sub-problems. From SVM's point of view, some data will become support vectors, so the boundary which is vital to SVM will be changed.

A two dimensional toy problem is depicted in Fig.1(a). The training data set of each class is randomly divided into two subsets. Four sub-problems shown in Figs.1(b) through 1(e) are generated by combining these two subsets. The response of SVMs (Rbf kernel with $\sigma = 0.5, C = 1$)corresponding to each of the four sub-problems are shown in Figs.2(a) through 2(d), and the response of the M³-SVM is shown in Fig. 2(e). We can see that the final decision boundary is not very smooth.

On the other hand, the training data set of each class is divided along the y-axis as shown in Fig.1(f). The response of SVMs corresponding to each of the four sub-problems are shown in Figs.3(a) through 3(d), respectively. The benefit of task decomposition using geometric relation is quite obvious. The function of each SVM is quite clear. The SVMs in Fig.3(b) and Fig.3(c) determine which part of the space the test data x belongs to. Training these two SVMs is very easy because the training data are linearly separable. The SVMs in Fig.3(a) and Fig.3(d) judge which class that test data x belongs to. The response of M³-SVM is shown in Fig.3(e), and is smoother than that in Fig.2(e). If the training data near to the y-axis are included in both the two subsets of each class, a smoother decision boundary will be obtained as shown in Fig.3(f).

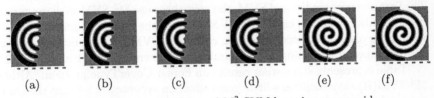

(a)   (b)   (c)   (d)   (e)   (f)

Fig. 3. The response of SVMs and M³-SVM by using space-grid

For more complicated high-dimensional problems, we divide the input space using the hyperplanes which are parallel with $z_1 + z_2, \ldots, +z_n = 0$. We never need to construct hyperplane explicitly by using a trick, so we still keep the advantage of M^3-SVM that don't need any prior or domain knowledge. Suppose we divide the training data set of class C_i to N_i subsets. Firstly, We compute the distance between each training sample x of class C_i and hyperplane H: $z_1 + z_2, \ldots, +z_n = 0$ as follows,

$$dist(x, H) = \frac{1 \times x_1 + 1 \times x_2, \ldots, +1 \times x_n}{\sqrt{1^2 + 1^2, \ldots, +1^2}} = \frac{x_1 + x_2, \ldots, +x_n}{\sqrt{n}} \qquad (6)$$

where x_i is the element of sample vector x. Then, we sort the training data according to the value of $dist(x, H)$, and divide the reordered sequence of training data to N_i parts equally to remain the size of sub-sets almost the same. Different from the toy problem, the hyperplanes used to divide the training data of different class is different for more general problems.

4   Experiments

In this section, we present experimental results on the Forest CoverType and banana data sets from UCI[5] to compare M^3-SVMs with traditional SVMs, as well as M^3-SVMs with different task decomposition methods.

Forest CoverType data set has seven classes, including 581012 samples, and the feature dimension is 54. We firstly normalize the original data in the range [0,1], and then randomly select 60% of the total data as training data, and the remainder as test data. Banana data set is a binary problem including 40000 training data and 490000 test data. All the simulations were done on an IBM p690.

Table 1. Three ways of decomposing the Forest CoverType problem

#	Num. of subsets			# classifier
	C_1	C_2	others	
A_1	2	2	1	27
A_2	3	2	1	41
A_3	4	3	1	57

Table 2. Three ways of decomposing the Banana problem

#	Num. of subsets		# classifier
	C_1	C_2	
B_1	2	1	2
B_2	2	2	4
B_3	3	2	6

We perform seven different experiments on the two data sets, respectively. The first one uses traditional SVM, the next three use M^3-SVM with random task decomposition method, called M^3-SVM(R), and the last three employ M^3-SVM with hyperplane task decomposition, called M^3-SVM(H). The original problems are decomposed according to the parameters shown in Table 1 and Table 2.

Table 3 presents simulation results on Forest CoverType problem. Although M^3-SVM(R) may sacrifice a little generalization accuracy, it reduces the training time in both serial and parallel ways. From Table 3, we can see M^3-SVM(H) can

Table 3. Results on Forest CoverType problem, where $\sigma = 0.25$ and $C = 128$

Method	# classifier	CPU Time(h.) parallel	serial	Speed up parallel	serial	Correct rate(%)	#SV
SVM	21	122.88	133.10	-	-	93.04	82719
M³-SVM(R)	27	61.29	128.12	2.00	1.04	92.57	98359
	41	19.94	120.75	6.16	1.10	92.48	118122
	57	10.40	117.32	11.82	1.13	92.35	133965
M³-SVM(H)	27	33.63	72.31	3.65	1.84	93.09	84272
	41	11.17	38.55	11.00	3.45	93.19	88049
	57	5.34	24.10	23.01	5.52	93.17	90579

get a better generalization performance in comparison with traditional SVMs, costs less training time, especially in a parallel way, and has less number of support vectors than M³-SVM(R). Table 4 lists the simulation results on banana problem. From Table 4, we can see that M³-SVM(H) is superior to traditional SVM and M³-SVM(R) in both generalization accuracy and training time.

Table 4. Results on Banana problem, where $\sigma = 1$ and $C = 361.2$

Method	# classifier	CPU Time(s.) parallel	serial	Speed up parallel	serial	Correct rate(%)	#SV
SVM	1	719.63	719.63	-	-	90.64	8430
M³-SVM(R)	2	322.34	644.36	2.23	1.12	89.79	8819
	4	155.28	594.55	4.63	1.21	90.67	9225
	6	102.58	577.80	7.02	1.25	90.20	8858
M³-SVM(H)	2	162.78	264.88	4.42	2.72	90.87	8419
	4	127.69	193.89	5.63	3.71	90.76	8404
	6	50.77	136.17	14.17	5.28	90.67	8539

Table 5 reports the performance of six sub-problems which are obtained by dividing training data sets belonging to class C_1 and class C_2. $T_{12}^{(i,j)}$ denotes a SVM corresponding to the sub-problem whose training data are from the ith sub-set of class C_1 and the jth sub-set of C_2, and T_{12} is combined from $T_{12}^{(i,j)}$ with three MIN units and one MAX unit. From Table 5, we see that the six problems in column "Random" are very similar in correct rate, training time, and the number of SVs. However, the other six problems in the column "Hyperplane" are quite different in the aspects of correct rate, training time, and the number of SVs. We can imagine that the 1st sub-set of class C_1 and the 2nd sub-set of class C_2 are located in different parts of the input space, and therefore it is easy to train $T_{12}^{(1,2)}$, and have a small number of SVs. Although the correct rate of each problem is quite low, the correct rate of whole T_{12} is high. This phenomenon can be explained as follows: each SVM in the "Hyperplane" column is responsible for one part of the input space and cooperates with each other to predict the whole input space.

Table 5. Subproblems of C_{12} of Forest CoverType problem

Task	#data		Random			Hyperplane		
	#pos.	#neg.	rate(%)	#SV	time(h.)	rate(%)	#SV	time(h.)
$T_{12}^{(1,1)}$	56660	63552	92.04	27027	19.00	67.04	19519	10.93
$T_{12}^{(1,2)}$	56660	63552	92.03	26704	18.32	56.42	2522	0.13
$T_{12}^{(2,1)}$	56660	63552	92.08	27163	19.03	66.81	14946	7.56
$T_{12}^{(2,2)}$	56660	63552	92.18	26977	18.52	69.35	9557	3.72
$T_{12}^{(3,1)}$	56660	63552	92.27	27406	19.34	49.47	2380	0.11
$T_{12}^{(3,2)}$	56660	63552	92.29	27214	19.94	64.36	18632	11.17
T_{12}	169980	127104	93.49	83996	114.15	94.19	61458	33.62

5   Conclusions

We have proposed a new task decomposition method using geometric relations among training data for M^3-SVM. We also have compared our method with random decomposition approach. From experiments results, we can draw the following conclusions. a) Although M^3-SVM have a few more number of support vectors, training M^3-SVM is faster than traditional SVM in both serial and parallel way. b) The proposed task decomposition method improves the performance of M^3-SVM in the aspects of training time, the number of support vectors, and generalization accuracy. It's worth noting that M^3-SVM with our proposed task decomposition method is superior to traditional SVM in both training time and generalization performance. As a future work, we will analyze the effectiveness of the decomposition method based on geometric relation theoretically.

References

1. Cortes, C., Vapinik, V.N.: Support-vector Network. Machine Learning, **20** (1995) 273-297
2. Lu, B.L., Ito, M.: Task Decomposition Based on Class Relations: a Modular Neural Network Architecture for Pattern Classification. In: Mira, J., Moreno-Diaz, R., Cabestany, J.(eds.), Biological and Artificial Computation: From Neuroscience to Technology, Lecture Notes in Computer Science. Vol. 1240. Springer (1997) 330-339
3. Lu, B.L., Ito, M.,: Task Decomposition and Module Combination Based on Class Relations: a Modular Neural Network for Pattern Classification. IEEE Transactions on Neural Networks, **10** (1999) 1244 -1256
4. Lu, B.L., Wang, K.A., Utiyama, M., Isahara, H.: A Part-versus-part Method for Massively Parallel Training of Support Vector Machines. In Proceedings of IJCNN'04, Budapast, **25-29** (2004) 735-740
5. Blake, C.L., Merz, C.J.: UCI Repository of Machine Learning Databases. In: ftp://ftp.ics.uci.edu/pub/machine-learning-databases(1998)

A Novel Ridgelet Kernel Regression Method

Shuyuan Yang, Min Wang, Licheng Jiao, and Qing Li

Institute of Intelligence Information Processing, Xidian University
Xi'an, Shaanxi 710071, China
syyang@xidian.edu.cn

Abstract. In this paper, a ridgelet kernel regression model is proposed for approximation of multivariate functions, especially those with certain kinds of spatial inhomogeneities. It is based on ridgelet theory, kernel and regularization technology from which we can deduce a regularized kernel regression form. Using the objective function solved by quadratic programming to define a fitness function, we adopt particle swarm optimization algorithm to optimize the directions of ridgelets. Theoretical analysis proves the superiority of ridgelet kernel regression for multivariate functions. Experiments in regression indicate that it not only outperforms support vector machine for a wide range of multivariate functions, but also is robust and quite competitive on training of time.

1 Introduction

In Machine Learning (ML), many problems can be reduced to the tasks of multivariate function approximation (MVFA) [1]. Depending on the community involved, it goes by different names including nonlinear regression, function learning etc. As we all know, approximation of a function from sparse samples is ill-posed, so one often assumes the function to be smooth to obtain a certain solution. This prior specification is reasonable in theory, but in practical applications such as industrial control, fault detection and system predicting, many systems are very complex MIMO (multi-input and multi-output) systems. Their equivalent models are not smooth MVFAs but MVFAs with spatial inhomogeneities, where classical mathematical methods fail in approximation.

Reconstructing a function by a superposition of some basis functions is a very inspiring idea on which many regressors are based, such as Fourier Transform (FT), Wavelet Transform (WT), Projection Pursuit Regression (PPR) [2] and Feed-forward Neural Network (FNN) [3]. FT and WT cannot avoid the 'curse of dimensionality' for using fixed basis in approximation, while PPR and FNN can bypass it by adaptively selecting basis. But PPR converges slowly and FNN has some disadvantages such as overfitting, slow convergence and too much reliance on experience.

For the task of MVFA, the most important thing is to find a good basis in high dimension. Though wavelet is the best basis in available basis functions and is able to process 1-D singularity efficiently, its good property cannot be extended to higher dimension. In 1996, *E.J. Candes* developed a new system to represent arbitrary multivariate functions by a superposition of specific ridge functions, the ridgelets [4]. Ridgelet proves to be a good basis in high dimension, and it has the optimal property for functions with spatial inhomogeneities. So ridgelet can serve as a new basis of

J. Wang, X. Liao, and Z. Yi (Eds.): ISNN 2005, LNCS 3496, pp. 893–899, 2005.

classical regressors to accomplish wider range of MVFAs, such as using ridgelet as the ridge function in PPR [5] or the activation function in FNN. However, they both adopt ERM principle, which leads to many shortcomings in learning including existence of local minimum, requirement of many samples and bad generalization etc [6].

Recently Kernel Machine (KM) has been a standard tool in ML, which has stricter mathematical foundation than NN and PPR for regression [7]. Its skill is to map the sample to a feature space using kernel functions, thus solve a linear approximation in the feature space. A main supervised algorithm in KM is support vector machine (SVM), which is established on the unique structure risk minimization (SRM) principle instead of empirical risk minimization (ERM) principle which only minimizes the training error. It is characteristic of good generalization, the absence of local minimum and sparse representation of solutions. In this paper, based on ridgelet and supervised KM, a ridgelet kernel regression model for MVFA is proposed. In the model, the minimum squared error (MSE) based on kernels and regularization technology, that is, the regularized kernel form of MSE, is adopted [8]. Employment of ridgelet can accomplish a wider range of MVFAs, and the regularized items in objective function are used to improve the generalization of solutions. Finally to get the directions of ridgelets, particle swarm optimization (PSO) is used in optimization of directions.

2   Ridgelet Kernel Regression Algorithm

2.1   Ridgelet Regression

As an extension of wavelet to higher dimension, ridgelet is a new harmonic analysis tool developed recently. It proves to be optimal for estimating multivariate regression surfaces with a speed rapider than FT and WT, especially for those exhibiting specific sorts of spatial inhomogeneities. The definition of ridgelet is as follows.

If $\psi : R^d \to R$ satisfies the condition of $K_\psi = \int (|\hat{\psi}(\xi)|^2 / |\xi|^d) d\xi < \infty$, then we call the functions $\psi_\tau (x) = a^{-1/2} \psi((u \cdot x - b)/a)$ as ridgelets. Parameter $\tau = (a, u, b)$ belongs to a space of neurons $\Gamma = \{\tau = (a, u, b), a, b \in R, a > 0, u \in S^{d-1}, \|u\| = 1\}$, where a, u, b are the scale, orientation and location of ridgelet. Denote the surface area of unit sphere S^{d-1} as σ_d; du is the uniform measure on S^{d-1}. Any multivariate function $f \in L^1 \cap L^2(R^d)$ can be expanded as a superposition of ridgelets:

$$f = c_\psi \int <f, \psi_\tau > \psi_\tau \mu(d\tau) \sigma_d da / a^{d+1} du db .$$

For function $Y=f(x):R^d \to R^m$, it can be divided into m mappings of $R^d \to R$. Selecting ridgelet as the basis, we get such an approximation equation:

$$\hat{y}_i = \sum_{j=1}^{N} c_{ij} \psi((u_j \cdot x - b_j)/a_j) \quad (\hat{Y} = [\hat{y}_1, \cdots, \hat{y}_m]; x, u_j \in R^d; \|u_j\|^2 = 1, i = 1, .., m)$$

where c_{ij} is the superposition coefficient of ridgelets. From above we can see ridgelet is a constant on spatial lines $t=u \cdot x$, which is responsible for its great capacity in dealing with line-like and hyperplane-like singularities. Moreover, this good characteristic can be extended to curvilinear singularity by a localized ridgelet.

2.2 Ridgelet Kernel Regression Model

SVM is a supervised KM method whose strength lies in its implement of SRM principle instead of ERM, and the equivalence of its training to solving a linear constrained protruding quadratic programming problem[9]. To get a ridgelet regressor with better generalization, in this section we used the SRM principle of SVM and proposed a kernel-based regressor with ridgelet being the kernel function.

Given a pair of sample set $S=\{(x_1, y_1),\ldots,(x_p, y_p)\}(x_i \in R^d, y_i \in R, i=1,..,P)$ generated from an unknown model $y = f(x)$, a regression task is to reconstruct f from **S**. Denote $X = [x_1,..,x_P], Y = [y_1,..,y_P]$ and ridgelet ψ_r. Let samples X firstly go through the directions of l ridgelets to get $R = [r_1,..,r_P]^T$ with each $r_i = [r_{i1},r_{i2},..,r_{il}]$ and $r_{ij} = u_j \cdot x_i$ $(i=1,..,P; j=1,..,l)$. Then the ridgelet regression of $R^d \rightarrow R$ becomes a $R^l \rightarrow R$ mapping using wavelet ψ. Considering this wavelet mapping, we construct such a linear function with weight w and threshold□for the linear regressor in feather space: $\hat{f}(r) = \sum w \cdot \psi(r) + \beta$. According to the reproducing kernel Hilbert space, the solution to this problem is in the space formed by the samples, i.e., w is a linear superposition of all the samples: $w = \sum_{i=1}^{l} a_i \psi(r_i)$. Denote the kernel function as $K(r_i, r_j) = \psi(r_i)\psi(r_j)$, the value of kernel equals to the inner product of two vectors r_i, r_j in the feather space $\psi(r_i)$ and $\psi(r_j)$. As shown in Fig.1, such an estimating equation in the feather space can be written as:

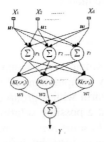

Fig. 1. Ridgelet kernel regression model.

$$\hat{f}(r) = \sum_{i=1}^{l} a_i \psi(r_i)\psi(r) + \beta = \sum_{i=1}^{l} a_i K(r_i, r) + \beta \qquad (1)$$

The elegance of using the kernel function is that one can deal with feather spaces of arbitrary dimension without having to compute the map $\psi(r)$ explicitly. The wavelet function $\psi(r) = \cos(1.75r)\exp(-r^2/2)$ proves to be a good kernel function[11]. Thus the constructed kernel function is: $K(r_i, r) = \cos(1.75 \times (r_i - r)/a)\exp(-\parallel r_i - r \parallel^2 /2a^2)$.

2.3 MSE Based Regularized Kernel Form

Vapnik showed that the key to get an effective solution is to control the complexity of the solution. In the context of statistical learning this leads to a new type of technique known as regularization networks or regularized kernel methods. Here we use the regularized kernel method to define a generalized estimator for the approximation. Then the solution is found by minimizing such an objective function E: $\min \ E = \sum V(y, \hat{f}(r)) + \lambda \|f\|_H^2 \ (\lambda > 0)$, where V is the loss function, H represents the Hilbert space of the hypotheses and λ is the regularization parameter. The squared loss function known as the minimum square error (MSE) is commonly used. The second term is a regularized term employed to get a specified solution with better generalization. Her we adopt such a form of $E = \frac{1}{2} \| y - \hat{f}(r) \|^2 + \frac{1}{2} \lambda \|w\|^2$ and let $e_k = y_k - w \cdot \psi(r_k) - \beta$. Exploiting this optimization problem and using the quadratic programming to represent the MSE based regularized kernel form:

$$\begin{cases} \min \quad E(w, \lambda, \beta) = \frac{1}{2} \sum_{k=1}^{l} \|e_k^2\| + \frac{\lambda}{2} \|w\|^2 \\ \text{s.t.} \quad y_k = w \cdot \psi(r_k) + \beta + e_k, \ k = 1,..,l \end{cases} \quad (2)$$

Finally define the *Lagrange* function as $L = E(w, \lambda, \beta) - \sum_{k=1}^{l} \mu_k [w \cdot \psi_\lambda(r_k) + \beta - y_k + e_k]$.

Then the optimal solution to this problem is:

$$\begin{aligned} & \partial L / \partial w = 0 \Rightarrow w = \sum_{k=1}^{l} \mu_k \psi(r_k), \quad \partial L / \partial \beta = 0 \Rightarrow \sum_{k=1}^{l} \mu_k = 0, \\ & \partial L / \partial e_k = 0 \Rightarrow \mu_k = \lambda e_k, \qquad \partial L / \partial \mu_k = 0 \Rightarrow w \cdot \psi(r_k) + \beta - y_k + e_k = 0 \end{aligned} \quad (3)$$

So as long as the directions of ridgelet functions are determined, this regression problem can be solved using the above model.

2.4 Optimization of Directions of Ridgelets

In this section, PSO algorithm is employed to obtain the directions of ridgelets. As we all know, PSO is an evolutionary computation technique inspired by social behavior of bird flocking[11]. It has a swarm with a group of particles, each of which has a speed and a position to represent a possible solution. In each iteration, the particles change their positions according to some recorded best individuals to make the whole swarm move towards the optimum little by little, so it is characteristic of rapid searching in solving an optimization problem. Here we applied it to our problem.

Firstly define the reciprocal value of object function $1/E$ as the fitness function. Since the directional vector of ridgelet $\|u\|=1$, the direction $u=[u_1,.., u_d]$ can be described using $(d-1)$ angles $\theta_1,...,\theta_{d-1}: u_1 = \cos\theta_1, u_2 = \sin\theta_1 \cos\theta_2,.., u_d = \sin\theta_1 \sin\theta_2...\sin\theta_{d-1}$. Each particle $\vec{p}^i = \{\theta_m^n\}^i \ (i=1,...,M, m=1,..,d-1, n=1,..l)$ records a group of angles. Our goal is to search for the optimal $l(d-1)$ angles which can maximum the fitness function. So the ridgelet kernel regression based on PSO algorithm (PSO-RKR) is:

Step 1: Init iteration times $t=0$, init a swarm P with M particles $P(0) = \{\vec{p}^1(0),...,\vec{p}^M(0)\}$;

Step 2: Compute the corresponding directions $U^i=[u_1,...u_l]^i$ for each particle $\vec{p}^i$;

Step 3: Each U^i determines the direction of l ridgelets; use quadratic programming to approximate the input samples; compute its fitness, i.e, the reciprocal value of the solved objective function $1/E$. Repeat step 2,3 for M times;

Step 4: Store the best particle in the swarm which has the maximum fitness, and judge the stop condition. If t exceeds the given maximum iterations T or the fitness function is large than $1/\varepsilon$, stop, else go on;

Step 5: Denote $\bar{p}$ as a particle with speed $\bar{v}$, $p_{groupbest}$ and $p_{selfbest}$ are the best particles met by the swarm and the particle. Update the swarm using: $\bar{p}(t+1) = \bar{p}(t) + \bar{v}(t)$, $\bar{v}(t+1) = c_1 \times \bar{v}(t) + c_2 \times r(0,1) \times (\bar{p}_{selfbest}(t) - \bar{p}(t)) + c_3 \times r(0,1) \times (\bar{p}_{groupbest}(t) - \bar{p}(t))$. Here r(0,1) is a random number in [0,1]. c1, c2, c3 represent the confidence of this particle on itself, its experience and its neighbors respectively, and they satisfiy c1+c2+c3=1.

Step 6: A new swarm P($t+1$) is thus obtained, $t=t+1$, go to step 2.

3 Simulation Experiments

This section shows some experimental results of our proposed PSO-RKR. Comparisons of the obtained result with SVM are also given.

1) 2-D function with pointlike singularity: A 2-D step function is firstly considered. The numbers of training and test sets are 25 and 400. Three models Gaussian kernel SVM (GSVM), wavelet kernel SVM(WSVM) and PSO-RKR are considered. The root mean squared error is used to estimate the approximation result. Fig.2 shows the approximation error reached by the methods under the same condition. To give a just result, PSO-RKR was run several times with different populations and an average result of 20 tests was given finally. As shown in Fig.2, WSVM performs better than GSVM, and PSO-RKR was superior to both of them.

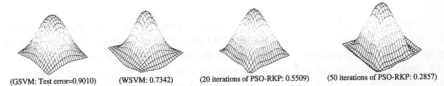

(GSVM: Test error=0.9010)   (WSVM: 0.7342)   (20 iterations of PSO-RKP: 0.5509)   (50 iterations of PSO-RKP: 0.2857)

Fig. 2. Comparison of the proposed PSO-RKR with GSVM, WSVM. The SVM package used was LIBSVM[12]. a=1, $\varepsilon =10^{-8}$, M=5, T=100, l=6,c1=0.1,c2=0.1,c3=0.8.

2) Functions with linear and curvilinear singularities: Then we have a test on other kinds of functions including smooth and singular functions. For F1-F3, the number of training sets is 49, while for D=5,D=8 of F4 are100 and 144. 400 samples are used for testing. The average error(A) and minimum error(M) of 20 tests are given in Table 1, where we see that PSO-RKR obtain more accurate results for all the functions. It should be pointed out that the additional iteration of PSO increases the consumed time in some sense. However, the direction of ridgelet maps the samples to low flow, which lows the computation complexity of the subsequent quadric programming remarkably in both time and space.

3) *Robustness to noise in the training data:* The above experiments have been performed with ideal training sets. Nevertheless, data are usually defected by some noise in practical. To give a further insight of PSO-RKR, a 5% of white noise was added to the training data of F1 and F3 in 2). The results in Table 2 show the robustness of the proposed method facing noisy data with different l. As the number of ridgelets l increases, the data are better modeled even in the existence of noise.

Table 1. Approximation results for seven functions F1-F4.

F	Error	GSVM	WSVM	PSO-RKR(A and M)		Expression
F1	Train	0.0009701	2.2764e-10	4.2470e-19	3.5520e-19	$f(x_1,x_2)=\begin{cases}4-x_1^2-x_2^2 & x_1+4x_2<1.2\\0 & otherwise\end{cases}$
	Test	1.5928	1.4641	1.1583	0.9214	
F2	Train	4.9328e-5	1.0865e-11	5.5366e-20	3.9046e-20	$f(x_1,x_2)=\begin{cases}\sqrt{x_1^2+(x_2^2+0.5)^2} & 3x_1+x_2>1\\1-0.2x_1^2-x_2^2 & x_1+2x_2<0.5\\0 & otherwise\end{cases}$
	Test	0.4074	0.3759	0.3271	0.2913	
F3	Train	1.552e-5	6.9517e-12	6.7489e-20	4.4481e-20	$f(x_1,x_2)=\begin{cases}e^{-(x_1^2+x_2^2)} & x_2\geq x_1^2\\0 & otherwise\end{cases}$
	Test	1.2543	1.2477	1.2232	0.9708	
F4 (D=5)	Train	3.7193e-1	1.0440e-2	5.3685e-4	8.3234e-5	$f=\sum_{i=1}^{D}\left[100(x_i^2-x_{i+1})+(x_i-1)^2\right]$
	Test	2.7495	2.0366	1.5153	1.0866	
(D=8)	Train	9.4860e-1	8.8913e-2	1.9503e-4	1.2311e-4	
	Test	3.0185	2.4565	1.4447	1.2057	

Table 2. Approximation results for noisy functions F1and F3. (Average result of 20 tests).

l	PSO-RKR for F1 (Train and test)		PSO-RKR for F3(Train and test)	
6	5.9772e-019	1.4548	9.3861e-020	0.8009
10	3.8687 e-019	1.0283	3.4752e-020	0.2937
15	7.5185e-020	0.9824	2.2823e-020	0.1888

4 Conclusions

Starting from the problem of MVFA, we propose a regression model based on ridgelet theory and kernel technology. It uses PSO algorithm to optimize the directions of ridgelets and can represent a wide range of functions with improved generalization. An important feather of the proposed approach is that it is able to efficiently approximate functions with certain kinds of spatial inhomogeneities with a robust and time-saving behavior, in addition to approximation of high dimensional functions.

References

1. Lorentz, G.G., Golitschek, M.V., Makovoz, Y.: Constructive Approximation, Advanced Problems. New York: Springer-Verlag. (1996)
2. Friedman, J.H.: Projection Pursuit Regression. J.Amer.Statist.Assoc, **76** (1981) 817-823
3. Cybenko, G.: Approximation by Superpositions of a Sigmoidal Function. Math. Control Signals Systems, **2** (1989) 303-314
4. Rakotomamonjy, A., Ridgelet Pursuit: Application to Regression Estimation, Technical Report, Perception Systèmes Information, ICANN (2001)
5. Candes, E. J.: Ridgelets: Theory and Applications. Dissertation, Stanford University (1998)
6. Vu, V.H.: On the Infeasibility of Training Neural Networks with Small Mean-squared error. IEEE Transactions on Information Theory, **44** (1998) 2892-2900

7. Gasser, T., Muller, H.G.: Estimating Regression Functions and Their Derivatives by the Kernel Method, Scandinavian Journal of Statistics,**11** (1984) 171-185
8. Xu, J.H,: Regularized Kernel Forms of Minimum Square Methods. ACTA AUTOMATIC SINICA, **30** (2004) 27-36
9. Vapnik, V.: Statistical Learning Theory. Wiley, New York (1988)
10. Zhang, L., Zhou, W.D., Jiao, L.C.: Wavelet Support Vector Machine. IEEE Trans. On Systems, Man, and Cybernetics. Part B: Cybernetics. **34** (2004)
11. Kennedy, J., Eberhart, R.C.: Particle Swarm Optimization. Proc. IEEE int'l Conf. on Neural Networks IV IEEE service center, Piscataway, NJ (1995) 1942-1948
12. Chang, C.C., Lin, C.J.: LIBSVM: A Library for Support Vector Machines. http://www.csie.ntu.edu.tw/~cjlin/libsvm (2001)

Designing Nonlinear Classifiers
Through Minimizing VC Dimension Bound

Jianhua Xu

Department of Computer Science, Nanjing Normal University, Nanjing 210097, China
xujianhua@njnu.edu.cn

Abstract. The VC dimension bound of the set of separating hyperplanes is evaluated by the ratio of squared radius of smallest sphere to squared margin. Choosing some kernel and its parameters means that the radius is fixed. In SVM with hard margin, the ratio is minimized through minimizing squared 2-norm of weight vector. In this paper, a bound for squared radius in the feature space is built, which depends on the scaling factor of RBF kernel and the squared radius bound in the input space. The squared 2-norm of weight vector is described as a quadratic form. Therefore, a simple VC dimension bound with RBF kernel is proposed for classification. Based on minimizing this bound, two constrained nonlinear programming problems are constructed for the linearly and nonlinearly separable cases. Through solving them, we can design the nonlinear classifiers with RBF kernel and determine the scaling factor of RBF kernel simultaneously.

1 Introduction

In statistical learning theory, the VC dimension plays an important role in various upper bounds of generalization ability or error [1], [2], [3]. These upper bounds consist of two terms, i.e. empirical risk and VC confidence. For fixed size of training set, as VC dimension increases, empirical risk decreases and VC confidence increases. Structural risk minimization principle [1], [2] suggests us to minimize some upper bound (sum of two terms) rather than empirical risk only to design an optimal learning machine for classification or regression. But it is very difficult to realize structural risk minimization principle in numerical computation.

For pattern classification, the VC dimension of the set of Δ-margin separating hyperplanes is bounded by $\min\left(\left[R^2/\Delta^2\right],m\right)+1$, where R is radius of smallest sphere containing all training data, Δ is margin which is the minimal distance between the hyperplane and the closest samples, and m denotes dimension of space.

When some kernel and its parameters are given previously, for given training set the radius calculated by a quadratic programming [2], [4] is fixed. In SVM with hard margin, $\Delta = 1/\|\mathbf{w}\|_2$ and the empirical risk is zero. Therefore VC confidence or VC dimension bound is minimized via minimizing 2-norm of weight vector only.

The success of SVM must depend on the optimal choice of several parameters that affect the generalization ability, i.e. kernel parameters and penalty parameter. In [2], [4], [5], the quantity $R^2\|\mathbf{w}\|_2^2$ is directly referred to as VC dimension bound or radius margin bound. Vapnik suggests us to choose kernel parameters for SVM according to

J. Wang, X. Liao, and Z. Yi (Eds.): ISNN 2005, LNCS 3496, pp. 900–905, 2005.

its minimum [2]. The optimal degree of polynomial kernel for USPS dataset is successfully estimated in a one-dimension search [6]. In [7], [8], [9], [10], this quantity and its modified versions are considered as objective functions to detect penalty parameter and/or kernel parameters (e.g., the width of RBF kernel). Furthermore they are minimized by gradient descent or quasi-Newton method. But a drawback in their work is that SVM and a quadratic programming for radius have to be solved repeatedly.

In this paper, a bound of squared radius in the feature space is built. It depends on the scaling factor of RBF kernel and the squared radius bound in the original input space. The squared 2-norm of weight vector is described as a quadratic form with kernel matrix. So a simple VC dimension bound based on RBF kernel for classification is built. For the linearly and nonlinearly separable cases, two constrained nonlinear programming problems are designed according to this bound. It is attractive that we can train the nonlinear classifiers and determine the scaling factor of RBF kernel simultaneously.

2 Radius and Weight Vector in the Feature Space

To estimate the VC dimension bound simply, in this section we derive a simple bound for R^2 and a simple expression for $\|\mathbf{w}\|_2^2$. Let the training sample set of two classes be $\{(\mathbf{x}_1, y_1),...,(\mathbf{x}_i, y_i),...,(\mathbf{x}_l, y_l)\}$, where $\mathbf{x}_i \in \mathfrak{R}^n$ and $y_i \in \{+1,-1\}$. In the feature space Vapnik [1] proposes the following formula to compute squared radius,

$$R^2 = \min_{\mathbf{a}} \; \max_{\mathbf{x}_i} \; (k(\mathbf{x}_i, \mathbf{x}_i) + k(\mathbf{a}, \mathbf{a}) - 2k(\mathbf{x}_i, \mathbf{a})), \tag{1}$$

where $\mathbf{a}$ is a vector in the original input space which corresponds to the center of sphere in the feature space and k is kernel function. In the original input space, it becomes,

$$R_0^2 = \min_{\mathbf{a}_0} \; \max_{\mathbf{x}_i} \; (\|\mathbf{x}_i - \mathbf{a}_0\|_2^2), \tag{2}$$

where $\|\cdot\|_2$ denotes 2-norm of vector and $\mathbf{a}_0$ is the center of sphere. If the center is fixed in advance, this squared radius is bounded by maximal Euclidean distance from the center to the far samples,

$$R_0^2 \leq \max_{\mathbf{x}_i} \; (\|\mathbf{x}_i - \mathbf{a}_0\|_2^2) = d^2. \tag{3}$$

In this paper, we define $\mathbf{a}_0 = (a_1, a_2,..., a_n)^T$ as the middle vector of training samples,

$$a_j = (\max_i(x_i^j) + \min_i(x_i^j))/2, \; i = 1,...,l, \tag{4}$$

where x_i^j implies the jth component of the ith sample. Thus the squared radius bound in the original input space can be calculated according to (3) and (4) easily.

We focus on the form of RBF kernel in this paper as follows,

$$k(x, y) = e^{-\gamma\|x-y\|_2^2}, \tag{5}$$

where $\gamma \geq 0$ is referred to as the scaling factor of RBF kernel. Thus the equation (1) can be rewritten as,

$$R^2 = \min_{\mathbf{a}} \ \max_{\mathbf{x}_i} \ (2(1 - e^{-\gamma \|\mathbf{x}_i - \mathbf{a}\|_2^2})). \tag{6}$$

Here we assume that the middle vector $\mathbf{a}_0$ corresponds to the center of sphere in the feature space, i.e. $\mathbf{a} = \mathbf{a}_0$. Then we have

$$R^2 \le \max_{\mathbf{x}_i} \ (2(1 - e^{-\gamma \|\mathbf{x}_i - \mathbf{a}_0\|_2^2})) = 2(1 - e^{-\gamma \max_{\mathbf{x}_i} \|\mathbf{x}_i - \mathbf{a}_0\|_2^2}) = 2(1 - e^{-\gamma d^2}). \tag{7}$$

In this case, we build an upper bound for squared radius, which only depends on the scaling factor of RBF kernel and the squared radius bound in the original input space.

In support vector machine and other kernel machines, the weight vector in the feature space can be represented as $\mathbf{w} = \sum_{i=1}^{l} \alpha_i \Phi(\mathbf{x}_i)$, where $\alpha_i \in \Re, i = 1, 2, \dots l$ are coefficients which describe significance of each sample in the weight vector, and $\Phi(\cdot)$ denotes some nonlinear transform form $\Re^n$ to some feature space. According to the definition of kernel function $k(\mathbf{x}_i, \mathbf{x}_j) = \Phi(\mathbf{x}_i)^{\mathrm{T}} \Phi(\mathbf{x}_j)$, the squared 2-norm of weight vector becomes

$$\|\mathbf{w}\|_2^2 = \sum_{i,j=1}^{l} \alpha_i \alpha_j k(\mathbf{x}_i, \mathbf{x}_j) = \sum_{i,j=1}^{l} \alpha_i \alpha_j e^{-\gamma \|\mathbf{x}_i - \mathbf{x}_j\|_2^2}. \tag{8}$$

3 Nonlinear Classifiers Based on Minimal VC Dimension Bound

In this section, we first build a simple VC dimension bound. For the linearly separable case in the feature space, when canonical form of constrains $y_j(\sum_{i=1}^{l} \alpha_i k(\mathbf{x}, \mathbf{x}_i) + \beta) \ge 1$ is used, we have $\Delta = \gamma_{\|\mathbf{w}\|_2}$. Thus using (7) and (8), the VC dimension bound can be described as

$$R^2 \|\mathbf{w}\|_2^2 \le 2(1 - e^{-\gamma d^2}) \sum_{i,j=1}^{l} \alpha_i \alpha_j e^{-\gamma \|\mathbf{x}_i - \mathbf{x}_j\|_2^2}. \tag{9}$$

It is noted that this bound includes the scaling factor of RBF kernel. According to this bound, we can construct a constrained nonlinear programming to design a nonlinear classifier and determine the scaling factor of RBF kernel, i.e.,

$$\begin{aligned} \min_{\alpha_i, \beta, \gamma} \ & 2(1 - e^{-\gamma d^2}) \sum_{i,j=1}^{l} \alpha_i \alpha_j e^{-\gamma \|\mathbf{x}_i - \mathbf{x}_j\|_2^2} \\ \text{s.t.} \ & y_j(\sum_{i=1}^{l} \alpha_i e^{-\gamma \|\mathbf{x}_i - \mathbf{x}_j\|_2^2} + \beta) \ge 1, \ \gamma \ge 0, \ j = 1, \dots, l \end{aligned} \tag{10}$$

To handle the nonlinearly separable case, a slightly modified form is considered by adding a penalty term for misclassified samples,

$$\begin{aligned} \min_{\alpha_i, \beta, \gamma} \ & 2(1 - e^{-\gamma d^2}) \sum_{i,j=1}^{l} \alpha_i \alpha_j e^{-\gamma \|\mathbf{x}_i - \mathbf{x}_j\|_2^2} + C \sum_{i=1}^{l} \xi_j^p \\ \text{s.t.} \ & y_j(\sum_{i=1}^{l} \alpha_i e^{-\gamma \|\mathbf{x}_i - \mathbf{x}_j\|_2^2} + \beta) \ge 1 - \xi_j, \xi_j \ge 0, \gamma \ge 0, j = 1, \dots, l \end{aligned} \tag{11}$$

where C is penalty parameter, which can control the tradeoff between the number of errors and the complexity of model, and p is a positive integer, which is taken 1 or 2 generally.

It's attractive that through solving (10) or (11) we can simultaneously determine all parameters $(\alpha_i, i = 1,...,l, \beta, \gamma)$ in the nonlinear classifier $f(\mathbf{x}) = \sum_{i=1}^{l} \alpha_i e^{-\gamma\|\mathbf{x}-\mathbf{x}_i\|_2^2} + \beta$.

Such learning machines could be referred to as nonlinear classifiers based on minimal VC dimension bound (or simply VCDC).

4   Experimental Results

The IRIS dataset is one of the most famous data sets used in statistics and machine learning. It contains fifty samples each of three types of plants (i.e., virginica, versilcolor and setosa). Each sample is described by four attributes (i.e., petal length and width, setal length and width). The setosa class is linearly separable from the other two classes. The last two attributes are used only in our experiment. We choose 135 samples for training set and the remaining 15 for test set. This classification problem is converted into three binary class problems via adjusting the class labels in order to examine our method.

In the original input space, the true squared radius 9.805 is obtained by a quadratic programming, while the squared radius bound is 9.913. For the linear case (setosa versus virginica and versicolor), we solve SVM and a quadratic programming for radius repeatedly, where the scaling factor is from 0.1 to 3.0 with a step 0.1. The VC dimension bound from SVM, true squared radius and squared radius bound are illustrated in Fig.1 (a). The minimum of VC dimension bound is $\gamma = 0.6$. The curve shapes of true squared radius and squared radius bound are similar. Fig.1 (b) and (c) show hyperplanes from SVM where $\gamma = 0.6$, and from our algorithm VCDC with $p = 1$ where the smaller factor $\gamma = 0.533$ is obtained automatically.

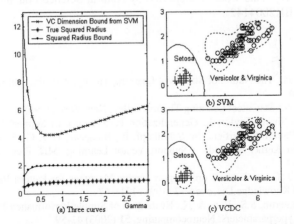

Fig. 1. (a) Three curves: VC dimension bound from SVM, true squared radius and squared radius bound versus scaling factor of RBF kernel. (b) Hyperplane from SVM corresponding to optimal scaling factor $\gamma = 0.6$. (c) Hyperplance from our classifier (VCDC) where $\gamma = 0.533$.

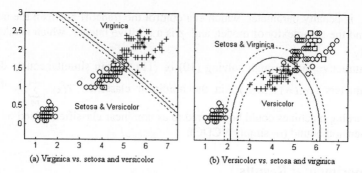

(a) Virginica vs. setosa and versicolor   (b) Versicolor vs. setosa and virginica

Fig. 2. The hyperplanes are obtained by our classifier (VCDC) for two nonlinearly separable cases. The stars and squares denote 15 test samples classified correctly.

For two nonlinearly separable cases (virginica versus setosa and versicolor, versicolor versus setosa and virginica), we obtain the hyperplanes illustrated in Fig.2. The samples denoted by stars and squares are test samples which all are classified correctly. These simple results show the efficiency of our classifiers initially.

5 Conclusions

In this paper, a simple VC dimension bound based on RBF kernel for classification is introduced, which depends on the parameter of nonlinear classifier and the scaling factor of RBF kernel. Based on this bound, we propose a constrained nonlinear optimization form for the linearly separable case. In order to handle the nonlinearly separable problem, a slightly modified version is presented by adding a penalty term for misclassified samples. Solving them, we can obtain nonlinear classifiers with RBF kernel and the scaling factor of RBF kernel simultaneously.

Our further work is to examine more benchmark databases and analyze our algorithms elaborately. This work is supported by Natural Science Foundation of Jiangsu Province (No. BK2004142).

References

1. Vapnik, V. N.: The Nature of Statistical Learning Theory. 2nd edn. Springer-Verlag, New York (1999)
2. Vapnik, V. N.: Statistical Learning Theory. Wiley, New York (1998)
3. Bartlett, P., Shawe-Taylor, J.: Generalization Performance of Support Vector Machines and Other Pattern Classifiers. In: Scholkopf, B., Buegesm, C. J. C., Smola, A. J. (eds.): Advances in Kernel Methods – Support Vector Learning. MIT Press, Cambridge MA (1998) 43-54
4. Burges, C. J. C.: A Tutorial on Support Vector Machines for Pattern Recognition. Data Mining and Knowledge Discovery, 2 (1998) 121-167
5. Duan, K., Keerthi, S.S., Poo, A. N.: Evaluation of Simple Performance Measure for Tuning SVM Hyperparameters. Neurocomputing, 51 (2003) 41-59
6. Scholkopf, B., Burges, C., Vapnik, V. N.: Extracting Support Data for a Given Task. In: Fayyad, U. M., Uthurusamy, R. (eds.): Proceedings of the First International Conference on Knowledge Discovery and Data Mining. AAAI Press, Menlo CA (1995) 252-257

7. Chapelle, O., Vapnik, V. N., Bousquet, O., Mukherjee, S.: Choosing Multiple Parameters for Support Vector Machines. Machine Learning, **46** (2002) 131-159
8. Keerthi, S. S.: Effective Tuning of SVM Hyperparameters Using Radius/Margin Bound and Iterative Algorithm. IEEE Transactions on Neural Networks, **13** (2002) 1225-1229
9. Tan, Y., Wang, J.: A Support Vector Machine with a Hybrid Kernel and Minimal Vapnik-Chervonenkis Dimension. IEEE Transactions on Knowledge and Data Engineering, **16** (2004) 385-395
10. Chung, K. M., Kao, W. C., Sun, C. L., Wang, L. L., Lin, C. J.: Radius Margin Bounds for Support Vector Machines with the RBF Kernel. Neural Computation, **15** (2003) 2643-2681

A Cascaded Mixture SVM Classifier for Object Detection

Zejian Yuan, Nanning Zheng, and Yuehu Liu

Institute of Artificial Intelligence and Robotics,
Xi'an Jiaotong University,
Xi'an 710049, China
{zjyuan,nnzheng,liuyh}@aiar.xjtu.edu.cn

Abstract. To solve the low sampling efficiency problem of negative samples in object detection and information retrieval, a cascaded mixture SVM classifier along with its learning method is proposed in this paper. The classifier is constructed by cascading one-class SVC and two-class SVC. In the learning method, first, 1SVC is trained by using the cluster features of the positive samples, then the 1SVC trained is used to collect the negative samples close to the positive samples and to eliminate the outlier positive samples, finally, the 2SVC is trained by using the positive samples and effective negative samples collected. The cascaded mixture SVM classifier integrates the merits of both 1SVC and 2SVC, and has the characters of higher detection rate and lower false positive rate, and is suitable for object detection and information retrieval. Experimental results show that the cascaded SVM classifier outperforms traditional classifiers.

1 Introduction

In recent years, many researchers have applied pattern recognition methods based on SVM to object detection and information retrieval, and have made important progress. These methods outperform traditional ones in image retrieval, document classification and spam detection [1–4]. In general,the existing methods are either two-class classifier or one-class classifier. In fact, neither of these two kinds of classifiers alone is capable of solving the object detection and information retrieval problem well.

One-class SVC (1SVC) for single class classification problem can find a decision boundary to decide whether the test samples fall inside the boundary or not, and it is not indispensable to construct a complex probability model for training samples[5, 6]. 1SVC is different from traditional reconstruction based method such as PCA and that which estimates the probability distribution of the object data based on a given model[7], but it outperforms traditional methods in such areas as object detection and media information retrieval, especially in pattern recognition in clutter background. Tax(1999) has also proposed a similar method named support vector data description(SVDD)[8]. This method has given better result in image retrieval. SVDD is not to find a super-plane, but to find a

J. Wang, X. Liao, and Z. Yi (Eds.): ISNN 2005, LNCS 3496, pp. 906–912, 2005.

super-ball with the smallest radial to contain most samples. SVDD is equivalent to one-clas SVC when the kernel function is a radial function. One-class classifiers only consider positive samples (object samples) and ignore negative samples (non-object samples). Therefor, detectors based on one-class classifier have no better performance for retrieving information in clutter background, and they would have high error rate while have high detection rate[5]. In the case of object detection and information retrieval, the positive samples and negative ones for training two-class classifier are seriously asymmetric or imbalanced data. Negative samples are richer and more abundant than positive ones, and are related to the background of positive ones. It is very difficult to describe or define negative samples, and to collect effective samples. In fact, we can not collect full and effective negative samples, which is important effect on two-class classifier. Obviously, the two-class classifier trained only using positive samples and non-full negative samples does not solve detection problem well despite that its learning method is very good.

To solve the low sampling efficiency problem of negative samples in object detection and information retrieval, a SVM cascaded mixture classifier and its learning method is proposed in this paper. The classifier is constructed by cascading one-class SVC and two-class SVC. The cascaded mixture SVM classifier integrates the merits of both 1SVC and 2SVC, and compared with the two, it has the characteristics of higher detection rate and lower false positive rate, and is suitable for object detection and information retrieval. At the same time, a fast implementation approach for 1SVC is also proposed, which improves the speed of the cascaded mixture SVM classifier.

The second part of this paper briefly introduces object detection problem in clutter background and its characteristic, and the third part introduces 1SVC and its fast implementation, and the forth part presents the structure of cascaded mixture SVM classifier and its learning method, and the last part is the analysis and discussion of the experiments.

2   Object Detection Problem

Assuming that objects exist in clutter background, object detection problem is to find the objects by using detector. Object detection can be represented as a pattern classification problem, i.e. the detector constructed by classifier can tell whether the input patterns are objects. Many detectors have been constructed by one-class classifier using only cluster features. Because one-class classifier ignores background information (see in the left of the Fig.1), the detector always has high false positive rate. In another case, the detectors can be constructed by using two-class classifier which is trained by using positive samples and non-full negative samples (see in the middle of the Fig.1). The performance of two-class classifier is significantly dependent on negative samples.

It is one effective approach to solve object detection in clutter background by combining cluster or representative features and discriminative features. So we can select one class classifier constructed by cluster features to remedy the

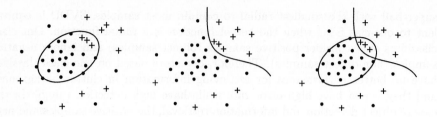

Fig. 1. Classifier for object detection: the left is one class classifier, the middle is two class classifier, and the right is a cascaded mixture classifier

shortcomings of two class classifier because of scarcity effective negative samples. The cascaded classifier constructed by 1SVC and 2SVC can integrate their merits (see in the right of the Fig.1), has higher detection rate and lower false positive rate, which will be discussed in the following section.

3   One-Class SVM Algorithm

3.1   SVM for One-Class Problem

One-Class SVC can find a super-plane in feature space to separate unlabeled data set from the origin with a maximum margin[5], which will have a good generalization[9]. In other words, we need to estimate a super-plane: $f_w(x) = \langle w, \Phi(x) \rangle$ and a threshold ρ to distinguish whether test sample satisfies $f_w(x) > \rho$, where $\Phi(\cdot)$ is the nonlinear mapping from original space to feature space. Optimal w and ρ can be obtained by solving the following quadratic program

$$\min_{w,\xi,\rho} \quad \tfrac{1}{2}\|w\|_2^2 - \rho + \tfrac{1}{vN}\sum_{i=1}^{N}\xi_i \tag{1}$$
$$s.t. \ \langle w, \Phi(x_i)\rangle \geq \rho - \xi_i, i = 1, 2, \cdots, N$$

where $\nu \in (0,1]$ is regularization parameter, which can balance complexity $\|w\|_2^2$ and the error, and ξ_i is a relaxation variables. From the optimal problem (1), the labels of training samples do not appear in inequality condition. It is different from the unsupervised learning methods.

If the dot product in feature space is defined as $\langle \Phi(x_i), \Phi(x_j)\rangle = K(x_i, x_j)$, and kernel function $K(\cdot, \cdot)$ satisfies Mercer's condition. Optimal decision function can be obtained by solving the dual optimal problem of the original problem

$$f(x) = \sum_{j \in sv} \alpha_j^* K(x_j, x) - \rho^* \tag{2}$$

Where $\alpha_j^* \geq 0$ is a optimal lagrange multiplier respondence to sample x_j, and sv is the set of index support vector.

3.2   The Fast Implementation of 1SVC

In the decision function (2) of 1SVC, owing to the positive definition of kernel function $K(\cdot,\cdot)$ and the non-negative characteristic of the lagrange multipliers, we have proposed a faster implementation method for 1SVC by using support vector one by one. In the algorithm, firstly, one support vector with biggest lagrange multiplier and threshold ρ^* are used, if the input pattern can not be classed as a object, then other support vectors will be added to distinguish input patterns. The algorithm is not finished until all support vectors are used. The implementation structure based on decision tree is shown in Fig.2.

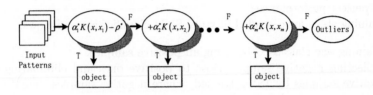

Fig. 2. The fast implementation structure of 1SVC

It is likely to classify input patterns as object by using a few support vectors, so can decrease some the kernel function computation and increase the computation efficient of 1SVC algorithm. While the dimension of input pattern is very high, the fast implementation method can improve largely the speed of 1SVC for detecting object.

4   Cascaded Mixture SVM Classifier

The cascaded mixture SVM classifier is made up of 1SVC and 2SVC by using decision tree structure. Its training process is given in Fig.3. The training process of 1SVC and 2SVC are independent in cascade mixture SVM classifier, and it has

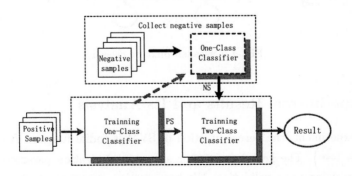

Fig. 3. The training process of cascaded mixture SVM classifier

a big freedom to select the kernel's type and its parameters in 1SVC and 2SVC because mixture classifier adopts the decision tree structure. 1SVC in the mixture classifier focuses mostly on the cluster feature of objects, and 2SVC focuses mostly on the discriminative feature between objects and non-objects. Negative samples close to positive ones can be collected from background by using the 1SVC trained by using positive samples, and these negative samples always contain important discriminative features and can improve the performances of 2SVC. The 2SVC can decrease the false positive rate of the 1SVC by removing the non-objects close to objects. At the same time, the 1SVC can also remedy the shortcomings of the 2SVC caused by non-full negative samples. Cascaded mixture classifier can combine cluster and discriminative features into a detector, so has perfect performances.

Training steps for cascaded mixture SVM classifier are as follows:

1. Training one-class SVC by using all positive samples.
2. Collecting negative samples close to positive ones and eliminating outlier positive samples by 1SVC trained, where negative samples should be from the background related to the objects.
3. Training 2SVC by using negative and positive samples collected by the 1SVC.
4. Output cascaded mixture SVM classifier.

Cascaded classifier based on decision tree is given as Fig.4. The first stage is the 1SVC which adopts the fast implementation method, and is used to receive most object patterns, and eliminate some outlier objects and most non-object patterns. The second stage is the 2SVC, and is used to remove one by one non-objects close to objects. The cascading 1SVC and 2SVC can not only improve the computation the computation efficiency of detector, but solve the problem caused by low negative sampling rate and non-full negative samples.

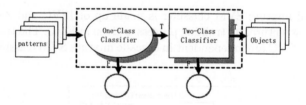

Fig. 4. Cascaded mixture classifier

5 Experimental Results and Its Analysis

In the experiment, the kernel function of SVM is radial function $K\left(x_i, x_j\right) = \exp\left(\frac{\|x_i - x_j\|_2^2}{s}\right)$. The kernel parameter s and regularization parameter are selected by using cross validation. We have carried out contrast experiment on 1SVC, 2SVC and CSVC with USPS and CBCL data set.

We select randomly 2000 training samples and 7000 test samples from USPS. 168 positive samples of digit "9" and 600 negative ones are collected from 2000 train samples. Experiment results are given is Table.1, where kernel parameter of 1SVC is $s = 11.3$, fraction is $\nu = 0.01$, the parameters of 2SVC are $s = 0.32$, regularization $C = 1$, and parameters in CSVC is $s = 0.20$, $C = 1$. Table.1 has given training time (TrTime), the number of support vector(SV), true positive rate (TPR), false positive rate (FPR) and test time (TsTime) while detecting digital '9'. The ROC of the classifiers is given in the left of Fig.5 while detecting the digit '9'. Digits detection experiment results show 1SVC can reject most non-digital '9' by only using 5 support vectors, and its rejection rate is 71.1%, true positive rate is 97.3%. At the same time, it has higher false positive rate (28.9%). To remedy the shortcomings of the 1SVC, cascaded classifier adopts 2SVC to distinguish the result of the 1SVC, and to remove non-objects. Because the numbers of patterns received by 1SVC are not quite large, 2SVC can complete object detection within a little time.

Table 1. The classifiers' performance on data set USPS and CBCL

Dataset	Classifier	TrTime(s)	SV	TPR(%)	FPR(%)	TsTime(s)
	1SVC	0.17	5	97.3	28.0	0.61
USPS	2SVC	2.25	167	93.9	2.1	2.39
	CSVC	1.20	5 + 161	95.2	1.4	1.69
	1SVC	0.29	31	92.3	19.5	2.03
CBCL	2SVC	5.25	500	74.2	0.27	12.84
	CSVC	2.50	31 + 468	76.8	0.13	5.39

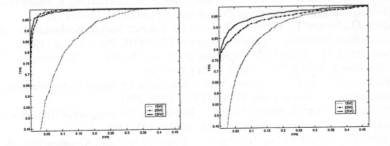

Fig. 5. The ROC of the classifiers on data set USPS (*left*) and CBCL (*right*)

We select 1000 positive samples and 1500 negative ones as training set from the train set of CBCL face data, and the remains are added into test data set. Face detection experiment results are given in Table.1, where the kernel parameters of 1SVC is $s = 13.4$, the fraction is $\nu = 0.001$, and the parameters of 2SVC is $s = 0.50$, and regularization is 1, and the parameters of the 2SVC in CSVC is $s = 0.27$ and regularization $C = 1$. The ROC of classifiers are given in the right of Fig.5. Experimental results of digital and face detection show the detector based on the cascaded mixture classifier has higher object detection rate, lower false positive rate and higher detection efficiency, especially, it have more advantages when the object samples' distribute is more complex or the background information is richer.

6 Conclusion

The cascaded mixture SVM classifier based on decision tree structure integrates the merits of 1SVC and 2SVC, and has high detection rate and low false positive rate in solving object detection,in the same time, and has higher detection efficiency. The learning method of CSVC can solve the low negative samples sampling efficiency and its non-fullness problem in object detection. Experimental results show CSVC has significant advantages in object detection and information retrieval. In future work, we will further discuss the applications of CSVC in document classification and web retrieval, and the problem of cascaded mixture one-class AdaBoost algorithm.

Acknowledgments

Z.J. Yuan was supported by National Innovation Group Foundation of China (No.60021302) and the NSF of China (No.60205001). The authors would like to thank professor X.C. Jia for contributing to the ideas in this paper.

References

1. Chen, Y.Q., Zhou X., Huang T.S.:A One-class SVM for Learning in Image Retrieval. IEEE Int Conference on Image Processing, Thessaloniki, Greece (2001)
2. Edgar, O., Robert F., Federico G.: Training Support Vector Machines: An Application to Face Detection. The proceedings of CVPR'97, Puerto Rico (1997)
3. Manevitz, L.M., Yousef, M.: One-class SVM for Document Classification. Journal of Machine Learning Research, 2 (2001) 139–154
4. Drucker, H., Wu H.D.: Support Vector Machines for Spam Categorization. IEEE Transaction on Neural Network, 10 (1999) 1048–1054
5. Scholkopf,B., Platt,J., Shawe-Taylor,J.: Estimating the Support of High-dimensional Distribution. Neural Computation, 13 (2001) 1443–1471
6. Gunnar,G., Sebastian,M., Bernhard,S., Klaus-Robert,M.: Constructing Boosting Algorithms from SVMs: An Application to One-class Classification. IEEE on Pattern Analysis and Machine Intelligence, 24 (2002) 1184–1198
7. Liu, C.J.: A Bayesian Discriminating Features Method for Face Detection. IEEE on PAMI, 25 (2003) 725–740
8. Tax,M.J., Robert,P.W., Messer Kieron: Image Database Retrieval with Support Vector Data Descriptoins. Technical Report, from http://www.boosting.org (1999)
9. Vapnik,V.: The Nature of Statistical Learning Theory. 2nd Edition, Springer-Verlag, New York (2000)

Radar High Range Resolution Profiles Feature Extraction Based on Kernel PCA and Kernel ICA*

Hongwei Liu, Hongtao Su, and Zheng Bao

National Lab of Radar Signal Processing, Xidian University,
Xi'an, Shaanxi 710071, China
{hwliu,htsu,zhbao}@xidian.edu.cn

Abstract. Kernel based nonlinear feature extraction approaches, kernel principal component analysis (KPCA) and kernel independent component analysis (KICA), are used for radar high range resolution profiles (HRRP) feature extraction. The time-shift uncertainty of HRRP is handled by a correlation kernel function, and the kernel basis vectors are chosen via a modified LBG algorithm. The classification performance of support vector machine (SVM) classifier based on KPCA and KICA features for measured data are evaluated, which shows that the KPCA and KICA based feature extraction approaches can achieve better classification performance and are more robust to noise as well, comparing with the adaptive Gaussian classifier (AGC).

1 Introduction

Radar automatic target recognition (ATR) from a returned signal has been received extensive attention from the radar technique community for decades. In particular, the wideband scattering behavior of radar target has been of considerable interest in the context of ATR. Among several kind of wideband radar target signatures, including target high range resolution profile (HRRP), synthetic aperture radar (SAR) images and inverse synthetic aperture radar (ISAR) images, target HRRP is more easier to be acquired, which makes the HRRP to be a promising signatures for radar ATR [1]-[4].

A number of radar HRRP recognition algorithms are proposed during past decades. Most of them are based on the traditional statistical modeling techniques. In the last couple of years, the theory of machine learning has developed a wide variety of novel classification algorithms. Several kernel based classifiers, including support vector machine (SVM), relevance vector machine (RVM) and kernel matching pursuit (KMP) classifier, are used for radar HRRP recognition recently [1],[2]. Besides using the kernel-based classifiers directly, to use the kernel-based approaches for feature extraction becomes a new research direction in pattern recognition application recently [5],[6]. In this paper, we will consider applying kernel principal component analysis (KPCA) and kernel independent component analysis (KICA) to radar HRRP feature extraction.

2 Radar HRRP Recognition Formulation

A HRRP is the coherent sum of the time returns from target scatterers located within a range resolution cell, which represents the distribution of the target scattering cen-

---

* This work was partially supported by the National NSFC under grant of 60302009.

J. Wang, X. Liao, and Z. Yi (Eds.): ISNN 2005, LNCS 3496, pp. 913–918, 2005.

ters along the radar line of sight. Three issues need to be considered when using HRRP for target recognition. First, the HRRP amplitude is a function of target distance, radar transmitter power, etc. Therefore, the HRRP amplitude should be normalized before performing recognition. Second, the target may exist at any position, generally the range between target and radar can't be measured exactly. To handle this problem, one can extract time-shift invariant feature from the original HRRP, or compensate the time-shift before performing classification. Finally, the targets of interest generally contain lot of scatterers with complex geometry structure, therefore, its scattered signal, HRRP is a function of target-radar orientation. A target model with the ability to describe the target scattering property of different target aspect is required.

A widely used HRRP recognition algorithm is the maximum correlation coefficient (MCC) classifier. Given a template HRRP data set contains M HRRPs denoted as $\{X_{Ti}(n), i=1,2,...,M, n=1,2,...,N\}$ and a test HRRP denoted as $X(n)$, assume they are 2-norm normalized, the MCC between the test HRRP and ith template is defined as

$$r_i = \max_\tau \int X_{Ti}(n)X(n-\tau)dn \cdot \tag{1}$$

A larger r_i means the test HRRP is more similar with the template. Generally, the average HRRPs associated with different target aspect sectors are generally used as templates in the above MCC classifier [3].

Another statistical classifier for radar HRRP recognition is the adaptive Gaussian classifier (AGC) [4]. In AGC, a Gaussian distribution model is used to represent the statistical property of HRRP and its discriminant function takes the form

$$y_i(X) = -\frac{1}{2}(X-\mu_i)^T \Sigma_i^{-1}(X-\mu_i) - \frac{1}{2}\ln|\Sigma_i| \cdot \tag{2}$$

where μ_i and Σ_i are the mean range profile and covariance matrix of ith target, respectively, which can be estimated from the training data. Generally, in order to represent the statistical property of HRRPs associated with different target aspect sectors, multiple Gaussian models are built for one target.

3 KPCA and KICA for Feature Extraction

3.1 Kernel PCA

Principal component analysis (PCA) is one of the most common statistical data analysis algorithms. However, it is difficult to apply PCA to HRRP feature extraction directly, due to the time-shift uncertainty problem, as described above. In addition, the traditional PCA is a linear algorithm, it can not extract the nonlinear structure in the data. The KPCA algorithm can realize nonlinear feature extraction via a nonlinear kernel function [5,6].

The KPCA algorithm first maps the data set $\{X_n, n=1,...,N\}$ into a high dimensional feature space via a function $\Phi(X)$, then calculates the covariance matrix

$$C = \frac{1}{N}\sum_{n=1}^{N}\Phi(X_n)\Phi(X_n)^T \cdot \tag{3}$$

The principal components are then computed by solving the eigenvalue problem as

$$\lambda V = CV = \frac{1}{N}\sum_{n=1}^{N}\Phi(X_n)^T V\Phi(X_n).\tag{4}$$

Note that V belongs to the space spanned by $\{\Phi(X_n), n = 1,..., N\}$, it can be represented as a linear combination of $\{\Phi(X_n), n = 1,..., N\}$ as $V = \sum_{n=1}^{N}\alpha_n\Phi(X_n).$ Substitute it into (4), it can be deduced that the eigenvalue problem is equivalent to $\lambda\alpha = K\alpha$, where the $(i,j)_{th}$ element of K is defined as $K_{ij} = \Phi(X_i)^T\Phi(X_j) = k(X_i, X_j).$ As in traditional PCA algorithm, the data need to be centered in the feature space. This can be done by simply substituting the kernel matrix K with $\overline{K} = K - 1_N K - K1_N + 1_N K1_N$, where 1_N is a N by N matrix with all elements equal to $1/N$. In addition, the solution should be normalized by imposing $\lambda_n(\alpha_n^T\alpha_n) = 1$ [5]. Thus the extracted feature vector can be obtained by projecting $\Phi(X)$ onto the space spanned by $\{V_m, m = 1,..., M\}$ as

$$F(X_n) = [V_1\ V_2...\ V_M]^T\Phi(X_n) = [\alpha_1\ \alpha_2...\ \alpha_M]^T[\overline{k}(X_1, X_N)\ \overline{k}(X_2, X_N)...\overline{k}(X_N, X_N)]^T.\tag{5}$$

3.2 Kernel ICA

Independent component analysis (ICA) is another statistical data analysis method. The aim of ICA is to linear transform the input data into uncorrelated components, along which the distribution of the data set is the least Gaussian. If the underlying data distribution matches with this assumption, it is expected a better classification performance can be achieved by ICA. Recently, the traditional nonlinear ICA has been extended to a nonlinear form, namely, the KICA [6]. The KICA can be realized by performing linear ICA based on the above KPCA features.

There are many iterative methods for performing ICA. In this paper, we use the FastICA algorithm proposed by Hyvarinen [7]. For the sake of simplicity, we denote the centered and whitened KPCA feature as $\{z_n, n = 1,..., N\}$. The optimal independent component direction $\mathbf{v}$ can be found by optimizing the below cost function

$$\eta(\mathbf{v}) = (E\{g(\mathbf{v}^T\mathbf{z})\} - E\{g(r)\})^2.\tag{6}$$

where $g(\bullet)$ is a nonlinear function and r is standard Gaussian variable. For radar HRRP feature extraction, we found using $g(v) = -\exp(-v^2/2)$ can achieve a better classification performance, which means that the underlying independent component of HRRP in the kernel induced feature space is more likely to be leptokuritic.

4 Application Considerations

Two issues need to be considered when applying above KPCA and KICA based feature extraction methods to radar HRRP feature extraction. The first is how to design the kernel function. The second is how to choose the kernel basis vectors for a large set of training data.

Almost all the existing kernel functions used in kernel based classifier or feature extraction are based on the inner product or Euclidean distance between input vectors and basis vectors, in other words, there is no time-shift variant problem involved. However, for the problem we concern herein, the kernel function should have ability to handle the time-shift uncertainty. We have defined a correlation kernel function in [1] and applied it in SVM classifier. In this paper, we extract HRRP features using the kernel-based methods based on the correlation kernel, which is defined as

$$k(X_i, X_j) = \exp\left(\frac{1 - \max_{\tau} \int X_i(n) X_j(n - \tau) dn}{\sigma^2}\right). \tag{7}$$

where σ is kernel function parameter. Note the value of $k(X_i, X_j)$ keeps invariant for any time-shifted X_i and X_j. Thus the KPCA and KICA features extracted based on the above kernel function is time-shift invariant.

As we have already seen, we need to solve an eigenproblem of $N \times N$ dimensional matrix when performing KPCA and KICA. It is undoubted that there will be computational and memory management problems if the amount of training data samples N is large. Fortunately, it is shown that a satisfactory classification performance can be achieved even a subset of the training data samples are used as the kernel basis vectors. . We proposed a modified LBG algorithm to automatic determine the kernel basis vectors, which is introduced as below.

LBG algorithm [8] is a well known vector quantization (VQ) and clustering algorithm. It can determine a set of codebook from a training data set by minimizing a particular objective function representing the quantization error. A key point involved in LBG algorithm designing is to define a distance operator to measure the quantization error. Due to the time-shift uncertainty of HRRP, we define a distance operator as

$$d(X_i, X_j) = 1 - \max_{\tau} \int X_i(n) X_j(n - \tau) dn. \tag{8}$$

Our simulation results show that the extracted KPCA and KICA features based on the kernel basis vectors determined by the above modified LBG algorithm can achieve better classification results than that of using the average profiles.

5   Example Results

The data used to evaluate the classification performance are measured from a C band radar with bandwidth of $400MHz$. The HRRP data of three airplanes, including An-26, Yark-42 and Cessna Citation S/II, are measured continuously when the target are flying. The measured data of each target are divided into several segments, the training data and test data are chosen from different data segment respectively. The SVM classifier is used to evaluate the classification performance of KPCA and KICA features. For each target, about 2000 HRRPs are used for training. Totally 150 kernel basis vectors are extracted from the training data set by using the modified LBG algorithm. Totally about 40000 HRRPs are used for test.

It is of interest to determine a suitable component number when performing KPCA and KICA. Shown in Fig.1 are the average classification rates vs. component number, for KPCA and KICA features respectively. From which we can see the best classification performance can be obtained if the component number set between 15 and 20, both less and larger than this range will decrease the classification rate. For smaller component number case, the average classification rate of KICA features are larger than that of KPCA features, but it decreases faster than that of KPCA features as the component number increasing. Note that KICA is to extract least Gaussian components from the data in kernel induced feature space, if the assumed component number larger than the actual component number of the data, the algorithm will force the extracted feature into non-Gaussian distribution, and unexpected feature will be extracted, thus decreases the classification performance. The classification performance of both KPCA and KICA features decrease as the number of component increases. This is because the data "noise" is also represented in the features if the component number is large enough.

For radar target recognition, the signal-noise-ratio (SNR) may vary in a large range in real system depends on the target-radar distance. Therefore, it is of interest to design a noise robust classifier. Here we compare the classification performance of AGC, SVM with KPCA and KICA features, under different SNR level. Shown in Tabel.1 are the confusion matrices of the three classifiers for the noise free case. The component number is set as 20. It shows that both KPCA and KICA feature can achieve a better classification performance than AGC. Among which KICA features achieve the best classification performance. Shown in Fig.2 are the classification rates of three classifiers vs. SNR. The classification performance of both KPCA and KICA features are much more robust to noise than that of AGC. This is not surprise because KPCA and KICA features only represent the structure information of the data in a *signal subspace*, the noise component outside this subspace is filtered out automatically during the feature extraction procedure.

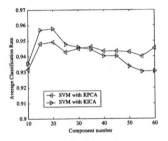
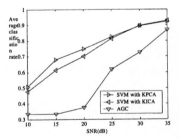

Fig. 1. The average classification rate of KPCA and KICA features vs. component number.

Fig. 2. The average classification rate vs. signal noise ratio.

6   Conclusions

KPCA and KICA are applied for radar HRRP feature extraction in this paper. A correlation kernel is used to handle the time-shift uncertainty problem of HRRP. A modified LBG algorithm is proposed to select the kernel basis vectors. The example results based on measured HRRP data show the KPCA and KICA features can achieve better classification performance and are more robust to noise than the AGC.

Table 1. Confusion Matrices of AGC, KPCA with SVM and KICA with SVM. The componenet number is set as 20.

	AGC			SVM with KPCA			SVM with KICA		
	Yark-42	An-26	Cessna	Yark-42	An-26	Cessna	Yark-42	An-26	Cessna
Yark-42	.9005	.0077	.0918	.9893	.0107	0	.9986	.0014	0
An-26	.0397	.7982	.1621	.0503	.8975	.0522	.0418	.8973	.0609
Cessna	0	.0017	.9983	0	.0421	.9579	0	.0253	.9747
Average	0.8990			0.9482			0.9569		

References

1. Liu, H. W., Bao, Z.: Radar HRR Profiles Recognition based on SVM with Power-Transformed-Correlation Kernel. Lecture Notes in Computer Science, Springer-Verlag, Berlin Heidelberg New York **3174** (2004) 531-536
2. Liao, X., Li, H., Krishnapuram, B.: An M-ary KMP Classifier for Multi-aspect Target Classification. Proceedings of IEEE ICASSP, **2** (2004) 61-64
3. Du, L., Liu, H. W., Bao, Z.: Radar HRRP Target Recognition Based on the High-order Spectra Features. IEEE Trans. on Signal Processing
4. Jacobs, S. P., O'sollivan, J. A.: Automatic Target Recognition Using Sequences of High Resolution Radar Range Profiles. IEEE Trans. on AES, **36** (2000) 364-380
5. Scholkopf, B., Smola, A. J., Muller, K. R.: Nonlinear Component Analysis as a Kernel Eigenvalue Problem. Neural Computing, **10** (1998) 1299-1319
6. Bach, F. R., Jordan, M. I.: Kernel Independent Component Analysis. Journal of Machine Learning Research, **3** (2002) 1-48
7. http://www.cis.hut.fi/projects/ica/fastica/
8. Linde, Y., Buzo, A., Gray, R.: An Algorithm for Vector Quantizer Design. IEEE Trans. on Communications, **28** (1980) 84-94

Controlling Chaotic Systems via Support Vector Machines Without Analytical Model

Meiying Ye

College of Mathematics and Physics, Zhejiang Normal University, China
ymy@zjnu.cn

Abstract. A controlling chaotic system method based on support vector machines (SVMs) is proposed. The method has been tested for controlling the Hénon map from arbitrary initial states to the desirable stationary point or function input without the need of an analytic model. We can see that its performance is very good in simulation studies. Even if there is additive noise, the proposed method is still effective.

1 Introduction

In the last decade, control of chaotic systems is a very active area of research due to its potential applications in diverse areas. The controller based on artificial neural networks (ANNs) have been widely used to control chaos in various systems. Although some of ANNs are developed in controlling unknown chaotic systems such as in Ref. [1], some inherent drawbacks, e.g., the multiple local minima problem, the choice of the number of hidden units and the danger of over fitting, etc., would make it difficult to put the ANNs into practice. These issues have long been a concern to researchers. Recently, Vapnik introduced support vector machines (SVMs) that are firmly based on the statistical learning theory [2]. Originally, it was designed to solve pattern recognition problems, where in order to find a decision rule with good generalization capability, a small subset of the training data, called the support vectors are selected. More importantly, the SVMs formulate these tasks in terms of convex optimization problems having a unique global optimum. This is in direct contrast to many versions of popular ANNs in which several local solutions exist. Further advantage of the SVM with respect to other neural is the application of Mercer's condition which allows the use of several possible kernel functions with less hidden layer parameters and no specification of the number of hidden units. In recent years, SVMs have been applied to various fields successfully and has become a hot topic of intensive study due to its successful application in classification and regression tasks. The SVMs have also been used for controlling chaotic systems by Kulkarni and co-workers [3]. In their method, the SVMs provide non-linearity compensation. It is applicable for systems, whose model equations can be decomposed into a sum of linear and non-linear parts. However, this means that the method proposed in Ref. [3] is based on the exact analytical model as the OGY method [4].

J. Wang, X. Liao, and Z. Yi (Eds.): ISNN 2005, LNCS 3496, pp. 919–924, 2005.

In this paper, a novel method that combines the advantages of several of these methods is present. Our method focus on ν-support vector machines (ν-SVMs) [5], a new class of SVM, and use it for controlling the chaotic system whose exact analytical model is unknown.

2 ν-Support Vector Machine Regression

In a number of instances, the chaotic system to be controlled is too complex and the physical processes in it are not fully understood. Hence, control design methods need to be augmented with an identification technique aimed at obtaining a better understanding of the chaotic system to be controlled. The ν-SVM regression can be used for identification to obtain an estimate model of the chaotic systems and using this model to design a controller. A simple description of the ν-SVM algorithm for regression is provided here, for more details please refer to Ref. [5].

The regression approximation addresses the problem of estimating a function based on a given set of data points $\{(\mathbf{x}_1, y_1), \cdots, (\mathbf{x}_l, y_l)\}$ ($\mathbf{x}_i \in R^n$ is an input and $y_i \in R^1$ is a desired output), which is produced from an unknown function. SVMs approximate the function in the following form:

$$y = \mathbf{w}^T \phi(\mathbf{x}) + b \,, \tag{1}$$

where $\phi(\mathbf{x})$ represents the high (maybe infinite) dimensional feature spaces, which is non-linearly mapped from the input space $\mathbf{x}$. The coefficients $\mathbf{w}$ and b are estimated by minimizing the regularized risk function

$$R(C) = \frac{1}{2}\mathbf{w}^T \mathbf{w} + C R^{\varepsilon}_{emp} \,, \tag{2}$$

$$R^{\varepsilon}_{emp} = \frac{1}{l}\sum_{i=1}^{l} \mid d_i - y_i \mid_{\varepsilon} \,, \quad \mid d - y \mid_{\varepsilon} = \begin{cases} \mid d - y \mid -\varepsilon \,, & \text{if } \mid d - y \mid \geq \varepsilon \,, \\ 0 \,, & \text{otherwise} \,. \end{cases} \tag{3}$$

In equation (2), the first term $(1/2)\mathbf{w}^T\mathbf{w}$ is called the regularized term. Minimizing this term will make a function as flat as possible, and the second term $C R^{\varepsilon}_{emp}$ is empirical error (risk) measured by the ε-insensitive loss function. This loss function provides the advantage of using sparse data points to represent the designed function (1). C is referred as regularized constant determining the trade off between the empirical error and the regularized term. ε is called the tube size of SVMs.

The parameter ε can be useful if the desired accuracy of the approximation can be specified beforehand. In some case, however, we just want the estimate to be as accurate as possible, without having to commit ourselves to a specific level of accuracy. Hence, Ref. [5] presented a modification of the SVM that automatically minimizes ε, thus adjusting the accuracy level to the data at hand. The Ref. [5] introduced a new parameter $\nu(0 \leq \nu \leq 1)$, which lets one control the number of support vectors and training errors. To be more precise, they proved that ν is an upper bound on the fraction of margin errors and a lower bound of the fraction of support vectors. We transform the equation (2) to the primal problem of ν-SVM regression. Then, by introducing Lagrange multipliers α_i, α_i^* and exploiting the optimality constraints, the regression estimative function (1) can be takes the following form:

$$y = \frac{1}{l} \sum_{i=1}^{l} (\alpha_i - \alpha_i^*) K(\mathbf{x}_i, \mathbf{x}_j) + b , \qquad (4)$$

where $K(\mathbf{x}_i, \mathbf{x}_j) = \phi(\mathbf{x}_i)^T \phi(\mathbf{x}_i)$ is the kernel function. Common kernel function is the Gaussian kernel

$$K(\mathbf{x}_i, \mathbf{x}_j) = exp(- \parallel \mathbf{x}_i - \mathbf{x}_j \parallel^2 /(2\sigma^2)) , \qquad (5)$$

where σ is the bandwidth of the Gaussian kernel.

It should be pointed out that training ν-SVMs is equivalent to solving a linearly constrained quadratic programming (QP) problem so that the solution of ν-SVMs is always unique and globally optimal, unlike training ANNs that requires nonlinear optimization with the danger of getting stuck into local minima.

3   Description of the Control Method

To control a general unknown chaotic system, a control input is applied to its right-hand side as follows

$$x(k + 1) = f(\mathbf{x}(k)) + u(k) , \qquad (6)$$

where $f(\cdot)$ is a chaotic system whose analytical model is unknown, $x \in R^n$ is the system state vector and u is a given control action. Let $n(k) = 0$ in equation (6), we obtain

$$x(k + 1) = f(\mathbf{x}(k)) . \qquad (7)$$

Equation (7) represents the dynamics of a chaotic system.

We assume that the input and output values of the chaotic system can be measured accurately. In the following, the ν-SVM will be trained using the input and output data of the chaotic system to obtain the estimate model with chaotic trajectories

$$\hat{x}(k + 1) = \hat{f}(\mathbf{x}(k)) . \qquad (8)$$

We define the tracking error by

$$e(k + 1) = x(k + 1) - x_d , \qquad (9)$$

where x_d is a desired trajectory.

If the state $x(k + 1)$ is expected to track a reference input $r = x_d$, the controller using the predicted state is chosen to be

$$u(k) = -\hat{f}(\mathbf{x}(k)) + x_d . \qquad (10)$$

The overall controlled chaotic system thus becomes

$$x(k+1) = -\hat{f}(\mathbf{x}(k)) + x_d + f(\mathbf{x}(k))$$
$$= x_d + [f(\mathbf{x}(k)) - \hat{f}(\mathbf{x}(k))] \qquad (11)$$
$$= x_d + \tilde{x}(k+1) \,,$$

where the approximation error $\tilde{x}(k+1) = f(x(k)) - \hat{f}(\mathbf{x}(k)) \approx 0$. This results in the controlled chaotic system dynamics with some small or weak residual non-linearity. If the prediction is accurate, then $\tilde{x}(k+1) \to 0$, $x(k+1) \to x_d$, and $e(k+1) \to 0$. The ν-SVM with the Gaussian function kernels is known to have very good approximating capabilities with excellent generalization properties on unknown function, so that the controlled chaotic system can track the reference input.

Briefly, the control method mentioned above can be described as follows: the observed input and output data pairs of the chaotic system are first obtained. Then, off-line identification (i.e., approximation) for the system using the ν-SVM is performed. Finally, the trained ν-SVM is incorporated into the system, to control the chaotic systems without exact analytical method.

4   Simulation Studies

In this investigation, the Gaussian function is used as the kernel function of the ν-SVM. The bandwidth of the Gaussian kernel, σ, is selected as 1.0. C and ν are the only two free parameters of ν-SVM if the kernel function has been considered. Both C and ν are chosen as 1000 and 0.5, respectively. The root mean squared error metric (RMSE) is used to evaluate the tracking performance of the proposed method. The numerical simulations have been carried out using MATLAB version 6.5 on WINDOWS XP machines.

To illustrate the effectiveness of using the ν-SVM regression for controlling chaotic systems, the two-dimensional Hénon map is used as evaluation of the forecasting power of ν-SVM regression. As it is well known, the Hénon map is described by the following equations:

$$x(k+1) = 1 - Ax(k)^2 + y(k) \,, \quad y(k+1) = Bx(k) \,, \qquad (12)$$

where A and B are the two bifurcation parameters fixed at 1.4 and 0.3.

We can also write it in the following form:

$$x(k+1) = 1 - Ax(k)^2 + Bx(k-1) \,, \qquad (13)$$

here take its x component.

The control u is added to the above system. We get the controlled chaotic system

$$x(k+1) = 1 - Ax(k)^2 + Bx(k-1) + u(k) \,. \qquad (14)$$

The ν-SVM is employed to learn the Hénon map. We select an initial state in $x(0) = -0.3$ and $y(0) = 0.1$, iterate 1000 times, and remove the data of the first 200 times to eliminate the effect of the initial value. Then we choose the data of the last 800 times as the training data.

The control objective is a stationary point $x_d(k) = 0.8$. In this simulation, control u is put into operation at the 200th step. The output of the controlled Hénon map is shown in Fig. 1(a). The proposed method can also control the system to track a function input. We choose the control objective as $x_d(k) = 0.4 + 0.5sin(k\pi/100 + \pi/6)$. The results are shown in Fig. 1(b). The fact that ν-SVM is known to have very good approximating capabilities with excellent generalization properties are well justified from the very low values of errors obtained. Two RMSE of tracking errors for the two control objectives, viz. the stationary point or the tracking function, are 2.25×10^{-4} where zero implies a perfect control.

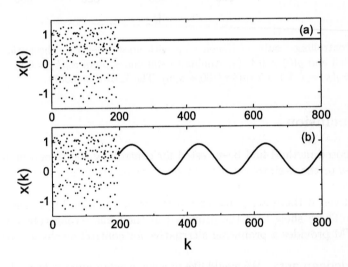

Fig. 1. Controlling results of Hénon map. The initial states are $x(0) = -0.3$ and $y(0) = 0.1$. (a) Aiming to stationary point $x_d(k) = 0.8$; (b) tracking function $x_d(k) = 0.4 + 0.5sin(k\pi/100 + \pi/6)$. The RMSE is 2.25×10^{-4}.

After above simulation with noise free data, we wish to test the control performance of the method in the presence of additive noise. The noisy data is generated by adding simulated random noise to the Hénon map data at each iteration. More specifically, MATLAB generates a normally distributed sequence of random numbers and superimposed these on the original clean Hénon map data. We use noise levels of 3%, (see Fig. 2), to investigate the control quality in comparison with the results of the noise free control mentioned above. Two RMSE of tracking errors is 0.0371. One should note that a high level of noise produces greater tracking errors, however, the control is also effective. The control objective is a stationary point $x_d(k) = 0.8$ in Fig. 2(a), a function input $x_d(k) = 0.4 + 0.5sin(k\pi/100 + \pi/6)$ in Fig. 2(b).

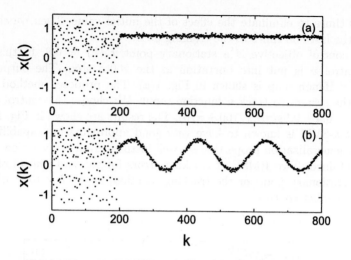

Fig. 2. Controlling results of Hénon map with noise level 3%. The initial states are $x(0) = -0.3$ and $y(0) = 0.1$. (a) Aiming to stationary point $x_d(k) = 0.8$; (b) tracking function $x_d(k) = 0.4 + 0.5sin(k\pi/100 + \pi/6)$. The RMSE is 0.0371.

5   Conclusion

The proposed method had been tested for controlling the Hénon map from initial states to the desirable stationary point or function input without the need of an analytic model and found the performance to be very good in simulation studies. Even if there are parameter perturbation and additive noise, it is still effective. These show that the proposed method has strong robustness. Thus, the ν-SVM provides a promising alternative for controlling chaotic systems.

Acknowledgements. We would like to acknowledge support from the Zhejiang Provincial Natural Science Foundation of China, Grant No. 602145.

References

1. Poznyak, A.S., Yu, W., Sanchez, E.N.: Identification and Control of Unknown Chaotic Systems via Dynamic Neural Networks. IEEE Transactions on Circuits and Systems I: Fundamental Theory and Applications. **46** (1999) 1491–1495
2. Vapnik, V.N.: The Nature of Statistical Learning Theory. Springer-Verlag, New York (1999)
3. Kulkarni, A., Jayaraman, V.K., Kulkarni, B.D.: Control of Chaotic Dynamical Systems Using Support Vector Machines. Phys. Lett. A. **317** (2003) 429–453
4. Ott, E., Grebogi, C., Yorke, J.A.: Controlling Chaos. Phys. Rev. Lett. **64** (1990) 1196–1199
5. Schölkopf, B., Smola, A., Williamson, R., Bartlett, P.L.: New Support Vector Algorithms. Neural Computation. **12** (2000) 1207–1245

Support Vector Regression for Software Reliability Growth Modeling and Prediction

Fei Xing[1,2] and Ping Guo[1,2]

[1] Department of Computer Science
Beijing Normal University, Beijing 100875, China
[2] Laboratory of Computer Science
Institute of Software, Chinese Academy of Sciences, Beijing 100080, China
xsoar@163.com, pguo@ieee.org

Abstract. In this work, we propose to apply support vector regression (SVR) to build software reliability growth model (SRGM). SRGM is an important aspect in software reliability engineering. Software reliability is the probability that a given software will be functioning without failure during a specified period of time in a specified environment. In order to obtain the better performance of SRGM, practical selection of parameter C for SVR is discussed in the experiments. Experimental results with the classical Sys1 and Sys3 SRGM data set show that the performance of the proposed SVR-based SRGM is better than conventional SRGMs and relative good prediction and generalization ability are achieved.

1 Introduction

Software reliability is one of key factors in software qualities. It is one of the most important problem facing the software industry. There exists an increasing demand of delivering reliable software products. Developing reliable software is a difficult problem because there are many factors impacting development such as resource limitations, unrealistic requirements, *etc*. It is also a hard problem to know whether or not the software being delivered is reliable. To solve the problem, many software reliability models have been proposed over past 30 years, they can provide quantitative measures of the reliability of software systems during software development processes [1–3].

Software reliability growth models (SRGMs) have been proven to be successful in estimating the software reliability and the number of errors remaining in the software [2]. Using SRGMs, people can assess the current reliability and predict the future reliability, and further more, conduct the software testing and debugging process. Now there have already existed many SRGMs such as Goel-Okumoto Model, Yamada Delayed S-Shaped Model, Yamada Weibull-Type Testing-Effort Function Model, *etc*. Most of SRGMs assume that the fault process follows the curve of specific type. Actually, this assumption may not be realistic in practice and these SRGMs are sometimes insufficient and inaccurate to analyze actual software failure data for reliability assessment.

J. Wang, X. Liao, and Z. Yi (Eds.): ISNN 2005, LNCS 3496, pp. 925–930, 2005.

In recent years, support vector machine (SVM) [4] is a new technique for solving pattern classification and universal approximation, it has been demonstrated to be very valuable for several real-world applications [5, 6]. SVM is known to generalize well in most cases and adapts at modeling nonlinear functional relationships which are difficult to model with other techniques. Consequently, we propose to apply support vector regression (SVR) to build SRGM and investigate the conditions which are typically encountered in software reliability engineering. We believe that all these characteristics are appropriate to SRGM.

2   Support Vector Regression

SVM was introduced by Vapnik in the late 1960s on the foundation of statistical learning theory [7]. It has originally been used for classification purposes but its principle can be extended easily to the task of regression by introducing an alternative loss function. The basic idea of SVR is to map the input data $\mathbf{x}$ into a higher dimensional feature space $\mathcal{F}$ via a nonlinear mapping ϕ and then a linear regression problem is obtained and solved in this feature space.

Given a training set of l examples $\{(\mathbf{x}_1, y_1), (\mathbf{x}_2, y_2), \ldots, (\mathbf{x}_l, y_l)\} \subset \mathbb{R}^n \times \mathbb{R}$, where $\mathbf{x}_i$ is the input vector of dimension n and y_i is the associated target. We want to estimate the following linear regression:

$$f(\mathbf{x}) = (\mathbf{w} \cdot \mathbf{x}) + b, \qquad \mathbf{w} \in \mathbb{R}^n, \ b \in \mathbb{R}, \tag{1}$$

Here we consider the special case of SVR problem with Vapnik's ϵ-insensitive loss function defined as:

$$L_\epsilon(y, f(\mathbf{x})) = \begin{cases} 0 & |y - f(\mathbf{x})| \le \epsilon \\ |y - f(\mathbf{x})| - \epsilon & |y - f(\mathbf{x})| > \epsilon \end{cases} \tag{2}$$

The best line is defined to be that line which minimizes the following cost function:

$$R(\mathbf{w}) = \frac{1}{2}\|\mathbf{w}\|^2 + C \sum_{i=1}^{l} L_\epsilon(f(y_i, \mathbf{x}_i)) \tag{3}$$

where C is a constant determining the trade-off between the training errors and the model complexity. By introducing the slack variables ξ_i, ξ_i^*, we can get the equivalent problem of Eq. 3. If the observed point is "above" the tube, ξ_i is the positive difference between the observed value and ϵ. Similar, if the observed point is "below" the tube, ξ_i^* is the negative difference between the observed value and $-\epsilon$. Written as a constrained optimization problem, it amounts to minimizing:

$$\frac{1}{2}\|\mathbf{w}\|^2 + C \sum_{i=1}^{l} (\xi_i + \xi_i^*) \tag{4}$$

subject to:

$$\begin{aligned} y_i - (\mathbf{w} \cdot \mathbf{x}_i) - b &\le \epsilon + \xi_i \\ (\mathbf{w} \cdot \mathbf{x}_i) + b - y_i &\le \epsilon + \xi_i^* \\ \xi_i, \xi_i^* &\ge 0 \end{aligned} \tag{5}$$

To generalize to non-linear regression, we replace the dot product with a kernel function $K(\cdot)$ which is defined as $K(\mathbf{x}_i, \mathbf{x}_j) = \phi(\mathbf{x}_i) \cdot \phi(\mathbf{x}_j)$. By introducing Lagrange multipliers α_i, α_i^* which are associated with each training vector to cope with both upper and lower accuracy constraints, respectively, we can obtain the dual problem which maximizes the following function:

$$\sum_{i=1}^{l}(\alpha_i - \alpha_i^*)y_i - \epsilon \sum_{i=1}^{l}(\alpha_i + \alpha_i^*) - \frac{1}{2}\sum_{i,j=1}^{l}(\alpha_i - \alpha_i^*)(\alpha_j - \alpha_j^*)K(\mathbf{x}_i, \mathbf{x}_j) \qquad (6)$$

subject to:

$$\sum_{i=1}^{l}(\alpha_i - \alpha_i^*) = 0$$
$$0 \le \alpha_i, \alpha_i^* \le C \qquad i = 1, 2, \ldots, l \qquad (7)$$

Finally, the estimate of the regression function at a given point $\mathbf{x}$ is then:

$$f(\mathbf{x}) = \sum_{i=1}^{l}(\alpha_i - \alpha_i^*)K(\mathbf{x}_i, \mathbf{x}) + b \qquad (8)$$

3 Modeling the Software Reliability Growth

In this section, we present real projects to which we apply SVR for software reliability growth generalization and prediction. The data sets are Sys1 and Sys3 software failure data applied for software reliability growth modeling in [2]. Sys1 data set contains 54 data pairs and Sys3 data set contains 278 data pairs. The data set are normalized to the range of [0,1] first. The normalized successive failure occurrence times is the input of SVR function and the normalized accumulated failure number is the output of SVR function. We denote the SVR-based software reliability growth model as SVRSRG.

Here we list the math expression of three conventional SRGMs refered in the experiments.

– Goel-Okumoto Model:

$$m(t) = a(1 - e^{rt}), \ a > 0, \ r > 0 \qquad (9)$$

– Yamada Delayed S-Shaped Model:

$$m(t) = a(1 - (1 + rt)e^{-rt}) \qquad (10)$$

– Yamada Weibull-Type Testing-Effort Function Model:

$$m(t) = a[1 - e^{-r\alpha(1 - e^{-\beta t^\gamma})}] \qquad (11)$$

The approach taken to perform the modeling and prediction includes following steps:

1. Modeling the reliability growth based on the raw failure data
2. Estimating the model parameters
3. Reliability prediction based on the established model

Three groups of experiments have been performed. Training error and testing error have been used as evaluation criteria. In the tables presented in this paper, the training error and the testing error are measured by sum-of-square $\sum_{i=1}^{l}(x_i - \hat{x}_i)^2$, where $x_i, \hat{x}_i$ are, respectively, the data set measurements and their prediction. In default case, SVR used in the experiment is ν-SVR and the parameters ν and C are optimized by cross-validation method.

In the experiment of generalization, we partition the data into two parts: training set and test set. Two thirds of the samples are randomly drawn from the original data set as training set and remaining one third of the samples as the testing set. This kind of training is called generalization training [8]. Fig. 1. (a) and (b) show the experimental result for software reliability growth modeling trained by using data set Sys1 and Sys3, respectively. It is obvious that SVRSRG gives a better performance of fitting the original data than the other models. From Table 1 we can find both the training error and the testing error of SVRSRG are smaller than the other classical SRGMs.

In the experiment of prediction, we will simulate the practical process of software reliability growth prediction. It is based on predicting future values by the way of time series prediction methods. Assuming software have been

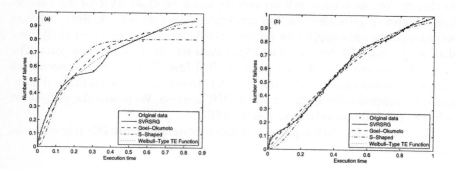

Fig. 1. (a) The generalization curve of four SRGMs trained with Sys1 data set (b) The generalization curve of four SRGMs trained with Sys3 data set

Table 1. The comparison of training error and testing error of four SRGMs for generalization

Data	Sys1		Sys3	
Models	Training error	Test error	Training error	Test error
Goel-Okumoto	0.1098	0.0576	0.1255	0.0672
S-Shaped	0.3527	0.1722	0.1792	0.0916
Weibull-type TE Function	0.0255	0.0137	0.1969	0.1040
SVRSRG	0.0065	0.0048	0.0147	0.0078

executed for time x_i and considering i data pairs $(x_1, y_1), (x_2, y_2), \ldots, (x_i, y_i)$, we calculate predicted number of failures y_{i+1} at time x_{i+1} in the following way. First, use first i data pairs to build model. Second, predict value at time x_{i+1} using current model. Experiment is processed at each time points and mean training error and mean testing error are reported. From Table 2, we can find that in practical process SVRSRG can achieve more remarkable performance than other four SRGMs.

Table 2. The comparison of training error and testing error of four SRGMs for prediction

Data	Sys1		Sys3	
Models	Training error	Test error	Training error	Test error
Goel-Okumoto	0.0259	0.0038	0.0864	0.0011
S-Shaped	0.0855	0.0110	0.1663	0.0015
Weibull-type TE Function	0.0127	0.0012	0.0761	0.0007
SVRSRG	0.0064	0.0021	0.0138	0.0001

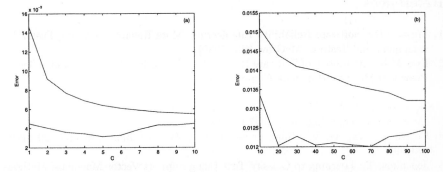

Fig. 2. The influence of parameter C on model generalization ability. Above line is training error, bottom line is testing error. (a) with Sys1 data set (b) with Sys3 data set

Parameter C in SVR is a regularized constant determining the tradeoff between the training error and the model flatness. The following experiment demonstrates the influence of parameter C on model generalization ability. It is conducted as experiment of prediction to simulate the practical process of software reliability growth prediction. The results are shown in Fig. 2. (a) and (b). We can see that with the increase of parameter C, the training error declines gradually, that is to say, the model fits the training data set better and better. However, as for testing, first the testing error declines gradually because the complexity of model is suitable for the need of testing data set more and more. And then the testing error raises because of overfitting problem. The problem of tradeoff between the training error and the model flatness can be solved by cross-validation technique which divides the training samples into two parts: one for training and another for validation to obtain satisfied generalization ability.

4   Conclusions

A new technique for software reliability growth modeling and prediction is proposed in this paper. SVR is adopted to build SRGM. From the experiments we can see that the proposed SVR-based SRGM has a good performance, no matter generalization ability or predictive ability, the SVRSRG is better than conventional SRGMs. Experimental results show that our approach offers a very promising technique in software reliability growth modeling and prediction.

Acknowledgements

This work was fully supported by a grant from the NSFC (Project No. 60275002) and the Project-sponsored by SRF for ROCS, SEM.

References

1. Musa, J.D.: Software Reliability Engineering: More Reliable Software, Faster Development and Testing. McGraw Hill (1999)
2. Lyu, M.R.: Handbook of Software Reliability Engineering. IEEE Computer Society Press and McGraw-Hill Book Company (1996)
3. Guo, P., Lyu, M.R.: A Pseudoinverse Learning Algorithm for Feedforward Neural Networks with Stacked Generalization Application to Software Reliability Growth Data. Neurocomputing, **56** (2004) 101–121
4. Cortes, C., Vapnik, V.: Support-vector Network. Machine Learning, **20** (1995) 273–297
5. Joachims, T.: Learning to Classify Text Using Support Vector Machines: Methods, Theory, and Algorithms. Kluwer (2002)
6. Xing, F., Guo, P.: Classification of Stellar Spectral Data Using SVM. In: International Symposium on Neural Networks (ISNN 2004). Lecture Notes in Computer Science. Vo. 3173. Springer-Verlag (2004) 616–621
7. Vapnik, V.N.: The Nature of Statistical Learning Theory. Springer-Verlag, New York (1995)
8. Karunanithi, N., Whitley, D., Malaiya, Y.: Prediction of Software Reliability Using Connectionist Models. IEEE Transactions on Software Engineering, **18** (1992) 563–574

SVM-Based Semantic Text Categorization
for Large Scale Web Information Organization

Peng Fu, Deyun Zhang, Zhaofeng Ma, and Hao Dong

Department of Computer Science and Technology, Xi'an Jiaotong University,
Xi'an, Shaanxi 710049, China
fupeng@lzu.edu.cn

Abstract. Traditional web information service can't meet the demand of users getting personalized information timely and properly, which can be think as a kind of passive information organization method. In this paper, an adaptive and active information organization model in complex Internet environment is proposed to provide personalized information service and to automatically retrieve timely, relevant information. An SVM-based Semantic text categorization method is adopted to implement adaptive and active information retrieval. Performance experiment based on a prototype retrieval system manifests the proposed schema is efficient and effective.

1 Introduction

In the case of traditional Web information service, the vast scale and scope of current online information sources make it difficult to find and process relevant personalized information, such as search engine and other many applications were widely used [1]. Locating relevant information is time consuming and expensive, push technology promises a proper way to relieve users from the drudgery of information searching [2,3]. Personalized recommender systems are an important part of knowledge management solutions on corporate intranets [4], but some important deficiency of which is that content-based or collaborative recommendation is difficult to provides kinds of users personalized information needs for users' feedback untimely, and dynamic perform in domains where there is not much content associated with items, or where the content is difficult for a computer to analyze-ideas, opinions etc. Those current approaches to delivering relevant information to users assume that user interests are relatively stable or static. In this paper, an adaptive and active information retrieval service model in complex Internet environment is proposed to provide users active and adaptive personalized information retrieval service timely in a proper way, which is a new approach centers information retrieval on a constantly changing model of user interests. In details, an SVM-based semantic text categorization method is introduced to implement incremental information retrieval and active delivery mechanism. Furthermore, the adaptive active information service model based on semantic retrieval, and the performance experiments are also discussed.

2 The Adaptive and Active Information Service Model

Usually traditional information retrieval system is composed of Retrieval-Enabled Information Source, Indexing Engine, Information Model, Searching Engine, User

J. Wang, X. Liao, and Z. Yi (Eds.): ISNN 2005, LNCS 3496, pp. 931–936, 2005.

Interface, entrance for users to retrieve for information, which includes the order option parameters.

Fig. 1. Pull-based Information Service Model.

Fig. 2. Push-based Information Service Model.

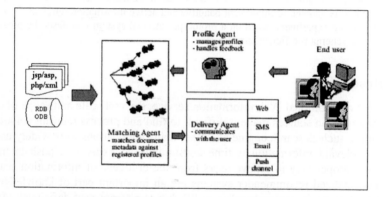

Fig. 3. Adaptive and Active Information Retrieval Model.

An adaptive and active information service for personalized information service is effective information retrieval framework for combining content and collaboration, which is user-centered, push personalized corresponding information service (fig.1, fig. 2) in a proper way, and can be generalized as the to automatically retrieve timely, relevant information timely and properly (fig. 3). The adaptive and active information service predict new items of interest for users, which can adjust the information content according to user's preference dynamically and adaptive to users interest in some degree, automatically retrieve and delivery related information to users with interaction.

The adaptive and active information service for user in a personalized way is a user correlative, context-sensitive push-based, event-driven information system, which is built on traditional information framework (fig. 3). This system for personalized information service is encapsulated on the traditional information system with adaptivity and activity, and can be viewed as the composition of user's interests or preference, information source, period that the system serve, trigger for the system serving, condition that which information source match which user(s). For a long-term user in information retrieval system, what one really needs in a relative period, is just a little part in the whole global information domain, The relation of information need in the information retrieval system can be described as: personalized information belong to domain information, and domain information belong to Global Information.

Considering the high-dimension of information system, usually the domains overlap each other especially in multi-subjects; the set of the above is n-dimension space.

Personalized Information Service over heterogeneous information source is a matching map from information to specific user with the specific condition and trigger during the specific period.

The information Model for the retrieval, which may be Boolean, Probable, or VSM modal, serve users adaptively and actively, the system must know users' needs; usually user profile decides the quality of adaptive and active service. Except for adaptively and activity, the adaptive and active information service also builds on the infrastructure of information indexing, information retrieval, information filtering.

3   SVM-Based Semantic Information Retrieval

We introduce the adaptive and active information service to manifest that our solution is active instead of passive, user-oriented but not system-oriented. The system is static and without adaptation for the users is sparse about single or group users. But with the growth of amount of users, the system then can serve users adaptively by their own access pattern or by group common usage mode, upon which the intelligent retrieval mode is called content-based filtering. Pure content-based retrieval adaptation can't account for community endorsements when little is known about a user, and collaborative adaptation is not proper when the group users is too sparse. In fact, integration of the two approaches is a reasonable trade-off, which can adapt user with fairly good stability and efficiency. In this paper we introduce Support Vector Machines (SVM) for integration of content-based and collaborative retrieval adaptation.

3.1   SVM-Based Semantic Retrieval

Text categorization is the task of assigning a Boolean value to each pair $<d_j, c_i> \in D \times C$, where D is a domain of documents and $C = \{ c_i, \ldots, c_{|C|}\}$ is a set of predefined categories. A value of T assigned to $<d_j, c_i>$ indicates a decision to file d_j under c_i, while a value of F indicates a decision not to file d_j under c_i. More formally, the task is to approximate the unknown target function $\ddot{O} : D \times C \rightarrow \{T, F\}$ (that describes how documents ought to be classified) by means of a function $\emptyset : D \times C \rightarrow \{T, F\}$ called the *classifier* (aka *rule*, or *hypothesis*, or *model*) such that $\ddot{O}$ and $\emptyset$ "coincide as much as possible". The goal of text categorization is the classification of documents into a fixed number of predefined categories. Each document d can be in multiple, exactly one, or no category at all. Using machine learning, the objective is to learn classifiers from examples, which do the category assignments automatically. To facilitate effective and efficient learning, each category is treated as a separate binary classification problem. Most efficient learning systems (SVM) are based on rather simple representations of documents such as term frequency - inverse document frequency (TF-IDF) that only take into account the frequencies of the terms in indexed documents and in the whole collection. Some study has investigated ways of introducing semantic knowledge into text classifiers and texts representation. The method we adopt [6, 8] for our adaptive and active retrieval relies on the a priori mapping of the input vectors into a space more semantic relationship. As the transformation of the data is linear, it comes to modify the Euclidean metric that allows comparing to features vectors.

(1) Semantic Retrieval

Once the training samples have been preprocessed in order to build an index of the most relevant terms, a proximity (symmetric) matrix S that reflects semantic relations between the index terms is built. Its dimension is thus the size of the index. Let us now consider the linear transformation defined by this matrix S.

Indeed terms that are semantically close to many others terms and are a priori strongly related to them, gain in importance (the corresponding feature value increases) while semantically isolated terms loose their importance. The new encoding of the documents is richer than standard TF-IDF encoding since, in addition to statistical information, it embodies grammatical/syntactical and semantic information. In a sense, this linear transformation smoothes the vectorial representation using a semantic bias. Let us now consider the metric induced by this transformation on the data. For two vectors α and β, ordinary Euclidean metric is transformed into the following metric where the positive matrix P is defined as $P = S^2$.

$$|| P(\alpha - \beta) ||^2 = (\alpha - \beta)^T P(\alpha - \beta) \tag{1}$$

This metric can be incorporated to the definition of a kernel in SVM.

(2) Support Vector Machines [5]

SVMs provides a framework between structural risk minimization (SRM) and VC-dimension theory for finite data sample, which build up the algorithm on the trade-off between the model's complexity and learning flexibility. Different symmetric functions can be used as kernels K: radial basis kernels and polynomial kernels for instance. Let us define the training sample for a binary classification problem. If the vector a and the scalar b are the parameters of the output hyperplane, and the generic decision hyperplane can be formulated as:

$$f(x) = \text{sgn}(\sum_{i=1.}^{l} y_i \alpha_i K(x_i \cdot x) + b) \tag{2}$$

The induction principal derived to determine weights of output hyperplane (and support data) are the maximization of the margin between the output hyperplane and the data encoded in the hidden layer of the network. To deal with non-separable data, the margin concept was softenized in order to accept as SV some points that are on the wrong side of the margin frontiers.

One of the most interesting characteristic of SVM is that they belong to the class of learning machines that implement Structural Risk Minimization principle: this brings a justification to their very good performance on all the real-life problems so far tested. However, as all algorithms based on relational coding, the accuracy of the SVM highly depends on the definition of the relation.

In this approach, the radial basis kernel is chosen that usually gets very good performance with few tuning: $K(x,y) = \tanh(-\gamma x \cdot y - \delta)$

After semantically smoothing the vectors: $K(x,y) = \tanh(-\gamma (x \cdot y)^T - \delta)$

Because there are fewer model parameters to optimize in the SVMs, reducing the possibility of over fitting the training data and therefore increasing the actual performance to a large degree, and thus have been widely used in classification, recognition and regression.

3.2 SVM-Based Semantic Retrieval Process

We integrate these two methods together to achieve much higher precision and recall. The adaptation is divided into two stages: at the first stage we apply SVMs to classify group interests according to the given user's profile, personalized information, domain information and retrieve from the information source(s) in a classification mode, and at the second stage, based on the returned retrieved results, then retrieve according to the PIS vector itself for content-based adaptation.

4 Performance Experiment

We have implemented the prototype system for active and adaptive information retrieval; the system is based on WebSphere (as Application Server). Based on this platform we have evaluated the performance in efficiency, incremental retrieval (precision / classification rate /recall in variant parameters, evaluation results manifest the system is efficient and feasible for active and adaptive information retrieval in distributed Internet environment. Our SVM semantic text Classifier is implemented based on SVMlight [7] .

The performance experiment results are tested under 100 user profiles, and using the breakeven point as the measurement of adaptive retrieval.

Table.1 showed that the SVM based semantic text categorization integration adaptive retrieval algorithm was more efficient than other classification algorithm. SVM-model parameters were fairly fewer and can learned from training data adaptively, moreover, SVM are more feasible to deal with high dimension data. This empirically shows that semantic smoothing is relevant for text categorization tasks. The SVM based semantic text categorization method is also proved to be more efficient than the pure SVM categorization method that only take into account the frequencies of the terms in indexed documents and in the whole collection.

Table 1. Average scores for accuracy, precision and recall.

Method	Classification Rate	Precision	Recall
SVM	83.24	82.36	85.68
K-NN	71.79	92.86	47.58
Bayes	79.22	81.05	72.36
SVM-Semantic	89.68	90.17	86.32

5 Conclusions

In our adaptive and active information service system, because SVM has been proved to be suited to cope with high-dimensional data very well, we have integrated the SVM-based semantic text categorization method to it, which is proved to be feasible, effective and efficient. Comparing with traditional retrieval system, our information service model provides a push-based, interest-driven, adaptive and active service mode, which serves users in an active style instead of passive mode. The adaptive and active information service is helpful for users to gain personalized information in time. Otherwise, other text categorization methods such as rough set or rough- neural network will be taken into account to improve our system performance in the future work.

References

1. Dominich, S.: Connectionist Interaction Information Retrieval. International Journal of Information Processing and Management, **39** (2003) 167-193
2. Underwood, G.M., Maglio, P.P., Barrett, R.: User-centered Push for Timely Information Delivery. Computer Networks and ISDN Systems, **30** (1998) 33-41
3. Chen, P.-M., Kuo, F.-C.: An Information Retrieval System based on a User Profile. Journal of Systems and Software. **54** (2000) 3-8
4. Yen, B., Kong, R.: Personalization of Information Access for Electronic Catalogs on the Web. Electronic Commerce Research and Applications, **1** (2003) 20-40.
5. Vapnik, V.: Statistical Learning Theory. Springer, Berlin, Heidelberg New York (1998)
6. Joachims, T.: Text Categorization with Support Vector Machines: Learning with Many Relevant Features. In: Proceedings of the European Conference on Machines Learning. Springer-Verlag, Berlin Heidelberg New York (1998)
7. Thorsten, J.: SVMlight, http://svmlight.joachims.org/
8. Georges, S. F.: Support Vector Machines Based on a Semantic Kernel for Text Categorization. IEEE IJCNN'00, Italy (2000)

Fuzzy Support Vector Machine and Its Application to Mechanical Condition Monitoring

Zhousuo Zhang, Qiao Hu, and Zhengjia He

School of Mechanical Engineering,
State Key Laboratory for Manufacturing Systems Engineering,
Xi'an Jiaotong University, Xi'an 710049, China
{zzs,hzj}@mail.xjtu.edu.cn, hq@mailst.xjtu.edu.cn

Abstract. Fuzzy support vector machine (FSVM) is applied in this paper, in order to resolve problem on bringing different loss for classification error to different fault type in mechanical fault diagnosis. Based on basic principle of FSVM, a method of determining numerical value range of fuzzy coefficient is proposed. Classification performance of FSVM is tested and verified by means of simulation data samples. A fuzzy fault classifier is constructed, and applied to condition monitoring of flue-gas turbine set. The results show that fuzzy coefficient can indicate importance degree of data sample, and classification error rate of important data sample can be decreased.

1 Introduction

It is assumed in algorithm of common support vector machine (SVM) that every training data sample has same importance. But importance of every training data sample usually is not same in practical classification problems. For example, importance of different classificatory fault samples is not same in condition monitoring and fault classification of machinery, because loss of misclassifying data sample of different fault type may have large difference. Therefore when data samples are classified, it is required to the greatest extent that correct rate of classifying important data samples should be higher than that of the other samples.

To overcome this shortcoming of common support vector machine, some research results have be obtained. Loss of misclassifying data samples and difference of selecting data samples are introduced into loss function of classification in reference [1], which shows that better results is got in simulation experiment. Loss difference of misclassifying training data sample of different types is treated by means of SVM algorithm based on loss function in reference [2], and the algorithm is applied to human face checking. Fuzzy support vector machine (FSVM) is proposed in reference [3]. FSVM modifies separating hyperplane by adding fuzzy coefficients to every training data sample, in order to indicate loss difference of misclassifying training data sample of different types.

First, basic principle of FSVM is introduced in this paper, and a method of determining numerical value range of fuzzy coefficient is proposed. Then classification performance of FSVM is tested and verified by means of simulation data samples that come from IDA database, and it is analyzed how importance of training data samples influences classification result. Finally, FSVM is applied to condition monitoring of flue-gas turbine set.

J. Wang, X. Liao, and Z. Yi (Eds.): ISNN 2005, LNCS 3496, pp. 937–942, 2005.

2 Basic Principle of Fuzzy Support Vector Machine

Fuzzy coefficients s_i, $0 < s_i \leq 1$, are introduced into training data samples to denote important degree of training data samples. Common SVM can be transformed into FSVM by means of fuzzy coefficients s_i [3]. Basic principle and algorithm of FSVM are researched as following.

Suppose two kinds of training data sample set are given as

$$(\mathbf{x}_i, y_i, s_i), \quad i = 1, 2, \cdots, n, \quad \mathbf{x} \in R^d, \quad y \in \{+1, -1\}. \tag{1}$$

Where $\mathbf{x}_i$ is training data samples, y_i is the class label (the value of one class equals +1, the value of another class equals -1), n is the number of the training samples, d is the dimension of each training sample.

Similar to algorithm of common SVM, the problem of seeking for the optimal separating hyperplane by FSVM algorithm can be transformed into the problem of seeking for maximum of following function

$$\text{Maximize} \qquad Q(\alpha) = \sum_{i=1}^{n} \alpha_i - \frac{1}{2} \sum_{i,j=1}^{n} \alpha_i \alpha_j y_i y_j K(\mathbf{x}_i \bullet \mathbf{x}_j).$$

$$\text{Subject to} \qquad \begin{cases} \sum_{i=1}^{n} y_i \alpha_i = 0 \\ 0 \leq \alpha_i \leq s_i C \end{cases} \qquad i = 1, 2, \cdots, n. \tag{2}$$

Where α_i is the optimizing coefficient, $K(\mathbf{x}_i, \mathbf{x}_j)$ is the kernel function, C is called punishing factor and controls the degree of punishing to samples misclassified.

After obtaining the α_i from the equation (2), toward testing samples $\mathbf{x}$, the classification function of the FSVM classifier is generally expressed as

$$f(\mathbf{x}) = sign\{\sum_{sv} \alpha_i y_i K(\mathbf{x}_i, \mathbf{x}) + b\} \tag{3}$$

Where **sign**{ } is the sign function, sv is support vector, b is the bias of the classification. So which class the testing sample $\mathbf{x}$ belongs to can be determined by means of the sign of classification function $f(\mathbf{x})$.

FSVM controls the importance of training data samples in training by means of the product of fuzzy coefficients s_i and punishing factor C. Fuzzy coefficient s_i denotes important degree of training sample $\mathbf{x}_i$, the bigger s_i is, the more important sample $\mathbf{x}_i$ is, the bigger weight of sample $\mathbf{x}_i$ in training is.

3 Determination of Numerical Value Range of Fuzzy Coefficient

The key of applying FSVM to practice is how fuzzy coefficient s_i of training data sample is determined according to the practical classification problem given. s_i is usually is determined according to the prior knowledge of practical classification problem. A method of determining numerical value range of fuzzy coefficient is proposed in this section.

According to the equation (2), the upper limit of optimizing coefficient α_i is controlled by the product of fuzzy coefficients s_i and punishing factor C. In fact, fuzzy coefficients s_i influence classification result by means of modifying the factor C of common SVM. In order to conveniently express, the maximum of α_i in common SVM is denoted by α_{max}, the maximum and minimum of s_i are denoted respectively by s_{max} and s_{min}. None but when $\alpha_{max} \leq s_{min} C$, or $s_{min} \geq \alpha_{max}/C$, fuzzy coefficients s_i may influence classification result of SVM. Based on these analyses, a method of determining numerical value range of fuzzy coefficient is given in this paper, and process of the method is in following.

1) Solve for α_{max}

Let $s_i = 1$, the optimizing coefficients α_i are obtained by the equation (2), and the maximum of α_i is α_{max}.

2) Determine s_{min}

The s_{min} is selected according to inequality $s_{min} \geq \alpha_{max}/C$.

3) Determine s_{max}

Based on s_{min}, according to important degree of training data samples or relation of s_i and other parameters, s_{max} is selected. Then numerical value range of s_i is determined by s_{min} and s_{max}.

4 Classification Performance of Fuzzy Support Vector Machine

Influence of fuzzy coefficients s_i on classification performance of FSVM is researched in this section by means of 32nd data set of Banana data sets in IDA database [4]. There are 400 training data samples in the 32nd data set, and 188 of them have positive class label, 212 of them have negative class label. There are 4900 testing data samples in the 32nd data set, and 2188 of them have positive class label, 2712 of them have negative class label. Dimension of the samples is 2. Kernel function of the classifier is radial base kernel function, and its parameter $\theta = 1$, punishing factor $C = 316.2$. Different fuzzy coefficients s_i are respectively given to two types of training data samples, in order to test their influence on classification performance of the classifier.

Fuzzy coefficients of samples with positive class label and samples with negative class label are respectively denoted by s_p and s_n. Let $s_n + s_p = 1$, but the ratio s_p/s_n changes from 0.1 to 9, classification error ratio of training samples and that of testing samples are shown in figure 1. The figure shows that the bigger ratio s_p/s_n is, the lower classification error ratio of samples with positive class label is and the higher classification error ratio of samples with negative class label. This indicates that fuzzy coefficient denotes important degree of data samples, the bigger fuzzy coefficient is, the more important the data sample is and the lower corresponding classification error ratio is.

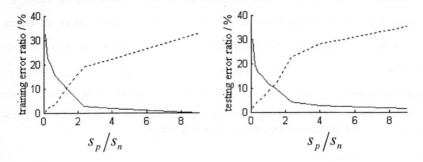

Fig. 1. Influence of fuzzy coefficients on classification performance. The line "——" denotes classification error ratio of samples with positive class label. The line "- - - - -" denotes classification error ratio of samples with negative class label.

5   Application of FSVM to Condition Monitoring of Flue-Gas Turbine Set

FSVM is applied to condition monitoring of flue-gas turbine set in this section. The important degree of training data samples is indicated by means of fuzzy coefficients, in order to explain the influence of importance of training data samples on classification results of FSVM.

When vibration sensors installed on a flue-gas turbine set are respectively in normal and looseness, signals of axis vibration of the turbine set in time domain are sampled as data samples. Sampling frequency of the data samples is 2000Hz, and every data sample contains 512 data points. Working frequency of the flue-gas turbine set is 97Hz. The data samples consist of 32 training data samples and 16 testing data samples. Data samples in condition of sensor normal have positive class label +1, and data samples in condition of sensor looseness have negative class label -1. Kernel function of the classifier is radial base kernel function, and its parameter $\theta = 1000$, punishing factor $C = 10$. Fuzzy coefficients of FSVM are $s_p = 0.1$, $s_n = 1$. Two fault classifiers based on common SVM and FSVM are respectively constructed, and their classification performance is tested and compared. The results of testing and comparing are shown in figure 2.

Half data samples sequence numbers of which are in front have positive class label in figure 2, and the other data samples have negative class label. Classification distance L indicates the distance from data sample point to separating hyperplane, and L can be computed by equation (4).

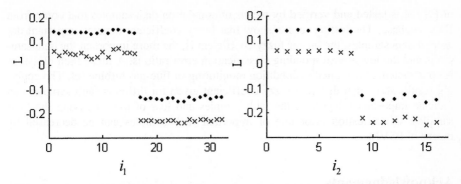

Fig. 2. Comparison of classification distance of common SVM with that of FSVM. The left of the figure shows classification distance of training data samples, and the right shows that of testing samples. The sign "•" indicates classification distance of common SVM, and the sign "×" indicates that of FSVM. The coordinates "L" denotes classification distance, and "i_1" denotes sequence number of training samples, and "i_2" denotes that of testing samples.

$$L(\mathbf{x}) = \frac{\sum_{sv} \alpha_i y_i K(\mathbf{x}_i, \mathbf{x}) + b}{\|\mathbf{w}\|} \tag{4}$$

Where $\mathbf{w}$ is weight coefficient vector of the separating hyperplane.

Figure 2 shows that classification distances of common SVM to samples with negative class label and samples with positive class label are basically symmetrical about zero, whether the samples are training samples or testing samples. This means that the separating hyperplane of common SVM is in the midst between samples with negative class label and samples with positive class label. However, classification distances of FSVM to samples with negative class label are about 0.2 bigger than that to samples with positive class label, which denotes the offset of the separating hyperplane toward samples with positive class label is about 0.1. Furthermore this offset means that classification error ratio of samples with positive class label increases and classification error ratio of samples with negative class label decreases. Obviously, this change of the classification error ratio is due to difference of fuzzy coefficients of two class samples, or s_n/s_p =10. Therefore it is indicated that fuzzy coefficients of training samples increases means classification distances to the training samples also increases and classification error ratio of the training samples decreases. In practical classifying of faults, it is usually requested that classification error ratio of the important fault samples misclassifying which results in bigger loss is lower. Hence the important degree of fault samples can be indicated by means of their fuzzy coefficients, and classification error ratio of the important fault samples can be decrease by increasing their fuzzy coefficients.

6 Conclusion

In basis of research on basic principle of FSVM, a method of determining numerical value range of fuzzy coefficient is proposed in this paper. Classification performance of FSVM is tested and verified by means of simulation data samples that come from IDA database. The testing results show that fuzzy coefficient denotes important degree of data samples, the bigger fuzzy coefficient is, the more important the data sample is and the lower corresponding classification error ratio is. A fuzzy fault classifier is constructed, and applied to condition monitoring of flue-gas turbine set. The applying results show that different fuzzy coefficient given for different fault samples can indicate importance degree of the fault samples and loss of misclassifying the fault samples. Classification error rate of important fault samples can be decreased by means of FSVM.

Acknowledgements

This work was supported by the project of National Natural Science Foundation of China (No.50335030, No.50175087).

References

1. Lin, Y., Lee, Y., Wahba, G.: Support Vector Machines for Classification in Nonstandard Situations. Machine Learning, **46** (2002) 199-202
2. Ma, Y., Ding, X.: Face Detection Based on Cost-sensitive Support Vector Machines. Lee, S.W., Verri, A. (eds.): Pattern Recognition with Support Vector Machines. Berlin: Springer-Verlag Press (2002) 260-267
3. Chun, F.C., Wang, S.D.: Fuzzy Support Vector Machines. IEEE Trans. On Neural Networks, **13** (2002) 464-471
4. IDA Benchmark Repository Used in Several Boosting, KFD and SVM papers, http://ida.first.gmd.de/~raetsch/data/benchmarks.htm

Guided GA-ICA Algorithms

Juan Manuel Górriz, Carlos García Puntonet,
Angel Manuel Gómez, and Oscar Pernía

Dept. Signal Theory. Facultad de Ciencias, Universidad de Granada
Fuentenueva s/n, 18071 Granada, Spain
gorriz@ugr.es

Abstract. In this paper we present a novel GA-ICA method which converges to the optimum. The new method for blindly separating unobservable independent components from their linear mixtures, uses genetic algorithms (GA) to find the separation matrices which minimize a cumulant based contrast function. We focuss our attention on theoretical analysis of convergence including a formal prove on the convergence of the well-known GA-ICA algorithms. In addition we introduce guiding operators, a new concept in the genetic algorithms scenario, which formalize elitist strategies. This approach is very useful in many fields such as biomedical applications i.e. EEG which usually use a high number of input signals. The Guided GA (GGA) presented in this work converges to uniform populations containing just one individual, the optimum.

1 Introduction

The starting point in the Independent Component Analysis (ICA) research can be found in [1] where a principle of redundancy reduction as a coding strategy in neurons was suggested, i.e. each neural unit was supposed to encode statistically independent features over a set of inputs. But it was in the 90´s when Bell and Sejnowski applied this theoretical concept to the blindly separation of the mixed sources (BSS) using a well known stochastic gradient learning rule [2] and originating a productive period of research in this area [3]. In this way ICA algorithms have been applied successfully to several fields such as biomedicine, speech, sonar and radar, signal processing, etc. and more recently also to time series forecasting [4].

In general, any abstract task to be accomplished can be viewed as a search through a space of potential solutions and whenever we work with large spaces, GAs are suitable artificial intelligence techniques for developing this optimization [4]. Such search requires balancing two goals: exploiting the best solutions and exploring the whole search space. In ICA scenario they work efficiently in the search of the separation matrix (i.e. EEG and scenarios with the BSS problem in higher dimension) proving the convergence to the optimum. We organize the essay as follows. In section 2 and 3 we give a brief overview of the basic ICA and GA theory. Then we introduce a set of genetic operators in sections 3 and 4 and prove convergence and conclude in section 5.

J. Wang, X. Liao, and Z. Yi (Eds.): ISNN 2005, LNCS 3496, pp. 943–948, 2005.

2   ICA in General

We define ICA using a statistical latent variables model (Jutten & Herault, 1991). Assuming the number of sources n is equal to the number of mixtures, the linear model can be expressed, using vector-matrix notation and defining a time series vector $\mathbf{x} = (x_1, \ldots, x_n)^T$, $\mathbf{s}$, $\tilde{\mathbf{s}}$ and the matrix $\mathbf{A} = \{a_{ij}\}$ and $\mathbf{B} = \{b_{ij}\}$ as:

$$\tilde{\mathbf{s}} = \mathbf{Bx} = \mathbf{BAs} = \mathbf{Gs} \tag{1}$$

where we define $\mathbf{G}$ as the overall transfer matrix. The estimated original sources will be, under some conditions included in Darmois-Skitovich theorem [5], a permuted and scaled version of the original ones.

The statistical independence of a set of random variables can be described in terms of their joint and individual probability distribution. This is equivalent to [6]:

$$\Pi = \sum_{\{\lambda, \lambda^*\}} \beta_\lambda \beta_{\lambda^*}^* \cdot \Gamma_{\lambda, \lambda^*} \qquad |\lambda| + |\lambda^*| < \tilde{\lambda} \tag{2}$$

where the expression defines a summation of cross cumulants [6] and is used as a fitness function in the GA. The latter function satisfies the definition of a contrast function Ψ defined in [7] as can be seen in the following generalized proposition given in [8].

Proposition 1. *The criterion of statistical independence based on cumulants defines a contrast function Ψ given by:*

$$\psi(\mathbf{G}) = \Pi - log|det(\mathbf{G})| - h(\mathbf{s}) \tag{3}$$

where $h(\mathbf{s})$ is the entropy of the sources and G is the overall transfer matrix.
Proof: To prove this proposition see Appendix A in [8] and apply the multi-linear property of the cumulants.

3   Genetic Algorithms: Definitions

Let $\mathcal{C}$ the set of all possible creatures in a given world and a function $f : \mathcal{C} \to R^+$, namely fitness function. Let $\Xi : \mathcal{C} \to \mathcal{V}_C$ a bijection from the creature space onto the free vector space over $\mathcal{A}^\ell$, where $\mathcal{A} = \{\bar{a}(i), \quad 0 \leq i \leq a - 1\}$ is the alphabet which can be identified by $\mathcal{V}_1$ the free vector space over $\mathcal{A}$. Then we can establish $\mathcal{V}_C = \otimes_{\lambda=1}^\ell \mathcal{V}_1$ and define the free vector space over populations $\mathcal{V}_\mathcal{P} = \otimes_{\sigma=1}^N \mathcal{V}_C$ with dimension $L = \ell \cdot N$ and a^L elements. Finally let $S \subset \mathcal{V}_\mathcal{P}$ be the set of probability distributions over $\mathcal{P}_N$ (populations with size N), that is the state which identifies populations with their probability value.

Definition 1. *Let $S \subset \mathcal{V}_\mathcal{P}$, $n, k \in \mathcal{N}$ and $\{P_c, P_m\}$ a variation schedule. A Genetic Algorithm is a product of stochastic matrices (mutation, selection, crossover, etc..) act by matrix multiplication from the left:*

$$\mathbf{G^n} = \mathbf{F_n} \cdot \mathbf{C_{P_c}^k} \cdot \mathbf{M_{P_m}} \tag{4}$$

where $\mathbf{F_n}$ is the selection operator, $\mathbf{C^k_{P_c^n}} = C(K, P_c)$ is the simple crossover operator and $\mathbf{M_{P_m^n}}$ is the local mutation operator (see [6] and [10])

In order to improve the convergence speed of the algorithm we have to included another mechanisms such as elitist strategy (a further discussion about elitism, can be found in [11]). Another possibility is:

4   Guided Genetic Algorithms

In order to include statistical information into the algorithm we define the hybrid statistical genetic operator based on reduction operators as follows (in standard notation acting on populations). The value of the probability measure from individual p_i to q_i depends on contrast functions (i.e. based on cumulants) as:

$$P(\xi_{n+1} = p_i \mid \xi_n = q_i) = \frac{1}{\aleph(T_n)} \exp\left(-\frac{||q_i - \mathbf{S^n} \cdot p_i||^2}{T_n}\right); \qquad p_i, q_i \in \mathbf{C} \tag{5}$$

where $\aleph(T_n)$ is the normalization constant depending on temperature T_n which follows a variation decreasing schedule, that is $T_{n+1} < T_n$ converging to zero, n is the iteration and $\mathbf{S^n}$ is the step matrix which contains statistical properties, i.e. it can be expressed using quasi-Newton algorithms as [3]: $\mathbf{S^n} = (\mathbf{I} - \mu^n(\mathbf{C}^{1,\beta}_{y,y}\mathbf{S}^\beta_y - \mathbf{I}))$; $\forall p_i \in \mathbf{C}$, where $\mathbf{C}^{1,\beta}_{y,y}$ is the cross-cumulant matrix whose elements are $[\mathbf{C}^{\alpha,\beta}_{y,y}]_{ij} = Cum(y_i, (\alpha times), y_i), y_j, \beta times, y_j)$ and $\mathbf{S}^\beta_y$ is the sign matrix of the output cumulants.

Proposition 2. *The guiding operator can be described using its associated transition probability function (t.p.f.) by column stochastic matrices $\mathbf{M^n_G}$, $n \in \mathcal{N}$ acting on populations.*

1. The components are determined as follows: Let p and $q \in \wp_N$, then we have

$$P^n_{\mathbf{M_G}}(x, B) \equiv \langle q, \mathbf{M^n_G} p \rangle = \frac{N!}{z_{0q}! z_{1q}! \dots z_{a^L-1q}!} \prod_{i=0}^{a^L-1} \{P(i)\}^{z_{iq}}; \qquad p, q \in \mathcal{P}_N \tag{6}$$

where z_{iq} is the number of occurrences of individual i on population q and $P(i)$ is the probability of producing individual i from population p given in the equation 5. The value of the guiding probability $P(i) = P(i, f)$ depends on the fitness function used, i.e. see equations 5[1]:

$$P(i) = \frac{z_{ip} \exp\left(-\frac{||q_i - \mathbf{S^n} \cdot p_i||^2}{T_n}\right)}{\sum_{i=0}^{a^L-1} z_{ip} \exp\left(-\frac{||q_i - \mathbf{S^n} \cdot p_i||^2}{T_n}\right)}$$

where the denominator is the normalization constant and z_{ip} is the number of occurrences of individual i on population p.

---

[1] The transition probability matrix $P(i, f)$ must converge to a positive constant as $n \to \infty$ (since we can always define a suitable normalization constant). The fitness function or selection method of individuals used within must be injective.

2. *For every permutation $\pi \in \Pi_N$, we have $\pi \mathbf{M_G^n} = \mathbf{M_G^n} = \mathbf{M_G^n}\pi$.*
3. *$\mathbf{M_G^n}$ is an identity map on $\mathbf{U}$ in the optimum, that is $\langle p, \mathbf{M_G^n}p \rangle = 1$ if p is uniform; and has strictly positive diagonals since $\langle p, \mathbf{M_G^n}p \rangle > 0 \quad \forall p \in \mathcal{P_N}$.*
4. *All the coefficients of a MC consisting of the product of stochastic matrices: the simple crossover $(\mathbf{C_{P_c}^k})$, the local multiple mutation $(\mathbf{M_{P_m}^n})$ and the guiding operator $(\mathbf{M_G^n})$ for all $n, k \in \mathcal{N}$ are uniformly bounded away from 0.*

Proof: (1) follows from the transition probability between states. (2) is obvious and (3) follows from [8] and checking how matrices act on populations. (4) follows from the fact that $\mathbf{M_{P_m}^n}$ is fully positive acting on any stochastic matrix $\mathbf{S}$. Note that the guiding operator $\mathbf{M_G^n}$ can be viewed as a suitable fitness selection. Of course it can be viewed as a certain Reduction Operator since it preserves the best individuals into the next generation using a non-heuristic rule unlike the majority of GAs.

4.1 Theoretical Convergence Analysis

The convergence and strong and weak ergodicity of the proposed algorithm can be proved using several ways. A Markov Chain (MC) modelling a canonical GA (CGA) has been proved to be strongly ergodic (hence weak ergodic, see [10]). So we have to focus our attention on the transition probability matrix that emerges when we apply the guiding operator. Using the equation 5 in binary notation we have:

$$\langle q, \mathbf{G^n}p \rangle = \sum_{v \in \wp_N} \langle q, \mathbf{M_G^n}v \rangle \langle v, \mathbf{C^n}p \rangle \tag{7}$$

where $\mathbf{C^n}$ is the stochastic matrix associated to the CGA and $\mathbf{M_G^n}$ is given by equation 6.

Proposition 3. Weak Ergodicity: *A MC modelling a GGA satisfies weak ergodicity if the t.p.f associated to guiding operators converges to uniform populations (populations with the same individual)*

Proof: If we define a GGA on CGAs, the ergodicity properties depends on the new defined operator since CGAs are strongly ergodic as we said before. To prove this proposition we just have to check the convergence of the t.p.f. of the guiding operator on uniform populations. If the following condition is satisfied:

$$\langle u, \mathbf{G^n}p \rangle \to 1 \quad u \in \mathbf{U} \tag{8}$$

Then we can find a series of numbers which satisfies:

$$\sum_{n=1}^{\infty} \min_{n,p}(\langle u, \mathbf{G^n}p \rangle) = \infty \leq \sum_{n=1}^{\infty} \min_{q,p} \sum_{v \in \wp_N} \min\left(\langle v, \mathbf{M_G^n}p \rangle \langle v, \mathbf{C^n}q \rangle\right) \tag{9}$$

which is equivalent to weak ergodicity [9].

Proposition 4. Strong Ergodicity: *Let* $\mathbf{M}_{\mathbf{P}_m}^n$ *describe multiple local mutation,* $\mathbf{C}_{\mathbf{P}_c}^k$ *describe a model for crossover and* $\mathbf{F}^n$ *describe the fitness selection. Let* $(P_m^n, P_c^n)_n \in \mathcal{N}$ *be a variation schedule and* $(\phi_n)_{n \in \mathcal{N}}$ *a fitness scaling sequence associated to* $\mathbf{M}_G^n$ *describing the guiding operator according to this scaling*[2]. *Let* $\mathbf{C}^n = \mathbf{F}^n \cdot \mathbf{M}_{\mathbf{P}_m}^n \cdot \mathbf{C}_{\mathbf{P}_c}^k$ *represent the first n steps of a CGA. In this situation,*

$$v_\infty = \lim_{n \to \infty} \mathbf{G}^n v_0 = \lim_{n \to \infty} (\mathbf{M}_G^\infty \mathbf{C}^\infty)^n v_0 \qquad (10)$$

exists and is independent of the choice of v_0, *the initial probability distribution. Furthermore, the coefficients* $\langle v_\infty, p \rangle$ *of the limit probability distribution are strictly positive for every population* $p \in \wp_N$.

Proof: The demonstration of this proposition is rather obvious using the results of Theorem 16 in [10] and point 4 in Proposition 2. In order to obtain the results of the latter theorem we only have to replace the canonical selection operator $\mathbf{F}_n$ *with our guiding selection operator* $\mathbf{M}_G^n$ *which has the same essential properties.*

Proposition 5. Convergence to the Optimum: *Under the same conditions of propositions 3, 4 the GGA converges to the optimum.*

Proof: To reach this result, we check the probability to go from any uniform population to the population containing only the optimum. It is equal to 1 when $n \to \infty$:

$$\lim_{n \to \infty} \langle p^*, \mathbf{G}^n u \rangle = 1 \qquad (11)$$

since the GGA is an strongly ergodic MC, hence any population tends to uniform in time. If we check this expression we finally have the equation 11. In addition we have to use point 3 in Proposition 2 to make sure the optimum is the convergence point. Thus any guiding operator following a simulated annealing law converges to the optimum uniform population in time.

Assembling all the properties and restrictions described it's easy to prove that:

Proposition 6. *Any contrast function* Ψ *can be used to build a transition matrix probability* $P(i)$ *in the guiding operator* $\mathbf{M}_G^n$ *satisfying proposition 5.*

5   Conclusions

A GGA-based ICA method has been developed to solve BSS problem from (in this case) the linear mixtures of independent sources. The proposed method obtains a good performance overcoming the local minima problem over multidimensional domains [6]. Extensive simulation results proved the ability to extract independent components of the proposed method [6] but no explanation about convergence was given. The experimental work on nonlinear ICA is on the way.

---

[2] A scaling sequence $\phi_n : (\mathcal{R}^+)^N \to (\mathcal{R}^+)^N$ is a sequence of functions connected with a injective fitness criterion f as $f_n(p) = \phi_n(f(p))$ $p \in \wp_N$ such that $\mathbf{M}_G^\infty = \lim_{n \to \infty} \mathbf{M}_G^n$ exist.

Finally we have introduced demonstrations of the GA-ICA methods used till now [6],[12], etc. We have proved the convergence of the proposed algorithm to the optimum unlike the ICA algorithms which usually suffer of local minima and non-convergent cases. Any injective contrast function can be used to build a guiding operator, as a elitist strategy i.e. the Simulated Annealing function defined in section 4.

References

1. Barlow, H.B.: Possible principles underlying transformation of Sensory messages. Sensory Communication, MIT Press, New York, U.S.A. (1961).
2. Bell,A.J., Sejnowski, T.J.: An Information-Maximization Approach to Blind Separation and Blind Deconvolution. Neural Computation, 7 (1995) 1129-1159.
3. Hyvärinen, A., Oja, E.: A fast fixed point algorithm for independent component analysis. Neural Computation, 9 (1997) 1483-1492
4. Górriz, J.M., Puntonet, C.G., Salmerón, S., González, J.: New Model for Time Series Forecasting using rbfs and Exogenous Data. Neural Computing and Applications, 13 (2004) 101-111
5. Cao, X.R., Liu W.R.: General Approach to Blind Source Separation. IEEE Transactions on Signal Processing, 44 (1996) 562-571
6. Górriz J.M., Puntonet C.G., Salmerón, Roja, F.: Hybridizing GAs with ICA in Higher dimension. Lecture Notes in Computer Science 3195 (2004) 414-421
7. Comon, P.: ICA, a new concept? Signal Processing 36 (1994) 287-314
8. Cruces, S., Castedo L., Cichocki A.: Robust blind Source separation algorithms using cumulants. Neurocomputing 49 (2002) 87-118
9. Isaacson, D.L., Madsen D.W.: Markoc Chains: Theory and Applications. Wiley (1985)
10. Schmitt, L.M., Nehaniv, C.L., Fujii, R.H.: Linear Analysis of Genetic Algorithms. Theoretical Computer Science, 200 (1998) 101-134
11. Rudolph, G.: Convergence Analysis of Canonical Genetic Algorithms. IEEE Transactions on Neural Networks, 5 (1994) 96-101
12. Tan, Y., Wang, J.: Nonlinear Blind Source Separation Using Higher order Statistics and a Genetic Algorithm. IEEE Transactions on Evolutionary Computation, 5 (2001)

A Cascaded Ensemble Learning
for Independent Component Analysis

Jian Cheng[1], Kongqiao Wang[1], and Yenwei Chen[2]

[1] Nokia Research Center, No.11 He Ping Li Dong Jie, Nokia H1, Beijing 100013, China
{ext-jian.2.cheng,kongqiao.wang}@nokia.com
[2] College of Information Science and Engineering, Ritsumeikan University, 525-8577, Japan
chen@is.ritsumei.ac.jp

Abstract. In case of application to high-dimensional pattern recognition task, Independent Component Analysis (ICA) often suffers from two challenging problems. One is the small sample size problem. The other is the choice of basis functions (or independent components). Both problems make ICA classifier unstable and biased. To address the two problems, we propose an enhanced ICA algorithm using a cascaded ensemble learning scheme, named as Random Independent Subspace (RIS). A random resampling technique is used to generate a set of low dimensional feature subspaces in the original feature space and the whiten feature space, respectively. One classifier is constructed in each feature subspace. Then these classifiers are combined into an ensemble classifier using a final decision rule. Extensive experimentations performed on the FERET database suggest that the proposed method can improve the performance of ICA classifier.

1 Introduction

As a generalization of Principal Component Analysis (PCA), Independent Component Analysis (ICA) has been applied in many areas, such as face representation, signal processing, medical image analysis, etc [1],[2]. However, ICA method often encounters two challenging problems in practical applications. One is the small sample size problem, i.e. there are only a small number of training samples available in practice. The training sample size is too small compared with the dimensionality of feature space. In case of small sample size, the ICA classifier constructed on the training sets is biased and has a large variance, which result in the unstable performance of ICA classifier. The other is the choice of basis functions (or independent components) problem. The most common approach is to throw away the eigenvectors corresponding to little eigenvalues in pre-processing stage. However, the choice criterion of this approach satisfies the least reconstruction error in PCA, but is not optimal for ICA.

Ensemble learning techniques have become extremely popular over the last years because they combine multiple classifiers using traditional machine learning algorithms into an ensemble classifier, which is often demonstrated significantly better performance than single classifier. Boosting [3], bagging [4], and the random subspace method [5] are three classic ensemble learning algorithms. In spirit to the random subspace method, we propose a cascaded ensemble learning scheme for ICA classifier to overcome the small sample size problem and the choice problem of basis

J. Wang, X. Liao, and Z. Yi (Eds.): ISNN 2005, LNCS 3496, pp. 949–954, 2005.

functions at the same time, called Random Independent Subspace (RIS). Firstly, a set of low dimensional feature subspaces is generated by resampling the original feature space with replacement. After whitened each feature subspace, we can obtain a set of orthogonal basis functions for ICA classifiers. Then selecting randomly a subset of orthogonal basis functions to design an ICA classifier. For a test sample, each ICA classifier gives a prediction. The final predictions are decided by aggregating all predictions using a final decision rule.

The rest of this paper is organized as follows. In section 2, we briefly introduce ICA algorithm. The proposed algorithm is given in section 3. Some experimental results are provided in section 4. Finally, we give concluding remarks in section 5.

2   Independent Component Analysis

Independent Component Analysis (ICA) technique is capable of finding a set of linear basis vectors to make the higher-order statistical independence besides the second-order statistical independence in PCA. Assume a set of training samples $X = [x_1, x_2, \cdots, x_n]$, where each column vector x_i represents a N-dimensional sample and the total number of training samples is n. The general model of ICA can be described as follows:

$$X = A * S . \tag{1}$$

where $S = [s_1, s_2, \cdots, s_n]$ is the coefficient, A is a mixing matrix and its column vectors are basis functions. The independent component analysis is to find a separating matrix W_I, so that

$$U_I = W_I * X . \tag{2}$$

approximates the independent component S, possibly permuted and rescaled. The components of S are as mutual independent as possible. A number of learning algorithms have been developed to learn approximate solutions for ICA. In this paper, we use Bell and Sejnowski's InfoMax algorithm [6], which maximize the mutual information between the inputs and outputs of a neural network.

ICA can be speeded up by a preprocessing operation W_P prior to learning, known as whitening or sphering. The transformed data is zero mean and decorrelated data:

$$W_P X X^T W_P^T = I . \tag{3}$$

This transformation can be accomplished by PCA (eigenvalue decomposition). In fact, when $W_P = \Lambda^{-\frac{1}{2}} V^T$, the Eq. (3) can be satisfied. Here, Λ and V are the eigenvalues matrix and eigenvectors matrix of covariance matrix of X, respectively.

3   Random Independent Subspace

ICA is demonstrated to be an effective method [1],[2],[6]. However, there are often only a small number of training samples are available in practice, i.e. the number of training samples is less than the dimensionality, which make the covariance matrix singular in ICA. It induces ICA classifier to be unstable. On the other hand, the choice of basis functions is still an open problem. The most common approach is to throw away the eigenvectors corresponding to the little eigenvalues in whitening stage. However, the choice criterion of this approach satisfies the least reconstruction error in PCA, but is not optimal for ICA.

In order to improve the performance and overcome the shortcoming of ICA classifier, we have attempted to design ICA classifier using ensemble learning approach [7]. In this paper, we proposed an enhanced ICA method adopting a cascaded ensemble learning scheme, named as **Random Independent Subspace (RIS)**. Let $X = [x_1, x_2, \cdots, x_n]$ be training set matrix, where n is the number of training samples. Each column $x_i = (x_{i1}, x_{i2}, \cdots, x_{iN})^T \in R^N$ is a N-dimensional feature representation for a training sample. Usually, N is very large. The cascaded ensemble learning scheme includes two level resampling. Firstly, We randomly resample K r-dimensional feature subspaces from the N-dimensional feature space, where $r < N$. The resampling is repeated with replacement. We name this resampling level 1. The new training set are $X^k = [x_1^k, x_2^k, \cdots, x_n^k]$, $k = 1, 2, \cdots, K$, where $x_i^k = (x_{i1}^k, x_{i2}^k, \cdots, x_{ir}^k)$. Then, each training set X^k is whitened. Denote the whitened training set $\widetilde{X}^k$. Secondly, we randomly resample s $(s < r)$ features from the whitened space to design ICA classifiers C^{kh}, $h = 1, \cdots, H$ which is named child classifier. This resampling is named level 2. For a new test sample y, it is firstly projected into each subspace and given a prediction by each ICA child classifier C^{kh}, $h = 1, \cdots, H$, in level 2. Then the predictions in level 2 are aggregated to decide the prediction of their parent classifier C^k in level 1. The final prediction is decided by aggregating all predictions in level 1. The pseudo codes are shown in Fig. 1:

Given a training set $X = [x_1, x_2, \cdots, x_n]$.

Step 1: Repeat for $k = 1, 2, \cdots, K$

    (a) Select randomly r features from the original N-dimensional feature space. The new training set $X^k = [x_1^k, x_2^k, \cdots, x_n^k]$, where $x_i^k = (x_{i1}^k, x_{i2}^k, \cdots, x_{ir}^k)$.

    (b) Perform whitening on the new training set X^k, and denote whitened training set $\widetilde{X}^k$.

    (c) Repeat for $h = 1, 2, \cdots, H$

        Select randomly s features from the whitened feature subspace for $\widetilde{X}^k$. The new training set $\widetilde{X}^{kh} = [\widetilde{x}_1^{kh}, \widetilde{x}_2^{kh}, \cdots, \widetilde{x}_n^{kh}]$, where $\widetilde{x}_i^{kh} = (\widetilde{x}_{i1}^{kh}, \widetilde{x}_{i2}^{kh}, \cdots, \widetilde{x}_{is}^{kh})$.

        Construct an ICA classifier C^{kh} on $\widetilde{X}^{kh}$.

    (d) Combine classifiers C^{kh}, $h = 1, 2, \cdots, H$, into an ensemble classifier C^k using the final decision rule.

Step 2: Combine classifiers C^k, $k = 1, 2, \cdots, K$, into an ensemble classifier C using the final decision rule.

Fig. 1. Random Independent Subspace (RIS).

In this paper, we use majority voting as the final decision rule, which simply assigns the test sample with the class label that appears most frequently in $C^k(y)$, $k = 1, 2, \cdots, K$.

4 Experiments

The experiments are performed on FERET face database [8]. We use 1002 front view images as training set. The gallery FA and probe FB contain 1195 persons. There is only one image for each person in FA and FB. All the images have been reduced to 48×54 by eye location. Histogram equalization is performed as preprocessing on all images. Few samples are shown in Fig.2.

Fig. 2. Upper are the original images and bottom are preprocessed image.

Face recognition performance is evaluated by the nearest neighbor algorithm using cosine similarity measure:

$$d(y_i, y_j) = 1 - \frac{y_i^T \cdot y_j}{\| y_i \| \cdot \| y_j \|}. \tag{4}$$

The first resampling in the original feature space (Level 1) gets rid of the shortcoming brought by the small sample size problem while the second resampling in the whitened feature space (Level 2) resolves the choice problem of basis functions. All child classifiers in Level 2 give voting to their parent classifiers in Level 1, and then all parent classifiers in Level 1 vote to the final prediction.

The total of features is $48 \times 54 = 2596$ in the original feature space. We randomly resample half of the total of features, i.e. 1298 in level 1. Then PCA technique is used to whiten these subspaces. Because zero and little eigenvalues usually correspond to noise, we only retain the eigenvectors corresponding to the 150 leading eigenvalues as features. In order to improve the accuracy of each classifier in level 2, we fix these features corresponding to the 20 largest eigenvalues, and the others are sampled randomly from the residual 130 features.

In most cases, the ensemble learning with more classifiers will achieve higher accuracy rate while more classifiers bring more computational expense. In practice the number of classifiers in each level is traded off in order to obtain better accuracy rate while require less computational expense. Fig. 3 shows that 10 times in the original feature space (level 1) and 5 times in the orthogonal whitened feature space (level 2) seem to be better choice.

As shown in Table 1, by aggregating 5 child classifiers in Level 2 (in whitened feature space), the parent classifiers in Level 1 are obvious better than those in

Level 2. For example, the first classifier in Level 1 get 80.42%, which is combined from 5 child classifiers (in Level 2) whose accuracies are 75.56%, 76.73%, 76.65%, 77.82% and 78.57%. Further, the 10 classifiers in Level 1 are aggregated into an ensemble classifier that obtains a best accuracy rate 85.36%. Fig.4 shows that the performance of RIS has a significant improvement compared with that of ICA. These experimental results suggest that the cascaded ensemble learning scheme can improve ICA classifier. We also consider the effect of the number of sampled features on accuracy rate. Experimental results show that RIS performs well as long as the number of sampled features more than 20%. For limited space, the detailed results are not described here.

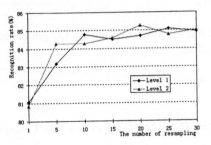

Fig. 3. Accuracy rate varies with the number of classifiers in level 1 and level 2, respectively.

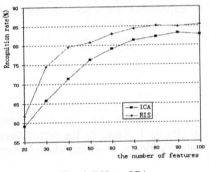

Fig. 4. RIS vs. ICA.

Table 1. Level 1 is the accuracy rate of parent classifiers. Level 2 is the accuracy rate of child classifiers constructed in sampled whitened feature subspace. The final is the accuracy rate of RIS aggregated from the 10 classifiers in Level 1.

Accuracy rate (%)		1	2	3	4	5	6	7	8	9	10
Level 2	1	75.56	78.91	76.48	77.48	77.82	77.99	77.99	78.99	77.48	77.99
	2	76.73	76.82	77.57	77.90	77.90	77.99	76.65	78.49	77.82	77.99
	3	76.65	78.91	77.32	77.90	76.48	77.07	76.56	77.40	76.31	78.07
	4	77.82	77.82	76.90	75.64	77.32	78.91	77.82	78.41	78.32	78.49
	5	78.57	76.56	75.56	78.82	79.24	78.07	77.40	78.66	77.57	77.99
Level 1		80.42	81.42	79.83	80.42	81.00	80.84	80.25	81.51	80.59	80.17
Final		85.36									

5 Conclusions

In pattern classification using ICA classifier, we often encounter two challenging problems: small sample size problem and the choice problem of basis functions, which result in the unstable performance of ICA classifier. In order to improve the performance of ICA classifier, we propose to design ICA classifier in cascaded ensemble learning scheme. The experimental results of face recognition demonstrate that the proposed method is effective.

Acknowledgements

The first author would like to sincerely thank Dr. Qingshan Liu and Prof. Hanqing Lu for their helpful suggestions.

References

1. Comon, P.: Independent Component Analysis – a New Concept?. Signal Processing, **36** (1994) 287-314
2. Bartlett, M.S., Lades, H.M., Sejnowski, T.J.: Independent Component Representations for Face Recognition. In the Proc. of SPIE, **3299** (1998) 528-539
3. Freund, Y., Schapire, E.: Experiments with a New Boosting Algorithm. Proc. of 13th Int'l Conf. on Machine Learning (1996) 148-156
4. Breiman, L.: Bagging Predictors. Machine Learning J., **24** (1996) 123-140
5. Ho, T.K.: The Random Subspace Method for Constructing Decision Forests. IEEE Trans. Pattern Analysis and Machine Intelligence, **20** (1998) 832-844
6. Bell, A.J., Sejnowski, T.J.: An Information-maximization Approach to Blind Separation and Blind Deconvolution. Neural Computation, **7** (1995) 1129-1159
7. Cheng, J., Liu, Q., Lu, H., Chen, Y.W.: Random Independent Subspace for Face Recognition. M.Gh.Negoita et al. (Eds.), KES 2004, LNAI **3214** (2004) 352-358
8. Philips, P.J., Wechsler, H., Huang, J., Rauss, P.: The FERET Database and Evaluation Procedure for Face Recognition Algorithms. Image and Vision Computing, **16** (1998) 295-306

A Step by Step Optimization Approach to Independent Component Analysis[*]

Dengpan Gao, Jinwen Ma[**], and Qiansheng Cheng

Department of Information Science, School of Mathematical Sciences
and LMAM, Peking University, Beijing 100871, China

Abstract. The independent component analysis (ICA) problem originates from many practical areas, but there has not been any mathematical theory to solve it completely. In this paper, we establish a mathematical theory to solve it under the condition that the number of super-Gaussian sources is known. According to this theory, a step by step optimization algorithm is proposed and demonstrated well on solving the ICA problem with both the super- and sub-Gaussian sources.

1 Introduction

Independent component analysis (ICA) [1] tries to blindly separate some independent sources $\mathbf{s}$ from their linear mixture $\mathbf{x} = \mathbf{As}$ via

$$\mathbf{y} = \mathbf{Wx}, \quad \mathbf{x} \in \mathbb{R}^m, \quad \mathbf{y} \in \mathbb{R}^n, \quad \mathbf{W} \in \mathbb{R}^{n \times m}, \tag{1}$$

where $\mathbf{A}$ is a mixing matrix, and $\mathbf{W}$ is the so-called de-mixing matrix to be estimated. This ICA problem is generally solved from the basic requirement that the recovered $\mathbf{y}$ should be component-wise independent so that it can be transformed into an optimization problem (e.g., [2]-[4]). Although several ICA algorithms were proposed and demonstrated well, they are generally heuristic and there has been no mathematical theory to ensure any algorithm to solve the ICA problem correctly. Essentially, the conditions on solving the ICA problem are not clear yet.

In this paper, we study the ICA problem from the point of view of step by step optimization on some related objective functions under the condition that the number of super-Gaussian (or sub-Gaussian) sources is known. By mathematical analysis, we prove that the step by step optimization (SBSO) process leads to a feasible solution. Accordingly, we propose a step by step optimization algorithm which is demonstrated well by the experiments.

2 The SBSO Theory

We begin to introduce some assumptions and definitions. In the ICA problem, we let $n = m$ and assume that the sources $\mathbf{s}$, the linearly mixing signals $\mathbf{x}$ and

---

[*] This work was supported by the Natural Science Foundation of China for Projects 60471054 and 40035010.

[**] The corresponding author, Email: jwma@math.pku.edu.cn.

J. Wang, X. Liao, and Z. Yi (Eds.): ISNN 2005, LNCS 3496, pp. 955–960, 2005.

the recovered signals $\mathbf{y}$ are all whitened so that the mixing matrix $\mathbf{A}$, the de-mixing matrix $\mathbf{W}$ and $\mathbf{R} = \mathbf{WA}$ are all orthogonal. From statistics, the kurtosis of a random variable z is defined as:

$$\kappa(z) = E\{z^4\} - 3E^2\{z^2\} \tag{2}$$

If $\kappa(z) > 0$, z or its probability distribution function (pdf) is called super-Gaussian; otherwise, it is called sub-Gaussian. If $z_1, \cdots, z_n$ are n independent random variables, we have

$$\kappa(\sum_{i=1}^{n} \alpha_i z_i) = \sum_{i=1}^{n} \alpha_i^4 \kappa(z_i) \tag{3}$$

where $\alpha_i(i = 1, \cdots, n)$ are constants. We now give a lemma for the SBSO theory as follows (see [5] for the proof).

Lemma 1. Let us denote by e_j the jth vector of the canonical basis of R^n. Consider the following function $f(r) = \frac{1}{4}\sum_{j=1}^{n} r_j^4 \kappa(s_j)$, where $\kappa(s_j)$ is the kurtosis of s_j and $r = (r_1, \cdots, r_n)^T \in R^n$ satisfying $\|r\|_2^2 = \sum_{i=1}^{n} r_i^2 = 1$ ($\|\cdot\|_2$ always represents the Euclidean norm). Moreover, we let p be a nonnegative number such that $\kappa(s_1) > 0, \cdots, \kappa(s_p) > 0$ and $\kappa(s_{p+1}) < 0, \cdots, \kappa(s_n) < 0$. We then have

(1) If $p \neq 0$ and $p \neq n$, the arguments of the local maxima of $f(r)$ on $\mathbb{S} = \{r \in R^n : \|r\|_2^2 = 1\}$ are the vectors e_j and $-e_j$ for $j = 1, \cdots, p$ and the arguments of the local minima of $f(r)$ are the vectors e_j and $-e_j$ for $j = p+1, \cdots, n$.

(2) If $p = 0$, the arguments of the local minima of $f(r)$ on $\mathbb{S}$ are the vectors e_j and $-e_j$ for $j = 1, \cdots, n$ and the arguments of the local maxima are reduced to the vectors r which satisfying $r_j^2 = \frac{\sigma}{\kappa(s_j)}$ for each j, where $\sigma = (\sum_j^n \frac{1}{\kappa(s_j)})^{-1}$.

(3) If $p = n$, the arguments of the local maxima of $f(r)$ on $\mathbb{S}$ are the vectors e_j and $-e_j$ for $j = 1, \cdots, n$ and the arguments of the local minima are reduced to the vectors r which satisfying $r_j^2 = \frac{\sigma}{\kappa(s_j)}$ for each j, where $\sigma = (\sum_j^n \frac{1}{\kappa(s_j)})^{-1}$.

With the above preparations, we have the following two theorems.

Theorem 1. Suppose that $W_i = (w_{i1}, \cdots, w_{in})$ is the ith row of $\mathbf{W}$, i.e., $\mathbf{W} = (W_1^T, \cdots, W_n^T)^T$. The following two optimization problems are equivalent.

$$(1). \max(or\ min)\quad f(W_i) = \kappa(W_i x)\quad s.t.\quad \|W_i\|_2^2 = 1;$$

$$(2). \max(or\ min)\quad f(R_i) = \kappa(R_i s) = \sum_{j=1}^{n} r_{ij}^4 \kappa(s_j)\quad s.t.\quad \|R_i\|_2^2 = 1,$$

where $R_i = (r_{i1}, \cdots, r_{in})$ is the ith row of $\mathbf{R}$.

Proof. According to $\mathbf{R} = \mathbf{WA}$ and the linearity properties of the kurtosis, we have for each i

$$\kappa(W_i x) \Longleftrightarrow \kappa(W_i As) \Longleftrightarrow \kappa(R_i s) \Longleftrightarrow \kappa(\sum_{j=1}^{n} r_{ij} s_j) \Longleftrightarrow \sum_{j=1}^{n} r_{ij}^4 \kappa(s_j) \tag{4}$$

According to the same orthogonality for both $\mathbf{W}$ and $\mathbf{R}$, we certainly have

$$\|R_i\|_2^2 = 1 \iff \|W_i\|_2^2 = 1. \tag{5}$$

Therefore, the two constrained optimization problems are equivalent. ∎

By Theorem 1, we get that if R_i^* is a maximum (or minimum) of the first problem, then the correspondence $W_i^* = R_i^* A^T$ is a maximum (or minimum) of the second problem. So, if we can solve R_i^* step by step to a permutation matrix $\mathbf{R}^*$, the corresponding $\mathbf{W}^*$ becomes a feasible solution of the ICA problem which can be solved step by step in the same way. Actually, when we solve the second problem step by step from $R_1^*, R_2^*, \cdots, R_n^*$ with the orthogonality constraint and the dimension reduction property, it can be easily proved via Lemma 1 that this step by step optimization process leads to a unique solution—a permutation matrix, which is summarized as the following theorem.

Theorem 2. The solution $R^* = (R_1^*, \cdots, R_n^*)$ of the the second problem by the step by step optimization process with the orthogonality constraint is a permutation matrix.

3   The SBSO Algorithm

According to Theorem 1&2, we can construct a step by step optimization (SBSO) algorithm for solving $\mathbf{W}$ correspondingly. In general, we have the following SBSO algorithm with p being the number of super-Gaussian sources given in advance.

Step1. Maximize the function $f(W_1) = \kappa(W_1 x)$ on $\mathbb{S}$ to get one maximum W_1^*;

Step2. At the number $i(1 < i \leq p)$, we have obtained the row vectors $W_1^*, \cdots, W_{i-1}^*$. We then maximize the function $f(W_i) = \kappa(W_i x)$ in $H^\perp(W_1, \cdots, W_{i-1})$ to get a maximum W_i^*, where $H^\perp(W_1, \cdots, W_{i-1})$ is the orthogonal complement space of the linear spanning subspace $\mathcal{L}(W_1, \cdots, W_i)$.

Step3. At the number $i(p + 1 \leq i \leq n)$, we minimize the function $f(W_i) = \kappa(W_i x)$ in $H^\perp(W_1, \cdots, W_{i-1})$ to get a minimum W_i^*.

Specifically, with $\mathbf{s}, \mathbf{x}$ and $\mathbf{y}$ whitened we can get each W_i^* by solving the constrained optimization problem as follows. When $i \leq p$, we take the maximization operation; otherwise, we take the minimization operation.

At the beginning time $i = 1$, we need to solve the following constrained optimization problem to get W_1^*:

$$\max(or\ min) \quad f(W_1) = \kappa(W_1 x) = E\{(W_1 x)^4\} - 3 \quad s.t. \quad \|W_1\|_2^2 = 1. \tag{6}$$

The normalization constraint can be naturally satisfied with the following substitution:

$$W_1 = \frac{\hat{W}_1}{\|\hat{W}_1\|_2}. \tag{7}$$

Via this substitution, the constrained optimization problem is turned into

$$\max(or\ min) \quad f(\hat{W}_1) = \kappa(\frac{\hat{W}_1}{\|\hat{W}_1\|_2} x) = E\{(\frac{\hat{W}_1}{\|\hat{W}_1\|_2} x)^4\} - 3. \tag{8}$$

With the following derivative of $f(\hat{W}_1)$ with respect to $\hat{W}_1$:

$$\frac{\partial f(\hat{W}_1)}{\partial \hat{W}_1} = 4E\{\frac{(\hat{W}_1 x)^3}{\|\hat{W}_1\|_2^4}x^T - \frac{(\hat{W}_1 x)^4}{\|\hat{W}_1\|_2^6}\hat{W}_1\}, \tag{9}$$

we can construct an adaptive gradient learning rule for W_1^* as follows.

$$\Delta\hat{W}_1 \propto (\frac{(\hat{W}_1 x)^3}{\|\hat{W}_1\|_2^4}x^T - \frac{(\hat{W}_1 x)^4}{\|\hat{W}_1\|_2^6}\hat{W}_1). \tag{10}$$

Here, the adaptive learning rule is just for the maximization problem. If the optimization problem is for the minimization, we just need to modify $\Delta\hat{W}_1$ in the opposite direction, i.e., to add the minus sign to the derivative at the right hand of Eq.(10). It will be the same in the following general case.

At the general time $i(1 < i \leq n)$, we have already obtained $W_1^*, \cdots, W_{i-1}^*$ in advance. We need to maximize (or minimize) the following constrained objective function to get W_i^*:

$$f(W_i) = \kappa(W_i x) = E\{(W_i x)^4\} - 3, \tag{11}$$

$$s.t. \quad \|W_i\|_2^2 = 1, W_i \perp W_k(k = 1, \cdots, i-1). \tag{12}$$

Since $W_1^*, \cdots, W_{i-1}^*$ are pairwise orthogonal vectors, we can expand a set of $n - i + 1$ normalized orthogonal vectors $W_i', \cdots, W_n'$ such that $W_1^*, \cdots, W_{i-1}^*$, $W_i', \cdots, W_n'$ form a canonical basis of $\mathbb{R}^n$. In this way, W_i^* should be a linear combination of the vectors $W_i', \cdots, W_n'$. Thus, we have

$$W_i^* = \sum_{j=i}^{n} \alpha_j W_j'. \tag{13}$$

With this expression, the constrained objective function can be represented by

$$f(\alpha) = \kappa(\frac{\hat{W}_i}{\|\hat{W}_i\|_2}X) = E\{(\frac{\hat{W}_i}{\|\hat{W}_i\|_2}x)^4\} - 3 \tag{14}$$

with the parameter $\alpha = (\alpha_i, \cdots, \alpha_n)$, where $\hat{W}_i = \sum_{j=i}^{n} \alpha_j W_j'$. Then, we can get the derivative of $f(\alpha)$ as follows.

$$\frac{\partial f(\alpha)}{\partial \alpha} = \frac{\partial f(\alpha)}{\partial \hat{W}_i}\frac{\partial \hat{W}_i}{\partial \alpha} = 4E\{[(\frac{(\hat{W}_i x)^3}{\|\hat{W}_i\|_2^4}x^T - \frac{(\hat{W}_i x)^4}{\|\hat{W}_i\|_2^6}\hat{W}_i)]W'\}, \tag{15}$$

where $W' = (W_i'^T, \cdots, W_n'^T)$.

According to this derivative, we can construct an adaptive gradient learning rule for α as follows.

$$\Delta\alpha \propto [(\frac{(\hat{W}_i X)^3}{\|\hat{W}_i\|_2^4}X^T - \frac{(\hat{W}_i X)^4}{\|\hat{W}_i\|_2^6}\hat{W}_i)]W'. \tag{16}$$

When α is obtained from the above learning rule, we finally get W_i^* from Eq.(13).

As a result, the SBSO algorithm can be implemented from a sample data set of the linearly mixed signals $\mathbf{x}$, which will be demonstrated in the following section.

4   Simulation Experiments

We conducted the experiments on the ICA problem of five independent sources in which there are three super-Gaussian sources generated from the the the Chisquare distribution $\chi^2(6)$, the F distribution $F(10, 50)$, and the Exponential distribution $E(0.5)$, respectively, and two sub-Gaussian sources generated from the β distribution $\beta(2, 2)$ and the Uniform distribution $U([0, 1])$, respectively. From each distribution, 100000 i.i.d. samples were generated to form a source. The linearly mixed signals were then generated from the five source signals in parallel via the following mixing matrix:

$$\mathbf{A} = \begin{bmatrix} 0.9943 & 0.3323 & 0.9538 & 0.7544 & 0.2482 \\ 0.3905 & 0.9552 & 0.4567 & 0.7628 & 0.1095 \\ 0.0449 & 0.7603 & 0.3382 & 0.0719 & 0.5296 \\ 0.7210 & 0.2491 & 0.7130 & 0.4943 & 0.2824 \\ 0.6461 & 0.2931 & 0.9977 & 0.5490 & 0.0969 \end{bmatrix}$$

The learning rate for each adaptive gradient learning rule in the implementation of the SBSO algorithm was selected as $\eta = 0.001$ by experience and these adaptive gradient learning algorithms were stopped when all the 100000 data points of the mixed signals had been passed only once.

The results of the SBSO algorithm are given in Table 1. As a feasible solution of the ICA problem in the whitened situation, the obtained $\mathbf{W}$ should make $\mathbf{WA} = \mathbf{P}$ be satisfied or approximately satisfied to a certain extent, where P is a permutation matrix.

Table 1. The result of $\mathbf{WA}$ in which $\mathbf{W}$ was obtained by the SBSEA algorithm.

$$\mathbf{WA} = \begin{bmatrix} \mathbf{0.9988} & 0.0292 & 0.0097 & 0.0292 & -0.0264 \\ 0.0264 & \mathbf{-0.9989} & -0.0276 & 0.0195 & -0.0150 \\ 0.0095 & 0.0279 & \mathbf{-0.9989} & -0.0329 & 0.0117 \\ 0.0262 & -0.0125 & 0.0128 & 0.0035 & \mathbf{0.9995} \\ -0.0293 & 0.0213 & -0.0345 & \mathbf{0.9988} & -0.0012 \end{bmatrix}$$

From Table 1, we can observe that the SBSO algorithm can really solve the ICA problem with both super- and sub-Gaussian sources as long as we know the number of super-Gaussian (or sub-Gaussian) sources. That is, the kurtosis signs of sources are sufficient to solve the general ICA problem. As compared with the natural gradient learning algorithm with Lee's switching criterion [6], the SBSO algorithm converges to a better solution. However, the SBSO algorithm generally needs more computation.

Theoretically, the SBSO theory and algorithm also provide a support to the one-bit-matching conjecture which states that "all the sources can be separated as long as there is a one-to-one same-sign-correspondence between the kurtosis signs of all source pdf's and the kurtosis signs of all model pdf's", which was summarized by Xu et al. [7] and theoretically developed in [8]-[9]. However, we don't need to select the model pdf's in the SBSO algorithm.

5   Conclusions and Further Works

We have investigated the ICA problem from a step by step optimization (SBSO) process and established an SBSO theory with the condition that the number of super-Gaussian sources is known. According to the theory, the SBSO algorithm is proposed with help of certain adaptive gradient learning rules in its implementation. It is demonstrated by the experiments that the SBSO algorithm can solve the ICA problem of both super- and sub-Gaussian sources with a good result.

In practice, the number p of super-Gaussian sources may not be available. In this situation, the SBSO algorithm cannot work. However, from its learning process we can find that $p + 1$ can be checked by the dependence of $y_i = W_i^* x$ to the previous observed outputs $y_1, \cdots, y_{i-1}$. With this checking step, we can conduct the SBSO algorithm without the information of p. This improvement will be done in our further works.

References

1. Comon, P.: Independent Component Analysis–a New Concept. Signal Processing, **36** (1994) 287-314
2. Amari, S. I., Cichocki, A., Yang, H.: A New Learning Algorithm for Blind Separation of Sources. Advances in Neural Information Processing, **8** (1996) 757-763
3. Attias, H.: Independent Factor Analysis. Neural Computation, **11** (1999) 803-851
4. Bell, A., Sejnowski, T.: An Information-maximization Approach to Blind Separation and Blind Deconvolution. Neural Computation. **7** (1995) 1129-1159
5. Delfosse, N., Loubaton, P.: Adaptive Blind Separation of Independent Sources: a Deflation Approach. Signal Processing, **45** (1995) 59-83
6. Lee, T. W., Girolami, M., Sejnowski, & T. J.: Independent Component Analysis Using an Extended Infomax Algorithm for Mixed Subgaussian and Supergaussian Sources. Neural Computation, **11** (1999) 417-441
7. Xu, L., Cheung, C. C., Amari, S. I.: Further Results on Nonlinearity and Separation Capability of a Liner Mixture ICA Method and Learned LPM. Proceedings of the I&ANN'98, (1998) 39-45
8. Liu, Z. Y., Chiu, K. C., Xu, L.: One-bit-matching Conjecture for Independent Component Analysis. Neural Computation, **16** (2004) 383-399
9. Ma, J., Liu, Z., Xu, L.: A Further Result on the ICA One-bit-matching Conjecture. Neural Computation, **17** (2005) 331-334

Self-adaptive FastICA
Based on Generalized Gaussian Model*

Gang Wang[1,2], Xin Xu[1,3], and Dewen Hu[1]

[1] Department of Automatic Control, National University of Defense Technology,
Changsha, Hunan 410073, China
dhu@nudt.edu.cn
[2] Telecommunication Engineering Institute, Air Force Engineering University,
Xi'an, Shanxi 710077, China
[3] School of Computer, National University of Defense Technology, Changsha,
Hunan 410073, China

Abstract. Activation function is a crucial factor in independent component analysis (ICA) and the best one is the score function defined on the probability density function (pdf) of the source. However, in FastICA, the activation function has to be selected from several predefined choices according to the prior knowledge of the sources, and the problem of how to select or optimize activation function has not been solved yet. In this paper, self-adaptive FastICA is presented based on the generalized Gaussian model (GGM). By combining the optimization of the GGM parameter and that of the demixing vector, a general framework for self-adaptive FastICA is proposed. Convergence and stability of the proposed algorithm are also addressed. Simulation results show that self-adaptive FastICA is effective in parameter optimization and has better accuracy than traditional FastICA.

1 Introduction

FastICA is a fast fixed-point algorithm for ICA firstly proposed by Hyvärinen and Oja in [1] based on negentropy. For its simplicity, robustness and fast convergence speed, it is now one of the most popular algorithms of ICA [2, 3].

In this popular algorithm, the activation function, which is a crucial factor for ICA, has to be selected from several predefined choices based on the prior knowledge of the sources. However, for a certain source, the best activation function is the score function defined on its probability distribution function (pdf) [4–6], and a suitable selection may become difficult and impractical. This problem in FastICA essentially belongs to self-adaptive blind source separation (BSS) [6]. Although in self-adaptive BSS, many models, such as the Pearson model, the Gaussian mixture [5, 6], have been proposed recently, the problem in FastICA has not been settled yet.

---

* Supported by National Natural Science Foundation of China (30370416, 60303012, & 60234030).

J. Wang, X. Liao, and Z. Yi (Eds.): ISNN 2005, LNCS 3496, pp. 961–966, 2005.

In this paper, the problem of selecting activation function adaptively for FastICA is addressed based on the generalized Gaussian model (GGM) [6, 7] as an extension of self-adaptive BSS [6]. The self-adaptive FastICA algorithm is presented where the optimization of the GGM parameter and that of the demixing vector are combined.

This paper is organized as follows. In section 2 some preliminaries about FastICA, GGM and score function are given. The iteration rule for score function is given in section 3 and a general framework for self-adaptive FastICA is presented as well. In section 4 performance analysis is given. Two simulations are shown in section 5 and in section 6 conclusions are drawn.

2   Preliminaries

FastICA and Activation Function FastICA [1] is a fast fixed-point algorithm based on negentropy for ICA. And the iteration for demixing vector is

$$\mathbf{w} \leftarrow E\{\mathbf{z}g(\mathbf{w}^T\mathbf{z})\} - E\{g'(\mathbf{w}^T\mathbf{z})\mathbf{w}\}, \mathbf{w} \leftarrow \mathbf{w}/\|\mathbf{w}\|. \tag{1}$$

where z are the centered and pre-whitened mixtures, $g(\cdot)$ is the activation function and $g'(\cdot)$ the derivative of $g(\cdot)$ [2]. The following predefined activation functions are available for FastICA

$$\begin{aligned}
g_1(u) &= \tanh(a_1 u), \\
g_2(u) &= u\exp(-a_2 u^2/2), \\
g_3(u) &= u^3,
\end{aligned} \tag{2}$$

where $1 \le a_1 \le 2$, $a_2 \approx 1$ are constants. And wherein $g_2(\cdot)$ is recommended for highly super-Gaussian source, $g_3(\cdot)$ for sub-Gaussian, and $g_1(\cdot)$ for general-purpose [2, 3].

Generalized Gaussian Distribution GGM is a conventional distribution model defined as

$$p_g(y, \theta, \sigma) = \left(2A(\theta, \sigma)\Gamma(1 + \frac{1}{\theta})\right)^{-1} \exp\left(-\left|\frac{y}{A(\theta, \sigma)}\right|^\theta\right), \tag{3}$$

where $A(\theta, \sigma) = \sqrt{\sigma^2 \Gamma(1/\theta)/\Gamma(3/\theta)}$ and $\Gamma(x) = \int_0^\infty \tau^{x-1} e^{-\tau} d\tau$ is the standard Gamma function [5, 6]. Parameter σ is the variance of variable y, and θ describes the sharpness of the distribution. A variety of pdfs can be obtained when θ changes, such as the Laplacian distribution ($p_g(y, 1, 1)$), Gaussian ($p_g(y, 2, 1)$) and nearly uniformly distribution ($p_g(y, 4, 1)$).

Score Function Many approaches to ICA, such as maximization Likelihood (ML), and Infomax may lead to the same cost function [3]

$$L(y, \theta, \mathbf{W}) = -\log(|\det(\mathbf{W})|) - \sum_{i=1}^n \log q_i(y_i, \theta_i), \tag{4}$$

where $q_i(y_i, \theta_i)$ is the model for the marginal pdf of y_i. And the score function is defined on the true pdf of y_i

$$\varphi(y_i) = -d\log p(y_i)/dy_i. \tag{5}$$

When $q_i(y_i, \theta_i)$ approaches $p(y_i)$, the best activation function is obtained [3, 6].

3   Framework for Self-adaptive FastICA

For the constraints on variables in FastICA, σ in (3) always equals 1. Denote $N(\theta) = -\log(2A(\theta))\Gamma(1 + 1/\theta))$, and the GGM can be simplified as

$$p_g(y, \theta) = \exp\left(-(y/A(\theta))^\theta + N(\theta)\right). \tag{6}$$

3.1   Score Function and Adaptive Rule

Similar to the derivation of the iteration rule to demixing matrix and GGM parameter for object function (4) in [6], object function for estimated signal y_i with demixing vector is defined as $l(y_i, \theta_i, \mathrm{w}) = -\log(p_g(y_i, \theta_i, \mathrm{w}))$ on its pdf (but not the whole demixing matrix and all the pdfs as addressed in [6]). And as an extension of self-adaptive BSS based on Cardoso's equivariant algorithm [6], the following adaptive rule can be obtained.

Score function to θ_i for recovered signal y_i can be expressed as

$$\frac{\partial l(y_i, \theta_i, \mathrm{w})}{\partial \theta_i} = -\left(\frac{y_i}{A(\theta_i)}\right)^{\theta_i}\left[\log\left(\frac{y_i}{A(\theta_i)}\right) - \frac{\theta_i A'(\theta_i)}{A(\theta_i)}\right] + N'(\theta_i). \tag{7}$$

And from (5) and (6), the score function to w is as follows

$$\frac{\partial l(y_i, \theta_i, \mathrm{w})}{\partial \mathrm{w}} = -\frac{\theta_i}{A(\theta_i)}\left(\frac{y_i}{A(\theta_i)}\right)^{\theta_i - 1} sign(y_i). \tag{8}$$

Then we can obtain the stochastic gradient algorithm [3] (or the natural gradient form) for the activation function of the ith component of the demixing model to θ_i

$$\triangle\theta_i = -\eta\frac{\partial l(y_i, \theta_i, \mathrm{w})}{\partial \theta_i}. \tag{9}$$

Substitute $g(\cdot)$ in (2) by (8) and the fast fixed-point iteration for w is

$$\mathrm{w} \leftarrow \mathrm{E}\left\{\mathrm{z} \cdot \frac{\partial l(y_i, \theta_i, \mathrm{w})}{\partial \mathrm{w}}\right\} - \mathrm{E}\left\{\frac{\partial^2 l(y_i, \theta_i, \mathrm{w})}{\partial^2 \mathrm{w}} \cdot \mathrm{w}\right\}, \mathrm{w} \leftarrow \mathrm{w}/\|\mathrm{w}\|. \tag{10}$$

For the direct iteration ((8.37) in [3]), a simpler form can also be obtained

$$\mathrm{w} \leftarrow \mathrm{E}\left\{\mathrm{z} \cdot \frac{\partial l(y_i, \theta_i, \mathrm{w})}{\partial \mathrm{w}}\right\}, \mathrm{w} \leftarrow \mathrm{w}/\|\mathrm{w}\|. \tag{11}$$

3.2   The Self-adaptive FastICA Algorithm

Based on the above rule, the self-adaptive FastICA algorithm is given as follows
   a) Center and pre-whiten the mixtures x;
   b) Initialize w randomly and $\theta_i = 2$;
   c) Iterate w and θ_i according to (11) and (9) respectively;
   d) Regulate w and θ_i;
   e) Rerun step c) and d) until both w and θ_i converge.
   It should be noted that, to ensure the efficiency and availability of θ_i, θ_i is initialized to 2 (corresponding to the Gaussian pdf) in step b), and in step d) constrained as

$$\theta_i(k) = \begin{cases} 1, & \theta_i(k) \leq 1 \\ \theta_i(k), & 1 < \theta_i(k) < 4 \\ 4, & \theta_i(k) \geq 4 \end{cases} \tag{12}$$

Obviously when θ_i varies from 1 to 4, GGM covers from Laplacian distribution to approximately uniform distribution correspondingly.

4   Performance Analysis

As to the performance of convergence and the stability for self-adaptive FastICA, similar analysis can be given as that has been made in section VI and VII of [6] based on Cardoso's equivariant algorithm. And the proof in detail is to be given in the future.

   A note should be made is the spurious equilibria in FastICA addressed in [8]. Here we apply the improvement by enhancing convergence constraint, and spurious independent components can be avoided.

5   Simulation

Two simulations are conducted to evaluate the performance of self-adaptive FastICA. In example 1 two sources are selected, in which one is super-Gaussian and the other sub-Gaussian. Three sources are employed in example 2, including two super-Gaussian and one Gaussian.

Example 1. Here s_1 is a speech signal from [9], and s_2 uniformly distributed. Mixing matrix is A=$\begin{bmatrix} 0.8 & 0.6 \\ 0.6 & 0.8 \end{bmatrix}$. Fig.1 shows the original sources (upper array), the recovered by self-adaptive FastICA (middle array) and those by FastICA (down array) where activation function is selected as recommended. And in Fig.2 the learning dynamics for parameter $\theta = (\theta_1, \theta_2)$ are given. The proposed algorithm's PI [9] (0.0116) is superior to that of the traditional FastICA (0.0242). It shows the accuracy of self-adaptive FastICA is better and the proposed algorithm is effective in parameter optimization.

Example 2. Among the three sources, s_1 and s_2 are speech signals from [9], and s_3 Gaussian distributed. Mixing matrix A is randomly chosen under the

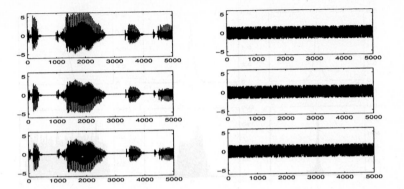

Fig. 1. Sources (upper array), the recovered signals by self-adaptive FastICA (middle array), and those by FastICA (down array)

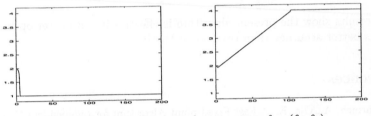

Fig. 2. Learning dynamics for parameter $\theta = (\theta_1, \theta_2)$

constraint of cond(A) < 50. Self-adaptive FastICA can also recover the sources well and shows its better accuracy. And in Fig.3 the pdfs of GGM for estimated parameters θ (upper array) and diagrams of corresponding estimated signals (down array) are given.

From the above two examples, we can see that self-adaptive FastICA is effective, even in the cases that different type sources are included, such as sub-Gaussian and super-Gaussian, or just one Gaussian source included. In estimating, the learning step-size for parameter θ is sensitive, and value between 0.01 and 0.001 is recommended.

Results in Fig.3 also show that even if the estimated θ_i is not the same as that of the original sources, self-adaptive FastICA is also effective. This behavior just coincides with the robustness of FastICA addressed in [1, 2], and the nonlinearity switching in extended infomax algorithm presented by T.-W Lee et al. in [7].

6   Conclusions

This paper focuses on the selection of activation function in FastICA. In the traditional FastICA, activation function has to be selected from predefined choices according to the prior knowledge of sources. Herein a self-adaptive FastICA algorithm is presented as the extension of self-adaptive BSS, where the optimization of score function based GMM and that of demixing vector are combined. Simu-

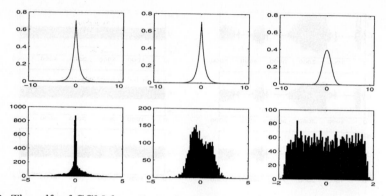

Fig. 3. The pdfs of GGM for estimated parameters θ (upper array) and diagrams of corresponding estimated signals (down array)

lation results show the present algorithm is effective in parameter optimization and has better accuracy than traditional FastICA.

References

1. Hyvärinen, A., Oja, E.: A Fast Fixed-point Algorithm for Independent Component Analysis. Neural Computation, **9** (1997) 1483-1492
2. Hyvärinen, A.: Fast and Robust Fixed-point Algorithms for Independent Component Analysis. IEEE Transactions on Neural Networks **10** (1999) 626-634
3. Hyvärinen, A., Karhunen, J., Oja, E.: Independent Component Analysis. John Wiley, New York (2001)
4. Lee, T.W., Girolami, M., Sejnowski, T.J.: Independent Component Analysis Using an Extended Infomax Algorithm for Mixed Sub-Gaussian and Super-Gaussian Sources. Neural Computation, **11** (1999) 417-441
5. Xu, L., Cheung, C.C., Amari, S.: Learned Parametric Mixture based ICA Algorithm. Neurocomputing, **22** (1998) 69-80
6. Zhang, L., Cichocki, A., Amari, S.: Self-adaptive Blind Source Separation Based on Activation Function Adaptation. IEEE Trans. on Neural Networks, **15** (2004) 233-244
7. Lee, T.-W., Lewicki, M.S.: The Generalized Gaussian Mixture Model Using ICA. International Workshop on Independent Component Analysis (ICA'00), Helsinki(2000) 239-244
8. Wang, G., Hu, D.: The Existence of Spurious Equilibrium in FastICA. The International Symposium on Neural Network (ISNN2004), Dalian, China, (2004) 708-713
9. Cichocki, A., Amari, S., et al.: ICALAB Toolboxes.
   http://www.bsp.brain.riken.jp/ICALAB

An Efficient Independent Component Analysis Algorithm for Sub-Gaussian Sources*

Zhilin Zhang and Zhang Yi

Computational Intelligence Laboratory,
School of Computer Science and Engineering,
University of Electronic Science and Technology of China
Chengdu 610054, China
zhangzl@vip.163.com, zhangyi@uestc.edu.cn

Abstract. A crucial problem for on-line independent component analysis (ICA) algorithm is the choice of step-size, which reflects a tradeoff between steady-state error and convergence speed. This paper proposes a novel ICA algorithm for sub-Gaussian sources, which converges fast while maintaining low steady-state error, since it adopts some techniques, such as the introduction of innovation, usage of skewness information and variable step-size for natural gradient. Simulations have verified these approaches.

1 Introduction

Independent component analysis (ICA) consists of separating unknown sources from their linear mixtures (or convolutional mixtures) using only the observed data [1]. It has drawn lots of attention in signal processing, biomedical engineering, communication and so on.

Many algorithms have been proposed for ICA [2–4]. One class are on-line algorithms. However, there exists a tradeoff when applying on-line algorithms: a large step-size leads to fast convergence, while resulting in large steady-state error; a small step-size leads to small steady-state error, but resulting in slow convergence. Many methods were proposed, attempting to solve this problem [6]. Until now almost all the work focus on how to select a suitable step-size. However there exist other approaches to accelerate convergence while maintaining lower steady-state error, which will be detailed later.

This paper proposes an ICA algorithm for sub-Gaussian sources. Since it adopts some novel techniques, it not only has a faster convergence speed but also maintains lower steady-state error, compared to other existed on-line algorithms. These techniques include usage of innovation extraction as preprocessing, exploitation of skewness information when encountering skewed sources, and adopting a special variable step-size technique for natural gradient.

* This work was supported by National Science Foundation of China under Grant 60471055 and Specialized Research Fund for the Doctoral Program of Higher Education under Grant 20040614017.

J. Wang, X. Liao, and Z. Yi (Eds.): ISNN 2005, LNCS 3496, pp. 967–972, 2005.

2   Framework of the Proposed Algorithm

The basic ICA model can be summarized as follows: assume that there exist mutually independent unknown sources $s_i(i = 1, \cdots, N)$, which have zero mean and unit variance. And also assume that the sources are linearly mixed with an unknown $M \times N(M \geq N)$ matrix $\mathbf{A}$:

$$\mathbf{x} = \mathbf{As}, \qquad (1)$$

where $\mathbf{s} = [s_1, s_2, \cdots, s_N]^T$ and $\mathbf{x} = [x_1, x_2, \cdots, x_M]^T$ are N-dimensional sources and M-dimensional mixed signals respectively. In independent component analysis, the basic goal is to find an $N \times M$ separating matrix $\mathbf{W}$ without knowing the mixing matrix $\mathbf{A}$, that is

$$\mathbf{y} = \mathbf{Wx}, \qquad (2)$$

such that $\mathbf{y} = [y_1, y_2, \cdots, y_N]^T$ is an estimate of $\mathbf{s}$ in the sense that each component of $\mathbf{s}$ may appear in any component of $\mathbf{y}$ with a scalar factor. For the sake of simplicity, we assume $M = N$.

Based on maximum likelihood estimation (MLE), we can easily derive the following algorithm [2]:

$$\mathbf{W}(t+1) = \mathbf{W}(t) + \mu[\mathbf{I} - \mathbf{g}(\mathbf{y}(t)) \cdot \mathbf{y}(t)^T]\mathbf{W}(t), \qquad (3)$$

where $\mathbf{y}(t) = \mathbf{W}(t)\mathbf{x}(t)$, and $\mathbf{g}(\mathbf{y}(t)) = [g_1(y_1(t)), \cdots, g_N(y_N(t))]^T$ is a component-wise vector function that consists of the function g_i defined as:

$$g_i = -(\log p_i)' = -\frac{p_i'}{p_i}, \qquad (4)$$

where p_i is the estimated density function of the ith unknown source.

This is an ordinary framework of on-line algorithm. However, through some improvements this algorithm will outperform many other on-line ones, which will be showed in section 5.

3   Novel and Efficient Techniques

3.1   Innovation Extraction

In the basic framework of ICA, the sources are considered as random variable, without time structure. It is an approximation to the real world. In fact, in the case of i.i.d., sources often correspond to independent physical processes that are mixed in similar autoregressive processes that give the sources. Thus the sources are less independent. What's more, in many cases, the sources are not strictly stationary. These factors will result in lower separation accuracy. Hyvärinen proposed [7] extracting innovation process from observation data as preprocessing for his FastICA [8]. Here, we point out that innovation extraction can also be used as preprocessing for on-line algorithms, that is,

$$\tilde{\mathbf{x}}(t) = \mathbf{x}(t) - \mathbf{x}(t-1), \qquad (5)$$

which is an approximation to innovation process of observation data $\mathbf{x}(t)$. The description of one iteration is as follows: at time t, we receive the observation data $\mathbf{x}(t)$. Then extract the current innovation $\tilde{\mathbf{x}}(t)$ according to (5) and get $\tilde{\mathbf{y}}(t) = \mathbf{W}(t)\tilde{\mathbf{x}}(t)$. Apply on-line ICA algorithm to $\tilde{\mathbf{y}}(t)$, and get the update of separating matrix $\mathbf{W}(t+1)$. Using $\mathbf{W}(t+1)$, we get the real output: $\mathbf{y}(t+1) = \mathbf{W}(t+1)\mathbf{x}(t+1)$.

It can be proved that the innovation process $\tilde{\mathbf{x}}(t)$ extracted from observed data $\mathbf{x}(t)$, and the innovation process $\tilde{\mathbf{s}}(t)$ extracted from original process $\mathbf{s}(t)$ still hold the ICA model (1) [7], that is to say, the coefficients of mixing matrix do not change. Since innovation are usually more independent from each other and more nongaussian than the original process, the technique above is expected to increase the accuracy of estimation of the ICA model, i.e., decrease the steady-state error.

3.2   Utilization of Skewness when Necessary

There are many asymmetrical sub-Gaussian sources, such as communication signals. Karvanen has pointed out [10] that skewness information can be used to improve the estimator needed in finding the independent components and consequently improve the quality of separation. Besides, we have found effective exploitation of sources' skewness can accelerate convergence.

In the maximum likelihood estimation of ICA model, an important issue is how to estimate the p.d.f. of unknown sources. The more accurate the estimated p.d.f., the lower the steady-state error. Thus, in order to exploit the skewness of sources, we use the following p.d.f. model to approximate asymmetrical sub-Gaussian distribution:

$$p(y) = (1 - a)N(1, 1) + aN(-1, 1), \tag{6}$$

where $N(1, 1)$ is gaussian density function with unit mean and unit variance. Parameter a serves to create levels of skewness in the distribution of y. Through tedious but straightforward calculation, we can obtain the skewness of the distribution:

$$skewness = \frac{8a(1 - a)(2a - 1)}{(-4a^2 + 4a + 1)^{3/2}}. \tag{7}$$

Obviously, if $a = 0.5$, the model therefore only serves for non-skewed sources, which is what the extended Infomax algorithm [2] adopts. If $0 < a < 0.5$, then $skewness < 0$; if $0.5 < a < 1$, then $skewness > 0$. Specially, we choose $a = 0.375$ for negative skewness, and $a = 0.625$ for positive skewness. Then the related non-linearities are

$$g(y) = -\frac{\frac{\partial p(y)}{\partial y}}{p(y)} = y - \frac{1 - 0.6exp(-2y)}{1 + 0.6exp(-2y)}, (skewness < 0), \tag{8}$$

$$g(y) = -\frac{\frac{\partial p(y)}{\partial y}}{p(y)} = y - \frac{1 - 1.67exp(-2y)}{1 + 1.67exp(-2y)}, (skewness > 0), \tag{9}$$

$$g(y) = -\frac{\frac{\partial p(y)}{\partial y}}{p(y)} = y - \tanh(y), (skewness = 0). \tag{10}$$

When applying the on-line algorithm (3), we choose the corresponding non-linearities according to the on-line calculated skewness.

3.3   Variable Step-Size for Natural Gradient

In order to accelerate convergence and maintain low steady-state error, some researchers adopted the idea of variable step-size from adaptive signal processing field [6]. However it maybe was not harmonious because their updating rules of step-size were based on stochastic gradient while those of separating matrix based on natural gradient [5]. Here, we derive an updating rule of step-size for natural gradient in the framework of MLE of ICA, which has simpler form. It is given by

$$\mu(t) = \mu(t-1) + \rho \nabla_{\mu(t-1)} \mathbf{J}(\mathbf{W}(t)), \tag{11}$$

where ρ is an enough small constant, and $\mathbf{J}(\mathbf{W}(t))$ is an instantaneous estimate of the cost function from which our MLE based algorithm is derived. Notice $\nabla_{\mu(t-1)} \mathbf{J}(\mathbf{W}(t))$ can be expressed by

$$\nabla_{\mu(t-1)} \mathbf{J}(\mathbf{W}(t)) = \left\langle \frac{\partial \mathbf{J}(\mathbf{W}(t))}{\partial \mathbf{W}(t)}, \frac{\partial \mathbf{W}(t)}{\partial \mu(t-1)} \right\rangle$$
$$= trace\left(\left[\frac{\partial \mathbf{J}(\mathbf{W}(t))}{\partial \mathbf{W}(t)}\right]^T \cdot \left[\frac{\partial \mathbf{W}(t)}{\partial \mu(t-1)}\right]\right), \tag{12}$$

where <> denotes inner product of matrixes.

On one hand, according to (3), we have

$$\frac{\partial \mathbf{W}(t)}{\partial \mu(t-1)} = \left[\mathbf{I} - \mathbf{g}(\mathbf{y}(t-1))\mathbf{y}^T(t-1)\right]\mathbf{W}(t-1) \triangleq \mathbf{H}(t-1). \tag{13}$$

On the other hand, $\mathbf{J}(\mathbf{W}(t))$ is an instantaneous estimate of the cost function based on natural gradient, so

$$\frac{\partial \mathbf{J}(\mathbf{W}(t))}{\partial \mathbf{W}(t)} = \left[\mathbf{I} - \mathbf{g}(\mathbf{y}(t))\mathbf{y}^T(t)\right]\mathbf{W}(t) \triangleq \mathbf{H}(t). \tag{14}$$

Substituting (13),(14) into (12), we obtain,

$$\nabla_{\mu(t-1)} \mathbf{J}(\mathbf{W}(t)) = trace(\mathbf{H}(t)^T \cdot \mathbf{H}(t-1)), \tag{15}$$

leading to the update of step-size as follows,

$$\mu(t) = \mu(t-1) + \rho \cdot trace(\mathbf{H}(t)^T \cdot \mathbf{H}(t-1)). \tag{16}$$

In order to prevent μ from becoming negative, a small positive bottom bound should be set. Also, in order to overcome the problem of numerical convergence, one can modify it as,

$$\mu(t) = (1-\delta)\mu(t-1) + \delta \cdot \rho \cdot trace(\mathbf{H}(t)^T \cdot \mathbf{H}(t-1)), \tag{17}$$

where δ is another small positive constant.

4   Algorithm Description

From the above discussion, the new algorithm can be summarized as follows:
(1). Extract innovation processes $\tilde{\mathbf{x}}(t)$ from the observed data $\mathbf{x}(t)$ according to formula (5), then obtain $\tilde{\mathbf{y}}(t) = \mathbf{W}(t)\tilde{\mathbf{x}}(t)$.
(2). Set a small positive threshold η, and on-line calculate skewness of $\tilde{\mathbf{y}}(t)$. If $skewness(\tilde{\mathbf{y}}(t)) > \eta$, choose the non-linearity (9); If $skewness(\tilde{\mathbf{y}}(t)) < -\eta$, choose the non-linearity (8); If $-\eta < skewness(\tilde{\mathbf{y}}(t)) < \eta$, choose the non-linearity (10).
(3). Apply the algorithm (3) on $\tilde{\mathbf{y}}(t)$, including updating the step-size using (16) or (17), and obtain $\mathbf{W}(t+1)$.
(4). Get the real (not innovation) output at time $t+1$: $\mathbf{y}(t+1) = \mathbf{W}(t+1)\mathbf{x}(t+1)$.
(5). Let $t \leftarrow t + 1$, return to step (1), starting next iteration.

5   Simulation Results

Because of the limit of space, we only show a simple simulation. Five skewed sub-Gaussian sources (6000 points) were generated using the method in [9]. The values of sources'skewness were fixed at 0.6. In Fig.1, the EASI algorithm [3], extended Infomax algorithm [2] and our new algorithm (using the three techniques) were compared in terms of performance index, i.e. the cross-talking error [4]. For EASI and extended Infomax algorithm, μ was fixed at 0.005,0.03, respectively, because via these parameters both algorithms obtained the best results (fastest convergence). For our proposed algorithm, μ was initialized at 0.02, $\rho = 0.00001$. From the figure, it is clear to see that our algorithm outperforms the other two, which had fastest convergence speed and lowest steady-state error.

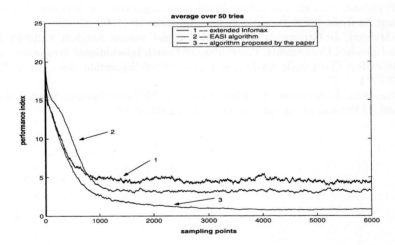

Fig. 1. Comparison of EASI, extended Infomax and our new algorithm in terms of performance index.

6   Conclusion

We propose an efficient on-line ICA algorithm for sub-Gaussian sources. By using some techniques, the algorithm has faster convergence speed, and at the meantime maintains lower steady-state error, compared to other similar on-line algorithms. It should be stressed that these techniques can be used in other algorithms to improve their performance. Simulation results show the validity of these techniques.

References

1. Comon, P.: Independent Component Analysis, a New Concept Signal Processing. **36** (1994) 287-314
2. Lee, T.-W., Girolami, M., Sejnowski, T.: Independent Component Analysis Using an Extended Infomax Algorithm for Mixed Sub-Gaussian and Supergaussian Sources. Neural Computation, **11(2)** (1999) 417-441
3. Cardoso, J.-F., Laheld, B.H.: Equivariant Adaptive Source Separation. IEEE Trans.On Signal Processing, **44** (1996) 3017-3030
4. Amari, S., Cichocki, A., Yang, H.: A New Learning Algorithm for Blind Source Separation. Advances in Neural Information Processing System. Vol. 8. MIT Press, Cambridge, MA (1996) 757-763
5. Amari, S.: Natural Gradient Works Efficiently in Learning. Neural Computation, **58** (1998) 251-276
6. Douglas, S.C., Cichocki, A.: Adaptive Step Size Techniques for Decorrelation and Blind Source Separation. in Proc. 32nd Asilomar Conf. Signals, Systems, Computers. Pacific Grove, CA. **2** (1998) 1191-1195
7. Hyvärinen, A.: Independent Component Analysis for Time-dependent Stochastic Processes. in Proc. Int. Conf. on Artificial Neural Networks (ICANN'98). Sweden. (1998) 541-546
8. Hyvärinen, A.: Fast and Robust Fixed-point Algorithm for Independent Component Analysis. IEEE Trans.On Neural Networks **10** (1999) 626-634
9. Karvanen, J.: Generation of Correlated Non-Gaussian Random Variables From Independent Components. Proceedings of Fourth International Symposium on Independent Component Analysis and Blind Signal Separation, Nara, Japan (2003) 769-774
10. Karvanen, J., Koivunen, V.: Blind Separation Methods Based on Pearson System and Its Extensions. Signal Processing, **82** (2002) 663-673

ICA and Committee Machine-Based Algorithm for Cursor Control in a BCI System

Jianzhao Qin[1], Yuanqing Li[1,2], and Andrzej Cichocki[3]

[1] Institute of Automation Science and Engineering,
South China University of Technology, Guangzhou, 510640, China
[2] Institute for Infocomm Research, Singapore 119613
[3] Laboratory for Advanced Brain Signal Processing, RIKEN Brain Science Institute
Wako shi, Saitama 3510198, Japan

Abstract. In recent years, brain-computer interface (BCI) technology has emerged very rapidly. Brain-computer interfaces (BCIs) bring us a new communication interface technology which can translate brain activities into control signals of devices like computers, robots. The preprocessing of electroencephalographic (EEG) signal and translation algorithms play an important role in EEG-based BCIs. In this study, we employed an independent component analysis (ICA)-based preprocessing method and a committee machine-based translation algorithm for the offline analysis of a cursor control experiment. The results show that ICA is an efficient preprocessing method and the committee machine is a good choice for translation algorithm.

1 Introduction

BCIs give their users a communication and control approach that does not depend on the brain's normal output channels (i.e. peripheral nerves and muscles). These new communication systems can improve the quality-of-life of those people with severe motor disabilities, and provide a new way for able-bodied people to control computers or other devices (e.g., robot arm).

EEG-based BCIs record EEG at the scalp to control cursor movement, to select letters or icons. Since the EEG signal includes some noise, such as eye movements, eye blinks and EMG, the BCIs should include a preprocessing procedure to separate the useful EEG signal from noise (including artifacts). A good preprocessing method can greatly improve the information transferring rate (ITR) of BCIs. ICA has been widely used in blind source separation [1], [2], [3], and biomedical signal analysis including EEG signal analysis [4]. In the offline analysis of a cursor control experiment, we used an ICA-based preprocessing method. The results show that the accuracy rate has improved dramatically after ICA preprocessing.

A translation algorithm transforms the EEG features derived by the signal preprocessing stage into actual device control commands. In the offline case without feedback, the translation algorithm primarily performs a pattern recognition task (We extract features from preprocessed EEG signal, then classify them into

J. Wang, X. Liao, and Z. Yi (Eds.): ISNN 2005, LNCS 3496, pp. 973–978, 2005.

several classes that indicate the users' different intentions). In supervised learning, if the size of training data is small (It is usual in BCIs), the overfitting problem may arise. A good transfer function should have a good generalization performance. In the analysis, we designed a simple and efficient committee machine as a transfer function to handle the overfitting problem.

2   Methods

In this section, we first describe the experiment data set and illustrate the framework of our offline analysis, then introduce the ICA preprocessing and the feature extraction. Finally, the structure of the committee machine and the classification procedure are presented.

2.1   Data Description

The EEG-based cursor control experiment was carried out in Wadsworth Center. The recorded data set was given in the BCI competition 2003. The data set and the details of this experiment are available on the web site

    http://ida.first.fraunhofer.de/projects/bci/competition.

The data set was recorded from three subjects (AA, BB, CC). The framework of our offline analysis is depicted as Fig. 1.

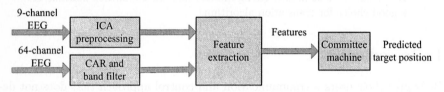

Fig. 1. Framework diagram

2.2   Independent Component Analysis

Independent component analysis is a method for solving the blind source separation problem [5]: A random source vector $\mathbf{S}(n)$ is defined by

$$\mathbf{S}(n) = [S_1(n), S_2(n), \ldots, S_m(n)]^T \tag{1}$$

where the m components are a set of independent sources. The argument n denotes discrete time. $\mathbf{A}$, a nonsingular m-by-m matrix, is called mixing matrix. The relation between $\mathbf{X}(n)$ and $\mathbf{S}(n)$ is as follows

$$\mathbf{X}(n) = \mathbf{AS}(n) \tag{2}$$

The source vector $\mathbf{S}(n)$ and the mixing matrix $\mathbf{A}$ are both unknown. The task of blind source separation is to find a demixing matrix $\mathbf{C}$ such that the original source vector $\mathbf{S}(n)$ can be recovered as below

$$\mathbf{Y}(n) = \mathbf{CX}(n) \tag{3}$$

The ICA method is based on the assumption that the original sources are statistically independent. The objective of an ICA algorithm is to find a demixing matrix $\mathbf{C}$, such that components of $\mathbf{Y}$ are statistically independent. We assume that the multichannel EEG can be modelled by (2), where $\mathbf{X}(n)$ is the recorded multichannel EEG at time n, $\mathbf{A}$ is the mixing matrix, and $\mathbf{S}(n)$ is the source vector at time n.

There are many algorithms to implement ICA. Bell and Sejnowski (1995) [6] proposed an infomax algorithm. Natural gradient (1995) was proposed and applied to ICA by Amari et al [7]. In the analysis, we applied a natural gradient-flexible ICA algorithm [8], which could separate mixtures of sub- and super-Gaussian source signals. We expected that ICA preprocessing can separate the useful EEG components from the noise (including artifacts).

2.3   Feature Extraction

In the analysis, we extracted and combined two kinds of features from the preprocessed EEG. One is the power feature, the other is the CSP feature.

The data includes 64 channels of EEG signal, but we only used 9 channels of EEG signal with channel number $[8, 9, 10, 15, 16, 17, 48, 49, 50]$ for the ICA preprocessing and power feature extraction. These 9 channels covered the left sensorimotor cortex, which is the most important part when the subject used his or her EEG to control the cursor in this experiment. During each trial with trial length 368 samples (subject AA and BB) or 304 samples (subject CC), we imagined that the position of the cursor was updated once in every time interval of 160 adjacent samples, and two subsequent time intervals were overlapped in 106 (Subject AA and BB) or 124 (subject CC) samples. Thus there were 5 updates of the position of the cursor in each trial. Only one best component, which had the best correct recognition rate in training sets (sessions 1–6), was used for power feature extraction. For each trial, the power feature is defined as,

$$\mathbf{PF} = [PF_1, PF_2, PF_3, PF_4, PF_5] \tag{4}$$

$$PF_n = \sum_{f \in [11,14]} P_n(f) * w_1 + \sum_{f \in [22,26]} P_n(f) * w_2 \tag{5}$$

where $P_n(f)$ is the power spectral of the $n-th$ time bin. The parameters w_1 and w_2 are determined by experiment. The criteria for choosing the two parameters is similar to that for choosing the best component.

CSP is a technique that has been applied to EEG analysis to find spatial structures of event-related (de-)synchronization [9]. Our CSP feature is defined as in [9]. The CSP analysis consists of calculating a matrix W and diagonal matrix D:

$$\mathbf{W}\Sigma_1\mathbf{W}^T = D \ and \ \mathbf{W}\Sigma_4\mathbf{W}^T = 1 - D \tag{6}$$

where Σ_1 and Σ_4 are the normalized covariance matrix of the trial-concatenated matrix of target 1 and 4, respectively. $\mathbf{W}$ can be obtained by jointed diagonalization method. Prior to calculating features by CSPs, common average reference

[10] was carried out, then the referenced EEG was filtered in 10–15Hz. The CSP feature for each trial consists of 6 most discriminating main diagonal elements of the transformed covariance matrix for a trial followed by a log-transformation [9].

3   Committee Machine-Based Translation Algorithm

Multi-layer perceptron is a strong tool in supervised-learning pattern recognition, but when the size of the training samples is relatively small compared with the number of network parameters, the overfitting problem may arise. In the session, based on the features mentioned above, we describe a committee machine consisting of several small-scale multi-layer perceptrons to solve the overfitting problem.

In our analysis, the data from sessions 1–6 (about 1200 samples) were used for training. A statistical theory on overfitting phenomenon [11] suggests that overfitting may occur when $N < 30W$, where N is the number of training samples, W denotes the number of network parameters. According to this theory, the maximum number of network parameters should be less than 40. In order to satisfy this requirement, we designed a committee machine to divide the task into 2 simple tasks, so the structure and training of each network in the committee machine can be simplified.

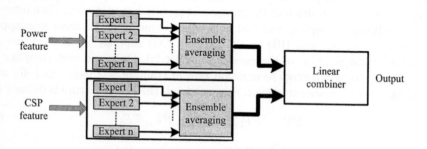

Fig. 2. The structure of a committee machine

The structure of the committee machine is depicted in Fig. 2. The units of this committee machine are several small-scale three-layer (including input layer) perceptrons with nonlinear activation function. We call these networks 'experts' which are divided into two groups. One group of experts make their decisions by using power features, while the other group's decision is from CSP features. The experts in the same group share common inputs, but are trained differently by varied initial values. Each network has four output neurons corresponding to four target positions. The final decision of a group is made by averaging all outputs of its experts, then the final outputs of the two groups are linearly combined to produce an overall output of the machine.

4 Result

We trained ICA on 250s (40000 samples) of EEG recording randomly chosen from session 1–6. All the trials in session 7–10 were used to test our method. For the purpose of comparison, we done feature extraction and classification under three conditions. 1) ICA was used for preprocessing, and committee machine was used for classification. 2) Without ICA preprocessing, the best channel of raw EEG signal was chosen for power feature extraction, and committee machine was used for classification. 3) ICA was used for preprocessing, while the committee machine was replaced by normal multiple-layer network for classification. The results for the three subjects are shown in Table 1, which were obtained under the three conditions.

Table 1. Accuracy rates (%) for the three subjects obtained under above three conditions

Subject	Condition	Session 7	Session 8	Session 9	Session 10	Average accuracy
AA	1	71.20	71.20	66.49	69.63	69.63
AA	2	68.06	68.06	65.45	68.59	67.54
AA	3	70.68	64.40	63.35	59.69	64.53
BB	1	63.87	62.30	47.12	54.97	57.07
BB	2	62.30	61.78	42.41	48.17	53.67
BB	3	62.30	57.07	46.07	46.60	53.01
CC	1	66.67	72.82	54.36	81.54	68.85
CC	2	63.59	70.26	50.77	72.82	64.36
CC	3	61.54	66.15	54.87	68.21	62.69

5 Discussion and Conclusion

Table 1 shows that the accuracy of offline analysis has been improved considerably by using the ICA preprocessing method. Furthermore, the committee machine is better in generalization performance than the normal multiple-layer network after comparing the results in conditions 1 and 3.

In the analysis, we used ICA as the preprocessing method for BCI. This method has some advantages. First, we think that the ICA preprocessing can separate useful source components from noise. Thus we can choose one or two components which contain more useful information for extracting power features than before preprocessing. Second, since we choose a smaller number of ICA components for feature extraction, the computation burden can be reduced. Furthermore, the dimensionality of the feature space can be reduced, as a consequence, not only the structure of the classifier can be simplified, but also the overfitting problem can be solved to some extent.

Meanwhile, a committee machine was used as a translation algorithm, which can also improve the performance of BCIs. The key point of committee machine is to divide a complex computational task into a number of simple tasks. Due to the simple network structure, the constituent experts of the machine are easy to be trained, and the generalization performance is improved.

Acknowledgements

This study was supported by the National Natural Science Foundation of China (No. 60475004, No. 60325310), Guangdong Province Science Foundation for Research Team Program (No. 04205789), and the Excellent Young Teachers Program of MOE, China.

References

1. Li, Y., Wang, J., Zurada, J. M.: Blind Extraction of Singularly Mixed Sources Signals. IEEE Trans. On Neural Networks, **11** (2000) 1413–2000
2. Li, Y., Wang, J.: Sequential Blind Extraction of Linearly Mixed Sources. IEEE Trans. on Signal Processing, **50** (2002) 997–1006
3. Cichocki, A., Amari, S.: Adaptive Blind Signal and Image Processing: Learning Algorithms and Applications. John Wiley, New York (2002).
4. Makeig, S., Bell, A.J., Jung, T.-P., Sejnowski, T.J.: Independent Component Analysis of Electroencephalographic Data. Adv Neural Info Processing Systems, **8** (1996) 145–151
5. Comon, P.: Independent Component Analysis - A New Concept Signal Procesing, **36** (1994) 287–314
6. Bell, A.J., Sejnowski, T.J.: An Information-maximization Approach to Blind Separation and Blind Deconvolution. Neural Computation, **7** (1995) 1129–1159
7. Amari, S., Chichocki, A., Yang, H.H.: A New Learning Algorithm for Blind Signal Separation. Advances in Neural Information Processing, **8** (1996) 757–763
8. Choi, S., Chichocki, A., Amari, S.: Flexible Independent Component Analysis. Proc. of the 1998 IEEE Workshop on NNSP, (1998) 83–92
9. Ramoser, H., Gerking, J.M., Pfurtscheller, G.: Optimal Spatial Filtering of Single Trial EEG during Imagined Hand Movement. IEEE Trans. Rehab. Eng. **8** (2000) 441–446
10. McFarland, D.J., McCane, L.M., David, S.V., Wolpaw, J.R.: Spatial Filter Selection for EEG-based Communication. Electroenc. Clin. Neurophys. **103** (1997) 386–394
11. Amari, S., Murata, N., Müller, K.-R, Finke, M., Yang, H.: Statistical Theory of Overtraining-Is Cross-Validation Asymptotically Effective Advances in Neural Information Processing Systems. Vol. 8. MIT Press Cambridge, MA (1996) 176–182

Fast Independent Component Analysis
for Face Feature Extraction

Yiqiong Xu, Bicheng Li, and Bo Wang

Department of Information Science, Information Engineering Institute,
Information Engineering University, Zhengzhou China
xuyiqiong@sina.com

Abstract. In this paper, Independent Component Analysis (ICA) is presented as an efficient face feature extraction method. In a task such as face recognition, important information may be contained in the high-order relationship among pixels. ICA is sensitive to high-order statistic in the data and finds not-necessarily orthogonal bases, so it may better identify and reconstruct high-dimensional face image data than Principle Component Analysis (PCA). ICA algorithms are time-consuming and sometimes converge difficultly. A modified FastICA algorithm is developed in this paper, which only need to compute Jacobian Matrix one time in once iteration and achieves the corresponding effect of FastICA. Finally a genetic algorithm is introduced to select optimal independent components (ICs). The experiment results show that modified FastICA algorithm quickens convergence and genetic algorithm optimizes recognition performance. ICA based features extraction method is robust to variations and promising for face recognition.

1 Introduction

ICA was originally developed to deal with the cocktail-party problem [1]. As an effective approach to the separation of blind signal, ICA has attracted broad attention and been successfully used in many other fields. One of the promising applications of ICA is feature extraction, where it extracts independent bases which are not necessarily orthogonal and sensitive to high-order statistics. In the task of face recognition, important information may be contained in the high-order relationship among pixels. So ICA seems reasonable to be a promising face feature extraction method. However ICA does not have advantages only. ICA algorithms are iterative, time-consuming and sometimes converge difficultly. A number of algorithms for performing ICA have been proposed and reviewed in [2][3]. Among these algorithms, the FastICA algorithm is a computationally highly efficient method for performing the estimation of ICA. It uses a fixed-point iteration scheme, which is introduced by Hyvärinen [4], but FastICA still time consuming, especially compared with PCA. A modified FastICA algorithm is introduced in this paper, with goal of fast convergence and computation reduction. Other difficulties of ICA are the ordering and selection of source vectors. In this paper, a genetic algorithm based feature selection method is carried out to obtain optimal ICs.

The remaining of this paper is organized as follows. Section 2 gives a brief review of ICA. Section 3 presents FastICA algorithm and its modified version. Section 4 introduces the face recognition based on the improved ICA algorithm. Finally, in the section 5, a conclusion is made and future work is put forward.

J. Wang, X. Liao, and Z. Yi (Eds.): ISNN 2005, LNCS 3496, pp. 979–984, 2005.

2   Review of ICA

The basic idea of ICA is to represent a set of random observed variables $x_1, x_2, \ldots, x_m$ using basis function $s_1, s_2, \ldots, s_n$, where the components are statistically independent as possible. We denote the observed variables x_i as a observed vector $X=(x_1, x_2, \ldots, x_m)^T$ and the component variables s_i as a vector $S= (s_1, s_2, \ldots, s_n)^T$. the relation between S and X can be modeled as

$$X = AS .\tag{1}$$

Where A is an unknown $m \times n$ matrix of full rank, called the mixing matrix. In feature extraction, the columns of A represent features, and s_i is the coefficient of the ith feature in the observed data vector X.

The current algorithms for ICA can be roughly divided into two categories. First category relies on minimizing or maximizing so-called contrast functions based on higher-order cumulants, which problem is that they required very complex matrix or tensor operations. The second category contains adaptive algorithms often based on stochastic gradient methods, which main problem is slow convergence. The FastICA algorithm is used for its good simulations and fast convergence.

3   FastICA Algorithm and Modified FastICA Algorithm

The basic idea of the FastICA algorithm is finding the local extreme of the kurtosis of a linear combination of the observed variables. Note that FastICA algorithms require a preliminary sphering of the data X, though also some versions for non-sphered data has been given. We assume that the X data is sphered in following sections. The FastICA learning rule finds a direction, i.e. a unit vector by the approximation of negentropy $J(w^T X)$ given as below

$$J(w^T X) \propto \left[E\{G(w^T X)\} - E\{G(v)\} \right]^2 ,\tag{2}$$

where G is nonquadratic function, v is a standard Gaussian variable. FastICA algorithm is based on fixed-point iteration scheme. The goal of the fixed-point algorithm is finding a maximum of the non-gaussianity of $w^T X$. We give the one-unit version of fixed-point algorithm. Firstly, the maxima of the negentropy of $w^T X$ are obtained at certain optima of $E\{G(w^T X)\}$, according to the Kuhn-Tucker conditions [4]. Under the constraint $E\{G(w^T X)\} = \|w\|^2 = 1$, the optima are obtained at points

$$E\{Xg(w^T X)\} - \beta w = 0 .\tag{3}$$

Where β is a constant which can easily evaluated as $\beta = E\{w_0^T Xg(w_0^T X)\}$, and w_0 is the value of w at the optimum. Newton's method [4] is used to solve this equation. Denoting the function on the left side of (3) by $F(w)$, its Jacobian matrix by $J(w)$ as

$$J(w) = E\{XX^T g'(w^T X)\} - \beta I .\tag{4}$$

Since the data is sphered, a reasonable approximation seems to be

$$E\{XX^T g'(w^T X)\} \approx E\{XX^T\} E\{g'(w^T X)\} = E\{g'(w^T X)\} I .\tag{5}$$

Thus the Jacobian matrix becomes diagonal, and can easily be inverted. Finally we obtain the following approximate Newton iteration

$$w_{k+1} = w_k - \left[E\{Xg(w_k^T X)\} - \beta w_k \right] / \left[E\{g'(w_k^T X)\} - \beta \right]. \tag{6}$$

Denote by g the derivative of G used in (2), the basic form of the FastICA algorithm is as follows, note that convergence means that the old and new values of w point in the same direction.

1. Initiate weight vector w
2. Let $w_{k+1} = w_k - \left[E\{xg(w_k^T x)\} - \beta w_k \right] / \left[E\{g'(w_k^T x)\} - \beta \right]$
3. Let $w_{k+1} = w_{k+1} / \|w_{k+1}\|$
4. If not converged, go back to 2.

By the introduction of FastICA, it is known that the time-consuming course is computing Jacobian matrix. Reducing the time of Jacobian matrix will improve the performance of FastICA algorithm. So a modified FastICA algorithm is developed, the basic form of the modified FastICA algorithm is as follows:

1. Initiate weight vector w
2. Let $w_{k+1} = w_k - F(w_k) / J(w_k)$
3. Let $w_{k+1} = w_{k+1} / \|w_{k+1}\|$
4. If not converged, go back to 2

In the modified FastICA, several iterations of FastICA are merged into one iteration, and, from (4), $J(w)$ need to be computed only once in one iteration. So computation is decreased and interaction speed is fast.

4 Face Recognition with Feature Extraction Based on ICA

4.1 Preprocessing for Face Images

Applying the ICA on face recognition, the random vectors will be the training face images. We organize each image in the database as a row vector, the dimension of which equals the number of pixels in the image. Before applying an ICA algorithm on the images, it is usually very useful and necessary to do some preprocessing.

The most basic preprocessing is to make X mean zero, by subtracting X by its mean. This course called "centering". This processing is made only to simplify the ICA algorithm. Another useful preprocessing in ICA is to make each image data's covariance one, this course called "whitening". In this processing, X is multiplied by $ED^{-1/2}E^T$. Where E is the orthogonal matrix of eigenvectors of $E\{XX^T\}$ and D is the diagonal matrix of its eigenvalue. By linearly transform of the observed matrix X, we obtain a new matrix $\tilde{X}$ which components are uncorrelated and their variances equal unity, as $E\{\tilde{X}\tilde{X}^T\} = I$. The utility of whitening resides in the fact that new mixing matrix $\tilde{A}$ is orthogonal [2]. We also reduce the number of parameters to be estimated by whitening. Assuming that the original matrix has n^2 parameters to be estimated, an orthogonal matrix contains only $n(n-1)/2$ parameters to be estimated. In the rest of this paper, we assume that all image data has been preprocessed by centering and whitening. For simplicity, X and A denote $\tilde{X}$ and $\tilde{A}$.

4.2 Face Feature Extraction Based on ICA

Before ICA applied to the image data, PCA step is carried out. We adopt PCA step approximately to Bartlett [5]. This step is not necessary, but useful to reduce high-dimension image data. We choose the first m principle component (PC) eigenvectors of the image set. The PCA vectors in the input did not throw away the high-order relationship because of the limitation of PCA. These relationships still existed. Let P_m denote the matrix containing the first m principle components. P_m^T contains the most power in X, we perform the modified FastICA on P_m^T, as follows

$$Y = WX = WP_m^T ,$$ (7)

$$P_m^T = W^{-1}Y ,$$ (8)

$$R_m = XP_m ,$$ (9)

$$X_{mse} = R_m P_m^T = XP_m P_m^T = XP_m W^{-1}Y = R_m W^{-1}Y .$$ (10)

Then, the row of $R_m W^{-1}$ contained the ICA coefficients, i.e., the coefficients for the linear combination of statistically independent face feature. For a face image to be tested x_t, the independent component representation is $x_t P_m W^{-1}$.

4.3 Face Feature Selection Based on Genetic Algorithm

In PCA, the eigenvectors rank according to the corresponding eigenvalues, the first ranked eigenvector corresponding to the largest eigenvalue. Relatively, one of the ambiguities of ICA is that we cannot determine the order of the independent components [2]. In order to select more suitable ICA face features, a genetic algorithm is presented in this section, which establishes the optimal feature subset through evolutionary computation.

Genetic algorithm uses selection and crossover operators to generate the best solution, mutation operator to avoid premature convergence. Firstly, we design the length of chromosome equal to m, whose each gene represents the feature selection, "1" denotes that the corresponding feature is selected, otherwise denotes rejection. The first population is generated with the probability of each gene equal to "1" is 0.9. Secondly, the fitness function is defined to represent the recognition performance on training data set, as

$$f(q) = \frac{1}{N}\sum_{i=1}^{N}\delta(x_i,q) .$$ (11)

Where if x_i are correctly recognized, $\delta(x_i,q)$ equals to 1, and if x_i are not correctly recognized, $\delta(x_i,q)$ equals to 0, q is chromosome of current population, N is the number of training data set. The roulette wheel strategy is used for selection operator, where each individual is represented by a space that proportionally corresponds to its fitness. The probability of each chromosome q_j which are chosen can be expressed as

$$P_{q_j} = f(q_j)/\sum_{k=1}^{n} f(q_k) .$$ (12)

Where n is the number of individuals. Thirdly, crossover operator is applied to randomly paired chromosomes based on randomly generated crossover template, where 0 and 1 represents non-exchange and exchange respectively. Then two new chromosomes are formed and the next population is generated. For this new population, the decoding and effectiveness of performed to verify whether it is satisfied. If yes, the optimization solution has been obtained, otherwise continue iteration to get next population.

4.4   Face Recognition Based on ICA

Face recognition was executed for the ICA coefficient vectors by the nearest neighbor algorithm, using Cosine similarity measures which were previous found to be effective for computational models of language and face recognition [6].

Table 1. Recognition performance of different face feature extraction algorithm with all Ics.

	PCA	FastICA	M-FastICA
The mean of FE time (s)	2.05	2.52	2.01
Recognition rate (%)	70.58	75.52	82.75

Table 2. Recognition performance of different feature selection.

	All ICs	20 ICs	15 ICs	10 ICs
Recognition rate (%)	82.75	90.32	87.80	83.45

The algorithms are tested in Yale face database, which consists of eleven persons and fifteen gray scale images per person. The image size is 98×116. Five images per person are used as ICA train set, and add five images per person as genetic algorithm train set. All images are used to test. In experiments, there are no step size parameters to choose, this means the modified FastICA algorithm is easy to use. Let FE be feature extraction and M-FastICA be the modified FastICA. Table 1 shows that, for feature extraction time, PCA based method consumes few time than FastICA based method. But M-FastICA consumes almost time as PCA. From the point of recognition performance, ICA based feature extraction method outperform PCA based method, especially M-FastICA based method. Table 2 shows that genetic algorithm retain optimal feature set, so face recognition turn more accurate and more robust to variations.

5   Conclusion and Future Work

In this paper, ICA as a subspace method is used for feature extraction from face images. ICA source vectors is independent and not necessarily orthogonal, they will be closer to natural features of images, and thus more suitable for face recognition. By modifying the kernel iterate course, the modified FastICA algorithm quicken convergence and achieve the corresponding effect of FastICA. Besides its own merits, the Modified FastICA algorithm inherits most of the advantages of FastICA algorithm: It is parallel, distributed, computational, simple, and requires little space. Genetic algorithm based feature selection method retains more suitable features and makes face recognition more robust. So ICA algorithm is improved in this paper by compensation its disadvantages.

Our future work may consist in replacing face recognition through simple distance comparisons by multi-dimension classification. Our related works are underway and the results will be reported in the near future.

References

1. Amari, S., Cichocki, A., Yang, H.H.: A New Learning Algorithm for Blind Signal separation. Advance in Neural Information Processing Systems, Cambridge, MA: MIT Press **8** (1996)
2. Hyvärinen, A., Karhunen, J., Oja, E.: Independent Component Analysis. Wiley, New-York (2001)
3. Lee T.-W.: Independent Component Analysis: Theory and Applications. Boston, MA: Kluwer (1998)
4. Hyvärinen A.: Fast and Robust Fixed-Point Algorithms for Independent Component Analysis. IEEE, Transaction on Neural Networks, **10** (1999) 626-634
5. Bartlett, M.S., Movellan, J.R., Sejnowski, T.J.: Face Recognition by Independent Component Analysis. IEEE, Transaction on Neural Networks, **13** (2002) 1450-1464
6. Toole A.O', Deffenbacher, K., Valentin, D., Abdi, H.: Structural Aspects of Face Recognition and the Other Race Effect. Memory Cognition, **22** (1994) 208-224

Affine Invariant Descriptors for Color Images Based on Independent Component Analysis*

Chengming Liu, Xuming Huang, and Liming Zhang

Dept. E.E, Fudan University, Shanghai 200433, China
{cmliu,022021035,lmzhang}@fudan.edu.cn

Abstract. In this paper we introduce a scheme to obtain affine invariant descriptors for color images using Independent Component Analysis (ICA), which is a further application of using ICA on contour-known objects. First, some feature points can be found by hue-histogram of the color images in HIS space. Then ICA is applied to extract an invariant descriptor between these corresponding points, which can be a representation of shape similarity between original image and its affine image. This proposed algorithm can also estimate affine motion parameters. Simulation results show that ICA method has better performance compared with Fourier methods.

1 Introduction

Independent Component Analysis (ICA) is a very useful method for blind signal separation, in which observed vector $X \in R^m$ is considered as a product of unknown matrix A and unknown independent sources S shown as follows:

$$X = AS, A \in R^{m \times n}, S \in R^n .$$ (1)

The ICA tries to separate the independent components from X in the case of unknown mixture matrix A and source signals S via

$$Y = WX = WAS \approx \hat{S} .$$ (2)

Here the matrix W can be obtained by unsupervised learning in the case of the components of vector Y as independent as possible. The independent components in vector Y can be changed in sign, scale and order, so $\hat{S}$ is used here.

Recently, we used ICA to recognize contour-known objects under affine transform and estimate affine motion parameters[3],[4].Experimental results show that the performance of the ICA method is better than other traditional invariant methods like moment or Fourier methods[2] using boundary images.

In practice, the contour of an object is always hard to extract, especially when color and texture are very abundant, there'll be many edges in the interior of the object and its contour is probably not close. In this paper, we extend affine invariant descriptors of contour-known objects based on ICA to color images. Some corresponding feature points of the original image and its affine image can be obtained via hue-histogram of them. And then ICA is used to extract affine invariant descriptors from these feature points. The proposed algorithm can also estimate affine motion parameters. Simula-

* This research is supported by the NSF(60171036) and the significant technology project (03DZ14015), Shanghai, China.

J. Wang, X. Liao, and Z. Yi (Eds.): ISNN 2005, LNCS 3496, pp. 985–990, 2005.

tion results indicate that the algorithm performs well in constructing invariants and motion estimates for color images.

2   Radical Idea of Affine Invariant Descriptors for Color Images Based on ICA

Firstly, recall the theory of affine transform based on object-contour: Suppose that a is sample point on contour of the object with its coordinate on x and y axes (x_a, y_a). The affine transformed coordinate (x_a', y_a') satisfy:

$$\begin{pmatrix} x_a' \\ y_a' \end{pmatrix} = \begin{pmatrix} z_{11} & z_{12} \\ z_{21} & z_{22} \end{pmatrix} \begin{pmatrix} x_a \\ y_a \end{pmatrix} + \begin{pmatrix} b_x \\ b_y \end{pmatrix} = ZX_a + B . \tag{3}$$

for any parameters of affine matrix Z and shift vector B. Consider the sequence of sample points on contour:

$$X_a(t)' = ZX_a(t) + B . \tag{4}$$

Here t is the index of sample points. Move the contour to central coordinate, the vector B can be omitted. If $x_a(t)$ and $x_a'(t)$ are random vector, via ICA we can get:

$$X_a(t)' = A'S_a(t); X_a(t) = AS_a(t) . \tag{5}$$

According to Eq. (4), omit the vector B, it has:

$$X_a(t)' = A'S_a(t) = ZAS_a(t); A' = ZA . \tag{6}$$

If two contours are transformed from the same object, they share the same independent vector $S_a(t)$. Because the independent vectors can make up a closed curve, the sampling start positions can be found by ring shifting method[4].

2.1   Color Image Affine Transform Representation

Suppose there is a random sample point q on color image with coordinate (x_q, y_q) and color (R_q, B_q, G_q). Under uniform illuminant, the color of the sample point remains unchanged or changes a scale factor α after affine transform. So the affine transform of color images can be represented as:

$$\begin{bmatrix} x'_q \\ y'_q \\ R'_q(x'_q, y'_q) \\ B'_q(x'_q, y'_q) \\ G'_q(x'_q, y'_q) \end{bmatrix} = \begin{bmatrix} z_{11} & z_{12} & 0 & 0 & 0 \\ z_{21} & z_{21} & 0 & 0 & 0 \\ 0 & 0 & \alpha(x_q, y_q) & 0 & 0 \\ 0 & 0 & 0 & \alpha(x_q, y_q) & 0 \\ 0 & 0 & 0 & 0 & \alpha(x_q, y_q) \end{bmatrix} \begin{bmatrix} x_q \\ y_q \\ R_q(x_q, y_q) \\ B_q(x_q, y_q) \\ G_q(x_q, y_q) \end{bmatrix} + \begin{bmatrix} b_x \\ b_y \\ 0 \\ 0 \\ 0 \end{bmatrix} \tag{7}$$

The last three rows of Eq. (7) can be simplified using a function $\Upsilon(x_q, y_q)$ of the three fundamental colors (R_q, B_q, G_q) to represent as:

$$\begin{bmatrix} x'_q \\ y'_q \\ \Upsilon(x'_q, y'_q) \end{bmatrix} = \begin{bmatrix} z_{11} & z_{12} & 0 \\ z_{21} & z_{22} & 0 \\ 0 & 0 & \alpha(x_q, y_q) \end{bmatrix} \begin{bmatrix} x_q \\ y_q \\ \Upsilon(x_q, y_q) \end{bmatrix} + \begin{bmatrix} b_x \\ b_y \\ 0 \end{bmatrix} \tag{8}$$

Further more we use HIS space instead of RGB, which contains three components: hue, H, intensity and saturation. Transformation from RGB to HIS is shown as follows:

$$H = \begin{cases} \theta & B \leq G \\ 360 - \theta & B > G \end{cases} \quad \text{with } \theta = \cos^{-1}(\frac{1/2[(R-G)+(R-B)]}{[(R-G)^2+(R-G)(G-B)]^{1/2}}) \qquad (9)$$

Divide H by $360°$, it can be normalized in the interval $[0,1]$. The color of sample point (x',y') in affine image satisfies: $R'=\alpha R$, $G'=\alpha G$, $B'=\alpha B$. According to Eq. (9), the θ' of sample point (x',y') is:

$$\theta'(x,y)=\cos^{-1}(\frac{1/2[(\alpha R-\alpha G)+(\alpha R-\alpha B)]}{[(\alpha R-\alpha G)^2+(\alpha R-\alpha G)(\alpha G-\alpha B)]^{1/2}})=\theta(x,y) \qquad (10)$$

Use Hue as the function Υ in Eq. (8). Since the function Υ remains the same after affine transform, the affine transform of color image can be simplified as the expression of coordinates like Eq. (3).

2.2 Choose Sample Sequence in Color Images

In HIS color space, the Hue of color images or objects remains unchanged after affine transform, so does Hue-histogram. Sample sequences of color images are obtained from the Hue histogram.

Suppose image C with $N \times M$ color pixels. Its hue on the kth pixel H_k, $k=1,2,\ldots$. $N \times M$ is normalized in the range $[0,1]$ that is divided by p equal parts. Compute the statistic number n_i of all pixels in hue value $[i/p,(i+1)/p)$, $i=0,\ldots,p-1$, here $\sum_{i=0}^{p-1} n_i = N \times M$. We can obtain hue-histogram in which the hue occupying the most proportion is defined as main-hue. Let n_{Max} and n_{Min} be the up-bound and low-bound thresholds in the hue-histogram. Only the intervals with the statistic number $n_{Min} \leq n_i \leq n_{Max}$ are considered as candidate intervals to extract observed sample points. Because the main-hue with the largest statistic number probably is the image's background, the smallest statistic number may be noise, so we discard them. Let the n_i pixels' positions in candidate intervals be (x_{ij},y_{ij}), $j=1,\ldots,n_i$. The average position of these pixels in interval i can be calculated as below:

$$\bar{x}_i = \frac{1}{n_i} \sum_{j=1}^{n_i} x_{ij} \text{ and } \bar{y}_i = \frac{1}{n_i} \sum_{j=1}^{n_i} y_{ij} \qquad (11)$$

As the same, for affine transformed image C', we can get the average positions $\bar{x}'_i$ $\bar{y}_i'$ of each interval i via its hue-histogram. Regard the coordinate sequences of these average positions in the candidate hue intervals as observed random vector. Since the lengths of two observed vectors must be the same in ICA, we choose L common intervals of both hue-histograms of C and C' as length of observed vector sequence, $L \leq p$. The observed random vectors are denoted as $X(t) = \{\bar{X}_t\}$, $X'(t) = \{\bar{X}'_t\}$,

$t=1,...,L$. As mentioned above, if the color is abundant and the intervals are small, the length of observed vector is enough. And index t in observed vector is sorted by hue value, so the corresponding positions are easy to be found. Furthermore, the sequence of average position of the points in hue interval is random, so ICA can be used to find independent components, and the affine invariant descriptors can be obtained.

3 Details of the Algorithm Using ICA to Extract Affine Invariant Descriptors for Color Images or Objects

Suppose two color images or objects $C1$ and $C2$ are affine transformed from the same one. Using Eq. (9), change the color spaces of $C1$ and $C2$ from RGB to HIS and get their hue-histograms and candidate intervals, then choose the common L intervals of the objects before and after affine transform. From Eq. (11), we have two random observed vectors $(X(1),X(2)...,X(L))$, $(X'(1),X'(2),...X'(L))$. Same as Eqs. (5), (6), that is:

$$X(t)=(X(1),X(2)...X(L))=A(S(1),S(2)...S(L))=AS(t),$$

$$X'(t)=(X'(1),X'(2)...X'(L))=ZX(t)=ZA(S(1),S(2)...S(L))^T=ZAS(t)=A'S(t).$$

(12)

where t is the index. It shows that the two random observed vectors X and X' are obtained from the same independent vector $S(t)$ if they have affine relation. The goal of ICA is to find a linear transformation matrix W so that the random variables in Y and Y' are as independent as possible. We have:

$$Y = WX = WAS = APS$$

$$Y' = W'X' = W'A'S = A'P'S.$$

(13)

where Λ is a diagonal matrix which changes the scale and sign of random variables, and P is a permutation matrix with only one element equal to 1 on each row and column which changes the order of elements in Y, $y_i, i = 1,2$. Here we adopt FastICA based on negentropy [5].Note that the Observed variables need to be whitened at first for the requirement of zero mean and unit variance, so the matrices Λ and Λ' in Eq. (13) are equal to identity matrices. And independent components are convergent after 2~10 iterations using FastICA. From Eq. (13) it has:

$$Y = (P'P^{-1})^{-1}Y' = MY'.$$

(14)

The permutation matrix P and P' only have eight cases, so we have $M=\begin{vmatrix} \pm1 & 0 \\ 0 & \pm1 \end{vmatrix}$ for the same order, and $M=\begin{vmatrix} 0 & \pm1 \\ \pm1 & 0 \end{vmatrix}$ for antithesis order. According to property of ICA, the independent components satisfy:

$$E\{YY^T\}=E\{Y'Y'^T\}=I.$$

(15)

where I is identity matrix. From Eqs. (14),(15) we have $M = YY'^T / L$. If the calculated M matrix is diagonal or anti-diagonal identity matrix with sign, the two images have a affine relationship. Let Y be the affine invariant descriptors. For any affine parameters Z and B, there is a M matrix which only changes the order or sign of independent components and we can get affine invariant descriptors using M matrix. Affine parameters can be further computed. From Eqs. (3),(13),(14), it has:

$$Z=W'^{-1}M^{-1}W \quad \text{and} \quad B=m_x'-Zm_x . \tag{16}$$

Where m_x, m_x' are means of X and X'.

4 Experiments and Results

Example 1: The hue-histogram is the same for two pictures in affine transform. Fig.1 (b) is the picture after affine transformation of Fig.1 (a), the rotated angle is $30°$ and illumination parameter α on RGB is 1.2.

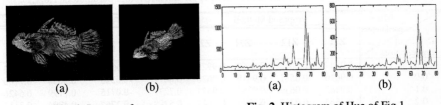

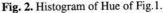

(a)	(b)
Fig. 1. Images of test.	**Fig. 2.** Histogram of Hue of Fig.1.

The hue-histograms of the two pictures are shown in Fig.2. The normalized H is quantified into 360 parts. There are only 80 color regions left if taking no account of the ones whose pixels are greater or less than the given thresholds. The two hue-histograms are very similar. Based on the hue-histogram and (11) we can get the sample sequences $x(t)$, $y(t)$ and $x'(t)$, $y'(t)$, $t=1,...,80$. The independent components separated by the ICA are y_x, y_y and y'_x, y'_y. Fig.3 shows the invariant description which we can get through the computation of alternate matrix M. Here the correlation coefficient of the x coordinate is 0.9971 while that of the y coordinate is 0.9879 for both pictures.

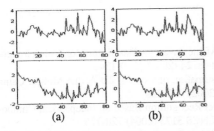

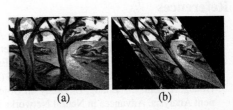

Fig. 3. The results after ICA: (a) Two random variables of Y; (b) Two random variables of MY'.

Fig. 4. Images of test: (a) original image; (b)its affine transform.

Example 2: We estimate affine motion parameters by proposed algorithm and Fourier method [1],[2]. Fig.4 contains two standard color images before and after affine transform. In order to compare both methods we choose the same example in [1].

The results for proposed method and Fourier method without noise are shown in Table 1. Under pepper noises, the results are shown in Table 2. It is obviously that the estimation precise of our method is better than Fourier method especially in noise case.

Table 1. Actual motion and estimated parameters.

Actual Affine Parameters				Proposed Method				Fourier Method			
Z11	Z12	Z21	Z22	Z11	Z12	Z21	Z22	Z11	Z12	Z21	Z22
1	0	1	1	1	0	1	1	1	0	1	1
1	0	0.3	1	0.973	0.019	0.342	1.075	1.057	0.101	0.399	1.184
0.9	0.2	0	1	0.965	0.229	0.067	0.950	1.017	0.390	0.092	0.860
0.9	0.2	0.2	1	0.970	0.251	0.265	0.981	1.063	0.455	0.281	0.884

Table 2. Affine parameters estimate under pepper noises;Actual parameters(1,0;1,1).

Affine parameters / Density of Pepper noises	Proposed Method				Fourier Method			
	Z11	Z12	Z21	Z22	Z11	Z12	Z21	Z22
0.02	0.9844	0.0013	0.9769	0.9923	0.9940	0.0037	0.9759	0.9753
0.05	0.9610	0.0581	0.9419	0.9620	0.8071	**-0.0228**	0.8034	0.7700
0.1	0.9467	0.0639	0.9232	0.9374	0.7277	**-0.0715**	0.7199	0.6426
0.2	0.8563	0.0782	0.7723	0.8012	0.5298	**-0.2767**	0.4794	0.3114

5 Conclusions

This paper proposes a scheme to extract affine invariant descriptors from color image or object based on ICA. Some preprocesses will help us to find some sorted random observed and corresponding points in the original image and its affine image. Experiments indicate that ICA can deal with not only sample points along contours but also corresponding discrete points. When estimating affine motion parameters, proposed algorithm is more robust on noise and more precise in the case of high skew compared with Fourier method in the color images.

References

1. Oirrak, A.E., Daoudi, M, Aboutajdine, D.: Affine Invariant Descriptors for Color Images Using Fourier Series. Pattern Recognition Letters, **24** (2003) 1339-1348
2. Oirrak, A.E., Daoudi, M, Aboutajdine, D.: Estimation of General 2D Affine Motion Using Fourier Descriptors. Pattern Recognition, **35** (2002) 223-228
3. Huang, X., Liu, C., Zhang, L.: Study on Object Recognition Based on Independent Component Analysis. Advances in Neural Networks LNCS **3173** (2004) 720-725
4. Zhang, L., Huang, X.: New Applications for Object Recognition and Affine Motion Estimation by Independent Component Analysis. Intelligent Data Engineering and Automated Learning, LNCS **3177** (2004) 25-27
5. Hyvärinen, A.: Fast and Robust Fixed-Point Algorithms for Independent Component Analysis. IEEE Trans. Neural Networks, **10** (1999) 626-634

A New Image Protection and Authentication Technique Based on ICA

Linhua Zhang[1,2], Shaojiang Deng[2], and Xuebing Wang[3]

[1] College of Science, Chongqing University, Chongqing 400044, China
Linzhang@cqu.edu.cn
[2] Department of Computer Science and Engineering, Chongqing University,
Chongqing 400044, China
sj_deng@cqu.edu.cn
[3] Network Center, Chongqing University, Chongqing 400044, China
Xuebingwang@cqu.edu.cn

Abstract. Based on chaotic communication and independent component analysis, an image protection and authentication technique is proposed. We modulate a watermark by chaotic signal and embed it in watermark host vector which is established in terms of the coefficients in DWT domain. When the author and the legal user want to authenticate digital image, they can detect and extract the watermark by using ICA and private key. Experimental results show that robustness of the watermarking satisfies both of their demands.

1 Introduction

With the fast development of computer network and multimedia technology, it is so easy to obtain digital image through information network and further process, reproduce and distribute it. Digital technology brings us much convenience, but it also makes it easy to infringe upon the rights of the author and the lawful user.

Although, with the help of cryptographic technology, digital image can be safely conveyed, if once decoded, no effective means can be taken to prevent the digital works from being illegally copied, distributed and used. On the other hand, the encoded digital image means nothing to others before decoding, which tampers the works from being appreciated by more individuals because encoded digital image lack the feature of direct perception and vividness. Therefore, the method of encryption is rather limited.

Digital watermarking technology employs characteristics of human audio-visual systems and embeds information into digital image. This method overcomes the drawbacks of the former and is regarded as the best protection means of digital image in the future.

Although watermarking technique has been popular for years, however, digital watermarking based on chaotic map became a hot topic in recent years. In the other hand, in 2002, by using ICA, a robust watermarking scheme was proposed firstly in [1], which can resist image scaling, cropping, low-pass filtering,

J. Wang, X. Liao, and Z. Yi (Eds.): ISNN 2005, LNCS 3496, pp. 991–996, 2005.

rotation, color quantization, noise addition, collusion and so on, but its key is too long to be practical. In [2], an improved scheme was put forward but it did not mention effects of mixing parameters. To sum up, the above schemes emphasize on robustness of watermark only. As a matter of fact, robustness of watermarking is crucial in protecting copyright of the digital image for the author, but for the legal user, fragility of watermark is required to examine whether the image was tampered with. In this paper, to meet demands of the author and the user, a new viewpoint is proposed, i.e, robustness of watermark can be controlled by the mixing matrix of ICA, and all of steps are elaborately devised from the author to the legal user.

2   The Proposed Scheme

We propose a new watermarking method based upon neural network. Firstly, we obtain the coefficients of DWT domain by using DWT and modulate the watermarking image by a chaotic signal. Secondly, we obtain the watermark host vector from DWT domain so that energy of watermark can spread through the original image. Thirdly, by choosing a mixing matrix to mix the modulated watermark and the host vector, we obtain a signal as a key which will be sent to legal user. Finally, the author or the legal user can compute the demixing matrix to authenticate watermarked image by using ICA. Fig.1 illustrates the block diagram of proposed scheme.

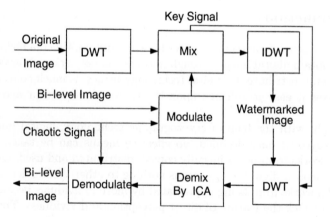

Fig. 1. Block diagram of the proposed scheme.

2.1   Chaotic Watermark Modulating

The chaotic signal with pseudo-random and broad-spectrum properties makes it suitable to secure communication and robust watermarking system. To modulate the watermark of gray image, bi-level image or other signal, some chaotic systems can be chosen to generate chaotic signals, such as Chen's system, Chua's system and Liao's system. Recent experimental results show that Liao's system can be realized by hardware, so it probably has a wide prospect of application. Hence, we choose Liao's system which is described by the expression [3]

$$\frac{dx(t)}{dt} = -x(t) + af[x(t) - bx(t-1) + c], t > 0 \tag{1}$$

$$f(x) = \sum_{i=0}^{n} \alpha_i[\tanh(x + k_i) - \tanh(x - k_i)] \tag{2}$$

We can obtain a distinct chaotic signal by virtue of control parameter b.

2.2   The Host Watermark Vector Constructing

Wavelet analysis is an efficient signal processing tool, especially in image processing. A 1-level 2-D wavelet image decomposition is composed of four wavelet channels: LL, LH, HL and HH, where LL corresponds to low frequencies of the original image and possesses a majority of energy. Experimental results indicate that LL is most stable against outside attacks [2]. In this letter, we divide approximate matrix into blocks of size 3×3 and draw out the central elements of them to construct a host watermark vector. In a sense, our scheme has properties of the popular spread spectrum [4].

2.3   ICA and Watermark Embedding

we can review the most classic application of ICA, such as blind source signals separation, noise in natural images reduction, hidden factors in financial data analysis and face recognition. In [5] and [6], a statistical "latent variable" model of ICA was described in detail. Compared with existing ICA algorithms, the well-known ICA implementation is Hyvärinen's Fastica, which has the advantages of neural algorithms: Parallel, distributed, but requires the user to select a contrast function according to the hypothetical probability density functions of the sources [6]. Recently, A novel nonparametric ICA algorithm(NICA) is introduced, which is non-parametric, data-driven, and does not require the definition of a specific model for the density functions [7]. Hence it is fit for watermark extraction.

In fact, NICA is an overall optimal problem, which can be posed as

$$\min \quad -\frac{1}{M} \sum_{i=1}^{N} \sum_{i=1}^{M} \log[\frac{1}{Mh} \sum_{m=1}^{M} \phi(\frac{w_i(x^{(k)} - x^{(m)})}{h})] - \log|detW| \tag{3}$$

$$s.t. \quad \|w_i\| = 1, i = 1, 2, ..., N$$

where ϕ is normal gaussian kernel and h is the kernel bandwidth, M is the size of sample data, N is the number of independent source signals. w_i is the i-th row of demixing matrix W. The quasi-Newton method can be used in the above problem to seek the overall optimal solution [8].

In this Letter, regarding the host watermark vector as a source signal, modulated watermark as the other source signal, we choose a matrix as mixing matrix. Since watermark was modulated, mixing matrix need not be random. In fact, one of mixed signals must be embedded in the host image, it is necessary that mixing matrix cannot be a random matrix.

Let X be the host watermark vector, M the modulated watermark, then ICA mixing model can be described as the expression

$$WI = X + aM \qquad (4)$$
$$KS = cX + bM$$

where WI will be used to change the coefficients of sub-band LL and KS key signal which will be transferred to legal user to detect the watermark. Therefore, parameter a should be small and b should be large, otherwise it is difficult to reconstruct the watermarked image and separate the modulated watermark by using ICA.

2.4 Watermark Detection and Extraction

It is crucial that the author protect the copyright of the watermarked image and that the legal user verify whether it is tampered with. The process of watermark detection and extraction is as follows:

- The author regards controlled parameter b in Eq.1 and key signal as a private key and send it to the legal user. There are many cryptosystems to be chosen, such as Diffie-Hellman key exchange, ElGamal encryption and digital signature, and all analogues in the elliptic curve case. By using ICA, the modulated parameters and mixing matrix is dispensable for the legal user.
- Regard WI and KS in Eq.4 as two mixed signals and extract two independent components. One of them must have the typical features of modulated watermark.
- Regard chaotic signal and the extracted modulated watermark as two mixing signals of ICA to acquire the original watermark.

3   Simulation Results and Performance Analysis

Now, we make some experiments with the proposed approach. The original image is a standard 256×256 gray-scale image of Woman, see Fig.2(d). The watermark shown in Fig.2(c) is a bi-level image of size 45×45, which looks like a seal. We transform the original image to DWT domain with 2-D discrete wavelet transformation, where a bi-orthogonal wavelet is used. In order to get the chaotic signal adopted in experiments, supposing that $a = 3, b = 4.5, c = 1, n = 2, \alpha_1 = 2, \alpha_2 = -1.5, k_1 = 1, k_2 = 4.0/3$, in Eq.1 and Eq.2, we get a lag-synchronization chaotic signal. Since $x(t)$ is continuous and predictable, we must sample and get a sub-signal whose waveform diagram is illustrated in Fig.2(b). The waveform of modulated watermark can be seen in Fig.2(a).

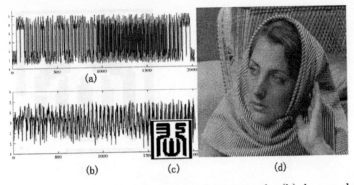

Fig. 2. Watermark embedding. (a) modulated watermark, (b) lag-synchronization chaotic signal, (c) watermarking image, (d) reconstructed image.

Let $c = 0.2987$ and $b = 1000$ in Eq.3, then the extracted watermarks from the watermarked image under many attacks are similar to the original watermark. It is easy to understand this result because the weight of the watermarking host vector is too small to affect one of the independent components, i.e., the modulated watermark. But the legal user can not realize whether the watermarked image was tampered with or replaced with an arbitrary image. Hence the ratio of parameter c to b must be proper in Eq.3. Meanwhile,we must take into account a in Eq.3, which corresponds to the embedding strength.

Let $c = 0.2987$, $a = 1$ and $b = 40$ in Eq.3. Table.1 indicates impacts of some attacks on the modulated watermark. These attacks include (a)Amplitude variation(rate=0.8), (b) image cropping(block of size 50×50), (c)gaussian noise addition(mean=0,variance=5), (d) salt&pepper noise addition(density=0.005), (e-g)JPEG compression(quality=80,30,10 respectively). Similarly, the correlation coefficients between the original watermark and the extracted watermark can be computed(omitted). In fact, we can take in account other attacks such as low-pass filtering, image scaling and rotating. Fig.3 shows the whole process of watermark extraction from the cropped image. As a whole, our scheme has strong robustness against JPEG compression and frailness against image cropping.

4 Conclusion

ICA is a classic method applied in signal processing. Chaotic Signal is selected to ensure security and to improve statistical independence of signals. Our watermarking scheme can be realized easily. By virtue of the mixing matrix of ICA,

Table 1. The correlation coefficients between the original modulated watermark and the extracted modulated watermark.

Attacks	(a)	(b)	(c)	(d)	(e)	(f)	(g)
SNR	13.96	1.923	24.00	14.16	18.36	14.78	11.66
Coefficients	0.998	0.600	0.989	-0.958	0.991	-0.941	-0.855

(a)   (b)

Fig. 3. Extracting the watermark from the cropped watermarked image. (a) the cropped watermarked image, (b) the extracted watermark.

robustness of watermark can be controlled efficiently so as to meet both the demands of the author and the legal user.

Acknowledgements

We would like to express our gratitude to prof. Xiaofeng Liao for unveiling to us many techniques of neural networks, and to the reviewers for their constructive comments.

References

1. Yu, D. and et al.: Water Detection and Extraction Using A Independent Component Analysis Method. EURASIP Journal on Applied Signal Precessing, **1** (2002) 92-104
2. Liu, J., Zhang, X.G, Sun, J.D.: A New Image Watermarking Scheme Based on DWT and ICA.IEEE Int. Conf. Neural Network&Signal Processing.(2003) 1489-1491
3. Liao, X.F., Wong, K.W.: Hopf Bifurcation and Chaos in A Single Delayed Neural Equation With Non-monotonic Activation Function. Chaos, Solitons&Fractals, **12** (2001) 1535-1547
4. Cox, I.J., Kilian, J.: Secure Spread Spetrum Watermarking For Multimedia. IEEE Trans. on Image Processing, **12** (1997) 1673-1687
5. Hykin, S.:Neural Networks: A Comprehensive Foundation. 2rd edn. Prentice Hall, New York(1999)
6. Hyvärinen, A., Oja, E.: Independent Component Analysis: Algorithms and Applications. Neural Networks, **13** (2000) 411-430
7. Boscolo, R., Pan, H.: Independent Component Analysis Based on Nonparametric Density Estimation. IEEE Trans. on Neurral Netwoks, **15** (2004) 55-64
8. Http://www.ee.ucla.edu/riccardo/ICA/npica.tar.gz

Locally Spatiotemporal Saliency Representation: The Role of Independent Component Analysis[*]

Tao Jiang[1,2] and Xingzhou Jiang[2]

[1] State Key Lab. of Intelligent Technology and Systems, Tsinghua University
Beijing 100084, China
[2] Department 5, Navy University of Engineering
Wuhan 430033, China
jt@s1000e.cs.tsinghua.edu.cn

Abstract. Locally spatiotemporal salience is defined as the combination of the local contrast salience from multiple paralleling independent spatiotemporal feature channels. The computational model proposed in this paper adopts independent component analysis (ICA) to model the spatiotemporal receptive filed of visual simple cells, then uses the learned independent filters for feature extraction. The ICA-based feature extraction for modelling locally spatiotemporal saliency representation (LSTSR) provides such benefits: (1) valid to use LSTSR directly for locally spatial saliency representation (LSSR) since it includes LSSR as one of its special case; (2) Plausible for space variant sampled dynamic scene; (3) Effective for motion-based scene segmentation.

1 Introduction

Psychologists believe that there exist underlying processes of visual attention that control primates visual behavior, e.g. saccadic eye movement, selective memory, etc. The concept of saliency is used to define *to which extend* the processed information feeding into the perceptual system *owns* some intrinsical properties to *cause attention* to focus on. The mechanisms of saliency representation may be divergent due to the divergence of processed information: local or global, spatial or temporal, see the psychological experiments result in [1]. In this paper, we only discuss the representation of salience of local contrast on spatiotemporal features and name the problem as locally spatiotemporal saliency representation (LSTSR). The current development of active robot vision systems also confronts the same task of using LSTSR to benefit their visual behaviors.

After Koch and Ullman in [2] firstly contributed a structural model on visual salience representation in 1985, and the origination of research on active vision for robots from the late of 1980s, many computation models have been developed for modelling primate visual attention selectivity or embedding robots with such capability. Most of the published computation models under the salience-map

---

[*] This research was funded by the State Key Lab. of Intelligent Technology and Systems, Tsinghua University, China, with the help of Prof. Peifa Jia.

J. Wang, X. Liao, and Z. Yi (Eds.): ISNN 2005, LNCS 3496, pp. 997–1003, 2005.

based approach mainly concentrate on locally spatial saliency representation (LSSR), except [3] and [4] adopting Reichardt energy detector and optical flow computation for motion feature extraction respectively. Few discusses the relationship between LSTSR and LSSR.

LSTSR is closely linked to the mechanism of visual motion perception. To meet some psychological experiment results on human visual motion search, Rosenholtz [5] proposed a simple model to estimate the ease or efficiency for searching target motion patterns among other distractors. To implement his model, issues as motion based object segmentation and motion parameter estimation (here, the velocity) must be involved. Due to the effect of illuminance change, noise and sensor shifting, whether optical flow based or feature based motion segmentation algorithms may cause severe error. Moreover, the approximation solution for parameter estimation may often be ill-posed.

Other facts should be noted before giving a model on spatiotemporal salience: the dichotomy of more dichromatic peripheral visual field and more chromatic foveal field of retina in primates, and the space variant sample manner at least adopted by the peripheral for imaging. See experiment result about dichromatic centrifugal motion salience in [6], and the space variant imaging in [7].

2   Computational Model

Based on upper motivations, we give an model for LSTSR by adopting ICA, an information-theoretic way, to learn feature extractors rather than adopt heuristic ways. We keep the elementary structural model of [8]. Moreover, our model includes space variant imaging before feature extraction as an optional process.

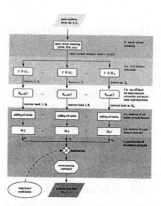

Fig. 1. The computational model of visual LSTSR.

a. The model structure
It is illustrated in fig. 1. Different to Itti *et al.*'s model, in order to provide higher efficiency on computation, some revision has been done. It consists of

three main steps after a space variant sampling operation on space uniform scene I. In the first step, the space variant scene I' is extracted by ICA filters (ICFs) $\{IC_i|i = 1, ..., m\}$ at each frame, then the contrast of each feature map and its multiscale representation B_i is obtained by applying an on-off bank filtering operation. The second step firstly adds all scales of the maps in the on-off bank of each feature channel, then gives a salience measurement S_i on the addition result for each channel. The last step only combines all the salience map $\{S_i|i = 1, ..., m\}$ from all channels by simply addition (i.e. the unbiased algebraic summation). Thus it gives the saliency representation S on I' at each frame. Permitting the top-down task-specific intentional modulation, it provides one way by adjusting the weighting value during the combination process. So the model links the stimulus-driven attentional mechanism and the task-specific intentional one together.

b. Inner processes
Some necessary inner processes in upper model that are not involved in Itti et $al.$'s model are briefly discussed in below. See [8] to understand the common normalization process.

(1) Space variant sampling
Methods for space variant sampling can be adopted as the the well-known log-polar mapping (LPM, [9]), or the so-called inverse polar mapping (IPM, [10]) with less data reduction ratio than LPM. The learned ICFs from Cartesian scene can be used on space variant sampled scene directly if the kept data resolution not too slow.

(2) Unsupervised feature extraction based on ICA learning
Applying ICA learning process, a set of independent filters can be get, then used as feature extractors on input scene to get feature maps in a parallel way. Suppose the set of feature extractors be $\{H_i|i = 1, 2, ..., m\}$, and the input scene $I(x, \tau)$, where $x \in \Theta$ and Θ is the spatial domain of each scene frame, $\tau \in [t-T, t]$ and T is the scope of temporal receptive field (RF) of ICFs, e.g. $T = 8/f$ for 9 frames as the temporal scope of RF of ICFs, f is the sample rate of frames, a set of feature maps $\{F_i|i = 1, 2, ..., m\}$ can be obtained by

$$F_i = I * H_i, i \in \{1, 2, ..., m\}, \tag{1}$$

where $*$ means the filter operation.

(3) Simultaneous feature contrast extraction and multiscale representation
We propose a so-called on-off bank filter for simultaneously doing spatial feature contrast extraction and multiscale representation. Each filter kernel f of it is defined by

$$\begin{cases} f_e(x, \sigma_e) = c_e g(x, \sigma_e); \\ f_i(x, \sigma_i) = c_i \cdot \frac{1}{2\pi\sigma_i^2} \exp\{-\frac{x'x-d^2}{2\sigma_i^2}\}; \\ f(x, \sigma_e, \sigma_i) = f_e(x, \sigma_e) - f_i(x, \sigma_i). \end{cases} \tag{2}$$

The parameter of c_e, c_i always are set as 0.75, 0.25 respectively, $\sigma_i = \sqrt{2}\sigma_e$, $d = \sqrt{8s \log \sqrt{2} \frac{2^s}{2^s-1}}\sigma_e$, where the quantity of s is equivalent to the difference of scale levels used in the cross-scale substraction operation in Itti *et al.*'s model. Then, the multiscale representation bank for each feature channel is obtained by

$$I_t(x) = |I(x) * f(x, \sigma_e(t), \sigma_I(t))|, t = 1, 2, ..., l. \tag{3}$$

(4) Getting salience map in each channel

Due to the assumption of linear time invariant property of visual system, to save the computation cost, we recommend to do one normalization process after adding all local contrast maps in the bank for each feature channel after applying on-off bank filtering. Our experiment result also reveals its validity and efficiency improvement.

3   Experiment and Result Analysis

In experiments, all the segments of traffic scenes are from [11]. Samples for ICA learning are from [11] and [12]. And we adopt FastICA [13] to learn the ICFs.

a. Spatiotemporal ICFs

In the left part of figure 2, 12 of the 125 learned spatiotemporal ICFs are shown. As an example, the right part of figure 2 shows 6 of its feature maps after filtering a traffic scene "bad" (frame 1 to frame 9) with ICFs.

Fig. 2. *Left.* Selected 12 learned ICA filters. The size is $11 \times 11 \times 9$ for height, width and temporal duration respectively. The frames for each ICF are shown from top to bottom in one column. *Right.* Selected 6 feature map after filtering scene "bad" (frame 1 to frame 9) with 6 ICFs.

b. LSTSR on Cartesian scene

The result on segment of scene "bad" is shown in figure 3. From the salience map, not limited in the illustrated set of result, we find that: (1) Main motion patterns and some salient spatiotemporal patterns (e.g. the related area at side of the street lamp pole when bus moving across it) are highlighted. Even small motion patterns are detected. (2) The salience map flow owns a constant latency to scene flow about 4 to 5 frames. It consists with the depth of filter block, which is 9 frames. It is plausible to the commonsense that motion only can be detected after its occurrence! (3) The synthesized flow (as shown in the fourth row) indeed reflects the temporal and spatial salience of the scene, e.g. all the occurrence of moving car into or out of the scene is quickly and correctly detected, and marked.

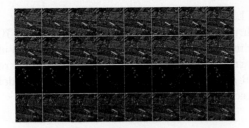

Fig. 3. The LSTSR on traffic scene segment of "bad". The first two rows are 16 frames of the source from left to right according their time sequence. The third row are 8 frames of the salience map for source frames 9 to 16 respectively. The fourth row are the synthesized images with marking salience maps on their source frames correspondingly.

c. Similar result between Cartesian imaging and space variant imaging
Adopting inverse polar mapping in experiments, it gives similar results between using LSTSR on scene after space variance imaging and on Cartesian stimulus directly, as shown in figure 4. It implies that in real implementation, we can directly use sensors sampling their peripheral visual field in a space variant manner. Here, based on our other result on comparing LSTSR on LPM sampled scene and Cartesian scene, we want to indicate that if the resolution reduction ratio after space variance sample is kept above some value, their similarity can be guaranteed.

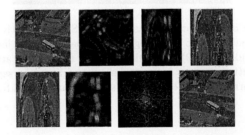

Fig. 4. The LSTSR on segment of traffic scene "bad". The first row is the result of doing LSTSR on Cartesian scene directly, the second on space variant sampled scene by inverse polar sampling. Illustration does not mean their true relative size. In the first row, from the left to the right are images for scene (frame 9) with indicating the most salient position, the salient map of it, the corresponding inverse polar sampled salient map, and the corresponding IPM sampled scene with indicating its most salient position based the IPM sampled salience map (the third image). In the second scene, from the left to the right, the four images are respectively the IPM sampled scene for LSTSR with the most salient position indicated, its salience map, the sampled frame illustrated in a Cartesian manner, and the original scene before sampling by indicating the most salient position after mapping the most salient position on salience map (the second image) inversely into the original scene.

d. Comparison of LSSR and LSTSR

Results from both using LSTSR (adopting spatiotemporal ICFs acquired from dynamic scenes or mixture of dynamic and static scenes) and using LSSR (adopting static ICFs learned on static scenes) shows quite similar results, see figure 5. We conclude LSSR a special case of LSTSR. Hence, the model of LSTSR can be used for LSSR directly without any modification.

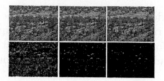

Fig. 5. Comparison of salience representation by LSTSR and LSSR on static image "Car-in-Forest" from [8]. From the left column to the right, the adopted ICFs are respectively learned from dynamic scenes (20,000 $11 \times 11 \times 9$ samples, 125 ICFs), mixture of static images and dynamic scenes (40,000 $11 \times 11 \times 9$ samples, 250 ICFs), static images (20,000 11×11 samples, 125 ICFs). In each column, the top image is the static scene indicated with the most salient position, and the bottom its salience map.

Fig. 6. Comparison of segmentation based on LSTSR on directly consecutive frames substraction of scene of traffic scene segment of "passat". The left are 2 consecutive frames (frame 29 and frame 30). The third image is the segmentation map by directly subtracting consecutive the two frames. The most right is the segmentation map by subtracting their corresponding consecutive salience map after LSTSR.

e. Using LSTSR for motion-based segmentation

Segmentation of static or dynamic stimulus always can be made only based on local feature contrast. Hence, LSSR and LSTSR are competent candidates for such segmentation process. Limiting the scene be always high dynamic, a background scene can not be acquired and keep stable even for a small moment. Below we compare the result of scene segmentation based on LSTSR with the one based on simple consecutive frames substraction, shown in figure 6. From experiments, we know that: (1) The result of substraction consecutive salience map sequences is better than direct on the scene, e.g. all moving objects or regions is segmented with high value, interestingly, the hidden moving cars among trees are also highlighted. (2) Segmentation based on salience map sequences can save it from noise, camera small shift, illuminance change to quite a large extend, not could direct substraction of scene. (3) Substraction on consecutive salience maps is better than on current map with an average map for a period.

4   Conclusion

Adopting spatiotemporal ICFs learned by ICA approaches, the proposed computational model gives a theoretic information oriented solution for modelling LSTSR, more effective and general than the heuristical feature space selection way. It indeed includes the modelling of LSSR as a subproblem of LSTSR. By embedding dichromatic periphery modelling and imaging it with space variant sampling, the model stands quite plausible ground for practical implementation. Our experiments with real-data shows its validity. Moreover, its efficiency for motion-based segmentation implies a competent segmentation nature of salience map on spatiotemporal stimulus. The inherent parallelism in the model, mainly due to the independence among ICFs-extracted feature channels, seems to guarantee its efficient real-time implementing. Future works may include to make more reasonable extension of ICA under constraints about structure information, and to adopt more principles than the ICA based redundancy reduction one for learning more plausible feature extractors.

References

1. Hans-Christoph Nothdurft: Salience from Feature Contrast: Temporal Properties of Saliency Mechanisms. Vision Research, **40** (2000) 2421 - 2435
2. Koch, C., Ullman, S.: Shifts in Selective Visual Attention: Towards the Underlying Neural Circuitry. Human Neurobiology, **4** (1985) 219 - 227
3. Itti, Dhavale, L.N., Pighin, F.: Realistic Avatar Eye and Head Animation Using a Neurobiological Model of Visual Attention, SPIE (2003)
4. Ouerhani, N., Hugli, H.: A Model of Dynamic Visual Attention for Object Tracking in Natural Image Sequences. LNCS **2686** (2003) 702 - 709
5. Rosenholtz, R.: A Simple Saliency Model Predicts a Number of Motion Popout Phenomena. Vision Research, **39** (1999) 3157 - 3163
6. Miura, K. et al.: Initiation of Smooth Pursuit in Humans Dependence on Target Saliency. Exp Brain Res, **141** (2001) 242 - 249
7. Wandell, B.A.: The Foundations of Vision. Stanford Press (1995)
8. Itti, L., Koch, C.: A Saliency-based Search Mechanism for Overt and Covert Shifts of Visual Attention. Vision Research, **40** (2000) 1489 - 1506
9. Bolduc, M., Levine, M.D.: A Review of Biologically Motivated Space-variant Data Reduction Models for Robotic Vision. Computer vision and image understanding, **69** (1998) 170 - 184
10. Jiang, T., et al.: Biologically Motivated Space Variant Sampling: First Step to Active Vision. Proceedings of Asian Conference on Computer Vision, (2004) 204 - 209
11. http://i21www.ira.uka.de/image_sequences, KOGS/IAKS Universität Karlsruhe
12. ftp://hlab.phys.rug.nl/pub.
13. Hyvärine, A.: Fast and Robust Fixed-Point Algorithms for Independent Component Analysis. IEEE Trans. on Neural Networks, **10** (1999) 626 - 634

A Multistage Decomposition Approach for Adaptive Principal Component Analysis

Dazheng Feng

National Lab. of Radar Signal Processing, Xidian University, Xi'an 710071, China
Dzfeng@rsp.xidian.edu.cn

Abstract. This paper devises a novel neural network model applied to finding the principal components of a N-dimensional data stream. This neural network consists of r ($\leq N$) neurons, where the i-th neuron has only $N - i + 1$ weights and a $N - i + 1$ dimensional input vector that is obtained by the multistage dimension-reduced processing (multistage decomposition) [7] for the input vector sequence and orthogonal to the space spanned by the first $i - 1$ principal components. All the neurons are trained by the conventional Oja's learning algorithms [2] so as to get a series of dimension-reduced principal components in which the dimension number of the i-th principal component is $N - i + 1$. By systematic reconstruction technique, we can recover all the principal components from a series of dimension-reduced ones. We study its global convergence and show its performance via some simulations. Its remarkable advantage is that its computational complexity is reduced and its weight storage is saved.

1 Introduction

The principal component analysis (PCA) is frequently applied to extract the important features and finds the signal subspace from a high dimensional data stream. In the "dimension-reduced" techniques, it helps eliminate information redundancy of the data transmitted through a band-limited channel. PCA identifies the most important features of a high-dimensional normal distribution in the sense that the projection error onto those feature subspaces is minimal. It can been shown [1] that the i-th principal component (PC) direction is along an eigenvector of the signal auto-correlation matrix, the PCA (signal) subspace can be interpreted as the maximizer of the projection variance of the stochastic signal. The optimal solution is the subspace spanned by the eigen-vectors of the signal auto-correlation matrix associated with the largest values. Interestingly, it has first been shown in [2] that Hebbian rule applied to train a single linear neuron can extract the first principal component of the input data streams, i.e. the weight vector of neuron converges to the eigenvector associated with the largest eigenvalue of the input auto-correlation matrix. Later, by proposing the subspace and the generalized Hebbian learning algorithms, many researchers [3]-[6] extended this method to find PC's or the subspace spanned by them, using multiple neurons. It is worth mentioning that any trained weight vector in these learning is the full length, i.e. its length is equal to the dimensional number of the input vector. The entire algorithms have explicitly or implicitly employed the deflation transformation to decorrelation weights from one another. In order to increase the computational efficiency of the adaptive PCA, this paper uses the reduced- dimension techniques.

J. Wang, X. Liao, and Z. Yi (Eds.): ISNN 2005, LNCS 3496, pp. 1004–1009, 2005.

2 Preliminaries

Suppose that a N-dimensional vector sequence $\mathbf{x}(k) = [x_1(k), \cdots, x_N(k)]^T$ ($k = 1, 2, \cdots$) is a stationary stochastic process with zero-mean and the covariance matrix $\mathbf{R} = E\{\mathbf{x}(k)\mathbf{x}^T(k)\} \in R^{N \times N}$. Let λ_i and $\mathbf{u}_i$ ($i = 1, \cdots, N$) denote the eigenvalues and the corresponding orthogonal eigenvectors of $\mathbf{R}$. We shall arrange the orthogonal eigenvectors $\mathbf{u}_1, \cdots, \mathbf{u}_N$ such that the corresponding eigenvectors are in a non-increasing order: $\lambda_1 \geq \lambda_2 \geq \cdots \geq \lambda_r > \lambda_{r+1} \geq \cdots \geq \lambda_N \geq 0$, where r denotes the principal component (PC) number. Furthermore, assume that $\mathbf{R}$ has the L distinct nonzero eigenvalues $\tilde{\lambda}_1 > \tilde{\lambda}_2 > \cdots > \tilde{\lambda}_L > 0$ with multiplicity $m_1, \cdots, m_L$ ($L \leq N$). Thus the eigenvalue decomposition (EVD) of $\mathbf{R}$ is represented as $\mathbf{R} = \sum_{j=1}^{N} \lambda_j \mathbf{u}_j \mathbf{u}_j^T = \mathbf{U}\Lambda\mathbf{U}^T$, where $\mathbf{U} = [\mathbf{u}_1, \cdots, \mathbf{u}_N]$ and $\Lambda = \text{diag}(\lambda_1, \cdots, \lambda_N)$. Our objective is adaptively to track all the eigenvalues and eigenvectors corresponding to the signal subspace of the $\mathbf{x}(k)$.

In the area of neural networks, Oja [2] showed that a normalized version of the Hebbian rule applied to a single linear unit can extract the principal component of the input vector sequence, i.e. it converges to the eigenvector associated with the largest eigenvalue of the input autocorrelation matrix. For the linear unit case, the realization of the Hebbian rule in its simplest form (just update each weight proportionally to the corresponding input-output product) is numerically unstable. Oja's algorithm [2] for tracking the single principal component is as follows:

$$y_1(k) = \mathbf{x}^T(k)\mathbf{w}_1(k) \tag{1a}$$

$$\mathbf{w}_1(k+1) = \mathbf{w}_1(k) + \eta_k[y_1(k)\mathbf{x}_1(k) - y_1^2(k)\mathbf{w}_1(k)] \tag{1b}$$

$$\lambda_1(k+1) = \alpha\lambda_1(k) + (1-\alpha)y_1^2(k) \tag{1c}$$

where $\mathbf{w}_1 = [w_{1,1}, \cdots, w_{1,N}]^T$ and $\mathbf{x}_1(k) = \mathbf{x}(k) = [x_{1,1}(k), \cdots, x_{1,N}(k)]^T$, respectively, are the connection weight vector and the input vector of neuron 1; λ_1 is the estimate of the first eigenvalue of the autocorrelation matrix $\mathbf{R}$, η_k is the time-step length and the learning rate and gain sequence; subscript "1" is associated with the first eigencomponent of $\mathbf{R}$ and neuron 1, which provides the convenience for considering the extraction of the PC's. Although Oja's algorithms can usually work satisfactorily, they are sometimes slow to converge. More importantly, both analytical and experimental studies show that convergence of these algorithms depends on appropriate selection of the gain sequence that relies on the input vector sequence. In order to improve the PCA algorithms, some appropriate selection methods for the learning rate sequence have been proposed. In fact, using some multistage representation [7], we can increase the computational efficiency of the PCA algorithms, especially in the case with the very high dimensional input vector.

3 Model

By using some multistage representation [7], we can get an efficient neural networks. The proposed neural network model is depicted in Fig.1 and made up of r linear neurons. There are r inputs $\mathbf{x}_1(k), \mathbf{x}_2(k), \cdots, \mathbf{x}_r(k)$, connected to r outputs $y_1(k), y_2(k), \cdots, y_r(k)$ through r feed-forward   weight   vectors $\mathbf{w}_1(k), \mathbf{w}_2(k), \cdots, \mathbf{w}_r(k)$. All the weights in the network will converge to the dimension-reduced principal eigenvectors $\widetilde{\mathbf{u}}_1(k), \widetilde{\mathbf{u}}_2(k), \cdots, \widetilde{\mathbf{u}}_r(k)$ with probability 1. The activation of each is linear function of its inputs

$$y_i(k) = \mathbf{w}_i^T(k)\mathbf{x}_i(k) \quad (i = 1, \cdots, r) \tag{2}$$

$$\mathbf{w}_i(k) = [w_{i,1}(k), \cdots, w_{i,N-i+1}]^T \tag{3a}$$

$$\mathbf{x}_1(k) = \mathbf{x}(k) \tag{3b}$$

$$\mathbf{x}_i(k) = \mathbf{T}_{i-1}(k)\mathbf{x}_{i-1}(k) \quad (i = 2, \cdots, r). \tag{3c}$$

In (3b), the dimension-reduced matrix is described by

$$\mathbf{T}_m(k) = [\mathbf{0}_{N-m+1}, \mathbf{I}_{N-m+1}] - \widetilde{\mathbf{w}}_m(k)(\mathbf{w}_m(k) + \mathbf{e})^T / (\|\mathbf{w}_m(k)\|^2 + w_{m,1}(k)) \tag{4}$$

$$\text{Where } \widetilde{\mathbf{w}}_i(k) = [w_{i,2}(k), \cdots, w_{i,N-i+1}]^T. \tag{5}$$

From (4) and (3c) we have

$$\mathbf{x}_i(k) = \widetilde{\mathbf{x}}_{i-1}(k) - \widetilde{\mathbf{w}}_{i-1}(k)(y_{i-1}(k) + x_{i-1,1}(k))/(\|\mathbf{w}_{i-1}(k)\|^2 + w_{i-1,1}) \tag{6}$$

$$\text{for } i = 2, \cdots, r)$$

$$\text{where } \widetilde{\mathbf{x}}_{i-1}(k) = [x_{i-1,2}(k), \cdots, x_{i-1,N-i+2}(k)]^T. \tag{7}$$

Only $\mathbf{w}_i(k)$ is trained in neuron i. The k-th iteration of algorithm is

$$\mathbf{x}_1(k) = \mathbf{x}(k) \tag{8a}$$

$$\mathbf{x}_i(k) = \widetilde{\mathbf{x}}_{i-1}(k) - \widetilde{\mathbf{w}}_{i-1}(k)(y_{i-1}(k) + x_{i-1,1}(k))/(\|\mathbf{w}_{i-1}(k)\|^2 + w_{i-1,1})$$
$$(i = 2, \cdots, r) \tag{8b}$$

$$y_j(k) = \mathbf{x}_j^T(k)\mathbf{w}_j(k) \tag{9a}$$

$$\mathbf{w}_j(k+1) = \mathbf{w}_j(k) + \eta_k[y_j(k)\mathbf{x}_j(k) - y_j^2(k)\mathbf{w}_j(k)] \tag{9b}$$

$$\lambda_j(k+1) = \alpha\lambda_j(k) + (1-\alpha)y_j^2(k) \tag{9c}$$

for $j = 1, \cdots, r$. Equations (9) are of Oja's type, however, the dimensional number of $\mathbf{w}_i(k)$ and $\mathbf{x}_i(k)$ is equal to $N - i + 1$ and less than N. In order to reduce the computational complexity, $\|\mathbf{w}_i(k)\|^2$ can been approximated as 1.

It can been shown by carefully computing that the above algorithm requires about $3N + (4N - 2r)(r-1)$ multiplications at each iteration. Oja's subspace algorithm [4] usually requires $3rN$ multiplications at each iteration. If $r > N/2$, then the

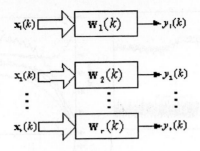

Fig. 1. The neural network model.

computational complexity of our algorithm is lower than that of Oja's subspace algorithm. Moreover, our algorithm only has $r(N-(r-1)/2)$ weights less than those of Oja's subspace algorithm, which significantly reduces the weight storage.

Once obtaining the $\tilde{\mathbf{u}}_i$ ($i = 2, \cdots, N$), the eigenvector matrix $\mathbf{U}$ can be reconstructed by the following equation:

$$\mathbf{U} = [\tilde{\mathbf{u}}_1, \mathbf{T}_1^T [\tilde{\mathbf{u}}_2, \mathbf{T}_2^T [\tilde{\mathbf{u}}_3, \cdots, \mathbf{T}_{N-1}^T \tilde{\mathbf{u}}_N]]] . \tag{10}$$

We can directly deduce the following back-forward recursive algorithm for computing (10):

$$\mathbf{U}_N = \tilde{\mathbf{u}}_N \tag{11a}$$

$$\mathbf{U}_{i-1} = [\tilde{\mathbf{u}}_{i-1}, \mathbf{T}_{i-1}^T \mathbf{U}_i] \ (i = N, N-1, \cdots, 2) \text{where } \mathbf{U} = \mathbf{U}_1 . \tag{11b}$$

4 Simulations

In this section we present two simulations that demonstrates the numerical performance of the proposed method. The error of the estimated principal at time k is measured by the "learning curve"

$$\text{dist}(k) = 20 \log_{10} [\|\mathbf{W}(k)\mathbf{W}^T(k) - \mathbf{U}_1 \mathbf{U}_1^T\|_F] .$$

In Example 1 and 2, $\mathbf{U}_1$ contains the first fifteen principal eigenvactors of the covariance matrix based on 200 data samples.

Example 1. Data are comprised of a spatial-colored random vector sequence with the dimension number 64 and the sample number 200. In order to insure stationary we have repeated the sequence, periodically, so that the convergence performance of the algorithm can be studied. Each period of the data is called a sweep and contains 200 data. In Fig.2, we show the convergence of the first 15 components using the proposed method and Oja's subspace algorithms. Results show that the convergence of these algorithms is exponential.

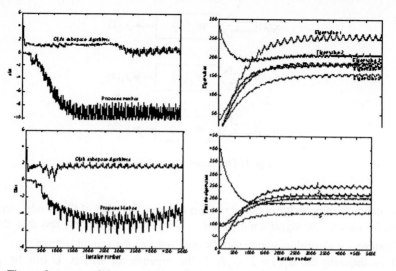

Fig. 2. The performance of the proposed method and Oja's subspace algorithms for PCA. We use 200 64-dimensional data vectors repeated in sweeps. In left figure, plot of the dist(k), between the components and the ones estimated using the proposed method and Oja's subspace algorithm. In right figure, the evolution of the estimation of the first five eigenvectors over each data sample.

Fig. 3. The performance of the proposed method and Oja's subspace algorithms for CPC. The same data used in Fig.2. In left figure, plot of the dist(k). In right figure, the evolution of the estimation of the first five eigenvectors over each data sample.

Example 2. We'll show that the proposed method has its ability to extract constrained principal components (CPC's) [6]. In CPC, we assume that there is an undesirable subspace $\Re$, spanned by the columns of an orthonormal constraint matrix $\mathbf{V}$. $\Re$ may stand for the zero-force subspace in adaptive beamformer or for the interference subspace in interference-cancellation. More formally, the CPC problem can be defined as follows: given an N-dimensional stationary stochastic input vector process $\{\mathbf{x}(k)\}$, extract the PC's of $\{\mathbf{x}(k)\}$ in the orthogonal complement of $\Re$. Thus we only require finding the PC's of $\overline{\mathbf{x}}(k) = (\mathbf{I}_N - \mathbf{V}\mathbf{V}^T)\mathbf{x}(k)$. In Fig.3, we show the convergence of the first 15 principal components using the proposed method and Oja's subspace algorithms. The data are artificially created and have dimension $N = 64$ and the 200 data samples. The eight column orthonormal constrain matrix $\mathbf{V}$ is randomly picked. The convergence is exponential as in the PCA case.

5   Conclusions

In this paper we propose a novel neural network model for extracting the multiple principal components.. Its superiority to older techniques is to save the weight storage under some cases and to reduce the computational complexity when the principal components are over the half of the signal order. We studied its global convergence and discussed some selected application examples.

References

1. Anderson, T.W.: An Introduction to Multivariate Statistical Analysis. 2nd ed., John Wiley & Sons (1984)
2. Oja, E.: A Simplified Neuron Model as A Principal Component Analyzer. J. Math. Biology, **15** (1982) 267-273
3. Cichocki, A., and Unbehauen, R.: Neural Network for Computing Eigenvalues and Eigenvectors. Biol. Cybern, **68** (1992) 155-159
4. Oja, E: Neural Networks, Principal Components, and Subspaces. Neural Systems, **1** (1989) 61-68
5. Sanger, T.D.: Optimal Unsupervised Learning in A Single-Layer Linear Feed-Forward Network. Neural Networks, **2** (1989) 459-473
6. Kung, S.Y., Diamantaras, K.I. and Taur, J.S.: Adaptive Principal Component Extraction (APEX) and Applications. IEEE Trans. Signal Processing, **42** (1994) 1202-1217
7. Goldstein, J.S., Reed, I.S., and Scharf, L.L.: A Multistage Representation of the Wiener Based on Orthogonal Projections. IEEE Trans. Information Theory, **44** (1998)

A New Kalman Filtering Algorithm
for Nonlinear Principal Component Analysis*

Xiaolong Zhu[1], Xianda Zhang[1], and Ying Jia[2]

[1] Department of Automation, Tsinghua University, Beijing 100084, China
{xlzhu_dau,zxd-dau}@tsinghua.edu.cn
[2] Intel China Research Center, Beijing 100080, China
ying.jia@intel.com

Abstract. This paper addresses the problem of blind source separation (BSS) based on nonlinear principal component analysis (NPCA),
and presents a new Kalman filtering algorithm, which applies a different
state-space representation from the one proposed recently by Lv et al.
It is shown that the new Kalman filtering algorithm can be simplified
greatly under certain conditions, and it includes the existing Kalman-
type NPCA algorithm as a special case. Comparisons are made with
several related algorithms and computer simulations on BSS are reported
to demonstrate the validity.

1 Introduction

Nonlinear principal component analysis (NPCA) makes the output signals mutually independent (or as independent as possible); hence it is closely related
to independent component analysis (ICA) and the problem of blind source separation (BSS) [1]. The noise-free instantaneous BSS problem is formulated as
follows. Given a sequence of observation vectors $\mathbf{x}_t = [x_{1,t}, \cdots, x_{m,t}]^T$, which
are generated according to

$$\mathbf{x}_t = \mathbf{A}\mathbf{s}_t \tag{1}$$

where $\mathbf{A}$ is an unknown $m \times n$ mixing matrix with full column rank $(m \geq n)$, and
$\mathbf{s}_t = [s_{1,t}, \cdots, s_{n,t}]^T$ is a vector of independent source signals, all but perhaps
one of them nongaussian. The objective is to process each $\mathbf{x}_t$ via an $n \times m$
separating matrix $\mathbf{B}$ such that the output vector

$$\mathbf{y}_t = \mathbf{B}\mathbf{x}_t \tag{2}$$

recovers the n source signals without crosstalk. Since this blind approach requires
no modeling for the mixing part of the model, it is useful in many applications of
signal processing and neural networks, in particular when the underlying physical
phenomena are difficult or impossible to be modeled accurately.

---

* This work was supported by the major program of the National Natural Science
Foundation of China (No. 60496311), by the Chinese Postdoctoral Science Foundation (No. 2004035061), and by the Foundation of Intel China Research Center.

J. Wang, X. Liao, and Z. Yi (Eds.): ISNN 2005, LNCS 3496, pp. 1010–1015, 2005.

A variety of algorithms have been proposed for BSS, see e.g. the textbook [1] for a review. We focus here on the NPCA, which has such implementations as the least-mean-square (LMS) algorithm [2], the recursive least-squares (RLS) algorithm [3], and the natural gradient-based RLS algorithm [4]. It is well known in adaptive filter theory [5] that the RLS filter is a special case of the Kalman filter, and the latter is noteworthy for its tracking ability. Following this clue, a Kalman filtering algorithm was proposed recently for BSS by Lv et al [6]. Different from the regular representations of a column state vector and a column observation vector, a state matrix and a row observation vector were designed, whereas the algorithm was obtained heuristically by using the correspondences between the Kalman variables and the devised state-space model. Although computer simulations on BSS have verified the effectiveness of the Kalman filtering algorithm of Lv et al [6], a rigorous theoretical justification remains to be given, which is just one of the aims of this paper.

2 A New Kalman Filtering Algorithm for NPCA

The separating matrix in (2) can be determined using either one-stage or two-stage separation systems. The first method updates $\mathbf{B}$ directly to optimize some contrast function for BSS. In the second approach, the observed data are first preprocessed by an $n \times m$ whitening matrix $\mathbf{U}$ and then an orthonormal matrix $\mathbf{W}$ is learned to achieve the source separation, yielding the total separating matrix $\mathbf{B} = \mathbf{WU}$.

There exist in the literature several contrast functions for the orthogonal matrix $\mathbf{W}$, of particular note is the NPCA criterion [2], [3], [4]

$$J(\mathbf{W}) = E\left\{\left\|\mathbf{v}_t - \mathbf{W}^T \varphi(\mathbf{W}\mathbf{v}_t)\right\|^2\right\} \tag{3}$$

where $E\{\cdot\}$ is the expectation operator, $\|\cdot\|$ represents the Euclidean norm, $\mathbf{v}_t = \mathbf{U}\mathbf{x}_t$ is the whitened vector satisfying $E\left\{\mathbf{v}_t\mathbf{v}_t^T\right\} = \mathbf{I}$, $\mathbf{y}_t = \mathbf{W}\mathbf{v}_t$ and $\varphi(\mathbf{y}_t) = [\varphi_1(y_{1,t}), \cdots, \varphi_n(y_{n,t})]^T$ denotes a vector of nonlinearly-modified output signals.

2.1 Development of Kalman Filtering Algorithms

Both the LMS-type [2] and the RLS-type [3], [4] algorithms can be derived from (3). On the other hand, the Kalman filter theory provides a unifying framework for the RLS filters. Therefore, it is intuitively affirmative that we can develop a Kalman filtering algorithm for NPCA. To proceed, we replace the term $\varphi(\mathbf{W}_t\mathbf{v}_t)$ in (3) with $\mathbf{z}_t = \varphi(\mathbf{W}_{t-1}\mathbf{v}_t)$ by applying the 'projection approximation' [7] and design the following process equation and measurement equation:

$$\overrightarrow{\mathbf{W}}_{t+1} = \overrightarrow{\mathbf{W}}_t, \quad \mathbf{v}_t = \left[\mathbf{I} \otimes \mathbf{z}_t^T\right] \overrightarrow{\mathbf{W}}_t + \mathbf{e}_t \tag{4}$$

where $\overrightarrow{\mathbf{W}}_t = vec(\mathbf{W}_t)$ is a vector obtained by stacking the columns of $\mathbf{W}_t$ one beneath the other, $\otimes$ stands for the Kronecker product, $\mathbf{I}$ denotes the $n \times n$

identity matrix, and $\mathbf{e}_t$ models the measurement noise. The process equation has a null process noise vector and the $n^2 \times n^2$ identity state transition matrix for the optimum weight matrix at equilibrium points is time-invariant. Clearly, (4) belongs to the family of standard state-space models of the Kalman filters. Denote the measurement matrix as $\mathbf{C}_t = \mathbf{I} \otimes \mathbf{z}_t^T$ and the covariance matrix of $\mathbf{e}_t$ as $\mathbf{Q}_t$, then the one-step prediction given in [5] can be used to formulate a new Kalman filtering algorithm for NPCA:

$$\mathbf{G}_t = \mathbf{K}_{t-1}\mathbf{C}_t^T \left[\mathbf{C}_t\mathbf{K}_{t-1}\mathbf{C}_t^T + \mathbf{Q}_t\right]^{-1}$$
$$\mathbf{K}_t = \mathbf{K}_{t-1} - \mathbf{G}_t\mathbf{C}_t\mathbf{K}_{t-1}$$
$$\overrightarrow{\mathbf{W}}_t = \overrightarrow{\mathbf{W}}_{t-1} + \mathbf{G}_t \left[\mathbf{v}_t - \mathbf{C}_t\overrightarrow{\mathbf{W}}_{t-1}\right] \tag{5}$$

where $\mathbf{K}_t = E\{(\overrightarrow{\mathbf{W}}_t - \overrightarrow{\mathbf{W}}_{opt})(\overrightarrow{\mathbf{W}}_t - \overrightarrow{\mathbf{W}}_{opt})^T\}$ is the state-error correlation matrix.

The algorithm (5) can perform the prewhitened BSS, as will be shown in Section 3. However, its application is inconvenient due to heavy computational load. To understand this point, we notice that $\mathbf{K}_t$ is an $n^2 \times n^2$ matrix, and the $n^2 \times n$ Kalman gain $\mathbf{G}_t$ involves the inversion of an $n \times n$ matrix.

Proposition 1. *Let* $\mathbf{H}_i$ *for* $i = 0, 1, \cdots$ *be* $n \times n$ *symmetric matrices, and* ρ_t *a scalar, the state-error correlation matrix at time t can be written as* $\mathbf{K}_t = \mathbf{I} \otimes \mathbf{H}_t$ *provided that* $\mathbf{Q}_t = \rho_t\mathbf{I}$ *and the initial matrix* $\mathbf{K}_0 = \mathbf{I} \otimes \mathbf{H}_0$ *is selected.*

Proof. We prove it by the mathematical induction and make the hypothesis that $\mathbf{K}_{t-1} = \mathbf{I} \otimes \mathbf{H}_{t-1}$. Using the properties of Kronecker product, we have

$$\mathbf{G}_t = (\mathbf{I} \otimes \mathbf{H}_{t-1})(\mathbf{I} \otimes \mathbf{z}_t)\left[(\mathbf{I} \otimes \mathbf{z}_t^T)(\mathbf{I} \otimes \mathbf{H}_{t-1})(\mathbf{I} \otimes \mathbf{z}_t) + \rho_t\mathbf{I}\right]^{-1}$$
$$= \mathbf{I} \otimes \mathbf{H}_{t-1}\mathbf{z}_t \Big/ \left(\mathbf{z}_t^T\mathbf{H}_{t-1}\mathbf{z}_t + \rho_t\right) \tag{6}$$

which implies

$$\mathbf{K}_t = \mathbf{I} \otimes \mathbf{H}_{t-1} - [\mathbf{I} \otimes \mathbf{H}_{t-1}\mathbf{z}_t \big/ \left(\mathbf{z}_t^T\mathbf{H}_{t-1}\mathbf{z}_t + \rho_t\right)](\mathbf{I} \otimes \mathbf{z}_t^T)(\mathbf{I} \otimes \mathbf{H}_{t-1})$$
$$= \mathbf{I} \otimes \left[\mathbf{H}_{t-1} - \mathbf{H}_{t-1}\mathbf{z}_t\mathbf{z}_t^T\mathbf{H}_{t-1} \big/ \left(\mathbf{z}_t^T\mathbf{H}_{t-1}\mathbf{z}_t + \rho_t\right)\right]. \tag{7}$$

Define $\mathbf{H}_t = \mathbf{H}_{t-1} - \mathbf{H}_{t-1}\mathbf{z}_t\mathbf{z}_t^T\mathbf{H}_{t-1} \big/ \left(\mathbf{z}_t^T\mathbf{H}_{t-1}\mathbf{z}_t + \rho_t\right)$, and it is obvious a symmetric matrix. This completes the proof.

Based on Proposition 1 together with its proof, we may save great computations by updating the $n \times n$ matrix $\mathbf{H}_t$ rather than the $n^2 \times n^2$ matrix $\mathbf{K}_t$. Applying the identity of $\mathbf{C}_t\overrightarrow{\mathbf{W}}_{t-1} = \mathbf{W}_{t-1}^T\mathbf{z}_t$, we can get a simplified and computationally efficient Kalman filtering algorithm for NPCA:

$$\mathbf{h}_t = \mathbf{H}_{t-1}\mathbf{z}_t, \mathbf{g}_t = \mathbf{h}_t \big/ \left(\mathbf{z}_t^T\mathbf{h}_t + \rho_t\right)$$
$$\mathbf{H}_t = \mathbf{H}_{t-1} - \mathbf{g}_t\mathbf{h}_t^T$$
$$\mathbf{W}_t = \mathbf{W}_{t-1} + \mathbf{g}_t\left[\mathbf{v}_t - \mathbf{W}_{t-1}^T\mathbf{z}_t\right]^T \tag{8}$$

where $\mathbf{y}_t = \mathbf{W}_{t-1}\mathbf{v}_t$ and $\mathbf{z}_t = \varphi(\mathbf{y}_t)$.

2.2   Comparisons with Related Algorithms

The simplified Kalman filtering algorithm (8) is identical in form with the one proposed recently by Lv et al [6], but they originate from different state-space models. To be specific, the latter owes to a state matrix and a row observation vector with the following pair of equations [6]:

$$\mathbf{W}_{t+1} = \mathbf{W}_t, \quad \mathbf{v}_t^T = \mathbf{z}_t^T \mathbf{W}_t + \mathbf{e}_t^T. \tag{9}$$

Since (9) is distinct from the standard state-space representation of the Kalman filter [5], the correspondences between the Kalman variables and (9) cannot be used directly without modification and thus the Kalman filtering algorithm in [6] is not so convincing. This paper gives the theoretical justification of (8) in that it is actually a simplified form of the complete Kalman filtering algorithm (5) under certain conditions specified by Proposition 1.

The algorithm (8) is also related to several NPCA algorithms stemmed from the criterion (3). The LMS-type NPCA subspace rule is given by [2]

$$\mathbf{W}_t = \mathbf{W}_{t-1} + \eta_t \mathbf{z}_t \left[\mathbf{v}_t - \mathbf{W}_{t-1}^T \mathbf{z}_t \right]^T \tag{10}$$

where η_t is a positive learning rate. The RLS-type NPCA update is [3]

$$\mathbf{h}_t = \mathbf{P}_{t-1} \mathbf{z}_t, \mathbf{g}_t = \mathbf{h}_t \big/ \left(\mathbf{z}_t^T \mathbf{h}_t + \lambda_t \right)$$
$$\mathbf{P}_t = Tri \left[\mathbf{P}_{t-1} - \mathbf{g}_t \mathbf{h}_t^T \right] \big/ \lambda_t$$
$$\mathbf{W}_t = \mathbf{W}_{t-1} + \mathbf{g}_t \left[\mathbf{v}_t - \mathbf{W}_{t-1}^T \mathbf{z}_t \right]^T \tag{11}$$

where λ_t is a forgetting factor ($0 < \lambda_t \leq 1$) and $Tri[\cdot]$ means that only the upper triangular of the argument is computed and its transpose is copied to the lower triangular part, making thus the matrix $\mathbf{P}_t$ symmetric.

The difference between (8) and (10) is straightforward, but (8) differs from (11) so slightly that they can be confused readily. In the RLS algorithm (11), $\mathbf{P}_t$ is indeed an approximation of the inverse of $E\{\mathbf{z}_t \mathbf{z}_t^T\}$ and it should converge to a positive definite diagonal matrix; therefore, the operator $Tri[\cdot]$ plays an important role otherwise the algorithm will diverge due to the increasing rounding error. In contrast, $\mathbf{H}_t$ in the Kalman algorithm (8) is related to the state-error correlation matrix, and it must converge to a null matrix (this is the reason why the symmetry operator is not used). Postmultiplying both sides of the second line of (8) by $\mathbf{z}_t$, we obtain

$$\mathbf{H}_t \mathbf{z}_t = \frac{\rho_t}{\mathbf{z}_t^T \mathbf{H}_{t-1} \mathbf{z}_t + \rho_t} \mathbf{H}_{t-1} \mathbf{z}_t. \tag{12}$$

Since the fraction term lies between zero and one, $\mathbf{H}_t \mathbf{z}_t$ approaches a null vector and thus $\mathbf{H}_t$ a null matrix as time goes to infinity. Finally, the parameters λ_t and ρ_t distinguish substantially from each other. In order to obtain satisfactory performance, the forgetting factor λ_t should increase gradually from a small number to unity whereas the noise-related parameter ρ_t should decrease gradually from a somewhat big value to zero.

3   Computer Simulations

The efficiency of the simplified Kalman filtering algorithm (8) has been confirmed by a number of experiments. In [6], it was used to separate several speech signals transferred through static or slow time-varying channels. Due to lack of space, we do not repeat the results but emphasize here the validity of the complete Kalman filtering algorithm (5) as well as the difference between the Kalman-type algorithm (8) and the RLS-type one (11). Consider adaptive separation of the six source signals given in [8] with the square mixing matrix $\mathbf{A}$ whose elements are randomly assigned in the range $[-1, +1]$.

For comparison, we run simultaneously the NPCA algorithms (5), (8), (10), (11) and the natural gradient-based RLS algorithm in [4] using the same non-linearity $\mathbf{z}_t = \tanh(\mathbf{y}_t)$. To isolate the influence of preprocessing, we obtain the

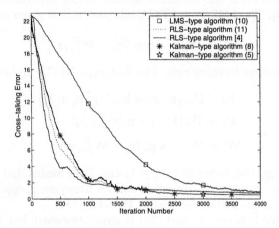

Fig. 1. The cross-talking errors averaged over 2000 independent runs

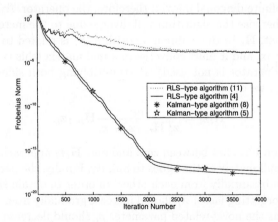

Fig. 2. The Frobenius norms of certain matrices averaged over 2000 runs

whitening matrix $\mathbf{U}$ via eigenvalue decomposition. All the matrices are initialized by the $n \times n$ identity matrix except for the $n^2 \times n^2$ matrix $\mathbf{K}_0$ which is a randomly generated symmetric positive definite matrix. The LMS-type algorithm (10) uses $\eta_t = 0.0125$. The two RLS-type algorithms adopt time-varying forgetting factors λ_t which increase linearly from 0.975 to 0.999 via 1500 (the algorithm (11)) and 800 (the algorithm in [4]) iterations, respectively. The two Kalman-type algorithms (5) and (8) apply the same time-varying noise parameter ρ_t which decreases exponentially with base 4 from 0.983 to 0.0001 via 550 iterations (In the former algorithm we take $\mathbf{Q}_t = \rho_t \mathbf{\Lambda}$ and the elements of the diagonal matrix $\mathbf{\Lambda}$ are randomly generated in $[0.5, 1.5]$).

Fig. 1 plots the cross-talking errors (which obtains the minimum of zero when perfect separation is achieved, see e.g. [1], [4] and [6] for its definition) of the five algorithms averaged over 2000 independent runs. Clearly, both the Kalman filtering algorithms (5) and (8) work satisfactorily. To further illustrate the difference between the Kalman-type algorithms and the RLS-type algorithms, the Frobenius norms of $\mathbf{K}_t$ in (5), $\mathbf{H}_t$ in (8), $\mathbf{P}_t - diag(\mathbf{P}_t)$ in (11) and $\mathbf{P}_t - diag(\mathbf{P}_t)$ in [4] are depicted in Fig. 2. They all converge to zero, which agrees with the theoretical analysis given in Section 2.2.

4   Conclusions

In this paper, we designed a column state equation and a column measurement equation to derive a new Kalman filtering algorithm for NPCA, which could be simplified under certain conditions to the one proposed recently by Lv et al [6]. Comparisons were made with the related NPCA algorithms by emphasizing the difference from the RLS-type algorithm [3] and computer simulations were given to validate the new algorithm and the theoretical analysis.

References

1. Hyvarinen, A., Karhunen, J., Oja, E.: Independent Component Analysis. Wiley, New York (2001)
2. Oja, E.: The Nonlinear PCA Learning in Independent Component Analysis. Neurocomputing, **17** (1997) 25-46
3. Pajunen, P., Karhunen, J.: Least-squares Methods for Blind Source Separation based on Nonlinear PCA. Int. J. of Neural Systems, **8** (1998) 601-612
4. Zhu, X.L., Zhang, X.D.: Adaptive RLS Algorithm for Blind Source Separation Using a Natural Gradient. IEEE Signal Processing Letters, **9** (2002) 432-435
5. Haykin, S.: Adaptive Filter Theory. 4th ed. Prentice-Hall, Englewood Cliffs New Jersey (2002)
6. Lv, Q., Zhang, X.D., Jia, Y.: Kalman Filtering Algorithm for Blind Source Separation. ICASSP2005 (to appear)
7. Yang, B.: Projection Approximation Subspace Tracking. IEEE Trans. Signal Processing, **43** (1995) 95-107
8. Ye, J.M., Zhu, X.L., Zhang, X.D.: Adaptive Blind Separation with an Unknown Number of Sources. Neural Computation, **16** (2004) 1641-1660

An Improvement on PCA Algorithm
for Face Recognition

Vo Dinh Minh Nhat and Sungyoung Lee

Kyung Hee University, South of Korea
{vdmnhat,sylee}@oslab.khu.ac.kr

Abstract. Principle Component Analysis (PCA) technique is an important and well-developed area of image recognition and to date many linear discrimination methods have been put forward. Despite these efforts, there persist in the traditional PCA some weaknesses. In this paper, we propose a new PCA-based method that can overcome one drawback existed in the traditional PCA method. In face recognition where the training data are labeled, a projection is often required to emphasize the discrimination between the clusters. PCA may fail to accomplish this, no matter how easy the task is, as they are unsupervised techniques. The directions that maximize the scatter of the data might not be as adequate to discriminate between clusters. So we proposed a new PCA-based scheme which can straightforwardly take into consideration data labeling, and makes the performance of recognition system better. Experiment results show our method achieves better performance in comparison with the traditional PCA method.

1 Introduction

Principal component analysis (PCA), also known as Karhunen-Loeve expansion, is a classical feature extraction and data representation technique widely used in the areas of pattern recognition and computer vision. Sirovich and Kirby [1], [2] first used PCA to efficiently represent pictures of human faces. They argued that any face image could be reconstructed approximately as a weighted sum of a small collection of images that define a facial basis (eigenimages), and a mean image of the face. Within this context, Turk and Pentland [3] presented the well-known Eigenfaces method for face recognition in 1991. Since then, PCA has been widely investigated and has become one of the most successful approaches in face recognition [4], [5], [6], [7]. However, Wiskott et al. [10] pointed out that PCA could not capture even the simplest invariance unless this information is explicitly provided in the training data. Recently, two PCA-related methods, independent component analysis (ICA) and kernel principal component analysis (Kernel PCA) have been of wide concern. Bartlett et al. [11] and Draper et al. [12] proposed using ICA for face representation and found that it was better than PCA when cosines were used as the similarity measure (however, their performance was not significantly different if the Euclidean distance is used). Yang [14] used Kernel PCA for face feature extraction and recognition and showed that the Kernel Eigenfaces method outperforms the classical Eigenfaces method. However, ICA and Kernel PCA are both computationally more expensive than PCA. The experimental results in [14] showed the ratio of the computation time required by ICA, Kernel PCA, and PCA is, on average, 8.7: 3.2: 1.0.

J. Wang, X. Liao, and Z. Yi (Eds.): ISNN 2005, LNCS 3496, pp. 1016–1021, 2005.

In face recognition where the data are labeled, a projection is often required to emphasize the discrimination between the clusters. PCA may fail to accomplish this, no matter how easy the task is, as they are unsupervised techniques. The directions that maximize the scatter of the data might not be as adequate to discriminate between clusters. In this paper, our proposed PCA scheme can straightforwardly take into consideration data labeling, which makes the performance of recognition system better. The remainder of this paper is organized as follows: In Section 2, the traditional PCA method is reviewed. The idea of the proposed method and its algorithm are described in Section 3. In Section 4, experimental results are presented on the ORL, and the Yale face image databases to demonstrate the effectiveness of our method. Finally, conclusions are presented in Section 5.

2 Principle Component Analysis

Let us consider a set of N sample images $\{x_1, x_2, ..., x_N\}$ taking values in an n-dimensional image space, and the matrix $A = [\overline{x_1} \overline{x_2} ... \overline{x_N}] \in \mathbb{R}^{n \times N}$ with $\overline{x_i} = x_i - \mu$ and $\mu \in \mathbb{R}^n$ is the mean image of all samples. Let us also consider a linear transformation mapping the original n-dimensional image space into an m-dimensional feature space, where $m < n$. The new feature vectors $y_k \in \mathbb{R}^m$ are defined by the following linear transformation:

$$y_k = W^T \overline{x_k} \text{ and } Y = W^T A \tag{1}$$

where $k = 1, 2, ..., N$ and $W \in \mathbb{R}^{n \times m}$ is a matrix with orthonormal columns.

If the total scatter matrix is defined as

$$S_T = AA^T = \sum_{k=1}^{N} (x_k - \mu)(x_k - \mu)^T \tag{2}$$

In PCA, the projection W_{opt} is chosen to maximize the determinant of the total scatter matrix of the projected samples, i.e.,

$$W_{opt} = \arg\max_W |W^T S_T W| = [w_1 w_2 ... w_m] \tag{3}$$

where $\{w_i | i = 1, 2, ..., m\}$ is the set of n-dimensional eigenvectors of S_T corresponding to the m largest eigenvalues.

3 Our Proposed PCA

In the following part, we show that PCA finds the projection that maximizes the sum of all squared pairwise distances between the projected data elements and we also propose our approach. Firstly we will take a look at some necessary background.

The *Laplacian* is a key entity for describing pairwise relationships between data elements. This is an symmetric positive-semidefinite matrix, characterized by having zero row and column sums. Let L be an NxN Laplacian and $z = [z_1 z_2 ... z_N]^T \in \mathbb{R}^N$ then we have

$$z^T L z = \sum_i L_{ii} z_i^2 + 2 \sum_{i<j} L_{ij} z_i z_j =$$

$$= \sum_{i<j} -L_{ij}(z_i^2 + z_j^2) + 2 \sum_{i<j} L_{ij} z_i z_j = \sum_{i<j} -L_{ij}(z_i - z_j)^2 \qquad (4)$$

Let $r_1, r_2, ..., r_m \in \mathbb{R}^N$ be m columns of the matrix Y^T, applying (4) we have

$$\sum_{k=1}^{m} r_k^T L r_k = \sum_{i<j} -L_{ij} \left(\sum_{k=1}^{m} ((r_k)_i - (r_k)_j)^2 \right) = \sum_{i<j} -L_{ij} d(y_i, y_j)^2 \qquad (5)$$

with $d(y_i, y_j)$ is the Euclidean distance. Now we turn into proving the following theorem, and develop it to our approach.

Theorem 1. *PCA computes the m-dimensional project that maximizes*

$$\sum_{i<j} d(y_i, y_j)^2 \qquad (6)$$

Before proving this Theorem, we define a NxN unit Laplacian, denoted by L^u, as $L^u = N\delta_{ij} - 1$. We have

$$A L^u A^T = A(N I_N - U) A^T = N S_T - A U A^T = N S_T \qquad (7)$$

with I_N is identity matrix and U is a matrix of all ones. The last equality is due to the fact that the coordinates are centered. By (5), we get

$$\sum_{i<j} d(y_i, y_j)^2 = \sum_{i=1}^{m} y_i^T L^u y_i = \sum_{i=1}^{m} w_i^T A L^u A^T w_i = N \sum_{i=1}^{m} w_i^T S_T w_i \qquad (8)$$

Formulating PCA as in (6) implies a straightforward generalization - simply replace the unit Laplacian with a general one in the target function. In the notation of Theorem 1, this means that the m-dimensional projection will maximize a weighted sum of squared distances, instead of an unweighted sum. Hence, it would be natural to call such a projection method by the name weighted PCA. Let us formalize this idea. Let be $\{wt_{ij}\}_{i,j=1}^{N}$ symmetric nonnegative pairwise weights, with measuring how important it is for us to place the data elements i and j further apart in the low dimen-

sional space. Let define NxN Laplacian $L_{ij}^w = \begin{cases} \sum_{i \neq j} wt_{ij} & i = j \\ -wt_{ij} & i \neq j \end{cases}$ and

$wt_{ij} = \begin{cases} 0 & x_i, x_j \in same \ class \\ 1/d(x_i, x_j) & other \end{cases}$. Generalizing (7), we have weighted PCA and it

seeks for the m-dimensional projection that maximizes $\sum_{i<j} wt_{ij} d(y_i, y_j)^2$. And this is

obtained by taking the m highest eigenvectors of the matrix $A L^w A^T$. The proof of this is the same as that of Theorem 1, just replace L'' by L''. Now, we still have one thing need solving. It is how to get the eigenvectors of $AL^wA^T \in \mathbb{R}^{nxn}$, because this is a very big matrix. And the other one is how to define wt_{ij}. Let D be the N eigenvalues diagonal matrix of $A^TAL^w \in \mathbb{R}^{NxN}$ and V be the matrix whose columns are the corresponding eigenvectors, we have

$$A^TAL^wV = VD \Leftrightarrow AL^wA^T(AL^wV) = (AL^wV)D \qquad (9)$$

From (9), we see that AL^wV is the matrix whose columns are the first N eigenvectors of AL^wA^T and D is the diagonal matrix of eigenvalues.

4   Experimental Results

This section evaluates the performance of our propoped algorithm compared with that of the original PCA algorithm and proposed algorithm (named WPCA) based on using ORL and Yale face image database. In our experiments, firstly we tested the recognition rates with different number of training samples. $k(k = 2,3,4,5)$ images of each subject are randomly selected from the database for training and the remaining images of each subject for testing. For each value of k, 30 runs are performed with different random partition between training set and testing set. And for each k training samples experiment, we tested the recognition rates with different number of dimensions, d, which are from 2 to 10. *Table 1& 2* shows the average recognition rates (%) with ORL database and Yale database respectively. In Fig. 1, we can see that our method achieves the better recognition rate compared to the traditional PCA.

Table 1. The recognition rates on ORL database.

d	2		4		6		8		10	
k	PCA	WPCA	PCA	WPCA	PCA	WPCA	PCA	WPCA	PCA	WPCA
2	39.69	**44.24**	61.56	**62.11**	69.69	**71.22**	78.13	**81.35**	78.49	**82.05**
3	40.36	**44.84**	66.79	**68.49**	70.00	**72.75**	78.21	**82.09**	80.36	**82.72**
4	38.75	**41.62**	63.75	**67.86**	78.33	**82.35**	83.75	**85.76**	86.25	**89.03**
5	37.00	**41.33**	68.00	**72.57**	79.50	**84.57**	85.50	**88.97**	89.00	**91.39**

Table 2. The recognition rates on Yale database.

d	2		4		6		8		10	
k	PCA	WPCA	PCA	WPCA	PCA	WPCA	PCA	WPCA	PCA	WPCA
2	40.56	**42.95**	58.33	**62.37**	66.48	**69.18**	70.93	**73.44**	76.11	**78.14**
3	42.50	**45.17**	74.17	**77.89**	78.33	**80.62**	81.67	**84.47**	86.67	**90.49**
4	43.10	**53.20**	71.67	**73.11**	83.10	**87.13**	88.81	**90.72**	90.71	**94.06**
5	57.22	**59.30**	72.78	**75.01**	83.89	**84.55**	87.22	**88.92**	88.33	**91.77**

5   Conclusions

A new PCA-based method for face recognition has been proposed in this paper. The proposed PCA-based method can overcome one drawback existed in the traditional PCA method. PCA may fail to emphasize the discrimination between the clusters, no

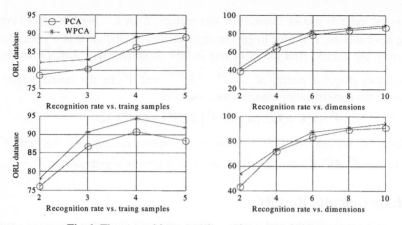

Fig. 1. The recognition rate (%) graphs on two databases.

matter how easy the task is, as they are unsupervised techniques. The directions that maximize the scatter of the data might not be as adequate to discriminate between clusters. So we proposed a new PCA-based scheme which can straightforwardly take into consideration data labeling, and makes the performance of recognition system better. The effectiveness of the proposed approach can be seen through our experiments based on ORL and Yale face databases. Perhaps, this approach is not a novel technique in face recognition, however it can improve the performance of traditional PCA approach whose complexity is less than LDA or ICA approaches.

Reference

1. Sirovich, L., Kirb, M.: Low-Dimensional Procedure for Characterization of Human Faces. J. Optical Soc. Am, **4** (1987) 519-524
2. Kirby, M., Sirovich, L.: Application of the KL Procedure for the Characterization of Human Faces. IEEE Trans. Pattern Analysis and Machine Intelligence, **12** (1990) 103-108
3. Turk, M., Pentland, A.: Eigenfaces for Recognition. J. Cognitive Neuroscience, **3** (1991) 71-86
4. Pentland, A.: Looking at People: Sensing for Ubiquitous and Wearable Computing. IEEE Trans. Pattern Analysis and Machine Intelligence, **22** (2000) 107-119
5. Grudin, M.A.: On Internal Representations in Face Recognition Systems. Pattern Recognition, **33** (2000) 1161-1177
6. Cottrell, G.W., Fleming, M.K.: Face Recognition Using Unsupervised Feature Extraction. Proc. Int'l Neural Network Conf., (1990) 322-325
7. Valentin, D., Abdi, H., O'Toole, A.J., Cottrell, G.W.: Connectionist Models of Face Processing: A Survey. Pattern Recognition, **27** (1994) 1209-1230
8. Penev, P.S., Sirovich, L.: The Global Dimensionality of Face Space. Proc. Fourth IEEE Int'l Conf. Automatic Face and Gesture Recognition, (2000) 264- 270
9. Zhao L., Yang, Y.: Theoretical Analysis of Illumination in PCA-Based Vision Systems. Pattern Recognition, **32** (1999) 547-564

10. Wiskott, L., Fellous, J.M., Kruger, N., Von Der Malsburg, C.: Face Recognition by Elastic Bunch Graph Matching. IEEE Trans. Pattern Analysis and Machine Intelligence, **19** (1997) 775-779
11. Bartlett, M.S., Movellan, J.R., Sejnowski, T.J.: Face Recognition by Independent Component Analysis. IEEE Trans. Neural Networks, **13** (2002) 1450-1464
12. Draper, B.A., Baek, K., Bartlett, M.S., Beveridge, J.R.: Recognizing Faces with PCA and ICA. Computer Vision and Image Understanding: Special Issue on FaceRecognition
13. Yuen, P.C., Lai, J.H.: Face Representation Using Independent Component Analysis. Pattern Recognition, **35** (2002) 1247-1257
14. Yang M.H.: Kernel Eigenfaces vs. Kernel Fisherfaces: Face Recognition Using Kernel Methods. Proc Fifth IEEE Int'l Conf. Automatic Face and Gesture Recognition (RGR'02), (2002) 215-220
15. Koren Y., Carmel L.: Robust linear dimensionality reduction. Visualization and Computer Graphics, IEEE Transactions, **10** (2004) 459-470

A Modified PCA Neural Network to Blind Estimation of the PN Sequence in Lower SNR DS-SS Signals

Tianqi Zhang[1], Xiaokang Lin[1], Zhengzhong Zhou[2], and Aiping Mu[1]

[1] Graduate School at Shenzhen, Tsinghua University, Shenzhen 518055, China
zhangtianqi@tsinghua.org.cn
[2] University of Electronic Science and Technology of China, Chengdu 610054, China

Abstract. A modified principal component analysis (PCA) neural network (NN) based on signal eigen-analysis is proposed to blind estimation of the pseudo noise (PN) sequence in lower signal to noise ratios (SNR) direct sequence spread spectrum (DS-SS) signals. The received signal is firstly sampled and divided into non-overlapping signal vectors according to a temporal window, which duration is two periods of PN sequence. Then an autocorrelation matrix is computed and accumulated by these signal vectors. The PN sequence can be estimated by the principal eigenvector of autocorrelation matrix in the end. Since the duration of temporal window is two periods of PN sequence, the PN sequence can be reconstructed by the first principal eigenvector only. Additionally, the eigen-analysis method becomes inefficiency when the estimated PN sequence becomes longer. We can use a PCA NN to realize the PN sequence estimation from lower SNR input DS-SS signals effectively.

1 Introduction

Since the direct sequence spread spectrum (DS-SS, DS) signals have the distinguished capability of anti-jamming and lower probability interception, the DS communication management and military scout become a key and challenging problem nowadays. Blind estimation of the PN sequence from the received DS signals is a vital step to DS communication management and military scout [1]. In [1], we also proposed an approach of neural network (NN) to realize the estimation. But the method did need to know a synchronous point between symbol waveform and observation window. Letter [2] proposed another method, which had an analogy to the method in [1] to discuss the interesting problem. But the methods both in [1] and [2] did need to estimate the synchronous point. It is still very hard to estimate the synchronous point from lower signal to noise ratios (SNR) DS signals. In [3], we used a multiple principal components analysis (PCA) NN to extract the PN sequence in lower SNR DS signals. We found that it is converged fast for the first principal component vector, but the convergence speed of the second principal component vector is very slow.

This paper proposes a modified approach to the PN sequence blind estimation. It doesn't need to search the synchronous point; it needs the first (or biggest) principal component vector only. Getting PN sequence, we can realize blind de-spread of DS signal. We assume that the signal is the same as [1],[2],[3]. For long code DS signals, we have another paper to discuss it.

J. Wang, X. Liao, and Z. Yi (Eds.): ISNN 2005, LNCS 3496, pp. 1022–1027, 2005.

2 Signal Model

The base band DS signal $x(t)$ corrupted by the white Gaussian noise $n(t)$ with the zero mean and σ_n^2 variance can be expressed as [3],[4],[5]

$$x(t) = s(t - T_x) + n(t) \tag{1}$$

Where $s(t) = d(t)p(t)$ is the DS signal, $p(t) = \sum_{j=-\infty}^{\infty} p_j q(t - jT_C)$, $p_j \in \{\pm 1\}$ is the periodic PN sequence, $d(t) = \sum_{k=-\infty}^{\infty} m_k q(t - kT_0)$, $m_k \in \{\pm 1\}$ is the symbol bits, uniformly distributed with $E[m_k m_l] = \delta(k - l)$, $\delta(\cdot)$ is the Dirac function, $q(t)$ denotes a pulse chip. Where $T_0 = NT_c$, N is the length of PN sequence, T_0 is the period of PN sequence, T_C is the chip duration, T_x is the random time delay and uniformly distributed on the $[0, T_0]$.

According to the above, the PN sequence and synchronization are required to despread the received DS signals. But in some cases, we only have the received DS signals. We must estimate the signal parameters firstly (We assume that T_0 and T_c had known in this paper), and then estimate the PN sequence and synchronization.

3 The Eigen-Analysis Based Estimator

The first case is the same as that in [3], the received DS signal is sampled and divided into non-ovelaping temporal windows, the duration of which is T_0. Then one of the received signal vector is

$$X(k) = s(k) + n(k), \quad k = 1, 2, 3, \cdots \tag{2}$$

Where $s(k)$ is the k-th vector of useful signal, $n(k)$ is the white Gaussian noise vector. The dimension of vector $X(k)$ is $N = T_0 / T_C$. If the random time-delay is T_x, $0 \le T_x < T_0$, $s(k)$ may contain two consecutive symbol bits, each modulated by a part of PN sequence, i.e.

$$s(k) = m_k \mathbf{p}_1 + m_{k+1} \mathbf{p}_2 \tag{3}$$

Where m_k and m_{k+1} are two consecutive symbol bits, $\mathbf{p}_1$ ($\mathbf{p}_2$) is the right (left) part of PN sequence waveform according to T_x.

According to signal eigen-analysis, we definite the $\mathbf{u}_i$ by $\mathbf{u}_i = \mathbf{p}_i / \|\mathbf{p}_i\|$

$$\mathbf{u}_i^T \mathbf{u}_j = \delta(i - j), \quad i, j = 1, 2 \tag{4}$$

Where $\mathbf{u}_i$ is the ortho-normal version of $\mathbf{p}_i$. From $\mathbf{u}_1$ and $\mathbf{u}_2$, we have

$$X(k) = m_k \|\mathbf{p}_1\| \mathbf{u}_1 + m_{k+1} \|\mathbf{p}_2\| \mathbf{u}_2 + n(k) \tag{5}$$

Assume $s(k)$ and $n(k)$ are mutually independent, we have

$$\mathbf{R}_X = E[\mathbf{X}\mathbf{X}^T] = \left[\sigma_n^2 \cdot \beta \cdot \left(\frac{T_0 - T_x}{T_C} \right) \right] \cdot \mathbf{u}_1 \mathbf{u}_1^T + \left(\sigma_n^2 \cdot \beta \cdot \frac{T_x}{T_C} \right) \cdot \mathbf{u}_2 \mathbf{u}_2^T + \sigma_n^2 \cdot \mathbf{I} \tag{6}$$

Where $\mathbf{I}$ is an identity matrix of dimension $N \times N$, the expectation of m_k is zero. The variance of m_k is σ_m^2, the symbol is uncorrelated from each other. The energy of PN sequence is $E_p \approx T_c \|\mathbf{p}\|^2$, the variance of $s(k)$ is $\sigma_s^2 = \sigma_m^2 E_p / T_0$, the β is $\beta = \sigma_s^2 / \sigma_n^2$. In a general way, we can reconstruct a whole period PN sequence from $\pm \mathbf{p} = \mathrm{sign}(\mathbf{u}_1 \pm \mathbf{u}_2)$, and solve the problem of PN sequence blind estimation.

The second case is as follows, the received DS signal is sampled and divided into non-overlapping temporal windows too, but the duration of which is $2T_0$. So $s(k)$ may contain three consecutive symbol bits m_k, m_{k+1} and m_{k+2}, i.e.

$$s(k) = m_k \mathbf{p}_1 + m_{k+1}\mathbf{p}_2 + m_{k+2}\mathbf{p}_3 \tag{7}$$

Where $\mathbf{p}_1$ ($\mathbf{p}_3$) is the right (left) part of PN sequence waveform according to T_x, $\mathbf{p}_2$ is a vector which contains a whole period of PN sequence.

In the same way in the first case, we have

$$\mathbf{X}(k) = m_k \|\mathbf{p}_1\| \mathbf{u}_1 + m_{k+1} \|\mathbf{p}_2\| \mathbf{u}_2 + m_{k+2} \|\mathbf{p}_3\| \mathbf{u}_3 + \mathbf{n}(k) \tag{8}$$

$$\mathbf{R}_X = \left(\sigma_n^2 \cdot \beta \cdot \frac{T_0}{T_C} \right) \cdot \mathbf{u}_2 \mathbf{u}_2^T + \left[\sigma_n^2 \cdot \beta \cdot \left(\frac{T_0 - T_x}{T_C} \right) \right] \cdot \mathbf{u}_1 \mathbf{u}_1^T + \left(\sigma_n^2 \cdot \beta \cdot \frac{T_x}{T_C} \right) \cdot \mathbf{u}_3 \mathbf{u}_3^T + \sigma_n^2 \cdot \mathbf{I} \tag{9}$$

Where $\mathbf{I}$ is an identity matrix of dimension $2N \times 2N$. From Eq.(9). We can reconstruct a whole period of PN sequence from $\pm \mathbf{p} = \mathrm{sign}(\mathbf{u}_2)$ in spite of T_x.

In this case, we can estimate the PN sequence by the first principal eigenvector of $\mathbf{R}_X$ only. It is easier to realize the PN sequence estimation. But the matrix eigenanalysis method becomes inefficient when the estimated PN sequence becomes longer. We'll use a PCA NN to overcome this shortcoming.

4   Implementation of the Neural Network

A two-layer PCA NN is used to estimate the PN sequence in DS signals [1],[3],[6]. The number of input neurons becomes $2N = 2T_0 / T_C$. One of the input vectors is

$$\mathbf{X}'(t) = \mathbf{X}(k) = \left[x_0'(t), x_1'(t), \cdots, x_{2N-1}'(t) \right]^T = \left[x(t), x(t-T_C), \cdots, x[t-(2N-1)T_C] \right]^T \tag{10}$$

Where $\{x_i'(t) = x(t - iT_C), i = 0,1,\cdots,2N-1\}$ are sampled by one point per chip.

The synaptic weight vector is

$$\mathbf{w}(t) = \left[w_0(t), w_1(t), \cdots, w_{2N-1}(t) \right]^T \tag{11}$$

Where the sign of $\{ w_i(t), i = 0,1,\cdots,2N-1 \}$ denotes the i-th bit of estimated PN sequence. Before $\mathbf{X}'(t)$ is put into the NN, we normalize it as follows.

$$\mathbf{X}(t) = \mathbf{X}'(t) / \|\mathbf{X}'(t)\| \tag{12}$$

Where $\|\mathbf{X}'(t)\| = \sqrt{\mathbf{X}'^T \mathbf{X}'}$. This will make the principal component vector to be estimated in a robust manner, hence yielding robust estimates of PN sequence.

The output layer of NN has only one neuron, its output is

$$y(t) = \mathbf{w}^T(t)\mathbf{X}(t) = \mathbf{X}^T(t)\mathbf{w}(t) \tag{13}$$

The original Hebbian algorithm of the NN is

$$\mathbf{w}(t+1) = \mathbf{w}(t) + \rho y(t)\mathbf{X}(t) \tag{14}$$

However, we will use the algorithm as follows here.

$$\mathbf{w}(t+1) = \mathbf{w}(t) + \rho y(t)\left[\mathbf{X}(t) - \mathbf{w}^T(t)\mathbf{X}(t)\mathbf{w}(t)\right] \tag{15}$$

Where ρ is a positive step-size parameter. In order to achieve good convergence performance, we express ρ as $\rho = 1/d_{t+1}$, and $d_{t+1} = Bd_t + y^2(t)$.

The analysis of the PCA NN is detailed in [3], [6]. We also discuss the principal component analysis algorithm, and give the following points:

(a) The probability that $\mathbf{w}(t) \to \pm\mathbf{C}_0$ (the first eigenvector of $\mathbf{R}_x$) is 1。 (b) When the NN has converged at its stable points, we will obtain $\lambda_0(\infty) = \lambda_0$ (the power of input signal), and $\lim_{t\to\infty}\mathbf{w}(t) = \pm\mathbf{C}_0$ ($\hat{\mathbf{P}}(t) = \text{sign}[\mathbf{w}(t)]$ denotes the estimated PN sequence). (c) Because $\pm\mathbf{C}_0$ have contained a whole period of PN sequence or its logical complement, we need not compute the second principal component vector, the convergence speed of NN can be very fast.

5 Simulations

The experiments mainly focus on the second case in section 3 and its NN implementation. We get principal eigenvectors and performance curves.

Fig.1 denotes the first, second and third principal component eigenvector with $N=100bit$ at $Tx=0.3T_0$. From them, we may estimate the T_x parameter and the PN sequence. Fig.2 shows the clearly distinguished first, second and third eigenvalues.

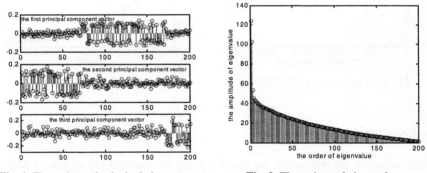

Fig. 1. The estimated principal eigenvectors.   Fig. 2. The estimated eigenvalues.

Fig.3 shows that the first, second and third eigenvalues are changed with T_x. It shows that λ_1 is the biggest eigenvalue in spite of T_x.

Fig.4 shows the learning curves of NN, where $N=1000bit, T_x=300Tc$. The curves from the left to the right corresponding to the noise standard deviation $\sigma_n = 1.0, 2.0, \cdots, 10.0$. The curves show that the convergence property of NN is very good. Fig.5 denotes the performance curves. It shows the time taken for the NN to perfectly estimate the PN sequence for lengths of *N=100bit, 300bit* and *1000bit* at $T_x/T_0=0.3$. Under the same condition, when the SNR is greater than −14dB, the longer the PN sequence is, some trivial worse the performance is, but when the SNR is less than −14dB, the longer the PN sequence is, the better the performance is.

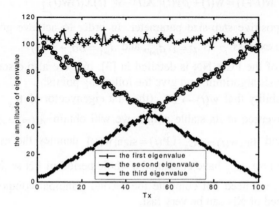

Fig. 3. λ_i, $i = 1,2,3$ versus T_x.

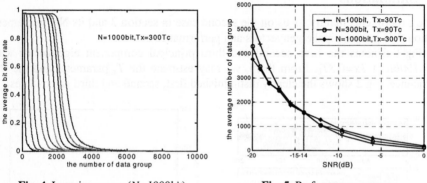

Fig. 4. Learning curves(*N=1000bit*).   **Fig. 5.** Performance curves.

6 Conclusions

Under common circumstances, we can de-spread the received DS signals of SNR of -20dB ∼-30dB if we know the PN sequence. In [4] Gardner used a method of time-delayed correlation to de-spread the DS signals blindly, and achieved −15dB of the SNR threshold. But on the same condition (the length of PN sequence $N = 100bit$), we can use the modified PCA NN to blindly estimate the PN sequence under threshold of $SNR = -20.0dB$ easily. So the performance of the methods in this paper is better. The modified PCA NN has the advantage of dealing with longer received signal vectors. The PN sequence blind estimation is very robust too. It can be further used in management and scout of DS communications.

References

1. Chang, T.Q., Guo, Z.X.: A Neural Networks to Estimation the PN Sequence of DS/SS Signals. the Ninth Telemetry and Telecommand Technology Annual Meeting of China, Haikou, China (1996) 535-537
2. Dminique, F., Reed, J.H.: Simple PN Code Sequence Estimation and Synchronization Technique Using the Constrained Hebbian Rule. Electronics Letters, **33** (1997) 37-38
3. Zhang, T.Q., Lin, X.K., Zhou, Z.Z.: Use PCA Neural Network to Extract the PN Sequence in Lower SNR DS-SS Signals. Lecture Notes in Computer Science, **3173** (2004) 780-785
4. French, C.A., Gardener, W.A.: Spread-Spectrum Despreading without the Code. IEEE Trans. Com, **34** (1986) 404-407
5. Dixon, R.C.: Spread Spectrum Systems. John Wiley & Sons, Inc. New York, USA (1984)
6. Haykin, S.: Neural Networks – A Comprehensive Foundation. Prentice Hall PTR, Upper Saddle River, NJ, USA (1999)

A Modified MCA EXIN Algorithm
and Its Convergence Analysis*

Dezhong Peng, Zhang Yi, and XiaoLin Xiang

Computational Intelligence Laboratory, School of Computer Science and Engineering,
University of Electronic Science and Technology of China, Chengdu 610054, China
dezhongpeng@yahoo.com.cn, zhangyi@uestc.edu.cn, xxllj@163.com

Abstract. The minor component is the eigenvector associated with the smallest eigenvalue of the covariance matrix of the input data. The minor component analysis (MCA) is a statistical method for extracting the minor component. Many neural networks have been proposed to solve MCA. However, there exists the problem of the divergence of the norm of the weight vector in these neural networks. In this paper, a modification to the well known MCA EXIN algorithm is presented by adjusting the learning rate. The modified MCA EXIN algorithm can guarantee that the norm of the weight vector of the neural network converges to a constant. Mathematical proofs and simulation results are given to show the convergence of the algorithm.

1 Introduction

MCA has many applications in diverse areas, such as total least squares (TLS) [1], curve and surface fitting [2], digital beamforming [3] and so on. To solve the task of MCA, many different neural algorithms are proposed. Among these algorithms, the MCA EXIN algorithm has better stability, speed and accuracy than other MCA algorithms [4]. But the algorithm has the same problem of divergence as other algorithms [5].

Consider a single linear neuron with the following input output relation:

$$y(k) = w^T(k)x(k), (k = 0, 1, 2 \ldots), \tag{1}$$

where $y(k)$ is the network output, the input sequence $\{x(k)|x(k) \in \mathbb{R}^n(k = 0, 1, 2 \ldots)\}$ is a zero mean stationary stochastic process, $w(k)$ is the weight vector of the neuron. The target of MCA is to get the minor component of the autocorrelation matrix $R = E[x(k)x^T(k)]$ by updating the weight vector of the neuron adaptively. The original MCA EXIN learning law has the following structure [4]:

$$w(k+1) = w(k) - \frac{\eta}{\|w(k)\|^2}\left[x(k)x^T(k)w(k) - \frac{w^T(k)x(k)x^T(k)w(k)}{\|w(k)\|^2}w(k)\right], \tag{2}$$

---

* This work was supported by National Science Foundation of China under Grant 60471055 and Specialized Research Fund for the Doctoral Program of Higher Education under Grant 20040614017.

J. Wang, X. Liao, and Z. Yi (Eds.): ISNN 2005, LNCS 3496, pp. 1028–1033, 2005.
© Springer-Verlag Berlin Heidelberg 2005

for $k \geq 0$, where $\eta > 0$ is the learning rate. In practice, the learning rate η is usually a constant. From the analysis of [5], if η is a constant, the norm of the weight vector is divergent. The reason of the divergence is simply analyzed as follows. Form (2), it holds that $w^T(k)[w(k+1) - w(k)] = 0$. So $w(k)$ is orthogonal to the update of the $w(k)$ in each iterate. Then $\|w(k+1)\|^2 = \|w(k)\|^2 + \|w(k+1) - w(k)\|^2$. Thus $\|w(k)\|$ is increasing and diverges.

In the paper, a modified MCA EXIN algorithm is derived in order to avoid the problem of the divergence. This paper is organized as follows: In Section 2, the modified MCA EXIN algorithm is presented. Section 3 gives the analysis of the convergence of the algorithm. Simulation results will be provided in Section 4. Finally, the conclusions will be drawn in Section 5.

2   Algorithm Description

The modified MCA EXIN algorithm has the same structure as the original MCA EXIN algorithm, but its learning rate is variant. It is described as :

$$w(k+1) = w(k) - \frac{\eta(k)}{\|w(k)\|^2}\left[x(k)x^T(k)w(k) - \frac{w^T(k)x(k)x^T(k)w(k)}{\|w(k)\|^2}w(k)\right], \quad (3)$$

for $k \geq 0$, where $\eta(k) > 0$ is the learning rate and

$$\eta(k+1) = \alpha\eta(k), \qquad (4)$$

where α is a constant $(0 < \alpha < 1)$.

It is difficult to study the dynamical behavior of the algorithm directly. Because the behavior of the conditional expectation of the weight vector can be studied by the deterministic discrete-time (DDT) algorithm, it is reasonable to analyze the behavior of the algorithm by its corresponding DDT algorithm indirectly [6]. By taking conditional expectation operator $E\{w(k+1)/w(0), x(i), i < k\}$ to (3) and identifying the conditional expected value as the next iterate, a corresponding DDT system can be obtained and given as:

$$w(k+1) = w(k) - \frac{\eta(k)}{\|w(k)\|^2}\left[Rw(k) - \frac{w^T(k)Rw(k)}{\|w(k)\|^2}w(k)\right], \qquad (5)$$

for $k \geq 0$, where $R = E[x(k)x^T(k)]$ is the correlation matrix. The main purpose of this paper is to study the convergence of the norm of the weight vector in (5).

Because the correlation matrix R is a symmetric nonnegative definite matrix, there exists an orthonormal basis of $\mathbb{R}^n$ composed of the eigenvectors of R. Let $\lambda_1, \ldots, \lambda_n$ be all the eigenvalues of R ordered by $\lambda_1 \geq \ldots \geq \lambda_n \geq 0$. Suppose that $\{v_i | i = 1, 2, \ldots, n\}$ is an orthonormal basis of $\mathbb{R}^n$ such that each v_i is a unit eigenvector of R associated with the eigenvalue λ_i. The weight vector $w(k)$ can be represented as :

$$w(k) = \sum_{i=1}^{n} z_i(k)v_i, \qquad (6)$$

where $z_i(k)(i = 1, 2, \ldots, n)$ are some constants, and then

$$w^T(k)Rw(k) = \sum_{i=1}^{n} \lambda_i z_i^2(k), \qquad (7)$$

and

$$w^T(k)RRw(k) = \sum_{i=1}^{n} \lambda_i^2 z_i^2(k). \qquad (8)$$

3   Convergence Analysis

Theorem 1. *If $\sum_{i=0}^{\infty} \|\Delta w(i)\|^2$ is convergent, then $\{\|w(k)\|^2\}$ converges .*

Proof. From (5), the update of the weight vector in each iterate is described as

$$\Delta w(k) = -\frac{\eta(k)}{\|w(k)\|^2}\left[Rw(k) - \frac{w^T(k)Rw(k)}{\|w(k)\|^2}w(k) \right]. \qquad (9)$$

Clearly, from (9), it holds that

$$w^T(k)\Delta w(k) = 0. \qquad (10)$$

So $w(k)$ and $\Delta w(k)$ is orthogonal. Then there exists the following relation among $w(k)$, $\Delta w(k)$, and $w(k+1)$:

$$\|w(k+1)\|^2 = \|w(k)\|^2 + \|\Delta w(k)\|^2 . \qquad (11)$$

$\|w(k)\|^2$ is increasing and described as:

$$\|w(k)\|^2 = \|w(0)\|^2 + \sum_{i=0}^{k-1} \|\Delta w(i)\|^2 . \qquad (12)$$

Thus,

$$\lim_{k \to \infty} \|w(k)\|^2 = \|w(0)\|^2 + \sum_{i=0}^{\infty} \|\Delta w(i)\|^2 . \qquad (13)$$

The proof is completed.

Theorem 2. *If*

$$\sum_{k=0}^{\infty} \frac{\eta(k)^2(\lambda_1^2 - \lambda_n^2)}{\|w(k)\|^2}$$

is convergent, then $\sum_{k=0}^{\infty} \|\Delta w(k)\|^2$ is convergent also.

Proof. From (5), (6), (7), (8) and (9), it follows that

$$
\begin{aligned}
\|\Delta w(k)\|^2 &= \Delta w^T(k)\Delta w(k) \\
&= \frac{\eta(k)^2}{\|w(k)\|^4}\left[w^T(k)RRw(k) - \frac{(w^T(k)Rw(k))^2}{\|w(k)\|^2}\right] \\
&= \frac{\eta(k)^2}{\|w(k)\|^4}\left[\sum_{i=1}^{n}\lambda_i^2 z_i^2(k) - \frac{[\sum_{i=1}^{n}\lambda_i z_i^2(k)]^2}{\|w(k)\|^2}\right] \\
&< \frac{\eta(k)^2}{\|w(k)\|^4}\left[\lambda_1^2\sum_{i=1}^{n} z_i^2(k) - \frac{[\lambda_n\sum_{i=1}^{n} z_i^2(k)]^2}{\|w(k)\|^2}\right] \\
&< \frac{\eta(k)^2}{\|w(k)\|^4}(\lambda_1^2\|w(k)\|^2 - \lambda_n^2\|w(k)\|^2) \\
&= \frac{\eta(k)^2}{\|w(k)\|^2}(\lambda_1^2 - \lambda_n^2).
\end{aligned}
$$

So $\|\Delta w(k)\|^2 < \eta(k)^2(\lambda_1^2 - \lambda_n^2)/\|w(k)\|^2$. It is clear that $\sum_{k=0}^{\infty}\|\Delta w(k)\|^2$ will converge if $\sum_{k=0}^{\infty}\eta(k)^2(\lambda_1^2 - \lambda_n^2)/\|w(k)\|^2$ converges. The proof is completed.

Theorem 3. *The series $\sum_{k=0}^{\infty}\eta(k)^2(\lambda_1^2 - \lambda_n^2)/\|w(k)\|^2$ is convergent.*

Proof. Let $u_k = \eta(k)^2(\lambda_1^2 - \lambda_n^2)/\|w(k)\|^2$, then,

$$
\begin{aligned}
\frac{u_k}{u_{k-1}} &= \frac{\eta(k)^2(\lambda_1^2 - \lambda_n^2)}{\|w(k)\|^2} \cdot \frac{\|w(k-1)\|^2}{\eta(k-1)^2(\lambda_1^2 - \lambda_n^2)} \\
&= \frac{\eta(k)^2\|w(k-1)\|^2}{\eta(k-1)^2\|w(k)\|^2} \\
&= \alpha^2\frac{\|w(k-1)\|^2}{\|w(k)\|^2}.
\end{aligned}
$$

Because $\|w(k)\|$ is increasing, then $\|w(k-1)\|^2 \le \|w(k)\|^2$ and $u_k/u_{k-1} \le \alpha^2 < 1$. Thus, $\sum_{k=0}^{\infty}u_k$ is convergent, i.e., $\sum_{k=0}^{\infty}\eta(k)^2(\lambda_1^2 - \lambda_n^2)/\|w(k)\|^2$ is convergent. The proof is completed.

By Theorem (1) (2) and (3), clearly, $\{\|w(k)\|^2\}$ is convergent. The next theorem can be easily obtained.

Theorem 4. *It holds that $\lim_{k\to\infty}\|w(k)\|^2 = S > 0$, where S is a constant.*

From the above theorems, it is clear that the norm of the weight vector is convergent in our modified MCA EXIN algorithm.

4   Simulation Results

The simulations presented in this section will show the convergence of the norm of the weight vector in the algorithm. We randomly generate a 5×5 correlation matrix as:

$$R = \begin{bmatrix} 0.1944 & 0.0861 & 0.0556 & 0.1322 & 0.1710 \\ 0.0861 & 0.2059 & 0.1656 & 0.1944 & 0.1467 \\ 0.0556 & 0.1656 & 0.2358 & 0.1948 & 0.1717 \\ 0.1322 & 0.1944 & 0.1948 & 0.3135 & 0.2927 \\ 0.1710 & 0.1467 & 0.1717 & 0.2927 & 0.3241 \end{bmatrix}.$$

In order to illustrate that the norm of the weight vector converges globally, we select different $\|w(0)\|$ and $\eta(0)$ for the simulations. In all the simulations, $\alpha = 0.999$.

In the first example, the initial weight vector is taken as:

$$w(0) = [0.3091, 0.3710, 0.1673, 0.9244, 0.9076]^T.$$

Clearly, $\|w(0)\| = 1.3926$. And $\eta(0) = 0.8, 1.8$ and 5. Fig1 shows the convergence of the norm.

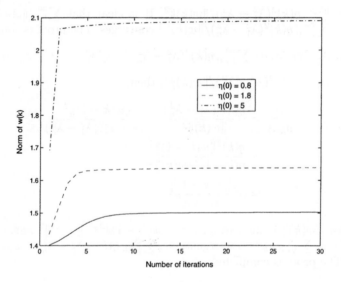

Fig. 1. Convergence of the norm in the modified EXIN algorithm($\|w(0)\| = 1.3926$)

In the second example, the initial weight vector is

$$w(0) = [0.3091, 0.3710, 0.1673, 0.9244, 0.9076]^T \times 100.$$

Clearly, $\|w(0)\| = 139.2633$. And $\eta(0) = 0.8, 1.8$ and 5. Fig2 shows the convergence of the norm.

Besides the above simulation results, further simulations with high dimensions persist to show that the modified MCA EXIN algorithm has a satisfactory results in the convergence of the norm of the weight vector.

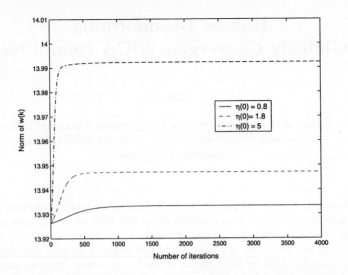

Fig. 2. Convergence of the norm in the modified EXIN algorithm($\|w(0)\| = 139.2633$)

5   Conclusion

In the paper, a modification to the MCA EXIN algorithm is given by adjusting the learning rate. The mathematical proof and the simulation results show that this algorithm can guarantee the global convergence of the norm of the weight vector.

References

1. Gao, K., Ahmad, M., O., Swamy, M., N.: Learning Algorithm for Total Least Squares Adaptive Signal Processing. Electron. Lett, **28** (1992) 430-432
2. Xu, L., Oja, E., Suen, C.: Modified Hebbian Learning for Curve and Surface Fitting. Neural Networks, **5** (1992) 441-457
3. Fiori, S., Piazza, F.: Neural MCA for Robust Beamforming. Proc of International Symposium on circuits and Systems (ISCAS'2000), **III** (2000) 614-617
4. Cirrincione, M., Cirrincione, G., Herault, J., Huffel, S., V.: The MCA EXIN Neuron for the Minor Component Analysis. IEEE Trans. Neural Networks, **13** (2002) 160-187
5. Taleb, A., Cirrincione, G.: Against the Convergence of the Minor Component Analysis Neurons. IEEE Trans. Neural Networks, **10** (1999) 207-210
6. Zufiria, P., J.: On the Discrete-time Dynamics of the Basic Hebbian Neural-network Nodes. IEEE Trans. Neural Networks, **13** (2002) 1342-1352

Robust Beamforming
by a Globally Convergent MCA Neural Network

Mao Ye

CI Lab, School of Computer Science and Engineering, University of Electronic
Science and Technology of China, Chengdu 610054, China
maoye@sina100.com

Abstract. Minor component analysis (MCA) by neural networks approach has attracted many attentions in the field of neural networks. Convergent learning algorithms of MCA neural networks are very important and useful for applications. In this paper, a globally convergent learning algorithm for MCA neural network is reviewed. Rigorous mathematical proof of global convergence is given and exponential convergence rate is obtained. Comparison experiments illustrate that this algorithm has good performance on beamforming problem.

1 Introduction

Minor component analysis (MCA) is a statistical method for extracting eigenvectors associated with the smallest eigenvalue of covariance matrix of input data. Neural networks method for MCA has been studied in recent years. Many algorithms have been proposed by many authors in recent years (for a review, see e.g.[2],[6]). And many applications for MCA neural network have been developed.

As denoted in [2],[5],[6],[9], if the learning rates are constants, MCA learning algorithms are either divergent or not globally convergent. Fiori applied OJA+ MCA neural network to robust beamforming problem, the good performance is obtained[4]. But OJA+ MCA learning algorithm is not globally convergent, the initial condition should be decided carefully.

A new MCA learning algorithm was proposed in [11]. The convergence of this algorithm is unknown. In this paper, we will present the further results based on [11]. The main contributions are the rigorous mathematical proof of the global convergence, and the exponential convergence rate of this learning algorithm. Since there are no comparison experiments on robust beamforming problem in [11], the comparison experiments will also be presented in the end, which confirm our theoretical results and show good performance of this algorithm.

2 The New MCA Learning Algorithm

Consider a linear neuron with the input $x(k) = [x_1(k), x_2(k), \cdots, x_n(k)]^T \in R^n$, the weight vector $w(k) \in R^n$ and output $y(k) = w^T(k)x(k)$ at time k.

J. Wang, X. Liao, and Z. Yi (Eds.): ISNN 2005, LNCS 3496, pp. 1034–1041, 2005.
© Springer-Verlag Berlin Heidelberg 2005

$x(k)(k = 0, 1, \cdots)$ is a zero-mean discrete-time stochastic process. Such process is constructed as a sequence $x(0), x(1), \cdots$ of independent and identically distributed samples upon a distribution of a random variable. And the weight vector $w(k)$ determines the relationship between the input $x(k)$ and the output $y(k)$ at time k.

Let the covariance matrix $C = E[x(k)x^T(k)]$. Obviously the eigenvalues of this matrix are nonnegative. Assume $\lambda_1, \lambda_2, \cdots, \lambda_n$ to be all the eigenvalues of C ordered by $\lambda_1 \geq \lambda_2 \geq \cdots > \lambda_n$. Since C is a symmetric matrix, then there exists an orthonormal basis of R^n composed by eigenvectors of C. Suppose that $\{v_i | i = 1, \cdots, n\}$ is an orthonormal basis in R^n and each v_i is a unit eigenvector of C corresponding to the eigenvalue λ_i. The goal of MCA by neural network is to make the $w(k)$ converges to $\pm v_n$, i.e., to the last principle vector, by adjusting the $w(k)$ on-line via the MCA learning law.

Assume that $\eta(k) > 0$ is the learning rate and $\| * \|$ stands for the Euclidean norm, we proposed the following MCA algorithm, and called it GMCA (global MCA) learning algorithm in [11]

$$w(k + 1) = w(k) - \eta(k) \cdot \left[y(k)x(k) - \frac{y(k)^2}{\|w(k)\|^2}w(k) \right.$$
$$\left. -\alpha \left(\frac{1}{\|w(k)\|} - 1 \right) w(k) \right], \tag{1}$$

for some constant $\alpha > 0$. For details about this algorithm, please refer to [11]. Taking the conditional expectation $E\{w(k+1)|w(0), x(i), i \leq k\}$ to both sides of the above equation and identifying the conditional expected value as the next iterate in the system, the GMCA DDT is the following

$$w(k + 1) = w(k) - \eta(k) \cdot \left[Cw(k) - \frac{w(k)^T Cw(k)}{\|w(k)\|^2}w(k) \right.$$
$$\left. - \alpha \left(\frac{1}{\|w(k)\|} - 1 \right) w(k) \right]. \tag{2}$$

Remark 1. If the initial vector is not zero, we will show $w(k) \neq 0$ for $k \geq 0$ in the next section. Thus algorithms (1) and (2) are reasonable.

Adding the term $\alpha \left(\frac{1}{\|w(k)\|} - 1 \right) w(k)$ to the algorithm is quite technical. For simplicity, we illustrate the idea in one dimensional case. Assume $C = 1$, $\eta(k) = \eta$ and $\eta\alpha < 1$, algorithm (2) in one dimension is the following

$$w(k + 1) = (1 - \eta\alpha)w(k) + \eta\alpha \frac{w(k)}{|w(k)|}.$$

If $w(0) > 0$, $w(k)$ will converge to $+1$; If $w(0) < 0$, $w(k)$ will converge to -1. Thus the norm of $w(k)$ will converge to one. However, the rigorous proof of convergence does not exist in [11], in this paper, we will give the rigorous mathematical proof. And the exponential convergence rate will be obtained.

3   Convergence Analysis

This section studies the discrete-time dynamical systems (1) and (2). For convenience, denote by σ the smallest eigenvalue of C. Immediately, $\sigma < \lambda_{n-1} \leq \cdots \leq \lambda_1$. Denote by V_σ the eigensubspace of the smallest eigenvalue σ. Clearly, $V_\sigma = span\{v_n\}$. Assume $\eta(k)$ is a constant, i.e., $\eta(k) = \eta$.

Lemma 1. *Suppose that there exists constant $\eta_0 > 0$ so that*

$$\eta_0 \leq \eta \leq \frac{1}{\lambda_1 + \alpha - \sigma},$$

and if $w(0)$ is not orthogonal to V_σ, we have $\|w(k)\| \geq \eta\alpha$ for $k \geq 1$.

Proof. Because $w(0)$ is not orthogonal to V_σ, so $\|w(0)\| \neq 0$. Since the vector set $\{v_1, \cdots, v_n\}$ forms an orthogonal basis of R^n, for each $k \geq 0$, $w(k)$ can be represented as

$$w(k) = z_n(k)v_n + \sum_{j=1}^{n-1} \epsilon_j(k)v_j \tag{3}$$

where $z_n(k)$ and $\epsilon_j(k)$ are constants. Because $w(0)$ is not orthogonal to V_σ, $z_n(0) \neq 0$.

Substitute (3) into (2), it follows that

$$z_n(k+1) = \left[1 + \eta\left(\beta(k) - \sigma + \frac{\alpha}{\|w(k)\|} - \alpha\right)\right] z_n(k), \tag{4}$$

and

$$\epsilon_j(k+1) = \left[1 + \eta\left(\beta(k) - \lambda_j + \frac{\alpha}{\|w(k)\|} - \alpha\right)\right] \epsilon_j(k), \tag{5}$$

for $1 \leq j \leq n-1$, where $k \geq 0$, and

$$\beta(k) = \frac{w^T(k)Cw(k)}{\|w(k)\|^2} \geq \sigma$$

for $k \geq 0$.

If the conditions of Lemma 1 are satisfied, we have

$$1 + \eta\left(\beta(k) - \sigma + \frac{\alpha}{\|w(k)\|} - \alpha\right) > 0, \tag{6}$$

and for each $j(1 \leq j \leq n-1)$,

$$1 + \eta\left(\beta(k) - \lambda_j + \frac{\alpha}{\|w(k)\|} - \alpha\right) \geq 0 \tag{7}$$

for $k \geq 0$.

From (6) and (7), it follows that

$$\|w(k+1)\|^2 \geq \left[1 + \eta\left(\sigma - \lambda_1 + \frac{\alpha}{\|w(k)\|} - \alpha\right)\right]^2 \cdot \|w(k)\|^2.$$

Thus by the condition of Lemma 1,

$$\|w(k+1)\| \geq [1 - \eta(\lambda_1 + \alpha - \sigma)] \cdot \|w(k)\| + \eta\alpha \geq \eta\alpha,$$

for $k \geq 0$.

Lemma 2. *Under the conditions of Lemma 1, the angle between the $w(k)$ of algorithm (2) and the eigenvector v_n will approach zero. Moreover,*

$$\beta(k) - \sigma \leq \sum_{j=1}^{n-1} (\lambda_j - \sigma) \cdot \theta^{k-1} \cdot \theta_0 \cdot \left[\frac{\epsilon_j(0)}{z_n(0)}\right]^2, \tag{8}$$

for $k > 0$, where

$$\theta = \left[1 - \frac{\eta(\lambda_{n-1} - \sigma)}{2 + \eta\lambda_1}\right]^2, \text{ and}$$

$$\theta_0 = \left[1 - \frac{\eta(\lambda_{n-1} - \sigma)}{1 + \eta\left(\lambda_1 + \frac{\alpha}{\|w(0)\|}\right)}\right]^2.$$

Proof. Clearly, $0 < \{\theta, \theta_0\} < 1$. From (4~7) and according to Lemma 1, by using the similar method in [9, 10], it follows that

$$\left[\frac{\epsilon_j(k+1)}{z_n(k+1)}\right]^2 \leq \theta^k \cdot \theta_0 \cdot \left[\frac{\epsilon_j(0)}{z_n(0)}\right]^2 \to 0, \quad (j = 1, \cdots, n-1)$$

as $k \to +\infty$. This means that the angle between $w(k)$ and v_n approaches to zero.
It follows that

$$\beta(k) - \sigma = \frac{\sum_{j=1}^{n-1}(\lambda_j - \sigma)\epsilon_j^2(k)}{z_n^2(k) + \sum_{j=1}^{n-1}\epsilon_j^2(k)} \leq \sum_{j=1}^{n-1}(\lambda_j - \sigma) \cdot \theta^{k-1} \cdot \theta_0 \cdot \left[\frac{\epsilon_j(0)}{z_n(0)}\right]^2$$

for $k > 0$.

Lemma 3. *If the conditions of Lemma 1 hold, and $\alpha > \lambda_1 - \sigma$, then, there exist constants Π_1, θ_1 and θ_2 so that*

$$|z_n(k+1) - 1| \leq k \cdot \Pi_1 \cdot \left[e^{-\theta_2(k+1)} + max\{e^{-\theta_1 k}, e^{-\theta_2 k}\}\right]$$

for $k > 0$, where

$$\theta_1 = -ln(max\{\theta, \theta_0\}), \theta_2 = -ln\delta, \delta = \max_k [1 + \eta(\beta(k) - \sigma) - \eta\alpha].$$

Clearly, $\theta_1, \theta_2 > 0$ and $\delta < 1$.

Proof. By Lemma 2, there exists a constant $M > 0$ so that

$$\beta(k) - \sigma \le Me^{-\theta_1 k} \tag{9}$$

for $k > 0$. From equation (4), it follows that

$$z_n(k+1) - 1 \le [1 + \eta\left(\beta(k) - \sigma\right) - \eta\alpha] \cdot (z_n(k) - 1)$$
$$+\eta\left(\beta(k) - \sigma\right) + \eta\alpha\left(\frac{z_n(k)}{\|w(k)\|} - 1\right)$$

for $k > 0$. Since $z_n(k) - \|w(k)\| \le 0$, by (9), it follows that

$$|z_n(k+1) - 1| \le \delta|z_n(k) - 1| + \eta Me^{-\theta_1 k}$$

for $k \ge 0$. Then,

$$|z_n(k+1) - 1| \le \delta^{k+1}|z_n(0) - 1| + \eta M \sum_{r=0}^{k}(\delta e^{\theta_1})^r e^{-\theta_1 k}$$
$$\le \delta^{k+1}|z_n(0) - 1| + k\eta M \cdot \max\left\{\delta^k, e^{-\theta_1 k}\right\}$$
$$\le k \cdot \Pi_1 \cdot \left[e^{-\theta_2(k+1)} + \max\left\{e^{-\theta_2 k}, e^{-\theta_1 k}\right\}\right]$$

where

$$\Pi_1 = \max\left\{|z_n(0) - 1|, \eta M\right\} > 0.$$

Theorem 1. *If the conditions of Lemma 1 and Lemma 3 are satisfied, the $w(k)$ of algorithm (2) will converge exponentially to the eigenvector v_n associated with the eigenvalue σ as $k \to +\infty$.*

Proof. By Lemma 3, clearly, $\lim_{k \to +\infty} z_n(k) = 1$. Then, by using Lemma 2,

$$\epsilon_j^2(k) = \left[\frac{\epsilon_j(k)}{z_n(k)}\right]^2 \cdot z_n^2(k) \to 0 \tag{10}$$

as $k \to +\infty$.

From (3) together with (10), it follows that $\lim_{k \to +\infty} w(k) = v_n \in V_\sigma$.

4   Experiments

A beamformer is an array of sensors which can do spatial filtering. The objective is to perform spatial filtering of a signal arriving from the desired direction in the presence of noise and other interfering signals. A beamformer can be realized by a neural unit with the input $x(k) = [x_1(k), x_2(k), \cdots, x_M(k)]^T \in \mathbb{C}^M$, the complex weight vector $w(k) \in \mathbb{C}^M$ and output $y(k) = w^H(k)x(k)$ at time k, where M is the number of sensors[4]. $x(k)(k = 0, 1, \cdots)$ are zero-mean discrete-time signals.

The optimization problem of training the beamforming neuron is the following [3, 4]:

$$\min w^H C w, \quad w^H w = \delta^{-2}, \quad \Gamma^H w = b, \tag{11}$$

where $C \equiv E[x(k)x^H(k)]$ is the complex correlation matrix of the input and $\delta \leq M$ is a constant that limits the white noise sensitivity. Large values of δ correspond to strong robustness, while small values of δ do not provide robust design[4]. The linear constrains $\Gamma^H w = b$ are used to improve the performances of the beamformer by accounting for prior knowledge[3],[4]. It is obvious that this problem is equivalent to finding the minor vector of matrix C with some constrains.

Define $P_c \equiv \Gamma(\Gamma^H \Gamma)^{-1}\Gamma^H$, $\tilde{P}_c = I - P_c$ and $w_c \equiv \Gamma(\Gamma^H \Gamma)^{-1}b$. Let the objective function $J(w) = E[|y|^2]$, Cox proposed an adaptive algorithm to solve the optimization problem (11) as the following [3]:

$$w(k+1) = w_c + \tilde{P}_c w(k) - \eta \tilde{P}_c \Delta w(k), \tag{12}$$

where η is the learning rate and $\Delta w = \dfrac{\partial J(w)}{\partial w}$ which is used to calculate the minor vector of matrix C. Since algorithm (12) only satisfies the linear constrains $\Gamma^H w = b$, we need to normalize the complex weight vector in each iteration so that $w^H w = \delta^{-2}$. Instead, by using the lagrange multiplier λ, Firori used the objective function $J(w) = 0.5E[|y|^2] + 0.5\lambda \left(w^H w - \delta^{-2} \right)$ which satisfies both constraints without additional mechanisms[4]. Thus,

$$\Delta w(k) \equiv \frac{\partial J(w)}{\partial w} = Cw - \delta^2(w^H Cw)w + \mu \left(w^H w - \delta^{-2} \right) w,$$

where $\mu > \lambda_1$ is a positive constant. This is OJA+ algorithm[4],[7]. Define

$$\Delta w(k) = Cw(k) - \frac{w^H(k)Cw(k)}{\|w(k)\|^2}w(k) - \alpha \left(\frac{1}{\|w(k)\|} - \delta \right) w(k),$$

which can be used to compute the minor vector with the constrain $w^H w = \delta^{-2}$, actually, this is our GMCA algorithm[11].

Since there are no comparison experiments in [11], in this paper, we will compare GMCA with OJA+ when $\delta = 1$. For the discussion of different values of δ, please refer to [4]. In our experiments, since OJA+ is only local convergence [5], the norm of initial weight vector of OJA+ algorithm should be small to avoid divergence.

In the simulation experiment, by restricting to planar geometry, the beamformer has $M = 6$ sensors placed in a linear array with the distance between the sensors $d = 0.26$m. The geometry is depicted in Fig.1. For this configuration the steering vector is

$$v(\theta) = \frac{1}{M} \left[1 \ e^{-i2\pi[(d\sin\theta)/\lambda]} \ \cdots \ e^{-i2\pi[(d\sin\theta)/\lambda](M-1)} \right],$$

where λ is the wavelength.

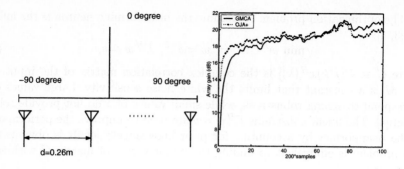

Fig. 1. Sensors array geometry(left) and array gain along θ_0(right)

The primary source signal s of unitary amplitude comes from $\theta_0 = \pi/8$ and a unitary disturbance i impinges the array from $\theta_d = -\pi/4$; a spatially uncorrelated noise of amplitude 0.01 is present. It follows that the input vectors are

$$x(k) = \sqrt{M} \cdot \left[v^H(\theta_0)s(k) + v^H(\theta_d)i(k) \right] + noise.$$

The linear constrain $\Gamma = v(\theta_0)$ and $b = 1$, i.e. the unit boresight response $v(\theta_0)^H b = 1$ which ensures no attenuation and zero phase shift in the direction of arrival of primary source. And the angular frequency of the signal $\omega = 2\pi \cdot 650$rad/s and speed $c \cong 340$m/s.

By using Theorem 1, we choose the learning rate $\eta = 0.00005$, $\mu = \alpha = 240$ and $\delta = 1$ for both algorithms. To evaluate the performances, define the array gain along direction θ as the following

$$G(\theta) = \frac{M|v^H(\theta)w|^2}{w^H C_{i+n} w},$$

where C_{i+n} is the covariance matrix of the disturbance and noise. The disturbance gain can be defined accordingly. The profile of array gain at different values of θ is called array beampattern. Fig.1-2 show the signal enhancement and disturbance rejection. From these figures, we can observe that the performances of GMCA is almost the same as that of OJA+. However, OJA+ is only locally convergent and the initial weight vector should be decided carefully.

5   Conclusions

A globally convergent learning algorithm for MCA neural network has been reviewed in this paper. The rigorous mathematical proof of global convergence is given and exponential convergence rate is obtained. In contrast to [11], comparison experiments were presented to confirm our theoretical results and illustrate the efficiency and effectiveness of the algorithm on beamforming problem.

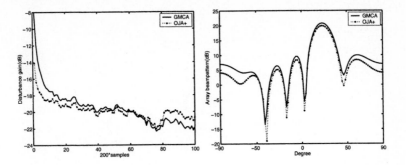

Fig. 2. Disturbance gain along θ_d(left) and array beampattern(right)

Acknowledgement

This work was supported in part by the National Science Foundation of China under grant numbers 60471055,10476006 and A0324638.

References

1. Anisse, T., Cirrincione, G.: Against the Convergence of the Minor Component Analysis Neurons. IEEE Trans. Neural Networks **10** (1999) 207-210
2. Cirrincione, G., Cirrincione, M., Herault, J., Huffel, S.V.: The MCA EXIN Neuron for the Minor Component Anlysis. IEEE Trans. Neural Networks **13** (2002) 160-187
3. Cox, H., Zeskind, R.M., Owen, M.M.: Robust Adaptive Beamforimg. IEEE Trans. on Acoustics, Speech, and Signal Processing **35** (1987) 1365-1376
4. Fiori, S., Piazza, F.: Neural MCA for Robust Beamforming. Proc. of International Symposium on Circuits and Systems(ISCAS'2000) **III** (2000) 614-617
5. Lv, J.C., Ye, M., Yi, Z.: Convergence Analysis of OJA+ Algorithm. In: Yin, F.L., Wang, J., Guo, C.G. (eds.): Advances in Neural Networks-ISNN2004. Lecture Notes in Computer Science, Vol. 3173. Springer-Verlag, Berlin Heidelberg New York (2004) 812-815
6. MöLLER, R.: A Self-stabilizing Learning Rule for Minor Component Analysis. International Journal of Neural Systems **14** (2004) 1-8
7. Oja, E.: Principal Components, Minor Components, and Linear Neural Networks. Neural Networks **5** (1992) 927-935
8. Xu, L., Oja, E., Suen, C.: Modified Hebbian Learning for Curve and Surface Fitting. Neural Networks **5** (1992) 441-457
9. Ye, M., Yi, Z.: On The Discrete Time Dynamics of the MCA Neural Networks. In: Yin, F.L., Wang, J., Guo, C.G. (eds.): Advances in Neural Networks-ISNN2004. Lecture Notes in Computer Science, Vol. 3173. Springer-Verlag, Berlin Heidelberg New York (2004) 815-821
10. Ye, M.: Global Convergence Analysis of a Self-stabilizing MCA Algorithm. Neurocomputing (to appear)
11. Zhang, X.S., Ye, M., Wang, Y.D., Li, Y.C.: Neural Networks for Adaptive Beamforming. Computer Science **31** (2004) 205-207

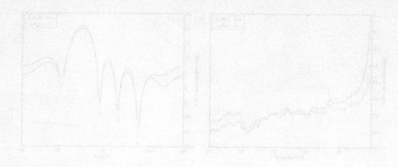

Author Index

Lecture Notes in Computer Science

For information about Vols. 1–3375

please contact your bookseller or Springer

Vol. 3429: E. Andres, G. Damiand, P. Lienhardt (Eds.), Discrete Geometry for Computer Imagery. X, 428 pages. 2005.

Vol. 3427: G. Kotsis, O. Spaniol (Eds.), Wireless Systems and Mobility in Next Generation Internet. VIII, 249 pages. 2005.

Vol. 3423: J.L. Fiadeiro, P.D. Mosses, F. Orejas (Eds.), Recent Trends in Algebraic Development Techniques. VIII, 271 pages. 2005.

Vol. 3422: R.T. Mittermeir (Ed.), From Computer Literacy to Informatics Fundamentals. X, 203 pages. 2005.

Vol. 3421: P. Lorenz, P. Dini (Eds.), Networking - ICN 2005, Part II. XXXV, 1153 pages. 2005.

Vol. 3420: P. Lorenz, P. Dini (Eds.), Networking - ICN 2005, Part I. XXXV, 933 pages. 2005.

Vol. 3419: B. Faltings, A. Petcu, F. Fages, F. Rossi (Eds.), Constraint Satisfaction and Constraint Logic Programming. X, 217 pages. 2005. (Subseries LNAI).

Vol. 3418: U. Brandes, T. Erlebach (Eds.), Network Analysis. XII, 471 pages. 2005.

Vol. 3416: M. Böhlen, J. Gamper, W. Polasek, M.A. Wimmer (Eds.), E-Government: Towards Electronic Democracy. XIII, 311 pages. 2005. (Subseries LNAI).

Vol. 3415: P. Davidsson, B. Logan, K. Takadama (Eds.), Multi-Agent and Multi-Agent-Based Simulation. X, 265 pages. 2005. (Subseries LNAI).

Vol. 3414: M. Morari, L. Thiele (Eds.), Hybrid Systems: Computation and Control. XII, 684 pages. 2005.

Vol. 3412: X. Franch, D. Port (Eds.), COTS-Based Software Systems. XVI, 312 pages. 2005.

Vol. 3411: S.H. Myaeng, M. Zhou, K.-F. Wong, H.-J. Zhang (Eds.), Information Retrieval Technology. XIII, 337 pages. 2005.

Vol. 3410: C.A. Coello Coello, A. Hernández Aguirre, E. Zitzler (Eds.), Evolutionary Multi-Criterion Optimization. XVI, 912 pages. 2005.

Vol. 3409: N. Guelfi, G. Reggio, A. Romanovsky (Eds.), Scientific Engineering of Distributed Java Applications. X, 127 pages. 2005.

Vol. 3408: D.E. Losada, J.M. Fernández-Luna (Eds.), Advances in Information Retrieval. XVII, 572 pages. 2005.

Vol. 3407: Z. Liu, K. Araki (Eds.), Theoretical Aspects of Computing - ICTAC 2004. XIV, 562 pages. 2005.

Vol. 3406: A. Gelbukh (Ed.), Computational Linguistics and Intelligent Text Processing. XVII, 829 pages. 2005.

Vol. 3404: V. Diekert, B. Durand (Eds.), STACS 2005. XVI, 706 pages. 2005.

Vol. 3403: B. Ganter, R. Godin (Eds.), Formal Concept Analysis. XI, 419 pages. 2005. (Subseries LNAI).

Vol. 3402: M. Daydé, J.J. Dongarra, V. Hernández, J.M.L.M. Palma (Eds.), High Performance Computing for Computational Science - VECPAR 2004. XI, 732 pages. 2005.

Vol. 3401: Z. Li, L.G. Vulkov, J. Waśniewski (Eds.), Numerical Analysis and Its Applications. XIII, 630 pages. 2005.

Vol. 3399: Y. Zhang, K. Tanaka, J.X. Yu, S. Wang, M. Li (Eds.), Web Technologies Research and Development - APWeb 2005. XXII, 1082 pages. 2005.

Vol. 3398: D.-K. Baik (Ed.), Systems Modeling and Simulation: Theory and Applications. XIV, 733 pages. 2005. (Subseries LNAI).

Vol. 3397: T.G. Kim (Ed.), Artificial Intelligence and Simulation. XV, 711 pages. 2005. (Subseries LNAI).

Vol. 3396: R.M. van Eijk, M.-P. Huget, F. Dignum (Eds.), Agent Communication. X, 261 pages. 2005. (Subseries LNAI).

Vol. 3395: J. Grabowski, B. Nielsen (Eds.), Formal Approaches to Software Testing. X, 225 pages. 2005.

Vol. 3394: D. Kudenko, D. Kazakov, E. Alonso (Eds.), Adaptive Agents and Multi-Agent Systems II. VIII, 313 pages. 2005. (Subseries LNAI).

Vol. 3393: H.-J. Kreowski, U. Montanari, F. Orejas, G. Rozenberg, G. Taentzer (Eds.), Formal Methods in Software and Systems Modeling. XXVII, 413 pages. 2005.

Vol. 3392: D. Seipel, M. Hanus, U. Geske, O. Bartenstein (Eds.), Applications of Declarative Programming and Knowledge Management. X, 309 pages. 2005. (Subseries LNAI).

Vol. 3391: C. Kim (Ed.), Information Networking. XVII, 936 pages. 2005.

Vol. 3390: R. Choren, A. Garcia, C. Lucena, A. Romanovsky (Eds.), Software Engineering for Multi-Agent Systems III. XII, 291 pages. 2005.

Vol. 3389: P. Van Roy (Ed.), Multiparadigm Programming in Mozart/Oz. XV, 329 pages. 2005.

Vol. 3388: J. Lagergren (Ed.), Comparative Genomics. VII, 133 pages. 2005. (Subseries LNBI).

Vol. 3387: J. Cardoso, A. Sheth (Eds.), Semantic Web Services and Web Process Composition. VIII, 147 pages. 2005.

Vol. 3386: S. Vaudenay (Ed.), Public Key Cryptography - PKC 2005. IX, 436 pages. 2005.

Vol. 3385: R. Cousot (Ed.), Verification, Model Checking, and Abstract Interpretation. XII, 483 pages. 2005.

Vol. 3383: J. Pach (Ed.), Graph Drawing. XII, 536 pages. 2005.

Vol. 3382: J. Odell, P. Giorgini, J.P. Müller (Eds.), Agent-Oriented Software Engineering V. X, 239 pages. 2005.

Vol. 3381: P. Vojtáš, M. Bieliková, B. Charron-Bost, O. Sýkora (Eds.), SOFSEM 2005: Theory and Practice of Computer Science. XV, 448 pages. 2005.

Vol. 3380: C. Priami (Ed.), Transactions on Computational Systems Biology I. IX, 111 pages. 2005. (Subseries LNBI).

Vol. 3379: M. Hemmje, C. Niederee, T. Risse (Eds.), From Integrated Publication and Information Systems to Information and Knowledge Environments. XXIV, 321 pages. 2005.

Vol. 3378: J. Kilian (Ed.), Theory of Cryptography. XII, 621 pages. 2005.

Vol. 3377: B. Goethals, A. Siebes (Eds.), Knowledge Discovery in Inductive Databases. VII, 190 pages. 2005.

Vol. 3376: A. Menezes (Ed.), Topics in Cryptology – CT-RSA 2005. X, 385 pages. 2005.

Vol. 3375: M.A. Marsan, G. Bianchi, M. Listanti, M. Meo (Eds.), Quality of Service in Multiservice IP Networks. XIII, 656 pages. 2005.